Ref Number:162
162. de Robertis, E.D.P. and de Robertis, Jr.,E.M.F. Cell and molecular biology, Philadelphia:Lea & Febiger, 1987. Ed. 8th

Cell and Molecular Biology

Cell and Molecular Biology

E.D.P. De ROBERTIS, M.D.

Emeritus Professor of Cell Biology
University of Buenos Aires
Buenos Aires, Argentina

E.M.F. De ROBERTIS, JR., M.D., Ph.D.

Norman Sprague Professor of Molecular Oncology
University of California
Los Angeles, California

EIGHTH EDITION

Lea & Febiger   *Philadelphia*

Lea & Febiger
600 Washington Square
Philadelphia, PA   19106-4198
U.S.A.
(215) 922-1330

The cover shows a giant nucleus of the *Xenopus* oocyte that has been manually isolated from the cytoplasm. The amplified nucleoli (bright spots on the periphery of the nucleoplasm) and the nuclear envelope are clearly visible in this darkfield image. Photo by E.M. De Robertis, Jr.

Library of Congress Cataloging-in-Publication Data

De Robertis, Eduardo D. P., 1913–
    Cell and molecular biology.

    Includes bibliographies and index.
    1. Cytology. 2. Molecular biology. I. De Robertis,
E. M. F. II. Title. [DNLM: 1. Cells. 2. Molecular
Biology. QH 581 D437c]
QH581.2.R613   1987   574.87   86-123
ISBN 0-8121-1012-9

PRINTED IN THE UNITED STATES OF AMERICA

Print number: 5   4   3   2

PREFACE

Life has an immense variety of forms that arose by the process of biological evolution, but all living organisms share a master plan of structural and functional organization. This book is about the building blocks—*cells* and *molecules*—that constitute the unity of the living world.

Progress in our field of science has been so rapid that during the lifetime of this book, first published in 1946, we have witnessed the most revolutionary discoveries, such as those involving the recognition of the ultrastructure and macromolecular organization of cell components, the uncovering of the DNA helix, and the molecular basis of the genetic code and gene expression. This tremendous progress necessitated making extensive revisions every five years and changing the original title, *General Cytology*, to *Cell Biology* in 1965 and to *Cell and Molecular Biology* in 1980.

The present, eighth, edition should be seen as a new book rather than as a revision because the changes have been so profound that the text has been entirely rewritten and more than half of the illustrations have been changed or added. We have tried to integrate the most recent advances in molecular biology with our knowledge of the structure and function of cells, while taking into account the work of classical cytologists—often forgotten these days—which laid the foundations of our understanding of the living cell.

The book has been organized to facilitate learning by proceeding from simple to more complex matters. For example, the first three chapters give an overview of the cell and its main structural components, a general introduction to the biochemistry of the cell, and an account of some of the current methods used in the study of cell biology.

Each of the following twenty-one chapters contains an introduction, stating the main objectives; sectional summaries that provide a synopsis of the essential points; and for further information and study, a list of numbered references and additional readings.

The specific fields of cell and molecular biology are presented dynamically, based on experimental data from which the student can draw his own conclusions. All the figures, many of them original drawings that represent classic experiments, are integrated in the text. A special feature of this edition is a glossary that defines words and concepts.

Cell and molecular biology have become the basic pillars on which all biological and medical sciences are based. Although primarily intended for courses at the college level, the book may be useful in more advanced courses, because of the wealth of up-to-date information contained in it. Much of the text should be of interest to those in the applied sciences such as medicine, veterinary, agronomy, and biotechnology.

By emphasizing that cell and molecular biology are in constant progress and much remains to be discovered, we hope to stimulate research among young scientists who want to dedicate their efforts to uncovering the basis of life on our planet.

Buenos Aires, Argentina   E. De Robertis
Los Angeles, California   E.M. De Robertis, Jr.

v

ACKNOWLEDGMENTS

We have been stimulated in our task by the good reception the book has received in previous English-language editions, as well as in the Spanish, Portuguese, Italian, French, Hungarian, Polish, Russian, Chinese, and Japanese translations. We especially want to recognize the contributions of Professors Richard Shivers, R. E. Hausman, James C. Tan, Ernest R. Vyse, Brower R. Burchill, Joseph F. Gennaro, Gerald Schatten, and William H. Beers, who have read the manuscript and have contributed valuable suggestions and criticisms.

Among the illustrations, we have included the original micrographs and designs of many experiments that led to major discoveries in cell and molecular biology. We hope that these experiments help the students to recognize the avenues by which scientific progress is made. We especially wish to express our gratitude to our colleagues who have contributed so generously to the illustrations of the present and past editions and whose names are acknowledged at each special instance.

We would like to thank the staff of Lea & Febiger who have contributed immensely to improving the presentation of this textbook.

We also want to acknowledge the collaboration of the staff of EL ATENEO, publishers of the Spanish edition, for their contribution of many valuable drawings to this book.

CONTENTS

THE CELL— STRUCTURAL ORGANIZATION

The study of the living world shows that evolution has produced an immense diversity of forms. There are about four million different species of bacteria, protozoa, plants, and animals that differ in their morphology, function, and behavior. We now know, however, that when living organisms are studied at the cellular and molecular levels there is a unique master plan of organization. The scope of *cell and molecular biology* is precisely that unifying organizational plan—in other words, the analysis of the cells and molecules that constitute the building blocks of all forms of life.

Ancient philosophers and naturalists, particularly Aristotle in antiquity and Paracelsus in the Renaissance, arrived at the conclusion that "All animals and plants, however complicated, are constituted of a few elements which are repeated in each of them." They were referring to the macroscopic structures of an organism, such as the roots, leaves, and flowers common to different plants, or segments and organs that are repeated in the animal kingdom. Many centuries later, the invention of magnifying lenses led to the discovery of the microscopic world, It was learned that a single cell could constitute an entire organism, as in the protozoa, or that it could be one of many cells grouped and differentiated into tissues and organs to form a multicellular organism.

The cell is thus a fundamental structural and functional unit of living organisms, just as the atom is the fundamental unit in chemical structures (Fig. 1–1). If cellular organization is destroyed by mechanical or other means, cellular function is likewise destroyed. Although some vital functions may persist (such as enzymatic activity), the cell becomes disorganized and dies.

Biochemical studies have demonstrated that

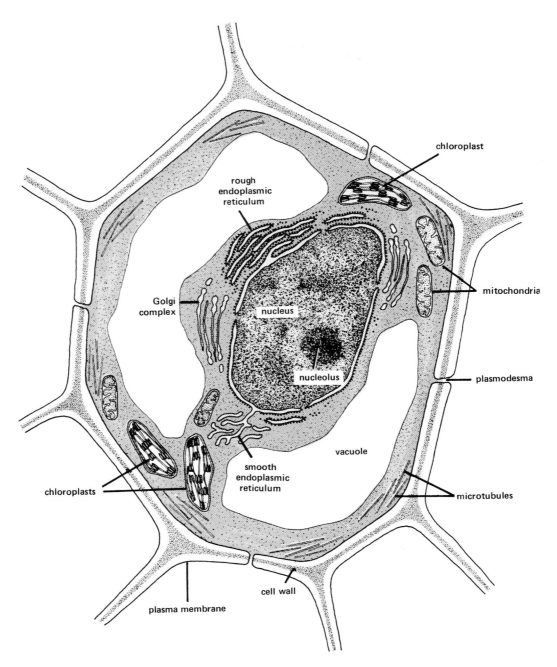

Fig. 1–1. Diagram of a plant cell showing the following cell components: the cell wall, the plasmodesmata, chloroplasts, and vacuoles.

living matter itself is composed of the same elements that comprise the inorganic world, although fundamental differences can exist in their organization. In the nonliving world there is a continuous tendency toward a thermodynamic equilibrium with a random distribution of matter and energy, whereas in a living organism there is a high degree of structure and function maintained by energy transformations based on the constant input and output of matter and energy. Biochemists have isolated from the complex mixture of cell constituents not only inorganic components, but also much more complex molecules such as proteins, fats, polysaccharides, and nucleic acids. Such biochemical studies have also demonstrated the underlying unity of the entire living world. Today it is known that the biochemical machinery in all organisms has essentially the same structure and function, and that all living organisms share the same genetic code.

In the last decade the use of the electron microscope (EM) has made it possible to obtain further knowledge about the structure of cells and to discover a whole world of subcellular organization that reaches the molecular level. On the other hand, the methods of cell fractionation have provided the way to separate the various subcellular organelles and to study them with the powerful techniques of biochemistry and molecular biology.

The main objectives of this chapter are to give an introduction to the study of the cell structure and to expose the student to the nomenclature of the cell components. These are of paramount importance in understanding the modern concepts of cell and molecular biology that are dealt with in the rest of the book. After a mention about the levels of organization in biology, some historical aspects will be considered with special reference to the *cell theory*. We shall describe the structural organization of the two main types of living organisms, prokaryotes and eukaryotes, putting emphasis on their similarities and differences. This chapter will also introduce the reader to the structure of the nucleus, the life cycle of the cell, and the processes of mitotic and meiotic division. It will end with a brief description of the ultrastructure of the cytoplasm and the main organelles. By the careful reading of this chapter the student will have an overview of the cell, as a base for studying the rest of the book.

1–1 LEVELS OF ORGANIZATION IN BIOLOGY

Biology at large deals with a complex hierarchy of organizational levels ranging from cells to populations and ecosystems. Modern studies of living matter demonstrate that a series of integrated *levels of organization* result in the vital manifestations of the organism. The concept of levels of organization as developed by Needham and others implies that in the entire universe—

in both the nonliving and living worlds—there are such various levels of different complexity that "The laws or rules that are encountered at one level may not appear at lower levels."

This concept can be applied to the different structural constituents of a cell or to the association of numerous cells in a tissue and of different tissues in an organism. For example, all vertebrates are composed of a limited number of cell types (e.g., muscle, skin, nerve cells) which are arranged in different proportions to produce the great variety of these organisms.

The diversity of the living world is ultimately dependent on a genetic program coded in the nucleic acids, which is executed through complex regulatory circuits that control the biochemical activities of cells. Molecular biology has demonstrated that the characteristics of the different cell types depend on special molecular components that are the result of the expression of genes (Jacob, 1977).

Levels of Organization and Instrumental Resolving Power

Table 1–1 shows the limits that separate the study of biological systems at particular dimension levels. The boundaries between levels of organization are imposed artificially by the *resolving power* of the instruments employed. The human eye cannot resolve (discriminate between) two points separated by less than 0.1 mm (100 μm). Most cells are much smaller than this and must be studied under the full resolving power of the light microscope (0.2 μm). Most cellular components are smaller still, and require the resolution of the EM (see Fig. 2–4). With this important tool, direct information can be obtained about structures ranging from 0.4 to 200 nm, thus extending our field of observation to the world of macromolecules. Results obtained by the application of electron microscopy have changed the field of cytology so much that a large part of the present book is devoted to discussions of the achievements made possible by this technique. Finally, in recent years great advances have been made in the detailed *x-ray diffraction* analysis of the molecular configuration of proteins, nucleic acids, and larger molecular complexes such as certain viruses.

In Figure 1–2 the sizes of different cells, bacteria, viruses, and molecules are indicated on a

TABLE 1–1.   VARIOUS LEVELS OF BIOLOGICAL STRUCTURE

Dimension	Field	Structures	Method
0.1 mm (100 μm) and larger	anatomy	organs	eye and simple lenses
100 μm to 10 μm	histology	tissues	various types of light microscopes,
10 μm to 0.2 μm (200 nm)	cytology	cell, bacteria }	x-ray microscopy
200 nm to 1 nm	submicroscopic morphology ultrastructure	cell components, viruses	polarization microscopy, electron microscopy
smaller than 1 nm	molecular and atomic structure	arrangement of atoms	x-ray diffraction

logarithmic scale and compared with the wavelengths of various radiations, as well as with the limits of resolution of the eye, the light microscope, and the EM. Note that the light microscope introduces a 500-fold increase in resolution over the eye, and the EM provides a 500-fold increase over the light microscope.

Table 1–2 shows the general relationships among some of the linear dimensions and weights used in different fields of chemical analysis of living matter. Familiarity with these relationships is essential to the study of cell and molecular biology. The weight of the important components of the cell is expressed in picograms

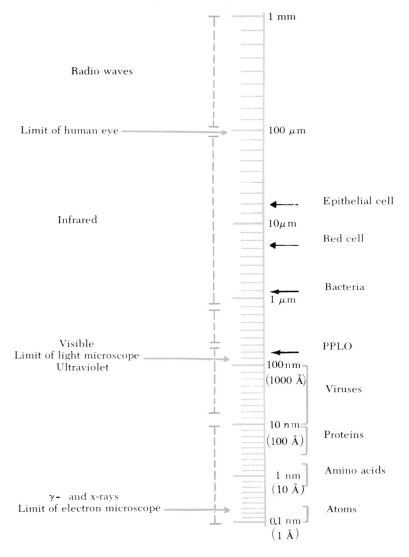

Fig. 1–2. Logarithmic scale of microscopic dimensions. Each main division represents a size ten times smaller than the division above. **To the left,** the position of the different wavelengths of the electromagnetic spectrum and the limits of the human eye, the light microscope, and the EM. **To the right,** the sizes of different cells, bacteria, PPLO (the smallest known living organism), viruses, molecules, and atoms. (Courtesy of M. Bessis.)

TABLE 1–2.   RELATIONSHIPS BETWEEN LINEAR DIMENSIONS AND WEIGHTS IN CYTOCHEMISTRY

Linear Dimension	Weight	Terminology	
1 cm	1 g	conventional biochemistry	
1 mm	1 mg or 10^{-3} g	microchemistry	
100 μm	1 μg or 10^{-6} g	histochemistry }	ultramicrochemistry
1 μm	1 μμg (or 1 picogram or 10^{-12} g)	cytochemistry }	

From Engström, A., and Finean, J.B.: *Biological Ultrastructure.* New York, Academic Press, Inc., 1958. Copyright 1958, Academic Press, Inc.

(1 pg = 1 μμg or 10^{-12} g), and that of molecules in *daltons*. The dalton is the unit of molecular weight (MW); one dalton equals one-twelfth the mass of carbon 12. For example, a water molecule weighs 18 daltons; a molecule of hemoglobin weighs 64,500 daltons.

SUMMARY:
Levels of Organization

Although there are about four million different species of living organisms, at the cellular and molecular levels there is a unique master plan of organization common to all. All organisms have the cell as their basic structural and functional unit, all have essentially the same biochemical machinery, and all share the same genetic code.

In living matter there are levels of organization that are interrelated in a complex manner and are maintained by energy transformations. The boundaries between these levels of organization are imposed by the resolving power of the eye (0.1 mm), the light microscope (0.2 μm), and the EM (~0.4 nm). These limits bring about the differentiation between the fields of anatomy, cytology, ultrastructure, and molecular biology. The EM provides direct information about structures ranging in size from 0.4 to 200 nm. It is important to correlate these boundaries with the spectrum of electromagnetic waves and the actual dimensions of cells, bacteria, viruses, and proteins (Fig. 1–2). The student should have a firm understanding of the relationships between linear dimensions and the weights used in cytochemistry. For example, the weight of cell components, such as nucleic acids, is expressed in picograms (10^{-12} g).

1–2   HISTORY OF CELL AND MOLECULAR BIOLOGY

The Development of the Cell Theory

The establishment of the *cell theory*, which essentially states that all living organisms are composed of cells and cell products, was the result of many investigations that started in the 17th century with the development of optical lenses and their combination in the compound microscope (Gr., *mikros*, small + *skopein*, to see, to look). The term *cell* (Gr., *kytos*, cell; L., *cella*, hollow space) was first used by Robert Hooke (1655) to describe his investigations on "the texture of cork by means of magnifying lenses." In these observations, repeated by Grew and Malpighi on different plants, only the cavities ("utricles" or "vesicles") of the cellulose wall were recognized. In the same century and at the beginning of the next, Leeuwenhock (1674) discovered free cells, as opposed to the "walled in" cells of Hooke and Grew. Leeuwenhock observed some organization within cells, particularly the nucleus in some red blood cells. For more than a century afterward, these observations were all that was known about the cell.

At the beginning of the 19th century, several discoveries were made about the structure of plant and animal tissues. These finally led the botanist Schleiden (1838) and the zoologist Schwann (1839) to establish the cell theory in a more definite form. After the discovery of the nucleus in all cells by Brown (1831) and the description of the cell content as *protoplasm*, the concept of the cell became that of a mass of protoplasm limited in space by a cell membrane and possessing a nucleus. The protoplasm surrounding the nucleus became known as the *cytoplasm* to distinguish it from the *karyoplasm*, the protoplasm of the nucleus.

A major expansion of the cell theory was expressed by Virchow in 1855 in his famous aphorism, "*Omnis cellulae e cellula*" (i.e., all cells arise from preexisting cells), which established cell division as the central phenomenon in the reproduction of organisms. Years later it was shown that cells ensure continuity between one generation and another by the mechanism of *mitosis* (Flemming, 1880) and the precise partitioning of the *chromosomes* (Waldeyer, 1890).

Another important discovery was that the development of an embryo starts with the fusion of two nuclei, one coming from an egg and the other from a sperm cell introduced during *fertilization* (Hertwig, 1875). Before the end of the century it was established that gametes (egg and sperm cells) are formed by a reductional division, later called *meiosis*, by which the number of chromosomes of a species remains constant from one generation to another.

All these discoveries led to the modern version of the cell theory, which states that (1) cells are the morphological and physiological units of all living organisms, (2) the properties of a given organism depend on those of its individual cells, (3) cells originate only from other cells, and continuity is maintained through the genetic material, and (4) the smallest unit of life is the cell.

The Development of Submicroscopic and Molecular Biology

In studying the limits and dimensions in biology (see Tables 1–1 and 1–2), the impact of instrumental analysis was mentioned and the modern fields of ultrastructure and molecular biology were delineated. These are the most progressive branches of biology, in which the merging of cytology with biochemistry, physicochemistry, and especially macromolecular and colloidal chemistry becomes increasingly intimate. Knowledge of the submicroscopic organization or ultrastructure of the cell is of fundamental importance, because practically all the functional and physicochemical transformations take place within the molecular architecture of the cell and at a supramolecular level.

The rapid development of cell and molecular biology in the present century can be attributed to two main factors: (1) the increased resolving power provided by electron microscopy and x-ray diffraction, and (2) the convergence of the field with other branches of biological research, especially genetics, physiology, and biochemistry. The convergence with genetics will be illustrative.

The fundamental laws of hereditary were discovered by Gregor Mendel in 1865, but at that time the cytological changes that take place in the sex cells were not sufficiently known to permit an interpretation of his findings. For this and other reasons, little attention was paid to Mendel's work until the botanists Correns, Tschermack, and De Vries independently rediscovered Mendel's laws in 1901. At that time cytology had advanced enough that the mechanism of distribution of the hereditary units postulated by Mendel could be understood and explained. The chromosome theory of heredity was finally established by Morgan and his collaborators, who assigned to the *genes* (Johanssen), or hereditary units, specific *loci* within the chromosomes. From this convergence of cytology and genetics, the study of *cytogenetics* has originated (see Chapter 18).

Within the past decade the study of genetics has also become linked to biochemistry, concerning events that take place at the molecular level. Although Miescher (1871) had already isolated "nuclein"—a substance now known as *deoxyribonucleic acid* or DNA—from white blood cells, its importance as genetic material was not recognized until the 1950s. Until that time the nuclear proteins had been considered the major constituents of genes. This concept was modified when nucleic acids were identified as the bearers of genetic information, especially in the light of fundamental work performed by Watson and Crick (1953), who proposed the *double helix model* of DNA. This model showed clearly how genes could duplicate and be transmitted from cell to cell (see Chapters 14 and 18). Since then the advances in molecular biology have been extraordinary. Another climax was reached with the deciphering of the *genetic code* (Nirenberg, Ochoa) and the discovery of the molecular mechanisms by which genes are transcribed and subsequently expressed as proteins. At the present time, the main focus of research in molecular biology is on the mechanisms that regulate gene expression.

SUMMARY:
Modern Cell Biology

In summary, it can be said that modern cell biology approaches the problems of the cell at all levels of organization—from molecular structure on. Cell biology is therefore the modern science in which genetics, physiology, and biochemistry converge. Cell biologists, without losing sight of the cell as a morphological and functional unit within the organism, must be prepared to study biological phenomena at all levels and to use all the methods, techniques, and concepts of the other sciences. This is a great challenge, but there is no other way to proceed if the life of the cell and of the organism is to be interpreted mechanistically, i.e., on the basis of combinations and associations of atoms and molecules.

1–3 GENERAL ORGANIZATION OF PROKARYOTIC CELLS

At the beginning of this chapter we mentioned that life is manifested as millions of different species that have their own special morphology and contain specific genetic information. The species can be arranged into progressively larger groups of organisms—genera, orders, families— up to the level of the classic kingdoms, plants, and animals. One of the most recent classification schemes, that of Whittaker, postulates a division into the five kingdoms monera, protista, fungi, plantae, and animalia, with their corresponding subdivisions (Table 1–3).

This complexity is simplified by examining living forms at the cellular level. Cells are identified as being of one of two recognizable types, prokaryotic or eukaryotic. Table 1–3 shows that only the monera (i.e., bacteria and blue-green algae) are prokaryotic cells, while all the other kingdoms consist of organisms made up of eukaryotic cells. The main difference between these two cell types is that prokaryotic cells (Gr., *karyon*, nucleus) lack a nuclear envelope. The prokaryotic chromosome occupies a space in the cell called a *nucleoid*, and is in direct contact with the rest of the protoplasm. Eukaryotic cells have a *true nucleus* with an elaborate nuclear envelope, through which the nucleocytoplasmic interchanges take place. Table 1–4 compares the structural organization in prokaryotes and eukaryotes, illustrating both the differences and similarities that exist between these two cell types.

From an evolutionary viewpoint, prokaryotes are considered to be ancestors of eukaryotes. Fossils three billion years old contain evidence of prokaryotes alone, whereas eukaryotes probably appeared one billion years ago. In spite of the differences between prokaryotes and eukaryotes, there are considerable homologies in their molecular organization and function. We shall see, for example, that all living organisms employ the same genetic code and a similar machinery for protein synthesis.

Cell Organization and the Energy Cycle

The sun is the ultimate source of energy for living organisms. The energy carried by photons of light is trapped by the pigment *chlorophyll*, present in the chloroplasts of green plants, and accumulates as chemical energy within the different foodstuffs consumed by other organisms.

All cells and organisms can be grouped into another two main classes, based on their mechanism of extracting energy for their own metabolism. Those of the first class, called *autotrophs* (e.g., green plants), use the process of *photosynthesis* to transform CO_2 and H_2O into the elementary organic molecule glucose, from which more complex molecules are produced. The second class of cells, called *heterotrophs* (e.g., animal cells), obtain energy from the various carbohydrates, fats, and proteins synthesized by autotrophic organisms. The energy contained in these organic molecules is released primarily by combustion with O_2 from the atmosphere (i.e., oxidation) in a process called *aerobic respiration*. The release of H_2O and CO_2 by heterotrophic organisms completes this energy cycle (Fig. 1–3).

These interrelated energy cycles have been maintained throughout evolution. Within the prokaryotic monera there are some species that are autotrophs and others that are heterotrophs. For example, photosynthetic bacteria and blue-green algae are autotrophs. Heterotrophic bacteria absorb soluble nutrients from the medium. Fungi and animals are heterotrophs, while all

TABLE 1–3. CLASSIFICATION OF LIVING ORGANISMS AND CELLS

Kingdom	Monera	Protista	Fungi	Plantae	Animalia
Representative Organisms	bacteria blue-green algae	protozoa chrysophytes	slime molds true fungi	green algae red algae brown algae bryophytes tracheophytes	metazoa
Cell Classification	prokaryotes			eukaryotes	

TABLE 1–4.   COMPARISON OF CELL ORGANIZATION IN PROKARYOTES AND EUKARYOTES

	Prokaryotic Cells Bacteria, blue-green algae, and mycoplasmas	Eukaryotic Cells Protozoa, other algae, metaphyta, and metazoa
Nuclear envelope	absent	present
DNA	naked	combined with proteins
Chromosomes	single	multiple
Nucleolus	absent	present
Division	amitosis	mitosis or meiois
Ribosomes	70S (50S + 30S)*	80S (60S + 40S)*
Endomembranes	absent	present
Mitochondria	respiratory and photo-synthetic enzymes in the plasma mem-brane	present
Chloroplast	absent	present in plant cells
Cell wall	non-cellulosic	cellulosic, only in plants
Exocytosis and endocytosis	absent	present
Locomotion	single fibril, flagellum	cilia and flagella

*S refers to the Svedberg sedimentation unit, which is a function of the size and shape of molecules.

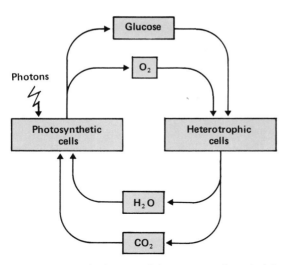

Fig. 1–3. Simple diagram of the energy cycle and of the interaction between photosynthetic and heterotrophic cells. (After A.L. Lehninger.)

plants (with the exception of a few species) are autotrophs.

Escherichia coli (E. coli)—The Most Studied Prokaryote

Although this book deals mainly with the more complex eukaryotic cells, it is important to know that much of our present knowledge of molecular biology stems from the study of viruses and bacteria. A bacterial cell such as *Escherichia coli (E. coli)* is easily cultured in an aqueous solution containing glucose and various inorganic ions. In this medium, at 37°C, the cell mass divides and doubles in about 60 minutes. This time—the *generation time*—can be reduced to 20 minutes if purines and pyrimidines (the precursors of nucleic acids) as well as amino acids are added to the medium.

As shown in the electron micrograph of Figure 1–4 A, the bacterium is surrounded by two definite membranes separated by the so-called *periplasmatic space*. The outer layer is rigid, serves for mechanical protection, and is designated as the *cell wall*. The chemical composition of the cell wall is rather complex; it contains polysaccharide, lipid, and protein molecules. One of the most abundant polypeptides, *porin*, forms channels that allow for the diffusion of solutes. In Figure 1–4 B, is shown a diagram of the cell wall of a bacterium with the possible disposition of porin channels made of six to eight subunits that span the full thickness of the outer membrane. At the bottom of the figure some proteins appear attached to the peptidoglycan grid present in the periplasmatic space. This space contains polysaccharides associated with proteins (i.e., *proteoglycans or mucoproteins*) forming a gel.

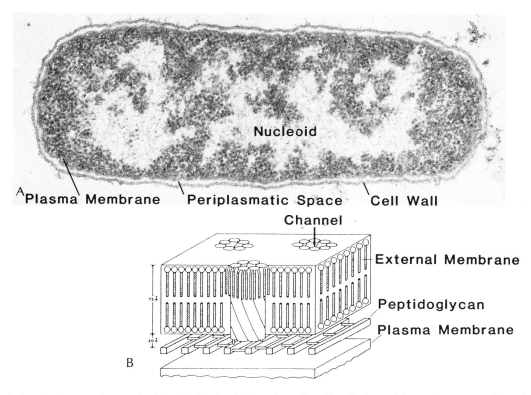

Fig. 1–4. *A,* electron micrograph of *E. coli* showing in its surface: the cell wall, the periplasmatic space, and the plasma membrane. The nucleoid appears as an irregular-shaped region of low electron density in which DNA filaments are observed. The rest of the protoplasm contains dense granules representing the ribosomes. (Courtesy of B. Menge, M. Wurtz, and E. Kellenberger.) *B,* diagram of the cell wall of a gram-negative bacterium. Observe the porin channels formed by 6 to 8 subunits traversing the lipid bilayer. Each subunit has 3 suspended hydrocarbon chains. The periplasmatic space contains a reticulum of peptidoglycans (see Table 5–1). A partial view of the plasma membrane is shown at the bottom.

The inner membrane, also called *plasma membrane*, is a lipoprotein structure serving as a molecular barrier with the surrounding medium. The plasma membrane, by controlling the entrance and exit of small molecules and ions, contributes to the establishment of a carefully regulated internal milieu for the protoplasm of the bacterium. It is interesting that the enzymes involved in the oxidation of metabolites (i.e., the respiratory chain), as well as the photosystems used in *photosynthesis*, may be present in the plasma membrane of prokaryotes.

Between the inner and outer membranes there are localized *regions of adhesion* or *junctions*, as can be demonstrated by the following experiment:

If bacteria of the gram-negative type (i.e., those that do not stain with Gram staining), like *E. coli* are exposed to hyperosmotic sucrose solution, the inner membrane pulls away from the cell wall by a process called *plasmolysis* and

zones of attachment become apparent. These adhesion zones contain receptors (i.e., recognition sites) for bacteriophages (Fig. 1–5), and for the attachment of flagella and pili. These regions are probably also the sites through which lipid and proteins are exported, across the inner membrane, to the periplasmatic space and the surrounding medium (see Fig. 8–12).

The bacterial chromosome is a single circular molecule of naked DNA tightly coiled within the *nucleoid*, which appears in the electron microscope as a lighter region of the protoplasm (Fig. 1–4). It is important to remember that the DNA of *E. coli*, which is about 1 mm long (10^6 nm) when uncoiled, contains all the genetic information of the organism. In this case there is sufficient information to code for 2000 to 3000 different proteins.

The single chromosome is circular and at one point it is attached to the plasma membrane. It is thought that this anchoring may help in the

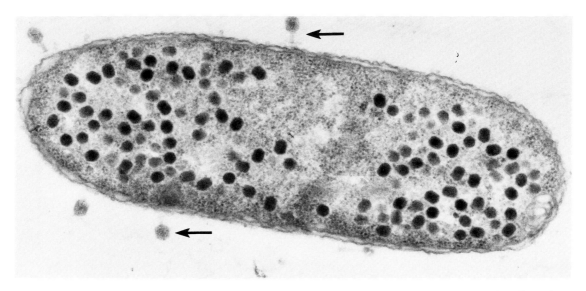

Fig. 1–5. *E. coli* infected with T$_4$ phage. This figure should be compared with a control in Figure 1–4. A few phage ghosts *(arrows)* are observed attached to the cell wall, after injection of the DNA. The nucleoid is no longer visible and the cell is filled with phage particles. (See text for details.) (Courtesy of B. Menge, M. Wurtz, and E. Kellenberger.)

separation of the two chromosomes after DNA replication. In fact, the growth of the intervening membrane may accomplish this separation.

In addition to a chromosome, certain bacteria contain a small extrachromosomal circular DNA called a *plasmid*. A plasmid may confer resistance to one or more antibiotics upon the bacterial cell. As will be mentioned in the chapters on molecular biology, plasmids can be separated and reincorporated; genes (specific pieces of DNA) can be inserted into plasmids, which are then transplanted into bacteria using the techniques of *genetic engineering* (see Chapter 19).

Surrounding the DNA in the dark region of the protoplasm are 20,000 to 30,000 particles, about 25 nm in diameter, called *ribosomes*. These particles are composed of *ribonucleic acid (RNA)* and proteins, and are the sites of protein synthesis. Ribosomes consist of a large and a small subunit and exist in groups called *polyribosomes* or *polysomes*. The remainder of the cell is filled with water, various RNAs, protein molecules (including enzymes), and various smaller molecules.

Certain motile bacteria have hair-like processes of variable length, called *flagella*, which are used for locomotion. In contrast with the cilia and flagella of eukaryotic cells, which contain several microtubules, each flagellum in bacteria is made of a single fibril (Table 1–4).

There is evidence that flagella rotate in a kind of "socket" located in the cell membrane, which consists of the basal body and the so-called "hook," which leads into the flagellum. The mechanism by which rotation occurs at the basal body is completely unknown.

SUMMARY:
General Organization of Prokaryotic Cells

Cells are identified as being of one of two types, prokaryotic or eukaryotic. The main difference between these types is that prokaryotic cells lack a nuclear envelope; their DNA occupies a space in the cell called a nucleoid. From an evolutionary standpoint, prokaryotes are considered to be ancestors of eukaryotes.

All organisms can be identified as either autotrophs or heterotrophs. Autotrophs can synthesize their own nutrients by the process of photosynthesis, using light energy from the sun to transform CO_2 and H_2O into glucose; heterotrophs must obtain their nutrients from among the products of autotrophs extracting energy from them by the process of aerobic respiration.

Escherichia coli is a common and representative prokaryote. It is easily cultured, has a short generation time, and is well suited to laboratory studies. A single bacterium *E. coli* (0.2×0.8 μm) is surrounded by a plasma membrane containing the respiratory enzymes. This membrane is surrounded by a more rigid cell wall. The DNA molecule is 1 mm long when uncoiled and is attached at one point to the plasma membrane. It contains information to code some 3000 protein molecules. The protoplasm contains some 20,000 to 30,000 ribosomes, principally as polysomes.

1–4 MYCOPLASMAS, VIRUSES, AND VIROIDS

Most prokaryotic cells are small, in the range of 1 to 10 μm, although some blue-green algae may reach 60 μm in diameter. From what has been said about *E. coli*, it is evident that there must be a minimum size limit for a cell. It must be large enough (1) to have a plasma membrane, (2) to contain sufficient genetic material to encode the various RNAs involved in protein synthesis, and (3) to contain the biosynthetic machinery that performs this synthesis.

Among living organisms that have the smallest mass, the best suited for study are small bacteria called *mycoplasmas* (Fig. 1–2), which produce infectious diseases in animals including humans and which can be cultured in vitro like any bacteria. These agents range in diameter from 0.25 to 0.1 μm; thus they correspond in size to some of the large viruses. These microbes are of biological interest because each is a thousand times smaller than the average bacterium and a million times smaller than a eukaryotic cell.

Viruses were first recognized by their property of being able to pass through the pores of porcelain filters and by the pathological changes they produced in cells. All viruses can now be visualized and identified morphologically with the EM and their macromolecular organization can be studied. Viruses are not considered true cells. Even though they share some cellular properties, such as autoreproduction, heredity, and mutation, viruses are dependent on the host's cells and are considered obligatory parasites.

It has been observed that many viruses display *icosahedral symmetry*. According to Caspar and Klug, this symmetry is due to the fact that the assembly of the protein subunits or *capsomeres* in this configuration enables the capsid of the virus to exist at a state of minimum energy. Such icosahedral symmetry has been found in a virus as small as φX174, which has only 12 capsomeres, and in one as large as adenovirus, which may have as many as 252 capsomeres.

Outside a living host cell viruses are inactive and may even be crystallized. Upon their intro-

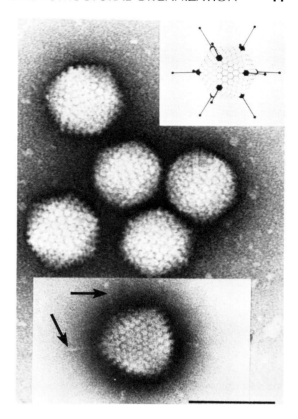

Fig. 1–6.   Electron micrographs of negatively stained adenoviruses showing the capsomeres. The projections *(arrows)* are observed attached to the pentons. In the drawing the hexons and pentons and the icosahedric shape of the virus are represented. (Courtesy of M. Wurtz.)

duction into the cell they become active and reproduce. From the point of view of their genetic constitution there are two kinds of viruses: those in which the chromosome is a molecule of RNA, such as tobacco mosaic virus, and those in which the chromosome is a DNA molecule, such as the bacterial viruses or bacteriophages. Viruses use their own genetic program for reproduction but rely on the biosynthetic machinery of the host (i.e., ribosomes, transfer RNA, enzymes) to produce the proteins of the coat, also called the *capsid*.

Viruses range in size between 30 and 300 nm, and their structures show different degrees of complexity. They are produced within a cell by a process of *macromolecular assembly*, which signifies that their components may be synthesized in different parts of the host cell and then assembled in a coordinated manner in another part.

Figure 1–6 shows *adenoviruses*, which tend

to form crystal aggregates inside the nucleus of the infected cell. This virus contains a linear double-stranded DNA molecule and the capsomeres are 240 hexons (i.e., having 6 neighbors) and 12 pentons (having 5 neighbors) situated at the vertex of the icosahedron. (In the diagram of Figure 1–6 the pentons are represented by black spheres and have filamentous projections that are visible in the electron micrograph at the bottom of the figure.)

Bacteriophages are viruses that use bacterial cells as hosts. The DNA is packed within the head of the phage and is injected into the bacterium by way of a tail that attaches to the host cell wall and functions as a syringe. The intracellular events that follow are rapid and start with disruption of the nucleoid and enzymatic hydrolysis of the host's DNA. The resulting nucleotides are used to synthesize phage DNA under the control of phage genes. The structural proteins of the phage are synthesized and it takes an average of 7 minutes to package and to assemble the mature phages inside the bacterium. As shown in Figure 1–5, after 30 minutes the *E. coli* is filled with many virus particles and is ready to break and release the bacteriophages. In this figure, on the surface of the cell wall, some ghosts of attached phages are still observed.

In eukaryotic cells infected with viruses the appearance of the new particles varies. For example, the DNA of *adenovirus* replicates in the host cell nucleus; the viral proteins are made on ribosomes in the cytoplasm and then the new virus particles are assembled in the nucleus. One special type is provided by the RNA tumor viruses in which maturation is produced by budding from the cell membrane. The complete virus contains a nucleocapsid, which in turn is wrapped by an envelope derived from the host cell membrane into which viral glycoproteins have become inserted (see Section 8–5).

Viroids are even simpler organisms than viruses. They are infectious agents that attack plant cells, and consist of a single RNA molecule that is not covered by a capsid of protein. Figure 1–7 shows a viroid that produces a disease in potatoes. Its circular RNA molecule consists of 359 nucleotides, and is able to multiply and infect other plants.

To conclude this short discussion about viruses let us compare them with true cells. We have defined a living cell as having the following characteristics: (1) A *specific genetic program* that permits the reproduction of new cells of the same type. (2) A *cell membrane* that establishes a boundary regulating all exchanges of matter and energy. (3) A *metabolic machinery* that can use energy trapped by the cell or obtained from foodstuffs. (4) A *biosynthetic machinery* for the synthesis of proteins. Viruses have only the first of these characteristics and lack all the others. For this reason they are not generally considered to be living organisms, despite the fact that they do contain a set of genetic blueprints from which a new virus can be made.

SUMMARY:
Mycoplasmas, Viruses, and Viroids

The smallest mass of living matter is represented by the *mycoplasma*, a small microbe 0.2 to 0.1 μm in diameter. At the minimum a living mass should contain (1) a plasma membrane, (2) the genetic material, and (3) the machinery for the synthesis of proteins.

Viruses are not considered to be true cells. Even though they share such cellular properties as heredity and mutations, they are dependent on the host's cells and are regarded as obligatory parasites. Viruses may contain either DNA or RNA, and they rely on the biosynthetic machinery of the host to produce capsid proteins from their genetic information. The capsids of many viruses display icosahedral symmetry, a configuration of the protein subunits that allows them to exist at a state of minimum energy. The entire virus particle is generally produced within the cell by a process of macromolecular assembly.

Viroids are even simpler than viruses, consisting of a single RNA molecule that is not covered by a protein capsid.

1–5   GENERAL ORGANIZATION OF EUKARYOTIC CELLS

Having studied the organization of prokaryotic cells it is useful to have another look at Table 1–4, in which the main points of comparison with eukaryotic cells are summarized. When comparing the organization of *E. coli* (Fig. 1–4) with that of a plant cell (Fig. 1–1) or an animal cell (Fig. 1–8), one is struck by the relative complexity of the eukaryotes. In a nondividing eukaryotic cell the *nucleus* exists as a separate compartment surrounded and limited by the nuclear envelope. Another and generally larger compartment is represented by the *cy-*

Fig. 1–7. Nucleotide sequence and secondary structure of potato spindle tuber viroid. The RNA is circular, and it has only 359 nucleotides and does not contain AUG codons (the signal for start of protein synthesis). Viroids are the simplest infectious agents known and are not covered with protein. (Reprinted by permission from Gross, J.J., et al.: Nature, 273:203, 1978. Copyright © 1978 Macmillan Journals.)

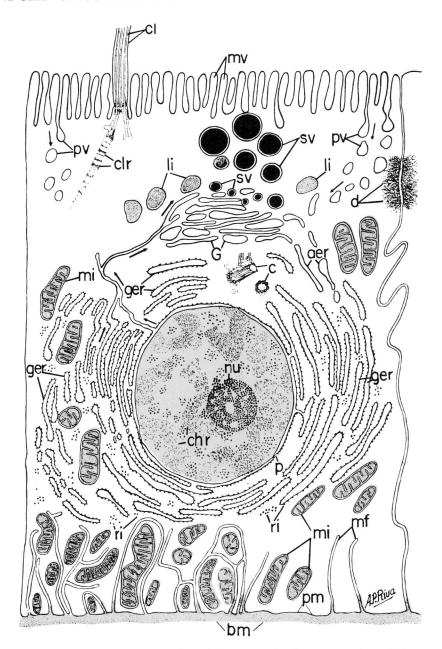

Fig. 1–8. General diagram of the ultrastructure of an idealized animal cell. *aer*, agranular endoplasmic reticulum; *bm*, basal membrane; *c*, centriole; *chr*, chromosome; *cl*, cilium; *clr*, cilium root; *d*, desmosome; *G*, Golgi complex; *ger*, granular endoplasmic reticulum; *li*, lysosome; *mf*, membrane fold; *mi*, mitochondria; *mv*, microvilli; *nu*, nucleolus; *p*, pore; *pm*, plasma membrane; *pv*, pinocytic vesicle; *ri*, ribosome; *sv*, secretion vesicle. (From E. De Robertis and A. Pellegrino de Iraldi, unpublished).

toplasm, and finally there is the *cell membrane* with its multiple infoldings and differentiations. Each of these three main components or compartments of the cell contains several subcomponents or subcompartments. Table 1–5 can be used as a guide to this complex organization since it lists the main morphological features of a eukaryotic cell and indicates the main functions of each. It will become apparent that this is an oversimplification when each of these components is discussed individually in later chapters of this book.

TABLE 1–5.   GENERAL ORGANIZATION OF THE EUKARYOTIC CELL

Main Components or Compartments	Subcomponents or Subcompartments	Main Function
Cell membrane	cell wall	protection
	cell coat	cell interaction
	plasma membrane	permeability, endocytosis and exocytosis
Nucleus	chromatin and chromo-somes	genetic information system
	nucleolus	synthesis of ribosomes
	nucleoplasm	
Cytoplasm		
Matrix, cytosol	soluble enzymes	glycolysis
Cytoskeleton	microfilaments	cell motility
	microtubules	cell shape and motility
	ribosomes	protein synthesis
Endomembrane system	nuclear envelope	nuclear permeability
	endoplasmic reticulum, rough and smooth	synthesis and transport of materials, secretion
	Golgi complex	secretion
Membrane organelles	mitochondria	cell respiration
	chloroplasts	photosynthesis
	lysosomes	digestion
	peroxisomes	peroxidation
Microtubular organelles	centrioles and spindle	cell division
	basal bodies, cilia, and flagella	cell motility

Morphological Diversity of Eukaryotic Cells

The cells of a multicellular organism vary in shape and structure and are differentiated according to their specific function in the various tissues and organs. This functional specialization causes cells to acquire special characteristics, but a common pattern of organization persists in all cells (Fig. 1–9).

Some cells, such as amebae and leukocytes, change their shape frequently. Others, such as nerve cells and most plant cells, always have a typical shape which is more or less fixed and specific for each cell type. The shape of a cell depends mainly on its functional adaptations and partly on the surface tension and viscosity of the protoplasm, the mechanical action exerted by adjoining cells, and the rigidity of the cell membrane. The orientation of microtubules within the cytosol also has a great influence on the shape of many cells (see Section 6–2).

The size of different cells ranges within broad limits. Some plant and animal cells are visible to the naked eye. Most cells, however, are visible only with a microscope, since they are only a few micrometers in diameter (Fig. 1–2). The smallest animal cells have a diameter of 4 μm.

In general, the volume of a cell is fairly constant for a particular cell type and is independent of the size of the organism. For example, kidney or liver cells are about the same size in the bull, horse, and mouse; the difference in the total mass of the organ depends on the number, not the volume, of cells.

The Cell Membrane

The structure that separates the cell content from the external environment is the *plasma membrane*. This is a thin film (6 to 10 nm thick) consisting of a continuous *lipid bilayer* with proteins intercalated in or adherent to both surfaces. The plasma membrane can be resolved only with the electron microscope, which reveals its numerous infoldings and differentiations as well as the different types of junctions that establish connections with neighboring cells (Fig. 1–8). The main function of the plasma membrane is to control selectively the entrance and exit of materials. This includes the entrance of water and large molecules by the process of *endocytosis* and the exit of cell products by *exocytosis* (Table 1–5). The plasma membrane is

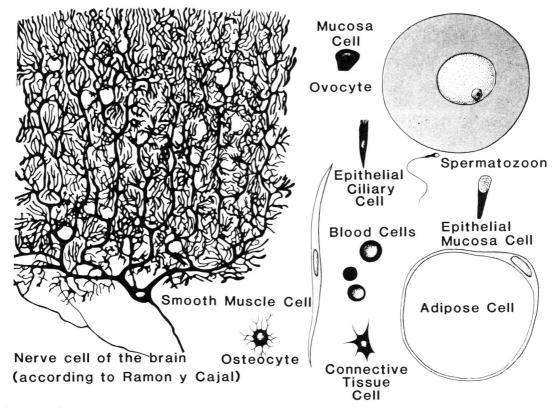

Fig. 1–9.   This composite figure illustrates the variety of cell types found in animal tissues. Observe the differences in shape and size of the cells.

covered and reinforced by the *cell wall* in plant cells and by the *cell coat* in animal cells.

The *cell wall* is responsible for the rigidity of most plant tissues. This structure consists mainly of cellulose fibers into which other substances (such as lignin, the main component of timber) may be incorporated. As shown in Figure 1–1, there are tunnels running through the cell wall called *plasmodesmata*, which allow communication with the other cells in a tissue.

In most animal cells the plasma membrane is covered by a *cell coat* made of glycoproteins, glycolipids, and polysaccharides that may extend the thickness of the cell membrane and continue far beyond it. The cell coat has several functions in addition to that of protection. It is involved in molecular recognition between cells, contains enzymes and antigens, and is fundamental in the association of cells in a tissue (see Chapter 5).

1–6 THE NUCLEUS AND THE CELL CYCLE

The shape of the nucleus is sometimes related to that of the cell, but it may be completely irregular. In spheroidal, cuboidal, or polyhedral cells, the nucleus is generally a spheroid. In cylindrical, prismatic, or fusiform cells, it tends to be an ellipsoid.

By 1905 Boveri had already noted that, in sea urchin larvae, the size of the nucleus was proportional to the chromosome number. In oocytes the nucleus (often called the *germinal vesicle*) is very active and may attain a large volume. In general it may be said that each somatic nucleus has a specific size that depends partly on its DNA content and mainly on its protein content, and that size is related to functional activity during the period of nondivision.

Almost all cells are *mononucleate*, but *binucleate* cells (some liver and cartilage cells) and *polynucleate* cells also exist. In the *syncytia*, which are large protoplasmic masses not divided into cellular territories, the nuclei may be extremely numerous. Such is the case with striated muscle fibers and certain algae, which may contain several hundred nuclei.

The growth and development of every living organism depends on the growth and multiplication of its cells. In unicellular organisms, cell division is the means of reproduction, and by this process two or more new individuals arise from the mother cell. In multicellular organisms, new individuals develop from a single primordial cell, the *zygote;* it is the multiplication of this cell and its descendants that determines the development and growth of the individual.

In many instances cells appear to grow to a certain size before division occurs. This process is repeated in the two daughter cells, so that the total volume eventually becomes four times that of the original cell. The growth of living material progresses rhythmically and according to a geometric progression that has been expressed as follows:

$$\frac{Mn}{Mc}, \frac{2Mn}{2Mc}, \frac{4Mn}{4Mc}, \frac{8Mn}{8Mc}, \text{ etc.}$$

where Mn is the *nuclear mass,* and Mc is the *cytoplasmic mass* of the cells. The two masses are in a state of optimum equilibrium, the so-called *nucleoplasmic index* (NP), which is expressed numerically as:

$$NP: \frac{Vn}{Vc - Vn}$$

where Vn is the *nuclear volume,* and Vc is the *cell volume.*

In general, every cell has two major periods in its life cycle; *interphase* (nondivision) and *division* (which produces two daughter cells). This cycle is repeated at each cell generation, but the length of the cycle varies considerably in different types of cells. The essential function of the nucleus is to store and make available to the cell the information present in its DNA molecule(s). This molecule duplicates during a special period of interphase called the *S phase* (or synthetic phase) in preparation for cell division (Fig. 1–10). During interphase the genetic information is also *transcribed* into different RNA molecules (messenger, ribosomal, and transfer RNAs) which, after passing into the cytoplasm, will *translate* the genetic information by facilitating the synthesis of specific proteins. Thus, the essential roles of the nucleus are (1) the storage of genetic information, (2) DNA duplication, and (3) transcription.

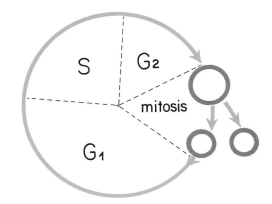

Fig. 1–10. Diagram of the life cycle of a cell indicating the mitotic and the interphase periods. Interphase is composed of the G_1, S, and G_2 phases. Duplication of DNA takes place during the synthetic or S phase.

In fixed and stained material the structure of the nucleus is distinguished by its complexity. The following structures are generally recognized in the interphase nucleus (see Fig. 1–8). A *nuclear envelope* composed of two membranes and perforated at intervals by the *nuclear pores.* (2) The *nucleoplasm* (or *nuclear sap*) that fills most of the nuclear space. This material represents regions of uncondensed *chromatin* (a substance consisting mainly of DNA and protein), the dispersed form assumed by the chromosomes in the nondividing cell. These regions correspond to the *euchromatin* (Gr., *eu*, true). (3) The *chromocenters* that (along with the twisted filaments of chromatin) represent parts of the chromosomes that remain condensed at interphase. These condensed regions of *heterochromatin* are frequently found near the nuclear envelope and are also attached to the *nucleolus.* (4) The *nucleoli,* which are generally spheroidal and very large in cells that are active in protein synthesis. The nucleoli are either single or multiple, and their role is to synthesize and to assemble the RNA molecules and numerous proteins that make up the ribosome before this organelle passes into the cytoplasm.

During cell division the nucleus undergoes a series of complex but remarkably regular and constant changes in which the nuclear envelope and the nucleolus disappear and the chromatin becomes condensed into dark-staining bodies, the *chromosomes* (Gr., *chroma,* color + *soma,* body). Chromosomes are always present in the nucleus even though they are not usually visible during interphase. A considerable amount is known about the structure of the chromatin fiber. Within the nucleus the DNA molecule is associated with basic proteins—the *histones*—to form very fine granular structures of about 8.5 nm, called *nucleosomes* (or nuclear bodies). Strings of nucleosomes constitute the thicker chromatin fiber, which has a mean diameter of 30 nm (see Chapter 13). Prior to cell division, as mentioned above, the chromatin fibers become coiled more tightly and can be seen under the light microscope as chromosomes (Fig. 1–11).

Mitosis and Meiosis—Essentials

It is important to introduce at this point the essentials of mitosis and meiosis, which are studied in detail in Chapters 15 and 16. We mentioned that all organisms that reproduce sexually develop from a single cell, the *zygote*, produced by the union of two cells, the *germ cells* or *ga-metes* (a *spermatozoon* from the male and an *ovum* from the female). The union of an egg and a sperm is called *fertilization*.

Every cell of an individual, with the exception of the *gametes*, contains the same number of chromosomes and hence the same amount of DNA and number of genes. In the somatic cells of a plant or an animal, chromosomes are paired, one member of each pair originally derived from one parent; the other member, from the other parent. Each member of a pair of chromosomes is called a homologue.

Man has 46 chromosomes or 23 pairs; the onion has 8 pairs; the toad, 11 pairs; the mosquito, 3 pairs; and so on. Homologues in each pair are alike, but the pairs are generally different from one pair to another. The original or *diploid* number of each somatic cell is preserved during successive cell divisions. Only in the gametes is the number reduced to half (the *haploid* number).

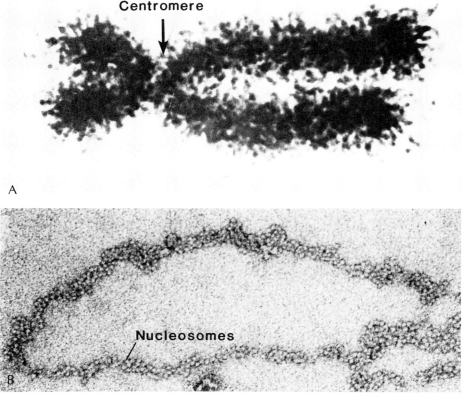

Fig. 1–11.   *A,* low power electron micrograph of a whole-mount of human chromosome 12 showing the two chromatids composed of thick (30 nm) chromatin fibers. The two chromatids are only joined at the centromere. (Courtesy of E.J. Dupraw.) *B,* higher power electron micrograph of a thick chromatin fiber from a lysed interphase nucleus. Observe that the fiber is made of closely packed nucleosomes. Negatively stained preparation. (Courtesy of J.B. Rattner and B.A. Hamkalo.)

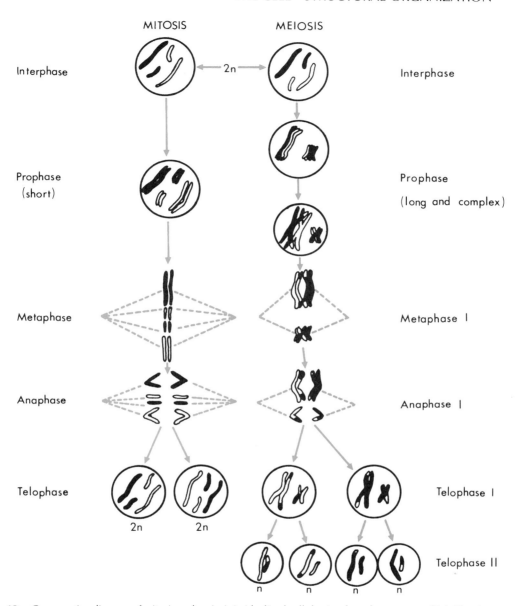

Fig. 1–12. Comparative diagram of mitosis and meiosis in idealized cells having four chromosomes (2n). The chromosomes belonging to each progenitor are represented in white and black. In mitosis the division is equational, and in meiosis it is reductional, the two divisions giving rise to four cells having only two chromosomes (n). In meiosis there is, in addition, an interchange of black and white segments of the chromosomes.

Mitosis—Maintenance of Chromosomal Continuity and Diploid Number

The continuity of the chromosomal set is maintained by a special type of *cell division*, which is called *mitosis*. At the time of cell division the nucleus becomes completely reorganized, as illustrated in Figure 1–12. Mitosis takes place in a series of consecutive stages known as prophase, metaphase, anaphase, and telophase. In a somatic cell the nucleus divides by mitosis in such a fashion that each of the two daughter cells receives exactly the same number and kind of chromosomes that the parent cell had.

Figure 1–12 represents two pairs of homologous chromosomes in a diploid nucleus. Each chromosome duplicates some time during *interphase* before the visible mitotic process begins. At this stage and at early *prophase* chromosomes appear as extended and slender threads. At late prophase chromosomes become short, compact rods by the packing of the nucleoprotein fibers. A spindle arises between the two centrioles and the chromosomes line up across the equatorial plane of the spindle at the *metaphase* plate. At this time each chromosome is made of two filaments called chromatids. At *anaphase* each chromatid separates, forming two daughter chromosomes, which go to opposite poles of the cell. Finally, at *telophase* the daughter chromosomes at each pole again become dispersed, and two daughter nuclei are formed.

In mitosis the original chromosome number is preserved during the successive nuclear divisions. Since the somatic cells are derived from the zygote by mitosis, they all contain the normal double set, or diploid number *(2n)*, of chromosomes.

Meiosis—Reduction of Chromosomal Number to Haploid Set

If the gametes (ovum and spermatozoon) were diploid, the resulting zygote would have twice the diploid chromosome number. To avoid this, each gamete undergoes a special type of cell division called *meiosis*, which reduces the normal diploid set of chromosomes to a single *(haploid)* set *(n)*. Thus, when the ovum and spermatozoon unite during fertilization, the resulting zygote is diploid. The meiotic process is characteristic of all plants and animals that reproduce sexually, and it takes place in the course of *gametogenesis* (Fig. 1–12).

Meiosis produces the reduction of the chromosome number by means of two nuclear divisions, the *first* and *second meiotic divisions*, that involve only a single division of the chromosomes.

The essential aspects of the process are simple. The homologous chromosomes pair longitudinally, forming a bivalent. Each chromosome is composed of two *chromatids*. The bivalent thus contains four chromatids and is also called a *tetrad*. In the tetrad, one chromatid of the homologue has a pairing partner. Portions of these paired chromatids may be exchanged from one homologue to the other, giving rise to cross-shaped figures, called *chiasmata*. The chiasma is a cytologic manifestation of an underlying genetic phenomenon called *crossing over*.

At metaphase I the bivalents arrange themselves on the spindle, and at anaphase I the homologous chromosomes and their two associated chromatids migrate to opposite poles. Thus, in the first meiotic division the homologous pair of chromosomes are segregated. After a short interphase the two chromatids of each homologue separate in the second meiotic division, so that the original four chromatids are distributed into each of the four gametes. The result is four nuclei with only a single set (haploid) of chromosomes (Fig. 1–12).

SUMMARY:
Essentials about Nucleus and Chromosomes

Multicellular organisms develop by the division of an initial diploid cell, the *zygote*, which is a product of the union of the haploid gametes (i.e., fertilization). Every cell has a life cycle, which is composed of a period of nondivision (interphase), and a period of division (generally, mitosis), The interphase period has a synthetic phase, S, in which the DNA duplicates. This S phase is preceded and followed by the G_1 and G_2 phases.

The size of the interphase nucleus is proportional to the ploidy of chromosomes. By observing a section of the liver, a student may recognize a few larger nuclei corresponding to tetra- and octoploid cells.

The following structures are recognized in the interphase nucleus: (1) the *nuclear envelope*; (2) the *nucleoplasm*, formed by uncondensed regions of chromatin (euchromatin); (3) *chromocenters* and other condensed portions of chromatin (heterochromatin); and (4) nucleoli. The chromosomes are always present in the nucleus, but during interphase they are dispersed, and they are generally not morphologically distinguishable. Each diploid cell has a number of homologous chromosome pairs (23 pairs in human). One chromosome of each pair comes from one parent.

The continuity of the diploid set of chromosomes is maintained by mitosis. Duplication of chromosomes occurs during the S phase of interphase; however, by means of *mitosis* (prophase, metaphase, anaphase, and telophase), the daughter chromosomes are distributed equally into the two daughter cells.

Meiosis is a special type of cell division found in germinal cells, by which the number of chromosomes is reduced to the haploid number. Meiosis involves two consecutive divisions without duplication of the chromosomes. The essential feature of meiosis is a long prophase during which the homologous chromosomes pair, forming bivalents. Each bivalent has four chromatids (tetrad). Parts of the paired homologous chromatids interchange at cross-points called *chiasmata*, which are manifestations of an underlying genetic recombination (crossing over). After the second meiotic division, the four resulting cells (gametes) are haploid.

1–7 THE ULTRASTRUCTURE OF THE CYTOPLASM

The cytoplasmic compartment of the eukaryotic cell has a very complex structural organization. Examination of this compartment in the electron miscroscope reveals a strikingly prodigious network of membranes. The *endomembrane system* pervades the ground cytoplasm, dividing it into numerous sections and subsections. This system is so polymorphous that it is difficult to describe and categorize. The cytoplasm is generally considered to have two parts, one contained within this membrane system and the other, the *cytoplasmic matrix proper*, remaining outside.

The most important constituents of the cytoplasm are in the matrix (ground cytoplasm), which lies outside the endomembrane system. This matrix constitutes the true internal milieu of the cell and contains all the principal structures involved in cell shape and movement, protein synthesis, and metabolic activity.

The Cytoskeleton—Microtubules, Microtubular Organelles, Microfilaments, and Intermediate Filaments

The microtubules and a variety of microfilaments and microtrabeculae present in the cytoplasmic matrix constitute a kind of dynamic and spongy cytoskeleton, providing a framework for soluble proteins, enzymes, and ribosomes (Fig. 1–13). This cytoskeleton is involved in the maintenance of cell shape, and it has a role in cell mobility and the colloidal changes that the cytoplasm may undergo.

Microtubules are thin, rather rigid tubular structures about 25 nm in diameter the walls of which are made of 13 individual filaments. The main protein comprising microtubules is *tubulin*. Cytoplasmic microtubules can change rapidly by a process of polymerization or depolymerization of the tubulin subunits (see Fig. 6–14). Microtubules facilitate changes of cell shape as well as the displacement of macromolecules and organelles throughout the cytoplasm. Microtubules can produce dynamic structures related to cell division, such as asters and the spindle, and even more complex organelles, such as centrioles, basal bodies, cilia, and flagella.

Microtubules are the main components of the *asters* and the *spindle*, which integrate the mitotic apparatus during cell division. These organelles appear and disappear at the appropriate stages of mitosis (prophase and telophase, respectively; see Chapter 15).

Centrioles are cylindrical structures of about 0.2×0.5 µm. They are open at both ends, and their walls contain nine groups of microtubular triplets arranged in a circle. Centrioles are generally double and disposed at right angles to one another. During mitosis, centrioles migrate to the poles of animal cells and are apparently involved in the formation of the spindle; plant cells lack centrioles, but the spindle is formed without their aid.

Basal bodies or *kinetosomes* are structurally similar to the centrioles but are located at the base of cilia and flagella. *Cilia* are short (3 to 10 µm) processes that extend into the surrounding medium. *Flagella* are longer than cilia but have the same diameter (0.5 µm) and structure. The axis of these organelles, the *axoneme*, is surrounded by the plasma membrane and has a characteristic microtubular structure consisting of nine doublets in a circle and two in the center. This $9 + 2$ organization differs slightly from that of the centrioles and basal bodies, which have nine triplets in a circle and none in the center (see Chapter 6). Cilia are used for locomotion in isolated cells, such as certain protozoa, or to move particles in the medium, as in air passages or the oviduct. Flagella are generally used for locomotion of cells, such as the spermatozoon.

Microfilaments are among the smallest structures observed with the EM. They are thin (4 to 6 nm) and composed mainly of the protein *actin*. These microfilaments may be associated with *myosin* and other proteins related to the process of contraction. It is now recognized that essentially all the manifestations of cell motility in nonmuscle cells involve the interaction of actin and myosin in the microfilaments.

The cytoskeleton also contains filaments of an intermediate size (10 nm) i.e., in between the size of microtubules and microfilaments. These *intermediate filaments* contain various fibrous proteins and their role is mainly mechanical.

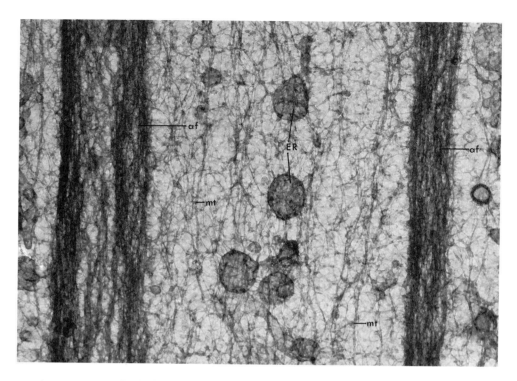

Fig. 1–13.   Electron micrograph of a thin section of a cultured cell made with the high-voltage EM at 10⁶ volts. Two bundles of actin filaments *(af)* run vertically. There are vesicles of the endoplasmic reticulum *(ER)* and a few microtubules *(mt)*. Observe the lattice of microtrabeculae that pervades the matrix × 40,000. (Courtesy of K.R. Porter. From Porter, K.R.: Introduction: Motility in cells. *In* Cell Motility. Vol. 1. Edited by R.D. Goldman, et al. Cold Spring Harbor, New York, Cold Spring Harbor Laboratory, 1976.)

The Endomembrane System—Nuclear Envelope, Endoplasmic Reticulum, and Golgi Complex

Figure 1–8 illustrates the possible continuities and functional interconnections of these different portions of the cytoplasmic membrane system.

The *nuclear envelope* is made of flattened sacs or *cisternae*, which consist of two membranes. These merge at the *pores*, openings that allow the transfer of materials between the nucleus and the cytoplasm. The inner membrane is in contact with the chromatin fibers, while the outer membrane is covered with ribosomes.

The *endoplasmic reticulum (ER)* constitutes the bulk of the endomembrane system (Fig. 1–14). It is made of tubules and flattened sacs. The outer surface of the *rough* or *granular* ER is covered with ribosomes which synthesize protein molecules that are delivered into the lumen, or cavity, of the reticulum. The *smooth* or *agranular* ER is in continuity with the rough

and is engaged, along with the former, in the transport of products within their cavities. Both parts of the ER are involved primarily in secretion.

The *Golgi complex* is a differentiated part of the endomembrane system. It is made up of dictyosomes, which are stacks of flattened sacs and vesicles (Fig. 1–14). This complex is involved in the processing and packaging of secretory products that come through the ER and, as secretory vesicles or granules, are released from the cell by exocytosis. The ER and the Golgi complex are also involved in the formation of lysosomes and peroxisomes.

Membrane Organelles—Mitochondria, Chloroplasts, Lysosomes, and Peroxisomes

As indicated in Table 1–4, the cytoplasm contains a variety of organelles (i.e., rather permanent structures with definite functions), several of which are membrane-bound.

Mitochondria are present in nearly all eukar-

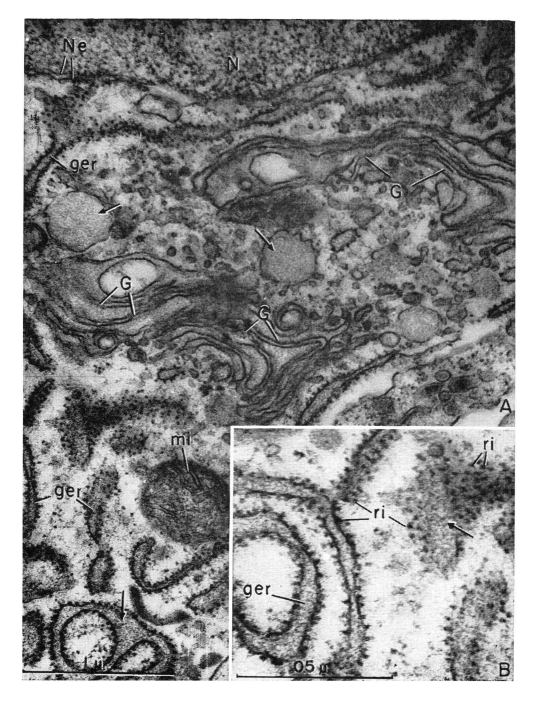

Fig. 1–14. Electron micrograph of a plasma cell showing near the nucleus *(N)* a large Golgi complex (G) formed of flat cisternae and small and large vesicles. Some of the large vesicles *(arrows)* are filled with material. Surrounding the Golgi complex is abundant granular endoplasmic reticulum *(ger)* having cisternae filled with amorphous material *(arrows)*. *mi,* mitochondrion; *Ne,* nuclear envelope; *ri* ribosomes. × 48,000; inset × 100,000. (From E. De Robertis and A. Pellegrino de Iraldi, unpublished).

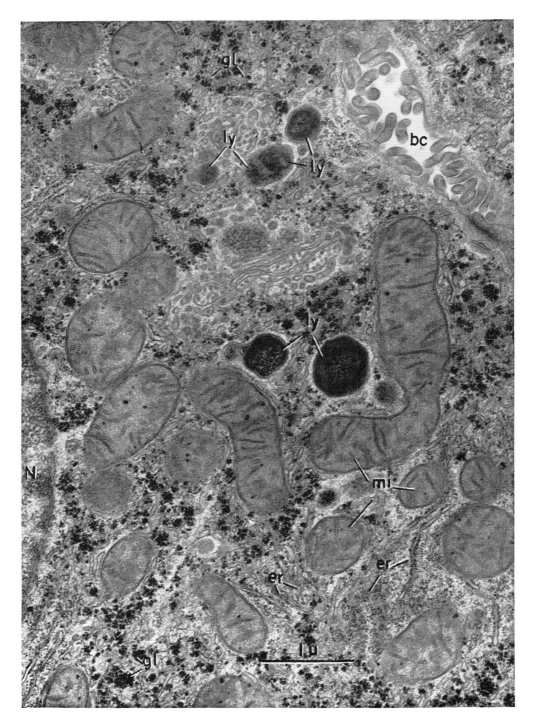

Fig. 1–15.   Peripheral region of a liver cell, showing a biliary capillary *(bc)* and several bodies interpreted as lysosomes *(ly)*, *er*, endoplasmic reticulum; *gl*, glycogen, *mi,* mitochondria; *N,* nucleus. × 31,000. (Courtesy of K.R. Porter.)

yotic cells in the form of cylindrical structures less than 1 μm in diameter. They consist of a double membrane. The inner membrane is folded into the *cristae* or *crests* (Fig. 1–15). These crests partially subdivide the inner chamber of the mitochondrion, and their inner surface is covered with mushroom-like projections that are related to phosphorylation, a portion of the process of cell respiration. Within the *matrix* of the mitochondria are numerous soluble enzymes that are involved in the *Krebs cycle*, the first step in the aerobic utilization of energy from nutrients.

Plant cells contain a variety of organelles; generally called *plastids*, which are not present in animal cells. Some of them, such as the leukoplasts, are colorless and participate mainly in the storage and metabolism of starch. Other plastids contain various pigments and are collectively called *chromoplasts*. The most important of these are the *chloroplasts*, which contain the green pigment *chlorophyll* (Fig. 1–1). The chloroplast has a double outer membrane, a *stroma* filled with many soluble enzymes, and a complex system of membrane-bound compartments. Chloroplasts are the site of photosynthesis, by which a plant is able to capture light energy and, using H_2O and CO_2, to synthesize a variety of compounds that are stored and eventually used as foodstuffs.

Both mitochondria and chloroplasts contain a kind of semiautonomous genetic system with its own DNA. They also contain ribosomes and RNA molecules, and are able to synthesize some of their own proteins.

Lysosomes are polymorphous organelles enclosed by a single membrane. They contain a vast array of hydrolytic enzymes (Fig. 1–15). These enzymes are responsible for the digestion of foreign substances that have been incorporated into the cell by endocytosis, and also for the digestion of parts of the cytoplasm. Lysosomes originate from the ER and the Golgi complex.

Peroxisomes are also bound by a single membrane. They contain enzymes related to the production and breakdown of peroxides (H_2O_2), and they fulfill a protective function in the cell.

SUMMARY:
Ultrastructure of the Cytoplasm

Eukaryotes are generally more complex than prokaryotes, consisting of three main components

(which are subdivided further): the nucleus, the cytoplasm, and the cell membrane. The shape, size, and volume of a eukaryotic cell can all vary considerably, but are ultimately determined by the specific function of the cell.

The cell or plasma membrane is a continuous lipid bilayer with proteins intercalated in or adherent to both of its surfaces. Its main function is to exercise selective control over the passage of materials in and out of the cell.

The cytoplasmic matrix is pervaded by a complex and polymorphic endomembrane system that is involved in the transport of materials within the cell. The principal structures responsible for the other activities of the cell are located within the matrix.

The cell is supported and mobilized by a variety of microtubules and microfilaments made up of the proteins tubulin, actin, and myosin. These structures are generally involved in dynamic processes such as changes of cell shape, the partitioning of chromosomes at cell division, and muscle contraction.

The endomembrane system consists of the nuclear envelope, the smooth and rough ER, and the Golgi complex. This system is engaged in the sequestering, packaging, and transport of substances produced either for use within the cell or for export to the medium.

The membrane organelles are generally involved in cellular metabolic processes. Mitochondria and chloroplasts are the sites of energy production; lysosomes are responsible for the digestion of substances within the cell; and peroxisomes fulfill a similar protective function.

Microtubules are the major structural components of several cell organelles including the asters, spindle, and centriole, which are involved in mitosis, and cilia and flagella, which are used for cellular locomotion or to move particles in the surrounding medium.

1–8 LITERARY SOURCES IN CELL AND MOLECULAR BIOLOGY

The preceding considerations of the present scope of cell biology demonstrate why the sources of literature are wide and multidisciplinary. Current studies are presented at scientific meetings and are published in specialized periodicals. Of the long list of literary sources that could be made, only a few of the more specific ones are mentioned: *Cell, Cell and Tissue Research, Chromosoma, Cytogenetics and Cell Genetics, Developmental Biology, Experimental Cell Research, Heredity, Hereditas, Journal of Cell Biology, Journal of Cell Science, Journal of Cellular Physiology, Journal of General Physiology, Journal of Molecular Biology, Journal of*

Supramolecular Structure, and Journal of Ultrastructure Research. Papers on cell biology are frequently published in more general periodicals, such as *American Scientist; Comptes Rendus Hebdomadaires des Seances de l'Academie des Sciences, Sciences Naturelles; Nature; Naturwissenchaften; Experientia; Proceedings of the National Academy of Sciences (Wash.); Proceedings of the Royal Society [London, Series B: Biological Sciences];* and *Science;* or even in such specialized publications as *Biochimica et Biophysica Acta, Biochemical Journal,* and *Journal of Biological Chemistry.*

Reviews of recent advances are found in the: *Advances in Cell and Molecular Biology, Advances in Genetics, Biological Reviews of the Cambridge Philosophical Society, Physiological Reviews, Plant Physiology, Protoplasma, Quarterly Review of Biology, International Review of Cytology, Scientific American, La Recherche,* and others.

Very useful in the compilation of a bibliography are special journals that give titles of papers or abstracts of the literature. These journals include: *Berichte über die Wissenschaftliche Biologie, Biological Abstracts, Chemical Abstracts, Excerpta Medica,* and *Index Medicus.* Journals such as *Current Contents and Bulletin Signalétique du Conseil des Recherches* publish titles of all papers that appear in various journals.

In addition there are many monographs, compendia, and textbooks that will be mentioned in later chapters of this book and which cover various specialized subjects of cell biology. The widest coverage is found in *The Cell: Biochemistry, Physiology, Morphology,* 6 volumes, edited by J. Brachet and A.E. Mirsky, 1959–1961); in *Handbook of Molecular Cytology,* edited by A. Lima-de-Faría (1969); and in *Cell Biology: A Comprehensive Treatise,* edited by Goldstein and Prescott (1977).

ADDITIONAL READINGS

Baumeister, W: Biological horizons in molecular microscopy. Cytobiologie, *17*:246, 1978.

Bernal, J.D., and Synge, A.: The origin of life. *In* Readings in Genetics and Evolution. London, Oxford University Press, 1973.

Beveridge, T.J.: Ultrastructure, chemistry and function of bacterial wall. Int. Rev. Cytol., *72*:229, 1981.

Brachet, J., and Mirsky, A.E.: The Cell: Biochemistry, Physiology, Morphology. New York, Academic Press, Inc., 1959–1961.

Claude, A.: The coming of age of the cell. Science, *189*:433, 1975.

De Robertis, E., and De Robertis, E.M.F., Jr.: Essentials of Cell and Molecular Biology. Philadelphia, Saunders College Pub., 1981.

Diener, T.O.: Viroids. Sci. Am., *244*:66, 1981.

Fawcett, D.W.: The Cell: Its Organelles and Inclusions. 2nd Ed. Philadelphia, W.B. Saunders Co., 1979.

Giese, A.C.: Cell Physiology. 5th Ed. Philadelphia, W.B. Saunders Co., 1979.

Goldstein, L., and Prescott, D.M.: Cell Biology: A Comprehensive Treatise. Vol. 1. New York, Academic Press, 1977.

Hayflick, L.: The cell biology of human aging. Sci. Am., *242*:58, 1980.

Hess, E.L.: Origins of molecular biology. Science, *168*:664, 1970.

Jacob, F.: Evolution and tinkering. Science, *196*:1161, 1977.

Johnson, J.E.: Aging and Cell Structure. Vol. 1. New York, Academic Press, 1982.

Karp, G.: Cell Biology. New York, McGraw-Hill Book Co., 1979.

Lima-de-Faría, A.: Handbook of Molecular Cytology. Amsterdam, North-Holland Pub. Co., 1969.

Margulis, L.: Symbiosis and evolution. Sci. Am., *225*:48, 1971.

Margulis, L., and Schwartz, K.V.: Five kingdoms. An illustrated guide to the phyla of life on earth. New York, W.H. Freeman & Co., 1982.

Monod, J.: Chance and Necessity. New York, Random House, Inc., 1971.

Schwartz, R., and Dayhoff, M.: Origins of prokaryotes, eukaryotes, mitochondria and chloroplasts. Science, *199*:395, 1978.

Swanson, C.P., and Webster, T.: The Cell. 4th Ed. Englewood Cliffs, New Jersey, Prentice-Hall, 1977.

Vidal, G.: The oldest eukaryotic cells. Sci. Am., *250*:48, 1984.

Watson, J.D.: Molecular Biology of the Gene. 3rd Ed. New York, W.A. Benjamin, Inc., 1975.

Woese, C.R., and Fox, G.E.: Phylogenetic structure of the prokaryotic domain: The primary kingdoms. Proc. Natl. Acad. Sci. USA, *74*:5088, 1977.

Wolfe, S.L.: Biology of the Cell. Belmont, California, Wadsworth Publishing Co., Inc., 1972.

MOLECULAR ORGANIZATION
OF THE CELL

The structure of the cell, which is visible with optical and electron microscopes, is the result of molecules arranged in a very precise order. Although there is still much to be learned, the general principles of the molecular organization of some cell structures, such as membranes, ribosomes, chromosomes, mitochondria, and chloroplasts are emerging, as will be shown in the corresponding chapters. The biology of cells is inseparable from that of molecules because, in the same way that cells are the building blocks of tissues and organisms, molecules are the building blocks of cells.

An early approach to the study of the chemical composition of the cell was the biochemical analysis of whole tissues, such as the liver, brain, skin, or plant meristem. This method had limited value because the material analyzed was generally a mixture of different cell types and, in addition, contained extracellular material. In recent years the development of cell fractionation methods and various micromethods has led to the isolation of different subcellular particles and thus to more precise information about the molecular architecture of the cell (see Chapter 3).

The chemical components of the cell can be classified as *inorganic* (water and mineral ions) and *organic* (proteins, carbohydrates, nucleic acids, and lipids).

Cells contain 75 to 85% water, 10 to 20% protein, and 2 to 3% inorganic salts. Among these molecules, the organic compounds derived from the chemistry of carbon atoms stand out as the molecules of life. Numerous cell structures are made up of large molecules called *macromolecules* or *polymers*, consisting of repeating units (called *monomers*), which are linked together by covalent bonds.

There are three important examples of polymers in living organisms. (1) *Nucleic acids* result from the repetition of four different units called *nucleotides*. The linear sequence of the four nucleotides in the DNA molecule is the ultimate source of genetic information. (2) *Polysaccharides* can be polymers of glucose, forming starch, cellulose, or glycogen, or may also involve the repetition of other molecules to form more complex polysaccharides. (3) *Proteins* or *polypeptides* are long chains made up of 20 different amino acids, linked together by *peptide bonds*. The order in which these 20 monomers can be linked gives rise to an astounding number of combinations in different protein molecules, and determines not only their specificity but also their biological activity.

In addition to emphasizing the stereochemical characteristics of these main constituents of the cell, in this chapter we will introduce the study of enzymes as molecular machines used by the cell to produce chemical transformations. We will also cover some of the mechanisms used to regulate their activity within the cell.

This study will be complemented with some notions about how macromolecules can be assembled and organized into more complex supramolecular structures, i.e., ones that are visible under the electron microscope (EM). These principles of molecular assembly were probably at work during the period of chemical and bio-

logical evolution that finally gave rise to the first living cell. For this reason, this introductory chapter will end with some speculations about the possible origin of prokaryotic and eukaryotic cells, a problem directly related to the origin of life on our planet.

This chapter should be considered as an elementary introduction to the understanding of cell and molecular biology. A comprehensive review of the field is the substance of biochemistry textbooks.

2–1 WATER, SALTS, IONS, AND TRACE ELEMENTS IN CELLS

With few exceptions, such as seeds, bone, and enamel, water is the most abundant component of cells. The water content of an organism is related to the organism's age and metabolic activity. For example, it is highest in the embryo (90 to 95%) and decreases progressively in the adult and in the aged. Water serves as a natural solvent for mineral ions and other substances and also as a dispersion medium of the colloid system of protoplasm. Water molecules also participate in many enzymatic reactions in the cell and can be formed as a result of metabolic processes.

Water exists in the cell in two forms: *free* and *bound*. *Free water* represents 95% of the total cellular water and is the principal part used as a solvent for solutes and as a colloid dispersion medium. *Bound water*, which represents only 4 to 5% of the total cellular water, is loosely held to the proteins by hydrogen bonds and other forces. Because of the asymmetrical distribution of charges a water molecule acts as a *dipole*, as shown in the following diagram.

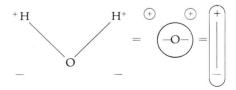

Because of this polarity, water can bind electrostatically to both positively and negatively charged groups in the protein. Thus, each amino group in a protein molecule is capable of binding 2.6 molecules of water.

Water is also used to eliminate substances from the cell and to absorb heat—by virtue of its high specific heat coefficient—thus preventing drastic temperature changes in the cell.

Another important property of water is its *ionization* into a hydroxide anion (OH⁻) and a hydrogen ion or proton (H⁺). At standard conditions of temperature and pressure only 1.0×10^{-7} mol/L of water molecules are dissociated. To this 10^{-7} concentration of H⁺ corresponds the neutral pH 7.

Salts dissociated into anions (e.g., Cl⁻) and cations (e.g., Na⁺ and K⁺) are important in maintaining *osmotic pressure* and the *acid-base equilibrium* of the cell. Retention of ions produces an increase in osmotic pressure and thus the entrance of water. Some of the inorganic ions, such as magnesium, are indispensable as cofactors in enzymatic activities; others, such as inorganic phosphate, form adenosine triphosphate (ATP), the chief supplier of chemical energy for the living processes of the cell, through oxidative phosphorylation.

The concentration of various ions in the intracellular fluid differs from that in the interstitial fluid (see Table 4–1). For example, the intracellular fluid has a high concentration of K⁺ and Mg⁺⁺, while Na⁺ and Cl⁻ are localized mainly in the interstitial fluid. The dominant anion inside cells is phosphate; some bicarbonate is also present.

Calcium ions are found in the circulating blood and in cells where they play important regulatory roles. In bone they combine with phosphate and carbonate ions to form a crystalline arrangement.

Phosphate occurs in the blood and tissue fluids as a free ion, but much of the phosphate of the body is bound in the form of phospholipids, nucleotides, phosphoproteins, and phosphorylated sugars. As primary phosphate ($H_2PO_4^-$) and secondary phosphate (HPO_4^{2-}), phosphate contributes to the acid-base equilibrium, thereby buffering the pH of the blood and tissue fluids.

Other ions found in tissues are sulfate, carbonate, bicarbonate, magnesium, and amino acids.

Certain *mineral ions* are also found as part of larger macromolecules. For example, *iron*, bound by metal-carbon linkages, is found in hemoglobin, ferritin, the cytochromes, and some enzymes (such as catalase and cytochrome oxidase). Traces of *manganese, copper, cobalt, iodine, selenium, nickel, molybdenum,* and *zinc*

TABLE 2–1. FUNCTIONS OF SOME INORGANIC IONS IN CELLS

Ions	Function in
Fe⁺⁺ or Fe⁺⁺⁺	Hemoglobin, cytochromes, peroxidases
Mg⁺⁺	Chlorophyll, phosphatases
Cu⁺⁺	Tyrosinase, ascorbic acid oxidase
Zn⁺⁺	Carbonic anhydrase, peptidase, alcohol dehydrogenase, Transcription Factor IIIA
Mn⁺⁺	Peptidases
Co⁺⁺	Peptidases
Mo	Nitrate reductase, xanthine oxidase
Ca⁺⁺	Calmodulin, actomyosin, ATPase

are indispensable for maintenance of normal cellular activities. The importance of some of these elements in enzymes is indicated in Table 2–1.

2–2 NUCLEIC ACIDS

Nucleic acids are macromolecules of the utmost biological importance. All living organisms contain nucleic acids in the form of deoxyribonucleic acid (DNA) and ribonucleic acid (RNA). Some viruses contain only DNA, while others have only RNA.

DNA is the major store of genetic information. This information is copied or *transcribed* into RNA molecules, the nucleotide sequences of which contain the "code" for specific amino acid sequences. Proteins are then synthesized in a process involving the *translation* of the RNA. The series of events just outlined is often referred to as the *central dogma* of molecular biology, and can be summarized in the form:

$$\text{DNA} \xrightarrow{\text{transcription}} \text{RNA} \xrightarrow{\text{translation}} \text{Protein}$$

The biological role of nucleic acids is discussed in detail in the chapters dealing with gene expression. However, the features of their chemical structure described here are essential for the understanding of the function of nucleic acids.

In higher cells, DNA is localized mainly in the nucleus, within the chromosome. A small amount of DNA is present in the cytoplasm and contained in mitochondria and chloroplasts. RNA is found both in the nucleus, where it is synthesized, and in the cytoplasm, where the synthesis of proteins takes place (Table 2–2).

TABLE 2–2. DNA AND RNA—STRUCTURE, REACTIONS, AND ROLE IN THE CELL

	Deoxyribonucleic Acid	Ribonucleic Acid
Localization	Primarily in nucleus; also in mitochondria and chloroplasts	In cytoplasm, nucleolus, and chromosomes
Pyrimidine bases	Cytosine	Cytosine
	Thymine	Uracil
Purine bases	Adenine	Adenine
	Guanine	Guanine
Pentose	Deoxyribose	Ribose
Cytochemical reaction	Feulgen	Basophilic dyes with ribonuclease treatment
Hydrolyzing enzyme	Deoxyribonuclease (DNase)	Ribonuclease (RNase)
Role in cell	Genetic information	Synthesis of proteins
^{3}H Precursor	^{3}H Thymidine	^{3}H Uridine

Nucleic Acids—A Pentose, Phosphate, and Four Bases

Nucleic acids consist of a sugar moiety (pentose), nitrogenous bases (purines and pyrimidines), and phosphoric acid.

A complete hydrolysis of DNA (or RNA) yields:

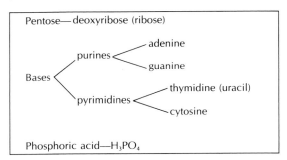

Pentose—deoxyribose (ribose)

Bases
  purines
    adenine
    guanine
  pyrimidines
    thymidine (uracil)
    cytosine

Phosphoric acid—H_3PO_4

A nucleic acid molecule is a linear polymer in which the monomers (nucleotides) are linked together by means of *phosphodiester* "bridges" or bonds (Fig. 2–1). These bonds link the 3' carbon in the pentose of one nucleotide to the 5' carbon in the pentose of the adjacent nucleotide. Thus the backbone of a nucleic acid consists of alternating phosphates and pentoses. The nitrogenous bases are attached to the sugars of this backbone.

As shown in Figure 2–1, the phosphoric acid uses two of its three acid groups in the 3', 5' diester links. The remaining group confers on the polynucleotide its acid properties and enables the molecule to form ionic bonds with basic proteins. (We mentioned in Chapter 1 that the DNA of eukaryotic cells is associated with basic proteins called histones, forming a nucleoprotein complex called chromatin.) This free acid group also causes nucleic acids to be highly *ba-*

sophilic; i.e., they stain readily with basic dyes (see Chapter 3).

A mild hydrolysis cleaves the nucleic acid into component nucleotides that result from the covalent bonding of a phosphate and a heterocyclic base to the pentose.

Pentoses are of two types: *ribose* in RNA, and *deoxyribose* in DNA. The only difference between these two sugars is that there is one less oxygen atom in deoxyribose (Fig. 2–1). A cytochemical reaction specific for the deoxyribose moiety, called the *Feulgen reaction*, can be used to visualize DNA under the microscope (see Chapter 3).

The *bases* found in nucleic acids are also of two types: *pyrimidines* and *purines*. Pyrimidines have a single heterocyclic ring, whereas purines have two fused rings. In DNA the pyrimidines are *thymine* (T) and *cytosine* (C); the purines are *adenine* (A) and *guanine* (G) (Fig. 2–1). RNA contains *uracil* (U) instead of thymine (Table 2–2).

It is useful to remember that there are two main differences between DNA and RNA: DNA has a deoxyribose and RNA a ribose moiety; DNA contains thymine and RNA uracil. The difference in pyrimidine bases has made it possible for cell biologists to use radioactive thymidine as a specific DNA label, and radioactive uridine to label RNA in living cells.

Heterocyclic bases absorb ultraviolet light at a wavelength (designated λ) of 260 nm. As shown in Figure 21–8, a cell photographed at λ = 260 nm, shows the nucleolus (containing RNA), the chromatin, and the RNA-containing regions of the cytoplasm, all of which absorb intensely. In fact, a simple way of determining the concentration of nucleic acid in solution is to measure the absorption at 260 nm.

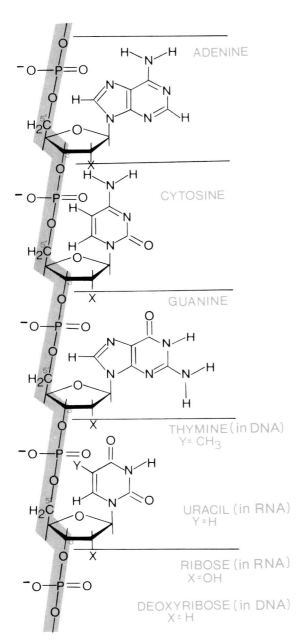

ADENINE

CYTOSINE

GUANINE

THYMINE (in DNA)
Y= CH₃

URACIL (in RNA)
Y = H

RIBOSE (in RNA)
X = OH

DEOXYRIBOSE (in DNA)
X = H

Fig. 2–1. A segment of a single hypothetical nucleic acid chain showing the nucleotides and their constituent parts. The pentose-phosphate backbone is indicated.

The combination of a base plus a pentose, minus the phosphate, constitutes a *nucleoside.* For example, *adenine* is a purine base; *adenosine* (adenine + ribose) is the corresponding nucleoside; while *adenosine monophosphate* (AMP), *adenosine diphosphate* (ADP), and

adenosine triphosphate (ATP) are nucleotides (Fig. 2–2).

In addition to functioning as the building blocks of nucleic acids, nucleotides are important because they are used to store and transfer chemical energy. Figure 2–2 shows that the two terminal phosphate bonds of ATP contain high energy. When these bonds are cleaved, the energy released can be used to drive a variety of cellular reactions. The high energy ~P bond enables the cell to accumulate a large quantity of energy in a small space and keep it ready for use when needed. In Figure 2–2B it is possible to appreciate how the energy stored in ATP can be used in different biosynthetic pathways. Other nucleotides, such as cytosine triphosphate (CTP), uridine triphosphate (UTP), and guanosine triphosphate (GTP), also have high energy bonds but the energy source for all of them is ultimately derived from ATP.

DNA Base Composition—A = T and G = C

DNA is present in living organisms as linear molecules of extremely high molecular weight. *E. coli,* for example, has a single circular DNA molecule of 3,400,000 base pairs and has a total length of 1.4 mm. In higher organisms the amount of DNA may be hundreds of times larger (700 times in the case of man); for example, the DNA in a single human diploid cell, if fully extended, would have a total length of 1.7 m.

All the genetic information of a living organism is stored in the linear sequence of these four bases. Therefore, a four-letter alphabet (A, T, G, C) must code for the primary structure (i.e., the sequence of the 20 amino acids) of all proteins. One of the most exciting discoveries in molecular biology was the elucidation of this code (see Chapter 20). One prelude to this discovery, having direct bearing on the understanding of DNA structure, was the finding that there were predictable regularities in base content. Between 1949 and 1953 Chargaff studied the base composition of DNA in great detail. He found that although the base composition varied from one species to another, in all cases the amount of adenine was equal to the amount of thymine (A = T). The number of cytosine and guanine was also found to be equal (C = G). Consequently, the total quantity of purines equals the total quantity of pyrimidines (i.e., A

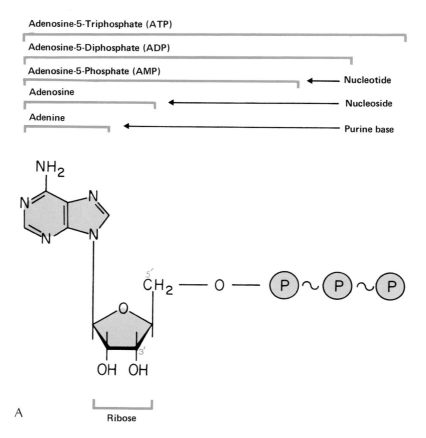

A

Fig. 2–2. *A,* structure of ATP and its components. Note the presence of two high energy phosphate bonds. *B,* channeling of phosphate bond energy by ATP into specific biosynthetic routes. (Courtesy of A.L. Lehninger.)

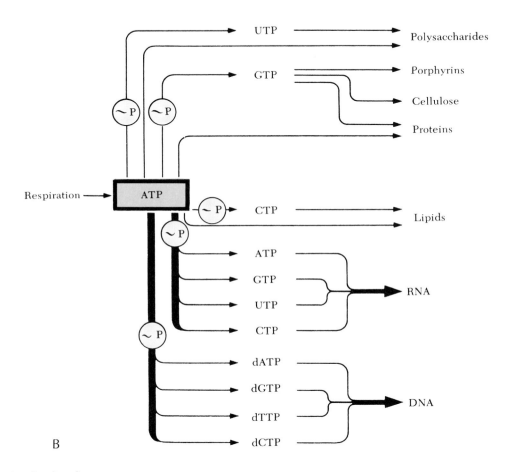

B

Fig. 2–2. *Continued*

+ G = C + T). On the other hand, the AT/ GC ratio varies considerably between species. For example in man the ratio is 1.52, while in *E. coli* the ratio is 0.93.

DNA Is A Double Helix

In 1953, based on the x-ray diffraction data of Wilkins and Franklin, Watson and Crick proposed a model for DNA structure that provided an explanation for its regularities in base composition and its biological properties, particularly its duplication in the cell. The structure of DNA is shown in Figure 2–3. It is composed of two right-handed helical polynucleotide chains that form a *double helix* around the same central axis. The two strands are antiparallel, meaning that their 3′, 5′ phosphodiester links run in opposite directions. The bases are stacked inside the helix in a plane perpendicular to the helical axis.

The two strands are held together by *hydrogen bonds* established between the two sugar moieties in the opposite strands, and only certain base pairs can fit into the structure. As shown in Figure 2–4, the only two pairs that are possible are AT and CG. It is important to note that two hydrogen bonds are formed between A and T, and three are formed between C and G, and that therefore a CG pair is more stable than an AT pair. In addition to hydrogen bonds, hydrophobic interactions established between the stacked bases are important in maintaining the double helical structure.

The *axial sequence* of bases along one polynucleotide chain may vary considerably, but on the other chain the sequence must be *complementary*, as in the following example:

First chain: 5′ T, G, C, T, G, T, G, G, T, 3′
                    ‖   ‖‖   ‖‖   ‖   ‖‖   ‖   ‖‖   ‖‖   ‖
Second chain: 3′ A, C, G, A, C, A, C, C, A, 5′

Because of this property, given an order of bases on one chain, the other chain is exactly complementary. During DNA duplication the two chains dissociate and each one serves as a template for the synthesis of a new complementary chain. In this way two double-stranded DNA molecules are produced, each having exactly the same molecular constitution.

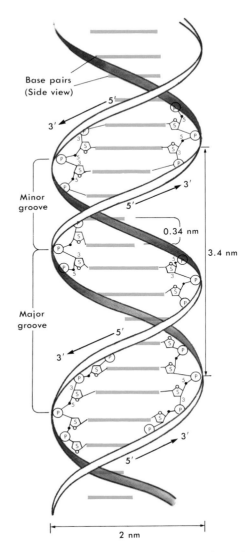

Fig. 2–3. The DNA double helix. The phosphate-ribose backbones are indicated as ribbons. The base pairs are flat structures stacked one on top of another perpendicular to the long axis of DNA, and they are therefore represented as horizontal lines in this side view. The base pairs can be seen from a top view in Figure 2–4. Note that the two strands are antiparallel and that the molecule has a minor and a major groove. The double helix gives one complete turn every ten base pairs (3.4 nm). *P*, phosphate group; *S*, sugar.

Right-Handed B- and A-DNA and Left-Handed Z-DNA

Most of the DNA in a cell is right-handed and of the *B-DNA form*, which is the most stable configuration described above. Some local regions, however, can form a slightly different right-handed DNA called *A-DNA*. Crystallographic studies on synthetic nucleotides con-

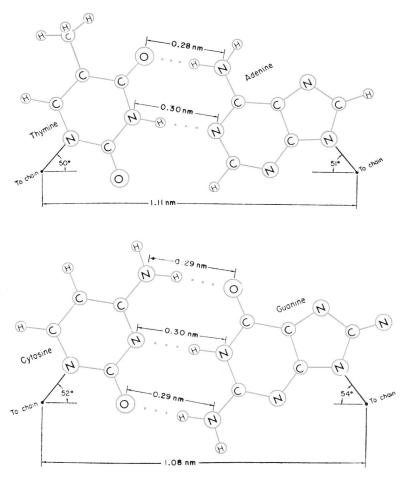

Fig. 2–4. The two base pairs in DNA. The complementary bases are thymine and adenine (T—A) and cytosine and guanine (C—G). Observe that between T—A there are two, and between C—G three, hydrogen bonds. (After studies of L. Pauling and R.B. Corey.)

sisting of alternating purines and pyrimidines have shown that left-handed double helices can also exist. This form has been called Z-DNA because of the zigzag array of the phosphate-ribose backbone, as can be observed in the space-filling models shown in Figure 2–5. In the Z-DNA configuration different parts of the molecule are exposed and this may have biological consequences. In fact, this alternate configuration suggests that DNA is a more flexible molecule than was previously thought and that it can adopt in the genome a variety of forms (Rich, 1980).

DNA Strands Can Be Separated and Annealed

Because the structure of the DNA double helix is preserved by weak interactions (i.e., hy-

drogen bonds and hydrophobic interactions established between the stacked bases), it is possible to separate the two strands by treatments involving heating, for example, or alkaline pH. This separation is called *melting* or *denaturation* of DNA. Since the temperature required to break the GC pairs (having three hydrogen bonds) is higher than that needed to break the AT pairs (having two hydrogen bonds), the temperature at which the DNA strands separate (the *melting point*) depends on the AT/GC ratio.

If the DNA is cooled slowly after denaturation, the complementary strands will base-pair in register, and the *native* (double helical) conformation will be restored. This process is called *renaturation* or *annealing*, and is a consequence of the base-pairing properties of nucleotides.

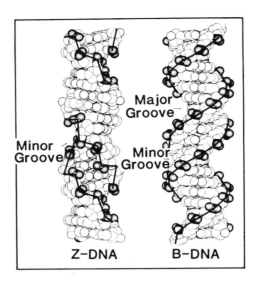

Fig. 2–5. Two molecular forms of the DNA double helix, right-handed B-DNA and left-handed Z-DNA. The heavy black line in each goes from phosphate group to phosphate group, indicating a smooth right-handed helix in B-DNA, but an irregular left-handed helix in Z-DNA. (Courtesy of A. Rich.)

Renaturation of DNA is a very useful tool in molecular biology. Figure 2–6 shows how DNA renaturation can be used to estimate the size (number of nucleotides) of the genome of a given organism. When DNA is renatured under standardized conditions, a large genome (e.g., calf) takes more time to reanneal than a small genome (e.g., *E. coli* or bacteriophage T$_4$). This is because the individual sequences take longer to find the correct partners (the larger the genome,

the more chances there are of incorrect molecular collisions).

Renaturation studies led to the discovery of *repeated sequences* in eukaryotic DNA. When certain DNA sequences are repeated many times, the rate of renaturation will be much faster than for sequences present as single copies. Some sequences (called *satellite* DNAs) can be repeated millions of times in the genome (see Section 22–2).

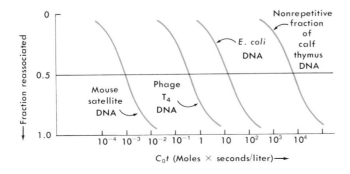

Fig. 2–6. Renaturation kinetics of various DNAs (C_0t curves). The velocity of renaturation is measured in the C_0t units required to obtain ½ renaturation. C_0t (short for Concentration × time) by convention is the initial concentration of DNA (in moles of nucleotides per liter) in the reaction mixture, multiplied by the time (in seconds) during which reannealing was allowed to proceed. A large genome renatures more slowly than a small one. (Redrawn from the classic paper by R.J. Britten and D.E. Kohne: Science, *161*:530, 1968.)

Single-stranded DNA will also anneal to complementary RNA, resulting in a hybrid molecule in which one strand is DNA and the other is RNA. *Molecular hybridization* is a very powerful method for characterizing RNAs because an RNA molecule will hybridize only to the DNA from which it was transcribed.

Labeled probes can be made by using an RNA molecule (e.g., globin messenger RNA), radioactive nucleosides and the enzyme *reverse transcriptase* that copies a complementary DNA strand (cDNA). This cDNA can then be used to localize the gene from which the RNA molecule arose. Furthermore, using the modern methods of genetic engineering described in Chapter 19, almost any gene can now be cloned and used as a probe for hybridization.

Denaturation done under carefully controlled conditions may be used for physical mapping of DNA in a technique known as *partial denaturation mapping*. This technique is based on the fact that the regions rich in AT separate more easily. Under the EM those regions are detected as single-stranded loops, and the distance between the loops and the end of the DNA molecule can be measured (Fig. 2–7).

Circular DNA—Supercoiled Conformation

Many viruses have covalently closed circular DNA molecules that when visualized under the EM show the twisted structure seen in Figure 2–8. This DNA is *supercoiled*, which indicates that the structure is under tension. Supercoiling arises from a relative lack of turns in the DNA double helix.

Let us consider a covalently closed circular DNA molecule with 5000 base pairs. In the *relaxed* conformation (smooth circle in Fig. 2–8), the DNA helix will have 500 helical turns (because the double helix gives one turn every 10 nucleotides). In a supercoiled molecule, however, the helix will have fewer turns (if it has 499, it will have one superhelical coil; if 490, it will have 10 coils). The molecule is under tension and folds upon itself in the supercoiled form. Since the molecule must be covalently closed, supercoils are always present in integral numbers; i.e., a molecule can lack one, two, or

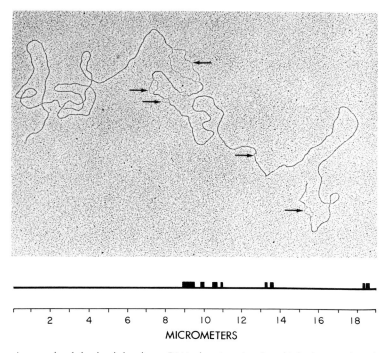

Fig. 2–7. Electron micrograph of the lambda phage DNA showing sites in which denaturation (i.e., separation of the strands) has resulted from the action of an alkaline medium (pH 11). A map of this partial denaturation is presented below the electron micrograph, showing at scale the two ends of the DNA molecules and indicating by rectangles the position and length of the denatured sites. (Courtesy of R.B. Inman. From Inman, R.B., and Schnös, M.G.: Mol. Biol., *49*:93, 1970.)

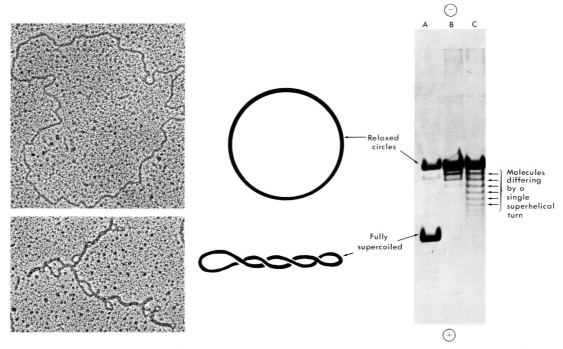

Fig. 2–8. Superhelical turns in circular DNA molecules detected by electron microscopy and gel electrophoresis. DNA from the virus SV 40 was electrophoresed in an agarose gel. The structure of the supercoiled molecules is more compact than the relaxed circles, and therefore it migrates faster. (A) SV 40 DNA sample, containing both relaxed and fully supercoiled molecules, (B) the same DNA was treated with *nicking-closing* enzyme that removes superhelical turns, (C) as in B, but treated with less nicking-closing enzyme so that some—but not all—of the superhelical turns are removed. Bands differing between themselves by only one superhelical turn can be clearly seen. (Electron micrographs courtesy of Sue Whytock, gel electrophoresis courtesy of R.A. Laskey.)

three turns, but not fractional amounts, such as 1.1, 1.2, or 1.3 (half-twists are not possible because the two DNA chains have opposite 5' to 3' polarities).

The number of 10 base pairs per turn of the helix is based on the x-ray diffraction pattern of B-form DNA fibers. In solution, however, recent studies have shown that the DNA helix gives a turn every 10.5 nucleotides instead.

Fully supercoiled DNA has about one superhelical turn every 200 base pairs. It is thought that supercoiling reflects the fact that in eukaryotic cells DNA is normally not free but bound to histone proteins. In vivo the DNA is not under tension and adopts the conformation most favorable for chromatin structure. When the histones are removed during purification of the DNA, the tension becomes apparent, and the molecules supercoil.

RNA Structure—Classes and Conformation

The primary structure of RNA is similar to that of DNA, except for the substitution of ribose for deoxyribose and uracil for thymine, as has already been discussed (see Table 2–2). The base composition of RNA does not follow Chargaff's rules described earlier in this section, since RNA molecules consist of only one chain.

As shown in Table 2–3, there are three major classes of RNA: *messenger* RNA (mRNA), *transfer* RNA (tRNA), and *ribosomal* RNA (rRNA). All are involved in protein synthesis—mRNA carries the genetic information for the sequence of amino acids; tRNA identifies and transports amino acid molecules to the ribosome, and rRNA represents 50% of the mass of ribosomes, the organelles that provide a molecular scaffold for the chemical reactions of polypeptide assembly. The structure and function of these RNAs will be described in detail in the chapters on gene expression.

Although each RNA molecule has only a single polynucleotide chain, RNA is not a simple, smooth, linear structure. RNA molecules have extensive regions of complementarity in which hydrogen bonds between AU and GC pairs are

TABLE 2–3.   MAJOR CLASSES OF RIBONUCLEIC ACIDS IN *E. COLI*

Type	Sedimentation Coefficient	Molecular Weight	Number of Nucleotide Residues	Percent of Total Cell RNA
mRNA	6S to 25S	25,000 to 1,000,000	75 to 3,000	~2
tRNA	~4S	23,000 to 30,000	75 to 90	16
rRNA	5S	~35,000	~100	
	16S	~550,000	~1,500	82
	23S	~1,100,000	~3,100	

formed between different regions of the same molecule. Figure 2–9 shows that as a result the molecule folds upon itself, forming structures called *hairpin loops*. In the base-paired regions the RNA molecule adopts a helical structure comparable to that of DNA. The compact structure of RNA molecules folded upon themselves has important biological consequences. For example, in bacteriophage MS2 a sequence that indicates the starting site for one of its proteins (called the polymerase) is in a part of the molecule inaccessible to the ribosomes, as shown in Figure 2–9, and can only be expressed when additional factors unfold the molecule. In Chapter 1 we discussed that *viroids* have no proteins and consist of a naked circular RNA molecule that can produce diseases in plants (see Fig. 1–7).

SUMMARY:

Nucleic Acids

Nucleic acids are the repositories of genetic information. In general, DNA is transcribed into RNA, and these are involved in the translation into proteins.

DNA → RNA → Protein

DNA is localized mainly in the nucleus (i.e., in the chromosomes); small amounts are in mitochondria and chloroplasts.

Nucleic acids are composed of a pentose, phosphoric acid, purine bases (adenine, guanine), and pyrimidine bases (thymine, cytosine, uracil). They are linear polymers of *nucleotides* linked by phosphate-diester bonds between the pentoses. Since one acid group is left free, nucleic acids are very acidic and bind to basic proteins. In eukaryotic cells DNA is bound to *histones*, forming a nucleoprotein structure called *chromatin*.

DNA differs from RNA in the pentose (deoxy-

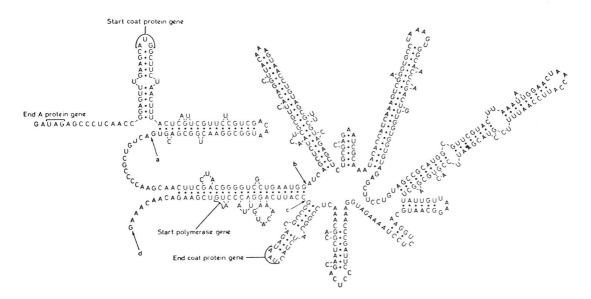

Fig. 2–9.   Part of the nucleotide sequence of the RNA bacteriophage MS2, a virus that infects *E. coli*. Note that the molecule folds back on itself, forming *hairpin loops*. The starting codon (AUG) for the coat protein is readily accessible to ribosomes. The start for the polymerase protein is blocked by the secondary structure and is accessible only when the structure is opened by the translation of the coat gene by ribosomes. (Reprinted by permission from Fiers, W., et al.: Nature, *260*:500, 1976. Copyright © 1976 Macmillan Journals Limited.)

ribose in DNA; ribose in RNA) and in one of the pyrimidine bases (thymine in DNA; uracil in RNA). Radioactive *thymidine* is used to label DNA; *uridine* is used for RNA. Genetic information is stored in the linear sequence of bases; the genetic alphabet consists of four letters: A, T, C, and G. The bases of DNA are in certain molar ratios; the amount of A always equals T, G equals C, and A + G = C + T. The ratio AT/GC varies among the different species.

Watson and Crick established the structure of the DNA molecule in 1953 on the basis of the x-ray diffraction studies of Wilkins and Franklin. The model is formed of two right-handed helical polynucleotide chains that are *antiparallel* and in which the bases are stacked perpendicularly. The two chains are held together by hydrogen bonds between the base pairs. Only two types of base pairs are possible within the double helix: AT and GC. This explains the regularities observed in the molar ratios of bases. AT pairs are held together by two hydrogen bonds, and GC pairs have three; for this reason GC pairs are more stable. The double helix has a major and a minor groove and completes one turn every 10 base pairs. A left-handed conformation (Z-DNA) has been found.

The axial sequence of bases along one chain is a complement of the other; DNA duplication takes place via a template mechanism, following the unwinding of the two strands. The sequence of bases in a long polymer provides an explanation for the immense number of combinations that carry different genetic information.

By an increase in temperature or other treatments, it is possible to denature (i.e., melt) the DNA. The two strands separate. More energy is required to break the G—C pair than is necessary to break the A—T pair.

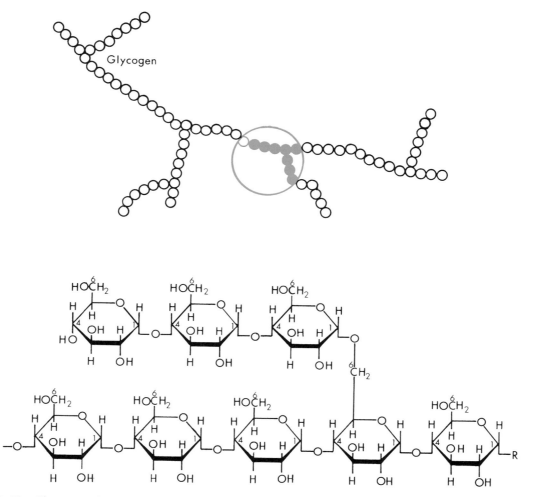

Fig. 2–10. Glycogen is a branched polymer formed by as many as 30,000 glucose units. The glycosidic bonds are established between carbons 1 and 4 of glucose, except at the branching points, which involve linkages between carbons 1 and 6. The top part of the figure shows a low resolution diagram; the circled area is enlarged on the bottom part.

Following denaturation, if the separated strands are kept at a suitable temperature, they can base-pair in register and restore the double helix *(renaturation* or *annealing)*. Renaturation is a very important tool in molecular biology, as it allows us to estimate the size of a genome, to detect *repeated sequences* in eukaryotic DNAs, and to *hybridize* RNA to the DNA from which it was transcribed.

RNA has only one chain, and it contains ribose and uracil. In RNA there are regions of secondary structure (hairpin loops) formed by hydrogen bonded AU or GC pairs. The major classes of RNA are ribosomal RNA, transfer RNA, and messenger RNA. The smallest infectious agents are called *viroids* and their only component is a single *circular* RNA molecule.

2–3 CARBOHYDRATES

Carbohydrates, composed of carbon, hydrogen, and oxygen, are the main source of cellular energy and are also important structural components of cell walls and intercellular materials. Carbohydrates are classified according to the number of monomers they contain.

Monosaccharides are simple sugars having the general formula $C_n(H_2O)_n$. Depending on the number of carbon atoms they contain, these are designated trioses, pentoses, or hexoses. The pentoses *ribose* and *deoxyribose* are components of nucleic acids. *Glucose*, a hexose, is the primary energy source for the cell. Other important hexoses are *galactose, fructose*, and *mannose*.

Disaccharides are sugars formed by the condensation of two hexose monomers with the loss of one molecule of water. Their formula is therefore $C_{12}H_{22}O_{11}$. The most important members of this group are *sucrose* (formed by glucose and fructose) and *lactose* (formed by galactose and glucose).

Polysaccharides result from the condensation of many hexose monomers, with a corresponding loss of water molecules. Their formula is $(C_6H_{10}O_5)_n$. Upon hydrolysis they yield molecules of simple sugars. The most important polysaccharides in living organisms are *starch* and *glycogen*, which are reserve substances in plant and animal cells, respectively, and *cellulose*, the most important structural element of the plant cell wall. These three substances are all polymers of glucose molecules, but differ in the way they are joined together. Figure 2–10 shows that glycogen is a branched molecule in which the glucose monomers can be joined together by two types of linkages.

Complex polysaccharides consist of hexoses plus nitrogen-containing compounds, such as glucosamine, that can also be acetylated or substituted with sulfuric or phosphoric acid. These polymers are important in molecular organization, particularly as intercellular substances. They are often found in combination with proteins or lipids.

Glycoproteins—Two-step Carbohydrate Addition

Glycoproteins are proteins that have covalently bound carbohydrates, as a short branch bound to asparagine, serine, or threonine. Several monosaccharides, such as galactose, mannose, and fucose, as well as N-acetyl-D-glucosamine and sialic acid, are bound to proteins. Glycoproteins can be divided into two major categories: cellular and secretory. Cellular glycoproteins are present mainly in the cell membrane, with the carbohydrate moiety protruding to the outside of the cell, and they have important functions in membrane interaction and recognition (see Chapter 5).

Most secreted proteins are glycoproteins and are produced by various cells: serum glycoproteins (i.e., seroalbumins), secreted by the liver; thyroglobulin, produced in the thyroid gland; immunoglobulins, secreted by plasma cells; ovalbumin, secreted by the hen oviduct; and ribonuclease and deoxyribonuclease, secreted by the pancreas. In most glycoproteins the protein is linked to the carbohydrate moiety by way of the amino acid asparagine (Asn).

A constant feature of asparagine-bound glycoproteins is a pentasaccharide "core" consisting of N-acetylglucosamine (GlcNAc) and mannose, to which are attached two side chains (R and R′) that differ in length and composition in the various proteins.

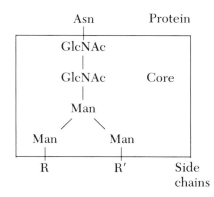

This complex is synthesized in two stages, each one taking place at a different intracellular location. The "core" is added to the protein inside the endoplasmic reticulum as the nascent peptides are being synthesized, while the side chains are added later in the Golgi complex by enzymes called *glycosyltransferases* (see Chapter 9). The oligosaccharide core is initially assembled on a lipid carrier (dolichol-phosphate) and then transferred to the protein molecule. The function of the lipid carrier is to enable the hydrophilic oligosaccharide to traverse the membrane of the endoplasmic reticulum.

Several exocrine glands, such as the salivary glands and the mucous glands of the digestive tract, secrete mucoproteins in which the linkage is between N-acetylglucosamine and the amino acids serine or threonine in the protein. Sialic acid and L-fucose are the terminal sugars in these compounds.

2–4   LIPIDS

The compounds in this group are characterized by their relative insolubility in water and solubility in organic solvents. The cause of this general property of lipids and related compounds is the predominance of long hydrocarbon chains or benzene rings. These structures are nonpolar and hydrophobic. In many lipids these chains may be attached at one end to a polar group, rendering it capable of binding water by hydrogen bonds. As explained below lipids may be simple or compound.

Triglycerides—Three Fatty Acids Bound to Glycerol

Simple lipids are alcohol esters of fatty acids. Among these are: (1) *Neutral fats* (glycerides) often called triglycerides; they are triesters of fatty acids and glycerol. Neutral fats accumulate in adipose tissue. (2) *Waxes*, which have a higher melting point than natural fats; they are esters of fatty acids with alcohols other than glycerol, such as in beeswax.

Fatty acids have long hydrocarbon chains with the general formula:

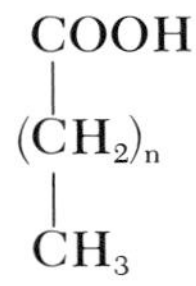

$$COOH$$
$$|$$
$$(CH_2)_n$$
$$|$$
$$CH_3$$

Fatty acids always have an even number of carbons because they are synthesized by joining two-carbon acetyl units. For example, palmitic acid has 16 carbons, and stearic acid has 18 carbons. Sometimes the hydrocarbon chain has double bonds ($-C{=}C-$), and in such cases the fatty acid is said to be nonsaturated. Double bonds are important because they increase the flexibility of the hydrocarbon chain, and thereby the fluidity of biological membranes.

The carboxyl groups of fatty acids react with the alcohol groups of glycerol in the following way:

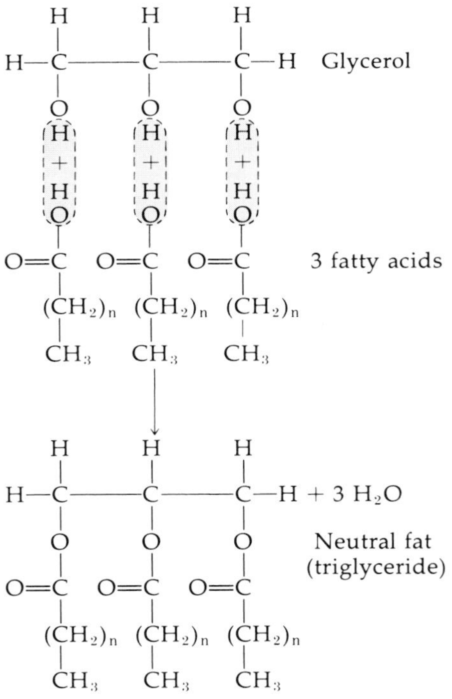

The repeating triglycerides, which accumulate in adipose tissue, are used by organisms for the convenient storage of spare energy. The oxidation state of the long hydrocarbon chains is very low, and they therefore liberate large amounts of energy (about twice as many calories per gram as carbohydrates and proteins) when oxidized to form CO_2 and H_2O in cells.

Phospholipids and Biological Membranes

Compound lipids are more complex and upon hydrolysis yield other substances in addition to fatty acids and alcohol. We mentioned earlier that some of these lipids are important structural components of the cell, in particular of cell membranes.

Phospholipids have only two fatty acids attached to the glycerol molecule. The third hydroxyl group of glycerol is esterified to phosphoric acid instead of to a fatty acid. This phosphate is also bound to a second alcohol molecule, which can be choline, ethanolamine, inositol, or serine, depending on the type of phospholipid:

From this formula it is possible to derive the main phospholipids such as *phosphatidyl choline (lecithin), phosphatidyl ethanolamine (cephalin), phosphatidyl serine* and *phosphatidyl inositol.*

As shown in Figure 2–11, phospholipids have two long hydrophobic fatty acid "tails" and a hydrophilic (polar) phosphate-containing "head." Phospholipids are thus *amphipathic* molecules (i.e., they contain a hydrophilic and a hydrophobic region), and their configuration accounts for many of the properties of biological membranes. Such membranes are bilayers of phospholipids with the hydrophilic heads (phosphate-containing regions) positioned at the water interface and the long hydrophobic tails arranged in the interior. When phospholipids are mixed with water, the bilayer arrangement—polar heads outside, nonpolar tails inside—is adopted spontaneously (see Chapter 4). This *principle of self-assembly*, in which the assembly of complex structures arises exclusively from the physiochemical properties of its molecular components, is characteristic of living systems. Viruses and ribosomes, for example, assemble in a similar manner.

Glycolipids and *sphingolipids* are characterized by the fact that glycerol is replaced by the amino alcohol *sphingosine*. To these groups belong the *sphingomyelins*, mainly in the myelin sheath of nerves; the *cerebrosides*, which are characterized by the presence of galactose or glucose in the molecule; the *sulfatides*, which contain sulfuric acid esterified to galactose; and the *gangliosides*.

The gangliosides deserve special mention because of their presence in cell membranes, their possible role as receptors of virus particles, and their influence on ion transport across membranes. A ganglioside is a complex molecule containing sphingosine, fatty acids, carbohydrates (lactose + galactosamine), and neuraminic acid. Gangliosides are long and highly polar molecules.

Steroids are lipids that derive from a complex ring structure. One of the best known is cholesterol, which is present in cell membranes (Fig. 2–12). Steroids have different functions depending on the groups attached to the basic structure. Several hormones such as estrogen, progesterone, corticosterone, and vitamin A are steroids.

2–5 PROTEINS

Proteins—Chains of Amino Acids Linked by Peptide Bonds

The building blocks of proteins are the *amino acids*. An amino acid is an organic acid in which the carbon next to the —COOH group (called an *alpha carbon*) is also bound to an —NH$_2$ group. In addition, the alpha carbon is bound to a *side-chain* (R), which is different in each amino acid.

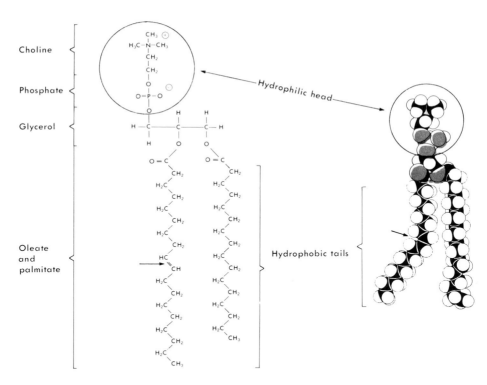

Fig. 2–11. A membrane phospholipid molecule has a hydrophilic head and two hydrophobic tails. The phospholipid represented here is palmitoyl-oleoyl-phosphatidyl-choline. Note that the double bond in the oleic acid produces a bend in the hydrocarbon chain (indicated by an arrow). Double bonds in the fatty acids increase cell membrane fluidity because unsaturated chains are more flexible (the rotation of the carbon-carbon single bonds on either side of the arrow is enhanced).

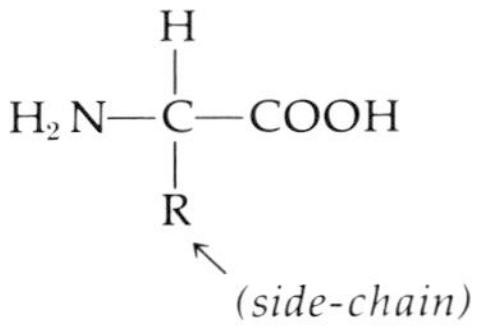

$$H_2N-\underset{R}{\overset{H}{\underset{|}{\overset{|}{C}}}}-COOH$$

(side-chain)

The amino acids differ from one another only in the side chain; for example, the R in alanine has one carbon, while in leucine it has four car-

Fig. 2–12. Chemical structure of cholesterol.

(b)

bons. Table 2–4 shows that the properties of the various amino acids depend on the chemical composition of their side chains; for example, lysine and arginine are basic because their side chains contain an extra amino group, and the acidic amino acids (glutamic and aspartic acids) contain an extra carboxyl group.

Because of the simultaneous presence of acidic (carboxyl) and basic (amino) groups, amino acids can have both positive and negative charges and are therefore *amphoteric molecules* or *zwitterions*.

The ionized form of an amino acid is:

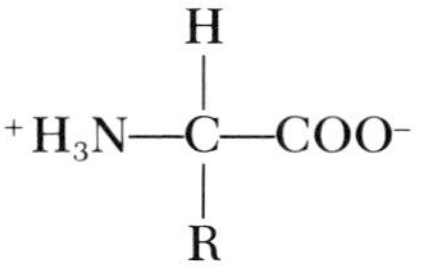

$$^+H_3N-\underset{R}{\overset{H}{\underset{|}{\overset{|}{C}}}}-COO^-$$

Figure 2–13 shows the structure of the 20 amino acids that are coded in biological systems. Of these, two are acidic, *aspartic acid* and *glutamic acid* (D and E in a one-letter nomenclature system that is sometimes used); three are basic,

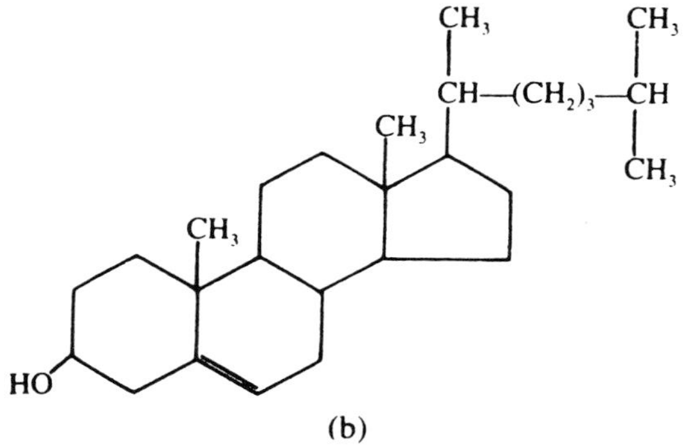

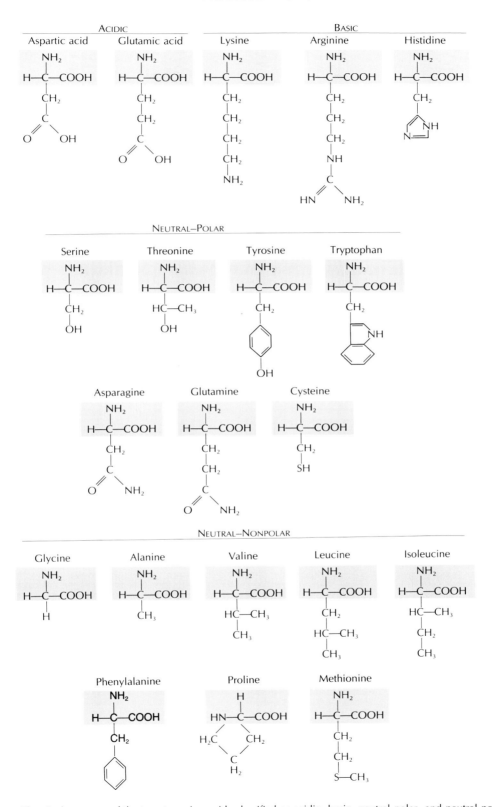

Fig. 2–13. Chemical structure of the twenty amino acids classified as acidic, basic, neutral polar, and neutral nonpolar. The structures below the conserved amino and carboxyl group are the R side chains.

lysine (L), *arginine* (R), and *histidine* (H); seven are neutral and at the same time polar (i.e., hydrophilic), *serine* (S), *threonine* (T), *tyrosine* (Y), *tryptophan* (W), *asparagine* (N), *glutamine* (Q), and *cysteine* (C); and eight are neutral nonpolar (i.e., hydrophobic), *glycine* (G), alanine (A), *valine* (V), *leucine* (L), *isoleucine* (I), *phenylalanine* (F), *proline* (P), and *methionine* (M). Note that two of these amino acids (Met and Cys) contain a sulfur atom. Between two cysteines a covalent *disulfide bridge* (—S—S—) can easily be formed because the H atoms of the —SH groups can be removed (see Fig. 2–14). The names of amino acids are usually abbreviated by using the first three letters of their name (Table 2–4).

The condensation of amino acids to form a protein molecule occurs in such a way that the acidic group of one amino acid combines with the basic group of the adjoining one, with the simultaneous loss of one molecule of water.

The linkage —NH—CO— is known as the *peptide linkage* or *peptide bond* (Fig. 2–15A). The formed molecule preserves its amphoteric character, since an acidic group is always at one end and a basic group is at the other, in addition to side chains that can be basic (Lys, Arg, His) or acidic (Asp, Glu). A combination of two amino acids is a *dipeptide;* of three, a *tripeptide.* When a few amino acids are linked together, the structure is an *oligopeptide* (Fig. 2–15B). A *polypeptide* consists of many (sometimes even 1000 or

more) amino acids. In the linear polymers there is always an amino (*N-terminal amino acid*) and a carboxyl group (*C-terminal amino acid*) at the ends (Fig. 2–15).

The distance between two peptide links is about 0.35 nm. A protein with a molecular weight of 30,000 consisting of 300 amino acid residues, if fully extended, should have a length of 100 nm, a width of 1.0 nm, and a thickness of 0.46 nm.

It is important to stress that the properties of proteins vary considerably. The side chains of the 20 amino acids have different chemical properties, and the number of different molecules possible by changing the linear sequence of amino acids is enormous. Much more diversity can be obtained in the chemical properties of proteins than in those of nucleic acids, which with their four component monomers are simple in comparison to proteins.

The *conjugated proteins* are attached to a nonprotein moiety, the so-called *prosthetic group.* To such a group belong the *glycoproteins*, the *lipoproteins* (e.g., blood lipoproteins), and the *chromoproteins*, which have a pigment as the prosthetic group, such as hemoglobin, hemocyanin, and the cytochromes. Hemoglobin and myoglobin (present in muscle) contain the prosthetic group *heme*, an iron-containing organic compound that combines with oxygen. Some other conjugated proteins are indicated in Table 2–5.

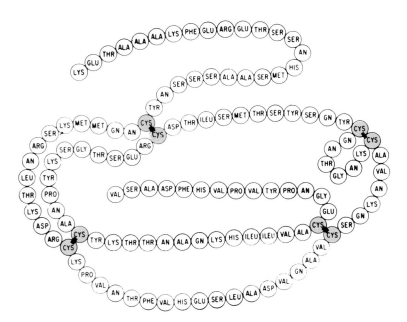

Fig. 2–14. The primary structure of bovine pancreatic ribonuclease. Notice the position of the four disulfide bridges between cystine residues. (From Anfinsen, C.B.: Principles that govern the folding of protein chains. Science, *181*:223, 1973.)

TABLE 2–4. THE TWENTY AMINO ACIDS

Type of Amino Acid	3-Letter Symbol	1-Letter Symbol
Hydrophobic		
(Aliphatic Side Chain)		
Glycine	Gly	G
Alanine	Ala	A
Valine	Val	V
Leucine	Leu	L
Isoleucine	Ile	I
Basic (Diamino)		
Arginine	Arg	R
Lysine	Lys	K
Acidic (Dicarboxylic)		
Glutamic acid	Glu	E
Aspartic acid	Asp	D
Amide-Containing		
Glutamine	Gln	Q
Asparagine	Asn	N
Hydroxyl-Containing		
Threonine	Thr	T
Serine	Ser	S
Sulfur-Containing		
Cysteine	Cys	C
Methionine	Met	M
Aromatic		
Phenylalanine	Phe	F
Tyrosine	Tyr	Y
Heterocyclic		
Tryptophan	Trp	W
Proline	Pro	P
Histidine	His	H

Four Levels of Structure in Proteins

Four levels of structure are commonly distinguished in proteins.

The *primary structure* is the sequence of amino acids, which form a chain connected by peptide bonds (Fig. 2–14). The amino acid sequence of a protein determines the higher levels of structure of the molecule. The biological importance of the amino acid sequence is exemplified by the human hereditary disease *sickle-cell anemia*, in which profound biological changes are produced by a single amino acid change in the hemoglobin molecule.

The *secondary structure* is the spatial arrangement of amino acids that are close to each other in the peptide chain. Some regions may display a rod-shaped structure, the *α-helix* (called alpha because it was the first structure deduced by Pauling and Corey in the early 1950s). In an α-helix the peptide chain is coiled around an imaginary cylinder (Fig. 2–16a) and stabilized by hydrogen bonds between the amino group of an amino acid and the carboxyl group of the amino acid situated four residues ahead in the *same* polypeptide chain. The protein α-helix has

3.6 amino acids per turn. In a *β-pleated sheet* (Fig. 2–16b) the amino acids adopt the conformation of a sheet of pleated paper, and the structure is stabilized by hydrogen bonds between the amino and carboxyl groups in *different* polypeptide strands. Other segments of the protein are not highly cross-linked and adopt a *random coil* configuration (Fig. 2–16c). This is partly because certain amino acids, such as proline, tend to disrupt helical structure.

The *tertiary structure* is the way in which helical and random coil regions fold with respect to each other (Fig. 2–17). That is, it refers to the three-dimensional relationship of amino acid segments that may be far apart from each other in the linear sequence.

The *quaternary structure* is the arrangement of protein subunits within complex proteins made up of two or more such subunits. For example, the hemoglobin molecule is composed of four polypeptide chains, two designated α and two β (Fig. 2–17). Separation and association of the subunits may occur spontaneously. Hemoglobin may be broken into two half-molecules (two α and two β) by urea. When urea is removed they reassemble, forming complete functional molecules.

Factors Involved in Determining Protein Structure

The spatial arrangement of a protein molecule is predetermined by its amino acid sequence (primary structure). This can be demonstrated by experiments in which protein *denaturation*, or disruption of tertiary structure, is brought about by nonphysiological conditions. Denaturation of a protein usually results in the loss of its biological activity.

Figure 2–18 shows that in the case of the enzyme ribonuclease, denaturation can be achieved by treatment with β-mercaptoethanol and high concentrations of urea. Mercaptoethanol is a reducing agent that can disrupt S—S bridges (disulfide bonds), reducing them to —SH groups, while urea disrupts other weak molecular interactions (in particular hydrogen bonds). The tertiary structure of ribonuclease is maintained by four disulfide bonds that are established between pairs of cysteines (an amino acid that contains an —SH group). After denaturation the enzyme can be *renatured*, or cor-

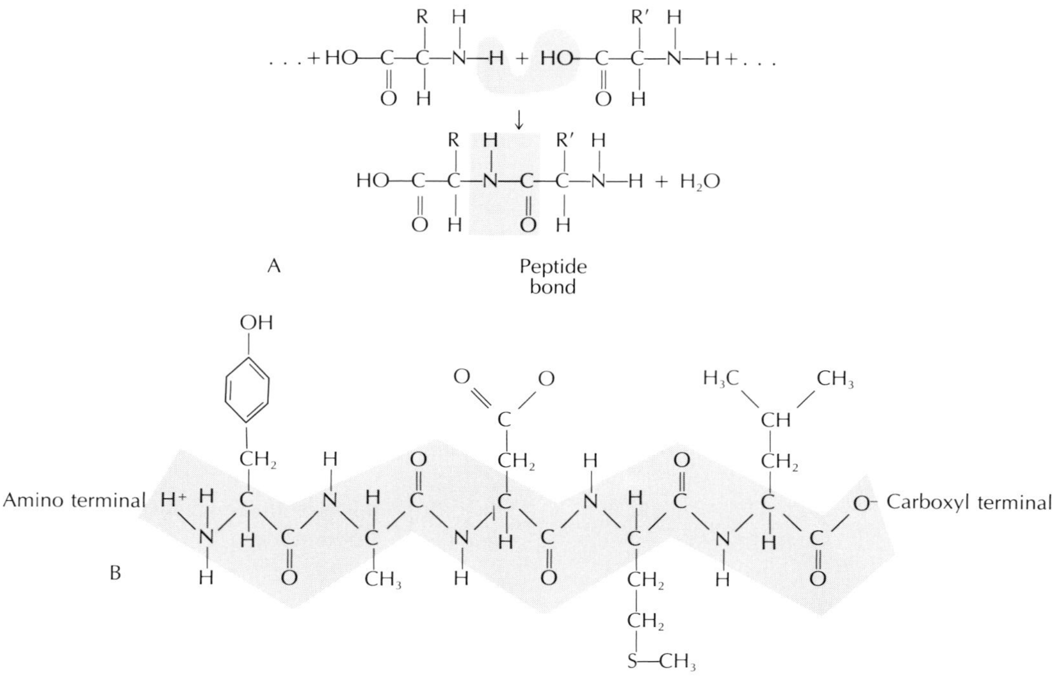

Fig. 2–15. *A,* formation of a peptide bond between two amino acids. *B,* a pentapeptide formed from the amino to the carboxyl terminus by tyrosine, alanine, aspartic acid, methionine, and leucine. The backbone of the peptide bonds is shaded. The acid and basic groups are in the ionized form as it occurs at physiological pH.

rectly refolded into its natural conformation, by gradually removing the urea and mercaptoethanol, after which the activity of the enzyme is recovered (Fig. 2–18). There are 105 possible combinations in which eight cysteines can pair to produce four disulfide bridges, but only the biologically active conformation is produced after careful renaturation because it is thermodynamically the most stable structure. This is clear evidence that all the information needed to produce the complex folding of a protein molecule is contained in its primary structure.

Several different types of bonds are involved in maintaining the four levels of protein structure. The *covalent bonds* in proteins are of two main types. The first is the *peptide bond* uniting amino acid monomers in the primary sequence. The second is the *disulfide bond* (S—S bridge) which, as we have just seen, is established between the —SH groups of two cysteine residues and is responsible for some aspects of secondary and tertiary structure.

Various kinds of *weak interactions* are impor-

tant in the establishment of secondary and tertiary structure. These weak bonds are all noncovalent; the main types (illustrated in Fig. 2–19) are as follows:

Ionic or *electrostatic bonds* result from the attractive force between ionized groups having opposite charges (Fig. 2–19a).

Hydrogen bonds result when a H^+ (proton) is shared between two neighboring electronegative atoms. The H^+ can be shared between nitrogen or oxygen atoms that are close to each other. Hydrogen bonds have many important biochemical functions. They are essential for the specific pairing between nucleic acid bases, thus providing the main force that holds the two DNA strands together as well as allowing the specific copying of DNA into RNA. Figure 2–4 shows hydrogen bonds in DNA, and Figure 2–19*b* shows them in a protein.

Hydrophobic interactions (Fig. 2–19c) involve the clustering of nonpolar groups, which associate with each other in such a way that they are not in contact with water. In globular pro-

TABLE 2–5.   A LIST OF CONJUGATED PROTEINS WITH THE COMPONENTS AND THE PERCENT WEIGHT THEY REPRESENT

Conjugated Protein	Prosthetic Component	% Weight
Nucleoprotein systems		
Ribosomes	RNA	50–60
Tobacco mosaic virus	RNA	5
Lipoproteins		
Plasma β_1-lipoproteins	Phospholipid, cholesterol, neutral lipid	79
Glycoproteins		
γ-Globulin	Hexosamine, galactose, mannose, sialic acid	2
Plasma orosomucoid	Galactose, mannose, N-acetylgalactosamine, N-acetylneuraminic acid	40
Phosphoproteins		
Casein (milk)	Phosphate esterified to serine residues	4
Hemoproteins		
Hemoglobin	Iron protoporphyrin	4
Cytochrome c	Iron protoporphyrin	4
Catalase	Iron protoporphyrin	3.1
Flavoproteins		
Succinate dehydrogenase	Flavin adenine dinucleotide	2
D-Amino acid oxidase	Flavin adenine dinucleotide	2
Metalloproteins		
Ferritin	$Fe(OH)_3$	23
Cytochrome oxidase	Fe and Cu	0.3
Alcohol dehydrogenase	Zn	0.3
Xanthine oxidase	Mo and Fe	0.4

From Lehninger, A.L.: Biochemistry: The Molecular Bases of Cell Structure & Function. 2nd Ed. New York, Worth, 1982.

teins the side chains of the most hydrophobic amino acids tend to aggregate inside the molecule, and the hydrophilic groups protrude from the surface of the structure. The hydrophobic residues tend to repel the water molecules that surround the protein, thereby causing the globular structure to be more compact.

Van der Waals interactions (Fig. 2–19d) occur only when two atoms come very close together. The closeness of two molecules can induce charge fluctuations, which may produce mutual attraction at very short range.

The essential difference between a covalent and a noncovalent bond is in the amount of energy needed to break the bond. For example, breaking a hydrogen bond requires only 4.5 kcal mole^{-1}, as compared with 110 kcal mole^{-1}, for the covalent O—H bond present in water. Although each individual bond is weak, large numbers of them can produce stable structures, as in the case of double-stranded DNA. Covalent bonds are generally broken by the intervention of enzymes, whereas noncovalent bonds are easily dissociated by physicochemical forces.

Many cellular proteins exist as complexes of multiple subunits that are held together by weak interactions. For analytical purposes, cell biologists sometimes find it desirable to dissociate them into their component polypeptides.

Electrical Charges of Proteins and the Isoelectric Point

In addition to the terminal —NH$_3$$^+$ and —COO$^-$ charged groups, proteins contain dicarboxylic- and diamino-amino acids (Table 2–4), which dissociate as follows:

1. The acidic groups lose protons and become negatively charged. For example, in aspartic and glutamic acids, the free carboxyl group dissociates into —COO$^-$ + H$^+$.

2. The basic groups, by gaining protons, become positively charged.

$$(-NH_2 + H^+ \rightarrow -NH_3{}^+)$$

This is found in amino acids with two basic groups, such as lysine or arginine.

(a)   (b)   (c)

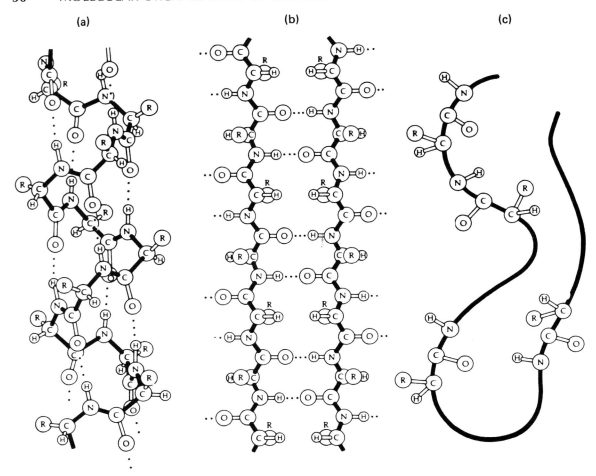

Fig. 2–16.   Representations of different secondary structures of proteins (a) α helix (note that it has 3.6 amino acids per turn), (b) β-pleated sheet, and (c) random coil configuration. (See text for further details.) (From Ayala, F.J., and Kiger, J.A., Jr.: Modern Genetics. Menlo Park, California, Benjamin-Cummings, 1980.)

The actual charge of a protein molecule is the result of the sum of all single charges. Because dissociation of the different acidic and basic groups takes place at different hydrogen ion concentrations of the medium, pH greatly influences the total charge of the molecule. Figure 2–20 shows that in an acid medium, amino groups capture hydrogen ions and react as bases ($-NH_2 + H^+ \rightarrow -NH_3^+$); in an alkaline medium the reverse takes place and carboxylic groups dissociate ($-COOH \rightarrow COO^- + H^+$). For every protein there is a definite pH at which the sum of positive and negative charges is zero (Fig. 2–20). This pH is called the *isoelectric point* (pI). At the isoelectric point, proteins placed in an electrical field do not migrate to either of the poles, whereas at a lower pH they migrate to the negative pole *(cathode)* and at a

higher pH, to the positive pole *(anode)*. This migration is called *electrophoresis*, and it provides a useful technique for the separation of cellular proteins.

Separation of Cell Proteins–Isoelectric Focusing and SDS Electrophoresis

Each protein has a characteristic isoelectric point, and this property can be used in the separation of proteins. In the technique called *isoelectric focusing*, proteins are subjected to electrophoresis on a pH gradient. Each protein moves until it reaches a pH equal to its individual isoelectric point. At that moment, migration in the electrical field stops because the net charge of the protein is zero.

The techniques of isoelectric focusing and

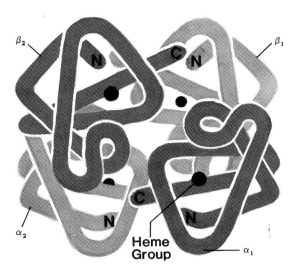

Fig. 2–17. Diagram representing the tertiary and quaternary structure of hemoglobin. This protein is composed of four subunits: two α and two β. The sites in which the four heme groups are located, as well as the amino (N) and carboxyl (C) termini of the polypeptide chains are indicated.

SDS polyacrylamide gel electrophoresis have been combined to produce two-dimensional separation of proteins. Figure 2–21 shows how several hundred cellular proteins can be resolved from one another. This technique is increasingly used in cell biology, and its great resolving power is due to the use of two independent properties of proteins. The proteins are first separated by isoelectric focusing (this is the first dimension), which separates proteins according to their charge (isoelectric point). The proteins are subsequently separated by electrophoresis (this is the second dimension) in polyacrylamide gels containing SDS, which

separates proteins according to their size (molecular weight). This technique results in a series of spots distributed throughout the polyacrylamide gel (if the same property of proteins had been used in both dimensions, the spots would be distributed along a diagonal).

When the detergent sodium dodecyl sulfate (SDS) is used in electrophoresis, the proteins are separated mainly according to their molecular weight. This is because SDS binds to the proteins, giving them large numbers of negative charges due to the sulfate. Thus, most of the protein charges will come from the SDS, minimizing the role of charge differences between individual proteins (differences which would otherwise affect electrophoretic mobility), and all the proteins migrate according to their size. The larger proteins move more slowly than the smaller ones because they encounter more resistance when traversing the molecular pores within the polyacrylamide gel used for electrophoresis. SDS electrophoresis is widely used as a method for determining molecular weights of proteins.

SUMMARY:
Molecular Components of the Cell

This chapter is an elementary survey of the molecular components of the cell. The protoplasm of a plant or animal cell contains 75 to 85% water, 10 to 20% protein, 2 to 3% lipid, and 1% carbohydrate. All living organisms contain DNA and RNA. The genetic information contained in the DNA is transcribed into RNA, which in turn is translated into protein; this series of events is often referred to as the central dogma.

Nucleic acids are linear polymers of nucleotides

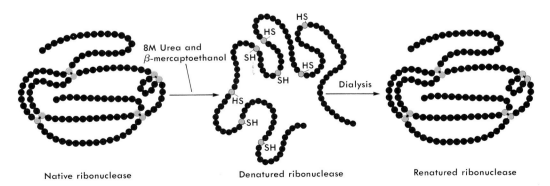

Native ribonuclease   Denatured ribonuclease   Renatured ribonuclease

Fig. 2–18. Denaturation and renaturation of ribonuclease. This experiment shows that the information for protein folding is contained in the amino acid sequence (primary structure) of proteins. (Cys residues are indicated with shaded beads. See Fig. 2–14.) Sulfhydryl groups, *SH, HS.*

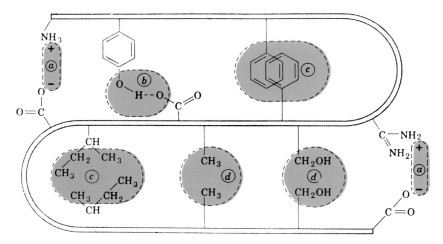

Fig. 2–19. Types of noncovalent bonds that stabilize protein structure. *a,* ionic bonds; *b,* hydrogen bonds; *c,* hydrophobic interactions; *d,* Van der Waals interactions. (From Anfinsen, C.B.: Principles that govern the folding of protein chains. *Science, 181*:223, 1973.)

linked together by phosphodiester bonds. The nucleotide monomers result from the covalent bonding of a phosphate and a base to a pentose moiety. The pentose is ribose in RNA and deoxyribose in DNA. The bases found in DNA are thymine (T) and cytosine (C), which are pyrimidines, and adenine (A) and guanine (G), which are purines. RNA contains uracil (U) instead of thymine. A nucleotide without its phosphate group is called a nucleoside. In addition to their role as nucleic acid constituents, nucleotides also have a major role in the storage and transfer of chemical energy.

All the genetic information of a living organism is stored in its linear sequence of the four bases. Although DNA base composition varies from one species to another, the amount of adenine always equals the amount of thymine (A = T), and the amounts of cytosine and guanine are also equal (C = G).

A DNA molecule is composed of two antiparallel polynucleotide chains that form a double helix around a central axis. The bases are stacked inside the helix in a plane perpendicular to its axis, and the two strands are held together by hydrogen bonds established between the base pairs. The only

pairs that occur are AT, held together by two hydrogen bonds, and GC, held together by three hydrogen bonds. The latter pair is the more stable. The pairing properties of the bases are such that, whatever the axial sequence on one strand may be, the sequence on the other strand must be exactly complementary to it.

If the two strands of a DNA molecule are denatured (separated) by physical or chemical treatments, they can subsequently reanneal as a consequence of their nucleotide base-pairing properties. Renaturation studies have led to the discovery of repeated sequences in eukaryotic DNA; similar hybridization studies have provided a powerful method for characterizing RNA molecules, which hybridize only to the DNA from which they were transcribed.

The three major classes of RNA are messenger, transfer, and ribosomal, all of which are involved in protein synthesis. Although each RNA molecule consists of only a single polynucleotide chain, the chain folds upon itself to make a more compact structure having such secondary structure characteristics as hairpin loops. These structures may have important biological consequences.

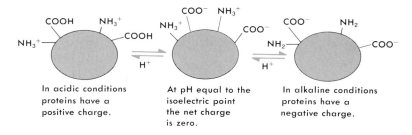

In acidic conditions proteins have a positive charge.

At pH equal to the isoelectric point the net charge is zero.

In alkaline conditions proteins have a negative charge.

Fig. 2–20. The ionization of proteins depends on the pH. This is of great importance in electrophoresis; in acidic conditions proteins migrate to the cathode (− pole), in alkaline conditions, to the anode (+ pole).

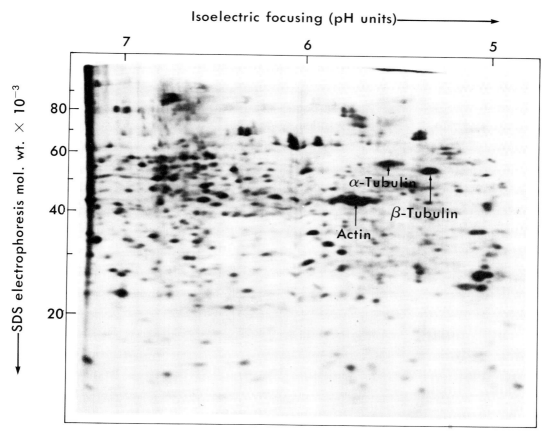

Fig. 2–21. Two-dimensional electrophoresis of the proteins of *Xenopus borealis* oocytes. The frog cells were labeled with S^{35}-methionine, homogenized, and the total proteins were separated by isoelectric focusing in the first dimension and by SDS (sodium dodecyl sulphate) electrophoresis in the second. The polyacrylamide gel was exposed against photographic film (autoradiography). Several hundred radioactive proteins can be seen. As in most other cell types, the most abundant protein is actin (3 to 5% of the total cell protein), followed by tubulin (which has α and β subunits, both of which form part of cell microtubules). Notice that the distance migrated by the proteins in the second dimension is proportional to the logarithm of the molecular weight. (From E.M. De Robertis.)

Carbohydrates serve as sources of energy or play a structural role (e.g., in cell walls). They are classified as mono-, di-, and polysaccharides. The main monosaccharides are pentoses (5 carbons) and hexoses (6 carbons). The main polysaccharides, which are composed of glucose monomers, are starch (plants) and glycogen (liver and muscle). Complex polysaccharides contain amino sugars (i.e., glucosamine), and sulfuric acid or phosphoric acid. The acidic mucopolysaccharides are strongly basophilic (i.e., heparin, chondroitin sulfate, hyaluronic acid). *Glycoproteins* are either *cellular*, present on the outside of cell membranes, or are *secretory* (e.g., seroalbumins, thyroglobulin, immunoglobulins, ribonuclease). In most glycoproteins the carbohydrate is linked to the amino acid asparagine by way of N-acetyl-D-glucosamine (GlcNAc). The carbohydrate moiety has a constant pentaoligosaccharide core of mannose of GlcNAc, which is added to the nascent peptides in the ER and two variable side chains, which are added later in the Golgi complex.

Lipids comprise a large group of different compounds characterized by their solubility in organic solvents. There are *simple* lipids (triglycerides), *steroids* (i.e., sex hormones, cholesterol), and *conjugated lipids* (i.e., phosphatides, glycolipids, cerebrosides, sulfatides, and gangliosides). Phospholipids are the main components of biological membranes. They have a hydrophilic (phosphate-containing) region and two hydrophobic (fatty-acid) tails.

Macromolecules are long *polymers* composed of *monomers*. Proteins are made of about 20 different monomers—the *amino acids*. These are *amphoteric* molecules because they carry an acidic (—COOH) and a basic group (—NH₂). Amino acids are linked by the *peptide bonds*, forming polypeptides. Essentially all basic functions of the cell depend on specific proteins.

The *primary structure* of the protein is the amino acid sequence. A change in a single amino acid may produce a profound change in the molecule (e.g., hemoglobins). The secondary structure of a protein may be of the *α-helix, β-pleated sheet,* or the random coil type. The *tertiary* structure of globular protein is complex and can be determined by x-ray diffraction. Proteins may contain parts with α-helix and β-configurations, or random coil. The disruption of the tertiary structure by high temperatures of other agents is called *denaturation.* The *quaternary* structure is characteristic of proteins having more than one subunit. Hemoglobin is a tetramer composed of two α- and two β-chains.

In proteins, noncovalent bonds, i.e., ionic, hydrogen, and hydrophobic interactions play an important role in determining the secondary and tertiary structure. The electrical charge of a protein is determined by the *ion-producing groups* (i.e., acidic or basic) and their degree of dissociation at different pHs. At the *isoelectric point* (pI) the net charge is 0, and the molecule will not migrate in an electrical field. This is the basis for *isoelectric focusing,* a technique used to separate proteins. Separation of proteins can be improved by the combination of isoelectric focusing and SDS polyacrylamide gel electrophoresis (Fig. 2–21).

2–6   ENZYMES AND THEIR REGULATION

The cell may be compared to a minute laboratory capable of carrying out the synthesis and breakdown of numerous substances. These processes are carried out by enzymes at normal body temperature, low ionic strength, low pressure, and a narrow range of pH.

Enzymes are biological catalysts. A *catalyst* is a substance that accelerates chemical reactions but that is not itself modified in the process, so that it can be used again and again. The vast majority of enzymes are proteins. In fact, until 1985 it was believed that all enzymes were proteins. We now know that some RNA molecules may also have catalytic activities and must thus be considered enzymes too (see below).

Enzymes are the largest and most specialized class of protein molecules. More than a thousand different enzymes have been identified; many of them have been obtained in pure, and even crystalline, condition. Enzymes represent one of the most important products of the genes contained in the DNA molecule. The complex network of chemical reactions which are involved in cell metabolism is directed by enzymes.

Enzymes *(E)* are proteins with one or more loci, called *active sites,* to which the *substrate*

(S) (i.e., the substance upon which the enzyme acts) attaches. The substrate is chemically modified and converted into one or more products *(P).* Since this is generally a reversible reaction, it may be expressed as follows:

$$E + S \rightleftarrows [ES] \rightleftarrows E + P \qquad (1)$$

where $[ES]$ is an intermediary enzyme-substrate complex. Enzymes accelerate the reaction until an equilibrium is reached. They are so efficient that the reaction may proceed from 10^8 to 10^{11} times faster than in the noncatalyzed condition.

A very important feature of enzyme activity is that it is *substrate-specific;* i.e., a particular enzyme will act only on a certain substrate. Some enzymes have nearly *absolute* specificity for a given substrate and will not act on even very closely related molecules, as, for example, stereoisomers of the same molecule. In Chapter 19 we will refer to the *restriction enzymes* that are able to cut DNA at very specific nucleotide sequences (see Fig. 19–10).

Some Enzymes Require Cofactors or Coenzymes

Some enzymes require the presence of a *cofactor* for their activity. This may be a metal as indicated in Tables 2–1 and 2–5 or a prosthetic group, in the case of conjugated proteins. Some enzymes require the presence of small molecules called *coenzymes.* For example, *dehydrogenases* require a nicotinamide-adenine dinucleotide (NAD$^+$) or an NADP$^+$ molecule (with an additional phosphate) to function. The reaction is as follows:

$$\text{Substrate + NAD + Enzyme} \rightarrow$$
$$\text{oxidized substrate + NADH and H}^+ + \text{Enzyme}$$

The two electrons gained by NADH can then be transferred to a second molecule, which will become reduced (i.e., it gains electrons).

In the cell the energy-producing enzymes use NAD as coenzyme; the synthetic processes, however, use NADPH as a hydrogen donor. In many coenzymes, as in NAD$^+$ and NADP$^+$ (which contain *nicotinamide*), the essential components are vitamins, particularly those of the B group. Some examples are *pantothenic acid* (vitamin B$_5$), which forms part of the important coenzyme A; *riboflavin* (vitamin B$_2$), incorporated into the molecules of flavin-adenine dinucleotide (FAD), and *pyridoxal* (vitamin B_6), a cofactor of transaminases and decarboxylases.

Substrates Bind to the Active Site

Enzymes have great specificity for their substrates and will frequently not accept related molecules of a slightly different shape. This can be explained by assuming that enzyme and substrate have a *lock-and-key* interaction. As shown in Figure 2–22 the enzyme has an *active site* complementary to the shape of the substrate. If a substrate has a different shape, it will not bind.

Although we can think of enzymes in terms of locks and keys, this does not mean that the active site is a rigid structure. In some enzymes the active site is precisely complementary to the substrate only *after* the substrate is bound, a phenomenon called *induced fit*. As shown in Figure 2–22, the binding of the substrate induces a conformational change in the protein, and only then will the chemical groups essential for catalysis come in close contact with the substrate.

The binding of the substrate to the active site involves forces of a noncovalent nature (ionic and hydrogen bonds, van der Waals forces), which are of very short range. This explains why the enzyme-substrate complex can be formed only if the enzyme has a site that is exactly comple-

mentary to the shape of the substrate. The active site is a three-dimensional entity; and because of the folding of the protein chain, the amino acid residues important in the function of the site can be far apart in the linear sequence of amino acids.

In molecular terms one can explain the function of an enzyme in terms of two steps: (a) the formation of the *specific complex* and (b) the *catalytic step proper*, in which the different mechanisms of catalysis, such as hydration, dehydration, and transfer of groups, are produced.

Essentials of Enzyme Kinetics—Km and Vmax Define Enzyme Behavior

The existence of an enzyme-substrate complex [ES] at the active site was postulated by Michaelis and Menten in 1913 on the basis of kinetic evidence. This concept has been of great importance in the understanding of the mechanism of the enzymatic reactions.

As was mentioned previously, the enzyme-substrate reaction proceeds in two steps (1). The first step can be written as follows:

$$E + S \underset{K_2}{\overset{K_1}{\rightleftarrows}} [ES] \qquad (2)$$

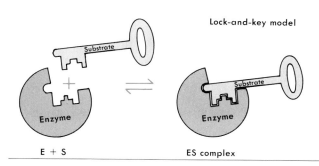

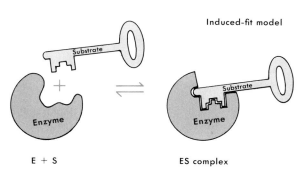

Fig. 2–22. Substrates interact with the active site in a precise way. Some enzymes have an induced-fit: the shape of the active site is complementary to the substrate only after the substrate is bound.

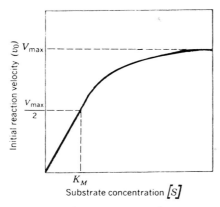

Fig. 2–23. Plot of the reaction rate of an enzyme at increasing substrate concentrations. The K_m, V_{max}, and $\frac{V_{max}}{2}$ are

described in the text. The curve is a hyperbola in which the first part follows first order kinetics (i.e., the reaction is proportional to the substrate concentration) and the second part corresponds to saturation (which has zero order kinetics as it no longer depends on substrate concentration).

In the second step the [ES] complex breaks down to form the product and the free enzyme, which will not be available for processing a new substrate molecule:

$$[ES] \underset{K_4}{\overset{K_3}{\rightleftarrows}} E + P \qquad (3)$$

(K_1, K_2, K_3, and K_4 are rate constants for the reactions.)

As shown in Figure 2–23 the velocity of the reaction depends on the substrate concentration [S]. At low concentrations the initial velocity (V) of the reaction describes a hyperbola. However, as [S] increases the reaction saturates and reaches a plateau. At this point, which corresponds to *Vmax*, all the enzyme is in the form of ES complexes. The equation of the curve is:

$$V = \frac{Vmax\ [S]}{Km\ +\ [S]} \qquad (4)$$

Km is the *Michaelis constant*, which experimentally may be defined as the [S] at which half of the enzyme molecules are forming ES complexes. The smaller the value of Km, the greater the *apparent affinity* of the enzyme for the substrate. Thus the kinetic behavior of an enzyme is defined by the values of Vmax and Km.

Enzyme Inhibitors Can Be Very Specific

Enzyme inhibition may be *reversible* or *irreversible*. There are two major types of reversible inhibition: *competitive* and *noncompetitive*. Competitive inhibition involves a compound similar in structure to the substrate, which forms a complex with the enzyme:

$$E + I \overset{K_i}{\rightleftarrows} [EI] \qquad (5)$$

where E is the enzyme, I the inhibitor, and K_i the association constant of the enzyme-inhibitor complex. Unlike the [ES] complex, the [EI] complex does not break down into the products of reaction and the free enzyme. The inhibition of succinic dehydrogenase by malonic acid, the molecular structure of which is similar to that of succinic acid, serves as an example of competitive inhibition:

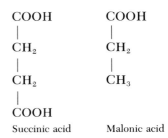

In this case, both the substrate (succinate) and the inhibitor (malonate) will compete for the active site, and the enzyme activity will be reduced. This inhibition can be reversed, nevertheless, by increasing the substrate concentration, so that the substrate molecules outnumber those of the inhibitor. Therefore, the *Vmax* is not changed by the competitive inhibition. This fact is easily observed using a Lineweaver-Burk plot as in Figure 2–24. Note that in competitive inhibition the Km increases, i.e., the apparent affinity of the enzyme for the substrate decreases.

In noncompetitive inhibition the inhibitor and the substrate are not structurally related, and the inhibitor binds to a different site than the substrate. Noncompetitive inhibition cannot be reversed by high concentrations of the substrate; therefore, the Km remains unchanged, but the Vmax is decreased (Fig. 2–24).

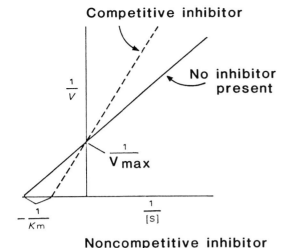

Competitive inhibitor

Noncompetitive inhibitor

Fig. 2–24. Lineweaver-Burk plot of the reactions. Here the plot is between the reciprocal of the velocity $\frac{1}{V}$ and the reciprocal of the substrate concentration $\frac{1}{[S]}$. In the presence of a competitive inhibitor V_{max} is unchanged, while with a noncompetitive inhibitor, K_m is unchanged.

Zymogens Are Inactive Forms of Enzymes

Many enzymes that are secreted by cells are synthesized in an inactive form and then converted into an active one. This occurs, for example, with the enzymes that are produced by the pancreas and used for digestion in the intestine. These enzymes like *trypsinogen* and *chymotrypsinogen* are stored in intracellular granules as inactive precursors called *zymogens* or *proenzymes*. These enzymes are activated only after they have been excreted into the digestive system. The proenzymes are converted into the active form (*trypsin, chymo-*

trypsin) by the action of proteases that cleave and remove a segment of the polypeptide chain, permitting the expression of the active site. In the proenzyme the active site is devoid of binding or catalytic activity, which is manifested only after the propeptide is broken. After such a break, there is a conformational change in the enzyme that activates the binding site, permitting the catalytic activity.

This mechanism is useful because it prevents the enzymes from digesting the pancreatic cells. A similar activation of proenzymes is found in the process of blood clotting. In Chapter 24 we discuss how nature uses proteolytic cleavage for the production of active neuropeptides from larger protein precursors. The signal for cleavage of protein precursors is usually provided by two basic amino acids (arg-arg; arg-lys; lys-lys) (see Fig. 24–23).

Isoenzymes

Isoenzymes are multiple forms of an enzyme that differ by minor variations in amino acid composition and sometimes in regulation. One of the best examples of an isoenzyme is lactic dehydrogenase (LDH), which catalyzes the conversion of pyruvate to lactate. There are five LDH isoenzymes that differ in their electrophoretic mobility in starch gels. LDH is a tetramer that can be formed by two types of subunits: M (predominant in muscle) and H (predominant in the heart). Each subunit is the product of a different gene. The five isoenzymes result from the five possible combinations of these subunits. (M_4, M_3H_1, M_2H_2, M_1H_3, and H_4). The relative proportions of the five isoenzymes are characteristic for each tissue and for each stage of embryonic development. For example the M_4 isoenzyme, having a low Km and a high Vmax, is prevalent during embryonic development. The M_4 is well adapted to tissues that are frequently low in oxygen and depend on anaerobic glycolysis for energy. On the other hand, the H_4 isoenzyme in the heart is adapted to aerobic conditions and has high Km and low Vmax.

The Cell Is Not Simply A Bag Full of Enzymes

Enzymes catalyze the thousands of chemical reactions that occur in cells. The enzymes of some biochemical pathways are in solution in the cytosol, and the substrates must diffuse freely from one enzyme to the next. In other cases, however, the enzymes involved in a chain of reactions are bound to one another and function together as a *multi-enzyme complex*. For example, the seven enzymes that synthesize fatty acids are tightly bound to one another. Similarly, the pyruvate-dehydrogenase complex is formed by three enzymes. Multi-enzyme systems facilitate complex reactions because they limit the distance through which the substrate molecules must diffuse during the sequence of reactions. The substrate is not released from the complex until all the reactions are completed.

The most complex multi-enzyme systems are associated with biological membranes and the ribosomes. For example, the respiratory enzymes necessary for electron transfer are arranged in a precise way in the two-dimensional scaffold provided by the inner membrane of mitochondria in eukaryotes (in prokaryotes they are arranged within the cell membrane). These multi-enzyme systems require this well-defined structure for activity; the enzymes become inactive when removed from the membrane. For this reason, the study of membrane biochemistry is especially difficult.

Enzyme distribution is never random. Some enzymes are packed into lysosomes, and others into secretion granules. Other enzymes, such as the RNA and DNA polymerases, are located in the nucleus and not in the cytoplasm. The mechanisms by which these proteins are segregated into the correct cellular compartments are emphasized throughout this book.

Some RNAs Have Enzymatic Activity— Ribozymes

In two instances it has been documented that pure RNA can catalyze biological reactions in the absence of any proteins. The first of these *ribozymes* is the intervening sequence of *Tetrahymena* (a protozoan) ribosomal DNA. As can be seen in Figure 20–16, this RNA can be spliced in the absence of proteins, i.e., a segment of RNA is excised (as a circle) and the molecule is religated. The driving force for this reaction is in the excised intervening sequence, and Zaug and Cech (1986) have shown that this molecule can also catalyze the cleavage and re-

ligation of added polycytidylic acid, an artificial oligonucleotide. Starting with poly C that is 5 nucleotides long, poly C up to 30 nucleotides long can be generated. Each ribozyme molecule can generate hundreds of elongated substrate molecules and is thus a classic enzyme. The reaction has hyperbolic kinetics and is specific for RNA, for deoxy poly C acts as a classic competitive inhibitor in reciprocal plots of the type shown in Figure 2–24.

The second example of catalytic RNA is provided by the enzyme ribonuclease P, which is involved in the maturation of tRNA precursors. As shown in Figure 21–31, ribonuclease P cuts off 5′ extensions precisely at the start of the mature molecule. Once purified, the enzyme was found to consist of an RNA molecule and a protein molecule. One molecule of the RNA component by itself, in the complete absence of protein, is able to cut many tRNA precursors precisely (Guerrier-Takada and Altman, 1984). The presence of the protein subunit makes this process faster but is not essential for catalysis.

Small nuclear RNPs (Table 20–3) contain RNAs (such as U1 and U2) which also seem good candidates to be ribozymes involved in the splicing of mRNA precursors, although a true catalytic activity has not yet been documented.

The fact that not all enzymes are proteins came as a surprise to most biochemists, and it gives a fresh perspective to speculations on the origin of life. As discussed at the end of this chapter, this suggests that RNA was the original molecule of life, probably capable of self-replication (one ribozyme that can elongate poly C molecules has already been found), from which the proteins, and later on DNA, may have evolved.

Cooperativity and Allosterism in Enzyme Kinetics

We mentioned above that when the velocity of an enzyme reaction is plotted as a function of increasing substrate concentration, many enzymes display the *hyperbolic curve* shown in Figure 2–23. As more substrate is added, more enzyme is found in the ES complex form, and the velocity of appearance of product increases.

$$E + S \rightleftharpoons [ES] \rightarrow P + E \qquad (6)$$

At high substrate concentrations essentially all

of the enzyme molecules are in the ES complex form, and the *maximal velocity* of the reaction is achieved. This type of kinetics is usually found in enzymes that are composed of a single polypeptide chain. Other enzymes consist of two, four, or more polypeptide chains. An extreme case is that of a huge enzyme involved in the metabolism of cytidine triphosphate, *aspartyl transcarbamylase* (ATCase), which consists of 12 polypeptide chains or *subunits*. We will see that in many cases such oligomeric enzymes exhibit *cooperativity*, do not obey the normal Michaelis-Menten kinetics, and are subject to *allosteric* regulation.

Allosteric enzymes have two functionally and topologically different sites. One is the regular binding site for the substrate *(active site)*; the other is called the *allosteric (allo,* other; *steric,* space) or *regulatory site*, which lacks catalytic activity but is able to bind to an *effector* molecule. The effector acting on the allosteric site may either inhibit *(negative effector)* or stimulate the enzyme activity *(positive effector)*. The allosteric site bound to its effector changes the conformation of the active site and makes it more or less active. Allosteric enzymes are composed of several subunits; the active and regulatory sites may be on separate subunits. A good example is provided by the ATCase, in which six subunits are catalytic and six regulatory. In allosteric enzymes there is a deviation in the usual Michaelis-Menten kinetics due to cooperative interactions between the different enzyme subunits. As shown in Figure 2–25, the enzyme ATCase gives a sigmoidal type of curve instead of a hyperbola (Fig. 2–23). This shape results from the fact that the first substrate molecule bound enhances the affinity for the binding of the second substrate molecule, the second substrate molecule enhances the affinity for the binding of the third substrate molecule, and so on. This is called cooperativity. As may be observed in Figure 2–25, there is a region of the curve in which a small increase in [S] causes a large increase in activity. The addition of ATP causes a positive effect and the curve tends to be more hyperbolic, while with CTP there is an inhibitory action that affects the ATCase activity and the curve becomes more sigmoidal. By shifting in this way the activity of enzymes at a given [S], the cell can regulate the utilization of its

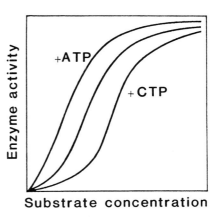

Fig. 2–25. Sigmoidal curve characteristic of a regulatory or allosteric enzyme. Observe the difference with a hyperbolic type of curve. Under the action of an activator (ATP) the curve becomes more hyperbolic, while with an inhibitor (CTP) it becomes more sigmoidal.

metabolic pathways, in this example by modifying ATP or CTP concentrations.

Enzyme Regulation at the Genetic and Catalytic Levels

The living cell seldom wastes energy synthesizing or degrading more material than necessary. Therefore, the thousands of chemical reactions that occur inside the cell must be carefully controlled. Enzyme activity is regulated by two major mechanisms: *genetic control* and *control of catalysis* (Fig. 2–26).

Genetic control implies a change in the total amount of enzyme molecules. The best known examples of this form of regulation are *enzyme induction* and *repression* in microorganisms in which enzyme synthesis is regulated at the gene level by the indirect action of certain small molecules. In Chapter 22 we will analyze in detail the cases of the *lactose operon* (induction) and the *tryptophan operon* (repression) of *E. coli*. We will see that gene expression is regulated by proteins called *repressors*, which bind to the DNA, and that small molecules such as the disaccharide lactose or the amino acid tryptophan can affect gene activity by binding to these repressors. As shown in Figure 2–26, in enzyme *induction* the availability of a substrate (e.g., lactose) induces the synthesis of the enzymes that degrade it, while in enzyme *repression* the accumulation of the end product of a metabolic chain (e.g., tryptophan) turns off production of

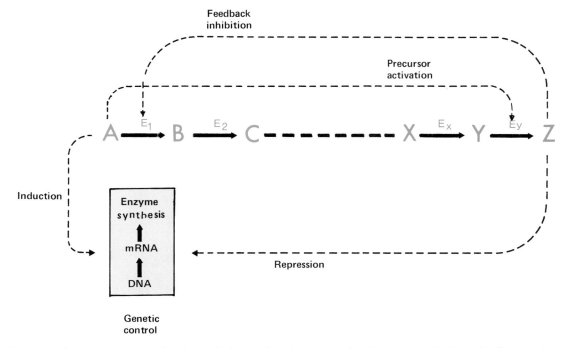

Fig. 2–26. Enzymes are regulated at the level of catalysis and at the gene level (enzyme synthesis). In *feedback inhibition* the end product (Z) of a metabolic chain acts as an allosteric inhibitor of the *first* enzyme of the pathway. In *precursor activation* the first metabolite (A) of a pathway is an allosteric activator of the final enzyme. The *genetic control* mechanisms modify enzyme synthesis according to cellular requirements. In *induction* (e.g., the lactose operon) the presence of a substrate (A) stimulates the synthesis of the enzymes that degrade it, while in *repression* (e.g., the tryptophan operon) the accumulation of the end product (Z) switches off enzyme production.

the enzymes involved in its synthesis. In both cases, the net result is that enzymes are synthesized only when required.

Control of catalysis involves a change in enzyme activity without a change in the total amount of enzyme synthesized. This is frequently produced in regulatory or allosteric enzymes by the action of allosteric activators or inhibitors. Two important mechanisms for this type of control are *feedback inhibition* and *precursor activation* (Fig. 2–26). In *feedback inhibition* the end product of a metabolic pathway acts as an *allosteric inhibitor* of the first enzyme of the metabolic chain. Thus, when enough product is synthesized, the entire chain can be shut off, and useless accumulation of metabolites is avoided. In *precursor activation* the first metabolite of a biosynthetic pathway acts as the *allosteric activator* of the last enzyme of the sequence (Fig. 2–26).

Another way in which metabolism may be regulated involves enzyme *interconversions*. Some enzymes may exist in two forms (active and inactive) that are interconvertible. Frequently the

mechanism of interconversion consists of *phosphorylation*, i.e., the covalent binding of a phosphate group that is provided by ATP.

Genetic control (induction, repression) is generally considered to be a coarse and relatively slow type of regulation, and feedback inhibition a finer, almost instantaneous way of ensuring that enzyme activity is adequate for cellular requirements.

Cyclic AMP—The Second Messenger in Hormone Action

Hormones are molecules that transfer information from one group of cells to another distant tissue. The molecular bases of hormonal action were very poorly understood until 1956, when E.W. Sutherland discovered *cyclic adenosine monophosphate* (cAMP), a cyclic nucleotide that has been found to regulate a large number of metabolic processes.

In 3′-5′ cAMP the phosphate group is covalently bound to the 3′ and 5′ carbons of the ribose ring. (Normally in AMP only the 5′ carbon is bound to the phosphate.) This cyclic nucleotide is synthesized by adenylate cyclase, an enzyme that is tightly bound to the cell membrane. Many hormones modify the activity of adenylate cyclase and therefore produce a change in the intracellular level of cAMP. In Sutherland's model of hormonal action (Fig. 2–27) the hormone is regarded as a *first messenger* that interacts with specific receptor sites located in the outer surface of the cell membrane. This hormone-receptor interaction results in a change in the activity of adenylate cyclase, which is mediated by an additional protein that binds GTP and is indicated as R in Figure 2–27. This GTP-binding protein is the product of the *c-ras* oncogene, which is frequently mutated in human tumors. The active site of adenylate cyclase is on the inner surface and using ATP as substrate, it produces cAMP in the cytoplasm. This nucleotide is considered as a *second messenger* that carries the information to the metabolic machinery of the cell. The effect of cAMP depends on the target organ; for example, an increase in cAMP will produce glycogen degradation in the liver and steroid production in the adrenal cortex.

Many hormones are known to act by way of specific receptors that stimulate adenylate cyclase. Among these hormones are: epinephrine (adrenalin), norepinephrine, glucagon, adrenocorticotropic hormone (ACTH), thyroid-stimulating hormone (TSH), melanocyte-stimulating hormone (MSH), parathyroid hormone, luteinizing hormone (LH), vasopressin, and thyroxine.

Each hormone listed has a particular receptor protein. However, all the receptors can interact with the same adenylate cyclase enzyme via the GTP-binding protein (Fig. 2–27). Some cells may have more than one receptor (e.g., liver cells respond both to epinephrine and to glucagon). As we will see in Section 4–2 the receptors are mobile and not permanently bound to adenylate cyclase; they can diffuse on the lipid bilayer, thus enabling the enzyme to interact with several receptors.

The main way in which cAMP affects metab-

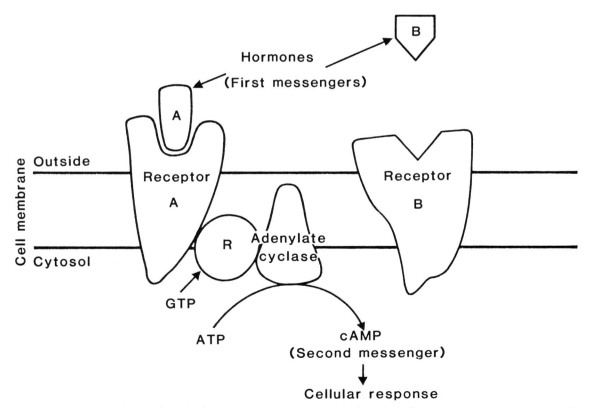

Fig. 2–27. Diagram showing how the hormones (first messengers), which are bound to specific receptors on the outside of the cell, can activate the enzyme adenylate that produces cAMP inside the cell (second messenger). The receptor molecules for various hormones can diffuse laterally in the lipid bilayer of the plasma membrane. (An additional regulatory (R) protein binds GTP, which is required for the receptor-cyclase interaction.)

olism is by stimulating the activity of *protein kinases*, a group of enzymes that catalyze the phosphorylation of proteins:

$$\boxed{\text{protein}} + \text{ATP} \longrightarrow \boxed{\text{protein}} \!-\! \textcircled{P} + \text{ADP} \quad (7)$$

Figure 2–28 shows the way in which cAMP activates protein kinase. The enzyme has two subunits, one catalytic, and one regulatory. The latter can bind cAMP. In the absence of cAMP the regulatory and catalytic subunits form a complex that is enzymatically inactive. When cAMP binds to the regulatory subunit it causes a conformational change, and the complex dissociates. The catalytic subunit thus freed is enzymatically active and will phosphorylate other enzymes.

The classic example of regulation produced by cAMP is the degradation and synthesis of glycogen. As shown in Figure 2–29 epinephrine (adrenalin) and glucagon induce the degradation of glycogen in the liver. These hormones, acting on the corresponding receptor, stimulate adenylate cyclase and raise the cAMP level. Cyclic AMP activates *protein kinase* by the mechanism shown in Figure 2–28. Protein kinase phosphorylates *phosphorylase b kinase*, thus converting it into its active form. This kinase is a specific enzyme that will, in turn, phosphorylate *phosphorylase a. Phosphorylase a* is the active form of the enzyme that degrades glycogen into its glucose units. By this cascade mechanism, the cell amplifies considerably the initial signal given by the hormones at the membrane level (Fig. 2–29).

Glycogen synthesis is also affected by the rise in cAMP. In this case, however, the phosphorylated form of *glycogen synthetase* is the *in-*

active form of the enzyme. As a result, an increase in cAMP increases glycogen degradation and coordinately decreases glycogen synthesis in the liver. This is the molecular mechanism by which epinephrine (which is liberated in conditions of alarm or stress) increases the amount of glucose available to the bloodstream.

Cyclic AMP is also involved in certain pathological conditions that are important in medicine. For example, a toxin of *Vibrio cholerae*, the bacterial agent in cholera, activates adenylate cyclase in the intestine. This stimulates salt and water secretion which may lead to a lethal diarrhea. Another aspect that is being actively investigated relates to the area of cancer research. It is known that many cancer cells have low levels of cAMP. Furthermore, it is known that certain abnormal features of cancer cells growing in vitro may be restored to normality by the addition of this cyclic nucleotide to the culture medium (see Fig. 6–24).

Cyclic GMP is a second type of cyclic nucleotide found in cells. Sometimes (although not always) the levels of GMP are in an inverse relationship to those of cAMP (when cAMP increases, cGMP decreases, and vice versa).

Calcium, Calmodulin, and Regulation of Cellular Functions

In addition to cAMP, calcium plays a fundamental regulatory role in a variety of cellular functions. Throughout this book, we shall see that Ca^{++} is involved in the regulation of contraction, secretion, endocytosis, transport across membranes, and in more general processes such as cell motility, cell growth, and cell division.

In a cell in resting condition, the intracellular

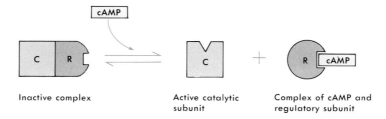

Fig. 2–28. Activation of a protein kinase by cAMP. The enzyme has a catalytic (C) and a regulatory (R) subunit, which form an inactive complex. Cyclic AMP binds to the regulatory subunit and induces a conformational change. The free catalytic subunit is the active form, which phosphorylates proteins in the reaction:

$$\boxed{\text{Protein}} + \text{ATP} \xrightarrow{\text{C}} \boxed{\text{Protein}} \!-\! \textcircled{P} + \text{ADP}$$

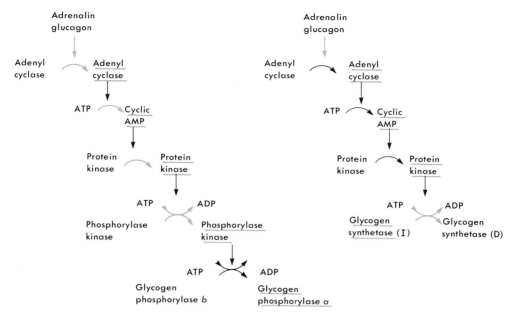

Fig. 2–29. Hormones regulate glycogen metabolism by inducing a cascade of reactions mediated by cAMP. The active forms are underlined. Note that phosphorylation activates glycogen phosphorylase (thus increasing glycogen degradation), but inhibits glycogen synthetase. The cascade mechanism greatly amplifies the hormonal signal.

Ca^{++} concentration is generally maintained at the low level of $10^{-7}M$ against a concentration gradient, the extracellular Ca^{++} being about $10^{-5}M$. In the stimulated state, for example during muscle contraction, there is Ca^{++} influx across the plasma membrane and the intracellular concentration rises. In Section 7–4 we shall see that restoration of the Ca^{++} level is obtained by the exit of this ion from the cell together with its sequestration into intracellular organelles.

In 1970 calcium was considered another intracellular messenger. It was later recognized that calcium acts mainly through a series of proteins that have the capacity to bind Ca^{++} and to serve as intermediaries in its regulatory action. Of these *calcium binding proteins*, the first to be extracted from muscle was *troponin C*. In Chapter 7 we shall learn about the important function played by this protein in contraction (see Fig. 7–10). A series of other proteins that exhibit affinity for Ca^{++} has been found, and of these the most widespread and best known is *calmodulin*. This protein was first recognized as a factor in the regulation of *phosphodiesterase*, the enzyme that hydrolyzes cAMP to 5'AMP, thus terminating the effect of the cyclic nucleotide (Cheung, 1982).

The interaction of calmodulin (CaM) with the enzyme (Enz) occurs in two steps, which can be described by the following equations:

(1) $CaM + nCa^{++} \rightleftharpoons CaM.Ca_n^{++} \rightleftharpoons CaM^*Ca_n^{++}$

(In which CaM* indicates that this protein has undergone a conformational change by the binding of Ca^{++}).

(2) $CaM^* Ca_n^{++} + Enz \rightleftharpoons (CaM^* Ca_n^{++}).\ Enz \rightleftharpoons (CaM^* Ca_n^{++}).\ Enz^*$

(In which Enz* indicates the activated state of the enzyme phosphodiesterase).

This example illustrates one of the various enzymes that are stimulated by the calcium-calmodulin complex. Others are brain adenylate cyclase, the Ca^{++} ATPase of erythrocytes, phosphorylase b kinase, and myosin kinase.

While several examples on the function of calmodulin in cell regulation will be given later in this book, we can now mention that the calcium-calmodulin system is frequently coupled to the cAMP system and that one of the important roles of calmodulin-Ca^{++} is precisely that of regulating the metabolism of cyclic nucleotides.

Calmodulin is a small acidic protein of about 17,000 daltons and 148 amino acids; its structure is well conserved during evolution, it is not species or tissue specific, and it influences a variety of cellular functions. The sequence shown in Figure 2–30 reveals the presence of four loops corresponding to Ca++ sites to which the ion binds with slightly different affinity. The binding of Ca++ to each of the sites produces a conformational change; therefore, one could envision the production of 16 different conformations for this flexible molecule.

The binding of Ca++ causes an increase in α helical content and thus a more compact conformation, with the exposure of hydrophobic regions. Some of these regions serve as binding sites for drugs, such as *trifluoperazine*, that inhibit calmodulin, or that may be used to bind other acceptor proteins, such as the recently described *calcimedins* (Moore and Dedman, 1982).

Calmodulin is present in the cytoplasm of all tissue cells, but it is excluded from the nucleus. In mitosis it becomes associated with the spindle (Section 15–2) and is related to the assembly and disassembly of microtubules (Section 6–2). In the brain, calmodulin is highly concentrated and probably plays an important role in the transmission of the nerve impulse and the phosphorylation of proteins in nerve terminals (Section 24–5).

SUMMARY:

Enzymes in the Cell

Enzymes are proteins that act as biological catalysts, accelerating chemical reactions. It has recently been found that some RNA molecules may also have catalytic activity. They contain a so-called *active site* to which the substrate attaches, forming a temporary [ES] complex; then the substrate is converted into one or more products, and the enzyme becomes free again (E + S ⇄ [ES] → E + P). The specificity for the substrate is considerable, and frequently analogues of a slightly different shape will not work at all.

The *lock-and-key theory* provides one explana-

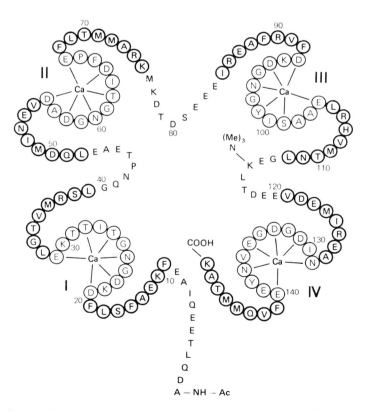

Fig. 2–30. Diagram showing the amino acid sequence of calmodulin from bovine brain. The amino acid residues are indicated by one letter code. Observe the four loops corresponding to the four Ca++ binding sites in which each ion is bound to six amino acids. The darker circles correspond to regions of α-helix. (From Klee, G.B., Crouch, T.H., and Richman, P.G.: Annu. Rev. Biochem., 49:489, 1980.)

tion for the specificity of the active site, which is dependent on the primary, secondary, and tertiary structure of the protein. The shape of the active site is complementary to that of the substrate. According to the *induced-fit theory*, the enzyme-substrate interaction may cause a conformational change in the protein. The binding of the substrate to the active site is by short range, noncovalent forces. After the ES complex is formed, the catalytic step, in which the substrate can undergo hydration, dehydration, oxidation, reduction, or transfer of chemical groups (among other processes), proceeds.

Some enzymes require *cofactors* for their activity. For example, the prosthetic group in cytochromes consists of a metalloporphyrin complex. Other enzymes use small nonprotein molecules, i.e., *coenzymes*, which become bound during the reaction to activate the enzyme (apoenzyme + coenzyme → holoenzyme). Important coenzymes are NAD^+ and NADP. Vitamins of the B group, such as nicotinamide, pantothenic acid, riboflavin, and pyridoxal, act as cofactors.

In the case of many enzyme-catalyzed reactions, the velocity of the reaction depends on the substrate concentration. The characteristic curve described in these reactions is a hyperbola that reaches a maximum velocity (Vmax) when all the enzyme active sites are saturated. The Km (Michaelis constant) is the substrate concentration at which the velocity is $\frac{Vmax}{2}$ (Fig. 2–23). The smaller the Km is, the greater is the affinity for the substrate.

Enzyme inhibition may be reversible or irreversible. Reversible inhibition may be *competitive* or *noncompetitive*. In competitive inhibition the inhibitor has a molecular structure similar to that of the substrate. In this case, Vmax is not changed, but the Km increases (Fig. 2–24). In noncompetitive inhibition the inhibitor is not structurally related to the substrate; Vmax is decreased and Km remains unchanged (Fig. 2–24).

Isoenzymes are multiple molecular forms of the same enzyme that differ in their electrophoretic mobility. Lactic dehydrogenase is a tetramer that has five isoenzymes derived from all the combinations possible between two types of subunits (M_4, M_3H_1, M_2H_2, M_1H_3, and H_4).

In some metabolic pathways the enzymes involved in a chain of reactions are bound to one another in a *multi-enzyme complex*. The substrate passes from one enzyme to the next, without needing to diffuse freely. The most complex multi-enzyme systems are associated with biological membranes, forming a precise two-dimensional array of enzymes.

RNA can act in a catalytic form in some circumstances. The intervening sequence of the ribosomal genes of the protozoon tetrahymena can cleave and rejoin oligonucleotides. Ribonuclease P, an enzyme that is involved in tRNA maturation, contains an RNA species that will leave 5' extensions on tRNA precursors. Thus some RNAs are enzymes. This gives a fresh perspective in the origin of life on earth, in which catalysis was perhaps first carried out by RNA before proteins arose.

Allosteric or regulatory enzymes have a sigmoidal V/[S] curve. Many of them are oligomers composed of two (dimer), four (tetramer), or more protein subunits. The binding of a substrate molecule to one subunit enhances the affinity for binding a second S molecule, the binding of the second substrate molecule enhances the affinity for the binding of the third substrate molecule, and so on. These enzymes have a great regulatory value since their activity can be changed by small modifications in the concentration of the substrate.

Regulatory enzymes are sensitive to *modifiers* or *modulators* that bind to a separate allosteric site which, in turn, influences the active site. In the case of aspartate transcarbamylase, the allosteric inhibitor CTP binds to a different subunit than the substrate.

The *regulation of enzyme activity* is by two major mechanisms: *genetic control* and *control of the catalytic activity*. In genetic control there is a change in the amount of enzyme as, for example, in *enzyme induction* and *repression* (see Chapter 22.) Control of catalysis consists of a change in the activity of the enzyme. This control may be achieved by *feedback inhibition*, *precursor activation*, and *enzyme interconversion* (see Fig. 2–26).

In feedback inhibition the *end product* of a metabolic pathway acts as an allosteric inhibitor of the *first* enzyme of the pathway, thus ensuring that the cells do not produce more metabolites than necessary.

Some enzymes are interconverted from a less active into a more active form. Frequently, this is done by phosphorylation of the enzyme. For example, *glycogen phosphorylase* (the enzyme that degrades glycogen) is more active when phosphorylated, while *glycogen synthetase* becomes less active when phosphorylated. Proteins are phosphorylated by enzymes, called *protein kinases*, that catalyze the reaction:

$$\text{Protein} + \text{ATP} \longrightarrow \text{Protein—}\textcircled{P} + \text{ADP} \qquad (8)$$

The activity of some protein kinases is stimulated by *cAMP* (Fig. 2–28). The level of cAMP in cells is regulated by *hormones*, which act on specific *receptors* on the outer cell surface, which in turn modify the activity of *adenylate cyclase*, the enzyme that synthesizes cAMP. The cyclic nucleotide is the *second messenger* in the action of many hormones, such as epinephrine, glucagon, ACTH, TSH, MSH, LH, and thyroxine. Figure 2–29 shows the way in which hormones affect glycogen metabolism, i.e., in a cascade of reactions that involve an increase in cAMP and the phosphorylation of enzymes.

Calcium regulates several basic cellular functions such as contraction, secretion, endocytosis, transport, motility, growth, and division. Ca^{++} acts by way of binding proteins such as troponin C and calmodulin. Calmodulin is a small acidic protein that has four Ca^{++} binding sites and the ability to undergo different conformational changes. It acts on several enzymes by a two step mechanism. Cyclic AMP and the calmodulin -Ca^{++} complex are frequently coupled in a variety of functions in cells.

2–7 THE ASSEMBLY OF MACROMOLECULES AND THE ORIGIN OF CELLS

Some of the molecular constituents of the cell can interact among themselves and become organized into supramolecular units. These units, in turn, are parts of structures recognizable within the cells by means of the EM.

When the molecules are associated linearly, the elementary units are primarily *unilinear* (fibrous); when the molecules are extended in two dimensions forming thin membranes, the units are *two-dimensional;* and when they are crystalline or amorphous particles, the units are *three-dimensional.*

In certain systems supramolecular structures may aggregate to form higher types of organization visible under the light microscope and even to the naked eye. In animal and plant tissues there are several series of components with this type of organization. They can be classified into three categories: *subcellular,* which comprise parts of cells, such as membranes, cilia, and chromosomes; *extracellular,* such as collagenous and elastic fibers, membranes of cellulose, or chitin situated outside the cells; and *supracellular,* which are macroscopic structures, such as hair, bone, and muscle, with a more complex supramolecular organization.

The function of these supramolecular structures in biological systems will be mentioned throughout this book. Several of these molecular systems are involved in *mechanical functions;* e.g., collagen fibers form tendons; fibrin fibers are used in blood clotting to prevent bleeding; and muscle proteins interact to produce shortening during contraction. Several of these supramolecular structures have *enzymatic properties* and may constitute *multi-enzymatic complexes.*

Proteins and Nucleic Acid May Self-Assemble

Earlier we mentioned that *self-assembly* can occur in oligomeric proteins such as hemoglobin

(Fig. 2–17). In self-assembly, the protein subunits contain the necessary information to produce the larger complex by means of secondary bonds. Complex macromolecules may be formed in the cell by the principle of self-assembly. One extraordinary example of a multienzyme is provided by the huge *pyruvate dehydrogenase complex* of *E. coli,* which contains three groups of enzymes and a total of 88 protein subunits.

In addition to *simple* self-assembly, in which no other component is involved, there is also *aided assembly,* in which certain enzymes may prepare the macromolecules for assembly. One example of aided assembly is that of blood clotting, in which fibrinogen must be activated by thrombin by the removal of a small peptide before the fibrin network can be formed. An excellent example of self-assembly is that of *viruses,* which are assembled from the genetic material contained either in DNA or RNA and the protein coat or *capsid.*

When outside the cell, viruses are metabolically inert and may even be crystallized. When they enter the host cell, they use their own genomes to program the replication of new virus particles, but they use the biosynthetic machinery (e.g., ribosomes) of the host cell to express the information that those viral genomes carry.

Tobacco mosaic virus (TMV), for example, is a particle 40×10^6 daltons in mass, with the form of a cylinder of 16×300 nm. This cylinder contains a single-stranded molecule of RNA consisting of 6500 nucleotides, forming a helix with a radius of 4.0 nm, and having a cylindrical cavity of 2.0 nm. Associated with this RNA helix and forming the protein coat are 2130 identical protein subunits of 18,000 daltons (Fig. 2–31).

In a classic series of experiments, Heinz Fraenkel-Conrat separated both components of the TMV virus and then reconstituted the active virus particles in the test tube. In this case, all the information needed for "rebuilding" the virus was inherent in its parts. Therefore, the self-assembly took place in solution, but at an extremely slow rate (6 hours). The mechanism of normal assembly in the living cell remained unclear, however, for the rate of this assembly was too slow. Furthermore, the viral RNA must be protected by the protein while the polynucleotide chain is growing to avoid degradation by ribonucleases. The solution of this problem

Fig. 2–31. Diagram of the molecular organization of the tobacco mosaic virus (TMV). In the center there is a spiral of RNA that is associated with protein subunits. There is one protein monomer for every three bases in the RNA chain. (From Caspar, D.L., and Klug, A.: Cold Spring Harbor Symp. Quant. Biol., 27:1, 1962.)

became evident when it was observed that the protein coat can, by itself, aggregate into disks made of about 18 subunits. As shown in Figure 2–32 the nucleation of TMV virus begins with the insertion of a hairpin loop of RNA into the central hole of the protein disk. Then the loop intercalates between two layers of subunits where the RNA strand becomes trapped. This is the start of the helical structure of the virus, which then grows by the addition of more disks and the elongation of the RNA molecule. When a new cell is infected, the virus particle must disassemble to liberate the viral RNA. This probably occurs with the removal of individual subunits from one of the ends of the TMV particle. Thus the mechanisms of assembly and disassembly are not simple reversals of one another (Butler and Klug, 1978).

The Origin of Cells

The way in which structures of higher and higher order are built in a cell is determined by the genetic information contained in DNA. This determines the primary sequence of the polypeptide chain in the protein, which in turn determines the secondary and tertiary structures, and finally, the formation of the oligomeric complexes. The interaction of different proteins with lipids and nucleic acids, on the other hand, results in the formation of molecular complexes, membranes, and structures of a higher order of complexity.

A fundamental problem is to determine by which mechanisms supramolecular organization originated on our planet and gave rise to the prokaryotic and eukaryotic cells. Any discussion of this problem should be considered highly speculative since it is directly related to the origin of life.

Although we do not know how cells were first formed, from fossil records it has been possible to establish that prokaryotic organisms preceded the eukaryotes and appeared between 3.5 and 3.0×10^9 years ago. Recent observations have shown that only one billion years after the earth was formed (i.e., about 3.5×10^9 years ago) there were bacteria-like organisms, whose fossil remnants were found in ancient Australian rocks (Bartusiak, 1981). Before that time there must have been a long period of *chemical evolution* in which carbon-containing molecules and macromolecular precursors, such as amino acids, sugar, and nucleic acid bases, appeared. Then, by polymerization, macromolecules and structures of a higher order of complexity were formed. It is possible that during this period the mechanisms of assembly of macromolecules discussed earlier were at work to give rise to the first self-reproducing supramolecular structures (Table 2–6).

Chemical Evolution Produced Carbon-containing Molecules

In *prebiotic times*, that is, before the origin of life, the atmosphere of the earth was devoid of molecular oxygen, and, as in the case of the major planets Jupiter and Saturn at present, it contained largely hydrogen, nitrogen, ammonia (NH_3), methane (CH_4), carbon monoxide (CO), and carbon dioxide (CO_2). There was also water, as vapor and as liquid covering parts of the earth's surface. Although normally these molecules do not react, they could have interacted

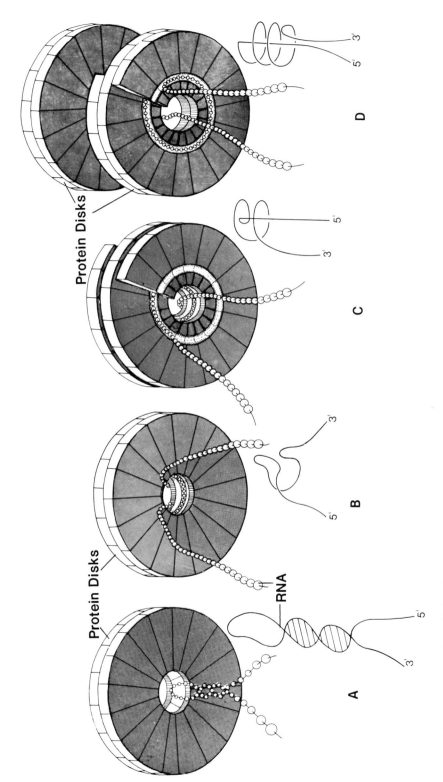

Fig. 2–32. From left to right this diagram depicts the nucleation of the tobacco mosaic virus (TMV), which starts with the insertion of an RNA hairpin loop into the disk of protein subunits (A), its intercalation between the two layers of subunits (B), and the trapping of the viral RNA (C). Finally, the RNA-disk complex provides for the start of the helix, with addition of disks and elongation of the RNA molecule (D). (Courtesy of A. Klug and P.J. Butler.)

TABLE 2–6. EVOLUTIONARY STEPS IN THE ORIGIN OF CELLS

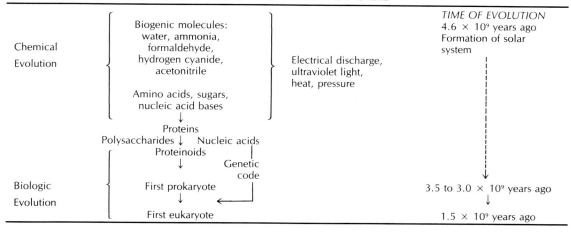

then because of the energy provided by ultraviolet radiation, heat, and electrical discharges (lightning). At that time the atmosphere also lacked a protective ozone layer, which acts as a filter. Therefore, ultraviolet rays could bathe the earth's surface with an intensity that would be lethal to modern animal life. This gave rise to highly reactive intermediary molecules, such as acetaldehyde, hydrogen cyanide, formaldehyde, and others, from which final products could have been synthesized (e.g., acetic acid, a simple fatty acid; or alanine, a simple amino acid).

By 1920 Oparin in Russia and Haldane in Britain postulated that these molecules, by polymerization, could have given origin to the proteins, nucleic acids, and carbohydrates found in living organisms. In 1953 Stanley Miller performed a fundamental experiment using conditions imitating those of the atmosphere of prebiotic times. He produced electrical discharges in a flask into which water vapor, H_2, CH_4, and NH_3 had been injected. In the condensed water he found that several amino acids, such as glycine, alanine, and glutamic and aspartic acids, had formed. So far 17 amino acids (of the 20 present in proteins) have been obtained in experiments simulating prebiotic conditions. Other compounds that have been formed are: sugars, fatty acids, and the bases that form part of the nucleic acids. For example, adenine, which is present in DNA, RNA, and ATP, has been produced in high yields.

Mechanisms of Assembly Were at Work to Form Primitive Proteinoids

We can speculate that the next step, *biological evolution*, was probably the polymerization of amino acids to form proteins. This could have been initiated by the catalytic action of clays. All the following prebiotic evolutionary processes probably occurred in a water medium (i.e., in ponds) in which the organic molecules were concentrated, forming a kind of *"primordial soup"* in which molecular interactions were favored. Once the first protein had been formed, the mechanisms of assembly described earlier in this chapter could have operated. The primary sequence of amino acids originates the secondary, tertiary, and—with the formation of oligomers—even the quaternary structure of proteins. In this way the enzymatic functions could have arisen. In the primordial soup the macromolecules probably formed larger complexes called *coacervates* or *"proteinoid"* droplets, which have a membrane-like outer surface and a fluid interior.

Primitive proteinoids could have exhibited enzymatic and transport activities, as in the case of the artificial membranes described in Chapter 4. These proteinoids, without nucleic acids to act as informational molecules, could not have had genetic continuity; thus, it is possible that extensive trial and error occurred, leading to short-lived structures.

Only after the origin of the genetic code, determined by the sequence of bases in nucleic acids, could a self-perpetuating organism have arisen in which the laws of natural selection had begun to operate. At that point, a first prokaryote with the minimum living mass was originated and life emerged on Earth.

It seems likely that RNA, and not DNA, was the primordial genetic material, so that from a chronological viewpoint macromolecules evolved as:

It was simpler to start with RNA because it can be used both as genetic material and as mRNA. Most key steps in the protein synthesis machinery depend on RNA-RNA interactions, such as mRNA—tRNA, rRNA—mRNA, and rRNA—tRNA. The replication of RNA is simpler than that of DNA, which requires a plethora of enzymes (see Chapter 14). Recombination is also much simpler in RNA, and perhaps the present RNA splicing mechanisms are a reflection of what started as a primitive mode of genetic recombination. Many important cofactors in biochemical reactions have nucleotides attached to them, such as NAD (nicotinamide-adenine dinucleotide) and UDPG (uridine diphosphate glucose, a sugar donor molecule), which might now be "biochemical fossils" of a time when chemical reactions were coupled by molecules whose "handles" were complementary to each other in the Watson-Crick sense (Reanney, 1979).

The recent finding that some RNA molecules have enzymatic activities (Zaug and Cech, 1986) suggests that in the beginning perhaps proteins were not required at all. If during early evolution RNA enzymes, also called ribozymes, had been able to replicate RNA, there would have been no need for protein molecules in the beginning of life. Today's introns can excise themselves (self-splicing, see Fig. 20–16); if this step were reversible, then the intron/exon structure may have been a way of recombining fragments of RNA, in a prebiotic version of sex. If RNA could catalyze its own replication, then splicing of some of the daughter molecules could have

given rise to new ribozyme activities (Gilbert, 1986). By using RNA cofactors such as NAD, an entire range of enzyme activities was then developed. Only with the appearance of a rudimentary genetic code would the first protein enzymes have been made, which turned out to be much better enzymes than their RNA counterparts and eventually dominated. Finally, DNA appears on the scene as a stable linear information storage molecule, which is error-correcting because of its double-stranded structure, but still able to undergo recombination and mutation. At this point RNA was relegated to the intermediate role it presently has today. Walter Gilbert (1986) suggests that the intron/exon structure of genes we see today is a relic of the basic mechanism of RNA recombination that was left imprinted in the DNA after RNA lost the center of the stage in early evolution.

A fundamental step came when specific aminoacyl-tRNAs appeared, permitting an orderly sequence for the addition of amino acids in protein synthesis. This then led to the elaboration of a genetic code. Once completed, the code became fixed because any changes in it would have introduced mutations in too many proteins simultaneously, with lethal effects. All living organisms have the same genetic code (see Chapter 19). (In Chapter 19 we shall see that the statement that there is a single genetic code is not completely correct because mitochondrial DNA has a slightly different code.)

The forces of evolution, selecting for favorable mutations in these primitive cells, subsequently led to the almost infinite variety of life forms on earth.

It is possible that the first prokaryotes were *heterotrophic* (i.e., they drew their nutrients from the organic molecules) and *anaerobic*, since there was no oxygen in the atmosphere. At a later time *autotrophic* prokaryotes, such as the blue-green algae, which have photosynthetic pigments, appeared. Because of photosynthesis, oxygen was produced, and it accumulated in the atmosphere, allowing the formation of *aerobic* prokaryotic cells.

It is only after the appearance of the autotrophic prokaryotes that the eukaryotic cell could have originated. From fossil records it is surmised that eukaryotic organisms appeared some 1.500 to 1.400 million years ago. A stable oxygen atmosphere had long been established and the organisms could then be fully aerobic. Thus far life had always been present in water and it was only later on that plants and animals colonized the earth. The invention of sex—the interchange of genetic information between individuals—some two billion years ago accelerated the evolution of living forms that until then was rather slow. Sex made possible by mutation and selection the immense number of different living forms now found on our planet.

SUMMARY:
Assembly of Macromolecules and the Origin of Life

The assembly of macromolecules may be brought about by simple *self-assembly*, in which no other component is involved; by *aided assembly*, in which an enzymatic process may prepare macromolecular parts for assembly, or by *directed assembly*, in which a template directs the formation. Self-assembly is found in some multi-enzyme complexes and in certain virus particles.

In the assembly of macromolecules, the size and shape of the protein subunits may play an important role. In some instances, such as with TMV (tobacco mosaic virus) and with ribosomes, the interaction between RNA and protein may be fundamental to guiding the assembly of the final structure.

From fossil records it has been determined that prokaryotic cells preceded eukaryotic and appeared on our planet between 3.5 and 3.0 $\times$ 10^9 years ago. In prebiotic times there was probably a long period of *chemical evolution* in which H_2, NH_3, CH_4, and CO_2 contained in the atmosphere interacted to form carbon-containing compounds. Experiments using conditions imitating those of the atmosphere at that time have yielded sugars, fatty acids, nucleic acid bases (e.g., adenine), and several amino acids. This step, chemical evolution, was probably followed by another, *biological evolution*, in which proteins were formed by polymerization, and aggregates appeared in the water medium. With the origin of the genetic code (nucleic acids), self-perpetuating cells could have arisen. Other schools of thought believe that RNA came first, providing the initial catalytic activities, and was only later followed by proteins, and even later by DNA. Heterotrophic anaerobic prokaryotes preceded the autotrophic ones, such as blue-green algae. The production of oxygen by photosynthesis allowed the reproduction of aerobic cells.

ADDITIONAL READINGS

Alberts, B., et al.: Molecular Biology of the Cell. New York, Garland, 1983.

Anfinsen, C.B.: Principles that govern the folding of protein chains. Science, *181*:223, 1973.

Attenborough, D.: Life on Earth. England, Collins, 1979.

Bernal, J.D., and Synge, A.: The origin of life. *In* Readings in Genetics and Evolution. Oxford, The Clarendon Press, 1973.

Britten, R., and Kohne, D.: Repeated sequences in DNA. Science, *161*:529, 1968.

Butler, P.J.G., and Klug, A.: The assembly of a virus. Sci. Am., *239*:62, 1978.

Cavalier-Smith, T.: The origin of nuclei and eukaryotic cells. Nature, *256*:463, 1975.

Cheung, W.Y.: Calmodulin. Sci. Am., *246*:48, 1982.

Cold Spring Harbor Symposium of Quantitative Biology: Structure and Function of Proteins at the Three-Dimensional Level. Cold Spring Harbor Laboratory, Cold Spring Harbor, New York, 1973.

Cohen, F.: The role of protein phosphorylation in neural and hormonal control of cellular activity. Nature, *296*:613, 1982.

Davidson, J.N.: The Biochemistry of the Nucleic Acids. 8th Ed. London, Chapman & Hall, 1976.

Dayhoff, M.E.: Atlas of Protein Sequence and Structure. Silver Spring, Maryland, National Biomedical Research Foundation, 1972.

Dickerson, R.E.: Chemical evolution and the origin of life. Sci. Am., *239*:70, 1978.

Doolittle, R.F.: Proteins. Sci. Am., *253* (No. 4):88, 1985.

Eigen, M.: Molecular self-organization and the early stages of evolution. Q. Rev. Biophys., *4*:149, 1971.

Fersht, A.: Enzyme Structure and Mechanism. San Francisco, W.H. Freeman & Co., 1977.

Frieden, E.: The chemical elements of life. Sci. Am., *227*:52, 1972.

Fox, S., and Dose, K.: Molecular Evolution and the Origin of Life. San Francisco, W.H. Freeman & Co., 1972.

Gilbert, W.: The RNA world. Nature, *319*:618, 1986.

Guerrier-Takada, C., and Altman, S.: Catalytic activity of an RNA molecule prepared by transcription in vitro. Science, *223*:285, 1985.

Hubbard, S.C., and Ivatt, R.J.: Synthesis and processing of asparagine-linked oligosaccharides. Annu. Rev. Biochem., *50*:555, 1981.

Jacob, F.: Evolution and Tinkering. Science, *106*:361, 1977.

Judson, H.F.: The Eighth Day of Creation: Makers of the Revolution in Biology. New York, Simon and Schuster, 1979. Highly recommended. The history of molecular biology as viewed from Cambridge.

Klee, G.B., Crouch, T.H., and Richman, P.G.: Calmodulin. Annu. Rev. Biochem., *49*:489, 1980.

Klug, A.: Assembly of tobacco mosaic virus. Fed. Proc., *31*:30, 1972.

Lehninger, A.L.: Biochemistry: The Molecular Bases of Cell Structure & Function. New York, Worth Publishers, Inc., 1982.

Monod, J.: Chance and Necessity. New York, Random House, Inc., 1971.

O'Farrell, P.H.: High resolution two-dimensional electrophoresis of proteins. J. Biol. Chem., *250*:4007, 1975.

Oparin, A.I.: The origin of life. Scientia, *113*:7, 1978.

Oparin, A.I.: Proceedings of the First International Symposium on the Origin of Life on the Earth. Oxford, Pergamon Press, 1969.

Perutz, M.: Hemoglobin structure and respiratory transport. Sci. Am., *239*:68, 1978.

Phillips, D.C., and North, A.C.T.: Protein Structure: Oxford Biology Readers. Vol. 34. Oxford, Oxford University Press, 1975.

Reanney, D.: RNA splicing and polynucleotide evolution. Nature, *277*:598, 1979.

Rossmann, M.G., and Argos, P.: Protein folding. Annu. Rev. Biochem., *50*:497, 1978.

Schopf, W.: Chemical evolution and the origin of life. Sci. Am., *239*:70, 1978.

Schopf, W.: The evolution of the earliest cells. Sci. Am., *239*:110, 1978.

Smith, E.L., Hill, R., Lehman, I., Lefkowitz, R., Handler, P., and White, A.: Principles of Biochemistry. New York, McGraw-Hill Book Co., 1983.

Spiegelman, S.: An approach to the experimental analysis of precellular evolution. Q. Rev. Biophys., *4*:213, 1971.

Stadel, J.M., De Lean, A., and Lefkowitz, R.J.: Molecular mechanism of coupling in hormone receptor adenylate cyclase systems. Adv. Enzymol., *53*:2, 1982.

Staneloni, R.J., and Leloir, L.F.: The biosynthetic pathway of the asparagine-linked oligosaccharides of glycoproteins. Trends Biochem. Sci., *4*:65, 1979.

Stryer, L.: Biochemistry. San Francisco, W.H. Freeman & Co., 1981.

Watson, J.D.: Molecular Biology of the Gene. 3rd Ed. London, W.A. Benjamin, Inc., 1976.

Watson, J.D.: The Double Helix. New York, Atheneum, 1968. The story of one of the most important scientific discoveries of this century.

Watson, J.D., and Crick, F.H.C.: Molecular structure of nucleic acid: A structure for deoxyribose nucleic acid. Nature, *171*:737, 1953.

Zaug, A.J., and Cech, T.R.: The intervening sequence RNA of Tetrahymena is an enzyme. Science, *231*:470, 1986.

TECHNIQUES IN CELL BIOLOGY

The observation of biological structures is made difficult by the fact that cells are, in general, very small and transparent to visible light and thus are not visible to the unaided eye. Furthermore, it is hard to uncover their molecular organization and to establish the way in which molecules interact within the structure to determine cell functions.

The extraordinary progress that cell and molecular biology have experienced in recent years is a result of the development of new methods for the study of the cell and its subcellular and molecular components. Such methods are the result of the application of biophysical and biochemical techniques to the cell and tissues. The number of methods and experimental approaches used is so large that it is impossible to encompass them within a single chapter; there-

fore, many of them, particularly those pertaining to the biochemical aspects, to molecular biology, and to genetic engineering will be dealt with throughout this book.

This chapter should be considered an introduction to the study of some general methods used for the instrumental analysis of biological structures (e.g., light and electron microscopy). The instrumentation discussion will be complemented by a brief presentation of a few of the main cytological and cytochemical methods for observation and experimentation of cells and tissues. Special sections will be dedicated to the study of the living cell, to the methods of immunochemistry, radioautography, and cell fractionation that, in recent years, have greatly contributed to the progress of cell biology.

3–1 VARIOUS TYPES OF LIGHT MICROSCOPY

In Chapter 1, we described the *limits of resolution* of the eye and of the different types of microscopes used (see Table 1–1). The student should now be aware that the human eye is also able to detect variations in wavelength and intensity of visible light. Throughout the years, developments in microscopy have tended to increase, on one hand, the *resolving power,* and on the other to counteract the transparency of the cell by increasing the *contrast* of cellular structures.

The majority of cell components are essentially transparent, except for some pigments (more frequent in plant cells) that absorb light at certain wavelengths (colored substances). The low light absorption of the living cell is caused largely by its high water content, but even after drying, cell components show little contrast.

One way of overcoming this limitation is to use dyes that selectively stain different cell components and thus introduce contrast by light absorption. In most cases, however, staining techniques cannot be used in the living cell. The tissue must be fixed, dehydrated, embedded, and sectioned prior to staining, and all these procedures may introduce morphological and chemical changes.

Light Microscope—Resolving Power

In the *light microscope,* as in any other type of microscope, the resolving power depends upon the wavelength (λ) and the numerical aperture (NA) of the objective lens (Fig. 3–1). The *limit of resolution,* defined as the minimum distance between two points that allows for their discrimination as two separate points is:

$$\text{Limit of resolution} = \frac{0.61\lambda}{\text{NA}} \qquad (1)$$

The numerical aperture is: NA $= n \times \sin \alpha$. Here, n is the refractive index of the medium and $\sin \alpha$ is the sine of the semiangle of aperture. Remember that the limit of resolution is inversely related to the resolving power; i.e., the higher the resolving power, the smaller the limit of resolution.

Since $\sin \alpha$ cannot exceed 1, and the refractive index of most optical material does not exceed 1.6, the maximal NA of lenses, using oil immersion, is about 1.4. With these parameters it is easy to calculate from formula (1) that the limit of resolution of the light microscope cannot surpass 170 nm (0.17 μm) using monochromatic light of $\lambda = 400$ nm (violet). With white light, the resolving power is about 250 nm (0.25 μm).

Phase and Interference Microscopy—Detect Small Differences in Refractive Index

In recent years remarkable advances have been made in the study of living cells by the development of special optical techniques, such as *phase contrast* and *interference microscopy.* These two techniques are based on the fact that although biological structures are highly transparent to visible light, they cause *phase changes or retardations* in transmitted radiations.

Figure 3–2 indicates the effects of a nonabsorbent transparent material (A) and an absorbent transparent material (C) on a light ray. In A, the wave impinges on a material that has a refractive index different from that of the medium. In passing through the object, the amplitude of the wave is not affected, but the velocity is changed. If the refractive index of the material is higher than that of the medium, there is a *delay* or *retardation.* After the wave emerges from the object, its original velocity is re-established, but retardation is maintained.

In the phase contrast microscope, the small phase differences are intensified. The most lateral light passing through the objective of the microscope is advanced or retarded by an *additional* $\frac{1}{4}$ wavelength ($\frac{1}{4}$ λ) with respect to the central light and by an annular phase plate that introduces a $\frac{1}{4}$ wavelength variation in the back focal plane of the objective. (The phase effect results from the interference between the direct geometric image given by the central part of the objective and the lateral image, which has been retarded or advanced to a total of $\frac{1}{2}$ wavelength.) In *bright,* or *negative,* contrast the two sets of rays are added, and the object appears brighter than the surroundings; in *dark,* or *positive,* contrast the two sets of rays are subtracted, making the image of the object darker than the surroundings (Fig. 3–3). Because of this interference, the minute phase changes within the ob-

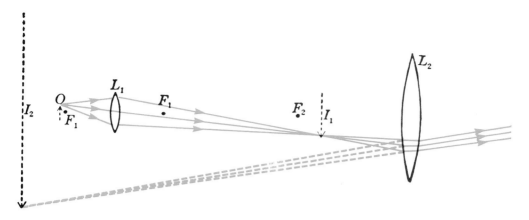

Fig. 3–1.   Light path in the ordinary light compound microscope. The group of ocular lenses is diagrammatically represented by L_2, the group of objective lenses by L_1. The object *(O)* on a microscope slide is placed just outside the principal focus of the objective lens *(L_1)*, which has a short focus. This lens produces a real image at I_1, which is formed inside the principal focus of the eyepiece lens *(L_2)*. *The eye, looking through the lens L_2, sees a magnified virtual image (I_2) of the image I_1. The eyepiece lens is thus used as a magnifying glass to view the real image (I_1).*

ject are amplified and translated into changes of amplitude (intensity).

A transparent object thus appears in various shades of gray, depending on its thickness and the difference between the refractive indexes of the object and the medium.

Phase microscopy is used routinely to observe living cells and tissues and is particularly valuable for observing cells cultured in vitro during mitosis (Fig. 3–3).

The *interference microscope* is based on principles similar to those of the phase microscope but has the advantage of giving quantitative data. Interference microscopy permits detection of small, continuous changes in refractive index, whereas the phase microscope reveals only sharp discontinuities. The variations of phase can be transformed into such vivid color changes that a living cell may resemble a stained preparation.

A special variation of the interference microscope is the so-called *Nomarski interference-contrast microscope*, in which the image obtained gives a characteristic relief effect and offers some advantage over ordinary phase contrast optics. It is particularly useful for the study of cells in mitosis.

Darkfield Microscopy—Based on Light Scattering at Cell Boundaries

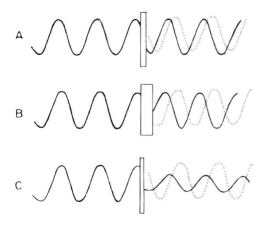

Fig. 3–2.   Diagram showing: *A,* the effect of a transparent and nonabsorbent material of higher refractive index than the medium, which introduces a phase change (retardation). *B,* the same, but thicker, object. The retardation or phase change is more pronounced. *C,* the effect of a transparent and absorbent object. There is a retardation, but also a decrease, in amplitude (intensity).

Darkfield microscopy, also called *ultramicroscopy*, is based on the fact that light is scattered at boundaries between regions having different refractive indexes. The instrument is a microscope in which the ordinary condenser is replaced by one that illuminates the object obliquely. With this darkfield condenser, no direct light enters the objective; therefore, the object appears bright because of the scattered light, and the background remains dark.

Under the darkfield microscope, objects smaller than those seen with the ordinary light microscope can be detected but not resolved.

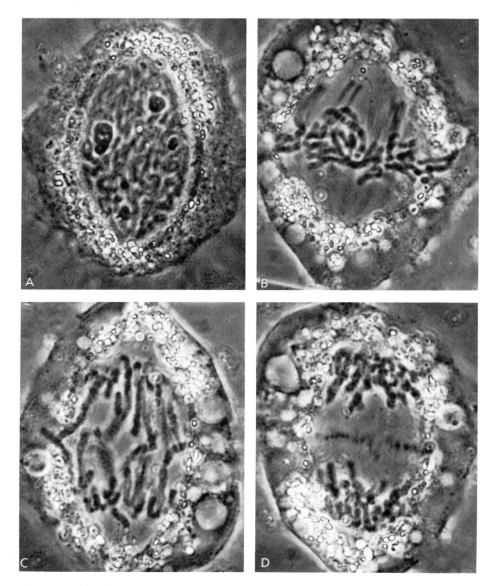

Fig. 3–3. Observation by phase contrast microscopy of mitotic cell division in a living cell of endosperm tissue of the plant *Haemanthus*. The same cell has been photographed at the following times: *A,* 10:32 hrs; *B,* 12:48 hrs; *C,* 13:12 hrs: and *D,* 13:21 hrs. *A,* late prophase showing the coiled chromosomes and the nucleoli within the nucleus; *B,* metaphase with chromosomes at the equatorial plane; *C,* anaphase; arid *D,* telophase showing the chromosomes at the poles and the formation of the phragmoplast at the equatorial plane. × 700. (Courtesy of A.S. Bajer.)

Polarization Microscopy—Detects Anisotropy with Polarized Light

This method is based on the behavior of certain components of cells and tissues when they are observed with polarized light. If the material is *isotropic*, polarized light is propagated through it with the same velocity, independent of the impinging direction. Such substances or structures are characterized by having the same *index of refraction* in all directions. On the other hand, in an *anisotropic* material the velocity of propagation of polarized light varies. Such material is also called *birefringent* because it presents two different indexes of refraction corresponding to the respective different velocities of transmission.

In biological fibers, birefringence is *positive* if the index of refraction is greater along the length of the fiber than in the perpendicular plane, and it is *negative* in the opposite case.

Birefringence (B) may be expressed quantitatively as the difference between the two indexes of refraction (N_e—N_o) associated with the fast and slow ray. In practice, the retardation (Γ) of the light polarized in one plane is measured relative to that of light polarized in another perpendicular plane with the polarizing microscope. The retardation depends on the thickness of the specimen (t) in this way:

$$B = N_e - N_o = \frac{\Gamma}{t} \qquad (2)$$

The *polarizing* microscope differs from the ordinary one in that two polarizing devices have been added: the *polarizer* and the analyzer, both of which can be made from a sheet of polaroid film or with Nicol prisms of calcite. The polarizer is mounted below the substage condenser and the analyzer is placed above the objective lens.

The polarizing as well as the interference microscope can be coupled to a video camera that enhances considerably the contrast and quality of the image (Inoué, 1981).

3–2 ELECTRON MICROSCOPY

The electron microscope (EM) permits a direct study of biological ultrastructure. Its resolving power is much greater than that of the light microscope. In the EM streams of electrons are deflected by an electrostatic or electromagnetic field in the same way that a beam of light is refracted when it crosses a lens. If a metal filament is placed in a vacuum tube and heated, it emits electrons that can be accelerated by an electrical potential and tend to follow a straight path with properties similar to those of light. Like light, the stream of electrons has a corpuscular and vibratory character, but the wavelength is much shorter (i.e., λ = 0.005 nm for electrons and 550 nm for light).

The filament or cathode of the EM emits the stream of electrons acting as a thermoionic gun. By means of a magnetic coil, which acts as a condenser, electrons are focused on the plane of the object and then are deflected by another magnetic coil, which acts as an objective lens and gives a magnified image of the object. This is received by a third magnetic "lens," which acts as an ocular or projection lens and magnifies the image from the objective. The final image

can be visualized on a fluorescent screen or recorded on a photographic plate (Fig. 3–4).

In spite of the apparent similarities there are great differences between the light and the EM; one of these is the mechanism of image formation.

While in the light microscope image depends on light absorption or phase changes, in the EM image is due principally to electron scattering. Electrons colliding against atomic nuclei in the object are often dispersed, so that they fall outside the aperture of the objective lens. The scattering may be elastic or inelastic. In *elastic dispersion* the image on the fluorescent screen results from the absence of those electrons blocked by the aperture. Dispersion may also be the result of multiple collisions, which diminish the energy of the passing electrons. In this case scattering is *inelastic*.

Electron dispersion is a function of the thickness and molecular packing of the object and depends especially on the atomic number of the atoms in the object. The higher the atomic number, the greater is the resultant dispersion. Most of the atoms that constitute biological structures are of low atomic number and contribute little to the image. For this reason, heavy atoms should be added to the molecular structure to increase contrast.

In the light microscope, magnification is largely determined by the objective, and a maximum magnification of 100 to 120× can be reached. Since the ocular lens can increase this image only 5 to 15 times, a total useful magnification of 500 to 1500× can be achieved.

In the EM the resolving power is so high that the image from the objective can be greatly enlarged. For example, with an initial magnification by the objective of 100×, the image can be magnified 200× with the projector coil, achieving a total magnification of 20,000×.

In the newer instruments a wide range of magnifications can be attained by introducing one or more intermediate lenses. Direct magnifications as high as 1,000,000× may thus be obtained, and the micrographs may be enlarged photographically to 10,000,000× or more, depending on the resolution achieved.

The light microscope differs from the EM in that the EM has a greater depth of focus, which has been used to develop three-dimensional electron microscopy of protein crystals and such

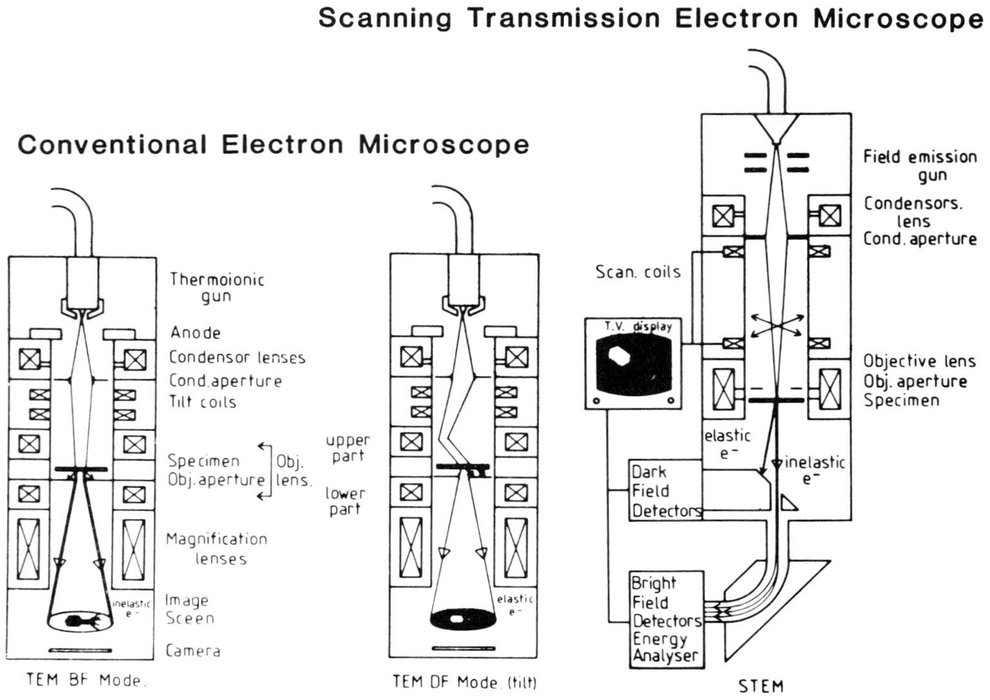

Scanning Transmission Electron Microscope

Conventional Electron Microscope

Fig. 3–4. Diagrams illustrating the basic principles of the conventional transmission electron microscope (TEM) and the scanning transmission electron microscope (STEM). For the TEM the bright field and the dark field modes are shown. See text for details. (Courtesy of E. Kellenberger.)

cell organelles as ribosomes (Hoppe, 1981; Amos et al., 1982).

Thin Specimens—Essential for EM Study

One limitation of the EM is the low penetration power of electrons. In current instruments, if the specimen is more than 500 nm (0.5 μm) thick, it appears almost totally opaque. The specimen is generally deposited on an extremely fine film (7.5 to 15 nm thick) of collodion, carbon, or other substance to support the specimen, and this film must be held up by a fine metal grid. For observation under the EM the specimen is usually dehydrated and then placed in a vacuum. Techniques for preparing specimens vary considerably. An important method for the study of macromolecules is the so-called monolayer technique of Kleinschmidt, in which the macromolecules are extended on an air-water interface before being collected on a film. This method has given excellent results in the demonstration of deoxyribonucleic acid (DNA) and ribonucleic acid (RNA) molecules from various sources (Fig. 3–5).

A similar technique, based on the disruptive action of an air-water interface, is used for the study of the structure of meiotic chromosomes (see Fig. 16–6). Another important method for the study of macromolecules, such as DNA and RNA, is the one developed by Oscar Miller, which is based on centrifugation, through appropriate media, directly onto a supporting film. This technique has provided the investigator with an important tool to observe minute changes in the DNA molecule and to follow, at the molecular level, the processes of transcription and translation. Numerous illustrations are provided in the corresponding chapters of this book.

Shadow Casting, Negative Staining, and Tracers—Increased Contrast

One technique, called *"shadow casting,"* consists of placing the specimen in an evacuated chamber and evaporating, at an angle, a heavy metal such as chromium, palladium, platinum, or uranium from a filament of incandescent tungsten. The material is thus deposited on one side

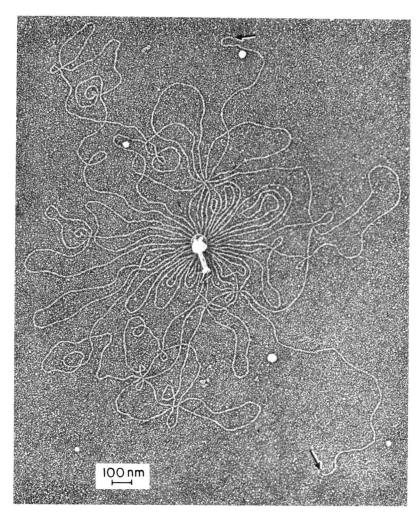

Fig. 3–5. Electron micrograph of a bacteriophage *(in the center)* that has undergone an osmotic shock. The DNA molecule that was contained in the "head" of the bacteriophage is now dispersed. Arrows indicate the extremes of the single, unbranched DNA molecule. Preparation shadowcast with platinum. × 76,000. (Courtesy of A.K. Kleinschmidt.)

of the surface of the elevated particles; on the other side a shadow forms, the length of which permits determination of the height of the particle. Photomicrographs made of such specimens have a three-dimensional appearance that is not found when other techniques are used (Fig. 3–5).

One of the most important techniques in the study of viruses and macromolecules is *"negative staining."* The specimen is embedded in a droplet of a dense material, such as phosphotungstate, which penetrates into all the empty spaces between the macromolecules (see Fig. 1–5).

A positive increase in contrast in biological structures has been obtained by the use of substances containing heavy atoms, such as osmium tetroxide, uranyl, and lead ions, which under certain conditions, act as *"electron stains."* These electron stains are comparable to histological stains, in that they combine with certain regions of the specimen.

For the study of certain biological processes, such as intercellular passages or the engulfment of macromolecules or particles by endocytosis, appropriate *tracers* having high electron opacity are used. These tracers do not stain the tissue but are able to detect the pathways through which a certain material is transported in between or within cells.

A variety of tracers are represented by enzymes able to produce a reaction that greatly enhances the contrast of biological specimens. For example, the presence of various *peroxidases* is demonstrated by their reaction with peroxide and 3,3'-diaminobenzidine. One of the smallest tracers of this kind is the so-called *microperoxidase*; which has a molecular weight of only 1900 daltons.

Freeze-Fracturing—Membranes Split at Cleavage Planes

The study of the fine structure of biological membranes has greatly benefited from the use of techniques involving the freezing and fracturing of cells. In general, the tissues are rapidly frozen in liquid nitrogen, fractured with a knife, and then subjected to a certain degree of water *sublimation* (etching) in a vacuum. This process, which involves the direct passage from ice to vapor, allows a three-dimensional (3-D) view of the structure. The effect of etching is thus to expose and to render visible the fine surface details.

As shown in Figure 3–6A, this step is followed by the deposit of a layer of evaporated metal (i.e., platinum), done at an angle, as in the previously described method of shadow casting. The metal, reinforced by a carbon layer also deposited by evaporation (but at a 90° angle), is detached from the specimen. This is the so-called *replica* of the object, which is then mounted on the EM grid for observation.

Freeze-fracturing has revealed a natural-looking representation of the surface of objects (Fig. 3–6B), and has allowed important discoveries on the molecular structure of different types of biological membranes. The fracture may disclose either the outer or the inner surface of a membrane but generally splits the membrane through the middle, thus revealing information about components passing through the lipid bilayer or present in one of the layers (see Fig. 4–3). Freeze-fracturing has been considerably improved by the use of methods in which the freezing is done very rapidly and is followed by a *deep etching*. In this way a three-dimensional replica of many cellular structures may be obtained. This technique gives three-dimensional images similar to those obtained with the scanning EM but with a higher resolution (Heuser, 1981). For further details on freeze-etching and on the nomenclature used for the splitting of the membranes see Sections 4–2 and 12–4.

Preparation of Thin Sections—Epoxy Resins and Ultramicrotomes

To satisfy the need for thinner sections, hard embedding media are used. Those most often used are epoxy resins that impregnate the tissue and are then polymerized by proper catalysts. Recently, water-miscible resins that can be infiltrated and polymerized at −35°C or −50°C have been developed. Such embedding reduces artifacts and permits the use of cytochemical methods on the section (Kellenberger et al., 1978).

To prepare the extremely thin sections, ultramicrotomes with specialized features have been developed. Several microtomes have been designed that have a thermal or a mechanical advancing device. With both types the thinnest sections that can be made are of the order of 20 nm. The limiting factors seem to be proper embedding and the sharpness of the cutting edge of the microtome. Diamond knives are now in general use. Thin sectioning can be performed at low temperature with simple embedding in gelatin.

High Voltage EM—Allows Study of Thicker Specimens

While most electron microscopes use accelerating voltages between 50 and 100 kV, there are instruments now in operation that greatly surpass this voltage and reach 500 to 3000 kV. The design of these newer instruments is essentially similar to the older models, but the construction is much more massive to permit higher acceleration, greater magnetic excitation for the lenses, and the shielding needed to protect against x-radiation. While its main application is in metallurgy, the high-voltage EM is being used increasingly in studies of biological material to examine thick sections (up to 5 μm) and whole cells, with less radiation damage resulting from ionization and temperature effects. With this instrument, theoretically, there is the possibility of examining living cells. The practice of obtaining stereomicrographs by tilting the specimen also gives a greater amount of three-dimensional information (see Fig. 1–13).

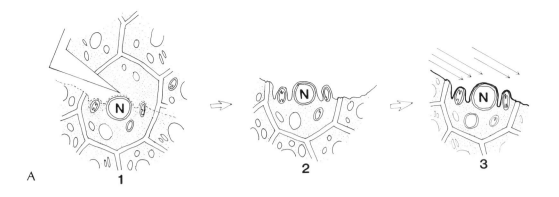

Fig. 3–6. *A,* diagram showing the method of freeze-etching. (1) The tissue kept frozen under vacuum is fractured with a knife. (2) By etching, part of the ice surface is removed. (3) The tissue is shadowed and the replica made is separated from the tissue and observed under the EM. *B,* observation of a replica of an onion root cell that has been submitted to the freeze-etching technique. The upper part of the figure corresponds to the nucleus (N) and shows the nuclear pore complexes (np) and the nuclear envelope (ne). In the cytoplasm, a Golgi complex (G) and a large vacuole (V) are observed. *cw,* cell wall. × 75,000. (Courtesy of D. Branton.)

Scanning EM and Scanning Transmission EM (STEM)

With the scanning EM a surface view of a specimen may be obtained by using the secondary electron emission that is ejected after the primary electron beam has interacted with the surface of a thick specimen (Fig. 3–7). In the scanning EM a thin beam of electrons moves back and forth across the specimen in the same way that the electron beam moves in a television tube. The secondary electrons are then collected by a photomultiplier tube, and an image is displayed on a television screen. In some cases, to increase the scattering power of the surface structures, infiltration with electron-contrasting chemicals or surface coating may be used.

Recently, a combination of the scanning and the transmission EM has been obtained in the so-called STEM (Scanning Transmission Electron Microscope). To understand the basic mechanism of the STEM we should refer again to Figure 3–4. The left side of this figure shows two uses of the conventional EM and, the right side, the STEM. In the most commonly used *bright field mode* of the EM, the image contrast is achieved by using an objective aperture that removes some of the scattered electrons. In this case, the image results from the interference between the nonscattered and the scattered electrons.

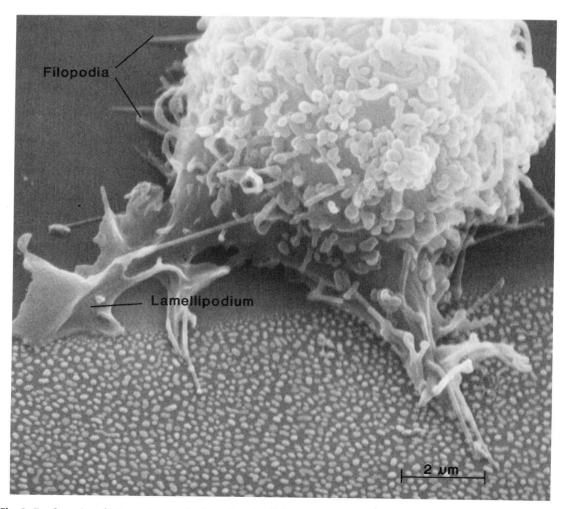

Fig. 3–7.   Scanning electron micrograph of a cultured cell that is in the process of respreading after cytokinesis. Notice the ruffling edge and the numerous microvilli. × 3260. (Courtesy of R.D. Goldman.)

In the *dark field mode* the electron beam is tilted and the image is made exclusively by the scattered electrons. Here a higher contrast is obtained, but more electrons impinge on the specimen producing more beam-induced destruction. To solve this problem the STEM uses a thin electron beam (a few Å in diameter) to scan the specimen. The electrons that are scattered are collected and analyzed by special detectors and the image obtained is displayed on a television screen (Kellenberger, 1978).

The STEM is particularly suited for the study of biological specimens. Due to the high contrast obtained it is possible to study unstained protein macromolecules and to obtain a more reliable image than those obtained by staining with heavy metals. For example with the STEM it is possible to make mass determination of proteins and to obtain finer and more reproducible details in viruses and biological membranes.

Image Reconstruction From Electron Micrographs

In the case of molecules or supramolecular structures that have a crystalline arrangement it is now possible to obtain more detailed information, even from rather fuzzy electron micrographs. This method, called *image reconstruction*, consists of placing the electron micrograph in the path of a beam of laser light to obtain an *optical diffraction pattern*. Using this pattern one can reconstruct the image of the individual molecules.

An example of an image reconstruction is given in Figure 5–6 in which, from a negatively stained array of macromolecules present in a gap junction, it was possible to obtain an image of the six protein molecules surrounding each hydrophilic channel. In this case the imaging method consisted of the computer processing of phases and amplitudes from a large surface area.

These methods reveal details that normally may be obscured by the random "noise" that is present in the image of each individual molecule. The resolution obtained may reach 0.7 nm (7 Å), which is good enough to determine the shape of the molecule. To reach a description of the molecular structure at the atomic level *X-diffraction methods* should be used, but their consideration is beyond the scope of this book.

SUMMARY:

Microscopy

Observation of biological structures is difficult because of their small size and lack of contrast. Optical instruments are specially designed to overcome both difficulties. In the *light microscope* the limit of resolution, $Lm = \dfrac{0.61\lambda}{NA}$, depends on the wavelength of the light and the numerical aperture. The resolving power is the inverse of the limit of resolution. The limit of resolution is, in general, $0.25\ \mu m$; with ultraviolet light it can be reduced to $0.1\ \mu m$.

Phase microscopy is used for the study of living cells, which are, in general, transparent to light. The principle on which the phase microscope is based is that the light passing through an object undergoes a retardation, or phase change, which normally is not detected. In this instrument, however, the phase difference is advanced or retarded one fourth of the wavelength (λ), and the small variations in phase produced by the various structures are thereby made visible. In phase microscopy the phase changes are translated into changes in light intensity.

Interference microscopy is based on similar principles.

The *Nomarski interference contrast* microscope, in particular, gives extraordinary images of living cells, with a relief effect.

Darkfield microscopy, also called ultramicroscopy, is based on the scattering of light and uses a darkfield condenser. With this microscope, as well as with phase contrast optics, objects smaller than the wavelength of light can be *detected*, although not *resolved*. Theoretically, a fiber of only 5.0 nm could be detected, provided it had enough contrast.

Polarization microscopy uses polarized light. A birefringent material (also called *anisotropic*) has two different indexes of refraction in two perpendicular directions. Birefringence is related to retardation and to the thickness of the object: $B = N_e - N_o = \dfrac{\Gamma}{t}$. In the polarizing microscope there is a polarizer and an analyzer (prisms or polaroid films) that are placed perpendicular to the specimen. The object is rotated 360° and if it is birefringent, it shows positions of maximum or minimum brightness. Most biological fibers show birefringence along the axis; a fiber of nucleic acid has negative birefringence. Crystalline, or intrinsic birefringence, is independent of the refractive index of the medium, whereas form birefringence changes when the refractive index of the medium varies. In general, both types are present. *Dichroism* is a special type of birefringence in which there is a change in absorption of polarized light with a change in the orientation of the object.

Electron microscopy is the best method for studying biological ultrastructure. Electrons are emitted and accelerated in a vacuum tube, and the electron beam is deflected by electromagnetic coils acting as condenser, objective, and projector lenses. The image formed on the fluorescent screen depends on the dispersion of electrons (electron scattering) by the atomic nuclei present in the object. This dispersion depends on the thickness of the object, the molecular packing, and, in particular, on the atomic number. Most atoms in biological objects do not scatter the electrons; heavy atoms are used as "electron stains" to increase the contrast. The resolving power depends on the wavelength (λ = 0.005 nm) and numerical aperture, as in the light microscope. The resolution reached is 0.3 to 0.5 nm, and the final magnification can be 10^6 times, or more.

The preparative techniques are of fundamental importance in the observation of biological material. Macromolecules such as DNA and RNA can be studied by the monolayer technique. Thick specimens can be studied via the technique of *thin sectioning*, in which the material is embedded in plastic and is cut with glass or diamond knives. The structure of membranes may be observed by freeze-fracturing and freeze-etching. The contrast can be increased by shadow casting, negative staining, or electron staining. Osmium tetroxide is used both as a fixative and an electron stain. Uranyl acetate and lead hydroxide are widely used for staining. Electron opaque substances, such as colloids and certain enzymes that give an opaque reaction, are used as tracers in electron microscopy for the study of several biological processes (e.g., pinocytosis). Thick sections and even living cells may be observed with *high-voltage electron microscopy*.

Scanning electron microscopy gives a surface view of structures by forming an image with the secondary electrons reflected by those structures. Special techniques and microscopes are used for this purpose.

Scanning transmission electron microscopy (STEM) uses a thin beam of electrons to scan the section and the scattered electrons are analyzed by special detectors. With STEM images of unstained macromolecules may be obtained.

3–3 STUDY OF THE LIVING CELL

One of the main methods for the study of the living cell is the use of *cell cultures*. Since 1912, when Alexis Carrel first succeeded in growing tissue explants for many cell generations, considerable progress has been made in the techniques of cell culture. At the present time these techniques represent some of the most powerful methods for the study of fundamental problems in cell biology. In early days of tissue culturing the technique consisted of explanting small por-

tions of different tissues (preferably embryonic) in a medium consisting of blood serum and embryo extract, plus saline solution. This system was extremely complex from the chemical viewpoint, and it was only after 1955 that the first chemically defined culture media became available. At present, the nutritional requirements of eukaryotic cells are well known, and most cells can be grown, with the addition of a small percentage of serum, in synthetic media.

Three main types of cultures can be distinguished: *primary*, *secondary*, and those using *established cell lines*. *Primary cultures* are those obtained directly from animal tissue. The organ is aseptically removed, cut into small fragments, and treated with trypsin. This proteolytic enzyme has the property of dissociating the cell aggregates into a suspension of single cells. without affecting viability. Thereafter, the cells are plated in sterile Petri dishes and grown in the appropriate culture medium. This culture can be trypsinized and replated in a fresh medium, resulting in a *secondary culture*.

The other major type of culture utilizes *established cell lines*, which have adapted to prolonged growth in vitro because of a cancerous transformation. Among the best known cell lines are the HeLa cells, obtained from a human carcinoma, the L and 3T3 cells from mouse embryo (Fig. 3–7), the BHK cells from baby hamster kidney, and the CHO cells from Chinese hamster ovary.

Normal mammalian cells do not survive indefinitely in culture and, after a variable time in vitro, they fail to divide and eventually die. Occasionally, some cells will survive and will grow permanently in culture. These established cell lines differ from normal cells in many respects: they grow more tightly packed; they have lower serum requirements; and they are usually *aneuploids*, i.e., their chromosome number varies from one cell to another. Despite these abnormalities, established cell lines are very useful as model systems for the study of cancer.

One of the major advances in cell culture was the obtaining of a *clone*, i.e., a population of cells derived from a single parent cell. After several days of incubation of cells plated at high dilutions, rounded colonies grow and adhere to the Petri dish. All the cells of this clone are derived from a single cell.

Microsurgery is another method that has contributed considerably to the knowledge of the living cell. Instruments such as micropipets, microneedles, microelectrodes, and microthermocouples are introduced into cells with the aid of a special apparatus that controls the movement of these instruments under the field of the microscope. Examples of microsurgical procedures are the dissection and extraction of parts of cells or tissues, the injection of substances, the measurement of electrical variables, and the grafting of parts from cell to cell. Laser beams are also used to damage special regions of the cell.

3–4 FIXATION AND STAINING

Fixation brings about the death of the cell in such a way that the structure of the living cell is preserved with a minimum of artifacts. Some fixation methods, at the same time, are useful in maintaining the chemical composition of the cell as intact as possible.

The choice of a suitable fixative is dictated by the type of analysis desired. For example, for studying the nucleus and chromosomes, *acid fixatives* are frequently used. Acetone, formaldehyde, and glutaraldehyde, which produce minimal denaturation and preserve some enzyme systems, are used for the study of enzyme cytochemistry.

Some fixing agents produce cross linkages between protein molecules. For example, aldehydes react with the amino, carboxyl, and indole groups of a protein and then produce methylene bridges with other protein molecules. The two-step reaction is shown below. Glutaraldehyde has two aldehyde groups ($HOC—CH_2—CH_2—CH_2—COH$) that can react with amino groups in two adjacent protein monomers.

$$
\underset{\substack{| \\ H \\ \text{amino group}}}{-N-H} \quad + \underset{\text{formaldehyde}}{HCHO} \rightarrow \underset{\text{methylol}}{—NH\cdot CH_2OH}
$$

$$
\underset{\text{methylol}}{—NH\cdot CH_2OH} + \underset{\substack{| \\ H \\ \text{amino group}}}{—N—H} \rightarrow \underset{\text{methylene bridge}}{—NH—CH_2—HN} + H_2O
$$

The preservation of a structure by fixation depends, to a great extent, on the degree of organization at the macromolecular level. In a well-organized structure, such as a chromosome, a mitochondrion, or a chloroplast, a great number of interacting forces hold the molecules together, and the action of the fixative is insufficient to break structural relationships. However, less organized regions of the cell, such as the cytoplasmic matrix, are more difficult to preserve, and the production of fixation artifacts is more likely to occur.

Freeze-drying and Freeze-substitution

This method consists of rapid freezing of the tissues, followed by dehydration in a vacuum at a low temperature. The initial freezing is generally accomplished by plunging small pieces of tissue in a bath cooled with liquid nitrogen to a temperature of -160 to $-190°C$. Fixation in liquid helium near absolute $0°$ (Kelvin) has also been used. The tissues are dried in a vacuum at -30 to $-40°C$. Under these conditions the ice in the tissues is changed directly into a gas, and dehydration is achieved.

The advantages of this method are obvious. The tissue does not shrink; fixation is homogeneous throughout; soluble substances are not extracted; the chemical composition is maintained; and the structure, in general, is preserved with very few modifications (Fig. 3–8). In addition, fixation takes place so rapidly that cell function can be arrested at critical moments, such as when kidney cells are excreting colored material.

In the freeze-substitution method, the tissue is kept frozen at a low temperature (-20 to $-60°C$) in a reagent that dissolves the ice crystals (e.g., ethanol, methanol, or acetone). The advantages of this method are somewhat similar to those just mentioned for fixation by freeze-drying.

Rapid freezing is generally used for preservation and storage of cultured cells for later recovery and plating. It is also frequently used for preservation and later recovery of viable sperm for *artificial insemination*, as well as fertilized eggs.

Microtomes and Embedding

Tissues should be conveniently sectioned before they are observed under the microscope. For this purpose, *freezing microtomes*, cooled with liquid carbon dioxide, are frequently used.

Instruments consisting of a microtome en-

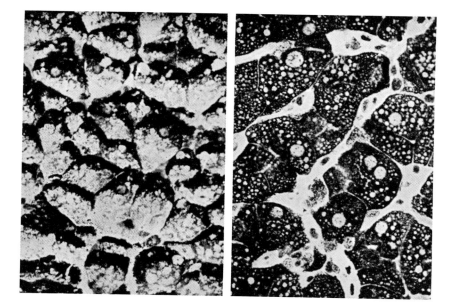

Fig. 3–8. *Left,* liver cells of *Ambystoma* fixed in Zenker-formol. The diffusion current produced by the chemical fixative (from the lower to the upper part of the figure) displaces the glycogen of the cell. *Right,* liver cells of *Ambystoma* fixed by freeze-drying. The glycogen appears to be distributed homogeneously in the cytoplasm. Spheroid nuclei and lipid droplets are distinguishable. Stain: Best's carmine. (Courtesy of I. Gersh.)

closed in a chamber at low temperature—the so-called *cryostat*—can make sections of fixed or fresh tissue for cytochemical purposes (Fig. 3–9). The use of the *vibratome,* in which a blade vibrates, allows the investigator to section fresh tissues for cytochemistry.

For the most frequently used sectioning techniques the tissue is embedded in a material that imparts the proper consistency to the section. For sections to be observed under the light microscope, paraffin or celloidin is generally used. The fixed tissue is dehydrated and then penetrated by the embedding material. This requires a proper intermediary solvent (e.g., xylene or toluene for paraffin; ethanol-ether for celloidin).

Chemical Basis of Staining

Most cytological stains are solutions or organic aromatic dyes. Since the time of the pioneer work of Ehrlich, two types of dyes have been recognized: basic and acid. In a basic dye the *chromophoric group,* which imparts the color, is basic (cationic). For example, methylene blue is a chlorhydrate of tetramethylthionine, in which the basic part carries the blue color. Eosin is generally used as potassium eosinate, in which the base is colorless. Sometimes the two com-

ponents of the salt are chromophoric, e.g., eosinate of methylene blue. The most frequently used chromophores for acid dyes contain nitro ($-NO_2$) and quinoid ($O=\!\!\!\!\bigcirc\!\!\!\!=O$) groups.

Basic chromophores contain azo ($-N\!=\!N-$) and indamin ($-N\!=\!$) groups. For example, picric acid has three nitro groups (chromophores) and one OH group, also called *auxochrome,* by which the dye combines with the tissue:

$$
\begin{array}{c}
\text{OH} \\
\text{NO}_2\!\!\!\bigcirc\!\!\!\text{NO}_2 \\
\text{NO}_2
\end{array}
$$

The properties that enable proteins, certain polysaccharides, and nucleic acids to ionize either as bases or acids should be noted. Acid ionization may be produced by carboxyl ($-COOH$), hydroxyl ($-OH$), sulfuric ($-HSO_4$), or phosphoric ($-H_2PO_4$) groups. Basic ionization results from amino ($-NH_2$) and other basic groups in the protein. At pH values above the isoelectric point, acid groups become ionized; below the isoelectric point, basic groups dissociate (see Fig. 2–20). Because of this property, at a pH above the isoelectric point, pro-

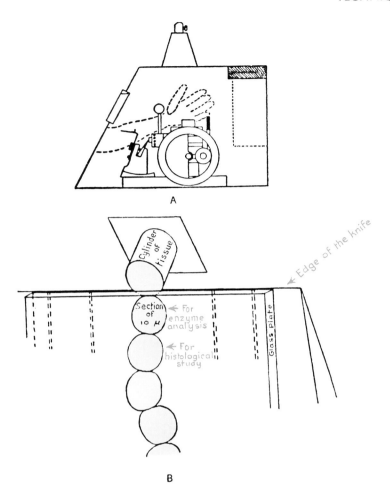

A

B

Fig. 3–9. *A*, microtome for frozen sectioning. The apparatus is kept in a refrigerated container that keeps the tissue and the sections frozen. *B*, scheme of sectioning with the freezing microtome. The cylinder of tissue is sectioned and the sections are collected so that each alternate piece is used for enzymatic analysis and the others for histological control. (Modified from Linderström-Lang.)

teins will react with basic dyes (e.g., methylene blue, crystal violet, or basic fuchsin) and below it, with acid dyes (e.g., orange G, eosin, or aniline blue).

The net charge of nucleic acids is determined primarily by the dissociation of the phosphoric acid groups, and the isoelectric point is very low (pH 2 or less). For this reason, staining with basic dyes (e.g., toluidine blue or azure B) at low pH values is selective for nucleic acids. Toluidine blue is frequently used to stain ribonucleic acid, and its specificity can be demonstrated by previous hydrolysis with ribonuclease (Fig. 3–10).

Metachromasia—A Change in Original Dye Color

Some basic dyes of the thiazine group, particularly thionine, azure A, and toluidine blue, stain certain cell components a different color than the original color of the dye. This property, called *metachromasia*, has interesting histochemical and physicochemical implications. The reaction occurs in mucopolysaccharides and, to a lesser extent, in nucleic acids and some acid lipids. This reaction is strong in cells that contain sulfate groups (such as chondroitin sulfate), e.g., in the cells of cartilage and connective tissue.

Some investigators believe that metachromasia depends on the formation of dimeric and polymeric molecular aggregates of dye on these high molecular weight compounds.

SUMMARY:
Observation of Living and Fixed Cells

Observation of cells can be made directly on living specimens or after cells have been fixed. *Tissue culture* consists of explanting a piece of tissue in a suitable medium (e.g., plasma, embryonic extract, or synthetic). The cells spread and divide, forming a zone of growth. Pure strains (i.e., clones)

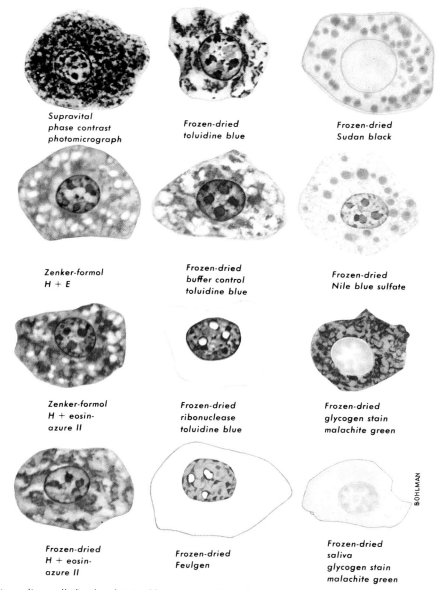

Fig. 3–10. Mouse liver cells fixed and stained by a variety of cytochemical procedures to show distribution of DNA, RNA, glycogen, and lipid droplets. Since these tests were all done on material fixed by freezing and drying, some of the sections are compared with similarly stained sections fixed by Zenker-formol. For further orientation, the fixed cell stained by hematoxylin and eosin and the unfixed cell photographed by phase contrast are also shown. × 1500. (From Bloom. W., and Fawcett, D.W.: Textbook of Histology. 10th Ed. Philadelphia, W.B. Saunders Co., 1975.)

of cells may be obtained after isolation of cells by trypsin. Organs, as well as tissues, can also be cultured. *Microsurgery* is a technique that may give information about the physicochemical properties of living cells. Through this method, transplants of subcellular parts—including the nucleus—can be made.

Fixation is brought about by chemicals that preserve cell structure. For nuclei and chromosomes, acid fixatives are preferred. To preserve the activity of certain enzymes, acetone, formaldehyde, and

glutaraldehyde are used. (Aldehydes react with amino groups of proteins, forming methylene bridges.) Glutaraldehyde has two aldehyde groups that can react with adjacent protein monomers. Fixatives penetrate by diffusion and produce a gradient of fixation. The preservation of cell structure depends on the tightness of macromolecular organization. Osmium tetroxide and glutaraldehyde are widely used in fixation for electron microscopy.

Freeze-drying consists of the rapid freezing of tissue using liquid nitrogen (-160 to $-190°C$), followed by dehydration at -30 to $-40°C$. The water is sublimed in a vacuum. In *freeze-substitution* the water of the frozen tissue is dissolved with alcohols or acetone. *Embedding* in paraffin or celloidin is used for light microscopy; special plastics are used for electron microscopy. Microtomes and ultramicrotomes are then used for *sectioning* the embedded tissues.

Cytological stains may be basic (e.g., methylene blue) or acid (e.g., eosin). Chromophores for acid dyes are nitro and quinoid groups; for basic dyes they are azo and indamin groups. The mechanism of staining is based on the ionization of acid groups (carboxyl, hydroxyl, sulfuric, phosphoric) or basic groups (amino) in proteins, polysaccharides, and nucleic acids. The staining of a protein is minimal at the isoelectric point; it stains with acid dyes below, and basic dyes above, the isoelectric point. Nucleic acids are stained selectively at low pH because their isoelectric point is at pH 2.

Metachromasia, the property by which a basic dye has a different color in tissue than in solution, is found in mucopolysaccharides, particularly those containing sulfate groups. This particular characteristic of the dye is probably due to dimerization or polymerization.

3–5 CYTOCHEMICAL METHODS

The immediate goal of *cytochemistry* is the identification and localization of the chemical components of the cell. As R.R. Bensley, one of the founders of modern cytology, once said: The aim of cytochemistry is the outlining "within the exiguous confines of the cell of that elusive and mysterious chemical pattern which is the basis of life." This aim is qualitative as well as quantitative; and once achieved, the next step is to study the dynamic changes in cytochemical organization taking place in different functional stages. In this way it is possible to discover the role of different cellular components in the metabolic processes of the cell.

Modern cytochemistry has followed two main methodological approaches. Of these, only one can be considered strictly *microscopic*, because it comprises a series of chemical and physical methods used to detect or measure different chemical components within the cell. The other method relies on *biochemical* techniques for the isolation and investigation of subcellular fractions, or it applies the techniques of *microchemistry* and *ultramicrochemistry* to the study of minute quantities of material.

For the cytochemical determination of a substance certain conditions must be fulfilled: (A)

The substance must be immobilized at its original location. (B) The substance must be identified by a procedure that is specific for it, or for the chemical group to which it belongs. This identification can be made by: (1) chemical reactions similar to those used in analytic chemistry, but adapted to tissues, (2) reactions that are specific for certain groups of substances, and (3) physical methods.

To demonstrate proteins, nucleic acids, polysaccharides, and lipids, some chromogenic agents that bind selectively to some specific groups of these substances may be used.

Schiff's Reagent—Detection of Aldehydes—Cytochemistry of Nucleic Acids

As an example of a cytochemical method we shall use *Schiff's reagent*. Deoxyribonucleic acid, certain carbohydrates, and lipids can be demonstrated with this reagent for aldehyde groups. Schiff's reagent is made by treating basic fuchsin, which contains parafuchsin (triamino-triphenyl-methane chloride), with sulfurous acid. Parafuchsin is converted into the colorless compound bis-N aminosulfonic acid (Schiff's reagent), which is then "recolored" by the aldehyde groups present in the tissue (Fig. 3–11).

In the histochemical tests involving Schiff's reagent, three types of aldehydes may be involved: (1) *free aldehydes*, which are naturally present in the tissue, such as those giving the plasmal reaction; (2) *aldehydes produced by selective oxidation* (which give the PAS reaction); and (3) *aldehydes produced by selective hydrolysis* (which give the Feulgen reaction).

Cytochemical staining methods for nucleic acids depend on the properties of the three components of the nucleotide (phosphoric acid, carbohydrate, and purine and pyrimidine bases).

Both DNA and RNA absorb ultraviolet light at 260 nm, because of the presence of nitrogenous bases. The deoxyribose present in DNA is responsible for the Feulgen reaction, which is specific for this type of nucleic acid (Table 3–1).

The phosphoric acid residue is responsible for the basophilic properties of both DNA and RNA. Among the basic stains, azure B gives a specific reaction with DNA and RNA. Another stain for DNA, based on the use of methyl green, also depends on the phosphoric acid residues.

Feulgen Reaction. DNA can be studied by

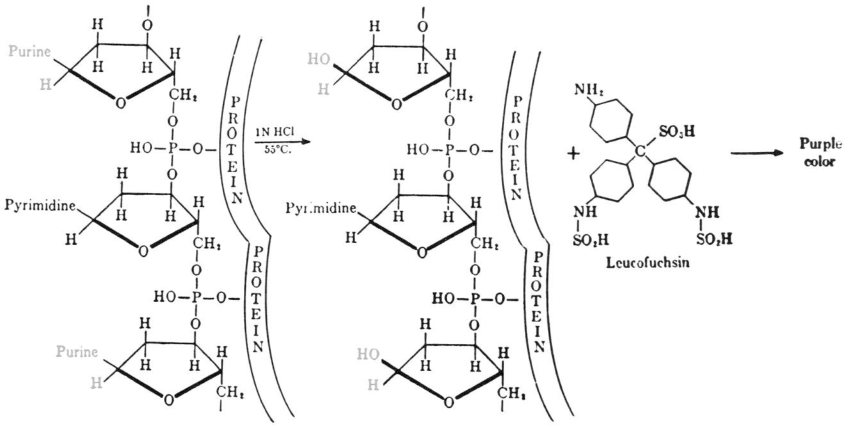

Fig. 3–11. Chemistry of the Feulgen reaction. Acid hydrolysis removes the purines and liberates the aldehyde groups, which react with leukofuchsin (Schiff's reagent), resulting in a purple color. In the diagram the size of deoxypentose is greatly exaggerated in relation to the protein. (From Lessler, M.A.: Int. Rev. Cytol., 2:231, 1953.)

means of the *nucleal reaction*, a technique developed in 1924 by Feulgen and Rossenbeck. Sections of fixed tissue are first submitted to a mild acid hydrolysis and then treated with Schiff's reagent. This hydrolysis is sufficient to remove RNA, but not DNA. The reaction takes place in the following stages: (1) The acid hydrolysis removes the purines at the level of the purine-deoxyribose glucosidic bond of DNA, thus unmasking the aldehyde groups of deoxyribose. (2) The free aldehyde groups react with Schiff's reagent (Fig. 3–11). The specificity of the reaction can be confirmed by treating the sections with deoxyribonuclease, which removes DNA.

The Feulgen reaction is positive in the nucleus and negative in the cytoplasm (Fig. 3–12). In the nucleus, the masses of condensed chromatin (i.e., heterochromatin) are intensely positive; the nucleolus is Feulgen-negative.

Periodic Acid-Schiff (PAS) Reaction. McManus devised a reaction based on the oxidation with periodic acid of the "1,2-glycol" group of polysaccharides with liberation of aldehyde groups, which give a positive Schiff reaction (Fig. 3–13). This test is done on plant cells for starch, cellulose, hemicellulose, and pectins; and on animal cells for mucin, mucoproteins, hyaluronic acid, and chitin.

Plasmal Reaction. Long-chain aliphatic aldehydes occurring in plasmalogens give the so-called *plasmal reaction* upon direct treatment of the tissue with Schiff's reagent. Since the substances giving the plasmal reaction are soluble in organic solvents, the tissue is not embedded in the usual way but is studied in frozen sections.

TABLE 3–1. SOME SPECIFIC REACTIONS USED IN CYTOPHOTOMETRIC ANALYSIS

Substance Tested For	Reaction or Test	Maximum Absorption (wavelength in nm)
Total nucleotides	Natural absorption of purines and pyrimidines	260
Soluble nucleotides	Natural absorption of purines and pyrimidines	260
Ribonucleic acid (RNA)	Natural absorption of purines and pyrimidines	260
Deoxyribonucleic acid (DNA)	Natural absorption of purines and pyrimidines	260
Deoxyribonucleic acid (DNA)	Feulgen nucleal reaction for deoxyribose	550 to 575
Deoxyribonucleic acid (DNA)	Methyl green	645
Nucleic acids (phosphoric acid groups)	Azure A	590 to 625
Protein (free basic groups)	Fast green	630
Protein (tyrosine)	Millon reaction	355
Polysaccharides with 1,2-glycol groupings	Periodic acid-Schiff reaction (PAS)	~550

From Moses, M.J.: Exp. Cell Res., 2 (Suppl.) 76, 1952.

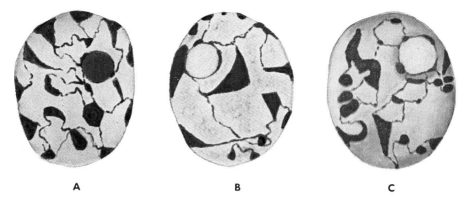

Fig. 3–12. Interphase nuclei of pancreatic cells fixed by freeze-drying. *A,* Azan staining: the nucleolus (in red), the chromonemic filaments with their enlarged portions (chromocenters), and the nuclear sap are visible. *B,* Feulgen reaction: the nucleolus gives a negative reaction; in the nuclear sap the reaction is slightly positive. *C,* action of ribonuclease and staining with Azan. The nucleolus does not stain, owing to the digestion of the ribonucleic acid. (From De Robertis, Montes de Oca, and Raffaele, Rev. Soc. Arg. Anat. Normal Patol., 7:106, 1945.)

The compounds are free aldehydes such as *palmitaldehyde*, $CH_3 (CH_2)_{14}CHO$, and *stearaldehyde*, $CH_3(CH_2)_{16}CHO$, corresponding to palmitic and stearic acids, respectively, which together constitute the so-called *plasmal*.

Lipids—Detection by Lipid-soluble Stains

Fat droplets can be demonstrated with osmium tetroxide, which stains them black by reacting with unsaturated fatty acids. Staining with Sudan III or Sudan IV (scarlet red) has a greater histochemical value. These stains act by a simple process of diffusion and solubility and accumulate in the interior of the lipid droplets. Sudan black B has the advantage of being dissolved also in phospholipids and cholesterol, and of producing greater contrast (Fig. 3–10).

Enzymes—Detection by Incubation with Substrates

To detect some enzymes, unfixed frozen sections are made in a cryostat; in other cases, the enzyme can withstand a brief fixation in cold

TABLE 3–2. SOME PHOSPHATASES STUDIED CYTOCHEMICALLY

Type	Substrate
Phosphomonoesterases	
Alkaline phosphatase	α- or β-glycerophosphate
Acid phosphatase	Naphthylphosphate
Adenosine triphosphatase (ATPase)	Adenosine triphosphate
5-Nucleotidase	5-Adenylic acid
Phosphamidase	Phosphocreatine
	Naphthyl phosphate acid diamines
Glucose-6-phosphatase	Glucose-6-phosphate
Thiamine pyrophosphatase	Thiamine pyrophosphate
Pyrophosphatase	Sodium pyrophosphate
	Dinaphthyl pyrophosphate
Phosphodiesterases	
Ribonuclease	Ribonucleic acid (RNA)
Deoxyribonuclease	Deoxyribonucleic acid (DNA)

acetone, formaldehyde, glutaraldehyde, or another dialdehyde.

Techniques for identifying and localizing enzymes are based on the incubation of the tissue sections with an appropriate substrate. For example, in the Gomori method for detecting alkaline phosphatase, phosphoric esters of glycerol are used as the substrate. The phosphate

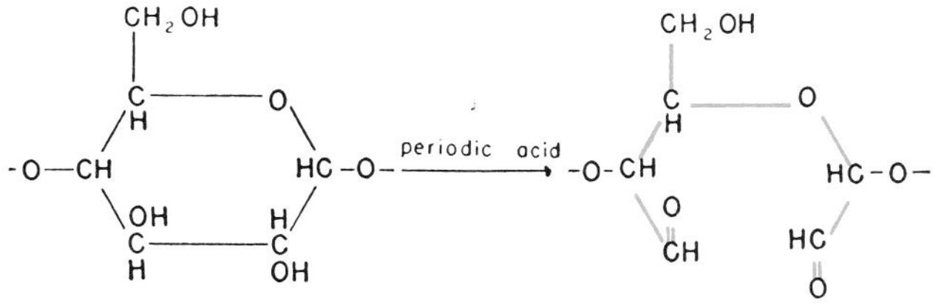

Fig. 3–13. Chemical diagram of a polysaccharide, showing the action site of periodic acid in the PAS reaction of McManus. The resulting aldehydes react with Schiff's reagent.

ion liberated by hydrolysis is converted into an insoluble metal salt (generally in the presence of Ca^{2+}), and the metal, in turn, is visualized by conversion into metallic silver, lead sulfide, cobalt sulfide, or other colored compounds.

Phosphatases. Phosphatases are enzymes that liberate phosphoric acid from many different substrates. A number of phosphatases differ with respect to substrate specificity, optimum pH, and the action of inactivators and inhibitors. The best known are the *phosphomonoesterases*, which hydrolyze simple esters held by P—O bonds, and the *phosphoamidases*, which hydrolyze P—N bonds. Table 3–2 indicates some of the most common enzymes studied cytochemically and some of the substrates used. An example of the alkaline phosphatase reaction is shown in Figure 3–14.

Esterases. Esterases are enzymes that catalyze the following reversible reaction:

$$—COOR + HOH \leftrightarrows COOH + R'OH$$

Esterases may be divided into *simple esterases (aliesterases)*, which hydrolyze short chain aliphatic esters; *lipases*, which attack esters with long carbon chains; and *cholinesterases*, which act on esters of choline.

Other *hydrolytic enzymes* studied cytochemically are β-D-glucuronidase, β-D-galactosidase, aryl sulfatase, and aminopeptidase.

Oxidases. Oxidases are enzymes that catalyze the transfer of electrons from a donor substrate to oxygen. They usually contain iron, e.g., peroxidase and catalase, or copper, e.g., tyrosinase and polyphenol oxidase.

Colorless substrates, such as benzidine, are used to detect peroxidases. These substrates are transformed into stained dyes by H_2O_2 in the presence of the enzyme. The reagent 3′,3′-diaminobenzidine (DAB) has allowed the study of peroxidase and cytochrome oxidase at the electron microscopic level.

Dehydrogenases. The pyridine nucleotide-linked dehydrogenases require the coenzymes NAD^+ or $NADP^+$. Among the best known NAD^+ enzymes are lactic acid dehydrogenase, which converts lactic acid into pyruvic acid, and malic acid dehydrogenase, which converts malic acid into oxaloacetic acid. Among the $NADP^+$ enzymes are isocitric acid dehydrogenase and the malic enzyme (malate → pyruvate + CO_2).

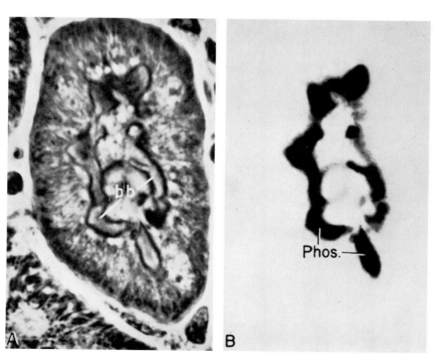

Fig. 3–14. Proximal convoluted tubule from a mouse kidney after freezing-substitution. *A,* observation in phase contrast with a medium of refractive index n = 1460; *bb,* brush border. *B,* observation with transmitted light; *Phos,* alkaline phosphatase reaction. (Courtesy of B.J. Davies, and L. Ornstein.)

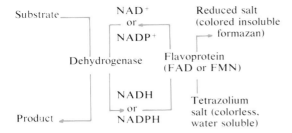

Fig. 3–15. Schematic representation of the transfer of electrons to tetrazolium salt. *FAD*, flavin adenine dinucleotide; *FMN*, flavin mononucleotide; *NAD⁺*, nicotinamide adenine dinucleotide; *NADP⁺*, nicotinamide adenine dinucleotide phosphate.

Figure 3–15 represents the mechanism of these histochemical reactions.

Cytophotometric Methods

Several cell components display a specificity in the way in which they absorb ultraviolet light. For example, the absorption range of nucleic acids is about 260 nm, whereas that of proteins is 280 nm. Some histochemical staining reactions give specific absorption in the visible spectrum and can be analyzed quantitatively with instruments called *cytophotometers.*

Table 3–1 indicates some of the histochemical reactions that can be analyzed by cytophotometric methods.

The specific ultraviolet absorption of nucleic acids is due to the presence of purine and pyrimidine bases, and, for this reason, is the same in DNA, RNA, and nucleotides (Table 3–1). By ultraviolet cytophotometry the two types of nucleic acids can be localized, but not distinguished. On the other hand, the Feulgen reaction shows the presence of DNA (Fig. 3–11). This reaction can be adapted to quantitative determinations of DNA in tissue sections. The Millon reaction, based on a nitrous-mercuric reagent that reacts with tyrosin groups, can be used to determine the protein content (Table 3–1).

Fluorescence Microscopy—Autofluorescence and Fluorochrome Dyes

In this method tissue sections are examined under ultraviolet light, near the visible spectrum, and the components are recognized by the fluorescence they emit in the visible spectrum. Two types of fluorescence may be studied; natural fluorescence *(autofluorescence)*, which is produced by substances normally present in the tissue, and *secondary fluorescence*, which is induced by staining with fluorescent dyes called *fluorochromes.*

Using *acridine orange* stain (after acetylation) on a tissue section it is possible to obtain a green fluorescence for DNA and a red one for RNA. Both fluorescences can be analyzed simultaneously using a special *cytofluorometer.*

Certain proteins can be tagged with fluorescent dyes, such as fluorescein isocyanate or rhodamine, without denaturing the molecule. These fluorescent proteins may then be injected into the animal and localized in sections within the cell or in the extracellular space.

Fluorescence often yields specific cytochemical information. Fluorescein and rhodamine as such are used to probe gap junctions between cells. Injection of the probe into one cell permits the study of transcellular permeability through these types of junctions (see Fig. 5–9).

The most important advantage of fluorescence microscopy is its great sensitivity. Fluorescence often yields specific cytochemical information because some of the normal components of the tissue have a typical fluorescent emission. Thus vitamin A, thiamine, riboflavin, and other substances can be detected. The cytochemical value of the method is increased considerably by spectrographic analysis of the radiation. Sometimes certain substances incorporated in cells, e.g., sulfonamides, can be localized.

An important application concerns the so-called lipogenic pigments, which are found in a great number of cells and which increase in number as the cell ages. It is thought that these pigments represent different degrees of oxidation and polymerization of unsaturated fatty acids. With fluorescence two types of pigments, the so-called *lipofuscin* and the *ceroid*, can be differentiated.

An important development has been the use of paraformaldehyde fixation, which, in freeze-dried tissues, produces condensation with catecholamines and indolamines, emitting a green and a yellow fluorescence, respectively.

In Chapter 19 the importance of certain fluorochromes, such as *quinacrine mustard*, in the study of human chromosomes will be mentioned.

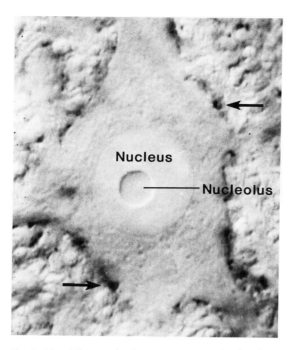

Nucleus

Nucleolus

Fig. 3–16. Micrograph of a motoneuron stained with an antiserotonin antibody using the PAP method. Because of the use of the Nomarski optics there is a three-dimensional view of the nucleus, nucleolus, and nerve endings on the neuron soma *(arrows)*. The other black points represent sections of serotoninergic axons. (Courtesy of J. Pecci Saavedra, P. Pasik, and T. Pasik.)

3–6   IMMUNOCYTOCHEMISTRY

Immunocytochemistry is based on the detection of *antigens* by *antibodies*. Antigens are, in general, large cellular molecules such as proteins, polysaccharides, and nucleic acids which, when injected into another animal, activate lymphocytes to produce specific antibodies. A small molecule, a so-called *hapten* (such as 5-hydroxytryptamine in Figure 3–16), when coupled to a larger molecular species can also produce antibodies. Basically each antibody is composed of four peptides; two identical *light chains* and two identical *heavy chains*. The specificity for the *antigenic determinants* depends on the variability of the amino acid sequence at the binding site. (For the molecular aspects of antibodies, see Section 23–4.) Most antibodies belong to the immunoglobulin G (IgG) class, present in the γ-globulin fraction of the serum. IgG has a molecular weight of 145,000 to 156,000 daltons and a Y shape, with two binding sites at the variable parts (Fab) and a constant region (Fc) (see Fig. 23–29).

In immunocytochemistry the antibody that interacts with the tissue antigen is known as the *primary antibody*. As shown in Figure 3–17, this first antigen-antibody complex can be revealed by suitable markers for observation with the light or the electron microscope.

In the early fifties Albert Coons conjugated the IgG with a fluorescent dye, such as *fluorescein*, which permitted the direct detection of the complex in a tissue section under the fluorescence microscope. In Figure 3–17A, two other *direct methods* of detection are illustrated. In one case the IgG was labeled with a radioactive marker (such as a tritiated amino acid added during the synthesis of the antibody) and then revealed by radioautography (see later discussion). In the other case an enzyme, such as *peroxidase*, was linked to the IgG and then reacted with diaminobenzidine in the presence of H_2O_2 (DAB reaction), giving an opaque deposit that can be visualized microscopically.

In addition to peroxidase other enzymes, such as alkaline phosphatase and β-galactosidase, are being used. This last enzyme is becoming that of choice because it gives low background and great sensitivity. The low background is due to the fact that the endogenous enzyme is negligible in animal cells.

Another direct method employs antibodies coupled to *ferritin*, an iron-containing protein, which is opaque to electrons.

Indirect Methods. Since direct methods have, in general, low sensitivity they make it difficult to detect the primary complex. Today, indirect methods are used most frequently (Fig. 3–17B and C). In this case the antigen-antibody interaction is further amplified by the introduction of a second labeled antibody. For example, if the primary antibody is an IgG made in a rabbit, an anti-rabbit IgG made in a goat or a sheep is introduced. Also in this case the reaction can be observed by fluorescence, autoradiography or by electron microscopy.

In the figure, besides the use of peroxidase or antibodies labeled with a fluorescent dye, tritium or peroxidase, more complex methods, to be described, are shown.

The Biotin-Avidin-Peroxidase Method. In this method the constant region (Fc) of the IgG is bound to *biotin* and then reacted with an *avidin*-peroxidase conjugate.

The enzyme most often coupled to the anti-

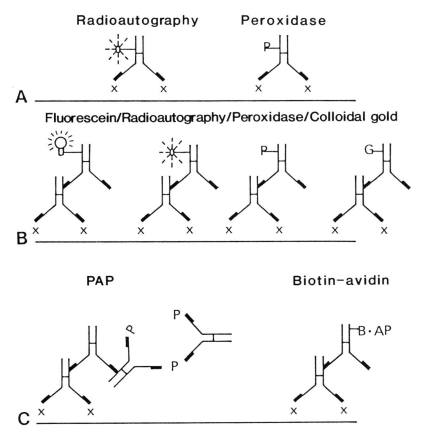

Fig. 3–17. Diagram representing different immunocytochemical procedures. *A*, direct methods in one step. *B* and *C*, different indirect procedures with several steps. (Courtesy of A.C. Cuello.)

IgG is *peroxidase*, which is reacted with 3',3'-diaminobenzidine (DAB) in the presence of H_2O_2. This reaction gives an electron-dense deposit that can be detected with the light microscope, as well as with the EM.

The PAP Method. In recent years, a further improvement has been introduced with the so-called *peroxidase-antiperoxidase* (PAP) method of Sternberger. As shown in Figure 3–18, after steps 1 and 2, in which unlabeled antibodies are used, step 3 introduces a peroxidase rabbit antiperoxidase complex, which also binds to the sheep antirabbit IgG. The peroxidase-antiperoxidase complex is made by interacting the enzyme with the corresponding IgG made in the rabbit. As in the case of the indirect method described above, the peroxidase is finally detected by the DAB method. Figure 3–16 shows nerve terminals on a motoneuron stained with the PAP method for 5-hydroxytryptamine. (For further details on the immunocytochemical

methods see Sternberger, 1979; Pearse, 1980; Cuello, 1983.)

The Protein A-Gold Technique. Another immunocytochemical technique, at the EM level, is based on the use of colloidal gold particles, which are very opaque to electrons. These particles are surrounded by protein A, extracted from *Staphylococcus aureus*, which has the property of interacting with IgG immunoglobulins. This interaction is with the Fc region of the IgG, and thus it does not affect the antibody binding site, which can react with the antigen (Fig. 3–19A). The protein A-gold complex can be used for any antigen, provided that it has interacted with the specific antibody. The gold particles are easily detectable under the EM and, because of its small size, the cellular structure containing the antigen can be clearly identified. Furthermore, it is possible to make a quantitative estimation of antigenic sites. In Figure 3–19B the localization of the enzyme amy-

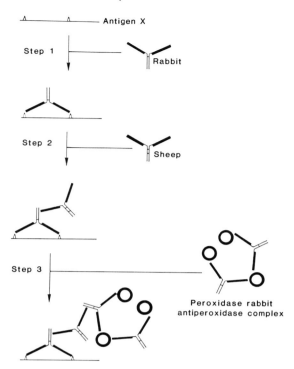

Fig. 3–18. Diagram illustrating the peroxidase-antiperoxidase (PAP) method. After steps 1 and 2, which are similar to steps I and II in Figure 3–17, there is step 3 in which the rabbit PAP complex is introduced. In a further step the enzyme is revealed by the DAB method: (diamino benzidine and H_2O_2). For electron microscopy this is followed by osmium tetroxide. (Courtesy of L.A. Sternberger.)

lase in a section of a pancreatic cell is observed (Roth, 1982).

Gold Anti-IgG Method. In addition to the protein A-gold technique previously described, colloidal gold may be used as a marker of an IgG for a tissue antigen, either in a direct way or by using a secondary antibody. Gold particles of 5 to 40 nm are available for tagging the IgG. The gold particles that are deposited at the antigen site can be enlarged by reacting them with silver, thus increasing the contrast. By using gold particles of different sizes associated to different IgG, or by associating the gold and PAP methods it is possible to tag two or more antigens in the same tissue (Pickel and Teitelman, 1984).

Monoclonal Antibodies. In Chapter 23 we refer to the recent developments in the production of monoclonal antibodies, by using cell fusion between lymphocytes and cells of myeloma (i.e., a tumor of the bone marrow). These monoclonal antibodies derive from a clone (i.e., a colony of a single lymphocyte), therefore, they are very pure and recognize a single immunogenic determinant in the protein. We also mention that these monoclonal antibodies have many important medical applications, and can be used advantageously in tissues for immunochemical methods (Milstein, 1980).

Throughout different chapters of this book examples are given of the important discoveries that the immunochemical methods have produced in cell and molecular biology.

3–7 RADIOAUTOGRAPHY

Radioautography is based on the ability to label cell components with radioisotopes, which can be then demonstrated by their capacity to interact with silver bromide crystals in a photographic emulsion.

In a *radioautography* the tissue section is put in contact with the emulsion for a certain period; then the radioautograph is developed like an ordinary photograph. By comparing the radioautograph with the cells in the tissues seen under a microscope, the radioisotope can be localized fairly accurately.

Of the various radioautographic techniques the one most used is that employing *liquid emulsions* (i.e., dipping radioautography). In this case the photographic emulsion, containing the silver halide grains, is in a gel that liquefies at 45°C. Essentially, the method consists of the following steps: (1) The tissue section, mounted on a glass slide or a grid, is immersed in the liquid emulsion at 45°C. (2) The section is removed, so that a film of gelatin covers it, and it is then left at room temperature (Fig. 3–20). (3) The specimens are kept in lightproof boxes for a period of days or weeks to allow the radiation to act on the film. (4) The photographic emulsion is developed. (5) The tissue is stained and observed. The silver grains stand out as black dots on the parts of the specimen where the isotope is localized (Fig. 3–20). For the EM a resolution of about 0.1 μm (100 nm) may be achieved.

Quantitative results are obtained by determining the density of the particles in the radioautographs by various optical methods, or by counting the grains.

Substance marked with the β-emitter ^{14}C have been widely used in radioautography. With such an emitter, *track radioautography* can also be used—a thick liquid emulsion is applied, and it

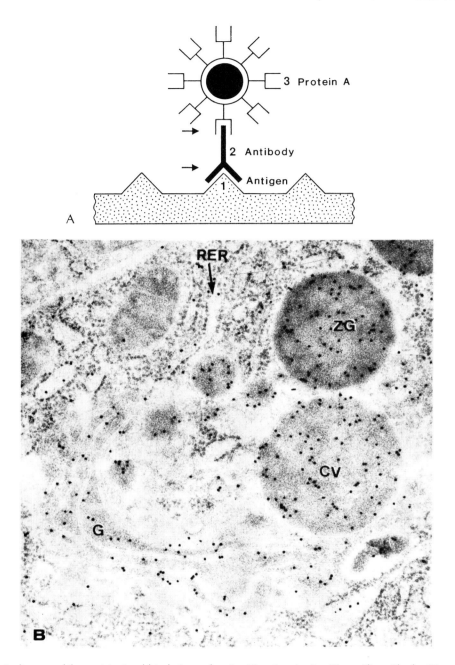

Fig. 3–19. *A,* diagram of the protein A-gold technique, showing (1) antigenic site, (2) specific antibody, (3) protein A-gold particle complex. The arrows indicate the two steps in the reaction at which the specificity of the binding can be tested. *B,* thin section of pancreas stained for amylase. Gold particles are present at the rough endoplasmic reticulum (RER), inside Golgi cisternae (G), in condensing vacuoles (CV), and in zymogen (ZG). (Courtesy of J. Roth.)

is possible to follow and count the single tracks of β-particles coming out from a definite area, thus providing a quantitative estimation. The resolution of this method is of the order of 1 to 2 μm.

Substances labeled with tritium (³H), a weak β-emitter, are most widely used in radioautography. For the study of deoxyribonucleic acid (DNA) metabolism of the cell, tritiated thymidine is used and is specific for this type of nucleic acid. To study RNA metabolism, ³H-uridine is used; and for examining the synthesis of pro-

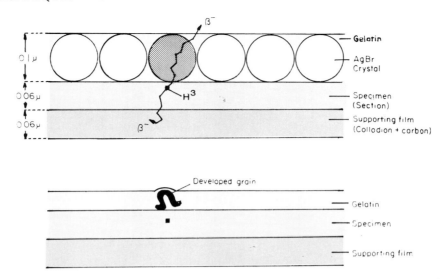

Fig. 3–20. Diagrammatic representation of an EM autoradiograph preparation. *Top, during exposure:* The silver halide crystals, embedded in a gelatin matrix, cover the section. A beta particle, from a tritium point source in the specimen, has hit a crystal (cross-hatched), causing the appearance of a latent image on the surface. *Bottom, during examination and after processing:* The exposed crystal has been developed into a filament of silver; the nonexposed crystals have been dissolved. The total thickness has decreased because the silver halide occupied approximately half the volume of the emulsion. (Courtesy of L.G. Caro.)

teins, various tritiated amino acids are used. For the polysaccharides and glycoproteins, tritiated monosaccharides, such as ^{3}H-mannose and ^{3}H-fucose, are employed.

Important investigations of the mechanism of DNA replication and RNA metabolism have been made using tritium labeling. In the example shown in Figure 3–21, the nuclei tagged with tritiated thymidine and initially present at the bottom of the intestinal crypts are found after 36 hours near the tip of the villus. This illustrates most graphically the life cycle of the cell—after a division at the bottom of the crypt, the cell ascends along the epithelium until, after a few hours, it is destroyed at the tip of the villus.

A two-emulsion radioautographic technique has been developed that can distinguish between the β-particles emitted from ^{14}C and those from ^{3}H atoms. With this method a tritiated precursor of DNA and a ^{14}C-labeled precursor of RNA or of a protein can be employed simultaneously (for further details see Pearse, 1980).

In Figures 9–10 and 9–11 an application of radioautography to the study of secretion is illustrated. In this case labeling was carried out with ^{3}H-leucine for a short period ("pulse"), followed by a "chase" with a nonradioactive amino acid. This procedure allows the sequential lo-

calization of the synthesized protein in the various cell compartments.

3–8 CELL FRACTIONATION

Cell fractionation methods involve, essentially, the homogenization or destruction of cell boundaries by different mechanical or chemical procedures, followed by the separation of the subcellular fractions according to mass, surface, and specific gravity.

Many different methods of cell fractionation are in use. Most of them are based on the homogenization of the cell in aqueous media—usually sucrose solutions in various concentrations.

A standard cell fractionation procedure is shown diagrammatically in Figure 3–22. The liver of an animal is first perfused with an ice-cold saline solution, followed by cold 0.25 M sucrose. The tissue is then forced through a perforated steel disk and homogenized in 0.25 M sucrose.

The classic type of cell fractionation is directed toward the subdivision of the cell components into four morphologically distinct fractions (nuclear, mitochondrial, microsomal, and soluble). In some glandular tissue a fifth fraction containing secretory granules may be obtained.

Note carefully that it is necessary to differ-

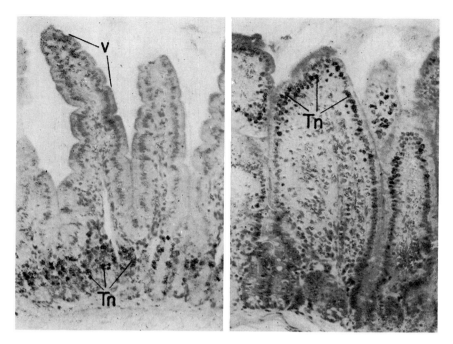

Fig. 3–21.   Section of intestine of a mouse injected with tritiated thymidine. *Left,* animal killed 8 hours after injection; *right,* 36 hours after injection. *Tn,* tagged nuclei; *V,* villus. (See text for details.) (Courtesy of C.P. Leblond.)

entiate between *cell fractions* and the parts of the cell *(organelles)* contained in the fraction. For example the mitochondrial fraction of the liver is composed principally of mitochondria, but the "mitochondrial fraction" of the brain is very heterogeneous and contains nerve endings and myelin in addition to free mitochondria. "Microsomes" do not exist, as such, in the cell; the "microsomal" fraction is composed mainly of broken parts of the endoplasmic reticulum, including the ribosomes, the Golgi complex, and other membranes (see Section 8–4).

Differential or Gradient Centrifugation

In the example just described (Fig. 3–22), the method used to separate the subcellular particles is called *differential centrifugation.* Depending on the strength of the centrifugal field needed, *standard centrifuges or preparative ultracentrifuges* are used.

At the initiation of the centrifugation all the particles are distributed homogeneously; as centrifugation proceeds, the particles settle according to their respective sedimentation ratios.

A similar principle may be used for separation of much smaller particles, such as viruses or macromolecules (i.e., nucleic acids and proteins), with the *analytical ultracentrifuge.* Using an analytical ultracentrifuge the sedimentation coefficient can be determined. This is expressed in Svedberg (S) units, which are related to the molecular weight of the particle (for example, transfer RNA, with 4S, has a molecular weight of 25,000 daltons).

Improvement in the technique of differential centrifugation may be achieved by using a density gradient, which may be either *discontinuous* or *continuous.* If it is discontinuous, the centrifuge tube is loaded with layers of varying densities (for example, with sucrose varying in molarity from 1.6 to 0.5 from the bottom to the top of the tube). However, mixing of two concentrations of sucrose produces a continuous gradient. Once the gradient is formed, the material is layered on the top and centrifuged until the particles reach equilibrium with the gradient. For this reason, this type of separation is also called *isopyknic* (equal density) centrifugation.

Improvements in this type of fractionation technique include the use of heavy water, cesium chloride, and media with different partition coefficients. To avoid drastic changes in osmotic pressure, macromolecular media such as glycogen Ficoll and Percoll are used.

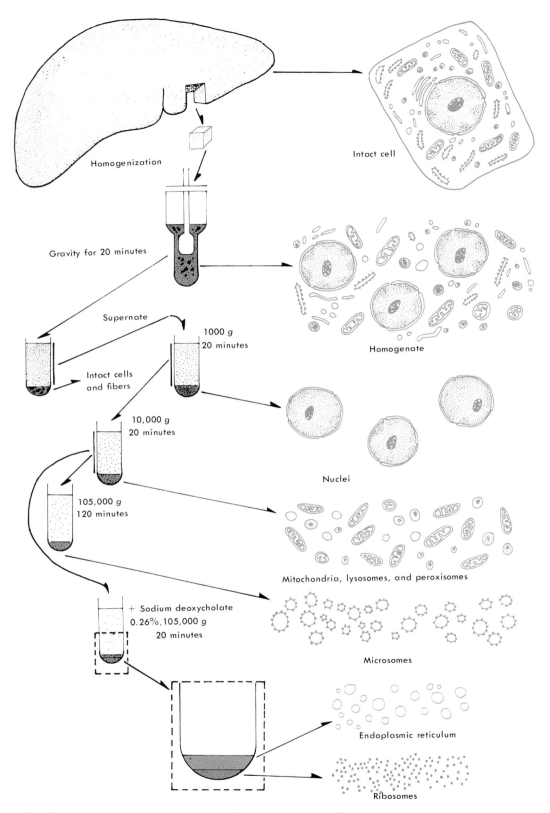

Homogenization

Intact cell

Gravity for 20 minutes

Homogenate

Supernate

1000 g
20 minutes

Intact cells
and fibers

Nuclei

10,000 g
20 minutes

105,000 g
120 minutes

Mitochondria, lysosomes, and peroxisomes

+ Sodium deoxycholate
0.26%, 105,000 g
20 minutes

Microsomes

Endoplasmic reticulum

Ribosomes

Fig. 3–22. Diagram showing the various steps used in the technique of differential centrifugation. A piece of tissue from liver is homogenized and then subjected to a series of centrifugations of increasing centrifugal force, as indicated on the left side of the figure. On the right side are diagrams of the various subfractions as they appear under the EM. (Modified from Bloom, W., and Fawcett, D.W.: Textbook of Histology. 10th Ed. Philadelphia, W.B. Saunders Co., 1975.)

With the use of the so-called *zonal rotors* the density gradient is formed while the rotor is spinning; then the sample is layered and centrifuged until the isopyknic zonal layering of the particles is reached.

Buoyant Density. Isopyknic centrifugation in preparative or zonal rotors permits the determination of the buoyant density of a macromolecule, i.e., the density at which it will reach an equilibrium with the suspending medium. This is important in studies of the molecular biology of nucleic acids (see Fig. 14–13).

Flow-sorting Cytometry

Methods have been developed to sort out whole cells, as well as metaphase chromosomes of a certain type. This can be achieved by the use of specially designed *sedimentation chambers* working at low gravity. In these chambers the cells or chromosomes are separated by size and density.

More useful has been the introduction of a *fluorescence activated cell sorter* or *cytometer*, which separates homogeneous groups of cells or chromosomes from a heterogeneous population, based on the amount of a fluorochrome bound to the DNA. The particles having a certain degree of fluorescence are displaced and separated by the cell sorter. Figure 14–4 shows an example of the application of the cell sorter to separate cells in different phases of the cell cycle.

Numerous applications of this technique are currently available. For example, isolated chromosomes can be used for gene mapping or to carry the transfer of genes to other cells by fusion (see Section 18–4).

SUMMARY:
Cytochemistry

Cytochemistry consists of the identification and localization of chemical components of the cell. Cytochemical (and histochemical) studies may be based on four main analytical techniques: (1) separation of cell fractions by conventional biochemical techniques; (2) isolation of minute amounts of tissues, and even single cells, by micro- and ultramicromethods; (3) direct detection of cell components in the cell by chemical staining; and (4) use of measurement of physical parameters.

Cytochemical staining is the method most often used in cytochemical studies. The substance to be investigated must be immobilized and then identified by chemical reactions. For the study of *proteins* the Millon and the diazonium reactions may be used. SH groups in protein may be determined by certain reagents.

Schiff's reagent, which is the leukobase of basic fuchsin, is recolored by *aldehyde* groups. This reagent is used in the *Feulgen reaction* for DNA. A prior mild acid hydrolysis removes RNA and unmasks aldehyde groups from deoxyribose linked to purine bases. The *periodic acid-Schiff (PAS) reaction* occurs in polysaccharides after oxidation of the 1,2-glycol groups by periodic acid. The *plasmal reaction* occurs in free aldehydes present in certain plasmalogens.

Detection of enzymes is accomplished by means of frozen sections made in a cryostat, or after light fixation with aldehydes. In general, the section is incubated with specific substrates, and the product is converted into a metal precipitate or a colored compound. Various phosphatases, esterases, cholinesterases, and other hydrolytic enzymes may be studied cytochemically. *Oxidases* and *dehydrogenases* are detected by special reagents, such as the Nadi reagent for cytochrome oxidase. The DAB reagent (3′,3′-diaminobenzidine) is widely used for peroxidases and other oxidases. Tetrazolium salts produce insoluble formazan dyes with some mitochondrial enzymes.

Several cytochemical methods are based on *physical determinations*. Ultraviolet absorption at 260 nm is characteristic of nucleic acids and results from the presence of purine and pyrimidine bases. Cytophotometric methods can be used in the ultraviolet or the visible spectrum to make quantitative determinations in cells and tissues. Thus, the Feulgen reaction permits the determination of the DNA content of the nucleus and even of single chromosomes.

Fluorescence microscopy, with ultraviolet light near the visible spectrum, is used to study the natural fluorescence *(autofluorescence)* of cell components such as vitamin A, thiamine, riboflavin, calcium, and porphyrins. More important is the use of fluorescent dyes *(secondary fluorescence)* in several cytochemical methods. Catechol and indolamines produce a fluorescent reaction with aldehydes.

Immunocytochemistry uses antibodies for the localization of macromolecules (antigens) in cells. (Antibodies are present in the γ-globulin fraction of serum.) In the direct reaction the antibody may be coupled with a fluorescent dye or an opaque molecule (e.g., ferritin) for the EM. In the *indirect* method the unlabeled antibody-antigen complex reacts with a labeled anti-γ-globulin antibody. This technique permits the study of antibody formation in plasmocytes and the localization of various proteins in different cells.

Radioautography uses substances labeled with radioisotopes. These radioisotopes are then detected using a photographic emulsion that is developed like an ordinary photograph. Most radioisotopes used are β-emitters and contain ^{14}C or tritium. Several techniques, such as the *track radioautography* in a thick liquid emulsion and the thin liquid emulsion for *electron microscope radioautography*, are employed. In most cases ^{3}H-thymidine is used for the study of DNA. ^{3}H-leucine, or other tritiated amino acids, is used for protein synthesis. Radioautography is one of the most frequently used techniques for following within the cell structure the mechanisms of DNA replication, DNA-RNA transcription, and their translation into protein.

ADDITIONAL READINGS

Amos, L.A., Henderson, R., and Unwind, P.N.T.: Three-dimensional structure determination by electron microscopy of two-dimensional crystals. Prog. Biophys. Mol. Biol., *39*:183, 1982.

Anderson, N.G.: The development of zonal centrifuges. Natl. Cancer Inst. Monogr., *21*, 1966.

Bahr, G.F.: Frontiers of quantitative cytochemistry. Anal. Quant. Cytol., *1*:1, 1979.

Beer, M., et al.: STEM studies of biological structure. Ultramicroscopy, 8:207, 1982.

Branton, D., and Kirchanski, S.: Interpreting the results of freeze-etching. J. Microsc. (Oxford), *111*:117, 1977.

Bullock, G.R., and Petrusz, P.: Techniques in Immunocytochemistry. Vol. 1. New York, Academic Press, 1982.

Carleman, E., and Kellenberg, E.: The reproducible observation of unstained cellular material in thin sections. EMBO J., *1*:63, 1982.

Cosslett, V.E.: Radiation damage in the high resolution electron microscopy of biological materials: A review. J. Microsc. (Oxford), *113*:113, 1978.

Coons, A.H.: Histochemistry with labeled antibody. Int. Rev. Cytol., *5*:1, 1956.

Crewe, A.V.: High resolution scanning transmission electron microscopy. Science, *221*:325, 1983.

Cuello, A.C.: Immunohistochemistry. Oxford, IBRO Publishers, 1983.

DeDuve, C.: Exploring cells with a centrifuge. Science, *189*:186, 1975.

Fujiwara, K., and Pollard, T.D.: Simultaneous localization of myosin and tubulin in human tissue culture cells by double antibody staining. J. Cell Biol., *77*:182, 1978.

Gabe, M.: Histological Techniques. Berlin, Springer-Verlag, 1976.

Heuser, J.: Quick-freeze, deep-etch preparation of samples for 3-D electron microscopy. Trends Biochem. Sci., *6*:64, 1981.

Hoppe, W.: Three-dimensional electron microscopy. Annu. Rev. Biophys. Bioeng., *10*:563, 1981.

Hopwood, D.: Theoretical and practical aspects of glutaraldehyde fixation. Histochem. J. (London), *4*:267, 1972.

Horne, R.W.: Special specimen preparation methods for image processing in transmission electron microscopy: A review. J. Microsc. (Oxford), *113*:241, 1978.

Inoué, S.: Video image processing greatly enhances contrast, quality and speed in polarization based microscopy. J. Cell Biol., *89*:346, 1981.

Kellenberger, E.: High resolution electromicroscopy. Trends Biochem. Sci., *3*:135, 1978.

Mazurkiewicz, J.E., and Nakane, P.K.: Light and electron microscopic localization of antigens in tissues embedded in polyethylene glycol with a peroxidase-labeled antibody method. J. Histochem. Cytochem., *20*:969, 1972.

Milstein, C.: Monoclonal antibodies. Sci. Am., *243*:66, 1980.

Pearse, A.G.: Histochemistry: Theoretical & Applied. Vol. 1. 4th Ed. London, Churchill, 1980.

Pickel, V.M., and Teitelman, G.: Light and electron microscope immunocytochemical localization of single and multiple antigens. *In* Histochemical and Ultrastructural Identification of Monoamine Neurons. Edited by J. Furness and M. Costa. New York, John Wiley & Sons, 1984.

Pollack, R.: Readings in Mammalian Cell Culture. Cold Spring Harbor, New York, Cold Spring Harbor Laboratory, 1973.

Pretlow, T.G., and Pretlow, T.P.: Cell Separation. Vol. 1. New York, Academic Press, 1982.

Puck, T.T.: The Mammalian Cell as a Microorganism. San Francisco, Holden-Day Inc., 1972.

Roddyn, D.B.: Subcellular Biochemistry. New York, Plenum Press, 1978.

Roth, J.: Applications of immunocolloids in light microscopy. J. Histochem. Cytochem., *30*:691, 1982.

Roth, J.: The protein A-Gold (pAG) technique, qualitative and quantitative approach for antigen localization on thin sections. *In* Techniques in Immunocytochemistry. Vol. 1. Edited by G.R. Bullock and P. Petrusz. London, Academic Press, 1981.

Sleytr, U.B., and Robards, A.W.: (1977) Freeze-fracturing: A review of methods and results. J. Microsc. (Oxford), *111*:77, 1977.

Sternberger, L.A.: Immunocytochemistry. 2nd Ed. New York, John Wiley & Sons, 1979.

Stumpf, W.E.: Autoradiography: Advances in methods and applications. J. Histochem. Cytochem., *29*:107, 1981.

CELL MEMBRANE AND PERMEABILITY

The cell has a different internal milieu from that of its external environment. For example, the ionic content of animal cells is quite dissimilar from that of the circulating blood. This difference is maintained throughout the life of the cell by the thin surface membrane, the *cell* or *plasma membrane*, which controls the entrance and exit of molecules and ions. The function of the plasma membrane of regulating this exchange between the cell and the medium is called *permeability.*

This membrane is so thin that it cannot be resolved with the light microscope, but in some cells it is covered by thicker protective layers that are within the limits of microscopic resolution. For example, most plant cells have a thick cellulose wall that covers and protects the true plasma membrane (see Fig. 1–1).

Most animal cells are surrounded by a *cell coat* or *external laminae* of cement-like substances that constitute visible membranes under the optical microscope. These cell coats, by including or excluding molecules according to their charge or size, play only a secondary role in permeability, but have other important functions as will be discussed in Chapter 5.

The objectives of this chapter are to study the plasma membrane from the viewpoint of its chemical composition and the ways in which the various molecular species are organized at the supramolecular level. This chapter also includes information about the assembly of artificial lipid bilayers and the important concepts of membrane fluidity and asymmetry. In addition, we will also cover the molecular structure of the plasma membrane in relation to its function, in the various types of cell permeability.

Cell surface research is currently one of the most active areas of cell and molecular biology, since it deals not only with the traffic of material between the cell and its environment, but also with many other functions related to the communication and interaction among cells in a tissue. These functions will constitute the substance of Chapter 5, which will deal mainly with the cell coat.

4–1 MOLECULAR ORGANIZATION OF THE CELL MEMBRANE

In the study of the molecular organization of the cell membrane, the first step is to isolate it from the rest of the cytoplasm in the purest form possible. The isolated membrane is then studied by biochemical and biophysical methods.

Several methods have been used to isolate plasma membranes from a variety of cells. In most cases the purity of the fraction has been controlled by electron microscopy, enzyme analysis, the study of surface antigens, and other criteria.

Plasma membranes are more easily isolated from erythrocytes subjected to hemolysis. The cells are treated with hypotonic solutions that produce swelling and then loss of the hemoglobin content (e.g., *hemolysis*). The resulting membrane is generally called a *red cell ghost*. Two main types of ghosts may be produced: *resealed ghosts* and *white ghosts*. The so-called resealed ghosts are produced when hemolysis is milder; the ghosts can be treated with substances that produce restoration of the permeability functions. White ghosts are formed if hemolysis is more drastic. There is complete removal of the hemoglobin, and the ghosts can no longer be resealed. These ghosts can be used for biochemical, but not for physiological, studies.

Figure 4–1 is meant to introduce to the student some of the concepts about the molecular organization of the red cell membrane, which will be dealt with in more detail later on. Free-floating erythrocytes and a red cell ghost cut in half are represented as seen under the light microscope; there is also a three-dimensional view of the membrane at the molecular level, in which the "fluid mosaic" model is reproduced.[1] It is important to observe that (1) the membrane is formed by a rather continuous bilayer of lipids into which protein complexes are embedded in a kind of "mosaic" arrangement, (2) there are other proteins peripheral to the bilayer and disposed on the inner surface (from (1) and (2) it can be concluded that the membrane is highly asymmetrical), and (3) the molecular asymmetry is further emphasized by the oligosaccharide chains of glycoproteins and glycolipids that protrude only at the outer surface of the membrane.

The Cell Membrane—Composed of Proteins, Lipids, and Carbohydrates

In the plasma membrane of the human red cell *protein* represents approximately 52% of its mass, *lipids* 40%, and *carbohydrates* 8%. Oligosaccharides are bound to lipids (i.e., *glycolipids*) and, mainly, to proteins (i.e., *glycoproteins*).

From the data shown in Table 4–1 it is evident that there is a wide variation in the lipid-protein ratio between different cell membranes. Myelin is an exception, in the sense that the lipid predominates; in the other cell membranes there is higher protein-lipid ratio. In myelin the area occupied by the protein is insufficient to cover that of the lipids, whereas in a red cell ghost the opposite situation is found.

Lipids are Asymmetrically Distributed within the Bilayer

The main lipid components of the plasma membrane are phospholipids, cholesterol, and galactolipids; their proportion varies in different cell membranes. The major proportion of membrane phospholipids is represented by phosphatidylcholine, phosphatidylethanolamine, and sphingomyelin, all of which have no net charge at neutral pH (i.e., *neutral phospholipids*) and tend to pack tightly in the bilayer. (This property is also shared by cholesterol.) Five to 20 per cent of the phospholipids are acidic, including phosphatidylinositol, phosphatidylserine, cardiolipin, phosphatidylglycerol, and sulfolipids. *Acidic phospholipids* are negatively charged and in the membrane are associated principally with proteins by way of lipid-protein interactions. Because of the many different arrangements of the fatty acids in the phospholipid molecule, it is understandable that a mammalian cell membrane may contain 100 or more species of phospholipids.[2]

One of the main characteristics of the molecular organization of the plasma membrane is the asymmetry of all of its chemical components. This refers to their nonuniform distribution between the two surfaces, i.e., the inner or *protoplasmic surface* (P_s), in contact with the ground cytoplasm, and the outer or *external surface* (E_s) in contact with the surrounding fluid medium (Fig. 4–1). The asymmetry is absolute

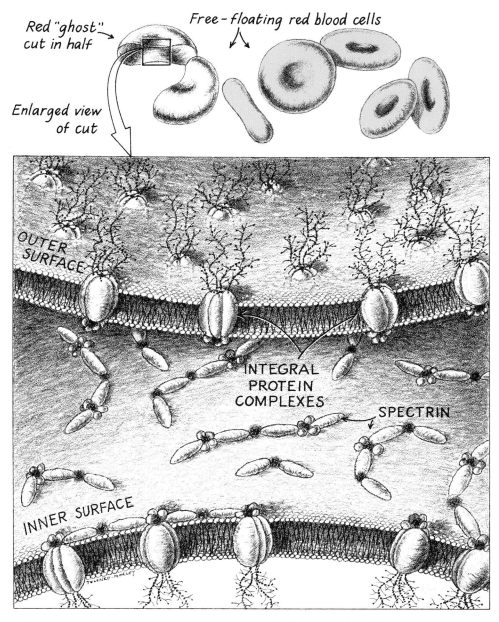

Fig. 4–1. Diagram showing red blood cells, a red cell ghost cut in half, and a view of the molecular organization of the membrane according to the "fluid mosaic" model. Integral protein complexes are represented with the oligosaccharide chains sticking out on the outer surface. At the inner surface the peripheral protein spectrin is represented. (Courtesy of G.L. Nicolson, 1978.)

for proteins and the carbohydrate moieties of glycoproteins and glycolipids, but it is only partial for the lipids.

Using reagents that do not cross the membrane and various *phospholipases* (i.e., enzymes that hydrolyze different parts of the phospholipid molecule), it has been demonstrated that the distribution of the *phospholipids* is partially asymmetrical. While the *outer layer* of the red cell membrane consists mainly of lecithin and sphingomyelin, the *inner layer* is composed mainly of phosphatidylethanolamine and phosphatidylserine. Furthermore, the *glycolipids* are mainly in the outer half of the bilayer. It is

TABLE 4–1.   LIPID AND PROTEIN RATIOS IN SOME CELL MEMBRANES

Species and Tissue		Protein (%)	Lipid (%)
Human	CNS myelin	20	79
Bovine	PNS myelin	23	76
Rat	Muscle (skeletal)	65	35
Rat	Liver	60	40
Human	Erythrocyte	60	40
Rat	Liver mitochondrion	70	27–29

	Molar ratio			Area ratio
	Amino Acid	Phospholipid	Cholesterol	Protein:Lipid
Myelin	264	111	75	0.43
Erythrocyte	500	31	31	2.0

From Triggle, D.J.: Neurotransmitter-Receptor Interactions. New York, Academic Press, Inc., 1971.

assumed that this asymmetry is rather stable and that there is little exchange of lipids across the bilayer.

Carbohydrates—In the Form of Glycolipids and Glycoproteins

The distribution of the oligosaccharides is also highly asymmetrical. In both *glycolipids* and *glycoproteins* they are confined exclusively to the external membrane surface. In erythrocyte membranes, hexose, hexosamine, fucose, and sialic acid are bound mainly to proteins. In fact all the proteins present at the outer surface are glycosylated.

Because of the presence of sialic acid residues, as well as carboxyl and phosphate groups, the outer surface of the membrane is negatively charged. Only a small amount of sialic acid exists in the form of *gangliosides* (i.e., glycolipids) in the plasma membrane of the liver (see Albers, 1981 in the Additional Readings at the end of this chapter).

Membrane Proteins—Peripheral or Integral

Proteins represent the main component of most biological membranes (Table 4–1). They play an important role, not only in the structure of the membrane, but also as carriers or channels, serving for transport; they may also be involved in regulatory or ligand-recognition properties. Numerous enzymes, antigens, and various kinds of receptor molecules are present in plasma membranes.

As explained in Section 2–5 one method usually used to separate the membrane polypep-tides is to dissolve the erythrocyte ghosts in sodium dodecyl sulfate (SDS) (an ionic detergent) and then to use polyacrylamide gel electrophoresis (PAGE). The gel can be stained for proteins (with Coomassie Blue) or for carbohydrates (with PAS reagent, see Section 3–5). Figure 4–2 shows a densitometric scan of the PAGE, which, by the position of the bands, permits estimation of the molecular weight of the polypeptides and, by the surface of the profile, enables determination of its relative mass (see Table 4–3). For example, *polypeptides 1* and *2* corresponding to *spectrin* represent about 30% of the total protein. *Polypeptide 3* is the major protein and represents about 25%.

Membrane proteins have been classified as *integral (intrinsic)* or *peripheral (extrinsic)* according to the degree of their association with the membrane and the methods by which they can be solubilized (Table 4–2).

Peripheral proteins are separated by mild treatment, are soluble in aqueous solutions, and are usually free of lipids. Examples include: *spectrin*, which may be removed from red cell ghosts by chelating agents, and cytochrome *c*, found in mitochondria, which is easily removed in high salt solutions.

Integral proteins represent more than 70% of the membrane proteins and require drastic procedures for isolation. Usually they are insoluble in water solutions, are associated with lipids, and need the presence of detergents to be maintained in a nonaggregated form. These proteins may be attached to oligosaccharides, thus forming glycoproteins.

To clearly understand the difference between integral and peripheral proteins, the student

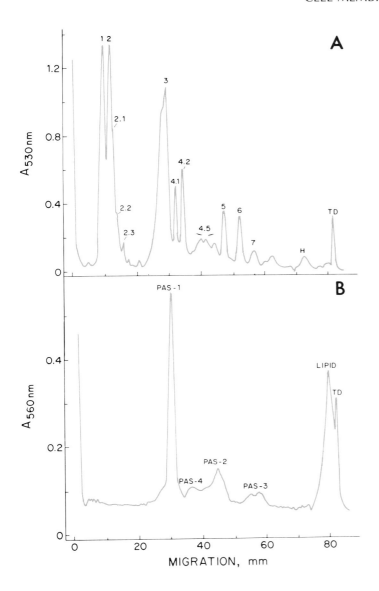

Fig. 4–2. Proteins and glycoproteins present in the erythrocyte membrane. *A,* densitometric scan of a gel stained for protein with Coomassie Blue. *B,* the same, stained with PAS reagent for carbohydrates. The nomenclature of the bands, the molecular weight of the proteins, and the nature of some of them are indicated in Table 4–3.

TABLE 4–2.   CRITERIA FOR DISTINGUISHING PERIPHERAL AND INTEGRAL MEMBRANE PROTEINS

Property	Peripheral Protein	Integral Protein
Requirements for dissociation from membrane	Mild treatments sufficient; high ionic strength, metal ion chelating agents	Hydrophobic bond-breaking agents required; detergents, organic solvents, chaotropic agents
Association with lipids when solubilized	Usually soluble—free of lipids	Usually associated with lipids when solubilized
Solubility after dissociation from membrane	Soluble and molecularly dispersed in neutral aqueous buffers	Usually insoluble or aggregated in neutral aqueous buffers
Examples	Cytochrome *c* of mitochondria; Spectrin of erythrocytes	Most membrane-bound enzymes; histocompatibility antigens; drug and hormone receptors

From Singer, S.J.: The molecular organization of biological membranes. *In* The Structure and Function of Biological Membranes. Edited by L.T. Rothfield. New York, Academic Press, Inc., 1971.

should remember that the primary structure of a protein involves domains of polar and nonpolar amino acids (see Section 2–2). In a peripheral protein, polar amino acids (i.e., hydrophilic) predominate at the surface, while nonpolar ones (i.e., hydrophobic) are buried in the interior. On the other hand, in an integral protein the non-polar amino acids are more exposed at the sur-face. In integral proteins the hydrophobic area may range between 20 and 60% of the total sur-face area that can be measured by using radio-active detergents.[3] The role of the hydrophobic surface is that of interacting with lipids serving as a kind of anchor or insertion into the bilayer.

One of the most hydrophobic proteins is *bac-teriorhodopsin*, which is present in the mem-brane of halophilic bacteria (i.e., bacteria living in high salt media). This protein functions as a proton (H^+) pump driven by light, and consists of a single polypeptide of 247 amino acids, which traverses 7 times the lipid bilayer. Each of the traversing segments is an α-helix with some 25 amino acid residues.[4]

Every Protein of the Cell Membrane Is Distributed Asymmetrically

The molecular organization of proteins is highly asymmetrical. This can be demonstrated by the use of reagents that are unable to cross the membrane. The reagent is first applied to the intact erythrocyte and then to the white ghost, in which there is access to the inner sur-face. The difference between the two prepara-tions may provide information about the position of a particular protein with respect to the outer or inner surface. Some specific labels can be detected on either side of the membrane by using cytochemical and electron microscopic techniques. For example, using antibodies against spectrin or actin (polypeptide 5), it is possible to show that each one is on the cyto-plasmic surface of the red cell membrane. On the other hand, acetylcholinesterase can be shown to be in the outer surface by its inacti-vation with proteolytic enzymes.

It can be said with certainty that every protein constituent is asymmetrically distributed in the red cell membrane. Most readily soluble poly-peptides are localized at the cytoplasmic surface, whereas those that are intrinsic to the mem-brane are more tightly bound to the lipid struc-ture of the membrane. A tentative model rep-resenting the position of the main polypeptides in the red cell membrane is shown in Figure 4–3A. This figure also shows how the membrane proteins split between the P and E leaflets (Fig. 4–3B).[5]

Major Polypeptides of the Red Cell Membrane and Cytoskeleton

In Table 4–3 are listed the major polypeptides found in a red cell ghost. Some of these poly-peptides are part of the membrane proper; oth-ers belong to the cytoskeleton, which in the erythrocyte has some special proteins. For ex-ample, α and β spectrin, as well as polypeptides 4.1 and 5 (actin), are considered part of the cy-toskeleton. These are the only proteins that re-main after the ghosts are extracted with a de-tergent that dissolves the membrane.

Spectrin and Ankryn. Spectrin represents the major mass of the cytoskeleton. It is composed of two nonidentical polypeptide chains (α of 240,000 daltons and β of 220,000 daltons) and may be present both as a dimer (α β) or as a tetramer (α β)2. Under the EM the spectrin molecule appears as two loosely intertwined and flexible strands of about 100 nm (the tetramer is 200 nm long). Spectrin binds to the protein called *ankryn* (band 2.1), which in turn is as-sociated with the integral protein of band 3. In this way the cytoskeletal spectrin becomes an-chored to the membrane. Spectrin also has bind-ing sites for band 4.1 and actin (band 5), which are present as oligomers of about 10 monomers (Fig. 4–4). Thus it can be said that the cytoskel-eton of the red cell is a meshwork composed of spectrin tetramers together with actin and band 4.1. (In Chapter 6 the cytoskeleton of other cells will be studied.)

The erythrocyte cytoskeleton can be observed under the EM after fixation with tannic acid-gluteraldehyde and also by scanning electron microscopy. This filamentous network provides stability to the plasma membrane and may also control the characteristic shape of the erythro-cyte (see Branton et al., 1981 in the Additional Readings at the end of this chapter).

There are several hereditary diseases in which there is an alteration of the typical biconcave shape of red cells. For example, erythrocytes may be spheroid (*spherocytosis*) or ellipsoid (*el-*

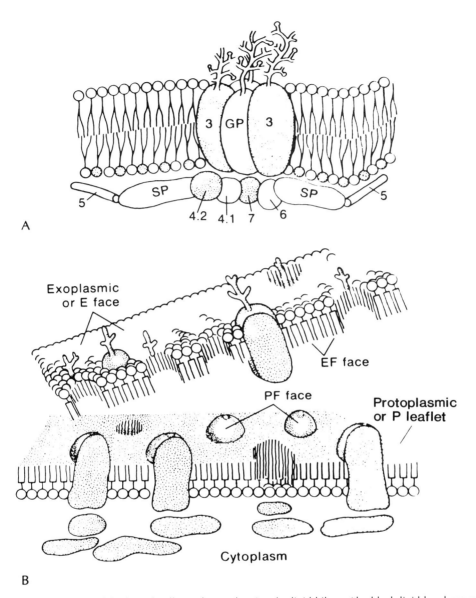

Fig. 4–3. *A,* hypothetical model of a red cell membrane showing the lipid bilayer (the black lipid head groups indicate the asymmetric distribution of phosphatidyl serine and phosphatidyl glycerol). *GP,* glycophorin (bands PS-1 and PS-2); *III,* band 3; *Sp,* spectrin (band 1); *IV a* and *IV b,* components 4.1 and 4.2; *V,* actin (band 5); *VI,* G3PD (band 6); *VII,* band 7. The nomenclature of the bands is as in Table 4–3. Peripheral proteins are in red. (Courtesy of G.L. Nicolson, 1977.) *B,* diagram of a membrane prepared using the freeze-fracturing method. Observe that the cleavage occurs along the contact between the two lipid leaflets. The PF face corresponds to the protoplasmic leaflet and shows more intrinsic proteins protruding on the surface. (See Fig. 4–1.) The EF corresponds to the exoplasmic face and shows pits corresponding to the particles in the other leaflet.

liptocytosis). In these diseases it has been found that there are alterations in the erythrocyte cytoskeleton. For example, in spherocytosis there is less ability for spectrin to undergo the dimer-tetramer transformation. On the other hand, in some cases of elliptocytosis there is a defect in the binding of band 4.1 to spectrin, or even a complete lack of 4.1 (see Cohen and Branton, 1981 and Bennett, 1982 in the Additional Readings at the end of this chapter).

Band 3. The major intrinsic protein *(band 3),* with a molecular weight of 93,000 daltons, spans the thickness of the membrane and has a small amount of carbohydrate on the pole at the outer

TABLE 4–3.   MAJOR ERYTHROCYTE MEMBRANE POLYPEPTIDES AND GLYCOPROTEINS

Polypeptide	Molecular Weight (daltons)	Copies per Ghost	Designation	
1	240,000	200,000	α-spectrin }	cytoskeleton
2	220,000	200,000	β-spectrin }	
2.1	200,000	100,000	ankryn	
3	93,000	1,200,000	anion channel	
4.1	82,000	200,000		cytoskeleton
4.2	76,000			
4.9	48,000	100,000		
5	43,000	500,000	actin	cytoskeleton
6	35,000	540,000	G 3PD	
7	28,000	403,000		
Glycoproteins				
PAS-1			} glycophorin	
PAS-2	55,000	500,000	} sialoglycoprotein	

Data from Steck, L.: J. Cell Biol., *62*:1, 1974 and Branton, D., et al., Cell, *24*:24, 1981.

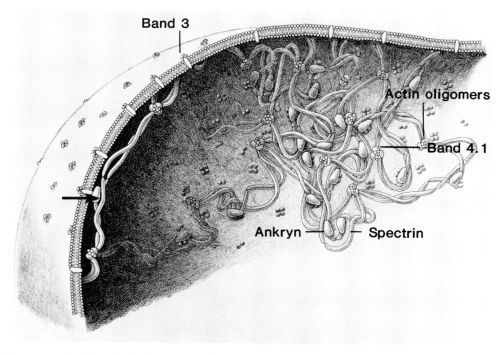

Fig. 4–4.   Diagram depicting the lipid bilayer with the intrinsic protein of band 3 and the main components of the erythrocyte cytoskeleton on the cytoplasmic side of the membrane. The long and flexible spectrin molecules are shown to bind to ankryn, which in turn binds to band 3 in the membrane. Spectrin also binds to band 4.1 and to actin oligomers. Observe that band 3 forms anionic channels across the membrane and binds to spectrin by way of ankryn *(arrow)*. (Courtesy of D. Branton).

surface (Fig. 4–3). This polypeptide is present in the membrane as a dimer held together by S–S bonds. There are between 500,000 and 600,000 such dimers per cell, enough to account for the 8 nm particles observed in freeze-fractured membranes (see Fig. 4–5). This dimeric protein appears to be involved in the transport *of anions* (i.e., chloride, bicarbonate) across the membrane.

Glycophorin. In the model shown in Figure 4–3 several glycoproteins, including band 3, are represented; however, the major one that spans the membrane is the so-called *glycophorin* (i.e., PAS-1 and PAS-2, Table 4–3). This protein has a molecular mass weight of 55,000, of which 60% is carbohydrate. Near the COOH end of the molecule there is a region that is very hydrophobic and which interacts with the lipids of the membrane. The COOH end is probably exposed to the interior of the red cell. The NH_2 end is more hydrophilic, is exposed to the external environment, and has the attached oligosaccharides that are at the outer surface of the membrane.

Glycophorin contains antigenic determinants for the ABO blood groups and others such as: the MN groups reacting with rabbit antisera, the influenza virus, phytohemagglutinin, and wheat germ agglutinin. It has been calculated that there are some 500,000 copies of glycophorin per human red cell. This protein accounts for 80% of the carbohydrate and 90% of the negatively charged sialic acid present in the cell surface.

Among the most hydrophobic integral proteins of the membrane, the so-called *proteolipids* are characterized by their strong association with lipids and the fact that they are soluble in organic solvent. First isolated by Folch and Lees from myelin, proteolipids are found in practically all cell membranes, and in many of them they represent receptor proteins for synaptic transmitters or form channels across those membranes (see De Robertis, 1975 in the Additional Readings at the end of this chapter).

Asymmetrical Distribution of Enzymes

More than 30 enzymes have been detected in isolated plasma membranes. Those most constantly found are 5'-nucleotidase, Mg^{2+} ATPase, $Na^+ - K^+$ ATPase, alkaline phosphatase, ad-

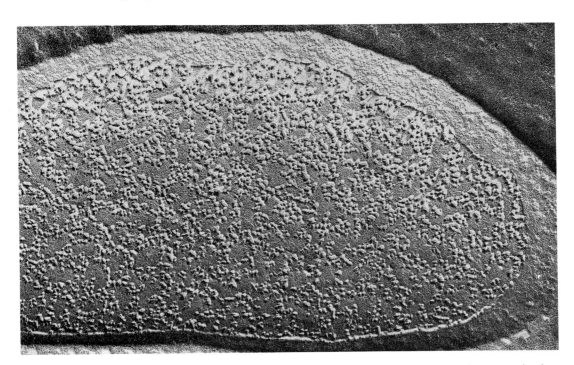

Fig. 4–5. Electron micrograph of a freeze-fractured and etched red cell ghost. Most of the surface shows particles that are intercalated in the plane of cleavage of the membrane. (P). The smooth area (S) corresponds to the true surface of the erythrocyte ghost that has been exposed by the etching. × 88,000. (Courtesy of D. Branton.)

enyl cyclase, acid phosphomonoesterase, and RNAse. A specific localization with a mosaic arrangement has been postulated for some of these enzymes. Disaccharidase forms 5 to 6 nm globular units coating the membrane of the microvilli.

Of all the membrane-associated enzymes, $Na^+ - K^+$ ATPase is one of the most important because of its role in ion transfer across the plasma membrane. This enzyme is dependent on the presence of lipids and is inactivated when all lipids are extracted.

Enzymes also show an *asymmetrical distribution*. For example, in the outer surface of erythrocytes there are acetylcholinesterase, nicotinamide-adenine dinucleotidase, and the ouabain binding site of the $Na^+ K^+$ ATPase. In the inner surface there is NADH-diaphorase, G3PD, adenylate cyclase, protein kinase, and Mg^{++} ATPase.

SUMMARY:
Molecular Organization of the Cell Membrane

The first step in the study of the molecular organization of the membrane consists of the isolation of the cell membrane. The easiest to isolate is the red cell membrane obtained after hemolysis. The red cell ghost contains 52% protein, 40% lipid, and 8% carbohydrate that is bound to lipid (glycolipids) or protein (glycoproteins). The lipids form a rather continuous bilayer in which the outer layer contains mainly lecithin and sphingomyelin, and the inner one, phosphatidylethanolamine and phosphatidylserine (lipid asymmetry). The oligosaccharides, present in glycolipids and glycoproteins, are located exclusively on the external surface.

Proteins play various roles: mechanical, transport, receptor, antigenic, and enzymatic. Classified by the degree of their association to the membrane, proteins are *peripheral* (extrinsic) or *integral* (intrinsic). The various proteins may be isolated by polyacrylamide gel electrophoresis (PAGE). Every protein in the red cell membrane is asymmetrically distributed. The peripheral proteins *spectrin* (polypeptides 1 and 2) and *actin* (polypeptide 5) form microfilaments associated with the inner surface of the membrane. Other peripheral proteins are glyceraldehyde-3-P dehydrogenase (G3PD) (polypeptide 6) and the protein of band 4.2.

The major intrinsic protein (band 3), with 93,000 daltons, spans the membrane and is present as a dimer. The major glycoprotein is *glycophorin*, with a molecular weight of 55,000, which also spans the membrane and has several antigenic sites. There are some 30 enzymes in the cell membrane, and these have an asymmetrical distribution.

4–2 MOLECULAR MODELS OF THE CELL MEMBRANE

The early theories on the molecular structure of the membrane were generally based on indirect information. Since substances soluble in lipid solvents penetrate the plasma membrane easily, in 1902 Overton postulated that the plasma membrane is composed of a thin layer of lipid. In 1926 Gorter and Grendell found that the lipid content of hemolyzed erythrocytes was sufficient to form a double layer of lipid molecules over the entire cell surface. This theory was also supported by electrical measurements that indicated a high impedance at the plasma membrane, which is due to the fact that it is difficult for ions to penetrate a lipid layer.

Other indirect information came from the study of the interfacial tension of different cells. Tension at a water-oil interface is about 10 to 15 dyn/cm, whereas surface tension of cells is almost nil. The low tension is due to the presence of protein among the lipid components. In fact, when a very small amount of protein is added to a model lipid-water system, the surface tension is lowered comparably.

To explain all these properties in 1935 Danielli and Davson proposed that the plasma membrane contained a *lipid bilayer*, with protein adhering to both lipid-aqueous interfaces.

Artificial Model Systems—Liposomes

The Danielli-Davson model was supported by the study of artificial lipid systems in which *monolayer* and *bilayer* films were formed. Recall that the various lipid molecules described in Chapter 2 are *dipoles*, containing both a polar (hydrophilic) group and a nonpolar (lipophilic) hydrocarbon chain (see Fig. 2–9). If a lipid dissolved in certain kinds of solvents is spread on the surface of water, it tends to form a *lipid monolayer* whose thickness depends on the number of carbons in the hydrocarbon chain. The monomolecular film can be deposited on the surface of a glass slide dipped into water, and bilayers or multilayers can be formed by the process of successive dippings.

Because of their many experimental applications, the *planar* and the *vesicular bilayers* are even more useful than the monolayers just described. By applying a droplet of lipid solubilized in an organic solvent to a small hole in a septum dividing two water chambers, it is possible to produce a lipid bilayer across which many biophysical properties (e.g., electrical resistance, ion permeability) can be studied. It is also possible to introduce into these membranes certain small polypeptides, proteins, and other substances capable of making channels for the passage of ions. These and other artificial systems have proved to be useful models for the study of some permeability mechanisms as well as the protein-lipid interactions within the membrane. Several types of phospholipid-water systems are illustrated in Figure 4–6.

In a partially hydrated phospholipid a hexagonal phase may be formed. This phase consists of a two-dimensional hexagonal lattice in which cylinders filled with water are embedded in a lipid matrix (Fig. 4–6A).[6] If a hexagonal phase, fixed with osmium tetroxide, is examined under the EM, dense dots corresponding to the water-filled cylinders are observed. This indicates that the osmium is bound to the polar groups of the lipids.

In the presence of increasing amounts of water, *lamellar phases*, generally called *myelin figures*, are formed. When studied at different temperatures, both the hexagonal and lamellar

systems exhibit special transitions that can be detected by X-ray diffraction, as well as by other physical methods. At these transitions the so-called *thermotropic mesomorphism* occurs. This phenomenon is apparently due to the change of the hydrocarbon chain from a liquid to a solid state. With a temperature above the transition point the mobility of the lipid molecules is increased.

As shown in Figure 4–7, the type of molecular arrangement depends on many factors, which may vary the shape of the molecule. For example, most neutral phospholipids have molecules with a cylindrical shape, which usually form bilayers. On the other hand, lysophospholipids (a phospholipid with one fatty acid removed) and some detergents, tend to produce micelles with the larger polar groups on the outside. Furthermore, phospholipids with unsaturated fatty acids, or acidic phospholipids have a conical-shaped molecule, with a smaller polar head that usually forms hexagonal phases. This last configuration is also promoted by raising the temperature and by the addition of Ca^{++}. The bilayer-hexagonal phase transition appears to be important in certain processes of membrane fusion and in the understanding of membrane biogenesis as well.

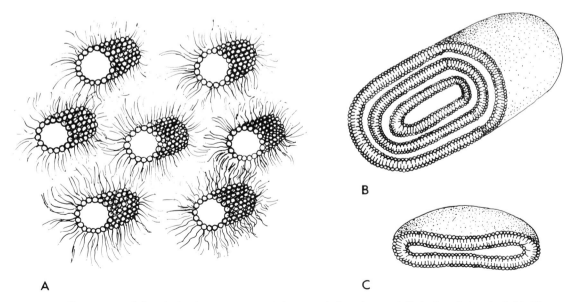

Fig. 4–6. Different types of phospholipid-water systems. *A*, hexagonal phase in a partially hydrated phospholipid. Observe that the water is contained within the cylinders, which are limited by polar groups; *B*, multilayered smectic mesophase corresponding to one liposome. Water is contained between lipid bilayers; *C*, phospholipid vesicle, single bilayered structure of 2 nm. (Courtesy of A.D. Bangham and D.A. Haydon. From Bangham, A.D., and Haydon, D.A.: Br. Med. Bull., 24:124, 1968.)

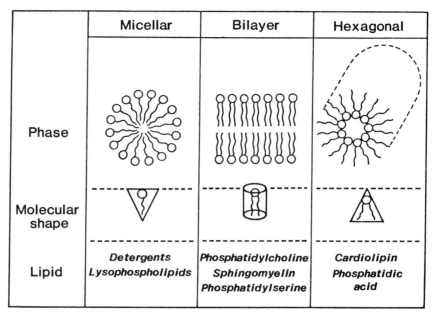

	Micellar	Bilayer	Hexagonal
Phase			
Molecular shape			
Lipid	*Detergents* *Lysophospholipids*	*Phosphatidylcholine* *Sphingomyelin* *Phosphatidylserine*	*Cardiolipin* *Phosphatidic acid*

Fig. 4–7. Diagram showing how the molecular shape of the lipids and the polar and nonpolar ends influences the type of phase produced, i.e., micellar, bilayer, and hexagonal.

In 1965 Bangham[7] defined a liposome as a special type of lamellar phase in which the water is self-contained and the lipid molecules are disposed in bimolecular layers attached by their nonpolar interfaces (Fig. 4–6*B*). Liposomes possess many properties akin to those of biological membranes: they are easy to manipulate; their lipid composition may be varied at will; and many substances may be trapped inside the interlamellar spaces. By sonicating a liposome suspension it is possible to obtain, under certain special conditions, single bilayer vesicles measuring about 2 nm (Fig. 4–6*C*).

In recent years important applications for liposomes have emerged in the medical field. Liposomes constitute excellent vehicles for carrying different molecules, which are protected within the lipid membrane, into cells and tissues. For example, a missing enzyme can be delivered into an organism that has a hereditary deficiency of that enzyme. Furthermore, drugs for cancer chemotherapy may be delivered more easily to the sites of attack; insulin may be given orally with liposomes;[8] and specific antibodies can be delivered to certain cell types.[9]

The Unit Membrane Model—Reevaluation of the Electron Microscopic Image

The EM showed, for the first time, that all cells are surrounded by a plasma membrane 6 to 10 nm thick. Early observations revealed a three-layered structure with two outer dense layers of 2.0 nm each and a middle one of about 3.5 nm (Fig. 4–8*A*). On the basis of these findings, Robertson (1959) postulated the *unit membrane model*.[10] In this model the electron microscopic image was interpreted along the lines of the Danielli-Davson model, with the clear central layer corresponding to the hydrocarbon chains of the lipids and the dense surrounding layers to the proteins on both sides. It was thought that many kinds of intracellular membranes were also constructed according to this pattern.

It is now known that the unit membrane model is an oversimplification. It does not, for example, account for the numerous protein molecules that traverse the membrane. In recent years the model has had to be re-evaluated on the basis of the following observations: (1) Extremely thin sections have revealed the existence of fine bridges that cross the bilayer (Fig. 4–8*C*). (2) Freeze-fracture through the erythrocyte membrane shows that numerous particles are intercalated in the plane of cleavage of the membrane (Fig. 4–5). (3) Fixation methods that avoid the removal of proteins tend to give a more granular appearance to the membrane in a cross-section. According to these findings the unit

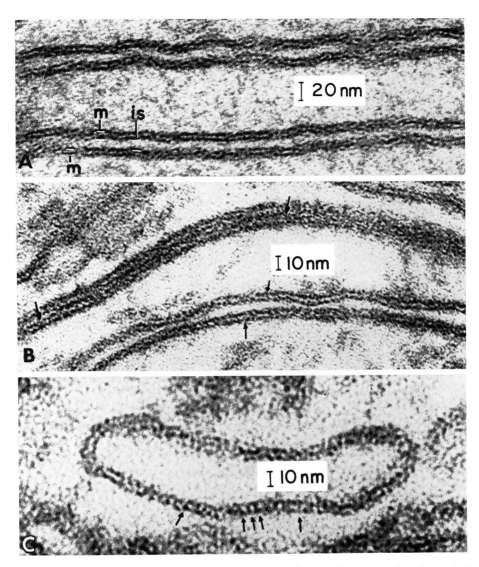

Fig. 4–8. *A,* electron micrograph of cell membranes of intestinal cells *(m),* showing the three-layered structure (unit membrane). *is,* intercellular space. × 240,000. *B,* cell membranes in the rat hypothalamus showing the unit membrane structure and, with arrows, some finer details across the membrane. The upper arrows indicate a region in which the two cell membranes are adherent *(tight junction)* and the intercellular space has disappeared. × 360,000. *C,* the same as *B,* showing fine bridges *(arrows)* across the unit membrane. × 380,000. (From E. De Robertis, unpublished.)

membrane is somewhat artifactual, perhaps signifying only the appearance of the three-layered structure after preparation for electron microscopy (see Luftig et al., 1977 in the Additional Readings at the end of this chapter).

The Fluid Mosaic Model—Generally Accepted

Present knowledge about the molecular organization of biological membranes comes mainly from an integration of the data on chemical analysis and from the application of several biophysical techniques.

The important concepts that have emerged are summarized in the so-called *fluid mosaic model* of membrane structure. This postulates: (1) that the lipid and integral proteins are disposed in a kind of mosaic arrangement, and (2) that biological membranes are quasi-fluid structures in which both the lipids and the integral proteins are able to perform translational movements within the bilayer.[1] The concept of fluidity implies that the main components of the membrane are held in place only by means of noncovalent interactions. In fact, the cohesive forces between lipids and with proteins are relatively weak interactions, i.e., ionic, hydrogen bonds, and mainly hydrophobic in character.

To better understand the molecular organization of the membrane, it is necessary to remember that, not only the lipids, but also many of the intrinsic proteins and glycoproteins of the membrane, are amphipathic molecules. The term *amphipathy*, coined by Hartley in 1936, refers to the presence, within the same molecule, of hydrophilic and hydrophobic groups.

In the fluid mosaic model represented in Figure 4–1 the integral proteins of the membrane are intercalated to a greater or lesser extent into a rather continuous lipid bilayer. These integral proteins are amphipathic, with polar regions protruding from the surface and nonpolar regions embedded in the hydrophobic interior of the membrane.

This arrangement may explain why different enzymes and glycoproteins may have their active sites exposed to the outer surface of the membrane. It is well recognized that a protein of appropriate size or a cluster of protein subunits may pass across the entire membrane (transmembrane proteins). Such traversing proteins may be in contact with the aqueous solvent on both sides of the membrane.

One of the major supports for this mosaic model of the membrane comes from the use of freeze-fracturing techniques in erythrocytes and other cell membranes (see Section 3–2). The red cell ghosts show a large number of particles, about 8 nm in diameter, that represent proteins embedded within the plane of cleavage, which passes through the middle of the lipid bilayer (Fig. 4–5).

There are between 500,000 and 600,000 such particles per cell, and more are attached to the inner half of the membrane (PF) than to the outer half (EF).

We mentioned earlier that these particles represent dimers of the polypeptide found in *band 3*, and that they may also correspond to anionic permeation channels. The degree of dispersion or aggregation of these particles within the membrane appears to be influenced by the *spectrin-actin system* (see earlier discussion). This suggests that some link between this system and

the dimers of band 3 exists, in the living membrane.

Membrane Fluidity and Membrane Fusion

The concept of membrane fluidity refers to the fact that both lipids and proteins may have considerable freedom of lateral movement within the bilayer. Vectorial movement across the membrane, however, is severely constrained (meaning that a lipid or protein in the outer half of the bilayer cannot pass into the inner half).

Membrane fluidity is essentially a property of the lipids. Normally these are fluid at body temperature and the main consideration is the degree of saturation of the hydrocarbon chains. Unsaturated fatty acids (i.e., those containing double and triple bonds) have a lower melting point than saturated ones, and in most biological membranes there are sufficient unsaturated lipids that the melting point of the lipid bilayer remains below physiological temperature. One factor that tends to increase rigidity is the concentration of cholesterol. Membrane fluidity may be subject to physiological regulations, for example, in hibernating animals, during the phase of hibernation, there is a change in fatty acids that increases fluidity.[11] Rapid changes in fluidity can be produced by methylation of phosphatidylethanolamine by methyltransferases present in the membrane; these in turn are regulated by receptors.[12]

The fluidity of the membrane can be studied with a series of techniques that can be classified as physical or biological.

The *physical techniques* are of two main types: (1) those that involve a minimal perturbation of the membrane, such as *X-ray diffraction* and *nuclear magnetic resonance (NMR) spectroscopy;* and (2) those that use certain added molecules to monitor specific sites of the membrane. Into this second class fall *fluorescence microscopy*, which uses fluorescent probes (see Section 3–6), and *electron spin resonance (ESR)* spectroscopy, which uses paramagnetic probes (e.g., nitroxide-containing amphipathic molecules) that are introduced into the lipid bilayer. The data obtained provide information on lipid-protein packing, lateral diffusion of lipids, lipid-protein interactions, the fluidity of the membrane, and the rate of transmembrane rotation (so-called flip-flop) of molecules across the bilayer. Detailed consideration of these techniques is beyond the scope of this book.

The *biological techniques* involve light and fluorescence microscopy and electron microscopy, including freeze-fracturing and radioisotope labeling methods. One of the simplest methods consists of binding gold or carbon particles to the cell surface and observing under the light microscope the movement of those particles on the surface. More information has come from studies in which different ligands, such as antibodies and plant lectins, interact with cell surface receptors. If these ligands are labeled with fluorescent dyes their movement can be followed by fluorescence microscopy, as illustrated in the following example: If lymphocytes are treated with fluorescent antibodies to certain membrane antigens, it is possible to observe the *capping phenomenon.* The antigens visibly displace in the membrane, forming patches (Fig. 4–9). Then they aggregate at one pole of the cell, producing a kind of cap that is highly fluorescent. At this point the membrane may invaginate into vesicles, which are internalized into the cytoplasm. This capping process can be inhibited by lowering the temperature so that the lipid bilayer solidifies.

A classic experiment involving these phenomena is that of Frye and Edidin,[13] in which two different cultured cells having different surface antigens are fused. *Cell fusion* is achieved by the use of an inactivated parainfluenza virus called Sendai (after a city in Japan). *Sendai virus* facilitates fusion of the cells' plasma membranes, merging the cytoplasms and thus producing a *heterokaryon* with two nuclei (Fig. 4–10, top). If the two cells are originally labeled with fluorescent antibodies of different colors, such as fluorescein (green) and rhodamine (red), it is possible at the onset of fusion to recognize the parts of the plasma membrane corresponding to each cell. However, intermixing occurs as the antigens are dispersed, and the two colors become less and less detectable. After 40 minutes the intermixing is complete and the two antigens can no longer be distinguished. Again, in this case, temperatures below 20° C impair the intermixing by causing solidification of the lipid bilayer.

Figure 4–10 (bottom) represents the possible molecular mechanism of this fusion. When the

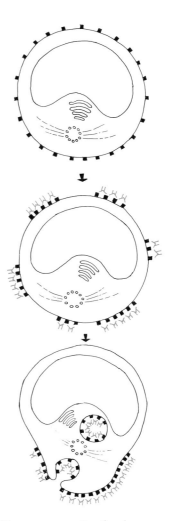

Fig. 4–9. Diagram representing the phenomenon of "capping" in lymphocytes. *Top,* corresponds to the normal distribution of antigens on the cell surface. *Middle,* the antigens are now clustered into patches after being cross-linked with a bivalent antibody. *Bottom,* a cap is formed by the active transport of the patches toward the pole that contains the centrosomal-Golgi area. Observe that the membrane is being internalized by endocytosis. Antibodies are in red.

two membranes approach each other and are under the influence of the *fusogen* (i.e., a fusion promoting factor, such as Sendai virus, lysophosphatides, oleic acid, and more recently the use of electric fields that induce cell fusion), the following phenomena may take place: (1) displacement of intrinsic proteins to produce protein free zones and contact between the two lipid bilayers; (2) formation of a hexagonal phase in between the two monolayers; (3) complete fusion and reorganization of the plasma membrane. The transition bilayer $\rightleftarrows$ hexagonal phase

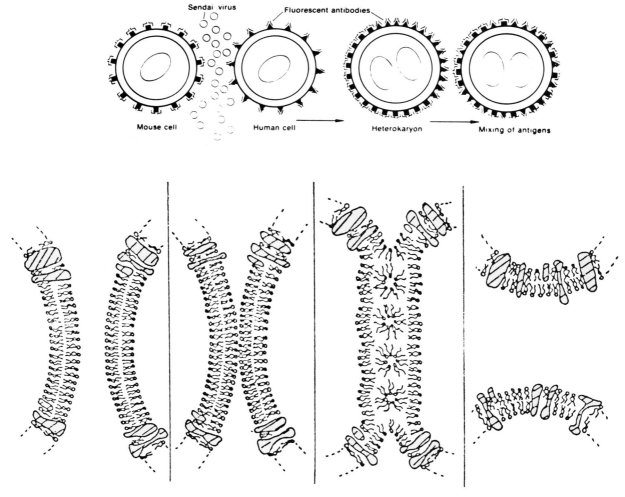

Fig. 4–10. *Top,* diagram representing the experiment of Frye and Edidin on cell fusion. A mouse and a human cell having different surface antigens labeled with fluorescent antibodies are fused by Sendai virus. The heterokaryon has each half of its surface covered with different types of antigens. After 40 minutes, mixing of the antigens occurs because of their movement in the plane of the membrane. *Bottom,* a possible interpretation of the molecular mechanism involved in cell fusion. (Redrawn from Schairer, H.V. and Overath, P.: J. Mol. Biol., *44:*209, 1969.)

seems to be essential in cell fusion, as well as in other transient processes, such as endo- and exocytosis, "blebbing" of the plasma membrane, and membrane biogenesis.

The Mobile Hypothesis of Receptors

Membrane receptors are macromolecules that have the dual function of recognizing a chemical signal and initiating a biological response. Several cellular regulatory agents (e.g., peptide hormones, neurotransmitters) are ligands that may act as chemical signals, which are recognized by such receptors. At the receptor site there is a specific type of binding that involves a *ligand-*

receptor interaction. Several kinds of receptors produce a response by interacting within the membrane with the enzyme *adenylate cyclase,* which as we have seen in Chapter 2 produces cyclic AMP (cAMP) from ATP. Several receptors that are specific for different ligands may act on a single type of adenylate cyclase. The important point is that the effect of the ligands is not additive; in other words, the system behaves as though several receptors were coupled to one molecule of enzyme. To explain this phenomenon, Cuatrecasas and others have postulated the *mobile hypothesis,* which is based on the fluidity of the membrane. According to this hypothesis, the lateral diffusion of receptors and enzyme within the plane of the lipid bilayer allows several receptors to couple with a single adenylate cyclase.[14]

This hypothesis has received experimental confirmation in fusion studies similar to those performed by Frye and Edidin. Cells having the β-adrenergic receptor in their membrane (and in which the adenylate cyclase was selectively inactivated by heat or N-ethylmaleimide prior to the fusion), were fused with others containing only adenylate cyclase (Fig. 4–11). The resulting hybrid cells contained both macromolecules, enabling coupling between the receptor and adenylate cyclase to occur. In these hybrid cells, under the influence of the ligand (in this case, isoproterenol), there was production of cAMP; the original unhybridized cells were unreactive to the ligand (see Schramm et al., 1977 in the Additional Readings at the end of this chapter).

SUMMARY:

Membrane Molecular Models

Several models of lipid-water and lipid-protein systems may be used to clarify our understanding of natural biological membranes. Lipid monolayers and bilayers, as well as myelin figures or cystalline phospholipid-water phases with a hexagonal configuration, may be produced and analyzed with polarization microscopy, x-ray diffraction, or electron microscopy. Such studies may give information about the relative position of polar and nonpolar groups and their association with proteins. Some inferences, based on studies with the EM, may be made about the electron density image.

Some special model systems of a lamellar type are provided by multilayered liposomes and the single, bilayered phospholipid vesicles. These systems may have important applications in biology and medicine.

To explain various properties of the cell membrane (e.g., permeability of lipid solvents, low surface tension) Danielli and Davson (1935) proposed

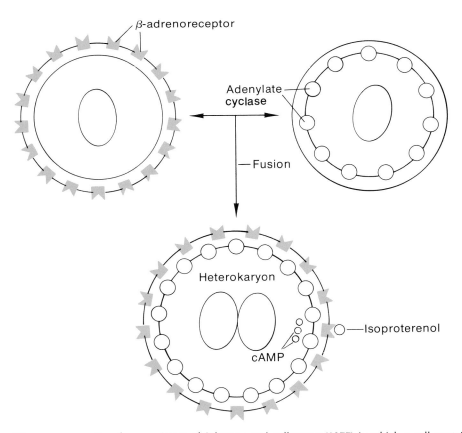

Fig. 4–11.   Diagram representing the experiment of Schramm and colleagues (1977) in which a cell containing the β-adrenergic receptor (with inactivation of adenylate cyclase) was fused with another carrying only adenylate cyclase. The heterokaryon contains both components and can now couple under the action of the specific ligand isoproterenol, resulting in the production of cAMP.

a model in which a lipid bilayer had protein adhering to both lipid-aqueous interfaces.

The observation of thin sections by electron microscopy has led to the concept of a trilayered plasma membrane, also called the "unit membrane." This is interpreted as indicating that the electron-dense outer layers correspond to the protein, and the less dense middle layer corresponds to the hydrocarbon chains of the lipids. The unit membrane concept, however, is certainly an oversimplification; numerous fine details suggest that the molecular organization of the membrane is much more complex. Recent observations indicate that the unit membrane image is mainly artifactual and does not account for the many proteins traversing the membrane.

The most favored model for the plasma membrane is the so-called fluid mosaic structure. According to this model, there is a rather continuous lipid bilayer into which the integral proteins of the membrane are intercalated; both these components are capable of translational diffusion within the overall bilayer. The mosaic model of the membrane is supported by the results of freeze-etching techniques in which protein particles are shown at the plane of cleavage of the bilayer.

There are between 500,000 and 600,000 particles per red cell, with the majority of these attached to the inner half of the membrane. Such particles represent the protein of band 3, which may form anionic permeation channels (see later discussion). The mosaic arrangement implies that: (1) the macromolecules have a characteristic *asymmetry*; (2) they are oriented for carrying information across the bilayer; and (3) they have considerable freedom of movement within the bilayer (*fluidity*). The fluidity of the lipids depends on the temperature and degree of saturation. Most lipids of the membrane are fluid at body temperature. With freeze-fracturing it is possible to demonstrate the movement of protein particles in different experimental conditions.

The fluidity of the lipids is supported by many indirect studies based on x-ray diffraction, differential thermal analysis, and electron spin techniques. The fluidity of the integral proteins is supported by experiments on cell fusion and on those of clustering and "capping" of surface antigens.

Membrane fluidity is used in the *mobile hypothesis* to explain the coupling between several membrane receptors and a single adenylate cyclase. This hypothesis is supported by experiments using cell fusion.

4–3   CELL PERMEABILITY

Permeability is fundamental to the functioning of the living cell and to the maintenance of satisfactory intracellular physiological conditions. This function determines which substances can enter the cell, many of which may be necessary to maintain its vital processes and the synthesis of living substances. It also regulates the outflow of excretory material and water from the cell.

The presence of a membrane establishes a net difference between the *intracellular* fluid and the *extracellular* fluid in which the cell is bathed. This may be fresh or salt water in unicellular organisms grown in ponds or the sea, but in multicellular organisms the internal fluid, i.e., the blood, the lymph, and especially the *interstitial* fluid, is in contact with the outer surface of the cell membrane.

One of the functions of the cell membrane is to maintain a balance between the osmotic pressure of the intracellular fluid and that of the interstitial fluid.

Passive Permeability—Concentration Gradient and Partition Coefficient

Permeability may be *passive* if it obeys only physical laws, as in the case of diffusion. It is common knowledge that if a concentrated solution of a soluble substance (e.g., sugar) is placed in water, there will be a net movement of the solute along the *concentration gradient* (i.e., from the region of high to that of low concentration). However, if a lipoprotein membrane (such as the plasma membrane) is interposed, the diffusion process is greatly modified and the membrane acts as a barrier to the passage of water-soluble molecules.

At the end of the last century, Overton demonstrated that substances that dissolve in lipids pass more easily into the cell, and Collander and Bärlund, in their classic experiments with the cells of the plant *Chara*, demonstrated that the rate at which substances penetrate depends on their solubility in lipids and the size of the molecule. The more soluble they are, the more rapidly they penetrate, and with equal solubility in lipids the smaller molecules penetrate at a faster rate (Fig. 4–12).

The permeability (P) of molecules across the membrane is:

$$P = \frac{KD}{t} \qquad (1)$$

with K, the partition coefficient; D, the diffusion

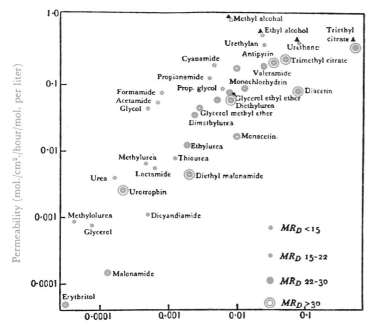

Fig. 4–12. Rate of penetration (permeability) in cells of *Chara ceratophylla* in relation to molecular volume (measured by molecular refraction, MR_D), and to the partition coefficient of the different molecules between oil and water. (After Collander, R., and Bärlund, H.: Acta Bot. Fenn., *11*:1, 1933.)

coefficient, which depends on the molecular weight; and t, the thickness of the membrane. The partition coefficient in most cell membranes is similar to that of olive oil and water.

The *partition coefficient* can be measured by mixing the solute with an oil-water mixture and then waiting until the phases are separated. The coefficient is the concentration of the solute in the oil, divided by the concentration in the aqueous phase. The *diffusion coefficient* can be easily determined by using radioactive solutes and measuring their rate of entry into the cytoplasm at various external concentrations. From the experiments shown in Figure 4–12 it is evident that many molecules may enter the cell by simple diffusion if the external concentration is higher than the internal. On the other hand, waste products accumulated in the cytoplasm may be eliminated by diffusion.

Equation (1) shows that molecules of equal size penetrate faster when their solubility in lipids is higher; in the case of molecules with equal lipid solubility, the smaller ones penetrate faster.

Passive Ionic Diffusion—Dependent on the Concentration and Electrical Gradients

In all cells there is a difference in ionic concentration with the extracellular medium and an

electrical potential exists across the membrane. These two properties are intimately related, since the electrical potential depends on an unequal distribution of the ions on both sides of the membrane.

As shown in Table 4–4, the interstitial fluid has a high concentration of Na^+ and Cl^-; and the intracellular fluid, a high concentration of K^+ and of large organic anions (A^-).

Using fine microelectrodes with a tip of 1 μm or less, investigators are able to penetrate through the membrane into a cell and also into the cell nucleus and to detect an *electrical potential* (also called the *resting*, or *steady*, potential), which is always negative inside. The values

TABLE 4–4. IONIC CONCENTRATION* AND STEADY POTENTIAL IN MUSCLE

		Interstitial Fluid	Intracellular Fluid
Cations	Na^+	145	12
	K^+	4	155
Anions	Cl^-	120	3.8
	HCO_3^-	27	8
	A^- and others	7	155
Potential		0	− 90 mv

Modified from Woodbury, J.W.: The cell membrane: Ionic and Potential gradients and active transport. *In* Neurophysiology. Edited by T.C. Ruch, H.D. Patton, J.W. Woodbury, and A.L. Towe. Philadelphia, W.B. Saunders Co., 1961.

of the membrane potential vary in different tissues between -20 and -100 mV. A more detailed consideration of the resting potential will be given in Chapter 24.

If we now consider the passive diffusion of ions across a membrane, we shall see that this process is more complex. Since ions are charged, their diffusion depends not only on the concentration gradient, but also on the electrical gradient.

In 1911, Donnan predicted that if a theoretical cell, having a nondiffusible negative charge inside, is put in a solution of KCl, K^+ will be driven into the cell by both the concentration and the electrical gradients; Cl^-, on the other hand, will be driven inside by the concentration gradient, but will be repelled by the electrical gradient. As shown by Donnan, the equilibrium concentrations will be exactly reciprocal:

$$\frac{[K^+_{in}]}{[K^+_{out}]} = \frac{[Cl^-_{out}]}{[Cl^-_{in}]} \qquad (2)$$

The relationship between the concentration gradient and the resting membrane potential is given by the Nernst equation:

$$E = RT \ln \frac{C_1}{C_2} \qquad (3)$$

where E is given in millivolts, R is the universal gas constant, and T is the absolute temperature.

From (2) and (3) the Donnan equilibrium for KCl can now be expressed as follows:

$$E = RT \ln \frac{[K^+_{in}]}{[K^-_{out}]} = RT \ln \frac{[Cl^-_{out}]}{[Cl^-_{in}]} \qquad (4)$$

According to (4) any increase in the membrane potential will cause an increase in the ion asymmetry across the membrane, and vice versa. While the first measurements of membrane potentials and ion concentration seemed to confirm this type of *passive* or *diffusion equilibrium*, more precise determinations in different cell types demonstrated that this was not the case. As mentioned in the next section, this discrepancy may be explained by the involvement of the active transport of ions.

Active Transport—The Sodium Pump

In addition to the diffusion or passive movement of neutral molecules and ions across membranes, cell permeability includes a series of mechanisms that require energy. These mechanisms are collectively described as *active transport processes*. Adenosine triphosphate, which is produced mainly by oxidative phosphorylation in mitochondria, is generally used as the energy source. For this reason, active transport is generally coupled to cell respiration.

When an ion is transported against an electrochemical gradient, extra consumption of oxygen is required. It has been calculated that 10% of the resting metabolism of a frog muscle is used for the transport of sodium ions. This consumption may increase to 50% in some experimental conditions in which the muscle is stimulated.

The resting membrane potential itself is maintained by active transport. This may be demonstrated in plant and animal cells that have been metabolically blocked by oxygen deprivation or specific poisons. In such cases leakage of K^+ occurs and the potential across the membrane may decrease to zero. Clearly, the active transport of ions is fundamental to the maintenance of cellular osmotic equilibrium. By regulating the specific concentrations of anions, cations, and other special ions needed for its metabolism, the cell keeps its osmotic pressure constant.

Potassium ions, which are concentrated inside the cell, must enter against a concentration gradient. This can be achieved by a pumping mechanism requiring energy. Sodium ions, which are continually exiting from the cell along with water, must also be transported by an active process, which has been named the *sodium pump*.

The diagram in Figure 4–13 summarizes the relationship existing between the transfer of K^+ and Na^+ by passive and active mechanisms and the resulting steady state potential. The passive (downhill) ionic fluxes are distinguished from the active (uphill) ionic fluxes. Notice that the active pumping out of Na^+ is the main mechanism for maintaining a negative potential of -50 mV inside the membrane. The diagram illustrates that the distribution of ions across the membrane depends on the summation of two distinct processes: (1) simple electrochemical diffusion forces

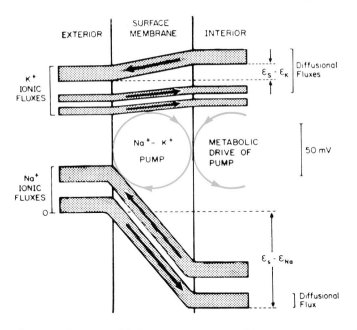

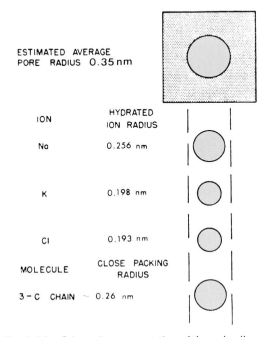

Fig. 4–13. Active and passive Na+ and K+ fluxes through the membrane in the steady state. The ordinate is the electrochemical potential of the ion ($\epsilon_s - \epsilon_k$ for K+, $\epsilon_s - \epsilon_{Na}$ for Na+). The abscissa is the distance in the vicinity of the membrane. The width of the band indicates the size of that particular one-way flux. Passive efflux of Na+ is negligible and is not shown. (After Eccles, J.C.: Physiology of Nerve Cells. Baltimore, Johns Hopkins Press, 1957.)

that tend to establish a Donnan equilibrium (i.e., passive transport), and (2) energy-dependent ion transport processes (i.e., active transport).

Ionic Transport through Charged Pores in the Membrane

The molecular machinery involved in ionic transport is located within the cell membrane. This has been demonstrated in two key materials. For example, if red blood cells are hemolyzed so that only the cell membrane remains, they can be filled again with appropriate solutions containing ions and ATP, and Na+ is transported and K+ is taken up as in a normal cell (i.e., a resealed ghost).

The giant axon of the squid, which has a diameter of about 0.5 mm, can be emptied of the axoplasm and then refilled with solutions of different electrolytes. The transport of ions against a concentration gradient, steady potentials, and even action potentials with the conduction of impulses can be obtained in this preparation in which most of the axoplasm is lacking and the excitable membrane left alone (see Section 24–1).

The use of radioisotopes demonstrated that ions can enter into the cell rapidly without obvious osmotic effects. It was then suggested that an ionic interchange across the membrane could take place through electrically charged pores.

Knowledge about the diameter of the different ions in the hydrated state is particularly pertinent. In this respect it is interesting to remember that the sodium ion, although smaller than K+ and Cl− in weight, is large in the hydrated condition and enters with more difficulty into the cell (see Fig. 4–14).

ESTIMATED AVERAGE
PORE RADIUS 0.35 nm

ION	HYDRATED ION RADIUS
Na	0.256 nm
K	0.198 nm
Cl	0.193 nm

MOLECULE	CLOSE PACKING RADIUS
3-C CHAIN	~ 0.26 nm

Fig. 4–14. Schematic representation of the red cell pore. Notice that the hydrated ion radius is larger for Na+ than for K+. (From Solomon, A.K.: J. Gen. Physiol., *43*:5, part 2, suppl. 1, p. 1, 1960.)

The total area of the pores in the red blood cell has been estimated to be on the order of 0.06% of the surface area. This means that a 0.7 nm pore would be surrounded by a nonporous square 20 × 20 nm. These findings indicate that the cell uses only a minute fraction of its surface area for ionic interchange. (Ionic channels of K^+, Na^+, and Ca^{++} will be covered in Chapter 24; see also Stefani and Chiarandini, 1982 in the Additional Readings at the end of this Chapter.)

Anion Transport in Erythrocytes Involves the Special Band-3 Polypeptide

The red cell membrane is endowed with an extremely active anion permeation mechanism to transport CO_2 from the tissues to the lung. This system can also exchange chloride and bicarbonate ions across the membrane. Using several chemical probes that inhibit anion permeability, knowledge has been gained about the molecular localization of this mechanism. Some of the probes can be transported by the anion system and, under certain conditions, can be fixed covalently to it. In this way it is possible to determine not only the number of permeation sites but also the polypeptide in which they are localized.

This approach has permitted the identification of the *polypeptide of band 3* with the site involved in the transport. This protein spans the membrane and may exist as a dimer (or tetramer). We saw earlier that it also corresponds to the 8 nm particles observed in freeze-fractured membranes (Fig. 4–5).

The model for anion transport that has been proposed is that of a continuous proteinaceous aqueous channel across the membrane having, near the outer surface, one anion binding site with three positive charges, and a hydrophobic barrier to limit the free diffusion of anions (Fig. 4–15). Passage through the barrier could occur only as a consequence of the binding of the anion.

In this model it is assumed that the segment containing the binding site could exist in two conformations, one facing outward, the other inward. In this way the segment could act as a gate in the proteinaceous channel, swinging between the two positions; this would permit the binding site to interact with anions coming from the outside or from the interior of the cell.

It has been proposed that, in the membranous portion of band 3, the hydrophilic residues are largely in the inner side forming an aqueous core through which the transport of anions occurs (see Rothstein and Ramjeesingh, 1982 in the Additional Readings at the end of this chapter).

Vectorial Function of Na^+K^+ATPase and Sodium Transport

We mentioned that Na^+, together with water, is eliminated from the cell by an active transport mechanism, generally called the *sodium pump*. This mechanism, discovered by Hodkin and Keynes in 1955, was soon associated by Skou with the enzyme *Na^+K^+ATPase*. This enzyme is able to couple the hydrolysis of ATP with the removal of Na^+ from the cytoplasm against an unfavorable electrochemical gradient.

$Na^+ K^+$ ATPase is an integral membrane protein that needs to be released and isolated. The best purifications have been obtained from tissues rich in this enzyme, such as electric eel organ, shark rectal gland, salt gland of marine birds and kidney of mammals. These purified preparations treated with negative staining show fragments that contain a granular structure on both sides of the membrane. Freeze-fracture particles of about 10 nm have smaller subunits in a quadripartite structure.[15] The molecular weight for the oligomer would be greater than 500,000 daltons.

Analysis by SDS gel electrophoresis has revealed the existence of two major polypeptides (α and β), ranging between 120,000 and 42,000 daltons, and a small proteolipid of about 12,000 that can be labeled with ouabain.[16,17] The α-chain contains the main ouabain-binding site and also the site for phosphorylation (see Siegel et al., 1981, and Zampighi et al., 1984 in the Additional Readings at the end of this chapter).

The $Na^+ K^+$ ATPase has been isolated in pure form from the kidney and reconstituted in phospholipid vesicles.[18] Localization of the $Na^+ K^+$ ATPase in tissues can be done by subcellular fractionation, autoradiography with ${}^3H$-ouabain, and the use of labeled antibodies.[19] With these cytochemical methods, it may be shown that the enzyme is localized almost exclusively at the basolateral side of Na^+ transporting epithelia, such as the thick limb of the kidney distal tubules. This localization suggests that the Na^+, which

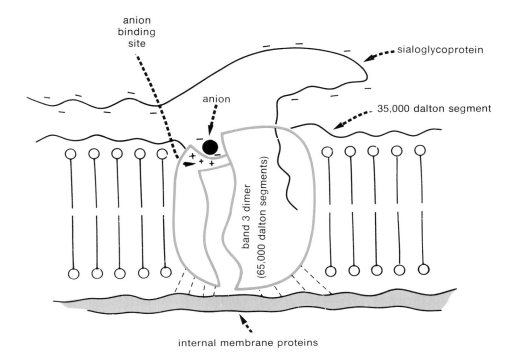

Fig. 4–15. Schematic model of the disposition of the band 3 protein and of the anion transport site in the erythrocyte membrane. (From Rothstein, A., Cabantchik, Z.I., and Knauf, P.: Fed. Proc., *35*:3, 1976.)

enters through the apex of the cell driven by the concentration gradient, is actively transported to the extracellular space at the basolateral surface of the cell.

In the case of the red blood cell it has been estimated that there are about 5000 ATPases, each of which may extrude 20 Na$^+$ ions per second. Figure 4–16 shows an idealized diagram of Na$^+$ K$^+$ ATPase within the red cell membrane. Note that the hydrolysis of one ATP provides the energy for the linked transport of two K$^+$ ions toward the inside and three Na$^+$ ions toward the outside of the cell. This diagram shows the vectorial characteristics of the enzyme, which are sensitive to ATP on the inside of the membrane but not on the outside. This ATPase is stimulated by a mixture of both Na$^+$ and K$^+$ and is inhibited by the cardiotonic glycoside *ouabain*. This specific inhibitor binds the enzyme only from the *extracellular surface*, and can be used as a marker to localize and measure the number of enzyme molecules in a membrane.[20]

Na$^+$ K$^+$ ATPase can also be inhibited by *vanadate* (analogous in structure to phosphate), but this inhibitor acts on the cytoplasmic side of the

enzyme (see Grantham and Glynn, 1979 in the Additional Readings at the end of this chapter).

The transport reaction takes place in two major steps. Step I consists of the formation of a *covalent phosphoenzyme intermediate* (E $\sim$ P) on the inner side of the membrane and in the presence of Na$^+$. This first reaction is inhibited

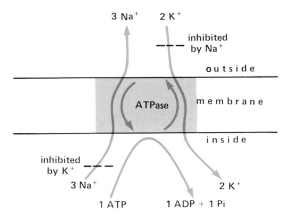

Fig. 4–16. Diagram of the Na$^+$-K$^+$ ATPase in the cell membrane. Observe that for each molecule of ATP hydrolyzed at the inner part of the membrane 3 Na$^+$ ions are transported outside and 2 K$^+$ ions are transported inside. (See text for details.)

by Ca^{2+}, but not by ouabain. In Step II the $[Na^+$ $E \sim P]$ complex is hydrolyzed to form free enzyme and Pi. This reaction requires K^+, added to the outside of the membrane, and is inhibited by ouabain, which competes with K^+. Steps I and II can be represented as follows:

Step I $Na^+_{in} + ATP + E \longrightarrow$ (5)
    $[Na^+ \cdot E \sim P] + ADP$
Step II $[Na^+ \cdot E \sim P] + K^+_{out} \longrightarrow$ (6)
    $Na^+_{out} + K^+_{in} + Pi + E$

It may be observed that Na^+ and K^+ are ultimately translocated in the opposite directions from which they were bound.

In the $Na^+K^+ATPase$ system, the transport of Na^+ outward is compensated by that of K^+ inward and electrical neutrality is achieved. However, there are other cases in which the pump becomes electrogenic, since the exit of Na^+ is not compensated by the entrance of K^+. The *electrogenic pump* generates a potential that may provide the driving force to transport other solutes. Several substances (glucose, amino acids) may enter the cell by means of the Na^+ pump.

Transport Proteins—Carrier and Fixed Pore Mechanisms

The transport of many different molecules across the membrane shows a high degree of specificity. That is, the permeability of a molecule is related to its chemical structure. While one molecule may readily enter the cell, another of the same size but a slightly different molecular structure may be completely excluded.

A clear example of this selectivity is provided by the transport of two isomers (glucose and galactose) into a bacterial cell. Although the only difference between them is in the position of an —OH group at carbon 4, these two sugars penetrate the membrane by different transport mechanisms. This type of selectivity is attributed to *transport proteins*, also called *carriers* or *permeases*. It is thought that a permease functions in a manner somewhat similar to an enzyme or a receptor in having a binding site able to recognize the molecule to be transported. Permeases accelerate the transport process, provide for special selectivity, and are recycled, meaning that they remain unchanged after having assisted in the entry or exit of a molecule.

Some permeases can transport only if there is a favorable concentration gradient, while others can do so against an unfavorable one. The first type of transport, being driven by a passive mechanism, is sometimes called *facilitated diffusion*. When the permease operates against the gradient, it is another case of active transport.

The details of the molecular mechanisms by which substances are selectively transported across the plasma membrane are largely unknown. Two general alternative hypotheses have been postulated: a carrier mechanism and a fixed-pore mechanism. A *carrier mechanism* implies that the molecule binds to the transport protein (carrier) at the outer surface of the cell, and that this complex rotates and translocates the molecule into the cytoplasm (Fig. 4–17). Rotation and translocation of the carrier are relatively unlikely, however, in view of what is known about the organization of the cell membrane. Rotation is thermodynamically difficult, and the translocation of a macromolecule from one half of the bilayer to the other is even less likely to occur. The *fixed-pore mechanism* seems more probable because it requires less expenditure of energy. In this mechanism the carrier is represented by integral proteins that traverse the membrane and which, once bound to the molecule to be transported, undergo conformational changes. A fixed pore or channel is generally thought to consist of several protein subunits (oligomers) having a hydrophilic lining in the middle (Fig. 4–17).

The vectorial function of $Na^+K^+ATPase$ can also be explained along the lines described for a carrier or fixed-pore mechanism. As diagrammed in Figure 4–16 the enzyme has binding sites for Na^+ and K^+ and a carrier mechanism coupled to the hydrolysis of ATP that generates the energy needed to move the ions against the concentration gradient.

A carrier mechanism has also been postulated to mediate the entrance of glucose into intestinal cells. This mechanism is activated by Na^+. The carrier in this model should have a binding site for Na^+ and another for glucose, and should exist in two conformational states, one with low affinity for glucose and one with high affinity. A carrier model can be postulated in which the glucose site is initially in a high affinity state and binds the glucose molecule at the surface of the cell. In a second step the carrier is translocated,

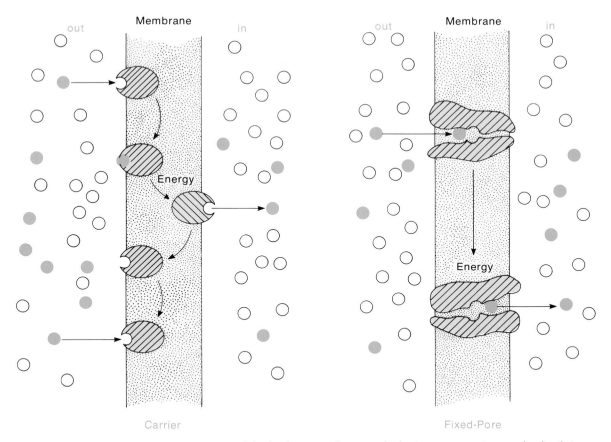

Fig. 4–17. Diagram representing the carrier and the fixed-pore mechanisms of selective transport. (See text for details.)

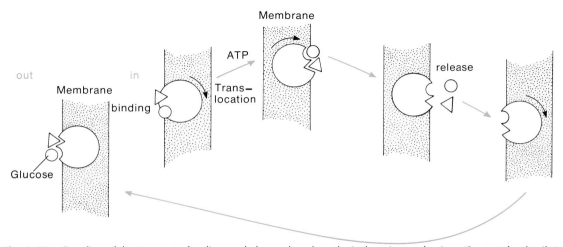

Fig. 4–18. Coupling of the transport of sodium and glucose by a hypothetical carrier mechanism. (See text for details.)

and the glucose site changes to a low affinity, releasing the molecule into the cell. The two affinities are modulated by the Na^+ concentration inside and outside the cell (Fig. 4–18).

SUMMARY:
Cell Permeability

The study of permeability requires knowledge of the chemical and molecular organization of the cell membrane. This membrane regulates the inflow of substances to the cell and the outflow of water ions and other materials. The cell membrane maintains a balance between the osmotic pressure of the intracellular and the interstitial fluid. Solutions may be isotonic, hypotonic, or hypertonic with respect to the intracellular fluid.

The *ionic concentration* in the intracellular fluid differs from that in the interstitial fluid. In the latter, Na^+ and Cl^- are high, and K^+ is low. Inside the cell, K^+ is high, Na^+ and Cl^- are low, and there is a large pool of organic anions (A^-) that do not cross the cell membrane. A *membrane potential* of -20 to -100 mV is detected in all cells. Cell membranes are polarized, being negatively charged inside and positively charged outside.

Movement of substances across the cell membrane may be by *passive diffusion* or by several mechanisms of *active transport*. *Diffusion* occurs when there is a concentration gradient and takes place from a high to a low concentration. *Diffusion of molecules* across membranes depends on molecular volume and the lipid solubility of the substance. *Diffusion of ions* depends on both the concentration and the electrical gradients across the membrane.

The distribution of K^+ and Cl^- generally follows a Donnan equilibrium that depends on the presence of nondiffusible anions inside the cell. The relationship between concentration gradient $\left(\dfrac{C_1}{C_2}\right)$ and membrane potential (E) is given by the Nernst equation:

$$E = RT \ln \frac{C_1}{C_2}$$

The *active transport* of neutral molecules and ions requires energy (ATP) and is generally coupled to the energy-yielding mechanisms of the cell. Active transport is blocked by cooling or by certain metabolic poisons of respiration and glycolysis. With complete energy block the membrane potential may decrease to zero and there is leakage of K^+. Active transport may be demonstrated by simple experiments using kidney tubules, frog skin, and toad bladder. Ionic transport is particularly intense in the tissues of the electric eel and in salt-secreting glands. By active transport Na^+ and water are pumped out of the cell and K^+ penetrates the cell. Both electrochemical diffusion (i.e., passive transport) and energy-dependent ion transport determine the distribution of ions and the membrane potential across the membrane. The passage of ions is thought to be across charged pores in the membrane. The size of the hydrated ion is important in the transport through the pores.

In a red blood cell 0.06% of the area may be occupied by pores. This membrane has an active anion permeation mechanism to transport chloride and bicarbonate. This mechanism has been localized to the polypeptide of band 3 that spans the membrane and to the 8 nm particles seen in *freeze-fractured membranes*. A channel-model for the transport of anions has been postulated.

Na^+ K^+ ATP*ase* is related to the active transport of Na^+ and K^+ in red blood cells, nerves, brain membranes, electric tissues, and others. The hydrolysis of 1 ATP $\rightarrow$ ADP $+$ Pi in the inner surface of the membrane is coupled with the transport of 3 Na^+ from the inside toward the outside of the membrane and with the transport of 2 K^+ in the opposite direction.

All these properties indicate that the enzyme has a vectorial orientation across the membrane. The Na^+ K^+ ATPase is specifically inactivated with *ouabain* from outside and vanadate from inside. The function of this enzyme is explained by a complex carrier mechanism that comprises: (1) binding sites for Na^+ and K^+, (2) a mechanism for translocation, and (3) a mechanism for the release of the ligand.

The production of a covalent phosphoenzyme intermediate has led to the concept that the transport is carried out in two steps as follows:

Step 1 $Na^+_{in} + ATP + E \longrightarrow$
$[Na^+ \cdot E \sim P] + ADP$

Step II $Na^+ \cdot E \sim P + K^+_{out} \longrightarrow$
$Na^+_{out} + K^+_{in} + Pi + E$

in which $[Na^+ \cdot E \sim P]$ is the phosphorylated complex of the enzyme. The enzyme has a large molecular weight ($>500,000$ daltons) and contains subunits α and β and a small proteolipid. Cytochemically it has been localized at the basolateral side of Na^+ transporting epithelia.

Transport of molecules across the membrane is highly specific. This specificity is attributed to transport proteins, also called carriers or permeases. These substances are thought to be similar to enzymes or receptors in having a binding site able to recognize the molecule to be transported. Permeases may be driven either by a passive mechanism (facilitated diffusion) or an active one (active transport). Carrier and fixed-pore mechanisms have been postulated to describe the means by which materials are selectively transported across the plasma membrane. The fixed-pore mechanism seems more probable because it requires less energy; it involves conformational changes in the molecules constituting the fixed pore. An example of the fixed-pore mechanism is the penetration of anions into red blood cells.

REFERENCES

1. Singer, S.J., and Nicolson, G.L.: Science, *175*:720, 1972.

2. Cullis, P.R., and De Kruijff, B.: Biochim. Biophys. Acta, *469*:99, 1979.
3. Rothman, J.E., and Lenard, J.: Science, *195*:743, 1977.
4. Henderson, R.: Annu. Rev. Biophys. Bioeng., *6*:87, 1977.
5. Nicolson, G.L.: Biochim. Biophys. Acta, *457*:57, 1976.
6. Luzzatti, V.: Biological Membranes. Edited by D. Chapman. New York, Academic Press, Inc., 1968, p. 71.
7. Banghman, A.D., Standish, M.M., and Watkins, J.C.: J. Mol. Biol., *13*:238, 1965.
8. Papahadjopoulos, D.: Ann. NY Acad. Sci., *308*:50, 1978.
9. Leserman, L.D., Machy, P., and Barbet, J.: Nature, *293*:226, 1981.
10. Robertson, J.D.: Biochem. Soc. Symp., *16*:3, 1959.
11. Goldman, S.S.: Am. J. Physiol., *228*:834, 1975.
12. Hirato, et al.: Receptors for Neurotransmitters and Peptide Hormones. Edited by G. Pepeu and H. Ladinsky. New York, Raven, 1980, p. 91.
13. Frye, L.D., and Edidin, M.: J. Cell Sci., *7*:319, 1970.
14. Cuatrecasas, P., and Hollenberg, M.D.: Adv. Protein Chem., *30*:251, 1976.
15. Haase, W., and Koepsell, H.: Pflugers, Arch., *381*:127, 1979.
16. Rivas, E., Lew, B., and De Robertis, E.: Biochim. Biophys. Acta, *290*:419, 1972.
17. Collins, J., et al.: Biochim. Biophys. Acta, *686*:7, 1982.
18. Forbush, B., Kaplan, J.H., and Hoffman, J.F.: Biochemistry, *17*:3667, 1978.
19. Di Bona, D.R., and Mills, J.W.: Fed. Proc., *38*:134, 1979.
20. Schaefer-Ridder, M., et al.: Science, *215*:166, 1982.

ADDITIONAL READINGS

Albers, R.W.: Biochemistry of cell membranes. *In* Basic Neurochemistry. Edited by G.J. Siegel and R.W. Albers. Boston, Little Brown & Co., 1981.

Bennett, V.: The molecular basis for membrane-cytoskeleton association in human erythrocytes. J. Cell. Biochem., *18*:49, 1982.

Branton, D., Chen, C.M., and Tyler, J.: Interaction of cytoskeletal proteins on the human erythrocyte membrane. Cell, *24*:24, 1981.

Carafoli, E.: Membrane Transport of Calcium. London, Academic Press, 1982.

Cohen, C.M., and Branton, D.: The normal and abnormal red cell cytoskeleton. Trends Biochem. Sci., *6*:266, 1981.

De Robertis, E.: Synaptic Receptors: Isolation and Molecular Biology. New York, Marcel Dekker, 1975.

Edidin, M.: Rotational and translational diffusion in membranes. Annu. Rev. Biophys. Bioeng., *3*:179, 1974.

Ellory, J.C., and Lew, V.L.: Membrane Transport in Red Cells. New York, Academic Press, 1977.

Grantham, J.J., and Glynn, I.M.: Renal Na$^+$ K$^+$ ATPase: Determinants of inhibition by vanadium. Am. J. Physiol., *236*:530, 1979.

Guidotti, G.: The structure of membrane transport systems. Trends Biochem. Sci., *1*:11, 1976.

Harrison, R., and Lunt, G.G.: Biological Membranes. 2nd Ed. Glasgow, Blackie & Son, 1980.

Jacobs, S., and Cuatrecasas, P.: The motile receptor hypothesis for cell membrane receptor action. Trends Biochem. Sci., *2*:280, 1977.

Keynes, R.D.: Ion channels in the nerve cell membrane. Sci. Am., *240*:126, 1979.

Knight, C.G.: Liposomes: From physical structure to therapeutic applications. Amsterdam, Holland Biomedical Press, 1981.

Lodish, H.F., and Rothman, J.E.: The assembly of cell membranes. Sci. Am., *240*:48, 1979.

Luftig, R.B., Wehrli, E., and McMillan, P.N.: The unit membrane image: A re-evaluation. Life Sci., *21*:285, 1977.

Nermut, M.V.: The cell monolayer technique in membrane research. Eur. J. Cell Biol., *28*:160, 1982.

Nicolau, C., and Poste, G.: Liposomes *in vivo*. Biol. Cell, *47*:1, 1983.

Poste, G., and Nicolson, G.L.: Membrane fusion. Vol. 5. Amsterdam, North-Holland, 1978.

Ralston, G.B.: The structure of spectrin and the shape of the red blood cell. Trends Biochem. Sci., *3*:195, 1978.

Rothstein, A., and Ramjeesingh, L.M.: The red cell band 3: Its role in anion permeability. Philos. Trans. R. Soc. Lond., *299*:497, 1982.

Schramm, M., Orly, J., Eimerl, S., and Korner, M.: Coupling of hormone receptors to adenylate cyclase of different cells by cell fusion. Nature, *268*:310, 1977.

Siegel, G.J., et al.: Ion transport. *In* Basic Neurochemistry. Edited by G.J. Siegel and R.W. Albers. Boston, Little Brown & Co., 1981.

Singer, S.J.: Thermodynamics, the structure of integral membrane protein and transport. J. Supramol. Struct., *6*:313, 1977.

Singer, S.J., and Nicolson, G.L.: The fluid mosaic model of the structure of cell membranes. Science, *175*:720, 1972.

Steck, T.L., and Hainfeld, J.F.: Protein ensembles in the human red-cell membrane. *In* International Cell Biology. Edited by B.R. Brinkley and K.R. Porter. New York, Rockefeller University Press, 1977.

Stefani, E., and Chiarandini, D.J.: Ionic channels in skeletal muscle. Annu. Rev. Physiol., *44*:357, 1982.

Zampighi, G., Kyte, J., and Freytag, W.: Structural organization of (Na$^+$ + K$^+$)–ATPase in purified membranes. J. Cell Biol., *98*:7851, 1984.

CELLULAR INTERACTIONS

While in Chapter 4 we described the structure and function of the plasma membrane, in this chapter we will cover the microdomains that originate in the fluid mosaic model. We will see that many cells have special regions that are adapted to certain functions. These regions are generally referred to as *differentiations of the cell surface*. One such differentiation, *the gap junction*, plays an important role in the electrical coupling and metabolic cooperation between adjacent cells in a tissue.

Another objective of this chapter is to study the cell coat and the mechanisms by which cells are able to "recognize" their partners in a tissue. It should be mentioned that changes in *intercellular communication* and *molecular recognition* are of paramount importance in the cancerous transformation of cells, since the cell surface is becoming one of the most fundamental fields in cell and molecular biology of both normal and transformed cells. This study will be continued in Chapter 6 in which we consider

how the cytoskeleton is related to the cell membrane and to the extracellular material, which plays a fundamental role in cell motility and cell adhesion.

5–1 DIFFERENTIATIONS OF THE CELL MEMBRANE

The differentiations of the cell membrane correspond to regions specially adapted to different functions such as absorption, secretion, fluid transport, mechanical attachment, or interactions with neighboring cells.

Figure 5–1 is a diagram of an idealized columnar cell in which various types of differentiations of the cell membrane may be observed. The apical surface is projected into slender processes called *microvilli*. At the base of the cell the plasma membrane is covered by a thick basement membrane or *basal lamina* of extracellular material in which infoldings of the plasma membrane may be observed.

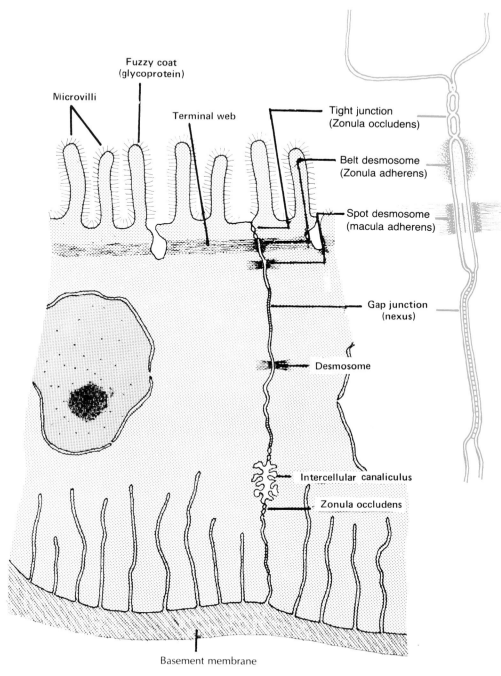

Fig. 5–1.   Diagram of an idealized columnar epithelial cell showing the main differentiations of the cell membrane. *To the right,* at a higher magnification, the series of differentiations found between two epithelial cells are indicated. (See text for details.)

In this figure the main intercellular junctions are also represented. There are four main types of intercellular junctions three of which, tight junctions, belt desmosomes, and spot desmosomes are involved in mechanical and sealing functions. The fourth represented by the gap junction is important in intercellular communications.

Microvilli—A Greatly Increased Cell Membrane Surface Area

In the intestinal epithelium microvilli are prominent and form a compact structure that appears under the light microscope as a *striated border*. These microvilli, which are 0.6 to 0.8 μm long and 0.1 μm in diameter, represent cytoplasmic processes covered by the plasma membrane. Within the cytoplasmic core fine microfilaments are observed which in the subjacent cytoplasm form a *terminal web*. As will be mentioned in Section 6–4, these filaments contain actin and are attached to the tips of the microvilli by α-actinin; their function is to produce contraction of the microvilli.

Microvilli from intestinal cells have been isolated and studied from the viewpoint of their biochemical organization. These studies have revealed that in the core, in addition to actin there are other proteins, such as *fimbrim* (68,000 daltons) and *villin* (95,000 daltons), that bind and cross-link the F-actin (see Section 6–4).[1]

In the *terminal web* there are microfilaments that in addition to actin contain myosin, α-actinin, tropomyosin, and spectrin. These web microfilaments undergo an ATP-dependent contraction that is perpendicular to the direction of the microvilli causing them to fan out. In vivo the function of the terminal web is that of preventing, rather than producing, movement by exerting a tension that is able to maintain the microvilli in an upright position (see Harris, 1983 in the Additional Readings at the end of this chapter).

The outer surface of the microvilli is covered by a coat of filamentous material (fuzzy coat) composed of glycoprotein macromolecules. Microvilli increase the effective surface of absorption. For example, a single cell may have as many as 3000 microvilli, and in a square millimeter of intestine there may be 200,000,000. The narrow spaces between the microvilli form a kind of sieve through which substances must pass during absorption. Numerous other cells have microvilli, although they are fewer in number. They have been found in mesothelial cells, in the epithelial cells of the gallbladder, uterus, and yolk sac, in hepatic cells, and so forth.

The *brush border* of the kidney tubule is similar to the striated border, although it is of larger dimensions. An amorphous substance between the microvilli gives a periodic acid-Schiff reaction for polysaccharides. Between the microvilli the cell membrane invaginates into the apical cytoplasm. These invaginations are apparently pathways by which large quantities of fluid enter by a process of pinocytosis. (Other specializations of the cell surface, such as cilia and flagella, are described in Section 6–3.)

Tight Junctions and the Sealing of Epithelia

Tight junctions (zonula occludens) are specially differentiated regions that seal the intercellular space, thus preventing the passage of fluid to and from the lumen. These junctions are situated just below the apical border of the cell. In a thin section, the two adjacent plasma membranes appear fused at a series of points (Fig. 5–1). If opaque tracers, such as ferritin, are put in the lumen of the epithelial cavity, they cannot penetrate the intercellular space because they are stopped at the level of the tight junctions. By this mechanism, a group of cells can maintain an intercellular environment different from that present at the apex.

The use of the freeze-fracture technique (Section 3–2) permits the study of the three-dimensional organization of tight junctions (Fig. 5–2). As shown in Figure 5–2 the tight junctions appear as a network of ridges on the cytoplasmic half of the membrane, with complementary grooves in the outer half (see Pinto da Silva and Kachar, 1982 in the Additional Readings at the end of this chapter). The ridges, in general, appear to be composed of two rows of particles, each one belonging to the adjacent cells. The lines of these particles produce the sealing, and for this reason have been named *sealing strands*.

More recent methods of rapid freezing have shown that the row of particles of the sealing strand results from a technical artifact. Under these conditions, the sealing strand appears as a pair of inverted cylinders. In the diagram of

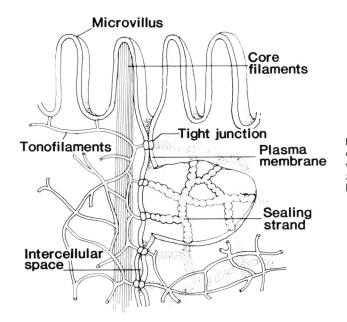

Microvillus

Core filaments

Tonofilaments

Tight junction

Plasma membrane

Sealing strand

Intercellular space

Fig. 5–2. Diagram of a tight junction. The adjacent cell membranes are held firmly at the sealing strand, which is composed of two rows of particles, as in a zipper. (See text for details.) (Redrawn from L.A. Staehelin and B. Hull, Sci. Amer., *238*:140, 1978.)

Figure 5–3, these cylinders are interpreted, at the molecular level, as lipid micelles situated between the line of fusion of the outer lipid leaflets of adjacent cells. It is possible that the previously used slow-freezing method resulted in the disruption of the cylindrical micelles into rows of beaded particles (Fig. 5–2).[2]

The tightness of an epithelium, which can be measured by the electrical resistance across it, is related to the abundance of the sealing

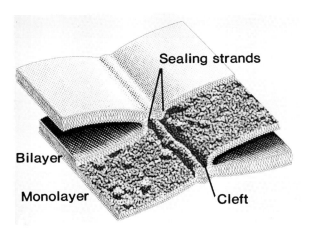

Sealing strands

Bilayer

Monolayer

Cleft

Fig. 5–3. Three-dimensional diagram representing sealing strands as observed after rapid freeze-fracture. The two sealing strands between adjacent rat Sertoli cells of the testicle are represented as cylindrical micelles made of lipid molecules. Observe that near one cylinder there is a cleft corresponding to the sealing strand of the adjacent cell. (Courtesy of B. Kachar and T.S. Reese.)

strands. These are highly flexible and adaptable to the physiological needs of the tissue.

Tight junctions are abundant in epithelia such as that of the collector tube of the kidney, in which the transport of water and ions occurs primarily across the apical and basal membranes and has a high electrical resistance. In other epithelia having low resistance and fewer tight junctions (e.g., gallbladder), the transport is mainly paracellular (i.e., occurs between the cells).[3]

In a polarized epithelium, one of the functions of the tight or occluding junctions is the maintenance of different intercellular environments in the apex and in the basolateral regions. Another function is the determination of a true asymmetry in the membrane of the epithelial cell. In other words, the tight junction could act as a barrier to the diffusion of macromolecules or lipids in the bilayer, so as to differentiate the apical from the basolateral portions of the cell membrane, each having their own particular compositions and physiological characteristics. In experiments using membrane-bound lectins (see Section 5–3), it has been observed that, when the lectins are applied either to the apical or basolateral side of epithelial cells, they are incapable of passing through the tight junctions and thus remain segregated. On the other hand, fluorescent lipid probes can pass this barrier if they are able to "flip-flop," i.e., to pass from the outer to the inner monolayer.[4]

An excellent material in which the function of the tight junction can be studied is represented by certain epithelioid cells of kidney origin, the so-called MDCK cells. Cultures of this cell line tend to form monolayers with functional properties similar to epithelia able to transport fluids and ions. These properties arise from the structural and functional polarity of the MDCK cell membrane, which is determined by the presence of tight junctions between the cells.[5]

In the MDCK cell monolayer, it has also been demonstrated that actin microfilaments are directly associated with the occluding junctions. In fact, treatment with cytochalasin B, which disrupts these microfilaments (see Section 6–4), produces abolition of the resistance across the monolayer and prevents the resealing of the tight junctions.[6]

Belt and Spot Desmosomes—Mechanical Function

Mechanical adhesion between cells is mainly held by the desmosomes, of which there are two types, belt and spot desmosomes (Fig. 5–1). The difference between them resides in their localization, extension, and relationship with the microfilaments.

Belt desmosomes (also called *zonula adherens*, *terminal bars* or intermediary junctions, in the old literature) are generally found at the interface between columnar cells, just below the region of the tight junctions (Fig. 5–1). They form a band that girdles the inner surface of the cell membrane. This band contains a web of 6 nm actin microfilaments and another group of interwoven intermediate filaments of 10 nm (see Section 6–6). Actin microfilaments are contractile and intermediate filaments play a structural role.

Spot desmosomes (macula adherens) appear as darkly stained bodies on the cell surface under the light microscope. They represent localized circular areas of contact about 0.5 μm in diameter (Fig. 5–1), in which the plasma membranes of the two adjacent cells are separated by a distance of 30 to 50 nm.

The electron microscope (EM) reveals a midline structure or intercellular core and two parallel desmosomal plasma membranes. Under each plasma membrane there is a discoidal intracellular plaque, toward which numerous fila-

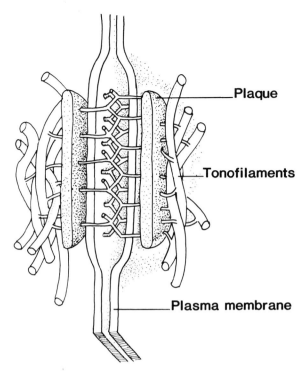

Fig. 5–4. Diagram of a spot desmosome. (See text for details.) (Redrawn from L.A. Staehelin and B. Hull, Sci. Amer., *238*:140, 1978.)

ments, called *tonofilaments* converge. These filaments describe a loop in a wide arc and course back into the cell (Fig. 5–4).

Tonofilaments are not contractile and belong to the 10 nm type of intermediate keratin filaments (Section 6–6) forming a structural framework for the cell cytoplasm. In addition, there are thinner filaments that arise from the dense plaques and traverse the cell membrane into the intercellular space forming the so-called "transmembrane linkers," which provide for mechanical coupling (Fig. 5–4).

The number of spot desmosomes is correlated with the degree of mechanical stress to which a tissue is subjected. For example, the epithelium of the vagina is rich in spot desmosomes.

Within the intercellular core a coating material, which sometimes forms a discontinuous middle dense line, may be observed. This extracellular material contains acid mucopolysaccharides and proteins. In fact, desmosomes are broken by trypsin, collagenase, and hyaluronidase and are sensitive to agents that chelate calcium. The dense intracellular attachment plaques are digested by proteolytic enzymes.

While the tonofilaments provide the intracellular mechanical support, cellular adhesion at the desmosome depends mainly on the extracellular coating material.

Desmosomes have been isolated from epidermal cells and from the desmosomes a number of glycosylated and nonglycosylated proteins have been separated. The major nonglycosylated proteins are called *desmoplakins I, II, and III;* the antibodies against desmoplakins have shown that these proteins are abundant in the dense region corresponding to the intracellular plaques. These antibodies beautifully stain the plaques in desmosomes.[7] The glycosylated proteins *(desmogleins I and II)* are probably transmembranal proteins with the carbohydrate moiety exposed toward the intercellular space. It has been observed that antibodies against these glycoproteins inhibit the adhesion of cultured epidermal cells (see Steinberg, 1984 in the Additional Readings at the end of this chapter).

Along the basal surface of some epithelial cells *hemidesmosomes* may be observed. These are similar to desmosomes in fine structure, but represent only half of them, the outer sides frequently being substituted with collagen fibrils.

SUMMARY:
Differentiations of the Cell Membrane

The cell membrane may present regional differentiations that are related to specialized functions such as absorption, fluid transport, and electrical coupling. Microvilli are found at the apical surface of the intestinal epithelium and form the brush border of the kidney tubules. They increase the absorption surface and are covered by a coat of glycoproteins.

Intercellular attachments comprise tight junctions, belt desmosomes, and spot desmosomes. Tight junctions serve to seal the intercellular spaces and to maintain the intercellular environment. They form a network of sealing strands below the apical regions of the cells.

Belt desmosomes have a system of actin and intermediate filaments and are situated below the zone of tight junctions. These belt desmosomes represent the terminal bars or intermediary junctions of the old literature. Spot desmosomes are localized circular areas of mechanical attachment having two dense plaques with keratin tonofilaments. The number of spot desmosomes is correlated with the degree of mechanical stress that the tissue has to support. *Hemidesmosomes* are found at the base of certain epithelia.

5–2 INTERCELLULAR COMMUNICATIONS AND GAP JUNCTIONS

The concept of the cell as the basic unit of all living matter should not lead us to disregard the fact that multicellular organisms are made of populations of cells that interact among themselves. Such *cellular interaction* is essential for the coordination of activities, and furthermore, the propagation between cells of signals for growth and differentiation is indispensable for development. It is now known that most cells in an organized tissue are interconnected by *junctional channels* and that they share a common pool of many small metabolites and ions that pass freely from one cell to another. Their individuality, however, is maintained by macromolecules that are not exchanged between cells.

Electrical Coupling between Cells Depends on Gap Junctions

One of the manifestations of cellular interaction is electrical coupling between cells. By introducing intracellular microelectrodes into adjacent cells of a tissue, investigators have been able to demonstrate that in many animal cells there are intercellular communications. In this case the cells are electrically coupled and have regions of low resistance in the membrane through which there is a rather free flow of electrical current carried by ions. The other parts of the cell membrane, which are not coupled, show a much higher resistance. This type of coupling, called *junctional communication*, is found extensively in embryonic cells. In adult tissue it is usually found in epithelia, cardiac cells, and liver cells. Skeletal muscle and most neurons do not show electrical coupling.

Electrical coupling seems to be related to the so-called *gap junctions* or *nexus* (also called *communicating junctions*) which are shown in the diagram of Figure 5–1. An excellent material for the study of gap junctions is the myocardial tissue, in which the action potential is transmitted from cell to cell by an electrical coupling.

The gap junction appears as a plaque-like contact in which the plasma membranes of adjacent cells are in close apposition, separated by a space of only 2 to 4 nm. This gap can be filled with electron opaque materials such as lanthanum and ruthenium red. In tangential sections gap junctions show a hexagonal array of 8- to 9-nm particles (Fig. 5–5A).

The electron-dense material is able to penetrate in between the particles, thus delineating their polygonal arrangement, and also into the central region of each particle. This central region is 1.5 to 2 nm in diameter and corresponds to the location of the channel.

With freeze-fracturing it is possible to split the junctions and to better define their internal structure. Upon the inner half-membrane of the fracture (PF face) the 8- to 9-nm particles are observed, while in the external half (EF face) a complementary array of pits or depressions appears.[8] Gap junctions are resistant to mechanical disruption, proteolysis, and removal of Ca^{++} and can be isolated from the cells.

At the gap junction the particles of one cell are in alignment with those of the other cell. After isolation they can be shown by electron microscopy and x-ray diffraction to have the same polygonal lattice structure, with 8- to 9-nm particles, as seen in the intact junction.

The Connexon—Opening and Closing of the Channel

The unit structure of the gap junction has been named the "connexon"; this structure is represented by a cylinder that spans the bilayer of each of the two connected membranes, as well as the gap in between, thus providing for intercellular communication (Fig. 5–5B). Figure 5–6 shows a surface view of a rat hepatocyte junction, isolated and negatively stained with uranyl acetate. The connexons appear as annuli arranged in a hexagonal lattice with periodicity of 8.5 nm. Each connexon has a ring-shaped structure made up of six identical protein subunits surrounding a hydrophilic channel. The inset of Figure 5–6 shows a densitometric and computer processing map carried out on electron micrographs. The features of this three-dimensional map suggest that the six protein subunits span the membrane and protrude from either side (see Unwin and Zamphigi, 1980 in the Addi-

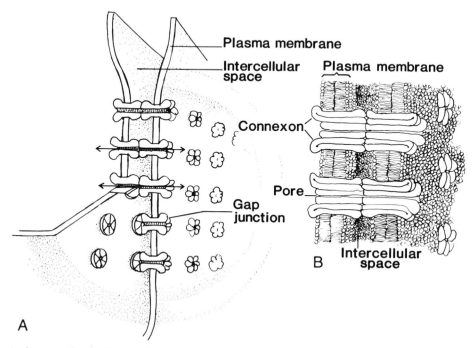

Fig. 5–5. A, diagram of a gap junction between two cells. The channels are made by particles in each membrane that traverse the intercellular space. The flow of fluid between the cells is indicated by arrows. (Redrawn from A. Staehelin and B. Hull, Sci. Amer., 238: 140, 1978). B, finer structure of the connexon or functional unit. The channel has a pore of about 2 nm and is formed by two hexamers (six subunits) traversing the lipid bilayer. (Redrawn from Makoski et al., 1977.) (Courtesy of W.R. Loewenstein.)

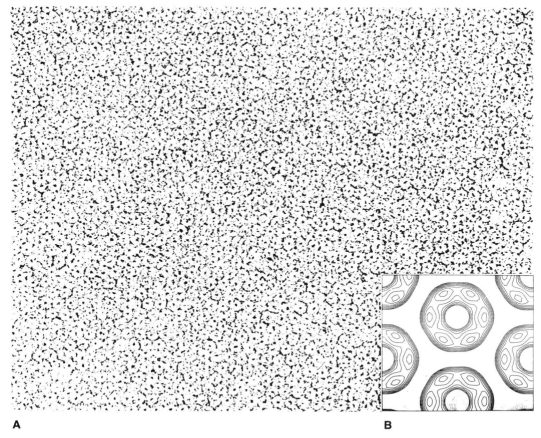

Fig. 5–6. *A,* surface view of a gap junction isolated from rat liver and negatively stained with uranyl acetate. Observe that each morphological unit has a ring-shaped appearance with a dot of stain in its center. Magnification $425,000\times$.

B, projection map obtained by computer processing of phases and amplitudes from an area of *A.* The unit cell dimensions are 8.5×8.5 nm. Each ring-shaped unit is made of six identical protein subunits surrounding a hydrophilic channel. (Courtesy of G. Zampighi.)

tional Readings at the end of this chapter [Fig. 5–5*B*]). A major polypeptide of 26,000 daltons has been found in the isolated gap junctions and it is suggested that the connexon may represent hexamers of this polypeptide.[9] This polypeptide is very hydrophobic and can be extracted with organic solvents as a proteolipid.

The three-dimensional map also suggests a mechanism for the opening and closing of the channel. As shown in Figure 5–7 closing is achieved by the sliding of the subunits against each other. In the diagram it may be observed that, by a rotation at the base, the inclination of the subunits is decreased and the closure occurs at the upper cytoplasmic face. This mechanism may be the one by which gap junctions regulate the passage of small molecules between living cells (see Unwin and Zamphighi, 1980 in the

Additional Readings at the end of this chapter). Improvements in the image of the connexon showing some asymmetric features are revealed by electron microscopy in which the radiation damage is kept low.[10]

In isolated and unfixed gap junctions embedded in a film of liquid and kept frozen (after image reconstruction) the connexon shows a star-shaped arrangement of the subunits on the cytoplasmic surface surrounding a channel of about 2 nm in diameter. The addition of Ca^{++} produces the closing of the channel by the displacement of the subunits with a reduction in tilt from 14 to 5° (see Unwin, 1984 in the Additional Readings at the end of this chapter).

The development of gap junctions has been observed in various cell types. The process consists of the appearance of plaques between cells

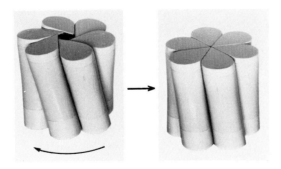

Fig. 5–7. Wooden model of the connexon, depicting the transition from the "open" to the "closed" configuration, as suggested by the maps in Figure 5–6B. It is proposed that the closure on the cytoplasmic face (uppermost) is achieved by the subunits sliding against each other, decreasing their inclination and hence rotating, in a clockwise sense, at the base. The darker shading on the side of the model indicates the portion that would be embedded in the membrane. (Courtesy of Unwin and G. Zampighi.)

having a reduction in the intercellular space and larger particles (probably precursors), and finally the arrangement of the particles in polygonal arrays.

When two cells are brought into direct contact, junctional communications may be formed in a matter of seconds, and this does not require new protein synthesis. It is therefore thought that the 8-nm protein particles are already present on the cell surface and that when cells come in contact the particles diffuse and interact with similar particles on the other membrane.

Gap or Communicating Junctions— Permeability to Ions and Small Molecules

The size of the channels present in gap junctions has been estimated by (1) electron microscopy (see earlier discussion), (2) by electrical measurements, and (3) by the diffusion of tracer substances.

By determining the conductance of single channels and the resistance of the cytoplasm and by assuming a cylindrical channel 20 nm long, the diameter can be calculated to be of the order of 1 nm, a size that is in agreement with morphological data.

The first tracer used was the fluorescent dye *fluorescein* (300 daltons), which when injected into a cell was shown to diffuse into adjacent cells. With the use of fluorescent peptides in cells of *Chironomus* salivary glands, a weight limitation of 1300 to 1900 daltons was reached,

which corresponds to a channel of about 1.4 nm.[11]

Regulation of the Gap Junction

The permeability of the channel is regulated by Ca^{++}. Normally the Ca^{++} content of the cell is low ($10^{-7}M$), and the permeability of the channel is high. If the intracellular Ca^{++} concentration increases, the permeability is reduced. The extracellular concentration of Ca^{++} is about $10^{-3}M$, so that any damage to the cell membrane or the junctional seals may increase the Ca^{++} inside the cells. When the Ca^{++} level reaches a certain limit the channels close,[12] and they open again if the cell eliminates the excess Ca^{++}.

It has been found that the junctional communications depend on the energy provided by oxidative phosphorylation (i.e., ATP). Treatments that inhibit cell metabolism, such as cooling at 8° C, dinitrophenol, cyanide, and oligomycin, produce uncoupling between the cells; this may be reversed by injection of ATP. It is thought that the uncoupling may be due to the influx of Ca^{++}, either from the medium or from mitochondria, during the action of the inhibitors.

The intracellular Ca^{++} concentration can be monitored by the injection of *aequorin*, a luminescent protein in which the light emission is proportional to the Ca^{++} concentration. With such a monitoring system and with the use of fluorescent peptides of various sizes, it has been possible to demonstrate that the permeability changes caused by Ca^{++} are *graded*. In other words, tracers of different molecular sizes can cross the channel according to the intracellular Ca^{++} concentration. As the Ca^{++} is elevated, the limit of permeation decreases gradually, and at $5 \times 10^{-5}M$ the channel is closed. It is thought that this control by Ca^{++} may provide a mechanism for regulation of intercellular communications (see Loewenstein, 1981 in the Additional Readings at the end of this chapter).

Closing of the channel may occur in one of the halves of the gap or in both, since each one has a shutter mechanism that can be activated independently by Ca^{++} and is situated on the cytoplasmic surface.

The closure of the channel, in certain cells, is also dependent on the membrane potential. De-polarizing the cells with high potassium may lead to shutting of the gaps. These findings suggest that a dipole sensitive to the electrical field may be associated with the channel (see Rose, 1984 in the Additional Readings at the end of this chapter).

In the formation of the gap junction cyclic AMP (cAMP) may play a regulatory role. The intracellular increase of this cyclic nucleotide, produced by different mechanisms, leads to formation of new channels, which can be demonstrated by electrical measurements and by the increase in the particles observed after freeze-fracturing. In neuroblastoma cells this effect may be produced by the addition of catecholamines acting on receptors. This type of regulation appears to be mediated by a cAMP dependent protein-kinase (see Lowenstein, 1984 in the Additional Readings at the end of this chapter).

Coupling between Cells Enables Metabolic Cooperation

Several functions may be ascribed to junctional communications. In cardiac muscle and in electrical synapses between certain neurons, the junctions are related to communication of electrical signals between these cells. Junctional interconnections are particularly widespread during embryonic development of stages in which cell differentiation takes place. Because diffusion of molecules can readily take place through them, these junctions are well suited for the dissemination of signals controlling cellular growth and differentiation at close range.

The coupling between cells is not only electrical but also *metabolic*. In addition to permitting free passage of electric current, junctional communications allow the passage of ions and small molecules such as nucleotides, sugars, vitamins, and other metabolites. This gives rise to the phenomenon of *metabolic cooperation* between cells. Figure 5–8 shows an experiment about the transfer of nucleotides between cells in culture. If a few cells prelabeled with a nucleotide are added to a culture of unlabeled cells; it can be seen that the nucleotide is transferred only between those cells that have direct contacts. Other experiments have shown that while nucleotides are easily transferred, macromolecules such as enzymes and RNA do not pass between cells.

Metabolic cooperation was first demonstrated in cell cultures of fibroblasts in which a mutation made the cells unable to incorporate exogenous ^{3}H-hypoxanthine into their nucleic acids. If such cells were grown in contact with wildtype fibroblasts (i.e., nonmutants), there was incorporation in both types of cells. These findings indicate that a metabolite (the nucleotide inosine monophosphate) from the wild-type fibroblast had penetrated the mutant fibroblast through intercellular channels.[13]

We can conclude that, in most tissues, the cells are coupled and have a metabolic cooperation that results in a coordinated control of enzyme activities. At the same time the different cells that compose a tissue retain their characteristic spectrum of macromolecules. The result is that the tissue has properties different from those of the constituent cells.[14]

Altered Coupling in Cancer Cells

Another important finding was that certain cancer cells show no intercellular coupling and that they fail to communicate with normal cells (Fig. 5–9). It is assumed that these cancer cells have a genetic defect that has interrupted the passage of growth-controlling molecules between them. Experiments in which heterokaryons have been produced between a cancer cell and a normal cell (i.e., cell fusion) have shown that it is possible to obtain a noncancerous hybrid. In this hybrid the intercellular communications, which were absent in the original cancerous cell, were now established between the hybrid cells.

Hybrid cells were made between human normal fibroblasts and mouse cancer cells that showed no intercellular communication. These hybrids were observed for several generations by studying the type of growth (i.e., cancerous or normal) and the presence or absence of channels. It has been found that a human factor probably linked to one chromosome can correct both for the growth and the channel defect.[15]

These experiments and others suggest that coupling and permanent uncoupling could be genetically determined. They also point to the possible regulatory role of molecules of small molecular weight—probably nucleotides—between cells controlling growth and differentiation.[15] Such regulatory agents are not transferred

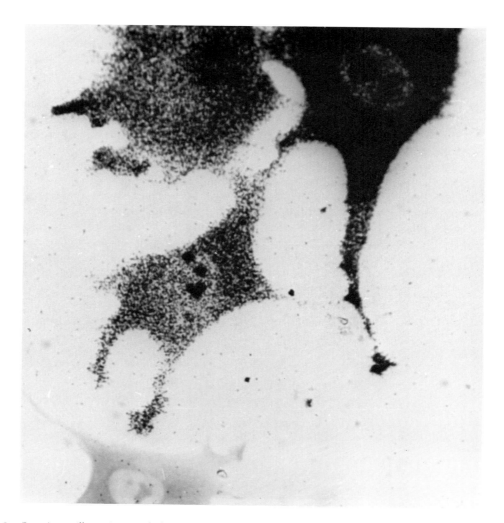

Fig. 5—8.   Experiment illustrating metabolic cooperation. Cultured BHK cells labeled with ³H-uridine were added to similar but unlabeled cells. (In this radioautograph the highly labeled cell on the top right was added to the culture.) It may be observed that the label was transferred to the two cells that make direct contact but not to another which is not in contact (bottom) × 1160. (Courtesy of J.D. Pitts.)

to the cancerous cells having no intercellular communication. It is evident that in most cells, as in a society, the exchange of information is essential to maintain normal health. (For further details on surface changes in cancer cells, see Section 5—4.)

SUMMARY:

Gap Junctions and Intercellular Communications

The so-called *gap junctions (nexus)* are essential in intercellular communications. They represent regions in which there are junctional channels through which ions and molecules can pass from one cell to another. Cells having gap junctions are electrically coupled; i.e., there is a free flow of

electrical current carried by ions. At the gap junction the membranes are separated by a space of only 2 to 4 nm, and there is a hexagonal array of 8- to 9-nm particles. At the center of each particle there is a channel 1.5 to 2 nm in diameter.

The macromolecular unit of the gap junction is called the connexon, which appears as an annulus of six subunits surrounding the channel. It is thought that the sliding of the subunits causes the channel to open and close.

Gap junctions provide direct intercellular communication by allowing the passage of molecules up to a limiting weight of 1300 to 1900 daltons (in *Chironomus* salivary glands). The permeability is regulated by Ca^{++}; if the intracellular Ca^{++} level increases, the permeability is reduced or abolished. Through the gap junction, metabolites (i.e., labeled nucleotides) can pass from one cell to an-

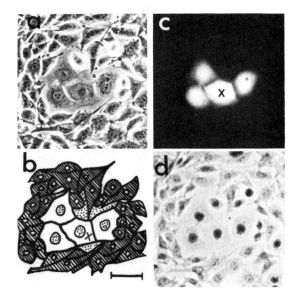

Fig. 5–9. Experiment that demonstrates the lack of coupling between normal cells and cancerous cells. *a,* phase contrast micrograph of a culture having four normal liver cells surrounded by cancerous cells; *b,* tracing of micrograph in *a; c,* the cell marked with X was injected with fluorescein; it is observed that the stain has diffused into the three other normal cells, but not into the cancerous ones; *d,* since the fluorescein is labeled with tritium, in the radioautograph only the four normal cells are ³H-labeled. (From Azarnia, R., and Loewenstein, W.R.: J. Membr. Biol., 6:368, 1971.)

other. In several strains of cancer cells there is no coupling as seen in normal cultured cells. Coupling is genetically determined, and probably genes linked to one chromosome can correct the cancerous growth and the channel defect. Junctional communication may convey electrical signals between certain neurons (i.e., electrical synapses) and between cardiac cells; however, most neurons and skeletal muscle lack electrical coupling. Gap junctions are also used in the transfer of substances that control growth and differentiation in cells.

5–3 CELL COAT AND CELL RECOGNITION

At the beginning of this chapter we mentioned that the plasma membrane is surrounded and protected by the *cell coat,* sometimes also called *glycocalix,* because it contains sugar units in glycoproteins and polysaccharides.

The cell coat is generally considered to be equivalent to the oligosaccharide side chains of glycolipids and glycoproteins that stick out from the cell surface (see Fig. 4–3) and are covalently attached to the protein moieties. In many cells,

however, there is a separate "fuzzy" layer, beyond the cell coat, which is composed mainly of carbohydrates and is secreted by the cell. The separation between the cell coat proper and the fuzzy coat is usually impossible to determine, because they are in continuity with one another and have the same staining properties.

The cell coat can be stained with PAS or Alcian blue for the light microscope, and with lanthanum or ruthenium red for the EM (Fig. 5–10).

Visualization of the carbohydrates may be made more specific by the use of *lectins.* These are proteins, in general, derived from plants, that bind to the cell surface and cause agglutination. The main property of lectins is the specific interaction with divalent or polyvalent carbohydrates. In this sense, they can be considered as recognition molecules for the sugar components of glycoproteins.

Among lectins *concanavalin A* isolated from jack beans is specific for glucose and mannose residues; and *germ agglutinin,* for N-acetylglucosamine. These lectins can be labeled with fluorescent dyes or with electron-dense materials for observation under the EM. (We shall see later that lectins are also present in animal tissues.)

The cell coat generally appears as a layer 10 to 20 nm thick that is in direct contact with the outer leaflet of the plasma membrane. In some cases, as in the ameba, the coat is formed by fine filaments 5 to 8 nm thick and 100 to 200 nm long.

The cell coat has negatively charged sialic acid termini, on both the glycoproteins and gangliosides, which may bind Ca^{++} and Na^+ ions. When the membranes are treated with *neuraminidase,* an enzyme that removes sialic acid, there is a reduction in the negative charge of the membrane.

The strength of the coat varies from cell to cell. For example, the coat of the intestinal epithelium resists vigorous mechanical and chemical attacks; in other cells the coat is labile and is depleted by washing or exposure to enzymes (see Luft, 1976 in the Additional Readings at the end of this chapter).

In Chapter 9 it will be mentioned that the biosynthesis of the glycoproteins forming the glycocalyx takes place in the ribosomes of the endoplasmic reticulum (ER) and that the final

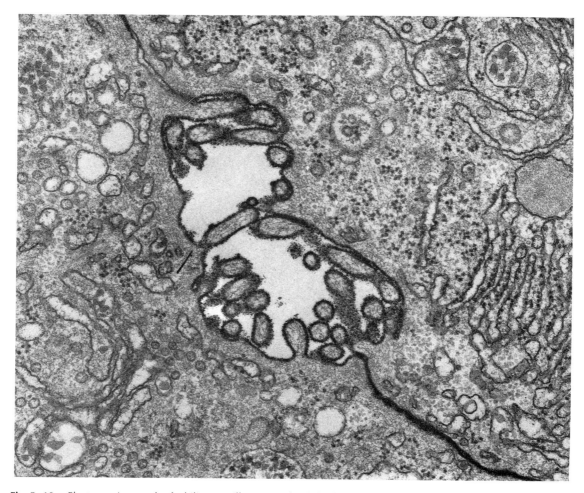

Fig. 5–10.   Electron micrograph of a biliary capillary stained with lanthanum nitrate to enhance the electron density of the surface coat of cells. Note the dark material between the liver cells and around the microvilli. × 38,400. (Courtesy of D. Ferreira.)

assembly with the oligosaccharide moiety is accomplished in the Golgi complex. The cell coat can be considered to be a secretion product of the cell that is incorporated into the cell surface and undergoes continuous renewal.

Glycosaminoglycans and Proteoglycans as Extracellular Materials

Extracellular materials lie outside the cell coat proper and the fuzzy layer of certain cells. This is apparent in the jelly coat of eggs of fishes and amphibians, the basal laminae of epithelia, the matrix material in which cartilage and bone cells are embedded, and the cell wall of plant cells (see Section 12–1). In these extracellular ma-terials the most prominent components are *collagens* (studied in Section 6–5) and *glycosaminoglycans*, (i.e., mucopolysaccharides). These are polysaccharides in which there is a repeating disaccharide unit that contains an amino sugar (either N-acetylglucosamine or N-acetylgalactosamine). Table 5–1 lists the main glycosaminoglycans with their repeating units. Because of the presence of carboxyl and sulfate groups, these polysaccharides are very acidic molecules. Frequently, they are associated with proteins forming *proteoglycans* (i.e., mucoproteins), which differ from glycoproteins in that the polysaccharide chain is much longer. Furthermore, proteoglycans are amorphous and form gels able to hold large amounts of water.

TABLE 5–1.   MAIN GLYCOSAMINOGLYCANS AND THEIR REPEATING DISACCHARIDE UNITS

Glycosaminoglycans	Disaccharide Unit
Hyaluronic acid	D-Glucuronic acid; N-acetyl-D-glucosamine
Chondroitin	D-Glucuronic acid; N-acetyl-D-galactosamine
Chondroitin sulfate A	D-Glucuronic acid; N-acetyl-D-galactosamine 4-sulfate
Chondroitin sulfate C	D-Glucuronic acid; N-acetyl-D-galactosamine 6-sulfate
Dermatan sulfate	L-Iduronic acid; N-acetyl-D-galactosamine 4-sulfate
Keratan sulfate	D-Galactose; N-acetyl-D-glucosamine 6-sulfate

Functions Attributed to the Cell Coat

In addition to the protection of the cell membrane numerous functions have been attributed to the cell coat. Although it is not absolutely necessary for the integrity of the cell and for the permeability of the plasma membrane, it nevertheless performs functions that are of great significance to studies in cell biology.

Filtration. The extracellular coats that surround many vertebrate capillaries, especially the kidney glomerulus, act as a filter and regulate the passage of molecules according to size.

Microenvironment. The glycocalyx may change the concentration of different substances at the cell surface by acting as a diffusion barrier. Because of its charge, the glycocalyx may act as a kind of exchange resin that changes the cationic environment of the cell. For example, a muscle cell with its excitable plasma membrane is surrounded by a glycocalyx that can trap sodium ions. Certain components, such as hyaluronate, can drastically change the electrical charge and pH at the cell surface.

Enzymes. Histochemical techniques have demonstrated alkaline phosphatase in the coat as well as on the surface of the intestinal microvilli. When these structures are isolated, practically all the enzymes involved in the terminal digestion of carbohydrates and proteins are found in them.

Blood Group Antigens. The classical division of the ABO blood groups is based on the occurrence of antigens on the red cell surface and of specific antibodies in the serum. The ABO antigens are found not only on the erythrocytes but in other tissues and also in secretions. Each one of the groups has a specific structure set by the terminal carbohydrates (see Harrison and Lunt, 1980 in the Additional Readings at the end of this chapter.)

Other Antigens. Among these antigens are those of histocompatibility, which permit the recognition of the cells of one organism and the rejection of other cells that are alien to it (e.g., the rejection of grafts from another organism). This function, is also related to molecular recognition.

The major sialoglycoprotein of the red cell membrane carries the M and N antigens that appear, infrequently, in man. It also contains the receptor sites for the influenza virus and various lectins.

Cell Adhesion. There are three main types of contact between cells which lead to their aggregation in a tissue.

1. *The aggregation may be formed by the inclusion of the cell in a common matrix, as it occurs in cartilage.* In most cases this extracellular matrix consists of the *coat* of the cells. The substances of the coat accumulate after the cells have made contact with each other.

2. *The aggregation of cells may have little intercellular material.* Electron microscopy shows the gap between cells to be frequently on the order of 15 nm. However, even in these small intercellular spaces there is some substance, possibly a mucoprotein, that accounts for the specificity of cell association. Furthermore, the enzymes that are more effective in separating these cells are proteases and mucases.

3. *The aggregation may imply the presence of intercellular channels.* In plant cells, cytoplasmic bridges or plasmodesmata (see Fig. 1–1) have been recognized for a long time. These provide narrow connections between the cells and across the cellulose walls and permit the free passage of ions and probably macromolecules.

Factors Mediating Cell-Self Recognition—The Aggregation Factor of Sponges

Other properties of the cell coat are related to the capacity of cells to "recognize" similar cells in a tissue, to adhere to one another, to dissociate and reassociate, and to produce an inhibition in the neighboring cells by the so-called *contact inhibition*. For molecular recognition, the oligosaccharides present in the cell coat may constitute a kind of molecular code for the cell surface. The number of permutations of their individual components (i.e., galactose, hexosamine, mannose, fucose, and sialic acid) makes for each cell type a special kind of *fingerprint* by which molecular recognition between cells could be established.

Molecular recognition reaches its maximum expression in the nervous system, where a neuron can make synaptic contacts with numerous other neurons, thereby forming specific neuronal circuits of immense complexity. Molecular recognition is one of the fields of cell biology and neurobiology that must be strongly developed in the future.

The history of cell recognition begins with the experiments done on sponges by H.V. Wilson in 1907. Forcing living sponges through a fine silk mesh disaggregated the cells; however, if the isolated cells were left standing for some time, they became aggregated into clusters and eventually new sponges were formed. This phenomenon depends on the motility of the cell and on *cell recognition* that allows for the *cell-self assembly*. In fact, if sponges of two species with different colors (e.g., one red, the other yellow) are dissociated, there is a process of *sorting out* by which only the cells of the same species aggregate to each other.

The study of sponges by Humprey, et al., and Jumblatt, et al., has more recently led to other important findings.[16,17] If the above experiments are carried out in artificial sea water, lacking Ca^{++} and Mg^{++}, the cells no longer reaggregate. In this case, the sponge cells have lost a huge proteoglycan (i.e., about 20 million daltons), which contains 50% carbohydrate and can be recovered in the soluble fluid. This proteoglycan, called the "aggregation factor," when added back to the system, is able to restore the property of self-recognition and to reproduce the aggregation.

Figure 5–11 illustrates a hypothetical model of the function of the aggregation factor. In Figure 5–11 the aggregation factor appears as two intercellular macromolecules that close the gap between cells and are joined by divalent calcium bridges. Ca^{++} is also indirectly related to the

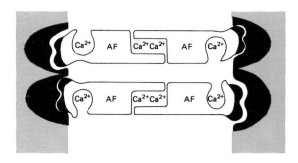

Fig. 5–11. A tentative model for specific cell to cell interaction in tissue construction. Two macromolecular aggregation factors (AF) are illustrated, each consisting of at least two subunits. The black termini at each pole carry the carbohydrates, which are recognized by the baseplate (BP) anchored in adjoining cell surfaces. (From Weinbaum, G., and Burger, M.M.: Nature, *244*:510, 1973.)

binding of the carbohydrate moiety to a recognition site (i.e., the so-called *base plate*) in the membrane. When the Ca^{++} is removed, the aggregation factor no longer bridges the two cells and these become separated. At the same time, the lack of Ca^{++} produces a conformational change in the factor by which these macromolecules are released into the medium.

Similar experiments were carried out by A.A. Moscona in the tissues of chick embryo. These tissues were dissociated by *trypsin*, an enzyme that removes the cell coat. If the cell suspension is allowed to stand and is gently rotated, reaggregation is produced. Similar cells (e.g., retinal, kidney, bone cells) attach to each other, forming aggregates that show the characteristics of the original tissue.[18] In contrast with what was observed in sponges, here the reaggregation is not species specific but organ specific. This can be demonstrated by the following experiment: If cells of the chick and mouse embryo are mixed, they reaggregate according to the tissue rather than the species. (The cells of a given species can be easily recognized by labeling them with radioactive thymidine and submitting the tissue sections to radioautography.)

An interesting phenomenon related to a cell-cell recognition is the so-called *homing of the lymphocytes*, in which lymphocytes leave the bloodstream, enter the lymphatic tissues, and spread to specific sites. Experimentally, the course of these cells has been followed by labeling lymphocytes with radioactive phosphate. Lymphocytes taken from the thoracic ducts of rats and injected into other rats, normally spread into the spleen and lymph nodes. However, treatment of lymphocytes with neuraminidase or with trypsin, which alters the cell coat, changes the normal homing, and causes the cells to accumulate in the liver.

In all of these phenomena of cell recognition the presence of specific carbohydrates seems essential. If the synthesis of carbohydrates is altered, the process of cell recognition may be impaired. For example, in the studies with embryonic cells in culture, it has been found that the presence of L-glutamine in the medium is important. This amino acid is involved in the synthesis of amino sugars, such as glucosamine, which are present in mucopolysaccharides (Table 5–1). In the absence of L-glutamine the cells no longer stick together; they remain separated from one another.

Animal Cell—Recognition Molecules

In the previously mentioned phenomena of cell recognition and adhesion two basic mechanisms may be present. One mechanism includes membrane-bound factors, such as glycophorin, histocompatibility (HLA) antigens, and others, that are *intrinsic* to the membrane. The other mechanism involves molecules that bind to the membrane by way of protein carbohydrate interactions. These *extrinsic* factors may be large proteoglycans, such as the association factor of sponges, or glycoproteins that specifically promote the reaggregation of dissociated embryonic cells. In recent years, a whole family of extrinsic protein factors of low molecular weight has been found in animal tissues. Because their behavior is similar to that of the plant lectins, they are sometimes referred to as *animal lectins*.

These substances are able to recognize saccharides on the cell surface and to cause specific hemoagglutination. Factors having this property have been extracted from embryonic and also from adult tissues. For example, there is an agglutinin capable of clustering together the erythroblastic cells around a nurse macrophage, forming the so-called *erythroblastic island* of the bone marrow. From the chick embryo, several substances that interact with the disaccharide *lactose* have been isolated. One of these substances is secreted by goblet cells into the intestinal lumen.[19] During development animal

lectins may mediate cell adhesion by way of a bridging mechanism (see Harrison and Chesterton, 1980 in the Additional Readings at the end of this chapter).

For cell-to-cell recognition and adhesion, two main theories have been proposed. One such theory, *chemoaffinity*, postulates that the cells that adhere have paired complementary genetic products that act as surface markers. This theory implies that there should be a large number of genetic products involved, a condition that would be terribly complex in the central nervous system (CNS) in which each neuronal type makes specific connections with other neurons (see Chapter 24).

At the present time, the more favored theory is one based on a small number of *cell adhesion molecules* that can undergo a process of modulation or change. For example, in the CNS of embryos there is a large *sialoglycoprotein* at the cell surface of neurons. This is an integral protein of the membrane that has a polypeptide portion (120,000 daltons) and contains 26 to 35% carbohydrate, most of which corresponds to *sialic acid* (the total molecular weight is 250,000 daltons). This protein appears to be the major neuron-neuron and also neuron-muscle adhesion molecule and is known as the *neural cell adhesion molecule* (N-CAM). The modulation of such a protein implies a change in the carbohydrate moiety, especially in its sialic acid content. For example, from the embryo to the adult, there is a considerable loss of sialic acid, while the amino acids and the neutral sugars remain constant (see Edelman, 1983 in the Additional Readings at the end of this chapter).

In addition to the N-CAM, a liver cell adhesion molecule (L-CAM) and a neuron-glia cell adhesion molecule (Ng-CAM) have been recognized. These proteins appear in the embryo in a definite sequence and in topological order. While L-CAM is observed in all three embryonic layers, N-CAM appears in ectoderm and mesoderm and Ng-CAM is found in ectoderm. All of these CAMs act as regulators of morphogenetic movements and may be a major factor in controlling embryonic induction. The use of antibodies against these adhesion molecules produces various perturbations of the normal development (see Edelman, 1984 in the Additional Readings at the end of this chapter).

Cellular Interactions and Cyclic AMP

Cells can also interact with each other through diffusible substances acting at a distance, or by short range actions at cell contacts. So far, we have studied mainly this second type of interaction. Long range interactions are better understood in the case of the cellular slime mold *Dictyostelium discoideum* (Fig. 5–12). This organism lives in forests, under layers of decomposing leaves. While food is plentiful (*Dictyostelium* feeds on soil bacteria), the organism lives as unicellular amebae, which are independent of each other and which multiply by mitosis. When food becomes scarce, some amebae start secreting a diffusible substance that attracts more amebae. A *nucleation center* is formed into which streams of amebae converge, until finally a *slug* or *pseudoplasmodium* is formed. The slug contains millions of cells and is able to migrate for considerable distances (towards light, where the spores will have a better chance to disperse). Eventually the slug differentiates, producing a stalk on top of which spores develop. By sporulating, *Dictyostelium* ensures survival during periods of low food supply.

The *Dictyostelium* life cycle has, thus, two distinct stages, one of unicellular life and a multicellular stage in which millions of cells move and differentiate in a coordinated way. This whole process is triggered by a substance that attracts the cells together. The attractant has been identified and is cAMP. Many cells use cAMP as an intracellular signal to regulate metabolism.

Cellular slime molds use cAMP as an *extracellular messenger* to communicate positional information to distant cells.

The biochemical machinery that controls the aggregation of cells and the differentiation of the *Dictyostelium* also involves a receptor for cAMP on the membrane, and the enzymes adenylate cyclase and phosphodiesterase, which are activated when pulses of cAMP are applied experimentally. This treatment results in aggregation of the amebae and formation of slugs (see Robertson and Grutsch, 1981 in the Additional Readings at the end of this chapter).

SUMMARY:
Cell Coats and Cell Recognition

Most cell membranes have a coat, sometimes referred to as the *glycocalyx*, made of glycoproteins or polysaccharides. The cell coat is negatively charged and may bind Na^+ and Ca^{++}. Several cytochemical techniques are used to reveal the cell coat (e.g., PAS and ruthenium red). The oligosaccharides may be visualized by the use of *lectins*.

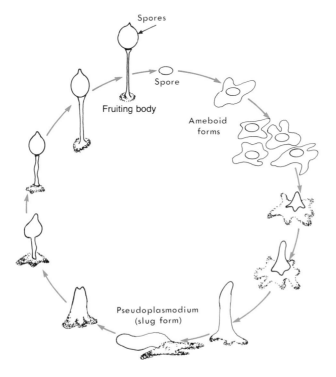

Fig. 5–12. Life cycle of *Dictyostelium discoideum*. The slime mold can exist as single-cell amebae which aggregate when food is scarce, attracted by cAMP. The slug (pseudoplasmodium) thus formed migrates and eventually sporulates.

The cell coat is a kind of secretion product that undergoes an active turnover.

Extracellular materials lie outside the cell coat proper and the fuzzy layer of certain cells. In these extracellular materials collagens and glycosaminoglycans are the main components. These are polysaccharides, such as hyaluronic acid and chondroitin sulfate, in which there is a repeating disaccharide unit. These acidic molecules are associated with proteins forming proteoglycans.

Many functions are attributed to the cell coat. It may act mechanically, protecting the membrane, and participate in the *filtration* and *diffusion* processes. The cell coat makes a kind of *microenvironment* for the cell. It contains enzymes involved in the digestion of carbohydrates and proteins.

Molecular recognition between cells may depend on a molecular code made up of the individual monosaccharides such as galactose, hexosamine, mannose, fucose, and sialic acid. The classical ABO blood groups are based on specific antigens of the red cell coat, which are specified by their terminal carbohydrates. Several other antigens are found on the cell coat. Molecular recognition reaches maximum expression in the nervous tissue. Cell adhesion and cell dissociation and reassociation are dependent on the coat. Cells are able to recognize similar cells in a tissue. This is dramatically shown by Wilson's experiments with sponges of different species and colors in which, after dissociation there is a process of sorting out. A huge proteoglycan, the association factor, and Ca^{++} are involved in this process. In the embryonic tissue of the chick the reaggregation is organ specific, not species specific, as in the case of the sponges.

In all cell-recognition phenomena the presence of specific carbohydrates at the membrane is essential. From embryonic and adult animal tissues, low molecular weight proteins that act as plant lectins have been isolated. These animal lectins recognize saccharides on the cell surface and cause a β-galactoside hemoagglutination. Cells can also interact through diffusible substances acting at a distance. One such example is the life cycle of *Dictyostelium discoideum*, a slime mold in which the single ameba may aggregate by the long range action of cAMP.

5–4 THE CELL SURFACE OF CANCER CELLS

In view of the many biomedical implications, the study of the cancer cell is of paramount importance in cell and molecular biology. This subject could have been treated in a holistic manner in a special chapter; however, we prefer to analyze the special characteristics of the cancer cell as they differ from the normal cell in their various cellular components. For this reason, considerations on the cancer cell will be found in those chapters related to the cell surface, the cytoskeleton, the nucleus, and the chromosomes. The molecular biology of *cell transformation* (i.e., the change from normal to cancerous) will be discussed in Section 22–3, however, in this section we will consider viruses as oncogenic (i.e., cancer producing) agents.

Cancer cells are characterized by an *uncontrolled cell growth*, *invasion of other tissues*, and *dissemination* to other sites of the organism producing secondary tumors. All of these characteristics suggest that cancer cells have escaped from the controls that regulate normal growth. Cancerous tumors are *monoclonal*, i.e., derive from the division of a single cell, which has been transformed into a cancer cell.

Cancer cells are structurally and biochemically different from normal cells. For example, the karyotype is frequently abnormal, with changes in chromosome number or with chromosomal alterations (see Sections 17–2 and 18–4). Malignant cells generally have a higher glycolytic activity. The cytoskeleton, which is composed of a network of microtubules, actin microfilaments, and intermediate filaments, is reduced or disorganized (see Section 6–5). In this section, we discuss the changes in the plasma membrane with special reference to the cell coat, which as we have seen, plays a prominent role in growth control and cell-to-cell interaction.

Surface Changes in Cancer Cells

Cancer cells undergo many changes in the plasma membrane, especially in the cell coat. In culture, they show more electrophoretic mobility, owing to the increased number of negative charges provided by a more abundant amount of glycosaminoglycans in the cell coat. Many cancer cells show electrical uncoupling owing to the disappearance of gap junctions. Another difference is demonstrated by the use of lectins that bind to glycoprotein receptors. In cancer cells, these receptors tend to diffuse more easily within the lipid bilayer than they do in normal cells. This greater diffusibility of receptors (which is also found in normal cells during mitosis) may be caused by disorganization of the microfilaments that normally are attached to the plasma membrane and that reduce the motility of receptors. Differences also exist in the *glycolipid* and *glycoprotein* content, in a reduction in the amount of *gangliosides* and in their enzymes of synthesis. In cells transformed by cancer certain proteins of the membrane disappear and *sugar uptake* increases. Cancer cells frequently release intracellular enzymes to the medium, indicating leakage from cell membranes.

A major protease secreted by cancer cells acts on *plasminogen* and converts it to *plasmin*. Plasmin is a proteolytic enzyme that dissolves blood clots and also removes exposed protein groups at the cell surface. If plasminogen is removed from the medium, the morphology of cancer cells returns to normal. Frequently, cancer cells carry *new antigens* not present in normal cells. For this reason, they can induce an immunological response that in certain favorable conditions may eliminate the cancer cells. The production of specific antibodies against such antigens could be a way of controlling some tumors.

Cancer Cells and Iron Transport

One important difference between normal cells and cancer cells is their method of metabolizing iron and trace metal ions. They transport and deliver iron to the cells through *transferrin*, a glycoprotein present in blood plasma that binds the metal and associates it with specific *transferrin receptors* on the cell membrane.[20] These receptors are glycoproteins consisting of two 95,000 dalton subunits bound by disulfide bonds (Trowbridge, 1984).

After its penetration inside the cell, the iron binds to other proteins, such as *ferritin*, and is deposited for use in many enzymatic systems (e.g., cytochromes). Transferrin is essential to the growth of normal cells in cultures and should be present in the medium to maintain the cell cycle. Normal cells probably adapt to iron deprivation by changing the transferrin binding capacity. Some cancerous cells, however, may adapt by an alternative mechanism. It has been found that cells transformed by an oncogenic virus secrete a low molecular weight agent—*the so-called siderophore-like growth factor*—which has a high capacity for binding iron (acts as a *chelating agent*) and transporting it inside the cell (acts as an ionophore or channel).

Figure 5–13 represents schematically the difference in metal ion transport between the normal cell and the transformed cell. Metal ions, such as Fe, Mn and Zn, play a role in regulatory growth mechanisms, which activate the cell division program (see Chapter 14). In a normal cell, receptors for the transferrin-metal complex are synthesized and integrated into the membrane. In a transformed cell, siderophore growth factors are delivered to the medium where they bind the metal, competing with transferrin. The siderophore-metal complex is then transported inside the cell. By this alternative mechanism cancer cells may decrease their dependence from transferrin and compete with normal cells for essential metal ions (see Fernandez-Pol, 1980 in the Additional Readings at the end of this chapter).

Fibronectin and the Cancer Cell

One of the major chemical components of the cell coat is *fibronectin*, a high molecular weight glycoprotein, which can be isolated from normal cultured fibroblasts by mild procedures, such as washing with a solution of urea. Together with other glycoproteins, fibronectin is found in the "footprints" that a moving cultured cell leaves on the substratum with which it comes in contact (e.g., a glass surface [see Section 6–5]).

Fibronectin occurs widely in connective tissues, among others, and it is thought to have a role in determining the distribution of cells within both embryonic and adult tissues. The presence of fibronectin increases cell adhesion to the substratum and to other cells and influences the morphology of the cell, as well as inducing locomotion and migration.

Fibronectin, originally called LETS (*large external transformation-sensitive*) protein, is absent or drastically reduced in cultured cells undergoing cancerous transformation. It is conceivable that the reduction of fibronectin, together with the uncoupling phenomenon mentioned above, could have adverse effects on the social behavior of transformed cells.

This change may explain the tendency of cancer cells to break up connections with other cells, to invade tissues locally, and to *metastasize* (i.e., to invade distant tissues through lymph and blood vessels).

It has been observed that the addition of fibronectin to cultured transformed cells pro-

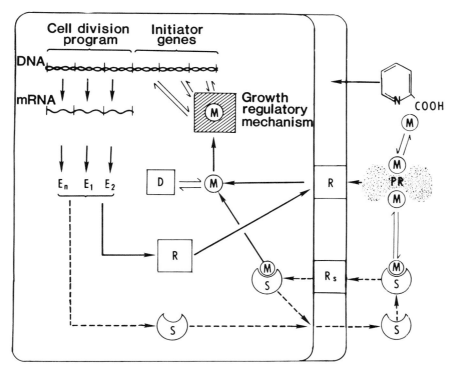

Fig. 5–13. Schematic representation of possible trace metal ion transport system in normal (———) and transformed (— — —) cells and its interaction with growth regulatory mechanism. Transformed cells may produce siderophores as their own growth factors which may be part of the cell iron transport system. M, metal ion (e.g., Fe, Ga); S, siderophore or other specific ionophore; Pr, protein (e.g., transferrin); R and Rs, plasma membrane receptors; E, enzyme; D, deposits (e.g., ferritin). (Courtesy of J.A. Fernandez-Pol.)

duces changes in their behavior, causing them to acquire a more normal appearance. For example, they attach more readily to the substrate, adopt a more flattened morphology, and tend to align with one another in a monolayer. The cytoskeleton of a transformed cell also becomes more organized under the influence of fibronectin.

Loss of Control of Growth in Cancer Cells

The previously mentioned changes in the cell surface of cancer cells are reflected in their behavior in culture. Normal cells growing in tissue culture tend to establish cell contacts by adhesion to neighboring cells. At the points of adhesion, some kind of electron-dense "plaque" is formed in both contacting cells. At the same time, there is a slowing down of the ameboid processes, which results in *contact inhibition of movement*. In contrast cancer cells are unable to form adhesive junctions and do not show this type of contact inhibition. Such cells tend to

have more blebs on the cell surface than normal cells. This difference is clearly shown in scanning electron micrographs of cancer cells (Fig. 5–14).

Experimentally, when normal cells have become completely surrounded by other cells, their motility stops and they form a monolayer (Fig. 5–15A). At the same time there is inhibition of growth, and the number of cells in the dish remains practically constant. On the other hand, cancer cells continue to multiply and pile up forming irregular masses several layers deep (Fig. 5–15B). The loss of *contact inhibition* is easily observed in normal cultured cells that have been "transformed" by oncogenic (i.e., cancer inducing) viruses.

As we mentioned in Section 3–3, the studies of Leonard Hayflick demonstrated that normal cells undergo a limited number of divisions (between 50 and 100 for human fibroblasts) in cultures that are repeatedly subcultured. Thereafter the cell ages and dies. Malignant cells, however, are *potentially immortal* and continue

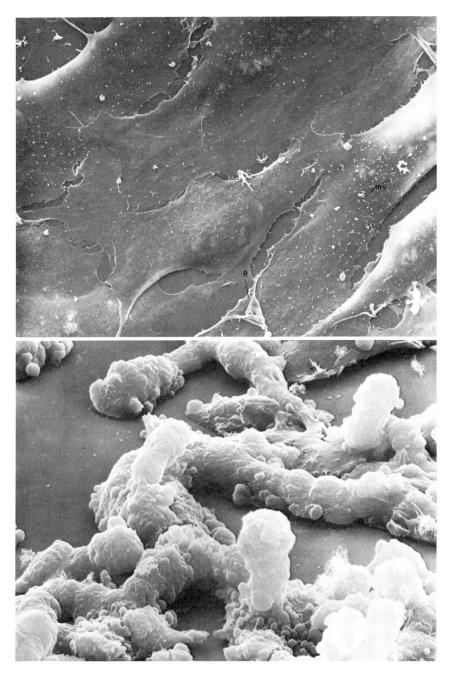

Fig. 5–14. Scanning electron micrographs of cultured cells. *Above,* well-spread normal hamster embryo cells. A few microvilli *(mv)* and ruffles *(R)* are observed. × 1620. *Below,* the same cell type after transformation by human adenovirus. Observe that the cells have blebs on the surface and tend to make several layers. × 3145. (Courtesy of R.D. Goldman.)

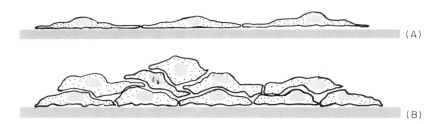

Fig. 5–15. Schematic representation of the growth pattern of normal *(A)* and transformed *(B)* cells cultured on a solid substratum. Normal cells are inhibited by contact and stop multiplying, forming a monolayer. Transformed cells continue to multiply, forming multilayers. (From Ringertz, N.R., and Savage, R.E.: *In* Cell Hybrids. New York, Academic Press, Inc., 1976.)

to divide indefinitely. In vivo, malignant cells act as if the body is a culture medium serving for their nutrition.

The growth in a normal cell culture is influenced by the action of different growth factors present in the surrounding fluid and by their interaction with the cell surface (see Section 14–1). In normal cultures, it is possible that growth factors become less available because of the packing of the cells and the tendency of mitosis to stop (see Holley, 1975 in the Additional Readings at the end of this chapter).

Transformed or malignant cells apparently escape from the inhibition control because they are less dependent on these external growth factors.

That *uncontrolled growth* is the main characteristic of a cancer cell can be illustrated by the following: If a large portion of the liver of a rat is removed, the portion that has remained regenerates. This occurs rapidly at first, growing more quickly than any malignant growth, but then it slows down. After a week, the liver has regenerated, and its cell divisions are as few as in the normal organ.

In a liver tumor (i.e., hepatoma), however, the growth is slower than occurs in regeneration, but it continues indefinitely because the regulation of cell proliferation has been lost.

Viruses as Oncogenic Agents

Numerous chemical, physical, and biological stimuli are able to convert a normal culture into a cancerous one, through a process called *transformation*. The study of these *oncogenic agents* is important because on it may hinge the solution to the problem of "spontaneous" cancer in man.

Throughout this book we refer to carcinogenic agents, all of which affect the DNA by changing its nucleotide sequence and by causing *mutations*. In Sections 17–2 and 19–2 we see that ionizing radiations, many alkylating drugs, and polycyclic aromatic hydrocarbons are carcinogenic. In this section, we refer to a biological oncogenic mechanism, i.e., one produced by certain viruses that can cause the cancerous transformation of normal cells in cultures.

In 1910 Peyton Rous demonstrated that the sarcoma of the chick could be reproduced by inoculating a cell-free extract. The oncogenic agent was later recognized to be a virus. It is now known that different viruses may produce the transformation. These may be viruses that contain either DNA or RNA. In both cases, the virus is able to infect and to transform the cell. The viral genome becomes integrated and hidden in the cell host DNA, and both of them undergo replication at each cell division. In addition, the viral genome is able to produce RNA (transcription) and special proteins (translation), which can be detected in the transformed cell.

Among the DNA viruses the most frequently studied are *polyoma* and *simian virus 40* (SV40). Both of these viruses are very small, of icosahedral shape (20 surfaces), and about 45 nm in diameter. The DNA molecule is circular and double-stranded. When added to a normal culture the virus may either kill the cell and fully reproduce new virus particles (case of *permissive cell*); or it may become integrated into the genome of the cell (case of a *nonpermissive cell*). The viral genome of SV40 has a few genes that code for the viral capsid and produce a protein called T (tumor) antigen, which accumulates in the cell nucleus. This protein T appears to be responsible for the transformation. In a transformed cell only the genes producing the T antigen are expressed and the viral capsid is not produced.

Other oncogenic viruses are the *adenovirus* (see Fig. 1–5), and the *Epstein-Barr virus*, which produces infectious mononucleosis and is thought to be involved in the human cancer called Burkitt lymphoma, which is endemic in Africa. This transformation occurs only in susceptible people having a specific genetic background.

Viruses that contain RNA belong to the so-called *retroviruses*, which have a single RNA chain. These viruses have an inner core of RNA with a few proteins. This inner core is surrounded by a shell and by an outer envelope, which are provided by the plasma membrane of the host cell. RNA viruses are natural agents in animal tumors, especially in rodents.

There is indication that the virus can infect the egg and thus pass from one generation to another. They may also pass to the offspring through the mother's milk. RNA viruses have been implicated in human malignancies, especially in breast cancer. In leukemias, the affected cells may contain viral information.

In the case of RNA viruses, the single RNA genome serves as a template for the production of a complementary DNA molecule by *reverse transcriptase* (see Section 14–2). Then, a double-stranded DNA of the viral genome is produced, which becomes integrated in the host DNA.

The use of oncogenic viruses in cancer research is of fundamental importance in understanding the molecular biology of cancer, an understanding that is a prerequisite to definite progress in its treatment (For a further discussion on the molecular biology of retroviruses and oncogenes, see Section 22–3).

SUMMARY:
Cell Surface in Cancer Cells

Cancer cells are characterized by uncontrolled growth, invasion, and dissemination (metastasis). In a tumor all cancer cells are monoclonal. They may show changes in karyotype, increased glycolysis, and disorganization of cytoskeleton.

In cancer cells there are many changes in the cell membrane and cell coat, such as the disappearance of gap junctions, loss of coupling, changes in glycolipids and glycoproteins, and a reduction in gangliosides. There is also more mobility of surface receptors, increased transport of sugars, and growth of new antigens.

While in normal cells the transport of iron and trace metal ions involves transferrin and transferrin receptors, in transformed cells there is an alternative mechanism. Transformed cells secrete siderophore-like growth factors (chelating agents that trap the metal ions), which compete with transferrin and transport the iron inside the cell.

In the cell coat, fibronectin, a large glycoprotein found in "footprints" of moving cultured cells, is reduced in cancer cells. A major characteristic of cancer cells is the combined loss of contact inhibition, motility, and growth control, which is characteristic of normal cells in culture. Cancer cells are "immortal" and tend to pile up, while normal cells die after a number of divisions and tend to form monolayers.

A normal cell can be converted into a cancerous cell by a number of transforming agents, all of which affect the DNA and cause mutations. Different DNA and RNA viruses can produce transformation; among the DNA viruses, polyoma and SV40 are extensively used in experiments. Adeno- and herpes virus (DNA) may also be involved. The RNA retroviruses are natural agents in animal tumors and can be transmitted to the offspring through the mother's milk.

In all of these cases the viral genome becomes integrated in certain regions of the host's genome and is expressed in the cancerous cell.

REFERENCES

1. Bretcher, A.: VII International Congress of Biophysics. Mexico, 1981, p. 49.
2. Kachar, B., and Reese, T.S.: Nature (Lond.), *296*:464, 1982.
3. Salas, P., and Moreno, J.H.: J. Membr. Biol., *64*:103, 1982.
4. Dragsten, P.R., Handler, J.S., and Blumenthal, R.: Fed. Proc., *41*:48, 1982.
5. Cerejido, M., Stefani, E., and Martinez Palomo, A.: J. Membr. Biol., *53*:19, 1980.
6. Meza, I., et al.: J. Cell Biol., *87*:746, 1980.
7. Mueller, H., and Franke, W.W.: J. Mol. Biol., *163*:647, 1983.
8. Goodenough, D.A., and Revel, J.P.: J. Cell Biol., *45*:272, 1970.
9. Henderson, D., Eibl, H., and Weber, K.: J. Mol. Biol., *132*:193, 1979.
10. Baker, T.S., Caspar, D.L.D., Hollingstead, C.J., and Goodenough, D.A.: J. Cell Biol., *96*:204, 1983.
11. Simpson, I., Rose, B., and Loewenstein, W.R.: Science, *195*:294, 1977.
12. Oliveira-Castro, G.M., and Loewenstein, W.R.: J. Membr. Biol., *5*:51, 1971.
13. Subak-Sharpe, H., Buck, P., and Pitts, T.D.: J. Cell Sci., *4*:353, 1969.
14. Pitts, J.D.: VII International Congress of Biophysics. Mexico, 1981, p. 209.
15. Azarnia, R., and Loewenstein, W.R.: J. Membr. Biol., *34*:1, 1977.
16. Humprey, T., Tonemoto, W., Humprey, S., and Anderson, D.: Biol. Bull., *149*:430, 1975.
17. Jumblatt, J.E., Schlup, V., and Burger, M.M.: Biochemistry, *19*:1028, 1980.

18. Hausman, R.E., and Moscona, A.A.: Exp. Cell Res., *119*:191, 1979.
19. Beyer, E.C., and Barondes, S.H.: J. Cell Biol., *92*:23, 1982.
20. Fernandez Pol, J.A., and Klos, D.J.: Biochemistry, *19*:3904, 1980.

ADDITIONAL READINGS

Bretcher, A., and Weber, K.: Purification of microvilli and analysis of protein components of microfilament core bundle. Exp. Cell Res., *116*:397, 1978.

Cameron, I.L., and Pool, T.B.: The Transformed Cell. New York, Academic Press, 1981.

Doljanski, F., and Kapeller, M.: Cell surface shedding— the phenomenon and its possible significance. J. Theor. Biol., *62*:253, 1976.

Dragstein, P.R., Handler, J.S., and Blumenthal, R.: Fluorescent membrane probes and the mechanism of maintenance of cellular asymmetry of epithelia. Fed. Proc., *41*:48, 1982.

Edelman, G.M.: Cell adhesion molecules. Science, *219*:450, 1983.

Edelman, G.M.: Cell adhesion molecules in morphogenesis. III International Congress of Cell Biology. Edited by S. Seno and Y. Okada. Japan, Academic Press, 1984, p. 170.

Edidin, M.: Rotational and translational diffusion in membranes. Annu. Rev. Biophys. Bioeng., *3*:179, 1974.

Fernandez-Pol, J.A.: Molecular bases of the regulation of iron-59 and gallium-67 transport in normal and simian virus 40 transformed cells. *In* Frontiers in Nuclear Medicine. Edited by W. Horst et al. Berlin, Springer-Verlag, 1980.

Geiger, B.: The molecular architecture of desmosomes. Trends Biochem. Sci. 6:R8, 1981.

Gilula, N.B.: Gap junctions and cell communication. *In* International Cell Biology. Edited by B.R. Brinkley and K.R. Porter. New York, The Rockefeller University Press, 1977.

Glaser, L.: Cell-cell recognition. Trends Biochem. Sci., *1*:84, 1976.

Harris, H.: Brush borders on the move. Nature, *302*:106, 1983.

Harrison, F.L., and Chesterton, C.J.: Factors mediating cell recognition and adhesion. FEBS Lett., *122*:157, 1980.

Harrison, R., and Lunt, G.G.: Biological Membranes, 2nd Ed. Glasgow, Blackie & Son, 1980.

Hertzberg, E.L., Lawrence, T.S., and Gilula, N.B.: Gap Junctional Communication. Annu. Rev. Physiol., *43*:479, 1981.

Holley, R.W.: Control of growth of mammalian cells in cell culture. Nature, *258*:487, 1975.

Junqueira, L.C.V., and Montes, G.S.: Biology of collagen-proteoglycan interaction. Arch. Histol. Jpn., *46*:589, 1983.

Loewenstein, W.R.: The cell-to-cell channel. III International Congress of Cell Biology. Edited by S. Seno and Y. Okada. Japan, Academic Press, 1984, p. 78.

Loewenstein, W.R.: Junctional intercellular communication: The cell-to-cell membrane channel. Physiol. Rev., *61*:829, 1981.

Loewenstein, W.R.: Junctional intercellular communication and the control of growth. Biochem. Biophys. Acta, *560*:1, 1979.

Luft, J.H.: The structure and properties of the cell surface coat. Int. Rev. Cytol., *45*:291, 1976.

Nicolson, G.L.: Cancer metastasis. Sci. Am., *240*:66, 1979.

Pastan, I., and Willingham, M.: Cellular transformation and the morphologic phenotype of transformed cells. Nature, *274*:645, 1978.

Pinto da Silva, P., and Kachar, B.: On tight junction structure. Cell, *28*:441, 1982.

Robertson, A.D.J., and Grutsch, J.F.: Aggregation in *Dictyostelium discoideum.* Cell, *24*:603, 1981.

Rose, B.: Cell-to-cell permeability and regulation. III International Congress of Cell Biology. Edited by S. Seno and Y. Okada. Japan, Academic Press, 1984, p. 79.

Staehelin, L.A., and Hull, B.E.: Junctions between living cells. Sci. Am., *238*:140, 1978.

Steinberg, M.S.: Molecular organization and interactions in the desmosome. III International Congress of Cell Biology. Edited by S. Seno and Y. Okada. Japan, Academic Press, 1984, p. 171.

Trowbridge, I.S.: Transferrin receptor and cell growth. III International Congress of Cell Biology. Edited by S. Seno and Y. Okada. Japan, Academic Press, 1984, p. 55.

Unwin, P.N.T.: Channel structure. III International Congress of Cell Biology. Edited by S. Seno and Y. Okada. Japan, Academic Press, 1984, p. 78.

Unwin, P.N.T., and Zampighi, G.: Structure of the junction between communicating cells. Nature, *283*:545, 1980.

Wurster, B.: On induction of cell differentiation by cyclic AMP pulses in *Dictyostelium discoideum.* Biophys. Struct. Mech. 9:137, 1982.

CHAPTER 6

THE CYTOSKELETON AND CELL MOTILITY—MICROTUBULES, MICROFILAMENTS, AND INTERMEDIATE FILAMENTS

In Chapter 1 we mentioned that there were notable similarities between prokaryotic and eukaryotic cells (see Table 1–3) in spite of their differences in structure. The cytosol, the *cytoplasmic matrix* or *ground cytoplasm*, of a eukaryotic cell contains the same components as a bacterium (e.g., ribosomes, RNA molecules, globular proteins, and enzymes). The new components that have evolved in higher cells are the many membranes that constitute the *endomembrane* or *vacuolar system* with its several portions (e.g., nuclear envelope, endoplasmic reticulum (ER), and Golgi apparatus) and the membrane-bound organelles (e.g., mitochondria, chloroplasts, lysosomes, peroxisomes, endosomes, and vacuoles). As a consequence of the endomembrane system, numerous compartments and subcompartments are formed in the cell.

In spite of the compartmentalization of the cell there is one continuum in the cytoplasm, the cytosol, which fills all the spaces of the cell and constitutes its true *internal milieu.* As will be studied in this Chapter, many fundamental functions of the cell take place in this part of the cytoplasm, which is particularly rich in differentiating cells (Fig. 6–1). The colloidal properties of the cell, such as those essential to sol-gel transformations, viscosity changes, intracellular motion (cyclosis), ameboid movement, spindle formation, and cell cleavage, depend, for the most part, on the cytoplasmic matrix.

In this chapter we show that within the cytoplasmic matrix there is a kind of cytoskeleton formed by microtubules, microfilaments, and intermediate filaments. Microtubules are dispersed throughout the cytoplasm, but are also found as part of organelles, such as cilia, flagella, and centrioles, that play a role in some types of cell motions.

Microfilaments represent the true contractile machinery of the cell cytoplasm, which involves the intervention of numerous proteins; some of these proteins are directly related to contraction while others play a regulatory or structural role. We show that cytoplasmic streaming in plant cells, and ameboid motion in Protozoa and animal cells are mediated by microfilaments, which frequently interact with the cell membrane and surface receptors. The importance of cell adhesion and of the extracellular proteins, fibronectin and collagen, in cell movement will be emphasized. A special section will be dedicated to the intermediate filaments and their role as mechanical integrators of cellular compartments.

6–1 CYTOSOL, ERGASTOPLASM, AND CYTOSKELETON

After the nuclear, mitochondrial, and microsomal fractions have been separated by cell fractionation (see Section 3–4), the remaining supernatant, or *soluble fraction* or *cytosol* (see Fig.

3–5), contains the soluble proteins and enzymes found in the cytoplasmic matrix. These proteins and enzymes constitute 20 to 25% of the total protein content of the cell. Among the important *soluble enzymes* present in the matrix are those involved in glycolysis and in the activation of amino acids for protein synthesis. The enzymes of many reactions that require ATP are found in the soluble fraction. Transfer RNA and all the machinery for protein synthesis, including the ribosomes, are also found in this part of the cytoplasm.

While it was initially thought that the ground cytoplasm was essentially amorphous, at the end of the 19th century it was discovered that portions of the cytoplasm in certain cells have a differential staining property. Because these areas stained with basic dyes, they were called *basophilic* or *chromidial cytoplasm*. The still common name *ergastoplasm* (Gr., *ergazomai*, to elaborate and transform) was coined by Garnier in 1887 to imply that biosynthesis is the fundamental role of this substance.

The ergastoplasm includes basophilic regions of the ground cytoplasm, such as the Nissl bodies of the nerve cells, the basal cytoplasm of serous cells (e.g., secretory cells of the pancreas and the parotid gland and chief cells of the stomach), and the basophilic clumps of liver cells. Caspersson, Brachet, and others demonstrated that the intense basophilic property of the ergastoplasm is due to the presence of RNA (Fig. 6–2).

It was shown that the ergastoplasm loses its staining properties if the cell is treated with ribonuclease, an enzyme that hydrolyzes RNA (Fig. 6–2). RNA is contained principally in the ribosomes, and as a result, a relationship between the ergastoplasm and protein synthesis was postulated.

The name *cytoskeleton* was coined many years ago, but it was abandoned because many of the structures observed in the cytoplasm with the light microscope were considered to be fixation artifacts. The existence of an organized fibrous array in the structure of the protoplasm was postulated in 1928 by Koltzoff. He concluded that "each cell is a system of liquid components and rigid skeletons, which generate the shape, and even though we rarely see the skeletal fibrils in living and fixed cells, that only means that these fibrils are very thin or that they are not distinguished by their refractive index from the sur-

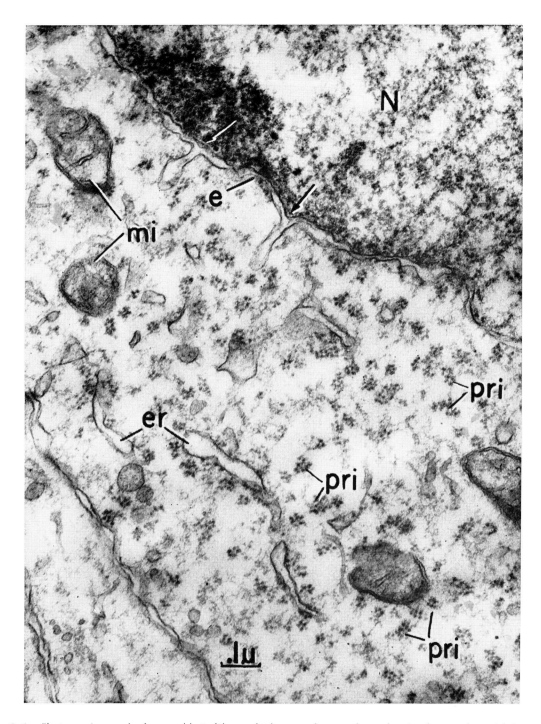

Fig. 6–1.   Electron micrograph of a neuroblast of the cerebral cortex of a rat embryo, showing the cytoplasm rich in matrix with numerous ribosomes and little development of the vacuolar system. *e,* nuclear envelope sending projections into the cytoplasm (arrows); *er,* endoplasmic reticulum; *mi,* mitochondria; *N,* nucleus; *pri,* polyribosomes (groups of ribosomes). ×45,000. (From E. De Robertis, unpublished.)

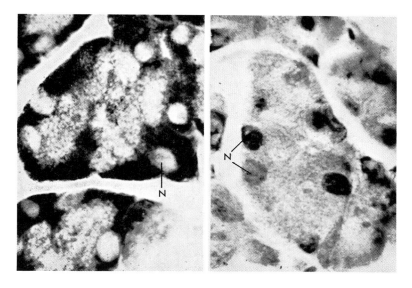

Fig. 6–2. *Left,* pancreatic acini frozen, dried, and stained with toluidine blue. The basophilic substance appears intensely stained. *Right,* the same tissue, but after digestion with ribonuclease; the basophilic substance has disappeared. *N,* nucleus. (From E. De Robertis.)

rounding colloidal solution." He thus conceived of a cytoskeleton that determines both the shape of the cell and the changes in its form.

Confirming this assumption, the electron microscope (EM) has revealed that the cytoplasm of most eukaryotic cells contains a cytoskeletal fabric formed of *microtubules, microfilaments,* and *intermediate filaments.*

Great progress has been made in the isolation of proteins that constitute these cytoskeletal structures, such as *tubulin, actin, myosin, tropomyosin,* and related molecules, and in the mechanism of their assembly (Table 6–1). In addition, by the production of specific antibodies against these proteins, it has been possible to examine under the light and the electron microscope the disposition of the microtubules and the microfilaments. The use of high-voltage electron microscopy on whole cells has also helped to demonstrate that there is a highly structured, three-dimensional lattice in the ground cytoplasm (see Fig. 1–14).[1]

A fundamental line of thought in the study of *cell motility* is that its various forms result from interaction among the several kinds of microfilaments and microtubules embedded in the ground cytoplasm. The most elaborated system of cell motility is represented by the *myofibril* of skeletal muscle, in which there is a highly differentiated macromolecular *mechanism for*

contractility. It is now evident, however, that nonmuscle cells use mechanisms essentially similar to those of muscle. While many of the concepts about molecular structure, including the biochemistry of the contractile proteins, apply to the material presented in this chapter, they will be treated in detail in Chapter 7.

SUMMARY:
The Cytoskeleton

Within the cell cytoplasm the *cytosol* or *cytoplasmic matrix* represents a continuum that constitutes the true internal milieu of the cell. In the cytosol many functions related to sol-gel transformations take place (i.e., cyclosis, ameboid movement, spindle formation, and cell cleavage). In the cytosol are contained the glycolytic enzymes and the whole machinery for protein synthesis. Basophilic regions of the cytosol, which are rich in RNA (ribosomes), were long ago named the *ergastoplasm.*

The name *cytoskeleton* can be applied to the fabric of microtubules, microfilaments, and intermediate filaments that pervade the cytosol and which are related to the various forms of cell motility. The main proteins that are present in the cytoskeleton are *tubulin* (in the microtubules), actin, myosin, tropomyosin, and others (in the microfilaments) which are also in muscle. Thus, the same proteins are involved in the contraction of nonmuscle cells as in that of muscle cells.

TABLE 6–1. CLASSIFICATION OF SOME ACTIN-BINDING PROTEINS

	Protein	Source	Mol Wt. × 10⁻³	Ca⁺ Sensitivity
I	*Cross-linking proteins*			
	Actin-binding protein	macrophage	2×270	No
	Filamin	fibroblast smooth muscle	2×250	No
	Spectrin	erythrocyte	2×240	Yes
		nonerythroid cells	2×260	
	α-Actinin	muscle HeLa cells	2×100	Yes
	Vinculin	smooth muscle HeLa cell	130	No
	Fimbrin	microvillus	68	No
	Villin	microvillus	95	Yes
II	*Severing and capping proteins*			
	β-actinin	muscle	$34 + 37$	No
	Gelsolin	macrophage	91	Yes
	Acumetin	macrophage	65	No
	Villin	microvillus	95	Yes
	Fragmin	*Physarum*	43	Yes
	Capping protein	*Acanthamoeba*	$29 + 31$	No
III	*Depolymerizing proteins*			
	Profilin	Mammalian tissues	$12 - 15$	No
	Brain depolymerizing factor	Brain	19	No

Data from Craig, S.W., and Pollard, T.D.: Actin-binding proteins. Trends Biochem. Sci., 7:88, 1982.

6–2 MICROTUBULES

Microtubules are structures universally present in the cytoplasm of eukaryotic cells and are characterized by their tubular appearance and their uniform properties in the different cell types. Most cytoplasmic microtubules are rather labile and do not resist the effects of fixatives such as osmium tetroxide; because of this, intensive studies began only after 1963, when glutaraldehyde fixation was introduced in electron microscopy.

The first observation of these tubular structures, in the axoplasm extruded from myelinated fibers, was made by De Robertis and Franchi in 1953 (see Fig. 24–2).[2] Here the so-called *neurotubules* appeared as elongated, unbranched, cylindrical elements 20 to 30 nm in diameter and of indefinite length. Microtubules were observed in a variety of animal cells studied in sections (Fig. 6–3).

Cytoplasmic microtubules are uniform in size and are remarkably straight. They are about 25 nm in outer diameter and several micrometers in length. In cross section they show an annular configuration with a dense wall about 6 nm thick and a light center. Each microtubule is surrounded by a zone of low electron density from which ribosomes or other particles are absent.

The wall of the microtubule consists of individual linear or spiraling filamentous structures about 5 nm in diameter, which, in turn, are composed of subunits. In cross section there are about 13 subunits with a center-to-center spacing of 4.5 nm (Fig. 6–4). Application of negative staining techniques has shown that microtubules have a lumen and a subunit structure in the wall. Occasionally, dense dots or rods have been detected in the center portion of some microtubules (see Fig. 24–3).

Although all the microtubules studied show approximately the same morphological characteristics, it is evident that they differ in other properties. For example, microtubules of cilia and flagella are much more resistant to various treatments. The microtubules forming the spindle fibers and the others present in the cytoplasm are, in general, labile and transitory structures. Cytoplasmic microtubules usually disappear if stored at 0° C or after treatment with colchicine.

Tubulin—The Main Protein of Microtubules

Microtubules are composed of protein subunits that are rather similar, even though they

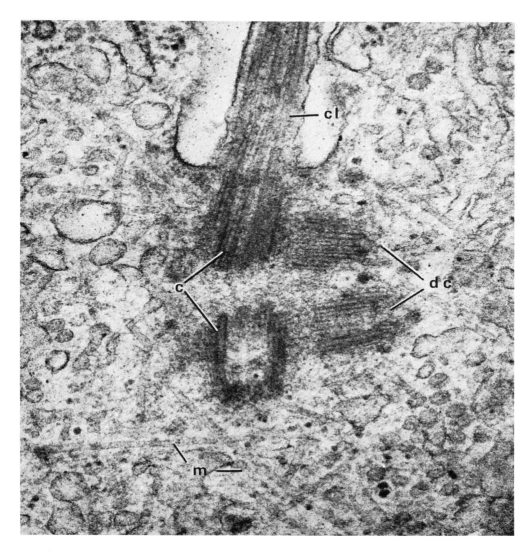

Fig. 6–3. Electron micrograph of the pancreas of a chick embryo showing cytoplasmic microtubules, and the replication of centrioles; *c*, the two centrioles; *dc*, daughter centrioles; *cl*, cilium; *m*, microtubules. ×50,000. (Courtesy of J. Andrè.)

are found in a variety of cell types. The term *tubulin*, used for the principal protein of cilia and flagella, is also used for the protein of cytoplasmic microtubules. Tubulin is a dimer of 110,000 to 120,000 daltons.

Two different monomers—tubulin A and B—are present in flagella. In most cases, tubulin is a heterodimer having two monomers of different kinds although they are quite similar in molecular weight (55,000 daltons).

The 8-nm spacing along the longitudinal axis of microtubules that is observed by electron microscopy probably reflects the pairing of the two types of tubulin monomers (Fig. 6–4). One

dimer of tubulin binds to a molecule of ^{3}H-colchicine, and this specific property is used for assay of this protein. Tubulin also binds to the Vinca alkaloid *vinblastine*, but at a site other than for colchicine. Vinblastine tends to produce crystal-like structures of tubulin in the cytoplasm, and in homogenates it produces precipitation of this protein, thus allowing rapid purification. The amino acid composition of tubulin from different sources shows little variation, although some differences are found between the monomers. Some enzymatic activities have been reported for tubulin (e.g., protein kinase activity). Some studies indicate that tubulin, in ad-

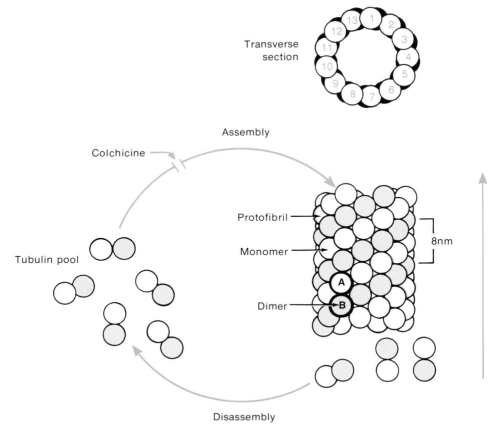

Fig. 6–4. Diagram of a microtubule and the process of assembly and disassembly of tubulin. *Top,* transverse section of a microtubule showing 13 protofibrils. *Bottom,* longitudinal view showing the helicoidal protofibrils made of tubulin dimers and the polarization of the microtubule. Observe that the microtubule is being disassembled at the bottom while being simultaneously assembled at the top. Colchicine, by blocking the assembly process, produces depolymerization of the microtubules.

dition to being present as a pool of free dimers and microtubules, may form an integral part of some membranes.

Microtubules—Assembly from Tubulin Dimers and Polarity

The assembly of microtubules from the tubulin dimers is a specifically oriented and programmed process. In the cell there are sites of orientation, i.e., centrioles, basal bodies of cilia, from which the polymerization is directed. These are the so-called *microtubule organization centers.* The quantity of polymerized tubulin is high at interphase (cytoplasmic microtubules) and metaphase (spindle microtubules), but low at prophase and anaphase.

Within the cell, microtubules are in equilibrium with free tubulin. Phosphorylation of the tubulin monomers by a cyclic AMP-dependent kinase favors the polymerization. In cultured epithelial cells cyclic AMP (cAMP) promotes the formation of microtubules, and the cell becomes elongated, with a fibroblastic appearance. Thus there is a relationship between cell shape, the number and direction of microtubules, and cAMP (see Section 6–5). The assembly and disassembly of tubulin constitute a polarized phenomenon. In a microtubule, the assembly of tubulin dimers takes place at one end while disassembly is prevalent at the other end (Fig. 6–4). If a cell is treated with colchicine, the assembly is inhibited, while the disassembly continues, leading to the disorganization of the microtubule. The assembly is accompanied by the hydrolysis of guanosine triphosphate (GTP) to guanosine diphosphate (GDP) and the lack of GTP stops the assembly.[3,4]

Microtubules can be assembled in vitro, but the presence of GTP and high concentrations of tubulin, as well as some associated proteins, is necessary. The polymerization is both temperature and pressure sensitive.[5]

When microtubules are disassembled using a cold technique, a mixture in the form of *rings* is produced. These rings are composed of tubulin dimers and associated proteins. The role of these rings in microtubule assembly is questionable because they are not observed in vivo and appear only in vitro.

The assembly-disassembly process is a dynamic equilibrium, as can be demonstrated by the use of ^{3}H-GTP as a marker for the addition or loss of tubulin dimers. Experimentally, using brain microtubules isolated in vitro, and a short pulse of ^{3}H-GTP, followed by cold GTP (chase), it is possible to demonstrate that the label is incorporated at the assembly end,- migrates along the microtubule, and is lost from the disassembly end. Thus, a unidirectional flux of tubulin from on end of the microtubule to the other, a process known as "treadmilling" occurs (see Fig. 6–4; see also Margolis and Wilson, 1981 in the Additional Readings at the end of this chapter). This flow of tubulin along the microtubule could be related to the function of these organelles. It is possible that they act as a "conveyor belt" for the intracellular translocation of cell components attached to their surface; however, this mechanism remains to be demonstrated in vivo.

In Section 15–2, when studying the mitotic spindle, we emphasize the problem of microtubular polarity. Microtubular polarity can be studied by adding tubulin, extracted from the brain, to lysed cells. Under the excess of tubulin, there is elongation of the microtubules at the growing end, where curved ribbons of protofilaments can be observed under the EM (see McIntosh, 1981 in the Additional Readings at the end of this Chapter).

The in vivo control of assembly and disassembly of tubulin involves Ca^{++} and the calcium-binding protein, *calmodulin*.[6] The addition of Ca^{++} inhibits polymerization to tubulin; this effect is enhanced by the addition of calmodulin (see Section 15–2).

Microtubular Associated Proteins. It was found that microtubules assemble at 37° C and disassemble on cooling to 5° C. However, the microtubules purified by several cycles of assembly-disassembly do not consist of tubulin only but contain about 15 to 20% other proteins, which have been called, generically, *microtubular associated proteins* (MAPs). A number of MAPs ranging in size between 300,000 and 55,000 daltons have been isolated. Although the particular role of each of the proteins is unclear, it is evident that MAPs are involved in microtubular assembly. These accessory factors appear to be involved stoichiometrically rather than catalytically in stimulating tubulin assembly. The most important MAPs are *tau* and the high molecular weight fractions MAP 1 and 2.

The small molecular weight MAPs are composed of a family of four proteins, collectively termed *tau*. A tau factor is a *calmodulin-binding protein* that is able to form a complex with calmodulin in the presence of small concentrations of Ca^{++}. This complex is reversible and contributes to the regulation of the assembly and disassembly of tubulin in the microtubule (see Kakiuchi and Sobue, 1983 in the Additional Readings at the end of this chapter).

Detection of Microtubules by Antibodies

The isolation of tubulin has permitted specific antibodies against this protein to be produced. These antibodies may be used as *immunofluorescent probes* for localizing microtubules in the cytoplasm of a wide variety of cultured cells.[7] Similar studies with antibodies against microfilament proteins (anti-actin antibodies and others) have permitted differentiation of the microtubules from the microfilaments in the same cell type (Fig. 6–5).

In a cultured cell, cytoplasmic microtubules often extend radially from the nucleus and appear as straight or curved filaments that seem to terminate near the cell surface. These filaments disappear when the cells are treated with *Colcemid*, a derivative of colchicine, or when cooled, and they reappear if the conditions are reversed. Such studies reveal that microtubules arise near the nucleus, from one or two focal points corresponding to the centrosomal region (i.e., the centrosphere).

In electron micrographs microtubules are often observed irradiating from the region containing the centrioles (see Fig. 6–3). In cells entering mitosis the cytoplasmic microtubules

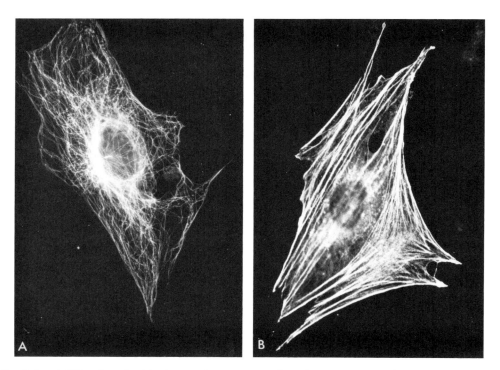

Fig. 6–5. Cultured 3T3 cells stained by immunofluorescence with antibodies against tubulin (A) and actin (B). ×600. (Courtesy of M. Osborn and K. Weber.)

disappear and are replaced by those integrating the spindle and asters.

The localization of microtubules in different cell types has also been carried out using antibodies against some of the associated proteins such as tau and the high molecular weight (HMW) fractions MAP 1 and 2. The anti-HMW, with the exception of cilia, labeled the microtubules of all mammalian cells. On the other hand, the anti-tau did not stain some cell types such as rat glial and pig kidney cells, suggesting that the tau protein could be absent in these cells.[8]

Functions of Cytoplasmic Microtubules

A rather long list of functions has been attributed to microtubules.

Mechanical Function. The shape of some cell processes or protuberances has been correlated to the orientation and distribution of microtubules. They are considered to be a framework that determines the shaping of the cell and redistributes its content. This is particularly evident in axons and dendrites of neurons.

Morphogenesis. Related to their mechanical function is the role that microtubules play in the shaping of the cell during *cell differentiation.* For example, the elongation of the cells during the induction of the lens placode in the eye is accompanied by the appearance of numerous microtubules. The morphogenetic changes that occur during spermiogenesis provide another interesting example. The enormous elongation that takes place in the nucleus of the spermatid is accompanied by the production of an orderly array of microtubules that are wrapped around the nucleus in a double helical arrangement.

Cellular Polarity and Motility. The determination of the intrinsic polarity of certain cells is also related to the microtubules. Treatment of various culture cells with Colcemid results in a change in motion. The following forms of movement have been observed to persist: (1) membrane ruffling, (2) endocytosis, (3) attachment to the surface, and (4) extension of microvilli. Only a saltatory movement of particles was found to be inhibited after the destruction of the microtubules. The directional gliding of the cell, however, is replaced by a random movement.

Circulation and Transport. Microtubules may also be involved in the *transport* of macromolecules in the cell's interior. A very interesting example is the protozoan *Actinosphaerium* (Heliozoa), which sends out long, thin pseudopodia within which cytoplasmic particles migrate back and forth. These pseudopodia contain as many as 500 microtubules disposed in a helical configuration. When these protozoans are exposed to cold, or high pressure, the pseudopodia are withdrawn and the microtubules depolymerize; however, the movement of granules continues.

Another example of the association between microtubules and transport of particulate material can be found in the melanocyte, in which the melanin granules move centrifugally and centripetally with different stimuli. The granules have been observed moving between channels created by the microtubules in the cytoplasmic matrix.

In the erythrophores found in fish scales the pigment granules may move at a speed of 25 to 30 μm per second between the microtubules. The role of microtubules in axoplasmic transport is discussed in Section 24–1.

SUMMARY:
Properties of Microtubules

Microtubules are found in all eukaryotic cells—either free in the cytoplasm or forming part of centrioles, cilia, and flagella. They are tubules 25 nm in diameter, several micrometers long, and with a wall 6 nm thick with 13 subunits. The stability of different microtubules varies. Cytoplasmic and spindle microtubules are rather labile, whereas those of cilia and flagella are more resistant to various treatments. The main component is a protein called *tubulin*. This heterodimer of 110,000 to 120,000 daltons, is formed of two different monomers (tubulin A and B) of the same molecular weight (i.e., 55,000). The monomers are 4 nm × 6 nm and correspond to the subunit lattice structure seen in the tubular wall. Tubulin binds colchicine and vinblastin at different binding sites.

The assembly of tubulin in the formation of microtubules is a specifically oriented and programmed process. Centrioles, basal bodies, and centromeres are sites of orientation for this assembly. Calcium and calmodulin may be regulating factors in the in vivo polymerization of tubulin. The level of polymerized tubulin is high at interphase and metaphase (i.e., spindle microtubules) and lower at prophase and anaphase.

Assembly involves the hydrolysis of GTP to GDP. The polarity of the assembly can be demonstrated, under the EM, by the addition of tubulin.

Specific anti-tubulin antibodies have permitted the localization of microtubules in cultured cells and study of the changes that occur during mitosis and after treatment with colchicine. Cell transformation (i.e., cancer) produces a disorganization of microtubules.

Several functions, some of which are related to the primitive forms of cell motility described in this chapter, have been attributed to microtubules. They play a *mechanical* function and the *shape* of the cell and cell processes is dependent on microtubules. This function is particularly apparent during *cell differentiation* of certain placodes, in the nerve cells, and in spermiogenesis. The *polarity* and directional gliding of cultured cells depend on microtubules. These structures are associated with *transport* of molecules, granules, and vesicles within the cell. They play a role in the contraction of the spindle and movement of chromosomes and centrioles, as well as in ciliary and flagellar *motion*. A possible role in *sensory transduction* has been postulated.

6–3   MICROTUBULAR ORGANELLES—
CILIA, FLAGELLA, AND CENTRIOLES

Several cell organelles are derived from special assemblies of microtubules. Cilia, flagella, basal bodies, and centrioles have, as one of their main components, groups of microtubules arranged in a special fashion. For this reason, and in spite of the complexity of their structure, these organelles will be considered at this point, along with some of their physiological aspects.

Ciliary and Flagellar Motions in Cells and in Tissues

Ciliary and flagellar cell motility is adapted to liquid media and is executed by minute, specially differentiated appendices that vary in size and number. They are called *flagella* if they are few and long, and *cilia*, if short and numerous. In Protozoa, especially the Infusoria, each cell has hundreds of thousands of minute cilia.

In some special regions of the Infusoria, several cilia fuse and form larger conical appendices, *the cirri*, or membranes known as *undulating membranes*.

One entire class of Protozoa, the Flagellata, is characterized by the presence of flagella. The spermatozoa of metazoans move as isolated cells by means of flagella. On the other hand, epithelial cells that possess vibratile cilia and constitute true ciliated sheets are relatively common. These may cover large areas of the external surface of the body and determine the motion of the animal. Such is the case with some Platyhelminthes and Nemertea and also with larvae of Echinodermata, Mollusca, and Annelida. More often, the ciliated epithelial sheets line cavities or internal tubes, such as the air passages of the respiratory system or various parts of the genital tract. In these organs all the cilia move simultaneously in the same direction, and fluid currents are thus produced. In some cases, the currents serve to eliminate solid particles in suspension (e.g., in the respiratory system). The eggs of amphibians and mammals are driven along the oviduct with the aid of vibratile cilia.

The Ciliary Apparatus—The Cilium, Basal Bodies, and Ciliary Rootlets

The essential components of the ciliary apparatus are: (1) the *cilium*, which is the slender cylindroid process that projects from the free surface of the cell, (2) the *basal body*, or granule, the intracellular organelle similar to the centriole from which it originates, and (3) in some cells fine fibrils—called *ciliary rootlets*—that arise from the basal granule and converge into a conical bundle, the pointed extremity of which ends at one side of the nucleus.

Various epithelia have appendices similar in shape to cilia, but immobile; these are called *stereocilia*. Examples are the processes of the epithelial cells of the epididymis. In the macula and crista of the inner ear, there are stereocilia in addition to motile cilia, or *kinocilia*.

Stereocilia of the hair cells of the inner ear are responsible for the transduction of sound (i.e., the motion of stereocilia caused by sound produces changes in membrane potential of the hair cells). These and other stereocilia do not contain microtubules; they do have about 3000 actin filaments, which are disposed longitudinally but have a definite polarity and a helical symmetry, with crossbridges around the filaments.[9]

The Axoneme Contains Microtubular Doublets

Axoneme is the term applied to the axial basic microtubular structure of cilia and flagella, and is the essential motile element. The axoneme may range from a few microns to 1 to 2 mm in length, but the outside diameter is about 0.2 μm. It is generally surrounded by an *outer ciliary membrane*, which is continuous with the plasma membrane. All the components of the axoneme are embedded within the *ciliary matrix* (Fig. 6–6).

Figure 6–7 shows a general diagram of the axoneme, with the fundamental 9 + 2 microtubular pattern. A plane perpendicular to the line joining the two central tubules divides the axoneme into a right and a left symmetrical half. It is generally admitted that the plane of the ciliary beat is perpendicular to this plane of symmetry. The two microtubules of the peripheral pairs are skewed so that one tubule, designated subfiber A lies closer to the axis than the other (i.e., subfiber B). The microtubule of subfiber A is smaller but complete, whereas that of subfiber B is larger and incomplete, since it lacks the wall adjacent to A. In fact, while A has 13 tubulin subunits, B has only 11.

Dynein Arms. Subfiber A has processes—the so-called *dynein arms*—that are oriented in the same direction in all microtubules. This orientation is clockwise when the axoneme is viewed from base to tip.

The use of quickly frozen deep-etch replicas of axonemes (see Section 3–2) has led to the recognition of finer details of structure in the outer dynein arm. These consist of an elliptical head, two spherical feet contacting the A microtubule, a stalk bridging the head with subfiber B of the next doublet, and an interdynein link.[10]

The arms contain dynein, a high molecular weight ATPase (300,000 to 400,000 daltons). Dynein is a Mg^{++} and Ca^{++}-activated enzyme, which after solubilization can recombine at the same position on the A microtubules. Two isoenzymatic forms of the ATPase; dynein I and II, have been separated by electrophoresis.[11]

Nexin Links. As shown in Figure 6–7 the doublets are linked by *interdoublet* or *nexin links;* another protein called *nexin* has been isolated from these links. This protein has a molecular weight of 150,000 to 160,000 daltons.[12] The function of these links is unknown, but they may serve as structures that maintain the integrity of the axoneme during the sliding motion.

Radial Spokes. There are radial bridges or links between the A subfiber and the sheath

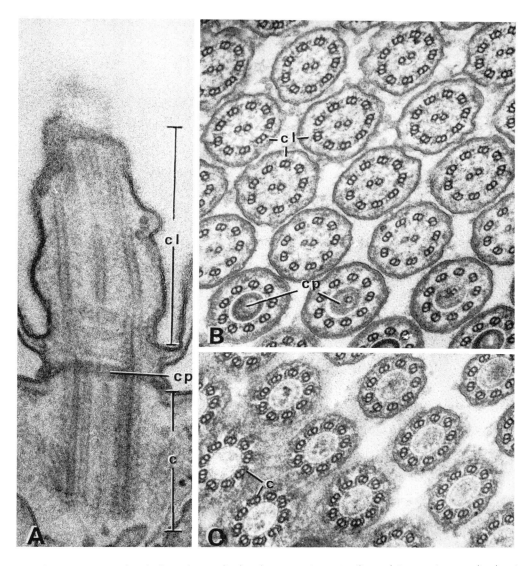

Fig. 6–6. Electron micrographs of cilia in longitudinal and cross sections. *A*, cilium of *Paramecium aurelia* showing the centriole or basal body *(c)*, the ciliary plate *(cp)*, and the cilium *(cl)* proper. *B* and *C*, cross sections through cilia of *Euplotes eurystomes. B*, the section passes through the cilium proper showing the typical structure and the ciliary plate *(cp)*. *C*, the section passes through the centriole or basal body *(c)*. Notice the absence of central tubules and the triple number of peripheral tubules. *A*, ×110,000; *B* and *C*, ×72,000. (Courtesy of J. Andrè and E. Fauret-Fremiet.)

containing the central microtubules. These spokes terminate in a dense knob or head, which may have a fork-like structure. Certain observations that the spokes are attached perpendicularly to the ciliary axis where it is straight and that they are relatively detached in bent or tilted regions of the axis has led to the hypothesis that they may be active in the conversion of active sliding between the outer doublets into local axial bending.[13]

Basal Bodies (Kinetosomes) and Centrioles Contain Microtubular Triplets

Since the classic works of Henneguy and Lenhossek in 1897, it has been suggested that basal bodies (or kinetosomes) of cilia and flagella are homologous with the centrioles found in mitotic spindles (see Fig. 15–1). In some cells a centriole engaged in mitosis could carry a cilium at the same time. Centrioles are cylinders that measure, on the average, $0.2\ \mu m \times 0.5\ \mu m$; at times

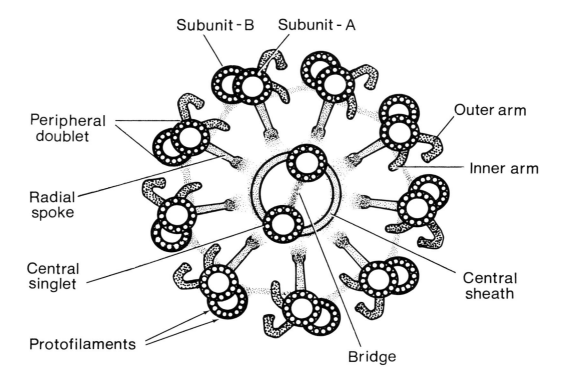

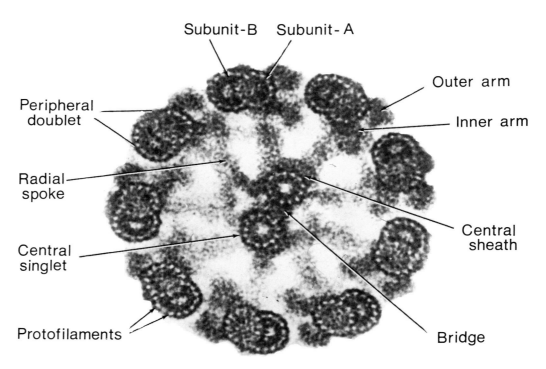

Fig. 6–7. *Above,* diagram of a cross section of the axoneme of a cilium showing the 9 + 2 microtubular structure. The view is from base to tip, with the dynein arms directed clockwise. Observe the tubulin subunits or protofilaments. The nexin links between the doublets are seldom visualized. *Below,* electron micrograph of a ciliary axoneme stained with tannic acid. (From Fawcett, D.: The Cell. 2nd Ed. Philadelphia, W.B. Saunders Co., 1981.)

they may be as long as 2 μm. This cylinder is open on both ends, unless it carries a cilium. In the latter case, it is separated from the cilium by a *ciliary plate* (Fig. 6–6A).

In freeze-fractured cilia it has been possible to observe the so-called *ciliary necklace*. This consists of two to six rows of membrane particles present just below the ciliary plate where the central pair of microtubules ends.

The wall of the centriole has nine groups of microtubules arranged in a circle. Each group is a *triplet* formed of three tubules (rather than two, as in cilia) that are skewed toward the center (Fig. 6–8). Each unit formed by the triplet has also been called *a blade*. Within each blade the tubules twist from one end to the other or describe a helical course. As shown in Figure 6–8 the tubules are designated A, B, and C from the center toward the periphery. Both A and B cross the ciliary plate and are continuous with the corresponding tubules in the axonemene.

In the formation of the axoneme of a cilium or flagellum, the centriole serves a template function and the polymerization of tubulin occurs at the distal end of tubules A and B, while the termination of tubule C occurs near the *ciliary* or *basal plate*. In this way the centriolar triplet is converted into the ciliary doublet.

There are no central microtubules in the centrioles and no special arms; however, they are linked by connectives. The proximal portion of the centriole has a cartwheel appearance, which provides the centriole with a structural and functional polarity. The growth of the centriole is from the distal end, and in the case of kinetosomes, it is from this end that the cilium is formed. Furthermore, the *procentrioles*, which are formed at right angles to the centriole, are located near the proximal end.

Ciliary Rootlets. In some cells, ciliary rootlets originate from the basal body. Most rootlets are striated, having a regular cross-banding with a repeating period of 55 to 70 nm.

The striated fibers are composed of parallel microfilaments; 3 to 7 nm in diameter, which in turn are formed of globular subunits. These fibers and filaments may serve a structural role such as anchoring the kinetosomes. By analogy with other microfilaments, a contractile role has been postulated for the ciliary rootlets.

Basal Feet and Satellites. Basal feet are dense processes that are arranged perpendicularly to the basal body. These processes impose a structural asymmetry on the basal body that has been related to the direction of the ciliary beat. The basal foot, which is composed of microfilaments that terminate in a dense bar, may be a focal point for the convergence of microtubules. *Satellites* or *pericentriolar bodies* are electron-dense structures lying near the centriole that are probably nucleating sites for microtubules.

Function of Centrioles and Basal Bodies. Besides a possible role in the formation of cilia, it has been suggested that these organelles are involved in ciliary and flagellar beat; mitosis, the organization of cytoplasmic microtubules, and the reception of optical, acoustic, and olfactory signals. In addition, it is interesting to mention that higher plant cells (which do not migrate) lack centrioles, while these organelles are present in lower migrating plant cells.

Recently, it has been suggested that centrioles could serve as devices for locating the directions of signal sources. This hypothetical function was conceived by comparing the geometric design of centrioles (with their disposition in pairs at the right angle and the ninefold symmetry) with human made devices, such as radar scanners, that detect directional signals (for more details on this interesting hypothesis, see Albrecht-Büehler, 1981 in the Additional Readings at the end of this chapter).

Ciliary Movement

Ciliary movement can be analyzed easily by scraping the pharyngeal epithelium of a frog or toad with a spatula and placing the scrapings in a drop of physiological salt solution between a slide and a coverglass. On the free surface of the epithelial cell, the rapid motion of the vibratile cilia can be seen. If a row of cilia is observed, the contraction is *metachronic* in the plane of the direction of motion; that is, it starts before or after the contraction of the next cilium. In this way true waves of contraction are formed.

On the other hand, in a plane perpendicular to the direction of motion, the contraction is *isochronic;* all the cilia are observed in the same phase of contraction at a given time. The rhythmical contraction of cilia has been interpreted in different ways. A two-step process that involves *intraciliary* excitation followed by *interciliary* conduction has been proposed to ex-

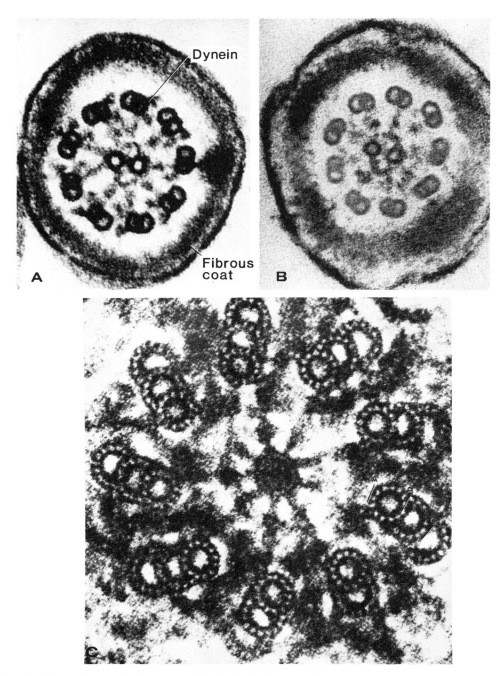

Fig. 6–8. *A,* electron micrograph of a cross section of the tail of a normal human spermatozoon. Note the attachment of dynein arms to microtubule A and the presence of a fibrous coat around the axoneme. *B,* similar cross section but from an individual with immotile ciliary syndrome. Note the lack of dynein arms. (Courtesy of B.A. Afzelius). *C,* cross section of a centriole stained with tannic acid showing the triplets with the tubulin protofilaments in negative staining. Observe that tubule A has 13 protofilaments while B and C have only 10. (Micrograph taken by Dr. Vitaus Kalnins from Fawcett, D.: The Cell. 2nd Ed. Philadelphia, W.B. Saunders Co., 1981.)

plain the metachronic rhythm of the ciliary beat. This mechanism persists even after the epithelium has been separated from the rest of the organism. However, if a cut is made in the row of cilia the waves of contraction of the two isolated pieces becomes uncoordinated.

The direction of the effective ciliary beat also depends on the underlying cytoplasm. If a piece of epithelium is removed from the pharynx of a frog and implanted with a reversed orientation, the movement is maintained but in the opposite direction.[11]

Ciliary contractions are generally rapid (10 to 17 per second in the pharynx of the frog). Analysis of the motion has been facilitated greatly by stroboscopic and ultrarapid microcinematography.

Ciliary movement may be pendulous, unciform (hook-like), infundibuliform, or undulant. In the pendulous movement, typical of the ciliated Protozoa, the cilium is rigid and the motion is carried out by a flexion at its base. On the other hand, in the unciform movement, the most common type in the Metazoa, the cilium is doubled upon contraction and takes the shape of a hook. In the infundibuliform movement, the cilium or flagellum rotates, describing a conical or funnel-shaped figure. In the undulant motion, characteristic of flagella, contraction waves proceed from the site of implantation and pass to the free border.

In Figure 6–7, the plane of beat passes through doublet 1 and between doublets 5 and 6. There are many indications, however, that there is a three-dimensional movement during the beat. Although the effective stroke is in a single plane, the cilium moves away from the original plane during the recovery phase. This movement, which can be beautifully demonstrated by scanning electron microscopy, implies that either the whole cilium rotates at its base or that the central microtubules rotate.

A Sliding of Microtubular Doublets that Involves Dynein

Ciliary and flagellar motion have been compared to muscle contraction. The key process in both appears to be the sliding of the microtubular doublets over each other, together with the associated making and breaking of crossbridges between adjacent doublets. The dis-

placement of the doublets can be easily observed because, as shown in Figure 6–9, the central microtubules extend to the very tip; whereas subfibers B terminate at a distance of 1 to 2 μm from the tip, and subfibers A terminate slightly beyond subfibers B. In Figure 6–9 doublets 1 and 6 are represented in the normal position, as well as during the effective and return strokes.

Observe that in the cross section of the normal position the 9 doublets are visible; whereas in the others, subfiber B is missing in certain doublets situated in the convex side of the cilium (an indication that there is sliding between the two subfibers).

Recent experimental work on ciliary motion has shown notable similarities with the sliding mechanism involved in the interaction of actin and myosin in muscle (see Section 7–2). The dynein arms attached to subfiber A (Fig. 6–7) have been compared with the crossbridges of myosin and it has been postulated that they form intermittent attachments, by which one doublet is able to push the adjacent one toward the tip of the axoneme (see Fig. 6–10). In Figure 6–10 two adjacent doublets (N and N + 1) are represented, together with the attachment of subfiber A of N to subfiber B of N + 1, by way of the dynein arm.

Under normal conditions, this attachment is not observed in an intact cilium. However, if the ciliary membrane is extracted with a detergent, the axoneme enters in a state of *rigor* in which the attachment is then produced (A). Addition of ATP to axonemes in the state of rigor restores motility and causes release of the dynein arm (B). This release is independent of the hydrolysis of ATP.

In this cycle the next step would be the reextension of the dynein arm (C) and its rebinding at an angle, with a new, more proximal, site on subfiber B (D). This step apparently needs the hydrolysis of ATP to ADP + Pi. In the last step, the arm returns to the rigor position and displacement of the doublets results (E). Thus the sliding of the microtubular doublets, which is at the base of ciliary motion, is apparently due to a mechanochemical cycle involving ATP and the ATPase present in the dynein arm (see Satir et al., 1981; Goodenough and Heuser, 1982 in the Additional Readings at the end of this chapter).

Experimentally, in glycerin-extracted cilia, flagella, or spermatozoon tails (or in cilia de-

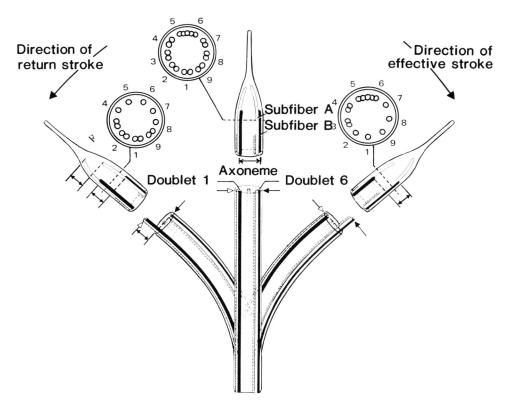

Fig. 6—9.   Diagram representing the sliding hypothesis in ciliary motion. The longitudinal section shows the tip of a cilium with doublets 1 and 6. The cross section shows the 9 doublets of microtubules in the normal position, and cilium bent in the direction of the effective stroke (right) and of the return stroke (left). Note that between subfiber A and B there is a displacement and that the cilium as a whole undergoes a certain rotation. These changes can be recognized by observing the number of microtubules in the cross section and the position of doublets 1 and 6. (Modified from P. Satir, from Fawcett, D.: The Cell. 2nd Ed. Philadelphia, W.B. Saunders Co., 1981.)

membranated by detergents), the addition of ATP restores rhythmical activity. On the other hand, extraction of dynein or the use of anti-dynein anti-serum leads to inhibition of movements.

In sperm flagella it has been observed that the beating is regulated by Ca^{++}, cAMP, and other factors that appear to act on the state of phosphorylation of certain proteins in the axoneme (see Gibbons and Gibbons, 1984 in the Additional Readings at the end of this chapter).

The Immotile Cilia Syndrome

Ciliary motion can be affected by many deficiencies in the protein composition of the organelle. In this respect the flagellar axonemes of *Chlamydomonas* have been a field of study of many mutations that may lead to paralysis of the flagellar function. Analysis of several mutants

has revealed that the defect is frequently in phosphoproteins (i.e., in proteins that can be phosphorylated and dephosphorylated) that act as regulators of flagellar motion (see Luck and Segal, 1984 in the Additional Readings at the end of this chapter).

From a medical point of view, the so-called *immotile cilia syndrome* is of special interest. In this syndrome cilia and flagella are immotile. This immotility leads to infertility (nonmotile sperm), chronic bronchitis, and sinusitis. It has been demonstrated that in most cases this syndrome is due to a mutation involving the dynein arms. In these patients *cilia and sperm are nonmotile because there is a lack or a deficiency of dynein arms, which contain ATPase.*

The immotile cilia syndrome is, however, a genetically heterogeneous disease, even though clinically it shows the same characteristics. In fact in some cases, there may be a deficiency of

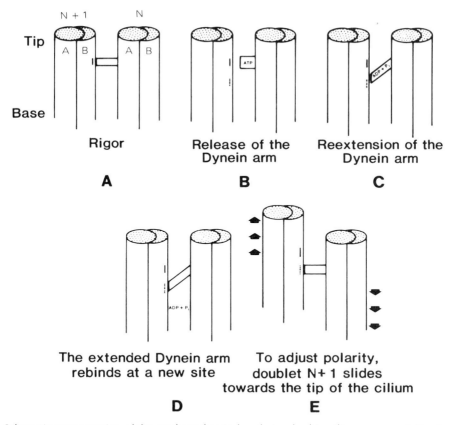

Fig. 6–10. Schematic representation of the mechanochemical cycle involved in ciliary movement. See text for details. (Courtesy of Satir, P., et al.: Cell Motility, *1*:303, 1981.)

the spokes, the nexin links, or of some other ciliary components (see Afzelius, 1985 in the Additional Readings at the end of this chapter).

Photoreceptors are Derived from Cilia

Studies on the structure of retinal rods and cones have shown that the short, fibrous connection found between the outer and inner segments is of a ciliary nature. Cross sections of this so-called connecting cilium have revealed nine pairs of filaments similar to those found in cilia (Fig. 6–11). In this structure, however, there are no central microtubules.

The outer segment of this cilium is composed of numerous double membrane rod sacs arranged like a stack of coins. The first stage in the development of a rod is a primitive cilium projecting from the bulge of cytoplasm that constitutes the primordium of the inner segment (Fig. 6–12). This cilium contains the nine pairs of filaments and the two basal centrioles.[14] The

apical end is filled with a vesicular material. In the second stage the apical region of the primitive cilium enlarges greatly, owing to the rapid building up of the vesicles and cisternae that constitute the primitive rod sacs. The proximal part of the primitive cilium remains undifferentiated and constitutes the connecting cilium of the adult (Fig. 6–12).

Cilia and Flagella Originate from Basal Bodies

The origin of centrioles and basal bodies is viewed from the standpoint that these are possibly semiautonomous cell organelles and in this way similar to mitochondria and chloroplasts. Isolation of kinetosomes from *Tetrahymena* has verified the presence of RNA and DNA, and this suggests that they are capable of some protein synthesis. More direct evidence of DNA in centrioles was obtained by use of the fluorescent dye acridine orange and by ^{3}H-thymidine incorporation followed by treatment with DNAse.

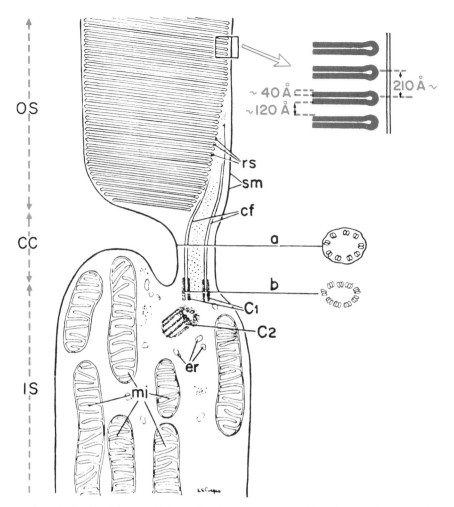

Fig. 6–11.   Diagram of a retinal rod cell in the rabbit showing the outer segment (OS) with the rod sacs (rs), the connecting cilium (CC), and the inner segment (IS). a and b correspond to cross sections through the connecting cilium and the centriole (C_1). C_1 and C_2, centrioles; cf, ciliary filaments; er, endoplasmic reticulum; mi, mitochondria; sm, surface membrane.

In most of these studies, however, the exact localization of DNA has not been determined.

Early cytologists realized that the development of cilia and flagella was directly related to the presence of centrioles. Flagella may be experimentally removed and their regeneration followed; thus, in *Chlamydomonas* it was found that the growth rate of a flagellum is about 0.2 μm per minute.

There are now several indications that centrioles may be generated by two different mechanisms.[15,16] Centrioles destined to form mitotic spindles or single cilia arise directly from the wall of the pre-existing centriole. The *daughter centrioles* appear first as annular structures (*procentrioles*), which lengthen into cylinders (see Fig. 6–3).

The groups of three tubules originate from single and double groups that first appear at the base of the procentriole. When they are half grown, the daughter centrioles are released into the cytoplasm to complete their maturation. (For a discussion on the centriolar cycle in mitosis, see Section 15–1.)

The other mechanism in the formation of centrioles is found in those destined to become kinetosomes, as in a ciliated epithelium. The centrioles are assembled progressively from a precursor fibrogranular material located in the apical cytoplasm. The newly formed centrioles become aligned in rows beneath the apical plasma membrane, and each centriole may then produce satellites from the side, a root from its base, and a cilium from its apex.

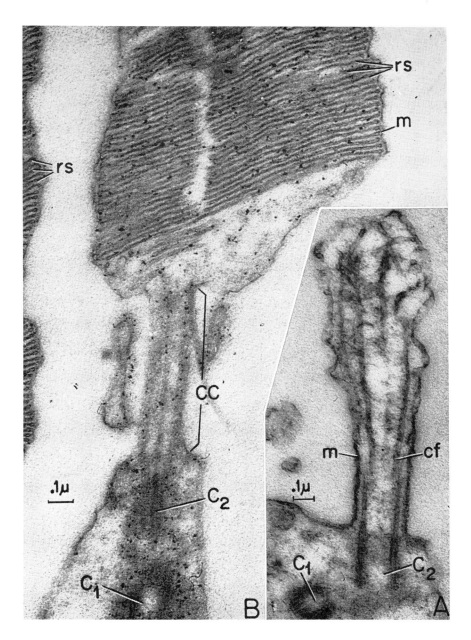

Fig. 6–12. *A,* electron micrograph of a primitive cilium in the retina of an 8-day-old mouse. C_1 and C_2, centrioles; *cf,* ciliary fibrils; *m,* ciliary membrane. ×72,000. *B,* electron micrograph of an adult rod, showing the outer segment with the rod sacs *(rs)* and the outer membrane *(m),* the connecting cilium (CC), and the basal centrioles (C_1 and C_2). ×62,000. (From E. De Robertis and A. Lasansky. *In* Structure of the Eye. New York, Academic Press, 1961.)

Development of the cilium begins with the appearance of a vesicle that becomes attached to the distal end of the centriole. The growing ciliary shaft invaginates the vesicular wall, which forms a temporary ciliary sheath until the permanent sheath is formed.

SUMMARY:

Structure, Motion, and Origin of Cilia and Flagella

Cilia and *flagella* are motile processes found in protozoa and in many animal cells; in plants, only the antherozoids have flagella. *Kinocilia* (i.e., motile cilia) should be differentiated from *stereocilia*, which are nonmotile and lack a microtubular structure. The ciliary apparatus comprises: (1) the *cilium*, (2) the *basal body*, and (3) the *ciliary rootlets*.

The *axoneme*, which is the motile element, is the basic microtubular structure of cilia and flagella. It is surrounded by the *ciliary membrane* and is embedded in the *ciliary matrix*. The fundamental structure is 9 + 2 (i.e., 9 pairs of peripheral microtubules and 2 single and central microtubules). The peripheral doublets can be numbered from 1 to 9, starting from the doublet cut by the plane perpendicular to the line joining the two central tubules. This plane coincides with the plane of the ciliary beat. In the doublet, *subfiber A* and *subfiber B* are distinguished by several morphological features. Subfiber A has two arms that are oriented in the same direction. These so-called *dynein* arms contain dynein, a high molecular weight ATPase. The interaction between tubulin and dynein is thought to be the basic mechanism of contraction in cilia and flagella. *Nexin* is another protein that links the microtubular doublets. *Radial spokes* connect each subfiber A with the central sheath that contains the central microtubules. These radial spokes, in longitudinal sections, are seen as periodically spaced processes.

Basal bodies or *kinetosomes* have the same structure as centrioles. A centriole is a cylinder (0.2 μm × 0.5 μm) open on both ends; the basal body at the distal end has a *ciliary plate* that separates it from the cilium. The centriolar wall contains triplets of microtubules. From the center to the periphery these tubules are designated A, B, and C. Tubules A and B traverse the ciliary plate and are continuous with the corresponding tubules in the cilium, whereas tubule C terminates near the plate. Centrioles lack the central microtubules. Centrioles are polarized structures; at the proximal end they have a cartwheel structure that connects with the triplets. Procentrioles are formed from the proximal end, whereas a cilium or flagellum may be formed from the distal end. Centrioles are generally in pairs and at right angles.

Basal bodies may be related to cross-banded fibers that constitute the *ciliary roots* and which serve a structural role. Other accessory structures are the basal *foot* and *satellites* or *pericentriolar bodies*.

Ciliary movement may be analyzed from a scraping of frog pharyngeal epithelium. The contraction of cilia is *metachronic* in the plane of motion and *isochronic* in a plane perpendicular to the direction of motion. The macromolecular mechanisms of ciliary motion are probably more complex than those taking place in muscle myofilaments. To account for the movement in cilia, a sliding mechanism is favored. Since the central microtubules end at the very tip of the cilium, and the peripheral ones end at a distance, it has been possible to measure the actual sliding of the doublets and to demonstrate that it increases with the angle of bend. Sperm flagella in which the membranes have been removed by detergents are reactivated by ATP. Extraction of dynein stops contraction, and this is restored by the addition of this protein. A mechanochemical cycle involving ATP and the ATPase of the dynein arm has been postulated in the sliding mechanism of ciliary motion. A human syndrome in which cilia and sperm are nonmotile (they lack dynein) is genetically determined.

The outer segments of the retinal rods and cones are connected to the inner segment by a *connecting cilium* that has a 9 + 0 pattern. The photoreceptor develops from a primitive cilium. Other ciliary derivatives are in the crown cells of the *saccus vasculosus* and in the pineal eye of certain lizards.

The *centrioles do not divide*. Those that later form mitotic spindles or single cilia arise directly from the wall of pre-existing centrioles. *Procentrioles* and *daughter centrioles* are formed at right angles. In the case of multiple centrioles, there is a fibrogranular material located in the apical cytoplasm from which the rows of basal bodies are formed. Cilia are formed from the distal end of the centriole.

6–4   MICROFILAMENTS

We mentioned earlier that the ground cytoplasm or cytosol contains a kind of *cytoskeleton* formed by microtubules and various types of microfilaments. From what was said about the function of cytoplasmic microtubules the reader may have gathered the impression that they play mainly a mechanical function, constituting a *passive portion* or *framework of the cytoskeleton*. This concept may not be completely accurate, since these microtubules could store energy by distortion, and during disassembly they could release elastic forces that might transport cytoplasmic structures. Furthermore, we have seen that the interaction between the tubulin of microtubules and dynein is the main mechanism of motion of cilia and flagella.

In recent years the *active* or *motile function of the cytoskeleton* has been attributed to some of the thin microfilaments that are observed by electron microscopy.

Microtrabecular Lattice—Revealed by High-Voltage Electron Microscopy

Because of its greater penetration, high-voltage electron microscopy has helped in identifying microfilaments in whole culture cells and in obtaining a three-dimensional view of the so-called *microtrabecular lattice* (MTL).

As shown in Figure 6–13 just beneath the membrane there are bundles of filaments of actin that are in continuity with a lattice of similar filaments that pervades the cytosol. The channels or spaces found in between this MTL are of the order of 50 to 100 nm and in the living cell may provide for the rapid diffusion of fluids and metabolites through all parts of the cytosol.

The filaments of the lattice are in contact with vesicles of the ER, with microtubules, and with polysomes, all of which seem to be contained or supported within the lattice (Fig. 6–13).

On the other hand, mitochondria appear to be free of the lattice, and this may explain the fact that they move rather freely in the cytosol.

The chromatophores, found in certain fish, are an excellent material to use when studying the structure and function of the cytoskeleton, since they may undergo rapid changes in color.

In these cells the pigment granules are suspended in a three-dimensional MTL. These granules can undergo very rapid movements. For example, they can expand to cover the entire cell or to contract to a central region, within a few seconds. These movements occur radially along pathways limited by microtubules, which irradiate from a complex cell center. The movement of pigment can be correlated with changes in the fine structure of the MTL. With contrac-

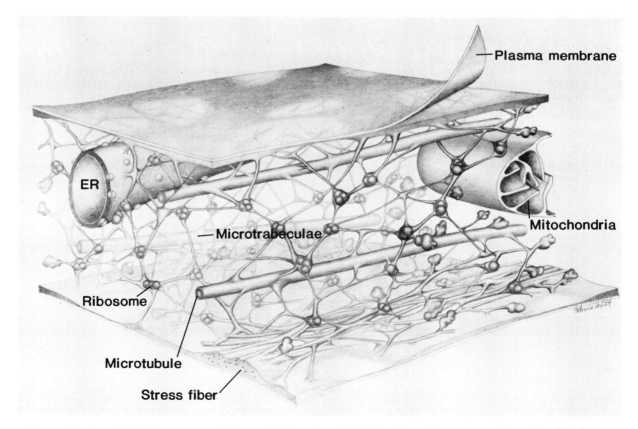

Fig. 6–13. A model of the structure of the cytoskeleton of a cell showing the various components contained in the cytoplasmic matrix. Below the plasma membrane bundles of actin microfilaments form a stress fiber. A microtubular lattice pervades the cytosol and is in contact with the ER, the microtubules, and the ribosomes. ×90,000. (Courtesy of K.R. Porter.)

tion, there is beading of the microtrabeculae. During expansion, it may be observed that each pigment granule returns to a fixed place in the cytoplasm, suggesting that it is embedded in a continuous structure. The movement of pigment granules is controlled by Ca^{++}; for the expansion of the lattice, ATP is required. Chromatophores are under the control of the autonomic nervous system. In tissue culture chromatophores contract with the addition of epinephrine and disperse under the action of cAMP (see Porter, 1984 in the Additional Readings at the end of this chapter).

Cytochalasin B Impairs Several Cellular Activities Involving Microfilaments

While in the study of microtubules the depolymerizing effect of colchicine has been of paramount importance, in the case of microfilaments the action of the drug *cytochalasin B* has had a decisive influence. Cytochalasin B has been found to impair numerous cell activities in which microfilaments are involved. For example, it inhibits smooth muscle contraction, beat of heart cells, migration of cells, cytokinesis, endocytosis, exocytosis, and other processes. Cytochalasin B disrupts the regular arrangement of microfilaments associated with some of these functions and affects the MTL. For example, the contractile ring of microfilaments observed during cell cleavage is altered by this drug (see Fig. 15–11).

Cytochalasin B binds preferentially to F-actin and, by blocking the assembly end of the microfilament, impairs its organization.[17] In addition to its effect on the polymerization of actin, cytochalasin B also inhibits the actin filament interactions that normally contribute to network formation.[18]

Microfilaments Detected by Specific Antibodies—The Cellular Geodome

Microfilaments can be detected by fluorescent antibodies against *actin, myosin, tropomyosin, α-actinin,* and other proteins. In culture cells actin has a typical pattern of straight and often parallel fibers that are predominantly localized on the lower, contacting side of the cell and that are in close contact with the plasma membrane (Fig. 6–14).

In these contacting regions actin microfilaments are arranged in parallel, forming bundles that in a living spreading cell may be visible as *stress fibers.* This pattern is completely disorganized by the action of cytochalasin B.

Staining of culture cells with anti-myosin antibodies produces a somewhat similar image to that of actin; however, the majority of the filaments show interruptions or striations (Fig. 6–14*B*). Staining with anti-α-actinin also reveals a periodic arrangement along the filament. These results suggest that α-actinin (and also *tropomyosin*) may be involved in the organization of the actin bundles of microfilaments.

An interesting case is that of the microvillus of the intestine, which contains a bundle of actin filaments that is attached to the plasma membrane by way of a dense material that stains with antibodies against α-actinin. This actin bundle has a polarization similar to that found in a muscle sarcomere (Fig. 6–15).

The localization of the various proteins can also be detected by the injection of fluorescently labeled structural proteins into the cell. Thus, *α-actinin* is deposited in a periodic pattern along the actin microfilaments and *vinculin* is found at the ends of actin filaments of the membrane.[19]

When anti-actin fluorescent antibodies were used on cultured cells in the process of flattening, it was observed that actin filaments around the nucleus could assume the shape of a regular network that resembles a *geodesic dome.* With antibodies against α-actinin, the fluorescence is localized mainly at the crossing points of the network, and with anti-tropomyosin it is found mainly in the short fibers connecting the vertexes (Fig. 6–16). Bundles of actin filaments (stress fibers) that originate from this geodome extend into the lamellipodia and filopodia. Geodesic domes are transitory structures that assemble and disassemble in relation to changes in the shape of the cell during movement.[20]

The Contractile Machinery in Nonmuscle Cells

Most of our knowledge about the molecular machinery involved in contraction was first derived from studies of skeletal muscle, which has a highly elaborate and regular sarcomeric structure. The movement of nonmuscle cells was puz-

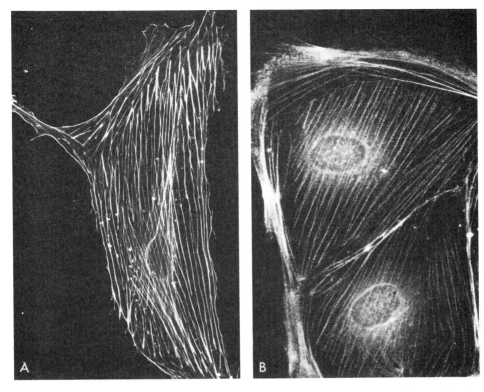

Fig. 6–14. *A,* cultured mammary rat cell stained with anti-actin antibodies. *B,* cultured rat cell stained with anti-myosin antibody. ×600. (*A,* courtesy of K. Weber; *B,* courtesy of K. Weber and U. Groeschel-Stewart.)

zling for a long time, until it was demonstrated that these cells contain the same contractile proteins—*actin* and *myosin*—as muscle cells. This suggested that the machinery for *nonsarcomeric contraction* in cells could be similar to that of muscle, and that the force-generation would be caused, as in muscle, by a sliding mechanism (see Section 7–3). In the case of nonmuscle cells, however, cycles of polymerization and depolymerization of actin may play a prominent role in cell movement.

In nonmuscle cells actin may appear in a variety of structural forms. For example, in intestinal microvilli actin microfilaments are well ordered and have regular polarization (Fig. 6–15). The *stress fibers,* observed in spreading moving cells in culture (Fig. 6–14), are also formed by bundles of actin microfilaments. These structures also contain other proteins such as myosin, tropomyosin, α-actinin, and filamin. In other parts of the cell (e.g., in pseudopodia, ruffling membranes, and in the bulk of the cytoplasm), actin filaments are disposed at random and form a cross-linked network. It is currently believed

that the changes in viscosity of the cytoplasm are due to rapid modifications in the actin network.

Actin. Actin comprises a large proportion of the cytoplasmic proteins of many cells; in developing nerve cells, it may constitute up to 20% of these proteins. Actin is present principally in its globular form (G-actin) with a molecular weight of 42,000 daltons, and it may quickly polymerize to form the microfilaments of fibrous actin (F-actin), which may measure several micrometers in length and contain thousands of monomers.

Cytoplasmic actin is very similar in structure to muscle actin and forms identical 6-nm wide microfilaments consisting of a double helical array of globular actin molecules (see Fig. 7–6). Actin filaments can be identified under the EM by virtue of the fact that they bind and become *decorated* with myosin (heavy meromyosin).

As shown in Figure 6–15, the actin-meromyosin complex has an arrow-like shape with a definite polarity. By comparison with an arrow it is possible to recognize a "pointed" and a

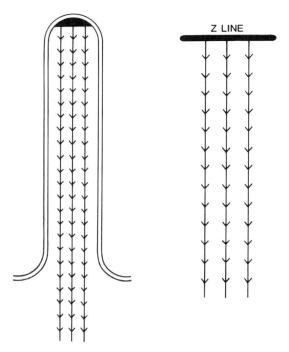

Z LINE

Fig. 6–15. A drawing comparing the polarity of the actin filaments in the microvillus with those in the skeletal muscle. (From Tilney, L.G.: J. Cell Biol., 67:737, 1975.)

"barbed" end. Although actin monomers can be added to both ends, the barbed one is preferred for assembly, while at the pointed end disassembly takes place. This type of polarization is somewhat similar to that observed in microtubules (Fig. 6–4) and in both cases the process is often called "treadmilling" (see Cleveland, 1982 in the Additional Readings at the end of this chapter).

Actin filaments result from the aggregation of a few monomers of G-actin by a process called *nucleation;* at these nucleation centers more monomers are added to the ends. Actin filaments can fragment spontaneously to produce new nucleation centers.[21]

Myosin. Compared with actin, myosin is found in a low concentration and is rather heterogenous in size and composition. In most cases it has a molecular weight of 500,000 daltons, similar to that of muscle myosin, but the two molecules contain different subunits. In *Acanthamoeba* there is a *"minimyosin"* with a molecular weight of 180,000 daltons. In spite of these differences, all myosins bind reversibly to actin filaments and contain a Ca^{++}-activated ATPase. As in muscle, both these functions are

localized in the globular heads of the molecule (see Section 7–2).

Identification of myosin filaments with the EM is less satisfactory. However, filaments that are 13 to 22 nm thick and 0.7 μm long, which are of probable myosin nature, have been observed.

Actin-Binding Proteins. Biochemical studies carried out on isolated cytoplasmic gels have revealed that, in addition to actin, these gels contain a variety of proteins that are designated generically as *actin-binding proteins.* Although this group of proteins is very complex and still under active investigation, some general concepts about their function are being defined (see Weeds, 1982; Craig and Pollard, 1982 in the Additional Readings at the end of this chapter).

Actin-binding proteins can be classified into three functional classes: (1) those that promote crosslinking and produce *gelation,* (2) those able to sever the long actin filaments by capping to short fragments, a process that results in *solation* (i.e., diminished viscosity), and (3) those proteins that tend to stabilize actin in the monomeric form (G-actin) also causing solation (Fig. 6–17).

Actin-Crosslinking Proteins. This group is characterized by their tendency to establish crosslinks between actin microfilaments, which results in the formation of a more viscous three-dimensional network. In some cases the density of crosslinks may be so high that it results in the formation of bundles of actin microfilaments. Some crosslinking proteins (Table 6–1), such as filamin, spectrin, and α-actinin are long and flexible rods, a shape that facilitates the attachment and crosslinking to two actin filaments.

In Figure 4–4 we illustrated the case of *spectrin* and its function in the cytoskeleton of the red blood cell. More recent evidence shows that spectrin is found not only in erythrocytes, but in many other cell types as well. The *nonerythroid spectrin* has a similar amino acid composition but differs in subunit structure. Spectrin is always concentrated in the cytoplasmic side of the plasma membrane. This suggests that by crosslinking actin filaments spectrin can contribute to the formation of a cortical network that can interact with membrane proteins. For example, in muscle, spectrin concentrates in the regions in which the Z and M lines terminate on the plasma membrane (see Chapter 7). Spec-

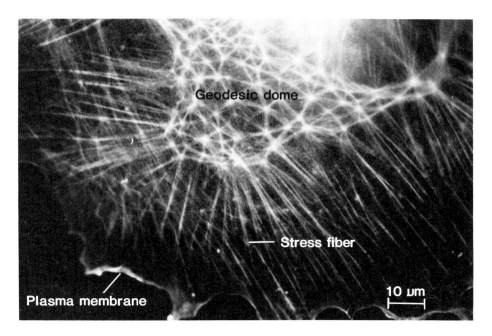

Fig. 6–16.   Geodesic dome arrangement of actin filaments around the nucleus of a cultured fibroblast that is extended and flattened. Actin is stained with the fluorescent anti-actin antibody. Observe that at the periphery of the cell the filaments have a straight radial disposition. (Courtesy of E. Lazarides.)

trin is also a *calmodulin-binding protein* and thus is sensitive to changes in Ca^{++} concentration (see Lazarides and Nelson, 1982 in the Additional Readings at the end of this chapter). It is possible that, like erythrocyte spectrin, this analogue can confer structural integrity to the plasma membrane and control the mobility of transmembrane proteins (including receptors) in the plane of the membrane (see Kakiuchi and Sobue, 1983 in the Additional Readings at the end of this chapter).

Control of gelation can be achieved by different mechanisms and calcium ions may regulate the valency (i.e., number of binding sites) or affinity of some of the crosslinking proteins. One special example is that of *villin*, a protein with a molecular weight of 95,000 daltons present in intestinal microvillae which, at low Ca^{++} concentrations, acts as a crosslinker and at 10^{-6} M Ca^{++} or more, has the opposite effect.[22] In some cases regulation of gelation involves a change in the association of the protein subunits. For ex-

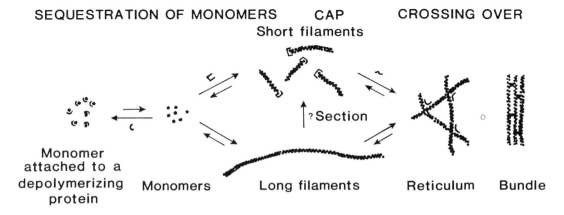

Fig. 6–17.   Diagram representing the actions of actin-binding proteins. Class I cross-links filaments into networks and bundles. Class II caps to one of the filaments and may sever preformed filaments. Class III inhibits polymerization by binding to actin monomers. (Courtesy of Susan Craig.)

ample, *filamin* may shift between monomer and dimer and spectrin between dimer and tetramer, in response to changes in ionic strength, pH, and temperature. In both cases it is only the oligomer that has the crosslinking function.

From *Dictyostelium discoideum* an actin binding protein that is rod-shaped and made of two subunits of 95,000 daltons has been isolated. This protein is regulated by Ca^{++} and pH and crosslinks actin filaments.

In studies on gels made with actin and actin-binding proteins, *cytochalasin B* decreases the viscosity by binding to the F-actin, causing depolymerization to G-actin. *Phalloidin*, another drug also of fungal origin, has the opposite effect, i.e., it stabilizes actin microfilaments and increases gel viscosity.[23]

Other crosslinking proteins, such as *α-actinin*, will be studied in Chapter 7 together with muscle.

Actin-Severing and Capping Proteins. This group of actin-binding proteins tends to shorten actin filaments by directly severing them or by capping to the elongation end, thus interfering with the polymerization process. The shortening of the actin filaments results in solation, i.e., in a lower viscosity of the gel. For example, the addition of *villin* (in the presence of Ca^{++}) causes fragmentation of actin filaments. This fragmentation may be observed under the EM. *Gelsolin*, a protein extracted from macrophages, causes a decrease in viscosity by capping to the barbed end of the actin microfilament. Gelsolin has two binding sites for Ca^{++}; this may account for the sensitivity of the cytoskeleton to minimal changes in Ca^{++} concentration.[24,25]

Acumetin is a protein extracted from macrophages that caps to the pointed end of the actin filaments and inhibits the addition of monomers. The action of acumetin increases the effect of gelsolin that caps to the barbed end and leads to complete control of the actin filament length in the cytoplasm.[26]

Actin-Depolymerizing Proteins. This group of proteins lowers the viscosity of actin by decreasing the process of polymerization, rather than by changing the length of the microfilaments. Depolymerizing proteins, such as *profilin*, bind to actin monomers thus preventing polymerization. In this case, it is assumed that there is a 1:1 ratio between profilin and G-actin. This is in contrast with the previous group of proteins

in which a small amount was enough to cause breakage of the actin filament. For example, in brain there is enough depolymerizing protein to account for the normal pool of G-actin (20 to 30% of the total).

Figure 6–17 summarizes the mechanism of action of the three classes of actin-binding proteins on the actin microfilament. (More information about some of these proteins is provided in Table 6–1).

In conclusion: The changes in viscosity of the cytosol, which are the basis of cell motility, result from the action of different regulatory factors on the actin microfilaments. These factors may either promote polymerization and crosslinking, thus increasing viscosity, or they may reduce crosslinking and produce depolymerization, thus decreasing cytoplasmic viscosity.

SUMMARY:
Microfilaments

Thin microfilaments ranging between 5 and 7 nm in width represent the active or motile part of the cytoskeleton and appear to play the major role in cyclosis and ameboid motion. With high-voltage electron microscopy a three-dimensional view of microfilaments has been obtained (i.e., an image of *microtrabecular lattice*). These microfilaments are sensitive to cytochalasin B, an alkaloid that also impairs many cell activities such as the beat of heart cells, cell migration, cytokinesis, endocytosis, and exocytosis. Applied to the alga *Nitella*, cytochalasin B produces a rapid halt in cyclosis. It is generally assumed that the cytochalasin-B-sensitive microfilaments are the contractile machinery of nonmuscle cells.

The contractile proteins *actin* and *myosin*, as well as tropomyosin, troponin-C, α-actinin, and other proteins found in muscle, are present in microfilaments. These proteins can be localized with specific antibodies. Actin filaments are made of globular actin molecules *(G-actin)*. In the presence of myosin they become "decorated," as can be seen with the EM.

Actin is one of the most abundant proteins in many cells. The globular (G-actin) fibrillar transition is the basis of the classical sol-gel transition in the cytoplasm of moving cells.

Myosin is found in amebae, blood platelets, and slime molds but in a much lower concentration than actin. It contains a Ca^{++}-activated ATPase as is found in muscle myosin. Identification of myosin filaments is more difficult.

Actin-binding proteins act: (1) to promote crosslinking and gelation of microfilaments (e.g., fimbrin, spectrin, α-actinin, filamin), (2) to sever long actin filaments and cap to shorter ones (e.g., gelsolin), and (3) to depolymerize actin filaments (e.g., profilin). (The possible mechanism of action of these factors is shown in Figure 6–17).

Table 6–1 lists some of the actin-binding proteins. This table is however incomplete, since new proteins are being added and knowledge in this field is rapidly increasing (see Pollard, 1984 in the Additional Readings at the end of this chapter).

6–5 MICROFILAMENTS AND CELL MOTILITY

We shall now consider two classic types of cell motility in which microfilaments and actin-myosin interactions are involved. One of them, called *cytoplasmic streaming* or *cyclosis*, is most remarkable in plant cells. The other, *ameboid motion*, is found mainly in certain protozoa and in animal cells.

Cytoplasmic Streaming (Cyclosis)—Observed in Large Plant Cells

Cytoplasmic streaming, or *cyclosis*, is easily observed in plant cells, in which the cytoplasm is generally reduced to a layer next to the cellulose wall and to fine trabeculae crossing the large central vacuole. Continuous currents can be seen that displace chloroplasts and other cytoplasmic granules.

In some plant cells the protoplasmic current can be initiated by chemicals (*chemodynesis*) or by light (*photodynesis*). Cyclosis is modified by temperature, by the action of ions, or by changes in pH. It is stopped by mechanical injuries, electrical shock, or some anesthetics. Some auxins (plant growth hormones) increase the rate of cyclosis. Cyclosis decreases progressively in cells subjected to increased hydrostatic pressure at the same time that the protoplasm becomes more liquid.

The classic experimental work on cyclosis has utilized the cylindroid cells of *Nitella*, which have a thin protoplasmic layer of about 15 μm surrounding a central vacuole of 0.5 mm by 10 cm. This protoplasmic layer is divided into a cortical region of structured cytoplasm in which are embedded nonmotile chloroplasts and a layer of less gelled cytoplasm (the endoplasm) where cyclosis takes place. The whole system is surrounded by the plasma membrane, under which is a single layer of microtubules. Bundles (cables) of actin filaments of about 0.2 μm are situated just beneath the chloroplasts in the cortical region and in a direction that is parallel to that of cyclosis. Each bundle is composed of about 100 microfilaments of 5 to 6 nm, all of them with the same polarity. Each bundle also has about five cables per row of chloroplasts. In addition to actin, the cytoplasm of *Nitella* contains myosin. It is assumed that the actin-myosin interaction, in the ectoplasm, could be the motive force for the movement of the endoplasm. Recent experiments have shed light on the mechanism of cyclosis and shown that it can be explained on the basis of the *sliding filament theory*, which is used to explain muscle contraction (see Chapter 7). The procedure consists of cutting open a cell of *Nitella* and observing it with a water immersion microscope. The endoplasm is washed away, and the ectoplasm with the chloroplast bed and the actin cables remains.

Then using a micropipette, fluorescent beads of about 0.7 μm in diameter, coated with skeletal muscle myosin (heavy meromyosin, HMM), are deposited on the endoplasm (Fig. 6–18A). In the presence of ATP and Mg, it is possible to observe for several hours a directed movement of the beads along the direction of the actin cables with an average velocity of 2.5 μm per second. As shown in Figure 6–18B, the beads move in opposite directions on either side of an indifferent zone, lacking actin cables. The direction is determined by the polarity of the actin filaments (see Sheetz and Spudich, 1983 in the Additional Readings at the end of this chapter).

This mechanism can explain many other motile phenomena in nonmuscle cells, for example, the axoplasmic flow in nerve axons (see Chapter 24).

In the slime mold *Physarum* the contraction of the cortical gel increases the internal pressure, and is responsible for the streaming of the endoplasm. Using excised strands of this slime mold it is possible to observe and to measure the contraction of the cortical gel. Cycles of tension and relaxation, which result from the concentration of Ca^{++} and ATP, can be detected in those strands, as well as in the living intact plasmodium.[27]

Ameboid Motion—Characteristic of Amebae and Many Free Cells

In ameboid motion the cell changes shape actively, sending forth cytoplasmic projections

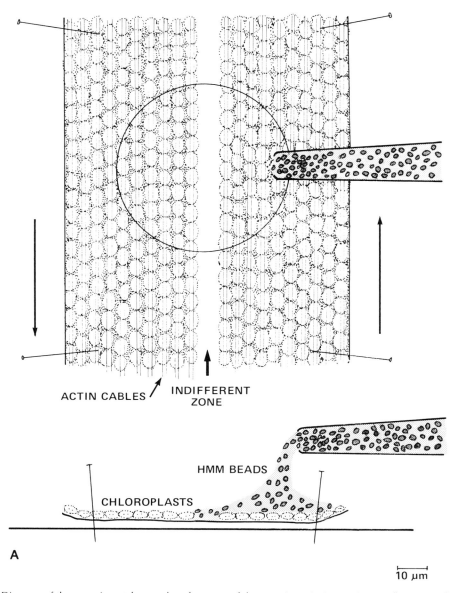

ACTIN CABLES / INDIFFERENT
ZONE

HMM BEADS

CHLOROPLASTS

A

10 μm

Fig. 6–18.   Diagram of the experimental procedure for assay of the myosin-actin interaction. *A,* fluorescent beads coated with HMM are added to a *Nitella* cell that has been dissected and pinned down at the edges and in which the endoplasm has been flushed away. The chloroplasts and the actin cables in the moving zone are represented, as well as the direction of movement. A circle shows the approximate size of the microscopic field.

called *pseudopodia,* into which the protoplasm flows. Although this special form of locomotion can be observed easily in amebae, it also occurs in numerous other types of cells.

Experimentally, one needs only to place a drop of blood between a slide and coverglass to see that the leukocytes, at first spheroidal, change their shape, emit pseudopodia, and move about. In tissue cultures, cells move out actively, forming the zone of migration. These changes also occur in vivo. For example, in epithelial repair, the cells free themselves and slide along actively toward the depth of the wound. In an inflammatory process, leukocytes wander out of the blood vessels *(diapedesis)* by active ameboid motion and progress toward the focus of infection.

Some amebae are predominantly *monopodial* (one pseudopodium), but others may be temporarily or permanently *polypoidal.* The shape

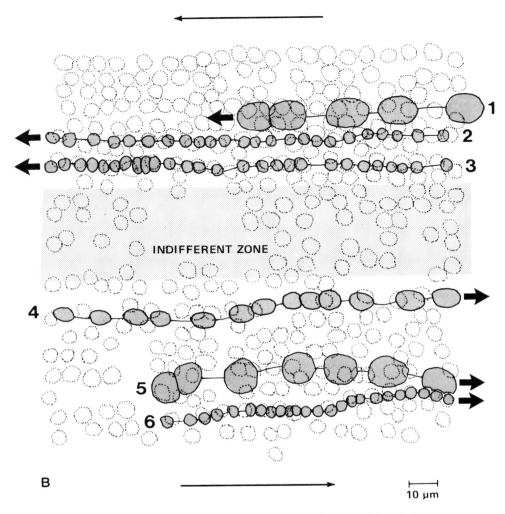

Fig. 6–18. *(continued.) B,* the movement of the beads is observed and determined directly from a video monitor. The chloroplasts remain fixed in position while the beads move in the directions indicated by the heavy arrows. The long thin arrows, as in A, indicate the direction of cyclosis. (From Sheetz, M.P., and Spudich, J.A.: Nature, *303*:31, 1983.)

of pseudopodia varies between a stout, almost cylindrical *lobopodium* and a fine filamentous or branching *filopodium.* Sometimes these fine-processes may be anastomosing *(reticulopodia),* as in Foraminifera.

Cell movements are best observed by time-lapse photography, as was done years ago by Warren H. Lewis. It was observed that, after a few hours of culture, the fibroblasts acquired a polygonal shape with sheet-like extensions of the cytoplasm or *lamellipodia.* These are also known as *ruffled membranes,* because they resemble the ruffles of a dress moving under the impulse of a breeze. These ruffling lamellipodia constitute the main locomotive processes of ameboid

cells and are well studied by scanning electron microscopy. Figure 6–19 shows a culture cell sending lamellipodia to a glass surface. Figure 6–19 also shows that ameboid motion is reversibly inhibited by cytochalasin B.[28] In addition to lamellipodia, fine thread-like processes called *filopodia* are sent out by the cell. These appear to have an exploratory function, i.e., "recognizing" the area in which the cell spreads. Filopodia are abundant in developing neurones, in which they may serve a sensory function prior to the formation of more permanent contacts with other cells.

The ameboid cell does not adhere to the solid support by the entire bottom surface, but only

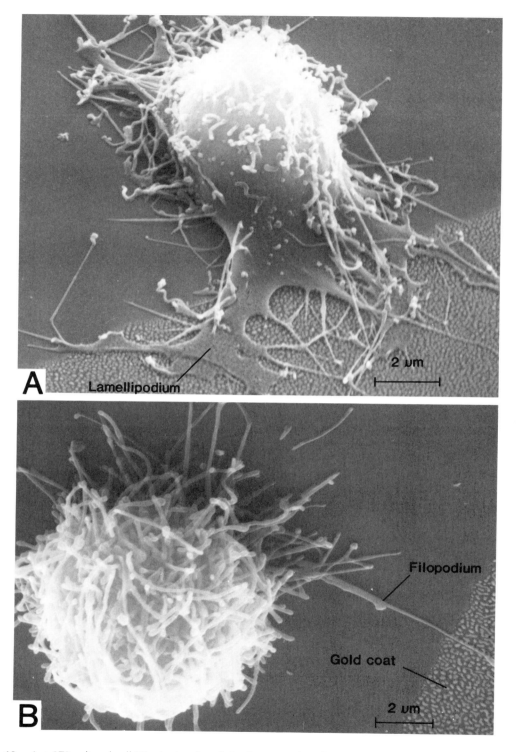

Fig. 6–19. *A,* a 3T3 cultured cell 30 minutes after plating in a normal medium. The cell has made attachments near the border line between the glass and an area covered with gold particles. Observe that a ruffling lamellipodium extends toward the gold plated area. There are small blebs on the dorsal surface. *B,* a similar cell but with the addition of cytochalasin B. The cell is covered with long filopodia (microvilli), but no lamellipodia have extended. This process is reversible if cytochalasin B is washed out. Electron micrographs made by scanning. Bars indicate 2 μm. (Courtesy of G. Albrecht-Büehler.)

by a small number of sites called *adhesion plaques* (Fig. 6–20). These focal sites of adhesion are continuously formed and broken as the cell moves along.

Another interesting experimental method for studying ameboid motion is putting the cells on a glass coated with gold particles. The displacement can be easily followed, because the migrating cell phagocytizes the particles producing a particle-free track. These tracks are called *phagokinetic* because they are the result of a combination of phagocytosis and locomotion (see Albrecht-Buehler, 1978 in the Additional Readings at the end of this chapter). This method permits the study of the movement of a large number of cells. Furthermore, quantitative determinations of such tracks can be used to study the effect of substances that may stimulate or inhibit the motion of migrating cells. One interesting observation is that after a cell division, the two daughter cells tend to follow symmetrical tracks that are mirror images of one another.

The directionality of the ameboid movement requires the integrity of the whole cell. This is revealed by the following experiment: By treating fibroblasts with cytochalasin B it is possible to produce tiny fragments of cytoplasm (microplasts), which show all known forms of ameboid movement (i.e., ruffles, filopodia, blebs), but are unable to move in a certain direction.[29]

In different amebae the rate of progression varies between 0.5 and 4.6 μm per second. In a leukocyte the rate of progression is about 0.6 μm per second. This rate is modified by temperature and other environmental factors.

There are substances that influence the motion by attracting or repelling the cells. This property, which is called *chemotaxis*, has great importance in defense mechanisms, especially during inflammation.

All that has been said thus far about microfilaments and cell motility applies to ameboid motion.

Calcium is required for this type of locomotion. Severe mechanical injury, electrical shock, or ultraviolet radiation causes retraction of pseudopodia and there seems no doubt that the actin-myosin interaction provides the actual motive force for ameboid motion. Nevertheless, the views are divided regarding the most likely site of contraction. Whereas some investigators regard the posterior region of the ectoplasmic tube as more active, others give more importance to the advancing end of the ameba.

The diagram of Figure 6–20 emphasizes this last viewpoint. From the velocity profiles indicated (v), it is evident that the streaming of the axial endoplasm increases toward the front, where the main region of contraction is indicated. The various portions of the cytoplasm are in states of contraction, relaxation, transition, or stabilized equilibrium, which can be recognized by the viscoelastic properties and the variations in birefringence revealed by the polarizing microscope. As the endoplasm streams foward, it becomes more rigid, assuming the viscosity of the ectoplasmic tube. In this model the sol-gel cycles described in the older literature are interpreted as contraction-relaxation cycles.

Interaction of Microfilaments with the Cell Membrane and Surface Receptors

Many motile events, such as cell spreading, shape changes, locomotion, membrane ruffling,

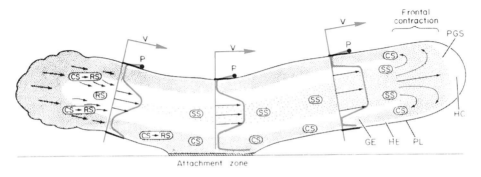

Fig. 6–20. A schematic diagram corresponding to a frontal contraction model for ameboid motion. *SS,* stabilized regions; *CS,* contracted and *RS,* relaxed regions; *V,* velocity profile; *HC,* hyaline cap; *PGS,* plasma gel sheet; *HL,* hyaline ectoplasm; *GE,* granular ectoplasm; *PL,* plasmalemma. (Courtesy of D.L. Taylor.)

cytokinesis, endocytosis, and the formation of pseudopods, require the tight coupling between the cell surface and the contractile apparatus provided by the actin microfilaments. In contrast to the sarcomeric contraction of muscle (see Chapter 7), in which the myofilaments are stable, in these other types of contractions, microfilaments are transitory structures and are assembled and disassembled within seconds (see Lazarides and Revel, 1979 in the Additional Readings at the end of this chapter).

In ameboid motion a fundamental role is played by the attachment of actin microfilaments to the plasma membrane. This binding is by ATP-stable linkages and special proteins such as *vinculin*. The cortical region and the pseudopods of ameboid cells (e.g., leukocytes) contain a three-dimensional network of actin filaments, crosslinked by myosin, α-actinin, and other actin crosslinking molecules. By using fluorescent antibodies, it is possible to observe that, in a polarized leukocyte, myosin and the other crosslinking proteins are concentrated in the anterior part of the pseudopod. This finding is consistent with a mechanism of contraction, in which the directionality of the movement results from localized changes in the structure of the actin gel. While the force-generation mechanism can be based on the sliding of actin and myosin filaments (see Section 7–2), the direction of the movement, in the isotropic actin lattice of the cell cortex, could derive from focal changes in crosslinking. The cortical cytoskeleton could then be converted into a "cytomusculature" whose contraction drives such events as cell locomotion, receptor redistribution, and endocytosis (see Condeelis, 1981a in the Additional Readings at the end of this chapter).

A good way to observe the interaction of the cytoskeleton with the cell membrane during locomotion is by achieving the fixation while an ameboid cell is being filmed. By using cultured fibroblast it is possible to follow three cytoskeletal components (microtubules, actin, and vinculin) during the process of ruffling. At the leading edge, in which there is ruffling activity, there is first an accumulation of actin, followed in a matter of seconds by vinculin and tubulin. The formation of microtubules at the leading end contributes to the stabilization of the ruffles and the subsequent sites of attachments to the substrate. On the contrary, the intermediate filaments are not directly involved in locomotion. (see Small et al., 1984 in the Additional Readings at the end of this chapter).

The *capping phenomenon*, which was studied in Section 4–2 (see Fig. 4–9), also involves the intervention of microfilaments. Such caps can be isolated and shown to contain high concentrations of actin, myosin, and other contractile proteins.

Figure 6–21 shows an experiment on an isolated cap in which concavalin A, (a lectin), coupled to the electron dense hemocyanin, is bound to the surface receptors. These appear as clusters on the membrane, while on the cytoplasmic side there are microfilaments of actin decorated with heavy meromyosin (see Fig. 7–9). These filaments show a polarity away from the membrane demonstrated by the pointed and barbed ends of the meromyosin. After the lectin becomes bound to the surface receptors, the amount of actin and myosin below the membrane increases considerably. The final result of the clustering of receptors is the internalization of the ligand-receptor complex (Fig. 4–9) (see Condeelis, 1981 in the Additional Readings at the end of this chapter).

In Section 10–3, we shall see that in the formation of *coated pits* and *coated vesicles*, as well as in other processes of endocytosis, actin microfilaments provide the motive force for the internalization of the vesicles. In the receptor mediated type of endocytosis, concentrations of *clathrin* (the protein of coated vesicles), actin, and *calmodulin* are found.

In conclusion: The actin cytoskeleton interacts with the plasma membrane, as well as with the organelles (including microtubules and intermediate filaments), thus providing for the dynamic integration of all regions of the cytoplasm.

Some of the previously mentioned phenomena (e.g., capping, coated pits and vesicles) suggest that *receptors* on the cell surface interact with microfilaments and exhibit some stable association with the cytoskeleton. One example, mentioned in Section 4–1, is the association between *band 3* of the red cell membrane and *spectrin*, which is part of the erythrocyte cytoskeleton (see Fig. 4–4). Biochemically, this interaction can be demonstrated by using a crosslinking agent that creates a tighter association between receptors and microfilaments.[30]

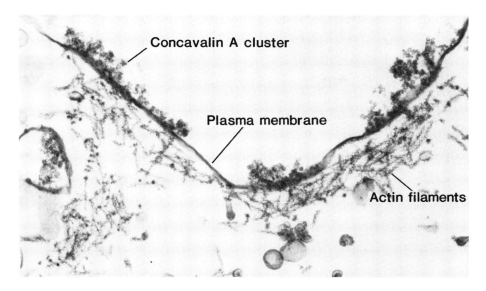

Fig. 6–21.   Cap isolated from an ameboid cell of *Dictyostelium* after lysis of the cell with a detergent. On the surface of the cell membrane clusters of concavalin A, tagged with hemocyanin, are observed. On the cytoplasmic side of the membrane, in the regions corresponding to the accumulations, actin filaments "decorated" with heavy meromyosin are visible. These filaments show a polarity with a direction pointing away from the membrane. (Courtesy of J. Condeelis.)

Cell Adhesion, Focal Contacts, Stress Fibers, Fibronexus, and Transmembrane Interaction

We mentioned that when cultured cells are grown on a solid support (e.g., plastic, glass) they tend to adhere and to spread, and that contacts between the ventral surface of the cell and the support are established at the so-called *adhesion plaques* or *focal contacts*. At these points, dramatic changes occur inside the cell (at the level of the cytoskeleton) and outside the plasma membrane (in the cell coat) (see Section 5–3).

Focal contacts are involved not only in attachment, but also in the movement of cells. These contacts are weakened and reduced when a cell undergoes transformation. We can also consider focal contacts as sites that mediate signals between the cytoplasm and the environment by way of transmembrane interaction.

As shown in Figure 6–22B, focal contacts can be observed in the living cell by using the interference reflection microscope under which they appear as darker areas of the cell membrane. In a freshly cultured cell these areas are radially arranged, whereas, as shown in Figure 6–22B, in a spreaded fibroblast they are aligned parallel to the direction of the movement. By using either the light or electron microscope, it may be observed that the focal contacts are the sites of implantation of actin-containing micro-

filament bundles, or *stress fibers*. By immunofluorescence, actin appears homogeneously distributed. On the other hand, α-actinin and myosin show a periodic arrangement that is reminiscent of the sarcomeres in muscle (see Section 7–1, see also Fig. 6–22C). The sarcomeres of the stress fibers are contractile and thus may be important in generating tensile forces between the cell and substratum.[31]

Vinculin. A specific protein present at the focal contacts is *vinculin* (Table 6–1).[32] Vinculin has a molecular weight of 130,000 daltons and is preferentially localized in the ventral region of cultured cells at the focal sites of adhesion, but has also been found in differentiations of the cell membranes such as the belt desmosomes and the intercalated discs of cardiac cells (see Geiger, 1981 in the Additional Readings at the end of this chapter). In vitro, it has been observed that one molecule of vinculin interacts with about 2000 actin monomers. This interaction probably occurs at the growing end of the actin microfilament; however, vinculin appears to play a bundling role at the tip end of the stress fiber, rather than directly linking the actin microfilaments to the membrane (Fig. 6–22C).

Outside the membrane of a moving cell numerous proteins and polysaccharides are secreted and deposited. When the cell moves or becomes detached this extracellular material

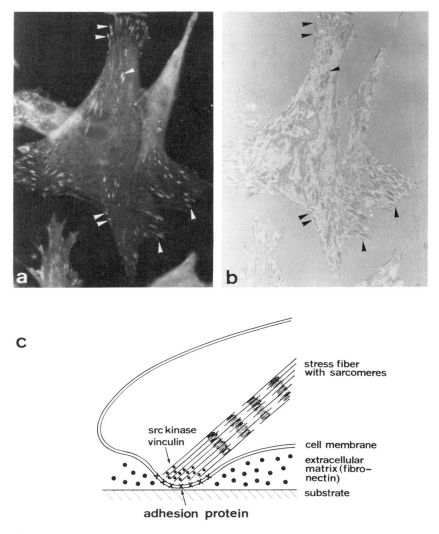

Fig. 6–22. *A* and *B* are images of focal contacts with the substrate of chick embryo fibroblasts observed with the optical microscope. *A,* immunofluorescence with a monoclonal antibody that inhibits the adhesion; the antigen is situated at the focal contacts. *B,* image by interference showing the focal contacts as dark bands. Arrows indicate some focal contacts in both *A* and *B. C,* diagram of the possible molecular organization of a focal contact and of the interaction, across the membrane, between intracellular and extracellular molecular components. (Courtesy of W. Birchmeier.)

leaves a "foot print" at the adhesion plaques. The abundance of this material is reflected in the fact that it contains 1 to 2% total cell proteins and 5 to 10% polysaccharides, most of which are of the glycosaminoglycan type (Table 5–1).[33]

Fibronectin. In various cell types, such as primitive mesenchymal cells, astroglia, fibroblasts, and some epithelial cells, a glycoprotein called *fibronectin* participates in cell adhesion. Fibronectin is a huge molecule of 440,000 daltons, which under the EM appears as a long, thin, and flexible strand. It has two subunits covalently linked by a single disulfide bond near the

-COOH end.[34] This protein is secreted in situ by the attached cells, probably by way of small vesicles observed under the EM in between focal contacts.

Fibronectin can be considered as a cell adhesion protein having numerous binding sites. It binds to collagen, fibrin, heparin, and other macromolecules, as well as to sites in the cell membrane. In addition to the cellular type, there is also a *soluble fibronectin* present in the blood plasma. Addition of high concentrations of the soluble fibronectin can inhibit the binding of the cellular type to the cell membrane. Small peptides have been produced that also bind to the cell recognition sites and displace fibronectin. Injection of such peptides in an amphibian gastrula leads to inhibition of cell migration and alterations in development (see Yamada and Kennedy, 1984 in the Additional Readings at the end of this chapter).

As mentioned in Section 5–3, fibronectin has received great attention because it is greatly reduced in cells undergoing cancerous transformation. Another point of interest is that some authors have found a certain transmembrane relationship between fibronectin and vinculin at the adhesion plaque. This relationship has been observed in cultured cells stained with antibodies against fibronectin and vinculin. It was found, at the ventral surface of the cells, that there is an alignment between the stress fibers (observed by interference reflection optics) and the sites of antivinculin and antifibronectin binding, stained with different fluorescent colors.[35]

The possible relationship between vinculin and fibronectin is disputed by some researchers who find fibronectin to be largely absent from focal contacts but concentrated around them (Fig. 6–22 C) (see Birchmeier, 1981 in the Additional Readings at the end of this Chapter). However, a direct relationship between the stress fibers inside the cell and fibronectin outside is apparently present in the so-called *fibronexus*.[36] Unlike the focal contacts that are established between the cells and substrate, the fibronexus appears to be involved in the binding of cells to extracellular matrices and other cells. Electron microscopic observations of fibronexus show that microfilaments are directly continued by lines of fibronectin fibers outside the plasma membrane.

In embryos it has been demonstrated that fibronectin is necessary for the migration of cells of the neural crests. In vitro experiments show that these cells prefer a glass strip coated with fibronectin to other substrates and that migration is inhibited by antibodies against the cell-binding region of fibronectin (see Rovasio et al., 1983 in the Additional Readings at the end of this chapter).

The release of fibronectin by the cell contributes to the deposition of extracellular matrices, such as the *basement membrane* or *basal laminae* found in many tissues. This membrane consists of collagenous and noncollagenous glyco-

proteins and polysaccharides, such as hyaluronic acid, heparin and chondroitin sulfate (see Table 5–1).

Laminin. Among the noncollagenous glycoproteins the best known is *laminin*, a huge molecule (950,000 daltons) that, as fibronectin, may serve as an adhesive material. Laminin has a rather peculiar cross-shaped structure with a long arm and three short arms joined by disulfide bonds and contains a fair amount of carbohydrates. It is an abundant component of all basement membranes in tissues, which also contain collagen Type IV and a heparan sulfate proteoglycan, to which laminin is tightly bound.

Laminin is the first extracellular matrix protein to appear in the embryo and can be detected as early as the morula stage (16 cells). In the kidney this protein acts as a major barrier to filtration, and antibodies against laminin can produce deposition in the glomerular basement membrane and severely affect kidney function. Laminin is increased in basement membranes of diabetic individuals, and antibodies against laminin are found in Chagas disease (see Timpl et al., 1983 in the Additional Readings at the end of this chapter).

As pointed out in this section, there should be an interaction by way of signals across the cell membrane; this may be essential for cell spreading, locomotion, and even for more complex functions such as contact inhibition of growth (see Section 5–4) and the integration of cells in a tissue. Later in this chapter we mention *cell transformation*, in which the transmembrane interaction appears to be profoundly affected (see Hymes, 1982 in the Additional Readings at the end of this chapter).

The problem of finding possible membrane proteins that could mediate such an interaction has been recently studied, using monoclonal antibodies that have an anti-attachment action. One of these antibodies, which inhibits the adhesion of fibroblasts to artificial substrates, is located exactly at the focal contacts (Fig. 6–22A and C). This antibody binds to a membrane protein of 60,000 daltons located on the exterior side, which is likely to be involved in the transmembrane interaction.

It has been assumed that a single fibroblast may have about 600,000 of these antigen molecules concentrated in about 200 focal contacts. This gives a total of a more than 1000 antigens per contact, a number that corresponds fairly well with that of actin microfilaments in a stress fiber attached to a single focal contact.[37] (The localization of this membrane antigen, which could represent a transmembrane adhesion protein, is illustrated in Figure 6–22C).

Cell Adhesion and Collagens

We mentioned in Section 5–3 that cell adhesion also involves proteins of the collagen type. Collagen represents the most abundant protein group found in the animal kingdom. It appears as an amorphous substance in the basement membranes of certain tissues, or, more frequently, constitutes the collagen and reticular fibers found in extracellular spaces. These fibers are made of fibrils that have a characteristic striation, with a 67 nm repeating period in the native state. These fibers serve as mechanical support for the tissue and represent surfaces on which cells may glide. Collagen is synthesized by almost every mesenchymal and epithelial cell of vertebrates with only few exceptions, but the connective tissue cells produce it in the largest quantities.

The basic molecular unit of collagen is *tropocollagen*, an elongated molecule about 300 nm long and 1.5 nm wide (Fig. 6–23). Tropocollagen consists of three polypeptides of about 95,000 daltons that are coiled together in a helical fashion. The largest portion of the molecule has an α-helix organization, with short nonhelical segments of 16 to 25 residues at both ends that are called *telopeptides*. The amino acid composition of collagen is rather simple: one-third is glycine, one-third is proline and hydroxyproline, and one-third contains the other amino acids.

There are at least five isotypes of collagen molecules, based on slight differences in the organization of the polypeptides and the association with other molecules (i.e., polysaccharides and glycoproteins). Type I is present in dermis, bone tendons, cornea, and dentin; Type II, mainly in cartilage; Type III, in fetal skin, the cardiovascular system, the uterus, and intestine; Type IV, in *basal laminae* or basement membranes; and Type V, in blood vessels, placenta, cornea, etc. While Types I, II, III, and V show typical striated fibrils, Type IV lacks a distinct fibrillar structure and is produced along with fibronectin and laminin.

Tropocollagen can be considered a macromolecular monomer because, by interaction, it is capable of forming different collagen structures. The tropocollagen molecule is polarized (i.e., it has a definite linear sequence of amino acid residues in the intramolecular strands), and it behaves as if it has a "head" and a "tail" (Fig. 6–23). The finding that tropocollagen, in the presence of ATP, produces short segments of only 300 nm in length, provides an explanation for the probable mechanism by which the 67-nm length is produced. The segments with long spacing result because tropocollagen molecules aggregate in parallel, with all the amino acids in register. In the case of the striated collagen, it is thought that overlapping occurs at about a quarter of its length (Fig. 6–24). This model has been modified slightly based on the fact that tropocollagen has a length equal to 4.4 D (in which D = 67 nm).

As shown in Figure 6–24, in each D period there is a *hole region* of 0.6 D and an *overlap region* of 0.4 D. The negative stain penetrates more in the hole region, producing the dark band, while it is excluded from the overlap region. In the native striated fiber, the molecules of tropocollagen are oriented in a single direction. However, when tropocollagen is mixed with glycoprotein, it can reconstitute *fibrils with long spacing*. In this case, it is thought that tropocollagen molecules are oriented in both directions (Fig. 6–23).

Collagen synthesis and fibrillogenesis is by a complex multistage mechanism that involves several intra- and extracellular steps. Collagen interactions, at a supramolecular level, begin inside the cell and continue in the extracellular space.

As previously mentioned, in basement membranes there is a tight relationship between collagen, fibronectin, laminin, and proteoglycans. A relationship between fibronectin and collagen fibers can be demonstrated experimentally in cell cultures on a collagen support. It has been found that fibronectin binds to special binding sites of the collagen molecule making a complex. Then the cell surface attaches to it and spreads making the focal contacts at the ventral side (for further details, see Kleinman et al., 1981 in the Additional Readings at the end of this Chapter.)

TROPOCOLLAGEN

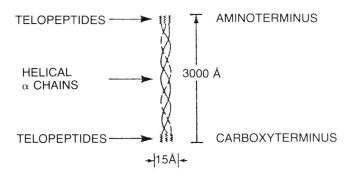

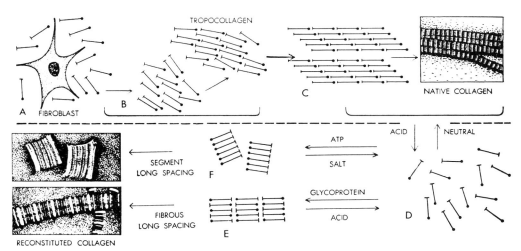

Fig. 6–23. *Top,* structure of a tropocollagen molecule with the three helical α-chains. *Bottom,* diagram of the formation and reconstitution of collagen. A fibroblast *(A)*, manufactures *tropocollagen* molecule *(B)*, which forms *native collagen (C).* Collagen fibrils are solubilized in acid *(D)* and the resulting tropocollagen, in the presence of glycoprotein, produces *fibrous long spacing (E)* and, with addition of ATP, *segment long spacing* collagen *(F)*. The long spacing of 280 nm results from the lateral aggregation of the tropocollagen molecules without overlapping. The 70 nm spacing of native collagen fibrils is due to the overlapping of the tropocollagen molecule. (Courtesy of J. Cross, 1961.)

Microtubules and Microfilaments in Cancer Transformation

In Section 5–4 we summarized the changes that the surface of the cell undergoes during cancerous transformation, and we also mentioned the possible influence of the diminution of fibronectin (and collagen) on the reduced adhesion of transformed cells.

Transformation also alters the cytoskeleton, and by immunofluorescence or electron microscopy, it can be seen that both microtubules and microfilaments are disorganized, producing considerable changes in cell shape.

Figure 6–25 shows that when normal fibroblasts, which appear star-shaped and flattened on the side, are transformed by a virus they become rounded (Fig. 6–25*B*). In these cells, the microtubules are very short and randomly oriented and the actin microfilaments, which are organized into stress fibers, have virtually disappeared. This figure shows another interesting phenomenon, i.e., that these changes can be reversed by treating the cell with dibutyl cAMP (Fig. 6–25*C*).[38]

Transformed cells also display *knobs* (or *blebs*) over the surface that can be easily observed by scanning electron microscopy. These knobs disappear under the influence of cAMP when the cell recovers its fibroblastic shape. The formation of the knobs has also been attributed to the disorganization of the cytoskeleton (microtubules and microfilaments) that occurs in transformation.[39]

A possible mechanism by which the transfor-

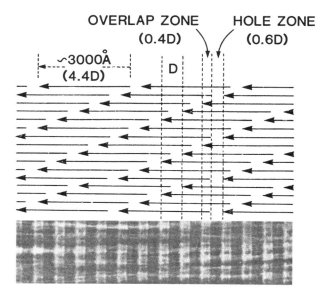

OVERLAP ZONE   HOLE ZONE
(0.4D)   (0.6D)

~3000Å
(4.4D)

D

Fig. 6–24. Two-dimensional model of the assembly of tropocollagen molecules (arrows) within the native collagen fiber. (See text for details.)

mation induced by a virus, e.g., the Rous Sarcoma virus (RSV), produces the disorganization of the cytoskeleton has been postulated. The malignant transformation of cells infected by the RSV depends on a single gene named *src* (for sarcoma formation). This gene produces a phosphoprotein of 60,000 daltons (pp60[src]) that functions as a protein kinase, and is able to phosphorylate several proteins including vinculin. An interesting finding is that the pp60[src] kinase is markedly enriched at the focal contacts where vinculin is located (Fig. 6–21C).[40] Unlike other cellular protein kinases, the amino acid that is phosphorylated by pp60[src] is *tyrosine*. As described earlier, vinculin is presented at the tips of the stress fibers near the plasma membrane, and serves as an excellent substrate for the action of pp60[src]. It is possible that a change in the vinculin molecule produced by phosphorylation could lead to the detachment and disorganization of the actin microfilaments. A consequence of this would be a more rounded morphology, cytoskeletal disorganization, and less adhesion to the substrate.[38,41] The fact that these changes are reversed by cAMP (Fig. 6–25C) suggests that this nucleotide could influence the phosphorylation of vinculin by the pp60[src]. In Chapter 23 the problem of cell transformation is considered and it is shown that phosphory-

lation of vinculin also occurs in normal cells, but at a rate 10 times lower than in transformed cells.

From these experimental observations it can be postulated that the organization of microfilaments is modulated by the phosphorylation of vinculin. The effect of pp60[src] would be that of tipping the balance toward the disorganization of the microfilaments with a profound effect on the cell shape.

SUMMARY:
Microfilaments and Cell Motility

The *mechanism of contraction* in nonmuscle tissue, which includes cyclosis and ameboid motion, is thought to involve the interaction of actin and myosin filaments (as in muscle) and the production of a shearing force. The random distribution of these filaments and the lower content of myosin may explain the slowness of this contraction, which requires ATP and the Ca^{++}-activated ATPase.

Cytoplasmic streaming or *cyclosis* is found in all kinds of cells and produces the displacement of various organelles. The alga *Nitella* is the organism most often used for this study. Cyclosis may be started by chemicals or by light. Various injuries and *cytochalasin* B stop cyclosis. Microtubules may provide the framework for cyclosis, but the actual motive force is provided by microfilaments.

Ameboid motion is observed in amebae, leukocytes, culture cells, and in healing wounds. The number and shape of pseudopodia vary (e.g., lobopodial, filopodial, and reticulopodial). In amebae there is an axial endoplasm separated by shear zone from the peripheral endoplasm. The ectoplasm forms a hyaline cap at the advancing end. In slime molds there is a pulsating movement that is inhibited by high pressure and is increased by ATP. Other important factors in ameboid motion are the levels of Ca^{++} and the adhesion to a solid support.

In ameboid motion of cultured cells ruffling lamellipodia are the main locomotive processes. Filopodia appear to have an exploratory function. The phagokinetic tracks allow the study of ameboid motion in many cells and of the effect of substances that stimulate or inhibit cell migration. In the cortical region and pseudopods of leukocytes there is a three-dimensional network of actin, crosslinked with myosin and actin-binding proteins. Directionality of movement may derive from focal changes in the crosslinking of actin.

In ameboid motion actin microfilaments are attached to the plasma membrane at the adhesion plaques. They are involved in the capping phenomenon and in the formation of coated pits

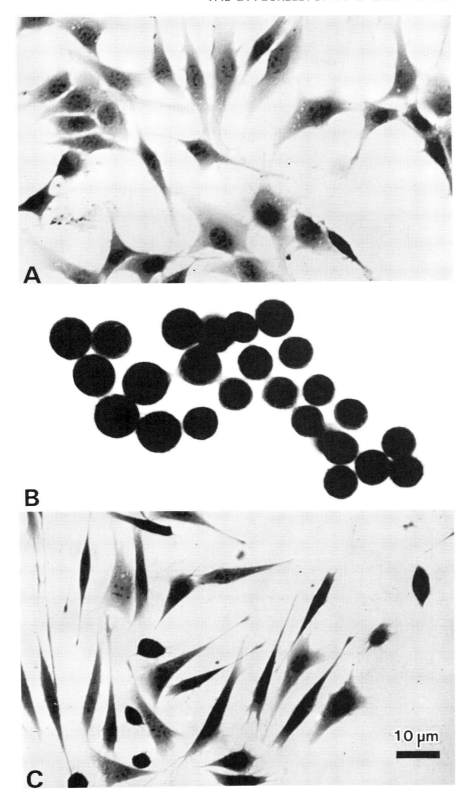

Fig. 6–25. *A,* normal cells in culture, showing the typical star shape after adhesion to the support. *B,* similar cells, but submitted to transformation by the avian sarcoma virus (ASV). Practically all cells are now dense and spherical. *C,* transformed cells as in *B,* but treated with dibutyl cAMP and testolactone for 96 hours. See text for details. (Courtesy of T.T. Puck.)

and coated vesicles. This attachment is by α-actinin and vinculin. On the solid support the ameboid cell secretes an extracellular material that leaves a "footprint." This material contains collagenous and noncollagenous glycoproteins such as fibronectin and laminin. Collagens are the most abundant proteins in animal tissues and are present in basement membranes and at the intercellular spaces as collagen and reticular fibers. Collagen results from the assembly of tropocollagen units of 280 nm. The characteristic length of 67 nm in collagen fibers is due to the overlap of the molecule of tropocollagen.

In transformed cells the cytoskeleton is disorganized. The Rous Sarcoma Virus produces transformation by a single gene whose product is a phosphokinase that phosphorylates vinculin.

6–6 INTERMEDIATE FILAMENTS— MECHANICAL INTEGRATORS OF CELLULAR COMPARTMENTS

In addition to the microtubules and the microfilaments, in the cytoskeleton of many cells, there is a group of 10 nm filaments called intermediate, because they range in thickness between the microtubules and the microfilaments. These *intermediate filaments* are very heterogeneous from the point of view of their biochemical properties, but by their morphology and localization can be grouped into four main types: keratin filaments, neurofilaments, glial filaments, and a more heterogenous type, which is composed of several proteins (see Lazarides, 1980; Davison, 1981; Steinert et al., 1985, in the Additional Readings at the end of this chapter).

In spite of their heterogeneity, intermediate filaments tend to show a similar structural organization. As seen with x-ray diffraction, all members of this family contain considerable amounts of α helix and, when examined under the EM after shadow casting, show a periodic structure with a length of 21 nm.[42] Intermediate filaments are insoluble in physiological solutions but dissolve at low and high pH. After dissolution they can be reconstituted to again form intermediate filaments.

Intermediate filaments form networks that interconnect the nucleus with the cell surface. In fibroblasts intermediate filaments attach to the nuclear envelope and form a cage around the nucleus; at the plasma membrane they sometimes appear associated with microfilaments at the adhesion plaques. The intermediate filaments that form the nuclear cage are continuous with other intermediate filaments that radiate through the cytoplasm (see Goldman et al., 1984 in the Additional Readings at the end of this chapter).

Intermediate filaments are formed by protein subunits of a multigene class that are expressed differentially in various cell types. However, the different intermediate filament proteins are organized in a similar way forming tetrameric complexes (see Franke and Jorcano, 1984 in the Additional Readings at the end of this chapter).

We will now briefly discuss the different types of intermediate filaments:

(1) *Keratin Filaments.* These filaments, also known as *tonofilaments*, prekeratin, or cytokeratin filaments are the most complex class of intermediate filaments (in man there are as many as 19 different cytokeratins).[43] As shown in Figure 5–4 these filaments are anchored to the cell surface and tend to converge upon the desmosomes. Mammalian cytokeratins are α-fibrous proteins that are synthesized in cells of the living layers of the epidermis, and form the bulk of the dead layers or *stratum corneum*. Cytokeratins are composed of multiple polypeptides ranging in size between 47,000 and 58,000 daltons and have a three-chain structural unit. The different cytokeratins are expressed during epithelial differentiation following different routes.

(2) *Neurofilaments.* These filaments, together with microtubules, are the main structural elements of axons (see Fig. 24–3), dendrites, and neuronal perikarya. They contain three polypeptides ranging between 200,000 and 68,000 daltons and are very sensitive to proteolysis in the presence of Ca^{++}.[44] Neurofilaments form a three-dimensional lattice that converts the axoplasm in a highly structured gel.[45]

(3) *Glial filaments.* These filaments are found throughout the cytoplasm of astrocytes and are composed of a protein of 51,000 daltons having very acidic properties. Antibodies against this glial protein do not cross-react with neurofilament protein or with desmin, vimentin, or synemin. It is interesting that only astrocytes, and not the oligodendrocytes of the brain, contain these intermediary filaments.

(4) *Heterogenous filaments.* Under this heading we include a group of intermediate filaments

that have a similar morphology and localization but which contain different proteins, such as desmin, vimentin, and synemin.

Desmin filaments are found predominantly in smooth, skeletal, and cardiac muscle. When studying the structure of skeletal muscle, we shall see that myofibrils are held by linkages present at the Z- and M-lines, which extend to the cell membrane (see Fig. 7–1). Such linkages are made of intermediate filaments composed mainly of desmin, vimentin, and synemin.

Desmin (G-link, bond) consists of two major polypeptides of 50,000 and 55,000 daltons that, after isolation, can reconstitute the 10 nm filaments. As will be shown in Section 7–2, desmin appears to be concentrated in the Z-lines of skeletal and cardiac muscle (see Fig. 7–4), together with vimentin, synemin, and α-actinin.[46] Also, in relation to the structure of muscle in Section 7–2, we shall mention the possible function of desmin filaments.

Vimentin filaments are found in cells of different origin. By immunofluorescence they show a wavy disposition, which explains their name (Latin, *vimentus* = wavy).[47] *Vimentin* has a molecular weight of 52,000 daltons and the antibodies show wide cross-reactivity among species of mammalian, avian, and amphibian origin. This suggests that this protein is preserved throughout evolution. In cells exposed to Colcemid, these filaments tend to form perinuclear caps which can be easily separated.[48]

From these caps both vimentin and desmin have been isolated. Vimentin filaments are very insoluble and tend to bind to the nuclear envelope and centrioles. These, and other intermediate filaments are resistant both to colchicine and cytochalasin B, which respectively depolymerize microtubules and actin microfilaments. Intermediate filaments, in general, are insoluble in physiological solutions but are susceptible to proteolysis. There is a Ca^{++}-dependent proteolytic mechanism that is probably involved in the disassembly of these filaments.

Synemin is a protein of 230,000 daltons, which is also present in the intermediate filaments of muscle, together with desmin and vimentin. Closely associated with these other proteins, synemin makes a network of interlinked rings within the plane of the Z-discs in skeletal muscle (Fig. 7–1).[49]

An excellent view of intermediate filaments containing vimentin and synemin is observed in chicken erythrocytes. A shown in Figure 6–26 there is a three-dimensional network of intermediate filaments that interlink the nucleus with the plasma membrane. Using antivimentin antiserum these filaments become uniformly decorated (Fig. 6–26A); whereas with antisynemin there is a periodic distribution of the antiserum (Fig. 6–26B). This suggests that the core of the filament is made up of vimentin; whereas synemin is at regularly spaced nodes.[50]

Intermediate Filaments During Mitosis

Mitosis of epithelial cultured cells show dramatic changes in the intermediate filaments of cytokeratin and vimentin. During prophase the 10 nm filaments unravel into threads of 2 to 4 nm and into spheroidal aggregates containing both proteins. At metaphase and anaphase most vimentin and cytokeratin appear as spheroid bodies, while at telophase the filamentous cytoskeleton becomes gradually reestablished. These observations suggest that the living cell contains factors that promote the reversible disintegration and restoration of intermediate filaments during mitosis.[51]

Recent studies have focused attention on the embryology and pathology of intermediate filaments. During oncogenesis keratin can be identified at the morula stage and is determined by multiple genes. Various tissues contain intermediate filaments of different nature that can be detected by the specific antibodies. While in some tissues (e.g., muscle) there are several intermediate filament proteins, in other tissues a single type is present. An important point to be stressed is that the tissue specificity is retained after cancerous transformation. For this reason intermediate filament antibodies are useful for identifying the origin of tumor metastasis when standard histological methods are not conclusive (see Lane and Anderton, 1982 in the Additional Readings at the end of this chapter). Such antibodies are also used to label the intermediate filament proteins during cell differentiation, since they appear at definite stages of development (see Anderton, 1983 in the Additional Readings at the end of this chapter).

In conclusion: Intermediate filaments are a distinct component of the cytoskeleton and are profoundly different from microtubules and

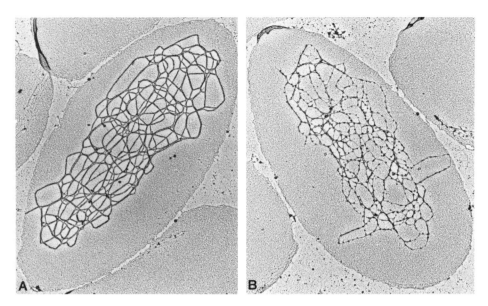

Fig. 6–26. Electron micrographs of hemolyzed chicken erythrocytes showing a network of intermediary filaments that interlinks the nucleus with the plasma membrane. In *A*, the filaments are uniformly decorated with antivimentin. In *B*, antisynemin shows a periodic structure on the filaments. × 8500. (Courtesy of B.L. Granger and E. Lazarides.)

actin filaments. Their biochemical properties, localization, and morphology suggest that all of them, in spite of their different composition, are involved in mechanical functions that determine the cell shape and the position of the various compartments within the cell.

SUMMARY:
Intermediate Filaments

Intermediate filaments have a mean thickness of 10 nm and represent a heterogeneous group from the viewpoint of chemical composition. They can be grouped into four main types:

(1) Keratin filaments (tonofilaments) are present in epithelial cells and contain the α-keratins. (2) Neurofilaments are present in axons, dendrites, and perikarya of neurons. (3) Glial filaments, containing a very acidic protein, are present in astrocytes. (4) Heterogenous filaments have a similar morphology and localization but contain a variety of proteins. This group consists of the desmin filaments which, in muscle, constitute a cytoskeleton that holds the myofibrils within the muscle cell. This cytoskeleton also contains vimentin and synemin. Vimentin filaments are wavy and in many cells tend to form perinuclear caps.

Intermediate filaments are resistant to colchicine and cytochalasin B and are sensitive to proteolysis. They are all involved in mechanical functions contributing to the determination of the cell shape and to the integration of cellular compartments. During mitosis of some epithelial cells, intermediate filaments undergo disintegration and restoration.

REFERENCES

1. Porter, K.R.: *In* Cell Motility. Vol. 1. Edited by R.D. Goldman, et al. Cold Spring Harbor, New York, Cold Spring Harbor Laboratory, 1976.
2. De Robertis, E., and Franchi, C.M.: J. Exp. Med., 98:269, 1953.
3. Margolis, R.L.: Eur. J. Cell Biol., 22(1):296, 1980.
4. Solomon, F.: Cell, 22:331, 1980.
5. Engelborghs, Y.: *In* Techniques and Applications of Fast Reactions in Solutions. Edited by W.J. Gettins and E. Wyn-Jones. D. Reidel Pub. Co., 1979.
6. Rehbun, L.H.: Eur. J. Cell Biol., 22(1):296, 1980.
7. Osborn, M., and Weber, K.: Proc. Natl. Acad. Sci. USA, 73:867, 1976.
8. Connolly, J.A., and Kalnins, V.I.: Exp. Cell Res., 127:341, 1980.
9. De Rosier, D.J., et al.: Nature, 287:291, 1980.
10. Goodenough, V.W., and Heuser, J.E.: J. Cell Biol., 95:798, 1982.
11. Ogawa, K., and Gibbons, R.: J. Biol. Chem., 251:5793, 1976.
12. Stephens, R.E.: *In* Cilia and Flagella. Edited by M.A. Sleigh. New York, Academic Press, Inc., 1974.
13. Warner, F.D., and Satir, P.: J. Cell Biol., 63:35, 1974.
14. De Robertis, E.: J. Biophys. Biochem. Cytol., 2:319, 1956.
15. Mizukanni, I., and Gall, J.G.: J. Cell Biol., 29:97, 1966.
16. Sorokin, S.P.: J. Cell Sci., 3:207, 1968.
17. Brown, S.S., and Sprudick, J.A.: J. Cell Biol., 88:487, 1981.
18. Mc-Lean-Fletcher, S., and Pollard, T.D.: Cell, 20:329, 1980.
19. Burridge, K., and Ferramisco, J.R.: Cell, 19:587, 1980.

20. Lazarides, E., and Revel, J.P.: Sci. Am., *240*:100, 1979.
21. Wegner, A.: Nature, *296*:266, 1982.
22. Matsudaira, P.I., and Burgess, D.P.: J. Cell Biol., *92*:648, 1982.
23. Condeelis, J., and Vahey, M.: J. Cell Biol., *94*:466, 1982.
24. Hartwig, J.H., and Stossel, T.P.: J. Mol. Biol., *134*:539, 1979.
25. Yin, H.L., and Stossel, T.P.: Nature, *281*:583, 1979.
26. Southwick, F.S., and Hartwig, J.H.: Nature, *297*:303, 1982.
27. Kamiya, N., et al.: In International Cell Biology. Edited by H.G. Schweiger. Berlin, Springer-Verlag, 1981.
28. Albrecht-Buehler, G.: J. Cell Biol., *69*:275, 1976.
29. Albrecht-Buehler, G.: Proc. Natl. Acad. Sci. USA. *77*:6639, 1980.
30. Kock, G.L.E.: In International Cell Biology. Edited by H.G. Schweiger. Berlin, Springer-Verlag, 1981.
31. Burridge, K., and McCullogh, A.: J. Supramol. Struct., *13*:53, 1980.
32. Geiger, B.: Cell, *18*:193, 1979.
33. Culp, L.A., et al.: J. Supramol. Struct., *11*:401, 1979.
34. Henderson, D., Geigler, N., and Weber, K.: J. Mol. Biol., *155*:173, 1982.
35. Singer, I.I.: J. Cell Biol., *92*:318, 1982.
36. Birchmeier, W.: Trends. Biochem. Sci., *6*:234, 1981.
37. Oesch, B., and Birchmeier, W.: Cell, *31*:671, 1982.
38. Puck, T.T., et al.: J. Cell Physiol., *107*:399, 1981.
39. Meek, W.D., and Puck, T.T.: J. Supramol. Struct. *12*:335, 1979.
40. Sefton, B.M., and Hunter, T.: Cell, *24*:165, 1981.
41. Barnekow, A., et al.: In International Cell Biology. Edited by H.G. Schweiger. Berlin, Springer-Verlag, 1981.
42. Milan, L., and Erickson, H.P.: J. Cell Biol., *94*:592, 1982.
43. Moell, R., Franke, W.W., and Schiller, D.L.: Cell, *31*:11, 1982.
44. Schlaepher, W.W.: Prog. Neuropathol., *4*:101, 1979.
45. Gilbert, D.: J. Physiol. (*Lond.*), *266*:81, 1976.
46. Granger, B.L., and Lazarides, E.: Cell, *18*:1053, 1979.
47. Franke, W.W., et al.: Proc. Natl. Acad. Sci. USA. *75*:5034, 1978.
48. Starger, A., et al.: J. Cell Biol., *78*:93, 1978.
49. Granger, B.L., and Lazarides, E.: Cell, *22*:727, 1980.
50. Granger, B.L., and Lazarides, E.: Cell, *30*:263, 1982.
51. Franke, W.W., et al.: Cell, *30*:103, 1982.

ADDITIONAL READINGS

Afzelius, B.A.: The immotile-cilia syndrome: A microtubule-associated defect. CRC Critical Rev. in Biochem., *19*:63, 1985.

Albrecht-Buehler, G.: Does the geometric design of centrioles imply their function? Cell Motil., *1*:237, 1981.

Albrecht-Buehler, G.: The tracks of moving cells. Sci. Am., *238*:68, 1978.

Anderton, B.: Cytoskeleton components. Nature, *302*:211, 1983.

Baccetti, B., and Gibbons, I.: International conference on development and function in cilia and sperm flagella. J. Submicrosc. Cytol., *15*:1, 1983.

Birchmeier, W.: Fibroblast's focal contacts. Trends Biochem. Sci., *6*:234, 1981.

Bretscher, A.: Fimbrin is a cytoskeletal protein that cross-links F-actin in vitro. Proc. Natl. Acad. Sci. USA, *78*:6849, 1981.

Cleveland, D.W.: Treadmilling of tubulin and actin. Cell, *28*:689, 1982.

Condeelis, J.S.: Reciprocal interaction between the actin lattice and cell membrane. Neurosci. Res. Prog. Bull., *19*:82, 1981a.

Condeelis, J.S.: Microfilament membrane interactions in cell shape and surface architecture. *In* International Cell Biology. Edited by H.G. Schweiger. Berlin, Springer-Verlag, 1981b.

Craig, S.W., and Pollard, T.D.: Actin-binding proteins. Trends Biochem. Sci., *7*:88, 1982.

Davison, P.F.: Intermediate filaments, intracellular diversity, and interspecies homologies. *In* International Cell Biology. Edited by H.G. Schweiger. Berlin, Springer-Verlag, 1981.

Dentler, W.L.: Microtubule-membrane interaction in cilia and flagella. Int. Rev. Cytol., *72*:1, 1981.

Dustin, P.: Microtubules. Sci. Am., *243*:58, 1980.

Franke, W.W., and Jorcano, J.L.: Organization and cell type-specific expression of the intermediate filament-desmosome complex. III Int. Congress Cell Biol. (S. Seno and Y. Okada, Eds.) Academic Press Japan Inc., 1984.

Gibbons, I.R.: Structure and function of flagellar microtubules. *In* International Cell Biology. Edited by B.R. Brinkley and K.R. Porter. New York, The Rockefeller University Press, 1977.

Gibbons, I.R., and Gibbons, B.: Regulation of sperm flagellar beating. III Int. Congress Cell Biol. (S. Seno and Y. Okada, Eds.) Academic Press Japan Inc., 1984.

Geiger, B.: Membrane-cytoskeleton interaction. Biochim. Biophys. Acta, *737*:305, 1983.

Geiger, B.: Transmembrane linkage and cell attachment: The role of vinculin. *In* International Cell Biology. Edited by H.G. Schweiger. Berlin, Springer-Verlag, 1981.

Goldman, R.D., et al.: Intermediate filaments may function as molecular connecting links between the nucleus and cell surface. III Int. Congress Cell Biol. (S. Seno and Y. Okada, Eds.) Academic Press Japan Inc., 1984.

Goodenough, V.W., and Heuser, J.E.: Substructure of the outer dynein arm. J. Cell Biol., *95*(3):798, 1982.

Hay, E.D.: Cell Biology of Extracellular Matrix. New York, Plenum, 1981.

Hymes, R.: Phosphorilation of vinculin by pp60[src] what might mean. Cell, *28*:437, 1982.

Kakiuchi, S., and Sobue, K.: Control of the cytoskeleton by calmodulin-binding proteins. Trends Biochem. Sci., *8*:59, 1983.

Kleinman, H.K., Klebe, R.J., and Martin, G.R.: Role of collagenous matrices in the adhesion and growth of cells. J. Cell Biol., *88*:473, 1981.

Lane, B., and Anderton, B.: Focus on filaments: Embryology and pathology. Nature, *298*:706, 1982.

Lazarides, E.: Intermediate filaments as mechanical integrators of cellular space. Nature, *283*:249, 1980.

Lazarides, E.: Intermediate filaments, a chemical heterogeneous developmentally regulated class of proteins. Ann. Rev. Biochem., *51*:219, 1982.

Lazarides, E., and Nelson, W.J.: Expression of spectrin in non-erythroid cells. Cell, *31*:505, 1982.

Lazarides, E., and Revel, J.P.: The molecular bases of cell movement. Sci. Am., *240*:100, 1979.

Luck, D.J.L., and Segal, R.: Protein phosphorilation as a regulator of flagellar function in *Chlamydomonas reinhardtii*. III Int. Congress Cell Biol. (S. Seno and Y. Okada, Eds.) Academic Press Japan Inc., 1984.

Margolis, R.L., and Wilson, L.: Microtubule treadmills—possible molecular machinery. Nature, *293*:705, 1981.

McIntosh, J.R.: Microtubule polarity and interaction in mitotic spindle formation. *In* International Cell Biology. Edited by H.G. Schweiger. Berlin, Springer-Verlag, 1981.

Pollard, T.D.: Cytoplasmic contractile proteins. J. Cell Biol., *91*:156, 1981.

Pollard, T.D.: Actin assembly and actin binding proteins from *Acanthamoeba*. III Int. Congress Cell Biol. (S. Seno and Y. Okada, Eds.) Academic Press Japan, Inc., 1984.

Porter, K.R.: Intracellular translocation. Lessons from chromatophores. III Int. Congress Cell Biol. (S. Seno and Y. Okada, Eds.) Academic Press Japan Inc., 1984.

Porter, K.R., and Tucker, J.B.: The ground substance of living cells. Sci. Am., *244*:56, 1981.

Poste, G., and Nicolson, G.L. (Eds.): Cytoskeletal Elements and Plasma Membrane Organization. Cell Surface Reviews. Vol. 7 (Amsterdam), North Holland Pub., 1982.

Rieder, C.L.: Ribonucleoprotein staining of centrioles and kinetochores in newt lung cell spindles. J. Cell Biol., *80*:1, 1979.

Sakai, H., Borisy, G.C., and Mohri, H. (Eds.): Biological Functions of Microtubules and Related Structures. New York, Academic Press, 1982.

Satir, P., Wais-Stlider, J., Lebduska, S., Nasr, A., and Avolio, J.: The mechanochemical cycle of the dynein arm. Cell Motility, *1*:303, 1981.

Sheetz, M.P., and Spudick, J.A.: Movement of myosin-coated fluorescent beads on actin cables in vitro. Nature, *303*:31, 1983.

Sherline, P., and Schiavone, K.: High molecular weight MAPs are part of the mitotic spindle. J. Cell Biol., 77:R9, 1978.

Small, J.V., et al.: Cytoarchitectural interactions in motile fibroblasts associated with substrate contact. III Int. Congress Cell Biol. (S. Seno and Y. Okada, Eds.) Academic Press Japan Inc., 1984.

Steinert, P.M., Steven, A.C., and Roop, D.R.: The molecular biology of intermediate filaments. Cell, *42*:411, 1985.

Tilney, L.G.: Actin: Its association with membranes and the regulation of its polymerization. *In* International Cell Biology. (Edited by B.R. Brinkley and K.R. Porter) New York, The Rockefeller University Press, 1977.

Timpl, R., Engel, J., and Martin, G.R.: Laminin: a multifunctional protein of basement membranes. TIBS, June:207, 1983.

Weeds, A.: Actin-binding proteins—regulators of cell architecture and motility. Nature, *296*:811, 1982.

Whealtly, D.N.: Centriole: A Central Enigma of Cell Biology. Amsterdam, Elsevier Biomedical Press, 1982.

Wiederhold, M.L.: Mechanosensory transduction in "sensory" and "motile" cilia. Annu. Rev. Biophys. Bioeng., 5:39, 1976.

Wilson, L. (Ed.): The cytoskeleton. Part A. Cytoskeletal proteins, isolation and characterization. *In* Methods in Cell Biol. *24*. New York, Academic Press, 1982.

Wilson, L. (Ed.): The cytoskeleton. Part B. Biological systems and in vitro models. *In* Methods in Cell Biol. *25*. New York, Academic Press, 1982.

Yamada, K.M., and Kennedy, D.W.: Dualistic nature of adhesive protein function. J. Cell Biol., 99:29, 1984.

CELLULAR AND MOLECULAR
BIOLOGY OF MUSCLE

In Chapter 6 we discussed the general motility of cells and emphasized that contraction results essentially from the interaction of the contractile proteins, actin and myosin, which are present in the microfilaments of the cytoskeleton. In addition to these force-generating proteins, there are regulatory and structural proteins such as tropomyosin, the troponins, α-actinin, desmin, and vimentin (see Table 6–1). All of these proteins are also present in muscle, where they have a specific organization. Muscle is adapted to the unidirectional shortening during contraction. Because of this, most muscle cells are elongated and spindle-shaped and the major part of the cytoplasm is occupied by contractile *myofibrils* (Fig. 7–1).

In smooth muscle, myofibrils are homogeneous and birefringent. In contrast, in cardiac and skeletal muscle, myofibrils are striated and have dark, birefringent (anisotropic) bands alternating with clear (isotropic) bands. In this chapter, we discuss the structural unit of the myofibril, i.e., the *sarcomere*. Since this structural unit is also a contractile unit, contraction in striated muscle is often called *sarcomeric*.

In muscle cells only a small part of the cytoplasm—the *sarcoplasm*—retains its embryonic characteristics. It lies between the myofibrils, particularly around the nucleus.

Some muscle cells are so highly differentiated that they are adapted to produce mechanical work equivalent to 1000 times their own weight and to contract 100 or more times per second.

The different types of muscle cells are included in histology textbooks and the special types of contraction are in physiology textbooks. Here, the emphasis is on the macromolecular organization of the striated skeletal muscle.

The study of the molecular biology of muscle is one of the most rewarding examples of the intimate association between structure and function and of the way in which chemical energy is transformed into mechanical work.

In this chapter, we also cover the coupling

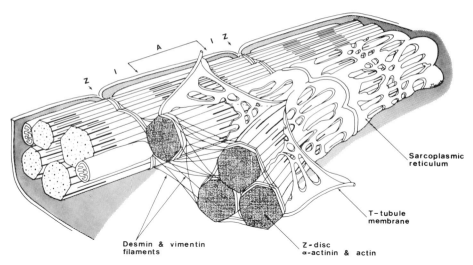

Fig. 7–1. Diagram of a striated muscle fiber showing the myofibrils with the sarcomeres, the sarcoplasmic reticulum, and the T-tubules joining the sarcolemma. The organization of the Z-discs containing α-actinin plus actin, and desmin plus vimentin is emphasized. The last two proteins, together with synemin, form intermediate filaments. Z, Z-disc; I, I-band; A, A-band. (Courtesy of E. Lazarides.)

between excitation and contraction, which is based on a complex intracellular conducting system that carries the action potential to the deeper portion of the muscle fiber, producing the functional synchronization of the myofibrils. In this functional coupling, sarcoplasmic reticulum (SR), calcium ions, and Ca^{++}-activated ATPase play an essential role. Muscle is an admirable example of physiological and structural integration at the macromolecular level comparable only to that of mitochondria and chloroplasts.

7–1 STRUCTURE OF THE STRIATED MUSCLE FIBER

Striated skeletal muscles are composed of multinucleate cylindrical fibers, 10 to 100 μm in diameter and several millimeters or centimeters long. These enormous structures arise in the embryo by the fusion of several primordial cells, the so-called *myoblasts*, which first form a *myotube*, and then a fully differentiated *muscle fiber*.

The entire fiber is surrounded by an electrically polarized membrane with an electrical potential of about −0.1 volt; the inner surface is negative with respect to the outer surface. This membrane, called the *sarcolemma*, becomes depolarized physiologically each time a nerve impulse that reaches the motor innervation of the

muscle *(end plate)* activates the membrane. The final result is a coordinated contraction of the entire muscle fiber. Three cytoplasmic components are highly differentiated in the muscle fiber. One is represented by the contractile machinery, which is essentially made of protein myofilaments and is formed embryonically within the cytoplasmic matrix. The myofilaments are disposed in parallel to form the larger fibrillar structures, the *myofibrils* (Fig. 7–1).

The arrangement of the myofilaments determines the different classes of muscle that are now recognized. For example, in striated skeletal and heart muscle of vertebrates the filaments are longitudinally oriented, and there is also a transverse repeating organization. In other types of nonstriated muscles the filaments are oriented in longitudinal or oblique arrays or have a more or less random distribution. The second component of striated muscle is a special differentiation of the endomembrane system, the so-called SR, which is involved with conduction inside the fiber and with coordination of the contractions of different myofibrils, in addition to being related to the relaxation of the muscle after a contraction (Fig. 7–1).

The third component is represented by numerous mitochondria, the so-called *sarcosomes*, which in some cases may attain large dimensions. The abundance of mitochondria may be related to the constancy with which the muscle contracts; for example, there is a greater number in steadily active muscles, such as the heart.

The Myofibril and Sarcomere are Structures Differentiated for Contraction

Myofibrils are long cylindrical structures about 1 μm in diameter that have transverse striations. These striations consist of the repetition of a fundamental unit, the sarcomere, which is limited by a dense line called the Z-line or Z-disc. This line is located in the center of a less dense zone known as the I-band, which corresponds to the relatively *isotropic* disc (for a definition of *isotropy* and *anisotropy*, see Section 3–1). The A-band, which is *anisotropic* under polarized light, has a greater density than the I-band. Under certain conditions, a less dense zone may be observed in the center of the A-band, subdividing it into two dark semi-

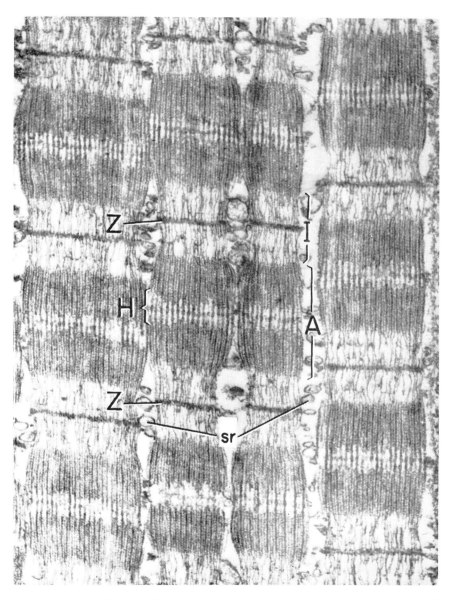

Fig. 7–2. Electron micrograph of four myofibrils, showing the alternating sarcomeres with the Z-discs and the H-, A-, and I-bands; *sr*, sarcoplasmic reticulum situated between the myofibrils. The finer structure of the myofibril represented by the thin and thick myofilaments is also observed. ×60,000. (Courtesy of H. Huxley.)

discs (Fig. 7–2). This zone constitutes the H-disc (Hensen's disc). In the middle of the H-disc an M-line can be observed.

In a relaxed mammalian muscle, the A-band is about 1.5 μm long, and the I-band, 0.8 μm. The striations of the myofibrils are the result of periodic variations in density, i.e., in concentration of the finer structures, the *myofilaments*, along the axis. These striations are in register in the different myofibrils, thus giving rise to the striation of the entire fiber.

Thick and Thin Myofilaments Are the Macromolecular Contractile Components

Morphologically, there are two kinds of myofilaments, *thick myofilaments* that are about 1.5 μm long and 10 nm wide and that are separated by a 40 nm space, and *thin myofilaments* that are about 1.0 μm long and 5 nm in diameter. Thick myofilaments are made of *myosin*, and thin myofilaments are composed of a more complex structure containing several proteins (i.e., actin, tropomyosin, and troponins) of which *actin* is the most important.

As shown in Figure 7–3, these two types of filaments are disposed in register and overlap to an extent that depends on the degree of contraction of the sarcomere. In a relaxed condition, the I-band contains only thin filaments; the H-band contains only thick filaments; and within the A-band the thick and thin filaments overlap. In a cross section through the A-band, the regular disposition of the two types of filaments can be observed best (Fig. 7–3). In vertebrate mus-

cle each thick filament is surrounded by six thin filaments, and each thin filament lies symmetrically among three thick ones (Fig. 7–3). As a consequence of this geometry there are twice as many thin filaments as thick ones.

A cross section through the H-band shows only thick myofilaments and through the I-band, only thin myofilaments.

Another interesting detail revealed by the electron microscope (EM) is that the two sets of filaments are linked together by a system of cross-bridges. These arise from the thick filaments at intervals of about 7 nm. Each bridge is situated along the axis with an angular difference of 60°. This means that the bridges form a helix about every 43 nm. As a result of this arrangement one thick filament joins the six adjacent thin ones every 43 nm.

A repeat period corresponding to this distance can be observed in the A-band after staining with uranyl acetate. There are 11 stripes in each half of an A-band starting from the M-line. Some of these are due to the myosin cross-bridges, but others, to the presence of C-protein.[1] The complex fine structure of the A-band can also be revealed by negative staining of ultrathin sections of frozen muscle (i.e., cryosections).[2] In cross section the M-line shows that the thick myofilaments are joined, forming a triangular lattice (Fig. 7–3).

The Z-disc Shows a Woven-basket Lattice and Contains α-Actinin, Desmin, Vimentin, and Synemin

In cross section the Z-disc shows a woven-basket lattice, which remains essentially un-

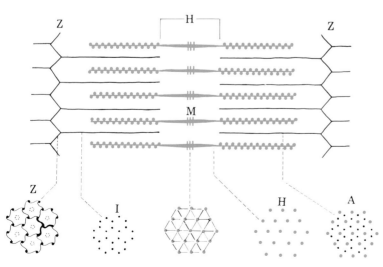

Fig. 7–3. Diagram representing the structure of one sarcomere in striated muscle in longitudinal and transverse sections. Observe that the I-band has only thin filaments, the H-band only thick ones, and the A-band both thick and thin filaments. Special structures are observed at the Z- and M-discs.

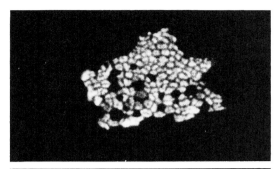

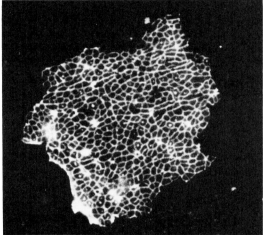

Fig. 7—4. Immunofluorescent staining of isolated Z-discs from skeletal muscle. *Top,* view of α-actinin localized in the central part of the disc. × 3250. *Bottom,* view of desmin localized at the periphery (similar results are obtained for vimentin or synemin). × 2180. (Courtesy of E. Lazarides and B.L. Granger.)

changed during contraction. This lattice is made of Z-filaments that are connected to the thin filaments of the I-band. It is presumed that as one thin filament enters the Z-disc, it is in continuity with three curved Z-filaments, which unite it with three other thin filaments of the same sarcomere. According to this model, the thin filaments of the opposite sarcomere are similarly arranged. The hexagonal lattice that is characteristic of the A-band (one thick filament surrounded by six thin ones) becomes compressed at the I-Z junction and is transformed into the square lattice characteristic of the woven-basket pattern (Fig. 7–3). This model, in which there is no interlooping of filaments from one sarcomere to the other, explains the splitting of the Z-disc that may occur under certain conditions.

The Z-discs can be isolated by homogenization

of the muscle, after extraction of actin and myosin. Under these conditions, the Z-discs of many myofibrils lie flat; therefore, it is possible to study the localization of the main protein constituents by immunofluorescence with specific antibodies. Figure 7–4 shows that *α-actinin* occupies the central domain of the disc, together with actin and an *85 K daltons protein*, while *desmin, vimetin,* and *synemin* are located at the periphery (see Table 6–1).[3]

As mentioned in Chapter 6, desmin, vimetin, and synemin form intermediate filaments. These filaments link the individual myofibrils at the Z-discs and all the myofibrils to other organelles and to the cell surface as indicated in Figure 7–1. This arrangement of the cytoskeleton of muscle ensures the alignment of the contractile apparatus during the contraction-relaxation cycle.

That the Z-disc has two differentiated portions poses the question of how they are assembled in cultures during myogenesis. Studies with fluorescent antibodies show that the first to appear is α-actinin with a punctate pattern along the actin myofilaments. Later, during myogenesis, vimentin and desmin become associated to α-actinin at the Z-disc (see Lazarides, 1981 in the Additional Readings at the end of this chapter).

A new high molecular weight protein called *paranemin* (280,000 daltons) has been identified in developing muscles. In myotubes, this protein, together with desmin, vimentin, and synemin, is localized in cytoplasmic myofilaments and later becomes associated with the Z-disc. With further development of the muscle, however, paranemin becomes gradually reduced and is no longer detected in the adult muscle.[4]

The Sarcomere I-Band Shortens and the Banding Inverts During Contraction

In a living muscle fiber, changes with contraction can be observed with the phase contrast and interference microscope. If this is done, one striking observation that can be made is that the A-band remains constant in a wide range of muscle lengths, whereas the I-band shortens in accordance with the contraction.

The shortening of the I-band is the result of the fact that the thin myofilaments slide farther and farther into the arrays of thick filaments (Fig. 7–5). With the progressing contraction,

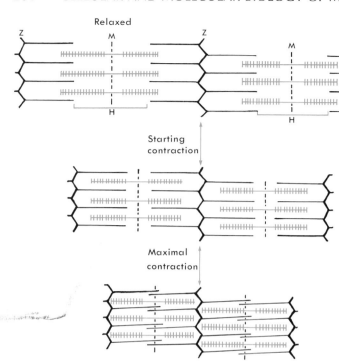

Fig. 7–5. Diagram showing the sliding model of contraction. In the relaxed condition the H-bands are wide and contain only thick filaments. With the beginning of contraction, the thin filaments slide toward the center of the sarcomere. With maximal contraction, the thin filaments penetrate into the H-band and produce inversion of the banding.

the thin filaments penetrate into the H-band and may even overlap, thereby producing a more dense band in the center of the sarcomere (*inversion of the banding*). Finally, the thick filaments make contact and are crumpled against the Z-discs.[5] These findings have been interpreted in the so-called *sliding filament mechanism* of contraction. The degree of contraction thus achieved can be measured by determining the length of the sarcomere (i.e., the distance between Z-discs) at rest and when it has shortened. Note that insect muscle, in general, shortens only slightly (about 12%), whereas the shortening in vertebrate muscle may be much greater (about 43%).

Smooth Muscles Lack the Z-disc

The EM has revealed that "smooth" muscles may have a varied macromolecular organization. In many cases they contain thin and thick myofilaments, as do striated muscles, but the difference lies in the absence of the Z-disc and the lack of periodicity. In mollusks and annelids there are muscles with a helical arrangement that have thin and thick myofilaments linked by cross bridges. In the adductor muscle of the oys-

ter, the so-called paramyosin muscle, each thick filament is surrounded by 12 thin filaments.

The smooth muscle of vertebrates apparently lacks these two types of myofilaments and even myofibrils are difficult to recognize. However, improvements in the preparative techniques have demonstrated coarse myofilaments. Furthermore, thick and thin myofilaments have been isolated. In smooth muscles the contraction is very slow, but extreme degrees of shortening may be achieved.

SUMMARY

The Structure of Muscle

Muscle cells are adapted to mechanical work by unidirectional contraction. The functional unit is the *myofibril*, which may be either striated or smooth. In skeletal muscle, myofibrils fill most of the large muscle fiber, leaving small amounts of sarcoplasm, which contains the nuclei, the SR, and large mitochondria, or *sarcosomes*. A *sarcolemma*, with -0.1 volt polarization, surrounds the fiber. This is activated through the *end plate*.

Myofibrils result from the repetition of *sarcomeres*. These are limited by the Z-lines and contain the I-bands, the A-bands, and the H-band. The M-line may be observed in the middle of the sarcomere. The myofibril is composed of thick (10 nm) and thin (5 nm) myofilaments. In the relaxed condition, the I-band contains only thin myofilaments, the H-band only thick, and the A-band, both thick and thin. Myofilaments are organized in a paracrystalline hexagonal array. From the thick filaments, at 7 nm intervals, there are cross-bridges extending toward the thin filaments. One thick filament joins to six adjacent thin ones every 43 nm.

In each half of an A-band there are about 11 transverse stripes, some of which are due to the cross-bridges, others to C-protein. The Z-disc shows a woven-basket square lattice made of Z-filaments of α-actinin, which are in continuity with the thin filaments of each sarcomere.

When the Z-discs are isolated they show that, while actin and α-actinin occupy the central domain; at the periphery, there are intermediate filaments of desmin, synemin, and vimentin that link all myofibrils, ensuring the alignment of them during the contraction-relaxation cycle.

During contraction the I-bands shorten as a result of the sliding of the thin filaments into the A- and H-bands. A further reversal of the band may be produced at the center of the sarcomere. In vertebrate muscle, the sarcomere may shorten up to 43%.

7–2 MOLECULAR ORGANIZATION OF THE CONTRACTILE SYSTEM

Knowledge about the molecular machinery involved in muscle contraction has become increasingly complex. To the classic force-generation proteins *myosin* and *actin* others having a regulatory function have been added, such as *tropomyosin* and the various *troponins*. Finally a series of other proteins, probably playing a structural role, such as *α-actinin* present in the Z-disc, the *M-disc proteins*, and the *C-protein*, have been described.

Here, we shall consider these proteins from the standpoint of their relation to the structure and function of the myofibrils.

The Thick Myofilament—Myosin Molecules; The Cross-Bridges—S₁ Subunits

Myosin can be extracted from muscles with a 0.3 M solution of KC1 and purified by precipitation at lower ionic strength. This large molecule comprises about half of the total protein of the myofibril and has a molecular weight of about 500,000 daltons. It contains 2 polypeptide chains of about 200,000 daltons each and 4 smaller ones in the range of 20,000 daltons. As shown in the diagram of Figure 7–6, the 2 heavy chains are coiled around each other, forming a double helical α-helix about 140 nm long and 2

nm in diameter. Only about half the heavy chain is rod-like, the rest, together with some of the smaller chains, is folded into two globular regions at one end of the molecule.

Myosin can be fragmented by the action of proteolytic enzymes; *trypsin* breaks it into the long, rod-like *light meromyosin (LMM)* and *heavy meromyosin* (HMM). This last portion can be further subdivided by *papain* into the globular *subunit* S_1 and the *helical rod* S_2, which joins S_1 to the light meromyosin. The most important part of the myosin molecule is HMM-S_1, since it contains the sites for the *ATPase* and for *binding to actin*. In fact it is the interaction between myosin and actin that results in contraction.

Myosin molecules are heterogeneous in different muscles, in the composition of heavy and light chains. For example, there are differences between fast and slow contracting muscles. In addition, after long-term denervation, the slow myosin disappears while the fast myosin persists. These findings suggest that the molecular organization of muscle is plastic and there is a continuous regulation by neural and nonneural mechanisms.[6]

At this point it is important to mention how the myosin molecules are integrated to constitute the *thick* or *myosin myofilament*. Each of these filaments has a smooth or bare portion in the middle (corresponding to the H-band) and two regions in which the surface is rough because of the presence of the *myosin cross-bridges*. It is now known that these cross-bridges represent the S_1 globular ends of myosin, while the shaft of the filament is formed by the rod-like portion of the molecule. As shown in Figure 7–7B, one half of the molecules in the myofilament are oriented in the opposite direction from the other half. In other words, there is a definite polarization of the S_1 ends in each half of the myofilament.

From the study of cross sections of the myosin filaments, which show a triangular shape, a model in which 12 parallel units are closely packed has been postulated.[7] Each one of the units, with a diameter of 4 nm, could be the result of the association of two myosin molecules.[8] Another interesting observation is that at the M-line a cross section of the myosin filament shows bridges that connect with the six neighboring thick filaments, forming a hexagonal lat-

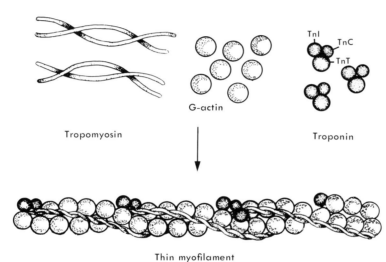

Fig. 7–6. Diagram indicating the molecular structure of the thick myofilaments (top) and thin myofilaments (bottom) in the myofibril. The myosin molecule is made of two polypeptide chains 140 nm long, having at one end the S_1 head (LMM-light meromyosin, HMM-S_2, heavy meromyosin-S_2 HMM-S_1, heavy meromyosin-S_1). Observe that the thick filament results from the assembly of myosin molecules. The thin filament is the result of the association of G-actin monomers, tropomyosin and the troponins (TnT, troponin-T; TnC, troponin-C; and TnI, troponin-I). Note that each tropomyosin molecule extends over seven actin molecules.

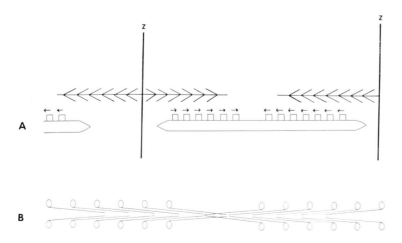

Fig. 7–7. *A,* diagram illustrating the polarity of cross-bridges in the myosin filaments and in the actin myofilaments. The sliding forces tend to move the actin filaments toward the center of the sarcomere. *B,* arrangement of myosin molecules within the thick filaments. Each molecule has a tadpole shape with a globular head and a tail. The axis of the filament is formed by the assembly of the tails. Observe that in each half of the myofilament the molecules are polarized. (From Huxley, H.E.: Proc. R. Inst. Gr. Br., *44*:274, 1970.)

tice (Fig. 7–3). Such bridges represent the attachment of the M-protein at the M-line of the sarcomere.[7]

Actin, Tropomyosin, and the Troponins Constitute the Thin Myofilament Structural Proteins

Actin needs high concentrations of KCl (0.6 M) to be extracted. It is the second major protein and represents about one quarter of the protein of the myofibril. The molecular weight of actin is 42,000 daltons. The actin myofilament is made up of two helical strands that cross over every 36 to 37 nm. Each of these cross-over repeats contains between 13 and 14 *globular* monomers of *G-actin* about 5 nm in diameter (Fig. 7–6). In muscle, actin is present mainly as fibrous or *F-actin*, which is the polymerized form of *G-actin*.

Tropomyosin. This protein represents between 5 and 10% of the total and is extracted with 1 M KCl or weak acids. It has a molecular weight of 64,000 daltons and is an elongated molecule of 40 nm made of two α-helical subunits. As shown in the diagram of Figure 7–6, tropomyosin stretches over seven actin monomers and forms two helixes that lie within the grooves of the actin double helix.

Troponins. Together with tropomyosin the troponins are a family of three small proteins that form a complex of about 80,000 daltons. Both proteins constitute a Ca^{++}-sensitive switch, within the thin filament, that regulates the contraction. Each of the troponin complexes binds to tropomyosin every 40 nm (i.e., every seven active monomers) as can be seen with antitroponin antibodies (Fig. 7–6). The three components present in equimolecular proportions are: (1) troponin-C (molecular weight 17,000 to 18,000 daltons), a small protein that specifically binds Ca^{++}; (2) troponin-I (molecular weight 22,000 to 24,000 daltons), a protein that binds to both actin and troponin-C and inhibits the ATPase of myosin;[9] and (3) troponin-T (molecular weight 37,000 to 40,000 daltons), the largest subunit that binds to tropomyosin and also interacts with the other troponins. Troponin has been considered a localized trigger that controls the motion of tropomyosin in the thin filament. Recent EM observations have revealed that the troponin complex has a tadpole shape with a globular domain composed mainly of troponin-C and -I, and a long tail that interacts with tropomyosin and corresponds mainly to troponin-I. It is suggested that this lengthwise interaction may be important in the regulation of the switching process[10] (see Ebashi, 1976 in the Additional Readings at the end of this chapter).

Among the structural proteins the following are considered:

α-Actinin. This protein is a rod-shaped molecule of about 95,000 daltons localized at the Z-disc (Fig. 7–4).

85 K Protein. This protein is present in the central region of the Z-disc and has a molecular weight of 85,000 daltons.

C Protein. This protein is present in the A-band along the middle portion of the myosin filament. With electron microscopy the anti-C-protein antibody is observed to form between 7 and 9 transverse stripes in each half of the A-band.[1,2]

M-Line Proteins. At the center of the H-band special proteins that form part of the M-line have been found. In addition to a creatine kinase, there is a structural protein called *myomesin*. This protein has a molecular weight of 165,000 daltons, can be detected in myoblasts, and has a high affinity for myosin. Myomesin acts as a specific marker during the differentiation of myoblasts into myotubes and muscle fibers. It is present in skeletal and heart muscle, but not in smooth muscle. Myomesin also plays a key role in the molecular organization of the thick myofilaments.[11]

Another M-line protein of 185,000 daltons was recently identified when monoclonal antibodies were used (see Grove et al., 1984 in the Additional Readings at the end of this chapter).

An Elastic Protein of Muscle. In recent years, a high molecular weight protein (one or more megadaltons) has been isolated from the myofibril and named *connectin* (see Kimura and Maruyama, 1983 in the Additional Readings at the end of this chapter). or *titin* (see Wang et al., 1984 in the Additional Readings at the end of this chapter). This protein can be isolated by using denaturing detergents such as sodium dodecyl sulfate. It represents 8 to 12% of the myofibrillar mass. Under the EM this protein appears as an extremely thin and long (more than 1 μm) strand.

Titin. It is extensible and flexible and shows

an axial periodicity. This protein interacts with both myosin and actin and forms nets that are concentrated at the A-I junction of the myofibril. Because of its elastic properties it is ideally suited to constitute a lattice or scaffold for the thick and thin myofilaments that contribute to the generation of passive tension when the muscle is stretched.

7–3 THE SLIDING MECHANISM OF MUSCLE CONTRACTION

We mentioned earlier that contraction is currently explained by a *sliding mechanism* in which the thin actin filaments are displaced with respect to the thick myosin filaments during each contraction-relaxation cycle.

The force generated is proportional to the degree of overlap between thick and thin filaments. This implies that the force is of a short range and is produced directly by the cross-bridges between the filaments. A resting muscle is rather plastic and extensible because the cross-bridges are not attached and can slide upon an applied external force (Fig. 7–8A).

Active sliding movement is thought to result from the repetitive interaction of the cross-bridges with the actin filaments. The following mechanism is postulated: (1) A cross-bridge binds to a specific site of the actin filament. (2) The cross-bridge undergoes a conformational change, which displaces the point of attachment toward the center of the A-band, thereby pulling the actin filament; at the same time a second cross-bridge becomes attached. (3) At the end of the cycle the first bridge returns to the starting configuration in preparation for a new cycle. According to this theory, the actin filaments from each half sarcomere are pulled as ropes toward the center of the sarcomere by the myosin arms that move to and fro (Fig. 7–8B through D).

Recently, support for this theory has been obtained from X-ray diffraction studies in living muscle, using high beam intensity and high speed time resolution. It was observed that a 14.3 nm reflection, believed to be produced by cross-bridges, changes in intensity with rapid stretches and releases. Such changes provide direct evidence that there is an active sliding mechanism operating during contraction.[12]

The energy for this interaction is provided by the splitting of ATP because of the ATPase pres-

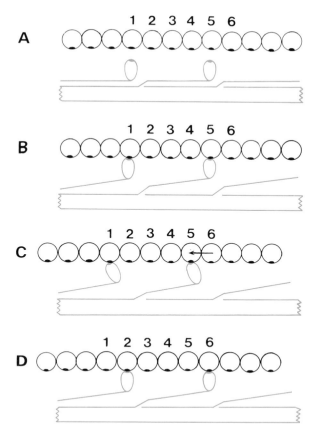

Fig. 7–8. A schematic interpretation of the mode of action of cross-bridges in the sliding of myofilaments. *A*, at relaxation there is no attachment of the S_1 cross-bridges to the actin filament; *B*, at the start of contraction interaction of S_1 and actin occurs; *C*, a conformational change in S_1 causes the actin myofilament to move toward the center of the A-band; *D*, at the end of the stroke S_1 detaches, returns to the original configuration, and re-attaches to the following G-actin.

ent in the cross-bridge. It is estimated that the splitting of one ATP accounts for a displacement of 5 to 10 nm for each cross-bridge.

Myosin and Actin Have a Definite Polarization within the Sarcomere

One of the most characteristic features of muscle is a molecular organization by which individual molecules may interact both spatially and temporally in a concerted fashion. Thus, the sliding mechanism depends on a very specific interaction between actin and myosin molecules and requires the existence of a *definite polarization* within the sarcomere. In each half sarcomere actin and myosin should have a different polarization; furthermore, the forces developed must have opposite directions (Fig. 7–7).

From electron microscopic studies of aggregates of myosin molecules, it is possible to obtain reconstituted myofilaments having the structure shown in the diagram of Figure 7–7B. The molecules of each half of the filament are in two antiparallel sets with a reversal of structural polarity in the center.

Muscle Contraction—A Cyclic Formation and Breakdown of Actin-Myosin Linkages

Since the classic experiments of Albert Szent-Györgyi, it has been known that the mixing of myosin and actin in the test tube results in the formation of the complex *actomyosin*, which contracts in the presence of ATP. Investigators following this interaction under the EM have observed that the myosin molecules bind to the F-actin with a directional orientation. As shown in Figure 7–9, the complex can also be produced by using F-actin and heavy meromyosin. The actin filaments become "decorated" with the myosin, and the complex shows an arrow-like

polarity. These findings demonstrate that in each filament of F-actin, the G-actin molecules are polarized with the same orientation.

Figure 7–10 shows a model illustrating the double helical filament of actin decorated with the S_1 myosin fragments.[13] The arrow-like configuration shown in the electron micrograph of Figure 7–9 results from the tilting and bending of the S_1 myosin along the helix of F_1 actin. The tilting of the S_1 myosin most likely corresponds to the configuration that the cross-bridge has at the end of the working stroke, when ADP and Pi have been released and no further force is being exerted. On the other hand, there is electron microscopic and x-ray diffraction evidence that in the relaxed muscle the bridges are approximately perpendicular to the axis of the myofilaments. (For further details and x-ray diffraction data, see Huxley, 1976 in the Additional Readings at the end of this chapter.)

In conclusion: The detailed ultrastructural and biochemical information obtained permits an interpretation of the macromolecular mechanisms involved in muscle contraction. It may be postulated that this is a cyclic event involving

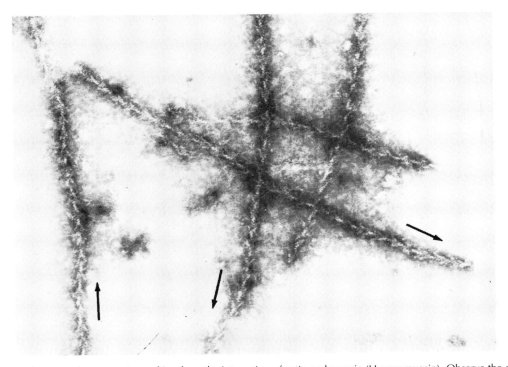

Fig. 7–9. Filaments of actomyosin resulting from the interaction of actin and myosin (H-meromyosin). Observe the arrow-like polarity of the actomyosin complex (arrows). Negatively stained; × 155,000. (Courtesy of H.E. Huxley.)

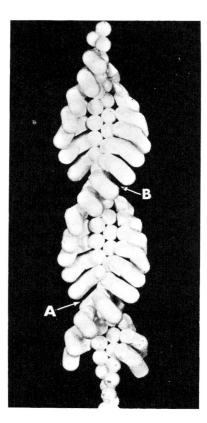

Fig. 7–10. Three-dimensional model of an actomyosin complex. In the axis is the double helical filament of F-actin (see Fig. 7–6) made of G-subunits. The actin has been treated with fragments of heavy meromyosin (fragment S_1, see Fig. 7–9). Points marked A and B are cross-over points in the actin helix. The arrowhead appearance is produced by the change in position of the point of attachment of the S_1 subunit to the actin molecules. (From Huxley, H.E.: Proc. R. Inst. Gr. Br., 44:274, 1970.)

the repetitive formation and breakdown of actin-myosin linkages at the bridges between thick and thin filaments. At each bridge the following sequence of events is probably produced: (1) formation of a perpendicular linkage between a heavy meromyosin head and one G-actin (globular) unit; (2) rupture of this linkage by one ATP molecule; (3) hydrolysis of the ATP by the Ca^{++}-activated ATPase of myosin; (4) formation of a new linkage between the same heavy meromyosin (bridge) and the next G-actin unit. The relative movement of the thin filament taking place in each sequence would be equivalent to the length of one-G-actin unit (5.3 nm).

SUMMARY:

Molecular Organization and the Sliding Mechanism

The molecular machinery involved in muscle contraction comprises (1) force-generating proteins (*myosin* and *actin*), (2) regulatory proteins (*tropomyosin* and *troponins*), and (3) structural proteins (α-actin in the Z-disc, M-disc proteins, and the C-protein).

Myosin represents 50% of the myofibril protein, and its molecular weight is about 500,000 daltons, with 2 polypeptides of 200,000 daltons and 4 of 20,000. Trypsin separates the rod-like light meromyosin from the heavy meromyosin, which can be divided further by papain into the subunits S_1 and S_2. S_1 contains ATPase and the binding site for actin. Myosin is in the thick myofilament, and S_1, in the cross-bridge. There is a definite polarization of both the myosin molecule and the S_1 in each half sarcomere (Fig. 7–7).

Actin represents about 25% of the myofibril protein, with a molecular weight of 42,000 daltons. The monomeric or globular actin (G-actin) of about 5 nm polymerizes into fibrous actin (F-actin) in a double helical strand with cross-overs every 13 to 14 monomers. The actin filament also contains tropomyosin and the troponins.

Tropomyosin (5 to 10%) is a molecule of 64,000 daltons and forms two elongated polypeptides 40 nm long that extend the length of 7 G-actins, within the grooves of the actin double helix.

Troponin has three components, troponin-C (binding Ca^{++}), troponin-T (binding tropomyosin), and troponin-I (inhibiting the ATPase). Troponin binds to tropomyosin every seven G-actins.

The *sliding mechanism of muscle contraction* postulates that the thin actin filaments are displaced with respect to the thick filaments at each contraction-relaxation cycle. Short-range forces are generated at the cross-bridges between the filaments. An S_1 portion of the myosin molecules attaches to a binding site of F-actin, and by a conformational change, pulls it toward the center of the sarcomere. The splitting of ATP by the Ca^{++}-activated ATPase provides the energy. One ATP may account for a 5 to 10 nm displacement of a cross-bridge. According to this theory, both the myosin and actin myofilaments should be polarized in each of the sarcomeres (Fig. 7–7). This polarization is actually demonstrated by the electron microscopic observation of reconstituted myosin filaments and of actin filaments treated with heavy meromyosin. The arrow-like configuration of these "decorated" filaments is the result of the F_1-actin helix and the tilting and binding of the S_1 heavy meromyosin.

7–4 REGULATION AND ENERGETICS OF CONTRACTION

In the *relaxed state* there is no attachment of the cross-bridges; the opposite condition could explain the state of *rigor* (i.e., of permanent contraction). In this state, the rigidity of the muscle could be due to the permanent attachment of the cross-bridges to the thin filaments. In order to have a contraction-relaxation cycle, a cyclic mechanism by which the cross-bridges become attached and detached must operate. It is now known that the transition between rest and activity is dependent on the *concentration of free calcium* in the vicinity of the contractile machinery (see Murray and Weber, 1974 in the Additional Readings at the end of this chapter).

The control of contraction by Ca^{++} requires the presence of the regulatory proteins tropomyosin and troponin. *Troponin-C* (Ca^{++}-binding) and *troponin-I* and *troponin-T* are placed along the actin filament at intervals of about 40 nm and such a periodicity corresponds to 7 G-actin monomers. Tropomyosin also extends along the actin filament for the same distance and is in close relationship with the troponin complex (Fig. 7–6).

In the regulation of the contraction-relaxation cycle by calcium, the troponin complex plays a fundamental role. At low Ca^{++} concentrations (i.e., 10^{-7} M) troponin-I inhibits the interaction between actin and myosin. When the Ca^{++} concentration increases above 10^{-6} M, this bivalent ion binds to troponin-C.[14] Remember that this protein has important similarities with the more diffuse Ca^{++}-binding protein calmodulin (Fig. 2–6). In fact, troponin-C has a 50% homology in amino acid composition with calmodulin and four calcium-binding sites. The binding of Ca^{++} to troponin-C relieves the inhibition by troponin-I and causes a conformational change that is transmitted by way of tropomyosin to the seven G-actin monomers to which it is associated. This chain of events leads to the contraction of the muscle.

Molecular Regulation—Displacement of Tropomyosin After Binding of Ca^{++} to Troponin-C

From electron microscopic and x-ray diffraction evidence a model has been constructed that may explain the regulatory action of the tropomyosin-troponin-Ca^{++} system.[14-16]

In this model (Fig. 7–11), an end-on view of the relative positions of actin, the S_1-myosin, and tropomyosin is presented. The tropomyosin

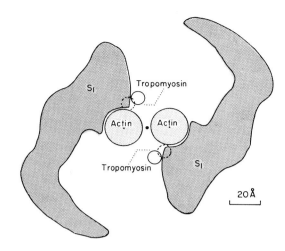

Fig. 7–11. A possible molecular model of regulation of contraction. The diagram shows an end-on view of the relative positions of actin and the S_1-myosin. Tropomyosin is shown with a line contour in the activated condition, in which there is interaction between actin and myosin. The dotted contour corresponds to the relaxed condition, in which tropomyosin is deep in the groove of F-actin and covers the actin binding site of S_1-myosin. (Courtesy of H. Huxley.)

molecule is shown in two positions, one corresponding to the activated state and the other to the relaxed state. According to the model in the latter condition (dotted contour) tropomyosin is deep in the groove of F-actin and covers the actin binding site of the S_1-myosin.

When Ca^{++} binds to troponin-C a displacement of tropomyosin away from the groove could be effected, exposing the actin binding site in S_1 for actin-myosin interaction.

Energy for Contraction Comes from Oxidative Phosphorylation and Glycolysis

Whereas the myofibrils constitute the mechanical machinery of the muscle, the fuel needed is produced mainly in the sarcoplasm. In all types of muscle, numerous mitochondria, called *sarcosomes*, provide the essential oxidative phosphorylation processes and the Krebs cycle system (see Chapter 11). These mitochondria are particularly prominent in size and number in heart muscle and in the flight muscles of birds and insects.

The sarcoplasmic matrix contains the glycolytic enzymes as well as other globular proteins, such as myoglobin, salts, and high phosphate compounds. Glycogen is present in the matrix as small granules or glycosomes observed under the EM. There are about 1% glycogen and 0.5% creatine phosphate as sources of energy in muscle. Glycogen disappear with contraction through glycolysis, and lactic acid is formed, which can be transformed into pyruvic acid to enter the Krebs cycle (see Chapter 11).

The initial energy source for contraction is ATP. The ADP produced after the initial contraction is again recharged to ATP by glycolysis or from creatine phosphate. Oxidative phosphorylation is the last and most important source of ATP.

SUMMARY:
Regulation and Energetics of Contraction

During the contraction-relaxation cycle there is a cyclic attachment and detachment of cross-bridges to actin filaments. (In *relaxation* there is no attachment, and in the state of *rigor* the attachment is permanent.) Regulation of this cycle depends on the Ca^{++} concentration and the presence of the regulatory proteins tropomyosin and the troponins. When Ca^{++} increases above 10^{-6} M in the cytosol, contraction is triggered by the binding of Ca^{++} to troponin-C. The influence of troponin is transmitted to seven G-actin monomers by way of tropomyosin. This regulatory action has been analyzed by electron microscopy and x-ray diffraction, and a model has been constructed. According to this, in the relaxed state tropomyosin is deep in the groove of F-actin and covers the actin binding site of S_1-myosin. When Ca^{++} binds to troponin-C, tropomyosin is displaced from the groove, and the binding site in S_1 is exposed for the actin-myosin interaction.

The initial source of energy for contraction is provided by ATP. This energy is furnished by oxidative phosphorylation in mitochondria *(sarcosomes)*, by glycolysis of glycogen, and by creatine phosphate.

7–5   EXCITATION-CONTRACTION COUPLING

Excitation-contraction coupling refers to the mechanism by which the electrical impulse is able to induce contraction in a muscle. It is known that with the arrival of the nerve impulse at the motor end plate a *chemical synaptic transmission* occurs (see Section 24–3). This in turn brings about a depolarization of the cell membrane, which is conducted along the muscle fiber and penetrates into it to induce the contraction.

The essential components in this excitation-contraction coupling mechanism are (1) the *T-system*, which conducts the action potential to the interior of the muscle fiber; (2) the release of Ca^{++} from the *SR;* (3) the induction of contraction by Ca^{++}; and (4) the reaccumulation of Ca^{++} into the SR by Ca^{++}-*ATPase*, which results in *muscle relaxation.* The study of this mechanism involves the knowledge of each one of the elements involved (see Ebashi, 1976; Hasselbach, 1977 in the Additional Readings at the end of this chapter).

The SR—A Longitudinal Component with Terminal Cisternae that Form Part of the Triad

The SR found in skeletal and cardiac muscle fibers is one of the most interesting specializations of the ER (see Chapter 8). It was discovered by Veratti in 1902 as a reticulum present in the sarcoplasm of the muscle fiber and extending in between the myofibrils. It was completely neglected until 1953 when Keith Porter and H. Stanley Bennett made the first electron micrographs of this structure.[17,18]

The SR is a membrane-limited reticular system whose organizational structure is regularly superimposed upon that of the sarcomeres.

As shown in Figures 7–1 and 7–12 the SR appears as an amplified form of SER formed by wide anastomosing tubules having a preferential longitudinal disposition. These tubules cover the surface of the sarcomere at the level of the A- and I-bands and terminate on the Z-discs by special *terminal cisternae.* In between the terminal cisternae of two consecutive sarcomeres a flattened cisterna, which is a direct extension of the plasma membrane of the muscle cell, is interposed and is called the *transverse or T-system.* The two apposed terminal cisternae and the interposed T-system together constitute the so-called *triad* (Fig. 7–12).

One such component has recently been isolated from skeletal muscle with retention of its structure, including the presence of osmiophilic processes or "feet" situated between the transverse tubule and the terminal cisternae.[19]

The T-system is in Continuity with the Plasma Membrane and Conducts Impulses Inward

The transverse component (i.e., the T-system) at certain points is continuous with the plasma

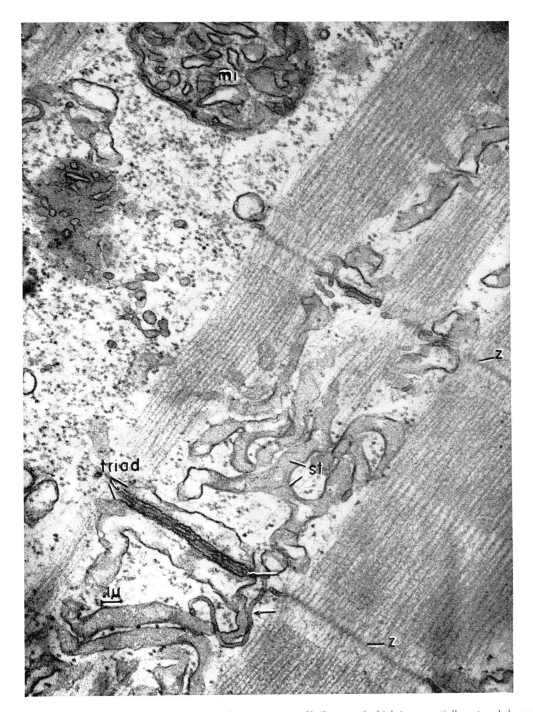

Fig. 7–12. Electron micrograph of striated muscle showing two myofibrils, one of which is tangentially cut and shows the arrangement of the SR. The two components of this system can be seen clearly. The transverse component is represented by the *triad* and especially by the central cisternae of the triad, which continues in special tubules (arrows). Notice the relationship of the triad to the Z-disc. The longitudinal component of the SR forms anastomosing tubules *(st)* on the surface of the sarcomere; *mi*, mitochondrion. (Courtesy of K. R. Porter.)

membrane of the sarcolemma and is the structure organized to conduct impulses from the fiber surface into the deepest portions of the muscle fiber.

The continuity between the T-system, the plasma membrane, and the extracellular space was first demonstrated indirectly. In frog muscle immersed for short periods in a solution containing ferritin molecules, it was found that the central vesicle of the triad had filled with these electron-opaque molecules. Ferritin was also found in certain tubules that were continuous with the central element of the triad. These findings suggested that at the plasma membrane there are a small number of openings communicating directly with the transverse system of the SR through fine tubules.

The function of the T-system is to transmit inward the electrical signal that will bring about the contraction of the individual myofibrils. The role of the system was suggested by experiments with microelectrodes in which stimulation of the sarcolemma, at the level of the Z-disc, produced a localized contraction of adjacent sarcomeres. Apparently, the T-system is not a passive channel but is actively involved in the conduction of action potentials. When a muscle is immersed in glycerol there is a lesion of the T-system (i.e., *detubulation*), and the mechanism of excitation-coupling is abolished.

The presence of this intracellular conducting system may explain the physiological paradox that all the myofibrils of a fiber between 50 and 100 μm in diameter may contract quickly and synchronously, once the activating action potential has passed over the surface.

Stimulation Releases Ca^{++} from the Terminal Cisternae

Within the muscle, Ca^{++} is stored mainly in the longitudinal components of the SR, so that the Ca^{++} concentration near the myofibrils is kept very low (i.e., about 10^{-7} M). Following stimulation, Ca^{++} is rapidly released, so that the concentration increases (up to 10^{-5} M) and induces contraction. It is now assumed that the triad plays an essential role in the release of Ca^{++}. It is thought that the action potential, arriving by way of the T-system, immediately opens Ca^{++} channels in the *terminal cisternae* of the SR, producing a localized release of Ca^{++}.

Thus, the triad is the most important site in the excitation-contraction coupling (see Ebashi, 1976 in the Additional Readings at the end of this chapter).

A Ca^{++}-activated ATPase is Present in the SR and Acts as a Ca^{++} Pump

We have just seen that during excitation Ca^{++} is liberated and that this causes the contraction. To produce relaxation the opposite phenomenon must take place. In other words, Ca^{++} must be rapidly removed, and this is done by a *Ca^{++}-activated ATPase* that is present in the membrane of the SR. This acts as a Ca^{++} pump that incorporates this cation under the action of ATP. The reactions involved are the following:

$$2 \text{ Ca}^{++} \text{ outside } + \text{ ATP } + \text{ E } \rightleftharpoons \\ \text{E-PCa}_2{}^{++} + \text{ ADP} \qquad (1)$$

$$\text{E-PCa}_2{}^{++} \rightleftharpoons \text{E-Pi} + 2\text{Ca}^{++} \text{ inside} \qquad (2)$$

The two equations show that the splitting of one molecule of ATP results in the translocation of two Ca^{++} by the ATPase *(E)*. This reaction is reversible (i.e., Ca^{++} inside can be used for the formation of ATP).[20]

The SR can be isolated by cell fractionation and vesicular structures may be obtained. The vesicles contain 40% lipid and two main proteins: the *Ca^{++} pump protein (ATPase)*, which, with a molecular weight of 105,000 daltons, represents about 80% of the protein of the membrane, and a *Ca^{++}-binding protein* (i.e., *calsequestrin)*, with a molecular weight of 65,000 daltons (see Meissner and Fleischer, 1974; Hasselbach, 1977 in the Additional Readings at the end of this chapter). The Ca^{++} ATPase is thought to be a highly asymmetric molecule spanning the membrane, which needs the presence of a certain amount of phospholipid to function (see Metcalfe and Warren, 1977 in the Additional Readings at the end of this chapter).

This membrane-bound ATPase can catalyze both the hydrolysis and the synthesis of ATP.[21]

The inward and outward movement of Ca^{++} from the SR is related to the phosphorylation and dephosphorylation of this Ca^{++} pump protein. In the presence of Ca^{++} and ATP this protein is rapidly phosphorylated and Ca^{++} is accumulated. In the absence of ATP this protein is dephosphorylated and calcium leaks from the SR (see De Meis, 1981 in the Additional Readings at the end of this chapter).

Since both the ATPase and the Ca^{++} uptake are inhibited by agents that block —SH groups, an electron microscopic study was conducted using a cytochemical reagent in which ferritin molecules were attached to an —SH blocking molecule. It was demonstrated that the active groups of the membrane-bound ATPase were localized in the outer surface of the isolated SR. All these findings suggest that the SR assumes the important role of relaxing the fiber after contraction and that this is accomplished by the binding of Ca^{++} at the outer surface and its transport within the sarcoplasmic vesicle. *Calsequestrin*, on the other hand, could be responsible for the binding of Ca^{++} within the SR.[22]

In summary, the series of events produced after the arrival of the electrical signal is as follows: (1) The signal is received at the level of the Z-line or the A-I junction by way of the T-system. (2) The coupling of the T-system with the terminal cisternae produces the release of Ca^{++} (this may occur in a matter of milliseconds). (3) Ca^{++} induces the contraction, affecting the regulatory proteins troponin and tropomyosin and thus enabling the interaction of actin and myosin. During this step, ATP is used as an energy source. (4) Ca^{++} is quickly removed and restored into the SR by way of the Ca^{++} pump. Ca^{++} uptake results in muscle relaxation.

Figure 7–13 is a diagram of how the Ca^{++}, liberated from the SR, is able to regulate muscle function. The intervention of troponin-C (TnC) is shown as is that of calmodulin (CaM) together with the coupling of energy with ATP, the involvement of the actomyosin ATPase, and the SR dephosphorylation driven by the interaction with calmodulin-Ca^{++}. The student should be well aware of these interconnected molecular events that control the contraction and relaxa-

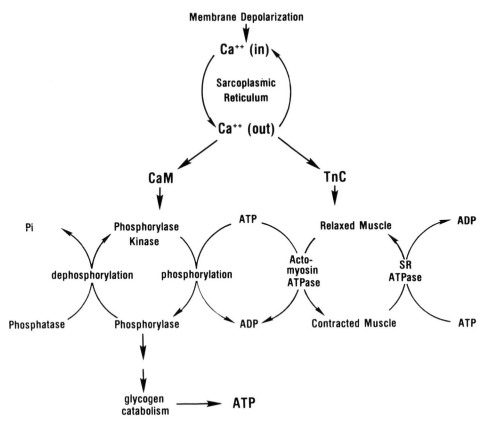

Fig. 7–13. General diagram showing the role of the calcium-binding proteins, calmodulin (CaM) and troponin-C (TnC), in muscle contraction and relaxation. The CaM is related to the phosphorylating system involved in glycogen catabolism. TnC plus Ca^{++} acts on the actomyosin ATPase for the contraction; while the relaxation is controlled by the ATPase of the sarcoplasmic reticulum (SR ATPase). (Courtesy of J.R. Dedman and P.B. Moore.)

tion of muscle (see Moore and Dedman, 1982 in the Additional Readings at the end of this chapter).

All these data, as well as those related to the sliding mechanism of contraction, can be put together in a molecular theory of muscular contraction. This is one of the best examples, so far studied, of a tight coupling between the processes furnishing energy and the actual machinery involved in contraction. In this case, structure and function are so intimately related in the realm of molecular organization that they are an inseparable unit.

SUMMARY:
Excitation-Contraction Coupling and the SR

The essential components of the excitation-contraction coupling are (1) the T-system, which conducts the action potential to the myofibril; (2) the release of Ca^{++} from the SR; (3) the induction of contraction by Ca^{++}; (4) the reaccumulation of Ca^{++} in the SR, which results in relaxation.

Within the muscle, Ca^{++} is stored mainly in the longitudinal components of the SR, so that the concentration in the cytosol is kept at about 10^{-7} M. Following stimulation, Ca^{++} is released and increases to 10^{-5} M, thus inducing contraction. This localized release occurs by the opening of Ca^{++} channels in the terminal cisternae of the triad. *Relaxation* is caused by the removal of *Ca^{++} by the Ca^{++}-activated ATPase*, which is present in the membrane of the SR, under the action of ATP. The splitting of one molecule of ATP results in the translocation of two molecules of Ca^{++}.

In the isolated SR the vesicles contain 40% lipids, and 80% of the protein consists of the Ca^{++}-ATPase (or Ca^{++} pump), which has a molecular weight of 105,000 daltons. There is also *calsequestrin*, a Ca^{++}-binding protein of 65,000 daltons. Muscle provides one of the best examples of the tight coupling between the processes furnishing energy and the actual machinery that produces contraction and relaxation. In muscle, structure and function constitute an inseparable unit.

REFERENCES

1. Craig, R.: J. Mol. Biol., *109*:69, 1977.
2. Sjöstrom, M., and Squire, J.M.: J. Mol. Biol., *109*:49, 1977.
3. Granger, B.L., and Lazarides, E.: Cell, *18*:1053, 1979.
4. Breckler, J., and Lazarides, E.: J. Cell Biol., *92*:795, 1982.
5. Huxley, H.E.: Proc. R. Inst. Gr. Br., *44*:274, 1970.
6. Carraro, V., et al.: Eur. J. Cell Biol., *22*(1):318, 1980.
7. Pepe, F.A.: J. Histochem. Cytochem., *23*:543, 1975.
8. Pepe, F.A., and Dowben, P.: J. Mol. Biol., *113*:199, 1977.
9. Potter, J.D., and Gergely, J.: Biochemistry, *13*:2697, 1974.
10. Flicker, P.F., Phillip, G.N., Jr., and Cohen, C.: J. Mol. Biol., *162*:495, 1982.
11. Eppenberger, H.M., Perriard, J.C., Rosenberg, W.B., and Strehler, E.E.: J. Cell Biol., *89*:185, 1981.
12. Huxley, H.E.: VII International Congress of Biophysics. México, 1981, p. 17.
13. Moore, P.B., Huxley, H.E., and De Rosier, D.J.: J. Mol. Biol., *50*:279, 1970.
14. Ebashi, S.E., and Endo, M.: Prog. Biophys. Mol. Biol., *18*:123, 1968.
15. Vibert, P.J., Halsegrove, J.C., Lowy, J., and Poulsen, F.R.: Nature, *236*:182, 1972.
16. Wakabayaski, T., Huxley, H.E., Amos, L., and Klug, A.: J. Mol. Biol., *93*:477, 1975.
17. Bennett, H.S., and Porter, K.R.: Am. J. Anat., *93*:1, 1953.
18. Porter, K.R.: J. Biophys. Biochem. Cytol., *10* (Suppl.):219, 1961.
19. Mitchell, R.D., Saite, A., Palade, P., and Fleicher, S.: J. Cell Biol., *96*:101, 1983.
20. Makinose, M., and Hasselbach, W.: FEBS Lett., *12*:271, 1971.
21. Hasselbach, W.: Eur. J. Cell Biol., *22* (1):473, 1980.
22. MacLennan, D.H., and Wong, P.T.S.: Proc. Natl. Acad. Sci. USA, *68*:1231, 1971.

ADDITIONAL READINGS

Chowraski, P.K., and Pepe, F.A.: The Z band- 85,000 dalton amorphin and α-actinin and their relation to structure. J. Cell Biol., *94*(3):565, 1982.
Curtin, N.A., and Woledge, R.C.: Energy changes and muscular contraction. Physiol. Rev., *58*:690, 1978.
De Meis, L.: Ca^{2+} transport and energy tranduction in the sarcoplasmic reticulum. VII International Congress of Biophysics. México, 1981, p. 293.
Dowen, R.M., and Shay, J.W.: Cell and Muscle Motility. New York, Plenum Pub. Corp., 1981.
Ebashi, S.E.: Excitation-contraction coupling. Annu. Rev. Physiol., *38*:293, 1976.
Grove, B.K., et al.: A new 185,000 dalton skeletal muscle protein detected by monoclonal antibodies. J. Cell Biol., *98*:518, 1984.
Hasselbach, W.: The sarcoplasmic reticulum pump. Biophys. Struct. Mech., *3*:43, 1977.
Huxley, H.E.: The structural basis of contraction and regulation in skeletal muscle. *In* Molecular Basis of Motility. Edited by Heilmeyer et al. Berlin, Springer-Verlag, 1976.
Kendrick-Jones J., and Scholey, J.M.: Myosin-linked regulatory systems. J. Muscle Res. Cell Motil., *2*:347, 1982.
Kimura, S., and Maruyama, K.: Interaction of native connectin with myosin and actin. Biomedical Res., *4*:607, 1983.
Lazarides, E.: Molecular morphogenesis of the Z-disc muscle cells. *In* International Cell Biology. Edited by H.G. Schweiger. Berlin, Springer-Verlag, 1981.
Lehman, W.: Phylogenetic diversity of proteins regulating muscular contraction. Int. Rev. Cytol., *44*:55, 1976.
Luther, P., and Squire J.: Three-dimensional structure of the vertebrate muscle M region. J. Mol. Biol., *125*:313, 1978.
Margosian, S.S., and Lowey, S.: Interaction of myosin subfragments with F-actin. Biochemistry, *17*:5431, 1978.

Martonosi, A., et al.: Development of sarcoplasmic reticulum in cultured chicken muscle. J. Biol. Chem., 252:318, 1977.

Meissner, G., and Fleischer, S.: Characterization, dissociation and reconstitution of sarcoplasmic reticulum. *In* Calcium Binding Proteins. Edited by W. Drabilowski. Amsterdam, Elsevier North-Holland Co., 1974.

Metcalfe, J.C., and Warren, G.B.: Lipid-protein interactions in a reconstituted calcium pump. *In* International Cell Biology. (Porter, K.R. ed.) New York, The Rockefeller University Press, 1977.

Moore, P.B., and Dedman, J.R.: Calcium binding proteins and cellular regulation. Life Sci., 31:2937, 1982.

Murray, J.M., and Weber, A.: The cooperative action of muscle protein. Sci. Am., 230(2):59, 1974.

Pepe, F.A., and Dowben, P.: The myosin filament. J. Mol. Biol., 113:199, 1977.

Seymour, J.: Unfolding troponin-C. Nature, 275:177, 1978.

Sjöstrom, M., and Squire, J.M.: Fine structure of the A-band in cryosections. J. Mol. Biol., 109:49, 1977.

Squire, J.: Contractile filament organization mechanics. Nature, 267:753, 1977.

Squire, J.: The Structural Basis of Muscular Contraction. New York, Plenum Publishing Corp., 1981.

Wang, K., Ramirez-Mitchell, R., and Palter, D.: Titin is an extraordinary long, flexible, and slender myofibrillar protein. Proc. Natl. Acad. Sci. USA, 81:3685, 1984.

CHAPTER 8

ENDOPLASMIC RETICULUM AND PROTEIN SEGREGATION

In Chapter 1 we mentioned that the eukaryotic cell has a high degree of ultrastructural organization. In both plant and animal cells the cytoplasm is permeated by a complex system of membrane-bound tubules, vesicles, and flattened sacs (cisternae) that, at many points, are intercommunicating. This endoplasmic reticulum (ER) may be interpreted three-dimensionally as a vast network that subdivides the cytoplasm into two main compartments, one enclosed within the membranes, the other situated outside and constituting the *cytoplasmic matrix* or *cytosol*.

The existence of the ER was discovered in 1945 after the introduction of electron microscopy applied to cultured cells.[1] The first micrographs showed a lace-like arrangement of tubules that did not reach the periphery of the cell, hence the term "endoplasmic" (Fig. 8–1). The use of high-voltage electron microscopy on such cells has rendered a clearer three-dimensional view of the ER (Fig. 7–3).[2] A more detailed analysis of the system became possible with the introduction of thin sectioning and freeze-fracturing of cells. Other advances were made by cell fractionation methods followed by biochemical analysis and the use of cytochemical techniques for the study of specific components—particularly enzymes. Recently, by the use of fluorescent dyes, it has been possible to observe the ER in living as well as in fixed cells (see Terasaki et al., 1984 in the Additional Readings at the end of this chapter).

A eukaryotic cell can no longer be considered as a bag containing enzymes, ribonucleic acid (RNA), deoxyribonucleic acid (DNA), and solutes surrounded by an outer membrane, as in the most primitive bacterium. As we shall see in this chapter, numerous membrane-bound compartments are responsible for vital cellular functions, among which are the separation and association of enzyme systems, the creation of diffusion barriers, the regulation of membrane potentials, ionic gradients, different intracellular pH values, and other manifestations of cellular heterogeneity. Furthermore, there is evidence that enzymes are spatially organized, forming multi-enzyme systems within the insoluble membrane framework of the cell.

One of the main functions of the endomembrane system is the segregation, within the lumen, of proteins that are synthesized by membrane-bound ribosomes. Many of these proteins are then processed and channeled for export. In Section 8–5, we discuss the possible mechanisms by which the thousands of proteins found in a cell are selectively translocated to their final destination, i.e., to the different cell compartments (e.g., nucleus, mitochondria, lysosomes, and chloroplasts) or membranes in which they function.

We also discuss the possibility that this selective traffic is due to the presence of special amino acid sequences in the protein and thus, genetically determined at the DNA-RNA level.

8–1 GENERAL MORPHOLOGY OF THE ENDOMEMBRANE SYSTEM

The ER is the main component of the endomembrane system, also called the cytoplasmic vacuolar system. This system comprises the following: (1) The nuclear envelope, consisting of two nonidentical membranes, one opposed to the nuclear chromatin and the other separated from the first membrane by perinuclear cisternae, the two being in contact at the nuclear pores; (2) the ER; and (3) the Golgi complex, which is a specialized region of the system, mainly related to some of the terminal processes of cell secretion (see Chapter 9).

As mentioned in Chapter 1, the ER comprises two parts differentiated by the presence or absence of ribosomes on the outer or cytoplasmic surface: the "rough" or "granular" endoplasmic reticulum (RER), and the "smooth" or "agranular" endoplasmic reticulum (SER).

In Figure 8–2 GERL (Golgi, *ER*, *L*ysosome) refers to a special region of the endomembrane system, which is more related to the Golgi complex and which has been implicated in the formation of lysosomes (see Section 10–3).

The fact that the entire system of endomembranes represents a barrier separating cytoplasmic compartments is clearly indicated in Figure 8–3. This diagram specially emphasizes the two faces of each membrane, (1) the so-called *cytoplasmic* or *protoplasmic face* and (2) the *luminal face*, also called the endoplasmic or extracellular face.[3]

The cytoplasmic face is directly opposed to the cytosol and in this sense is equivalent to the inner face of the plasma membrane and to the surfaces of the outer mitochondrial membranes and of the other intracellular organelles, such as the Golgi complex, the lysosomes, the peroxisomes, and the secretory granules (Fig. 8–3).

The luminal face borders the perinuclear cisternae, the cavities of the RER and the SER, and the Golgi elements. This surface also corresponds to the interior of the secretory granules, the lysosomes, and the peroxisomes and to the faces of the mitochondria confronting the outer chamber. It is important to remember that after freeze-fracturing, the membrane halves containing the cytoplasmic (A or P) faces show many more intramembranous particles than those consisting of the luminal (B or E) faces.[4]

As in the case of the plasma membrane, we may consider that in the endomembrane system there is a lipid bilayer of 5 to 6 nm with peripheral and integral proteins, some of which may be exposed only toward either the cytoplasmic or the luminal face, and in addition this bilayer may have integral proteins that have a transmembrane disposition (see Fig. 5–1).

The size of the ER varies considerably in different cell types and is related to their functions. It is often small and relatively undeveloped in eggs and in embryonic or undifferentiated cells but increases in size and complexity with differentiation. A simple SER found in cells engaged in lipid metabolism, such as adipose, brown fat, and adrenocortical cells. In the reticulocytes, which produce only proteins to be retained in the cytosol (hemoglobin), the ER is poorly developed or nonexistent, although the cell may contain many ribosomes.

The RER—Ribosomes and Protein Synthesis

The RER is especially well-developed in cells actively engaged in protein synthesis. In gen-

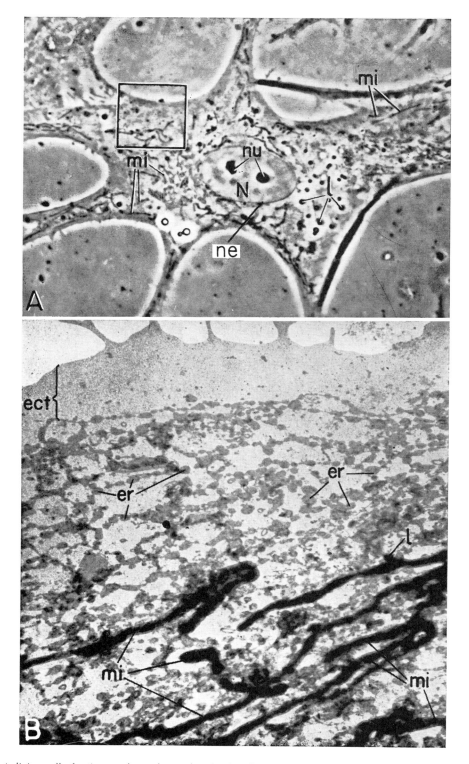

Fig. 8–1. *A,* living cell of a tissue culture observed under the phase contrast microscope: *l,* lipid; *mi,* mitochondria; *ne,* nuclear envelope; *nu,* nucleoli. The region indicated in the inset is similar to B. (Courtesy of D.W. Fawcett.) *B,* electron micrograph of the marginal region of a mouse fibrocyte in tissue culture: *er,* endoplasmic reticulum; *mi,* filamentous mitochondria; *l,* lipid. The peripheral region *(ect)* is homogeneous. ×7000. (Courtesy of K.R. Porter.)

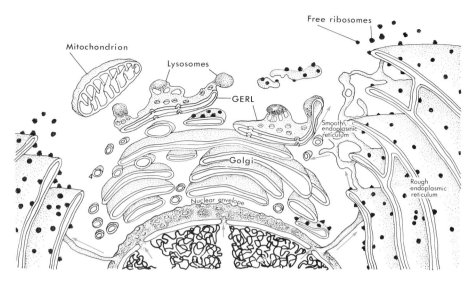

Fig. 8–2. Three-dimensional diagram of the endomembrane system of the cell. The nucleus with its chromosomal fibrils shows interchromatin channels (arrows) leading to nuclear pores. Note the double-membrane organization of the nuclear envelope. Cisternae of rough endoplasmic reticulum are interconnected and have ribosomes attached to the outer surface. Some of these cisternae are extended by tubules of smooth endoplasmic reticulum. The Golgi apparatus shows the GERL region. The large arrows indicate the probable dynamic relationship of the portions of the endomembrane system.

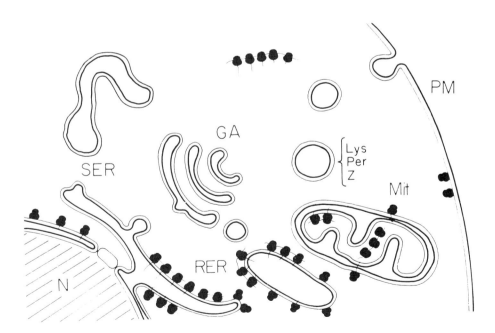

Fig. 8–3. Diagram of cellular membranes and their relationship to compartments containing the ribosomes. In each membrane the luminal faces are shown in thick lines, while the cytoplasmic faces are depicted by thin lines. Ribosomes are always on the cytoplasmic (or matrix) side. N, nucleus; GA, Golgi apparatus; Lys, lysosome; Mit, mitochondria; RER, rough endoplasmic reticulum; Per, peroxisome; PM, plasma membrane; SER, smooth endoplasmic reticulum; Z, zymogen granule. (Courtesy of D.D. Sabatini and G. Kreibich. From Sabatini, D.D., and Kreibich, G.: Functional specialization of membrane-bound ribosomes in eukaryotic cells. In The Enzymes of Biological Membranes. Vol. 2. Edited by A. Martonosi. New York, Plenum Press, 1976, p. 432.)

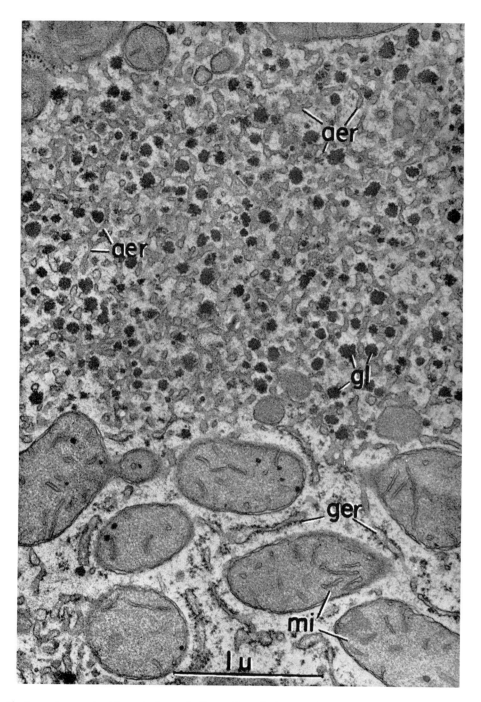

Fig. 8–4. Electron micrograph of the cytoplasm of a liver cell. At the bottom, the rough or granular endoplasmic reticulum *(ger)* and mitochondria *(mi)*; at the top, the smooth or agranular endoplasmic reticulum *(aer)* mixed with glycogen particles *(gl)*. ×45,000. (Courtesy of G.E. Palade.)

TABLE 8–1. RELATIVE CYTOPLASMIC VOLUMES AND MEMBRANE SURFACE AREAS OF SECRETORY COMPARTMENTS IN RESTING GUINEA PIG PANCREATIC EXOCRINE CELLS

Compartment	Relative Cytoplasmic Volume %	Membrane Surface Area µm²/Cell
RER	~20	~8000
Golgi complex	~8	~1300
Condensing vacuoles	~2	~150
Secretory granules	~20	~900
Apical plasmalemma		~30
Basolateral plasmalemma		~600

Data originally from Bolender, 1974; and Amsterdam and Jamieson, 1974; (From Jamieson, 1977).

eral, it occupies the regions of the cytoplasm that appear to be basophilic (corresponding to the *ergastoplasm*) when viewed under the light microscope. As mentioned in Section 6–1, this property is due to the presence of the RNA-containing ribosomes.

In rapidly dividing cells, such as those of plant and animal embryos, as well as those of cancers, the cytoplasm may be strongly basophilic, but the ER is poorly developed. In cells engaged in the production of large amounts of proteins for export (enzymes), such as the cells of the pancreas, the RER is highly developed and consists of parallel stacks of large flattened cisternae occupying the cytoplasm of the base and the lateral regions of the cell.

Table 8–1 indicates the relative volumes and membrane areas of the main secretory compartments of the pancreatic cell and shows the large structural amplification of the RER. In liver cells the RER occurs as groups of cisternae interspersed with regions of SER (Fig. 8–4).

The cavity of the RER is sometimes very narrow, with the two membranes closely apposed; but more frequently there is a true space between the membranes that may be filled with a material of varying opacity. This space is much distended in certain cells actively engaged in protein synthesis, such as the plasma cells and goblet cells. In these cases, a dense macromolecular material can be observed inside the cisternae. In the pancreas, intercisternal secretion granules, smaller than the zymogen granules, may be observed.

The total surface of the ER contained in 1 ml

of liver tissue has been calculated to be about 11 square meters, two thirds being of the granular, or rough type.[5]

Ribosomal Binding to the ER—60S Subunit and Ribophorins Involved

The main characteristic of the RER is the presence of attached ribosomes on the outer surface. These are present as polysomes held together by mRNA (see Section 21–1) and are often arranged in typical "rosettes" or spirals (Fig. 8–5).

Another characteristic, which is shown in Figures 8–3 and 8–6, is that the ribosomes appear to be attached to the membrane by the large 60S subunit. It should be stressed, however, that the free ribosomes in the cytosol, and the ER-bound ribosomes are able to exchange their subunits with each cycle of protein synthesis (see Section 22–2).

Several mechanisms have been proposed to explain the selective binding of ribosomes to the RER. It has been found that the RER, when stripped of ribosomes, is capable of high-affinity binding to the ribosomes, while the SER lacks this property. Besides a strong ionic interaction, the emerging polypeptide from the ribosomes establishes a functional link with the ER membranes. Furthermore, the presence of special *receptor sites* for ribosomes is postulated.

Isolated RER contains two transmembrane glycoproteins (*ribophorins I and II* of 65,000 and 64,000 daltons, respectively) that are not found in the SER. If the microsomes are treated with a crosslinking agent (i.e., an agent that establishes covalent links or bridges), such as glutaraldehyde, the two ribophorins become attached to the ribosomes and segregate with them. These observations, as well as the fact that the concentration of these proteins is stoichiometrically related to the number of ribosomes, suggest that ribophorins are involved in mediating the attachment of ribosomes to the membrane (see Kreibich et al., 1981 in the Additional Readings at the end of this chapter).

Ribophorins tend to interact with each other, and they may form an intramembranous network, which could control the distribution of ribosome binding sites in the plane of the ER membrane, thereby determining the limits between RER and SER. The typical "rosette" arrangement of ribosomes (Fig. 8–5) can also be

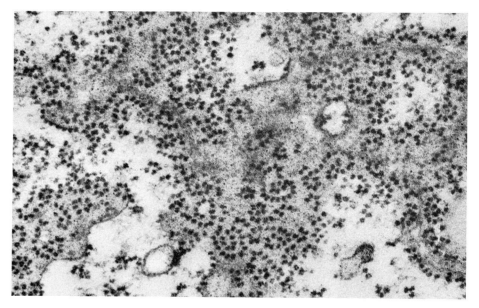

Fig. 8–5. Electron micrograph of a root hair from an epidermal cell of the radish. The tangential section through the membrane of the endoplasmic reticulum shows groups of ribosomes (i.e., polyribosomes) disposed in recurrent patterns. × 57,000. (Courtesy of H.T. Bonnett, Jr., and E.H. Newcomb.)

determined by the ribophorin network. Using antibodies against ribophorins I and II, the ribosomes become displaced and aggregated resulting in the formation of inverted vesicles in which the ribosomes are in the inner side of the vesicle instead of the outside.

The SER Lacks Ribosomes—Glycosomes

Although the SER forms a continuous system with the RER, it has a different morphology. In the liver cell it consists of a tubular network that pervades large regions of the cytoplasmic matrix. These fine tubules are present in regions rich in glycogen and can be observed as dense particles in the matrix (Fig. 8–4).

These glycogen particles or *glycosomes* are spheroidal in shape, measure 50 to 200 nm, and show an internal structure made of smaller particles of about 20 to 30 nm. The glycosomes observed under the electron microscope (EM) do not correspond directly to glycogen, since this polysaccharide does not stain with the usual electron stains. Glycosomes represent the staining of the protein moiety that corresponds to the enzymes of synthesis of glycogen.[6] Many of the glycosomes attached to SER are observed in the liver and in the conduction fibers of the heart.

SUMMARY:

The Endoplasmic Reticulum

The cytoplasm of animal and plant cells is traversed by a complex system of membranes that form closed compartments in structural continuity. The main components of the system are the ER with its three portions (the nuclear envelope, the RER and the SER) and the Golgi complex. The RER has ribosomes attached to its outer surface and is particularly well developed in the basophilic regions of the cytoplasm (i.e., the ergastoplasm). The tubular and cisternal cavities of the ER may be closed, but more often contain material that has been synthesized on the ribosomes (e.g., protein) or by the enzymes present in the membranes (e.g., lipids, oligosaccharides). The SER is devoid of ribosomes. Frequently, it forms a tubular network, and in the liver it is related to glycogen deposits (glycosomes) and peroxisomes.

The structure of the ER membrane is similar to that of the cell membrane in the sense that there is a lipid bilayer with peripheral and integral proteins. The ribosomes are bound by their 60S (large) subunits. In the RER are two proteins (ribophorins I and II) that are absent in the SER. These are transmembrane glycoproteins that may correspond to the ribosomal binding sites.

8–2   MICROSOMES—BIOCHEMICAL STUDIES

In Chapter 2 the concept of the "microsomal fraction" that can be isolated by differential cen-

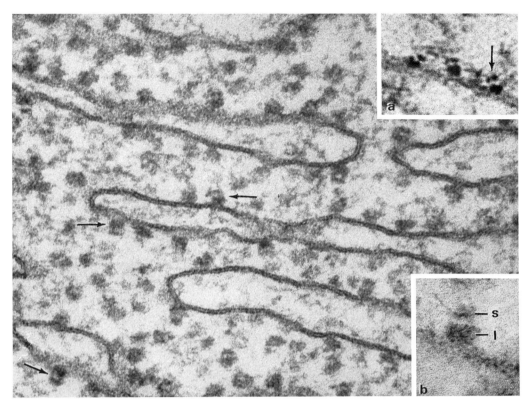

Fig. 8–6.   Electron micrograph showing ribosomes attached to the membranes of the endoplasmic reticulum. Observe the "unit membrane" structure. Arrows indicate ribosomes where the attachment of the large (60S) subunit is best observed. × 208,000. (Courtesy of G.E. Palade.) Inset *a*, attachment of the large and small subunits forming a "cap," which is subdivided into two portions (arrow). Inset *b*, at higher magnification, the small *(s)* and large *(l)* subunits appear to be separated by a clear cleft. *a*, × 200,000; *b*, × 410,000. (Both insets courtesy of N.T. Florendo.)

trifugation following the separation of the nuclear and mitochondrial fractions was introduced. Electron microscopic studies have revealed that such a fraction is rather complex. In addition to fragments of the plasma membrane, microsomes include various parts of the vacuolar system, i.e., the RER and the SER and the Golgi complex (Fig. 8–7).

Microsomes, as discrete entities, are not found in the intact cell, but are the result of the fragmentation of most of the cytoplasmic membranous components.

During cell disruption the cisternae of the ER break but immediately reseal, maintaining intact the topological relationships between the membranes, the lumen, and the ribosomes. Because of the difference in density, the smooth and the rough microsomes may be separated. It is also possible to subfractionate the microsomes into their components. Thus, the membranes may be stripped of the ribosomes by high salts and treatment with puromycin, which stops the growth of the polypeptide chain.

The ribosomes may be separated by treatment with desoxycholate, a surface-active agent that solubilizes the membrane (Fig. 8–7). Finally, it is possible to extract the content of the lumen, consisting mainly of precursors of secretory proteins (albumin and other serum proteins in the case of the liver), by the use of sonication or very low concentrations of detergents that make holes in the membranes, enabling the exit of the luminal content.

Microsomal Membranes—A Complex Lipid and Protein Composition

Microsomes constitute about 15 to 20% of the total mass of the liver cell. They contain 50 to

Endomembrane System

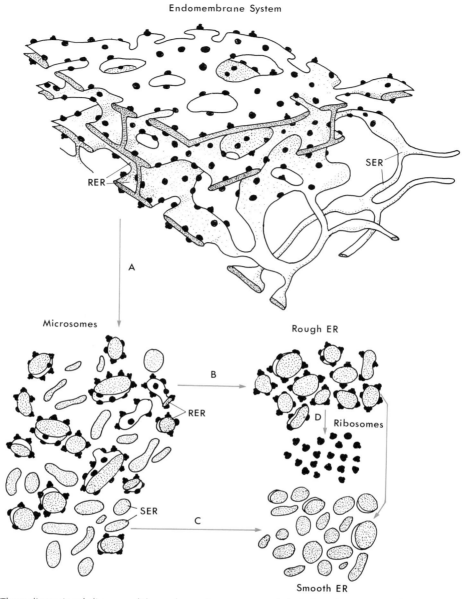

Fig. 8–7.   Three-dimensional diagram of the endomembrane system. *A,* isolation of microsomes by homogenization and differential centrifugation; *B* and *C,* separation of rough and smooth vesicles of the endoplasmic reticulum; *D,* separation of ribosomes from the RER. *RER,* rough endoplasmic reticulum; *SER,* smooth endoplasmic reticulum.

60% of the cellular liver RNA, which is due to the presence of ribosomes. Microsomal membranes have a high lipid content that includes: phospholipids, lipids, phosphatidylinositol, plasmalogens, and some gangliosides. There is more lipid in relation to proteins in the SER and Golgi membranes than in the RER. Microsomal membranes also have some special proteins that serve as markers for the recognition of this particular fraction and to differentiate it from Golgi membranes, secretory granules, or plasma membranes.

The membrane proteins of the ER have been separated and as many as 30 polypeptide bands ranging from 15,000 to 150,000 daltons have been identified (see Fig. 11–7).

TABLE 8–2. SOME MICROSOMAL ENZYME ACTIVITIES

Synthesis of glycerides:
    Triglycerides
    Phosphatides
    Glycolipids and plasmalogens
Metabolism of plasmalogens
Fatty acid synthesis
Steroid biosynthesis:
    Cholesterol biosynthesis
    Steroid hydrogenation of unsaturated bonds
$NADPH_2 + O_2$-requiring steroid transformations:
    Aromatization
    Hydroxylation
$NADPH_2 + O_2$-requiring drug detoxification:
    Aromatic hydroxylation
    Side-chain oxidation
    Deamination
    Thio-ether oxidation
    Desulfuration
L-Ascorbic acid synthesis
UDP-uronic acid metabolism
UDP-glucose dephosphorylation
Aryl- and steroid-sulfatase

Modified from Rothschild, J.: The isolation of microsomal membranes. Biochem. Soc. Symp., *22*:4, 1963.

We mentioned earlier the existence of the so-called ribophorins I and II—present in the RER and absent in the SER.

Two Microsomal Electron Transport Systems—Flavoproteins and Cytochromes b₅ and P-450 Involved

The ER contains many of the enzymes utilized in the synthesis of triglycerides, phospholipids, and cholesterol.

Table 8–2 indicates some of enzymatic activities of microsomes. Many of the metabolic functions are carried out by two electron transport chains present in these membranes. Microsomes contain at least two flavoproteins (*NADH-cytochrome-c-reductase* and *NADH-cytochrome-b₅-reductase*) and two hemoproteins (*cytochrome b₅* and *cytochrome P-450*).

Study of this system has centered especially on the role of cytochrome P-450, which is characterized by a molecular weight of 50,000 daltons and absorption at 450 nm.

Cytochrome P-450 functions as a terminal oxidase and in the liver is used to detoxify or inactivate many drugs by oxidation, and to hydroxylate steroid hormones (see Table 8–2).

The administration of barbiturates to a rat produces a rapid increase in SER and a great increase in cytochrome P-450.

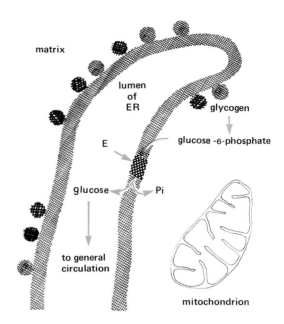

Fig. 8–8. Diagram of the intervention of the smooth endoplasmic reticulum in glucogenolysis with the consequent release of glucose. The enzyme *(E)*, glucose-6-phosphatase, is present in the membrane and has a vectorial disposition by which it receives the glucose-6-phosphate from the matrix surface. The product-glucose penetrates the lumen of the ER.

It has been calculated that in these conditions cytochrome P-450 may increase to represent 10% of the total protein of the microsome and that one molecule of NADPH-cytochrome-c-reductase should interact with 20 to 30 molecules of cytochrome P-450.

A cluster arrangement of the electron transport carriers has been postulated.[7] In this model the flavoprotein is located at the core of the complex, surrounded by molecules of cytochrome b₅ and cytochrome P-450.

The second electron transport chain in microsomes is composed of *cytochrome b₅, NADH-cytochrome-b₅-reductase*, and *fatty acyl CoA desaturase*. This system uses electrons from NADH to desaturate fatty acids.

Microsomal Enzymes—Glycosidation and Hydroxylation of Amino Acids

In the luminal face of the ER there are enzymes that can modify compounds released into the cavity. For example, *peptidases* may cut out portions of protein precursors, and other enzymes may modify the nascent polypeptides by *glycosidation* or *hydroxylation* of amino acid residues.

Other important enzymes are Mg^{++}-activated ATPase and glucose-6-phosphatase. In the liver this last enzyme has the important function of splitting off the phosphate from glucose-6-phosphate, thus allowing glucose to be liberated. As shown in Figure 8–8 glucose-6-phosphate originates from the degradation of glycogen in the cytoplasmic matrix, and the enzyme, which is embedded in the membrane, probably acts in a vectorial manner sending glucose into the lumen of the ER.

As in other membranes the main enzymes of the ER show a definite asymmetry across the membrane. Among the proteins present on the cytoplasmic surface are cytochrome b_5, NADH-cytochrome-b_5-reductase; NADH-cytochrome-c-reductase, cytochrome P-450, and 5-nucleotidase. Among those in the luminal surface are glucose-6-phosphatase, nucleoside diphosphatase, and β-glucuronidase. Several of the enzymes of the former group, particularly cytochrome b_5, are bound to the membrane by a hydrophobic tail that penetrates into the lipid bilayer.

SUMMARY:

Microsomes

By differential centrifugation most of the components of the endomembrane system can be isolated in the so-called microsomal fraction. Further separation of the RER and the SER and Golgi complexes may be achieved by gradient centrifugation; the membranes, ribosomes, and luminal contents may also be separated.

The ER contains enzymes for the synthesis of triglycerides, phospholipids, and cholesterol. There are two electron transport systems that contain two flavoproteins (NADH-cytochrome-c-reductase and NADH-cytochrome-b_5-reductase) and two hemoproteins (cytochrome b_5 and cytochrome P-450). P-450 absorbs at 450 nm, is concentrated in the liver, and increases with inducing agents (i.e., barbiturates, polycyclic hydrocarbons, and others). P-450 is used to detoxify certain drugs by oxidation, but it may also transform a drug into a carcinogen. A cluster arrangement within the membrane is postulated to explain the relation of one molecule of NADH-cytochrome-c-reductase to 20 to 30 molecules of cytochrome P-450. The second electron transport chain contains NADH; cytochrome-b_5-reductase, cytochrome b_5, and fatty acid acyl CoA desaturase. Other enzymes of the ER are peptidases, glycosyl transferases, and hydroxylases; these modify the nascent polypeptides.

Glucose-6-phosphatase produces the degradation of glycogen in the smooth ER. The various enzymes have different topologies with respect to the luminal or cytoplasmic surfaces of the ER.

8–3 BIOGENESIS AND FUNCTIONS OF THE ER

The great complexity of the ER stems from the fact that a large portion of it is associated with the ribosomes, and thus it plays a fundamental role in the storage and processing of proteins that are destined for export from the cell. This is done, in many cases, through an elaborate series of steps that lead to *cell secretion* (see Chapter 9). Other functions are more related to the fact that the ER represents a fundamental part of the endomembrane system, which divides the cytoplasm into two definite compartments. Furthermore, these membranes contain many enzyme systems that are able to carry out various functions, including the biogenesis of some of the membrane components.

Membrane Biogenesis Involves a Multi-Step Mechanism

The origin of the ER membrane is not definitely known. Some electron microscopic observations of differentiating cells suggest that it may develop by evagination from the nuclear envelope (see Fig. 6–1). At telophase, however, the nuclear envelope is re-formed by vesicles of the ER. The close relationship between these two portions of the system is also suggested by cytochemical studies.

The relationship between RER and SER may be studied in differentiating cells. In rat liver cells before birth there is a preferential increase of the rough type, whereas after birth the growth is mainly of the smooth type. Experimental studies using the protein or lipid precursors, ^{14}C-leucine and ^{14}C-glycerol, have shown that in the period of rapid growth of the ER, the incorporation into proteins and lipids is greater in the rough than in the smooth type. This finding suggests that the synthesis of membranes follows the direction RER → SER.

In addition to the *structural* and *functional continuity* between the various membrane compartments of the vacuolar system, there is also a *temporal continuity*. This refers to the fact that one cell receives a full set of membranes from its ancestor cell.

Current concepts of membrane biogenesis generally assume a multi-step mechanism involving, first, the synthesis of a basic membrane of lipids and intrinsic proteins and, thereafter, the addition, in a sequential manner, of other constituents such as enzymes, specific sugars, or lipids. The process by which a membrane is modified chemically and structurally may be regarded as *membrane differentiation.*

Available evidence suggests that lipids and proteins may be added to the various cellular membranes independently of each other. In other words, the growth of the membrane results from the insertion of individual lipid and protein molecules. The ER (especially the smooth part) is the organelle containing the main phospholipid synthesizing enzymes. In contrast with the plasma membrane, in which the transbilayer movement of phospholipids is very slow, in the ER there is a rapid distribution of the phospholipids between the two monolayers (i.e., there is an intense flip-flop of the lipid components).

The insertion of proteins into the ER membrane is independent of that of the lipids. Most proteins are formed on membrane-bound ribosomes. It is now evident, however, that some proteins of the ER are formed by free ribosomes in the cytosol and then inserted into the membrane. This is the case of NADH-cytochrome-b_5-reductase, which after being synthesized in the cytosol, becomes incorporated in various parts of the endomembrane system (i.e., RER and SER, Golgi complex) and in the outer mitochondrial membrane.[8]

ER-Membrane Fluidity and Flow through the Cytoplasm

The ER may act as a *circulatory system* for intracellular circulation of various substances. Membrane flow may also be an important mechanism for carrying particles, molecules, and ions into and out of the cells. The continuities observed in some cases between the ER and the nuclear envelope suggest that the membrane flow may also be active at this point. This flow would provide one of the several mechanisms for export of RNA and nucleoproteins from the nucleus to the cytoplasm.

We may consider the membrane of the ER at body temperature to be highly dynamic and also fluid. There is evidence from freeze-fracturing studies that the bound ribosomes are mobile on the membrane and that their mobility is controlled by its fluidity.[9] The concept of flow refers not only to the content of the ER but to the membrane proper. Different studies suggest that the transfer of secretory proteins is accompanied by a flow of newly synthesized membrane proteins that are incorporated into the RER.

Ions and Small Molecules—Transport Across ER Membranes

The ER, along with the cytoplasmic matrix, may participate in many of the mechanical functions of the cell. By dividing the fluid content of the cell into compartments the ER provides supplementary *mechanical support* for the colloidal structure of the cytoplasm.

The enormous internal surface of the ER gives an idea of the importance of the exchanges taking place between the matrix and the inner compartment.

It is known that in the cell the endomembrane system has *osmotic properties.* After isolation, microsomes expand or shrink according to the osmotic pressure of the fluid. Diffusion and active transport may take place across the membranes. As in the plasma membrane, the presence of *carriers* and *permeases* that are involved in active transport across the membrane has been postulated. The existence of an endomembrane system makes possible the existence of ionic gradients and electrical potentials across these intracellular membranes. This concept has been applied especially to the sarcoplasmic reticulum (SR), a specialized form of ER found in striated muscle fibers, which is considered as an intracellular conducting system (see section 7.5).

Special Functions of the SER—Detoxification, Lipid Synthesis, and Glycogenolysis

In addition to the functions just mentioned that apply to both parts of the ER the following are some that predominate in the smooth portion:

Detoxification. As mentioned earlier, drugs such as phenobarbital, administered to an animal, result in increased activity of enzymes related to detoxification, as well as a considerable hypertrophy of the SER. This mechanism for detoxification also applies to endogenous or administered steroid hormones. Carcinogens such as 3-methylcholanthrene and 3,4-benzopyrene are among the most potent inducers of drug-metabolizing enzymes.

The enzymes involved in the detoxification of aromatic hydrocarbons are *aryl hydroxylases*. It is now known that benzopyrene (found in charcoal-broiled meat) is not carcinogenic (i.e., an inducer of cancer) but under the action of aryl hydrolases in the liver becomes converted to 5,6-epoxide, which is a powerful carcinogen.

The effect of these drugs is to produce a true induction of the enzymes of the ER. In other words, there is an increased synthesis of the enzymes, which can be prevented by inhibitors of protein synthesis, such as puromycin.

Synthesis of Lipids. Phospholipid biosynthesis is largely confined to the membranes of the ER. Studies with radioactive precursors indicate that the newly synthesized phospholipids are rapidly transferred to other cellular membranes. How this transfer occurs is not yet known, but it may be caused by the intervention of *phospholipid exchange proteins*, which are found in the cytosol and are often specific for a phospholipid class.

Glycogenolysis. In fasted animals it has been observed that the residual glycogen remains associated with the tubules and vesicles of the ER. When feeding is resumed, there is an increase in SER, which maintained its association with the accumulating glycogen in the form of glycosomes (Fig. 8–4).

The enzyme UDPG-glycogen transferase, which is directly involved in the synthesis of glycogen by addition of uridine diphosphate glucose (UDPG) to primer glycogen, is bound to the glycogen particle rather than to the membranous component. This suggests that the ER (Fig. 8–8) is related to glycogenolysis, but not glycogenesis.

In prenatal liver cells, just before birth, the amount of glycogen increases and then decreases simultaneously with an increased amount of glucose-6-phosphatase. This depletion of glycogen is accompanied by an increase in SER.

SUMMARY:
Biogenesis and Functions of the ER

Between the various membrane compartments there is not only a structural and functional continuity but also a temporal continuity. A multi-step mechanism involving the synthesis of a basic membrane with the addition of other molecular components is postulated. Some proteins, such as NADH-cytochrome b_5-reductase, are made on free ribosomes and are then inserted into the ER membrane.

Numerous functions are attributed to the ER. It contributes to the *mechanical support* of the cytoplasm, has *osmotic properties*, and is involved in intracellular *exchanges* between the matrix and the internal cavity. These exchanges are brought about by diffusion or *active transport*, in which carriers and permeases may be involved. In liver there are about 11 square meters of membrane per milliliter available for exchange. *Ionic gradients* and electrical potential may be generated across these membranes, and *conduction of intracellular impulses* has been postulated in the case of the SR. The ER serves as a *circulatory system* for the transport of various substances. Although slower, the flow of membranes may be effective in various types of intracellular transport. The *synthesis of proteins* for export is one of the main functions of the RER. The membrane of the ER is highly dynamic and fluid (the bound ribosomes are mobile), and the transfer of secretory proteins is accompanied by a flow of newly synthesized membrane proteins. The *synthesis of lipids and lipoproteins* is associated with the RER and the SER. The *synthesis of glycogen* is accomplished in the cytoplasmic matrix, but the smooth ER is involved in *glucogenolysis* through the action of glucose-6-phosphatase. Another important function of the SER is in the *detoxification* of many endogenous and exogenous compounds. The prolonged administration of certain drugs produces an increase in the SER and the induction of the specific enzymes.

8–4   THE ER AND SYNTHESIS OF EXPORTABLE PROTEINS

Although the process of protein synthesis takes place in the cytosol through the interaction of the polysomes, the mRNA, and the tRNAs, which carry the various amino acids (see Section 22–2), it is now well established that the polypeptide chain grows through a "groove" or "tunnel" in the large subunit that is linked directly to a channel in the ER membrane. In this way, at some stage of the elongation, the polypeptide chain begins to emerge and is finally deposited into the lumen of the ER.

Figure 8–9 shows an experiment using pro-

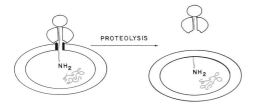

Fig. 8–9. Diagram showing the location of the nascent polypeptide chains within free and attached ribosomes. By proteolysis, it is possible to demonstrate the existence of a protected segment in both types of ribosomes and also in those attached to the ER. (Courtesy of G. Blobel, from D.D. Sabatini and G. Blobel, 1970.)

teolytic enzymes on free and membrane-bound ribosomes. This experiment demonstrated the presence, in both free and membrane-bound ribosomes, of a protected polypeptide chain of about 40 amino acids at the —COOH end of the protein. Furthermore, in bound ribosomes, the rest of the chain, with the —NH$_2$ end, was found to be protected inside the cisternal space where the protein is segregated.

On the basis of these findings, it was thought that the polypeptide chain moves through the large ribosomal subunit, becomes attached to the ER membrane, and then through a channel spanning the membrane enters the ER lumen.[10]

The Signal for Secretory Proteins Resides in the Nascent Polypeptide: The Signal Theory

Because free and membrane-bound ribosomes are continuously interchanged and show no differences between them (see Section 22–2), what determines whether or not a ribosome binds to the ER? The present view, named the *signal theory*, postulates that the information resides in the N-terminus of the polypeptide chain. According to Blobel and Sabatini,[10] the signal should be present in the messenger RNA (mRNA) at its 5′ end. At this end, there should be a set of *signal* codons localized immediately after the initiation codon AUG. As shown in Figure 8–10 protein synthesis starts on free ribosomes, and it is only when the special *signal peptide* emerges that the large ribosomal sub-

unit of the ribosomes becomes attached to the ER membrane.

The signal peptide is thought to recognize and to interact with special receptor proteins on the membrane, which form a "tunnel" through which the polypeptide penetrates into the ER lumen.[11,12]

The Hydrophobic Signal Peptide Is Removed by a Signal Peptidase

An important clue to the signal theory was provided by Milstein and colleagues in 1972.[13] They conducted an experiment in which the synthesis of the kappa light chain of immunoglobulin from mouse myeloma cells was followed in vitro on two different systems: (1) reticulocytes (i.e., young erythrocytes) containing no ER; (2) ascitic cell lysate containing microsomes. The analysis of the synthesized proteins showed that in system 1, the polypeptide was about 15 amino acids longer at the N-terminal end; whereas in system 2, the protein was shorter, and of the same size as the mature protein secreted by the myeloma cell.

They concluded that: (a) the N-terminal sequence functions as a signal for the ribosomal attachment to the ER, and (b) this extra sequence is removed by an enzyme in the ER. Their experiment demonstrated that the signal peptide is separated by a processing protease, generally called the signal peptidase. This protease is associated with the membrane of the ER and its active site faces the interior of this organelle. The signal peptidase processes the precursor polypeptide while it is being synthesized and translocated. This protease can be solubilized by the use of detergents and purified by various chromatographic procedures.[14] It is not species-specific, and even a signal peptidase from bacteria can act on eukaryotic cells.

The student should find it interesting to calculate the length of the polypeptide before it is hydrolyzed by the peptidase. For example, to emerge from the ribosome and to become inserted into the ER membrane the nascent peptide must be longer than 30 to 40 amino acids. Since some 20 more are needed to cross the membrane, and the signal peptide is 18 to 30 residues long, it can be said that, for the signal peptidase to act, the polypeptide must be at least 70 to 90 amino acids long.

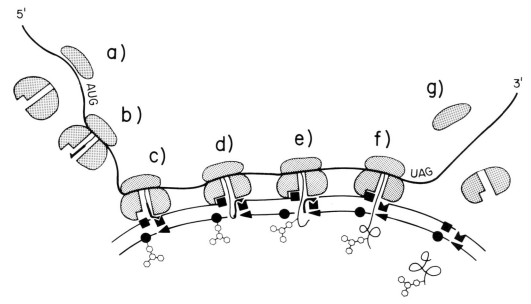

Fig. 8–10.   Diagram indicating the main molecular events taking place during the synthesis and translocation of a polypeptide across the ER membrane.

(a) After the initiation of the translation the signal peptide (thicker line) emerges from the 60S ribosomal subunit (b) and recognizes a receptor (translocator) on the membrane (c). The association of the ribosome to the membrane is reinforced by ionic linkages and the ribophorins. The nascent chain initially in a loop configuration crosses the ER membrane (d).

Other *cotranslational modifications* of the growing chain are the *proteolytic processing* by the signal peptidase (e) at the luminal surface; the *glycosidation* to asparagine residues from dolichol pyrophosphate oligosaccharide intermediates (●○) (f). After termination of protein synthesis (g) the secretory protein is segregated into the ER lumen and the ribosomal subunits are dissociated and detached. (Courtesy of D.D. Sabatini and G. Kreibich.)

The now "processed" polypeptide continues to grow until it is completely segregated into the lumen. Figure 8–10 is a diagram of some of the major molecular events that occur in the synthesis and translocation of the polypeptide into the ER lumen. These events include the translation of the signal peptide, the association of the ribosome to the ER membrane, and the recognition of the signal peptide by a possible receptor, which acts as a translocator. Other events are the clipping of the signal peptide by the peptidase, the first glycosidation of the protein by dolichol pyrophosphate oligosaccharide intermediates, the termination of protein synthesis, and the detachment of the ribosomal subunits.

The signal theory is supported by abundant evidence. Messenger RNAs for several secreted proteins have been translated on free ribosomes (in the absence of microsomal membranes) and shown to produce larger polypeptide chains. These have been called *preproteins* to differentiate them from the *proproteins*, such as proinsulin, proalbumin, proparathyroid hormone, and others, which are larger precursors

of the secreted proteins (see Fig. 9–15). For example, a mRNA for albumin first produces a preproalbumin; this, after removal of the signal peptide, is converted into proalbumin, and later on it is again cleaved at a specific secretory stage to produce the final product, *albumin*. Signal peptides from different proteins differ in their amino acid sequence; however, they resemble in that they usually contain a series of 10 to 14 highly hydrophobic residues in the middle region and one or two charged groups near the amino terminal end. Table 8–3 shows the signal sequences of a series of preproteins and the number of amino acids that are removed from the N-terminal end by the signal peptidase.[14] While in most proteins the signal peptide is removed, there are some exceptions. For example, it has been found that there is a signal peptide in ovalbumin, but this is not removed by cleavage.[15]

The signal theory also applies to secretory proteins produced by plant cells such as zein from corn, glycinin from soybean, the chymotrypsin inhibitor from tomato leaves, and the seed reserve protein from the common bean.[16]

TABLE 8–3.   SIGNAL SEQUENCES PRESENT IN SECRETED PROTEINS

Precursor	Amino Acid Sequence from N-Terminus
Pre-ovomucoid	M A M A G V F V L F S F V L C G F L P D A A F G A E V D (↓24)
Pre-lysozyme	M R S L L I L V L C F L P L A A L S K V F X (18)
Pre-proinsulin	M A L W M R F L P L L A L L V L W E P K P A Q A F V K Q (24)
Pre-growth hormone	M A A D S Q T P W L L T F S L L C L L W P Q E A G A L P A M (26)
Pre-lactalbumin	M M S F V S L L L L V G I L F X A T Q A E Q L T (19)
Pre-opiocortin	M P R L C S S R S G A L L L A L L L Q A S M E V R G W C L E (26)
Pre-penicillinase	M S I Q H F R V A L I P F F A A F C L P V F A H P E T (23)

The arrow points to the site of action of the signal peptidase, the number indicates the length of the signal peptide separated by the peptidase. For amino acid nomenclature see Table 2–4. Hydrophobic amino acids are underlined.

The Role of the Signal Recognition Particle (SRP)

One of the most interesting questions posed by the signal theory is that of the mechanism by which the nascent polypeptide, starting on a free ribosome in the cytosol, is later geared to become attached to ER membrane. In recent years, it has become evident that the hydrophobic signal peptide does not possess the needed information to ensure its transfer through the membrane; this implies that some other factors should be present in the system. This was first suggested by experiments in which microsomes were either extracted with high salt solutions or were mildly treated with proteolytic enzymes. In both cases, the translocation was prevented and it was restored by adding back the extract. The factor removed by this treatment is a large complex of proteins and RNA of about 250,000 daltons, which acts on the translocation.[17] This complex, which consists of 6 polypeptides[18] and contains one RNA molecule with 140 nucleotides belonging to the 7S cytoplasmic RNA type,[19] is the so-called *signal recognition particle* (SRP) (see Fig. 20–28). The SRP aids in the decoding of the information carried by the signal peptide for secretory proteins, as well as that present in the signal peptides for lysosomal and certain membrane proteins.[20]

In an in vitro system, in which protein synthesis is going on without the presence of microsomal membranes, it can be demonstrated that the SRP specifically arrests the synthesis as soon as the signal peptide emerges from the large ribosomal subunit. At this stage, the polypeptide chain has about 70 amino acids (i.e., 40 required to span the ribosomal subunit plus about 30 amino acids corresponding to the signal peptide). It should be readily apparent to the student why ribosomes carrying cytosolic and other proteins that do not have signal peptides cannot be recognized by the SRP and remain in the cytosol.

Figure 8–11 *A* through *E* represents a proposed model for the function of the SRP. At first, this particle is free in the cytosol and in a "dynamic equilibrium" with SRPs bound to the ER membrane (SRP cycle, *A-E*). The free SRP (*A*) binds to a ribosome with low affinity (*B*); later on, when the translation of the signal peptide starts, the affinity increases (*C*). With the emergence of the signal peptide from the ribosome the SRP binds to it and the synthesis is arrested (*D*). This situation continues until the ribosome attaches to the surface of the ER. The attachment occurs by means of ribosome and SRP receptors present in the ER membrane (*E*). When the final assembly of the ribosome with the ER membrane is produced, the block of protein synthesis is released, and the polypeptide chain resumes its elongation and is translocated through a pore-like structure.[21]

In the removal of the block imposed by the SRP a membrane component, corresponding to the SRP receptor (Fig. 8–11), is involved. By using mild proteolysis and high salt extraction on membranes of the RER, a 60,000 dalton protein that is capable of counteracting the block caused by the SRP was separated.[22,23] This pro-

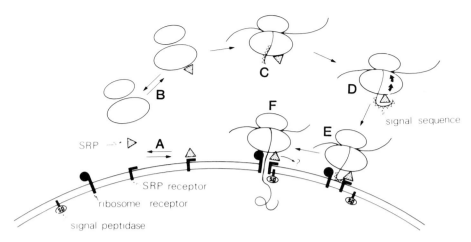

Fig. 8–11. Model suggesting the function of the signal recognition protein (SRP) in the process of translocation of a secretory protein. The SRP receptor could correspond to the so-called docking protein (see text for details). (Courtesy of P. Walter and G. Blobel. Reproduced from the Journal of Cell Biology, 91:557–561, 1981. Copyright permission of the Rockefeller University Press.)

tein, which corresponds to the SRP receptor, is a fragment from a 72,000 dalton protein known as the "docking protein" (see Meyer, 1982 in the Additional Readings at the end of this chapter). The SRP receptor has been isolated, purified, and shown to mediate the release of the arrest of protein synthesis imposed by the SRP particle.[24]

It was suggested that the interaction ribosome SRP-SRP receptor may be transient and mainly involved in the targeting of the arrested ribosome to a specific site on the ER membrane. (This interpretation is based partly on the fact that there are more ribosomes attached to the ER than SRP receptors.) Later on, this interaction might be replaced by a direct interaction with the ribophorins I and II, which are in stoichiometric amounts with respect to the membrane-bound ribosomes.

In conclusion: The main role of the SRP is that of stopping the synthesis of protein after the signal peptide has been exposed from the ribosome. By this mechanism, the ribosomes making secretory, lysosomal, and membrane proteins, such as albumin, acid phosphatase, and glycophorin, are recognized and have time to find the site of attachment to the membrane necessary for the translocation and the further processing of the polypeptide into the lumen of the ER.

The SRP appears to be an indispensable component of the machinery for protein synthesis (see Chapter 21) that functions as an adapter

between the cytoplasmic translation of a protein and its translocation across the ER membrane (see Walter and Blobel, 1982 in the Additional Readings at the end of this chapter).

Protein Secretion in Bacteria Uses a Signal Peptide

Proteins that are secreted by bacteria follow a pattern similar to that of eukaryotic cells. Although bacteria lack an ER, they are able to secrete proteins that attach to the periplasmatic space and the cell wall, or that diffuse into the environment (see Fig. 1–4). For example, bacteria that are resistant to penicillin secrete the enzyme *penicillinase* and *Corynebacterium diphtheriae* export *diphtheria toxin.* The synthesis of these and other proteins (including those that reside in the membrane) occurs on ribosomes attached to the plasma membrane.

Because of the high concentration of ribosomes in a section of a bacterium, it is impossible to detect the membrane-bound ribosomes by electron microscopy (see Fig. 1–4). If bacteria are disrupted by sonication, however, a certain percentage of ribosomes remains attached to the membrane. This fraction is enriched in the mRNA of the proteins that are secreted by the bacterium. As in eukaryotic cells, the proteins produced on the membrane-bound ribosomes are larger and contain a hydrophobic signal peptide that facilitates the translocation through the plasma membrane; this peptide is then removed

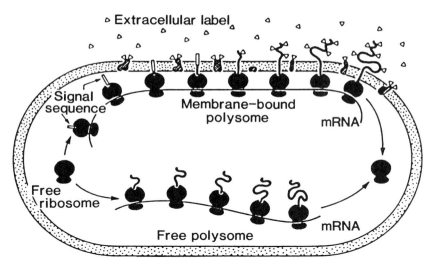

Fig. 8–12. Diagram showing the synthesis of proteins in bacteria by the use of free and membrane-bound ribosomes. Notice that in the latter case a signal sequence is removed. The nascent polypeptide chain traverses the plasma membrane and can be labeled by the use of extracellular labels that bind to —NH₂ groups or to certain amino acids.

by a signal peptidase (see Davis and Tai, 1980 in the Additional Readings at the end of this chapter).

Figure 8–12 illustrates the difference between protein synthesis on free or membrane-bound polysomes in bacteria. By the use of different labels (i.e., drugs that bind covalently) that do not penetrate into the bacterium, it is possible to demonstrate that the proteins that are secreted become labeled while the ribosomes are still attached to the membrane. On the other hand, the proteins produced on free ribosomes remain unlabeled.

Bacteria offer the possibility of studying mutants in which changes in the amino acid sequence of the peptide are produced (see Robertson, 1984 in the Additional Readings at the end of this chapter). The substitution of a simple hydrophobic residue may be enough to abolish the translocation of the protein. As expected, transport was most affected when hydrophobic amino acids in the signal peptide were substituted by charged amino acids.[25]

SUMMARY:

Synthesis of Exportable Proteins—The Signal Hypothesis

The molecular mechanism involved in the synthesis of proteins will be studied in the chapters on molecular biology. Although this process occurs in the cytosol, the polypeptide chain of the ex-

portable protein grows into a "groove" or "tunnel" in the 60S ribosomal subunit and penetrates, through a channel, into the lumen of the ER.

The signal hypothesis tries to explain the mechanism by which the mRNA is able to recognize free or bound ribosomes. It is postulated that the mRNA for secretory proteins contains a set of *special signal codons* localized after the initial codon AUG. Once the ribosome "recognizes" the signal the ribosome becomes attached to the membrane, and the polypeptide penetrates. It is also postulated that at the luminal surface there is a *signal peptidase* that removes the signal peptide. Thus the mRNA produces a *preprotein* of larger molecular weight than the final protein. The signal peptide has between 15 and 30 amino acids, which are generally hydrophobic. Such a signal peptide probably establishes the initial association of the ribosome with the membrane, but some protein factors are involved. An SRP complex binds to the nascent signal peptide and stops the translation until it reaches the ER membrane. It has been postulated that a docking protein removes the SRP block, allowing for the translocation of the polypeptide. A similar mechanism of protein secretion by a signal peptide is found in bacteria that secrete penicillinase or diphtheria toxin to the medium. In mutants, the exchange of one amino acid of charged nature for a hydrophobic one affects the secretion.

8–5 · PROTEIN SEGREGATION

In an elementary way, Figure 8–13 illustrates the different types of protein segregation. For example, histone and other nuclear proteins must cross the nuclear envelope to reach the

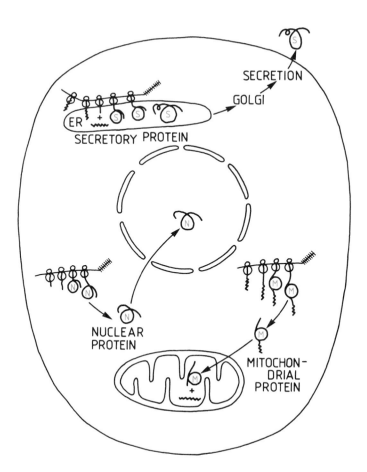

Fig. 8–13. Mechanisms of protein segregation. The correct segregation of secretory, mitochondrial, and nuclear proteins is achieved by different mechanisms.

nucleoplasm; enzymes of the Krebs cycle should traverse the two membranes of the mitochondrion to enter the matrix; some peroxisomal proteins, such as catalase, are produced in the cytoplasm and, to enter the peroxisome, must go through its membrane. On the other hand, some of the enzymes of the respiratory chain in mitochondria, or the photosystem complexes in chloroplasts, become integrated into the inner membranes of both organelles.

In a cell, about 10,000 different proteins are synthesized, either on free ribosomes in the cytosol or on ribosomes attached to membranes of the ER. The final fate of these proteins is (1) to be secreted outside the cell; (2) to be incorporated into various intracellular compartments; or (3) to be attached or integrated into the membranes.

The understanding of the mechanism involved in protein segregation requires identification of all of the steps by which a polypeptide is transferred from the site of synthesis to the site where it will perform its function. For sorting out the various subpopulations of proteins, special mechanisms should be involved. In a generalized hypothesis, it is postulated that the information for the selective traffic of proteins resides in discrete sequences in the polypeptide chain (see Blobel, 1980 in the Additional Readings at the end of this chapter). These sequences can be compared with specially addressed labels or postal codes on a letter which help it reach its destination. Because the number of codes is relatively small compared with the number of proteins, it is conceivable that each subpopulation shares a similar type of sequence. (In Section 8–4 we described the case of secretory proteins having the *signal peptide*, which has a sequence of hydrophobic amino acids at the amino terminal.)

Ultimately, the information contained in the various sequences must be determined genetically, i.e., it must reside in the DNA molecule. This is clearly shown in the following experiment:

Xenopus oocytes are microinjected with different types of mRNA and, after some time, the proteins that are translated are studied in their location within the cell. It is demonstrated that, if the mRNA codes for a secretory protein (e.g., immunoglobulin, albumin) the product will accumulate within membrane-bound vesicles. If the mRNA codes for a cytosolic protein, such as globin, it accumulates in the cytosol.

Cotranslational and Postranslational Translocation of Proteins—Segregation Sequences

There are two distinct mechanisms by which proteins may be translocated to their destination: cotranslational and postranslational translocation (Fig. 8–13).

In the first case, exemplified by the secretory proteins, the translocation is coupled with the translation (i.e., it occurs simultaneously with protein synthesis). In the second case, exemplified by some of the proteins of the mitochondrial or chloroplastic matrices (see Sections 11–2 and 12–3); the proteins are translocated after they have been fully synthesized on free ribosomes in the cytosol. To reach their destination these proteins should be translocated through the membranes of these organelles. In both types of translocation, in addition to a special polypeptide sequence, a *receptor* or *translocator* in the membrane should recognize (i.e., to decode) the corresponding signal (Fig. 8–10).

In addition to the *signal sequences* for secretory proteins, the existence of two more types of sequences are postulated. In proteins that are integrated in the membrane there should be *stop or halt transfer sequences (or signals)*. Such sequences will allow the anchorage of the polypeptide chain while passing through the hydrophobic portion of the membrane.

Sorting out sequences exist in the proteins that go through membranes to reach a certain compartment. These proteins are synthesized on free ribosomes in the cytosol and then enter various organelles such as peroxisomes, mitochondria, or chloroplasts (Fig. 8–13).

Sometimes, after the proteins enter such compartments, the sorting out sequence may be removed by proteolysis. Examples of various proteins segregated into cell organelles are given in the corresponding chapters.[26,27]

Protein entering the nucleus probably use a different mechanism; they selectively accumulate after completion of their synthesis in the cytosol.[28] The segregation property apparently resides in the mature protein (see Section 13–1). The segregation of proteins is achieved by several mechanisms that differ in their transfer sequences and probably in the receptor. This is not at all surprising in view of the structural and functional complexity of the living cell.

Synthesis and Assembly of Membrane Proteins—Stop Transfer Sequences

To study the problem of how membrane proteins are synthesized and inserted, the student should recall the fluid model analyzed in Section 4–1. As shown in Figure 8–14, proteins may be exposed to the cytoplasmic side *(endoproteins)* or to the extracytoplasmic side *(ectoproteins)*. In addition, an important group of proteins traverses the membrane and is exposed on both sides. What are the mechanisms of synthesis for these various types of proteins?

Endoproteins (both peripheral and integral) could be synthesized either on free or bound ribosomes localized in the same cell compartment. For example, in reticulocytes, which do not contain ER, all proteins on the cytoplasmic face are produced by free ribosomes. Ectoproteins (both peripheral and intrinsic) appear to be synthesized exclusively on ribosomes attached to the ER membranes. Such proteins are then transferred into the lumen. Some of these proteins may be secreted, but others may remain as permanent residents in the ER. This is the case with *calsequestrin* a calcium-binding protein present inside the sarcoplasmic reticulum of muscle (see Section 7–5).

One special case is that of transmembrane proteins in which the synthesis also occurs on membrane-bound ribosomes, but the insertion is caused by an incomplete vectorial transfer of

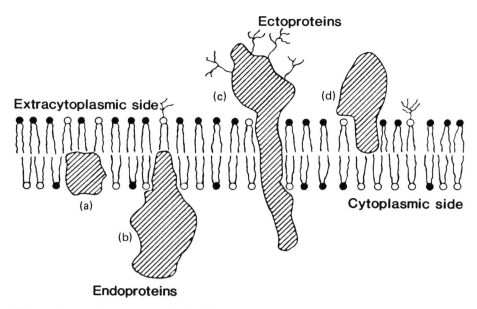

Fig. 8–14. Diagram showing the position in the lipid bilayer of endoproteins *(a,b)* and ectoproteins *(c,d).* Observe that *(c)* is also a transmembrane protein.

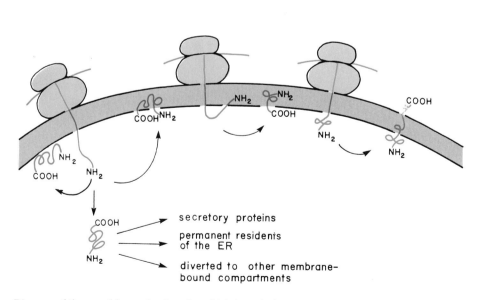

Fig. 8–15. Diagram of the possible mechanisms by which bound ribosomes may give rise to proteins that are deposited into the lumen of the ER or are incorporated either as peripheral or integral proteins at the luminal face, or as transmembrane proteins. (Courtesy of D.D. Sabatini and G. Kreibich. From Sabatini, D.D., and Kreibich, G.: Functional specialization of membrane-bound ribosomes. *In* Enzymes of Biological Membranes. Vol. 2. Edited by A. Martonosi. New York, Plenum Press, 1976, p. 431.)

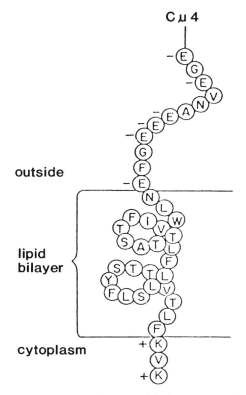

Fig. 8–16. Diagram of immunoglobulin in the membrane of a lymphocyte. The μ-chain has a sequence of 26 uncharged residues spanning the membrane, which is followed, on the cytoplasmic side, by lys-val-lys. It is also shown that the transmembrane chain is flanked by positive and negative charged amino acids.

the polypeptide chain. We shall see that transmembrane proteins have a hydrophobic segment somewhat similar to the signal peptide, but followed by a hydrophilic sequence that functions as a *stop transfer signal*. In all cases studied so far, these stop transfer sequences appear to consist of a hydrophobic stretch of 11 to 25 amino acid residues that are followed by several closely spaced basic amino acid residues, which may help to fix the transmembrane protein into the membrane (Fig. 8–15). While at first glance there are similarities between the signal peptides for secretory proteins and those for membrane proteins (in the sense that both consist of a stretch of hydrophobic amino acid residues), there are evident differences.

One interesting example is provided by the differences in the IgM that is secreted into the blood plasma and the surface immunoglobulin IgM, that functions as a surface receptor in lymphocytes. They have slight differences in one of

the chains (μ-chain) at the C-terminus. The secreted μ-chain has a hydrophilic region of 20 amino acids after a constant (Cμ4) region. The membrane-bound μ-chain, however, has 41 residues at the C-terminus; among these is a 26-hydrophobic sequence, presumably transmembranous, which is followed by a stop transfer sequence of lys-val-lys exposed to the cytoplasmic sides (Fig. 8–16). This example shows that a given polypeptide can be targeted to two different compartments by a change in the signal sequence.

Figure 8–15 summarizes the various mechanisms by which bound ribosomes may give rise to proteins that lie in the lumen or are incorporated into the membrane as peripheral, integral, or transmembranal proteins. In the last type, there is a special orientation of the protein, with the $—NH_2$ end facing the lumen and the —COOH terminal at the cytoplasmic face.

SUMMARY:
Protein Segregation

The several thousand proteins made by a cell are synthesized either on free or membrane-bound ribosomes. The final fate of the proteins is (1) to be exported, (2) to be incorporated into different intracellular compartments (e.g., nucleoplasm, peroxisome, or mitochondria), or (3) to be integrated into membranes. The sorting out of protein subpopulations is attributed to special sequences in the polypeptide chain. Ultimately, the information resides in a genetic mechanism involving DNA-mRNA. In addition to the transfer sequence, the presence of specific receptors or translocators in the membrane is postulated.

Translocation may occur simultaneously with translation (cotranslation) as in the case of secreted proteins, or may occur after completion of protein synthesis (postranslation). Several types of transfer sequences are postulated: (1) *Signal peptides* used for secretory proteins, (2) *stop transfer sequences* involved in the integration of proteins into membranes, (3) *sorting* sequences that determine the posttranslational translocation.

An interesting problem is posed by the mechanism of synthesis of the various types of proteins that are present in the membrane. A special case is that of the proteins that traverse the membrane. It is thought that they are synthesized on ribosomes attached to the membrane, with a more hydrophobic segment in the middle that remains embedded in the membrane. The possible mechanisms of synthesis for the various types of proteins are indicated in Figure 8–15.

REFERENCES

1. Porter, K.R., Claude, A., and Fullan, E.F.: J. Exp. Med., *81*:233, 1945.
2. Buckley, I.K., and Porter, K.R.: J. Microsc. (Oxford), *104*:107, 1975.
3. Sabatini, D.D., and Kreibich, G.: *In* Enzymes of Biological Membranes. Vol. 2. Edited by A. Martonosi. New York, Plenum Press, 1976, p. 531.
4. Branton, D., et al.: Science, *190*:54, 1975.
5. Weibel, E.R., Stäbli, W., Gnagi, R., and Hess, E.A.: J. Cell Biol., *42*:68, 1969.
6. Rybicka, K.: J. Histochem. Cytochem., *29*:4, 1981.
7. Estabrook, R.W., et al.: *In* The Structural Basis of Membrane Function. Edited by Y. Hatefi, D. Javadi, and L. Ohaniance. New York, Academic Press, 1976, p. 429.
8. Borghese, N., and Gaetani, S.: Eur. J. Cell Biol., *22* (1):154, 1980.
9. Ojakian, G.K., Kreibich, G., and Sabatini, D.D.: J. Cell Biol., *72*:530, 1977.
10. Blobel, G., and Sabatini, D.D.: *In* Biomembranes. Vol. 2. Edited by L.A. Manson. New York, Plenum, 1971, p. 193.
11. Blobel, G., and Dobberstein, B.: J. Cell Biol., *67*:835 and 852, 1975.
12. Blobel, G.: *In* International Cell Biology. Edited by B.R. Brinkley and K.R. Porter. New York, The Rockefeller University Press, 1977, p. 318.
13. Milstein, C., et al.: Nature New Biol., *239*:117, 1972.
14. Jackson, R.C., and Blobel, G.: Proc. Natl. Acad. Sci. USA., *76*:1795, 1977.
15. Palmiter, R.D., et al.: Proc. Natl. Sci. USA., *75*:94, 1978.
16. Bollini, R., Vitale, A., and Chrispeels, M.J.: J. Cell Biol., *96*:999, 1983.
17. Dobberstein, B.: Z. Physiol. Chem., *359*:1469, 1978.
18. Walter, P., and Blobel, G.: Proc. Natl. Acad. Sci. USA, *77*:7112, 1980.
19. Walter, P., and Blobel, G.: Nature, *299*:691, 1982.
20. Anderson, D.J., Walter, P., and Blobel, G.: J. Cell Biol., *93*:501, 1982.
21. Walter, P., and Blobel, G.: J. Cell Biol., *91*:551 and 557, 1981.
22. Meyer, D.I., Krause, E., and Dobberstein, B.: Nature, *297*:647, 1982.
23. Meyer, D.I., Loward, D., and Dobberstein, B.: J. Cell Biol., *92*:579, 1982.
24. Gilmore, R., Blobel, G., and Walter, P.: J. Cell Biol., *95*:463 and 470, 1982.
25. Bassford, P., and Beckwith, J.: Nature, *277*:538, 1979.
26. Highfield, P.E., and Ellis, R.J.: Nature, *271*:420, 1978.
27. Maccechini, M.L., et al.: Proc. Natl. Acad. Sci. USA., *76*:343, 1979.
28. De Robertis, E.M., Jr., Longthorne, R.F., and Gurdon, J.B.: Nature, *272*:254, 1978.

ADDITIONAL READINGS

Autori, A., Svensson, H., and Dallner, G.: Biogenesis of microsomal membrane glycoproteins in rat liver: I, II, and III. J. Cell Biol., *67*:687, 1975.

Blobel, G.: Intracellular topogenesis. Proc. Natl. Acad. Sci. USA., *77*:1496, 1980.

Blobel, G.: Synthesis and segregation of secretory proteins, the signal hypothesis. *In* International Cell Biology. Edited by B.R. Brinkley and K.R. Porter. New York, The Rockefeller University Press, 1977, p. 318.

Davis, B.D., and Tai, P.C.: The mechanism of protein secretion across membranes. Nature, *283*:433, 1980.

Eriksson, L.C., De Pierre, J.W., and Dallner, G.: Preparation and properties of microsomal fractions. *In* Pharmacology and Therapeutics. Vol. 2. Edited by W.C. Bowman. New York, Pergamon Press, 1978, p. 281.

Haguenau, F.: The ergastoplasm: Its history, ultrastructure, and biochemistry. Int. Rev. Cytol., *7*:425, 1958.

Kreibich, G., et al.: Components of microsomal membranes involved in the insertion and co-translation processing of proteins made on bound ribosomes. *In* International Cell Biology. Edited by H.G. Schweiger. Berlin, Springer-Verlag, 1981, p. 579.

Lingappa, V.R., Katz, F.N., Lodish, H.F., and Blobel, G.: A signal sequence for insertion of transmembrane glycoprotein. J. Biol. Chem., *253*:8867, 1978.

Meyer, D.I.: The signal hypothesis: A working model. Trends Biochem. Sci., *7*:320, 1982.

Novikoff, A.B.: The endoplasmic reticulum: A cytochemist's view (a review). Proc. Natl. Acad. Sci. USA, *73*:2781, 1976.

Robertson, N.: Membrane traffic and the problem of protein secretion. Nature, *307*:594, 1984.

Sabatini, D.D., and Kreibich, G.: Functional specialization of membrane-bound ribosomes in eukaryotic cells. *In* The Enzymes of Biological Membranes. Vol. 2. Edited by A. Martonosi. New York, Plenum Press, 1976, p. 531.

Terasaki, M., et al.: Localization of endoplasmic reticulum in living and glutaraldehyde—fixed with fluorescent dyes. Cell, *38*:101, 1984.

Walter, P., and Blobel, G.: Signal recognition particle contains a 7S RNA essential for protein translocation across the ER. Nature, *292*:691, 1982.

GOLGI COMPLEX AND
CELL SECRETION

In Chapter 8 we mentioned that the *Golgi apparatus* or *Golgi complex* is a differentiated portion of the endomembrane system found in both animal and plant cells. This membranous component is spatially and temporally related to the endoplasmic reticulum (ER) on one side and, by way of secretory vesicles, may fuse with specific portions of the plasma membrane. Because of the important functions that this structure plays as an intermediary in secretory processes, it will be studied in this chapter together with *cell secretion.*

In 1898 by means of a silver staining method Camillo Golgi discovered a reticular structure in the cytoplasm of nerve cells (Fig. 9–1). Since its refractive index is similar to that of cytosol, the Golgi complex in the living cell was difficult to observe with the light microscope, and this led to many controversies regarding its true nature. However, the use of electron microscopy provided a distinct image of this organelle, and its structure could be studied in detail. Later on, the study of thick sections with the high-voltage electron microscope,[1] the use of

freeze-etching (Fig. 3–9), and the observation of isolated Golgi membranes by negative staining contributed to the study of this structure. Other advances came with the use of autoradiography with special radioactive precursors and cell fractionation, a technique that enabled isolation of purified fractions and performance of biochemical studies. Through all these advances, the role of this organelle in cellular functions has been, to a great extent, elucidated (see Morré, 1977; Farquhar and Palade, 1981 in the Additional Readings at the end of this chapter).

In this chapter, we emphasize not only the morphology and cytochemistry of the Golgi complex, but also the role of this organelle in the glycosidation of lipids (glycolipids) and proteins (glycoproteins) and in the packaging and release of secretory products. We show that, in some glandular cells, secretion takes place in consecutive steps, several of which occur in the Golgi complex.

The Golgi apparatus is also related to the lysosomes and can be considered as a sorting center that is able to discriminate between the pro-

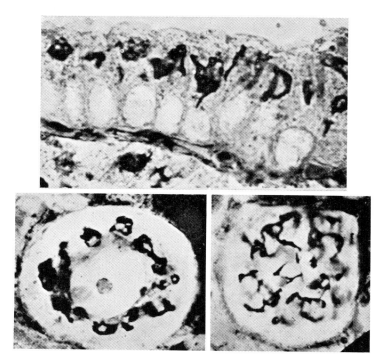

Fig. 9–1. *Above,* Golgi complex (Golgi apparatus) in cells of the thyroid gland of the guinea pig, apical position. Osmic impregnation. *Below left,* ganglion cell, perinuclear Golgi apparatus. *Below right,* same structure as at left, optical section tangential with respect to the nucleus. Silver impregnation. (From E. De Robertis, unpublished.)

teins that are to be secreted and those that are to be delivered to the lysosomes (see Tartakoff, 1982 in the Additional Readings at the end of this chapter).

9–1 MORPHOLOGY OF THE GOLGI COMPLEX (DICTYOSOMES)

The Golgi complex is morphologically very similar in both plant and animal cells (Fig. 9–2). It consists of *dictyosome* units formed by stacks of flattened disc-shaped cisternae and associated secretory vesicles.

In cells that have a polarized structure the Golgi complex is, in general, a single large structure and occupies a definite position between the nucleus and the pole of the cell in which the release or secretion takes place. This is the case, for example, in thyroid cells (Fig. 9–1) in the exocrine pancreas, or in the mucous cells of the intestinal epithelium. In nerve (Fig. 9–1) and liver cells (Fig. 9–3) and in most plant cells there are many dictyosomes that do not show a special polarity. In liver cells there are some 50 *dictyosomes* per cell. These structures represent some

2% of the total cytoplasmic volume. Although the localization, size, and development of these organelles vary from one cell to another and also with the physiological state of the cell, they show morphological characteristics that permit their differentiation from the other parts of the endomembrane system.

One such quality is a lack of attached ribosomes; in fact most of the Golgi complex appears to be surrounded by a zone from which most ribosomes, glycogen, and mitochondria are absent; the so-called *zone of exclusion*.[2] Some free polysomes, present at the periphery of the Golgi complex, have been observed, however, and the possibility of specific protein synthesis that may lead to membrane modification of the Golgi has been recognized.

In general, three membranous components are recognized under the electron microscope (EM): (1) flattened sacs (i.e., cisternae); (2) clusters of tubules and vesicles of about 60 nm; and (3) larger vacuoles filled with an amorphous or granular content. The Golgi cisternae are arranged in parallel and are separated by a space or 20 to 30 nm, which may contain rod-like ele-

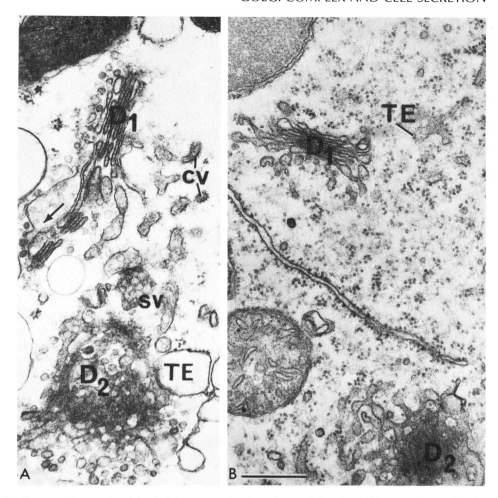

Fig. 9–2. Electron micrographs of the Golgi apparatus in: *A,* rat liver and *B,* onion *(Allium cepa)* stem. D_1, dictyosome in cross section showing the stacked cisternae; D_2, dictyosome in tangential section showing a face view of cisternae with a central plate-like region and a fenestrated margin; *TE,* transition element of the ER adjacent to the Golgi; *cv,* coated vesicles; *sv,* secretory vesicles. An arrow shows a connection between the SER and a secretory vesicle. ×35,000. (Courtesy of D.J. Morré. From Morré, D.J.: Membrane differentiation and the control of secretion. A comparison of plant and animal Golgi apparatus. *In* International Cell Biology. Edited by B.R. Brinkley and K.R. Porter. New York, Rockefeller University Press, 1977, p. 293.)

ments or fibers. Often the cisternae are arranged concentrically with a convex and a concave face. There may be from three to seven of these structures in most animal and plant cells. In certain algae, however, there may be as many as 10 or 20 cisternae.

Dictyosomes—A Forming Face, and a Maturing Face Near the GERL

Each stack of cisternae forming a dictyosome is a polarized structure having a *proximal* or *forming* face generally convex and closer to the nuclear envelope or the ER and a *distal* or *maturing* face of concave shape that encloses a re-

gion containing large secretory vesicles (Fig. 9–2). This polarization is often referred to as the *cis-trans axis* of the Golgi complex. The cis or forming face is characterized by the presence of small transition vesicles or tubules that converge upon the Golgi cisternae, forming a kind of fenestrated plate (Fig. 9–2). These *transition vesicles* are thought to form as blebs from the ER and to migrate to the Golgi where, by coalescence, they form new cisternae. As will be mentioned later, a mechanism of membrane flow is postulated in which new cisternae are formed at the proximal end and thus compensate for the loss at the maturing face that occurs with the release of secretory vesicles.

Associated with the trans or maturing face

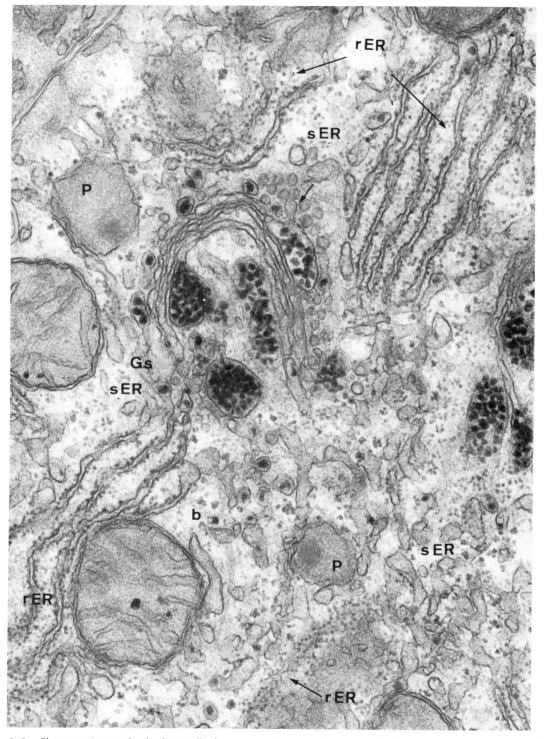

Fig. 9–3. Electron micrograph of a liver cell of an animal having a diet rich in fat. The synthesis and transport of the lipoprotein granules is observed. The rough endoplasmic reticulum *(rER)* is observed as two stacks of lamellae converging toward a Golgi complex having Golgi sacs *(Gs)* on its convex or "forming face." There are portions of smooth endoplasmic reticulum *(sER)* connecting both parts of the vacuolar system. Mitochondria and peroxisomes *(P)* are observed. × 56,000. (Courtesy of A. Claude.)

there is often a saccular structure that is rich in acid phosphatase and has been called the GERL (see Fig. 8–2). This denomination indicates that it has been interpreted as a region of *smooth endoplasmic reticulum (SER), near the Golgi,* which is involved in the production *of lysosomes.* More recent work relates the GERL to *Golgi condensing vacuoles* or *presecretory granules.*[3]

Polarization of Dictyosomes and Membrane Differentiation

The polarization of the dictyosomes is also expressed in terms of what may be called *membrane differentiation.* As shown in Table 9–1 the thickness of the membranes increases progressively from the ER to the Golgi and the plasma membrane, and even within the various cisternae of the dictyosome. There are also differences in staining properties; thus the cisternae at the forming face stain more strongly with osmium, and a stain for glycoproteins, such as periodic acid-methenamine, appears in the maturing cisternae. This is particularly evident in the Golgi complex of certain algae in which scales of cellulose are being produced and secreted (see Brown and Willison, 1977 in the Additional Readings at the end of this chapter).

The mucin-producing goblet cells of the intestine provide another chance to observe the polarization and differentiation of the Golgi apparatus. In Figure 9–4, the N-acetyl-D-galactosamine residues are detected by means of a lectin-gold complex. By observing this figure, the student finds that although the reaction is absent in the rough endoplasmic reticulum (RER), it starts heavily in the Golgi stacks. The mucin granules and the plasma membrane also show signs of glycosidation. This finding suggests that the corresponding glycosyl-transferase is found in the Golgi but not in the ER. Other examples of the polarization and heterogeneity of the Golgi apparatus are discussed at the end of this chapter, when we study the possible functional compartmentalization of this organelle (see Figs. 9–13 and 9–14). We show that thiamine-pyrophosphatase is found in the *trans* region of the Golgi; whereas acid phosphatase appears even more distally, in the region of the GERL, which is related to the lysosomes (see Fig. 8–2).[4]

The transitions between the various parts of the Golgi complex can also be observed in liver cells under certain experimental conditions (e.g., a diet rich in fat) that make it possible to trace the synthesis and transport of lipoproteins, which appear as discrete dense granules of about 40 nm. As shown in Figure 9–3 these granules appear first within tubules of the SER and then enter the outer fenestrated cisternae, in the cis region of the Golgi. After longer periods they accumulate in the large vacuoles that are formed by dilatation of the edge of the sacs and, finally, they are detached.

SUMMARY:
Morphology of the Golgi Complex

The Golgi apparatus or complex is a differentiated portion of the endomembrane system, which is morphologically very similar in animal and plant cells. It is spatially and temporally related to the

TABLE 9–1. MEMBRANE DIFFERENTIATION IN THE GOLGI COMPLEX OF ANIMALS AND PLANTS

Membrane Type	Membrane Thickness (nm)			
	Rat Liver	Rat Mammary Gland	Onion Stem	Soybean Hypocotyl
Nuclear envelope	65 ⎱	60	56	56
Endoplasmic reticulum	65 ⎰		53	56
Golgi complex:				
Cisterna 1	65 ⎫		53	56
Cisterna 2	68 ⎪	70	60	38
Cisterna 3	72 ⎬		65	61
Cisternae 4 and 5	80 ⎭		75	69
Secretory vesicle	83	85	88	78
Plasma membrane	85	97	93	88

From Morré, D.J.: Membrane differentiation and the control of secretion: A comparison of plant and animal Golgi apparatus. *In* International Cell Biology. Edited by B.R. Brinkley and K.R. Porter. New York, The Rockefeller University Press, 1977, p. 293.

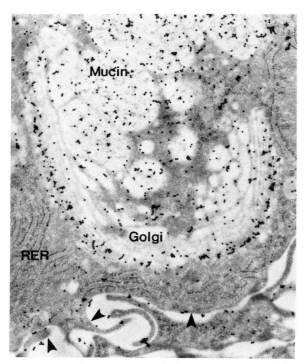

Fig. 9–4. Electron micrograph of a mucin-producing goblet cell from the colon of a rat. The N-acetyl-D-galactosamine residues are detected by a lectin from *Helix pomatia*, complexed to gold particles. See the text for details. ×8,000. (Courtesy of J. Roth.)

ER on one side, and to secretory vesicles leading to the cell membrane on the other.

The Golgi complex consists of *dictyosomes* formed by stacks of curved cisternae, associated tubules, and secretory vesicles. The cisternae lack ribosomes and are surrounded by a zone in which organelles are excluded. However, some free ribosomes may be at the periphery. Each dictyosome is polarized, with a proximal or *forming face* (convex and close to the nucleus) and a distal or *maturing face* (concave). At the forming face are vesicles and tubules that converge, forming a fenestrated plate. There is a continuous mechanism of membrane flow to compensate for the loss of secretory vesicles and also a process of differentiation of membrane revealed by an increase in its thickness and by its staining for glycoproteins. The interrelationship between the various parts of the Golgi and between it and the ER and secretory vesicles is clearly seen in the secretion of lipoprotein granules by the liver.

9–2   CYTOCHEMISTRY OF THE GOLGI COMPLEX

Golgi complex isolation has been achieved in numerous plant and animal cells.[5,6] The methods used are based on gentle homogenization, which tends to preserve the stacks of cisternae, thus allowing large fragments of Golgi complexes to be obtained by differential and gradient centrifugation (see Section 3–4). The Golgi complexes have a lower specific density than the ER or the mitochondria and they are equilibrated in a band having a density of 1.16. Under the EM the stacked cisternae appear to be bordered by an extensive system of tubules and vesicles (the secretory products remain within these vesicles) (Fig. 9–5). Washing the Golgi complexes in distilled water results in further purification with a consequent loss of the secretory products.

A subfractionation of the Golgi complex has been achieved after unstacking of the cisternae of the dictyosome by the use of proteolytic enzymes. Several subfractions, some richer in secretory vesicles, others containing mainly cisternae plates and other portions of the Golgi have been separated by gradient centrifugation and were characterized by some biochemical parameters.[7]

Chemical Composition of the Golgi Complex—Intermediate Between Those of the ER and the Plasma Membrane

Information about the biochemical makeup of the different endomembranes has come from studies in which highly purified fractions of Golgi complex have been compared with others from the ER and plasma membranes isolated in parallel.

The Golgi complex isolated from rat liver consists of about 60% protein and 40% lipid. By the use of gel electrophoresis it was found that the Golgi complex and the ER contain some proteins in common, but the former has fewer protein bands. In the plasma membrane fraction, however, fewer proteins were observed.

Studies carried out on isolated Golgi complex from different plant and animal cells show marked differences in protein and enzyme content. The Golgi complex has a phospholipid composition that is intermediate between those of the ER and plasma membrane. In contrast to rat liver, in which the most abundant phospholipid is phosphatidylcholine, plants have large amounts of phosphatidic acid (30 to 40%) and phosphatidylglycerol (see Morré, 1977 in the Additional Readings at the end of this chapter).

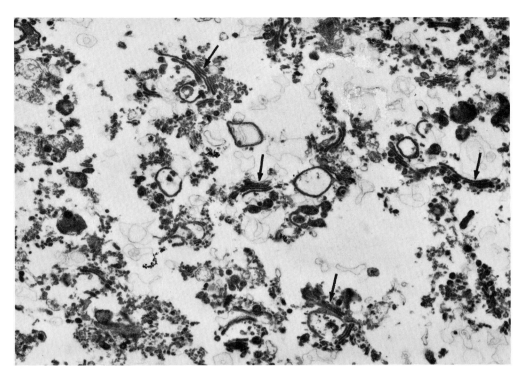

Fig. 9–5. Isolated Golgi complexes from liver cells. The complexes that best show the stacks of cisternae are indicated by arrows. (Courtesy of D.J. Morré.)

Plant membranes, including those of the Golgi, lack sialic acid; on the other hand, this sugar is very high in rat liver, and glycosphingolipids are abundant. Both animal and plant cells have some carbohydrate components in common (i.e., glucosamine, galactose, glucose, mannose, and fucose), but plants also have pentoses (i.e., xylose and arabinose) and other special carbohydrates.

Glycosyl Transferases are Concentrated in the Golgi

In the isolation of subcellular organelles it is of great importance to determine the degree of contamination between the various fractions. Table 9–2 shows some enzyme markers used for plasma membranes (5′-nucleotidase), for the ER (glucose-6-phosphatase), for mitochondria (cytochrome oxidase), and for peroxisomes (uric acid oxidase). It has been calculated that a fraction of Golgi membranes purified by washing contains small percentages of contamination.

The enzymes that are concentrated in the Golgi fraction are *thiamine-pyrophosphatase*

and several *glycosyl transferases*. The Golgi fraction also contains acid phosphatase and other lysosomal enzymes, presumably in relation to primary lysosomes (see Chapter 11).

The most characteristic enzymes of the Golgi fraction are those related to the transfer of oligosaccharides to proteins (i.e., glycosyl transferases) with the resulting formation of glycoproteins. Table 9–3 shows the specific activity of four glycosyl transferases that are involved in the transfer of CMP-neuraminic acid (CMP-NANA), UDP-galactose, and UDP-acetylglucosamine to protein acceptors. (All of these transferases are highly concentrated in the Golgi fraction of the rat liver.)

The isolated Golgi complex can be subfractionated into fractions of different densities. In rat liver, the subfractionation is facilitated if ethanol is given before sacrifice. This treatment results in the accumulation of low-density lipoproteins inside the Golgi vacuoles when accompanied by a diet rich in fat (see Fig. 9–3).

This experimental condition facilitates the separation of a *light or GF₁ fraction* containing large vesicles filled with lipoproteins; an inter-

TABLE 9–2.   DISTRIBUTION OF ENZYME MARKERS IN THE CORRESPONDING FRACTION AND IN THE GOLGI COMPLEX FROM RAT LIVER

Enzyme	Fraction	Specific Activity in Corresponding Fraction	Specific Activity in Golgi Apparatus	Ratio
5'-Nucleotidase	Plasma membrane	41.9	5.8	0.14
Glucose-6-phosphatase	Endoplasmic reticulum	11.4	1.1	0.10
Cytochrome oxidase	Mitochondria	67.8	1.8	0.03
Uric acid oxidase	Peroxisomes	50.0	0.5	0.01

From Morré, D.J., Keenan, T.W., and Mollenhauer, H.H.: Golgi apparatus function in membrane transformations and product compartmentalization: Studies with cell fractions isolated from rat liver. In Advances in Cytopharmacology. Vol. I. First International Symposium on Cell Biology and Cytopharmacology. Edited by F. Clementi and B. Ceccarelli. New York, Raven Press, 1971.

mediary or GF_2 fraction, similar to GF_1 but with some cisternal and smaller vesicles; and a *GF_3 or heavy fraction*, in which Golgi cisternae and small empty vesicles predominate. These fractions follow the above mentioned cis-trans-polarization of the complex in the following order: $GF_3 \rightarrow GF_2 \rightarrow GF_1$.[8]

The terminal glycosyl transferases occur throughout the Golgi complex in a fairly constant proportion, but their concentration increases in the cis $\rightarrow$ trans direction. Golgi membranes also contain other enzymes in common with the ER, for example, NADH-cytochrome-b_5-reductase, NADH-cytochrome-c-reductase, and 5'-nucleotidase.

SUMMARY:
Cytochemistry of the Golgi Complex

The Golgi complexes of many plant and animal cells can be separated by gentle homogenization, followed by differential and gradient centrifugation. Both the ER and mitochondria have a higher specific density than the Golgi complex (density: 1.16). The Golgi from liver contains 60% protein and 40% lipids. (The pattern of protein is less complex than in the ER and more complex than in the plasma membrane.)

Various subcellular fractions can now be characterized by enzyme markers, and the degree of contamination of the Golgi fraction by mitochondria, plasma membranes, ER, and peroxisomes may be determined. The enzymes that appear most concentrated in the Golgi fraction are thiamine-pyrophosphatase and several glycosyl transferases that transfer oligosaccharides to the glycoproteins. The isolated Golgi complex can be subfractionated.

9–3   FUNCTIONS OF THE GOLGI COMPLEX

The major functions of the Golgi complex appear to be related to the fact that it represents a special membranous compartment interposed between the ER and the extracellular space. Through this compartment there is a continuous traffic of substances, which may have been synthesized elsewhere, but which are modified and transformed while transported. This process also involves the flow and differentiation of membranes.

It is now well established that most cytoplasmic membranes of the eukaryotic cell arise from the RER (exceptions are the inner membranes of mitochondria and chloroplasts). This

TABLE 9–3.   GLYCOSYL TRANSFERASE ACTIVITY IN GOLGI COMPLEX FRACTION OF RAT LIVER

Glycosyl Transferase	Specific Activity* in Total Homogenate	Specific Activity* in Golgi Complex	% of Total Activity in Golgi Complex
Sialyl	50	422	44
Galactosyl	11	128	42
N-acetylglucosaminyl	24	219	43
Galactosyl-N-acetylglucosamine	6	64	40

Modified from Morré, D.J., Keenan, T.W., and Mollenhauer, H.H.: Golgi apparatus function in membrane transformations and product compartmentalization: Studies with cell fractions isolated from rat liver. In Advances in Cytopharmacology. Vol. 1. First International Symposium on Cell Biology and Cytopharmacology. Edited by F. Clementi and B. Ceccarelli. New York, Raven Press, 1971; Unpublished data of H. Schachter.
*Specific activity of the four glycosyl transferases is expressed in mμ moles/hrs/mg protein of sugar nucleotide bound to the protein acceptor.

mechanism involves loss of attached ribosomes to generate SER, pinching off of vesicles, which will be fused with the *cis*-face of the Golgi complex, modification of proteins within the Golgi, and production of secretory vesicles from the *trans*-face. After this process of membrane flow and differentiation, the vesicles can finally fuse with the plasma membrane.

Within the walls of the Golgi complex a vast diversity of materials may be found, varying from simple fluids to macromolecules and even preformed wall units, as in the case of certain algae. Virtually every major class of macromolecule is transported through the Golgi and secreted, and this implies a continuous and fast turnover of the Golgi membranes. Thus, dictyosomes in algae and liver cells are estimated to turn over every 20 to 40 minutes, and cisternae are released at the rate of 1 every few minutes.

Synthesis of Glycosphingolipids and Glycoproteins—A Major Role of the Golgi

The Golgi complex plays a major role in the glycosidation of lipids and proteins to produce *glycosphingolipids* and *glycoproteins* (see Chapter 2). The carbohydrate prosthetic groups of both these complex substances are added in a sequential manner in the Golgi complex by the action of the various glycosyl transferases (Table 9–3).

The peptide backbone of all glycoproteins is synthesized on the membrane-bound ribosomes of the RER. Then the polypeptide is glycosidated in a sequential way, as it passes through the secretory pathway (i.e., ER → Golgi complex → secretory granule). As mentioned in Section 2–5 glycoproteins have a branched oligosaccharide chain attached to the polypeptide. Most of them contain a common core consisting of five saccharide residues of mannose and N-acetyl-glucosamine (see diagram on page 41). This core is linked to an asparaginyl residue in the protein, and sometimes to a serine residue. Besides the core, there are terminal side chains that contain galactose, fucose, and sialic acid, which are added in the Golgi complex in a stepwise manner by the various glycosyl transferases.

Glycosyl Transferases. As shown in Figure 8–10 the core is synthesized as dolichol pyro-

phosphate, a lipid intermediate that transfers the sugar residue to the nascent polypeptide inside the ER (see Staneloni and Leloir, 1979 in the Additional Readings at the end of this chapter). Dolichol is a polyisoprenol with 17 to 21 isoprenes. The series of reactions are the following: dolichol-P reacts with UDP-acetylglucosamine to form dolichol-P-P-ClcNac (a reaction that is blocked by tunicamycin, i.e., an inhibitor or glycoprotein synthesis), which then transfers the sugar under the action of the glycosyl transferase. The terminal side chains are added to the common core by the corresponding sialyl-, galactosyl- and N-acetylglucosamine transferases.

Different patterns of incorporation may be observed in experiments using radioactive precursors of various carbohydrates (Table 9–4).[9] When D-mannose is used, the RER is the first to be marked and then the other membrane compartments: Golgi cmplex → secretory granules → extracellular space or plasma membrane. When sialic acid, L-fucose or D-galactose is used the Golgi complex is the first to be labeled. With D-glucosamine both the RER and the Golgi complex are simultaneously labeled.

For cytological studies using electron microscopic radioauography, ^{3}H-fucose is preferred because it does not enter into mucopolysaccharides, and in glycoproteins fucose is present only in the terminal group. In intestinal cells, 2 minutes after the injection of ^{3}H-fucose the labeling is almost exclusively in the Golgi complex. Then there is migration of the glycoproteins to the sides of the cell and into the apical cell membranes, where it is concentrated 4 hours after injection. The intracellular migration of the glycoproteins is carried out by small vesicles formed in the Golgi region.

The Golgi appears to be also involved in the addition of sulfate to the carbohydrate moiety of the glycoproteins. In cartilage cells, mucopolysaccharides as well as glycoproteins are synthesized in the Golgi complex.[10,11]

The diagram of Figure 9–6 is an idealized representation of the synthesis and secretion of glycoproteins and the intervention of the various segments of the endomembrane system: RER amd SER → Golgi membranes → secretory vesicles and plasma membrane.

In addition to the glycoproteins that are secreted and those that are incorporated into the plasma membrane there are others that become

TABLE 9–4. INCORPORATION OF VARIOUS PRECURSORS INTO GLYCOPROTEINS*

Pattern Type	Radioactive Precursor	Interpretation of Time Course Experiments
A	Leucine ↑ D-mannose	RER → Golgi → SG, EC, or PM ↑ Label
B	Sialic acid L-fucose D-galactose	RER → Golgi → SG, EC, or PM ↑ Label
C	D-glucosamine	RER → Golgi → SG, EC, or PM ↑ ↑ Label Label

*Abbreviations: *RER*, rough endoplasmic reticulum; *SG*, secretory granule; *EC*, extracellular space; *PM*, plasma membrane. This table represents, in abbreviated form, experiments by various groups of investigators. The experimental work involved the injecting of whole animals, perfused tissues, or tissue slices with the precursors and the fractionating of the tissue at various time intervals. (Modified from Schachter, H. J. Biol. Chem., *248*:974–976, 1973.)

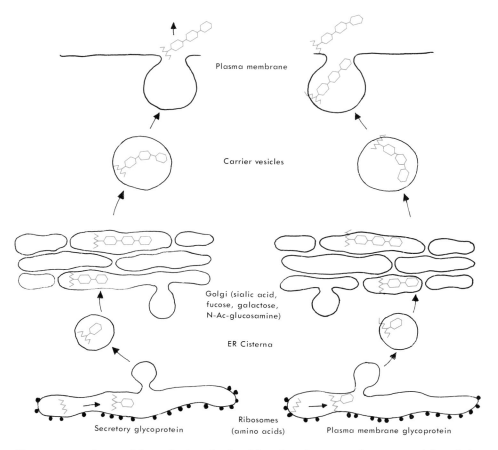

Fig. 9–6. Diagram comparing possible mechanisms for the elaboration of secretory glycoproteins (left) and plasma membrane glycoproteins (right). Both proteins are first synthesized by ribosomes. The secretory proteins are released into the lumen, while the plasma membrane glycoproteins remain inserted in the wall of the ER and are transported by membrane flow. The various oligosaccharides are added in a sequential manner in the ER and the Golgi apparatus (see Table 9–4). The glycoproteins are transported from the Golgi by carrier vesicles and released by exocytosis at the plasma membrane. (Modified from Schachter, 1974. Courtesy of C. P. Leblond and G. Bennett. From Leblond, C.P., and Bennett, G.: Role of the Golgi apparatus in terminal glycosylation. *In* International Cell Biology. Edited by B.K. Brinkley and K.R. Porter. New York, Rockefeller University Press, 1977, p. 326.)

incorporated into lysosomes. For example, in liver cells ^{3}H-fucose is first incorporated into the Golgi and later on is found in lysosomes, in the blood plasma, and on the cell membranes (see Leblond and Bennett, 1977 in the Additional Readings at the end of this chapter).

In conclusion: The polypeptide backbone of the glycoprotein is synthesized on membrane-bound ribosomes and penetrates into the ER by the signal peptide mechanism mentioned in Section 8–4. Within the ER the oligosaccharide core is transferred to the nascent protein by a dolichol pyrophosphate oligosaccharide precursor. After reaching the Golgi complex, the terminal side chains of galactose, fucose, and sialic acid are added by the corresponding transferases. These enzymes are probably organized in close proximity to one another, so as to permit the transfer of the acceptors, from enzyme to enzyme, without the release of the intermediate into the lumen of the Golgi complex.

Translocation of precursors for glycosidation and sulfatation through Golgi membranes. The substrates for the above mentioned glycosyl and sulfate transferases are not the sugar moieties (fucose, mannose, galactose, glucose N-acetyl-glucosamine, neuraminic acid) or the sulfate used in radiolabeling experiments. The true donor molecules are sugar nucleotides, rich in energy, such as guanosine-5'diphosphate fucose or mannose (GDP-F or GDP-M), uridine-5'diphosphate glucose, galactose, or N-acetyl-glucosamine (UDP-glu, UDP-gal or UDP-NAcGlc). On the other hand, the adenosine 3'phosphate, 5'phosphosulfate (PAPS) is the donor of the sulfate moiety. All these compounds are synthesized in the cytosol and traverse the membrane to become substrates of the enzymes, which have their active centers at the luminal side of the Golgi vesicle. These precursors are molecules of rather high molecular weight (>1000 daltons), very hydrophilic and negatively charged at physiological pH so that they cannot be translocated by simple diffusion. Furthermore, when these molecules enter the Golgi vesicle and are hydrolyzed by the transferases, the nucleotides that are liberated should be able to exit to the cytoplasm. In fact, if this were not the case, the accumulation of nucleotides could inhibit the transferases and sulfatases and be harmful to the cell.

The two problems—the translocation of precursors and the exit of nucleotides—have apparently been solved by a mechanism of *facilitated diffusion* through specific transporting proteins present in the Golgi membrane. These proteins bind the sugar nucleotides at the cytoplasmic surface of the Golgi membrane and transfer them to the interior to be used as substrates. At the same time these proteins are able to function as *antiports*, binding the remaining nucleotides at the luminal surface and translocating them to the cytoplasm.

In the diagram of Figure 9–7 are indicated the translocators for GDP-F, UDP-NAcGlc, CMP-Ac Nem (Cytosine 5'-monophosphate neuramimic acid) and PAPS. It may be observed that GDP-F binds to the translocator by way of the nucleotide and penetrates into the Golgi vesicle to be used by the fucosyltransferase (Ft) in the fucosylation of glycoproteins or glycolipids. The generated GDP can be converted into GMP to be interchanged for GDP-F at the antiport protein. This mechanism applies also to the other sugar nucleotides and to PAPS. In all cases there is a stoichiometric equimolecular relationship, in the sense that one molecule of nucleotide exits for one molecule of sugar nucleotide that enters the Golgi vesicle. This is a typical example of facilitated diffusion through antiport proteins in which there is no expenditure of energy (see Section 4–3).

Cancer Cells—Changes in Glycosidation of Lipids and Proteins

The Golgi complex plays a central role in the biosynthesis of *gangliosides* and other glycosphingolipids. Several transferases involved in the glycosidation of glycosphingolipids have been found to be concentrated in the Golgi complex and to a lesser extent in the ER. As in the case of the glycoproteins the synthesis of glycosphingolipids occurs by a stepwise addition of sugars (from sugar nucleotides) to ceramide or to growing glycolipid acceptors. There is evidence that in cancer cells there are surface membrane changes that involve a loss of glycosphingolipids, and these alterations are due, in part, to the reduction of one or more glycosyltransferases present in the Golgi complex. The possibility that changes in surface glycoproteins and glycosphingolipids may be related to the development of the malignant properties of cells is at

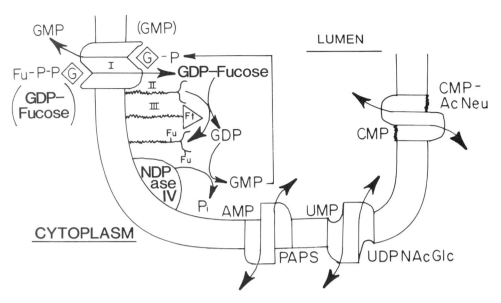

Fig. 9–7. Diagram of the mechanism of translocation of sugar nucleotides and PAPS across the membrane of the Golgi apparatus. The GDP-F binds to the cytoplasmic side of the antiport protein (I) at the Golgi membrane. This precursor is translocated intact into the lumen where it serves as substrate for the fucosyltransferases (Ft), together with endogenous glycoproteins or glycolipids (II). The diagram also shows that the GDP generated can be converted into GMP by a phosphatase (NDPase), and this nucleotide is then interchanged for GDP-F at antiport (I) with an equimolecular stoichiometry. Similar antiport proteins are responsible for the translocation of PAPS, CMP-AcNeu, UDP-NAcGlc and other sugar nucleotide precursors. (Courtesy of J.M. Capasso. From Capasso, J.M. and Hirschberg, C.B.: Proc. Natl. Acad. Sci. USA, *81*:7051, 1984.)

present receiving wide attention.[12,13] Some of these changes were mentioned in Section 5–4.

9–4   SECRETION—THE MAIN FUNCTION OF THE GOLGI COMPLEX

The study of cell secretion in relation to the ER and the Golgi system is justified at this point because these organelles are the ones more directly involved in the synthesis, transport, and release of macromolecules from the cell. The relationship between the Golgi complex and cell secretion was postulated by Cajal in 1914 in his study of goblet cells. Cell secretion is not confined to animal cells alone; in fact plant cells secrete polysaccharides and proteins to make their cell walls. Furthermore, the enzymes present *in lysosomes* are produced by a kind of secretory process (see Section 10–1).

Generalizing the concept of cell secretion, it may be said that such activity appears even in the prokaryotic cells, in relation to the production of the cell wall of bacteria and to the secretion of a variety of enzymes to the medium. In certain protozoa, vacuoles similar to the Golgi

complex are found, which by their contraction expel water into the medium.

The Secretory Cycle—Continuous or Discontinuous

Secretion involves a continuous change that can be best interpreted by studying the cell throughout the different stages of cellular activity. In some secretory cells secretion is continuous; the product is discharged as soon as it is elaborated. In the secretion of glycoprotein by the liver cell and in the case of the antibodies produced by the plasma cells, the secretion produced is not accumulated into special storage granules, and the release of secretory materials is more or less simultaneous with the synthesis and intracellular transport of these substances.

In other cells the secretory cycle is *discontinuous*: it is specially timed so that the synthesis and intracellular transport are followed by the accumulation of the secretion product in special storage granules that are finally released to the extracellular space. In these cells the secretory cycle has extremely variable cytological expressions, but it is generally characterized by prod-

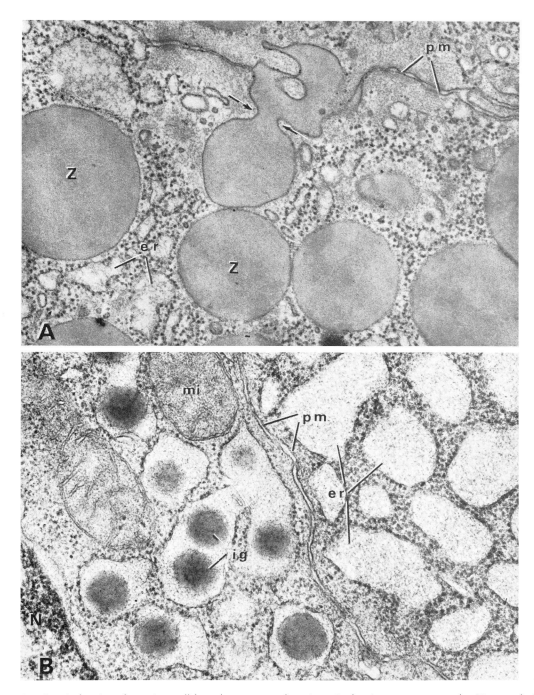

Fig. 9–8.   *A,* apical region of an acinar cell from the pancreas of a guinea pig showing zymogen granules *(Z),* one of which is being expelled into the lumen by exocytosis followed by membrane fusion; *er,* granular endoplasmic reticulum; *pm,* plasma membrane. × 30,000. (Courtesy of G.E. Palade.) *B,* the same as above, but from the basal portion showing the enlarged cisternae of the endoplasmic reticulum *(er),* some of which contain intracisternal granules *(ig); mi,* mitochondria; *N,* nucleus; *pm,* plasma membrane. × 30,000. (Courtesy of D. Zambrano.)

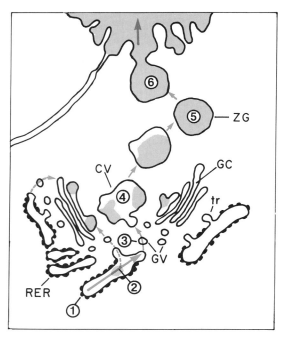

Fig. 9–9. Diagram of the processing of secretory proteins in a typical glandular cell. Steps 1 through 6 are described in the text. *RER,* rough endoplasmic reticulum; *tr,* transitional vesicles; *GC,* Golgi cisternae; *CV,* condensing vacuole; *ZG,* zymogen granules. (Courtesy of J.D. Jamieson and G.E. Palade. From Jamieson, J.D., and Palade, G.E.: Production of secretory proteins in animal cells. *In* International Cell Biology. Edited by B.R. Brinkley, and K.R. Porter. New York, Rockefeller University Press, 1977, p. 308.)

ucts, visible with the microscope, that accumulate in the cell, and then are ultimately eliminated. These products may be dense and refractile granules, vacuoles, droplets, or other structures having a definite location in the cell and, at times, characteristic histochemical reactions. The dense secretory granules containing enzymes, generally in an inactive form (proenzyme), are called the *zymogen granules* (Fig. 9–8A).

The Secretory Process in the Pancreas—Six Consecutive Steps

In a glandular exocrine cell, such as that of the pancreas, the following six steps that involve the endomembrane system may be recognized (Fig. 9–9) (see Jamieson and Palade, 1977 in the Additional Readings at the end of this chapter).

Step 1 or The Ribosomal Stage. The initial stage involves the *synthesis of proteins* in direct contact with the polysomes attached to the RER. This step, which is actually begun in the cytosol, was studied in Chapter 8 (for protein synthesis, see also Section 21–2).

Step 2 or The Cisternal Stage. This corresponds to the *vectorial transport* of synthesized proteins into the cisternae of the ER and was also studied in Chapter 8. The student should bear in mind here the mechanism of the binding of ribosomes to the membrane, the *signal hypothesis* with the signal codons in the mRNA, the signal peptide, the ribosomal receptor proteins; the signal peptidase, and the mechanisms by which the protein is processed and stored within the ER. The protein material appears in most cases as a dilute solution of macromolecules, but in certain conditions the ER may have small *intracisternal granules* (Fig. 9–8B).

Step 3 or Intracellular Transport. In this stage the secretory proteins are transported through the RER and enter the so-called *transitional elements*, which are located at the boundary between the ER and the Golgi complex. These transitional components have ribosomes attached to most of their surface except in the regions facing the cis or forming face of the Golgi, which are smooth and bud off into vesicles similar to the Golgi peripheral vesicles.

According to the studies of Palade and co-workers these vesicles reach the next station of the secretory pathway and fuse with the large condensing vacuoles of the Golgi that are present at the maturing face.

The intracellular transport of the secretory protein has been followed cytochemically by the use of *radiolabeled precursors,* such as ^{3}H-leucine, either injected into the animal or applied to tissue slices.

The kinetics of the labeling of the secretory protein can be studied by the use of a *short pulse* followed by a so-called *chase.* For example, slices of pancreas are exposed first to a leucine-free medium (i.e., a saline solution containing all amino acids except leucine), which produces depletion of the leucine-endogenous pool. After a short time the tissue is submitted to a medium containing ^{3}H-leucine for a few minutes, and this procedure is then followed by the chase, which consists of washing and incubation with a solution containing excess nonlabeled leucine. In the fixed tissue, it is possible to observe that after a few minutes the isotope is localized in the ER

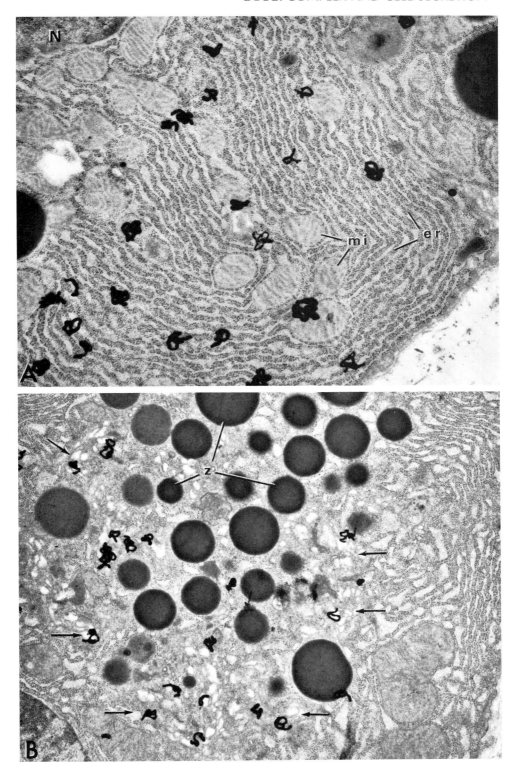

Fig. 9–10. Electron microscopic radioautograph of acinar cells from the pancreas of a guinea pig. *A*, three minutes after pulse labeling with ^{3}H-leucine. The radioautographed grains are located almost exclusively on the granular endoplasmic reticulum *(er)*; *mi*, mitochondria; *N*, nucleus. ×17,000. *B*, the same as above, but incubated for seven minutes after pulse labeling. The label is now in the region of the Golgi complex (arrows); *z*, zymogen granules. ×17,000. (Courtesy of J.D. Jamieson and G.E. Palade.)

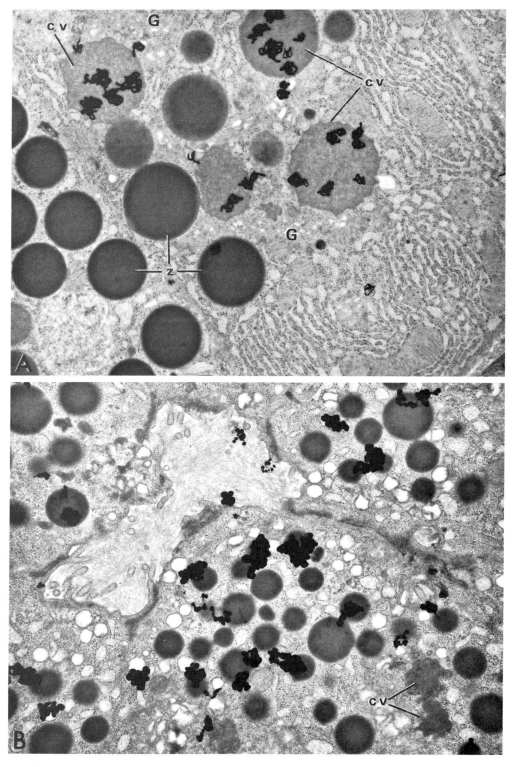

Fig. 9–11.   *A,* the same experiment as in Figure 9–10, but 37 minutes after pulse labeling. The label is now concentrated in the condensing vacuoles *(cv)* of the Golgi complex *(G).* The zymogen granules *(z)* are unlabeled. ×13,000. *B,* the same as above, but incubated for 117 minutes after pulse labeling. The radioautographed grains are now localized primarily over the zymogen granules, while the condensing vacuoles *(cv)* are devoid of label. Some grains are in the lumen of the acinus, indicating the secretion. ×13,000. (Courtesy of J.D. Jamieson and G.E. Palade.)

of the basal region. Later, the newly synthesized protein passes into the Golgi complex. In this region apparently it becomes progressively concentrated into prozymogen granules or condensing vacuoles surrounded by a membrane. After a longer time, the label is found principally in the zymogen granules and in the lumen of the acinus. Figures 9–10 and 9–11 show electron microscopic radioautographs of such sequences of events in sections of pancreas.

Quantitative date may be obtained by measuring the number of radioautographic grains over the various cell components. Immediately after the pulse labeling there is a sudden increase in radioactivity in the RER with a tendency to decline rapidly. The radioactivity then becomes high in the Golgi complex, while it increases more slowly in both the immature and the mature secretory granules. In the parotid gland, there is a wave-like movement of the pulse-labeled secretory protein through the various intracellular compartments in the following order: RER → Golgi complex → immature granules (i.e., condensing vacuoles and prozymogen) → mature secretory (zymogen) granules. In the rat pancreas the total life span of a zymogen granule has been estimated at 52.4 minutes.[14]

With small variations, the transport of the secretory protein from the ER to the Golgi is the same in all protein-producing cells.

Since the intracellular transport is made against an apparent concentration gradient, it is of interest to determine what its possible metabolic requirements are. The transport of the newly synthesized polypeptide chain from the ribosome into the cisternae of the ER does not require additional energy and seems to be controlled mainly by the structural relationship of the large ribosomal subunit with the membrane of the reticulum (see Section 8–1). When the process of protein synthesis is inhibited by puromycin, the incomplete peptides formed are transported into the vacuolar cavity.[15]

Transport from ER to the condensing vacuole is blocked by inhibitors of cell respiration (nitrogen, cyanide, antimycin A) or of oxidative phosphorylation (dinitrophenol, oligomycin). An important conclusion to be drawn from these studies is that at the periphery of the Golgi complex, in transitional elements between the ER and the small vesicles of the Golgi complex, there is an *energy-dependent lock* in the transport. Such a lock may regulate the flow of secretory proteins.

Recent studies show the bioenergetics of the transit of secretion products through the Golgi complex. Isolated Golgi organelles contain high concentrations of an ATPase, which functions as a proton pump producing acidification of the secretion (see Zhang and Schneider, 1983 in the Additional Readings at the end of this chapter). In Chapter 10 we show that there is also a proton pump in lysosomes and it is possible that all compartments of the endomembrane system may require an ATP-dependent acidification.

Step 4 or Concentration of the Secretory Protein. During this period the condensing vacuoles are converted into zymogen granules by the progressive filling, and concentration of their content, and finally acquire their characteristic electron-opaque content. This conversion is not dependent on a supply of metabolic energy, since it continues after inhibition of glycolysis or respiration.

These findings provide no explanation for the mechanism by which the secretory protein is condensed. However, the fact that ^{35}S-sulfate is incorporated into the Golgi elements and into the zymogen granules has suggested possible mechanisms.[16,17] Recently it has been observed that sulfatation (as well as glycosylation) occurs in the Golgi in which there is translocation of sugar nucleotides and adenosine 3'-phosphate 5'-phosphosulfate by specific carrier proteins (see Capasso and Hirschberg, 1984 in the Additional Readings at the end of this chapter). A sulfated peptidoglycan that acts as a large polyanion interacting with the basic secretory proteins has been found in the Golgi complex and condensing vacuoles. It is thought at present that the formation of osmotically inactive aggregates may result in a passive flow of water from the secretion granule to the relatively hyperosmotic cytosol.[18]

Step 5 or Intracellular Storage. The previous step culminates with the storage of the secretory product into *secretory granules*, which are released when the appropriate stimulus (a hormone or neurotransmitter) acts on the cell. This storage provides a mechanism by which the cell may cope with a demand for secretory product that exceeds its rate of synthesis. This mechanism of storage is utilized not only in the protein-secreting cells but also in others producing smaller molecules such as peptides and amines. For example, the *catecholamine-containing granules* of the adrenal medulla contain several proteins (the so-called chromogranins) and ATP, with which the amines are complexed.[19] These granules also contain opioid-like materials similar to those of the enkephalins (see Section 24–4).[20] All of these components are released upon exocytosis.

Step 6 or Exocytosis. The discharge of the secretory granule is effected by a process of exocytosis, which involves its movement toward the apical region of the cell and fusion between its membrane and the luminal plasma membrane. As a result of the fusion of the membranes and the consequent fission, with elimination of the layers in between, an orifice is made through which the secretory product is discharged (Fig. 9–8A).

The mechanism of exocytosis was first described in the adrenal medulla after electrical stimulation of the splanchnic nerves.[21] As shown in Figure 9–12, it was found that the catechol-containing granules (or vesicles) first attach themselves to the plasma membrane, then they swell, and finally their content is evacuated, leaving empty membranes. Biochemically, exocytosis requires the intracellular elevation of Ca^{++} and the production of ATP. Thus, exocytosis is an energy-dependent process and signifies a second energy-requiring step in the entire secretory process. Presumably part of this energy requirement is related to the process of membrane fusion-fission taking place during exocytosis, but it is possible that energy could be consumed partly in the propulsion of the granule to the cell apex.

The process of exocytosis involves the pres-

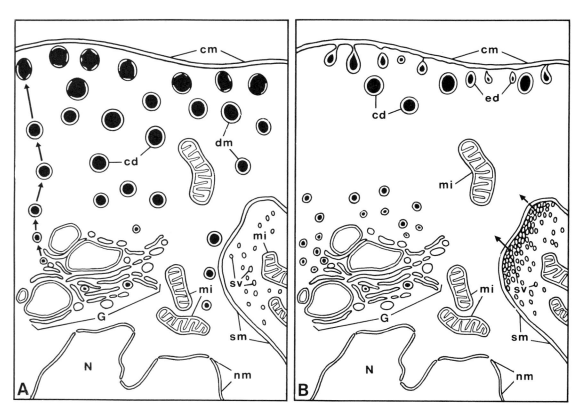

Fig. 9–12. Diagrammatic interpretation of the mechanism of secretion in the chromaffin cell. *A*, cell in the resting stage, showing the storage of mature catechol droplets in the outer cytoplasm. Near the nucleus within the Golgi complex new secretion is being formed at a slow rate. At the right, a portion of a nerve terminal, showing the synaptic vesicles *(sv)* and mitochondria *(mi)*; cd, catechol droplets; cm, cell membrane; dm, droplet membrane; ed, evacuated droplets; G, Golgi complex; N, nucleus; nm, nuclear membrane; sm, surface membrane. *B*, cell after strong electrical stimulation by way of the splanchnic nerve. Most of the catechol droplets have disappeared; the few that remain can be seen in different stages of excretion into the intercellular cleft. The Golgi complex is now forming new droplets at a higher rate. The nerve ending shows an increase of synaptic vesicles with accumulation at "active points" on the synaptic membrane. (From E. De Robertis and D.D. Sabatini, Fed. Proc., *19:*70, 1960.)

ence of a protein of 47,000 daltons called *synexin*, which could regulate the Ca^{++}-dependent molecular events involved in membrane fusion and release of catecholamines.[22] (For a review on the role of Ca^{++} in cell secretion, see Case, 1984 in the Additional Readings at the end of this chapter.)

In the parotid gland stimulated by adrenergic drugs acting on beta receptors, the zymogen granules are discharged by exocytosis in a sequential manner, i.e., in the order in which they are present at the apical region. It has been postulated that cyclic 3'-5' AMP may be implicated in the process of exocytosis. In fact, the secretion can be induced directly by the use of dibutyryl cyclic AMP.[23]

Endocytosis and Recycling of Membranes of Secretory Granules

A pancreatic cell in which protein synthesis has been inhibited can go through an entire secretory cycle of transport, concentration, storage, and release. This fact suggests that within the period of time of one cycle (i.e., 60 to 90 minutes) the synthesis of new membrane proteins, serving as containers of the secretion, is not required. In other words, intracellular membranes have a much longer half-life and are re-utilized extensively during the secretory process.

The membrane of the zymogen granules may be regarded as a vacuole that shuttles between the Golgi apparatus and the cell surface. In the removal of excess membrane from the apical region of the cell it has been suggested that patches of membranes are invaginated from the surface as small vesicles that move back into the Golgi region, to be re-utilized in the packing of more secretion. Thus, the process of exocytosis is coupled with that of endocytosis.

The specificity of this recycling process appears to depend on the formation of *coated pits* and *vesicles*, which separate the membrane patches from the plasma membrane by a mechanism of *selective endocytosis* (see Section 10–5).

Figure 9–13 is a diagram of two mechanisms by which the membranes of the secretion granules are recycled and retrieved by endocytosis. In *A*, illustrated by an exocrine gland cell, the membranes are removed by endocytosis, using coated vesicles, and are then delivered to the Golgi region, to be reused in another secretory cycle. In *B*, illustrated by an endocrine cell of the thyroid gland, the endocytic vesicles are internalized and transferred to lysosomes (see Herzog, 1981 in the Additional Readings at the end of this chapter).

The GERL Region and Lysosome Formation

In addition to serving as the site for packaging the secretory products and providing a limiting membrane to zymogen granules, the Golgi membranes are involved in the formation of the primary lysosomes. This process appears to have the same sequence as that described above, i.e., synthesis, aggregation, transport, and packaging of the enzymes.[24] Since many lysosomal enzymes are glycoproteins, the Golgi has also been implicated in their glycosidation.

We mentioned earlier that in the maturing face of the Golgi is the GERL region, in which acid phosphatase, a characteristic lysosomal enzyme, is present. This region has been implicated in the formation of primary lysosomes (see Section 10–1).

According to Novikoff the GERL of the Golgi complex appears also to be involved in the formation of melanin granules, in the processing and packaging of secretory material in endo- and exocrine cells, and in lipid metabolism.[3]

In Chapter 10 we again refer to the GERL (also known as the *trans-reticular Golgi*) as the site of the cell that is involved in the sorting out of proteins. The relationship of this region of the Golgi with endosomes and lysosomes will then be emphasized.

Compartments of the Golgi Apparatus

At the beginning of this chapter we mentioned some of the morphological and histochemical evidences that indicate a possible heterogeneity of function within the Golgi apparatus. For example, as shown in Figure 9–14, *galactose residues*, labeled by a specific lectin complexed to gold particles, are localized in the *trans* region of the Golgi apparatus (a similar distribution is also found in the enzyme *galactosyl-transferase*), while the *cis* region is unlabeled (see Roth, 1984 in the Additional Readings at the end of this chapter). In the same cells thiamin-pyrophosphatase is localized in similar stacks of Golgi

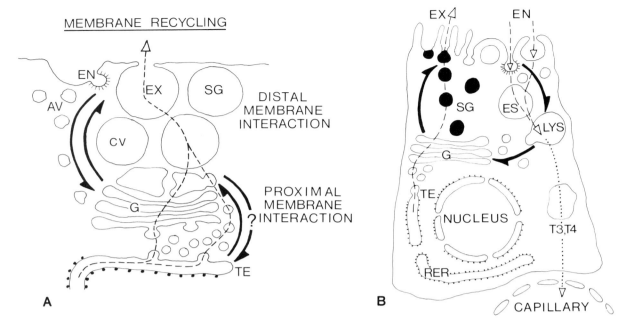

Fig. 9–13. *A,* diagram illustrating two possible mechanisms of membrane retrieval after exocytosis of the secretory granule *(SG).* In *A,,* illustrated by an exocrine gland cell, two membrane interactions are shown. The proximal one shows the fusion of transitional elements (TE) of the ER that are transferred to the Golgi complex *(G).* In the distal membrane interaction after exocytosis *(EX),* patches of the SG membrane are endocytosed *(EN)* by coated vesicles and then, as uncoated vesicles *(AV),* are transferred back to the Golgi. The question mark indicates a still unknown route. In *B,* illustrated by an endocrine cell of the thyroid gland, secretion of thyroglobulin from a thyroid cell is transported from the RER by way of TE to the Golgi, in which SG are formed and secreted. Thyroglobulin is endocytosed and goes to the lysosomes *(lys)* where it is hydrolyzed, and the thyroid hormones, triiodothyroxine *(T3)* and thyroxin *(T4)* are secreted at the basal pole into the capillary. (Courtesy of V. Herzog.)

membranes. The codistribution of these two enzymes suggests the existence of functional compartmentalization in the Golgi apparatus, and favors a model of glycosidation in which the two enzymes act in concert.[25] Recently, using monoclonal antibodies and electron microscopy, it has been found that N-acetylglucosamine transferase 1 (the enzyme that initiates the conversion of asparagine-linked oligosaccharides to the complex type) occurs in medial cisternae of the Golgi (see Dunphy et al., 1985 in the Additional Readings at the end of this chapter).

Additional evidence substantiating the compartmentalization of the Golgi apparatus is provided by the presence of a gradient of cholesterol from the *cis* to the *trans* stacks. In Figure 9–15, cultured rat hepatocytes were treated with filipin, a drug that complexes with cholesterol. These complexes can be detected under the EM after freeze-fracture. Observe that the filipin-sterol complexes are absent in the *cis* region, start in the *medial* cisternae, and increase in the *trans* region.

Monensin, a drug that acts as an ionophore in membranes has also been used recently to verify the compartmentalization of the Golgi. By the action of monensin a cell gains sodium and loses potassium thereby producing dramatic changes in the Golgi. In many cells, a few minutes after the addition of monensin, the Golgi cisternae became greatly dilated while the ER cisternae remain unaltered and protein synthesis continues. Under the action of monensin the transport of proteins, destined either to the plasma membrane, to the lysosomes, or to be secreted, is slowed down within the Golgi complex. This approach suggests that all three types of proteins reach a proximal subcompartment that is responsible for receiving the materials produced by the RER. The block in the transport by monensin occurs between this receiving subcompartment and a distal one in which the terminal sugars are added and the sorting out occurs normally (see Tartakoff, 1982 in the Additional Readings at the end of this chapter).

The previously mentioned studies suggest

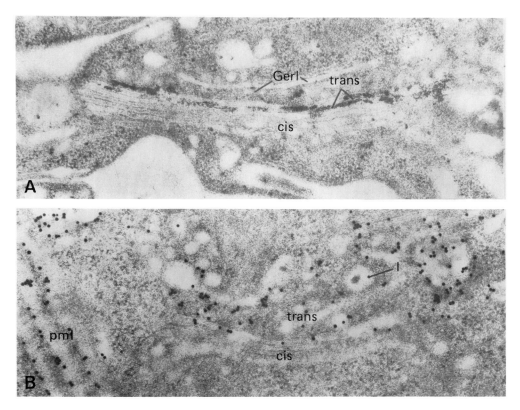

Fig. 9–14. Electron micrograph demonstrating the compartmentalization of the Golgi apparatus in intestinal epithelium. A, thiamine-pyrophosphatase activity is present in a few Golgi stacks in the *trans* side, while it is absent in the *cis* part and also in the GERL region. × 35,000. *B,* histochemical reaction for galactose residues using a lectin from *Ricinus communis* complexed to gold particles. Observe that the complexes are not only in the *trans* side of the Golgi apparatus but also in lysosomal bodies (l) and in the lateral plasma membrane *(pml).* × 50,000. (Courtesy of J. Roth.)

that the Golgi complex may have at least three distinct sets of cisternae, besides the GERL region. If each cisterna has a specific function then the traffic between the *cis* and *trans* side would be effected by vesicular carriers that operated from the dilated rims of the cisternae.

According to this hypothesis, the Golgi membrane proteins and specific enzymes are in fixed positions within the Golgi complex, but those proteins that enter the cisternae may diffuse throughout the entire complex by this carrier mechanism (see Palade and Farquhar, 1984 in the Additional Readings at the end of this chapter). The proteins transported by small vesicles from the ER are received at the *cis* cisternae and they exit from the *trans* cisternae at the opposite end of the stack. An intervening or *medial* cisternal compartment is also considered. The follow-up of the glycosidation pathway of glycoproteins has helped in the elucidation of such compartmentalization by showing that the

glycosyl-transferases are localized in the cisternae according to their order of use.

Another recent approach stems from the experiments of cell fusion between mutant cells that differ in a special marker of the Golgi (see Rothman and Leonard, 1984 in the Additional Readings at the end of this chapter). It was found that there was a rapid transport between the two Golgi complexes, producing a complete randomization of the marker protein. This finding suggests that between the two Golgi complexes in the hybrid cell a transfer may have occurred by carrier vesicles. These vesicles probably originate at the rim of the Golgi cisternae of one cell complex and become associated with cisternae of the other cell Golgi complex. It is possible that a similar mechanism is normally at work within a cell. Instead of a model in which there is a *cis-trans* progression and differentiation, as suggested at the beginning of this chapter, a *dissociative model,* based on vesicular carriers

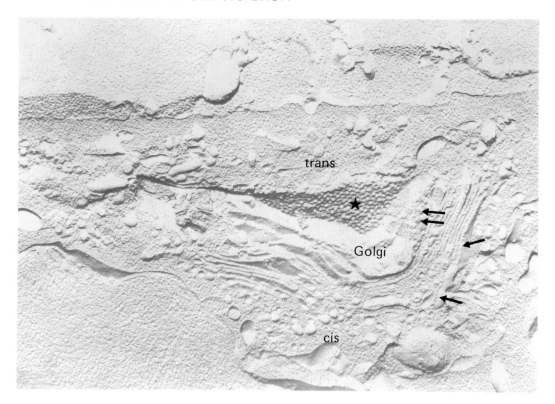

Fig. 9–15. Freeze-fracture electron micrograph of cultured rat hepatocytes showing the differential distribution of filipin-sterol complexes in different regions of the Golgi stacks. Membrane fracture faces of *cis* cisternae (arrows) are devoid of filipin-sterol complexes, whereas these occur in *medial* cisternae (double arrows) and increases in *trans* cisternae (asterisk). ×55,000. (Courtesy of J. Roth.)

between the stacked cisternae, is favored at present (see Rothman and Leonard, 1984 in the Additional Readings at the end of this chapter). These findings emphasize the importance of the many small vesicles (some of the coated type) that normally surround the Golgi region (Fig. 9–2).

Based on the possible functional subcompartmentalization of the Golgi complex, Rothman suggested that the successive Golgi stacks of membranes from the *cis* to the *trans* face could act as a molecular purification device.[26] For example, in order to enrich the stacks of proteins destined to the plasma membrane, those proteins characteristic of the ER could be "extracted" and returned back to the ER through small vesicles coming out from the edges of the cisternae. By observing Figure 9–16, it is evident that at each stack purification of some proteins could be achieved by a cross-current mechanism. This figure also illustrates that the proteins destined to be exported, become progressively more concentrated from the *cis* to the *trans* face of the Golgi apparatus.

In conclusion: In the function of the Golgi complex, subcompartmentalization with a division of labor is proposed between the *cis* region (in which proteins of the ER are sorted and returned back), and the *trans* region, in which the most refined proteins are further separated for their delivery to the various cell compartments (e.g., plasma membrane, secretion granules, lysosomes) (see Rothman, 1981 in the Additional Readings at the end of this chapter).

Insulin Biosynthesis—A Good Example of the Molecular Processing of Secretion

It is well known that many secreted proteins are first synthesized as biologically inactive precursors, which are activated later. Activation consists essentially of the removal of a portion of the polypeptide chain and may occur at different sites. For example, the zymogens secreted by the pancreas are activated extracellularly, i.e., after the release of the secretion. Various polypeptide hormones are produced as

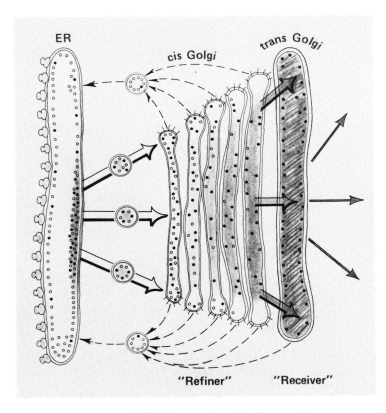

Fig. 9–16. Diagram illustrating the hypothetical dual function played by the cis and trans portions of the Golgi apparatus. The density of shading portrays the progressive transport and accumulation of exported proteins. The closed circles represent some of these proteins, while the open circles correspond to ER proteins that are removed from the rims of the cis Golgi cisternae and return to the ER (dashed arrows) by a cross current. The thick arrows represent steps in the transport into the cis- and the trans-Golgi compartments working in tandem. (Courtesy of J.E. Rothman).

inactive *prohormones*, which are then activated intracellularly by the *converting* (i.e., proteolytic) *enzymes* presumably present in the Golgi apparatus.

Examples of such *proproteins* are: proalbumin, proparathyroid hormone, proglucagon, progastrin, and the well-known case of *proinsulin*. The hormone insulin, produced by the beta cells of the pancreatic islands, has a molecular weight of only 12,000 daltons, and two chains: the A chain of 21 amino acids and the B chain of 30 amino acids; the chains are linked by two —S—S bonds.

In Figure 9–17 is represented the insulin mRNA with its B and A regions (cistrons), which code for the B and A chains. In addition the mRNA contains a preregion that codes for the *signal peptide* (see Section 8–1), a C segment, and a poly A tail. This mRNA was isolated and

found to contain information coding for a polypeptide chain of 330 amino acids. The complete translation of this mRNA on free ribosomes gives rise to *pre-proinsulin*. However, as described in Chapter 8, in the ER, the preregion (i.e., a signal peptide 23 amino acids long) is removed by the *signal peptidase*.[27] The completed *proinsulin chain* is then activated in the Golgi through the removal of the C peptide by the converting enzyme and is packaged as active hormone within the secretion granule (see Chan and Steiner, 1977 in the Additional Readings at the end of this chapter).

A possible mechanism linking the processing of the proproteins with exocytosis has been reported for proalbumin.[28] In this case the conversion of proalbumin into albumin depends on the fusion of small Golgi vesicles containing a protease. The general problem of molecular

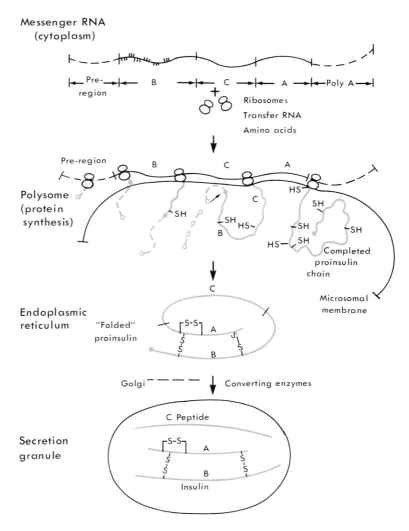

Fig. 9–17.   Diagram indicating steps in the molecular processing of insulin secretion. Observe that the pre-region that forms part of pre-proinsulin is separated within the endoplasmic reticulum. The converting enzyme present in the Golgi transforms the proinsulin into insulin by removal of the C peptide. (From Chan, S.J., and Steiner, D.F.: Trends Biochem. Sci., 2:254, 1977.)

processing of secretory proteins is at present of great importance and is related to the possibility that cleavage products may have different functions. For example, in pituitary cells a single precursor molecule contains the hormone ACTH (adrenocorticotrophin) and β-endorphin, a peptide with opiate-like activity[29,30] (see Fig. 24–23).

SUMMARY:

Functions of the Golgi Complex

The Golgi complex may be considered as an intermediary compartment interposed between the

ER and the extracellular space, through which there is a continuous traffic of substances (i.e., fluids, macromolecules, cell wall units, and other cell constituents). This traffic also involves the flow and differentiation of membranes and a rapid turnover (every 20 to 40 minutes in dictyosomes of the liver).

One of the major functions of the Golgi is the *glycosidation* of *lipids* and *proteins* to produce glycolipids and glycoproteins. The sugar residues are transferred in a sequential order by the glycosyltransferases. In glycoproteins, after the synthesis of the protein backbone in the ER, the oligosaccharides of the core are added and then those of the terminal group (Table 9–4). Glycoproteins may be secreted or incorporated into the cell coat or the lysosomes. The glycosidation of lipids, which leads to the synthesis of gangliosides and glycosphingolipids, may be altered in cancerous cells.

The main function of the Golgi complex is *cell secretion*, not only of exportable proteins but also of the enzymes present in lysosomes and peroxisomes (see Chapter 10). Secretion may be continuous, the secretion products being discharged without storage (i.e., glycoproteins by the liver cells, antibodies by the plasma cells), or it may be discontinuous, with storage in secretory or zymogen granules (e.g., in the pancreas, parotid gland, and others).

The following six steps of secretion can be recognized in the pancreas and other zymogen-secreting glands: (1) *The ribosomal stage*. The synthesis of proteins by polysomes attached to the RER. (2) *The cisternal stage*. This step corresponds to the vectorial transport of the proteins into the lumen of the ER. (The student should remember here the signal hypothesis described in Section 8–4.) The secreted material may be a dilute solution, but may sometimes have intracisternal granules. (3) *Intracellular transport*. The secreted proteins enter the *transitional* tubules and vesicles that lead to the Golgi complex, in which they fuse with the large *condensing vacuoles* present at the maturing face of the Golgi. The transport may be followed by the use of labeled amino acid precursors with subsequent study by radioautography or cell fractionation. In the parotid gland the total life span of a zymogen granule has been estimated to be 52.4 minutes. The transport from the ER to the condensing vacuole requires the use of energy (ATP). (4) *Concentration of the secretion*. By a process of concentration the condensing vacuole is converted into the zymogen granule. This conversion does not require energy and is probably due to the formation of osmotically inactive aggregates of sulfated peptidoglycans, with loss of water to the cytosol. (5) *Intracellular Storage*. Formation of the secretory granules. (6) *Exocytosis*. The discharge of the secretory granule involves its movement toward the apical region and fusion of the membranes (of the granule and the cell). Exocytosis requires Ca^{++} and energy (ATP). The energy is presumably related to the process of fusion-fission of membranes.

At the apical region there is a mechanism of removal and recycling of the excess membrane. Cyclic 3′-5′ AMP may also be implicated in the process of exocytosis. A dual function at the cis and trans faces has been postulated.

The Golgi apparatus is involved in the formation of primary lysosomes. A number of lysosomal enzymes are glycoproteins and are glycosidated at the Golgi complex.

An excellent example of the molecular processing of secreted proteins is provided by the hormone insulin, which has two precursors: preproinsulin and proinsulin.

REFERENCES

1. Rambourg, A., Clermont, Y., and Marraud, A.: Am. J. Anat., *140*:27, 1974.
2. Morré, D.J.: Membrane differentiation and the control of secretion: A comparison of plant and animal Golgi apparatus. *In* International Cell Biology. Edited by B.R. Brinkley and K.R. Porter. New York, The Rockefeller University Press, 1977, p. 293.
3. Novikoff, A.B.: Proc. Natl. Acad. Sci. USA, 73:2781, 1976.
4. Friend, D.S.: J. Cell Biol., *41*:269, 1969.
5. Morré, D.J., Cheetham, R., and Yunghans, W.: J. Cell Biol., *39*:961, 1968.
6. Fleischer, B.: Fed. Proc., *28*:404, 1969.
7. Ovtracht, L., Morré, D.J., Cheetham, R.D., and Mollenhauer, H.H.: J. Microsc. (Paris), *18*:87, 1973.
8. Bretz, R., Bretz, H., and Palade, G.E.: J. Cell Biol., *84*:87, 1980.
9. Schachter, H., et al.: VII International Congress of Biophysics. Mexico, 1981, p. 192.
10. Neutra, M., and Leblond, C.P.: Sci. Am., *220*:100, 1969.
11. Young, R.W.: J. Cell Biol., *57*:195, 1973.
12. Richardson, C.L., et al.: Biochim. Biophys. Acta, *417*:175, 1975.
13. Cook, G.M.W.: The Golgi apparatus. *In* Oxford Biology Reader. Vol. 77. Oxford, Oxford University Press, 1975.
14. Warshawsky, H., Leblond, C.P., and Droz, B.: J. Cell Biol., *16*:1, 1973.
15. Zambrano, D., and De Robertis, E.: Z. Zellforsch., 76:458, 1968.
16. Berg, N.B., and Young, R.W.: J. Cell Biol., *50*:469, 1971.
17. Reggio, H., and Palade, G.E.: J. Cell Biol., *70*:360, 1976.
18. Reggio, H.A., and Palade, G.E.: J. Cell Biol., *77*:288, 1978.
19. Blaschko, H., et al.: Biochem. J., *103*:300, 1967.
20. Creutz, C.E.: J. Cell Biol., *91*:247, 1981.
21. De Robertis, E., and Vaz Ferreira, A.: J. Biophys. Biochem. Cytol., 3:611. 1957.
22. Viveros, O.H., Diliberto, E.J., Hazun, E., and Chang, K.J.: Adv. Biochem. Psychopharmacol., *22*:191, 1980.
23. Schramm, M., et al.: Nature, *240*:203, 1972.
24. Baiton, D.F., Nichols, B.A., and Farqhuar, M.G.: *In* Lysosomes in Biology and Pathology. Vol. 5. Edited by J.T. Dingle and R.T. Dean. Amsterdam, North-Holland Pub. Co., 1976, p. 3.
25. Roth, J., and Berger, E.C.: J. Cell Biol., *93*:233, 1982.
26. Rothman, J.E., and Fries, E.: J. Cell Biol., *89*:162, 1981.
27. Steiner, D.F.: Diabetes, *26*:322, 1977.
28. Judah, J.D., and Quim, P.S.: Nature, *271*:384, 1978.
29. Mains, R.E., Eipper, B.A., and Ling, N.: Proc. Natl. Acad. Sci., USA, 74:3014, 1977.
30. Mains, R.E., and Eipper, B.A.: J. Biol. Chem., *253*:651, 1978.

ADDITIONAL READINGS

Brown, R.M., Jr., and Willison, J.H.M.: Golgi apparatus and plasma membrane involvement in secretion and cell surface deposition. *In* International Cell Biology. Edited by B.R. Brinkley and K.R. Porter. New York, The Rockefeller University Press, 1977, p. 267.
Capasso, J.M., and Hirschberg, C.B.: Mechanism of gly-

cosidation and sulfatation in the Golgi apparatus. Proc. Natl. Acad. Sci. USA, *81*:7051, 1984.

Case, R.M.: The role of Ca^{2+} stores in secretion. Cell Calcium, 5:89, 1984.

Chan, S.J., and Steiner, D.F.: Preproinsulin, a new precursor in insulin biosynthesis. Trends Biochem. Sci., 2:254, 1977.

Conn, P.M.: Cellular regulation of secretion and release. *In* Cell Biology Series. New York, Academic Press, 1982.

Dunphy, W.G., et al.: Attachment of terminal N-acetylglucosamine to asparagine-linked oligosaccharides occurs in central cisternae of the Golgi stack. Cell, *40*:463, 1985.

Farquhar, M.G., and Palade, G.E.: The Golgi apparatus (Complex) (1954–1981): From artifact to center stage. J. Cell Biol., *91* (No. 3, Pt. 2): 77S, 1981.

Goldfisher, S.: The internal reticular apparatus of Camilo Golgi. J. Histochem. Cytochem., *30*:717, 1982.

Herzog, V.: Pathways of endocytosis in secretory cells. Trends Biochem. Sci., 6:319, 1981.

Jamieson, J.D., and Palade, G.E.: Production of secretory proteins in animal cells. *In* International Cell Biology. Edited by B.R. Brinkley and K.R. Porter. New York, The Rockefeller University Press, 1977, p. 308.

Leblond, C.P., and Bennett, G.: Role of the Golgi apparatus in terminal glycosylation. *In* International Cell Biology. Edited by B.R. Brinkley and K.R. Porter. New York, The Rockefeller University Press, 1977, p. 326.

Membrane Recycling. Ciba. Found. Symp., 98, 1982.

Morré, D.J.: Membrane differentiation and the control of secretion: A comparison of plant and animal Golgi apparatus. *In* International Cell Biology. Edited by B.R. Brinkley and K.R. Porter. New York, The Rockefeller University Press, 1977, p. 293.

Morré, D.J., and Mollenhauer, H.H.: The endomembrane concept: A functional integration of endoplasmic reticulum and Golgi apparatus. *In* Dynamic Aspects of Plant Ultrastructure. Edited by A.W. Robards. New York, McGraw-Hill Book Co., 1974, p. 84.

Neutra, M., and Leblond, C.P.: The Golgi apparatus. Sci. Am., *220*:100, 1969.

Northcote, D.H.: The Golgi complex. *In* Cell Biology in Medicine. Edited by E.E. Bittar. New York, John Wiley & Sons, 1973.

Palade, G.E.: Intercellular aspects of the process of protein secretion. Science, *189*:347, 1975.

Palade, G.E., and Farquhar, M.G.: The Golgi complex: A current overview. *In* III International Cell Biology Congress. Edited by S. Seno and Y. Okada. Japan, Japan Inc., 1984, p. 96.

Parodi, A., and Leloir, L.: Lipid intermediates in protein glycosidation. Trends Biochem. Sci., *1*:58, 1976.

Richardson, C.L., Baker, S.R., Morré, D.J., and Keenan, T.W.: Biochim. Biophys. Acta, *417*:175, 1975.

Roth, J.: Cytochemical localization of terminal N-acetyl-D-galactosamine residues in cellular compartments of intestinal goblet cells. J. Cell Biol., 98:399, 1984.

Rothman, J.: The Golgi apparatus: Two organelles in tandem. Science, *213*:1212, 1981.

Rothman, J., and Leonard, J.: Membrane traffic in animal cells. Trends Biochem. Sci., 9:176, 1984.

Rothman, J., Miller, R.L., and Urbani, L.J.: Intercompartmental transport of Golgi complex is a dissociative process. J. Cell Biol., 99:260, 1984.

Schachter, H.: The subcellular sites of glycosylation. Biochem. Soc. Symp., 40:57, 1974.

Staneloni, A., and Leloir, L.F.: The biosynthetic pathway of the asparagine-linked oligosaccharides of glycoproteins. Trends Biochem. Sci., 4:65, 1979.

Tartakoff, A.M.: Simplifying the complex Golgi. Trends Biochem. Sci., 7:174, 1982.

Whaley, W.G.: The Golgi apparatus. *In* Cell Biology Monographs. Vol. 2. Berlin, Springer-Verlag, 1975, p. 1.

Winkler, H.: The biogenesis of adrenal chromaffin granules. Neuroscience, 2:657, 1977.

Zhang, F., and Schneider, D.L.: The bioenergetics of Golgi apparatus function: Evidence for an ATP-dependent proton pump. Biochem. Biophys. Res. Commun., *114*:620, 1983.

CHAPTER 10

LYSOSOMES, ENDOCYTOSIS, COATED VESICLES, ENDOSOMES, AND PEROXISOMES

Every eukaryotic cell has a group of cytoplasmic organelles, *the lysosomes*, of which the main function is intracellular or extracellular digestion. They may be distinguished from other organelles by their morphology and especially by the following functions they perform: (1) digestion of food or various materials taken by *phagocytosis* or *pinocytosis*, (2) digestion of parts of the cell by a process called *autophagy*, and (3) breakdown of extracellular material by the release of enzymes into the surrounding medium.

In this chapter we show that lysosomes contain numerous hydrolytic enzymes involved in the mechanisms of cellular digestion. When some of these enzymes are genetically altered, important modifications of the cell functions occur, and the lysosomes play an important role in pathology and medicine.

The lysosome is directly related to the process

of *endocytosis*, which comprises phagocytosis of solid material and pinocytosis of liquid. There is also a special type of selective endocytosis and transfer of intracellular membranes by a receptor-mediated mechanism. In this case, the *coated or clathrin vesicles* play an essential role.

Another cell compartment, the so-called *endosome*, is characterized by its acidic content and its function in the selective intracellular sorting of receptors and ligands.

The other organelle studied in this chapter, the *peroxisome*, is related to the *production and decomposition of H_2O_2* and to the *β-oxidation of fatty acids*. In plants, peroxisomes are also involved in photorespiration. A related plant organelle, the *glyoxysome*, is discussed in Chapter 12.

10–1 MAJOR CHARACTERISTICS OF LYSOSOMES

The concept of the lysosome originated from the development of cell fractionation techniques, by which different subcellular components were isolated. By 1949, a class of particles having centrifugal properties somewhat intermediate between those of mitochondria and microsomes was isolated by De Duve and found to have a high content of acid phosphatase and other hydrolytic enzymes. Because of their enzymatic properties, in 1955 these particles were named lysosomes (Gr., *lysis*, dissolution; *soma*, body).[1]

At present some 50 lysosomal hydrolases are known (Table 10–1), which are able to digest most of the biological substances.[2] Lysosomes have been found both in animal and plant cells and in *Protozoa*. In bacteria there are no lysosomes, but the so-called *periplasmatic space*, found between the plasma membrane and the cell wall, may play a role similar to that of the lysosomes.[3]

One important property of lysosomes is their stability in the living cell. The enzymes are enclosed by a membrane and are not readily available to the substrate. After isolation by mild methods of homogenization, the amount of enzyme that can be measured is small. The yield increases considerably if the particles are treated with hypotonic solutions or surface-active agents. This so-called *latency* of the lysosomal enzymes is due to the presence of the membrane.

The membrane is resistant to the enzymes that it encloses, and the entire process of digestion is carried out within the lysosome. In this way it protects the rest of the cell from the destructive effect of the enzymes, and its stability is of fundamental importance to the normal function of the cell. In fact, pathological conditions are known in which this membrane becomes more labile and permits the exit of the enzymes with catastrophic consequences to the cell. Another interesting point is that most of the lysosomal enzymes act in an acid medium.

Various Cytochemical Procedures for Identifying Lysosomes

While the concept of lysosome was, at the beginning, based on purely biochemical grounds, its identification soon became possible with the use of the electron microscope (EM) and various cytochemical techniques. The most widely used procedure is the Gomori stain for *acid phosphatase* (see Section 3–5) (Fig. 10–1). Cytochemical reactions for *β-glucuronidase, aryl sulfatase, N-acetyl-β-glucosaminidase*, and *5-bromo-4-chloroindoleacetate-esterase* can also stain the lysosomes. Certain substances that are taken up by lysosomes may be used in their identification. Thus, *peroxidase* ingested by the cell may be detected cytochemically. Certain dyes added to living cells may accumulate in lysosomes and then be demonstrated by fluorescence microscopy. For example, neutral red stain, some anti-malarial drugs, and vitamin A are taken up by lysosomes.

We mentioned that lysosomes represent an acidic compartment of the cell and later on we show that endosomes also have an acidic content, but lack the hydrolases. It is possible to determine the intralysosomal pH, which is about 5, by the incorporation of a fluorescein and dextran complex into the cell; the fluorescence spectrum of the complex changes with the H^+ concentration. Acidification depends on an ATP-dependent *proton pump*, present in the membrane of the lysosome, by which H^+ is accumulated (see Reijngoud, 1978 in the Additional Readings at the end of this chapter).

It is interesting to note that when certain amines, such as the anti-malarial compound *chloroquine*, are given they accumulate in the lysosomes because of their weak-base proper-

TABLE 10–1. LYSOSOMAL ENZYMES

Hydrolases acting on ester bonds

Arylesterase	Acid phosphatase	Sphingomyelin phosphodiesterase
Triglycerol lipase	Phosphodiesterase I	Arylsulfatases A and B
Phospholipase A_1 and A_2	Deoxyribonuclease II	Chondroitinsulfatase
Cholesterol esterase	Ribonuclease II	

Hydrolases acting on glycosyl compounds

Lysozyme	Hyaluronoglucosidase	α-L-fucosidase
Neuraminidase	α- and β-mannosidase	L-iduronidase
α- and β-glucosidase	α- and β-N-acetyl-glucosaminidase	
	α- and β-N-acetyl-galactosaminidase	
α- and β-galactosidase		

Hydrolases acting on peptide bonds

Carboxypeptidase A, B, and C	Neutral proteinase	Renin
Dipeptidase	Plasminogen activator	Cathepsin E
Dipeptidylpeptidase	Cathepsin B	Cathepsin G
Kininogen	Cathepsin D	Collagenase
Elastase		

Hydrolases acting on other carbon-nitrogen bonds

Aspartylglucosaminidase	Amino acid naphthylamidase	Benzoyl arginine naphthylamidase

Hydrolases acting on acid anhydrides
Inorganic pyrophosphatase

Hydrolases acting on phosphorus-nitrogen bonds
Phosphoamidase

Hydrolases acting on sulfur-nitrogen bonds
Heparin sulfamidase

From Barrett, A.J., and Dean, R.T.: Lysosomal Enzymes. Data Book on Cell Biology. Bethesda, MD, Federation of American Societies for Experimental Biology, 1976.

ties. Chloroquine reacts with the acid interior of the organelle increasing the pH. This leads to water uptake and vacuolization and decrease in lysosomal enzymes.

Lysosomes Are Very Polymorphic

The most remarkable morphological characteristic of the lysosome is its polymorphism, particularly regarding the size of the particle and the irregularities of its internal structure. This polymorphism, which suggests that these organelles are extremely dynamic, can be recognized after isolation of the lysosomes (Fig. 10–2). Within the cell, lysosomes are surrounded by *multivesicular bodies*, smooth and *coated vesicles*, and dense bodies; the images observed under the EM suggest the possibility of fusion and fission events between these structures.

Bodies with similar morphological characteristics were observed in intact liver cells and first named "pericanalicular dense bodies" because of their preferential location along the fine bile canaliculi (see Fig. 1–16).

According to the current interpretation, the polymorphism is the result of the association of primary lysosomes with the different materials that are phagocytized by the cell. A summary of these concepts is presented in Figure 10–3.

Primary Lysosomes and Three Types of Secondary Lysosomes

At present four types of lysosomes are recognized, of which only the first is the *primary lysosome*; the other three may be grouped together as *secondary lysosomes*.

(1) The *primary lysosome* (i.e., *storage granule*) is a small body whose enzymatic content is synthesized by the ribosomes and accumulated in the endoplasmic reticulum (ER). From there the enzymes penetrate into the Golgi region, in which the first acid-phosphatase reaction takes place[4] (Fig. 10–3). The primary lysosome may be charged preferentially with one type of enzyme or another; it is only in the secondary lysosome that the full complement of acid hydrolases is present. The formation of primary lysosomes may be followed in cultures of monocytes, which become transformed into macrophages. In a short time there is considerable synthesis of hydrolytic enzymes, which may be

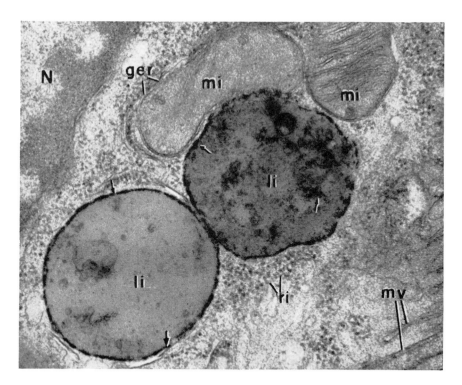

Fig. 10–1.   Electron micrograph of a proximal convoluted tubule cell of mouse kidney, 2 hours after injection of crystalline ox hemoglobin. Two absorption droplets (phagosomes, lysosomes) have formed at the apical region, and the acid phosphatase reaction becomes positive at the surface (arrows) and penetrates inside the lysosome. *ger*, granular endoplasmic reticulum; *li*, lysosomes; *mi*, mitochondria; *mv*, microvilli; *N*, nucleus; *ri*, ribosomes. × 60,000. (Courtesy of F. Miller.)

blocked by the protein synthesis inhibitor puromycin. In these activated cells, using ³H-leucine and radioautography at the electron microscopic level, the transfer of protein was observed in the following sequence: ER → Golgi complex → lysosomes.

(2) The *heterophagosome* or *digestive vacuole* results from the phagocytosis or pinocytosis of foreign material by the cell. This body, which contains the engulfed material within a membrane, shows a positive phosphatase reaction, which may be due to the association with a primary lysosome (Fig. 10–3). The engulfed material is progressively digested by the hydrolytic enzymes, which have been incorporated into the lysosome. Under ideal conditions digestion leads to products of low molecular weight that pass through the lysosomal membrane and are incorporated into the cell to be used again in many metabolic pathways.

(3) *Residual bodies* are formed if the digestion is incomplete. In some cells, such as ameba and other protozoa, these residual bodies are eliminated by *defecation* (Fig. 10–3). In other cells they may remain for a long time and may load the cell. For example, the pigment inclusions found in nerve cells of old animals may be a result of this type of process.

(4) The *autophagic vacuole, cytolysosome,* or *autophagosome* is a special case, found in normal cells, in which the lysosome contains a part of the cell in the process of digestion (e.g., a mitochondrion or portions of the ER) (Fig. 10–3).

Lysosomes regularly engulf bits of cytosol, which is degraded by a mechanism called *microautophagy.* During starvation the liver cell shows numerous autophagic vacuoles; mitochondrial remnants can be found in some of these. This is a mechanism by which the cell can achieve the degradation of its own constituents without irreparable damage. In the liver, autophagy may be induced by injection of the pancreatic hormone glucagon. This treatment produces a considerable increase in cytolysosomes while the small primary lysosomes diminish in number.[5]

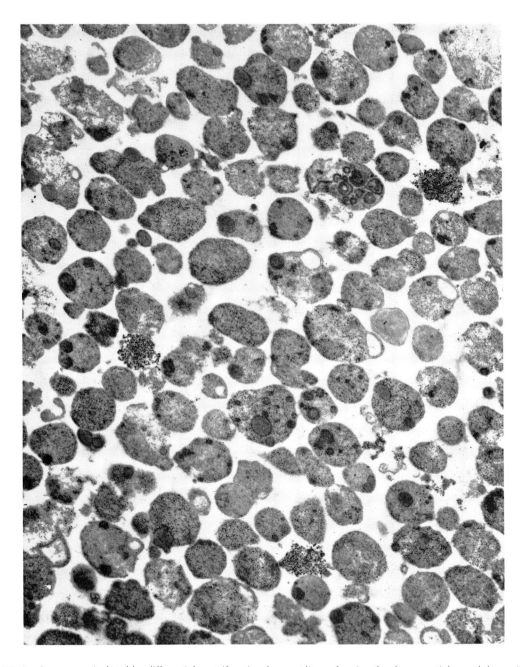

Fig. 10–2. Lysosomes isolated by differential centrifugation from rat liver, showing the dense particles and the variety of other dense material contained within the single membrane of the lysosome. ×60,000. (Courtesy of C. De Duve.)

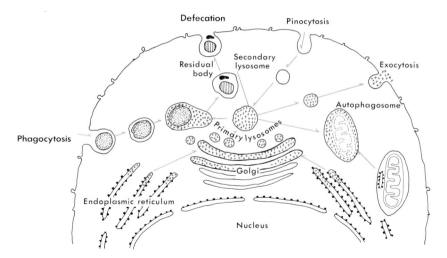

Fig. 10–3. Diagram representing the dynamic aspects of the lysosome system. Observe the relationships between the processes of phagocytosis, pinocytosis, exocytosis, and autophagy.

Lysosomal Enzymes—Synthesized in the ER, Packaged in the Golgi—The Mannose-6-Phosphate Receptor

According to current concepts, primary lysosomes are thought to be secretion products of the cell which, like other secretions (see Section 9–4), are synthesized by ribosomes, enter the ER, and reach the Golgi region for final packaging. Since by this mechanism a cell may produce different types of lysosomes, and many other secretion products, it is likely that there is a topological specificity in the ER-Golgi system. In other words, the different products must be dispersed through different channels of the intracellular membrane system.

Studies on the biosynthesis of the lysosomal enzymes cathepsin and β-glucuronidase favor the biogenesis of lysosomal enzymes on ribosomes bound to the ER.[6,7] These studies demonstrate that lysosomal enzymes are discharged into the ER lumen using a signal peptide mechanism similar to that of secretory proteins (see Section 8–5).

How the lysosomal enzymes are sorted out and find their way into lysosomes remains in question. An increasing body of evidence indicates that these enzymes contain terminal mannose-6-phosphate (man-6-P) residues that interact with man-6-P receptors in intracellular membranes. This interaction would be instrumental in the traffic of lysosomal enzymes.

After being secreted into the ER lumen the enzymes are glycosylated.[8] Later on, the addition of the man-6-P residue is catalyzed by enzymes (N-acetylglucosamine 1-phosphotransferase and N-acetylglycosamyl phosphodiesterase) that are associated with the Golgi complex. Newly synthesized lysosomal enzymes, bearing the man-6-P recognition marker, bind to specific *man-6-P receptors* located in Golgi membranes and are then removed from the Golgi cisternae and delivered to the GERL region and to the lysosomes by a vesicular carrier. After reaching its final destination the terminal man-6-P residue is cleaved off (see Sly and Fisher, 1982 in the Additional Readings at the end of this chapter).

Recently, the localization of the man-6-P receptor has been followed by immunochemical methods. It has been found that these receptors are concentrated in Golgi stacks on the *cis* side of the complex and also in coated vesicles, endosomes, and lysosomes; all of these structures may serve as stations in the delivery route of lysosomal enzymes (see Brown and Farquhar, 1984; Geuze et al., 1984 in the Additional Readings at the end of this chapter) (Fig. 10–4).

The possible "routing" of most lysosomal enzymes by a man-6-P receptor mechanism may explain the so-called "*I-diseases*" in which lysosomal enzymes are not retained by the cell but are secreted in massive amounts. In these cases there is a deficiency in the N-acetylglucosamine 1-phosphotransferase, and the man-6-P residue is not synthesized, thus preventing

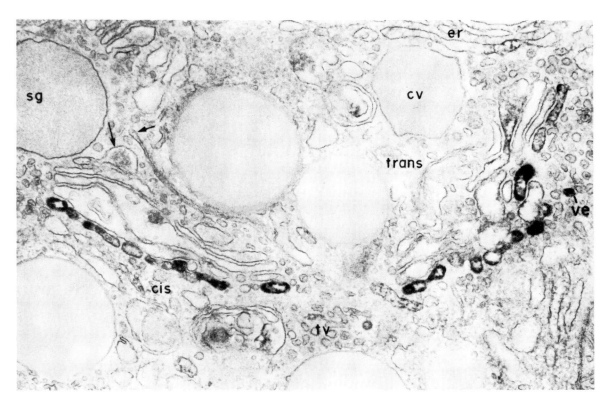

Fig. 10–4.   Electron micrograph of a pancreatic acinar cell showing the localization of the G-P-mannose receptor using an antibody. Observe the neat polarization of the Golgi complex where the immune reaction is strictly localized in the *cis* cisternae of the complex; this is missing in the *trans* region. A few vesicles *(ve)* also give a positive reaction; *tv*, transporting vesicles; *cv*, condensing vesicles; *sg*, secretion granules; *er*, endoplasmic reticulum. (Courtesy of M.G. Farquhar.)

the interaction with the man-6-P receptor and the normal routing. Similar alterations occur in culture cells that are deficient in the receptor and also in cells treated with *tunicamycin*, which prevents glycosidation of the lysosomal enzymes. The administration of certain amines, like *chloroquine*, also causes the secretion of large amounts of lysosomal enzymes because these amines prevent maturation of the protein.

All of these findings suggest that special carbohydrates may act as sorters in the intracellular traffic of proteins synthesized in the ribosomes attached to the ER.

SUMMARY:
Lysosomal Morphology and Cytochemistry

Lysosomes are cytoplasmic organelles that contain numerous (about 50) hydrolytic enzymes and in which the main functions are intracellular and extracellular digestion (Table 10–1). These structures digest materials taken in by endocytosis (phagocytosis and pinocytosis), parts of the cell (by *autophagy*), and extracellular substances. Lyso-

somes are separated as a fraction that is intermediate between mitochondria and microsomes (see Section 2–1). The lysosomal enzymes are enclosed within a membrane (accounting for their *latency*) and generally act at acid pH. Under the EM they can be recognized by using different cytochemical reactions (e.g., acid-phosphatase).

Lysosomes show considerable polymorphism. The *primary lysosomes* (i.e., storage granules) are dense particles of about 0.4 μm surrounded by a single membrane. Their enzymatic content is synthesized by ribosomes in the ER and appears in the Golgi region. The formation of primary lysosomes can be blocked by puromycin. *Secondary lysosomes* (i.e., digestive vacuoles) result from the association of primary lysosomes with vacuoles containing phagocytized material. The so-called *phagosome* fuses with lysosomes (i.e., *heterophagosome*) and is digested by hydrolytic enzymes. Sometimes *residual bodies* containing undigested material are formed. These structures may be eliminated, but in most cases they remain in the cell as pigment inclusions and may be related to the aging process. The autophagic vacuole or cytolysosome is a special case in which parts of the cell are digested. This normal process is stimulated during starvation and by the pancreatic hormone glucagon.

Lysosomal enzymes are synthesized in the ER and then packaged at the GERL region of the Golgi complex to form the primary lysosomes. The mechanism by which these proteins are sorted out and routed toward the lysosomes is as yet unknown. However, the terminal phosphorylation of mannose and attachment to a receptor inside the ER may play an important role.

10–2   FUNCTIONS OF LYSOSOMES— INTRACELLULAR DIGESTION

Within the secondary lysosome the ingested materials or those resulting from autophagy are subjected to the action of the many hydrolases already present in the lysosome (Table 10–1). The majority of these enzymes have acid pH optima, and secondary lysosomes may indeed be acidic (pH about 4.0). *Digestion of proteins* usually ends at the level of the dipeptide, which can pass through the membrane and be further degraded into amino acids.

Carbohydrates are usually hydrolyzed to monosaccharides, which are easily released. Disaccharides or polysaccharides (cellobiose, inulin, or dextran), however, are not digested and remain within the lysosome. Sucrose may be taken into macrophages by pinocytosis; however, it is not hydrolyzed and remains trapped in secondary lysosomes that become swollen by the increase in osmotic pressure; such cells are intensely vacuolated. These vacuoles correspond to the lysosomes overloaded with sucrose. It is interesting to note that if at this point the enzyme *sucrase* is given to the cell, it penetrates into the lysosomes and normality is restored. This experiment points out one way in which certain lysosomal diseases could be treated.

Autophagy by Lysosomes—The Renovation and Turnover of Cell Components

Many cellular components, such as mitochondria, are constantly being removed from the cell by the lysosomal system. Cytoplasmic organelles become surrounded by membranes of smooth endoplasmic reticulum (SER), then the lysosomal enzymes are discharged into the autophagic vacuoles, and the organelles are digested. Autophagy is a general property of eukaryotic cells and is related to the normal renovation and turnover of cellular components. Autophagy may be greatly stimulated in certain conditions; for ex-

ample, in protozoa deprived of nutrients or in the liver of a starving animal, many autophagic vacuoles appear. This is a mechanism by which parts of the cell are broken down to facilitate survival.

Lysosomal Removal of Cells and Extracellular Material and Developmental Processes

Many developmental processes involve the shedding or remodeling of tissues, with removal of whole cells and extracellular material.

During metamorphosis of amphibians there is considerable destruction of cells, and this is accomplished by lysosomal enzymes. For example, the degeneration of the tadpole tail is produced by the action of cathepsins (i.e., proteolytic enzymes) contained in the lysosomes.

In organs undergoing regression, such as wolffian ducts in the female embryo and Müller's ducts in the male, there is a considerable increase in lysosomal enzymes. Some tissues that undergo regression after a period of activity may do so by the action of lysosomes. For example, the human uterus just after birth weighs 2 kg and in 9 days returns to its former size, weighing 50 gm! During this process there is a large infiltration of phagocytic cells carrying lysosomes which digest all the debris, extracellular material, and parts of the endometrium. Autophagy is also prominent in the mammary gland after lactation.

Release of Lysosomal Enzymes to the Medium for Extracellular Action

We mentioned earlier that the contents of primary lysosomes may be released to the medium by a process of exocytosis somewhat similar to that of cell secretion (Fig. 10–3). The so-called *osteoclasts*, the multi-nucleated cells that remove bone, may do so by the release of lysosomal enzymes that degrade the organic matrix. This process is activated by the parathyroid hormone and inhibited by *calcitonin*, a hormone secreted by the C cells of the thyroid gland. In bone cultured in vitro, lysosomal enzymes may be activated by vitamin A. Vitamin A intoxication may cause spontaneous fractures in animals. Release of lysosomal enzymes also occurs when cells are treated with anti-cellular antibodies in the presence of serum complement. Such a process may occur in rheumatoid arthritis, a human disease in which the cartilage of the joints may be eroded by lysosomes.

Lysosomal Enzyme Involvement in Thyroid Hormone Release and in Crinophagy

Thyroid hormones (thyroxine and triiodothyronine) are released from a large protein molecule (thyroglobulin) stored within the follicles. In 1941 De Robertis found that this release was due to the activation of a proteolytic enzyme.[9] It is now known that this enzymatic mechanism involves lysosomal proteases.

The name *crinophagy* has been applied to a mechanism by which secretory granules produced in excess of physiological needs may be removed. This mechanism, which is a special case of autophagy, was first observed in the pituitary gland, but it is likely that it is present in many types of endocrine and exocrine glands.

Leukocyte Granules Are of a Lysosomal Nature

All leukocytes of vertebrates contain granules of a lysosomal nature. In the *polymorphonuclear leukocytes* are various types of granules, demonstrable by electron microscopy and gradient centrifugation, that contain several of the known lysosomal enzymes (see Bainton et al., 1976 in the Additional Reading at the end of this chapter). *Monocytes* have few lysosomes, but when they enter the tissues and are transformed into *macrophages*, they gain many lysosomes.

Lysosomes Are Important in Germ Cells and Fertilization

The *acrosome* of the spermatozoon, which develops from the Golgi region and covers the anterior end of the nucleus, can be considered as a special lysosome. Indeed, it contains *protease* and *hyaluronidase* (an enzyme that digests cell coats containing hyaluronic acid) and abundant *acid phosphatase*. During fertilization of the oocyte, hyaluronidase disperses the cells around it *(cumulus oöphorus)* and protease digests the zona pellucida, making a channel through which the sperm nucleus penetrates. In *eggs*, lyso-

somes play a role in the digestion of the stored reserve materials.

Lysosomal Involvement in Human Diseases and Syndromes

Lysosomes are of particular importance in medicine, since they are involved in many diseases and syndromes. In certain pathological conditions, such as *rheumatoid arthritis*, *silicosis* and *asbestosis* (diseases produced by the inhalation of silica or asbestos particles), and *gout* (in which crystals of urate accumulate in the joints), there is a release of lysosomal enzymes from the macrophages and an acute inflammation of the tissues that may lead to an increase in collagen synthesis (fibrosis).

An acute release of lysosomal enzymes occurs in states of anoxia, acidosis, and shock, and results in increased amounts of enzymes in the blood. A common example is that of the myocardial infarct. In regard to all these diseases it is important to remember the nature of the membrane of the lysosome. This structure may be made labile by many substances. For example, all the *liposoluble vitamins* (A, K, D, and E) and the steroid sex hormones tend to make the membrane more susceptible to rupture. On the other hand, cortisone, hydrocortisone, and other drugs having an anti-inflammatory action tend to stabilize the lysosomal membrane.

We have seen that lysosomes of *leukocytes* and *macrophages* are essential to the defense of the organism against bacteria and viruses. There are diseases, such as leprosy and tuberculosis, however, in which bacteria are resistant to the lysosomal enzymes (because of special components of the cell wall) and survive within the macrophages. The same is true of certain mycoses and parasitoses, e.g., toxoplasmosis (see De Duve, 1974 in the Additional Readings at the end of this chapter).

Storage Diseases—Caused by Mutations That Affect Lysosomal Enzymes

Several congenital diseases have been found in which the main alteration consists of the accumulation within the cells of substances such as glycogen or various glycolipids. These are also called *storage diseases* and are produced by a mutation that affects one of the lysosomic enzymes involved in the catabolism of a certain substance. For example, in *glycogenosis* type II, the liver and muscle appear filled with glycogen within membrane-bound organelles. In this disease, *α-glucosidase*, the enzyme that degrades glycogen to glucose, is absent (see Hers and Van Hoof, 1973 in the Additional Readings at the end of this chapter). Some 20 congenital diseases involving lysosomes are presently known. Most of them involve glycolipids and mucopolysaccharides that accumulate in the tissues. These diseases are also due to the lack of certain lysosomal enzymes, such as β-glucosidase and sulfatidase (see Table 10–2).

Many of the diseases involving the *glycosphingolipids* affect mainly the brain, because these lipids are principal components of the myelin sheath (e.g., Tay-Sachs disease and Niemann-Pick disease). In these patients the synthesis of sphingolipids is normal, but they can not be degraded and accumulate in the lysosomes.

Lysosomes in Plant Cells and Their Role in Seed Germination

Membrane-bound organelles containing digestive enzymes have been identified in plant cells. In corn seedlings the large vacuoles of parenchymatous cells may show some characteristic lysosomal enzymes, e.g., protease, carboxypeptidase, DNAse, RNAse, β-amylase, or α-glucosidase. In Chapter 12 we discuss the so-called protein or aleurone bodies, the starch granules present in seeds, and the changes they undergo during germination. These changes are accomplished by the release of various lysosomal enzymes (e.g., α-amylase) that are able to hydro-

TABLE 10–2. STORAGE DISEASES RESULTING FROM INABILITY TO DEGRADE BODY CONSTITUENTS IN LYSOSOMES

Disorder	Enzyme deficiency	Main metabolite affected
Mucopolysaccharidoses		
Hurler's and Scheie syndromes	α-L-iduronidase	Dermatan sulfate, heparan sulfate
Hunter's syndrome	Iduronate sulfatase	Dermatan sulfate, heparan sulfate
Sanfilippo syndrome A	Heparan-N sulfatase	Heparan sulfate
Sanfilippo syndrome B	N-acetyl-α-glucosaminidase	Heparan sulfate
Maroteaux-Lamy syndrome	N-acetylgalactosamine sulfatase	Dermatan sulfate
β-glucuronidase deficiency	β-glucuronidase	Dermatan sulfate, heparan sulfate
Morquio syndrome	Uncertain	Keratan sulfate
Sphingolipidoses		
GM$_1$ Gangliosidosis	β-Galactosidase	GM$_1$ Ganglioside, glycoprotein fragments
Krabbe's disease	β-Galactosidase	Galactosylceramide
Lactosylceramidosis	β-Galactosidase	Lactosylceramide
Tay-Sachs disease	Hexosaminidase A	GM$_2$ ganglioside
Sandhoff's disease	Hexosaminidases A and B	GM$_2$ ganglioside, globoside
Gaucher's disease	β-Glucosidase	Glucosylceramide
Fabry's disease	α-Galactosidase	Tubexosylceramide
Metachromatic leukodystrophy	Arylsufatase A	Sulfatide
Nieman-Pick disease	Sphingomyelinase	Sphingomyelin
Farber's disease	Ceramidase	Ceramide
Disorders of glycoprotein metabolism		
Fucosidosis	α-L-Fucosidase	Glycoprotein and glycolipid fragments
Mannosidosis	α-Mannoside	Glycoprotein fragments
Aspartylglycosaminuria	Amidase	Aspartyl-2-deoxy-2-acetamido glucosylamine
Other disorders with single enzyme defect		
Pompe's disease	α-Glucosidase	Glycogen
Wolman's disease	Acid lipase	Cholesterol esters, triglyceride
Acid phosphatase deficiency	Acid phosphatase	Phosphate esters
I-cell disease	Almost all lysosomal enzymes, present extracellularly	Mucopolysaccharides, glycolipids

From Neufeld, E.F., Lim, T.W., and Shapiro, L.J.: Inherited disorders of lysosomal metabolism. Annu. Rev. Biochem. *44*, 357–363, 1975.

lyze the stored material to be used by the growing plant. Thus, lysosomes in plants may be involved in intracellular and extracellular digestion and also in the process of development (see Allison, 1974 in the Additional Readings at the end of this chapter).

SUMMARY:
Functions of Lysosomes

Lysosomal enzymes digest proteins into dipeptides and carbohydrates into monosaccharides. Some disaccharides (sucrose) and polysaccharides (inulin, dextran) are not digested and remain in the lysosomes. Through the process of autophagy, lysosomes are involved in the renovation and turnover of cellular components. In development, lysosomes are active in the remodeling of tissues (e.g., removal of the tadpole tail, regression of wolffian and Müller's ducts, and like processes). Digestion of extracellular material involves the release of primary lysosomes by exocytosis (e.g., osteoclasts). Degradation of bone is activated by vitamin A and parathyroid hormone. In rheumatoid arthritis lysosomal enzymes erode cartilage. Crinophagy is a process by which excess secretory granules are removed. The acrosome of the spermatozoon, which develops in the Golgi, is a special lysosome.

The study of lysosomes is particularly important in medicine. For example, they act in rheumatoid arthritis, silicosis, asbestosis, and gout. Lysosomes of leukocytes (specific granules) and monocytes are essential in defense against bacteria and viruses. There are about 20 congenital diseases called *storage diseases*, in which there is accumulation of substances (i.e., glycogen, glycolipids) in lysosomes. These diseases are due to the lack of certain lysosomal enzymes (Table 10–2).

Lysosomes are found in plant cells. In seedlings they are involved in the hydrolysis and removal of protein and starch during germination.

10–3 ENDOCYTOSIS

The intracellular secretion of primary lysosomes is in some way coupled to another system of extracellular origin that is formed by the process of *endocytosis*, which is related to the activity of the plasma membrane. Endocytosis includes the processes of *phagocytosis*, *pinocytosis*, and *micropinocytosis*, by which solid or fluid materials are ingested in bulk by the cell (Fig. 10–3). (We saw in Chapter 9 that *exocytosis* is the reverse process, by which membrane-lined products are released at the plasma membrane.)

Phagocytosis—The Process of Cellular Ingestion of Solid Material

Phagocytosis (Gr., *phagein*, to eat) is found in a large number of protozoa and among certain cells of the metazoa in which it is, in general, a means of defense against particles that are foreign to the organism, such as bacteria, dust, and various colloids.

Phagocytosis is highly developed in granular leukocytes (first described by Metschnikoff at the end of the 19th century), and also in the cells of mesoblastic origin ordinarily grouped under the common term *macrophagic* or *reticuloendothelial system*. The cells belonging to this group include the histiocytes of connective tissue, the reticular cells of the hematopoietic organs, and those endothelial cells lining the capillary sinusoids of the liver, adrenal gland, and hypophysis. All these cells can ingest not only bacteria, protozoa, and cell debris, but also smaller colloidal particles. In this instance phagocytosis is called *ultraphagocytosis*.

In protozoa, phagocytosis is intimately linked to ameboid motion. An ameba ingests large particles, including microorganisms, by surrounding them with pseudopodia to form a food vacuole within which the digestion of food takes place.

Analyzing the process of phagocytosis, one may distinguish two distinct phenomena. First the particle *adheres* (is *adsorbed*) to the mass of the protoplasm, and then the particle actually penetrates the cell. In some cases it has been possible to dissociate these two phases of phagocytosis. For example, at low temperature, bacteria may adhere to the cytoplasm of a leukocyte without being ingested.

Pinocytosis—The Ingestion of Fluids

The uptake of fluid vesicles by the living cell, first observed by Edwards in amebae and by Lewis in cultured cells, has been called pinocytosis (Gr., *pinein*, to drink). As can be readily seen in Lewis's motion pictures, the uptake of fluids is accompanied by vigorous cytoplasmic motion at the edge of the cell, as if vesicles of fluid were being surrounded and engulfed by clasping folds of cytoplasm. Vacuoles taken up at the edge of the cell are then transported to other portions of the cell.

Pinocytosis may be easily demonstrated with proteins labeled with fluorescent dyes. The presence of the protein seems to act as a stimulus to pinocytosis, and the uptake is surprisingly high. The ameba practically "drinks" the protein solution, and with it the organism may absorb other substances that normally do not penetrate.

Pinocytosis is induced not only by proteins but also by amino acids and certain ions. In ameba the pinocytotic activity, once started, is kept going for about 30 minutes, during which time some 100 fluid channels are formed. Then the process comes to a stop, and the ameba has to wait for 2 to 3 hours before starting another pinocytotic cycle. This has been interpreted as an indication that the surface membrane available for invagination is exhausted.

That phagocytosis and pinocytosis are essentially similar phenomena can be demonstrated by allowing the ameba to phagocytize some ciliated cells first and then inducing pinocytosis. The number of channels formed is much lower under these conditions. In the reverse experiment it has been found that an ameba can ingest much fewer ciliates for food.

Pinocytosis is related to the cell coat that covers the plasma membrane. A fluorescent protein may be concentrated 50 times in the cell coat before the ameba starts to engulf it.

The binding to the coat explains why the concentration of the incorporated protein may reach enormous figures.

Micropinocytosis and the Ingestion and Transport of Fluids

The use of the EM demonstrated that the plasma membrane of numerous cells could invaginate, forming small vesicles of about 65 nm.

These vesicles were first found in endothelial cells lining capillaries in which they concentrate in the region adjacent to the inner and outer membranes. The presence of vesicles opening on both surfaces and of others traversing the cytoplasm suggested a possible transfer of fluid across the cell. A similar component was then observed in numerous cell types, particularly in macrophages, muscle cells, and reticular cells.

The transport of fluid across the capillary endothelium is assumed to occur in quantal amounts; this process implies the invagination of the plasma membrane, the formation of a closed vesicle by membrane fission, the movement of the vesicle across the endothelial cytoplasm, the fusion of the vesicle with the opposite plasma membrane, and the discharge of the vesicular content.

The transendothelial passage of fluids can be demonstrated by using opaque tracers such as peroxidase. This process represents a short distance shuttle that bypasses intracellular compartments, and permits the passage of fluid across the cell (transcytosis).[10]

Micropinocytosis is a rare event compared to the receptor-mediated endocytosis by the coated pit-vesicle system, which we discuss in Section 10–4.

Endocytosis—An Active Mechanism Involving the Contraction of Microfilaments

Phagocytosis and pinocytosis are active mechanisms in the sense that the cell requires energy for their operation. During phagocytosis by leukocytes oxygen consumption, glucose uptake, and glycogen breakdown all increase significantly. Induction of phagocytosis also produces an increased synthesis of phosphatidic acid and phosphatidylinositol. In cultured cells, addition of ATP increases the rate of endocytosis, and this is inhibited by respiratory and other metabolic poisons.

Various amines, like *chloroquine*, inhibit some types of endocytosis and this may have physiopathological and therapeutic implications. In endocytosis the attachment of the particles to the plasma membrane apparently leads to some kind of change beneath the attachment site, which in turn may be related to the microfilaments of actin inserted in the cytoplasmic side of the membrane.

The mechanism of endocytosis involves the contraction of microfilaments of actin and myosin present in the peripheral cytoplasm, which causes the plasma membrane to invaginate and form the endocytotic vacuole. Involvement of actin microfilaments is demonstrated by the action of the drug *cytochalasin B*, which inhibits endocytosis and disorganizes actin microfilaments (see Section 6–5). After entering the cell, the endocytotic vacuole or the phagosome moves toward the Golgi region where primary lysosomes attach to it, fuse, and liberate their contents, forming secondary lysosomes (*heterophagosomes*) (Fig. 10–3). (For a review on endocytosis, see Steinman et al., 1983; Bretscher, 1984 in the Additional Readings at the end of this chapter.)

SUMMARY:
Endocytosis and Lysosomes

Primary lysosomes are coupled with an extracellular system that is the result of the activity of the cell membrane (endocytosis).

Bulk ingestion of solid or fluid material by the cell is called, respectively, *phagocytosis* and *pinocytosis*. Both processes show similarities and are included within the more general concept of *endocytosis*.

Phagocytosis is found in many protozoa and in certain cells of metazoa where, principally, it plays a defense role against bacteria and colloidal particles. In the ameba, phagocytosis is related to ameboid motion. The particles are surrounded by pseudopodia and are digested inside food vacuoles. In phagocytosis two distinct phases may be distinguished: adsorption of the particle and actual penetration. The macrophagic system in metazoa accumulates negatively charged vital dyes and colloids, concentrating them into vacuoles. Pinocytosis is easily observed in the ameba as the formation of fluid channels that break up inside the cell into vacuoles. This process is induced by addition of negatively charged proteins, amino acids, and ions. In a fluorescent protein solution the cell coat of an ameba concentrates the protein 50 times or more; then the protein is taken up by invagination of the cell membrane. The relation between pinocytosis and the cell coat is best observed using colloidal gold and the EM. With this instrument it is observed that pinocytosis is widely present in endothelial and other cells as small vesicles of 65 nm (i.e., *micropinocytosis*). These vesicles are engaged in transcellular fluid transport. An important case is found in the erythroblasts, in which the ferritin molecules used to manufacture hemoglobin are transferred into the cell by micropinocytosis. Endocytosis, in general, is an active process that requires energy. Large amounts of plasma membrane may be internalized by endocytosis. In this process microfilaments of actin are involved. Endocytosis may be inhibited by cytochalasin B. After entering the cell the endocytotic vacuole fuses with primary lysosomes to form heterophagosomes.

10–4 COATED VESICLES AND RECEPTOR-MEDIATED SELECTIVE TRANSPORT

Coated vesicles, whose function is related to endocytosis and the intracellular transport of membranes and soluble proteins, can be observed under the EM in the cytoplasm of most eukaryotic cells. These vesicles vary in size between 50 and 250 nm, and are characterized by the presence of a *coat* made of tiny, regularly spaced bristles that cover the cytoplasmic side of the vesicle. On the plasma membrane, invaginating regions (so-called *coated pits*), which are dynamically related to the coated vesicles, may be observed. The coated pit-coated vesicle system is illustrated in Figure 10–5.

The Coat is Made of Clathrin

Coated vesicles can be isolated and purified from brain and other tissues (see Pearse, 1980 in the Additional Readings at the end of this chapter). With negative staining a lattice of pentagons and hexagons, which constitute the coat, is observed (Fig. 10–6). This coat is mainly composed of a major protein of 180,000 daltons named *clathrin*.[11] It has been calculated that a 200-nm vesicle contains about 1000 clathrin molecules. This protein forms a flexible cage or basket that participates in the invagination of the coated pit. The *clathrin cage* rapidly assembles and disassembles as required by the functional cycle of the vesicle (Fig. 10–7).

Using a freeze-etching method, which permits a three-dimensional view, it has been observed that at the beginning of its formation the clathrin cage of the *coated pit* is made of hexagons. Afterward, when the pit starts to invaginate, pentagons appear in the coat. This hexagon to pentagon transformation may be the force-generating mechanism for the coated vesicle formation.[12]

The assembly of the clathrin coat may be observed in vitro under the EM. The unit of assembly is a trimer (also called a *triskelion*) with three extended projections of 44.5 nm that are bent in the same direction.[13,14] The trimer occurs at each vertex of the hexagon or pentagon of the clathrin coat, and the projections extend along two edges (Fig. 10–7). The extensive fibrous contact of triskelions confers mechanical strength and flexibility to the clathrin coat. It should be emphasized that clathrin is a multifunctional protein capable not only of a rapid assembly and disassembly, but of binding to different cellular membranes.[15] Triskelions are flexible molecules that can associate either as pentagons or hexagons and form surfaces of various curvatures to cover vesicles of different size.

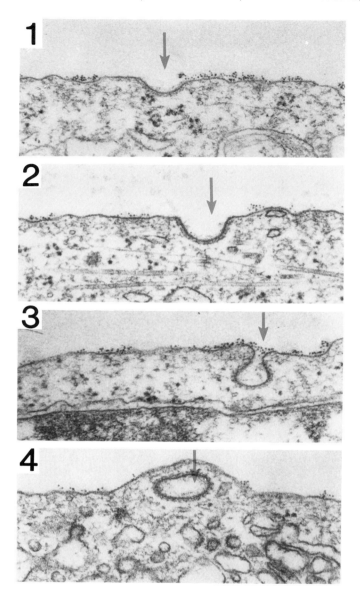

Fig. 10–5. Sequence of events in the formation of a coated vesicle in the surface of a mouse fibroblast. (1) and (2) coated pits; (3) invaginating pit; (4) a typical coated vesicle. The cell surface was labeled with a ferritin conjugate that binds the sites of an antigen. Observe that the regions of the cell surface labeled with ferritin are excluded from the coated pits and vesicles. (Courtesy of M.S. Bretscher.)

The clathrin cage also contains some minor proteins, one of which—of small molecular weight—serves to "glue" the triskelions together (see Slayter, 1982 in the Additional Readings at the end of this chapter).

Functions of Coated Vesicles—Receptor-Mediated Endocytosis

At least ten different types of membranes with characteristic lipid and protein composition occur in a cell. If one considers that membranes are essentially fluid and that there is an extensive transport between those membranes, how can a cell maintain this diversity? One way to solve this problem is to have microdomains in which the lateral diffusion of the membrane components is somewhat restricted; this might be the function of coated pits and vesicles.

Coated vesicles are observed in many cells in which selective endocytosis of bits of membranes and macromolecules from the plasma membrane occurs (Fig. 10–5). This cellular "traffic" also runs in the opposite direction (i.e., toward the membrane) and between different intracellular membrane compartments (e.g., ER and Golgi complex). The function of coated vesicles appears to be related directly to the intracellular traffic of membranes (Fig. 10–8).

Unlike bulk endocytosis (Section 10–3), in

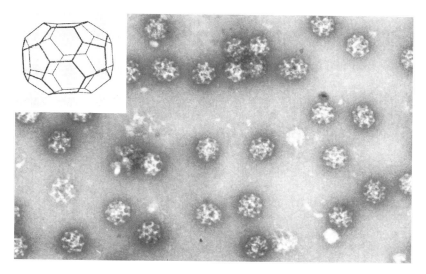

Fig. 10–6. Isolated negative-stained coated vesicles. Inset, model showing the pentagonal and hexagonal structure of the clathrin network. ×67,500. (Courtesy of B. Pearse.)

which the transport is rather nonspecific, the coated-vesicle system is highly selective. Clathrin cages encase special regions of the membrane and maintain them as a unit or microdomain while being transferred to other regions of the cell (Fig. 10–8). At the plasma membrane, coated pits act as molecular filters in selecting some membrane regions to be endocytosed, while others are excluded. In Figure 10–5, we can observe that the clathrin pits exclude regions

of the plasma membrane that have been labeled with ferritin to locate a special antigen.[16] It should be emphasized that, together with the protein, the lipid component of the membrane is also invaginated to form the coated vesicle.

At the present time, many functions are attributed to the coated vesicles. All of them involve the traffic of membranes and material bound to special receptors. Some examples are:

(1) *Uptake of extracellular yolk into chicken*

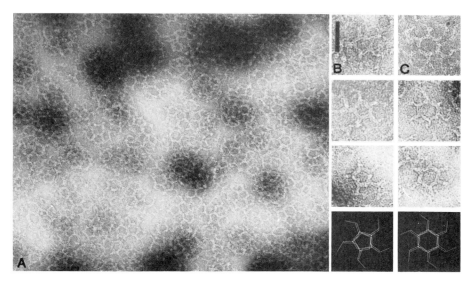

Fig. 10–7. Isolated small fragments of clathrin cages observed by negative staining during the process of assembly. *(A)* A general field containing pentagons and hexagons is observed. ×105,000. Some selected pentagons *(B)* and hexagons *(C)* are shown at higher magnification. ×195,000. The diagrams show the disposition of the individual trimers (or triskelions) that build up the pentagons and hexagons. (Courtesy of R.A. Crowther and B.M.F. Pearse.)

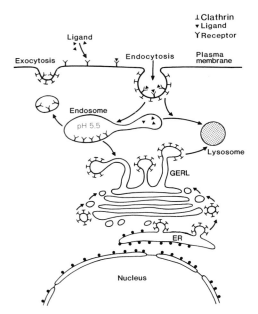

Fig. 10–8. Diagram showing the mechanism of receptor-directed endocytosis involving coated vesicles and endosomes. The ligand interacts with a surface receptor and both are incorporated by vesicles having a clathrin coat. The passage of the vesicle to the endosome is characterized by a drop in pH to 5.5 and the separation of the ligand from the receptor. While the ligand is transferred to the lysosome, the receptor is recycled back to the plasma membrane by clathrin vesicles and exocytosis. The function of coated vesicles in the traffic between the ER and the Golgi, as well as between the GERL region and the endosomal and lysosomal compartments, is also indicated. (Courtesy of I. Pastan, redrawn.)

oocytes. It is well known that in the hen's egg the main yolk protein, *vitellogenin*, is transported from the blood plasma into the oocyte through coated vesicles. There are about 10^{14} receptors on the oocyte surface, which cover a large part of the membrane, and as much as 1 gram of protein can be incorporated per day. The yolk proteins are stored until embryonic development and then phagocytosed by the yolk sac and the catabolic products used by the embryo (see Roth et al., 1984 in the Additional Readings at the end of this chapter).

(2) *Transfer of maternal immunoglobulins (IgG) through the placenta into the fetus.* In most higher animals the transport of IgG provides the newborn with a set of antibodies that help to protect it in early times, when the immune system is still not well developed. This protection is complemented by the endocytosis of antibodies from the mother's milk into the

infant gut. This explains the importance of breast feeding.

(3) *Incorporation into cells of macromolecules and protein bound nutrients* (i.e., cholesterol, iron, and vitamin B_{12}). This function also implies the receptor-mediated endocytosis of protein hormones and growth factors such as insulin, gonadotrophins, epidermal growth factor, and the iron-transporting protein *transferrin*. The same mechanism is also active in the penetration of toxins (such as diphtheria toxin) and certain viruses into cells.

(4) *The recycling of membranes from exocytosed secretory granules and synaptic vesicles.* In these examples and in others, *special receptors* are present at coated pits. These receptors recognize and bind to the molecules that are to be transferred by this type of *receptor-mediated endocytosis.*

To study the function of the coated vesicles more specifically, let us consider the uptake of *low density lipoproteins* (LDL) by human fibroblasts. The LDL is a large particle that transports cholesterol; it originates in the liver and circulates in the plasma. In a cultured fibroblast, about 10,000 surface receptors can be recognized by using an LDL bound to ferritin. If the conjugate is added at 4°C, we see binding but not endocytosis, and the LDL is found on coated pits that occupy about 2% of the cell surface. If the cell is then warmed to 37°C, in a few minutes the LDL-ferritin complex disappears from the surface and is found first in coated vesicles and later in endosomes and lysosomes. Within the lysosomes the LDL is degraded and the cholesterol that it transports is made available for membrane synthesis.

A special mention should be made of the LDL receptor, which has been purified and sequenced. Information about the coding of this receptor has been facilitated by the finding of genetic defects that cause a common human disease called *familial hypercholesterolemia*, in which atherosclerosis occurs early. These mutations block the movement of receptors from the ER to the Golgi apparatus and from the plasma membrane to the coated vesicles. The result is the lack of degradation of the LDL by the lysosomes[17] (see Brown and Goldstein, 1984 in the Additional Readings at the end of this chapter).

The use of anticlathrin antibodies provides another means to study the function of coated pits and coated vesicles. In cultured fibroblasts it has been observed that this protein is concentrated in the perinuclear region. As shown in Figure 10–9, it also appears as dots on the cell surface and in regions that can be identified as coated pits and that cover a certain proportion of the plasma membrane.[18] The process of coated vesicle formation is fast; it has been estimated that in less than 1 hour it can endocytose the equivalent of the whole plasma membrane. This implies a rapid recycling of clathrin and receptor molecules back to the plasma membrane. As mentioned in Section 6–5, the internalization of the coated pits and vesicles also involves the function of actin microfilaments attached to this endocytotic system.

Coated Vesicles and Intracellular Traffic

In addition to selective endocytosis, coated vesicles are involved in other transport mechanisms between intracellular membranes. One of the most interesting is the transport of the coat protein of the *vesicular stomatitis virus* (VSV) from the ER to the plasma membrane.[19] This so-called protein G (a glycoprotein glycosidated at two sites) is produced in the ER and transported to the Golgi and then to the plasma membrane (Fig. 10–10). The function of the coated vesicles is that of sorting out and transporting the G-protein to its final destination. At the cell surface, the patches containing the G-protein become integrated with the *virion* (containing RNA and other proteins), forming the mature virus, which buds out from the cell surface.[20] The intimate relationship between the transport of the G-protein and coated vesicles is evident by the fact that cells infected with VSV contain G-protein and clathrin in nearly stoichiometric amounts.

Experimentally, the study of the transit of the G-protein, from the site of synthesis in the ER to the final destination at the plasma membrane, is facilitated by the fact that the protein is glycosidated first in the ER, producing a G1-protein, and subsequently is terminally glycosidated at the Golgi complex, producing a G2-protein. (The difference between G1 and G2 is that the first is sensitive to an endoglycosidase and the second is resistant.)

The following experiment illustrates the kinetics of the G-protein synthesis: If cells are infected with the VSV and the coated vesicles are isolated, at different time intervals, it is found that there are two waves: (1) one corresponding to G1 (in the first 15 minutes), probably representing coated vesicles going from the ER to the Golgi, and (2) a second wave (starting later), presumably associated with coated vesicles that transport G2-protein from the Golgi to the plasma membrane. All the steps involving

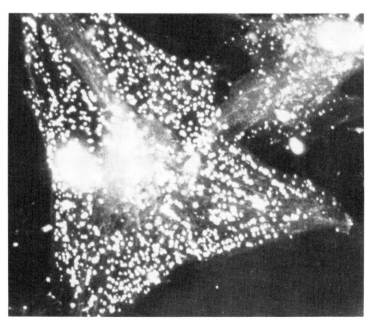

Fig. 10–9. Fibroblast stained with clathrin antiserum by indirect immunofluorescence. Note the presence of a characteristic punctuated pattern, which is due to the staining of coated pits on the cell surface or nearby coated vesicle. ×2000. (Courtesy of E.M. Merisko, M.G. Farquhar, and G.E. Palade.)

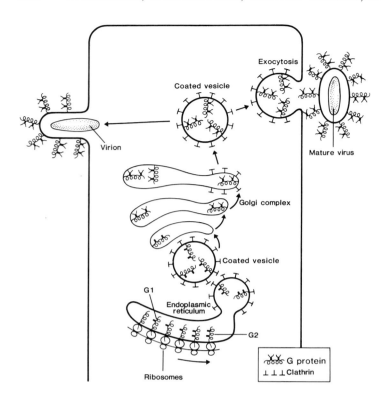

Fig. 10–10. Diagram showing the transport of the coat glycoprotein G of the vesicular stomatitis virus from the site of synthesis at the ER to the plasma membrane. This selective transport is mediated by clathrin coated vesicles. Observe that protein G is glycosidated at one (G_1) and two sites (G_2). The patches of membranes containing G_2 move into the Golgi complex and then, by means of coated vesicles, are integrated by exocytosis with the plasma membrane. At the later stage these patches are integrated with the virion (containing RNA and proteins), which buds out from the cell surface. (Drawing based on the work of J.R. Rothman and R.E. Fine: Proc. Natl. Acad. Sci. USA, 77:780, 1980.)

the transport by coated vesicles are energy dependent and can be stopped by inhibitors of respiration.

As a general conclusion, it can be said that by the use of coated vesicles the cell solves the problem of sorting out and of transporting different proteins and lipids. Thus, it can be considered that these vesicles are the key intermediaries in most sorting and transporting events within the eukaryotic cells.

A polarized epithelial cell may pose special problems to the intracellular traffic. In this case, membrane proteins may either be segregated toward the luminal or the basolateral domains of the cell surface. As mentioned in Chapter 5, in polarized epithelial cells tight junctions are important in establishing such membrane domains. These domains can be demonstrated by the use of two viruses that have different budding polarities. For example, while vesicular stomatitis virus buds and the G-glycoprotein are released at the basolateral side (Fig. 10–10), the *influenza hemagglutinin virus* is transported directly to the luminal surface (see Sabatini et al., 1984 in the Additional Readings at the end of this chapter). These results suggest that glycoproteins may exit from the Golgi and are deliv-

ered by carrier vesicles to specific cell membrane domains (see Bretscher, 1984; Rothman and Lenard, 1984 in the Additional Readings at the end of this chapter).

The Endosome and the Sorting of Ligands and Receptors

In recent years, there has been increasing evidence about the presence of another organelle, the *endosome* or *receptosome*, related to the intracellular traffic and the sorting of ligands and receptors.

Endocytosis by clathrin coated vesicles is, in many cases, followed by the presence of uncoated vesicles and tubules of 200 to 400 nm that move by saltatory motion and reach the Golgi region where they fuse with tubular elements present in the *trans* region of this complex. This region also called the *transreticular Golgi system* (TR-Golgi), may correspond to the GERL region (see Fig. 8–2).

One important characteristic of endosomes is their rapid acidification with a pH between 5 and 5.5 (thus in the cell there are two acidic compartments, the endosomes and the lysosomes). Acidification of endosomes can be studied by the endocytosis of fluorescein bound to dextran. The intravesicular pH is then determined by following the changes in fluorescence spectrum. It can be demonstrated that, as soon as the coated vesicles are transformed into endosomes, they become acidified, a process that is ATP-dependent (see Sly and Doisy, 1984 in the Additional Readings at the end of this chapter).

In the endocytosis of LDL, it can be demonstrated that while the ligand finds its way to the lysosome and is degraded, the LDL receptor is spared from proteolysis and is recycled to the plasma membrane, to be used again and again in the transport of LDL. The rescue of the receptor is carried out in the endosome. It is thought that with the lowering of the pH, the LDL becomes unbound from the receptor, which remains in the membrane, being transported back to the cell membrane. The content of the vesicle with the LDL is transferred by fusion into the lysosomes (Fig. 10–8).

The major role of the endosome would be that of permitting the entrance of ligands into the cell and avoiding early fusion with the lysosome. The material that is endocytosed by a nonreceptor mechanism, however, would quickly reach the lysosomal compartment (see Willingham et al., 1981; Pastan et al., 1984 in the Additional Readings at the end of this chapter). It should be emphasized here that endosomes also differ from lysosomes in that they lack degradative enzymes.

The study of the endocytosis of several macromolecules has revealed that the mechanism of degradation may differ in every case. For example, both the ligand and the receptor of *epidermic growth factor* are degraded in lysosomes. On the other hand, *transferrin*, the iron transporting protein is not degraded after its incorporation by endocytosis and its receptor is spared from degradation. In this case, only the iron is separated from the transferrin to be incorporated into the cytosol.

One of the best examples of receptor-mediated endocytosis is provided by the penetration of asialoglycoproteins (ASGP) from the blood into the liver cells. While the degradation of the ASGP is rapid (it occurs in a matter of minutes), the half life of the receptor is 40 hours or longer. The explanation for this discrepancy has been obtained from follow-up EM studies

of the ligand and the receptor labeled with different immunochemical markers. The ASGP binds to clathrin-coated pits that contain the receptor. Then it is taken up by endocytosis into coated vesicles, followed by tubular vesicle structures that become uncoated. Such organelles have been recently called *CURL* (i.e., compartments for uncoupling receptors and ligands).[20] In these, the vesicular part contains the ligand and fuses with lysosomes, in which the ASGP is degraded. The tubular portions (which may correspond to endosomes) are enriched with receptor, and probably function as intermediates in recycling of the receptor to the cell surface.

In conclusion: The endosome represents a special compartment intercalated in the intracellular traffic, which is characterized by the lack of clathrin and degradative enzymes and the acid pH. This compartment enters in connection with the transreticular Golgi region and is involved in the recycling of certain receptors and ligands.

SUMMARY:
Coated Vesicles and Selective Transport

Coated vesicles of 50 to 250 nm are observed in the cytoplasm of many cells, and are characterized by the presence of a coat of clathrin. Coated pits are also found at the plasma membrane. Coated vesicles appear to be involved in the selective transport of material to and from the plasma membrane, as well as between intracellular compartments (e.g., ER, Golgi). Clathrin pits and vesicles function in the uptake of yolk, transfer of immunoglobulins, recycling of membranes, and the entrance of growth factors, insulin, and LDL. In all of these mechanisms special receptors are present at the coated pit-vesicle system. Besides selective endocytosis, coated vesicles intervene in transport between endomembranes. A special example is provided by protein G, which makes the coat of the VSV. This protein is transported from the ER to the Golgi and then to the plasma membrane by clathrin vesicles. Transport by coated vesicles is energy dependent.

10–5 PEROXISOMES

The use of cell fractionation methods has been instrumental in the isolation of another group of particles (in addition to lysosomes) from liver cells and other tissues. Such particles, isolated by De Duve and co-workers in 1965, were enriched with some oxidative enzymes, such as peroxidase, catalase, D-amino oxidase, and urate oxidase. The name *peroxisome* was applied because this organelle is specifically involved in the formation and decomposition of hydrogen peroxide (H_2O_2). A second function of peroxisomes, discovered in 1976, was the β-oxidation of fatty acids (see Lazarow, 1981 in the Additional Readings at the end of this chapter).

The morphological characterization of isolated peroxisomes led to their being related to the microbodies found earlier by electron microscopy. Peroxisomes have also been found in protozoa, yeast, and many cell types of higher plants.

In plant cells, some organelles show morphological similarities to the peroxisomes of animal cells, but their enzymatic make-up includes the enzymes of the glyoxylate cycle, hence their name *glyoxysomes*. As it will be described in Section 12–2, glyoxysomes are related to the metabolism of triglycerides. The glyoxylate cycle allows fungi, protozoa, and plants to convert fat into carbohydrates. Glyoxysomes also contain enzymes for the oxidation of alcohols and alkanes, which may be induced by the substrate. For example, in yeast grown in methanol, as much as 80% of the cell volume may be occupied by peroxisomes. These and other observations demonstrate that peroxisomes are multi-purpose organelles, often containing inducible enzymes, that are of great importance in plants, fungi, and protozoa.

Morphology of Peroxisomes and Microperoxisomes

Peroxisomes are ovoid granules limited by a single membrane; they contain a fine, granular substance that may condense in the center, forming an opaque and homogeneous core (Fig. 9–3). In a quantitative study on rat liver cells the average diameter of peroxisomes was shown to be 0.6 to 0.7 μm. The number of peroxisomes per cell varied between 70 and 100, whereas 15 to 20 lysosomes were found per liver cell. In many tissues peroxisomes show a crystal-like body made of tubular subunits. The number of organelles having these bodies is sometimes correlated with the content of urate oxidase.

In contrast to the core-containing peroxisomes found in liver and kidney are others that are smaller and lack a core. These *microperoxisomes* are found in all cells.

Both peroxisomes and microperoxisomes can be demonstrated under the EM using the histochemical reaction for peroxidase, in which the oxidation of 3'-3' diaminobenzidine (DAB) is carried out in the presence of H_2O_2.[23] A more specific method consists of the localization of *catalase* by inmunocytochemistry. Using anti-catalase antibodies, labeled with horseradish peroxidase or ferritin, it is possible to detect the sites in which catalase is deposited.[24]

This enzyme has been found in the matrix of the peroxisomes, but not in the cisternae of the ER or the Golgi complex (Fig. 10–11). These findings support the view that catalase is not synthesized in the ER but in the cytosol.

Biogenesis of Peroxisomes—A Mixed Model

The mechanism of biogenesis of peroxisomes is complex and not completely known. It is possible that the peroxisomal membrane proteins are synthesized in the ER, but both organelles have different protein composition. In the past it was considered that peroxisomes were formed as dilatations or "buds" from the ER, which became swollen and filled with electron dense material. However, most peroxisomes exist without connections to the ER but with transient interconnections between themselves (see Lazarow, 1981; Fujiki et al., 1984 in the Additional Readings at the end of this chapter).

Peroxisomal, as well as glyoxysomal, enzymes are synthesized in the cytosol on free ribosomes and are then incorporated, post-transcriptionally, into these organelles.

The main peroxisomal enzyme, *catalase* is made as a monomeric precursor (apoenzyme) which, after being in the cytosol with a half life of 14 minutes, is translocated into the peroxisome. Once inside the organelle, the *heme*, carrying the iron, is added and the final enzyme (holoenzyme), with its tetrameric form, becomes active. The half life of catalase has been calculated as 36 hours.[25] It is assumed that the peroxisome as a whole grows slowly and is destroyed, probably by autophagy, with a life span of 5 to 6 days.[26]

The process by which catalase (and also urate oxidase) is translocated into peroxisomes corre-

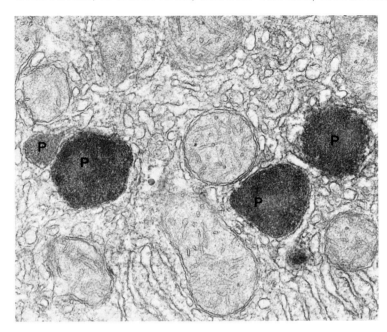

Fig. 10–11. Electron-micrograph of a liver cell immunochemically stained for the localization of catalase in peroxisomes. Note the close association of perixosomes (P) with the ER and the lack of diffuse cytoplasmic staining. Several mitochondria are observed. × 30,000. (Courtesy of H.F. Fahimi and S. Yokota.)

sponds to that of a post-translational transfer of protein (see Section 8–5).

These studies have led to a *mixed model of peroxisome biogenesis*, in which: (1) their membrane proteins are mainly synthesized on membrane-bound ribosomes; (2) the peroxisomal enzymes are made in the cytosol on free ribosomes and are translocated into the organelle.[27]

Peroxisomes Contain Enzymes Related to the Metabolisms of H_2O_2 and β-Oxidation

Isolation of liver peroxisomes has demonstrated that these organelles contain four enzymes related to the metabolism of H_2O_2. In fact, three of them— urate oxidase, D-amino oxidase, and α-hydroxylic acid oxidase—produce peroxide (H_2O_2), and catalase destroys it. *Catalase* is in the matrix of liver peroxisomes and represents up to 40% of the total protein. The enzyme *urate oxidase* and two other enzymes present in amphibian and avian peroxisomes are related to the catabolism of purines.

Another function of peroxisomes is the β-oxidation of fatty acids, which is also carried out in mitochondria.

The peroxisomal β-oxidation cycle comprises several enzymes (i.e., acyl-CoA oxidase,[28] enoyl-CoA hydratase, 3-hydroxyacyl-CoA and 3-ketoacyl-CoA thiolase) that serve to activate and oxidize fatty acids and to produce acetyl-CoA for anabolic reactions. Because of these functions peroxisomes play a role in *thermogenesis* and are concentrated in brown fat, which is abundant in hibernating animals.

Several drugs, including several hypolipemic compounds (which reduce lipids in the blood) such as *clofibrate*, and certain phthalate esters, used as plasticizers, have strong proliferative properties in hepatic peroxisomes. The percent volume of peroxisomes may increase from 2 to 17%, and catalase and especially the β-oxidation enzymes may increase considerably. There is also an increase in the amount of messenger RNAs, indicating that the action of these drugs is held at the transcriptional level (see Osumi and Hashimoto, 1984; Reddy, 1984 in the Additional Readings at the end of this chapter).

As in catalase, the β-oxidation enzymes are synthesized on free ribosomes, and they are packaged into peroxisomes without undergoing a proteolytic processing.

In addition to their role in β-oxidation of fatty acids, it has recently been found that perixosomal enzymes are involved in the initial steps of alkoxy-phospholipid biosynthesis, which leads to the production of plasmalogens; these are important components of biological membranes. (For a recent discussion on peroxisomes and lipid metabolism see Masters and Crane, 1984; Osumi and Hashimoto, 1984 in the Additional Readings at the end of this chapter).

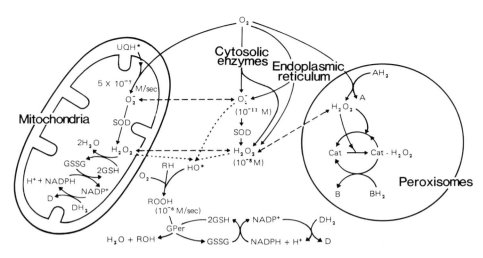

Fig. 10–12. General overview of the pathways for H_2O_2 and O_2^- (superoxide anion) inside the cell. It is observed that the peroxisome is the main source of H_2O_2 and that this is degraded by catalase *(Cat.)*. *GSH,* reduced glutathione; *GSSG,* oxidized glutathione; *G per,* glutathione peroxidase; *B and BH_2,* hydrogen donors; *SOD,* superoxide dismutase; *UQH*,* ubiquinone radical; *DH_2 and D,* a non-specific NADP reducing system. See the text for details. (Courtesy of B. Chance and A. Boveris.)

The enzymatic composition of peroxisomes varies according to the tissue, and the steps in their metabolic pathways are interrelated with other steps taking place in the cytosol or in the mitochondria. Figure 10–12 gives an overview of the pathways involved in the production of *hydrogen peroxide,* as well as of *superoxide anion* (O_2^-), which are normal metabolites in aerobic cells. The peroxisome is the main source of H_2O_2; however, only a fraction of this H_2O_2 diffuses to the cytosol, and most of it is degraded by the intraorganellar catalase.

Most cytosolic H_2O_2 is produced by mitochondria and membranes of the ER, although there are also H_2O_2-producing enzymes localized in the cytoplasmic matrix. Catalase acts as a "safety valve" for dealing with the large amounts of H_2O_2 generated by peroxisomes; however, other enzymes such as *glutathione peroxidase,* are capable of metabolizing organic hydroperoxides, and also H_2O_2, in the cytosol and mitochondria. The production of superoxide anion in mitochondria and cytosol is regulated mainly by the enzyme *superoxide dismutase.* All of these protective enzymes are present in high levels in aerobic tissues.[29]

In recent years the possible relationship between peroxides and free radicals (such as O_2^-), with the process of cell aging is being emphasized. Such radicals acting on DNA could produce mutations, altering the transcription into mRNA and the translation into proteins. In addition free radicals and peroxides can affect the membranes by causing peroxidation of lipids and proteins. For these reasons reducing compounds such as vitamin E or enzymes like superoxide dismutase could play a role in keeping the healthy state of a cell.

As in the case of lysosomes certain familial diseases may throw light on the study of peroxisomes. In the *cerebro-hepato-renal syndrome* of Zellweger, a rare disease that produces death within a year after birth, the liver and kidney lack peroxisomes. The finding of many metabolic changes in this syndrome, including alterations in the synthesis of bile acid and cholesterol esters, may lead to a more complete knowledge of the function of peroxisomes. In fact, inborn errors of metabolism are useful tools to analyze many aspects of the function of cellular organelles (see Borst, 1983 in the Additional Readings at the end of this chapter).

Plant Peroxisomes Are Involved in Photorespiration

In green leaves there are peroxisomes that carry out a process called *photorespiration.* In this process, *glycolic acid,* a two-carbon product of photosynthesis that is released from chloroplasts, is oxidized by *glycolic acid oxidase,* an enzyme present in peroxisomes. This oxidation, carried out by oxygen, produces hydrogen peroxide, which is then decomposed by catalase inside the peroxisome. Photorespiration is so called because light induces the synthesis of glycolic acid in chloroplasts. The entire process involves the intervention of two basic organelles: chloroplasts and peroxisomes.

Catalase, a typical peroxisomal enzyme, has been found in microbodies isolated from certain fungi zoospores, as well as *malate synthetase* and *isocitrate lyase*, two typical enzymes of glyoxysomes (see Powell, 1976 in the Additional Readings at the end of this chapter). *Glyoxysomes*, which are special plant cell organelles involved in the metabolism of stored lipids, will be studied in Chapter 12, together with the plant cell from which they originate (see Tolbert and Essner, 1981 in the Additional Readings at the end of this chapter).

SUMMARY:
Peroxisomes

Peroxisomes are organelles rich in peroxidase, catalase, D-amino-acid oxidase, and urate oxidase. They are abundant in the liver, kidney, and in many cell types of animals and plants. They have 0.6 to 0.7 μm granules with a single membrane and a dense matrix. Frequently, a crystal-like condensation is observed.

Many cells contain microperoxisomes that can be demonstrated by the DAB reaction of peroxidases. The use of a specific immunochemical method for catalase also serves to detect peroxisomes and microperoxisomes. The enzymes of this organelle are synthesized in the cytosol on free ribosomes and are then transferred by a post-translational mechanism. A mixed model for the biogenesis of peroxisomes has been postulated.

Three of the enzymes—urate oxidase, D-amino oxidase, and α-hydroxylic acid oxidase—produce H_2O_2 and catalase decomposes it. These organelles are also involved in β-oxidation of fatty acids and play a role in thermogenesis. The metabolic pathways of peroxisomes are interrelated with steps taking place in the cytosol and mitochondria (Fig. 10–12). Catalase acts a "safety valve" to deal with peroxides that are dangerous to the cell.

In plants peroxisomes carry out the process of photorespiration, which involves the cooperation of chloroplasts and peroxisomes. *Glyoxysomes* are special plant organelles involved in the metabolism of stored lipids (see Chapter 12).

REFERENCES

1. De Duve, C.: General properties of lysosomes. *In* Lysosomes. Ciba Foundation Symposium. London, J. & A. Churchill, 1963, p. 1.
2. Allison, A.C.: Lysosomes. *In* Oxford Biology Readers. Vol. 58. Oxford, Oxford University Press, 1974.
3. De Duve, C.: Les lysosomes. La Recherche, 5:815, 1974.
4. Essner, E., and Novikoff, A.B.: J. Cell Biol., 15:289, 1962.
5. Deter, R.L., Baudhuin, P., and De Duve, C.: J. Cell Biol., 35:11C, 1967.
6. Erickson, A., and Blobel, G.: J. Biol. Chem., 254:1771, 1979.
7. Rosenfeld, M., et al.: J. Cell Biol., 87:308a, 1980.
8. Fischer, H.D., et al.: J. Biol. Chem., 255:9608, 1980.
9. De Robertis, E.: Anat. Rec., 80:219, 1941.
10. Simionescu, N.: Eur. J. Cell Biol., 22 (1):180, 1980.
11. Pearse, B.M.F.: Proc. Natl. Acad. Sci. USA, 73:1255, 1976.
12. Heuser, J.: J. Cell Biol., 84:560, 1980.
13. Ungewickell, E., and Branton, D.: Nature, 289:420, 1981.
14. Uname, E., Ungewickell, E., and Branton, D.: Cell, 26:439, 1981.
15. Crowther, R.A., and Pearse, B.M.F.: J. Cell Biol., 91:790, 1981.
16. Bretscher, M.S., Thompson, J.N., and Pearse, B.M.F.: Proc. Natl. Acad. Sci. USA, 77:4156, 1980.
17. Goldstein, J.L., Anderson, R.G., and Brown, M.S.: Nature, 279:679, 1979.
18. Merisko, E.M., Farquhar, M.G., and Palade, G.E.: J. Cell Biol., 92:846, 1982.
19. Willingham, M.C., and Pastan, I.H.: Cell, 21:67, 1980.
20. Schmid, S.L., Matsumoto, A.K., and Rothman, J.E.: Proc. Natl. Acad. Sci. USA, 79:91, 1982.
21. Geuze, H.J., et al.: Cell, 32:277, 1983.
22. Rothman, J.E., and Fine, R.E.: Proc. Natl. Acad. Sci. USA, 77:780, 1980.
23. Yokota, S., and Fahimi, H.D.: J. Histochem. Cytochem., 28:613, 1981.
24. Fahimi, H.D., and Yokota, S.: *In* International Cell Biology. Edited by H. Schweiger. Berlin, Springer-Verlag, 1981, p. 640.
25. De Duve, C., and Baudhuin, P.: Physiol. Rev., 238:3952, 1966.
26. Poole, B., Leighton, F., and De Duve, C.: J. Cell Biol., 41:536, 1970.
27. Goldman, B.M., and Blobel, G.: Proc. Natl. Acad. Sci. USA, 75:5065, 1978.
28. Inestrosa, N.C., Bronfman, M., and Leighton, F.: Eur. J. Cell Biol., 22 (1):166, 1980.
29. Chance, B., Sied, H., and Boveris, A.: Physiol. Rev., 59:527, 1979.

ADDITIONAL READINGS

Allison, A.C.: Lysosomes. *In* Oxford Biology Readers. Vol. 58. Oxford, Oxford University Press, 1974.

Bainton, D.F., Nichols, B.A., and Farquhar, M.G.: Primary lysosomes of blood leukocytes. *In* Lysosomes in Biology and Pathology. Vol. 5. Edited by J.T. Dingle. Amsterdam, North-Holland Publishing Co., 1976, p. 3–32.

Borst, P.: Animal peroxisomes (microbodies) lipid biosynthesis and the Zellweger syndrome. Trends Biochem. Sci., 8:269, 1983.

Bretscher, M.S.: Endocytosis: Relation to capping and cell locomotion. Science, 224:681, 1984.

Brown, M.S., and Goldstein, J.L.: How LDL receptors influence cholesterol and atherosclerosis. Sci. Am., 251:58, 1984.

Brown, W.J., and Farquhar, M.G.: The mannose-6-phosphate receptor for lysosomal enzymes is concentrated in Cis Golgi cisternae. Cell, 36:295, 1984.

Callaham, J.W., and Lowden, J.A.: Lysosomes and Lysosomal Storage Diseases. New York, Raven Press, 1981.

De Duve, C.: Microbodies in the living cell. Sci. Am., 242:74, 1983.

De Duve, C.: Exploring the cell with a centrifuge. Science, 189:186, 1975.

De Duve, C.: Les lysosomes. La Recherche, 5:815, 1974.

Dingle, J.T., and Shaw, J.I.H.: Lysosomes in Applied Biology and Therapeutics. Vol. 6. Amsterdam, North-Holland Publishing Co., 1979.

Fujiki, T., Rachukinski, R.A., and Lazarow, P.B.: Synthesis of a major integral membrane polypeptide of rat liver peroxisomes on free polysomes. Proc. Natl. Acad. Sci. USA, 81:7127, 1984.

Geisow, M.: Lysosome proton pump identified. Nature, 298:511, 1982.

Geuze, H.J., et al.: Ultrastructural localization of the mannose-6-phosphate receptor in rat liver. J. Cell Biol., 98:2047, 1984.

Hers, H.G., and Van Hoof, F.: Lysosomes and Storage Diseases. New York, Academic Press, 1973.

Kindl, H. The biosynthesis of microbodies (Peroxisomes, glyoxysomes). Int. Rev. Cytol., 80:193, 1982.

Lazarow, P.B.: Functions and biogenesis of peroxisomes. In International Cell Biology. Edited by H.C. Schweiger. Berlin, Springer-Verlag, 1981, p. 633.

Masters, C., and Crane, D.: The role of peroxisomes in lipid metabolism. Trends Biochem. Sci., 9:315, 1984.

Masters, C., and Holmes, R.: Peroxisomes: New aspects of cell physiology and biochemistry. Physiol. Rev., 57:816, 1977.

Novikoff, A.B.: Golgi apparatus and lysosomes. Fed. Proc., 23:1010, 1964.

Novikoff, A.B., and Novikoff, P.M.: Microperoxisomes. J. Histochem. Cytochem., 21:963, 1973.

Ohkuma, S., Moriyama, Y., and Takano, T.: Identification and characterization of proton pump on lysosomes by fluorescence in isothiocyanate dextran fluorescence. Proc. Natl. Acad. Sci. USA, 79:758, 1982.

Osumi, T., and Hashimoto, T.: The inducible fatty acid oxidation system in mammalian peroxisomes. Trends Biochem. Sci., 9:317, 1984.

Pastan, I., and Willingham, M.C.: Journey to the center of the cell: Role of the receptosome. Science, 214:504, 1981.

Pastan, I., et al.: Internalization and intracellular sorting of ligands and receptors. In III International Congress

of Cell Biology. Edited by S. Seno and Y. Okada. Japan, Academic Press Inc., 1984, p. 69.

Pearse, B.M.F.: Coated vesicles. Trends. Biochem. Sci., 5:13, 1980.

Powell, M.J.: Ultrastructure and isolation of glyoxysomes (microbodies) in zoospores of the fungus. Entophlyctis sp. Protoplasma, 89:1, 1976.

Reddy, J.K.: Induction of peroxisome proliferation in liver by hypolipemic drugs and phthalate ester plasticizers. In III International Congress of Cell Biology. Edited by S. Seno and Y. Okada. Japan, Academic Press Inc., 1984, p. 103.

Reijngoud, D.J.: The pH and transport of protons in lysosomes. Trends Biochem. Sci., 3:178, 1978.

Roth, T.F., et al.: Coated vesicles structure and function in maternal-fetal transport. In III International Congress of Cell Biology. Edited by S. Seno and Y. Okada. Japan, Academic Press Inc., 1984, p. 66.

Rothman, J., and Lenard, J.: Membrane traffic in animal cells. Trends Biochem. Sci., 9:176, 1984.

Sabatini, E., et al.: Biogenesis of plasma membranes in polarized epithelial cells. In III International Congress of Cell Biology. Edited by S. Seno and Y. Okada. Japan, Academic Press Inc., 1984, p. 93.

Schneider, Y.J., Octave, J.N., Limet, J.N., and Trouet, A.: Functional relationship between cell surface and lysosomes during pinocytosis. In International Cell Biology. Edited by H.C. Schweiger. Berlin, Springer-Verlag, 1981, p. 590.

Silverstein, S.C., Steinman, R.M., and Cohn, Z.A.: Endocytosis. Annu. Rev. Biochem., 46:669, 1977.

Slayter, A.: Clathrin lock up vesicle structure. Nature, 298:221, 1982.

Sly, W.S., and Doisy, E.A.: ATP-dependent acidification of endosomes. In III International Congress of Cell Biology. Edited by S. Seno and Y. Okada. Japan, Academic Press Inc., 1984, p. 66.

Sly, W.S., and Fisher, H.D.: The phosphomannosyl recognition system. J. Cell Biochem., 11:67, 1982.

Steinman, R.M., Mellman, I.S., Muller, W.A., and Cohn, Z.A.: Endocytosis and the recycling of plasma membrane. J. Cell Biol., 96:1, 1983.

Tolbert, N.E., and Essner, E.: Microbodies: Peroxisomes and glyoxysomes. J. Cell Biol., 92:846, 1981.

Vigil, E.L.: Plant microbodies, J. Histochem. Cytochem., 21:958, 1973.

Walters, M.N.I., and Papdimitrion, J.M.: Phagocytosis: A review. CRC Crit. Rev. Toxicol., 5:377, 1978.

Willingham, M.C., Haigler, H.T., Dickson, R.B., and Pastan, I.H.: Receptor-mediated endocytosis in cultured cells; coated pits, receptosomes and lysosomes. In International Cell Biology. Edited by H.C. Schweiger. Berlin, Springer-Verlag, 1981, p. 63.

MITOCHONDRIA AND OXIDATIVE PHOSPHORYLATION

We mentioned in Chapter 1 that all cells and organisms can be grouped into two main classes that differ in the mechanism of extracting energy for their own metabolism. In the first class, called *autotrophs*, CO_2 and H_2O are transformed by the process of *photosynthesis* into the elementary organic molecule *glucose*, from which more complex molecules are made (Fig. 1–3). The second class of cells, called *heterotrophs*, obtains energy from the different nutrients (carbohydrates, fats, and proteins) synthesized by autotrophic organisms. The energy contained in these organic molecules is released mainly by combustion with O_2 from the atmosphere in a process called *aerobic respiration*. The release of H_2O and CO_2 by heterotrophic organisms completes this cycle of energy.

Energy transformation in cells takes place

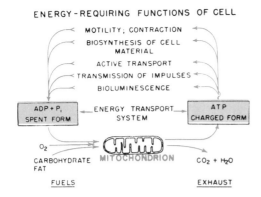

Fig. 11–1. Diagram showing that the mitochondrion constitutes the central "power plant" of the cell. The ATP produced is used in the various functions that are indicated. (From Lehninger, A.L.: Physiol. Rev., *42*:3, 467, 1962.)

with the intervention of two main *transducing systems* (i.e., systems that produce energy transformations) represented by *mitochondria* and *chloroplasts*. These two organelles in some respects function in opposite directions. Chloroplasts, present only in plant cells, are specially adapted to capture light energy and to transduce it into chemical energy, which is stored in covalent bonds between atoms in the different nutrients or *fuel molecules*. On the other hand, mitochondria are the "power plants" that, by oxidation, release the energy contained in the fuel molecules and make other forms of chemical energy (Fig. 11–1). The main function of chloroplasts is *photosynthesis*, while that of mitochondria is *oxidative phosphorylation*. Table 11–1 shows some of the basic differences between these two transducing systems. Note that photosynthesis is an *endergonic* reaction, which means that it captures energy; oxidative phosphorylation is an *exergonic* reaction, meaning that it releases energy.

TABLE 11–1. DIFFERENCES BETWEEN PHOTOSYNTHESIS AND OXIDATIVE PHOSPHORYLATION

Photosynthesis	Oxidative Phosphorylation
Only in presence of light; thus periodic	Independent of light; thus continuous
Uses H_2O and CO_2	Uses molecular O_2
Liberates O_2	Liberates CO_2
Hydrolyzes water	Forms water
Endergonic reaction	Exergonic reaction
$CO_2 + H_2O$ + energy $\rightarrow$ foodstuff	Foodstuff + $O_2 \rightarrow CO_2$ + H_2O + energy
In chloroplasts	In mitochondria

Mitochondria (Gr., *mito-*, thread + *chondrion*, granule), granular or filamentous organelles present in the cytoplasm of protozoa and animal and plant cells, are characterized by a series of morphological, biochemical, and functional properties. Among these are their size and shape, visibility in vivo, special staining properties, specific structural organization, lipoprotein composition, and content of a large "battery" of enzymes and coenzymes that interact to produce cellular energy transformations.

First observed at the end of the 19th century and described as "bioblasts" by Altmann (1894), these structures were called "mitochondria" by Benda (1897). Altmann predicted the relationship between mitochondria and cellular oxidation, and Warburg (1913) observed that respiratory enzymes were associated with cytoplasmic particles.

A most important advance was the first isolation of liver mitochondria by Bensley and Hoerr in 1934. This established the possibility of a direct study of the organelles by biochemical methods. The final demonstration that the mitochondrion was indeed the site of cellular respiration was made in 1948 by Hogeboom and co-workers.[1] In recent years important advances in the study of its ultrastructural organization have been made with the aid of the electron microscope (EM). The study of this organelle is particularly thrilling because the mitochondrion is one of the best known examples of structural-functional integration within the cell.

The objectives of this chapter are: (1) to demonstrate the close relationship between molecular organization and respiratory function; (2) to study the ultrastructure and compartmentalization of mitochondria, as well as the isolation of mitochondrial compartments; and (3) to explain the coupling of oxidation and phosphorylation on the basis of the localization of the enzymes and the asymmetry of the inner mitochondrial membrane. That both mitochondria and chloroplasts contain DNA and ribosomes, and perform a certain amount of local protein synthesis will also be discussed. These observations have given rise to the *symbiont hypothesis* for these organelles.

For many years, we believed that the mitochondrion was a completely closed organelle surrounded by two distinct membranes, and had a high degree of autonomy within the cell. This model of structure and function has recently been revised, however, and many observations suggest that mitochondria are dynamic and may interact extensively with other cellular structures. Furthermore, most of the several hundred mitochondrial proteins are coded by nuclear genes and transported into mitochondria (see Yaffe and Schatz, 1984 in the Additional Readings at the end of this chapter).

11–1 MORPHOLOGY OF MITOCHONDRIA

Although the examination of mitochondria in living cells is somewhat difficult because of their low refractive index, they can be observed easily in cells cultured in vitro, particularly under darkfield illumination and phase contrast (Fig. 8–1A). This examination has been greatly facilitated by coloration with *Janus green*. The resultant greenish blue stain is due to the action of the cytochrome oxidase system present in mitochondria, which maintains the vital dye in its oxidized (colored) form. In the surrounding cytoplasm the dye is reduced to a colorless leukobase.

Fluorescent dyes, which are more sensitive, have been used in isolated mitochondria and in intact cultured cells. Such dyes are suitable for metabolic studies of mitochondria in situ.[2]

Micromanipulation has demonstrated that mitochondria are relatively stable. The *specific gravity* is greater than that of the cytoplasm. By ultracentrifugation of living cells mitochondria are deposited intact at the centrifugal pole.

In cultured fibroblasts, continuous and sometimes rhythmic changes in volume, shape, and distribution of mitochondria can be observed.

The two main types of mitchondrial motion are *displacement* from one part of the cell to another, mainly by cytoplasmic streaming, and *bending* and *stretching cycles*. It has been found that the zones of bending are constant and correspond to regions in which the mitochondrial crests are vesicular.[3]

Active changes in volume and shape of mitochondria may be caused by chemical, osmotic, and mechanochemical changes. In living cells, low amplitude contraction cycles associated with oxidative phosphorylation have been observed.

The *shape* of mitochondria is variable, but in general these organelles are *filamentous* or *gran-*

ular (Fig. 11–2). During certain functional stages, other derived forms may be seen. For example, a long mitochondrion may swell at one end to assume the form of *a club* or be hollowed out to take the form of a *tennis racket*. At other times mitochondria may become *vesicular* by the appearance of a central clear zone. The morphology of mitochondria varies from one cell to another, but it is more or less constant in cells of a similar type or in those performing the same function.

The *size* of mitochondria is also variable; however, in most cells the width is relatively constant (about 0.5 μm), and the length is variable, reaching a maximum of 7 μm. Mitochondria are, in general, uniformly distributed throughout the cytoplasm, but there are many exceptions to this rule. In some cases, they accumulate preferentially around the nucleus or in the peripheral cytoplasm. Overloading with inclusions, such as glycogen and fat, displaces these organelles. During mitosis, mitochondria are concentrated near the spindle, and upon division of the cell they are distributed in approximately equal number between the daughter cells.

The distribution of mitochondria within the cytoplasm should be considered in relation to their function as energy suppliers. In some cells they can move freely, carrying ATP where needed, but in others they are located permanently near the region of the cell where presumably more energy is needed. For example, in certain muscle cells mitochondria are grouped like rings or braces around the I-band of the myofibril. The basal mitochondria of the kidney tubule are intimately related to the infoldings of the plasma membrane in this region of the cell. It is assumed that this close relationship with the membrane is related to the supply of energy for the active transport of water and solutes.

Mitochondria may have a more or less definite orientation. For example, in cylindrical cells they are generally oriented in the basal-apical direction, parallel to the main axis. In leukocytes, mitochondria are arranged radially with respect to the centrioles. The mitochondrial content of a cell is difficult to determine, but, in general, it varies with the cell type and functional stage. It is estimated that in liver, mitochondria constitute 30 to 35% of the total protein content of the cell, and in kidney, 20%.

A normal liver cell contains between 1000 and

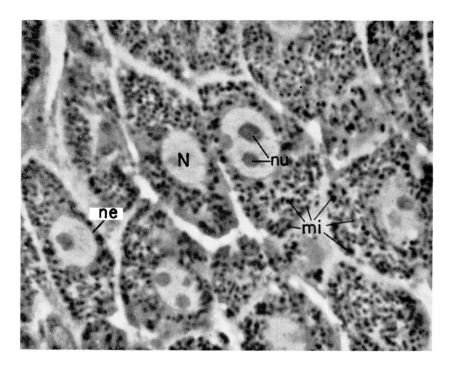

Fig. 11–2.   Liver cells of the rat fixed at −180°C and dehydrated in acetone at −40°C. Photomicrograph made with phase contrast microscope in a medium of n = 1.460. *mi*, mitochondria; *N*, nucleus; *ne*, nuclear envelope, *nu*, nucleoli. (Courtesy of S. Koulish.)

TABLE 11–2.   MEASUREMENTS IN RAT LIVER MITOCHONDRIA

	Peripheral Cells	Midzonal Cells	Central Cells
Cytoplasmic volume (%)	19.8	19.1	12.9
Number per cell	1060	1300	1600
Diameter (μm)	0.56	0.47	0.32
Length (μm)	3.85	4.32	5.04

From Loud, A.V.: J. Cell Biol., *37*:27, 1968.

1600 mitochondria (Table 11–2), but this number diminishes during regeneration and also in cancerous tissue.[4] This last observation may be related to decreased oxidation that accompanies the increase to anaerobic glycolysis in cancer. Another interesting finding is that there is an increase in the number of mitochondria in the muscle after repeated administration of the thyroid hormone, thyroxin. An increased number of mitochondria has also been found in human hyperthyroidism.

Some oocytes contain as many as 300,000 mitochondria. There are fewer mitochondria in green plant cells than in animal cells, since some of their functions are taken over by chloroplasts.

SUMMARY:
Mitochondrial Morphology

Mitochondria are organelles present in the cytoplasm of all eukaryotic cells. They provide an energy-transducing system by which the chemical energy contained in foodstuffs is converted, by oxidative phosphorylation, into high-energy phosphate bonds (ATP). Mitochondria may be observed in the living cell; visibility is increased by the vital stain, *Janus green*. They display passive and active motion and show changes in volume and shape that are related to their function. Swelling of mitochondria can be induced by Ca^{++}, various hormones, and certain drugs.

The morphology of mitochondria is best studied after fixation. In general, they are rod-shaped, with a diameter of about 0.5 μm and a variable length that may range up to 7 μm. There are 1000 to 1600 mitochondria in a liver cell and 300,000 in some oocytes. Green plants contain fewer mitochondria than animal cells. The distribution of mitochondria may be related to their function as suppliers of energy. Their orientation in the cell may be influenced by the organization of the cytoplasmic matrix and vacuolar system.

11–2   MITOCHONDRIAL STRUCTURE

As indicated in Figure 11–3, a mitochondrion consists of two membranes and two compart-

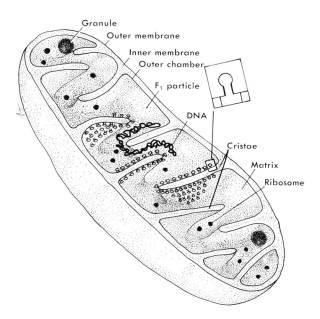

Fig. 11–3. Three-dimensional diagram of a mitochondrion cut longitudinally. The main features are shown. Observe that the cristae are folds of the inner membrane and that on their matrix side they have the F_1 particles. The inset shows an F_1 particle with the head piece and stalk.

ments. An outer limiting membrane, about 6 nm thick, surrounds the mitochondrion. Within this membrane, and separated from it by a space of about 6 to 8 nm, is an inner membrane that projects into the mitochondrial cavity complex infoldings called *mitochondrial crests*. This inner membrane divides the mitochondrion into two chambers or compartments: (1) the outer chamber contained between the two membranes and in the core of the crests and (2) the inner chamber filled with a relatively dense proteinaceous material usually called the *mitochondrial matrix*. This is generally homogeneous, but in some cases it may contain a finely filamentous material, or small, highly dense granules (see Figure 11–4).

These granules contain phospholipids, which give them an affinity for calcium when this is added to the fixative. There is no evidence, however, for the normal presence of calcium phosphate in the granules.

The mitochondrial crests are, in general, incomplete septa or ridges that do not interrupt the continuity of the inner chamber; thus, the matrix is continuous within the mitochondrion.

Within the mitochondrial matrix are small ribosomes (Fig. 11–4) and a circular DNA. The outer and inner membranes and the crests can be considered to be fluid molecular films with a compact molecular structure; the matrix is gel-like and contains a high concentration of soluble proteins and smaller molecules. This double (solid-liquid) structure is important in providing an explanation for some of the mechanical properties of mitochondria (e.g., deformation and swelling under physiological or experimental conditions).

F_1 Particles—On the M Side of the Inner Mitochondrial Membrane

The use of negative staining has enabled recogition of other details of mitochondrial structure. If a mitochondrion is allowed to swell and break in a hypotonic solution and is then immersed in phosphotungstate, the inner membrane in the crest appears covered by particles of 8.5 nm that have a stem linking each with the membrane (Fig. 11–5). These so-called "elementary" or "F_1" particles,[5] are regularly spaced at intervals of 10 nm on the inner surface of these membranes. According to some estimates, there are between 10^4 and 10^5 elementary particles per mitochondrion. These particles correspond to a special ATPase involved in the coupling of oxidation and phosphorylation.[6]

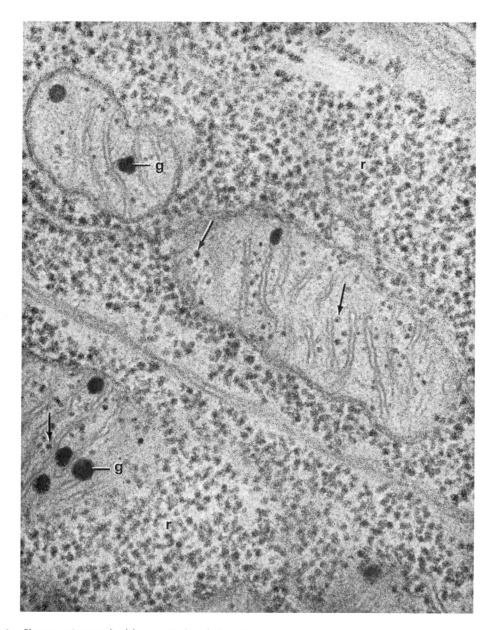

Fig. 11–4. Electron micrograph of the intestinal epithelium showing a large accumulation of ribosomes *(r)* in the cytoplasm. In mitochondria, dense granules *(g)* and ribosomes (arrows) are observed. ×95,000. (Courtesy of G.E. Palade.)

The presence of such F_1 particles on the matrix side (M side) confers to the inner mitochondrial membrane a characteristic asymmetry that is of fundamental importance to its function. It is interesting to recall here that bacterial plasma membranes contain F_1 particles with a similar asymmetric distribution (i.e., toward the bacterial matrix). On the other hand, the opposite orientation is found in chloroplasts (see Section 12–4), where there are F_1 particles on the outer or stromal side of the thylakoid membrane. As will be shown later, the asymmetric location of the F_1 particles in the membranes is related to the direction of the proton pump (i.e., the flow of H^+).

Mitochondrial Structural Variations in Different Cell Types

Some investigators assume that a common pattern of mitochondrial structure developed at

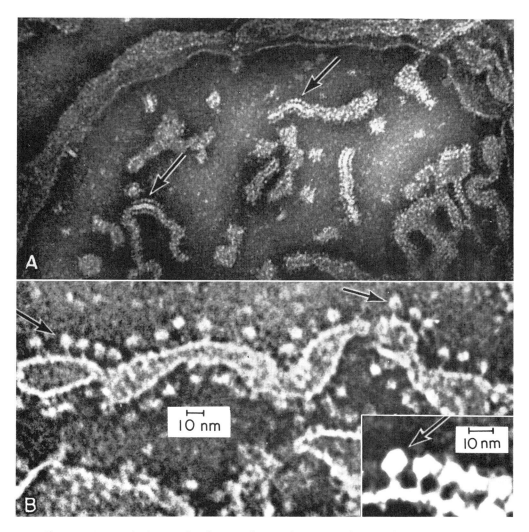

Fig. 11–5.   Electron micrograph of a mitochondrion swollen in a hypotonic solution and negatively stained with phospho-tungstate. *A,* at low power; isolated crests can be observed in the middle of the swollen matrix. Arrows point to some of these crests. *B,* at higher magnification (×500,000), a mitochondrial crest showing the so-called "elementary particles" on the surface adjacent to the matrix. Inset at ×650,000, showing the elementary particles with a polygonal shape and the fine attachment to the crest. (Courtesy of H. Fernández-Morán.)

an early stage of evolution and was subsequently transmitted, without considerable modifications, from protozoa to mammals and from algae to flowering plants. Detailed structural variations can be observed, however. The crests may be arranged longitudinally (e.g., in nerve and striated muscle) or they may be simple or branched. In protozoa, insects, and adrenal cells of the glomerular zone, the infoldings may be tubular, and the tubules may be packed in a regular fashion.[7]

The number of crests per unit volume of a mitochondrion is also variable. Mitochondria in liver and germinal cells have few crests and an abundant matrix, whereas those in certain muscle cells have numerous crests and little matrix. In some cases the crests are so numerous that they may have a quasi-crystalline disposition. The greatest concentration of crests is found in the flight muscle of insects. In general, there seems to be a correlation between the number of crests and the oxidative activity of the mitochondrion.

Relationship to Lipids. Various authors since the time of Altmann have observed that the *disposition of lipids* may be related to mitochondrial activity. In pancreas and liver cells, for example, after a short period of starvation, the mitochondria come into contact with lipid droplets. The relationship may be so tight that only the inner mitochondrial membrane can be seen adjacent to the lipid. The images suggest that an active process of fat utilization takes place under the action of the fatty acid oxidases present in mitochondria.

Intramitochondrial Inclusions. An accumulation of pigment derived from hemoglobin has been observed in mitochondria of amphibians. Ferritin molecules accumulate within mitochondria in subjects suffering from Cooley's hereditary anemia. Another example is the transformation of mitochondria into yolk bodies in eggs of mollusks. In these, masses of protein molecules may assume a regular crystalline disposition. In amphibian oocytes, crystalline yolk bodies also form within mitochondria.

Mitochondrial Sensitivity to Cell Injury and Resultant Degeneration

Mitochondria are one of the most sensitive indicators of injury to the cell. Although mitochondria may be readily altered by the action of various agents, the changes are, within certain limits, reversible. If the alteration reaches a certain critical point, however, it becomes irreversible, and this is generally considered degeneration of the mitochondria. Essentially there are three types of change: (1) fragmentation into granules, followed by lysis and dispersion, (2) intense swelling with transformation into large vacuoles, and (3) a great accumulation of materials with transformation of mitochondria into hyalin granules. This last change is characteristic of the so-called cloudy swelling and hyalin degeneration that frequently results in cellular death.

Another type of degeneration is the fusion of mitochondria to form large bodies called *chondriospheres*. This degeneration has been found in patients with scurvy and seems to be normal in the adrenal gland of the hamster.[8]

SUMMARY:
Structure of Mitochondria

The mitochondrion contains two compartments—an inner one filled with the mitochondrial matrix and limited by an inner membrane, and an outer one located between the inner and outer membranes. Both membranes have a trilaminar (unit membrane) structure; however, by negative staining the outer membrane is smooth but the inner membrane shows, on its inner surface, particles of 8.5 nm linked to the membrane (F_1 particles). It will be shown later that these particles contain a special ATPase. Complex infoldings of the inner membrane, called mitochondrial crests, project into the matrix. The shape and disposition of these crests vary in different cells and their number is related to the oxidative activity of the mitochondrion. Mitochondria may show a close relationship to lipid droplets. They may accumulate iron-containing pigments or protein molecules, thereby forming yolk bodies. Injury to mitochondria may produce degenerative changes consisting of fragmentation, intense swelling, or accumulation of material. Degenerating mitochondria may be found forming cytolysosomes or large chondriospheres.

11–3   ISOLATION OF MITOCHONDRIAL MEMBRANES

In recent years the two mitochondrial membranes and the compartments they limit have been separated.[9,10]

A so-called *mitoplast*, which includes the inner membrane and matrix, has been produced by separating these two elements from the outer membrane with digitonin (Fig. 11–6). The mitoplast has pseudopodic processes and is able to carry out oxidative phosphorylation.

Outer and Inner Membranes—Structural and Chemical Differences

The outer membrane fraction has a 40% lipid content (compared to 20% in the inner membrane), contains more cholesterol, and is higher in phosphatidyl inositol; on the other hand, it is lower in cardiolipin. These figures demonstrate that a fundamental difference between the outer and inner membrane is in the lipid/protein ratio (i.e., about 0.8 in the outer membrane and about 0.3 in the inner membrane). This low lipid/protein ratio indicates that in the inner membrane there is a greater degree of intercalation of the protein within the lipid bilayer. This fact has been confirmed by electron microscopic observations with freeze-fracturing methods.[11]

The outer membrane contains a major intrinsic protein of 29,000 daltons that is resistant to

Liver mitochondrion

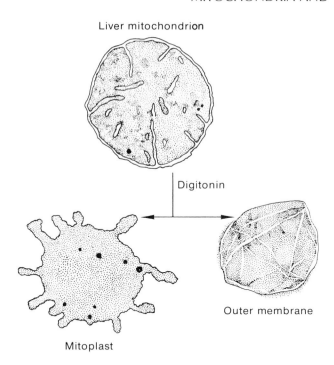

Digitonin

Mitoplast

Outer membrane

Fig. 11–6. Diagram showing an intact liver mitochondrion and its dissection by the action of digitonin. The mitoplast comprises the inner membrane with the finger-like unfolded mitochondrial crests and the matrix. The outer membrane revealed by negative staining shows a "folded-bag" appearance.

tryptic digestion and appears to form channels for the passage of solutes. Because of similarities in function with the *porin* present in the outer membrane of Gram-negative bacteria (see Fig. 1–4*B*), the same name has been given to this mitochondrial protein.[12] Porin appears to be synthesized on free ribosomes and then integrated, post-transcriptionally, into the outer mitochondrial membrane.[13]

As shown in Figure 11–7, in yeast the pattern formed by the polypeptides of the outer mitochondrial membrane is different from that of the inner mitochondrial membrane and of the rough endoplasmic reticulum (RER). The very prominent band of 29,000 daltons, porin is clearly observed in this SDS gel electrophoresis (see Reizman et al., 1983 in the Additional Readings at the end of this chapter).

From the morphological viewpoint the outer membrane lacks the elementary particles that are prominent in the inner membrane (Fig. 11–5). As shown in Figure 11–6 the outer membrane has a characteristic "folded bag" appearance in negatively stained preparations.

Mitochondrial Enzymes Are Highly Compartmentalized

Table 11–3 lists some of the enzymes present in the various mitochondrial fractions. The ac-

tivities of some enzymes that may be considered as markers of the fractions are presented in Table 11–4.

The *outer membrane* contains, as in the endoplasmic reticulum (ER), an NADH-cytochrome-c-reductase system that consists of a flavoprotein and cytochrome b_5. The most specific enzyme system of the outer membrane is *monoamine oxidase*, which may serve as an enzyme marker (Table 11–4).[10] This membrane also contains kynurenine hydroxylase, fatty acid coenzyme A ligase, a phospholipase, and various enzymes of the phospholipid metabolism. The supernatant, obtained after treatment with digitonin is considered to originate from the outer compartment. In this compartment are found adenylate kinase, nucleoside diphosphokinase; DNAse I, and 5′-endonuclease.

The *mitoplast fraction* contains the components of the respiratory chain and oxidative phosphorylation and the soluble enzymes present in the matrix.

The *inner membrane* carries all the components of the respiratory chain and the oxidative phosphorylation system. It also has several *carriers or translocators* for the permeation of phosphate, glutamate, aspartate, ADP, and ATP. It is calculated that the respiratory chain represents about 20% and the phosphorylating

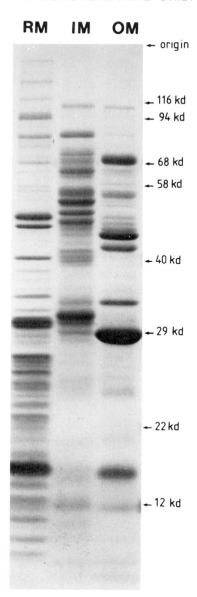

RM IM OM

← origin

← 116 kd
← 94 kd

← 68 kd

← 58 kd

← 40 kd

← 29 kd

← 22 kd

← 12 kd

Fig. 11–7. SDS polyacrylamide gel showing the different polypeptides present in the rough ER (RM), inner mitochondrial membrane (IM), and outer mitochondrial membrane (OM). Observe that the bands in the OM differ from those in the other membranes. The 29 Kd porin is the major protein in the OM. (Courtesy of H. Reizman, G. Schatz, and colleagues.)

TABLE 11–3. ENZYME DISTRIBUTION IN MITOCHONDRIA

Outer membrane
  Monoamine oxidase
  Rotenone-insensitive NADH-cytochrome-c
    reductase
  Kynurenine hydroxylase
  Fatty acid CoA ligase

Space between outer and inner membranes
  Adenylate kinase
  Nucleoside diphosphokinase

Inner membrane
  Respiratory chain enzymes
  ATP synthetase
  Succinate dehydrogenase
  β-Hydroxybutyrate dehydrogenase
  Carnitine fatty acid acyl transferase

Matrix
  Malate and isocitrate dehydrogenases
  Fumarase and aconitase
  Citrate synthetase
  α-Keto acid dehydrogenases
  β-Oxidation enzymes

Courtesy of A.L. Lehninger.

all systems involving electron transport. The *inner compartment* or *mitochondrial matrix* contains all the soluble enzymes of the citric acid or Krebs cycle and those involved in the oxidation of fatty acids. The matrix also contains DNA, ribosomes, and other RNA species and enzymes involved in the synthesis of protein.

The numerous proteins present in the mitochondrial matrix are densely packed,[14] thus it is possible that the enzymes of the Krebs cycle may have specific interactions between them and also with components of the inner membrane. For example, in a heart mitochondrion it is calculated that there are about 20,000 Krebs cycle assemblies and many of them may be in contact with the inner membrane surface and may interact with components of the electron transport system. These interactions might be important in the control of aerobic oxidation and mitochondrial biogenesis.[15]

The Inner Membrane—Regional Structural and Enzymatic Differences

Various studies have led to the suggestion that the inner mitochondrial membrane has two major portions, one corresponding to the *mitochondrial crests* and the other to the region that lines the *outer membrane*. For example, it has

system about 15% of the total protein content of the membrane. These impressive figures give an idea of the importance of these systems in the inner membrane and mitochondrial crests.

The inner membrane contains a remarkably high concentration of cardiolipin (polyglycero phosphatides), which appear to be important in

TABLE 11–4. ENZYME ACTIVITY OF VARIOUS MITOCHONDRIAL COMPONENTS

Component	Inner Membrane	Outer Membrane	Matrix	Outer Chamber
Enzyme marker	Cytochrome oxidase	Monoamine oxidase	Malate dehy-drogenase	Adenylate kinase
Specific activity of marker in component	9315	551	3895	6690
Specific activity in whole mitochondria	1980	22	2608	421
% Protein in component	21.3	4.0	66.9	6.3

From Schnaitman, C.A., and Greenawalt, J.W.: J. Cell Biol., *38*:158, 1968.

been reported that the F_1 particles and the coupling factors are present only in the cristae (Fig. 11–3). Furthermore, the activities of several enzymes predominate in the cristae.[16]

Studying the density and distribution of anionic sites, it was found that while the crests have few sites, there are patches of high density in the regions lining the outer membrane. Another regional difference of the inner membrane involves its association with the outer membrane at the *contact zones*. These are observed especially when the mitochondria are in the condensed state (see Fig. 11–17B); there are about 100 such contacts per liver mitochondrion. At each of these points is a concentration of cytoplasmic ribosomes, which has led to the suggestion that the contact zones could be the sites through which cytoplasmic polypeptides enter the mitochondrion.

SUMMARY:
Isolation of Mitochondrial Membranes

Mitochondria, as well as their membranes and compartments, may be separated by subcellular fractionation. The so-called mitoplast is a mitochondrion from which the outer membrane has been stripped by osmotic action or by digitonin. The outer membrane comes off in a lighter fraction because it has a much higher lipid-protein ratio than the inner membrane. By negative staining it shows a smooth surface and a "folded bag" appearance.

The mitochondrial enzymes show a definite compartmentalization. The outer membrane contains NADH-cytochrome-c-reductase, which consists of a flavoprotein and cytochrome b_5. Monoamine oxidase is the specific enzyme marker of this membrane. The *outer compartment* contains adenylate kinase and other soluble enzymes. The *inner membrane* carries all the components of the respiratory chain and of oxidative phosphorylation. Taken together, these components represent 35% of the protein of the membrane. In addition to other bound enzymes the inner membrane contains several specific carriers or translocation proteins involved in the permeation of metabolites. The *mitochondrial matrix* contains soluble enzymes of the Krebs cycle, DNA, RNA, and other components of the machinery for protein synthesis of the mitochondrion.

11–4 BIOENERGETICS: MOLECULAR ORGANIZATION AND FUNCTION OF MITOCHONDRIA

The molecular organization of the mitochondria is very complex. Within the realm of this organelle there are more than 70 enzymes and coenzymes, as well as numerous cofactors and metals that operate in an orderly and integrated fashion. The whole metabolic machinery of the mitochondria needs only the entrance of phosphate, ADP, and acetylcoenzyme A (acetyl-CoA) to produce ATP, CO_2 and H_2O.

To understand mitochondrial function, besides some knowledge of the enzymatic mechanisms, we need to know how energy is transformed and how the chemical energy contained in foodstuffs is released.

The energy that the cell has at its disposal exists as chemical energy primarily locked in high energy bonds. The cell uses only part of the *total energy* (H), also called *enthalpy*, contained in a chemical compound. This portion of the total energy, the *free energy* (G), does not dissipate as heat. Expressed as an energy change:

$$\Delta H = \Delta G + T\Delta S$$

The equation shows that the change in total energy (ΔH) is equal to the change in free or available energy (ΔG) plus the unavailable energy ($T\Delta S$), which dissipates as heat. (In this equation, T is the temperature in degrees Kelvin, and S is the *entropy* of the system.)

Entropy is Related to the Degree of Molecular Order

It is important to have a clear idea of the role of entropy in biological systems. In the preceding equation ΔS is a measure of the irreversibility of a reaction. As the entropy increases, more energy ($T\Delta S$) becomes unavailable, and the process becomes less reversible.

According to the Second Law of Thermodynamics, the entropy of an isolated system of reactions tends to increase to a maximum, at which point an equilibrium is reached and the reaction stops. The concept of entropy is related to the ideas of "order" and "randomness." When there is an orderly arrangement of atoms in a molecule, the entropy is low. The entropy of the system increases when, during a chemical reaction, there is a tendency toward molecular disorder. Thermodynamically, it is well established that the flow of energy proceeds from a higher to a lower level, a phenomenon accompanied by increased entropy.

In any protein molecule the sequence of amino acids is precisely determined. Therefore, the molecule shows a high degree of order and low entropy. The synthesis of such a molecule from the individual amino acids requires considerable amounts of energy or "work" (an *endergonic reaction*). On the other hand, the breakdown of specific proteins into amino acids or into carbon dioxide and water is a highly irreversible process that gives up considerable energy (an *exergonic reaction*).

When energy is forced to flow in a reverse direction (from a lower to a higher level), the entropy decreases. Such processes are thermodynamically unlikely unless they are connected with another system in which the entropy increases accordingly, thus compensating for the decrease. In the plant cell, synthesis of glucose from carbon dioxide and water, simultaneously locking energy derived from the sun into the molecule, is accompanied by a decrease in entropy. On the other hand, in the oxidation of glucose in the animal cell there is a considerable increase in entropy. The interaction of the two systems thus satisfies the second law of ther-

modynamics, i.e., that entropy must always increase.

These concepts are of great importance in biological systems, since cells are characterized by a high degree of order expressed in their molecular and subcellular structure. When a cell dies, disintegration begins, and entropy increases.

Chemical Energy and ATP

The *chemical* or *potential* energy of foodstuffs is locked in the covalent bonds between the atoms of a molecule. For example, during the hydrolysis of typical chemical bonds, about 3000 calories per mole are liberated. In a mole ($\sim$ 180 g) of glucose, between the various atoms of C, H, and O, there are about 686,000 calories of potential energy that can be liberated by combustion, as in the following reaction:

$$C_6H_{12}O_6 + 6\ O_2 \rightarrow$$
$$6\ H_2O + 6\ CO_2 + 686,000\ \text{calories} \quad (1)$$

Within the living cell this enormous amount of energy is not released suddenly, as in combustion by a flame. It is made available in a stepwise and controlled manner, requiring a great number of enzymes that finally convert the fuel into CO_2 and H_2O.

The liberated energy may be used by the cell (1) to synthesize new molecules (i.e., proteins, carbohydrates, and lipids), by means of endergonic reactions, which can then be used to replace others or for the normal growth and metabolism of the cell; (2) to perform mechanical work such as cell division, cyclosis, or muscle contraction; (3) to carry out active transport against an osmotic or ionic gradient; (4) to maintain membrane potentials, as in nerve conduction and transmission, or to produce electrical discharges (e.g., in certain fish); (5) to perform cell secretion; or (6) to produce radiant energy (e.g., in bioluminescence). Only in the reactions of group (1) is the energy provided by nutrients transformed into chemical bond energy. In all the other types of reactions, chemical energy is transformed into other forms of energy.

The common link in all these transformations is the compound ATP. ATP is found in all cells. Its most significant chemical characteristic is that it has two terminal bonds with a potential energy much higher than that of all the other chemical bonds. ATP is composed of the purine base adenine, a ribose moiety, and three molecules of phosphoric acid. (Adenine plus ribose forms the nucleoside *adenosine*.) ATP and the closely related molecule ADP are the most important compounds in energy transformation. If phosphate is represented by P, the simplified formula of ATP and its transformation into ADP or AMP is as follows:

$$\boxed{\text{Adenosine}} \text{---} \textcircled{P}\sim\textcircled{P}\sim\textcircled{P} \ ; \rightleftharpoons \boxed{\text{Adenosine}}$$
$$\text{---} \textcircled{P}\sim\textcircled{P} \ + \ \text{Pi} \ + \ 7300 \text{ calories} \qquad (2)$$

or:

$$\boxed{\text{Adenosine}} \text{---} \textcircled{P}\sim\textcircled{P}\sim\textcircled{P} \rightleftharpoons \boxed{\text{Adenosine}}$$
$$\text{---} \textcircled{P} \ + \ \text{PPi} \ + \ 7300 \text{ calories} \qquad (3)$$

Note that the release of any one of the two terminal phosphates of ATP yields about 7300 calories per mole, instead of the 3000 calories from common chemical bonds. The high energy $\sim$P bond enables the cell to accumulate a great quantity of energy in a very small space and keep it ready for use whenever it is needed.

Other nucleotides having high energy bonds, such as cytosine triphosphate (CTP), uridine triphosphate (UTP), and guanosine triphosphate (GTP), are involved in biosynthetic reactions. The energy source for these nucleotide triphosphates derives ultimately from ATP, however.

Glycolysis—First Step in the Release of Energy

To study how the energy contained in foodstuffs is liberated, we must consider the glucose molecule. Under anaerobic conditions (i.e., in the absence of oxygen) glucose is degraded into lactate by a process called glycolysis (lysis or splitting of glucose). If glycolysis is carried out under aerobic conditions the final products are pyruvate and coenzyme NADH (Fig. 11–8). Glycolysis is achieved by a series of 10 enzymes, all of which are located in the cytosol. As shown in Figure 11–8, in this chain of reactions, the product of one enzyme serves as a substrate for the next reaction. To facilitate its analysis the sequence can be divided as follows: (1) in reactions 1 through 4 the glucose molecule is converted into two (3-carbon) *glyceraldehyde 3-phosphates*. This step requires the use of the

enzymes hexokinase, phosphoglucose isomerase, phosphofructokinase, and aldolase and the investment of the two ATP molecules; (2) in reactions 5 and 6 the aldehyde group of glyceraldehyde phosphate is oxidized to the carboxylic acid and one ATP molecule is produced (the enzymes involved in this step are triosephosphate isomerase and glyceraldehyde phosphate dehydrogenase); (3) in reactions 7,8,9, and 10 pyruvate is made and two more ATP molecules are synthesized. (The enzymes used are: phosphoglycerate kinase, phosphoglyceromutase, enolase, and pyruvate kinase.) As shown in Figure 11–8, the net energy yield of this chain of reactions is the production of two ATP molecules from one molecule of glucose.

Under aerobic conditions, the products are *pyruvate and coenzyme NADH*. NADH carries two electrons, taken from glyceraldehyde-3 phosphate, and contains little energy.

Under anaerobic conditions, pyruvate is used as a hydrogen acceptor and converted into lactate. In this case, the following equation represents the overall reaction of glycolysis:

$$C_6H_{12}O_6 \ + \ \text{Pi} \ + \ 2 \text{ ADP} \rightarrow$$
glucose
$$2 \ C_3H_6O_3 \ + \ 2 \text{ ATP} \ + \ 2 \text{ H}_2O \qquad (4)$$
lactate

Since pyruvate still contains a large amount of energy, it must undergo further degradation, but this time inside the mitochondrion. This is done in two consecutive steps: *the Krebs or tricarboxylic acid cycle* and *oxidative phosphorylation*. Pyruvate directly enters the mitochondrial matrix and is converted into *acetyl-CoA* by the huge enzyme *pyruvate dehydrogenase*. NADH cannot penetrate directly and their electrons are transferred to dihydroxyacetone phosphate, which shuttles them into the mitochondrion.

For anaerobic organisms and for tissues such as skeletal muscle, which can function for some time under anaerobic conditions, glycolysis is a major source of energy. In these conditions, pyruvate remains in the cytosol and can be converted into ethanol (as in yeast) or into lactate (muscle).

Coenzymes Play a Central Role in Mitochondria

Before considering the Krebs cycle and oxidative phosphorylation we should mention the

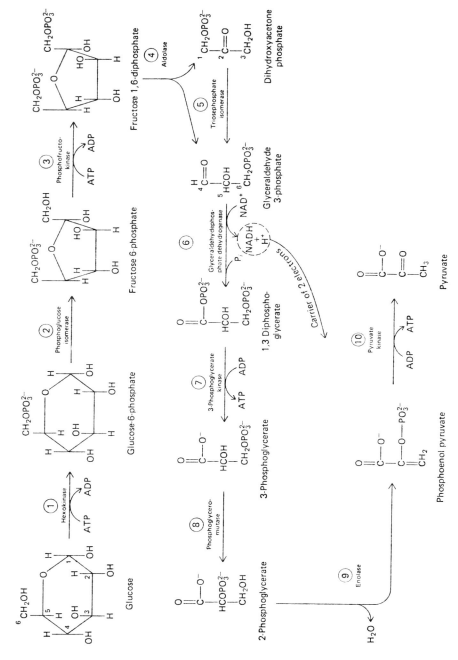

Fig. 11–8.   Diagram of anaerobic glycolysis—from glucose to pyruvate. (See text for details.)

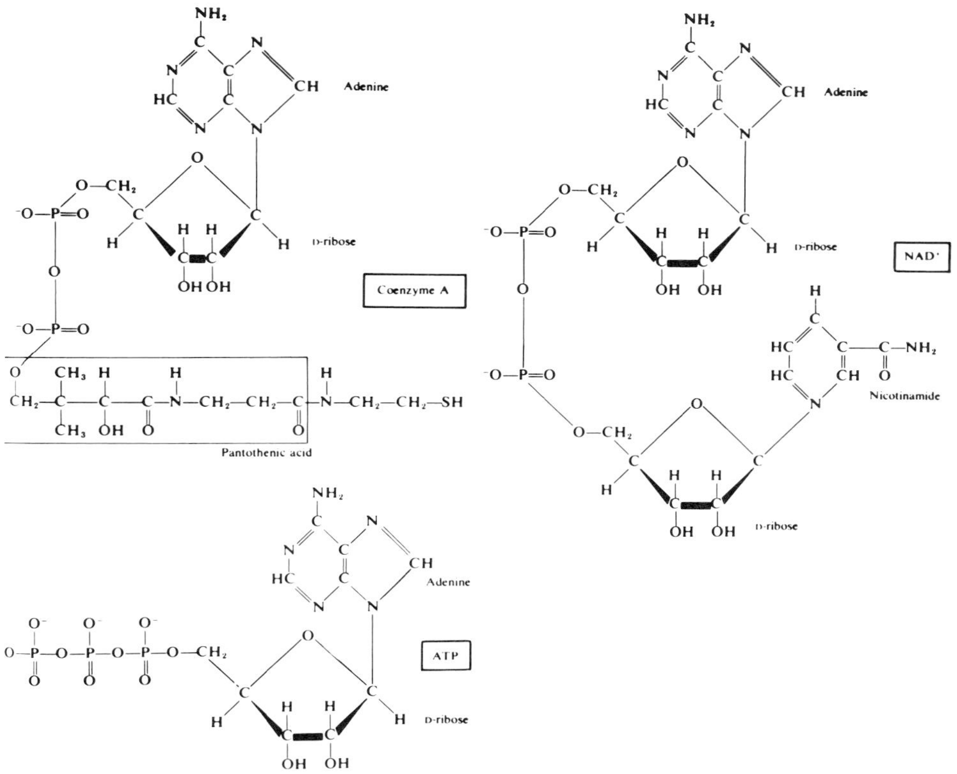

Fig. 11–9. Chemical structure of coenzymes A, NAD⁺, and ATP showing the similarity in nucleoside structure. Observe the presence of pantothenic acid in coenzyme A and nicotinamide in NAD⁺.

role that some coenzymes play in mitochondrial function. *Coenzyme A* (CoA) is part of a group that is derived from a nucleoside (adenine-D-ribose) and contains pantothenic acid (a vitamin of the B complex) linked to the ribose by pyrophosphoric acid (Fig. 11–9). CoA can be easily transformed into an ester at the thiol end (—SH) by acetyl groups making acetyl-CoA.

Other mitochondrial coenzymes are *nicotine amide dinucleotide* (NAD +), which contains the vitamin nicotinic acid; and *flavin mononucleotide* (FMN) and *flavin adenine dinucleotide* (FAD), which both contain riboflavin. NAD⁺, FMN, and FAD are important coenzymes, not only in mitochondria but also in chloroplasts. In Figure 11–9, the similarities between CoA, NAD⁺ and ATP may be observed.

The Krebs Cycle—A Common Pathway in the Degradation of Fuel Molecules

The *Krebs cycle*, also called the *tricarboxylic acid cycle*, or citric acid cycle takes place in the mitochondrial matrix. It serves as the first step in a common pathway for the degradation of fuel molecules such as carbohydrates, fatty acids, and amino acids. In the cell cytoplasm these substances are first acted upon metabolically to produce acetyl groups, which enter the mitochondrion and undergo the two-stage transformation diagrammed in Figure 11–10. In the first stage the acetyl groups, linked to acetyl-CoA, are taken into the Krebs cycle. Among the products of this cycle are CO_2 molecules and protons (H^+). In the second stage the H^+ are taken up by the *respiratory chain* and eventually combined with O_2 to yield H_2O. This process also generates ATP by the phosphorylation of ADP.

As illustrated in Figure 11–11, the first step of the Krebs cycle involves the condensation of the acetyl group (2 carbons) of acetyl-CoA with oxaloacetate (4 carbons) to make *citric acid*, (6 carbons). This step is directed by the enzyme *citrate synthase*. From citric acid, H_2O is released twice by aconitase to produce *isocitric acid*. This is followed by a decarboxylation (loss

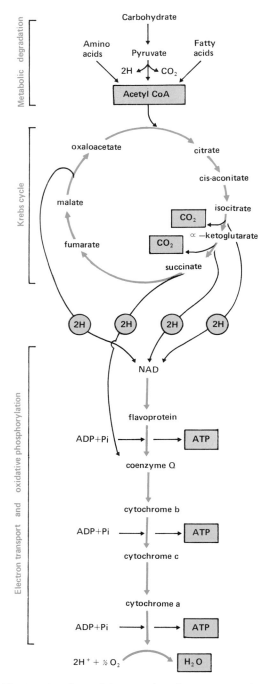

Fig. 11–10. General diagram of aerobic respiration showing the Krebs cycle, the respiratory chain, and its coupling with oxidative phosphorylation. (From Lehninger, A.L.: Biochemistry. New York, Worth Publishers, Inc., 1975.)

of CO_2) by isocitrate dehydrogenase, producing the 5-carbon *α-ketoglutaric acid*. CO_2 released by α-ketoglutarate dehydrogenase in the presence of CoA leads to *succinyl CoA* and by succinyl kinase to the 4-carbon compound *succinic acid*. The next enzyme succinate dehydrogenase produces *fumaric acid*, and then fumarase produces *malic acid*. The intervention of malate dehydrogenase produces oxaloacetic acid, and thereby closes the cycle.

An important consequence of the Krebs cycle is that at each turn four pairs of hydrogen atoms are removed from the substrate intermediates by enzymatic dehydrogenation and two CO_2 are released. These hydrogen atoms (or equivalent pairs of electrons) enter the respiratory chain, being accepted by either NAD^+ or FAD. Three pairs of hydrogen molecules are accepted by NAD^+, reducing it to NADH, and one pair by FAD, reducing it to $FADH_2$ (this last pair comes directly from the succinic dehydrogenase reaction). Since it takes two turns of the cycle to metabolize the two acetate molecules that are produced by glycolysis from one molecule of glucose, a total of six molecules of NADH and two of $FADH_2$ are formed at the starting points of the respiratory chain.

Electrons Flow Along a Cytochrome Chain Having a Gradient of Redox Potential

The early work of Warburg and Keilin using spectrophotometric techniques led to the concept that cell oxidations are brought about by electron carriers arranged in a chain of *oxidation-reduction* or *redox potential*. Keilin employed a simple hand spectroscope to study insect muscle and found that there were special pigments that underwent changes with oxidation and reduction during respiration. Such pigments, which had special absorption bands in the reduced state, were called *cytochromes*.

The cytochromes are iron-containing proteins in which the iron atom is contained within a chemical structure called a *porphyrin ring*. This portion of the molecule is a *heme* that is bonded to the protein by sulfur bridges. (A similar heme group is present in molecules of hemoglobin.) During the electron transfer, the iron atom passes from the ferrous to the ferric state, releasing an electron. This reaction is the basis of all cellular oxidation-reduction processes.

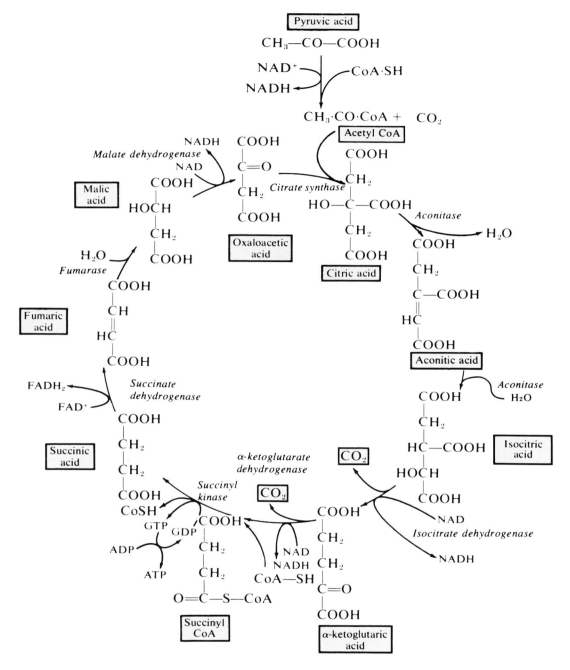

Fig. 11–11. Diagram of the Krebs or tricarboxylic acid cycle in mitochondria. The different enzymes and the steps of the reactions involved are represented.

$$Fe^{2+} \rightarrow Fe^{3+} + e^- \qquad (5)$$

The inner mitochondrial membrane contains the cytochromes b, c, c_1, a, and a_3, aligned to form a chain in which the redox potential becomes more and more positive. At each cyto-chrome the iron may be in the reduced or oxidized state. In these processes an electron, a pair of electrons, or hydrogen atoms are transferred from one cytochrome to another along a gradient of *redox potential.*

As shown in Figure 11–12, the couple $NAD^+/$

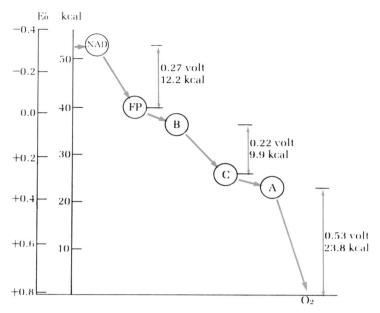

Fig. 11–12. Diagram showing the decline of free energy as electron pairs flow down the respiratory chain from NAD to O_2. E_0', oxidoreduction potential; *kcal*, free energy; *FP*, flavoprotein; *B,C,A*, are cytochromes. At the three points indicated there is enough energy drop to generate a molecule of ATP from ADP and phosphate. (Courtesy of A.L. Lehninger. From Lehninger, A.L.: Biochemistry. 2nd Ed. New York, Worth Publishers, 1975, p. 516.)

NADH has the most negative redox potential (designated E_0, in this case -0.32 volts) of the chain and the highest free energy (designated ΔG, in this case -52.7 kcal/mole). This free energy is released upon oxidation. The reversible NAD^+/NADH system is followed in the chain by a *flavin mononucleotide* (FMN)- and FAD-containing flavoprotein. A lipid-soluble protein called coenzyme Q then acts as a kind of shuttling system between the flavoproteins and the series of cytochromes beginning with cytochrome b. The cytochrome series ends with O_2, which has the most positive redox potential ($E_0 = 0.82$) and no free energy to be utilized ($\Delta G = O$). Thus the most reduced members of the electron transport chain are found at the beginning and the most oxidized members at the end.

This sequence of electron transfer reactions is consistent with the redox potentials of the various electron carriers. Sensitive spectrophotometric methods developed by Britton Chance have confirmed the functional nature of this sequence, showing that NAD^+ is the most reduced member of the chain and that cytochrome a and a_3 (which form part of *cytochrome oxidase*) are the most oxidized.

In recent years it has been found that the electron transport chain is extremely complex and that there are additional components. Of particular interest is *coenzyme Q* (CoQ) or *ubiquinone* (because it is ubiquitous in cells), which is a lipid-soluble benzoquinone with a long side-chain of ten isoprenoid units (thus the name CoQ_{10}). In addition, the mitochondrial respiratory chain contains enzyme-bound copper and iron-sulfur protein. According to Slater in the respiratory chain there are about 30 reaction centers able to accept electrons.

The Respiratory Chain—Four Molecular Complexes

All the previously discussed components of the respiratory chain, as well as those involved in the mechanism of phosphorylation, are integrated within the molecular structure of the inner mitochondrial membrane.

The components of the respiratory chain can be separated into multi-molecular complexes by the use of mild detergents. Such complexes have a considerable amount of lipid, which is essential for their activity.

Extraction of phospholipids by acetone inactivates some catalytic functions of the membrane; however, restoration of function can be

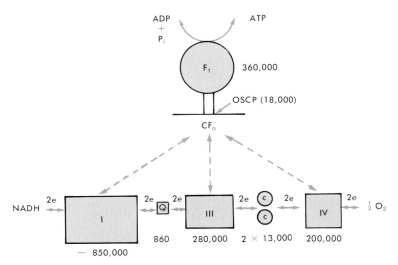

Fig. 11–13. Diagram of the electron-transferring complexes (i.e., respiratory chain) and the ATPase complex (F_1) present in the mitochondrial inner membrane; *complex I*, NADH-Q-reductase; *complex III*, QH$_2$-cytochrome-c-reductase; *complex IV*; cytochrome-c-oxidase; CF$_0$, F$_0$ coupling factor; OSCP, oligomycin sensitive protein; Q, ubiquinone; c, cytochrome. (See text for details.) (Courtesy of E.C. Slater.)

achieved by addition of the lipids and CoQ$_{10}$. David Green and associates have recognized four main complexes (I to IV) which, if mixed in correct stoichiometric ratios, can reconstitute to form the electron transport chain. Three of these complexes are indicated in Figure 11–13 (*i.e.,*, I, III, and IV). Complex II corresponds to succinate-Q-reductase (i.e., succinate dehydrogenase), which can transfer electrons directly from succinate to CoQ.[17] In Table 11–5 the main characteristics of these complexes are indicated. Listed there are the molecular weight, the number of protein subunits, the various prosthetic groups in the flavoproteins (FMM and FAD), and the iron-sulfur centers and cytochromes (hemes) present in each of the complexes. The table also gives the topology of the active sites in relation to the matrix (M side) or cytosol (C side) (see Papa, 1976; De Pierre and Ernster, 1977 in the Additional Readings at the end of this chapter).

Complex I (NADH-Q-reductase). This is the largest complex, with a structure consisting of 15 subunits. It contains *FMN* and six iron-sulfur centers. Complex I spans the inner mitochondrial membrane and is able to translocate protons across it, from the M side to the C side.

Complex II (Succinate-Q-reductase). This complex is composed of two polypeptides. It contains *FAD* and three iron-sulfur centers. In contrast to complex I, succinate-Q-reductase ap-

parently is unable to translocate protons across the membrane.

Complex III (QH$_2$-cytochrome-c-reductase). This complex contains a number of subunits, cytochromes b, cytochrome c$_1$, and iron-sulfur protein. In the topology of this complex the Q-site may be in the middle of the membrane in the hydrophobic area, and the cytochrome c site, on the C side.

Complex IV (Cytochrome-c-oxidase). Cytochrome oxidase is a large complex consisting of several polypeptides and having two cytochromes (a and a$_3$) and two copper atoms. This complex is thought to traverse the mitochondrial membrane, protruding on both surfaces. Such a *transmembranal orientation* is associated with the *vectorial transport of protons* across the membrane.

In yeast mitochondria it has been shown that cytochrome oxidase is made of seven subunits that can be resolved by gel electrophoresis and that range between 40,000 and 5000 daltons. It is of great biological interest that the three larger polypeptides are very hydrophobic and that their synthesis occurs in mitochondrial ribosomes, and that the four small subunits are relatively hydrophilic and originate in the cytoplasm.[18]

The seven subunits are arranged in the membrane in a functional sequence, being in contact with cytochrome c on the C side. The electrons

TABLE 11–5. COMPOSITION AND TOPOLOGY OF THE MITOCHONDRIAL RESPIRATORY CHAIN

Enzyme, Complex	Number of Subunits	Prosthetic Groups	Association with Membrane	Topology of Catalytic Sites
NADH-Q-reductase (Complex I) ~550,000	16	1 FMN 16–24 FeS (5–6 centers)	Integral	NADH site: M side; Q site: Middle
Succinate-Q-reductase (Complex II) 97,000	2	1 FAD (covalently bound) 8 FeS (3 centers)	Integral	Succinate site: M side Q site: Middle
QH$_2$-cytochrome-c-reductase (Complex III) 280,000	6–8	2 b-hemes 1 c-heme (c$_1$) 2 FeS	Integral	Q site: Middle Cytochrome-c-site: C side
Cytochrome c 13,000	1	1 c-heme	Peripheral	Cytochrome-c$_1$ site: C side Cytochrome-a site: C side
Cytochrome-c-oxidase (Complex IV) 200,000	6–7	2 a-hemes (a, a$_3$) 2 Cu	Integral	Cytochrome-c site: C side O$_2$ site: M side (?)

From De Pierre, J.W., and Ernster, L.: Annu. Rev. Biochem., 46:201, 1977.

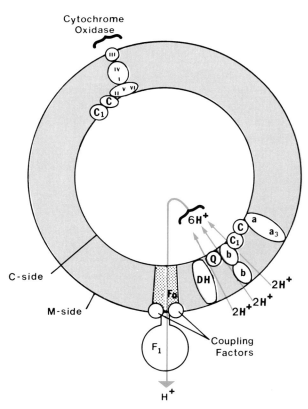

Fig. 11–14. The topography of a vesicle originated from the inner membrane of mitochondrion; the transmembrane distribution of the various complexes integrating the respiratory chain, and ATPase are indicated. (Courtesy of E. Racker, 1977.)

then pass to cytochrome a, then to Cu^{++}; and finally to cytochrome a_3 and oxygen at the M side (Fig. 11–14).

Using antibodies against cytochrome oxidase, it was found that this complex was inhibited when the antibodies were applied to either side of the inner membrane. Such a finding is in agreement with the traverse topography of complex IV.

Electron Transport—Coupled to the Phosphorylation at Three Points

The electron transport system of mitochondria is *coupled* at three points with the phosphorylating system (Figs. 11–13 and 11–14). The protons (H^+) originating from electron transfer are translocated by the respiratory chain across the membrane from the M side to the C side. According to the *chemiosmotic hypothesis* of Peter Mitchell, this translocation creates a pH difference and a membrane potential. Both constitute the proton-motive force that tends to move H^+ from the C side back to the M side of the membrane. Since the inner mitochondrial membrane is highly impermeable to H^+ ions, these can only reach the M side through the "proton channel" of the ATPase. When H^+ moves from the C side to the M side, the F_1-ATPase, operating in reverse, catalyzes ATP synthesis. Conversely, when the F_1-ATPase hydrolyzes ATP, it functions as a "proton pump" and ejects H^+ from the M side to the C side. Figure 11–14 indicates that for each NADH that is oxidized, six protons (H^+) are translocated through the inner membrane; these six H^+, when returning to the M side through the F_1-ATPase, give rise to three molecules of ATP.

Mitochondrial ATPase—A Structurally Complex Proton Pump

The mitochondrial ATPase molecule is indeed a multi-polypeptide complex. Three parts can be distinguished: (1) The F_1 *particle* or soluble ATPase, which is in the head piece of the projection observed on the matrix side of the mitochondrial crests (Fig. 11–5). F_1 contains five associated subunits. (2) A complex of very hydrophobic proteins (i.e., proteolipids) localized in the membrane bilayer and containing the proton translocating mechanism (this corresponds

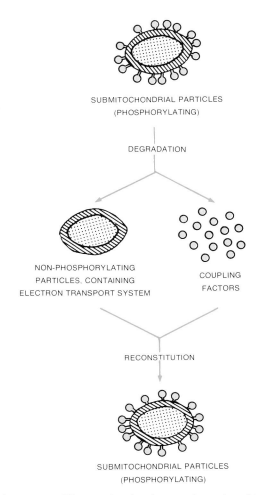

SUBMITOCHONDRIAL PARTICLES
(PHOSPHORYLATING)

DEGRADATION

NON-PHOSPHORYLATING
PARTICLES. CONTAINING
ELECTRON TRANSPORT SYSTEM

COUPLING
FACTORS

RECONSTITUTION

SUBMITOCHONDRIAL PARTICLES
(PHOSPHORYLATING)

Fig. 11–15. Diagram showing the experiment by which the submitochondrial particles, corresponding to the inner mitochondrial membranes, are submitted to urea to remove the coupling factor F_1, thus leaving non-phosphorylating particles. The lower part of the figure shows the subsequent reconstitution of a phosphorylating submitochondrial particle. (From Racker, E.: The membrane of the mitochondrion. Sci. Am., *218*:32, 1968. Copyright © 1968 by Scientific American, Inc. All rights reserved.)

to F_0 in Fig. 11–14). (3) A protein stalk that connects the two. This portion corresponds to the *oligomycin-sensitive conferring protein* (OSCP) and to another protein needed to bind F_1 to the membrane (Fig. 11–13). (For further details on this complex system see De Pierre and Ernster, 1977 in the Additional Readings at the end of this chapter.)

The function of the phosphorylating system is best demonstrated by the elegant experiment of Efraim Racker, shown in Figure 11–15.

If isolated mitochondria are subjected to ultrasonic fragmentation, vesicles are formed that show a reverse organization ("inside-out" vesicles), with the F_1 particles situated on the outer surface and with the C side facing the interior of the vesicle. These vesicles are able not only to respire but also to phosphorylate.[19]

After treatment with urea, which removes the F_1 coupling factor, the membranes no longer phosphorylate. Reconstitution of the complete phosphorylating system can be achieved by mixing the non-phosphorylating membranes with F_1 particles. After it is isolated, the F_1 particle causes the hydrolysis of ATP to ADP and phosphate; however, when it is attached to the membrane it functions as a synthesizing enzyme (Fig. 11–13).[20]

Topological Organization of the Respiratory Chain and the Phosphorylation System

All that was just said about the various complexes of the respiratory chain and the phosphorylating system indicates that the inner mitochondrial membrane is highly asymmetrical in its molecular organization. In other words, there is a special "sidedness" of the most important components, which is summarized in Table 11–5 and in the diagram of Figure 11–14. Under the EM the sidedness has been demonstrated by the use of antibodies and also by reagents that produce electron-dense deposits in the mitochrondrial crests.

By using both a special tetrazole, which is converted to formazan by succinate dehydrogenase (see Fig. 3–15) and 3',3'-diaminobenzidine (DAB), which is oxidized by cytochrome c, it has been shown that the dense products accumulate on the outside surface of the mitochondrial crest.

From experimental evidence it can be demonstrated that the electron transport system is accessible to NADH and succinate only from the inner matrix side, while cytochrome c is reached from the outer surface of the membrane.

In addition to this transverse topology there is also a lateral topology of the assemblies involved in the respiratory chain and oxidative phosphorylation.

The number of assemblies varies according to the tissue. A mitochondrion from liver has about 15,000 assemblies, while one from the flight muscle of an insect may have as many as 100,000. Within this assembly or *unit of phosphorylation*

TABLE 11–6. STOICHIOMETRY OF ELECTRON-TRANSFER AND ATP-SYNTHESIZING COMPONENTS IN HEART MITOCHONDRIA

Component*	Stoichiometry	Molecular Weight ($\times 10^{-3}$)
I	1	$1 \times 550 = 550$
III	4	$4 \times 230 = 920$
c	8	$8 \times 12 = 96$
IV	8	$8 \times 200 = 1600$
ATPase	4	$4 \times 360 = 1440$
OSCP	4	$4 \times 18 = 72$
	29	4678
CFo	4	
Q	64	

Modified from Slater, E.C.: Electron transfer and energy conservation. 9th International Congress on Biochemistry. Stockholm, 1973.

*Components: I, Complex I; III, Complex III; C, cytochrome c; IV, Complex IV; OSCP, oligomycin sensitive protein; CF_o, F_o coupling factor; and Q ubiquinone or coenzyme.

the main enzymes are in equimolecular quantities (see Table 11–6). This finding has resulted from the use of ultra-rapid spectrophotometric studies and from the determination of binding sites of drugs that specifically affect the various components of the system (Fig. 11–13). (For example, *aurovertin* binds to the ATPase; *oligomycin* acts at the level of the binding of ATPase to the membrane; *antimycin* is an inhibitor acting on the QH_2 cytochrome-c-reductase; *rotenone* acts on NADH-Q-reductase; and *cyanide* has its well-known poisoning effect on cytochrome-c-oxidase.)

Topology and Membrane Fluidity. While it was previously thought that in the inner mitochondrial membrane there was a compact mosaic of macromolecules, representing the various respiratory and oxidative phosphorylation complexes, in recent years the problem of membrane fluidity has been taken into consideration. Although the inner mitochondrial membrane has a high protein content (75% against 25% of lipids), the lipids are virtually devoid of cholesterol and contain a high concentration of unsaturated fatty acids. This explains the low viscosity of this membrane, which is one of the most fluid membranes.

Different studies demonstrate that the various electron-transport and oxidative-phosphorylation complexes contain polypeptides that are not embedded in the membrane and are situated externally on the bilayer. In other words, these complexes occupy only a certain proportion of the bilayer. Studies using freeze-fracture electron microscopy show that the transmembrane protein occupies only 40 to 50% of the total membrane area. This can be illustrated by submitting the membrane to an electrophoretic field, which results in the packing of the proteins to one side. After removing the field of force, the proteins again become randomized because of diffusion in the fluid bilayer. These and other studies, in which the lipid area was increased by fusion with liposomes, suggest that the inner mitochondrial membrane is not structurally ordered or densely packed. Thus, the electron transfer, between the different components of the electron-transport chain could be mediated by lateral diffusion and collisions in the plane of the membrane (see Hackenbrock, 1981 in the Additional Readings at the end of this chapter).

The Chemiosmotic Theory—An Electrochemical Link Between Respiration and Phosphorylation

The previously mentioned studies have demonstrated that for each pair of electrons received from the Krebs cycle and transported by the electron transfer system there is one molecule of ATP synthesized by the ATPase. The fundamental problem is that of establishing the nature of the link between these two systems (represented by the dotted lines between the respiratory chain and the F_1 coupling factor in Fig. 11–13).

Several hypotheses have been proposed to explain the mechanism of this link. For the most part, the *chemiosmotic coupling* theory (in which the link is essentially electrochemical in nature) proposed by Mitchell[21] is currently in favor.

According to this theory, the inner membrane of the mitochondrion acts as a transducer converting the energy, provided by an electrochemical gradient, into the chemical energy of ATP. In this model, the membrane is impermeable to both H^+ and OH^- ions. For this reason, if pH differences are established across the membrane, they act as energy-rich gradients. The electron transport system is organized in "redox loops" within the membrane, and the electrons are passed from one carrier to another on the respiratory chain. At the same time, protons (H^+) are ejected toward the cytoplasmic side (C side), while OH^- remain on the matrix side (M

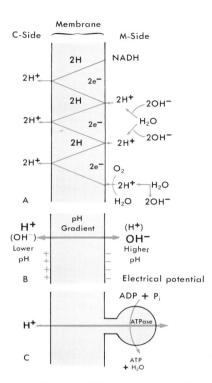

Fig. 11–16. Diagram of chemiosmotic coupling according to P. Mitchell. *A*, During electron transport H^+ ions are driven to the C side of the inner mitochondrial membrane. *B*, This process produces a pH gradient and an electrical potential across the membrane. *C*, This gradient drives the proton pump of the ATPase, and ATP is synthesized from ADP and Pi.

side) (Fig. 11–16A). This vectorial movement of protons creates a difference in pH (i.e., lower pH on the C side and higher on the M side), which results in an electrical potential (Fig. 11–16B).

Calling ΔpH the *pH gradient* and $\Delta\psi$ the resulting *electrical gradient* in volts, the energy produced ΔP is the *proton motive force:*

$$\Delta P = \Delta\psi + 2.3 \frac{RT}{F} \Delta pH \qquad (6)$$

(where R is the universal gas constant, T the absolute temperature, and F the Faraday constant).

The chemiosmotic theory postulates that the primary transformations occurring in the respiratory chain guide the *osmotic work* needed to accumulate ions. The energy generated by electron transport is conserved in the energy-rich form of a H^+-ion gradient. As indicated in Figure 11–16C, this gradient, through the action of the proton pump of the ATPase, drives the oxidative phosphorylation of ADP to form ATP, by which mechanism free energy is conserved.

$$ADP + Pi \rightleftharpoons ATP + H_2O \qquad (7)$$

In this reaction H_2O is also formed because of the dehydration, which leads to the removal of H^+ and OH^- ions.

A special feature of this theory is that both the respiratory chain and the ATPase may operate independently. Since coupling by electrical fields can operate at *long-range* distances there is no need for a direct contact between the macromolecules involved. In recent years, much experimental evidence has supported the validity of Mitchell's chemiosmotic theory, and the mechanism of the redox proton pump has been elucidated by experiments in which various components of the electron transport and phosphorylating system were reconstituted into vesicular bilayers, the so-called liposomes (see Racker, 1975 in the Additional Readings at the end of this chapter).

The Chemical-Conformational Hypothesis Involves Short-Range Interactions

An alternate explanation of the coupling between respiration and phosphorylation—the chemical-conformational hypothesis—postulates that the electron transport and phosphorylating systems are in molecular contact and that information is transmitted by short-range interactions that may, in part, be electrostatic in nature. It is postulated that upon acceptance of the two electrons, each electron transfer protein undergoes a conformational change that is then transmitted to the ATPase by the short-range interactions.

According to some recent views the chemical-conformational and chemiosmotic mechanisms of electron transport linkage to phosphorylation do not represent mutually exclusive alternatives but may indeed be two aspects of the same mechanism (see Ernster, 1977 in the Additional Readings at the end of this chapter).

Overall Energy Balance—Glucose and the P:O Ratio

At this point it is interesting to mention the overall energy balance that results from the function of mitochondria.

Of the 686,000 calories contained in a mole of glucose, less than 10% (i.e., 58,000) can be released by anaerobic glycolysis. Much more of this energy is released by oxidative phosphorylation.

Since the formation of ATP occurs in three steps, in the electron chain, the equation for this process can be written as follows:

$$NADH + H^+ + 3\,ADP + 3\,Pi \\ + 1/2\,O_2 \rightarrow NAD^+ + 4\,H_2O + 3\,ATP \qquad (8)$$

One way of indicating the ATP yield from oxidative phosphorylation is the *P/O ratio*, which is expressed as the moles of inorganic phosphate (Pi) used per oxygen atom consumed. [In Equation (8) the P/O ratio is 3 because 3 Pi and 1/2 O_2 are used.]

The energy balance of aerobic respiration shows that 36 ATP molecules are produced from each glucose molecule. The overall equation can be written as follows:

$$C_6H_{12}O_6 + 36\,Pi + 36\,ADP \rightarrow 6\,CO_2 \\ + 36\,ATP + 42\,H_2O \qquad (9)$$

The cell stores 40% of the chemical energy liberated by the combustion of glucose in the form of ATP. The rest of the energy is dissipated as heat or used for other cell functions.

SUMMARY:
Molecular Organization and Function

Mitochondria function as energy-transducing organelles into which the major degradation products of cell metabolism penetrate and are converted into chemical energy (ATP) to be used in the various activities of the cell. This entire process requires the entrance of O_2, ADP, and phosphate, and it brings about the exit of ATP, H_2O, and CO_2.

The chemical energy of nutrients is stored in the covalent bonds between the atoms of each molecule. This energy is not released suddenly, but is made available in a stepwise and controlled manner. It can then be used to synthesize new molecules, perform mechanical work, carry out active transport, maintain membrane potentials, or produce radiant energy. The common link in all these processes is the compound ATP, which is distinguished by having two terminal bonds of very high potential energy.

The processes of energy transformation that occur in mitochondria are based on three coordinated steps: (1) the Krebs cycle, carried out by a series of soluble enzymes present in the mitochondrial matrix, which produces CO_2 by decarboxylation and removes electrons from the metabolites; (2) the *respiratory chain* or *electron transport system*, which captures the pairs of electrons and transfers them through a series of electron carriers, which finally leads by combination with activated oxygen to the formation of H_2O; (3) a *phosphorylating system*, tightly coupled in the respiratory chain, which at three points gives rise to ATP molecules.

The Krebs or carboxylic acid cycle, which takes place in the mitochondrial matrix, is the first step in a common pathway for the degradation of fuel molecules. These molecules are acted upon metabolically (in the cytoplasm) to produce acetyl groups, which are taken into the cycle by acetyl coenzyme A. Each turn of the cycle generates two CO_2 molecules and four pairs of hydrogen atoms. The hydrogen atoms then enter the respiratory chain: three pairs are accepted by NAD^+, reducing it to NADH, and one pair is accepted by FAD, reducing it to $FADH_2$. It takes two turns of the cycle to metabolize the two acetate molecules that are produced by glycolysis from one molecule of glucose.

The electron transport system starts with NAD^+, which is reduced to NADH. NAD^+/NADH has the most negative redox potential ($E_0 = -0.32$ volts) and highest free energy ($\Delta G = -52.7$ kcal/mol). This is followed by a FMN- or a FAD-flavoprotein, coenzyme Q_{10}, and a series of cytochromes, ending with O_2 ($E_0 = +0.82$ and $\Delta G = 0$) (Fig. 11–12). The respiratory chain also contains enzyme-bound copper and iron-sulfur proteins. The respiratory chain and the phosphorylating system are in the inner mitochondrial membrane. This is rich in proteins (60 to 70%), most of which are of the integral type.

The components of the respiratory chain may be separated into multi-molecular complexes in which lipids are essential for the activity. Four main complexes have been isolated (Table 11–5).

Complex I or *NADH-Q-reductase* is the largest (molecular weight >500,000) and has 15 subunits. It contains FMN and six iron-sulfur centers. The NADH reaction site is at the M side (matrix).

Complex II or *succinate-Q-reductase* contains FAD and three iron-sulfur centers. This complex transfers electrons from succinate to CoQ.

Complex III or *QH_2-cytochrome-c-reductase* contains cytochromes b_1, c_1, and an iron-sulfur protein. Cytochrome-c is on the C side (cytosol), and CoQ is in the hydrophobic region of the membrane.

Complex IV or *cytochrome-c-oxidase* is a large complex with cytochromes a and a_3 and two copper atoms. As do other complexes, it has a transmembranal orientation that is associated with the vectorial transfer of protons. In yeast mitochondria, complex IV has seven subunits of which the three

larger ones are hydrophobic and synthesized on mitoribosomes.

The *phosphorylating* system is represented by the F_1-ATPase. This is a multipeptide complex with three main parts: (1) the F_1 particle of soluble ATPase, (2) a complex of very hydrophobic proteins (i.e., proteolipids), which represent the proton translocating portion (Fo in Fig. 11–14), and (3) a protein stalk that connects the two and contains the coupling factors. By proper methods it is possible to separate the F_1 particle and then to reconstitute the system.

The respiratory chain and phosphorylating system have a fine topology that runs transversely and laterally in a mosaic arrangement. The entire assembly may be 5×10^6 daltons and may cover a surface of 20×20 nm.

The nature of the link between the respiratory chain and the ATPase is unknown, but there are two favored hypotheses: the *chemiosmotic* and the *chemical-conformational*. According to the chemiosmotic theory, the ATPase synthesizes ATP under the influence of the vectorial field of the proton motive force generated by the transfer of the electrons through the respiratory chain and the release of protons on the C side. This effect operates at long range and direct molecular contact is not needed. The chemical-conformational hypothesis postulates that the transfer of electrons originates a conformational change in the proteins of the respiratory chain which, by short-range interaction, transmits the signal to the ATPase that is needed for the synthesis of ATP.

To follow the overall energy balance the fate of a molecule of glucose is illustrative. This molecule is first degraded by anaerobic glycolysis in the cytosol to lactic acid, producing two ATP. Of the 686,000 calories contained in a mole of glucose, only 58,000 are relesed by glycolysis. In oxidative phosphorylation much more energy is generated. The energy balance of aerobic respiration can be written as follows:

$$C_6H_{12}O_6 + 36\ Pi + 36\ ADP \rightarrow 6\ CO_2 + 36\ ATP + 42\ H_2O$$

The 36 ATPs produced represent 40% of the total energy contained in glucose. The P:O ratio expresses the moles of Pi (inorganic phosphate) used per $1/2\ O_2$ consumed. The normal value of P:O is about 3.

11–5 PERMEABILITY OF MITOCHONDRIA

Since the most important metabolic activities of mitochondria take place within the inner mitochondrial compartment, there should be a rapid and active flow of certain metabolites across the two membranes. The products of extramitochondrial metabolism, for example, must reach the mitochondrial matrix in order to undergo oxidation, and ADP and phosphate must enter to form ATP. Simultaneously, end products such as H_2O, ATP, urea, and ammonia must leave the mitochondrion.

There are important differences in permeability between the mitochondrial membranes. The outer membrane is freely permeable to electrolytes, water, sucrose, and molecules as large as 10,000 daltons. The inner membrane, on the other hand, is normally impermeable to ions, as well as to sucrose.

ADP, ATP, and Pi—Transport by Specific Carriers

The inner mitochondrial membrane uses specific *carriers* or *permeases* for the translocation of various substances.

In mitochondria the existence of specific carriers for ATP (or ADP), phosphate, succinate (or malate), isocitrate, glutamate, aspartate, and bicarbonate has been suggested. A carrier mechanism has been postulated for Ca^{++}, Mn^{++}, or Sr^{++}.

Of all these the most important carriers are those involved in the *transport of ADP, ATP, and Pi* through the inner mitochondrial membrane. Such mechanisms provide for the entrance of ADP and Pi into the mitochondrial matrix and for the exit of ATP to the cytoplasm. Study of the ADP-ATP translocation is facilitated by the fact that there are specific inhibitors of this transport, the experimental use of which has pointed to the *gated pore* as the most plausible of the models (see Section 4–3 for the anion-gated-pore model). Such an ADP-ATP carrier would be asymmetrically disposed through the membrane, with two conformations, one facing the C side, and the other the M side. The same carrier could be used for the reverse translocation of ADP and ATP. Isolation of the carrier protein has been achieved by the use of [35]S *carboxyatractylate* (^{35}S CAT), which inhibits transport at the C side, followed by the use of detergents. In this way, a ^{35}S CAT-carrier complex having a molecular weight of 29,000 daltons has been separated from mitochondria. Two of these subunits probably constitute the carrier (see Klingenberg et al., 1976 in the Additional Readings at the end of this chapter).

Mitochondrial Conformation—Changes with Stages of Oxidative Phosphorylation

At the beginning of this chapter it was noted that low amplitude contraction cycles that could be associated with stages in oxidative phosphorylation were observed in living cells. Similar observations have been made in isolated mitochondria by using absorbancy or light scattering to study them. Such changes were interpreted as being the result of small variations in volume due to an energy linked swelling contraction phenomenon. The changes in absorbancy and light scattering, however, may also reflect a rearrangement of the internal structure of the mitochondrion. By means of a quick sampling method, which permits the fixation of isolated mitochondria at different stages of their metabolism, reversible ultrastructural changes were observed.[22]

Mitochondria may alter their internal conformation between the two extreme states shown in Figure 11–17. One is the so-called *orthodox* state that is usually observed in intact tissues. The other corresponds to the *condensed* state, in which there is a dramatic contraction of the inner compartment of the mitochondria accompanied by accumulation of fluid in the outer compartment. In the orthodox conformation the inner membrane shows the characteristic crests; the matrix fills practically the entire volume of the mitochondrion and has a reticular or granular aspect (Fig. 11–17A). In the condensed conformation the inner membrane is folded and the matrix, now more homogeneous, represents only about 50% of the mitochondrial volume (Fig. 11–17B). In this state, at certain points, it is possible to see contact zones between the inner and outer membranes of the mitochondrion.

By freeze-fracturing it has been observed that the number of such contacts increases in the condensed conformation and decreases in the orthodox. A regulatory function has been proposed, which could be involved in the ATP/ADP exchange across the two mitochondrial membranes.[23]

The electron transport system is required for the change from condensed to orthodox conformation to take place. Inhibition of the respiratory chain by cyanide, antimycin A, or Amytal impairs this transformation. The orthodox state

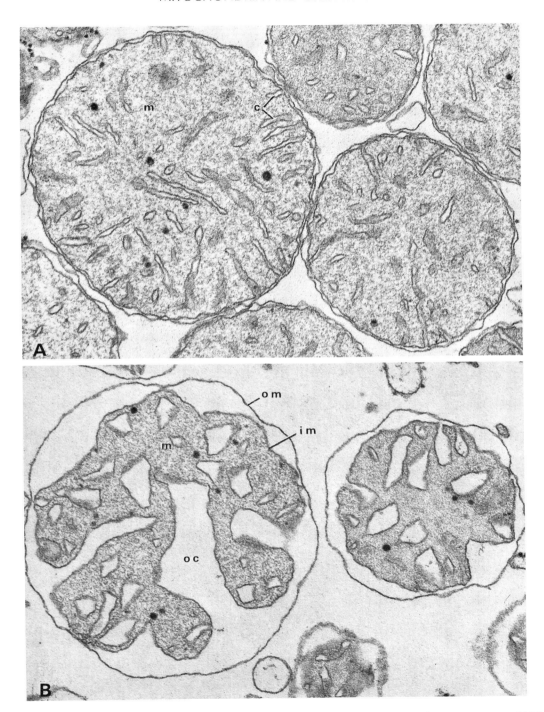

Fig. 11–17.   Electron micrographs of isolated mitochondria from rat liver in two extreme conformation states. ×110,000. A, the *orthodox* conformation. The inner membrane is organized into crests *(c)*, and the matrix *(m)* fills the entire mitochondrion. B, the *condensed* conformation. Mitochondrial crests are not observed, and the outer chamber *(oc)* represents about 50% of the volume; *om*, outer membrane; *im*, inner membrane. (Courtesy of C.R. Hackenbrock.)

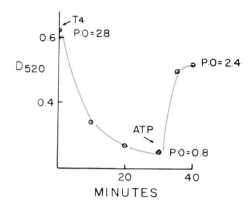

Fig. 11–18. Swelling of rat liver mitochondria in the presence of thyroxine and later contraction by ATP. The decrease in optical density corresponds to an increase in water content, and vice versa. Note that the P:O ratio declines during swelling, but is restored again during the contraction stage. *P:O*, ratio between the passage of inorganic phosphorus to ADP and the O_2 consumed. (Courtesy of A.L. Lehninger.)

is induced when the external ADP becomes low and there is none left to be phosphorylated. If at this time ADP is added, respiration is rapidly enhanced, and the contraction of the inner membranes takes place. It is thought that during the transition from the orthodox to the condensed stage there is a change both in the inner membrane and in the matrix. The inner membrane is believed to contain the contractile elements or "mechano-enzymes," but the possible role of the matrix in this contraction should also be considered.

Mitochondrial Swelling and Contraction—Agent-Induced

During swelling the mitochondrial volume may increase three to five times its normal value in the absence of ADP. With the addition of ATP, or the restoration of respiration, the mitochondrion may regain its original size. Studies on the phenomenon of swelling demonstrate that mitochondria have an important function in the uptake and extrusion of intracellular fluid.

Agents known to induce swelling are phosphate, Ca^{++}, reduced glutathione, and, in particular, the thyroid hormone thyroxine. Thyroxine is the most effective swelling agent and is capable of producing swelling at physiological concentrations.

Figure 11–18 shows an experiment in which the swelling effect of thyroxine as measured by

light absorption, and the reversible contraction produced by the addition of ATP. This experiment also shows that the P:O ratio is lowered by thyroxine and returned to normal range by ATP. The water uptake is associated with the uncoupling of oxidative phosphorylation. Mitochondria can actively squeeze out water and small molecules in the proportion of several hundred per each ATP molecule split.

Mitochondrial Accumulation of Ca^{++} and Phosphate

Although respiration and oxidative phosphorylation are the most important functions of mitochondria, another related function is the accumulation of cations. Ca^{++} can be concentrated in isolated mitochondria up to several hundred times normal values. Phosphate enters the mitochondria along with Ca^{++}. The amounts of Ca^{++} and phosphate accumulated may be so great that the dry weight may increase by 25%, and microcrystalline, electron-dense deposits may become visible within the mitochondria. This process usually occurs in the osteoblasts present in tissues undergoing calcification.

In the presence of Ca^{++}, mitochondria no longer phosphorylate but, instead, accumulate Ca^{++} and phosphate. Both oxidative phosphorylation and accumulation of Ca^{++} depend on the maintenance of cell respiration (i.e., the Krebs cycle). The accumulation of Ca^{++} and other cations inside the mitochondrion is accompanied by loss of H^+, so as to maintain an electrical equilibrium. The matrix becomes more alkaline because of OH^- accumulation, while acid is simultaneously released to the outside. Interestingly enough, chloroplasts behave precisely in the reverse direction by pumping out OH^- and accumulating H^+.

SUMMARY:
Mitochondrial Permeability

Since most metabolic processes in mitochondria occur in the inner compartment, an active flow of metabolites across both membranes takes place. The outer membrane is freely permeable, but the inner one is rather impermeable to ions and metabolites, and must use specific carriers or translocators to achieve this goal. Carriers for ATP (or ADP), phosphate, succinate or malate, isocitrate, glutamate, aspartate, bicarbonate, and Ca^{++} have been postulated. In isolated mitochondria cycles of contractions associated with oxidative phosphorylation have been observed. Physiologically, the mitochondrion passes from a conformational stage called *orthodox* into that called *contraction*. In the orthodox conformation, found at low external ADP, the matrix fills the entire volume of the inner compartment, and the crests may be seen clearly. In this stage the organelle does not phosphorylate. If ADP is added, the contraction stage is brought about, the matrix becomes condensed, and about 50% of the water diffuses into the outer compartment.

Mitochondrial swelling is induced by several agents, such as phosphate, Ca^{++}, and especially thyroxine. This hormone increases the volume of the mitochondrion by water uptake at the same time that it uncouples phosphorylation. With ATP the swelling is reversed; Ca^{++} and phosphate are accumulated in large amounts. Mitochondria tend to pump out H^+ and to retain OH^-. Chloroplasts act in a reverse manner and tend to pump out OH^- and retain H^+.

11–6 BIOGENESIS OF MITOCHONDRIA

Two main mechanisms for the biogenesis of mitochondria have been postulated: origination by division from parent mitochondria or origination *de novo* from simpler building blocks. Mitochondria are distributed between the daughter cells during mitosis, and their number increases during interphase. It has been observed with time-lapse cinematography that mitochondria gradually elongate and then fragment into smaller mitochondria. This observation has been verified in *Neurospora*.[24] In an experiment in which a choline deficient mutant of *Neurospora* was labeled with radioactive choline, the radioactivity was followed in the mitochondria of the second and third generations. By radioautography it was found that all mitochondria of the original progeny were labeled. The mitochondria of each daughter cell were also labeled but contained about half the radioactivity. This seemed to indicate that mitochondria had divided and grown by the addition of new lecithin molecules to the existing mitochondrial framework.

Yeast cells grown anaerobically lack a complete respiratory chain (cytochromes b and a are absent); under the EM they show no typical mitochondria. When the yeast cells are placed in air, however, the mitochondrial membranes present fuse, unfold, and form true mitochondria that contain the cytochromes. This observation is somewhat in agreement with the *de novo* synthesis theory of the biogenesis of mitochondria.

The differences between the outer and the inner membrane should be considered in relation to the origin of mitochondria. The chemical composition of the outer membrane is similar to that of the ER and very different from that of the inner membrane. Furthermore, continuities between the outer membrane of the mitochondrion and the ER, and even with the axon membrane, have also been observed.[25]

Mitochondria are Semiautonomous Organelles

In recent years the study of mitochondrial and chloroplast biogenesis became of great interest because it was demonstrated that these organelles contain DNA, as well as ribosomes, and are able to synthesize proteins. The term *semiautonomous organelles* was applied to the two structures in recognition of these findings. At the same time, however, the name indicated that the biogenesis was highly dependent on the nuclear genome and the biosynthetic activity of the ground cytoplasm. It is now known that the mitochondrial mass grows by the integrated activity of both genetic systems, which cooperate in time and space to synthesize the main components. The mitochondrial DNA codes for the mitochondrial, ribosomal, and transfer RNA and for a few proteins of the inner membrane. Most of the proteins of the mitochondrion, however, result from the activity of the nuclear genes and are synthesized on ribosomes of the cytosol (Fig. 11–19).

Mitochondrial DNA—Circular and Localized in the Matrix

In 1963 Margit and Sylvan Nass[26] observed filaments within mitochondria that they interpreted as DNA molecules. This finding was fully confirmed in cell sections, as well as in DNA extracted and studied by the surface spreading technique (Fig. 11–20). The mitochondrial DNA (mtDNA) is localized in the matrix and is probably attached to the inner membrane at the point where DNA duplication starts. This duplication is under nuclear control and the enzymes used (i.e., polymerases) are imported

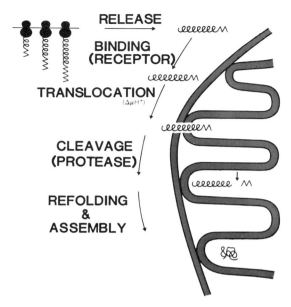

Fig. 11–19. Diagram with the proposed sequence of events in the import of a protein into the mitochondrial matrix. (See text for details; Courtesy of G. Schatz.)

from the cytosol (Fig. 11–19). A single mitochondrion may contain one or more DNA molecules, depending on its size; i.e., the larger the mitochondrion, the more DNA molecules present. Thus, a normal eukaryotic cell has at least as many copies of mtDNA as it has mitochondria. It has been calculated that in an adult man there may be 10^{17} mtDNA molecules. So far the results obtained suggest that these mtDNA molecules are identical and that genes are represented in single copies.

In most animals and plants the mtDNA is circular, although it may be highly twisted (Fig. 11–20). In most animal cells mtDNA is only 5.5 μm long, while in yeast it may be longer. Because of its higher guanine-cytosine content, mtDNA has a higher buoyant density and denatures at a higher temperature than nuclear DNA. It is relatively easy to separate the strands of mtDNA and to localize genes after hybridization with RNA molecules. Genetic maps have been constructed for mtDNAs of yeast and mammalian cells (see Borst, 1981; Attardi et al., 1981 in the Additional Readings at the end of this chapter).

Mitochondrial DNA from human cells is a circular DNA molecule of about 15,000 base pairs. A significant proportion of this DNA is devoted to coding for the protein synthesizing machinery. MtDNA codes for the mitochondrial ribosomal RNAs (12S and 16S), for about 19 tRNAs, and for the mRNAs of about 12 proteins. The mRNAs transcribed from the mitochondrial DNA code mostly for highly hydrophobic proteins.

Mitochondrial DNA duplicates by a mechanism, which will be described in Chapter 14. Incorporation of ^{3}H-thymidine into DNA has been observed in mitochondria of *Tetrahymena*. All mitochondria incorporate ^{3}H-thymidine within a population doubling time. In synchronized cultured human cells labeled with ^{3}H-thymidine, it was found that mitochondrial DNA was synthesized during a period extending from the G_2 phase to cytokinesis. In Chapter 19 the special genetic code present in mitochondria, which differs from the "universal code" is mentioned (see Borst, 1981 in the Additional Readings at the end of this chapter).

Mitochondrial Ribosomes—Smaller than Cytoplasmic Ribosomes

Mitochondria contain ribosomes and polyribosomes (see Chapter 21) that can be demonstrated, in many cases, by electron microscopy (Fig. 11–4). In yeast and *Neurospora*, ribosomes have been assigned to a 70S class similar to that of bacteria; in mammalian cells, however, mitochondrial ribosomes are definitely smaller and have a total sedimentation coefficient of 55S, with subunits of 35 and 25S.[27] Ribosomes in mitochondria appear to be tightly associated with the membrane.

In spite of their lower sedimentation coefficient, animal ribosomes are slightly larger than the 70S bacterial ribosomes; this is because they contain more protein. All the ribosomal proteins are imported from the cytosol (Table 11–7). (For a review on mitoribosomes see Curgy, 1985 in the Additional Readings at the end of this chapter.)

Mitochondria Synthesize Mainly Hydrophobic Proteins

We mentioned earlier that mitochondria can synthesize about 12 different proteins, which are incorporated into the inner mitochondrial membrane. These proteins are very hydrophobic

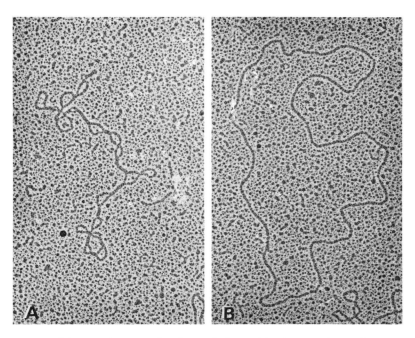

Fig. 11–20. DNA extracted from rat liver mitochondria and observed by the spreading technique. *A*, configuration in twisted circle ("super-coiled"); *B*, configuration in open circle. (Courtesy of B. Stevens.)

(i.e., they are proteolipids). It has been recognized, however, that the three largest subunits of cytochrome oxidase and one protein subunit of the cytochrome b-c$_1$ complex are made on mitoribosomes. Furthermore, in yeast, several subunits of the ATPase and probably a few hydrophobic proteins of the membrane are also made on *mitoribosomes.*

Table 11–7 shows a list of proteins originating in the cytosol and in the mitochondria and clearly demonstrates the cooperation of the two genetic systems in the synthesis of important mitochondrial protein complexes. One of the best known differences between the two mechanisms of protein synthesis is in the effect of some inhibitors. The mitochondrial system is in-

while synthesis in the cytosol is not affected. On the other hand, *cycloheximide* has the reverse effect.

As shown in Table 11–7, most mitochondrial proteins are coded by nuclear genes and are synthesized on free ribosomes in the cytosol. The import of the polypeptide seems to involve several of the steps mentioned in Section 8–5. These include the following: (1) the synthesis of a precursor polypeptide; (2) the binding of a precursor polypeptide to a receptor on the mitochondrial surface; and (3) the translocation of a precursor polypeptide into or across the mitochondrial membranes (see Schatz and Butow, 1983 in the Additional Readings at the end of this chapter). There are some differences, how-

TABLE 11–7.   BIOSYNTHESIS OF MAJOR MITOCHONDRIAL ENZYME COMPLEXES IN YEAST

		Number of Subunits	
Enzyme Complex	**Total**	**Made in Cytosol Ribosomes**	**Made on Mito- chondrial Ribosomes**
Cytochrome oxidase	7	4	3
Cytochrome b-c$_1$ complex	7	6	1
ATPase (oligomycin-sensitive)	9	5	4
Large ribosomal subunit	30	30	0
Small ribosomal subunit	22	21	1

From Borst, P.: Structure and function of mitochondrial DNA. *In* International Cell Biology. New York, The Rockefeller University Press, 1977, p. 237.

into the inner membrane and matrix compartment, the intermembrane space, or the outer mitochondrial membrane.

Figure 11–19 shows a suggested sequence of events in the import of a protein into the mitochondrial matrix (in this figure, the zigzag portion corresponds to the NH_2 terminal region, while the spiral portion corresponds to the mature sequence):

(1) In general, the polypeptide that is synthesized is longer than the mature protein and frequently has a NH_2 *terminal extension* that is longer than the signal peptide of secreted proteins (see Table 8–3).[28]

(2) In a second step, the precursor polypeptide attaches to a receptor at the mitochondrial surface. There are probably different specific receptors for the various proteins to be imported. It is interesting to note that these receptors are usually located at the sites in which the two mitochondrial membranes are in close contact (Fig. 11–17). Furthermore, the attachment of some cytoplasmic ribosomes to these regions has frequently been observed.

(3) The actual translocation across the membranes requires an electrochemical gradient[29,30] ($\Delta\bar{\mu}\,H^+$ in Fig. 11–19), which is produced by a post-transscriptional mechanism (i.e., it occurs after the molecule is fully synthesized, see Section 8–5). In fact, by using precursors that were synthesized in vitro and put in contact with isolated mitochondria, it has been shown that the precursor polypeptides reach the correct intramitochondrial location.[31]

(4) After the translocation, the NH_2^- terminal extension is cleaved by a protease located in the matrix. This enzyme, which has been partially purified in yeast, is also imported into the mitochondria as a larger precursor. How this enzyme is processed, however, still remains unclear (see Hay et al., 1984 in the Additional Readings at the end of this chapter).

(5) Once inside the matrix, the mature protein undergoes a process of refolding and in some cases a process of assembly with other proteins takes place.

The import of intrinsic proteins into the inner mitochondrial membrane is by a mechanism similar to that described in Figure 11–19 for a matrix protein. The transfer of proteins into the outer membrane is apparently done without proteolytic processing and does not need an electrochemical gradient across the intermembrane.[32]

A special case is posed by some proteins imported into the intermembrane space such as cytochrome b_2, which in its mature form carries a flavin and a heme group and is located in the intermembrane space in yeast mitochondria. In this case, however, a more complex two-step proteolytic process occurs (see Schatz and Butow, 1983 in the Additional Readings at the end of this chapter).

The Symbiont Hypothesis—Mitochondria and Chloroplasts are Intracellular Prokaryotic Parasites

At the end of the last century, cytologists like Altmann and Schimper speculated on purely morphological grounds that mitochondria and chloroplasts might be intracellular parasites that had established a *symbiotic* relationship with the eukaryotic cell. Bacteria were thought to have originated the mitochondria, and blue-green algae, the chloroplasts. In modern cell biology, with the recognition that these organelles have a certain degree of autonomy, this symbiont hypothesis has been reframed. According to the revised theory, the original host cell is conceived of as an anaerobic organism deriving its energy from glycolysis, a process that occurs in the cytoplasmic matrix, and the parasite contains reactants of the Krebs cycle and the respiratory chain and is able to carry on respiration and oxidative phosphorylation. The symbiont hypothesis is even more plausible in the case of plant cells, since the parasite would be the chloroplast, an autotrophic microorganism able to transform energy from light (see Margulis, 1971; Bogorald, 1975; Mahler and Raff, 1975; Saccone and Quagliarello, 1975; Bucher et al., 1977 in the Additional Readings at the end of this chapter).

The symbiont hypothesis is based on the many similarities between mitochondria, chloroplasts, and prokaryotes, which when considered from an evolutionary viewpoint, may be more than circumstantial. There are similarities in the *localization of the respiratory chain*. In bacteria this is situated in the plasma membrane, which also has the F_1-ATPase projecting into the matrix.

Certain bacteria have membranous projections extending from the plasma membrane, forming the so-called *mesosomes*. Such membranous projections, which are comparable to mitochondrial crests, have been separated and shown to contain the respiratory chain. The inner mitochondrial membrane and the matrix may represent the original symbiont enclosed within a membrane of cellular origin.

The *mitochondrial DNA* is circular, as it frequently is in chromosomes of prokaryons.

Protein synthesis in mitochondria and in bacteria is inhibited by chloramphenicol, whereas the extramitochondrial protein synthesis of the higher cell is not affected.

In mitochondria there is evidence of a DNA-dependent RNA synthesis, which indicates partial autonomy of this organelle. The amount of information carried by the mitochondrial DNA, however, is insufficient for an autonomous biogenesis.

In support of the possible prokaryotic origin of mitochondria and chloroplast is the fact that *intracellular symbiosis* may be found in nature. Thus, *paramecia* may contain certain bacteria, and blue-green algae may occur in simple animals. The endosymbiosis of a photosynthetic prokaryote confers upon the host the ability to capture light energy to synthesize various products. On the other hand, this provides for the prokaryote a constant environment in which to grow and reproduce. In evolutionary terms it is possible to conceive that a symbiotic relationship could have evolved into the present situation, in which the organelles have only a certain degree of autonomy and depend on the nucleus and cytosol of the cell for the synthesis of most of their specific components.

SUMMARY:
Mitochondrial Biogenesis

Mitochondrial and chloroplast biogenesis became of great interest with the demonstration that these organelles contain DNA and ribosomes and are able to synthesize proteins. They are called *semiautonomous* organelles because their biogenesis is the result of the cooperation of two genetic systems (i.e., mitochondrial and nuclear). A mitochondrion has one or more circular DNA molecules about 5.5 μm long localized in the matrix. This DNA is able to code for ribosomal RNA, transfer RNA, and a few proteins of the inner membrane. Mitoribosomes are in general smaller than cytoplasmic ribosome (55S against 80S). The proteins of these ribosomes are imported from the cytosol.

Mitochondria can synthesize some 12 proteins, which are very hydrophobic (i.e., they are proteolipids). Mitochondrial protein synthesis is inhibited by chloramphenicol (as in bacteria), while cycloheximide inhibits that of the cytosol. From Table 11–7 it is evident that the two genetic systems cooperate to make the mitochondrial proteins.

About 90% of the proteins are coded by nuclear genes and synthesized in the cytosol. Then they are transferred into the mitochondrion by a posttranscriptional mechanism.

The symbiont hypothesis postulates that mitochondria (and chloroplasts) originated from the symbiosis of a prokaryotic organism with a host cell that was anaerobic and derived its energy only from glycolysis. Mitochondria could be the result of a bacterium parasite, and chloroplasts, the result of a blue-green algae (having chlorophyll). In evolutionary terms it is possible that a symbiotic relationship could have evolved into the present situation, in which these organelles have only a certain degree of autonomy.

REFERENCES

1. Hogeboom, G.H., Schneider, W., and Palade, G.: J. Biol. Chem., *172*:619, 1948.
2. Bereiter-Hahn, J.: Biochim. Biophys. Acta, *423*:1, 1976.
3. Bereiter-Hahn, J.: Cytobiologie, *12*:429, 1976.
4. Loud, A.V.: J. Cell Biol., *37*:27, 1968.
5. Fernández-Morán, H.: Science, *140*:381, 1963.
6. Racker, E.: Fed. Proc., *26*:1335, 1967.
7. Sabatini, D.D., De Robertis, E., and Bleichmar, H.: Endocrinology, *70*:390, 1962.
8. De Robertis, E., and Sabatini, D.D.: J. Biophys. Biochem. Cytol., *4*:667, 1958.
9. Levy, M., Toury, R., and André, J.: Biochim. Biophys. Acta, *135*:599, 1967.
10. Schnaitman, C.V., Erwin, V.G., and Greenawalt, J.W.: J. Cell Biol., *32*:719, 1967.
11. Melnick, R.L., and Packer, L.: Biochim. Biophys. Acta, *253*:503, 1971.
12. Zalman, L.S., Nikaido, H., and Kagawa, Y.: J. Biol. Chem., *255*:1771, 1980.
13. Mihara, K., Blobel, G., and Sato, R.: Proc. Natl. Acad. Sci. USA *79*:7102, 1982.
14. Srere, P.A.: Trends Biochem. Sci., *6*:4, 1981.
15. Srere, P.A.: Trends Biochem. Sci., *7*:1, 1982.
16. Werner, S., and Neupert, W.: Eur. J. Biochem., *25*:369, 1972.
17. Green, D.E., and Goldberger, R.F.: *In* Molecular Insights into the Living Process. New York, Academic Press, Inc., 1966.
18. Schatz, G.: *In* The Structural Basis of Membrane Function. Edited by Y. Hatefi and L.D. Javadi-Ohaniance. New York, Academic Press, Inc., 1976.
19. Racker, E.: Sci. Am., *218*:32, 1968.
20. Racker, E.: Biochem. Soc. Trans., *3*:27, 1975.
21. Mitchell, P.: Fed. Proc., *26*:137, 1967.
22. Hackenbrock, C.R.: J. Cell Biol., *37*:345, 1968.
23. Knoll, G., and Brdiczka, D.: Eur. J. Cell Biol., *22*(1):281, 1980.
24. Luck, D.J.L.: J. Cell Biol., *24*:461, 1965.
25. De Robertis, E., and Bleichmar, H.B.: Z. Zellforsch, *57*:572, 1962.
26. Nass, M.M.K., and Nass, S.: J. Cell Biol., *19*:593, 1963.

27. Attardi, B., Attardi, G., and Aloni, Y.: J. Mol. Biol., 55:231, 251, and 271, 1971.
28. Neupert, W., and Schatz, G.: Trends Biochem. Sci., 6:1, 1981.
29. Gasser, S.M. et al.: Proc. Natl. Acad. Sci. USA, 79:267, 1982.
30. Schleyer, M., Schmidt, B., and Neupert, W.: Eur. J. Biochem., 125:109, 1982.
31. Zimmermann, R., and Neupert, W.: Eur. J. Cell Biol., 22 (1):152, 1980.
32. Freitag, H., et al.: Eur. J. Biochem., 126:197, 1982.

ADDITIONAL READINGS

Attardi, G., et al.: Organization and expression of genetic information in human mitochondrial DNA. In International Cell Biology. Edited by H.G. Schweiger. Berlin, Springer-Verlag, 1981, p. 225.
Bogorald, L.: Evolution of organelles and eukaryotic genomes. Science, 188:891, 1975.
Borst, P.: The biogenesis of mitochondria in yeast and other primitive eukaryotes. In International Cell Biology. Edited by H.G. Schweiger. Berlin, Springer-Verlag, 1981.
Boyer, P.D., et al.: Oxidation phosphorylation and photophosphorylation. Annu. Rev. Biochem., 46:995, 1977.
Bücher, T., Neupert, W., Sebald, W., and Werners, S.: Genetics and Biogenesis of Chloroplasts and Mitochondria. Amsterdam, North-Holland Publishing Co., 1977.
Bygrave, F.L.: Mitochondria and the control of intracellular calcium. Biol. Rev., 53:43, 1978.
Capaldi, R.A.: Arrangement of proteins in the mitochondrial inner membrane. Biochem. Biophys. Acta, 694:291, 1982.
Chance, B.: The nature of electron transfer and energy coupling reactions. FEBS Lett., 23:1, 1972.
Curgy, J.J.: The mitoribosome. Biol. Cell. 54:1, 1985.
De Pierre, J.W., and Ernster, L.: Enzyme topology of intracellular membranes. Annu. Rev. Biochem., 46:201, 1977.
Erecinska, M., and Wilson, D.F.: Cytochrome-c-oxidase: A synopsis. Arch. Biochem. Biophys., 188:1, 1978.
Ernster, L.: Chemical and chemiosmotic aspects of electron transport-linked phosphorylation. Annu. Rev. Biochem., 46:981, 1977.
Frederick, J.F.: Origin and evolution of eukaryotic intracellular organelles. Ann. NY Acad. Sci., 361, 1982.
Gillham, N.W.: Organelle Heredity. New York, Raven Press, 1978.
Hackenbrock, C.L.R.: Lateral diffusion and electron-transfer in the mitochondrial inner membrane. Trends Biochem. Sci., 6:151, 1981.

Hay, R., Böhni, P., and Gasser, S.: How mitochondria import proteins. Biochem. Biophys. Acta, 179:65, 1984.
Hinkle, P.C., and McCarty, R.E.: How cells make ATP. Sci. Am., 238:104, 1978.
Klingenberg, M., et al.: Mechanism of carrier transport and ADP, ATP carrier. In The Structural Basis of Membrane Function. Edited by Y. Hatefi and L. Djavadi-Ohaniance. New York, Academic Press, Inc., 1976, p. 293.
Lehninger, A.L.: Bioenergetics. New York, W.A. Benjamin, Inc., 1965.
Lehninger, A.L.: The Mitochondrion. New York, W.A. Benjamin, Inc., 1964.
Linnane, A.W., and Nagley, P.: Structural mapping of mitochondrial DNA. Arch. Biochem. Biophys., 187:277, 1978.
Mahler, H.R., and Raff, R.A.: The evolutionary origin of the mitochondrion, a nonsymbiotic model. Int. Rev. Cytol., 43:2, 1975.
Margulis, L.: Cell organelles such as mitochondria may have once been free living organisms. Sci. Am., 225:48, 1971.
Mitchell, P.: A commentary on alternative hypotheses of protonic coupling. FEBS Lett., 78:1, 1977.
Neupert, W., and Schatz, G.: How proteins are transported into mitochondria. Trends Biochem. Sci., 6:1, 1981.
O'Brien, T.W.: Transcription and translation in mitochondria. In International Cell Biology. New York, Rockefeller University Press, 1977, p. 245.
Papa, S.: Proton translocation reactions in the respiratory chains. Biochem. Biophys. Acta, 456:39, 1976.
Racker, E.: Reconstitution, Mechanism of Action, and Control of Ion Pumps. Biochem. Soc. Trans., 3:27, 1975.
Racker, E.: Membranes of Mitochondria and Chloroplasts. New York, Van Nostrand Reinhold. Co., 1970.
Saccone, C., and Kroon, A.M.: The Genetic Function of Mitochondrial DNA. Amsterdam, North-Holland Publishing Co., 1976, p. 354.
Saccone, C., and Quagliarello, E.: Biochemical studies on mitochondrial transcription and translation. Int. Rev. Cytol., 43:125, 1975.
Schatz, G., and Butow, R.A.: How are proteins imported into mitochondria? Cell, 32:316, 1983.
Sjöstrand, F.S.: The structure of mitochondrial membranes: A new concept. J. Ultrastruct. Res., 64:217, 1978.
Slater, E.C.: Mechanism of energy conservation. In Mitochondrial Biomembranes. Amsterdam, North-Holland Publishing Co., 1972, p. 133.
Tager, J.M., et al.: Control of mitochondrial respiration. FEBS Lett., 151:1, 1983.
Tzagoloff, A.: Mitochondria cellular organelles. New York. Plenum Publishing Corp., 1982.
Yaffe, M., and Schatz, G.: The future of mitochondrial research. Trends Biochem. Sci., 9:179, 1984.

PLANT CELL—CHLOROPLAST AND PHOTOSYNTHESIS

The cell biology of plants is essentially similar to that of animals; therefore throughout the book we use examples from the Plant Kingdom to illustrate a particular subject. In this way, we demonstrate the unity of structure and function of the eukaryotic cell. A separate chapter on the plant cell refers to some special topics, such as (1) the thick *cell wall* that covers the plasma membrane; (2) the *development* and *germination* of seeds; (3) some special organelles, like the *glyoxysomes* involved in lipid metabolism, and the *plastids*, which are related to the synthesis and storage of substances found in these cells. Of the plastids, the most important are the *chloroplasts*, which, along with the mitochondria,

are biochemical machines that produce energy transformations. In a chloroplast the electromagnetic energy contained in light is trapped and converted into chemical energy by the process of *photosynthesis*.

12–1 THE CELL WALL OF PLANT CELLS

The cell wall constitutes a kind of *exoskeleton* that provides protection and mechanical support for the plant cell. This includes the maintenance of a balance between the osmotic pressure of the intracellular fluid and the tendency of water to penetrate the cell. When plant cells are placed in a solution that has an osmotic pressure similar to that of the intracellular fluid, the cytoplasm remains adherent to the cellulose wall. When the solution of the medium is more concentrated than that of the cell, it loses water, and the cytoplasm retracts from the rigid cell wall. On the other hand, when the solution of the medium is less concentrated than that of the intracellular fluid, the cell swells and eventually bursts.

The *growth* and differentiation of plant cells—such as the formation, from cambial cells, of *xylem vessels* (which become lignified) or of *phloem tissue* (used to transport to the rest of the plant the material that has been photosynthesized in the leaves)—is mainly the result of the special synthesis and assembly of the cell wall.

The Cell Wall—A Network of Cellulose Microfibrils and a Matrix

The cell wall, in some respects, can be compared to a piece of plastic reinforced by glass fibers. The wall consists of a *microfibrillar* network lying in a gel-like *matrix* of interlinked molecules. The microfibrils are mostly *cellulose* (the most abundant biological product on earth) consisting of straight polysaccharide chains made of glucose units linked by 1–4 β-bonds. These are the *glucan* chains, which by intra- and intermolecular hydrogen bonding produce the structural unit known as the *microfibril*. Each microfibril is 25 nm in diameter and is composed of about 2000 glucan chains. Microfibrils in turn associate among themselves and with non-cellulosic polysaccharides and proteins to form the cell wall (Fig. 12–1).

Cellulose is synthesized by a wide variety of cells that include bacteria (e.g., *acetobacter*, *agrobacter*, and *rhizobium*), algae, fungi, cryptogams, and seed plants.

In addition to protein, the gel matrix contains some *polysaccharides* and *lignin*. The major polysaccharide fractions are (1) *pectin substances* (containing galactose, arabinose, and galacturonic acid), which are soluble in water, and (2) *hemicelluloses* (composed of glucose, xylose, mannose, and glucuronic acid), which are extracted with alkali.

Lignin is found only in mature cell walls and is made of an insoluble aromatic polymer resulting from the polymerization of phenolic alcohols. Certain cell walls may have *cuticular*

substances (*cutin waxes*) and *mineral* deposits in the form of calcium and magnesium carbonates and silicates. In fungi and yeasts the cell wall is composed of *chitin*, a polymer of glucosamine.

Development of Primary and Secondary Walls and Plant Cell Differentiation

We mentioned that cell walls are complex and highly differentiated in certain plant tissues. In fact, to a great extent the cell wall determines the *shape* of cells and serves in the classification of plant tissues. (Consideration of specialized plant tissues is beyond the scope of this book, for further information see Preston, 1974; Albersheim, 1975 in the Additional Readings at the end of this chapter.) In general, *primary* and *secondary* walls develop in a special sequence and are differentiated by the disposition of the cellulose microfibrils and by the composition of the matrix.

As we describe in Section 15–2, the *cell plate*, which is formed mainly from vesicles of the Golgi complex that become aligned at the equatorial plane (Fig. 12–2), forms the intercellular layer or *middle lamella* of the more mature cell wall; this layer contains pectin substances. After cell division each daughter cell deposits other layers that constitute the *primary cell wall*, which is composed of pectin, hemicellulose, and a loose network of the cellulose microfibrils, most of which are oriented transverse to the long axis of the cell.

When the cell has enlarged to its mature size, usually with the development of large vacuoles (Fig. 1–1), the *secondary wall* appears. This consists of material that is added to the inner surface of the primary wall, either as a *homogeneous thickening*, as in cells forming sieve tubes (*phloem*), or as *localized thickenings*, as in xylem vessels. In both cases this material consists mainly of cellulose and hemicellulose, with little pectin.

In xylem development a further differentiation consists of the penetration of *lignin* from the outside into the secondary thickenings. This hydrophobic polymer replaces water and finally encrusts all the microfibrils and the matrix. At this stage the wall is lignified and the cell dies.

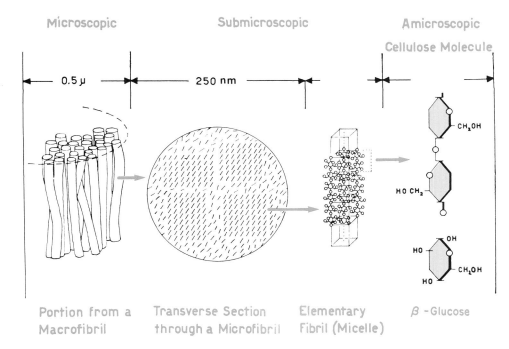

Fig. 12–1. Structural elements of cellulose at different levels of organization. (From Mühlethaler, K.: Plant cell walls. *In* The Cell. Vol. 2. Edited by J. Brachet, and A.E. Mirsky. New York, Academic Press, Inc., 1961.)

Cell Wall Components—Synthesis Associated with the Golgi or the Plasma Membrane

In the biogenesis of cellulose and other cell wall components two main pathways have been described. One involves the *Golgi complex*, and the other appears to be directly associated with the *plasma membrane*.

Involvement of the Golgi is very evident in some scale-bearing algae. These scales consist of a radial microfibrillar network associated with a spiral one, and an amorphous wall. The Golgi membranes with their *glycosyl transferase* content are able to polymerize glucan chains into cellulosic microfibrils. The whole process of synthesis, assembly, and release of the scales at the cell surface has been followed under the electron microscope (EM) (see Brown and Romanovicz, 1976 in the Additional Readings at the end of this chapter).

The plasma membrane is the most common site of cellulose synthesis in plant cells. One must not disregard, however, the essential functions that are probably carried out by the endoplasmic reticulum (ER) and the Golgi complex (see Chapters 8 and 9). It is possible that the glycosyl transferases synthesized in the ER could be transferred to the Golgi, and from there to the plasma membrane, where they may be-come active in the synthesis of microfibrils.[1] Using freeze-fracturing, globular complexes situated at the end of growing microfibrils have been observed in the secondary walls of cotton fibers. These globules have been interpreted as enzyme complexes involved in the synthesis of the microfibrils. Such globules, which are mobile, may be guided by cytoplasmic microtubules. (For further information on cellulose assembly and deposition see Brown, 1979 in the Additional Readings at the end of this chapter.)

We mentioned that in fungi the cell is mainly made of a polymer of glucosamine, known as *chitin*. This polysaccharide is synthesized by the enzyme *chitin synthetase* in the presence of the appropriate substrate (UDP-acetylglucosamine). This enzyme is activated by proteolysis, and light accelerates both the enzyme and the biosynthesis of chitin. Chitin synthetase has been found in vesicular organelles of 40 to 70 nm, the so-called *chitosomes*, which appear to be the main vehicle in the delivery of chitin synthetase to the sites of wall synthesis at the cell surface.[2]

Plasmodesmata Establish Communication Between Adjacent Cells

A characteristic of most plant cells is the presence of bridges of cytoplasmic material that es-

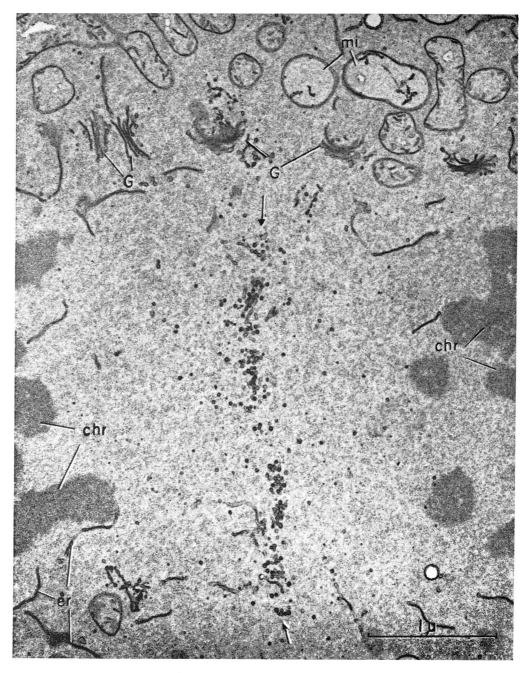

Fig. 12–2.   Electron micrograph of root cells of *Zea mays* at telophase. This region corresponds to the cell plate. Note at the top the marginal mitochondria *(mi)* and the Golgi complex *(G)* (dictyosomes). Between the arrows the vesicles are aligned to form the first evidence of a cell plate. *chr,* telophase chromosomes in the two daughter cells; *er,* endoplasmic reticulum. ×45,000. (Courtesy of W. Gordon Whaley and H.H. Mollenhauer.)

tablish a continuity between adjacent cells. These bridges, called *plasmodesmata*, pass through the thickness of the pectocellulose membrane. The presence of plasmodesmata permits the free circulation of fluid, which is essential to the maintenance of plant cell tonicity, and probably also allows passage of solutes and even of macromolecules. According to these concepts, cell walls do not represent complete partitions between cells, but constitute a vast syncytium supported by a skeleton formed by the pectocellulose membranes.

The formation of plasmodesmata is related to the *cell plate*, mentioned earlier, which appears at the equator of dividing cells during telophase. At this time the cell plate is crossed by vesicles and tubules of the ER that determine the location of the plasmodesmata (see Section 15–2).[3,4]

It has been suggested that plasmodesmata may play a role in cell differentiation.[5] In a cell that undergoes elongation the number of plasmodesmata is reduced along the axis and increased in the transverse walls.

SUMMARY:
The Plant Cell Wall

The plant cell wall represents a kind of exoskeleton that protects and provides mechanical support, as well as serving to balance the intracellular osmotic pressure with that of the medium. Differentiation of cambial cells into xylem vessels and phloem tissue is also dependent on the cell wall. The wall consists of a microfibrillar network (mainly of cellulose) and a gel-like matrix (containing pectin, hemicellulose, and lignin).

Development of the cell wall starts with the middle lamellae formed by the cell plate after cell division. Then the primary cell wall is made by the daughter cells. After growth of the cell, the secondary wall is added. In lignified tissues the secondary wall becomes encrusted with lignin. Biogenesis of cell wall components either involves the Golgi apparatus or is associated with the cell membrane. The formation of scales in algae occurs in the Golgi. The transfer of glycosyl transferases from the ER of the Golgi and the plasma membrane aids in the production of the cell wall.

Plasmodesmata are bridges of cytoplasm that establish communications between cells across the cell wall, and they may play a role in cell differentiation.

12–2 PLANT CELL CYTOPLASM

In Chapter 1, we discussed the basic similarities between the cytoplasm of animal and plant cells. In meristematic cells the membranes of the cytoplasmic vacuolar system are relatively scanty and are masked by the numerous ribosomes that fill the cytoplasmic matrix. Indeed, in undifferentiated cells most of these particles are not attached to the membranes, but are free in the matrix (Fig. 12–3).

In this section, we list some special features of the cytoplasm of plant cells.

Microtubules are preferentially localized below the plasma membrane and are oriented tangentially (see Fig. 1–1) in the regions of cytoplasm that underlie the points at which the microfibrils of the secondary wall are being deposited.[6] Thus, microtubules appear to play an important morphogenetic role in providing precise positional information for the deposition of the cell wall. This in turn determines cell division, cell shape, and other aspects of cell differentiation.[7]

The ER plays a role in the biosynthesis of proteins, lipids, and polysaccharides and in the formation of certain organelles, such as *protein bodies*, *glyoxysomes*, and vacuoles. Plasmodesmata are frequently crossed by tubules of smooth endoplasmic reticulum (SER). In cells that differentiate to form sieve tubes, pads of *callose*, a polysaccharide composed of glucose units linked by 1–3 β-bonds, is deposited over the ER. Thus, this organelle is also related to the formation of sieve pores (see Northcote, 1974 in the Additional Readings at the end of this chapter).

The great development of the SER during cell differentiation is related to the intense hydration of the cytoplasm. This process may give rise to huge vacuoles that are filled with liquid and may be confluent. As a result, the cytoplasm may become compressed in a thin layer against the cellulose membrane and may show cytoplasmic movements, called *cyclosis* (see Section 6–4). As in animal cells, the dictyosomes of some plant cells (e.g., root cap cells of maize) are directly related to secretion. The Golgi cisternae become filled with secretion products, which are then concentrated and discharged (Fig. 12–2).

Dictyosomes and their associated vesicles are numerous in cells involved in the synthesis of mucilage.

In developing cotyledons of *Vicia faba*, dictyosomes participate in the intracellular pathway of storage proteins such as *vicilin* and *legumin*.

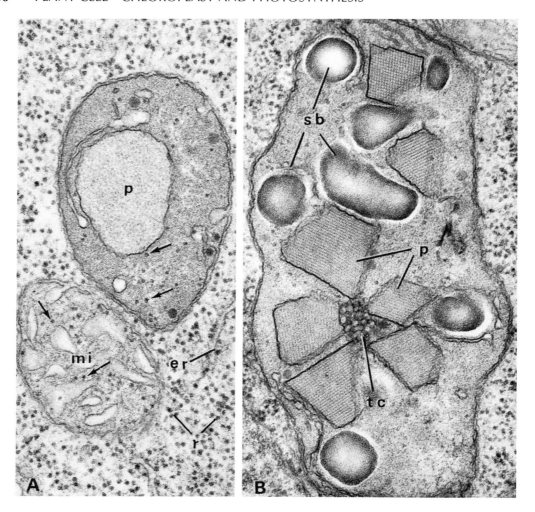

Fig. 12–3. Electron micrograph of plastids in bean root tips. *A,* a young plastid and a mitochondrion *(mi)* are observed in a meristematic cell. The plastid contains a protein body *(p).* The stroma is dense and contains granules and ribosomes (arrows). Numerous ribosomes *(r)* are in the cytoplasm. *B,* a plastid containing several crystalline protein bodies *(p)* and starch bodies *(sb).* Near the center it has a system of tubules called the tubular complex *(tc). A,* ×67,000; *B,* ×77,000. (Courtesy of E. H. Newcomb.)

Using antibodies and the protein A-colloidal gold technique, the different stages in the synthesis and transport of these proteins have been studied. Some of these proteins are glycosidated in the rough endoplasmic reticulum (RER) while others do so in the dictyosomes, which are involved in the transfer of legume storage proteins to the protein bodies, the final accumulation site (see Nieden et al., 1984 in the Additional Readings at the end of this chapter).

Mitochondria Can Be Distinguished from Proplastids

Mitochondria of plant cells have a structure essentially similar to that of those in animal cells.

In meristems, mitochondria have relatively few crests and an abundant matrix. During differentiation this internal structure may vary. In cells engaged in photosynthesis (leaf cells), mitochondria show an increased number of crests; in cells containing starch granules (amyloplasts), mitochondria remain undifferentiated, as in meristems (Fig. 12–3A).

One important point is the relationship between mitochondria and chloroplasts. Guillermond postulated that in early stages there are two types of organelles in leaf meristems. Although early plastids, also called proplastids, in some ways resemble mitochondria, they are readily distinguishable because of fewer projections of the inner membrane, their large size, and the presence of dense granules in the matrix (Fig. 12–3A).

Seed Development—Cell Division, Cell Expansion, and a Drying Phase

The seed can be considered as a resting stage of the plant in which the cells are in a highly dehydrated state. In addition to the primordia of the root and shoot, the seed is composed of food stores in the form of starch, fat, and proteins.

Seed development involves the following three main stages: (1) A *division phase*, in which the cells of the embryo and the cotyledons are formed, (2) A *phase of cell expansion*, in which proteins and other reserves are synthesized and stored in cotyledons, and (3) A *drying phase*, in which there is loss of water with further deposition of the reserves.

Many of the proteins (i.e., *zein* in corn endosperm, *vicilin* and *legumin* in legume cotyledons) are localized in special organelles called *protein bodies* or *aleurone grains* (Fig. 12–3A and B). During the expansion phase, the cells have an extensive RER, which is involved in the biosynthesis of stored proteins.[8] *Zein*, the main protein of corn, is synthesized by mRNA that is specifically associated with ribosomes attached to the ER.[9,10]

Seed Germination Involves Synthesis of Hydrolytic Enzymes

Seed germination is accompanied by the metabolism of the protein bodies of cotyledons. An *endopeptidase* synthesized in the seedlings penetrates into the protein bodies, causing hydrolysis of the stored proteins.

When treated with the hormone *gibberellic acid*, the aleurone cells of cereal endosperm produce α-amylase and other hydrolytic enzymes, which are released and attack the starch reserves. The synthesis of these enzymes is correlated with an extensive proliferation of the RER in these aleurone cells (see Chrispeels, 1977 in the Additional Readings at the end of this chapter).

Glyoxysomes—Organelles Related to Triglyceride Metabolism

The germination of the castor bean *(Ricinus)* is accompanied by the breakdown of a large amount of fat stored in the endosperm. This process is accompanied by the development of *glyoxysomes*, special organelles involved in triglyceride metabolism. Glyoxysomes consist of an amorphous protein matrix surrounded by a limiting membrane.

As in peroxisomes (see Section 10–4), glyoxysomal enzymes are made on free ribosomes in the cytosol, and then transferred into the organelle. The membrane originates from the ER.[11,12] The enzymes of the glyoxysome are used to transform the fat stores of the seed into carbohydrates by way of the *glyoxylate cycle*, which is a modification of the Krebs cycle. The overall equation of this cycle is as follows: 2 acetyl CoA → succinate + $2H^+$ + 2 CoA. The difference between the glyoxylate cycle and the Krebs cycle is that the former uses two auxiliary enzymes, *isocitrate lyase* and *malate synthetase*, and requires two molecules of acetyl CoA, instead of one. The other three enzymes of the cycle (*citrate synthetase*, *aconitase*, and *malate dehydrogenase*) are the same as those present in the Krebs cycle (Fig. 12–4 *Top*).

All the enzymes of the glyoxylate cycle have been found in isolated glyoxysomes of the castor bean endosperm. When seeds germinate, considerable change is produced involving the synthesis of glyoxysomic enzymes. If the endosperm is fractionated at different stages of germination, it is possible to observe that the marker enzyme malate synthetase shifts from fractions corresponding to the ER to others representing the glyoxysomes (Fig. 12–4 *Bottom*). These findings support the concept that these organelles originate from the ER, although the enzymes are made in the cytosol (see Chrispeels, 1977 in the Additional Readings at the end of this chapter).

SUMMARY:
Plant Cell Cytoplasm

Microtubules are preferentially localized under the plasma membrane, and they may play a role in guiding the enzymes involved in cell wall biosynthesis.

Mitochondria of plant cells do not differ from those of animal cells and can be distinguished from early plastids or proplastids.

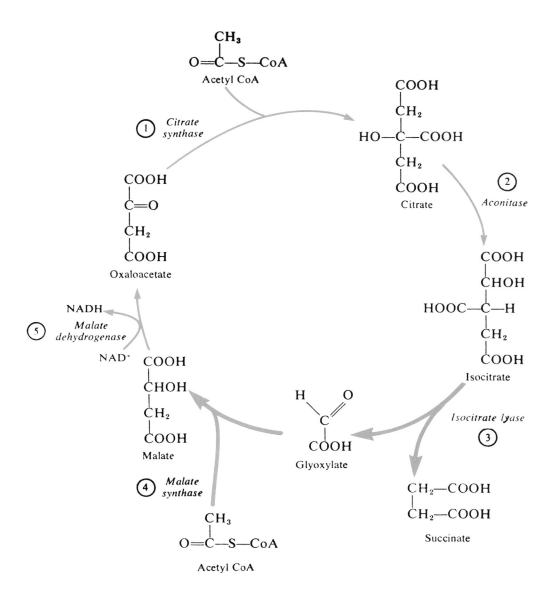

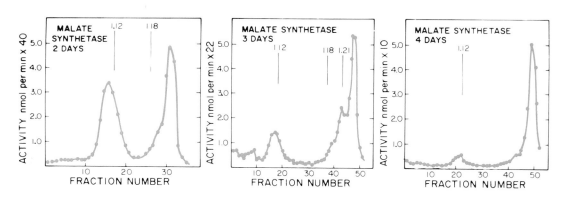

Fig. 12–4. The glyoxylate cycle in castor bean endosperm and an experiment showing the localization of enzyme synthesis. *Top*, the glyoxylate cycle, showing the special steps (heavy arrows) that differentiate it from the Krebs cycle of mitochondria. *Bottom*, fractionation of the enzyme malate synthetase during germination of the castor bean endosperm. Germination was carried out for 2, 3, and 4 days, and the endosperm homogenate was centrifuged on a continuous gradient. Observe that at 2 days there are two peaks of enzyme activity corresponding to the ER and the glyoxysomes; with time, the first peak is reduced and the second becomes predominant. (From Gonzalez, E., and Beevers, H.: Plant Physiol., *57*:406, 1976.)

The ER (rough and smooth) is involved in the formation of protein bodies, glyoxysomes, and vacuoles. A seed can be considered as the plant primordium, in a highly dehydrated state, together with food stores of starch, fat, and proteins. Many of the proteins (zein, vicilin, legumin) are in protein or aleurone bodies. During seed germination an endopeptidase is synthesized and penetrates into the protein bodies and hydrolyzes the proteins. In cereal endosperm treated with gibberellic acid (plant hormone), α-amylase is synthesized and released to attack the starch stores. In *Ricinus* germination results in the development of *glyoxysomes*, which become involved in triglyceride metabolism. Fat stores are transformed into carbohydrates by the glyoxylate cycle.

Two special enzymes of glyoxysomes are *isocitrate lyase* and *malate synthetase* (Fig. 12–4). Golgi complexes in the form of dictyosomes are dispersed throughout the cytoplasm and are involved in secretion (e.g., of mucilage).

12–3 THE CHLOROPLAST AND OTHER PLASTIDS

In 1883 Schimper first used the term "plastid" for special cytoplasmic organelles present in eukaryotic plant cells. The most important of all, the *chloroplasts*, are characterized by the presence of pigments such as *chlorophyll* and *carotenoids* and by their fundamental role in *photosynthesis*.

In addition to chloroplasts, other colored plastids may be observed. These are grouped under the name *chromoplasts*. Yellow or orange chromoplasts occur in petals, fruits, and roots of certain higher plants. In general, they have a reduced chlorophyll content and are thus less active photosynthetically. The red color of ripe tomatoes is the result of chromoplasts that contain the red pigment *lycopene*, a member of the carotenoid family. Chromoplasts containing various pigments (e.g., *phycoerythrin* and *phycocyanin*) are found in algae.

Colorless plastids or leukoplasts are found in embryonic and germ cells. Leukoplasts in certain differentiated zones of the root produce starch granules, called *amyloplasts* (Fig. 12–3*B*). Leukoplasts are also found in meristematic cells and in those regions of the plant not receiving light.

Plastids located in the cotyledon and the primordium of the stem are colorless at first but eventually become filled with chlorophyll and acquire the characteristic green color of chloroplasts. True leukoplasts, however, are found in fully differentiated cells and never become green. Recently, the leukoplasts have also been characterized by the absence of thylakoids and ribosomes (see Carde, 1984 in the Additional Readings at the end of this chapter).

We mentioned earlier the protein bodies or *proteinoplasts;* others contain starch (*amyloplasts*, fat (*elaioplasts or oleosomes*), or essential oils.[13]

Chloroplasts are the most common of the plastids and also those with the greatest biological importance. By the process of *photosynthesis*, they produce oxygen and most of the chemical energy used on our planet by living organisms. Life is essentially maintained by the chloroplasts; without them there would be no plants or animals, since the latter feed on the foodstuffs produced by plants. From a human point of view we may say that each molecule of oxygen used in respiration and each carbon atom in our bodies has at one time passed through a chloroplast. Chloroplasts are localized mainly in the cells of the leaves of higher plants and in algae.

From the evolutionary viewpoint the first living organisms were anaerobic and appeared at least one billion years before photosynthesis. This process appeared first in prokaryotes and then in eukaryotes. With the release of oxygen, a by-product of photosynthesis, the atmosphere favored aerobic organisms, and anaerobic organisms became a small fraction of all living forms.

Chloroplast Morphology Varies in Different Cells

The *shape* and *size* of chloroplasts may vary in different cells within a species, but these organelles are relatively constant within cells of the same tissue. In leaves of higher plants, each cell contains a large number of spheroid, ovoid, or discoid chloroplasts. Chloroplasts are frequently vesicular, with colorless centers. The presence of starch granules in these organelles is detected by the characteristic blue iodine reaction.

The average diameter in higher plants is 4 to 6 μm. This is constant for a given cell type, but sexual and genetic differences are found. For instance, chloroplasts in polyploid cells are larger than those in the corresponding diploid cells. In general, chloroplasts of plants grown in the shade are larger and contain more chlorophyll than those of plants grown in sunlight.

The *number* of chloroplasts is relatively constant in the different plants. Algae, such as *chlamydomonas*, often possess a single huge chloroplast. In higher plants there are 20 to 40 chloroplasts per cell. It has been calculated that the leaf of *Ricinus communis* contains about 400,000 chloroplasts per square millimeter of surface area. When the number of chloroplasts is insufficient, it is increased by division; when excessive, it is reduced by degeneration. Chloroplasts are motile organelles that have passive and active movements.

Changes in shape and volume caused by the presence of light have been observed in chloroplasts isolated from spinach. The volume decreases considerably after the chloroplasts are struck by light and photophosphorylation is initiated; this effect is reversible. Chloroplasts apparently multiply by division—elongation of the plastid and constriction of the central portion.

Chloroplast Structure—Envelope, Stroma, and Thylakoids

The light microscope had already demonstrated that many chloroplasts have a heterogeneous structure made up of small granules called *grana*, which are embedded within the stroma, or matrix. The size of the grana varies from 0.3 to 1.7 μm depending on the species.

The EM has revealed the true structure of the chloroplast, with its three main components: the envelope, the stroma, and the thylakoids (Fig. 12–5).

The *envelope* is made of a double limiting membrane across which the molecular interchange with the cytosol occurs. In contrast to mitochondria, the inner membrane of mature chloroplasts is not in continuity with the thylakoids (Fig. 12–6). Isolated membranes have a yellow color due to the presence of small amounts of carotenoids. These areas, however, lack chlorophyll and cytochromes and contain only 1 to 2% of the protein of the chloroplast.

The *stroma* fills most of the volume of the chloroplasts and is a kind of gel-fluid phase that surrounds the thylakoids. This component contains about 50% of the chloroplast proteins and most of these are of the soluble type. It has ribosomes and also DNA, both of which are in-

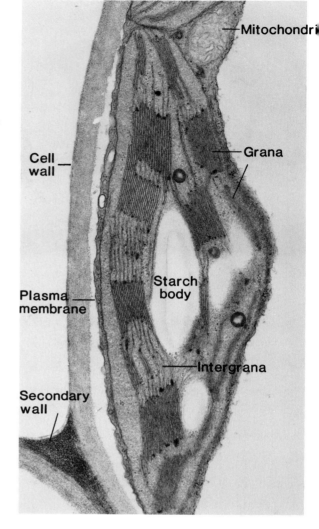

Fig. 12–5. Longitudinal section of a chloroplast of *Linaria vulgaris* showing the grana and the intergranal thylakoids. (Courtesy of M. Cresti and M. Wurtz.)

volved in the synthesis of some of the structural proteins of the chloroplast. The stroma is where CO_2 fixation occurs and where the synthesis of sugars, starch, fatty acids, and some proteins takes place.

The *thylakoids* consist of flattened vesicles arranged as a membranous network. The outer surface of the thylakoid is in contact with the stroma, and its inner surface encloses an *intrathylakoid space*. Thylakoids may be *stacked* like a pile of coins, forming the grana (Fig. 12–6B), or they may be *unstacked (stroma thylakoids)*, forming a system of anastomosing tubules that

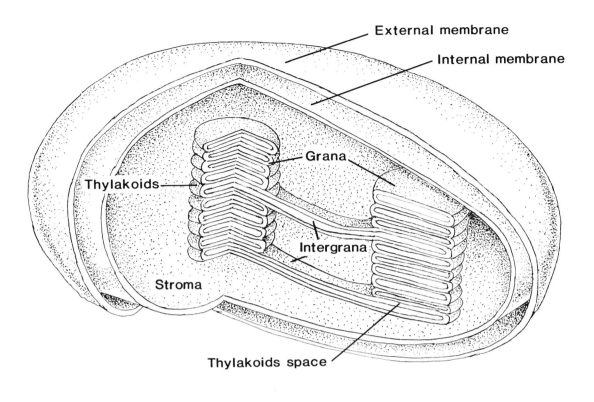

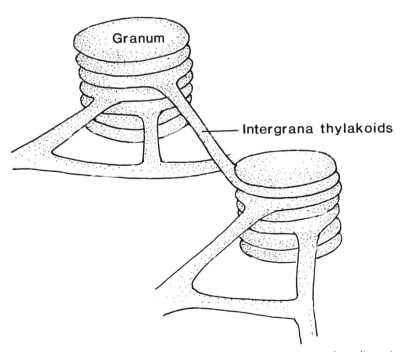

Fig. 12–6. *Top,* diagram of a chloroplast showing the main structural components. *Bottom,* three-dimensional diagram of two grana with the stacked thylakoids and the unstacked ones that cross through the stroma of the chloroplast.

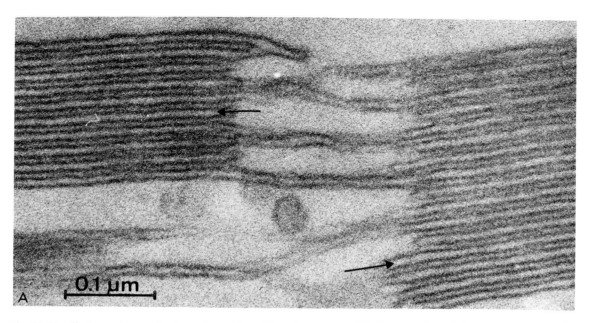

Fig. 12–7.   Electron micrograph of a section of two chloroplast grana and their connecting tubules (see Fig. 12–6). The arrows indicate the cavity of the membranous compartment of the grana (i.e., thylakoid). ×240,000. (Courtesy of I. Nir and D.C. Pease.)

are joined to the *grana thylakoids* (Fig. 12–7). The number of thylakoids per granum may vary from a few to 50 or more. The thylakoids contain about 50% of the protein and all the components involved in the essential steps of photosynthesis.

12–4   MOLECULAR ORGANIZATION OF THYLAKOIDS

Modern ideas about the molecular organization of thylakoids are largely based on the fluid lipid-protein mosaic model of the membrane proposed by Singer and Nicolson in 1972 (see Section 4–2). This basic concept includes several fundamental properties: (1) *fluidity*, which allows for the free lateral movement of components in the thylakoid membrane; (2) *asymmetry*, in the sense that the crossing of molecules from one half of the bilayer to the other is restricted, permitting the separation of different components; and (3) *economy*, since placing the components in a thin membrane eliminates random movements in the third dimension.[14,15]

Lipids represent about 50% of the thylakoid membrane; this includes those directly involved in photosynthesis, such as *chlorophylls*, *carotenoids*, and *plastoquinones*. There are also *structural* lipids, which are mainly glycolipids,

sulfolipids, and a few phospholipids. Most of the structural lipids are highly unsaturated, which confers to the thylakoid membrane a high degree of fluidity.

Chlorophyll is an asymmetrical molecule having a hydrophilic head made of four pyrrole rings bound to each other and forming a porphyrin (Fig. 12–8). This part of the molecule is similar to some animal pigments, such as hemoglobin and cytochromes. In chlorophyll, however, there is a Mg atom forming a complex with the four rings. In animal pigments the Mg is replaced by Fe. Chlorophyll has a long hydrophobic chain (phytol chain) attached to one of the rings.

In higher plants there are two types of chlorophyll—*a* and *b*. In chlorophyll *b* there is a —CHO group in place of the —CH$_3$ group, indicated by a circle in Figure 12–8.

A small amount of pigment absorbing at a wavelength of 700 nm has been found. This pigment is called P700 because it is bleached at 700 nm. There is also a P680 that bleaches when absorbing light at 680 nm.

Pigments that belong to the group called *carotenoids* are masked by the green color of chlorophyll. In autumn, the amount of chlorophyll decreases and the other pigments become ap-

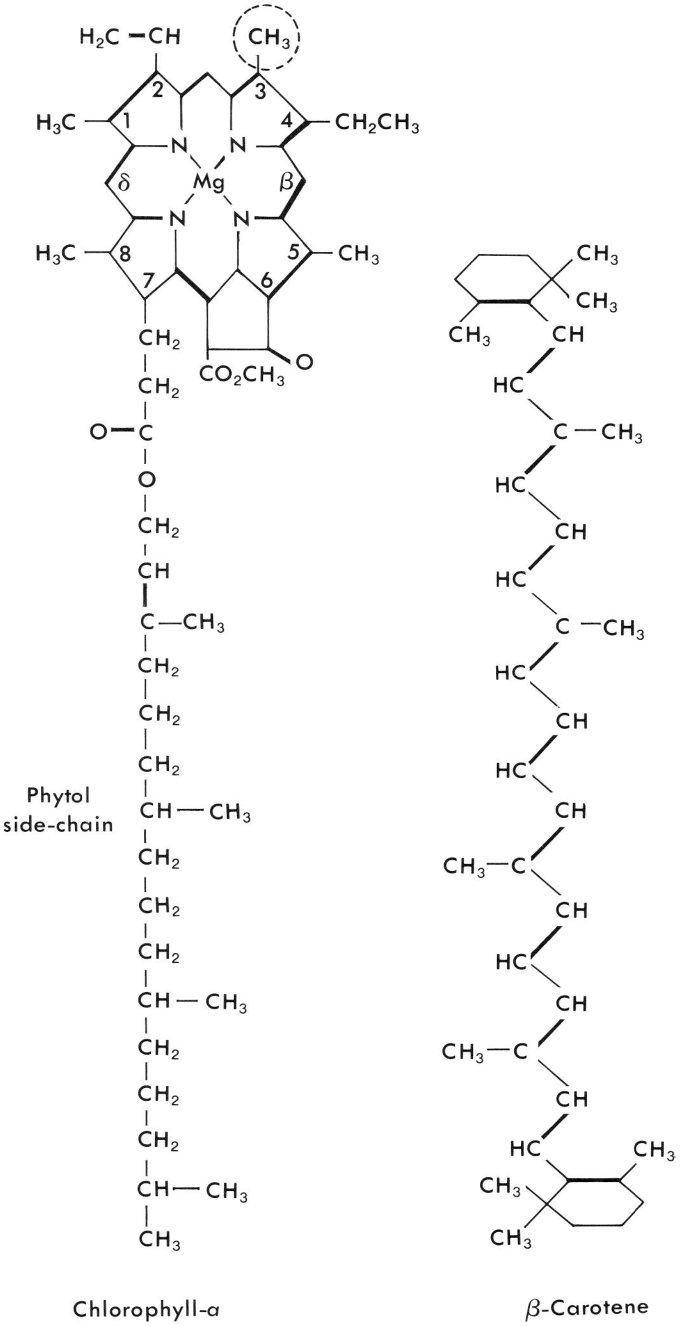

Fig. 12–8. Structural formula of chlorophyll-a and β-carotene.

Chlorophyll-*a*

β-Carotene

parent. These belong to the *carotenes* and *xanthophylls*, which are both related to vitamin A. Carotenes are characterized chemically by the presence of a short chain of unsaturated hydrocarbon, which makes them completely hydrophobic (Fig. 12–8). Xanthophylls, on the contrary, have several hydroxyl groups.

Several Chlorophyll-Protein Complexes in Thylakoid Membranes

The protein components of the thylakoid membrane are represented by 30 to 50 polypeptides disposed in the following five major supramolecular complexes, which can be isolated with mild detergents[16,17] (Table 12–1) (see Staehelin and De Witt, 1984 in the Additional Readings at the end of this chapter).

(A) *Photosystem I (PS I)*. This complex contains a reactive center composed of P700, several polypeptides, a lower chlorophyll *a/b* ratio than that found in PS II, and β-carotene. It also acts as a light trap and is present in unstacked membranes.

(B) *Photosystem II (PS II)*. This complex is comprised of two intrinsic proteins that bind to a reaction center composed of chlorophyll P680. It contains a high ratio of chlorophyll *a* over *b* and β-carotene. Frequently, the PS IIs are associated with the light-harvesting complex. PS II works as a light trap in photosynthesis and is mainly present in the stacked membranes of grana.

(C) *Cytochrome b/f*. This complex contains one cytochrome f, two cytochromes of b563, one FeS center, and a polypeptide. It is uniformly distributed in the grana.

As shown in Figure 12–9, these three complexes are related to the electron transport and are linked by mobile electron carriers (i.e., plastoquinone, plastocyanin, and ferredoxin). Electron transport through PS II and PS I finally results in the reduction of the coenzyme $NADP^+$. Simultaneously, the transfer of protons from the outside to the inside of the thylakoid membrane occurs.

(D) *ATP synthetase*. As in mitochondria, this complex consists of a CF_0 hydrophobic portion, a proteolipid that makes a proton channel, and a CF_1 or *coupling factor* that synthesizes ATP from ADP and Pi, using the proton gradient provided by the electron transport (Fig. 12–9).

(E) *Light-Harvesting Complex* (LHC). The main function of the LHC is to capture solar energy. It contains two main polypeptides and both chlorophyll *a* and *b*. As shown in Figure 12–9 the LHC is mainly associated with PS II, but may also be related to PS I.[14] LHC is mainly localized in stacked membranes and lacks photochemical activity (Table 12–1).

Freeze-fracturing Best Reveals the Fine Structure of the Thylakoid Membrane

Detailed information about the macromolecular organization of the thylakoid membrane has been gained by the use of the technique of freeze-fracturing (see Section 3–2). The diagram of Figure 12–10 shows the nomenclature that has been adopted to interpret the images produced by freeze-fracturing.[18] It is based on the fact that all biological membranes consist of two leaflets, a protoplasmic (P) and an exoplasmic (E).

Each leaflet has a true surface (S) and a fracture face (F). Thus, the designation PF refers to a fracture showing the protoplasmic leaflet, and ES corresponds to the true surface of the exoplasmic leaflet (in this case corresponding to the intrathylakoid space). The nomenclature also contains the subscripts *s* and *u*, referring to stacked or unstacked membranes. (Thus PFu refers to a fracture showing the protoplasmic face of an unstacked thylakoid membrane.)

In experiments using freeze-fracturing, the thylakoid membrane appears as a smooth continuum, representing one half of the lipid bilayer, with particles of various sizes believed to be protein macromolecules or aggregates that

TABLE 12–1. CHLOROPHYLL-PROTEIN COMPLEXES IN THYLAKOID MEMBRANES

Complex	Chlorophyll Content	Localization in Grana	Function
Photosystem II	a/b ratio > 25 contains P680	Stacked membranes (~85%)	Light induced release of O_2 from H_2O
Photosystem I	a/b ratio > 7.5 contains P700	Unstacked membranes (>85%)	Light induced reduction of $NADP^+$
Cytochrome b/f	traces	Uniform distribution	Electron carrier
ATP Synthetase	none	Unstacked membranes (100%)	Proton translocation synthesis of ATP
Light Harvesting	a/b ratio 1.3–1.5	Stacked membranes (70–90%)	Binds to PS II, lacks photochemical activity

Data from Staehelin, L.A., and De Wit, M.: Correlation of structure and function of chloroplast membranes at the supramolecular level. J. Cell. Biochem., 24:261, 1984.

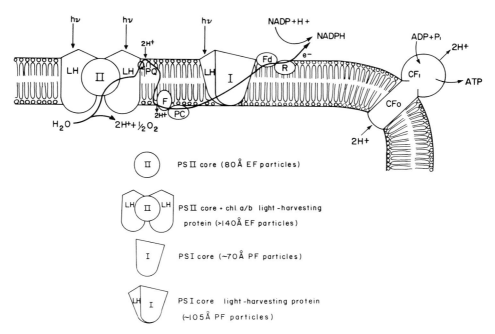

Fig. 12–9. Diagram representing the molecular structure of the thylakoid membrane. The continuous line indicates the flow of electrons through the various complexes. The light energy (hv) is trapped by photosystems II and I (PS II and PS I). Photolysis of water is produced at the inner side of the thylakoid; electrons are passed to plasquinone (PQ), which picks 2 protons (H+) from the stroma. The electron is passed to cytochrome f (F) and the 2 protons are discharged into the lumen. Electrons are then passed to plastocyanin (PC) and to the core of PS I, which absorbs light primarily through the associated light-harvesting proteins (LH). Then the electrons are passed to ferredoxin (Fd) and by way of a reductase (R) are able to reduce NADP+ to NADPH. The accumulation of protons inside the thylakoid makes a gradient relative to the stroma that induces the production of ATP by the ATP synthetase (CF$_1$). CF$_0$ acts as a proton channel. The various complexes and the corresponding size particle that may be observed in freeze-fracture electron micrographs are indicated below. See Figure 12–13 for more details on the electron flow. (Courtesy of L. Bogorad.)

are asymmetrically localized within the lipid matrix. The freeze-cleaving of the membrane results in a selective segregation of the particles in either the PF or the EF leaflet.[18] Observation of Figure 12–10 shows that PFu and PFs contain a large number of small particles, while in EFs the particles are larger and less numerous. Very few of these larger particles are found in the EFu face (corresponding to the unstacked intra-thylakoid face).

Table 12–2 shows the distribution of particles in the PF and EF leaflets of the thylakoid membranes. The numbers indicate that the small PF particles are distributed similarly between stromal (PFu) and granal (PFs) membranes, whereas, the large particles of the EF leaflet are more concentrated in the granal regions (EFs).

Several studies identify the large EF particle with PS II, together with the associated light-harvesting complex, i.e., the complete PS II. As indicated in Table 12–2, the proportion of PS II in both the stacked and unstacked regions of the

thylakoid is the same as that of EF particles. These concepts are summarized in the model of thylakoid membrane shown in Figure 12–11, in which the various complexes are embedded in a lipid continuum.

The described distribution of particles is representative of the heterogeneous organization of the photosystems and LHC complexes in chloroplasts that contain stacks of grana. In chloroplasts having a single thylakoid (e.g., during development) a random distribution of all the complexes occurs along the plane of the membrane. It is only when the thylakoid membranes adhere (or are appressed) that the heterogeneity of macromolecular organization appears (Fig. 12–11). Some evidence suggests that the adhesion in the grana region is mediated by the complete PS II, with LHC. For example, if thylakoid membranes become unstacked when a low salt solution is used, the distribution of the EF particles becomes random. On treatment with a high salt solution, an opposite reaction takes

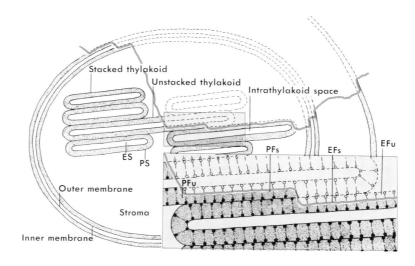

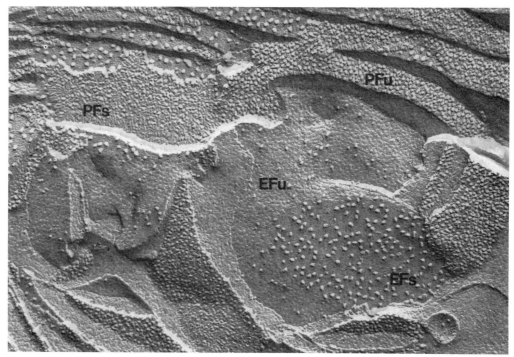

Fig. 12–10. *Top,* nomenclature of freeze-fracturing technique as applied to the chloroplast structure. The red line traces the course of the hypothetical fracture prior to etching. In the inset note that the cleavage plane tends to follow the hydrophobic region of the membrane. *P,* protoplasmic leaflet; *E,* exoplasmic leaflet; *S,* the membrane surface; *F,* fracture; *s,* stacked thylakoid; *u,* unstacked thylakoid. (Modified from Branton, D., et al.: Science, *190*:54, 1975. Copyright 1975 by the American Association for the Advancement of Science.)

Bottom, electron micrograph of a freeze-fractured chloroplast of *Pisum sativum l.* showing the large particles observed in EFs and the smaller particles in PFu and PSs. ×76,000. (From Armond, P.A., Staehelin, L.A., and Arntzen, C.J.: J. Cell Biol., *73*:400, 1977.)

TABLE 12–2.   DISTRIBUTION OF PARTICLES IN THYLAKOIDS

	Stromal Thylakoids	Granal Thylakoids
Small PF particles per μm^2	3409	3620
Large EF particles per μm^2	574	1495
Proportion of EF particles	20%	80%
Proportion of photosystem II	20%	80%

Data from freeze-fracture electronmicroscopy by Arntzen, C.J.: Dynamic structural features of chloroplast lamellae. *In* Current Topics in Bioenergetics. Vol. 8. Edited by D.R. Sanadi and L.P. Vernon. New York, Academic Press Inc., 1978, p. 111.

place, i.e., the membranes become restacked, and segregation of the EF particles occurs (see Anderson, 1982 in the Additional Readings at the end of this chapter). (For the arrangement of polypeptides in the thylakoid membrane, see von Wettstein, 1981 in the Additional Readings at the end of this chapter.)

SUMMARY:

Structure and Molecular Organization of Chloroplasts

Eukaryotic plant cells have specialized organelles—the *plastids*—which contain pigments and may synthesize and accumulate various substances. *Leukoplasts* are colorless plastids; *amyloplasts* produce starch; *proteinoplasts* accumulate protein; and *elaioplasts* produce fats and essential oils. *Chromoplasts* are colored plastids that contain less chlorophyll than the chloroplasts, but more carotenoid pigments, such as lycopene. Some plastids may store starch and protein at the same time.

Chloroplasts are the most important and most common plastids. Their shape, size, number, and distribution vary in different cells but are fairly constant for a given tissue. In higher plants they are discoid, 4 to 6 μm in diameter, and number about 20 to 40 per cell. The size and number are genetically controlled; in polyploids, therefore, they are larger. Chloroplasts multiply by division. The quantity of light available causes chloroplasts to undergo changes in shape and volume caused by contraction or swelling.

Many chloroplasts, under the light microscope, show small granules, called grana, of 0.3 to 1.7 μm embedded within the matrix. The EM reveals their three main components: the envelope, stroma, and thylakoids. The envelope is made of two membranes, as in mitochondria; however, in the grana there is no continuity between the inner membrane and the thylakoids.

The stroma is a gel-fluid phase that contains 50% of the chloroplast proteins (most are soluble). It has ribosomes and DNA. The thylakoids are flattened

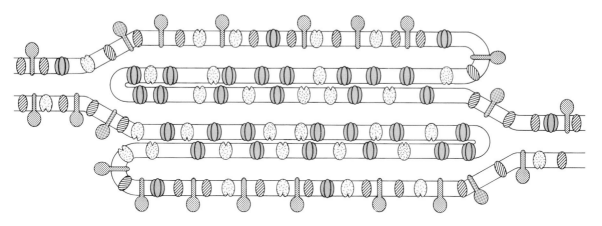

$\oslash$ PS 1 complex — LHC$_1$ ATP synthetase

$\oslash$ PS 2 complex — LHC$_2$ Cytochrome <u>b</u> <u>f</u> complex

Fig. 12–11. Diagram showing the distribution of the main complexes within the thylakoid membrane both in the granal (stacked) and stromal (unstacked) regions. The different complexes are indicated at the bottom of the figure. Observe that the ATP synthetase is only present in the stromal surface of the grana and the stromal thylakoids. The stacked regions contain the other three complexes. (Courtesy of J.M. Anderson and B. Anderson.)

vesicles forming a membranous network. Within grana, thylakoids are stacked; those that are unstacked (stromal thylakoids) form a system of anastomosing tubules that connect the granal thylakoids. Thylakoids contain about 50% of the chloroplast protein and all the components essential to photosynthesis.

The fine structure of the thylakoid membrane is best revealed by freeze-fracturing (Fig. 12–10), after which a smooth continuum corresponding to half of the lipid bilayer and containing particles of various sizes can be observed. In the PFu and PFs (protoplasmic leaflets, unstacked and stacked), there are numerous small particles, and in the EFs (exoplasmic leaflet, stacked) the particles are larger and less numerous. In the exoplasmic unstacked leaflet (EFu), there are few of the large particles. Molecular organization of thylakoids is based on the fluid-mosaic model of the membrane, of which the main characteristics are fluidity, asymmetry, and economy (i.e., lack of movement in the third dimension).

In thylakoids, chlorophyll, carotenoid molecules, and a reaction center are assembled, forming two photosystems (I and II). Each photosystem is associated with an electron transport system and with structural proteins. Lipids represent 50% of the thylakoid membrane. These include those involved in photosynthetsis (i.e., chlorophyll, carotenoids, and plastoquinone) and structural lipids.

Chlorophyll, the main pigment, is an asymmetrical molecule with a porphyrin head composed of four pyrrole rings and forming a complex with a Mg atom. The molecule also has a long hydrophobic phytol chain. There are several types of chlorophylls (a, b, c, d, and e). Types a and b are found in higher plants. This green pigment absorbs at 663 nm in the isolated condition; another pigment called P700 is bleached at 700 nm.

Chlorophyll molecules and structural proteins are assembled in the following complexes: PS I complex, which has the reaction center molecule P700 and a high a/b ratio; the PS II complex, which has a low a/b ratio and is present mainly in grana thylakoids, and the light-harvesting chlorophyll-protein complex (LHCP), which contains 40 to 60% of the chlorophyll in the thylakoid, but is photochemically inactive.

Chloroplasts have a photophosphorylation coupling factor (CF_1) similar to that of mitochondria. The CF_1 particles are localized mainly in the outer surface of stromal thylakoids. Both PS I and PS II have a vectorial arrangement across the membrane.

Several molecular complexes may be isolated by solubilizing thylakoid membranes with non-ionic detergents. In addition to PS I and PS II and the LHC are the cytochrome f-b_6 complex and the CF_0 or HF_0 (the hydrophobic component of ATPase). The CF_0 complex has six polypeptides and is capable of proton transfer. The CF_1 is similar to the F_1 of mitochondria and has five polypeptides. Based on the existence of these complexes and the

data from freeze-fracturing, a functional model of the thylakoid membrane has been proposed. The large EF particles are identified as PS II systems associated with LHCs. The small PF particles are more difficult to interpret, since they may correspond to the other protein complexes mentioned in Table 12–1.

12–5   PHOTOSYNTHESIS

Photosynthesis is one of the most fundamental biological functions. By means of the chlorophyll contained in the chloroplasts green plants trap the energy of sunlight emitted as photons (i.e., excitons) and transform it into chemical energy. This energy is stored in the chemical bonds that are produced during the synthesis of various foodstuffs.

In Chapter 11, we emphasized how mitochondria can utilize and transform the energy contained in the foodstuffs by oxidative phosphorylation. Photosynthesis is somewhat the reverse process (see Table 11–1). Chloroplasts and mitochondria have many structural and functional similarities, but several differences also exist.

The overall reaction of photosynthesis is:

$$nCO_2 + nH_2O \xrightarrow[\text{chlorophyll}]{\text{light}} (CH_2O)_n + nO_2 \tag{1}$$

This indicates that, essentially, photosynthesis is the combining of carbon dioxide and water to form various carbohydrates with release of oxygen.

It has been calculated that each CO_2 molecule from the atmosphere is incorporated into a plant every 200 years, and that all the oxygen in the atmosphere is renewed by plants every 2000 years. Without plants there would be no oxygen in the atmosphere, and life would be almost impossible.

In 1771, Joseph Priestley observed that a plant put in a closed glass chamber not only survived, but was able to sustain a candle flame or a living mouse. With this experiment demonstrating the production of oxygen by the plant, he found the fundamental relationship between plants and animals.

At this point, it is important to emphasize that in reaction (1) water is the donor of both H_2 and O_2. This can be demonstrated by an experiment in which either water or carbon dioxide is labeled with an isotope of oxygen (^{18}O). The results obtained can be written as follows:

$$nCO_2 + nH_2{}^{18}O \xrightarrow{\text{light}} [CH_2O]_n + n^{18}O_2 \tag{2}$$

$$nC^{18}O_2 + nH_2O \xrightarrow{\text{light}} [CH_2{}^{18}O]_n + nO_2 \tag{3}$$

From these two experiments, the student can immediately recognize that oxygen comes from water and not from carbon dioxide. In these reactions, CO_2 acts as an acceptor (A) of electrons and H_2O as a donor (D) of electrons. Thus, a more general equation for photosynthesis would be:

$$H_2D + A \xrightarrow{\text{light}} H_2A + D \tag{4}$$

(where H_2D is the donor and A the acceptor of H and electrons).

The carbohydrates first formed by photosynthesis are soluble sugars; these can be stored as granules of starch or other polysaccharides inside the chloroplasts (Fig. 12–5) or, more usually, inside the amyloplasts. After several steps involving different types of plastids and enzymatic systems, the photosynthesized material is either stored as a reserve product or used as a structural part of the plant (e.g., cellulose).

Pigment Molecules Are Photoexcited

The visible spectrum corresponds to a small portion of the total electromagnetic radiation and comprises the region between 400 and 700 nm. The energy contained in these wavelengths is made of discrete packets called *photons*. A photon contains one quantum of energy. This is expressed mathematically by the equation derived in 1900 by Max Planck:

$$E = \frac{hC}{\lambda} \tag{5}$$

where h is Planck's constant (i.e., 1.585×10^{-34} cal-sec), C is the speed of light (3×10^{10} cm/sec), and λ is the wavelength of the radiation.

This equation shows that photons with shorter wavelengths have higher energy.

We mentioned that certain pigments, such as chlorophyll and the reaction centers P680 and P700, present in the photosystems (see Table 12–1) are specially adapted to be excited by light. The photons that are absorbed produce excitation by displacing electrons from a *ground state* to a higher energy level or *excited state* in the orbitals within the atoms. This excitation by photons may be dissipated as heat or emitted as radiation (fluorescence), but it may also be converted into chemical energy.

The most essential phenomenon in photosynthesis is the excitation by photons in the photochemical reaction.

Primary Reaction of Photosynthesis— Photochemical Reaction

Photosynthesis involves a complex series of reactions, some of which take place only in the presence of light while others can also be carried out in the dark. Hence the names of *light or photochemical reactions* and *dark reactions*.

In 1937, Robert Hill was the first to provide evidence that the photochemical reactions of plant cells took place in chloroplasts. Hill's experiment demonstrated that leaves ground in water, to which a hydrogen acceptor (e.g., quinone) was added, gave off O_2 when exposed to light. This reaction took place without involving the synthesis of carbohydrates and without requiring the presence of CO_2. The *Hill reaction* followed the general equation mentioned in reaction (4):

$$2 H_2O + 2A \xrightarrow{\text{light}} 2AH_2 + O_2 \tag{6}$$

where A is the acceptor and AH_2 is the reduced form of the acceptor. Later it was found that in photosynthesis the oxidized form of coenzyme NADP+ was the normal hydrogen acceptor of the photochemical reaction, producing NADPH.

Another important finding was made by Arnon in 1954, when he showed that, under the action of light, chloroplasts could make ATP from ADP and Pi. This is the process called *photophosphorylation, which is similar to that of oxidative phosphorylation* in mitochondria.

When studying photochemical reactions the student should recall the process of oxidative phosphorylation in mitochondria, in which the flow of electrons is from NADH to O_2, following the oxidation-reduction potential, i.e., from -0.6 to $+0.82$ volts (see Fig. 11–12). In photosynthesis the opposite phenomenon occurs, and the electrons flow from H_2O to NADPH. The photochemical reaction takes place in the thylakoid membranes. When these are illuminated, there is a transfer of electrons from water ($E'_0 = +0.82$ volts) to the final acceptor ($E'_0 = -0.6$ volts). Transfer against this negative electrochemical gradient uses the energy provided by the photons of light. Furthermore, the transfer of electrons is accomplished by a chain of *electron carriers* and is coupled with the *phosphorylation* of ADP to ATP. Within the thylakoid membrane, the energy of light is collected by the two *photosystems* (PS I and PS II), which act as collector antennae.

According to the diagram of Figure 12–12, the reaction centers of both photosystems (P 680 and P 700) are excited by photons. At these centers are special molecules of chlorophyll *a* with a higher absorption maximum, in which the photochemical reactions take place. The excitation produced facilitates the transfer of electrons (e^-) from a *donor* to an *acceptor* molecule.

Both photosystems operate in a sequential manner. In *PS II*, the absorbed quanta of light produce the removal of electrons from H_2O and boost them to a higher energy level.

The *primary acceptor of PS II* is not well known but could be a semiquinonic form of plastiquinone. Using short pulses of light in the near ultraviolet spectrum, a compound X_{320} was detected, which could be the primary acceptor. Another compound C_{550} (in Fig. 12–12), which seems to be also in the region of the primary acceptor, has been recognized to be β carotene. It is thought that, because of its hydrophobicity, this acceptor of PS II is localized in the lipid interior of the membrane (Fig. 12–9).

The *primary acceptor of PS I* is a complex in which ferrodoxin (a protein that contains iron, but lacks a porphyrin ring) is associated to a sulfurferroprotein and, because of its absorption at 430 nm, is called P_{430}. The reduction of NADP$^+$ to NADPH, which is the terminal acceptor of electrons transported by ferrodoxin, is catalyzed by the enzyme *ferrodoxin-NADP$^+$-reductase*, a flavoprotein that uses NADP$^+$ as an acceptor. Different studies suggest that these components of the acceptor of PS I (i.e., ferrodoxin, ferrodoxin-NADP$^+$-reductase and a structural component) are exposed to the hydrophilic side of the chloroplastic matrix, where the enzymes responsible for the dark assimilation of CO_2 are also present (Fig. 12–9).

The *donor of electrons for PS II* in green plants is water. The mechanism by which H_2O is oxidized to liberate O_2 is not completely known. However, two components, Cl$^-$ and Mn^{++}, are essential. Apparently, 6 Mn atoms per 400 chlorophyll molecules are needed for optimal functioning of the PS II. Both Cl$^-$ and Mn^{++} are also needed for plant growth. The donor of PS II is located in the lipid part of the membrane.

The *donor of electrons in PS I is plastocyanine*, a copper-containing protein that is related to two quinones, i.e., vitamin K_1 and plastoquinone. *Plastoquinones*, similar to ubiquinone in mitochondria, are highly mobile lipophilic molecules that could establish a dynamic link between the two photosystems by flowing within the membrane. The donor of electrons of PS I is also located in the interior of the membrane.

Electron Transport Chain

The transport of electrons permits the connection of the photolysis of water in one extreme with the reduction of NADP$^+$ in the other (Fig. 12–12). This is done by way of the two photosystems, which are connected by a transport chain. This transport comprises cytochromes b 563 and f, as well as plastoquinone.

In summary: As shown in Figure 12–12, the excitation of the reaction center of PS II will cause the release of O_2 and electrons. These are boosted from the reducing potential of water ($+0.82$ volts) to a higher potential (0 volts) to reach acceptor X_{320}. Then the electrons move in a downhill sequence to reach PS I. Here the electrons are again boosted to a higher reducing level (-0.6 volts), from which they finally reach NADP$^+$, reducing it to NADPH.

Topology of the Electron Transport and Photosystems

As in mitochondria, there is evidence of a vectorial disposition of the photosystems and electron carriers across the membrane of the chloroplast (Fig. 12–9). We already mentioned the possible location of the donor and acceptor sites of both photosystems. Recent findings suggest that the LHC complexes also span the membrane.[19] Regarding the lateral organization, the electron microscopic observations of Miller and Staehelin in 1976[20] demonstrated that the coupling factor CF_1 was only present on the outer surface exposed toward the stroma. On the other hand, the CF_1 is lacking in the stacked grana (Fig. 12–11).

Subfractionation of chloroplasts has demonstrated that the grana stacks of membranes are enriched in PS II and LHC, while there is little ATP synthetase. On the other hand, a fraction of stroma thylakoids is rich in PS I and ATPase and poor in PS II and LHC. This contrasts with the uniform distribution of the cytochrome b/f complex in all thylakoid regions.

Such a lateral organization suggests that the electron transport system in chloroplasts is not a fixed structure and that some distant connection must exist between the two photosystems. These connections may occur through shuttle molecules such as plastoquinone and plastocyanine, which are able to diffuse rapidly within the bilayer. This flexible organization permits a

way of controlling photosynthetic rates in various membranes. This can be illustrated by comparing the structure of chloroplasts of plants that grow in plain sunshine to those that develop in the shade.

Shade plants have chloroplasts with giant grana stacks and few stroma thylakoids. This implies that they contain more PS II and LHC than ATP synthetase and thus have a lower rate of photosynthesis. The opposite (high PS I and ATP synthetase) occurs in the stroma rich chloroplasts of sun plants in which the rate of photosynthesis is high (see Anderson and Anderson, 1982 in the Additional Readings at the end of this chapter).

Photophosphorylation Involves a CF_1 ATPase—The Chemiosmotic Coupling

The electron transport system of chloroplasts is coupled to the phosphorylation of ADP to

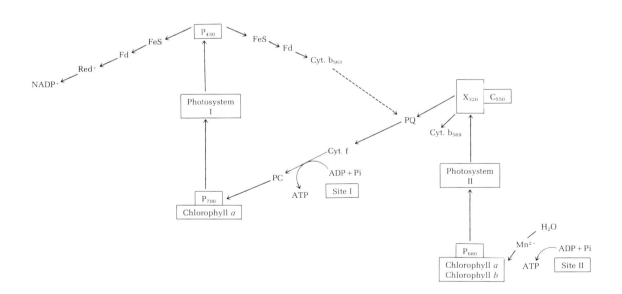

Fig. 12–12. Diagram showing the possible electron transport in chloroplasts. P 700 and P 680 are the photoexcited pigments of photosystems I and II. P_{430} and X_{320} are the primary acceptors of both photosystems; C_{550} is β carotene; PQ, plastoquinone; Cyt., cytochromes; PC, plastocyanine; FeS, sulfurferroprotein; Fd, Red^+-$NADP^+$-reductase. Sites I and II indicate the places where ATP is synthesized (see the description in the text). (Taken from C.S. Andreo, J.J. Cazzulo and R.H. Vallejos, *in* Bioquímica General. H.N. Torres et al., ed. El Ateneo. Buenos Aires, 1983.)

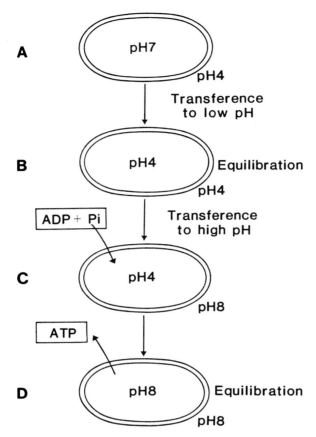

Fig. 12–13. Diagram of Jagendorf's experiment in which a suspension of spinach chloroplasts is submitted to changes in pH of the medium. In the conditions indicated there is entrance of ADP and Pi and synthesis of ATP, with increase of pH inside the chloroplast.

ATP. This coupling is effected by an ATP synthetase consisting of the CF_1 coupling factor, situated on the outer surface of the thylakoid membrane, and of the CF_0 hydrophobic segment that acts as a proton channel (Fig. 12–9).

Of the several theories proposed to explain the coupling between the photosystems and photophosphorylation, as in mitochondria, the *chemiosmotic theory* of Mitchell is favored (see Section 11–4). According to this theory, coupling is effected by a link of electrochemical nature, in which a gradient of potential and of pH across the membrane is produced.

The experiment done by Jagendorf[21] strongly supports this theory (Fig. 12–13). A suspension of spinach chloroplasts with an initial pH 7 was equilibrated in an acid medium of pH 4. ADP and Pi were then added and the chloroplasts brought to pH 8. Under these conditions ATP

synthesis occurred, driven by the pH difference between the chloroplast and the medium, exactly as postulated by Mitchell.

In contrast to the oxidative phosphorylation of mitochondria, O_2 is not used in photophosphorylation of chloroplasts (see Table 11–1). Green plants can produce 30 times as much ATP by photophosphorylation as by oxidative phosphorylation in their own mitochondria. In addition, these plants contain many more chloroplasts than mitochondria.

In the overall photosynthetic pathways for every one molecule of O_2 liberated eight light quanta are needed, and two molecules of $NADPH_2$ are formed; the efficiency is close to 25%. Along the electron transport there is the synthesis of three ATP molecules.

The Photosynthetic Carbon-Reduction (PCR) Cycle—Dark Reactions

The synthesis of carbohydrates, with the reduction of CO_2, can take place after the plant that had been exposed to light is placed in the dark. This is why reactions involved in this cycle are called *"dark reactions."* In the PCR cycle the molecules of ATP and NADPH produced by the photochemical reaction provide the energy needed to fix the CO_2 and to synthesize carbohydrates (Fig. 12–14).

The PCR cycle takes place in the stroma of the chloroplast by the intervention of numerous enzymes that work in a cycle, the sequence of which was solved mainly by Melvin Calvin and associates (i.e., *Calvin cycle*) with the use of $^{14}CO_2$. The reactions involved are so rapid that they occur 1 second or less after the addition of the radioactive CO_2.

As shown in Figure 12–14, the initial reaction in which CO_2 and H_2O enter the cycle is carried out by the enzyme *ribulose-1,5-diphosphate carboxylase (i.e., carboxydismutase)*. This is a huge enzyme (500,000 daltons) that represents about half of the stromal proteins. Through its action, *ribulose-1,5-diphosphate* (a pentose) is integrated with CO_2 and H_2O to produce two molecules of *3-phosphoglycerate*. These triose sugars are phosphorylated by ATP and form an activated molecule that is able to accept hydrogens (and electrons) from NADPH. *3-phosphoglyceraldehyde* is then reduced to form more complex carbohydrates and ribulose-1,5-diphos-

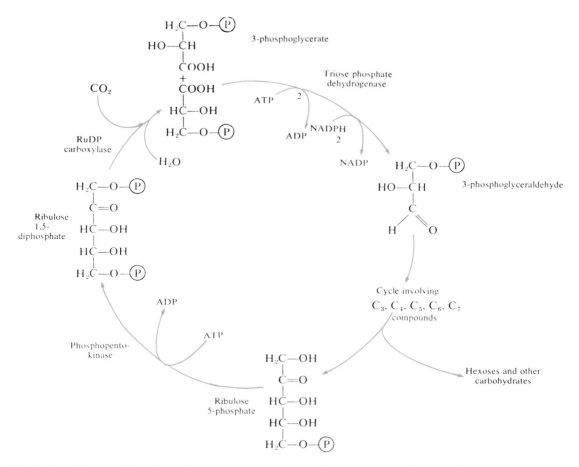

Fig. 12–14.   Diagram of the Calvin or C_3 cycle of photosynthesis in which CO_2 is reduced and carbohydrates are synthesized.

phate is regenerated. (Among the intermediates formed, one of the most important is fructose 6-phosphate, which eventually leads to the formation of glucose.)

The energy balance of photosynthesis is:

$$6\ CO_2\ +\ 12\ H_2O\ \xrightarrow{\text{light}}\ C_6H_{12}O_6$$
$$+\ 6\ CO_2\ +\ 6\ H_2O\quad(7)$$

which represents a storage of 686,000 calories per mole. This energy is provided by a total of 12 NADPH and 18 ATP molecules, which represent 750,000 calories. The efficiency reached by the PCR cycle is thus as high as 90% (686/750 × 100 = 90%).

In summary: Photons absorbed by chlorophyll and other light-sensitive pigments are first converted into chemical energy as NADPH and ATP. During this photochemical reaction, water is oxidized and molecular oxygen is released into the atmosphere as a by-product. The reduction of CO_2 takes place even in the dark, provided that NADPH and ATP are present. The initial product of this cycle is a triose, which is finally processed to *fructose 6-phosphate*, glucose, and other carbohydrates.

If the student compares all of the actions that result in the synthesis of hexoses (Fig. 12–14) with those involved in their breakdown by glycolysis (Fig. 11–8) and by aerobic respiration (Fig. 11–11), it is evident that they follow entirely different pathways.

The C_4 Pathway in Angiosperms

The PCR cycle we described is most commonly found in higher plants. In 1966, however, Hatch and Slack described an alternate pathway of CO_2 fixation found in many *Angiosperm* spe-

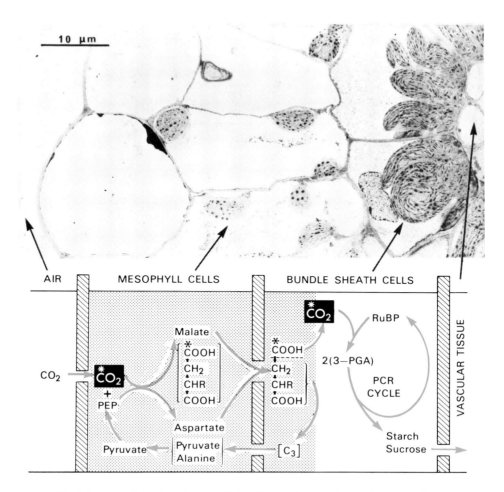

Fig. 12–15. Simplified diagram of the C_4-pathway and the photosynthetic carbon reduction cycle in angiosperms. *Top,* electron micrograph taken by S. Craig of *Panicum miliaceum* leaf. See the text for details. (From Hatch, M.D.: Trends Biochem. Sci., 2:199, 1977.)

cies. Using ^{14}C labeled CO_2, as in the experiment of Calvin, they found that the product of the cycle was not 3-phosphoglycerate but a 4-carbon acid, oxaloacetate. The initial enzyme is *phosphoenolpyruvate carboxylase* and the first reaction consists of the reduction of CO_2 by phosphoenolpyruvate (PEP) (Fig. 12–15).

From the point of view of cell biology it is interesting that the PCR cycle and the C_4 pathway are integrated in different cells of the same plant (see Hatch, 1977 in the Additional Readings at the end of this chapter).

As shown in Figure 12–15, in the mesophyll cells the CO_2 is assimilated by carboxylation of PEP, which gives rise to C_4 acids, such as malate and aspartate. These are then transferred to the bundle sheath cells, probably by diffusion.

Within these cells the CO_2 that is released by decarboxylation enters the PCR cycle, giving origin to 3-phosphoglyceric acid, and the C_3 products may go back to the mesophyll cell to enter the C_4 pathway again.

Ion Fluxes and Conformational Changes—Caused by Light and Darkness

Structural changes in chloroplasts caused by the action of light or darkness have been observed both in vivo and in vitro. In many plant species chloroplasts flatten in the light and become more spherical in the dark.[22] Such changes appear to be mediated by ion fluxes across the membranes of individual thylakoids.[23] Changes of this type are blocked by drugs that affect ion transport. Similar findings have been observed in vitro in isolated chloroplasts.

Using freeze-fracturing it has been demonstrated that during illumination there are changes in the size and density of the particles within the membrane.[24] A more dramatic change can be observed if chloroplasts are suspended in a solution low in cations. There is unstacking of the membranes coupled with a dispersion of the EF particles. This change is reversible, and upon readdition of cations, stacking is re-established and the EF particles become concentrated in the grana.[25] These observations suggest that the protein complexes have mobility within the fluid membrane.

It was mentioned in Chapter 11 that mitochondria eject H^+ ions during the electron transport and that the interior becomes more alkaline. In chloroplasts the opposite reactions occur; under the action of light, H^+ ions are absorbed, and the interior becomes acidified. At the same time, K^+ and Mg^{++} are ejected into the medium. In the dark, H^+ is ejected and K^+ and Mg^{++} are reabsorbed by the chloroplasts. Thus, the chloroplast membrane has a vectorial organization that is the reverse of the mitochondrial membrane. Photophosphorylation requires the separation of positive and negative charges into specific pathways of electron flow.

As in mitochondria, according to the *chemiosmotic hypothesis* of Mitchell, the proton gradient may serve as a driving force for the formation of ATP at the CF_1 ATPase.

SUMMARY:
Photosynthesis and the Chloroplast

Photosynthesis is the process by which chloroplasts trap light quanta (i.e., photons, excitons) and transform them into chemical energy. Photosynthesis is, in some ways, the reverse of oxidative phosphorylation in mitochondria. The overall reaction of photosynthesis is:

$$2nH_2O + nCO_2 \rightarrow nH_2O + nO_2 + (CH_2O)n$$

Thus, using H_2O as hydrogen donor and CO_2 from the atmosphere, carbohydrates are synthesized, and O_2 is released.

Photosynthesis consists of a *photochemical reaction* (Hill reaction), which occurs in the presence of light, and a *dark* or *thermochemical reaction*. In the first reaction, O_2 is released when the chloroplasts are exposed to light; in the second, CO_2 is fixed and carbohydrates are formed. In the photochemical reaction electrons flow from H_2O to

$NADPH_2$ (i.e., from $+0.81$ to -0.6 volts of redox potential) because electrons are boosted to high energy levels by the absorbed light. In photosynthesis there are two photosystems (I and II) that are excited at different wavelengths. *Photosystem I* (excited at 680 to 700 nm) comprises 200 molecules of chlorophyll a, 50 of carotenoids, and one of P700. *Photosystem II* (excited at 650 nm) contains chlorophyll a and b, only this second system is associated with the release of O_2 from H_2O. Both photosystems operate in a sequential and interrelated fashion (Fig. 12–12). In photosystem I, two quanta of light, trapped by P700, boost electrons that reduce *ferredoxin*, a protein containing iron and sulfur. In turn, ferredoxin transfers the electrons to $NADP^+$, reducing it to $NADPH_2$. The electrons in PS I are restored by the second system. Here the light quanta remove the electrons from the hydrogen of water, releasing O_2. The electrons are transferred to P700 by a transport system that comprises several cytochromes and *plastocyanine* (a copper-containing protein). Two quinones–*vitamin K*, and *plastoquinone*—are also included in this electron transport system. As in mitochondria, the transfer of electrons is coupled with the phosphorylation of ATP at the level of PS II. The dark, or thermochemical, reaction involves many steps that start with the uptake of CO_2 and its reduction by $NADPH_2$ to form the various carbohydrates (Fig. 12–14). In several steps of this complex cycle of reactions, ATP is also used as an energy source. A dominating compound is 3-phosphoglyceric acid, which gives rise to glucose phosphate, from which come the various disaccharides and polysaccharides. The formation of phosphoglyceric acid depends on the initial enzyme, *carboxydismutase*. As in mitochondria, there is a strict structural-functional correlation in chloroplasts. All the enzymes of the dark reactions (like those of the Krebs cycle) are soluble and are contained in the stroma of the chloroplast. The photosystems, including electron transport and photophosphorylation, form part of the membranes of the chloroplasts.

Under the influence of light, chloroplasts undergo conformational changes which are mediated by ion fluxes. Chloroplasts flatten in the light, and the intrathylakoid space is reduced. There are also changes in the size and density of the particles. In a solution low in cations there is unstacking of the membranes and dispersion of the EF particles. In the dark, the chloroplasts swell and the cavity of the thylakoid expands; H^+ is ejected, and K^+ and Mg^{++} are reabsorbed. Under the action of light, the interior of the chloroplasts becomes more acid because of the uptake of H^+. At the same time, K^+ and Mg^{++} are released and the cavity of the thylakoid becomes smaller.

12–6 BIOGENESIS OF CHLOROPLASTS

Since the classical work of Schimper and Meyer (1883) it has been accepted that chloroplasts multiply by fission, a process that implies the growth of the daughter organelles. This is easily observed in the alga *Nitella*, which contains a single huge chloroplast. In *Nitella* a division cycle of 18 hours has been recorded cinematographically.

During the development of the chloroplast the first structure to appear is the so-called *proplastid*, which has a double membrane.

The following stages of development are illustrated in Figure 12–16:

(A) In the presence of light, the inner membrane grows and gives off vesicles into the matrix that are transformed into discs. These intrachloroplastic membranes are the thylakoids which, in certain regions, pile closely to form the grana. In the mature chloroplast the thylakoids are no longer connected to the inner membrane, but the grana remain united by intergranal thylakoids.

(B) If a plant is put under low light intensity, a reverse sequence of changes takes place. This is the process called *etiolation*, in which the leaves lose their green pigment and the chloroplast membranes become disorganized.

The chloroplasts are transformed into *etioplasts*, in which there is a paracrystalline arrangement of tubules forming the so-called *prolamellar body*. Attached to these bodies are young thylakoid membranes that lack photosynthetic activity. The three-dimensional organization of the etioplast is most apparent in freeze-fractured preparations observed with a high-resolution scanning EM.

(C) The regular crystal lattice of two prolamellar bodies surrounded by young thylakoid membranes is observed (see Osumi et al., 1984 in the Additional Readings at the end of this chapter). If etiolated plants are re-exposed to light, thylakoids are re-formed and the prolamellar material is used for assembly.

Chloroplasts as Semiautonomous Organelles

Chloroplasts, like mitochondria, are very special organelles in the way in which they originate and develop. Both chloroplasts and mitochondria are formed from *pre-existing organelles*. These organelles are thus semiautonomous and result from the cooperation between two genetic systems (i.e., one of their own, the other belonging to the rest of the cell).

Chloroplasts have their own DNA, RNA, ribosomes, and other components involved in protein synthesis, which produce a certain number of proteins. However, many more proteins are coming from the cytosol by the action of the nuclear genes.

These concepts have been elaborated in the so-called *symbiont hypothesis* (see Section 11–6). After the general laws of inheritance were established by Mendel, Morgan, and others (Chapter 17), it was found that in certain plants the results of crosses did not agree with Mendelian predictions. These findings referred to cases of *leaf variegations* (i.e., mutants in which the leaves appeared spotted or striped) that were due to the alternation of pigmented and non-pigmented chloroplasts. In these cases, the trait was inherited in a strictly maternal non-Mendelian fashion. For example, when a variegated male was crossed with a wild type female all of the offspring were wild type. On the contrary, if the female was variegated and the male wild type, the progeny was variegated. This finding suggested that variegation was transmitted by the female cytoplasm. It should be kept in mind that the sperm contributes very little cytoplasm and no chloroplasts to the zygote.

The work carried out later on various organisms (yeasts, *Neurospora*, *Chlamydomonas*) led to the conclusion that this exceptional behavior is due to the presence of *extra chromosomal genes* in the cytoplasm. This behavior has been shown experimentally:

When the unicellular alga *Euglena* is grown in the dark it contains small, colorless plastids. Under these conditions, irradiation of the nucleus with an ultraviolet microbeam that destroys nuclear genes does not interfere with the formation of chloroplasts, and these organelles appear if the alga is brought back to the light. On the other hand, if the nucleus is shielded, and the cytoplasm is irradiated, colonies lacking chloroplasts (colorless) develop. These results suggest the presence of a cytoplasmic DNA in this alga.

Chloroplasts contain a distinct genetic system based on a double helical DNA circle with an average length of 45 μm (about 135,000 base pairs). The replication of chloroplast DNA has been followed with ^{3}H-thymidine. Maps of the location of genes have been made in several chloroplast DNAs using restriction enzymes (see

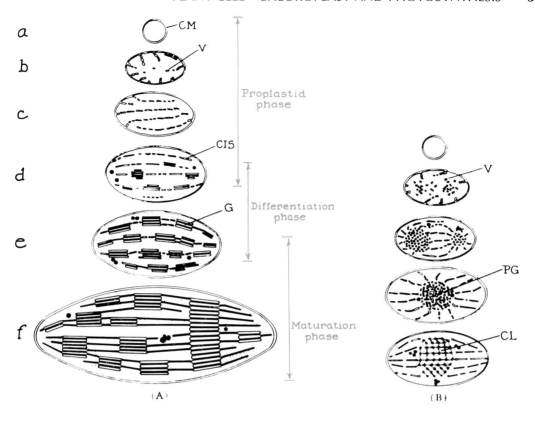

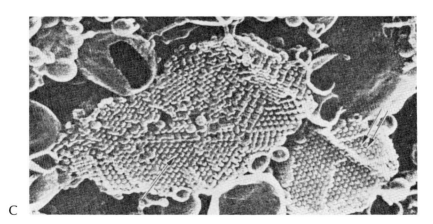

Fig. 12–16. *A,* phases in the development of a proplastid into a chloroplast in the presence of light. *B,* same, but in the dark, showing the formation of the primary granum (PG), or prolamellar body. *CIS,* flattened cisternae; *CL,* crystal lattice; *CM,* double chloroplast membrane; *G,* granum; *V,* vesicles. (Modified from von Wettstein, D.J.: Ultrastruct. Res., *3*:235, 1959.) *C,* scanning electron micrograph of an etioplast showing the crystalline structure of the prolamellar body. (Courtesy of M. Osumi.)

Chapter 19). Furthermore, the gene for the large subunit of carboxydismutase has been fully sequenced (1425 nucleotides).

Chloroplasts contain 70 S ribosomes, which account for up to 50% of the total ribosomes in photosynthetic cells. The presence of ribosomes within chloroplasts is easily demonstrated by electron microscopy. The presence of such ribosomes provides evidence that these organelles contain a specific protein synthesizing system. This protein synthesis is preferentially inhibited by chloramphenicol concentrations that do not affect protein synthesis in the cytoplasm.

The involvement of the two types of protein synthesis in the assembly of chloroplasts may also be studied by using cycloheximide to inhibit synthesis by cytoplasmic ribosomes. Although some protein synthesis (about 10%) occurs within the chloroplast, most proteins are made in the cytosol and are translocated into this organelle. Of the 30 known thylakoid polypeptides that function in photosynthesis, so far 9 have been demonstrated to be synthesized on chloroplastic ribosomes and 9 are coded by nuclear genes and synthesized on cytoplasmic ribosomes (see von Wettstein, 1981 in the Additional Readings at the end of this chapter).

In spite of the large potential coding capacity of chloroplastic DNA many chloroplast polypeptides are synthesized on cytoplasmic ribosomes. This is the case with the enzymes of the Calvin cycle, which are present in the stroma, and the protein that binds chlorophyll a/b in thylakoids. In addition, several structural polypeptides have their genes localized in the nuclear genome. Analysis of protein synthesis by chloroplasts has been facilitated by the fact that these organelles, under the action of light, can synthesize RNA molecules and proteins. Of the proteins, a major one is associated with the ATP synthetase complex, present in thylakoid membranes. Another major protein is soluble, and has been identified as the large subunit of *ribulose-diphosphate carboxylase* or *carboxydismutase*.

As we mentioned earlier this huge enzyme represents about 50% of total soluble proteins in chloroplasts and thus may be the most abundant protein in nature. Carboxydismutase comprises 16 subunits: 8 of high molecular weight (55,000 daltons) and 8 of much smaller molecular weight (14,000 daltons). The large subunit is coded by genes present in the chloroplastic

DNA, while the small subunit is produced by nuclear genes. This last polypeptide (P 20) is synthesized as a precursor weighing 20,000 daltons on free ribosomes; it then enters *post-translationally* into the stroma to be cleaved to attain its final size.[26] It is postulated that the chloroplastic envelope has receptor sites that recognize the proteins that are to be incorporated into the organelle. The extra sequence (or signal) that is present in P 20 is composed of acidic amino acids, in contrast to the hydrophobic ones in the signal sequence of secretory proteins (see Section 8–4). After entering the chloroplast the extra sequences are removed by a protease, present in the envelope, and the small subunit is released into the stroma. Figure 12–17 summarizes a current model for the synthesis of carboxydismutase (CDase), which can be applied both to chloroplasts and mitochondria. This model suggests that the small subunit is also able to control the synthesis of the large subunit with which it will be associated to make the final active enzyme (see Ellis, 1981 in the Additional Readings at the end of this chapter).

In summary: Chloroplasts have been described as having many of the characteristics of a semiautonomous or symbiotic organism living within the plant cells. They divide, grow, and differentiate; they contain DNA, ribosomal RNA, and messenger RNA, and are able to conduct protein synthesis. It has been suggested that chloroplasts may have resulted from a symbiotic relationship between an autotrophic microorganism, one able to transform energy from light, and a heterotrophic host cell.

It is evident, however, that chloroplasts have in their DNA the genetic information to code for only a limited number of proteins and the elements necessary to establish a DNA-RNA-directed protein synthesis (i.e., ribosomes and tRNA).

In chloroplasts protein may be synthesized (1) by an exclusive chloroplastic mechanism, (2) by a mechanism involving nuclear genes and chloroplastic ribosomes, and (3) by nuclear genes and cytoplasmic ribosomes. Although the symbiont hypothesis is attractive; it is evident that synthesis and assembly of all the components integrating a chloroplast depend to a great extent on nuclear genes and on cytoplasmically directed protein synthesis (for further discus-

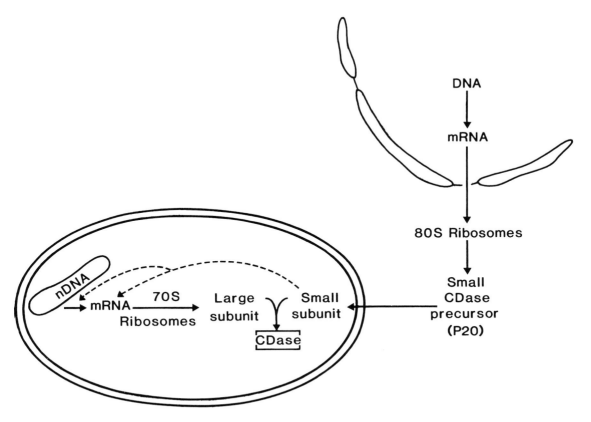

Fig. 12–17.   Postulated model for the synthesis of the enzyme carboxydismutase (CDase) in higher plants. DNA stands for nuclear and chloroplastic DNA. Observe that the small subunit of CDase is made on free 80S ribosomes as a precursor (P_{20}). The large subunit is synthesized by chloroplastic DNA and mRNA. (Redrawn from Highfield, P.E., and Ellis, R.J.: Nature, *271*:420, 1978.)

sions, see Bogorad, 1981 in the Additional Readings at the end of this chapter; see also Section 11–6).

SUMMARY:

Biogenesis of Chloroplasts

Like mitochondria, chloroplasts exhibit a certain degree of functional autonomy. They undergo division and contain genetic information. They have DNA, ribosomes, and all the machinery for protein synthesis, which is inhibited by chloramphenicol. Protein in chloroplasts may be synthesized (1) by an exclusive chloroplastic mechanism, (2) by nuclear genes and chloroplastic ribosomes, and (3) by nuclear genes and cytoplasmic ribosomes. The symbiont hypothesis has also been applied to chloroplasts (see Section 11–6).

REFERENCES

 1. Kiermayer, O., and Dobberstein, B.: Protoplasma, 77:437, 1973.

 2. Bartnicki-García, S., Bracker, Ch.E., and Ruiz-Herrera, J.: Eur. J. Cell Biol., 22 (1):461, 1980.
 3. Frey-Wyssling, A., López-Saez, J.E., and Mühlethaler, K.: J. Ultrastruct. Res., 10:422, 1964.
 4. Hepler, P.K., and Newcomb, E.H.: Ultrastruct. Res., 10:497, 1967.
 5. Juniper, B.E.: Theor. Biol., 66:583, 1977.
 6. Porter, K.R.: *In* Principles of Biomolecular Organization. Ciba Foundation Symposium. Edited by G.E.W. Wolstenholme. London, J. & A. Churchill Ltd., 1966, p. 308.
 7. Newcomb, E.H.: Annu. Rev. Plant Physiol., 20:253, 1969.
 8. Bailey, C.J., Cobb, A., and Boulter, D.: Planta (Berlin), 95:103, 1970.
 9. Larkins, B.A., and Dalby, A.: Biochem. Biophys. Res. Commun., 66:1048, 1975.
10. Burr, B., and Burr, F.A.: Proc. Natl. Acad. Sci. USA, 73:515, 1976.
11. Kagawa, T., Lord, J.M., and Beevers, H.: Plant Physiol., 51:61, 1973.
12. González, B., and Beevers, H.: Plant Physiol., 57:406, 1976.
13. Amelunxen, F., and Groman, G.: Pflanzen Physiol., 60:156, 1969.
14. Anderson, J.M.: Biochim. Biophys. Acta, 416:191, 1975.
15. Anderson, J.M.: *In* International Cell Biology. Edited by B.R. Brinkley and K.R. Porter. New York, The Rockefeller University Press, 1977, p. 183.

16. Vernon, L.P., and Klein, S.M.: Ann. NY Acad. Sci., *244*:281, 1975.
17. Wessel, J.S.C., and Borchert, M.T.: *In* Third International Congress on Photosynthesis. Vol. 1. Edited by M. Avron. Amsterdam, Elsevier Publishing Company, 1975, p. 473.
18. Branton, D., et al.: Science, *190*:54, 1975.
19. Anderson, B., Anderson, J.M., and Kyrie, I.J.: Eur. J. Biochem., *123*:465, 1982.
20. Miller, K.R., and Staehelin, A.: J. Cell Biol., *68*:30, 1976.
21. Jagendorf, A.T.: Fed. Proc., *34*:1718, 1975.
22. Miller, M.M., and Nobel, P.S.: Plant Physiol., *49*:535, 1972.
23. Murakami, S., Torres-Pereira, J., and Packer, L.: *In* Bioenergetics of Photosynthesis. Edited by Govinjee. New York, Academic Press, 1975.
24. Wang, A.V., and Packer, L.: Biochim. Biophys. Acta, *347*:134, 1973.
25. Staehelin, L.A.: J. Cell Biol., *71*:136, 1976.
26. Highfield, P.E., and Ellis, R.J.: Nature, *271*:420, 1978.

ADDITIONAL READINGS

Albersheim, P.: The walls of growing plant cells. Sci. Am., *232*:80, 1975.
Anderson, J.M.: The role of chlorophyll-protein complexes in the function and structure of chloroplasts thylakoids. Mol. Cell. Biochem., *46*:161, 1982.
Anderson, J.M., and Anderson, B.: The architecture of photosynthetic membranes. Trends Biochem. Sci., *7*:288, 1982.
Armond, P.A., and Arntzen, C.J.: Localization and characterization of photosystem II in grana and stroma lamellae. Plant Physiol., *59*:398, 1977.
Avron, M.: Energy transduction in photophosphorylation. FEBS Lett., *96*:223, 1978.
Avron, M.: Energy transduction in chloroplasts. Annu. Rev. Biochem., *46*:143, 1977.
Barber, J.: The Intact Chloroplast. Amsterdam, Elsevier Publishing Company, 1976.
Barber, J.: Cation control in photosynthesis. Trends Biochem. Sci., *1*:33, 1976.
Bogorad, L.: Chloroplasts. J. Cell Biol., *91* N3 Pt 2:256s, 1981.
Brown, R.M., Jr.: Cellulose and Other Natural Polymer Systems. New York, Plenum Publishing Corp., 1982.
Brown, R.M., Jr.: Biogenesis of natural polymer systems with special reference to cellulose assembly and deposition. *In* Structure and Biochemistry of Natural Biological Systems. Edited by E.M. Walk. New York, Philip Morris Inc., 1979, p. 51.
Brown, R.M., Jr., and Romanovicz, D.K.: Biogenesis and structure of Golgi-derived cellulosic scales in *Pleurochrysis*. Appl. Polym. Symp., *28*:537, 1976.
Carde, J.-P.: Leucoplasts: A distinct kind of organelles lacking typical 70S ribosomes and free thylakoids. Eur. J. Cell Biol., *34*:18, 1984.
Chrispeels, M.J.: The role of the endoplasmic reticulum in the biosynthesis and transport of macromolecules in plant cells. *In* International Cell Biology. New York, The Rockefeller University Press, 1977, p. 284.
Crease, L.L.: Cellular and Subcellular Localization in Plant Metabolism. New York, Plenum Publishing Corp., 1982.
Douce, R., and Joyard, J.: Le chloroplaste. La Recherche, *8*:527, 1977.
Ellis, R.J.: Chloroplast proteins: Synthesis, transport and assembly. Annu. Rev. Plant Physiol., *32*:111, 1981.
Govindjee: Photosynthesis: Energy conversion by plants and bacteria. Vol. 1. New York, Academic Press, 1982.
Hatch, M.D.: C_4 pathway photosynthesis: Mechanism and physiological function. Trends Biochem. Sci., *2*:199, 1977.
Joliot, P. Photosynthesis. La Recherche, *9*:331, 1978.
Juniper, B.E.: Some feature of secretory systems in plants. Histochem. J., *9*:659, 1977.
Kirk, J.T.O., and Tilney-Basset, R.A.E.: The plastids: Their chemistry, structure, growth and inheritance. Amsterdam, Elsevier North-Holland Publishing Co., 1978.
Ledbetter, M.C., and Porter, K.R.: Introduction to the Fine Structure of Plant Cells. New York, Springer-Verlag, 1970.
Lehninger, A.L.: Biochemistry: The Molecular Basis of Cell Structure and Function. 2nd Ed. New York, Worth Publishers, Inc., 1975.
Loewus, F.: Biogenesis of Plant Cell Wall Polysaccharides. New York, Academic Press, 1973.
Marchant, H.J.: Microtubules, cell wall deposition and determination of plant cell shape. Nature, *278*:167, 1979.
Nieden, V., et al.: Dictyosomes participate in the intracellular pathway of storage proteins in developing *Vicia taba* cotyledons. Eur. J. Cell Biol., *34*:9, 1984.
Northcote, D.H.: Differentiation in higher plants. Oxford, Oxford Biology Readers, No. 44, 1974.
Ohad, I.: Ontogeny and assembly of chloroplast membrane polypeptides in *Chlamydomonas reinhardtii*. *In* International Cell Biology. New York, The Rockefeller University Press, 1977, p. 193.
Osmuni, M., et al.: Three-dimensional observation of the prolamellar bodies in etioplasts of Squash. Scan. Electron Micros., I:111, 1984.
Preston, R.D.: The Physical Biology of Plant Cell Walls. London, Chapman & Hall, Ltd., 1974, p. 491.
Robards, A.W.: Dynamic Aspects of Plant Ultrastructure. New York, McGraw-Hill Book Co., 1974.
Staehelin, L.A.: Freeze-fracture studies of green plants and prochloron thylakoids. Proceedings 5th International Congress on Photosynthesis. Greece, 1980, p. 1.
Staehelin, L.A., and De Wit, M.: Correlation of structure and function of chloroplast membranes at the supramolecular level. J. Cell. Biochem., *24*:261, 1984.
von Wettstein, D.: Chloroplast and nucleus. *In* International Cell Biology 1980–1981. Edited by H.G. Schweiger. Berlin, Springer-Verlag, 1981, p. 250.
Willison, J.H.M., and Brown, R.M., Jr.: Cell wall structure and deposition in *glaucosystis*. J. Cell Biol., *77*:103, 1978.

CHAPTER 13

THE INTERPHASE NUCLEUS, CHROMATIN, AND THE CHROMOSOME

Over a century ago, in 1876, Balbiani described rod-like structures that were formed in the nucleus before cell division. In 1879 Flemming used the word "chromatin" (Gr., *chroma*, color) to describe the substance that stained intensely with basic dyes in interphase nuclei. He suggested that the affinity of chromatin for basic dyes was due to their content of "nuclein," a phosphorus-containing compound isolated initially from pus cells by Meischer in 1871 and later on from salmon sperm, a substance that is now called DNA. In 1888 Waldeyer used the word *chromosome* to emphasize the continuity between the chromatin of interphase nuclei and the rod-like objects observed during mitosis. In this chapter we analyze the molecular organization of chromatin and of the chromosomes, as revealed by more recent studies. But first we discuss the general structure of the interphase nucleus and the nuclear envelope.

13–1 THE INTERPHASE NUCLEUS

In eukaryotic cells the DNA, which contains the genes, is separated from the cytoplasm by the nuclear envelope. Large amounts of DNA are packed in one region of the cell, which has important consequences, because the site of the process of transcription of RNA from the DNA is physically separated from the site of protein synthesis, which takes place in the cytoplasm. This has enabled eukaryotes to evolve highly complicated pathways of RNA processing. The precursor messenger RNAs (mRNAs) transcribed from the DNA in the nucleus frequently undergo extensive processing, with segments cut out and others spliced together (see Chapter 20), before a mature cytoplasmic mRNA molecule is produced. Only the fully matured mRNA molecules exit through the nuclear envelope. Conversely, the nuclear envelope prevents the entrance into the nucleus of functional ribosomes and other components required for protein synthesis, thus ensuring that the immature precursor mRNA molecules do not translate into protein. In prokaryotes, protein synthesis is not separated from the DNA by a nuclear membrane, thus the RNA copied from the genes is immediately available for translation into protein, which results in a different type of gene regulation. (For further information on gene regulation see Chapter 20.)

Ultrastructure of the Interphase Nucleus

The light microscope reveals that the interphase nucleus is surrounded by a distinct envelope, although it does not permit distinguishing among the envelope's components. Experiments in which a variety of basic stains were used revealed that chromatin can be present in a condensed state (heterochromatin) or in a dispersed form (euchromatin) in interphase nuclei. Condensed chromatin tends to attach to the internal side of the nuclear envelope. In addi-

tion, one or several large bodies, called *nucleoli* can be observed inside the nucleus. The nucleolus has a high concentration of RNA and proteins, is not surrounded by a membrane, and is the site of transcription of ribosomal RNAs and of assembly of ribosomes, as we discuss in Chapter 21.

Figure 13–1 shows a nucleus visualized by electron microscopy. Osmium tetroxide, the most commonly used fixative for electron microscopy, is not bound by nucleic acids, but when sections are stained with uranyl acetate this displays DNA strongly and RNA slightly less so. The condensed chromatin is bound to the inner side of the nuclear envelope *(c)*, except at the nuclear pores, which leads to the formation of interchromatin "channels" (arrows). Large clumps of chromatin are also associated with the external part of the nucleolus *(ac)*. The rest of the nucleus is occupied by the apparently structureless nucleoplasm, which contains numerous electron-dense granules, some of which contain RNA. At the electron microscopy level, it is possible to distinguish RNA-containing structures from those containing DNA by "bleaching" the DNA stain with EDTA (ethylenediaminetetraacetic acid).[1] In Figure 13–2 the nucleus was also stained with uranyl acetate (as in Figure 13–1), but it was then treated with EDTA. EDTA is a chelating (metal-binding) agent that preferentially displaces the UO_2^{++} ions bound to DNA. A number of granules stained by uranyl acetate now become apparent in the nucleoplasm. That these granules contain RNA can be shown by digestion of the sections with ribonuclease.

These nucleoplasmic granules are ribonucleoprotein particles (RNP) that probably contain mRNA precursors. Some RNPs are small and heterogeneous in size ("*interchromatin granules*" in Fig. 13–2) while others are more compact, larger (45 nm), and usually surrounded by a clear halo ("*perichromatin granules*," arrowed in Fig. 13–2). The nature of these perichromatin granules is still unknown.[2]

Although the nucleoplasm is amorphous in electron microscopy, it probably has considerable structure. Isolated nuclei extracted biochemically with drastic methods (such as 1 Molar salt and various detergents) and digested extensively with DNAse and RNAse still retain some of the structures observed morphologically in electron microscopic sections.[3,4] This suggests

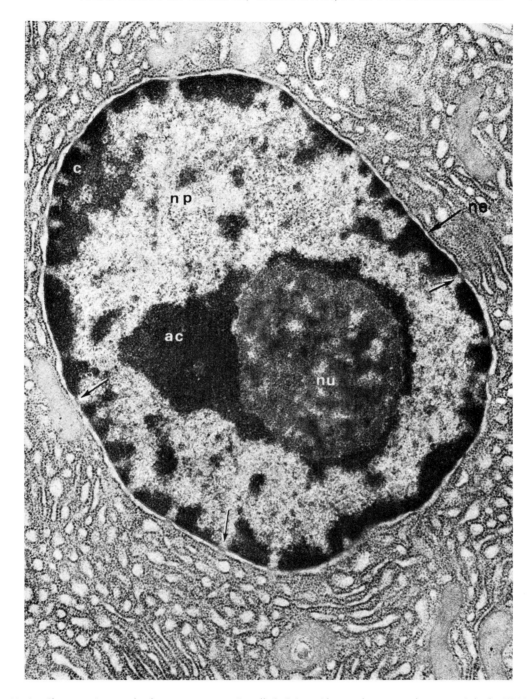

Fig. 13–1.   Electron micrograph of a mouse pancreatic cell. Staining with uranyl acetate enhances mainly the DNA- and RNA-containing parts of the cell. The chromatin is mainly associated with the internal part of the nuclear membrane *(c)*, except at the nuclear pores. The interchromatin channels are indicated by arrows; *ac*, indicates chromatin associated with the nucleolus *(nu)*; *np*, nucleoplasm. Note that the nuclear envelope *(ne)* is a double membrane with pores, with the external membrane covered by ribosomes on the cytoplasmic side. The nuclear envelope can be considered a cisterna of the ER. The cytoplasm of this pancreatic cell contains abundant RER because pancreatic cells are involved in the secretion of a number of digestive enzymes (trypsin, chymotrypsin, amylase, RNAse, and DNAse). ×24,000. (Courtesy of J. André.)

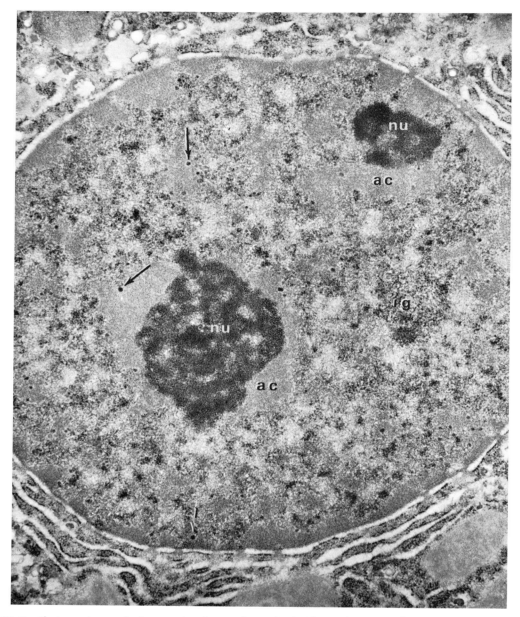

Fig. 13–2.   Electron micrograph of a normal rat liver nucleus stained with uranyl acetate and treated with EDTA to visualize ribonucleoprotein components; *nu*, nucleolus; *ac*, associated chromatin; *ig*, interchromatin granules; perichromatin granules (arrows). Chromatin appears as pale areas because of the EDTA treatment (compare with Fig. 13–1). The RNA-containing granules probably represent precursor mRNAs at different stages of processing. ×25,000. (Courtesy of W. Bernhard.)

that the nucleoplasm may contain a nuclear matrix or nuclear skeleton of protein nature. The best characterized part of the nuclear skeleton, however, is the *nuclear lamina*, which is attached to the internal part of the nuclear envelope.

13–2   THE NUCLEAR ENVELOPE

The light microscope provides little information about the *nuclear envelope*. The electron microscope, however, clearly shows that the nuclear envelope is a special perinuclear cisterna of the cell endomembrane system with an inner

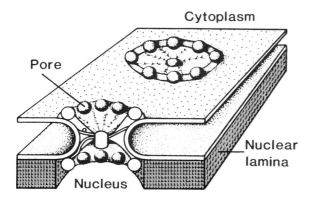

Fig. 13–3. Diagram of the nuclear envelope, which consists of two lipid bilayers with 80 nm pores occupied by pore complexes of octagonal radial symmetry. On the inner side, a nuclear *lamina* 50 to 80 nm thick covers the membrane, except at the nuclear pores. The lamina proteins bind chromatin, which thus becomes attached to the nuclear envelope. (Modified from Franke, W.W., et al.: J. Cell Biol., 91:39, 1981.)

and outer membrane enclosing a lumen and traversed by pores.

Nuclear Pores Are Not Wide-Open Channels

The nuclear envelope consists of two concentric membranes separated by a perinuclear space 10 to 15 nm in width. These membranes have a bilayer structure similar to that of other biological membranes, having ribosomes only on the outer surface. The lipid constitution of these membranes is similar to that of the rough endoplasmic reticulum (RER), and direct continuities with the RER can frequently be observed branching from the nuclear membrane (Fig. 1–1).

As shown in Figure 13–3, at certain points the nuclear envelope is interrupted by structures called *pores*.[5,6] Around the margins of these nuclear pores both membranes are in continuity. At their nuclear side the pores are generally aligned with channels of nucleoplasm (Fig. 13–1, arrows) situated between more condensed lumps of chromatin, which attach to a *fibrous lamina* of proteins 50 to 80 nm thick that is attached to the inner membrane, but not to the nuclear pores (Fig. 13–3).

The nuclear pores are large, 80 nm in diameter. In nuclei of *Mammalia* it has been calculated that nuclear pores account for 5 to 15% of the surface area of the nuclear membrane. In amphibian oocytes, certain plant cells, and pro-

tozoa the surface occupied by the pores may be as high as 20 to 36%. Nuclear pores, however, are not wide-open channels, and the electron microscope has revealed that they are occluded by an electron-dense material. The pores are enclosed by circular structures called *annuli*. The pores and annuli are together designated the *pore complex*.

Negative staining techniques have demonstrated that pore complexes have an eight-fold symmetry.[6] More recently, computerized image-processing techniques[7] have shown that the pore complex consists of two "rings" (or annuli) with an inside diameter of 80 nm, large particles that form a central plug *(c)*, and eight radial "spokes" *(s)* that extend from the plug to the rings, as shown in Figure 13–4. Sometimes, but not always, we can see particles attached to the cytoplasmic side of the ring *(p* in Fig. 13–4) that are also octagonally arranged. These particles might be inactive ribosomes attached to the periphery of the pore complex.

The number of pores in the nuclear membrane seems to correlate with the transcriptional activity of the cell. In the frog *Xenopus laevis*, oocytes (which are very active in transcription) have 60 pores/μm^2 (and up to 30 million pore complexes per nucleus), whereas mature erythrocytes (inactive in transcription) have only about 3 pores/μm^2 (and a total of only 150 to 300 pores per nucleus).[8]

Two proteins have been found associated to the nuclear envelope. One is an integral membrane protein, a glycoprotein (which binds the lectin concanavalin A) of 120,000 daltons that may anchor the annuli to the lipid bilayers (Gerace et al., 1982). The second protein is a 63,000 dalton protein (that has covalently bound acetylneuraminic acid, and binds the lectin wheat germ aglutinin), located on the cytoplasmic side of the electron-dense material that occludes the nuclear pores (Davis and Blobel, 1986) and which may be involved in the transport of materials through the nuclear pore. The isolation of the nuclear pore proteins has been difficult because they tend to be proteolytically degraded during isolation and because they usually are isolated together with the nuclear lamina components which tend to mask the less abundant pore components.

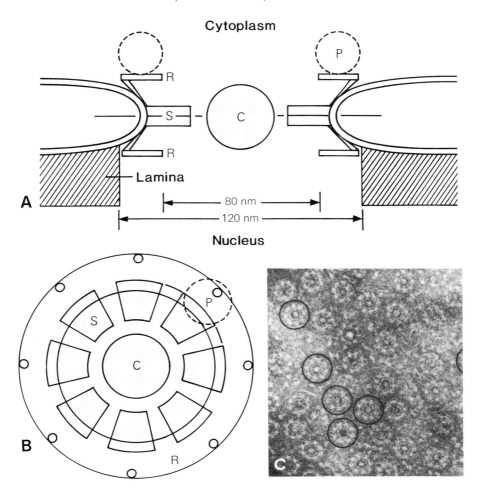

Fig. 13–4. Diagram of the pore complexes as visualized by computerized image reconstruction techniques. *A,* cross section; *B,* top view; *c,* nuclear envelope from *Xenopus* oocytes, which was negatively stained with gold thioglucose, manually isolated, and spread on an EM specimen holder. The data for the diagram was derived from the electron micrograph. Note the two peripheral rings *(R),* the central spokes *(S),* the central plug *(C),* and the nuclear lamina. The cytoplasmic particles *(P)* sometimes found attached to the cytoplasmic ring are the same size as ribosomes. (Redrawn from Unwin, P.T.N., and Mulligan, R.A.: J. Cell Biol., *93*:63, 1982.)

We can expect that in the next few years the identification of other pore components, and the study of their assembly and disassembly during the cell cycle, will lead to insights on the function of this large octagonal macromolecular machine in regulating exchanges between nucleus and cytoplasm.

Annulate Lamellae—Cytoplasmic Stores of Pore Complexes

In developing oocytes and spermatocytes, as well as in certain embryonic and tumor cells, the so-called *annulate lamellae* may be found in the cytoplasm. As shown in Figure 13–5, these membranous structures appear as stacks of cis-ternae of the endoplasmic reticulum (ER), which at regular intervals are traversed by pore complexes that are similar in morphology and structure to those found in the nuclear envelope.[9] These are closely spaced and may occupy up to 50% of the surface of the membrane (Fig. 13–5). Thus, pore complexes are not exclusive to the nuclear envelope. Annulate lamellae are usually associated with rapidly proliferating cells and can be considered as a storage form of preassembled pore complexes.

The Nuclear Lamina Depolymerizes During Mitosis

In most eukaryotic nuclei, a proteinaceous layer 50 to 80 nm thick separates the peripheral

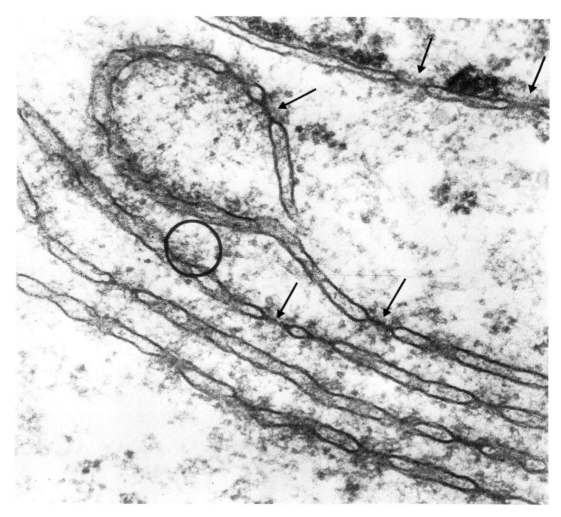

Fig. 13–5. Electron micrograph of annulate lamellae observed in a human melanoma cell cultured in vitro. *Upper right,* the nucleus, with the nuclear envelope, showing two pore complexes (arrows). Similar complexes are observed in the lamellae present in the cytoplasm. ×80,000. (Courtesy of G.G. Maul.)

heterochromatin from the inner nuclear membrane,[10] as can be seen clearly in Figure 13–6. When nuclear membranes isolated by cell fractionation procedures are treated with detergents, the lipids are removed, but the nuclear pore complexes remain attached to each other by way of the meshlike nuclear *fibrous lamina*.[11,12] These preparations are enriched in three polypeptides of 74,000, 72,000, and 62,000 daltons called *lamins* A, B, and C. Antibodies have been made against the three proteins and they have been found to be antigenically related.[13,14]

The mRNAs coding for the lamin proteins have been cloned and sequenced (see McKeon et al., 1986 in the Additional Readings at the end of this chapter). The protein sequence de-

duced from this has shown that lamins, surprisingly, have stretches of extensive amino acid sequence homology with α helical regions of another family of cytoskeletal proteins, the intermediate filament proteins (such as vimentin, desmin, and keratins). Lamins A and C result from differential RNA processing of the same gene and both proteins differ only at the COOH terminus. The α helices conserved with the intermediate filament proteins are present in the part common to lamins A and C, and they are presumably responsible for the assembly of these proteins into a fibrous network (see McKeon et al., 1986 in the Additional Readings at the end of this chapter). Not all cell types have identical lamin proteins, and specialized

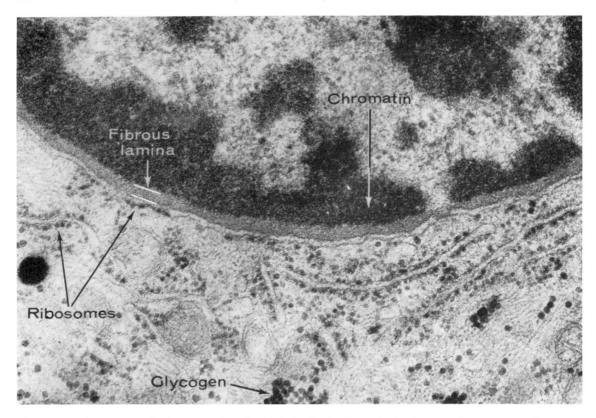

Fig. 13–6. Ultrastructure of a human epidermal epithelial cell. The fibrous lamina is unusually conspicuous because it stains less intensely than the peripheral heterochromatin. Note that the lamina does not extend over the nuclear pore at the right of the figure. (From Fawcett, D.W.: The Cell. Philadelphia, W.B. Saunders, 1981.)

lamins appear in some cell types and at different stages of development (see Benavente et al., 1985 in the Additional Readings at the end of this chapter).

Immunofluorescence studies have shown that in interphase cells the lamins are localized in the inner part of the nuclear envelope and that at the onset of prophase the lamins start to depolymerize and appear in the cytoplasm. During mitosis, the lamins are distributed throughout the cell. At telophase, they re-polymerize around the nuclear periphery.[13–15] Biochemical studies[16] have shown that the depolymerization of the lamina during mitosis is due to phosphorylation of the lamins. Lamin B seems to remain associated with membrane vesicles during mitosis, and these vesicles might in turn remain as a distinct subset of membrane components from which the nuclear envelope is reassembled at telophase. Lamins A and C become entirely soluble during mitosis, and at telophase they become dephosphorylated again, and polymerize around the chromatin. The nuclear pore octagonal structures are also disassembled during mitosis, because they can no longer be detected by electron microscopy. Because of our scant knowledge on the pore complex components, we do not yet know if any remain associated with membrane vesicles during mitosis (see Blobel, 1985 in the Additional Readings at the end of this chapter). Understanding what happens to the nuclear envelope components during mitosis will no doubt help us to understand the mechanism of nuclear assembly.

Chromatin binds strongly to the inner part of the nuclear lamina, which could conceivably interfere with chromosome condensation. During meiotic chromosome condensation, the nuclear lamina completely disappears by the pachytene stage of prophase[17] and reappears later during diplotene in oocytes, but does not reappear at all in spermatocytes (see Fig. 16–12).

The lamins are the most abundant DNA-binding proteins in most eukaryotic nuclei[18] and can bind DNA even after denaturation by SDS-electrophoresis. It also appears that the nuclear lamina interacts strongly with the inner nuclear membrane. Thus, the lamins may play a crucial role in the assembly of interphase nuclei. In fact, when cells are left for a long time in colchicine (which arrests cells in metaphase), the lamins assemble around individual chromosomes,[15] which then become surrounded by nuclear envelopes giving rise to micronuclei containing only one chromosome. A similar phenomenon takes place during normal amphibian development. In the first few cleavages of amphibian development, the nuclear envelope initially forms around individual chromosomes, forming several vesicles that then fuse together to form a single nucleus. This suggests that chromatin is the nucleating center for the deposition of a nuclear lamina and envelope.

Interphase Nuclei Can Be Assembled from Bacteriophage λ DNA

The frog egg is a cell that is programmed to rapidly assemble nuclei. After fertilization, the egg divides rapidly and in only 6 hours it makes 4000 new cells, each one with a nucleus. Thus, *Xenopus* eggs appear to be an ideal system to assemble nuclei. Forbes and colleagues (1984) microinjected DNA from bacteriophage λ into activated eggs. The injected DNA bound to histone proteins forming chromatin and about 1 hour later they could see spherical bodies budding off from the main mass of bacteriophage DNA. The rounded bodies were indistinguishable from normal nuclei by electron microscopy. Recently, it has been possible to assemble nuclei in vitro using concentrated extracts from *Xenopus* eggs and either demembranated sperm nuclei or pure bacteriophage λ DNA (see Lohka and Masui, 1984 and Newmeyer et al., 1986 in the Additional Readings at the end of this chapter).

These experiments have shown that the initial requirement for assembling a nucleus is DNA. This DNA binds first to histones and other proteins producing chromatin, which in turn binds to lamins and eventually becomes surrounded by membranes. The in vitro assembly systems have shown that small membrane vesicles, which have relatively electron dense contents bind directly to the chromatin, and then flatten

out to form the nuclear envelope (see Lohka and Masui, 1984 in the Additional Readings at the end of this chapter). In addition, it has been shown that the process of membrane assembly is different from pore complex insertion, because under certain conditions nuclei can be obtained which have few or no pores, which are however surrounded entirely by a double membrane (see Newmeyer et al., 1986 in the Additional Readings at the end of this chapter). Because artificial nuclei can be made from a variety of prokaryotic DNAs, it appears that no specific sequence information is necessary to trigger assembly of a nucleus. Artificial nuclei can also undergo chromatin condensation and breakdown of the nuclear membrane during mitosis. The availability of such a system might hopefully lead to understanding the molecular mechanisms by which an interphase nucleus is assembled.

Nuclear Pores Are a Selective Diffusion Barrier Between Nucleus and Cytoplasm

The materials exchanged between nucleus and cytoplasm must traverse the nuclear pore complexes. This exchange is very selective and allows passage of only certain molecules, of either low or very high molecular weight. One of the main functions of the nuclear envelope is to prevent the entrance of active ribosomes into the nucleus. Ribosomes and other cytoplasmic components (Fig. 21–26) are selectively excluded from the nucleus.

The permeability of the nuclear membrane with respect to small ions is reflected in some of the electrochemical properties of this structure, which can be investigated with fine microelectrodes.[18] As shown in Figure 13–7, when giant cells from the salivary gland of *Drosophila* are penetrated with a microelectrode, there is an abrupt change in potential at the plasma membrane (-12 mV); then, as the microelectrode enters the nucleus, there is another drop in negative potential at the nuclear membrane (-13 mV). These results suggest that the nuclear envelope may constitute a diffusion barrier for ions as small as K^+, Na^+, or Cl^- (or that other mechanisms, such as Donnan's equilibrium, [see Chapter 4] are utilized).

On the other hand, very large structures such as ribosomal subunits, which are assembled in

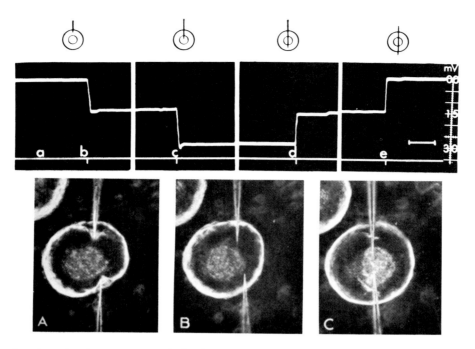

Fig. 13–7. Experiment in microsurgery to study the electric potential of the nuclear envelope in giant nuclei of the salivary gland of *Drosophila*. *Above,* a diagram of the penetration of the microelectrode into the cell is shown along with the membrane or steady potentials registered at each position. *Below,* photomicrographs of cells penetrated by two microelectrodes. *A,* penetration of the membrane; *B,* into the cytoplasm; *C,* into the nucleus. (Courtesy of W.R. Loewenstein and Y. Kanno.)

the nucleolus, are able to leave the nucleus. This can be visualized in Figure 21–4, where ribonucleoprotein particles are seen traversing the nuclear pore complexes. Frequently, dumbbell-like structures can be seen traversing the nuclear pores in cells that are very active in transcription. Interestingly, only a small (10–15 nm) part of the central region of the pore complex is used in this process, as shown in Figure 13–8. Not all RNAs can exit the nucleus, however, heterogeneous nuclear RNA (hnRNA, a large precursor of mRNA which contains intervening sequences—see Chapter 20) is never found in the cytoplasm, and the mRNA sequences can traverse the pores only after extensive processing has occurred within the nucleus. Other RNAs, such as the so-called small nuclear RNAs, also accumulate specifically inside the nucleus in their snRNP form (Fig. 20–24).

Xenopus oocytes are particularly useful for studying nucleocytoplasmic exchanges because these cells are very large (up to 1.2 mm diam-

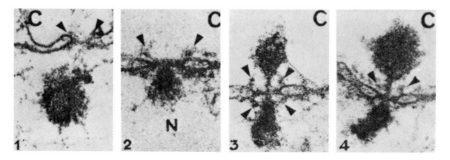

Fig. 13–8. Electron-dense material (probably ribonucleoproteins) being extruded through nuclear pores in salamander oocytes. *(N),* nucleus at the bottom; *(C),* cytoplasm at the top. Note that in panel 3 the material is extruded through the center of the pore, occupying only the middle 10–15 nm. (Courtesy of Uli Scheer.)

eter), and substances can be easily microinjected into the cytoplasm. One can then find out whether or not these substances entered the nucleus by manually isolating the giant oocyte nucleus (also called the germinal vesicle), which measures up to 0.4 mm in diameter. Studies involving the microinjection of dextrans, colloidal gold particles, and labeled proteins suggest that the oocyte nuclear membrane has functional pores with a radius of about 4.5 nm.[19,20] It was found that small proteins (such as lysozyme, molecular weight 14,000; and ovalbumin, molecular weight 44,000), were able to enter the nucleus, that bovine serum albumin (molecular weight 62,500) entered extremely slowly, and that larger proteins did not enter at all. This upper limit for diffusion applies only to proteins that are not normally nuclear proteins for, as we see in the following section, nuclear-specific proteins behave differently.

Some contradiction exists between the measurement of a pore size of 4.5 nm by microinjection studies and the finding that the electrical potential of the nucleus is different from the cytoplasm. With pores so large, there should be no effective barrier to the diffusion of ions. This contradiction probably merely reflects the fact that we do not yet understand how the nuclear pore complexes function as molecular filters.

Nuclear Proteins Accumulate Selectively in the Nucleus

The cell nucleus contains a specific subset of proteins. For example, RNA polymerase, DNA polymerase, and histones are located specifically inside the nucleus. All these proteins, however, are synthesized in the cytoplasm and must be subsequently transported into the nucleus. Furthermore, during each cell division the nucleus is disassembled. Figure 13–9 shows that nuclear proteins (in this case a protein that binds to hnRNA) become dispersed throughout the cell during mitosis and do not remain associated with the mitotic apparatus or the condensed chromosomes (with the exception of histones and other proteins that are directly bound to DNA). These proteins reenter the nucleus at each telophase when the nucleus is reassembled.[21,22] Clearly, selective mechanisms must be required to achieve this sorting out of nuclear components in the cell.

The following experiments show that nuclear proteins contain, within the structure of the mature protein, a signal (or property) that enables them to accumulate in the nucleus after microinjection into the cytoplasm as first proposed in 1978.[23]

Figure 13–10 shows the three ways in which proteins can behave after microinjection into the cytoplasm of frog oocytes. Some proteins (such as actin) distribute almost evenly through the nucleus and cytoplasm; others (such as tubulin) remain predominantly cytoplasmic; and others (also called karyophilic proteins) migrate and accumulate in the cell nucleus. A large variety of nuclear migrating proteins have been found.[23,24] These experiments have shown that essentially all proteins initially isolated from nuclei have the ability to migrate and accumulate in the nucleus, while cytoplasmic proteins do not. Even naked RNAs microinjected into the oocyte can migrate to and accumulate in the nucleus when they are of nuclear origin (such as snRNAs; Fig. 20–24); it is thought that this intracellular sorting out is due to the binding of these RNAs to karyophilic proteins.[25] (For further information see Chapter 20.)

The great degree of selectivity of protein accumulation in nuclei is illustrated by the experiment shown in Figure 13–11. When nuclear proteins of normal oocytes are examined by two-dimensional gel electrophoresis, two types of proteins can be distinguished: (1) nucleus-specific proteins, and (2) proteins common to *both* nucleus and cytoplasm. Actin is the most conspicuous member of the "common-to-both" proteins, and it is one of the major constituents of the oocyte nucleoplasm. A labeled preparation of *Xenopus* nucleoplasmic proteins (which contained actin and nuclear-specific proteins, such as the protein indicated as N1—see Figure 13–11A) was injected into the cytoplasm of living oocytes.[23] After 24 hours the oocytes were manually separated into nucleus and cytoplasm, and the proteins were analyzed by two-dimensional gels (Fig. 13–11B and C). It was found that nucleus-specific proteins were greatly concentrated in the nucleus (100-fold in the case of N1), while microinjected actin was found in both nucleus and the cytoplasm (it is the major protein in Fig. 13–11C). This is as expected, because actin is normally present in both cell compartments.

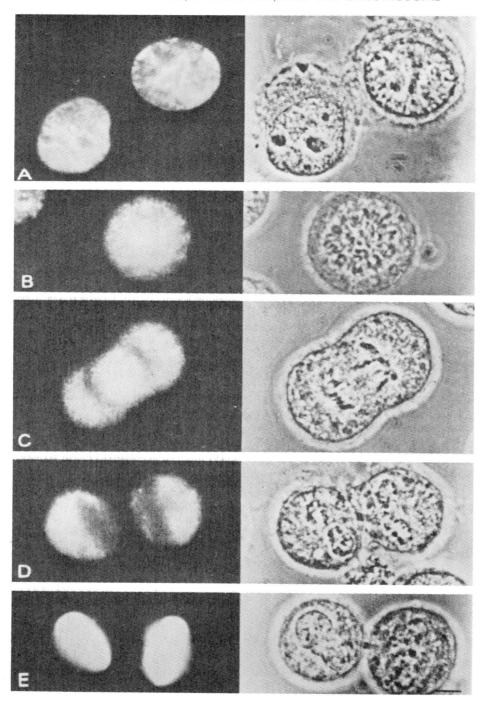

Fig. 13–9. Immunofluorescent localization of the main protein that binds to pre-messenger RNA (hnRNA) during the cell cycle. The fluorescent antibody is shown on the left-hand side (ultraviolet illumination). The morphology of the same cells with phase contrast can be seen on the right-hand side. *A,* interphase cells, the antigen is exclusively nuclear; *B,* metaphase cell, the antigen is distributed throughout the cell; *C,* anaphase, the antigen is everywhere, except on the chromosomes; *D,* early telophase, the nuclei are starting to form but do not yet concentrate the protein; *E,* late telophase, the antigen re-enters the newly assembled nuclei, and the daughter cells are still held together by the midbody. (Courtesy of Dr. Terry Martin.)

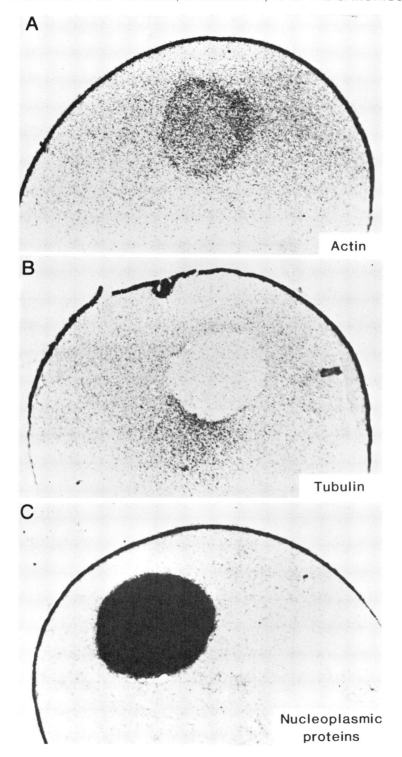

Fig. 13–10. Autoradiograms of histological sections of oocytes microinjected with ^{125}I-actin *(A)*, ^{125}I-tubulin *(B)*, and a soluble preparation of oocyte ^{35}S-labeled nucleoplasmic proteins *(C)*. Note the differential behavior of the three proteins: (1) Actin distributes throughout both cell compartments, (2) tubulin remains in the cytoplasm, and (3) proteins of nuclear origin strongly accumulate in the nucleus. The black ring at the periphery of the oocyte is melanin pigment, not autoradiographic grains. (From De Robertis, E.M.: *Cell*, 32:1021, 1983.)

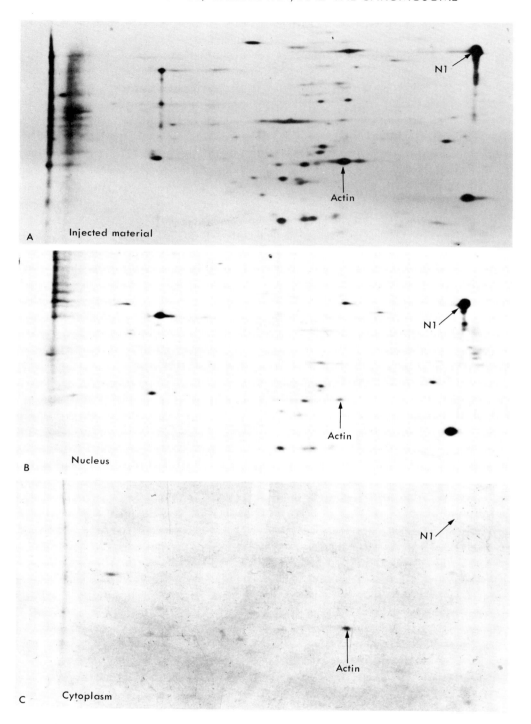

Fig. 13–11.   Selective migration of nuclear proteins. Two-dimensional gel analysis of ^{35}S-methionine-labeled nucleoplasmic proteins microinjected into *Xenopus* oocytes. *A*, a preparation of [35]S-*Xenopus* nucleoplasmic proteins that was injected into oocytes. Note that it contains actin and many nuclear proteins. *B*, the nuclear fraction reisolated from unlabeled oocytes injected 24 hours earlier into the cytoplasm with the ^{35}S proteins shown in *A*. Note that the nuclear proteins re-enter the nucleus, while actin decreases in relative amount. *C*, cytoplasmic fraction from the same microinjected oocytes. Note that actin is the most prominent protein, while the nuclear proteins are not detectable in the cytoplasm. (From E. M. De Robertis.)

The nuclear-specific proteins have in their molecular structure some type of signal that enables them to accumulate selectively in the nucleus.[23] This karyophilic signal must be absent in proteins that do not accumulate in the nucleus, such as actin. Some of the nuclear proteins are quite large (N1 has a molecular weight of 120,000 daltons) but are still able to re-enter the nucleus rapidly; it therefore seems that when proteins are nucleus-specific, factors other than size determine nuclear pore permeability.

In an important experiment, Feldherr et al. (1984) coated colloidal gold particles with *nucleoplasmin*, a protein which is the most abundant protein in *Xenopus* oocyte nuclei and which is strongly nuclear-migrating. The coated gold particles were microinjected into *Xenopus* oocytes, which were then fixed, sectioned, and analyzed by electron microscopy. The gold particles were able to traverse the nuclear pores, being transported into the nucleus through a smaller (20 nm) channel in the central region of the pore complex (as in Fig. 13–8). This experiment proved formally that nucleocytoplasmic exchanges take place through the nuclear pores. Since large gold particles, of up to 16 nm, could be transported as easily as smaller particles, it suggests that having a karophylic signal somehow opens up the nuclear pore to passage of macromolecules. One can imagine that the nuclear pore might function like a sphincter that might open for the entrance or exit of macromolecules. Whichever the mechanism, it is clear that the way in which the pore functions is unknown, even though this enormous (diameter 80–120 nm, Fig. 13–4) proteinaceous octagonal structure has been conserved since the appearance of the first eukaryotic cells. Understanding how nuclear pores work will be one of the challenges for cell biologists in the next few years.

The nuclear concentration is a property of the mature protein because even proteins labeled in vitro by iodination are able to migrate.[26] When *nucleoplasmin* was cleaved in half, only one part had the property of migrating to and accumulating in nuclei.[27] Recently, it has been possible to identify the karyophilic signal on proteins using gene manipulation techniques, by progressively deleting parts of genes. In a detailed study it was found that the T antigen protein of SV40 virus requires the sequence Thr-Lys-Lys-Lys-Arg-Lys-Pro for mi-

gration into the cell nucleus (see Kalderon et al., 1984 *a* and *b* in the Additional Readings at the end of this chapter). Nucleoplasmin also requires a similar sequence (T. Burglin and E. De Robertis, in preparation). Thus it seems that the karyophilic signal in proteins is extremely simple; a series of basic amino acids (Lys or Arg) flanked by one or more amino acids (Pro or Gly) that disturb the formation of α-helices (perhaps giving more flexibility to this part of the protein). This simple signal, or variations of it, is all that is required, as demonstrated by experiments in which a synthetic peptide containing the SV40 karyophilic signal was cross-linked to antibody molecules. Antibody molecules are large and cannot enter nuclei after microinjection into the cytoplasm, but will readily do so if a signal peptide is attached to them (see Lanford et al., 1986 in the Additional Readings at the end of this chapter).

We also know little about the mechanism by which some proteins are kept within the nucleus against a concentration gradient. Two main mechanisms for nuclear accumulation are envisaged: (1) active transport of the karyophilic proteins through the nuclear pores, and (2) binding of karyophilic protein to a non-diffusible nucleoplasmic component (reviewed by DeRobertis, 1983). Experiments involving damage of the nuclear membrane[28] have shown that nuclear accumulation can be obtained in the absence of a functional nuclear envelope; this favors the binding mechanism. On the other hand, we were able to show that in vitro assembled nuclei, which can efficiently transport nucleoplasmin, are unable to do so if ATP is depleted from the system. This ATP requirement for nuclear entry seems to favor active pumping through the pore (see Newmeyer et al., 1986 in the Additional Readings at the end of this chapter). However, it may well be that both mechanisms coexist: an active, selective, opening of the nuclear pore, followed by binding to a nucleoplasmic component. If so, probably both mechanisms for accumulation utilize the same karyophilic signal.

Cells Use Multiple Mechanisms to Segregate Proteins into Cellular Compartments

Cells have the capacity to place specific proteins into the corresponding cell compartments, as occurs with the mitochondria or chloroplast proteins that are synthesized in the cytoplasm, or with proteins that are exported from the cell. The mechanisms by which proteins are segregated into the ER (see Chapter 8) or into mitochondria or chloroplasts (see Chapter 11) are different from the one used for the accumulation of proteins inside the nucleus. In secretory and membrane proteins, the hydrophobic "signal peptide" enables the proteins to traverse the ER as they are being synthesized, and it is rapidly cleaved off thereafter (Fig. 8–10). In mitochondrial proteins, synthesis is completed with an extra signal segment, the protein then penetrates the mitochondrion and the signal sequence is cleaved off after the final destination has been achieved, making the translocation irreversible. In nuclear proteins, nuclear migration and accumulation occur independently of protein synthesis, and do not involve transient precursor proteins. The karyophilic signal is contained in the mature nuclear proteins, which is in line with the biological need of disassembling and reassembling the cell nucleus at each cell division.

SUMMARY:
The Nuclear Envelope

The nuclear envelope, a differentiation of the RER, is composed of two membranes and a perinuclear space. Nuclear pores represent openings in this envelope at sites where the two membranes are in contact. The pores are octagonal orifices about 80 nm in diameter. They are not, however, freely communicating openings, but are plugged by a cylinder of protein materials. In rapidly growing cells (oocytes, embryonic cells, and cancer cells) preassembled pore complexes can be stored in the cytoplasm in membranous structures called *annulate* lamellae.

The inner membrane of the nuclear membrane is attached to a layer of fibrous proteins 50 to 80 nm thick that is known as the *nuclear lamina*. During mitosis, the lamina proteins become phosphorylated and the nuclear envelope disassembles. At telophase, the lamina proteins bind strongly to chromatin and are thought to be important in reassembling the nucleus.

The nuclear envelope regulates the passage of ions and macromolecules. By way of the pore complex, the envelope may have an important role in the transfer of macromolecules between the nucleus and the cytoplasm—and vice versa. The passage of very large ribonucleoprotein particles (ribosomal subunits, mRNA and others) may be seen by electron microscopy.

Nuclear proteins are able to accumulate inside

TABLE 13–1. DNA CONTENT AND CHROMOSOME COMPLEMENT

Cells	Mean DNA-Feulgen Content	Presumed Chromosome Set
Spermatid	1.68	haploid (n)
Liver	3.16	diploid (2n)
Liver	6.30	tetraploid (4n)
Liver	12.80	octoploid (8n)

From Pollister, A.W., Swift, H., and Alfert, M: J. Cell. Comp. Physiol., *38* (Suppl. 1):101, 1951.

the nucleus after microinjection. Nuclear proteins contain in their mature molecular structure a signal that enables them to accumulate inside the nucleus. This signal is absent in those proteins (such as actin) that are normally distributed in both the nucleus and the cytoplasm. The mechanism involved in this selective distribution of proteins is very different from that involved in the intracellular segregation of export proteins.

13–3 CHROMATIN

DNA is the main genetic constituent of cells, carrying information in a coded form from cell to cell and from organism to organism. Within cells, DNA is not free but is complexed with proteins in a structure called chromatin.

The C-Value Paradox

The demonstration that nuclei contain a constant amount of DNA[29,30] was a landmark in cell biology because it suggested that DNA was the genetic material and that genetic information is not lost during the differentiation of the various somatic tissues. The DNA in nuclei was stained using the Feulgen reactions, and the amount of stain in single nuclei was measured using a special microscope (cytophotometer). All the cells in an organism contain the same DNA content (2C) provided that they are diploid. Gametes are haploid and therefore have half the DNA content (1C). Some tissues, like liver, contain occasional cells that are polyploid and their nuclei have a correspondingly higher DNA content (4C or 8C), as seen in Table 13–1.

Each species has a characteristic content of DNA, which is *constant* in all the individuals of that species and has thus been called the *C-value*. Eukaryotes vary greatly in DNA content but always contain much more DNA than prokaryotes. Lower eukaryotes in general have less

DNA, such as the nematode *C. elegans*, which has only 20 times more DNA than *Escherichia coli*, or the fruit fly *Drosophila melanogaster*, which has 40 times more (0.18 pg per haploid genome). Vertebrates have greater DNA content (about 3 pg), in general about 700 times more than *E. coli*. One of the highest DNA contents is that of the salamander *Amphiuma*, which has 84 pg of DNA. Man has about 3 pg of DNA per haploid genome, or 3×10^9 base pairs, i.e., the human genome could accommodate about 3 million average-sized proteins if all the DNA were coding. But then, if this were true, salamanders would have 30 times more genes than man. This is the so-called *C-value paradox*. Quite early on it was found that there was little connection between the morphological complexity of eukaryotic organisms and their DNA content. *E. coli* (3,400,000 base pairs) has about 3000 genes. Although it is difficult to estimate how many different genes exist in the human genome, there are probably not more than 20,000 to 30,000, and there is no reason to believe that salamanders should have any more. Thus, most of the DNA in eukaryotic genomes must be of a non-coding nature. In Chapter 22, we show that a larger part of the eukaryotic genome is made up of repetitive, non-coding DNA. There are presumably few selective pressures to keep the eukaryotic genome streamlined, which results in a large amount of DNA that must be tightly packed within the nucleus.

Considering that 1 pg of DNA is equivalent to 31 cm of DNA, it is possible to calculate that there are about 174 cm of DNA in human diploid cells. The DNA content in the 46 human chromosomes has been estimated by cytophotometry, and from these measurements it appears that the DNA content is proportional to the size of the chromosome. The largest chromosome (1), which is 10 μm long, should accommodate about 7.2 cm of DNA in a tightly packed form; i.e., the DNA is compacted about 7000-fold in a metaphase chromosome. In the rest of this chapter we analyze how this folding is achieved.

Chromatin Is a Complex of DNA and Histones

Chromatin can be isolated biochemically by purifying nuclei and then lysing them in hypotonic solutions. When prepared in this way,

TABLE 13–2. COMPOSITION OF CHROMATIN
(Data expressed as % dry weight)

Component	Liver	Pea Embryo
DNA	31	31
RNA	5	17.5
Histone	36	33
Non-histone proteins	28	18

chromatin appears as a viscous, gelatinous substance that contains DNA, RNA, basic proteins called histones, and non-histone (more acidic) proteins (Table 13–2). Since the early 1960s it was known that the content of RNA and non-histone protein is variable between different chromatin preparations, but histone and DNA are always present in a fixed ratio of about 1:1 (Table 13–2). The non-histone proteins are very heterogeneous; they vary in different tissues and include RNA and DNA polymerases, among other enzymes, and putative regulatory proteins. On the other hand, Figure 13–12 shows that there are only five types of histones, each one present in large amounts. This suggests that histones could have a structural role.

Table 13–3 shows some of the properties of histones. Histones are small proteins that are basic because they have a high content (10 to 20%) of the basic amino acids arginine and lysine. Being basic, histones bind tightly to DNA, which is acid. The four main histones, H2A, H2B, H3, and H4, are very similar in different species, being among the most conserved proteins known. For example, the sequence of histone H3 from the rat differs only in two amino acids from that of peas, out of 102 total amino acids. These four "core" histones are present in equimolar amounts (2 of each every 200 base pairs of DNA).

Histone H1 is not conserved between species and has tissue-specific forms; is present only once per 200 base pairs of DNA; is rather loosely associated with chromatin (it can be eluted from DNA by adding low concentrations of salt and may be easily lost during the biochemical preparation of chromatin) and is involved in the maintenance of a higher-order folding of chromatin.

Electron Microscopy of Chromatin Spreads Revealed a Beaded Structure

When eukaryotic nuclei were *spread* on EM specimen holders (instead of prepared as thin

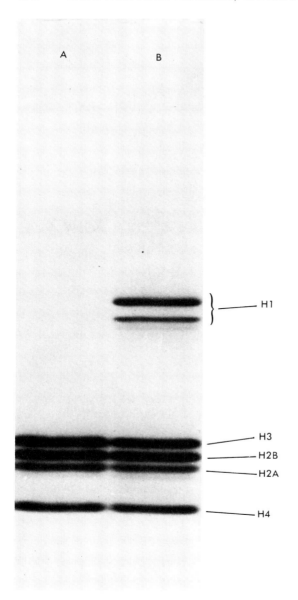

Fig. 13–12.   Histone proteins after electrophoresis in polyacrylamide gels containing sodium dodecyl sulphate. *A,* histones from nucleosome "cores" that do not contain H1. *B,* histones from total chromatin. Note that the four fundamental histones are present in about equal amounts. (Courtesy of Jean O. Thomas.)

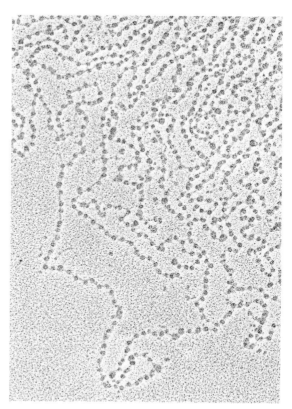

Fig. 13–13.   A *Drosophila melanogaster* embryonic nucleus spread for electron microscopy. Most of the DNA, if not all, has a beaded structure. This beads-on-a-string appearance is seen when histone H1 is lost during preparation. × 50,000. (Courtesy of O.L. Miller, Jr.)

pared, which results in a loss of histone H1. With more mild treatments (which do not remove H1), or in thin electron microscopic sections of fixed nuclei, the beads are not stretched but can be observed as a 10 nm fiber, with the beads touching each other. The 10 nm fiber therefore represents the first level of organization of chromatin within cells.

The Nucleosome—An Octamer of Four Histones (H2A, H2B, H3, and H4) Complexed with 200 Base Pairs of DNA

The existence of a repeating unit of chromatin—called the *nucleosome*—was also predicted from biochemical studies independent of the electron microscopic results described above.[32]

Studies of digestion of chromatin by an enzyme called micrococcal nuclease showed that the DNA was cut into multiples of a unit size, which was later found to be about 200 base pairs

sections), it was found that chromatin has a repeating structure of beads about 10 nm in diameter connected by a string of DNA.[31] Most, if not all, of the DNA is present in this form, as shown in Figure 13–13.

The beads-on-a-string appearance is not the true structure of chromatin but is an artifact arising from the way in which the sample is pre-

TABLE 13–3.   SOME PROPERTIES OF HISTONES

Histone	Molecular Weight	Amino Acid Composition	Species Variation	Number of Molecules per 200 Base Pairs of DNA
H1	20,000	Lysine-rich	Wide	1
H2A	13,700 ⎫	Moderately lysine-rich	Fairly well-conserved	2
H2B	13,700 ⎭			2
H3	15,700 ⎫	Arginine-rich	Highly conserved	2
H4	11,200 ⎭			2

of DNA in length;[33] i.e., the DNA was cut at intervals of 200, 600, 800 base pairs and so forth. Other studies showed that histones H3 and H4 tend to associate in solution, forming tetramers consisting of two of each. Because the four histones are in equimolar amounts (2 of each per 200 base pairs of DNA—see Table 13–3), in 1974 Roger Kornberg proposed a model for the nucleosome, in which the four histones H2A, H2B, H3, and H4 are arranged in octamers containing two of each of them, every 200 base pairs of DNA.[32] The histone octamers are in close contact (constituting the 10 nm fiber), and the DNA is coiled on the *outside* of the nucleosome.

That the biochemical subunit is identical to the beads observed in the EM was demonstrated by the so-called *"one, two, three, four"* classical experiment shown in Figure 13–14. After partial digestion of chromatin with the enzyme micrococcal nuclease, monomer, dimer, trimer, and tetramer fractions were isolated on a sucrose gradient (Fig. 13–14A). When analyzed by electrophoresis, these fractions contained DNA fragments 200, 400, 600, and 800 base pairs long respectively (Fig. 13–14B) and one, two, three, and four beads long respectively under the EM (Fig. 13–14).

The diagram in Figure 13–15 explains why DNA in chromatin is cleaved into discrete fragments by micrococcal nuclease. Neighboring nucleosomes are connected by *linker* DNA, which is more exposed to the enzyme. Brief digestion gives rise to *nucleosomes*, which contain a histone octamer, 200 base pairs of DNA, and one molecule of histone H1. The nucleosome is a flat disc-shaped particle 11 nm in diameter and 5.7 nm in height. The DNA makes two complete turns around the histone octamers and these two turns are sealed off by an H1 molecule. Extensive digestion (Fig. 13–15C) generates a nucleosome *core particle* that has lost the linker DNA and contains only 146 base

pairs of DNA. As shown in Figure 13–12A, nucleosome cores also lose histone H1.

The DNA enters and exits the nucleosome at sites close to each other, and the two turns of DNA are stablized or "sealed off" by histone H1. As a result, when chromatin *containing H1* is spread for electron microscopy at very low salt concentrations a typical zigzag pattern is produced (Fig. 13–16A) because the nucleosomes, being flat, lie on their sides on the specimen holder, connected by the linker DNA.

Chromatin lacking H1 produces the beads-on-a-string appearance (Fig. 13–13) in which the DNA enters and leaves the nucleosomes at random places. Histone H1 can also interact with the H1 moiety of other neighboring nucleosomes, and this leads to the further folding of the fiber.

The length of DNA per nucleosome can vary in different tissues. For example, chicken erythrocytes have a repeat length of 210 base pairs, and chicken liver has about 195 base pairs per repeat. The reason for this is not clear, but it appears to be related to the type of histone H1 in each tissue. The variability is restricted to the length of the linker segment, and all tissues have a constant amount of DNA in the nucleosome core.

When nucleosomes are in close apposition in the 10 nm filaments, the packing of DNA is about five-to seven-fold, i.e., five to seven times more compact than free DNA. This is still 1000-fold lower than the packing ratio of DNA in metaphase chromosomes, which is achieved by further folding the 10 nm fiber.

The 30 nm Fiber—A Nucleosome Solenoid

Electron microscopic studies of chromatin fibers in interphase nuclei and mitotic chromosomes[34] have revealed a *"thick fiber"* of a diameter that varies from 200 to 30 nm and which probably represents the structure of in-

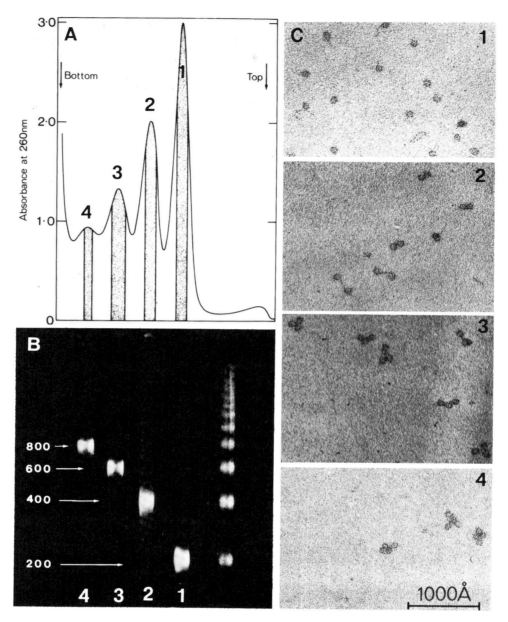

Fig. 13–14. The biochemical nucleosome corresponds to the beads observed with the EM. Chromatin was partially digested with micrococcal nuclease and centrifuged in a sucrose gradient. The fractions corresponding to the monomer, dimer, trimer, and tetramer peaks in *A* were analyzed by electrophoresis in agarose after extraction of the DNA in *B;* or examined directly under the EM in *C.* (From Finch, J.T., Noll, M., and Kornberg, R.D.: Proc. Natl. Acad. Sci. USA, *72:*3320, 1975.)

active chromatin. Figure 13–16*C* shows that the 30 nm fiber consists of closely packed nucleosomes.

The 30 nm fiber probably arises from the folding of the nucleosome chain into a solenoidal structure having about six nucleosomes per turn.[35,36] In chromatin containing H1 the chromatin fiber can be folded into such structures

simply by increasing the salt concentration, as shown in Figure 13–17. At 1 mM salt the usual zigzag appearance is observed, at 5 mM the nucleosomes come closer together (Fig. 13–16*B*), and at concentrations over 60 mM a 30 nm thick fiber is found.[36] The whole structure is stabilized by interactions between H1 molecules in neighboring nucleosomes. The H1 molecules can in-

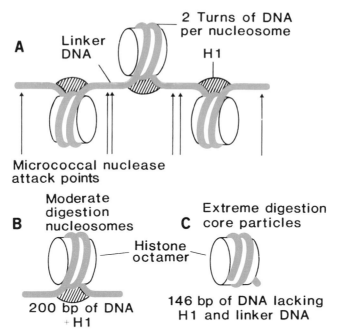

Fig. 13–15. Diagram illustrating the effect of nucleases on chromatin. *A*, intact chromatin, note that the nucleosome is flat and that each histone octamer has two turns of DNA sealed off by histone H1. In living cells, the nucleosomes would be touching each other forming a 10 nm fiber and not stretched out as shown here. *B*, nucleosomes released by moderate digestion, containing 200 base pairs of DNA. *C*, core particles obtained by extensive digestion with only 146 base pairs of DNA.

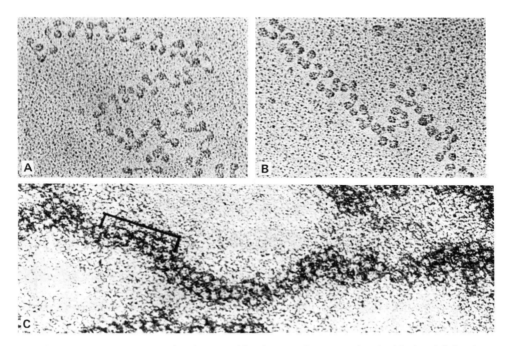

Fig. 13–16. Chromatin (containing H1 molecules) spread for electron microscopy. *A*, at 1 mM salt a definite zigzag structure can be seen. *B*, at 5 mM salt the nucleosomes are closer together and the linker DNA cannot be distinguished clearly. *C*, 30 nm fiber; the nucleosomes can be seen closely packed within it. (Panels A and B courtesy of T. Koller; Panel C courtesy of B. Hamkalo.)

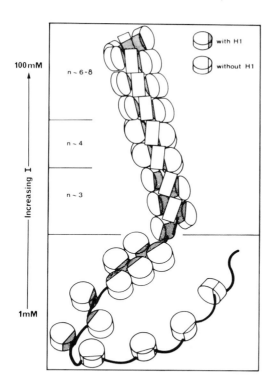

Fig. 13–17. Idealized drawing of the solenoid superstructures found by chromatin containing H1 when exposed to increasing ionic strength. The open zigzag of nucleosomes (bottom left) closes up to form helices of increasing numbers of nucleosomes per turn (*n*). When H1 is absent (pictured at bottom right), no zigzags or definite higher-order structures are found. The data for this diagram was derived from the electron micrographs in Figure 13–16. (From Thoma, F., Koller, T., and Klug, A.: J. Cell Biol., 83:403, 1979.)

teract with each other because they are located in the central "hole" of the solenoid. Histone H1 is thought to play a fundamental role in this condensation because solenoids can not be formed when histone H1-depleted chromatin is treated in the same way. Although other structural arrangements have been proposed for the 30 nm fiber ("superbeads," "double ribbons"), the solenoid model remains the most likely form of chromatin folding (see Felsenfeld and Mc-Ghee, 1986; Widom and Klug, 1985 in the Additional Readings at the end of this chapter).

The DNA of a 30 nm solenoid has a packing that is about 40-fold. The DNA of a metaphase chromosome, however, is packed between 5000 and 10,000 times, i.e., the 25 nm fiber must be further folded more than 100-fold during mitosis.

Bacterial Chromosomes Are Not Complexed with Histones

Prokaryotes do not have histones, and the free phosphate groups on the DNA do not seem to be neutralized by basic proteins but rather by other substances, such as polyamines. On the other hand, all higher organisms have histones, the amino acid sequence of which has been greatly conserved through evolution, suggesting that histones play an essential role in the life of eukaryotic cells. Perhaps eukaryotes require chromatin because they contain large amounts of DNA, which must be tightly packed during cell division. In higher organisms, the chromatin undergoes condensation-decondensation cycles during cell proliferation, whereas bacterial DNA divides without changing its condensation state.[37] Chromatin can therefore be seen as a reversible DNA condensation mechanism evolved to handle the high eukaryotic DNA contents.[37]

Chromatin Structure of Active Genes

Biochemical studies have shown that active genes are more accessible to nuclease digestion than are regions of the genome that are inactive. Digestion is of isolated nuclei by the enzyme DNAse I.[38,39] Globin genes and the region surrounding them are ten times more accessible to digestion than the bulk of the transcriptionally inactive chromatin. In tissues such as oviduct in which the globin genes are inactive, the genes are not preferentially digested. Conversely, ovalbumin genes and their surrounding regions are DNAse-sensitive in oviduct nuclei, but not in red blood cells. This suggests that the chromatin of active genes is less condensed than that of inactive chromatin, which is presumably under the form of a 30 nm fiber.

What keeps active chromatin in this more open conformation? This is not known, but several possibilities have been implicated. For example, modifications of the histones, such as phosphorylation and acetylation, might be responsible. Acetylation of lysine groups of histones is more prevalent in active chromatin. Another modification that histones may have is the covalent joining of a small protein called *ubiquitin* (Table 13–4) by way of its COOH end to the free amino group of one of lysine 119 of

TABLE 13–4.   SOME ABUNDANT NON-HISTONE PROTEINS FROM CALF THYMUS CHROMATIN

	Number of Amino Acids	Chemical Properties	Location in Chromatin
Ubiquitin	74	Can bind covalently to a lysine side chain of histone H2A	U-H2 in nucleosome core
HMG-17	89	Soluble in 10% trichloroacetic acid	Nucleosome cores?
HMG-14	100		
HMG-1	270	Soluble in 2% trichloroacetic acid	Internucleosomal linker DNA?
HMG-2	270		

histone H2A. This results in a branched protein, called U-H2A. This bond, although covalent, is reversible, being cleaved during mitosis and rejoined in interphase.

Other important components of chromatin are the abundant non-histone proteins known as HMGs (for High Mobility Group proteins because they migrate fast during some types of electrophoresis). HMGs are present in relatively high amounts in chromatin, one tenth of all nucleosomes in the cell are associated with HMGs. Table 13–4 lists some characteristics of HMG proteins. Although their function is as yet unknown, HMGs and ubiquitin are known to associate preferentially with active chromatin.[40,41,42]

The vast majority of the DNA is covered by nucleosomes that do not have any DNA sequence specificity, but it also contains punctuation marks provided by gene-controlling proteins tightly bound at specific sites. In Chapter 20 we discuss a protein that binds tightly to 5S genes, and hundreds of other similar proteins must indeed exist. One of these specific DNA-binding proteins, called T *antigen*, binds to a defined region of the virus SV40 genome, and this region appears devoid of nucleosomes when probed with nucleases or examined under the EM. Similarly, a "DNAse I hypersensitive region" of a few hundred nucleotides, devoid of nucleosomes, is found at the beginning of many protein-coding genes,[43,44] perhaps reflecting the presence of specific DNA-binding proteins at these sites.

SUMMARY:

Chromatin

The cells of eukaryotic organisms have a constant DNA content (C value) that is characteristic for each species and much higher than in bacteria.

When packed in a metaphase chromosome, the DNA is compacted 5000- to 10,000-fold. Most, if not all, of the DNA is present in chromatin, which is a complex of DNA with an equal weight of basic proteins called histones.

Histones are small proteins that are basic because they have a high content of arginine or lysine. There are only five histones. The four fundamental histones, H2A, H2B, H3, and H4, are present twice every 200 base pairs of DNA. Histones H3 and H4 were almost completely conserved during evolution; histone H3 from rat liver differs in only two amino acids from that of pea embryos. The fifth histone, H1, is present only once per 200 base pairs of DNA, and it varies considerably between species and even within tissues of the same species.

Chromatin is formed by a series of repeating units called *nucleosomes*. Under the EM, nucleosomes can be visualized as beads about 10 nm in diameter. During spreading of the preparation for electron microscopy, chromatin frequently becomes stretched, showing a beads-on-a-string configuration (Fig. 13–13). In the living cell, however, the nucleosomes touch each other, forming a 10 nm fiber.

Each nucleosome contains a histone octamer consisting of two of each of the four histones H2A, H2B, H3, and H4, with about 200 pairs of DNA coiled on the outside of the nucleosome. The whole structure has the shape of a flattened disc with the DNA entering and leaving from the same side, and histone H1 sealing off the two turns of DNA (Fig. 13–15). The chains of nucleosomes can be folded into a 30 nm fiber, probably by forming a solenoidal structure having six nucleosomes per turn (Fig. 13–17). This structure is stabilized by interactions between different H1 molecules. The 30 nm fiber can be observed in metaphase chromosomes and in interphase nuclei, and it probably represents the natural conformation of transcriptionally inactive chromatin.

The packing of DNA in a chain of nucleosomes is about five to seven times more compact than free DNA, and in a 30 nm solenoid it is packed about 40-fold. This is still more than 100-fold less than the compaction of metaphase chromosomes (5000- to 10,000-fold).

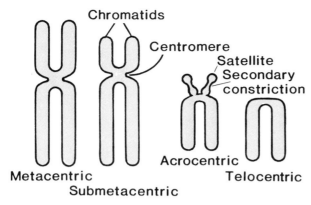

Fig. 13–18. Chromosomes are classified according to the position of the centromere.

13–4   THE CHROMOSOMES

At the time of cell division, chromatin becomes condensed into the chromosomes. Because chromosomes can be clearly visualized with the optical microscope, they were the subject of much study soon after their discovery in 1876, and by 1910 it became clear that genetic phenomena could be explained in terms of chromosome behavior. For this same reason, the study of chromosomes remains of the utmost importance even today. In this section we analyze the general morphology and composition of chromosomes, and their behavior during mitosis and meiosis is treated in Chapters 15 and 16.

Chromosomal Shape is Determined by the Position of the Centromere

Chromosomes can be studied in tissue sections, but they are best visualized in *squash* preparations. Fragments of tissues are stained with basic dyes (e.g., orcein or Giemsa) and then squashed between slide and coverslip by gentle pressure. Sometimes hypotonic solutions are used prior to squashing to produce swelling of the nucleus and a better separation of the individual chromosomes. The morphology of the chromosomes is best studied during metaphase and anaphase, which are the periods of maximal contraction.

Chromosomes are classified into four types according to their shape, which, in turn, is determined by the position of the centromere (point of attachment to the mitotic spindle). As seen in Figure 13–18, *telocentric* chromosomes have the centromere located on one end; *ac-rocentric* chromosomes have a very small or even imperceptible short arm; *submetacentric* chromosomes have arms of unequal length; and *metacentric* chromosomes have equal or almost equal arms. During anaphase movements the chromosomes bend at the centromere, so that metacentric chromosomes are V-shaped, and ac-rocentric chromosomes are rod-shaped.

Chromosomal Nomenclature—Chromatid, Centromere, Telomere, Satellite, Secondary Constriction, Nucleolar Organizer

Classical cytogeneticists produced many names to describe the various components that can be visualized in mitotic chromosomes. The student may find this terminology initially confusing, but it is important to become familiar with these names, for they not only define particular structures but also specific properties of the chromosomes.

Chromatid. At metaphase each chromosome consists of two symmetrical structures; the *chromatids* (Fig. 13–18), each one of which contains a single DNA molecule (see later discussion). The chromatids are attached to each other only by the centromere (Fig. 13–18) and become separated at the start of anaphase, when the sister chromatids migrate to opposite poles. Therefore, *anaphase chromosomes have only one chromatid, while metaphase chromosomes have two.*

Chromonema(ta). During prophase (and sometimes during interphase) the chromosomal material becomes visible as very thin filaments, which are called chromonemata and which represent chromatids in early stages of condensation. "Chromatid" and "chromonema," therefore, are two names for the same structure: a single linear DNA molecule with its associated proteins.

Chromomeres. These components are bead-like accumulations of chromatin material that are sometimes visible along interphase chromosomes. Chromomeres are especially obvious in polytene chromosomes, where they become aligned side by side, constituting the chromosome bands. These tightly folded regions of DNA are of considerable interest, because they may correspond to the units of genetic function in the chromosome (Chapter 22). At metaphase the chromosome is tightly coiled, and the chromomeres are no longer visible.

Centromere or Kinetochore. This is the region of the chromosome that becomes attached to the mitotic spindle. The centromere lies within a thinner segment of the chromosome, the *primary constriction*. The regions flanking the centromere frequently contain highly repetitive DNA and may stain more intensely with basic dyes (heterochromatin). Centromeres contain specific DNA sequences (which have been cloned, see below) with special proteins bound to them, forming a disc-shaped structure. Sometimes this disc, to which microtubules bind, is called the *kinetochore*. In thin electron microscopic sections the kinetochore shows a trilaminar structure, with a dense outer proteinaceous layer, a middle layer of low density, and a dense inner layer tightly bound to the centromeric DNA (Fig. 15–7). Between 4 and 40 microtubules become attached to the kinetochore and provide the force for chromosomal movement during mitosis. The function of the kinetochore is to provide a center of assembly for microtubules; when isolated metaphase chromosomes are incubated in a solution of tubulin with an appropriate buffer, microtubules are reassembled at the kinetochores of all the chromosomes.[45]

The vast majority of chromosomes have only one kinetochore *(monocentric chromosomes)*. Some species have diffuse kinetochores, with microtubules attached along the length of the chromosome *(holocentric chromosomes)*. In some chromosomal abnormalities (induced for example by x-rays), chromosomes may break and fuse with other ones, producing chromosomes without kinetochores *(acentric)* or with two kinetochores *(dicentric)*. Both types of aberration are unstable; one because it cannot attach to the mitotic spindle and remains in the cytoplasm, the other because the two centromeres tend to migrate to opposite poles, thus leading to chromosomal fragmentation.

Telomere. This term applies to the tips of the chromosomes. Cytologists have recognized for a long time that telomeres have special properties. When chromosomes are broken by x-rays, the free ends without telomeres become "sticky" and fuse with other broken chromosomes; they do not, however, fuse with a normal telomere. Stable deletions in chromosomes are always interstitial and preserve the original telomere. Telomeres contain the ends of the long linear DNA molecule contained in each chromatid, and therefore have an unusual DNA structure, which will be discussed later.

Secondary Constrictions. Other morphological characteristics of chromosomes are the *secondary constrictions*. Constant in their position and extent, these constrictions are useful in identifying particular chromosomes in a set. Secondary constrictions are distinguished from the primary constriction by the absence of marked angular deviations of the chromosomal segments during anaphase.

Nucleolar Organizers. These areas are certain secondary constrictions that contain the genes coding for 18S and 28S ribosomal RNA and that induce the formation of nucleoli. The secondary constriction may arise because the rRNA genes are transcribed very actively, interfering with chromosomal condensation. In man, the nucleolar organizers are located in the secondary constrictions of chromosomes 13, 14, 15, 21, and 22, all of which are acrocentrics and have satellites (see Chapter 18).

Satellites. Another morphological element present in certain chromosomes is the *satellite*. This is a rounded body separated from the rest of the chromosomes by a secondary constriction (Fig. 13–18). The satellite and the constriction are constant in shape and size for each particular chromosome.

Chromosome satellites are a morphological entity and should not be confused with satellite DNAs, which are highly repeated DNA sequences (see Chapter 22).

Karyotype—All the Characteristics of a Particular Chromosomal Set

The name *karyotype* is given to the whole group of characteristics that allows the identification of a particular chromosomal set; i.e., the number of chromosomes, relative size, position of the centromere, length of the arms, secondary constrictions, and satellites. In Chapter 18 we will see that by the use of banding techniques, other morphological criteria are available that have immensely increased the possibilities for identification of chromosomes and regions of chromosomes in the human and other species.

The karyotype is characterististic of an individual, species, genus, or larger grouping, and may be represented by a diagram in which the pairs of homologues are ordered in a series of decreasing size (Fig. 18–1). Some species may have special characteristics; for example, the mouse has acrocentric chromosomes, many amphibia have only metacentric chromosomes, and plants frequently have heterochromatic regions at the telomeres. Some species are particularly favorable for cytogenetic analysis. Salamanders and grasshoppers have high DNA content and very large chromosomes that produce very beautiful meiotic preparations; among plants, broad beans (*Vicia fava*) and onions (*Allium cepa*) have particularly large chromosomes that produce nice preparations of mitotic cells from root tips.

Each Chromatid Has a Single DNA Molecule

Each chromatid represents a single linear DNA molecule with its associated proteins. This concept, which is essential in understanding chromosome behavior, was formerly known as the *unineme theory*. It took much experimental evidence to convince the scientific community that chromosomes could in fact be so simple (for a fine review, see Gall, 1981, in the Additional Readings at the end of this chapter). As we explain in Chapter 14, the replication of chromatids is consistent with the semiconservative replication of a single DNA molecule (Fig. 14–13). When cells are labeled with ^{3}H-thymidine or with bromodeoxyuridine (an analogue that becomes incorporated into DNA and prevents the staining of the chromatid with certain dyes) and then grown for two cycles in an unlabeled medium, only one of the granddaughter chromatids is labeled (Fig. 13–19), as expected for a single DNA molecule. Furthermore, exchanges between sister chromatids can be induced experimentally, and in this case every change in a chromatid is accompanied by a reciprocal change in the other (Fig. 13–19), a fact that can be explained only if one assumes that there is only one DNA molecule per chromatid. Additional evidence comes with the finding that G_1 chromosomes (which have not yet replicated their DNA) have only one chromatid, and that G_2 chromosomes have two (see Fig. 14–6). Perhaps the most direct demonstration of the unineme organization of the chromatid was the isolation of *Drosophila* DNA molecules that are long

enough to contain all the DNA from a single chromatid. The length of these giant molecules was estimated from viscoelastic measurements.[46]

Metaphase Chromosomes Are Symmetrical

The two sister chromatids are mirror images one of the other. The morphological characteristics of one chromatid, such as chromomeres, secondary constrictions, and satellites, always have their counterparts in the other (Fig. 13–19). The reason for this symmetry is that sister chromatids contain identical DNA molecules, and the morphological features of chromosomes are ultimately determined by the DNA sequence they contain.

Centromeric DNA Has Conserved Sequences

The new recombinant DNA techniques have allowed the isolation of yeast centromeric DNA, using plasmid vectors that can grow in yeast cells. These plasmids, however, are relatively unstable in yeast in the absence of selective pressure. The presence of a yeast centromere in the plasmid confers mitotic stability, and thus provides a functional assay for centromere activity.[47] Fragments that are only 500 base pairs long have centromere function.[48] Yeast chromosomes are peculiar in that each chromosome has a single microtubule bound to it, while higher eukaryotes have 20 to 40 microtubules. Thus, it seems likely that in higher organisms the centromere sequences will turn out to be repeated several times.

Sequence comparisons between cloned centromeres from three different yeast chromosomes have shown that they share certain conserved features (see Fig. 13–20). Two conserved sequences of 14 and 11 base pairs are separated by a segment of 90 base pairs of highly A T-rich DNA. An additional conserved sequence of 10 base pairs has been found 250 nucleotides downstream.[47] These regions are thought to bind specialized proteins because digestion of yeast chromatin revealed a 250 base pair region that is resistant to nucleases (Fig. 13–20).[49] Presumably, the proteins bound to these conserved regions provide a binding site for microtubules. The availability of cloned centromere DNA will permit the isolation of proteins that bind preferentially to it.

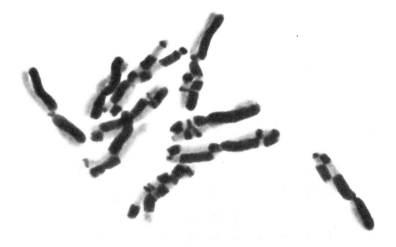

Fig. 13–19. Chromosomes of onion root tips that had been treated two generations earlier with bromodeoxyuridine and that had experimentally induced sister chromatid exchanges. Note that each chromosome has two chromatids. The DNA molecule that incorporated bromodeoxyuridine does not stain with Giemsa. Observe that each change in one chromatid is accompanied by a reciprocal one in the other. These images are consistent with the idea that each chromatid contains a single DNA molecule. (Courtesy of J.B. Schvartzman.)

Chromosome Ends Can Be Cloned in Linear DNA Vectors in Yeast

Because chromatids are long linear molecules of DNA, their ends present special problems. As mentioned earlier, DNA ends broken artificially (e.g., by x-ray irradiation) are very sticky and unstable within cells,[50] thus, DNA in telomeres must be somehow "blocked" or protected. An additional problem is posed by DNA replication,[51] which always starts with an RNA primer as we will discuss in Chapter 14. Because a primer is required, this would mean that at each replication a small part of the chromosome end would be lost during each replication cycle, if special measures were not taken. How cells solve this problem can be answered by cloning telomeres and analyzing their structure, but this presents formidable problems because all bacterial vectors are circular DNA molecules (or bacteriophages that have circular intermediates at some stage of their life cycle), and telomeres are by definition linear molecules. This problem was solved by Szostak and Blackburn, who devised linear plasmid vectors that could replicate in yeast, a microorganism that is convenient for

genetic manipulations. They cut open a circular vector, and placed on one of its ends the terminus of a linear DNA molecule that occurs normally in the ciliate protozoan *Tetrahymena*. The other end could then be used to clone yeast telomeres (Fig. 13–21*B*).

Tetrahymena contains extrachromosomal linear ribosomal DNA molecules, each of which contains only two rDNA genes. The ends of these linear molecules have been analyzed[52] and found to contain a variable number of repeats 20 to 70 of six nucleotides (CCCCAA). The end is not free to react and must be somehow cross-linked, in all probability forming a hairpin loop as shown in Figure 13–21. The C_4A_2 strand contains several single-strand interruptions and the opposite strand (G_4T_2) also has at least one interruption (Fig. 13–21). This sequence can work as a telomere in yeast, suggesting that the mechanisms involved in chromosome end replication have been strongly conserved through evolution.

Linear plasmids with one tetrahymena and one yeast telomere (Fig. 13–21) can multiply in yeast and are fundamental in analyzing chromosome ends. Some features are already emerg-

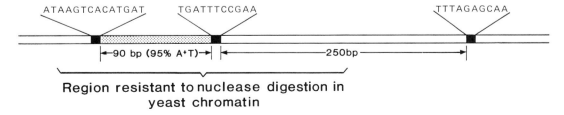

Region resistant to nuclease digestion in yeast chromatin

Fig. 13–20. DNA sequences conserved in centromeres of *Saccharomyces cerevisiae* (bakers' yeast).

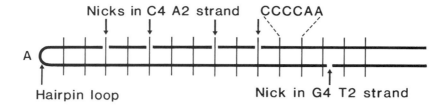

Fig. 13–21. Diagram of the procedures used for cloning telomeres. *A,* end of a tetrahymena rDNA molecule. Note that it contains many C_4A_2 repeats and single strand interruptions. *B,* linear plasmid for cloning telomeres. It contains an rDNA end and a restriction site for cloning other chromosome ends. The plasmid also contains a Leucine 2 gene required for the biosynthesis of leucine. Plasmids are selected by transformation into yeast cells that have an inactive Leucine 2 gene. In the absence of leucine in the culture medium only cells containing the linear plasmid grow.

ing. Telomeres of all yeast chromosomes have similar sequences, as determined by nucleic acid hybridization. Furthermore, the cloned yeast telomeres also have a variable number of both repeats rich in C + A and nicks in their terminal region, thus resembling the tetrahymena rDNA ends.[53,54] An additional intriguing feature of telomeres is that the number of C + A terminal repeats increases in each cell division, until at a later stage a deletion of the extra repeats occurs and the telomeres resume their lengthening cycle once again (see Blackburn, 1984 in the Additional Readings at the end of this chapter). Furthermore, when the tetrahymena telomeres were examined after many generations of growth in yeast cells, it was found that their C + A repeats had the sequence corresponding to yeast telomeres and had lost the original tetrahymena $C_4 A_2$ repeats. This occurs because new repeats are added constantly at telomeres, by a yeast enzyme (see Greider and Blackburn, 1985 in the Additional Readings at the end of this chapter). This novel enzyme is a terminal transferase able to add synthetic TTGGGG deoxyoligonucleotides (the opposite strand of $C_4 A_2$) to telomere ends.

Interphase Chromosomal Arrangement— Random or Non-random?

The condensation and decondensation cycles of the chromatin fibers during mitosis and meiosis could be facilitated if each chromosome were maintained in a definite space within the nuclear cavity during interphase, so as to avoid entanglements of the chromatin fibers. Some evidence exists in plant cells to indicate that during interphase the chromosomes retain the same spatial order they had at telophase. This was shown to be true in onion chromosomes, which have characteristic heterochromatic regions both at the centromeres and at the telomeres. Throughout interphase the centromeres remain on one side of the nucleus, whereas the telomeres are on the opposite side[55] in the same orientation that occurs during telophase. A similar "telophase configuration" has been observed in Drosophila embryonic and polytene nuclei (see Foe and Alberts, 1984, Mathog et al., 1984 in the Additional Readings at the end of this chapter).

It is even possible that chromosomes may be arranged in a fixed pattern with respect to each other, as is suggested by the classic work of D. P. Costello, who analyzed the first embryonic division in a turbellarian that has chromosomes that can be recognized by particular morphological features. The chromosomes were arranged in a precise order during metaphase, as shown in Figure 13–22. The probability that such an arrangement would arise at random is less than 1 in 2.[48]

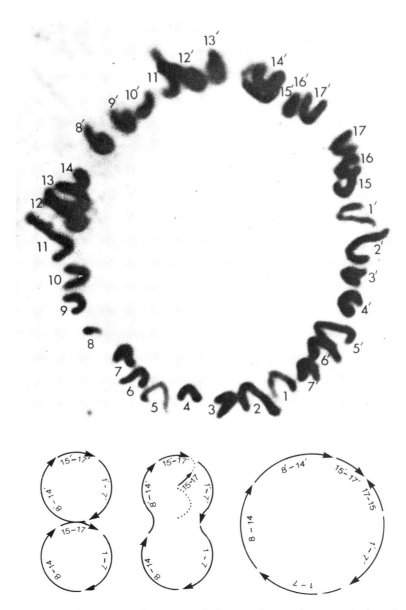

Fig. 13–22. Section of a first cleavage metaphase in a *Turbellarian* embryo (polar view). The homologous chromosomes can be recognized by their particular morphology and are indicated by numbers and prime numbers. The chromosomes show an ordered pattern, which might have arisen in the way shown in the lower diagram, which assumes that the chromosomes of both gametes were in an identical linear order. (From Costello, D.P.: Proc. Natl. Acad. Sci. USA, 67:1951, 1970.)

Chromosomes Are Made of Radial Loops of 30 nm Chromatin Fibers

As discussed earlier, a metaphase chromosome is 5000 to 10,000 times shorter than the equivalent amount of free DNA. By comparison, the DNA in the head of bacteriophage T_4 is packed at a ratio of 520:1. The DNA in mitotic chromosomes is contained in nucleosome chains, which are in turn coiled into 30 nm chromatin fibers (Fig. 13–17); this represents a packing ratio of 40:1. How the higher levels of packaging are achieved is not known, but some ideas about this are emerging from structural studies on metaphase chromosomes.

Structural non-histone proteins could be in-

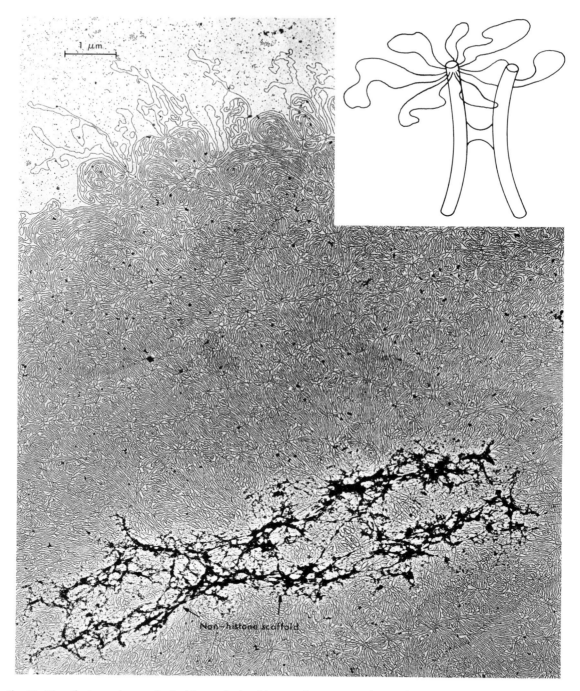

Fig. 13–23. Electron micrograph of a histone-depleted human chromosome. The non-histone proteins form two scaffolds, one per chromatid, which are joined at the centromere. The scaffold retains the shape of an intact chromosome, while the naked DNA fibers form a halo around it. The inset shows an interpretation of these and other results. In this model, chromosomes are organized in loops of DNA that emerge from the non-histone protein scaffold.

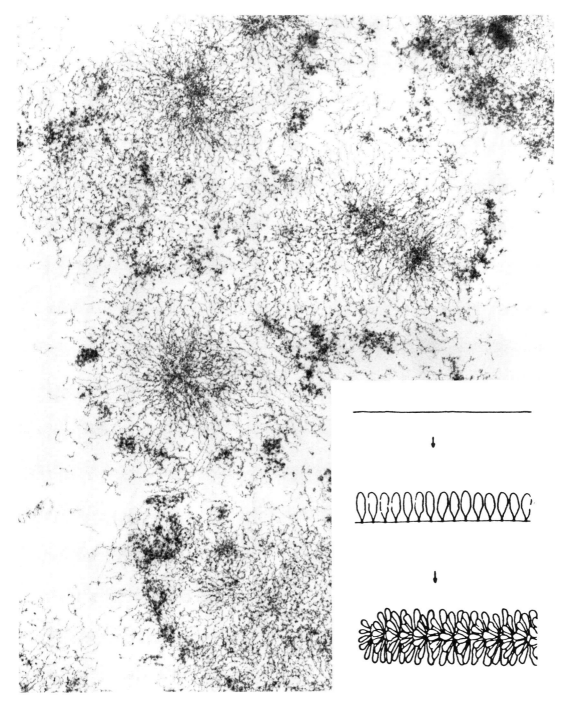

Fig. 13–24. Thin section showing the radial fiber arrangement of human chromosomes treated with EDTA prior to fixation and embedding. The nucleoprotein fiber is in the 100 Å configuration. Several cross sections of chromatids appear in the micrograph. The inset shows Laemmli's radial loop model of chromosome structure. (From Marsden, M., and Laemmli, U.K.: Cell, *17*:849, 1979.)

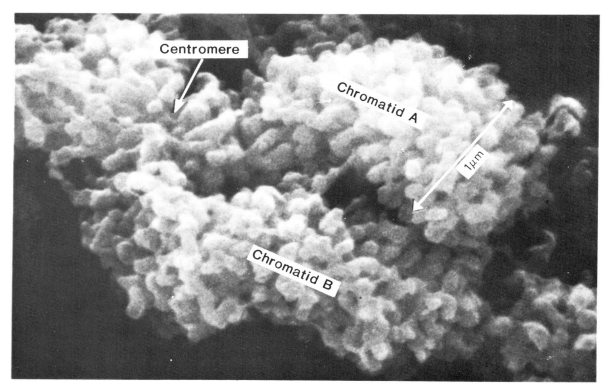

Fig. 13–25. Scanning electron micrograph of a human metaphase chromosome. Note the compact projections, which are thought to be the condensed loops of the thick chromatin fiber. (Courtesy of U. Laemmli.)

volved in organizing the 30 nm chromatin fibers into loops. It is possible to remove the histones from metaphase chromosomes (by adding the polyanion dextran sulfate, which competes with the DNA for the basic histones).[56] As shown in Figure 13–23, histone-depleted chromosomes have a central core or *scaffold*, surrounded by a halo made of hoops of DNA. The scaffold is made of non-histone proteins and retains the general shape of the metaphase chromosome. Each chromosome has two scaffolds, one for each chromatid, and they are connected together at the centromere region. When the histones are removed, the DNA, which was packed about 40-fold in the 30 nm chromatin fiber, becomes extended and produces loops with an average length of 25 μm (75,000 base pairs). In each loop the DNA exits from the scaffold and returns to an adjacent point (Fig. 13–23, inset). On the basis of these observations a model of chromosome structure has been proposed in which the DNA is arranged in loops anchored to the non-histone scaffold.[57] Because the lateral loops have 25 μm of DNA, after contracting 40-

fold in the 30 nm fiber, they would be only about 0.6 μm long, a length consistent with the diameter of metaphase chromosomes (1 μm). With milder treatments (low magnesium) that do not remove histones but unfold the 30 nm fiber into the 10 nm fiber (chain of nucleosomes), a similar pattern of loops can be observed; however, these loops are now shorter (3–4 μm) than those of naked DNA. Figure 13–24 is a micrograph of several cross sections of chromatids, showing a central dense core from which loops of 10 nm fibers emerge. When fully condensed chromosomes are observed with the scanning EM, compact chromatin projections can be seen protruding radially on the chromatid surface, as shown in Figure 13–25. Each projection is thought to be a loop of the 30 nm thick chromatin fiber.

The inset of Figure 13–24 shows how the chromatin is arranged in loops which, during metaphase, become arranged so that the base of the loops forms a scaffold in the center of the chromatid. The base of the loops might be arranged on a helical coiled path. In some favorable organisms, such as the plants *Trillium* and *Tra-*

descantia, helically coiled chromatids can be observed. In human chromosomes, coiled chromatids can be seen only after certain pretreatments that partially open the tightly packed chromosomes.[58]

Other work, using different methodology, has suggested that it is possible that even in interphase nuclei the chromatin might be organized in "domains" or loops containing about 75,000 base pairs of DNA.[59] Furthermore, the average replication unit of mammalian cells (replicons) has a similar length as well. This suggests that the loops observed in metaphase chromosomes might persist throughout the cell cycle. The loop organization of chromatin could have profound functional significance in eukaryotic gene regulation.

13–5   HETEROCHROMATIN

Heterochromatin—Chromosomal Regions That Do Not Decondense During Interphase

In 1928 Heitz defined *heterochromatin* as those regions of the chromosome that remain condensed during interphase and early prophase and form the so-called *chromocenters*. The rest of the chromosome, which is in a non-condensed state, was called *euchromatin* (Gr., *eu*, true). Heitz followed cells throughout the cell cycle and found chromosomal segments that do not decondense.

The heterochromatic segments tend to show preferential localizations in the pericentromeric region of most plants and animals, at the telomeres (especially in plants), or adjacent to the nucleolar organizers. In other cases, whole chromosomes become heterochromatic.

The heterochromatic regions can be visualized in condensed chromosomes as regions that stain more strongly or more weakly than the euchromatic regions, showing what is called a *positive* or a *negative heteropyknosis* of the chromosomes (Gr., *hetero* + *pyknosis*, different staining).

It is thought that in heterochromatin the DNA remains tightly packed in the 30 nm fiber, which probably represents the configuration of transcriptionally inactive chromatin.

Heterochromatin Can Be Facultative or Constitutive

Two types of heterochromatin are generally recognized: *constitutive* heterochromatin, which is permanently condensed in all types of cells, and *facultative* heterochromatin; which is condensed only in certain cell types or at special stages of development. Frequently, in *facultative* heterochromatin one chromosome of the pair becomes either totally or partially heterochromatic. The best known case is that of the X chromosomes in the mammalian female, one of which is active and remains euchromatic, whereas the other is inactive and forms the sex chromatin, or Barr body, at interphase (Fig. 18–5).

Constitutive heterochromatin is the most common type of heterochromatin. Most chromosomes contain large blocks of heterochromatin flanking the centromeres. This type of heterochromatin contains highly repeated DNA sequences, called satellite DNA, which might have a structural role in chromosomes.

A property common to all types of heterochromatin is *late replication*. When cells are given a brief pulse of ^{3}H-thymidine during the late S phase, the label is incorporated only into the heterochromatic segments, indicating that they replicate after the bulk of the DNA. The other common property is that heterochromatin is not transcribed.

Heterochromatin Is Genetically Inactive

It is generally agreed that condensed chromatin is inactive in RNA synthesis. For example, in Chapter 14 we see that condensed mitotic chromosomes do not synthesize RNA (Fig. 14–7). There is good genetic evidence indicating that genes contained in heterochromatic segments are not expressed. Here we examine three of these lines of evidence.

Some cats have a striking spotted black and yellow coloring of the coat. Because of the peculiar colored patches of the coat, these cats are known as *tortoiseshells* (Fig. 13–26). Tortoiseshell cats are always female, never male. The reason for this became clear when Mary Lyon put forward the hypothesis that the patchy pigmentation is produced by a gene contained in the X chromosome, which becomes heterochro-

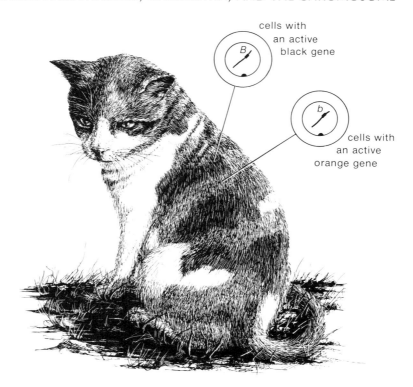

cells with
an active
black gene

cells with
an active
orange gene

Fig. 13–26.   A tortoiseshell cat. This cat has a white background and spots that have black or orange patches. The genes for the spot color are located in the X chromosomes, and this cat is heterozygous (one X chromosome has the orange allele; the other has the black allele). The appearance is patchy because one of the chromosomes becomes heterochromatic at random during early development, only one of the X chromosomes is active in any given cell. An orange patch represents the cells descended from a single cell in which the X chromosome containing the gene for black color became heterochromatic. Tortoiseshell cats are always female (XX). Male cats (XY) can have either orange or black coats, but never display the tortoiseshell patches. (From Mange, A.P., and Mange, E.J.: Genetics: Human Aspects. Saunders College Publishing, Philadelphia, 1980, p. 150.)

matic and inactive in some groups of cells but not in others. One of the X chromosomes in mammalian females (including humans) becomes inactivated early in embryogenesis on a random basis, so that adult animals are mosaics in which 50% of the cells have the paternal X chromosome euchromatic and active, and the other 50% have the maternal X chromosome in an active state.[60] The inactivation is random only in the cells destined to give rise to the body, for the embryonic cells that give rise to the placenta in mammals always inactivate the paternal X chromosome.

In some mouse strains segments from other chromosomes sometimes become translocated into the X chromosome. In such cases, the translocated genes also become inactivated on a random basis, suggesting that the heterochromatinization can spread along the whole chromosome, even to sequences that are not normally located in the X chromosome.

Another striking example of facultative heterochromatin is found in the mealy bug *(Planococcus citri),* an insect in which the males have a haploid set of chromosomes that are entirely heterochromatic. The heterochromatic set is the paternal one. The genes in these paternal chromosomes are switched off, as demonstrated by the fact that it is not possible to induce lethal mutations in them, even by high doses of radiation (administered to the father), whereas dominant lethals can be readily obtained by irradiation of the euchromatic set (administered to the mother). In female mealy bugs both sets of chromosomes are euchromatic and both are genetically active; i.e., lethal mutations can be induced in both the paternal and maternal set (see Brown, 1966). The third and final example comes from genetic studies that have shown that genes can be inactivated by translocation into heterochromatic regions of the chromosomes. This was first shown in *Drosophila melanogaster* in which a gene coding for eye color (*w*) can be inactivated by translocating it close to the centromeric heterochromatin. This is called a "position effect," and many examples of genes that become inactivated by insertion close to heterochromatin since the new techniques for introducing cloned genes into the Drosophila genome (Fig. 22–20) became available.

The biological significance of the switching off of genes in condensed chromatin might be more universal than indicated in the limited examples we discussed. Multi-cellular organisms have many specialized tissues, and in each one of them a particular set of genes is active and others are inactive. For example, globin genes are expressed in red blood cells but are not expressed in other tissues. We do not know how entire sets of genes are switched off, but many believe that variable degrees of condensation of chromatin domains could be involved in the maintenance of the differentiated state.

Studies have been conducted to understand the molecular mechanisms by which heterochromatin becomes inactive. Most of these studies have concentrated on the X chromosome inactivation; unfortunately, definitive answers have not yet emerged. It appears that DNA is changed in inactive X chromosomes. This is deduced from the finding that DNA taken from inactive X chromosomes does not transform mammalian cells deficient in an enzyme (called hypoxanthine-guanine phosphoribosyl transferase or HGPRT; see Figure 18–16) the activity of which is readily detected and which is coded by the X chromosome, whereas DNA taken from an active X chromosome does transform the same cells.[61] The change most likely to affect DNA is methylation. Eukaryotic DNA contains 5-methylcytosine, usually in Cs that are followed by a G (CpG) (see Chapter 23). A hint that DNA methylation might be involved in X chromosome inactivation comes from the reactivation of the HGPRT gene in cells that have been grown in the presence of 5-azacytidine, an analogue of cytosine and a powerful inhibitor of DNA methylation.[62] In these cells, however, the X chromosome remains on the whole heterochromatic and presumably only limited regions become reactivated. The methylation pattern of individual X chromosome genes, however, has not yet shown extensive differences between heterochromatic and euchromatic X chromosomes.[63]

Constitutive Heterochromatin Contains Repetitive DNA Sequences

Constitutive heterochromatin can be found in regions of the genome that contain short repetitive sequences of DNA called *satellite* DNAs. The best known one is the mouse satellite DNA, a 240 base pair sequence that is repeated about 1,000,000 times in the mouse genome (Fig. 22–7), constituting 10% of the total mouse DNA.

Figure 13–27 shows the classic experiment of Pardue and Gall[64] in which they took purified mouse satellite DNA, copied it into ³H-labeled RNA with RNA polymerase, and annealed it directly on a microscope slide to chromosomes whose DNA had been denatured. They found that this hybridization created the constitutive heterochromatin present in the pericentromeric region of mouse chromosomes. This experiment started the in situ hybridization technique, which has allowed the location of many DNA sequences on chromosomes.

During the course of their experiments, Pardue and Gall also noted that after the denaturation procedure heterochromatin became strongly stained by Giemsa stain while the rest of the chromosome was bleached. Simply omitting the hybridization step from the in situ hybridization procedure, Pardue and Gall created the C-banding (for constitutive heterochromatin) technique. C-banding and the other banding techniques that followed allow us to recognize many finer details of chromosome morphology, which have been tremendously useful in human cytogenetics (Chapter 18). The C-banding technique has proven to be specific to the extent that, if one sees such constitutive heterochromatin on a chromosome, one can predict that this region contains a repetitive simple sequence of DNA.

Why constitutive heterochromatin is associated so frequently with centromeres and telomeres remains unexplained. No a priori reason exists for it to be so abundant. For example, *Drosophila melanogaster* has 0.18 pg of DNA per haploid genome, whereas the similar fly *Drosophila virilis* has 0.36 pg. All the extra DNA in *D. virilis* is in the pericentromeric heterochromatin, which occupies 40% of the chromosomes. Clearly, much work remains to be done before we understand the paradoxes of eukaryotic genomes.

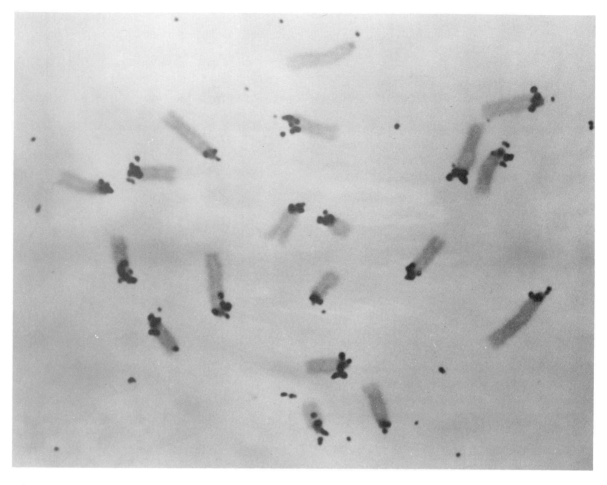

Fig. 13–27. Mouse chromosomes hybridized in situ with ³H-labeled RNA complementary to mouse satellite DNA. After hybridization, the slide was dipped in photographic emulsion, and the autoradiography was developed after several days of exposure. Note that the autoradiographic grains are over the pericentromeric regions. All of the chromosomes of the mouse are acrocentric (Fig. 13–18) and thus hybridization occurs close to one end. (Courtesy of J. Gall.)

SUMMARY:
The Chromosomes and Heterochromatin

The study of the chromosomes is of the utmost importance in biology, because it allows one to follow the behavior of DNA molecules and genes in a visual way. The morphology of chromosomes can be best studied during metaphase and anaphase. Chromosomes are of four types: (1) telocentric; (2) acrocentric; (3) submetacentric, and (4) metacentric (Fig. 13–18), depending on the position of the centromere, which is at the primary constriction. Other morphological characteristics include the secondary constrictions, the telomeres, the satellites, and the nucleolar organizers.

Each metaphase chromosome has two chromatids, which are attached to each other only by the centromere (the region in which the chromosome attaches to the mitotic spindle). Each chromatid consists of a single linear DNA molecule with its associated proteins (unineme theory). Sister chromatids are symmetrical in all respects, because they contain identical DNA molecules.

Mitotic chromosomes are made of chromatin fibers of 30 nm, which represent a condensation of 40 times, far less than the packing of DNA in metaphase chromosomes, which is about 7000-fold (compared with only 520-fold for the packaging of DNA in bacteriophage heads). Structural nonhistone proteins could be involved in this higher-order coiling of chromatin fibers. Chromosomes are believed to have a central scaffold of non-histone proteins that anchors loops of DNA of about 25 μm (75,000 base pairs). The DNA loops emerge radially from this core (Fig. 13–24).

Some regions of the chromosome remain condensed during interphase and are stained differentially by basic dyes. These heterochromatic regions are late-replicating and genetically inert, and they probably consist of 30 nm chromatin fibers.

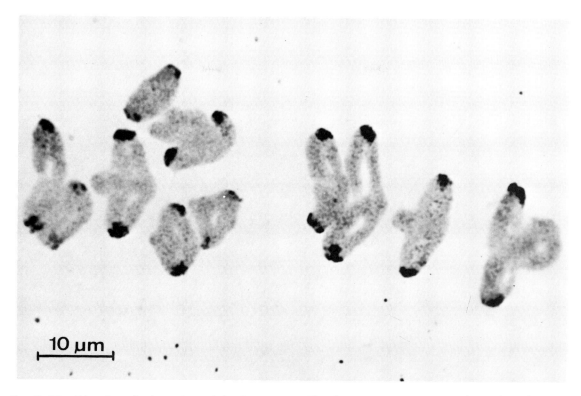

Fig. 13–28. C-banding of salamander meiotic chromosomes. The chromosomes were processed exactly as for in situ hybridization (which includes a DNA denaturation step) and then stained with Giemsa. The centromeric constitutive heterochromatin is stained, probably because the highly repeated DNA sequences have reannealed, while the rest of the genome remains denatured and does not stain with Giemsa. These chromosomes are in meiotic metaphase I. This can be diagnosed because the chromosome pairs are still held together by chiasmata, while the centromeres tend to be pulled toward the poles by the spindle microtubules. (Courtesy of Herbert McGregor.)

Two types of heterochromatin can be recognized: *constitutive* and *facultative*. *Constitutive heterochromatin* is permanently condensed in all types of cells, and is related to highly repetitive DNAs (satellite DNAs). *Facultative heterochromatin* is condensed in certain cell types or at special stages of development. The genes contained in facultative heterochromatin are not expressed, as shown in the cases of X chromosome inactivation in mammals (Fig. 13–26) of the inactive paternal chromosomal set of male mealy bugs, and in certain translocations in *D. melanogaster*. The switching-off of genes in condensed chromatin provides a mechanism that might explain the regulation of genes in cell differentiation.

REFERENCES

1. Monneron, A., and Bernhard, W.: J. Ultrastruct. Res., 27:266, 1969.
2. Daskal, Y.: The Cell Nucleus. 8:117, 1981.
3. Capco, D.G., and Penman, S.: J. Cell Biol., 96:896, 1983.
4. Fisher, P., Berrios, M., and Blobel, G.: J. Cell Biol., 92:674, 1982.
5. Callan, H.G., Randall, J.R., and Tomlin, S.G.: Nature, 163:280, 1949.
6. Gall, J.G.: J. Cell Biol., 32:391, 1967.
7. Unwin, P.T.N., and Mulligan, R.A.: J. Cell Biol., 93:63, 1982.
8. Scheer, U.: Dev. Biol., 30:13, 1973.
9. Maul, G.G.: Int. Rev. Cytol., 6 (Suppl.):75, 1977.
10. Fawcett, D.: The Cell. Philadelphia, W.B. Saunders Co., 1981.
11. Aaronson, R.P., and Blobel, G.: Proc. Natl. Acad. Sci. USA, 72:1007, 1975.
12. Dwyer, J. and Blobel, G.: J. Cell Biol., 70:581, 1974.
13. Gerace, L., Blum, A., and Blobel, G.: J. Cell Biol., 79:546, 1978.
14. Krohne, G., et al.: Cytologie, 18:22, 1978.
15. Jost, E., and Johnson, R.T.: J. Cell Sci., 47:25, 1981.
16. Gerace, L., and Blobel, G.: Cell, 19:277, 1980.
17. Stick, R., and Schwartz, H.: Cell Differ., 11:235, 1982.
18. Lebrowsky, J.S., and Laemmli, U.K.: J. Mol. Biol., 156:325, 1982.
18. Loewenstein, W.R., Karno, Y., and Ito, S.: Ann. NY Acad. Sci., 137:708, 1966.
19. Bonner, W.M.: J. Cell Biol., 64:42, 1975.
20. Paine, P.L., Moore, L.C., and Horowitz, S.B.: Nature, 254:109, 1975.
21. Beck, J.S.: Exp. Cell Res., 28:406, 1962.
22. Martin, T.E., and Okamura, C.S.: *In* International Cell

Biology. Edited by H.O. Schweiger. Berlin, Springer-Verlag, 1981, p. 77.

23. De Robertis, E.M., Longthorne, R.F., and Gurdon, J.B.: Nature, 272:254, 1978.
24. Debauvalle, M.C., and Franke, W.W.: Proc. Natl. Acad. Sci. USA, 79:5302, 1982.
25. De Robertis, E.M., Lienhard, S., and Parisot, R.F.: Nature, 295:572, 1982.
26. Mills, A.D., Laskey, R.A., Black, P., and De Robertis, E.M.: J. Mol. Biol., 139:561, 1980.
27. Dingwall, C., Sharnick, S.V., and Laskey, R.A.: Cell, 30:449, 1982.
28. Feldherr, C.M., and Ogburn, J.A.: J. Cell Biol., 87:589, 1980.
29. Mirsky, A.E., and Ris, H.: J. Gen. Physiol., 31:1, 1948.
30. Swift, H.: Int. Rev. Cytol., 2:1, 1953.
31. Olins, A.L., and Olins, D.E.: Science, 183:330, 1974.
32. Kornberg, R.D.: Science, 184:868, 1974.
33. Noll, M.: Nature, 251:249, 1974.
34. Ris, H., and Kubai, D.: Annu. Rev. Genet., 4:263, 1970.
35. Finch, J.T., and Klug, A.: Proc. Natl. Acad. Sci. USA, 73:4382, 1976.
36. Thoma, F., Koller, T., and Klug, A.: J. Cell Biol., 83:403, 1979.
37. Kellenberger, E.: Experientia, 36:267, 1980.
38. Weintraub, H., and Groudine, M.: Science, 193:848, 1976.
39. Garel, A., and Axel, R.: Proc. Natl. Acad. Sci. USA, 73:3966, 1976.
40. Levi, B., and Dixon, G.: Proc. Natl. Acad. Sci. USA, 76:1682, 1979.
41. Weisbrod, S., Groudine, M., and Weintraub, H.: Cell, 14:289, 1980.
42. Johns, E.W.: The HMG Chromosomal Proteins. New York, Academic Press, 1983.
43. Wu, C.: Nature, 286:854, 1980.
44. Wu, C., and Gilbert, W.: Proc. Natl. Acad. Sci. USA, 78:1577, 1981.
45. Telzer, B.R., Moses, M.J., and Rosenbaum, J.L.: Proc. Natl. Acad. Sci. USA, 72:4023, 1975.
46. Kavenoff, R., and Zimm, B.H.: Chromosoma, 41:1, 1973.
47. Fitzgerald-Hayes, M., Clarke, L., and Carbon, J.: Cell, 29:235, 1982.
48. Panzeri, L., and Philippsen, P.: EMBO J., 1:1605, 1982.
49. Bloom, K., and Carbon, J.: Cell, 29:305, 1982.
50. Orr-Weaver, T.L., Szostak, J.W., and Rothstein, R.J.: Proc. Natl. Acad. Sci. USA, 78:6354, 1981.
51. Cavalier-Smith, T.: Nature, 301:112, 1983.
52. Blackburn, E.H., and Gall, J.G.: J. Mol. Biol., 120:33, 1978.
53. Szostak, J.W., and Blackburn, E.H.: Cell, 29:245, 1982.
54. Lustig, A.L., and Petes, T.D.: Proc. Natl. Acad. Sci. USA 83:1398, 1986.
55. Fussel, C.P.: Chromosoma, 50:201, 1975.
56. Paulson, J.R., and Laemmli, U.K.: Cell, 12:817, 1977.
57. Mirkovitch, J., Mirault, M.E., and Laemmli, U.K.: Cell, 39:223, 1984.
58. Ohnuki, Y.: Chromosoma, 25:401, 1968.
59. Benyajati, C., and Worcel, A.: Cell, 9:393, 1976.
60. Lyon, M.: Proc. R. Soc. Lond. (Biol.), 187:243, 1974.
61. Liskay, R.M., and Evans, R.J.: Proc. Natl. Acad. Sci. USA, 77:4895, 1982.
62. Mohandas, T., Sparkes, R.S., and Shapiro, L.J.: Science, 211:393, 1981.
63. Wolf, M., and Migeon, B.: Nature, 295:667, 1982.
64. Pardue, M.L., and Gall, J.G.: Science, 168:1356, 1970.

ADDITIONAL READINGS

Benavente, R., Krohne, G., and Franke, W.W.: Cell type-specific expression of nuclear lamina proteins during development of Xenopus laevis. Cell, 41:177, 1985.
Blackburn, E.H.: Telomeres: Do the ends justify the means? Cell, 37:7, 1984.
Blobel, G.: Gene gating: A hypothesis. Proc. Natl. Acad. Sci. USA, 82:8527, 1985.
Bonner, W.: Protein migration and accumulation in nuclei. In The Cell Nucleus. Vol. 4. Edited by H. Busch. New York, Academic Press, 1978, p. 97.
Bostock, C.J., and Summer, A.T.: The Eukaryotic Chromosome. Amsterdam, North-Holland Publishing Co., 1978.
Brown, S.W.: Heterochromatin. Science, 151:417, 1966. (A classic paper. Highly recommended.)
Chambon, P.: The molecular biology of the eukaryotic genome is coming of age. Cold Spring Harbor Symp. Quant. Biol., 42:1209, 1977. (Highly recommended.)
Davis, L.I., and Blobel, G.: Identification of a nuclear pore complex protein. Cell, 45:699, 1986.
De Robertis, E.M.: Nucleocytoplasmic segregation of proteins and RNAs. Cell, 32:1021, 1983.
De Robertis, E.M., Longthorne, R.F., and Gurdon, J.B.: Intracellular migration of nuclear proteins in Xenopus oocytes. Nature, 272:254, 1978.
Feldherr, C.M., Kallenbach, E., and Schultz, N.: Movement of a karyophilic protein through the nuclear pores of oocytes. J. Cell Biol., 99:2216, 1984.
Felsenfeld, G.: DNA. Sci. Am., 253: No. 4, p. 58, 1985.
Felsenfeld, G., and McGhee, J.D.: Structure of the 30 nm chromatin fiber. Cell, 44:375, 1986.
Fitzgerald-Hayes, M., Clarke, L., and Carbon, J.: Nucleotide sequence comparisons and functional analysis of yeast centromere DNAs. Cell, 29:235, 1982.
Foe, V., and Alberts, B.: Reversible chromosome condensation induced in Drosophila embryos by anoxia: visualization of interphase nuclear organization. J. Cell Biol., 100:1623–1636, 1984.
Forbes, D.J., Kirschner, M.W., and Newport, J.: Spontaneous formation of nucleus-like structures around bacteriophage DNA microinjected into Xenopus eggs. Cell, 34:13, 1983.
Franke, W.W., Scheer, U., Krohne, G., and Jarasch, E.D.: The nuclear envelope and the architecture of the nuclear periphery. J. Cell Biol., 91:39s, 1981.
Gall, J.G.: Chromosome structure and the C-value paradox. J. Cell Biol., 91:3s, 1981. (A fine historical review by one of the greatest cell biologists of our time.)
Gerace, L., and Blobel, G.: The nuclear envelope lamina is reversibly depolymerized during mitosis. Cell, 19:277, 1980.
Gerace, L., Ottaviano, Y., and Kondor-Koch, C.: Identification of a major polypeptide of the nuclear pore complex. J. Cell Biol., 95:826, 1982.
Grieder, C.W., and Blackburn, E.H.: Identification of a specific telomere terminal transferase activity in Tetrahymena extracts. Cell, 43:405, 1985.
Hieter, P., et al.: Functional selection and analysis of yeast centromere DNA. Cell, 42:913, 1985.
Kalderon, D., Richardson, W.D., Markham, A.F., and Smith, A.E.: Sequence requirements for nuclear location of simian-virus 40 T/antigen. Nature, 311:499, 1984a.
Kalderon, D., Roberts, B.L., Richardson, W.D., and Smith, A.E.: A short amino acid sequence able to specify nuclear location. Cell, 39:499, 1984b.

Kornberg, R.D.: The location of nucleosomes in chromatin: Specific or statistical? Nature, *292*:579, 1981.

Kornberg, R.D., and Klug, A.: The nucleosome. Sci. Am., *244*(2):52, 1981.

Laemmli, U.K., et al.: Metaphase chromosome structure: The role of nonhistone proteins. Cold Spring Harbor Symp. Quant. Biol., *42*:351, 1977.

Lanford, R.E., Kanda, P., and Kennedy, R.C.: Induction of nuclear transport with a synthetic peptide homologous to SV40 T antigen. Cell, *46*:575, 1986.

Lohka, M.J., and Masui, Y.: Roles of cytosol and cytoplasmic particles in nuclear envelope assembly and sperm pronuclear formation in cell-free preparations from amphibian eggs. J. Cell Biol., *98*:1222, 1984.

Marsden, M.P.F., and Laemmli, U.K.: Metaphase chromosome structure: Evidence for a radial loop model. Cell, *17*:849, 1979.

Mathog, D., et al.: Characteristic folding pattern of polytene chromosomes in Drosophila salivary gland nuclei. Nature, *308*:414, 1984.

McGhee, J.D., and Felsenfeld, G.: Nucleosome structure. Annu. Rev. Biochem., *49*:1115, 1980.

McKeon, F.D., Kirschner, M.W., and Caput, D.: Homologies in both primary and secondary structure between nuclear envelope and intermediate filament proteins. Nature, *319*:463, 1986.

Newmeyer, D.D., Lucocq, J.M., Burglin, T.R., and De Robertis, E.M.: Assembly in vitro of nuclei active in nuclear protein transport: ATP is required for nucleoplasmin accumulation. The EMBO J., *5*:501, 1986.

Ris, H., and Witt, P.L.: Structure of the mammalian kinetochore. Chromosoma, *82*:153, 1981.

Stick, R., and Schwarz, H.: Disappearance and reformation of the nuclear lamina structure during specific stages of meiosis in oocytes. Cell, *33*:949, 1983.

Swift, H.: The organization of genetic material in eukaryotes: Progress and prospects. Cold Spring Harbor Symp. Quant. Biol., *38*:963, 1973.

Szostak, J.W., and Blackburn, E.H.: Cloning yeast telomeres on linear plasmid vectors. Cell, *29*:245, 1982.

Weisbrod, S.: Active chromatin. Nature, *297*:289, 1982.

Widom, J., and Klug, A.: Structure of the 300 Å chromatin filament: x-ray diffraction from oriented samples. Cell, *43*:207, 1985.

THE CELL CYCLE

"An unbroken series of cell divisions extends backwards from our own day throughout the past history of life. The individual is but a passing eddy in the flow which vanishes and leaves no trace, while the general stream of life goes forwards."

E.B. Wilson, 1927

The ability to reproduce is a fundamental property of cells. The magnitude of cell multiplication can be appreciated by realizing that an adult person is formed by 10^{14} cells, all derived from a single cell, the fertilized egg. Even in fully grown adults, the amount of cell multiplication is impressive. A man contains 2.5×10^{13} red blood cells (5 liters of blood, with 5,000,000 red blood cells/mm³), and the average life span of a red blood cell is 120 days (10^7 seconds). Therefore, to maintain a constant blood supply, 2.5×10^{13} cells must be produced every 10^7 seconds; i.e., 2.5 million new cells are required per second. Furthermore, cell reproduction is precisely regulated so that the production of new cells compensates exactly the loss of cells in adult tissues. In this chapter, we analyze the life cycle of eukaryotic cells.

14–1   THE CELL CYCLE

A growing cell undergoes a cell cycle that is comprised of essentially two periods; (1) *interphase* (the period of non-apparent division), and (2) the period of *division*. In eukaryotes, division generally takes place by mitosis or meiosis.

For many years, cytologists were concerned mainly with the period of division in which dramatic changes visible under the light microscope could be observed, and interphase was considered a "resting" phase. Cells, however, spend most of their life span in interphase, which is a period of intense biosynthetic activity in which the cell doubles in size and duplicates precisely its chromosome complement. Some differentiated cell types divide only rarely, and in mammals nerve cells do not divide at all after birth. Thus, for a human neuron the interphase period lasts the entire lifetime of a person.

The cell cycle can be considered as the complex series of phenomena by which cellular material is divided equally between daughter cells. Cell division is only the final and microscopically visible phase of an underlying change that has occurred at the molecular level. Before the cell divides by mitosis, its main molecular components have already been duplicated. In this respect, cell division can be considered as the final separation of the already duplicated molecular units.

Interphase—The G_1, S, and G_2 Phases

The introduction of cytochemical methods, such as the Feulgen stain, followed by a cytophotometric quantitative assay, first suggested that the doubling of DNA takes place during interphase. The studies done by autoradiography with labeled thymidine, however, were the most important in determining the exact period in which DNA replication takes place in a eukaryotic cell. These studies demonstrated that the synthesis occurs only in a restricted portion of the interphase—the so-called S period (*i.e.*, *synthetic period*), which is preceded and followed by two "gap" periods of interphase (G_1 and G_2) in which there is no DNA synthesis. This led Howard and Pelc[1] to divide the cell cycle into four successive intervals: G_1, S phase, G_2, and mitosis (Fig. 14–1); G_1 is the period between the end of mitosis and the start of DNA synthesis, S is the period of DNA synthesis, and G_2, the interval between the end of DNA synthesis and the start of mitosis.

During G_2 a cell contains two times (4C) the amount of DNA present in the original diploid cell (2C). Following mitosis the daughter cells again enter the G_1 period and have a DNA content equivalent to 2C (Fig. 14–2).

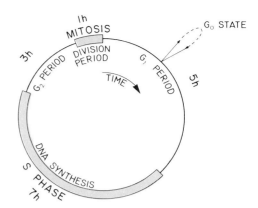

Fig. 14–1. The cell cycle. The duration of each phase corresponds to a mammalian cell growing with a generation time of 16 hours. (After D.M. Prescott.)

G_1—The Most Variable Period of the Cell Cycle

The duration of the cell cycle varies greatly from one cell to another. For a mammalian cell growing in culture with a generation time of 16 hours, the different periods would be as follows: G_1 = 5 hours, S = 7 hours, G_2 = 3 hours, and mitosis = 1 hour. Generally speaking, the S, G_2, and mitotic periods are relatively constant in the cells of the same organism. The G_1 period is the most variable in length. Depending on the physiological condition of the cells, it may last days, months, or years. Those tissues that normally do not divide (such as nerve cells or skeletal muscle), or that divide rarely (e.g., circulating lymphocytes), contain the amount of DNA present in the G_1 period. Cultured cells that stop multiplying because of density-dependent inhibition of growth (or contact inhibition—see Fig. 5–15) also stop at G_1.

The regulation of the duration of the cell cycle occurs primarily by arresting it at a specific point of G_1, and the cell in the arrested condition is said to be in the G_0 state.[2] In the G_0 state the

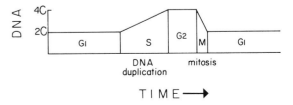

Fig. 14–2. Life cycle of a cell showing the changes in DNA content during the various periods as a function of time. 2C corresponds to the diploid content of DNA; 4C corresponds to the tetraploid content.

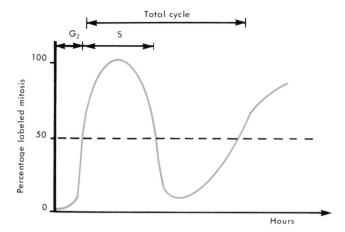

Fig. 14–3. Determination of the duration of the cell cycle stages by the percentage of mitotic cells labeled after a ³H-thymidine pulse. See the text for details. The length of each stage can be determined from the points at which the curve transects the 50% line.

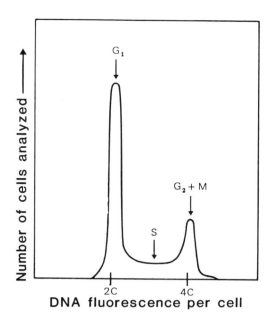

Fig. 14–4. Measurement of the DNA content of individual cells stained with a fluorescent DNA-binding dye.

cell may be considered to be withdrawn from the cell cycle (Fig. 14–1); when conditions change and growth is resumed, the cell re-enters the G_1 period (Fig. 14–1).

The different stages of the cell cycle were initially discovered by autoradiographic pulse-chase experiments.[1] In a typical experiment, cultured cells are exposed for 10 minutes to ³H-thymidine, a nucleoside that labels only DNA. The culture is then washed thoroughly and cultured in unlabeled medium; samples are fixed for autoradiographic studies at 1 to 2 hour intervals for 24 hours. The percentage of *mitotic* cells that are labeled is then used to measure the cell cycle stages (Fig. 14–3). The first labeled mitotic chromosomes appear about 3 hours later, indicating that the G_2 period is about that long. The S phase spans the time between the increase in the labeled mitosis percentage and its decrease (Fig. 14–3). Eventually a new cell cycle starts, and the percentage of labeled mitosis rises again. The total duration of the cell cycle can be estimated from the interval between the two ascending slopes of the curve (or measured directly by counting the number of cells in the culture). The length of mitosis (usually about 1 hour) can be determined by microscopic observation (percentage of cells in mitosis) if the generation time of the culture is known. The length of G_1 can then be calculated by subtracting the lengths of G_2, S, and mitosis from the total duration of the cell cycle.

This rather cumbersome method of analysis can presently be circumvented by using a fluorescence-activated cell analyzer.[3,4] This machine is a microfluorometer; when a suspension of cells that have been treated with a DNA-finding fluorescent stain is passed through a fine nozzle, the intensity of fluorescence of each cell can be measured. When plotted as in Figure 14–4, the proportions of cells in G_1, S, or G_2 can be easily calculated. (Other more sophisticated machines based on the same principle, called *cell sorters*, are able to separate cells according to their degree of fluorescence and have been useful in immunological studies in which surface antigens of cells are marked with fluorescent antibodies.)

Cell cycle studies can be greatly simplified by using *synchronized* cells, in which all the cells in the population are at the same stage of the cell cycle. Many methods exist for synchronizing cells (see Prescott, 1976; John, 1981 in the Additional Readings at the end of this chapter). The simplest one is the *mitotic selection* of cultured cells, which takes advantage of the fact that cells in mitosis become rounded up and loosely attached to the surface of the culture flask, whereas cells in interphase are flat and firmly attached to the culture vessel. By vigorously shaking the culture, it is possible to preferen-

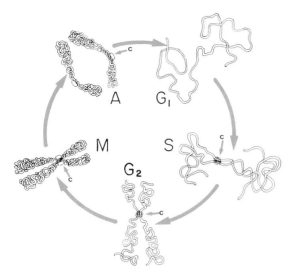

Fig. 14–5. The condensation-decondensation cycle of chromosomes. G_1, chromosomes are completely dispersed; S, duplication occurs; and G_2, condensation starts. At metaphase, *M*, and anaphase, *A*, the condensation is maximal and the two centromeres are clearly visible.

tially detach the mitotic cells, while the interphase ones remain attached, thus obtaining preparations in which up to 99% of the cells are in mitosis.

Visualization of Chromosomes During G_1, S, and G_2 by Premature Condensation

Eukaryotic chromosomes undergo condensation-decondensation cycles at cell division (Fig. 14–5), whereas the DNA of prokaryotes is never cycled in this way. This constitutes a specific difference between eukaryotes and prokaryotes, as a result of which the DNA of the latter can be replicated continuously in fast growing cultures (see Fig. 14–18).

During interphase the chromosomes are decondensed and cannot be distinguished under the microscope. Using an experimental trick, however, it is possible to induce the condensation of chromosomes in all three stages of interphase.[5,6] Mitotic cells fused to interphase cells (using inactivated Sendai virus—see Fig. 23–1), are able to induce *premature chromosome condensation* in the interphase nuclei. As can be clearly seen in Figure 14–6, the prematurely condensed chromosomes from G_1 nuclei show only one chromatid (Fig. 14–6A), whereas the ones from G_2 nuclei have two chromatids (Fig. 14–6C). This shows visually that

the G_1 phase corresponds to the period of interphase prior to DNA replication and that G_2 is the period of interphase that follows DNA replication.

When S phase nuclei are fused to mitotic cells, a more complex pattern is observed (Fig. 14–6B) in which the chromosomes adopt a "pulverized" configuration. It is not known whether the replicating chromosomes actually break during premature condensation, or if the condensed segments are joined by threads of DNA that cannot be distinguished under the microscope.

Condensed Chromosomes do not Synthesize RNA

RNA synthesis occurs throughout interphase. This can be shown simply by incubating cells for a short period (pulse) in ^{3}H-uridine, a precursor of RNA, and then studying them by autoradiography. All of the interphase nuclei become labeled (unlike what happens with ^{3}H-thymidine, in which only S phase nuclei are labeled). RNA synthesis, however, stops during mitosis, as is shown dramatically in Figure 14–7. The rate of RNA synthesis declines rapidly in late prophase and stops in metaphase and anaphase. As occurs in other instances (Section 13–5), highly condensed chromatin cannot be transcribed, perhaps because the DNA cannot be reached by RNA polymerase.

Several Molecular Events Occur at Defined Stages of the Cell Cycle

Of the biochemical events that occur at defined stages of the cell cycle, the most noticeable one is DNA synthesis. S phase cells contain a factor that induces DNA synthesis. This has been shown by cell fusion experiments, in which the onset of DNA replication in G_1 nuclei can be accelerated by fusion with S phase cells.[6] Interestingly, G_2 nuclei do not respond to this factor, indicating that some mechanism must exist to block reinitiation of DNA synthesis within the same cell cycle. Additional evidence for a factor inducing DNA synthesis comes from multi-nucleated cells, such as the plasmodium of the slime mold *Physarum polycephalum*, in which all the nuclei start DNA synthesis simultaneously.

S phase lasts for several hours, during which

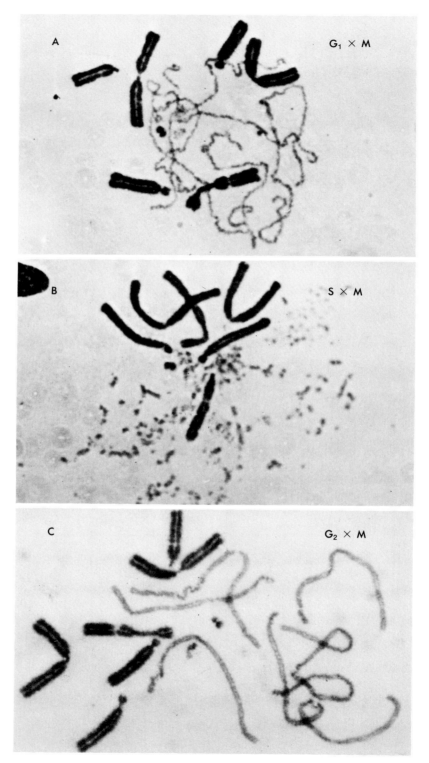

Fig. 14–6. Premature chromosome condensation of G_1, S, and G_2 chromosomes, induced by fusion to mitotic cells. All cells are from the Indian muntjac, a deer that has a small number of chromosomes. The thick metaphase chromosomes are from the mitotic cells. *A*, $G_1 \times M$, note that the G_1 chromosomes have a single chromatid. *B*, S $\times$ M, the S phase chromosomes have a fragmented appearance. *C*, $G_2 \times M$, the G_2 chromosomes have two chromatids. (Courtesy of D. Röhme.)

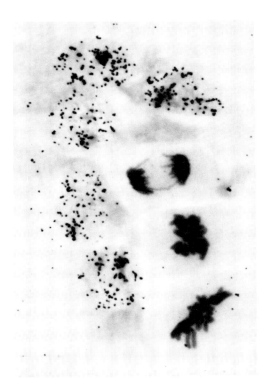

Fig. 14–7. RNA synthesis stops during mitosis. Chinese hamster ovary (CHO) cells were labeled for 15 minutes with ³H-uridine and then subjected to autoradiography. (From Prescott, D.M.: Reproduction of Eukaryotic Cells. New York, Academic Press, Inc., 1976.)

many units of replication are sequentially activated. In all cells, the more condensed, heterochromatic regions of the chromosomes replicate late during S phase. Autoradiographic studies have shown that the centromeric heterochromatin (which contains satellite DNAs, see Section 13–5) replicates later than the rest of the chromosome. Similarly, the X chromosome that becomes inactivated in mammalian females is heterochromatic and late replicating, whereas the one that remains activated is euchromatic and replicates earlier. Figure 14–8 shows other molecular events linked to the cell cycle, including the synthesis of histones during S phase; the decrease in protein synthesis that occurs during mitosis (of about 75%); the decrease in cAMP levels during mitosis; phosphorylation of histones (especially H1) during chromatin condensation, and many other events.

The most important point in the regulation of the cell cycle occurs in the G_1 phase, during which it must decide whether the cell will start a new cell cycle or become arrested in the G_0

state. Once this G_1 checkpoint has been passed, the cell goes on to complete a new cycle. Unfortunately, we still know little about the regulation of this fundamental step in cell proliferation.

A Cytoplasmic Clock Times the Cell Cycle in Early Embryogenesis

Our account of the cell cycle thus far could have misled the reader into thinking that the nucleus plays the most important role in the cell cycle (because it is where DNA and RNA synthesis as well as chromosome condensation take place). The great importance of the cytoplasm in the timing of the cell cycle has been demonstrated directly by experiment during the early embryonic divisions of the frog *Xenopus laevis*.[7]

After fertilization, the first division takes place in 90 minutes and thereafter the embryos divide every 35 minutes until the 12th cleavage (or 4000-cell stage) is reached (Fig. 23–16). This extraordinarily rapid cell cycle (all the DNA must replicate in 30 minutes) is also found during cleavage in embryos of other species. Before each division the cortex of the cells contracts, and the cells become tense and turgid. This can be best seen in eggs in which cell division is stopped with colchicine or vinblastine (inhibitors of microtubules) but that still undergo the cortical contraction (which requires functional actin). Figure 14–9 shows a similar phenomenon in an egg that was activated by pricking with a fine needle. The egg does not divide because it lacks a functional centriole, which is normally provided by the sperm. One can observe, however, that the egg increases in height (due to the cortical contraction) 90 minutes after pricking and every 30 minutes thereafter (Fig. 14–9, panels *c, e, g, i, k*). This contraction, which has the same periodicity as the cell cycle, does not require the presence of a cell nucleus. This was demonstrated experimentally by constricting a recently fertilized egg into two halves with a loop of thin hair. As shown in Figure 14–10, the fusing male and female pronuclei are trapped on one half, which continues to divide, while the other half, which lacks nuclei and centrioles, is unable to divide. When the changes in height were measured in the non-nucleated half of the embryo (Fig. 14–11), the same periodic con-

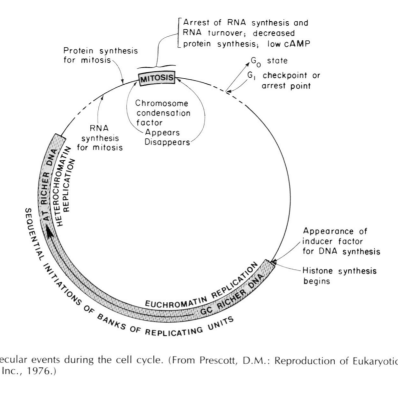

Fig. 14—8. Molecular events during the cell cycle. (From Prescott, D.M.: Reproduction of Eukaryotic Cells. New York, Academic Press, Inc., 1976.)

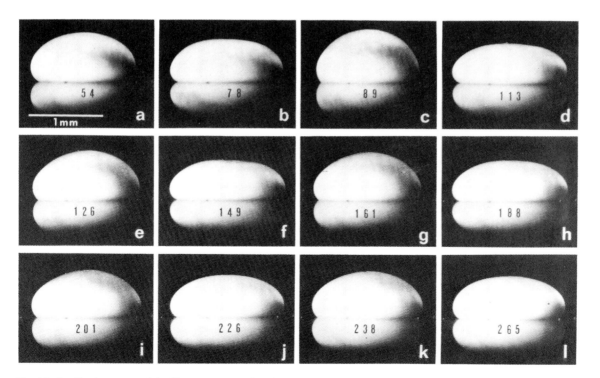

Fig. 14—9. Twelve sequential still pictures reproduced from a 16-mm time-lapse film, showing periodic changes in the height of an unfertilized *Xenopus* egg activated by pricking. Numerals indicate time (minutes) after activation. Each picture includes reflection of the egg in the supporting glass surface. The contractions have the same timing as the cell cycle in these cells. (From Hara, K., Tydeman, P., and Kirschner, M.: Proc. Natl. Acad. Sci. USA, 77:464, 1980.)

Fig. 14–10. Series of still pictures showing constriction of a fertilized egg with newborn human hair and cleavages in one of the two halves (25°C). (a) At 25 minutes after fertilization: vitelline membrane is still intact, and sperm entrance spot (SES) is clear. (b) At 38 minutes: halfway constriction after removal of vitelline membrane. (c) At 46 minutes: constriction is completed and two halves are separated. (d) At 64 minutes: beginning of first cleavage (arrow). (e) At 90 minutes: second cleavage is started. (f) At 116 minutes: third cleavage is started. The non-dividing half does not contain nuclei, but is nevertheless able to undergo periodic contractions (Fig. 14–11). (From Hara, K., Tydeman, P., and Kirschner, M.: Proc. Natl. Acad. Sci. USA, 77:465, 1980.)

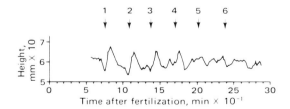

Fig. 14–11. Periodic surface activities in a non-cleaving (non-nucleated) fragment of a fertilized egg prepared by constriction, expressed as changes in the height of the non-cleaving fragment. Arrows 1–6 indicate times of onset of cleavage in the cleaving (nucleated) partner fragment (21°C). This shows that changes of the same periodicity as the cell cycle can be observed in the absence of a nucleus. (From Hara, K., Tydeman, P., and Kirschner, M.: Proc. Natl. Acad. Sci. USA, 77:462, 1980.)

tractions were observed. These results demonstrate that events of the same timing as the cell cycle can take place in frog eggs in the complete absence of a cell nucleus. A biological clock of undetermined nature exists in the cytoplasm of these cells.

Polypeptide Growth Factors Control Cell Proliferation

Cultured cells can presently be grown in entirely defined media, provided the proper nutrients, hormones, and growth factors are added.[8] In the past, it was always necessary to add 5 to 10% serum to the culture medium. The addition of serum provides essential protein growth factors without which cells cannot proliferate.

In the last 10 years synthetic culture media

have been developed which permit cell growth in the absence of serum. These defined media have permitted us to assay the effect of growth factors on cell proliferation. In multicellular organisms, these tissue-specific growth factors are important in coordinating cell proliferation. Furthermore, cancer cells have lowered requirements of these growth factors.

In recent years numerous growth factors have been isolated. Among these are:

Nerve growth factor (NGF). Composed of 118 amino acids, this polypeptide is required for the development of sympathetic neurons (see Chapter 24). Positive evidence exists that this factor is necessary for normal development because when anti-NGF antibodies are injected into newborn mice, selective death of sympathetic neurons is induced.[9,10]

Epidermal growth factor (EGF). A 6000 dalton polypeptide whose mode of action is the best understood.

Fibroblast growth factor (FGF). This growth factor stimulates cells of mesenchymal origin.

Platelet derived growth factor (PDGF). A growth factor that is liberated from blood platelets after stimulation with thrombin; it promotes proliferation of smooth muscle cells in arterial walls. One of the genes that may become activated in cancer cells, the *sis* oncogene, was found to be identical to the PDGF gene. Overproduction of this growth factor can produce cancer.

Lymphokines. Soluble protein mediators that regulate proliferation of lymphocytes in the immune system. For example, the so-called *helper T-lymphocytes* (derived from the thymus) do not secrete antibodies, but their presence ("help") is necessary for the immune response. One cell growth factor activates the function of macrophages, another one (interleukin-2) stimulates the proliferation of antigen-activated T-lymphocytes.[11] Many different growth factors are involved in the cell interactions of the immune response, and most are active only at short distances from the cell that produces it.

We now consider in a little more detail EGF, which is a potent mitogen in a variety of cells of ectodermic and mesodermic origin. It is active at concentrations (1 μg/ml) even lower than those required for the action of many hormones. EGF stimulates keratinization of epithelia and, when injected into newborn rats, accelerates the

opening of the eyelids and the eruption of teeth. It acts on cells that are related to basement membranes and type IV collagen (see Section 6–5). Among those cells of mesodermal origin, it stimulates cell division in chondrocytes, fibroblasts, smooth muscle, and endothelial and corneal cells.[12] The ability of EGF to trigger epithelial cell proliferation has raised the question whether this factor could act as a *fetal growth hormone*, for the development of certain epithelial territories of the embryo. For example, it has been found that EGF stimulates lung development and differentiation, and it may be of fundamental importance in its maturation (see James and Bradshaw, 1984; Heldin and Westermark, 1984 in the Additional Readings at the end of this chapter). The relationship between growth factors and normal embryonic development has been highlighted by the finding that a Drosophila gene called *notch*, whose function is necessary for the normal differentiation of ectoderm into neurons, is a protein whose sequence is very related to EGF (see Wharton et al., 1985 in the Additional Readings at the end of this chapter).

In intact organisms, the production of a growth factor induces multiplication at a short distance only, stimulating the cell that secretes it and the nearby cells. This is sometimes called *autocrine* secretion, in order to distinguish it from *endocrine* secretion, in which hormones act at long distances through the bloodstream. These short range interactions are possibly involved in the many induction events observed during embryogenesis (Chapter 25).

When EGF reaches the membrane of a target cell it binds to a specific receptor protein of 170,000 daltons and the complex is subsequently internalized.[13,14] The receptor protein becomes phosphorylated during this process and, interestingly, the phosphorus binds to tyrosine. Phosphorylation of tyrosine is rare in proteins (normally one finds mostly phosphoserine and phosphothreonine) except in cells that have been transformed by retroviruses and become cancerous (see Sections 5–4 and 22–3). One of the oncogenes (called *erb* B) has been found to be EGF-receptor, which in an oncogenic virus has its protein truncated and produces a more active protein (see Ulrich et al., 1984 in the Additional Readings at the end of this chapter). Rous sarcoma virus carries an oncogene that is a protein kinase that also specifically phosphorylates tyrosine. Many other oncogenes are also tyrone kinases, and are presumed to be receptors for other growth factors, because their amino acid sequence resembles that of EGF-receptor (Chapter 22–3).

Both in EGF and in viral transformation phosphorylation of tyrosine results in increased cell proliferation. Cells start dividing only after many hours of adding the growth factors, and what happens during this period remains a black box. In other words, we remain ignorant of what induces a cell to leave the G_0 state and to re-enter the cell cycle. On the other hand, great progress has been made with the finding that the great majority of the oncogenes are genes belonging to the growth factor pathway. Some oncogenes (discussed in detail in section 22–3) are growth factor genes, others are the receptors for these peptides which affect cell growth by phosphorylating specific cellular proteins in tyrosines. Subversion of the growth factor signalling pathway is the most frequently occurring mechanism of carcinogenesis. Since nearly 20 oncogenes have been found to be normally involved in this growth control pathway, there is every hope that cell proliferation will be understood in the next few years, and that understanding this will help control the abnormal growth of cancer cells (see Varmus, 1984 and Bishop, 1985 in the Additional Readings at the end of this chapter).

A variety of growth inhibitors, called *chalones* (*Gr.*, to slow down), have been isolated from tissues but none of them have been completely purified and characterized. Nevertheless, it may turn out that growth inhibitors will be as numerous and important as growth factors have already been shown to be.

SUMMARY:
The Cell Cycle

Mitosis represents only a small part of the life cycle of a cell (about 1 hour in most cells). The cell spends most of its lifetime in interphase, the period during which it doubles in size and replicates its DNA. The cell cycle can be divided into four periods: G_1, S, G_2, and mitosis. G_1 is the time *"gap"* between the end of mitosis and the start of DNA synthesis, S is the period of DNA synthesis, and G_2 is the time *"gap"* between the end of DNA synthesis and the beginning of mitosis (Fig. 14–1). For a mammalian cell with a generation time of 16 hours, the different periods, for example, could be:

G_1 = 5 hours, S = 7 hours, G_2 = 3 hours, and mitosis = 1 hour. The most variable period is G_1; depending on the physiological conditions of the cell, it may last days, months, or years. Cells that stop proliferating become arrested at a specific point of G_1 and remain withdrawn from the cell cycle in the G_0 state.

When interphase cells are fused to mitotic cells, the chromosomes become prematurely condensed, and it can be observed that G_1 chromosomes (unreplicated) have only one chromatid and that G_2 chromosomes (which have already replicated) have two (Fig. 14–6).

During mitosis RNA synthesis stops in the condensed chromosomes (Fig. 14–7), and the rate of protein synthesis decreases. During the S phase heterochromatin is late replicating. Sometimes whole chromosomes are late replicating (heterochromatic X chromosomes); and sometimes only part of a chromosome is late replicating, as occurs with centromeric heterochromatin, which contains satellite DNA. The most important point in the regulation of cell proliferation occurs during G_1, when the crucial decision of whether the cell undergoes a new division cycle or enters the G_0 state is taken, but we do not know how this is achieved.

The cytoplasm may play an important role in timing the cell cycle, as shown by experiments with enucleated frog egg fragments, in which actin contracts with the same periodicity as the normal cell cycle (Fig. 14–11).

Cell proliferation in the organism is controlled by a myriad of specific growth factors (NGF, EGF, FGF, and lymphokines) that can induce cell proliferation at extremely small concentrations and in a tissue-specific way. These factors play an important role in fetal development. A lower requirement of growth factors is related to the increased proliferation of cancer cells. Several oncogenes have been found to be genes coding for growth factors and growth factor receptors.

14–2   DNA REPLICATION

The mechanism of DNA replication may be considered a direct consequence of the DNA structure presented in the molecular model proposed by Watson and Crick in 1953, with which one should be familiar in order to understand this section (see Chapter 2–1). As shown in Figure 14–12, the two strands of the double helix can be separated because they are joined by relatively weak hydrogen bonds. Each polynucleotide chain serves as a template for the synthesis of a new DNA molecule. DNA replication follows the base pairing rules by which A pairs only with T, and G with C, and as a consequence of this, each daughter molecule is an exact replica of the parental molecule (Fig. 14–12).

DNA Replication is Semiconservative

The Watson-Crick model also suggested that replication was semiconservative, which means that half of the DNA is conserved (i.e., only one strand is synthesized; the other half of the original DNA is retained) (Fig. 14–13). This has been verified by several demonstrations. In their classic experiment Meselson and Stahl made use of the heavy isotope ^{15}N. The DNA containing ^{15}N (heavy-heavy, or HH DNA) is more dense than the DNA containing ^{14}N. *Escherichia coli* was grown in a medium containing

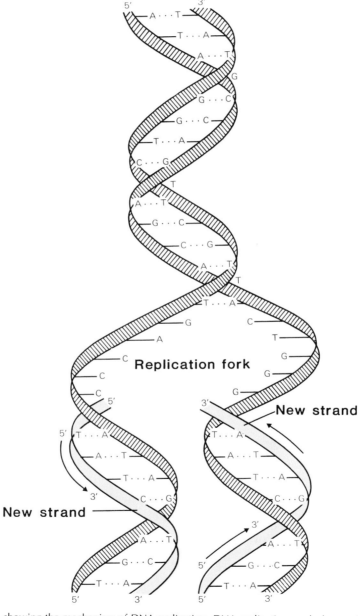

Fig. 14–12. Diagram showing the mechanism of DNA replication. DNA replication works by unwinding the two strands of the double helix and then using each strand as a template for the newly-synthesized molecules, following the base pairing rules that A pairs with T and G with C. Note that replication takes place only in the 5′ → 3′ direction, which has important consequences in DNA replication.

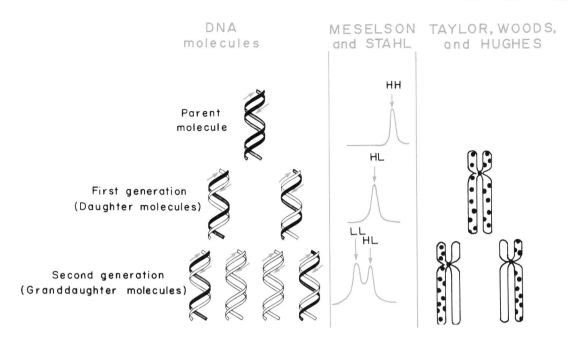

DNA molecules

MESELSON and STAHL

TAYLOR, WOODS, and HUGHES

Parent molecule

First generation (Daughter molecules)

Second generation (Granddaughter molecules)

HH

HL

LL HL

Fig. 14–13. Semiconservative replication of DNA in *E. coli* and in eukaryotic chromosomes. Meselson and Stahl labeled the two chains of the parent DNA molecule with the heavy isotope ^{15}N and analyzed the results by CsCl density gradients, while Taylor and colleagues used ^{3}H-thymidine and radioautography. (See the text for details.)

^{15}N for many generations, and was then passed to another medium containing normal ^{14}N. The DNA was isolated and its density was determined by ultracentrifugation on a cesium chloride gradient. It was found that after the first division cycle there is only one DNA peak corresponding to the heavy-light (HL) hybrid molecule (i.e., one strand is labeled with ^{14}N, the other with ^{15}N). At the second generation two peaks of DNA appear—one in which the two DNA strands contain ^{14}N (LL) and the other still corresponding to hybrid molecules (Fig. 14–13).[15]

Semiconservative DNA replication can also be demonstrated in higher organisms, because as explained in Chapter 13, each chromatid in a metaphase chromosome represents a single DNA molecule. In their classic experiment Taylor and colleagues[16] labeled *Vicia fava* (broad bean) root tip cells with ^{3}H-thymidine and then allowed them to grow in unlabeled medium as shown in Figure 14–13. When metaphase chromosomes were analyzed by autoradiography one or two generations later it was found that both chromatids were labeled after one generation, but only one was radioactive after two cell

cycles, as expected in semiconservative replication.

More recently, techniques have been developed that allow visualization of semiconservative chromosome replication without autoradiography. These methods are based on the use of 5-bromodeoxyuridine (BrdU), an analogue that is incorporated into chromosomes in place of thymidine. DNA that contains BrdU does not stain with a fluorescent dye (33258 Hoechst) or with certain modifications of the Giemsa staining technique. When cells labeled with BrdU are subsequently grown for two generations in a medium without the analogue, the metaphase chromsomes have the appearance shown in Figure 14–14.

DNA Synthesis is Discontinuous on One Strand

DNA is synthesized by enzymes called DNA polymerases, which are required not only for DNA replication but also for DNA repair. The enzymology of DNA replication is not considered in detail here (for further information see Kornberg, 1980 in the Additional Readings at the end of this chapter). DNA polymerases of

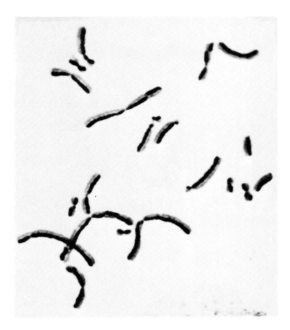

Fig. 14–14. Semiconservative replication of DNA. Chinese hamster ovary cells (CHO) were labeled with bromodeoxyuridine for several generations and were then allowed to grow for two generations in unlabeled medium. Only one of the chromatids stains normally; the chromatid that contains BudR does not stain with Giemsa. Some sister chromatid exchanges, which arise from the experimental treatment, are apparent. (Courtesy of S. Latt.)

both prokaryotic and eukaryotic origin require for their function the presence of a *template DNA*, a *primer* (RNA or DNA, but only RNA is used in vivo), and the four deoxynucleoside triphosphates (dATP, dTTP, dGTP, and dCTP), which are the building blocks of DNA. As shown in Figure 14–15, DNA polymerases can only add nucleotides to the 3' OH group of a preexistent primer, and DNA synthesis can only proceed in the 5' → 3' direction. Because both strands of a DNA molecule are antiparallel, i.e., run in opposite directions, (see Section 2–1) this creates a problem.

As shown in Figure 14–16, when the two strands unwind at the replication fork the *leading* strand faces the DNA polymerase in the correct 5' to 3' direction, so that the synthesis of a long continuous complementary strand takes place. On the *lagging* strand this is not possible; therefore, replication proceeds in a *discontinuous* way, synthesizing short segments of DNA (always in the 5' → 3' direction). These segments are then joined together by the action of a *DNA ligase* (Fig. 14–16).

Okasaki Fragments Start with an RNA Primer

The short discontinuous DNA segments were discovered by T. Okasaki who exposed bacteria to ^{3}H-thymidine for a few seconds and found that fragments of DNA 1000 to 2000 nucleotides long became labeled. These pieces, called *"Okasaki fragments"* after their discover, are only 200 nucleotides long in eukaryotes. The shorter Okasaki fragments and slower rate of advancement of the replication fork in eukaryotes (Table 14–1) might be connected to the fact that eukaryotic DNA is packed into nucleosomes.

DNA polymerases can only elongate a primer molecule; they cannot start a new chain by themselves. This poses a problem when initiating each Okasaki fragment, which is solved by synthesizing a short segment of RNA (Fig. 14–16; Table 14–1) that acts as the primer. This RNA segment is subsequently removed by repair enzymes, the gap is filled by *DNA polymerase*, and then joined to the neighboring Okasaki fragment by a *DNA ligase* (Fig. 14–16).

Why are these RNA primers necessary? It would seem much simpler for DNA polymerase to start on its own without any primers. The answer lies in the extreme fidelity with which DNA replication must proceed to maintain the genome constant. Only one erroneous nucleotide is introduced every 10^9 nucleotides that are replicated. (Mammals have a haploid genome of 3×10^9 nucleotides, so even this low rate means 3 mutations per round of replication). Yet, DNA polymerases introduce incorrect nucleotides at a rate of one every 10^5 nucleotides when copying a template. This discrepancy is due to the *proofreading properties* of DNA polymerase.[17] When the last nucleotide incorporated does not base-pair correctly with the template DNA, synthesis is cut short (because it must have a correctly matching primer) and the DNA polymerase turns back on its track, removes the incorrect nucleotide (by a 3'-5' *exonuclease* activity), and then proceeds to replace it with the correct one, continuing with synthesis in the 5' → 3' direction. This proofreading by DNA polymerase can occur only because the enzyme requires a primer. Thus, each Okasaki fragment starts with an RNA segment. RNA is used (instead of DNA) because it immediately provides a "tag" indicating that this part of the molecule has not yet

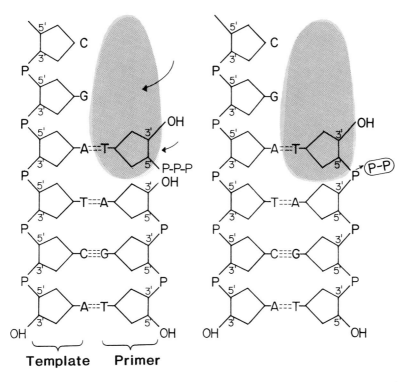

Template   Primer

Fig. 14–15. Diagram showing the action of DNA polymerase in DNA duplication. *A*, the template chain to the left and a piece of primer DNA to the right. A nucleotide (i.e., thymidine triphosphate) is being put in place along with the enzyme DNA polymerase. *B*, the enzyme has produced the linking of the nucleotide with release of P—P. Observe that the enzyme can add only nucleotides in the direction of 5′ → 3′.

been subjected to the proofreading mechanism and must be replaced.

Helix-Destabilizing Proteins and DNA Topoisomerases are Required for Replication

The DNA replication machinery consists of many proteins that must act in a concerted fashion on the replication fork. The following are some examples of these proteins:

Helix-destabilizing proteins. These proteins bind to single-stranded DNA and thus help to open up the double helix at the replication fork. This help is necessary, for although the DNA strands are held together by weak hydrogen bonds, high temperatures (~80°C) are required to separate strands of naked DNA in the laboratory.

DNA topoisomerases. These enzymes can change the topological form (or shape) of DNA. Some of these enzymes can introduce supercoils, unwind (Fig. 2–8), or untangle DNA molecules. The simplest DNA topoisomerases (type

I) introduce a nick (or cut) on only one of the DNA strands. This allows the molecule to rotate around the phosphodiester bond on the opposite strand as if it were a swivel. This is important during DNA replication, because as the two strands of DNA are separated at the replication fork (Fig. 14–12) the rest of the DNA tends to "wind up" and accumulate tension. Figure 2–8B shows the action of such an enzyme (also called *nicking-closing* enzyme) on circular supercoiled DNA. *Type I* topoisomerases are remarkable in that they do not require ATP to work; they bind covalently to the cut DNA strand storing the chemical energy of the phosphodiester bridge, which they then use to reseal the strand after the tension has been removed from the DNA.

Type II DNA topoisomerases can cut (and subsequently reseal) both strands of a DNA molecule.[18] This action can be used to prevent the problems that arise when DNA molecules become tangled during replication. Figure 14–17 illustrates how two interlocked DNA circles can be untangled by such an enzyme, which requires ATP.

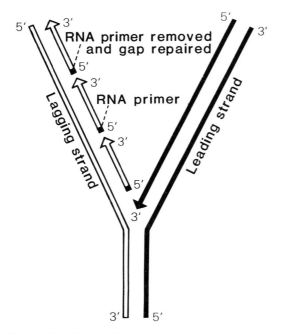

Fig. 14–16. DNA replication is continuous on one strand, and discontinuous on the other. The lagging strand is made in short Okasaki fragments that start with an RNA primer. The RNA primer is subsequently removed by a repair mechanism, the gap is filled in, and the molecule joined by a DNA ligase.

Reverse Transcriptase can Copy RNA into DNA

The central dogma of molecular biology is:

$$\text{DNA} \xrightarrow[\text{transcription}]{} \text{RNA} \xrightarrow[\text{translation}]{} \text{Protein.}$$

However, there is an exception to this strict unidirectional flow of information, which is that RNA can sometimes be copied into DNA. In other words, *RNA may be transcribed into DNA.* This is done by *reverse transcriptase*, an *RNA-dependent DNA polymerase* that is able to synthesize DNA on an RNA template.

$$\text{DNA} \rightleftharpoons \text{RNA} \longrightarrow \text{Protein}$$

Tumor-producing viruses containing an RNA genome, called retroviruses, (see Chap. 22) have been found to act as templates for the synthesis of DNA.[19] The tumor virion contains the reverse transcriptase by which an RNA/DNA hybrid molecule can be produced. In this way viral genes can be integrated into the genome of the host cell. Most higher eukaryotes contain retroviruses within their genome and reverse transcriptase, which can copy normal cellular RNAs into DNAs that can be subsequently integrated in the genome. The human genome is full of these *pseudogene* sequences, which result from copying mRNA or snRNA and reinserting them into the DNA of germ cells (see Chaper 22).

Replication of the *E. coli* Chromosome is Bidirectional

Much of what we know about DNA replication comes from the study of *E. coli*. As mentioned in Chapter 2, this prokaryote has a circular chromosome, which upon unfolding, has a length of about 1.1 mm (about 3.3×10^6 base pairs). The *E. coli* chromosome is made of a single molecule of DNA and is attached at one point to the cell membrane.

In *E. coli* (as in eukaryotes) DNA replication proceeds bidirectionally[20,21] starting from a fixed origin of replication. The entire replication of the *E. coli* chromosome takes 30 minutes. If the culture medium is enriched, however, the cells can divide in less time (every 20 minutes). This paradox is explained by the appearance of new replication forks even before the replication of

TABLE 14–1. SOME DIFFERENCES BETWEEN PROKARYOTIC AND EUKARYOTIC DNA REPLICATION

	Prokaryotes	Eukaryotes
RNA primer length	~50 nucleotides	9 nucleotides
Okasaki fragments of lagging strand	1000–2000 nucleotides	~200 nucleotides (due to nucleosomes?)
Fork movement rate	~500 nucleotides/sec	50 nucleotides/sec
Number of origins	1	thousands, developmentally regulated
Block of reinitiation in same cycle	no	yes

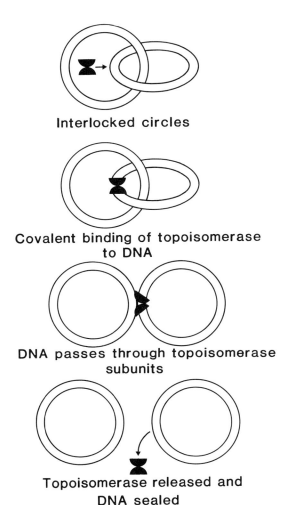

Interlocked circles

Covalent binding of topoisomerase
to DNA

DNA passes through topoisomerase
subunits

Topoisomerase released and
DNA sealed

Fig. 14–17. Diagram showing the action of a type II topoisomerase in separating two interlocked DNA circles by passing one strand through the enzyme subunits. (The action of a type I DNA topoisomerase was shown in Figure 2–8.)

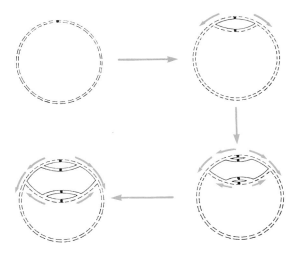

Fig. 14–18. Diagram of a circular bacterial chromosome containing multiple replication forks that start from a single origin of replication. The small arrows indicate the directions of movement of replication forks. (From Wake, R.G.: J. Mol. Biol., 68:501, 1972.)

human chromosomes have 7 cm. If these huge molecules were replicated from a single origin of replication, as in *E. coli* (in which the chromosome is only 1.1 mm), the S phase would be exceedingly long. Eukaryotic cells solve this problem by having multiple replication initiation sites in each chromosome (Fig. 14–19).

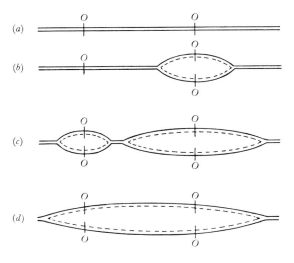

Fig. 14–19. Bidirectional model of DNA replication in eukaryotes. Each pair of horizontal lines represents a segment of a double helical DNA molecule with two strands. The newly formed chains are indicated by broken lines. (O), indicates sites of origin of replication; (a), two adjacent replicons prior to replication; (b), replication started in the unit to the right; (c) replication started in the unit to the left; and (d) replication completed in both units. (From Huberman, J.A., and Riggs, A.D.: J. Mol. Biol., 32:327, 1968.)

the molecule is completed. Figure 14–18 shows the interpretation of these findings. From the initiation point of the circular chromosome, two replicating forks proceed bidirectionally; at a certain stage—also starting from the same point of origin—two new replication forks are initiated, thus producing four forks that advance bidirectionally.

Eukaryotic Chromosomes Have Multiple Origins of Replication

Eukaryotic chromosomes contain a large amount of DNA. The largest chromosomes, from the salamander *Amphiuma*, contain 2.5 meters of DNA, all of it in a single molecule! The large

Eukaryotic DNA replication is best studied by the technique of DNA fiber autoradiography.[22] In this technique cells are labeled with ^{3}H-thymidine and placed in a dialysis chamber of which the walls are formed by nitrocellulose filters. The cells are gently lysed by adding a detergent and treated with pronase to digest the proteins, and the chamber is punctured and drained. Some of the DNA fibers adhere to the nitrocellulose filters. After autoradiography, the DNA molecules can be seen as parallel tracks of grains oriented in the direction in which the chamber was drained (Fig. 14–20).

In 1968 Huberman and Riggs[23] demonstrated that eukaryotic DNA has tandem units of replication (called replicons). The average replicons are 30 μm long, and there are about 30,000 of them per haploid mammalian genome. Each chromosome may have several thousand origins of replication. Each replication fork progresses at a speed of 50 nucleotides per second, which is 10 times slower than in *E. coli*. As seen in Figure 14–19 replication forks proceed in opposite directions from each origin (O) until they reach the terminal point at which they meet the neighboring replicon. The points of termination are probably not fixed, and they correspond simply to the point at which two growing forks converge.

Eukaryotic DNA Synthesis is Bidirectional

As in *E. coli*, replication in eukaryotic chromosomes proceeds bidirectionally. This was demonstrated by experiments in which cells were labeled with ^{3}H-thymidine of two different specific activities. As shown in Figure 14–21 cells grown in highly radioactive thymidine and then transferred to less radioactive thymidine show two "tails" of diminishing grain density after DNA fiber autoradiography. Conversely, when cells grown in moderately labeled thymidine are then placed in highly labeled thymidine, the two tails show increased grain density (Fig. 14–21).

The Number of Replicons is Developmentally Regulated

The cell cycle is exceptionally rapid during the first cleavages of most egg cells, in which the blastomeres divide without any intervening

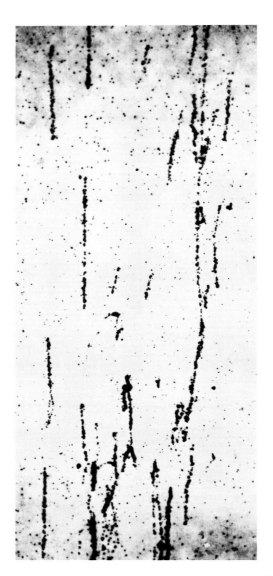

Fig. 14–20. Radioautograph of DNA replication in a cell grown in tissue culture. The various segments that are being duplicated are indicated by the parallel lines of silver grains of ^{3}H-thymidine. The replicating regions are seen as single lines (instead of two replicating forks) because sister strand separation is not apparent in DNA fiber radioautography until at least 50 μm of DNA have replicated. See the text for details. (Courtesy of H.G. Callan.)

growth of the cell. The S period is considerably shortened during cleavage. Cases in which the S phase is longer than usual also occur; the S phase immediately preceding meiosis is exceptionally prolonged in all organisms studied so far. A good example is provided by the newt *Triturus cristatus*, which has an S phase that is 1 hour long in blastula cells, about 20 hours long in adult somatic cells, and 200 hours long in the

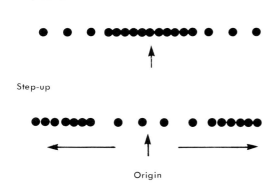

Step-down

Step-up

Origin

Fig. 14–21. Diagram showing the bidirectionality of eukaryotic DNA synthesis, as revealed by DNA fiber autoradiography. Cells were incubated first with highly labeled ³H-thymidine and then with moderately labeled thymidine (step-down), and vice versa (step-up), as described in the text.

premeiotic S phase of spermatogonia.[24] The differences in S phase duration are due to changes in the number of replication initiation sites. Embryonic cells have many more origins of replication, and the number of replicons is greatly reduced in the premeiotic chromosomes.[25]

A similar case occurs in *Drosophila*, in which the S period of embryonic nuclei lasts 3 to 4 minutes, while it lasts 600 minutes in adult somatic cells. The cleavage nuclei have origins of replication every 2 to 3 μm of DNA, and the distances are much longer (12 μm) in adult cells (see Blumenthal, Kriegstein and Hogness, 1974 in the Additional Readings at the end of this chapter). Clearly, some of the replication origins active in embryonic cells are not utilized in adult cells.

Eukaryotic Origins of Replication and the ARSes of Yeast

Although it is clear that viruses that infect eukaryotic cells (e.g., SV40) contain fixed origins of replication, the molecular nature of the chromosomal origins is still poorly characterized and continues to be a topic under intense investigation (see Laskey and Harland, 1981 in the Additional Readings at the end of this chapter). Some complexities are far from being understood, e.g., during the S phase clusters of contiguous replicons are activated at different times. This is particularly apparent in polytene chromosomes, in which entire chromosome bands replicate at different times during S phase.[26] This might be related to the degree of chromatin condensation, for it is well known that heterochromatin replicates late during S phase (Fig. 14–8).

Some evidence indicates that DNA replication starts at invariant specific chromosomal locations in mammalian cells,[27] but the best studied system so far is that of baker's yeast. It is possible to force yeast cells to take up foreign DNA from the culture medium (called transformation) and a variety of selectable plasmid vectors have been developed for this purpose. Bacterial plasmids are unable to multiply in yeast, but when certain sequences of yeast chromosomal DNA are introduced into them the plasmids can replicate.[28,29] These sequences are called *A*utonomous *R*eplication *S*equences or ARSes, which shows that humor exists among molecular biologists too. ARSes increase DNA replication in vitro using yeast extracts,[30] whereas in vivo different ARS elements in the genome replicate at specific times during the yeast S phase. It has also been shown that plasmid-borne ARSes are replicated at the same time as their chromosomal counterparts.[31] Although indirect, the evidence suggests that ARSes might be true chromosomal origins of replication.

Figure 14–22 shows the sequence of two short segments, 57 and 49 base pairs long, of yeast DNA that suffice to confer autonomous replication to plasmids.[32] Only a short consensus sequence (shared with many other ARS elements as well) is conserved, and the rest of the sequences, although required for function, do not show striking homologies, except that the region is A·T-rich. The homologies observed are too short to expect cross-hybridization to occur between different ARSes in the genome.

Reinitiation Within the Same Cycle is Prevented in Eukaryotic Cells

Each chromosome set has 30,000 origins of replication that are successively activated in the course of several hours in S phase. However, to duplicate the DNA accurately, the cells must ensure that no initiation site is used more than once in each cell cycle. Prevention of reinitiation must therefore be a crucial feature of eukaryotic replication,[33] and constitutes an important difference between prokaryotes and eukaryotes

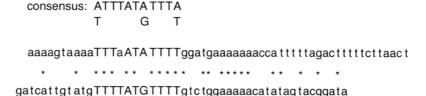

consensus: ATTTATATTTA
T   G   T

aaaagtaaaaTTTaATATTTTggatgaaaaaaacca ttt t tagact ttt tct taac t

*   *   *** ** ***** ** *****   ** * * *

gatcat tgt atgTTTTATGTTTTgtc tggaaaaacat a tag tacggata

Fig. 14—22. Sequence homology between two yeast ARS elements. Asterisks indicate homologous positions, and capital letters signify the best homology to the yeast ARS consensus sequence. This "consensus" sequence was identified by comparing nucleotide sequences from many ARS elements. (From Kearsey, S.: EMBO J., 2:1571, 1983.)

(Table 14–1). We do not know how this is achieved, but some mechanism must exist to distinguish newly replicated chromatin from the unreplicated material; for example, different methylations in the bases of newly replicated DNA or specific modifications in the histones of the newly assembled nucleosomes could be utilized. The molecular mechanism for reinitiation blocking remains unknown in DNA replication.

New Nucleosomes are Assembled Simultaneously with DNA Replication

Figure 14–23 shows a replication unit of *Drosophila melanogaster* spread for electron microscopy under conditions that retain the chromosomal proteins. It can be observed that the newly replicated chromatin becomes rapidly associated with histones, giving a nucleosome configuration on both daughter molecules.[34] In eukaryotic cells we must therefore think in terms of replication of chromatin rather than of naked DNA (see De Pamphilis and Wasserman, 1980 in the Additional Readings at the end of this chapter). Figure 14–24 shows a close-up model

of an eukaryotic replication fork as it is currently viewed.

Large amounts of histones are newly synthesized during S phase and become rapidly associated with the newly replicated DNA. Density-label experiments indicate that once histone octamers are bound to DNA they do not exchange histones. In other words, in each cell cycle, half of the resulting nucleosomes are made up of newly synthesized histones and half consist solely of old histones. Disagreement exists, however, over how the old and new nucleosomes segregate between the daughter strands. When new protein synthesis is inhibited with cycloheximide and examined under the EM, long stretches of naked DNA can be observed on the lagging strand, while the old nucleosomes are found on the leading strand.[35] An interpretation of this result is shown in Figure 14–25, in which all the new nucleosomes become incorporated into only one of the daughter strands. If this model of nucleosome segregation turned out to be correct it could have important biological consequences, particularly during the differentiation of stem cells (see Fig. 23–33).

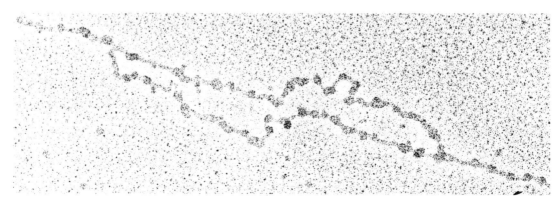

Fig. 14—23. Replicating chromatin. (Courtesy of S.L. McKnight and O.L. Miller, Jr.)

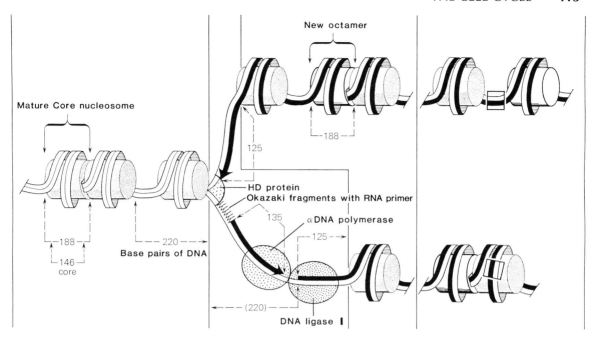

Fig. 14–24. Replication fork of an eukaryotic chromosome. Several components are indicated: RNA-primed Okasaki fragment, helix destabilizing (HD) protein, DNA polymerase (α pol, which is the main enzyme involved in eukaryotic DNA replication), and DNA ligase I. The numbers indicate distances in nucleotides. (Reproduced, with permission, from the Annual Review of Biochemistry, Vol. 49. © 1980 by Annual Reviews Inc. From DePamphilis, M., and Wasserman, P.: Annu. Rev. Biochem., 49:627, 1980.)

RNA Synthesis Continues During DNA Replication

Figure 14–26 shows a replication unit of *D. melanogaster* that is being transcribed. It is apparent that RNA synthesis can be initiated almost immediately after replication and that therefore transcription continues during the S phase.[34] The nascent RNA chains can be ob- served on both sister chromatids in a rather sym- metrical arrangement.

Electron microscopy of macromolecules spread by the technique devised by Oscar Miller is an enormously powerful tool for cell biologists, as can be readily realized from Figures 14–23 and 14–26, and from others throughout this book. The direct visualization of molecules can

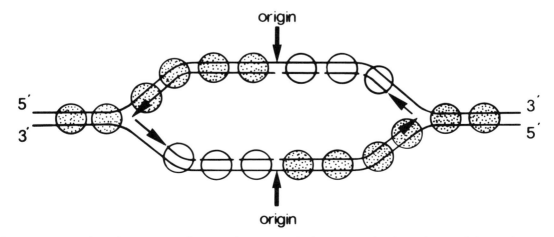

Fig. 14–25. Hypothetical segregation of new nucleosomes to the lagging strand at the replication fork. See the text for details. (Courtesy of R. Laskey.)

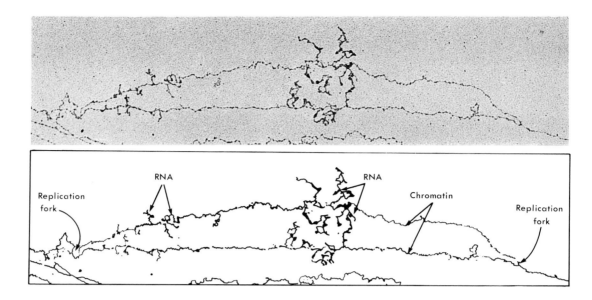

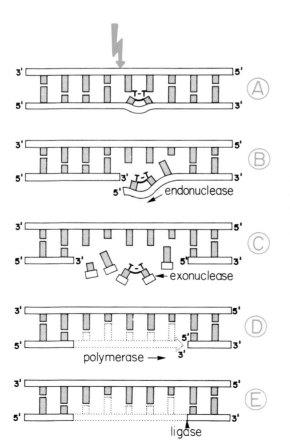

Fig. 14–27. Diagram showing the effect of ultraviolet light on the DNA molecules and the mechanism of DNA repair. *A,* under the action of ultraviolet light a dimer of T—T in produced; *B,* the affected DNA strand is recognized and incised by a molecule of endonuclease; *C,* the strand segment is excised by a molecule of exonuclease; *D,* the gap is filled by DNA polymerase; and *E,* the synthesized segment is joined by a DNA ligase.

sometimes settle, in a single stroke, years of biochemical controversy.

DNA Repair Enzymes Remove Thymine Dimers Induced by Ultraviolet Light— *Xeroderma Pigmentosum*

As discussed in the previous section the cell takes great care to introduce as few errors as possible during DNA replication. Mutations can also be introduced after replication has been completed, by exposure to elements such as chemical carcinogens (e.g., cigarette smoke) or ultraviolet light. Cells have DNA repair enzyme systems that constantly survey the DNA for such changes. In *E. coli* the production of such enzymes is stimulated by exposure of the cells to mutagens (see Howard-Flanders, 1981 in the Additional Readings at the end of this chapter).

Mutations can be considerably increased by exposure of *E. coli* to ultraviolet light. This treatment tends to produce *dimerization* of adjacent pyrimidine bases; predominantly T—T but also C—C and C—T *dimers* are produced within a single strand of DNA. Dimerization reduces the distance between nucleotides and impairs the mechanism of DNA duplication. In most cases this type of mutation is corrected by a mechanism of *DNA repair* that involves (Fig. 14–27): (1) the recognition and incision of the affected DNA strand by an endonuclease, (2) excision and broadening of the gap by an exonuclease (e.g., DNA polymerase), (3) filling of the gap by repair replication (this is also done by DNA polymerase), and (4) covalent joining of the polynucleotides by the ligase. Similar repair mechanisms exist in human cells, and their importance is demonstrated by the disease *Xeroderma pigmentosum*.

Xeroderma pigmentosum is a hereditary disease characterized clinically by severe intolerance to sunlight. The sun-exposed areas (face and hands) develop fibrosis, pigmentations, and more importantly, multiple skin cancers. The disease is due to a defect in the enzymes that repair damage due to ultraviolet light. The defective excision and repair of thymine dimers can be demonstrated cytologically by irradiating

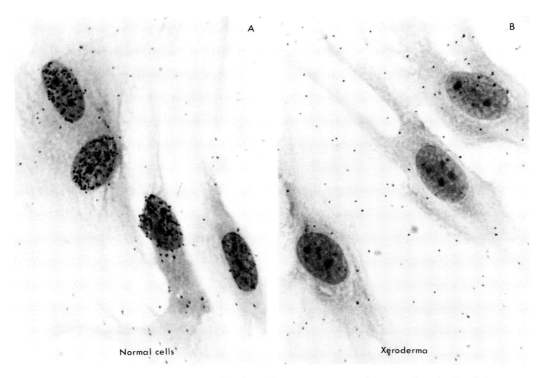

Fig. 14–28. Repair of thymine dimers in normal and xeroderma pigmentosum human cultured cells. Cells were treated with ultraviolet light and then with ³H-thymidine for 1 hour. The normal cells incorporate low amounts of thymidine in their nuclei, while xeroderma cells do not. None of the cells shown here was in S phase. (Courtesy of Y. Okada.)

cultured fibroblasts with ultraviolet light and then incubating them with ³H-thymidine for 1 to 2 hours. As can be seen in Figure 14–28A, ultraviolet treatment induces normal cells to incorporate low levels of thymidine (when not replicating their DNA). This incorporation represents the repair of the excised thymine dimers. Fibroblasts from xeroderma pigmentosum patients are unable to repair the ultraviolet damage and therefore do not incorporate thymidine after the same treatment (Fig. 14–28B). This disease clearly links human cancer with somatic mutations and shows the benefit of having mechanisms to repair DNA defects.

SUMMARY:
DNA Replication

The mechanism of DNA replication is implicit in the Watson-Crick model of DNA (see Chapter 2). Each strand acts as a template for the production of two identical daughter molecules, according to the base pairing rules (A = T; G = C). DNA synthesis is semiconservative (i.e., one of the parental DNA strands is conserved), as shown in Figure 14–13.

DNA polymerases proceed only in the 5′ to 3′ direction, and for this reason DNA synthesis is discontinuous in the lagging strand (see Fig. 14–16). Short pieces of replicating DNA, called *Okasaki* fragments, have been isolated (Table 14–1). These segments start with an RNA primer that is subseqently removed, the gap is filled by DNA polymerase, and the DNA strands joined together by a DNA ligase. This complicated procedure is necessary to maintain the proofreading properties of DNA polymerase, which ensure that only one error is made every 10^9 nucleotides that are replicated.

Mammalian cells have about 30,000 origins of replication per haploid genome. From each origin two replication forks advance bidirectionally until they meet the neighboring replication fork. Yeast ARSes 50 to 60 nucleotides long are required for replication of plasmids in yeast, and could correspond to the chromosomal origins of replication.

E. coli has a small (1.1 mm or 3,400,000 base pairs) circular chromosome, which replicates bidirectionally from a single origin of replication in about 30 minutes. In rapidly growing cells DNA synthesis can restart before the previous replication is completed (Fig. 14–9). In eukaryotic cells, however, reinitiation within the same cell cycle is prevented, to avoid differential replication of some replicons.

The number of replication origins can be developmentally regulated. In the newt *Triturus*, S phase is 1 hour long in cleaving embryos, 20 hours long in adult somatic cells, and 200 hours long in the premeiotic S phase. These differences can be explained by variations in the number of replication units. The newly synthesized DNA is assembled into chromatin simultaneously with replication (Fig. 14–23), and RNA synthesis can be resumed immediately after replication (Fig. 14–26).

REFERENCES

1. Howard, A., and Pelc, S.R.: Heredity [London], 6 (Suppl):261, 1953.
2. Lajtha, L.G.: J. Cell. Comp. Physiol., 62:143, 1963.
3. Van Dilla, M.A., Trujillo, T.T., Mullaney, P.F., and Coulter, J.R.: Science, 163:1213, 1969.
4. Herzenberg, L.A., Sweet, R.G., and Herzenberg, A.: Sci. Am., 234:(3):108, 1976.
5. Johnson, R.T., and Rao, R.N.: Nature, 266:717, 1970.
6. Johnson, R.T., and Rao, R.N.: Nature, 265:159, 1970.
7. Hara, K., Tydeman, P., and Kirschner, M.: Proc. Natl. Acad. Sci. USA, 77:462, 1980.
8. Bottenstein, J.A., and Sato, G.H.: Proc. Natl. Acad. Sci. USA, 76:514, 1979.
9. Levi-Montalcini, R., and Booker, B.: Proc. Natl. Acad. Sci. USA, 46:384, 1960.
10. Levi-Montalcini, R., and Calissano, P.: Sci. Am., 240(6):68, 1979.
11. Cohen, S., Pick, E., and Oppenheim, J.: Biology of the Lymphokines. New York, Academic Press, 1979.
12. Carpenter, G., and Cohen, S.: Annu. Rev. Biochem., 48:193, 1979.
13. Fernandez-Pol, J.A.: J. Biol. Chem., 256:9742, 1981.
14. Rubin, R.A., O'Keefe, E.J., and Earp, H.S.: Proc. Natl. Acad. Sci. USA, 79:776, 1982.
15. Meselson, M., and Stahl, F.W.: Proc. Natl. Acad. Sci. USA, 44:671, 1958.
16. Taylor, J.H., Woods, P.S., and Hughes, W.L.: Proc. Natl. Acad. Sci. USA, 43:122, 1957.
17. Fersht, A.R.: Trends Biochem. Sci., 5:262, 1980.
18. Gellert, M.: Annu. Rev. Biochem., 50:879, 1981.
19. Temin, H.: Sci. Am., 226:24, 1972.
20. Prescott, D.M., and Kuempel, P.: Proc. Natl. Acad. Sci. USA, 69:2842, 1972.
21. Wake, R.G.: J. Mol. Biol., 68:501, 1972.
22. Cairns, J.: J. Mol. Biol., 15:372, 1966.
23. Huberman, J.A., and Riggs, A.D.: J. Mol. Biol., 32:327, 1968.
24. Callan, H.G., and Taylor, J.H.: J. Cell Sci., 3:615, 1968.
25. Callan, H.G.: Cold Spring Harbor Symp. Quant. Biol., 38:195, 1974.
26. Holmquist, A., et al.: Cell, 31:131, 1982.
27. Heintz, N.H., and Hamlin, J.L.: Proc. Natl. Acad. Sci. USA, 79:4083, 1982.
28. Stinchcomb, D.T., Struhl, K., and Davis, R.W.: Nature, 282:39, 1979.
29. Stinchcomb, D.T., et al.: Proc. Natl. Acad. Sci. USA, 77:4559, 1980.
30. Celniker, S.E., and Campbell, J.L.: Cell, 31:201, 1982.
31. Fangman, W.L., Hice, R.H., and Chlebowicz, E.: Cell, 32:831, 1983.
32. Kearsey, S.: EMBO J., 2:1571, 1983.
33. Harland, R.M., and Laskey, R.A.: Cell, 21:761, 1980.
34. McKnight, S.L., and Miller, O.L., Jr.: Cell, 12:795, 1977.
35. Riley, D., and Weintraub, H.: Proc. Natl. Acad. Sci. USA, 76:328, 1979.

ADDITIONAL READINGS

Alberts, B., and Sternglanz, R.: Recent excitement in the DNA replication problem. Nature, 269:655, 1977.

Alberts, B., et al.: Molecular biology of the cell. New York, Garland Publishing Inc., 1983.

Bishop, J.M.: Viral oncogenes. Cell, 42:23, 1985.

Blumenthal, A., Kriegstein, J., and Hogness, D.: The units of DNA replication in *Drosophila melanogaster*. Cold Spring Harbor Symp. Quant. Biol., 38:205, 1974. (A classic.)

Callan, H.: Replication of DNA in eukaryotic chromosomes. Br. Med. Bull., 29:192, 1973. (Highly recommended.)

DNA: Replication and recombination. Cold Spring Harbor Symp. Quant. Biol., 43: 1978.

DePamphilis, M.L., and Wasserman, P.M.: Replication of eukaryotic chromosomes: A close-up of the replication fork. Annu. Rev. Biochem., 49:627, 1980.

Hara, K., Tydeman, P., and Kirschner, M.: A cytoplasmic clock with the same period as the division cycle in *Xenopus* eggs. Proc. Natl. Acad. Sci. USA, 77:462, 1980. (A beautiful paper.)

Hartwell, L.H., Culotti, J., Pringle, J.R., and Reid, B.J.: Genetic control of the cell division cycle in yeast. Science, 183:46, 1974.

Heldin, C., and Westermark, B.: Growth factors: Mechanism of action and relation to oncogenes. Cell, 37:9, 1984.

Holley, R.W.: Control of animal cell proliferation. J. Supramolec. Struct., 13:191, 1980.

Howard-Flanders, P.: Inducible repair of DNA. Sci. Am., 245(5):72, 1981.

Huberman, J., and Riggs, A.: On the mechanism of DNA replication in mammalian chromosomes. J. Mol. Biol., 32:327, 1968.

James, R., and Bradshaw, R.A.: Polypeptide growth factors. Ann. Rev. Biochem., 53:259, 1984.

John, P.C.L.: The Cell Cycle. Cambridge, England, Cambridge University Press, 1981.

Johnson, R.T., and Rao, P.N.: Mammalian cell fusion: Induction of premature chromosome condensation in interphase nuclei. Nature, 226:717, 1970.

Kornberg, A.: DNA Replication. San Francisco, W.H. Freeman Co., 1980.

Laskey, R.A., and Harland, R.M.: Replication origins in the eukaryotic chromosome. Cell, 24:283, 1981.

Lewin, B.: Genes. New York, John Wiley & Sons, 1985.

McKnight, S.L., and Miller, O.L., Jr.: Post-replicative non-ribosomal transcription units in *D. melanogaster* embryos. Cell, 17:551, 1979.

Prescott, D.M.: Reproduction of Eukaryotic Cells. New York, Academic Press, Inc., 1976. (Highly recommended.)

Taylor, J.H.: Units of DNA replication in chromosomes of eukaryotes. Int. Rev. Cytol., 37:1, 1974.

Ullrich, A., et al.: Human epidermal growth factor receptor. cDNA sequence and aberrant expression in epidermoid carcinoma cells. Nature, 309:418, 1984.

Varmus, H.E.: The molecular genetics of cellular oncogenes. Ann. Rev. Genet., 18:553, 1984.

Wang, J.C.: DNA topoisomerases. Sci. Am., 247(1):94, 1982.

Wharton, K.A., Johansen, K.M., Xu, T., and Artavanis Tsakonas, S.: Nucleotide sequence from the neurogenic locus Notch implies a gene product that shares homology with proteins containing EGF-like repeats. Cell. 43:567, 1985.

MITOSIS AND CELL DIVISION

Cell division is the complex phenomenon by which cellular material is divided equally between daughter cells. This process is the final, and microscopically visible, phase of an underlying change that has occurred at molecular and biochemical levels. Before the cell divides by mitosis, its fundamental components—including those involved in hereditary transmission—have duplicated. Thus cell division can be considered the final separation of the already duplicated macromolecular units.

The essential features of cell division by mitosis were considered in Chapter 1. In this chapter a more detailed analysis is made. Mitosis involves a series of complex changes in both the nucleus and the cytoplasm.

Special consideration is given to the *mitotic spindle*, the fibrous component made of microtubules (see Section 6–2), which is assembled every time the cell begins to divide and is disassembled at the end of mitosis. We show that the mitotic spindle functions as a structural framework and force-generating system that sets the chromosomes in position and distributes them between the daughter cells. In our discussion of mitosis we also deal with some of the structural differentiations of chromosomes, such as the *kinetochores* to which the spindle micro-

tubules are anchored, and the general changes in fine structure that chromosomes undergo during the mitotic cycle. The study of mitosis also comprises the question of the continuity of chromosomes as entities capable of autoduplication and of maintaining their morphological characteristics through successive cell divisions. Finally, we discuss *cytokinesis*, or separation of the cytoplasm of the two daughter cells, in animal as well as in plant cells.

15–1 A GENERAL DESCRIPTION OF MITOSIS

Despite the similarity in the overall process of mitosis in all eukaryotic cells, variations exist between different organisms—particularly between animal and plant cells—that make it difficult to give a general description. For example, in higher plants the spindle has no *centrioles or asters* at the poles, as in animal cells and lower plants, and, furthermore, the separation of the daughter cells is a more complex phenomenon. Although plant *meristems* (e.g., young root tips) are particularly advantageous for the study of cell division, because 10 to 15% of the cells are in mitosis, our description is based mainly on animal cells such as those in tissue culture or dividing eggs.

Figure 15–1 shows the stages of mitosis. The cycle is considered to begin at the end of the intermitotic period *(interphase)* and end at the beginning of a new interphase. The main divisions of this cycle are: *prophase, metaphase, anaphase,* and *telophase. Cytokinesis,* a process of separation of the two cytoplasmic territories, is simultaneous with anaphase or telophase, or it can occur at a later stage.

The various phases of mitosis are described in a way that should give an idea of the sequence of events that occur in the nucleus and the cytoplasm. Some of the underlying processes are considered in terms of their fine structure, bio-

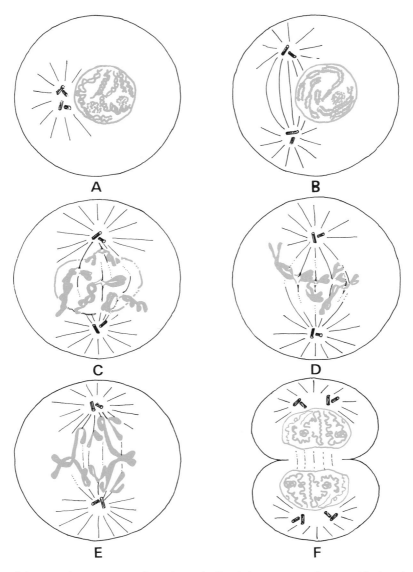

Fig. 15–1.   General diagram of mitosis. *A, prophase,* the nucleoli and chromosomes, shown as thin threads; in the cytoplasm the aster with the pairs of centrioles are shown. *B, prophase,* a more advanced stage of this phase in which the chromosomes have shortened. The primary constriction with the kinetochore is shown; in the cytoplasm the spindle is formed between the asters. *C, late prophase* or *prometaphase,* the nuclear envelope disintegrates and the chromosomes become attached to the spindle fibers. *D, metaphase,* the chromosomes are arranged along the equatorial plane. *E, anaphase,* the daughter chromosomes, preceded by the kinetochores, are moving toward the poles. *F, telophase,* the daughter nuclei are in the process of reconstitution; cell cleavage has started.

chemistry, and physiological significance in the second part of this chapter.

Prophase—Chromatid Coiling, Nucleolar Disintegration, and Spindle Formation

The beginning of *prophase* is indicated by the appearance of the chromosomes as thin threads inside the nucleus. In fact, the word "mitosis" (Gr., *mitos*, thread) is an expression of this phenomenon, which becomes more evident as the chromosomes start to condense (Fig. 15–1A). The condensation occurs by a process of folding of the chromatin fibers (see Section 14–2). At the same time, the cell becomes spheroid, more refractile, and viscous.

Each prophase chromosome is composed of two coiled filaments, the *chromatids*, which are a result of the replication of the DNA during the S phase. As prophase progresses, the chromatids become shorter and thicker, and the primary constrictions (defined in Section 13–4), which contain the kinetochores, become clearly visible (Fig. 15–1B). During early prophase, the chromosomes are evenly distributed in the nuclear cavity; as prophase progresses, the chromosomes approach the nuclear envelope, causing the central space of the nucleus to become empty. At this time each chromosome is composed of two cylindrical, parallel elements that are in close proximity. Not only the primary constrictions, but also the secondary constrictions along some chromosomes may be observed (Fig. 15–1B). Other changes are the reduction in size of the nucleoli, and their disintegration within the nucleoplasm. At the end of prophase, with the rapid fragmentation and disappearance of the nuclear envelope, the nucleolar material is released into the cytoplasm.

In the cytoplasm the most conspicuous change is the formation of the spindle. In early prophase there are two pairs of centrioles, each one surrounded by the so-called *aster*, composed of microtubules that radiate in all directions. (The name "aster" refers to the star-like aspect of this structure.) The two pairs of centrioles migrate along with the asters, describing a circular path toward the poles, while the spindle lengthens between. The migration of the asters continues until they become situated in antipodal positions (Fig. 15–1C).

Metaphase—Chromosomal Orientation at the Equatorial Plate

Sometimes the transition between prophase and metaphase is called *prometaphase* (Gr., *meta*, between). This is a short period in which the nuclear envelope disintegrates and the chromosomes are in apparent disorder (Fig. 15–1C). After that the spindle fibers invade the central area, and their microtubules extend between the pole. The chromosomes become attached by the kinetochores to some of the spindle fibers and oscillate until they become radially oriented in the equatorial plane and form the *equatorial plate* (Fig. 15–2). Those fibers of the spindle that connect to the chromosomes are generally called the *chromosomal fibers;* those that extend without interruption from one pole to the other are the *continuous fibers*.

Mitosis in which the spindle has centrioles and asters is called *astral* or *amphiastral* and is found in animal cells and some lower plants. Mitosis in which centrioles and asters are absent is called *anastral* and is found in higher plants, including all angiosperms and most gymnosperms (Fig. 15–2). Thus, centrioles and asters are not indispensable to the formation of the spindle and in a certain way, the formation of the spindle in *astral mitosis* is a mechanism that leads to the distribution of the centrioles between the two daughter cells.

Anaphase—Movement of the Daughter Chromosomes toward the Poles

At anaphase (Gr., *ana*, back) the equilibrium of forces that characterizes metaphase is broken by the separation of the kinetochores—a process that is carried out simultaneously in all the chromosomes. The kinetochores move apart, and the chromatids separate and begin their migration toward the poles (Fig. 15–1E). The kinetochore always leads the rest of the chromatid or *daughter chromosome*, as if it is being pulled by the chromosomal fibers of the spindle. The chromosome may assume the shape of a V with equal arms if it is metacentric, or with unequal arms if it is submetacentric. During anaphase the microtubules of the chromosomal fibers of the spindle shorten one-third to one-fifth of the original length. Simultaneously, the microtubules of the continuous fibers increase in length. Some of

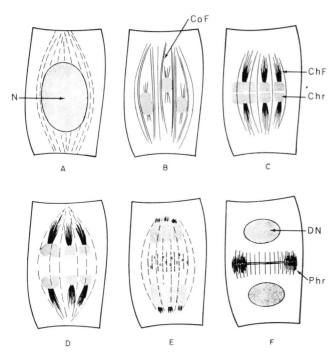

Fig. 15–2. Diagram of mitosis in a plant cell showing the changes in birefringence of the spindle fibers. Note the absence of centrioles and asters. Abbreviations: *CoF*, continuous fibers; *ChF*, chromosomal fibers; *Chr*, chromosome; *DN*, daughter nucleus; *N*, nucleus; *Phr*, phragmoplast. (Courtesy of S. Inoué.)

these stretched spindle fibers now constitute the so-called interzonal fibers.

Telophase—Reconstruction of the Daughter Nuclei

The end of the polar migration of the daughter chromosomes marks the beginning of telophase (Gr., *telo*, end). The chromosomes start to unfold and become less and less condensed by a process that in some ways recapitulates prophase, but in the reverse direction. At the same time, the chromosomes gather into masses of chromatin that become surrounded by discontinuous segments of nuclear envelope made by the endoplasmic reticulum (ER). Such segments fuse to make the two complete nuclear envelopes of the daughter nuclei. During the final stages the nucleoli reappear at the sites of the nucleolar organizers.

Simultaneous with the unfolding of the chromosomes and the formation of the nuclear envelope, *cytokinesis* occurs. In animal cells the cytoplasm constricts in the equatorial region, and this constriction is accentuated and deepened until the cell divides (Fig. 15–1F).

In cells of higher animals the period of cytokinesis is marked by active movements at the cell surface that are best described as "bub-

bling." Some investigators suggest that bubbling may reflect the activity of a rapidly expanding membrane. In ameboid cells at telophase both daughter cells have active movements, which appear to pull them apart. This is best observed in films of dividing cells and in scanning electron micrographs (Fig. 15–3). The high viscosity of the cytoplasm, characteristic of metaphase and anaphase, decreases during telophase. In most cases asters become reduced and tend to disappear. During cytokinesis the cytoplasmic components are distributed, including the mitochondria and the Golgi complex. Cytokinesis in plant cells is considered later on in this chapter.

Cycle of the Centrioles

In Section 6–3 we discussed the ultrastructure of centrioles as microtubular organelles having a triplet organization. At the end of interphase, two pairs of centrioles exist. These centrioles are generally found in a specially differentiated region—the *centrosome, cell center* or *centrosphere*. The centrosome is juxtanuclear and firmly attached to the nuclear envelope. The centrosome's position often determines the polarity of the cell, with the *cell axis* passing through it and the nucleus.

Although the centrioles were once thought to

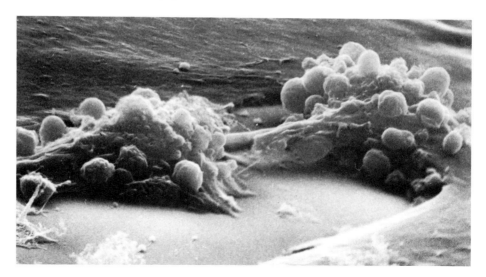

Fig. 15–3. Scanning electron micrograph of cultured normal hamster kidney cells. Two cells showing many blebs on the surface area in late cytokinesis, while the interphase cells remain flattened. × 1820. (Courtesy of R.D. Goldman.)

play a role in forming the spindle, this is not the case. We have seen that in plant cells the spindle is produced in the absence of centrioles (Fig. 15–2). The electron microscope (EM) has revealed that the microtubules of the spindle converge toward the centrioles but do not come in contact with them. As shown in Figure 15–4 these microtubules terminate in dense granules called *centriolar satellites* that surround the centrioles and are part of the centrosphere. It is from this pericentriolar material that the microtubules polymerize, as observed in isolated centrospheres incubated in vitro with tubulin.[1]

Centrioles replicate during interphase, generally in the S phase. At the beginning of prophase, a single aster surrounds the two pairs of centrioles. One of the pairs remains in position, with half the original aster, while the other pair, with the other half aster, migrates about 180° around the periphery of the nucleus to reach the opposite pole. Each of the two daughter cells receives a pair of orthogonally oriented centrioles (Fig. 15–1F). During the G_1 phase of the following cycle, the two centrioles lose their orthogonal disposition (*disorientation*).

At late G_1 or the S phase, a *procentriole* is produced by *nucleation*. This procentriole is oriented perpendicularly to the parent centriole (see Fig. 6–3), but not in contact with it. The parent centriole appears to function as a nucleation induction site in which the assembly of precursor material gives rise to the new cen-

triole. During the S phase and G_2, the procentriole grows (*elongation*) slowly; it attains full size at prophase. At this time, the two pairs of centrioles are again ready to start the spindle formation (Fig. 15–1A). Disorientation, nucleation, and elongation are typical changes occurring during the centriolar cycle.[2]

SUMMARY:
Events in Mitosis

Mitosis is a mechanism by which the cell distributes, in equivalent amounts, the components that have been duplicated during interphase. Prophase, metaphase, anaphase, and telophase are characterized by morphological changes that take place in the nucleus and the cytoplasm (Fig. 15–1).

In *prophase* chromosomes appear as thin threads that condense by coiling and folding. Each chromosome contains two *chromatids* that will be the future *daughter chromosomes*. With condensation each chromatid shows the *centromere* and *kinetochore*. The nucleolus tends to disintegrate and disappears at the end of prophase. In the cytoplasm the spindle is formed between the asters (and centrioles), which move toward the poles. Centrioles replicate at interphase during the S phase. A procentriole, oriented perpendicularly to the parent centriole, is produced by nucleation, which is followed by elongation.

At the beginning of *metaphase* (prometaphase) the nuclear envelope disintegrates, and mixing of the nucleoplasm with the cytoplasm occurs. Chromosomes become attached to the microtubules of the spindle and are oriented at the *equatorial* plate. The spindle has both continuous and chromosomal

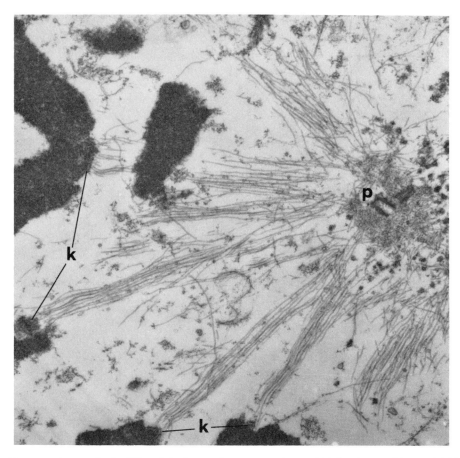

Fig. 15–4. Electron micrograph of a PtK₁ cell in tissue culture at metaphase which has been cold treated (4°C for 1 hour). Because of this treatment mainly kinetochoric microtubules remain. These are implanted at kinetochores (*k*). *P,* pole showing the two centrioles and dense material of the centrosphere. × 12,000. (Courtesy of V. Euteneur and J.R. McIntosh.)

microtubules. Animal cells have the type of spindle shown in Figure 15–1 *(astral mitosis).* In higher plant cells centrioles and asters are absent *(anastral mitosis)* (Fig. 15–2).

In *anaphase* the daughter chromosomes, led by the kinetochore, move toward the poles. The spindle fibers shorten one-third to one-fifth the original length.

In *telophase* chromosomes again unfold; the nuclear envelope is re-formed from the ER; and the nucleolus reappears.

Cytokinesis is the process of separation of the cytoplasm. In animal cells there is a constriction at the equator that finally results in the separation of the daughter cells.

15–2   MOLECULAR ORGANIZATION AND FUNCTIONAL ROLE OF THE MITOTIC APPARATUS

The term "mitotic apparatus" has been applied to the *asters* that surround the centrioles together with the *mitotic spindle.*[3]

We mentioned previously that the spindle has the so-called chromosomal fibers, joining the chromosomes to the poles; the continuous fibers, extending pole to pole; and the interzonal fibers, observed between the daughter chromosomes and nuclei in anaphase and telophase (Fig. 15–2). All these fibers, including those of the aster, are composed of microtubules.

The fine structure of the mitotic apparatus is studied principally with the EM. In Figure 15–4, because of the cold treatment of the cell, mainly the microtubules attached to the kinetochores remain. Interesting observations may also be made using polarization microscopy and the Nomarski interference-contrast microscope.[4] Because of their positive birefringence, spindle and aster fibers are readily observed in living cells. In plant cells, which are devoid of centrioles and asters, the first spindle fibers appear at prophase in a clear zone surrounding the nucleus (Fig. 15–2*A*). Birefringence is strongest near the kinetochores but becomes weaker toward the poles (Fig. 15–2*B–D*). During anaphase, the chromosomes are led by intensely birefringent chromosomal spindle fibers (Fig. 15–2*C* and *D*). The continuous fibers, in which birefringence is low in early anaphase, become more conspicuous in late anaphase and telophase. In animal cells such fibers form a bundle that maintains a connection between the two daughter cells for some time. Under the EM this bundle of microtubules appears surrounded by a dense material and some vesicles that together constitute a structure called the *midbody*[5] (see Fig. 15–13).

The use of antibodies against tubulin has permitted interesting observations on the changes in the mitotic apparatus during mitosis in cultured cells. The degree of fluorescence is related to the number and condensation of the microtubules in the aster and in the various regions of the spindle (see Fig. 15–5).

The use of the immunochemical methods based on the protein A-gold technique (see Fig. 3–19) permits a clear detection of the microtubules. As shown in Figure 15–6, during anaphase in a plant cell it is possible to differentiate the microtubules attached to the kinetochores of the chromosomes from those forming the continuous and interzonal fibers.

Centromere or Kinetochore—Microtubule Implantation Sites

Mitotic chromosomes of most plant and animal cells have a special region to which some spindle microtubules are attached. Usually, this region coincides with the *primary constriction* (see Section 13–4). In the older literature this region of the chromosome was referred to as the *centromere* or *kinetochore* (the terms were used interchangeably). In this region the sister chromatids of the duplicated chromosome are joined and at the same time the chromosomal microtubules are attached. With more recent developments the term "kinetochore" is preferred by cell biologists to define the site of implantation of the microtubules, whereas "centromere" is still used by geneticists. The chromosomal or kinetochoric spindle microtubules play an im-

portant role in chromosomal movements during mitosis.

The kinetochore may be detected during prophase, before the microtubules become attached to it (Fig. 15–1*B*). Under the EM the kinetochore appears as a plate or cup-like disc, 0.20 to 0.25 nm in diameter situated upon the primary constriction (Fig. 15–7).

In a cross section the kinetochore consists of the following three layers: (1) an *outer* layer of dense material 10 nm thick, (2) a *middle* layer of lower density, and (3) an *inner* dense layer in contact with the underlying chromatin fibers. Emanating from the convex surface of the outer layer, in addition to the microtubules, a "corona" or "collar" of fine filaments has been observed (Fig. 15–7).

The chemical nature of the kinetochore is largely unknown. Most authors believe that the inner layer of the kinetochore consists of chromatin, but it has often been claimed that the outer layer is composed of a non-chromatin material. It has also been maintained that the disc structure is the result of a special arrangement of the chromatin fibers, which are held together in a plane; these fibers probably lack nucleosomes.[6]

The antibodies found in the human disease scleroderma—the so-called CREST autoantibodies—have proved an excellent marker to study the kinetochore. The preferential binding of the antibody to this region of the chromosome can be seen by immunofluorescence and by electron microscopy in intact cells, as well as in isolated chromosomes (Fig. 15–8). Using this technique, one can trace the kinetochore not only in mitotic chromosomes but also at interphase, during the G_1, S, and G_2 phases of the cell cycle. At interphase kinetochores appear in the nucleus as subunits that are synthesized in the S phase. The kinetochores can be isolated by the CREST technique as follows: First, purified metaphase chromosomes are digested with nucleases; next the histones are removed. The remaining scaffold material (see Fig. 13–23) is dissociated by detergent and fractionated, and the pellets are stained with CREST. In a highly enriched fraction of kinetochores several proteins were recognized, and structures resembling the kinetochore plates were observed by electron microscopy (Fig. 15–4). These studies reveal that special proteins, characterized by the

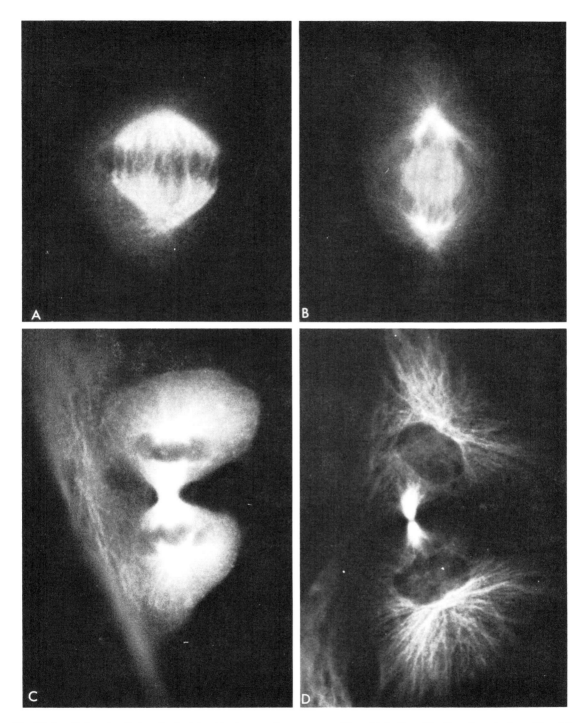

Fig. 15–5. Cultured cells in mitosis stained by immunofluorescence with anti-tubulin antibodies. *A, metaphase.* The stain is concentrated in the spindle fibers. *B, anaphase.* Spindle fibers are apparent, but there is staining at the asters and the midzone of the spindle. *C, telophase.* The asters and the midzone of the spindle are more heavily stained. *D, after telophase.* The tubulin fibers radiate from the asters, and the cytoplasmic bridge (*midbody*) between the two daughter cells is heavily stained. (Courtesy of B.R. Brinkley.)

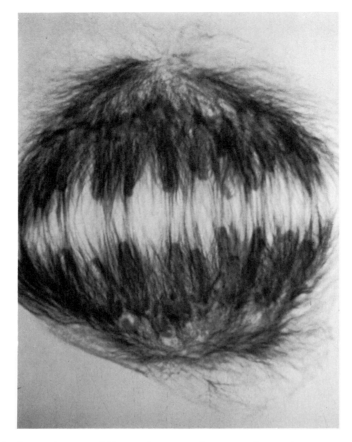

Fig. 15–6. Micrograph of an endosperm of *Haemanthus* at anaphase stained using the protein A-gold technique with an anti-tubulin antiserum and a blue stain for chromosomes. The fibers attached to the kinetochores of the chromosomes and those formed by continual and interzonal microtubules are clearly differentiated. (Courtesy of A. Bajer.)

antibody, are found in this region of chromosome that represents the point of attachment of microtubules (see Earnshaw et al., 1984; Brinkley et al., 1984; Rattner and Lin, 1985, in the Additional Readings at the end of this chapter).

The main function of kinetochores seems to be related to the attachment of the chromosomal microtubules. They may also serve as nucleation centers for the polymerization of tubulin.

Studies with the electron microscope using anti-tubulin antibodies have revealed that tubulin is found not only in the spindle and centrioles (both of which contain microtubules) but also in kinetochores. This suggests that they contain the nucleating intermediates necessary for the assembly of tubulin monomers into microtubules.[7] We shall see that the kinetochore is essential for the separation of the daughter chromosomes in anaphase.

Spindle Microtubules—Kinetochoric, Polar and Free

A precise study of the number of microtubules in the various regions of the spindle has been carried out in dividing animal cells.[5] There may be as few as a single microtubule per chromosome in the spindle of yeast cells and as many as 5000 in the spindle of a higher plant cell.

In a cultured cell line it has been found that about 34 ± 5 microtubules end at each kinetochore (Fig. 15–4). These *kinetochore tubules* correspond to the chromosomal fibers mentioned in the older literature. The microtubules point toward the poles, but not all of them are long enough to reach the pole. Quantitative measurements have shown that the so-called continuous fibers are not actually so because only a few microtubules may be so long as to

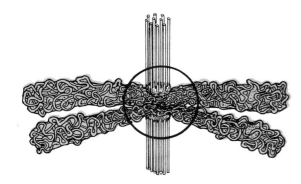

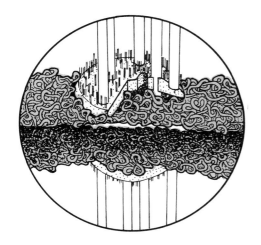

Fig. 15–7. *Top,* diagram of a metaphase chromosome showing the folded-fiber structure and the kinetochore with implanted microtubules. *Bottom,* an inset showing the convex electron-dense layer and the fibrillar material forming the "corona" of the kinetochore. Several microtubules of the spindle are shown penetrating the various layers of the kinetochore.

span the two poles. As better names, in addition to kinetochore microtubules, the terms *polar* and *free* microtubules have been proposed.[8]

Assembly and Polarity of Spindle Microtubules

The cyclic changes in the birefringence of the spindle are interpreted as reflecting the systematic assembly and disassembly of the microtubules.

Studies in living cells reveal that the spindle fibers represent a very dynamic structure. Their birefringence is abolished by low temperature, but after return to normal temperature, the cell recovers in a few minutes, with continuation of the arrested mitosis. Intense hydrostatic pres-

sure, microbeam ultraviolet irradiation, and certain drugs, such as colchicine, Colcemid, and others, also induce disappearance of the birefringence and of the microtubules. One interesting change is produced with heavy water (D_2O). When dividing sea urchin oocytes are placed in 45% D_2O, the birefringence increases twofold, and the volume of the spindle increases about tenfold in 1 to 2 minutes. After the eggs are returned to H_2O, the birefringence reverts to normal within a few minutes.[9] These experiments imply that in the cytoplasm there is a great excess of building material for the mitotic apparatus.

In Vitro Studies. By using gentle methods of cell fractionation on sea urchin eggs, Mazia and Dan (1952)[3] first isolated the mitotic apparatus.

A new approach has been used based on the discovery of the conditions under which brain tubulin can be polymerized in vitro.[10] The technique consists of preserving the isolated spindle in a medium containing exogenous tubulin.[8,11,12] With this method, some of the spindle functions are preserved. For example, in cells lysed at anaphase the chromosomes have been observed to move in normal fashion.

Control of the Assembly. Some observations suggest that the assembly of microtubules is "*controlled*" by the *poles* and also by the *kinetochores.* The *lateral interaction* between the spindle microtubules may also be involved. Although the exact mechanism of the assembly is unknown, it is thought to involve first the polymerization of tubulin subunits and then the formation of microtubules by secondary bonds.

Normally in a cell there is a large *pool of tubulin monomers*, which are in a dynamic equilibrium with the polymerized microtubules. An equilibrium is also established with the *cytoplasmic microtubules* (see Section 6–2). When cells enter prophase, these microtubules become depolymerized and are replaced by the mitotic spindle. At metaphase, only the spindle microtubules are present; at anaphase with the movement of the chromosomes, the spindle becomes depolymerized, and at telophase the daughter cells are held by the *midbody*, and the cytoplasmic microtubules reappear. (All these stages are clearly seen in the sequence shown in Fig. 15–5.[13])

Action of Ca^{++} and Calmodulin. The local concentration of Ca^{++} and of the Ca^{++}-binding

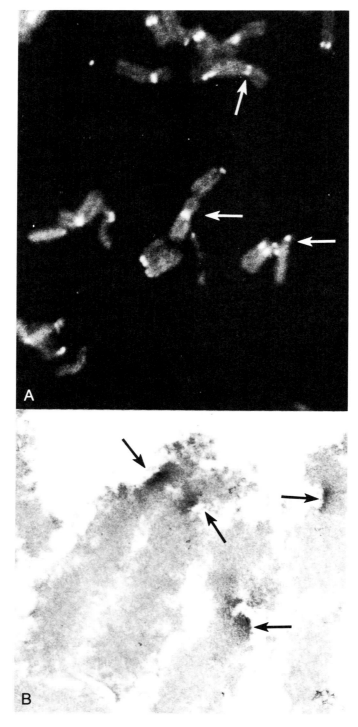

Fig. 15–8. Isolated chromosomes from CHO cultured cells stained by the CREST antibodies of human scleroderma. Such antibodies preferentially stain the kinetochores, (*top*) by immunofluorescence and (*bottom*) under the EM. The *arrows* point to kinetochores. (Courtesy of B.R. Brinkley and M. Valdivia.)

protein *calmodulin* appears to have a controlling role in the assembly and disassembly of spindle microtubules. This is shown in the following experiment: If $CaCl_2$ is injected into the mitotic apparatus of dividing eggs, there is a loss of birefringence, which is reversible.[14]

The level of Ca^{++} in the cytosol may be regulated through various mechanisms by which this cation is bound or sequestered in the cell.

The EM has revealed, around the spindle, the presence of vesicles of smooth endoplasmic reticulum (SER), that are able to take up labeled Ca^{++} by an ATP-dependent mechanism. These vesicles may serve to regulate the Ca^{++} concentration and, in this way, could be involved in the control of the assembly and disassembly of microtubules in the mitotic apparatus.[15]

The action of *calmodulin* may also be involved in this control. Using a fluorescent antibody, it is possible to demonstrate that this protein is concentrated in the mitotic apparatus. At prophase, it is found near the centrioles; at metaphase, it is prevalent in the polar regions; at anaphase, it is present in each half spindle but absent in the interzone; and at telophase it is absent in the contractile ring but appears in the midbody fibers.[16] Using the PAP method (see Fig. 3–18), at the electron microscopic level, it is possible to demonstrate that calmodulin is associated with the spindle microtubules and is highest in places in which these structures are more packed or have more extensive lateral interactions (see De Mey et al., 1980; Dedman et al., 1982; Petzelt and Hafner, 1986 in the Additional Readings at the end of this chapter).

In Section 5–2 we mentioned that the *tau factor*, a low molecular weight microtubule associated protein (MAP) present in microtubules, is capable of forming a reversible complex with calmodulin in the presence of low concentrations of Ca^{++}. This complex is dissociated when the Ca^{++} concentration increases. In this way the tau factor acts as a Ca^{++}-dependent switch between Ca^{++}, calmodulin, and the microtubule.[17]

Polarity of Spindle Microtubules. In Section 6–2 we mentioned that microtubules have distinct polarity with a *fast growing* or *plus end* and a *slow growing* or *minus end* (see Fig. 6–4). This polarity can be demonstrated under the EM by the following experiment: If an excess of tubulin is added to dividing cells that have been made permeable by a detergent, the microtubules become decorated with curved sheets (or hooks) of tubulin protofilaments. The direction of the hooks indicates the polarity. The results obtained in cells undergoing metaphase or anaphase (Fig. 15–9) is that all microtubules situated between the poles and the kinetochores have the same polarity, i.e., with the fast growing ends distal to the poles. This finding implies that in kinetochore microtubules the plus ends are also distal to the pole or, in other words, near the kinetochore.[18]

This observation suggests that the kinetochores are able to "capture" some microtubules initiated at the poles. In dividing plant cells the plus ends of the microtubules are also directed toward the phragmoplast (Fig. 15–2F). This observation suggests that both kinetochore and phragmoplast may serve as positioning sites for microtubules initiated elsewhere.[19]

Some work on diatoms also favors the idea that, during mitosis, kinetochores interact with and "capture" the microtubules that emanate from the poles. A system of fine filaments that forms the so-called *corona* or *collar* (Fig. 15–7) and is implanted in the kinetochore appears to be important in this interaction with the microtubules (see Pickett-Heaps et al., 1982 in the Additional Readings at the end of this chapter).

Metaphase Chromosomal Motion—Caused by Kinetochore and Polar Microtubular Interaction

Current hypotheses about the role of the mitotic apparatus are based on the idea that the microtubules can generate some sort of mechanical force by "pushing" or "pulling" the other cell components. Pushing is accomplished by the elongation and pulling by the shortening of the microtubules. In prophase, the migration of centrioles toward the poles is probably caused by the "pushing," which is a result of the elongation of the continuous fibers between the two poles. In fibroblasts, separation of centrioles occurs at a rate of 0.8 to 2.4 μm/min. As soon as the nuclear envelope begins to disintegrate, the nuclear region is invaded by microtubules that establish pole-kinetochore attachments before the chromosomes move toward the metaphasic plate.

The motion of chromosomes toward the metaphase plate can be explained as the result of

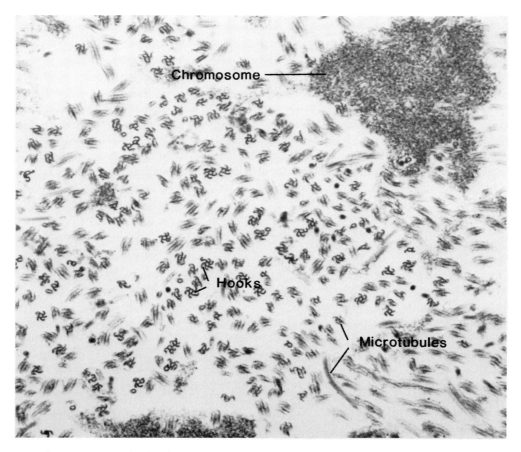

Fig. 15–9. Electron micrograph of a thin section of a PtK$_1$ cell in tissue culture at anaphase. This culture was made permeable and treated with an excess of tubulin. Observe that there are hooks of polymerized tubulin (most of which have a counterclockwise curve) surrounding the microtubules. × 48,000. (Courtesy of V. Euteneur and J.R. McIntosh.)

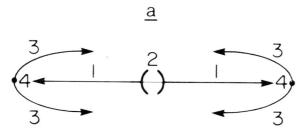

Fig. 15–10. Diagram indicating the equilibrium of forces at metaphase: *1*, pole-directed forces that pull sister kinetochores to opposite poles; *2*, forces that link the two kinetochores until anaphase; *3*, forces that pull the poles inward; *4*, resistance to the inward force. (From McIntosh, J.R., Cande, Z. W., and Snyder, J.: *In* Molecules and Cell Movement. Edited by S. Inoué and R.E. Stephens. New York, Raven Press, 1975, p. 31.)

lateral interaction between polar and kinetochoric microtubules.

During metaphase an equilibrium of forces is established, as indicated in the diagram of Figure 15–10. There are pole-directed forces that tend to pull the sister kinetochores toward the poles; however, because these are linked until the onset of anaphase, the opposing forces tend to balance each other.[8] It is possible that the polar microtubules exert a counteracting force that tends to pull the poles inward. (Such an equilibrium of forces could also explain the bowed shape of the spindle and the fact that the chromosomes remain stationary at metaphase.)

Anaphase Chromosomal Movement

With the separation of the sister chromatids at anaphase, this equilibrium is broken. The chromosomes usually move toward the poles at the rate of about 1 μm/min.

The characteristic shapes assumed by the chromosomes in metaphase and anaphase and the anaphase bridges that may result from the stretching of dicentric chromosomes (see Section 17–2) suggest that the forces responsible for the pulling of the chromosomes toward the poles are transmitted to the kinetochores. These forces may be great enough to produce the rupture of dicentric chromosomes. By hydrodynamic analysis it has been calculated that to move a chromosome, a force of about 10^{-11} dynes is needed, and that the entire displacement—from the equator to the pole of a chromosome—may require the use of about 30 ATP molecules. Recent experiments have permitted a direct measurement of the force necessary to move a single chromosome and have demonstrated that it is at least 10,000 times higher than was thought previously. The force has been measured by introducing a flexible microneedle into dividing cells and, with the tip, generating a force opposed to that exerted by the spindle. Because the deflection of the needle was calibrated in dynes, it was possible to calculate the force needed to counteract the chromosome movement. The opposing force needed to stop a single chromosome from moving toward the pole was calculated to be approximately 7×10^{-5} dynes. This finding may be relevant to possible mechanisms involved in anaphase movements (see Nicklas, 1983 in the Additional Readings at the end of this chapter).

A view of the spindle microtubules at different stages of mitosis has been obtained in thick sections of cultured cells by high-voltage electron microscopy (1000 kvolts). The resultant images suggest that at anaphase two distinct events occur: (1) the motion of chromosomes toward the poles on each half spindle and (2) the separation of the two half spindles by elongation of the free microtubules at the interzonal region. At the midregion of the interzonal fibers it is possible to identify a dense zone classically known as the *stem body.* Electron microscopic observations have shown that in this region there is a high concentration of interdigitating microtubules.

The importance of microtubules in the movement of chromosomes at anaphase is revealed by experiments of *micromanipulation.* Spindle

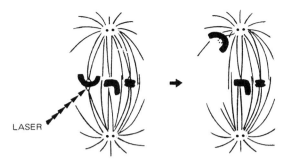

Fig. 15–11. Diagram of an experiment in which one kinetochore is inactivated by laser microbeam irradiation at metaphase. The two daughter chromosomes immediately move to the pole to which the other kinetochore is attached. In this way an anaphase movement can occur during metaphase. (From McNeill, P.A., and Berns, M.W.: J. Cell Biol., 88:543, 1981.)

fibers, within certain limits, resist extension by microneedles and maintain mechanical integrity. Movement of chromosomes toward the poles is not affected by removal of the centrosomal region at metaphase or anaphase. If the centrosomal region is removed earlier (i.e., at prophase or prometaphase), movements are abnormal. If the interzonal fibers are cut, there is no elongation of the spindle, but the polar movement of chromosomes is scarcely altered. These findings suggest that the centrosomes and the polar and interzonal regions are not indispensable for anaphase movements and that the motive force is probably located near the kinetochore region.[20] This interpretation was also supported by another experiment along these lines, in which one of the kinetochores was inactivated by laser microirradiation at metaphase. The result was that the two daughter chromosomes were transported to the pole corresponding to the intact kinetochore[21] (Fig. 15–11).

Anaphase movement of chromosomes comprises two distinct mechanisms that have been called A and B. In *anaphase A* the movement of chromosomes to the poles is associated with the shortening of the kinetochore microtubules. In *anaphase B*, the spindle elongates until it nearly duplicates the length it had at metaphase. In experiments with cells that have been made permeable with polyethylene glycol, it has been found that ATP is needed only for anaphase B.[22] The elongation can be inhibited with *vanadate*, which acts on an ATPase. These two findings, action of ATP and inhibition by vanadate, are consistent with the view that, in anaphase B, a dynein-like ATPase (see Section 6–3) may be active in spindle elongation and involved in force generation. On the other hand, anaphase A does not require ATP and is insensitive to vanadate, suggesting a different mechanism of force generation.

Hypotheses about the Mechanisms of Anaphase Movements

Hypotheses about the role of the mitotic apparatus in mitosis have been influenced by the discovery of a cyclic mechanism in muscle and cilia in which there is a direct relationship between the hydrolysis of ATP and mechanical events. Such a coupling is based on an elaborate and rather stable molecular organization of these systems (see Chapters 6 and 7). On the other hand, the mitotic apparatus is very labile and is assembled and disassembled at every mitotic cycle.

Several hypotheses have attempted to interpret the molecular mechanism responsible for the movement of chromosomes.

The *dynamic equilibrium hypothesis* first proposed in 1949 by Ostergren has been established mainly by Inoué and Sato[9] as a consequence of their studies with the polarization microscope. This hypothesis postulates that the polymerization-depolymerization of microtubules is directly responsible for the movement of the chromosomes. While it is rather easy to see that the microtubules could have a "cytoskeletal" role, serving as a guide for maintaining the shape of the cell and for "pushing" the centrioles, it is more difficult to understand how their depolymerization could provide a motive force. However, calculations of the cohesive forces that could be involved in assembling microtubules have led some authors to conclude that the microtubules contain enough energy to move chromosomes through the cytoplasm.[23]

A variation of the dynamic equilibrium hypothesis postulates that microtubules are firmly anchored at kinetochores, while the polar ends are free for disassembly. In this *anchorage-translocation* system the motive force would be provided by the polarized assembly-disassembly of the microtubules (see Fig. 6–4).[24]

According to the *sliding hypothesis*, the spindle microtubules assemble and disassemble, but do not generate directly the driving force to move the chromosomes. Rather, they are considered the "railroad tracks" on which the chromosomes move. This hypothesis postulates the involvement of lateral interactions of molecules with microtubules that could play a role analogous to *dynein* in flagella (see Section 6–2).

The contractile protein *actin* has been identified as a component of the spindle by immunofluorescence and by electron microscopy using "decoration" with myosin fragments (Fig. 7–9). Although the bulk of actin is not attached to microtubules, some may be associated with them. Experimentally, it has been observed that by mixing isolated microtubules with actin in solution, a gel can be produced, provided that the MAPs are also present (Section 6–2). *Myosin* has also been localized in the spindle but with a diffuse distribution. The protein *dynein* (Section 6–3), which has ATPase activity, is also present in the spindle.[25]

Recently a possible regulatory role of proteins of the MAP has been studied. Some proteins of high molecular weight have been isolated. These are present in the interphase nucleus and then, after breakdown of the nuclear membrane, are released in the cytoplasm and become attached to the spindle microtubules. A monoclonal antibody against a high molecular weight MAP has been used to stain cytoplasmic and spindle microtubules. Such an antibody, when injected into a living cell, interferes with normal mitosis, suggesting that this MAP may contribute to the function of the spindle.[26]

These findings suggest the possibility that the interaction of spindle actin with myosin, or dynein with microtubules, could provide the motive force in chromosomal movement. In this context mitosis could be interpreted as a special case of the more general mechanisms involved in cell motility described in Chapter 6.

According to the *microtubular lattice hypothesis*, such a portion of the cytoskeleton (Section 6–4) produces the motive force, while the direction is given by the microtubules.[27] This model has been developed on the basis of other intracellular movements such as those found in erythrophores (e.g., pigment cells found in scales of fishes).

In these last two hypotheses the interaction of actin with myosin is in some way implicated. Against this interpretation is the fact that several of the movements involved in mitosis, however, are insensitive to cytochalasin B, which depolymerizes actin. Furthermore, it has been found that the injection of an anti-myosin antibody, which completely inhibits cytokinesis in vivo, has no effect on the rate or extent of anaphase chromosome movements (Fig. 15–12). These experiments suggest that myosin is not involved in chromosome movement during mitosis.[28]

In conclusion the mechanism of chromosome anaphase motion has not been clarified and even the two phases A and B cannot be explained by a single mechanism (see Pickett-Heaps, 1982; McIntosh, 1984 in the Additional Readings at the end of this chapter).

The Nuclear Envelope and the Lamina Are Re-Formed at Telophase

In Chapter 13 we studied the nuclear envelope with the two membranes in continuity with the ER, the pore complexes, and the lamina with its three polypeptides or *lamins*. We mentioned that at the end of prophase the nuclear envelope breaks up into vesicles indistinguishable from the ER, most of the pore complexes become attached to the chromosomes, and the lamina is disrupted by the disassembly of the lamins, which is apparently related to the phospho-

rylation of these proteins (see Khrone and Benavente, 1986).

At telophase, vesicles of the ER become associated with the surface of the chromosomes and then fuse to reconstruct the nuclear envelope. At the same time, the pore complexes are reinserted into the envelope and the lamins are dephosphorylated and repolymerized, re-forming the nuclear lamina. At this point, the nucleolus reappears, and the RNA synthesis becomes active while the chromatin becomes dispersed as in the interphase conformation. In Section 16–3 we discuss the changes in the nuclear lamina during meiosis (see Fig. 16–13).

Cytokinesis in Animal Cells—A Contractile Ring of Actin and Myosin

The separation of the daughter nuclei and cytokinesis or *cell cleavage* may be two separate processes. For example, the eggs of most insects undergo division of the nucleus to form a multinucleate *plasmodium* without separation of the cytoplasmic territories. Cleavage of animal cells has been studied mainly in tissue culture and dividing eggs and is produced by the furrowing of the cell. The first visible changes consist of the appearance of a dense material around the microtubules at the equator of the spindle at

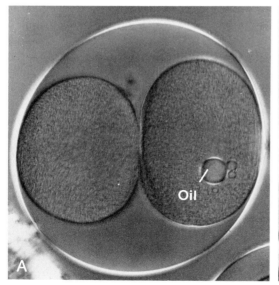
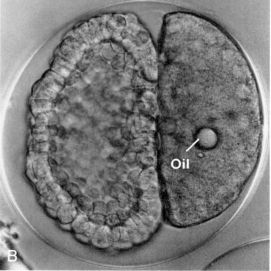

Fig. 15–12.   Experiment in which anti-myosin antibody is injected into *Asteria* eggs. In *A*, the right blastomere, receiving the injection at the two-cell stage, is marked with oil drops. In *B*, 10 hours later, the left blastomere has divided normally, producing a blastula; whereas in the right blastomere cytokinesis is blocked. (Courtesy of D.P. Kierhart, I. Mabuchi, and S. Inoué.)

either mid- or late anaphase. Then, while most spindle microtubules tend to become disorganized and disappear during telophase, some may persist, and even increase in number at the equator, frequently being intermingled with a row of vesicles and dense material; the entire structure is called the *midbody* (Fig. 15–13). Simultaneously, a depression appears on the cell surface—"a constriction" that deepens gradually until reaching the midbody. With the completion of the furrowing the separation of the cell is concluded.

An important point is that the position of the cleavage is apparently determined by the spindle poles. In fact, the cleavage furrow is positioned at the intersection of the sets of astral rays coming from the poles. This can be observed best in the initial divisions of sea urchin eggs in which microtubules from the two poles intercept at the surface where the contractile mechanism is activated. A signal that moves from the poles toward the surface, at a rate of 2 μm/min, has been postulated.

Cleavage by furrowing appears to be the result of a contractile ring present at the cell cortex and can continue in a cell lysed by detergents in the presence of ATP and Ca^{++}. Under the EM, a system of microfilaments can be observed in the region of the furrow (Fig. 15–14), and by the use of labeled antibodies, both actin and myosin have been detected in the cortical layer. Another indication of an actin-myosin mechanism is provided by the action of *cytochalasin B*. Cleavage is inhibited by this drug, while anaphasic movements are not affected. We mentioned earlier that cell cleavage is completely inhibited by an injection of an anti-myosin antibody, while anaphase movements are not impaired (Fig. 15–12). These results suggest that cleavage and anaphase movements use different force-generating mechanisms. While in cleavage an actin-myosin mechanism appears to be active, this is apparently not the case for the anaphasic movements.[28]

At the beginning of the chapter we mentioned that in cell division all the nuclear and cytoplasmic components are equally divided between the two daughter cells. Cell organelles like mitochondria, chloroplasts, or centrioles separate at cytokinesis. An exception is the Golgi apparatus, which becomes fragmented and thus difficult to identify during mitosis. Recently this task has been facilitated by the use of a monoclonal antibody against a specific protein of the Golgi. At prophase the entire Golgi complex becomes dispersed into small vesicles that reassociate at telophase to reconstruct the organelle.[29]

Cytokinesis in Plant Cells—The Phragmoplast and Cell Plate

In plant cells at anaphase the interzonal region of the spindle becomes transformed into the *phragmoplast*, which is similar to the midbody of animal cells (Fig. 15–2E and F).

Under the EM it is possible to observe that the interzonal microtubules at the equator plane have scattered patches of vesicles and of dense material applied to their surface. The vesicles are derived from the Golgi complexes, which are found in the regions adjacent to the phragmoplast and which migrate into the equatorial region to be clustered around the microtubules. Although the phragmoplast initially is found in the mid-region of the cell, with time it grows centrifugally by addition of microtubules and vesicles until it extends across the entire equatorial plane (Fig. 15–E and F). The vesicles increase in size and then fuse until the two daughter cells become separated by fairly continuous plasma membranes. At this time the phragmoplast has been transformed into the *cell plate*.

The cytoplasmic connections—the *plasmodesmata*—traverse the cell plate and remain in place for communication between the adjacent daughter cells. (In Section 12–1 we described that the formation of the cell plate leads to the synthesis of the *cell wall*.) The *Golgi* vesicles in the phragmoplast are already filled with a secretory material consisting mainly of *pectin* (see Fig. 12–2). The fusion of the vesicles results in the combining of the pectin in the extracellular space between the daughter cells, thereby forming the main body of the *primary cell wall*. (As mentioned in Section 12–1 pectin is an amorphous polymer made of galacturonic acid.) Later on, microfibrils of cellulose are laid down in a semicrystalline lattice on the two surfaces of the facing daughter cells. (For a review see Gunning and Wick, 1985, in the Additional Readings at the end of this chapter.)

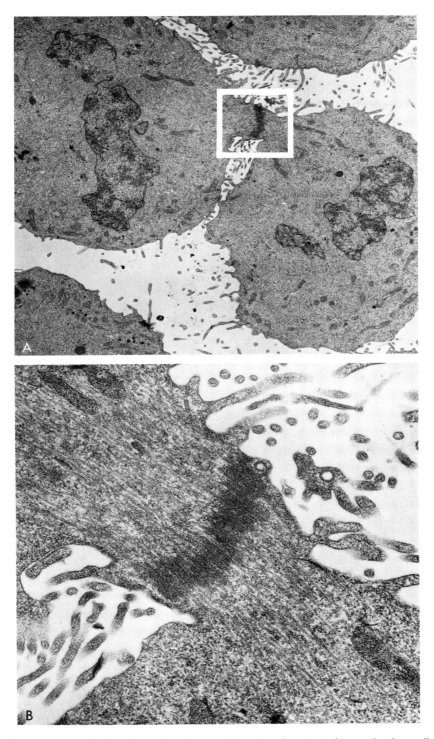

Fig. 15–13. Electron micrograph of a HeLa cell at the completion of cytokinesis. *A,* the two daughter cells are still joined by a small bridge, which contains the interzonal microtubules and the electron-dense midbody. *B,* an inset at higher magnification. *A,* ×10,000; *B,* ×30,000. (Courtesy of B.R. Brinkley.)

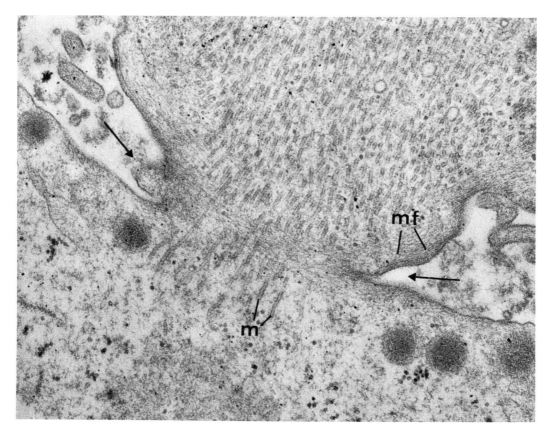

Fig. 15–14.   Electron micrograph of the advancing furrow (*arrows*) in a cleaving rat egg. Microtubules (*m*) of the interzonal fibers of the spindle are observed in the bridge between the daughter cells. Below the plasma membrane a network of the microfilaments (*mf*) is observed. (See the text for details.) ×34,000. (Courtesy of D. Szollosi.)

SUMMARY:

The Mitotic Apparatus

The *mitotic apparatus* comprises the spindle and the asters, which surround the centrioles. The spindle is made of the *chromosomal* fibers, the *continuous* fibers, and the *interzonal* fibers; the interzonal fibers are observed at anaphase and telophase between the daughter chromosomes.

The mitotic apparatus may be observed in the living cell through polarization microscopy. Anti-tubulin antibodies and fluorescence microscopy may provide information about the relative number and condensation of microtubules.

The *kinetochore* is the site of implantation of the microtubules in the chromosome. There is usually one kinetochore per chromatid. It appears as a cup-like disc 0.20 to 0.25 μm in diameter. It has three layers: an outer dense layer, an inner dense layer, and a middle layer of lower density. Kinetochores also contain tubulin, and they act in the assembly of microtubules. About 34 microtubules attach in a single kinetochore. The fibers of the spindle are formed by microtubules which may be kineto-

choric, polar, or free. Kinetochoric microtubules correspond to the chromosomal spindle fiber. Cross-bridges have been observed between microtubules and also microfilaments which may correspond to actin.

Microtubules are assembled and disassembled during the various phases of mitosis. The mitotic apparatus may be isolated, and best preservation is obtained in media containing exogenous tubulin.

The assembly of tubulin is controlled by the poles and by the kinetochore. There is a large pool of tubulin monomers in equilibrium with the microtubules. Ca^{++} inhibits the polymerization of tubulin.

A special ER may regulate the Ca^{++} concentration, and the relationship between Ca^{++}, calmodulin, and microtubules may play a role in the assembly and disassembly of the mitotic apparatus.

Elongation and shortening of microtubules seem to be the two major mechanisms by which chromosomes are moved toward the poles. At metaphase there is an equilibrium of forces (Fig. 15–10). In anaphase, chromosomes move at a rate of 1 μm/min. The motive force is transmitted by way of the kinetochores. There is also an elongation of the spindle shortening of the kinetochore microtubules. In the mid-region of the spindle, numerous microtubules appear.

Several hypotheses have been devised to interpret the molecular mechanisms of chromosome movement.

The dynamic equilibrium hypothesis postulates that the polymerization-depolymerization of microtubules is directly responsible for the movement. The sliding hypothesis postulates that the driving force may originate from lateral interactions (as in the case of dynein in flagella) or by interaction with actin microfilaments. Recent experiments, however, tend to disprove actin and myosin involvement.

Cytokinesis or cell cleavage differs considerably in animal and plant cells. In the former, separation of daughter cells is produced by an equatorial constriction which involves a contractile mechanism at the cell cortex. This is achieved by a system of actin-like microfilaments. A dense structure called the *midbody* may be formed.

In plant cells cytokinesis starts with the formation of the *phragmoplast*, which comprises the interzonal microtubules and Golgi vesicles. This structure is transformed into the *cell plate*, which separates the territories of the daughter cells. Within the cell plate the primary cell wall is produced by a secretory mechanism consisting mainly of the production of pectin, which is contained in Golgi vesicles (see Chapter 12).

REFERENCES

1. Gould, R.R., and Borisy, G.G.: Cell Biol., *73*:601, 1977.
2. Kuriyama, R., and Borisy, G.G.: J. Cell Biol., *91*:814, 1981.
3. Mazia, D., and Dan, K.: Proc. Natl. Acad. Sci. USA, 38:826, 1952.
4. Bajer, A., and Allen, R.D.: Science, *151*:572, 1966.
5. Brinkley, B.R., and Cartwright, J.: J. Cell Biol., *50*:416, 1971.
6. Ris, H., and Witt, P.L.: Chromosoma, *82*:153, 1981.
7. Pepper, D.A., and Brinkley, B.R.: Chromosoma, *60*:223, 1977.
8. McIntosh, J.R., Cande, Z.W., and Snyder, I.: *In* Molecules and Cell Movement. Edited by S. Inoué and R.E. Stephens. New York, Raven Press, 1975, p. 31.
9. Inoué, S., and Sato, H.: J. Gen. Physiol., *50*:259, 1967.
10. Weisenberg, R.C.: Science, *177*:1104, 1972.
11. Inoué, S., Borisy, G.G., and Kierhart, D.P.: J Cell Biol., *62*:175, 1974.
12. Rebhun, I.I., Rosenbaum, J., Lefebvre, P., and Smith, G.: Nature, *249*:113, 1974.
13. Fuller, G.M., and Brinkley, B.R.: J. Supramol. Struct., 5:497, 1976.
14. Kierhart, D.P.: J. Cell Biol., 88:604, 1981.
15. Silver, et al.: Cell, *19*:505, 1980.
16. Welsh, M.J.: J. Cell Biol., *81*:624, 1979.
17. Kakiuchi, S., and Sobue, K.: Trends Biochem. Sci., 8:59, 1983.
18. Heideman, S.R., and McIntosh, J.R.: Nature, *286*:517, 1980.
19. Enteneuer, V., and McIntosh, J.R.: J. Cell Biol., 89:338, 1981.
20. Hiramoto, Y., and Shöje, Y.: Eur. J. Cell Biol., *22*(1):311, 1980.
21. McNeill, P.A., and Berns, M.W.: J. Cell Biol., 88:543, 1981.
22. Cande, W.Z.: Cell, 28:15, 1982.
23. Inoué, S., and Ritter, H., Jr.: *In* Molecules and Cell Movement. Edited by S. Inoué and R.E. Stephens. New York, Raven Press, 1975.
24. Wilson, L., and Margolis, R.L.: Symp. Mol. Cell. Biol., *12*:241, 1978.
25. Pollard, T.D.: Eur. J. Cell Biol., *22*(1):289, 1980.
26. Izant, J.G., Weatherbee, J.A., and McIntosh, J.R.: J. Cell Biol., *96*:424, 1983.
27. Zieve, G., and Solomon, F.: Cell, 28:233, 1982.
28. Kierhart, D.P., Mabuchi, I., and Inoué, S.: J. Cell Biol., *94*:165, 1982.
29. Burke, B., et al.: EMBO J., *1*:1621, 1982.

ADDITIONAL READINGS

Bajer, A.S., and Molé-Bajer, J.: Spindle dynamics and chromosome movements. *In* International Review of Cytology, suppl. 3. New York, Academic Press, 1972.
Beams, H.W., and Kessel, R.G.: Cytokinesis. Am. Sci., *63*:279, 1976.
Brinkley, R.B., Brenner, S.L., Tousson, A., and Valdivia, M.M.: Microtubular organizing centers: Evolution and biochemical characterization of centromeres, kinetochores of mammalian chromosomes. III International Cell Biology Congress. Edited by S. Seno and Y. Okada. Japan, Academic Press, 1984, p. 117.
Cande, W.Z., Lazarides, E., and McIntosh, R.: A comparison of the distribution of actin and tubulin in the mitotic spindle. J. Cell Biol., *72*:552, 1977.
Dedman, J.R., et al.: Localization of calmodulin in tissue culture cells. Calcium and Cell Function, *3*:455, 1982.
De Mey, Z., et al.: *In* Microtubules and Microtubule Inhibitors. Edited by M. De Brabander and J. De Mey. Amsterdam, Elsevier North-Holland, 1980, p. 227.
Earnshaw, W.C., Halligan, N., Cooke, C., and Rothfield, N.: The kinetochore is part of the metaphase chromosome scaffold. J. Cell Biol., *98*:352, 1984.
Fuller, G.M., and Brinkley, B.R.: Structure and control of assembly of cytoplasmic microtubules in normal and transformed cells. J. Supramol. Struct., *5*:497, 1976.
Gunning, B.E.S., and Wick, S.M.: Preprophase bands, phragmoplasts, and spatial control of cytokinesis. J. Cell Sci. Suppl. *2*:157, 1985.
Hartwell, J.H.: Cell division from a genetic perspective. J. Cell Biol., *77*:627, 1978.
Inoué, S.: Cell division and the mitotic spindle. J. Cell Biol., *91*:3, 1981.
John, B., and Lewis, K.R.: The chromosome cycle. Protoplasmatologia, *6b*:1, 1969.
Krohne, G., and Benavente, R.: The nuclear lamins. Exp. Cell Res. *162*:1, 1986.
Kubai, D.F.: The evolution of the mitotic spindle. Int. Rev. Cytol., *43*:167, 1975.
Mazia, D.: The cell cycle. Sci. Am., *230*(1):54, 1974.
McIntosh, J.R.: Mechanism of mitosis. Trends Biochem. Sci., *9*:195, 1984.
Nicklas, R.B.: Chromosome movement, facts and hypoth-

esis. *In* Mitosis, Facts and Questions. Edited by M. Little, et al. Berlin, Springer-Verlag, 1977.

Nicklas, R.B.: Measurement of the force produced by the mitotic spindle in anaphase. J. Cell Biol., 97:542, 1983.

Paweletz, N.: Membranes in the mitotic apparatus. Cell Biol. Int. Rep., 5:323, 1981.

Petzelt, C., and Hafner, M.: Visualization of the Ca^{2+} transport system of the mitotic apparatus of sea urchin eggs with a monoclonal antibody. Proc. Natl. Acad. Sci. USA, 83:1719, 1986.

Pickett-Heaps, J.D., Tippit, D.H., and Porter, K.R.: Rethinking mitosis. Cell, 29:729, 1982.

Porter, K.R.: Motility in cells. *In* Cell Motility. Edited by C.R. Goldman, et al. New York, Cold Spring Harbor, 1976, p. 1.

Prescott, D.M.: Reproduction of Eukaryotic Cells. New York, Academic Press, 1976.

Rattner, J.B., and Lin, C.C.: Centromere organization in chromosomes of the mouse. Chromosoma (Berl.) 92:325, 1985.

Sherline, P., and Schiavone, K.: High molecular weight MAPs are part of the mitotic spindle. J. Cell Biol., 77:R9, 1978.

Zimmerman, A., and Forer, A.: Mitosis—Cytokinesis. New York, Academic Press, 1981.

MEIOSIS AND SEXUAL REPRODUCTION

Meiosis is a special type of cell division found in organisms in which there is *sexual reproduction*. In many protozoa, algae, and fungi reproduction is asexual, i.e., by simple cell division or *mitosis*. In this form of reproduction all the descendant individuals have *uniparental inheritance*. On the other hand, in most multicellular organisms—animals or plants—the reproduction is sexual. This means that during their lifetime they produce *sexual cells* or *gametes (eggs and sperms)* which by their fusion in the zygote will originate the new organism. Through a process of nuclear fusion called *fertilization*, the descendant individuals have *biparental inheritance*.

To understand the essentials of meiosis let us take the case of humans, illustrated in Figure 16–1. The human karyotype, which will be studied in Chapter 18, has 46 chromosomes, 44 + XY in the male, and 44 + XX in the female.

(The pairs XY and XX are the sex chromosomes.) If mitosis were the only type of cell division, then each gamete would have 46 chromosomes and the zygote, 92. This process would be repeated in the next generation, and the number of chromosomes would increase according to the square of the number of generations!

Meiosis (Gr., meioum, to diminish) is the mechanism that prevents this from happening. By a series of two divisions, the number of chromosomes is reduced by half—a process that gives rise to four *haploid* cells (four spermatozoa in the male, and one ovum and three polar bodies in the female) (Fig. 16–1). The processes by which the gametes are produced are called *spermatogenesis* and *oogenesis*, respectively, and they take place in the gonads (i.e., testicle and ovary). The type of meiosis found in humans is characteristic of all animals and a few lower plants; it is called *terminal* or *gametic* because

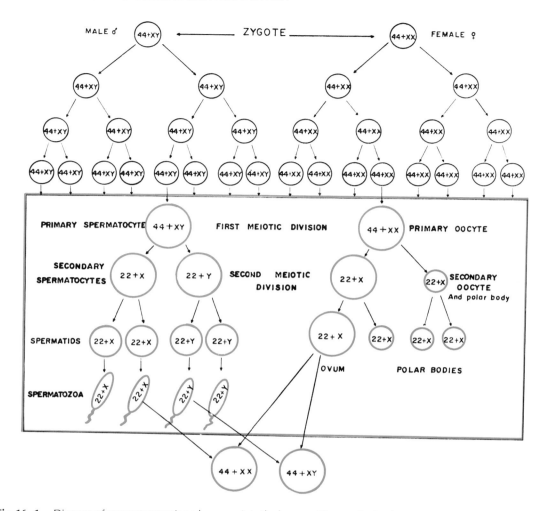

Fig. 16–1. Diagram of spermatogenesis and oogenesis in the human. *Above,* mitosis of spermatogonia and oogonia. *Middle* (within the box), the meiotic divisions. *Below,* fertilization and zygotes. Notice the 44 autosomes and the XX and XY sex chromosomes.

the meiotic divisions occur just before the formation of the gametes. We shall see later that in most plants *meiosis* is *intermediary* or *sporic* and occurs at some time between fertilization and the formation of gametes. In other words, in this case meiosis is preceded and followed by mitosis. Cells undergoing meiosis are often called *meiocytes.*

In this chapter, we describe the essential processes of meiosis: *reduction* in the number of chromosomes, *recombination* with interchange of chromosome segments (i.e., blocks of genes) by way of *crossing-over,* and *random assortment* of homologous chromosomes.

To understand cytogenetics the student should have a clear understanding of the morphology and biochemistry of meiosis.

16–1 A COMPARISON OF MITOSIS AND MEIOSIS

In Chapter 1 (Fig. 1–13) we compared the two types of cell division. Many of the events we have studied in mitosis are also found in meiosis—for example, the sequence of changes in the nucleus and cytoplasm; the stages of prophase, metaphase, anaphase, and telophase; the formation of the mitotic apparatus with the asters and spindle; the condensation cycle of the chromosomes; and the structure and function of the kinetochores. There are, however, essential differences between the two processes, some of which are the following:

1. Mitosis occurs in all *somatic cells* of an

individual, but meiosis is limited to the *germinal cells*.

2. In mitosis each replication cycle of DNA is followed by one cell division. The resulting daughter cells have a *diploid number* of chromosomes and the same amount of DNA as the parent cell. In meiosis one replication cycle of DNA is followed by two divisions, and the four daughter cells are *haploid* and contain half the amount of DNA as each parent cell.

3. In mitosis DNA synthesis occurs in the S phase, which is followed by a G_2 phase before the onset of division. In meiosis there is *premeiotic DNA* synthesis, which is much longer than that in mitosis[1,2] and which is followed immediately by meiosis. In other words, the G_2 phase is short or nonexistent.

4. In mitosis every chromosome behaves independently; in meiosis the *homologous chromosomes* become mechanically related during the first meiotic division (i.e., *meiotic pairing*).

5. While mitosis is rather brief (1 or 2 hours), *meiosis is a long process*. For example, in the human male it may last 24 days, and in the female it may go on for several years.

6. A fundamental difference between the two types of cell division is that in mitosis the genetic material remains constant (i.e., with only rare mutations or chromosomal aberrations), while *genetic variability* is one of the main consequences of meiosis.

16–2 A GENERAL DESCRIPTION OF MEIOSIS

As shown in the diagram of Figure 16–1, after several mitotic divisions of the spermatogonia (or oogonia), meiotic division starts. A "switch" shifts these cells from mitosis to meiosis (see Riley and Flavell, 1977 in the Additional Readings at the end of this chapter). Some experiments in cultured cells of lily anthers show that this change probably takes place at the beginning of G_2, but its exact nature is still unknown. In these cells it is possible experimentally to reverse the switch back to mitosis before the process becomes irreversible.

Premeiotic cells have either two times (2n) or four times (4n) the haploid (n) amount of DNA.

In the primary spermatocyte as well as in the oocyte there is a tetraploid (4n) amount of DNA. After the first meiotic division, the DNA content in the resulting cells is diploid (2n), while after the second meiotic division only the haploid amount of DNA is present. The DNA content during meiosis was measured by Pollister and collaborators more than 30 years ago by the use of the Feulgen reaction followed by cytophotometry (Table 16–1).

We mentioned earlier in this chapter that meiosis comprises two cell divisions (I and II). From the morphological viewpoint the first meiotic division is characterized by a long prophase during which homologous chromosomes pair closely and interchange hereditary material. The classic stages of mitosis do not suffice to describe the complex movements of the chromosomes in meiosis. The successive meiotic stages are the following:

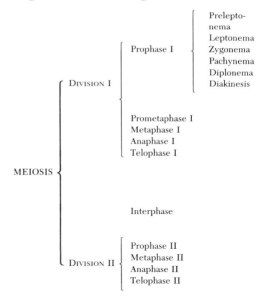

Leptotene Chromosomes Appear to Be Single and Have Chromomeres

Preleptonema is the early prophase of meiosis. Chromosomes are extremely thin and difficult to observe. Only the sex chromosomes may stand out as compact heteropyknotic bodies.

Leptonema (Gr., *leptos*, thin + *nema*, thread) (Fig. 16–2) is a period in which the nucleus has increased in size and the chromosomes have become more apparent. The leptotene chromosomes differ in the following two ways from the

Cell Type		DNA Content in picograms
Premeiotic	Class 2n	3.28 ± 0.07
	Class 4n	5.96 ± 0.07
Primary spermatocyte		6.28 ± 0.07
Secondary spermatocyte		3.35 ± 0.04
Spermatid		1.68 ± 0.02

From Pollister, A.W., Swift, H., and Alfert, M.: J. Cell Comp. Physiol., *38*(Suppl. 1):101, 1951.

prophase mitotic ones: (1) Although DNA duplication has already occurred and they have two chromatids, leptotene chromosomes look single rather than double, and (2) these chromosomes show bead-like thickenings, the so-called *chromomeres*, which appear at irregular intervals along their length (Fig. 16–3A). Because these beads are characteristic in size, number, and position for a particular chromosome, they may be used as landmarks to identify a specific chromosome of an organism. With the further contraction of chromosomes during zygonema and pachynema, chromomeres become larger and fewer in number.

Chromomeres were once thought to represent the genes. They are too few in number, however (1500 to 2500 in the lily), to include all the genes.

Frequently, leptonemic chromosomes have a definite polarization and form loops whose ends are attached to the nuclear envelope at points near the centrioles, contained within an aster. This peculiar arrangement is often called the "bouquet."

Zygonema—Pairing of Homologous Chromosomes and Synaptonemal Complex Formation

During *zygonema* (Gr., *zygon*, adjoining), the first essential phenomenon of meiosis occurs. The homologous chromosomes become aligned and undergo pairing in a process often called *synapsis of the chromosomes* (Fig. 16–2). Pairing is highly specific and involves the formation of a special structure that can be observed under the electron microscope (EM) and is called the *synaptonemal complex* (SC).[3]

The pairing does not have a special starting point. Sometimes the chromosomes unite at their polarized ends and continue pairing toward the antipodal extremity; in other cases, fusion occurs simultaneously at various places along the length of the thread. Polarization seems to favor regularity in pairing. Pairing is remarkably exact and specific; it takes place point for point, and chromomere for chromomere, in each homologue. The two homologues do not fuse during pairing, but remain separated by a space of about 0.15 to 0.2 μm, which is occupied by the SC.

First described by Moses in 1956, this complex is composed of two *lateral components* or arms and a *central* or *medial* element which are interposed between the two pairing homologues (Fig. 16–4). It should be emphasized that each lateral component is shared by the two sister chromatids of a homologue.

The morphology of the SC is very similar in plant and animal meiocytes. In cross section it can be observed that the SC is a flattened, ribbon-like structure. The lateral arms vary in width from 20 to 80 nm in various species. They are formed of electron-dense coarse granules or fibers. These arms are joined to the adjacent chromosomes by fine fibrils.

In most plants and animals—with the exception of the insects—the two lateral arms are separated by an axial space of lower density (Fig. 16–4). In insects the central component may be very complex. In this case, it has the aspect of a ladder, with three dense parallel lines and bridges crossing at intervals of 20 to 30 nm. These bridges are formed by fine fibrils that span the central and lateral components and are arranged perpendicularly to them (Fig. 16–5).

The SC can be considered the structural basis for pairing and synapsis of meiotic chromosomes. At the end of leptonema the lateral elements of the SC have already appeared in the space between the two chromatids, while the medial component appears, with the pairing, at zygonema.[4]

When pairing starts, the packing of the chromosomes is already very high, on the order of at least 300/1 (i.e., the DNA is 300 times longer than the length of the chromosomes). In other words, only 0.3% of the DNA of the homologous chromosomes is matched along the length of the

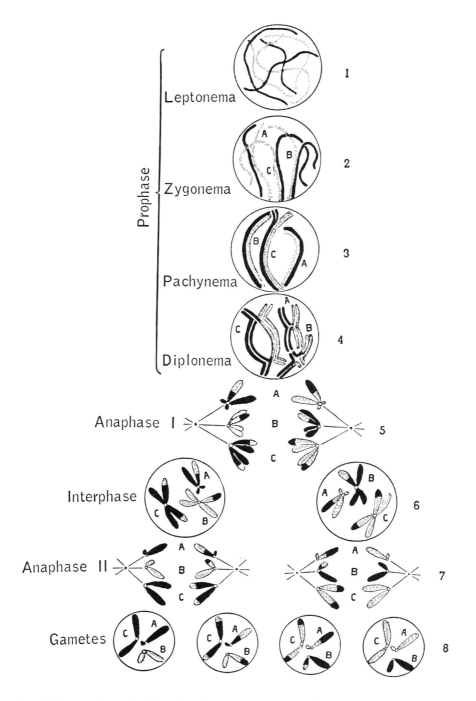

Fig. 16–2.   General diagram of meiosis, illustrating the union, separation, and distribution of the chromosomes.

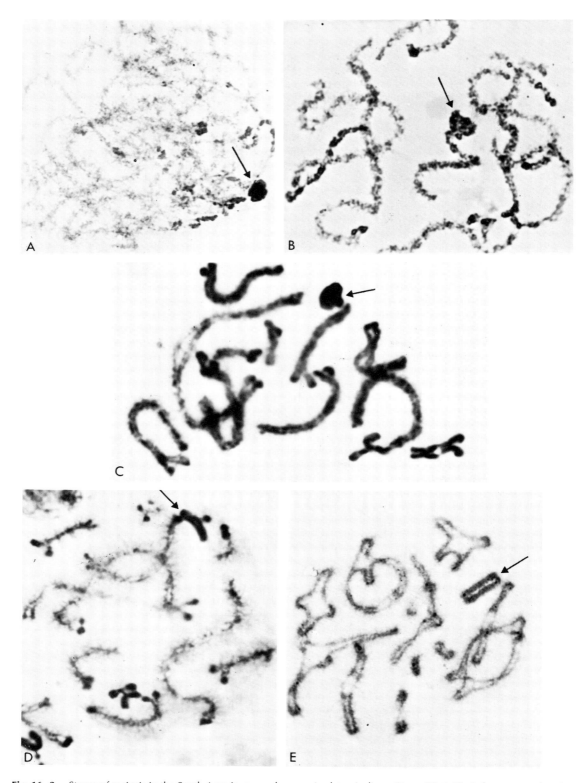

Fig. 16–3. Stages of meiosis in the South American grasshopper, *Laplatacris dispar* (2n = 22 + X). *A, leptonema,* showing the long, thin filaments with the chromomeres. (The X chromosome is indicated by an arrow in this and subsequent micrographs.) *B, pachynema,* showing thick filaments in which the homologous chromosomes have paired. *C, early diplonema,* in which the homologous chromosomes have shortened considerably and have begun to separate. *D, mid-diplonema,* showing the configuration of the bivalents with the chiasmata. *E, late diplonema,* showing the chiasmata in distinct form. (Microphotographs taken by the late professor and former coauthor F.A. Saez.)

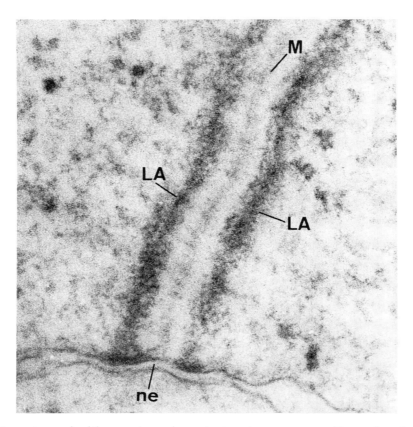

Fig. 16–4. Electron micrograph of the synaptonemal complex in a dog spermatocyte. The two lateral arms (*LA*), corresponding to the homologous chromosomes, are shown parallel to each other and extending into the nuclear envelope (*ne*). The medial element (*M*) is simpler than in certain invertebrates. ×125,000. (Courtesy of J.R. Sotelo.)

lateral components of the SC. Prezygotene condensation is a highly ordered process that follows a similar pattern in each homologue, so that the synapsis becomes specific for the corresponding set of genes.

The alignment of the homologous chromosomes is made easier by the fact that the telomeres of the chromosomes are frequently attached to the nuclear envelope. At these points of attachment the SCs are linked by an electron-dense thickening, the so-called *fixation plate*. The fixation of the telomeres to the nuclear envelope is accompanied by a redistribution of anuli during meiotic prophase. The anuli of the nuclear pores (Section 13–1) are no longer homogeneously distributed but tend to accumulate at the sites of fixation of the telomeres.

Methods to Study Synaptonemal Complexes

To visualize the SC two methods have been used. In Figure 16–5, a three-dimensional *re-construction* has been achieved from serial sections of an oocyte from the *Bombix mori* (silkworm). This method is rather time-consuming but gives excellent results. It has been found that, at leptonema, homologous chromosomes are distributed at random (i.e., they are not pre-aligned). In the *Bombix* cells the SC starts at the telomeres and progresses to the rest of the chromosome, being highly specific for homologous regions (Fig. 16–5). During zygonema most nuclei contain one or more bivalents that are interlocked; however, during pachynema, these interlockings are resolved.[5]

The second and easier technique consists in the spreading of the nuclei at a water-air interface. This produces the dispersion of all the bivalents with the corresponding SC, which can then be recognized without overlappings.[6] *Microspreading* permits one to observe all the SCs present in a karyotype and even to recognize a variety of chromosomal aberrations, such as tan-

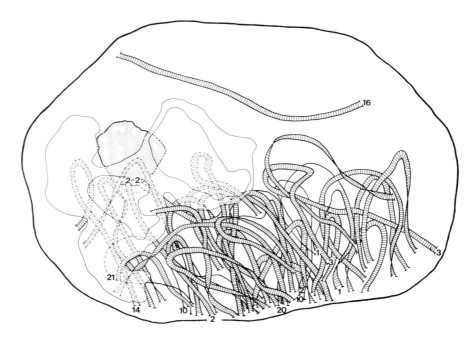

Fig. 16–5. Reconstruction of the late zygonema stage of meiosis in the *Bombyx mori* (silkworm) female. All homologous chromosomes are paired, with the exception of pairs 1, 2, 5, 10, and 17. In those that are paired a full synaptonemal complex is present; in those with unpaired regions, the lateral arms of the *SC* are present. Note that most of the telomeres are attached to a region of the nuclear envelope, forming a *bouquet* figure. ×11,500. (Courtesy of S.W. Rasmussen.)

dem duplications and inversions (see Section 17–2). These appear as loops or buckles in the corresponding SC.[7,8]

Figure 16–6 shows all the SCs of a human karyotype. The figure, made from a spermatocyte in the pachytene stage, shows the 22 pairs of homologues and the XY pair. Special features that the student should recognize in this figure are: (1) the *kinetochores*, which appear as stained prominences in the lateral elements of the SCs; (2) the nucleolar organizers, in pairs 13, 14, and 21; and (3) the XY pair, with a short region that has synapsed, and a long portion of SC of the X chromosome that has not synapsed (see Solari, 1980 in the Additional Readings at the end of this chapter). The chemical nature of the SC has been studied through the use of deoxyribonuclease to digest DNA and the heavy metal indium to stain nucleic acids. The results suggest that the main component of the SC is protein in nature. This protein is basic and probably similar to the histones. The fine fibers that cross between the two lateral arms and connect with the chromosomes probably contain DNA.

The Synaptonemal Complex and Homologue Alignment

We have already mentioned that the lateral arms of the SC appear in leptonema, before pair-

ing, and that completion of the SC occurs during zygonema and becomes more conspicuous at pachynema. After this stage, during diplonema, the SC disintegrates and usually disappears.

One function of the SC appears to be to stabilize the pairing of the homologues; another, to facilitate recombination. Because of the deposition of new protein molecules in the lateral arms of the SC, the matching segments of DNA are placed in a way that allows interchange at the molecular level. The SC can be interpreted as a protein framework that permits not only the proper alignment of the homologues but also their recombination.

Organization of Chromatin in Early Meiotic Prophase

Before we continue with the description of meiosis, it is important that the student remember what was said about the macromolecular organization of chromatin in the interphase nucleus of mitotic cells (Section 13–2). In cells that undergo meiosis the basic structure is the same; that is, the thin chromatin fiber is composed of a tandem array of nucleosomes (Fig. 13–9), folded to make the 30-nm fiber (Fig. 13–12).

In early meiotic prophase, the chromatin fi-

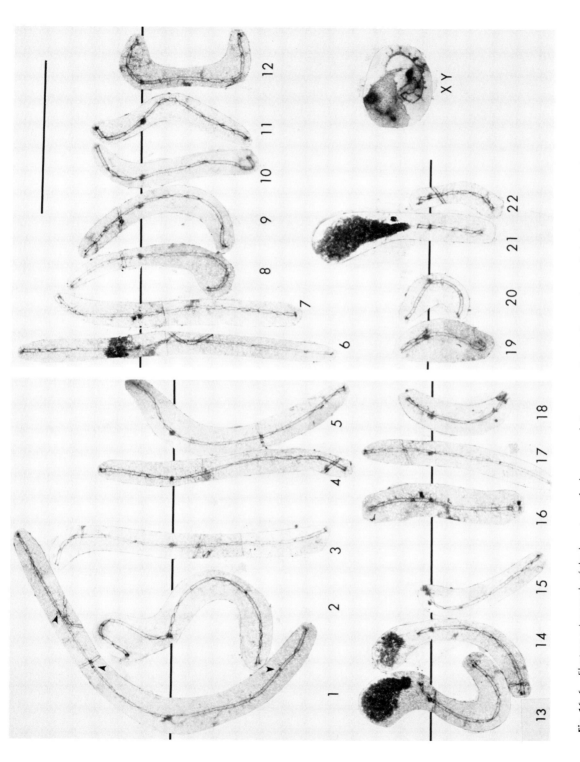

Fig. 16–6.   Electron micrograph of the karyotype of a human pachytene spermatocyte. Preparation made by the microspreading technique, followed by staining with phosphotungstic acid. The 22 pairs of autosomes and the XY pairs are shown in decreasing order of size. The lines pass through the kinetochores. The SCs of chromosomes 13, 14, and 21 show the nucleolar organizers. In 1, recombination nodules are marked with arrowheads. Bar = 10 μm. (Courtesy of A.J. Solari.)

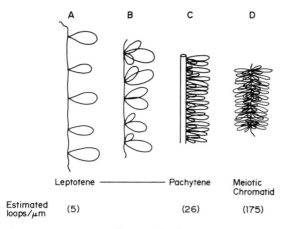

Fig. 16–7. Diagram showing the loop organization of chromatin in meiotic prophase. From leptotene to pachytene there is an increase in the concentration of loops. Observe that, at pachytene, the lateral element of the SC has appeared. (Courtesy of J.B. Rattner, M. Goldsmith, and B.A. Hamkalo.)

bers are in the form of discrete loops, each of which contains between 5 and 25 μm of DNA. As shown in the diagram of Figure 16–7A, during leptonema the loops of chromatin are arranged as loose linear arrays. Later on, the loops tend to concentrate at certain regions or foci that have electron-dense centers. At such foci, during pachynema the lateral arms of SC are deposited (Fig. 16–7B–C). This figure clearly shows that during meiotic prophase there is a progressive packing of the loops.[9]

At pachynema, the homologues, observed under the EM, show a brush-like appearance, with the loops of chromatin inserted into the lateral elements of the SC. Figure 16–8 shows these details clearly and also shows that there are regions of the loops that are undergoing transcription into RNA. The transcription is revealed by the presence of nascent ribonucleoprotein fibrils inserted into some chromatin loops[10,11] (see Hamkalo, 1985 in the Additional Readings at the end of this chapter).

The loop organization of the chromatin fiber of both mitotic and meiotic chromosomes suggests that this is a universal mechanism by which these fibers are folded into higher-order structures. As in mitotic chromosomes (Fig. 13–20), it is possible that in meiotic ones a non-histone protein scaffold also exists.[12,13] In meiosis, however, there is the further addition of a protein structure that constitutes the SC.

Pachynema—Crossing-Over and Recombination between Homologous Chromatids

During pachynema (Gr., *pachus*, thick), the pairing of the chromosomes reaches completion (Figs. 16–2 and 16–3B). The chromosomes contract longitudinally so that the threads become shorter and thicker. By middle pachynema because of the pairing of the homologues, the nucleus contains half the number of chromosomes. Each unit is a *bivalent* or *tetrad* composed of two homologous chromosomes in close longitudinal union and four chromatids.

Each chromatid has its own kinetochore. Thus in a tetrad there are four kinetochores, two homologous and two sister kinetochores (Fig. 16–9,2). (The two chromatids of each homologue are called *sister chromatids*.) During the first meiotic division the kinetochores of the two homologous chromatids behave as a functional unit (Fig. 16–9,3 through 5). In late pachytene a line of separation perpendicular to the plane of pairing appears, and the four chromatids become visible. During pachynema, the space occupied by the SC is maintained, and all the homologous chromosomes have finished the pairing process.

Experimental evidence suggests that during pachynema two of the chromatids of the homologues exchange segments (i.e., recombine). It is thought that transverse breaks occur at the same level on each of the chromatids and that this event is followed by interchange and final fusion of the chromatid segments (Fig. 16–9).

Pachynema may last for days, weeks, or even years, whereas leptonema and zygonema may last only a few hours.

The Recombination Nodule in the SC is Probably Related to Crossing-Over

Electron microscopic studies of meiosis at the pachytene stage have revealed the presence of dense nodules of about 100 nm in intimate association with the SC. The number of these nodules, called *recombination nodules*, and their location along the bivalents are related to the number and distribution of the genetic exchanges (cross-overs). It has been suggested that these nodules may be involved in the recombination process during pachytene (see Carpenter, 1975; Zickler, 1977 in the Additional Readings at the end of this chapter).

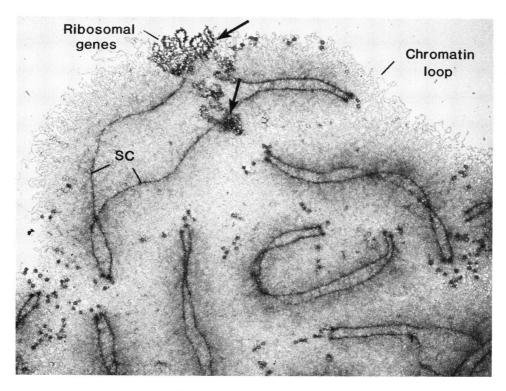

Fig. 16—8.   Meiotic prophase from a testis of the *Bombyx mori* (silkworm) showing several bivalents with the SCs and the chromatin attached in loops. At the points indicated with arrows, there is an active transcription of ribosomal genes, with the formation of ribonucleoproteins. (Courtesy of J.B. Rattner, M. Goldsmith, and B.A. Hamkalo.)

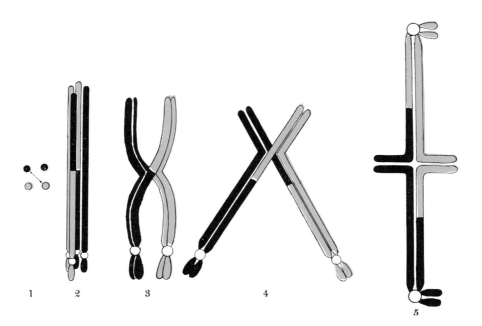

Fig. 16—9.   *1* and *2,* diagrams showing the process of crossing-over; *3,* formation of a chiasma; *4,* terminalization; *5,* rotation of the chromatids of one bivalent.

In human chromosomes these structures appear as electron-dense *bars* located across the two lateral arms of the SC, or they are longitudinally disposed between them (Fig. 16–6). The number of bars is in the range of the number of chiasmata or crossings (i.e., about 50) observed at diplonema (see below). This is a strong indication that the recombination nodule is the site at which crossing-over takes place.[14]

At the molecular level, for the recombination to occur, the DNA fibers of both homologous chromatids must reach a distance of about 1 nm in the central component of the SC. It is thought that the contact between chromatids may be achieved by the transverse bridges, which connect the homologue chromatids with the lateral arms and the central component of the SC (Fig. 16–4). At these points homologous sequences of nucleotides could search for each other to finally exchange DNA segments. The recombination nodule or bar could represent a morphological expression of such an exchange.

The nodule has also been interpreted as the site of a large multi-enzyme complex that could be involved in the bringing together of the maternal and paternal chromatids across the 100-nm width of the SC. (For the molecular mechanism of recombination see Fig. 16–18.)

Diplonema—Separation of the Paired Chromosomes Except at the Chiasmata

At *diplonema* the intimately paired chromosomes repel each other and begin to separate (Fig. 16–3C through E). However, this separation is not complete, because the homologous chromosomes remain united by their points of interchange, or *chiasmata* (Gr., *chiasma*, crosspiece). Chiasmata are generally regarded as the sites where the phenomenon of *crossing-over*, or *recombination*, takes place, by which chromosomal segments with blocks of genes are exchanged between homologous members of the pairs. With few exceptions, chiasmata are found in all plants and animals. At least one chiasma is formed for each bivalent. Their number is variable; some chromosomes have one chiasma and others have several. During diplonema the four chromatids of the tetrad become visible and the SC can no longer be observed (Fig. 16–10).

Diplonema is a long-lasting period. In the fifth month of prenatal life, for example, human

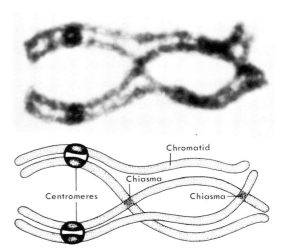

Fig. 16–10. *Above,* photomicrograph of a single bivalent at diplonema in a salamander spermatocyte. *Below,* diagram interpreting the photomicrograph. The four chromatids are clearly apparent. Two chiasmata and the positions of the centromeres are also seen. Observe that the two sister centromeres are side by side. (Courtesy of J. Kezer.)

oocytes have reached the stage of diplonema and remain in it until many years later, when ovulation occurs. In most species there is uncoiling of the chromosomes during diplonema. In fish, amphibian, reptilian, and avian oocytes the uncoiling becomes so marked that the greatly enlarged nucleus assumes an interphase appearance. In these cases the bivalent chromosomes may attain a special configuration known as the *lampbrush chromosome,* in which the chromonema uncoils into loops that converge upon a more coiled axis. It will be shown later that the presence of these lampbrush chromosomes is related to an intensive RNA synthesis and to the enormous growth of the oocyte (see Section 22–5).

Diakinesis—Reduction in the Number of Chiasmata

In diakinesis (Gr., *dia*, across) the chromosomes again contract (Fig. 16–11B). The tetrads are more evenly distributed in the nucleus, and the nucleolus disappears. During this period the number of chiasmata diminishes. By the end of diakinesis, in general the homologues are held together only at their ends (Fig. 16–9,4 and 5), a process that has been named *terminalization.*

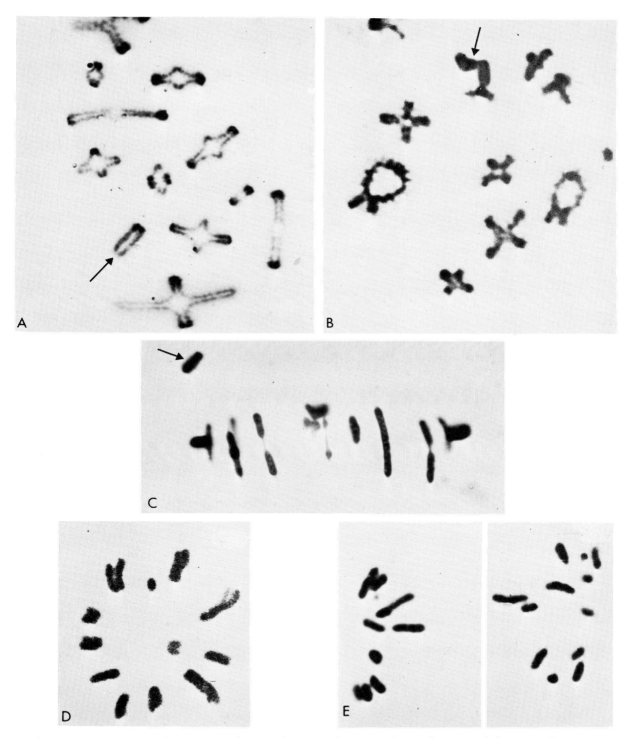

Fig. 16–11. Continuation of Figure 16–3, showing other stages of meiosis in the grasshopper. *A, diplonema,* with Giemsa stain showing the constitutive heterochromatin localized at the centromere regions (C bands). *B, diakinesis,* showing the condensation of the bivalents. *C, metaphase I,* side view. The arrows point to the X chromosome in A, B; in C it points to the X-chromosome that is advancing toward the pole; all the autosomes are still in the equatorial plane. *D, metaphase II,* polar view. *E, anaphase II,* side view. At this moment each chromosome consists of a single chromatid, and each daughter cell has the haploid number of chromosomes. (Micrographs taken by the late professor and former coauthor F.A. Saez.)

Meiotic Division I—Separation of the Homologous Kinetochores

In *prometaphase I* condensation of the chromosomes reaches its maximum. The nuclear envelope breaks down, and the spindle microtubules become attached to the kinetochores. Each homologue is attached to one of the poles by the homologous kinetochore; thus the two sister chromatids behave as a functional unit.

In *metaphase I* the chromosomes become arranged at the equator. If the bivalent is long, it presents a series of anular apertures between the chiasmata in perpendicularly alternating planes. If the chromosomes are short, they have a single anular aperture. Meiotic metaphase I can be distinguished by the fact that the homologues are still attached by chiasmata at their ends while the kinetochores are being pulled toward the poles (Fig. 16–11*B* and *C*).

In *anaphase I* the sister chromatids of each homologue, united by the kinetochore, move toward their respective poles. The short chromosomes, generally connected by a terminal chiasma, separate rapidly. Separation of the long chromosomes, which have interstitial and unterminalized chiasmata, is delayed.

It should be recalled that by recombination, segments were transposed between two of the chromatids of each homologue. Thus, when the homologous paternal and maternal chromosomes separate in anaphase, two of their chromatids are mixed; the other two maintain their initial nature (Figs. 16–2 and 16–9).

Telophase I begins when the anaphase groups arrive at their respective poles. Chromosomes may remain in a condensed state for some time.

Following telophase is a short *interphase*, which has characteristics similar to mitotic interphase.

The result of the first meiotic division is the formation of the daughter nuclei, which in animals are called secondary spermatocytes (in the male) and oocyte plus the first polar body (in the female).

At the interphase between the two meiotic divisions *there is no replication of the chromosomes*. These are now haploid in number, although each one consists of two chromatids.

Meiotic Division II—Separation of the Sister Kinetochores

The short prophase II is followed by the formation of the spindle, which marks the beginning of metaphase II. At metaphase II (Fig. 16–11*D*) chromosomes become arranged on the equatorial plane, and the sister kinetochores separate; the two sister chromatids go toward the opposite poles during anaphase II (Fig. 16–11*E*).

Each of the four nuclei of telophase II has one chromatid, and each nucleus has a haploid number of chromosomes.

Chromosome Distribution in Mitosis and Meiosis Is Dependent on Kinetochore Orientation

We have just said that in meiotic division I there is a separation of the homologous kinetochores (and chromatids), while in meiotic division II, as in mitosis, separation of the sister kinetochores (and chromatids) occurs. The differences observed are of great importance in explaining the general mechanism of chromosome distribution in mitosis and meiosis.

Based on early observations by Schrader in 1936, Ostergren in 1951 postulated that the differences in chromosome distribution in mitosis and meiosis resulted from the different orientation of the kinetochores. In fact, in mitosis the kinetochores of the sister chromatids are arranged back to back (Fig. 15–7); in meiotic division I the kinetochores of the sister chromatids lie side by side (Fig. 16–10).

The factor that determines the distribution of the chromosomes is their initial interaction with the spindle. This interaction occurs by way of spindle microtubules that run from the kinetochore toward one of the poles and that determine the particular movement of the given chromosome.

The problem of chromosome distribution has been studied in experiments using microsurgery in grasshopper spermatocytes. Meiocytes in the I and II divisions were fused by mechanical means to produce one cell containing the two spindles—one spindle with bivalents, the other with unpaired chromosomes. In each spindle the chromosome behaved as in the original cells. Furthermore, it was possible to detach a bivalent from division I and place it in the division II spindle. In this arrangement the attached bivalent also divided in the original fashion (i.e., the two sister chromatids moved together to each pole); while in all the other chromosomes, single sister chromatids moved to each pole (see Nicklas, 1977 in the Additional Readings at the end of this chapter). These and other experiments suggest that in the mechanism of chromosomal distribution the key role is played by the orientation of the kinetochore, which determines the nucleation and preferential polarization of microtubules.

SUMMARY:
Events in Meiosis

Meiosis is a special type of cell division present in germ cells of sexually reproducing organisms. Sexual reproduction occurs by way of sexual cells or gametes (eggs and sperms), which after fertilization constitute the zygote. Meiosis consists of a single duplication of the chromosomes followed by two consecutive divisions. In animals and lower plants meiosis is *terminal or gametic* (i.e., it occurs before the formation of the gametes). In the male, four haploid sperms are produced; in the female, one ovum and three polar bodies are produced. In most plants meiosis is *intermediary* or *sporic* (i.e., it occurs sometime between fertilization and the formation of gametes). Cells in meiosis are called *meiocytes*. The essential differences between mitosis and meiosis are: (1) Mitosis occurs in somatic cells, and meiosis is limited to germinal cells. (2) In mitosis, one replication cycle of DNA is followed by one division, resulting in diploid cells. In meiosis, one replication of DNA is followed by two divisions, resulting in haploid cells. (3) In mitosis DNA replication occurs in the S phase, which is followed by G_2. In meiosis there is a premeiotic DNA synthesis, which is very long and followed immediately by meiosis. (4) In mitosis each chromosome behaves independently, and in meiosis there is pairing of homologous chromosomes. (5) Mitosis lasts 1 to 2 hours, and meiosis lasts longer; in the male it may last 24 days; in the female, several years. (6) In mitosis the genetic material remains constant, and in meiosis there is genetic variability.

The essential processes of meiosis are (1) pairing of homologous chromosomes, (2) formation of chiasmata with underlying genetic recombination, and (3) the segregation of the homologous chromosomes.

Meiosis is divided into divisions I and II. In division I there is a long prophase of which the stages are preleptonema and leptonema, zygonema, pachynema, diplonema, and diakinesis. Leptotene chromosomes look single (in spite of their two chromatids), and they have chromomeres that are characteristically placed for each chromosome. The unit fiber of chromatin is folded at the chromomeres. Sometimes leptotenes are polarized, forming the so-called "bouquet."

During *zygonema*, pairing and synapsis of the homologues occur. Pairing involves the formation of the SC. This is composed of two lateral arms (which appear in each homologue at the end of leptonema) and a medial element. At the time of pairing the packing of DNA is 300/1 and there is only 0.3% matching between homologous DNA. Pairing starts at random, but telomeres are generally inserted at the nuclear envelope. The 0.2 µm space between homologous chromosomes is occupied by the SC. This may have the appearance of a ladder, with bridges crossing the medial element. The main component of this complex is protein.

Three-dimensional reconstructions of all bivalents have been made by serial sections (Fig. 16–5) and by a special microspreading technique that permits observation of all the SCs in a karyotype (Fig. 16–6).

The macromolecular organization of meiotic chromosomes is similar to that of mitotic ones with a 20- to 30-nm chromatin fiber composed of nucleosomes. In early prophase, the fibers make discrete loops that coalesce in dense foci, where the lateral arms of the SC are deposited. The packing of these loops increases throughout meiotic prophase, and some RNA transcription may be seen in certain loops corresponding to active nucleolar genes.

During *pachynema*, pairing is complete and chromosomes become shorter and thicker. The number of chromosomes has been halved (i.e., bivalents or tetrads). Each tetrad has four kinetochores (two homologous and two sister). During pachynema, two homologous chromatids exchange segments at a molecular level (recombination). The SC appears to stabilize the pairing, thus enabling the interchange.

At pachynema the SC may show the recombination nodules or bars, which may be the actual sites of crossing over.

During *diplonema*, the paired chromosomes begin to separate, but they are held together at the *chiasmata* (points of interchange or *crossing over*). There is at least one chiasma per bivalent chromosome. At diplonema, the SC is shed from the bivalents. At each chiasma there is a piece of SC that ultimately disappears and is replaced by a chromatin bridge. Diplonema may last for months or years.

During *diakinesis* there is a reduction in the number of chiasmata and further contraction. This is followed by prometaphase I, metaphase I, anaphase I, and telophase I. In anaphase I the sister chromatids of each homologue move to the respective poles (segregation of the homologues).

During the interphase between the two meiotic

divisions there is no replication of the chromosomes, and division II is very similar to mitosis, by the end of which each nucleus has a haploid number of chromosomes composed of single chromatids.

The fact that in meiotic division I there is separation of the homologue kinetochores (and chromatids) and that in division II the sister kinetochores (and chromatids) are separated can be explained by the orientation of the kinetochores, which determines nucleation and preferential polarization of microtubules.

16–3 GENETIC CONSEQUENCES OF MEIOSIS AND TYPES OF MEIOSIS

We said at the beginning of this chapter that the essential processes of meiosis were (1) *pairing* or *synapsis*, (2) formation of *chiasmata* and *recombination*, and (3) *segregation of the homologous chromosomes*. Meiosis not only results in the halving of the chromosome number but also segregates each member (i.e., chromatid) of the homologous pair into four different nuclei. One point of special interest is that the homologous kinetochores separate in anaphase I, and the sister kinetochores, in anaphase II; but because crossing-over mixes the homologous chromatids of the bivalents, both meiotic divisions are needed to segregate the genes contained in the chromatids. (The student can follow this process by studying closely the distribution of chromosome *a* in Fig. 16–12.)

We shall now see that segregation of the homologues and recombination have important genetic consequences. From the genetic viewpoint, meiosis is a mechanism for distributing the genes between the gametes which permits their recombination and random segregation.

Figure 16–12 is a diagram in which the genetic consequences of the meiosis of three pairs of chromosomes having one, two, and three chiasmata can be observed. The chromosomes are followed through diplonema, metaphase I, and anaphase I, and then into anaphase II. It is obvious that each of the four gametes has a different genetic constitution. As a result of the crossing-over, each chromosome does not consist solely of maternal or paternal material but of alternating segments of each.

The random assortment of the genes is due not only to crossing-over but also to the random distribution of the chromosomes in the first and second division: (In Figure 16–12 to simplify the matter, the three paternal and maternal homologues are shown separating into different cells.) However, since this separation is a random process, the resulting cells will contain eight (i.e., 2^3) different chromosomal combinations, even in the absence of crossing-over. In a human (with 23 pairs of chromosomes), the possible chromosomal combinations in gametes will be an immense number, 2^{23} or 8,388,608. Meiosis is the mechanism by which genetic variation is brought about, and knowledge of its mechanism is a prereqisite for the understanding of the chromosomal basis of genetics.

Meiosis Is Intermediary or Sporic in Plants

We mentioned earlier in this chapter that intermediary or sporic meiosis is characteristic of flowering plants. As shown in Figure 16–13 meiosis takes place some time between fertilization and the formation of gametes.

In higher plants the reproductive organs—anthers in male and ovary or pistil in female—produce microspores and megaspores, respectively. The cells that undergo meiosis to produce megaspores are called megasporocytes. Microspores are produced by microsporocytes (pollen mother cells). Each microsporocyte gives rise, by meiosis, to four functional microspores. Each megasporocyte produces four megaspores by meiosis, of which three degenerate. The remaining megaspore develops into the female gametophyte, which gives rise to the egg cell.

In plants, microspores and megaspores are not the final gametes. Before fertilization, they undergo two mitotic divisions in the anther (not shown in Fig. 16–13) or three in the ovary to produce the male and female gametophytes, respectively.

Fertilization in plants is a complex phenomenon. Each pollen grain or *microspore* (haploid) carries two sperm nuclei. One of them fertilizes the egg nucleus to give a *diploid zygote*, which will eventually form the embryo of a new plant (i.e., *sporophyte*). The other sperm nucleus fuses with two polar nuclei to form a *triploid endosperm nucleus*, which by mitotic divisions will give rise to the *endosperm*, which contains the nutritive material for the embryo. Thus the seed is a mosaic of tissues consisting of the diploid zygote, the triploid endosperm, and the dip-

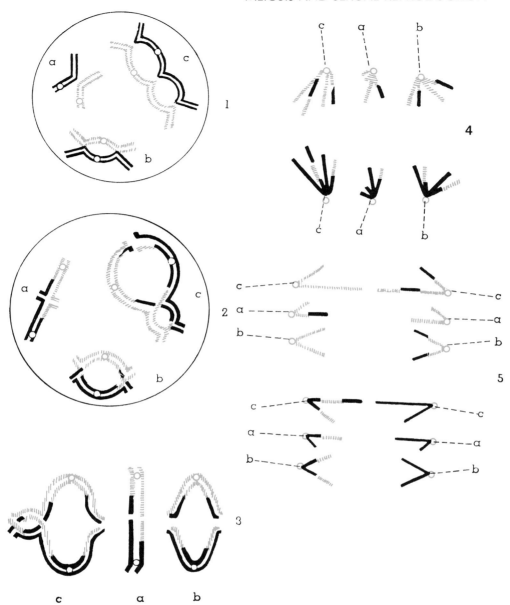

Fig. 16–12. Diagram showing the genetic consequences of the meiosis of three pairs of chromosomes with (a) one chiasma, (b) two chiasmata, and (c) three chiasmata; *1*, diplonema; *2*, advanced diplonema showing the process of terminalization; *3*, metaphase I; *4*, anaphase I; *5*, anaphase II, showing the distribution of the chromosomes in the four nuclei formed. *Solid line*, the paternal chromosomes; *dashed line*, the maternal chromosomes. The kinetochore is represented by a *circle*.

loid integuments, which are of maternal origin (Fig. 16–13).

Meiosis May Last for 50 Years in the Human Female

The primary female germ cells (i.e., *gono-cytes*) appear in the human embryo in the wall of the yolk sac at about 20 days and during the fifth week migrate to the gonadal ridges. By mitosis, they form oogonia and become surrounded by the follicular cells, forming the *primary follicles*. At the end of the third month of prenatal development, oogonia enter meiosis, becoming oocytes I, and are then arrested at the stage of diplonema until sexual maturity at about 12

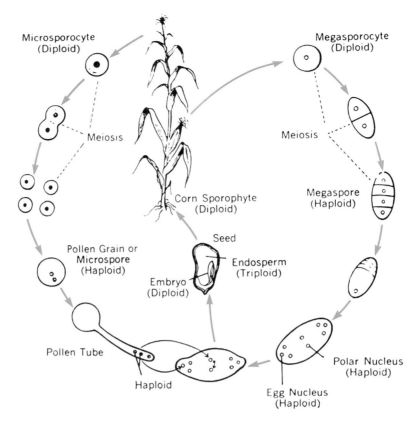

Fig. 16–13.   Life cycle of a plant. (Modified from Sinnott, E.W., Dunn, L.C., and Dobzhansky, T.: Principles of Genetics. 5th Ed. New York, McGraw-Hill Book Co., 1958.)

years. During this long phase, the chromosomes remain in a rather uncondensed state and resemble the *lampbrush chromosomes* (see Section 22–5).

The number of oocytes in a newborn female has been estimated at about one million; however, most of these degenerate. By the age of 7 years, there are some 300,000 oocytes, but only about 400 reach maturity between 12 and 50 years. Thus meiosis may last as long as 50 years. This may explain the increase in the incidence of chromosomal aberrations with the increasing age of the mother (see Chapter 18).

When the ovum is released into the *oviduct*, the first meiotic division occurs, giving one polar body (Fig. 16–1). It is only when the ovum is fertilized by the spermatozoon that the second meiotic division takes place, giving the second polar body. As a result of oogenesis, only one viable egg is produced; the haploid nucleus is known as the *female pronucleus* (Fig. 16–1). The polar bodies degenerate and do not take part in embryogenesis.

Meiosis Starts after Puberty in the Human Male

In the human male the primitive gonad also starts with the migration of the primary gonocytes into the gonadal ridges, and later these are incorporated into the *seminiferous tubules*. These contain spermatogonia that enter meiosis at puberty. In contrast to oogenesis in the female, spermatogenesis continues in the male until advanced age. Completion of this process takes about 24 days, and the meiotic prophase I occurs in 13 to 14 days. Four viable spermatozoa are formed from each meiotic cycle (Fig. 16–1). The transformation of spermatids into mature sperm cells occurs by a complex process called spermiogenesis, which can be studied in histology textbooks.

Cell Biology of Fertilization

Fertilization of the egg by the sperm is the first step in the process of embryonic development and for this reason is dealt with mostly in embryology textbooks. Here we shall present an overview of the cellular and molecular aspects of fertilization.

In fertilization there are two coordinated phenomena. First, by self propulsion from its flagellum, the sperm reaches the egg surface, achieves contact, and, after membrane fusion, becomes incorporated into the cytoplasm of the egg. Second, the two haploid nuclei fuse in a process called *syngamy* and a diploid nucleus results, which soon starts the first division. An important point to remember is that fertilization is species-specific, in the sense that the sperm from one species cannot fertilize the egg of another. This is the main reason for the stability of a species (see Section 17–3).

The two actors are the male gamete represented by the spermatozoon and the unfertilized egg, of which more will be said in Chapter 23. The sperm is a polarized cell having an undulating tail or flagellum, whose motility is similar to that of cilia (Section 6–3), a head that contains the compact nucleus, and the *acrosome*, a derivate from the Golgi apparatus that plays an important role in fertilization. In this region there is also a pool of monomeric actin that is activated and polymerized into microfilaments at a certain moment. Under the influence of egg-associated factors the sperm undergoes the so-called *acrosome reaction* by which there is a secretion of material from the acrosomal vesicle and elongation of the acrosomal process by the microfilaments; the elongated acrosome establishes initial contact with the egg surface. In mammals the acrosome reaction is prevented by factors at the surface of the epididymis that render the sperm incompetent to react. These factors are removed within the oviduct tract, and the sperms become competent to have the acrosomal reaction; this is a process called *capacitation* of the sperms.

In sea urchins the acrosomal reaction results in the liberation of a species-specific egg-binding protein, *bindin*, and a protease called *acrosomin*. This protease is apparently responsible for the initial digestion of the vitelline layer that covers the unfertilized egg; bindin may play a role in causing adhesion of sperms to eggs of the same species.

Figure 16–14 summarizes the steps that can be recognized in fertilization by microscopic methods, including scanning electron microscopy. In Figure 16–14A, two spermatozoa are attached to the egg surface, and their tails are moving. In *B–C*, the tail motility stops in the sperm undergoing fusion (this may occur within 13 seconds in a sea urchin). In *D*, there is a rapid cortical contraction irradiating from the point of fusion and the elevation of a fertilization coat by the discharge, *by exocytosis of cortical granules* into the space between the vitelline layer and the plasma membrane. This coat impedes the penetration of other sperms into the egg, thus preventing polyspermia. In *E–F*, penetration of the sperm is produced by the formation of a *fertilization cone* at the site of fusion. This cone results from the elongation of microvilli (having a core of microfilaments) that surround the sperm. In *G–H*, after its penetration into the egg cortex, the sperm glides and rotates so that the centrioles move closer to the female pronucleus. In *J–K*, the pair of centrioles, serving as basal body for the sperm tail, originate an aster, and the male pronucleus is moved centripetally. In *L–M*, the two pronuclei are interconnected by the aster microtubules and are moved toward the center of the egg. At this point the unfertilized egg does not contain morphologically apparent centrioles; for the development of a normal bipolar mitotic apparatus, the egg must use the pair of centrioles contributed by the sperm. In *N*, syngamy occurs at the egg center after disassembly of the aster microtubules. In *O*, the zygote nucleus is distorted by a streak of microtubules. In *P*, the streak is disassembled and breakdown of the nuclear membrane occurs. Finally in *Q*, the spindle is formed and cleavage occurs. Figure 16–15 shows scanning electron micrographs of different stages of the penetration of the sperm into the egg.

The complex series of movements described demonstrate that the cytoskeleton, with its two main components, microtubules and microfilaments, is involved in all the steps of fertilization. Studies using video microscopic data and microelectrodes to record bioelectrical responses have shed some light on the initial phenomenon that activates the egg. Without going into much detail, it appears that after the acrosome reaction there is an elevation of Ca^{++} and an increase in pH inside the sperm and that fusion produces a sudden and localized increase in cytoplasmic Ca^{++}, sufficient to trigger the exocytosis of the

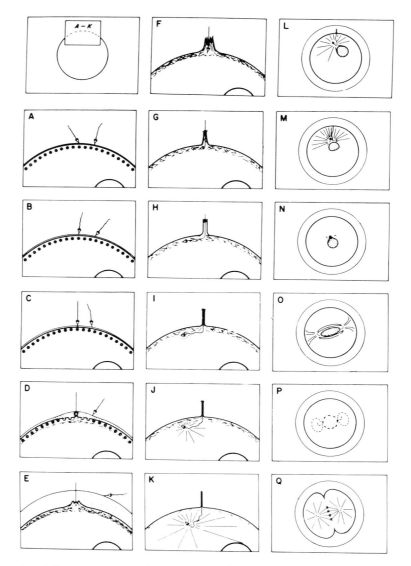

Fig. 16–14. This series of diagrams represents the main steps in fertilization: the attachment of two sperms; the fusion of one; the cortical reaction with the exocytosis of the cortical granules; the formation of the fertilization cone, made of microvilli, that cause the penetration of the sperm; the formation of the aster; and the syngamy and final cell division of the zygote nucleus. See the text for details. (Courtesy of G. Schatten.)

cortical granules. This is followed by a propagating wave of Ca^{++} released from intracellular stores and by an increase in cytoplasmic pH. These two components (Ca^{++} and pH) appear to be the regulators that trigger the various cytoskeletal changes occurring in fertilization. In addition to protons and Ca^{++}, calmodulin and possibly cyclic nucleotides are also involved in these complex molecular mechanisms.

Centrosomes and fertilization: Since the early studies of Boveri in 1904, the question of the origin of centrosomes has been raised. In his

studies in sea urchin eggs he proposed a paternal contribution with the introduction of the centrosome at fertilization, together with the sperm centriole. However, the situation in mammalian oocytes seems to be different and, at least in the mouse, the centrosome appears to be of maternal origin.

This problem has recently been studied using an anticentrosomal antiserum derived from patients with the disease scleroderma that specifically stains the centrosomes. With this technique it has been possible to confirm that while

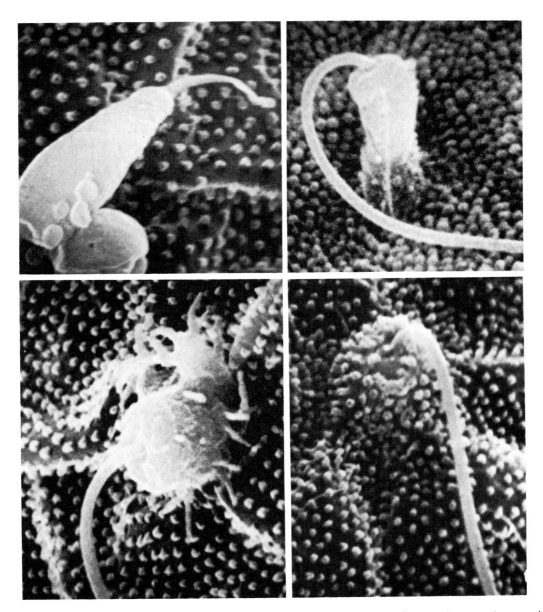

Fig. 16–15. Four stages in the penetration of a sperm into the egg studied by scanning electron microscopy in sea urchin. *Top left,* the acrosome of the sperm elongates and establishes the initial contact with the egg surface. *Top right,* the sperm head starts to penetrate the egg; the membrane from the egg now surrounds the anterior portion of the sperm. *Bottom left,* further penetration; microvilli elongate around the sperm, which is being drawn in. *Bottom right,* the sperm head has completed its entry, and the underlying microvilli have withdrawn. On the surface of all the electron micrographs observe the tips of the papillae of the vitelline sheath. (Courtesy of G. Schatten.)

in the sea urchin the centrosome is from paternal origin, in the mouse oocyte it comes from maternal line. Furthermore, strong parallels between the chromosome and centrosome cycles have been observed. (For further readings on this interesting problem see Schatten et al., 1986 in the Additional Readings at the end of this chapter).

Changes in the Nuclear Lamina in Meiosis

In Chapter 13 we mentioned that the interphase chromatin is probably held, in a three-dimensional ordered state, by its association with the nuclear lamina, a proteinaceous layer under the nuclear envelope. Because of the movements that the chromosomes must undergo during mitosis and meiosis it is understandable that the *lamins* (the proteins of the lamina) undergo depolymerization and disaggregation. While in mitosis this occurs simultaneously with the removal of the nuclear envelope, to reappear together with the nuclear envelope at telophase, in meiosis the lamina disappears during the long prophase and is completely absent in spermatocytes and spermatids (see Stick and Schwartz, 1982 in the Additional Readings at the end of this chapter). In contrast, in the female meiotic cell the lamina disappears early in prophase but then reappears at the diplotene stage. This is clearly shown in Figure 16–16 in which two oocytes in the pachytene stage are completely devoid of lamins; on the other hand lamins are abundant in the surrounding somatic nuclei. In a diplotene nucleus the immunological stain again reappears. It should be emphasized that while these changes occur in the lamina, the nuclear envelope, including the pore complexes, remains unchanged (see Stick and Schwartz, 1983 in the Additional Readings at the end of this chapter).

SUMMARY:
More about Meiosis

Meiosis not only results in the halving of the number of chromosomes but also in the segregation of each chromatid into a different nucleus. Meiosis is a mechanism for distributing the genes between gametes, enabling their recombination and random segregation. In this process, genetic variation is produced.

The study of meiosis is a prerequisite for understanding the chromosomal basis of genetics. Only after the process of meiosis is understood will its significance in hereditary phenomena become apparent (see Chapter 17).

In the human female, gonocytes appear in the embryo at 20 days and migrate into the gonadal ridges. At the third month, the oogonia enter meiosis, which is arrested at diplonema until sexual maturity (at about 12 years). The chromosomes remain uncondensed resembling lampbrush chromosomes. Most oocytes degenerate, and only about 400 reach maturity between 12 and 50 years. Thus meiosis may last for 50 years! When the ovum is released from the follicle, it produces the first polar body. The second meiotic division occurs only with fertilization.

In the human male, the primary gonocytes become incorporated into seminiferous tubules.

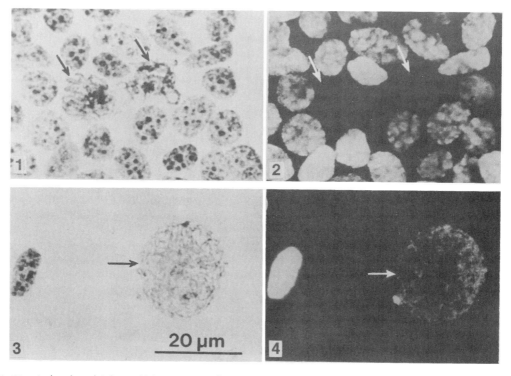

Fig. 16–16. Isolated nuclei from chick ovary. *1* and *3* are stained with orcein; and *2* and *4* with a fluorescent antibody against proteins of the nuclear lamina. Comparing *1* and *2* it is observed that, while somatic nuclei give a positive lamina reaction, two oocytes in pachynema lack the lamina. In *3* and *4*, one nucleus in diplonema gives a faint reaction. (Courtesy of R. Stick and H. Schwartz.)

Meiosis starts at puberty and ends in old age. Meiosis is completed in 24 days, and prophase I in 13 to 14 days.

Fertilization comprises the incorporation of the sperm into the egg cytoplasm and the fusion of the two haploid nuclei, which is called syngamy. In the vicinity of the egg the sperm undergoes an acrosome reaction by which bindin and acrosomin are liberated. In the egg a fertilization coat is produced by exocytosis of cortical granules. Penetration of the sperm involves the formation of the fertilization cone, which is made of microvilli. Microtubules and microfilaments undergo changes during fertilization.

The nuclear lamina is depolymerized during meiotic prophase in spermatogenesis and disappears in spermatocytes and spermatids. In oogenesis the lamina disappears until the pachytene stage but reappears in the diplotene stage.

16–4 DNA Metabolism in Meiosis

During meiotic prophase there are some special molecular events related to the pairing and crossing-over of chromosomes. The isolation of meiocytes at a particular stage of meiosis has facilitated the study of such events. This isolation has been achieved in the anthers of *Lillium* and also in male gonads of mammals.[15] In both cases three consecutive stages of meiosis were studied in relation to the DNA replication.

1. *Premeiotic DNA replication.* We mentioned in Section 16–1 that after several mitotic divisions in gonial cells (Fig. 16–1) the DNA metabolism changes. At this point, after a G_2 phase, the cell enters meiotic prophase, and immediately there is a second S phase, without previous division. This S phase is 100 to 200 times longer than usual, primarily because of the reduced frequency in initiation points of replication (see Section 14–2). Another important difference is that the DNA duplication is not complete at this time, and *0.3 to 0.4% of the DNA remains unreplicated.*

2. *Zygonemal DNA (Zy-DNA) replication.* During zygonema most of the unreplicated DNA is synthesized. This Zy-DNA is produced in apparent coordination with chromosome pairing and is distributed among all chromosomes. Zy-DNA has a higher buoyant density than other DNA (Fig. 16–17), is rich in guanine-cytosine (GC) bases, and contains highly repeated sequences of nucleotides. Using radioautog-

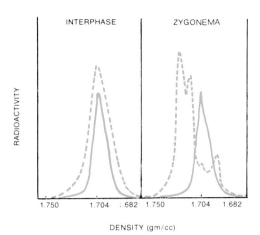

Fig. 16–17. Ultracentrifugation patterns of DNA from meiotic cells of *Lillium* after equilibration in a CsCl gradient. Observe that the buoyant density of the DNA synthesized at zygonema (broken line) is different from that at interphase (solid line absorption at 260 nm). (Courtesy of H. Stern and Y. Hotta.)

raphy at the electron microscopic level, it has been found that the Zy-DNA is localized in the region of the SC; this finding has led to speculation about the possible role of Zy-DNA in chromosome pairing.

3. *Pachynemal DNA replication (P-DNA).* In the period from zygonema to pachynema, there is a third wave of DNA synthesis. This is of the DNA repair type (Section 14–2) and is related to the process of crossing-over or recombination. In this period an *endonuclease* is activated which produces nicks (i.e., cuts) in the two DNA molecules, which are aligned for recombination (Fig. 16–18A). Next the DNA strands unravel, the strands are rejoined in the opposite chromatids, the gaps are filled by a process similar to DNA repair (Fig. 16–18B–D), and the remaining overlapping nucleotides are excised. (In Fig. 16–18A–D the student can follow these stages of P-DNA replication.)

One interesting finding is that the number of nicks produced during DNA repair is much larger than the number of recombinations; nevertheless the two phenomena are related, as the following experiment shows: In an *achiasmatic* hybrid of *Lillium* (i.e., lacking chiasmata at diplotene), only 10% of the normal frequency of chiasmata is found and the P-DNA synthesis is proportionally reduced.

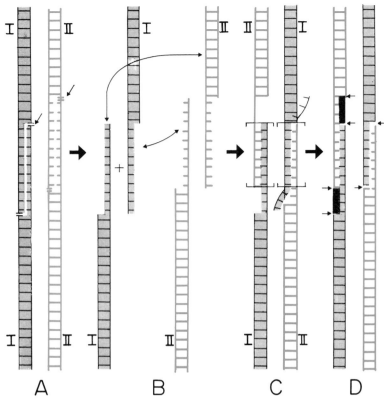

Fig. 16–18. Molecular model of recombination during meiotic prophase. *A*, the DNA of the pairing chromatids (I and II) undergoing the effect of endonucleases, producing "nicks" on each of the strands; *B*, unraveling of the strands; *C*, rejoining of the opposite chromatids; and *D*, elimination of excess pieces of DNA and filling of gaps by a process similar to that of DNA repair. (Modified from a diagram of C.A. Thomas, Jr.)

P-DNA sequences are of the middle repeat type (Section 22–2) and are identified by the fact that they are labeled with thymidine during pachynema. It is supposed that the sequences of P-DNA are localized at the preferred sites where recombination occurs (i.e., at potential regions for crossing-over).

Recently it has been observed that a particular class of small RNA molecules having only 125 nucleotides is able to modify the chromatin structure at the specific regions in which the nicking occurs. These small RNA molecules are synthesized during zygotene and pachytene and apparently become localized in the regions of chromatin containing P-DNA. These RNA molecules apparently render chromatin more accessible to the action of endonucleases and DNase II, which produces the nicking.[16]

Proteins Involved in Meiotic DNA Metabolism

During meiosis there are cyclic changes in some specific proteins which are related to the metabolism of Zy-DNA and P-DNA. These proteins increase during leptonema, reach a peak between zygonema and early pachynema, and then return to the original level at the end of pachynema.

Among these specific proteins are the following:

1. A β-*DNA polymerase*, responsible for the repair synthesis at pachynema.
2. A meiotic specific *endonuclease*, responsible for the nicks in the DNA at pachynema.
3. An ATP-dependent *DNA unwinding protein*, which unwinds several hundred base pairs, making the DNA strands available for duplex formation between the homologous chromosomes (Fig. 16–18B).
4. A *reassociation of R-protein*, which catalyzes the reassociation of single-stranded DNA (Fig. 16–18C). It has been suggested that the R-protein may play a role in the alignment of homologous stretches of DNA during recombination.

It is worth mentioning that the changes in DNA metabolism during meiosis are similar in plants and animals. This fact suggests that certain features of meiotic organization are highly conserved across the phylogenetic spectrum.[17]

SUMMARY:
DNA and Protein in Meiosis

Biochemical studies have been carried out in meiocytes of *Lillium* at different stages of meiosis.

At prezygonema there is a premeiotic DNA replication (S phase) which is 100 to 200 times longer than that in somatic cells. However, 0.3 to 0.4% of the DNA remains unreplicated. During zygonema, replication of the previously unreplicated DNA occurs (Zy-DNA), in coordination with pairing. Zy-DNA has a higher buoyant density than other DNA and is made of highly repeated nucleotide sequences.

During *pachynema* there is also a DNA synthesis (P-DNA) of small magnitude which is probably related to the process of recombination.

An endonuclease produces nicks in the two DNA molecules. The DNA strands unwind, the gaps are filled, and overlaps are excised (Fig. 16–18). Several proteins—a DNA polymerase-β responsible for DNA repair, an endonuclease, an unwinding protein, and a reassociation protein—undergo cyclic changes and are involved in the metabolism of DNA during meiosis.

REFERENCES

1. Bennett, M.D., Chapman, V., and Riley, R.: Proc. R. Soc. Lond. (Biol.), *178*:259, 1971.
2. Callan, H.G.: Proc. R. Soc. Lond. (Biol.)., *181*:19, 1972.
3. Moses, M.J.: Annu. Rev. Genet., 2:363, 1968.
4. Westergaad, M., and Von Wettstein, D.: Annu. Rev. Genet., 6:71, 1972.
5. Holm, P.B., and Rasmussen, S.W.: Eur. J. Cell Biol., 22 (1):23, 1980.
6. Moses, M., Counces, D., and Poulson, D.: Science, *187*:363, 1975.
7. Poorman, P.A., Moses, M.J., Rusell, L.B., and Cacheiro, N.L.A.: Chromosoma, *81*:507, 1981.
8. Moses, M.J., and Poorman, P.A.: Chromosoma, *81*:519, 1981.
9. Rattner, J.B., Goldsmith, M., and Hamkalo, B.A.: Chromosoma, 79:215, 1980.
10. Kierszerbaum, A.L., and Tres, L.L.: J. Cell Biol., 63:293, 1974.
11. Rattner, J.B., Goldsmith, M.R., and Hamkalo, B.A.: Chromosoma, 82:341, 1981.
12. Adolph, K.W., et al.: Cell, *12*:805, 1977.
13. Marsden, M.P.F., and Laemli, U.K.: Cell, *17*:849, 1979.
14. Solari, A.J.: Chromosoma, *81*:315, 1980.
15. Stern, H., and Hotta, Y.: Philos. Trans. R. Soc. Lond. (Biol.), 277:277, 1977.
16. Stern, H., and Hotta, Y.: Mol. Cell. Biochem., *29*:145, 1980.
17. Hotta, Y., and Stern, H.: Cell, 27:304, 1981.

ADDITIONAL READINGS

Callan, G.H.: Biochemical activities of chromosomes during the prophase of meiosis. *In* Handbook of Molecular Cytology. Edited by A. Lima-de-Faría. Amsterdam, North-Holland, 1969, p. 540.

Carpenter, A.T.C.: Electron microscopy of meiosis in *D. melanogaster* females. Proc. Natl. Acad. Sci. USA, 72:3186, 1975.

Chandley, A.C., Hotta, Y., and Stern, H.: Biochemical analysis of meiosis in the male mouse. Chromosoma, 62:243, 1977.

Cotton, R.W., Manes, C., and Hamkalo, B.A.: Electron microscopic analysis of RNA transcription in preimplantation rabbit embryos. Chromosoma, 79:169, 1980.

Gillies, C.B.: Synaptonemal complex and chromosome structure. Annu. Rev. Genet., 9:91, 1975.

Grossman, L.: Enzymes involved in the repair of damaged DNA. Arch. Biochem. Biophys., *211*:511, 1981.

Hamkalo, B.A.: Visualizing transcription in chromosomes. Trends in Genetics *1*:1, 1985.

Moens, P.B.: Ultrastructural studies of chiasma distribution. Annu. Rev. Genet., *12*:433, 1978.

Moses, M.: Synaptonemal complex karyotyping: I, II and III. Chromosoma, *60*:99, 127, 345, 1977.

Moses, M., Counces, D., and Poulson, D.: Synaptonemal complex complement of man in spreads of spermatocytes. Science, *187*:363, 1975.

Nicklas, R.B.: Chromosome distribution. Experiments on cell hybrids and in vitro. Philos. Trans. R. Soc. Lond. (Biol.), *277*:267, 1977.

Riley, R., and Flavell, R.B.: A first view of the meiotic process. Philos. Trans. R. Soc. Lond., 277:191, 1977.

Schatten, H., Schatten, G., Mazia, D., Balczon, R., and Simerby, C.: Behavior of centrosomes during fertilization and cell division in mouse oocytes and sea urchin eggs. Proc. Natl. Acad. Sci. USA, 83:105, 1986.

Solari, A.J.: Synaptonemal complexes and associated structures in microspread human spermatocytes. Chromosoma, *81*:315, 1980.

Sotelo, J.R.: Ultrastructure of chromosomes at meiosis. *In* Handbook of Molecular Cytology. Edited by A. Lima-de-Faría. Amsterdam,. North-Holland, 1969, p. 412.

Stern, H., and Hotta, Y.: Biochemistry of meiosis. Philos. Trans. R. Soc. Lond. (Biol.), 277:277, 1977.

Stick, R., and Schwartz, H.: The disappearance of the nuclear lamina during spermatogenesis. Cell Differ., *11*:235, 1982.

Stick, R., and Schwartz, H.: Disappearance and reformation of the nuclear lamina during specific stages of meiosis in oocytes. Cell, 33:949, 1983.

Westergaad, M., and von Wettstein, D.: The synaptosomal complex. Annu. Rev. Genet., 6:71, 1972.

Zickler, D.: The synaptonemal complex and recombination nodules. Chromosoma, 61:289, 1977.

CYTOGENETICS—CHROMOSOMES AND HEREDITY

In Chapter 1 we mentioned the fundamental discoveries that have established a link between cell and molecular biology and genetics and have led to the emergence of *cytogenetics*. This discipline is concerned with the chromosomal and molecular basis of heredity and deals with important issues with applications in agriculture and medicine. In Chapter 18 we consider some recent advances in the application of cytogenetics to the human species, and in Chapter 19 we study some molecular aspects of genetics. However, because all these subjects and related problems (i.e., genetic variation, mutation, phylogeny, morphogenesis, and evolution of organisms) can be studied in textbooks on genetics and general biology, we shall consider here only those subjects that concern cell biology, particularly the relationship between chromosomes and heredity.

The main objective of this chapter is to give a general overview of the chromosomal bases on which are founded the principles of heredity discovered by Mendel, Morgan, and others. Part of this chapter is devoted to the study of linkage

and recombination. We also consider, in an elementary way, the various chromosomal aberrations that can be produced, either spontaneously or by the action of radiation or chemical agents. Very briefly we mention the chromosomal aspects of evolution.

17–1 MENDELIAN LAWS OF INHERITANCE

The transmission of hereditary characters is governed by the behavior of the chromosomes during meiosis. However, in 1865 when Gregor Johann Mendel discovered the fundamental laws of inheritance, nothing was known about the chromosomes and meiosis. His discovery was based on the precise quantitation of experimental crosses and exceptional abstract thinking. He studied crosses between peas (*Pisum sativum*) that had pairs of differential or contrasting characteristics. For example, he used plants that have white and red flowers, smooth and rough seeds, yellow and green seeds, long and short stems, and so forth. After crossing the parental generation (P_1), he observed the resulting *hybrids* of the first filial generation, F_1. Then he crossed the hybrids (F_1) among themselves and studied the result in the second filial generation, F_2.

The Law of Segregation—Genes Are Distributed without Mixing

In a cross between parents with yellow seeds and parents with green seeds, in the first generation Mendel found that all the hybrids had yellow seeds and, thus, the characteristic of only one parent. In the second cross (F_2), the characteristics of both parents reappeared in the proportion of 75% to 25%, or 3:1.

Mendel postulated that the color of the seeds was controlled by a "factor" that was transmitted to the offspring by means of the gametes. This hereditary factor, which is now called the *gene*, could be transmitted without mixing with other genes. He postulated that the gene could be segregated in the hybrid into different gametes to be *distributed* in the offspring of the hybrid. For this reason this is called the *law* or *principle of segregation of the genes*. Mendel found that the plants with yellow seeds in F_2, despite showing the yellow color, had different genetic constitutions. One-third of this group, when crossed with each other, always gave yellow seeds, but the other two-thirds of the yellow-seeded plants of the F_2 generation, when crossed with each other, produced plants with yellow seeds and plants with green seeds in the ratio of 3:1. When the 25% of plants in F_2 with green seeds were crossed among themselves, they always produced green seeds. Thus they were a pure strain for this character. If we represent the genes in the crossing by letters, designating by A the gene with yellow character and by a the gene with green character, we have the following:

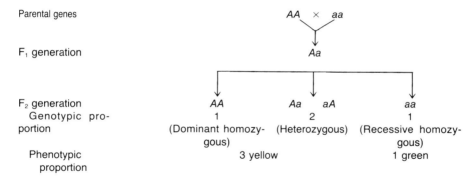

In the first generation (F_1) both A and a genes are present, but only A is revealed because it is *dominant;* gene a remains hidden and is called *recessive*. In the hybrid F_1 each gene is segregated and enters a different gamete. Half of the gametes will have the gene A, and the other half, a. Because each individual produces two types of gametes in each sex, there are four possible combinations in F_2. This gives as a result the proportion 1:2:1, corresponding to 25% of plants with pure yellow seeds (AA), 50% with hybrid yellow seeds (Aa), and 25% with pure green seeds (aa).

Mendel's results can now be explained in terms of the behavior of chromosomes and genes. The genes present in the chromosomes are found in pairs called *alleles*. In each homologous chromosome the gene for each trait occurs at a particular point called a *locus* (plural *loci*).

In the case illustrated in Figure 17–1, the gray mouse will have two GG genes, one in each homologue. Because the two homologues pair and then separate at meiosis, the two GG genes must also separate to enter the gametes. The mechanism is the same in a dominant (GG) as in a recessive (gg) white mouse. In the hybrid F_1, one chromosome bears gene G and the homologous chromosome bears gene g. When hybrids are self-fertilized, the gametes unite in the combinations shown by the checkerboard method illustrated in Figure 17–1. The individuals having two like alleles are called *homozygous* (i.e., GG or gg); those with different alleles, *heterozygous* (i.e., Gg). It is observed that there is one gray mouse that is homozygous dominant, two mice heterozygous and one mouse homozygous recessive.

Genotype and Phenotype. In 1911 Johanssen

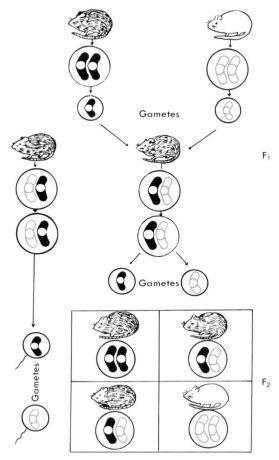

Fig. 17–1. A monohybrid cross between a gray mouse (dominant) and a white mouse (recessive). The parallelism between distribution of genes and chromosomes is indicated, as well as the resulting phenotypes in the F_1 and F_2 generations.

proposed the term *genotype* for the genetic constitution and *phenotype* for the visible characteristics shown by the individual. For example, in the case of the peas with green or yellow seeds there are two phenotypes in F_2: yellow seeds and green seeds in the ratio of 3:1. According to the genetic constitution, however, there are three genotypes: *AA*, *Aa*, and *aa*, in the ratio of 1:2:1.

The phenotype includes all the characteristics of the individual that are an expression of gene activity. For example, in humans phenotypic characteristics include eye color, baldness, the hemoglobins, the blood groups, and the ability to taste thiourea.

In crossings of certain plants that have white and red flowers, such as *Mirabilis jalapa*, it is

possible to find in F_2 three phenotypes (red, pink, and white flowers), which correspond to the three genotypes. This is due to incomplete dominance. The rule of dominance and recessiveness is not always accomplished completely; dominance may be complete in most cases, but incomplete in others. In this case there is a mixture of characteristics, called *intermediary heredity*.

Independent Assortment—Genes in Different Chromosomes Are Distributed Independently during Meiosis

Whereas the law of segregation applies to the behavior of a single pair of genes, the *law of independent assortment* describes the simultaneous behavior of two or more pairs of genes located in different pairs of chromosomes. Genes that lie in separate chromosomes are independently distributed during meiosis. The resulting offspring is a hybrid (also called a dihybrid) at two loci.

Figure 17–2 diagrams the cross between a black, short-haired guinea pig *(BBSS)* and a brown, long-haired guinea pig *(bbss)*. The BBSS individual produces only *BS* gametes; the bbss guinea pig produces only *bs* gametes. At F_1 the offspring are heterozygous for hair color and hair length. Phenotypically they are all black and short-haired. When two of the F_1 dihybrids are mated, however, each produces four types of gametes *(BS, Bs, bS, bs)*, which by fertilization result in 16 zygotic combinations. As shown in F_2 there are nine black, short-haired individuals; three black, long-haired; three brown, short-haired, and only one brown, long-haired individual. This phenotypic proportion (9:3:3:1) is characteristic of the second generation of a cross between individuals having two allelic pairs of genes.

Genes Are Linked When They Are on the Same Chromosome

All these examples of genetic crosses illustrate the fact that during meiosis there is a random distribution of the chromosomes which leads to the segregation of the genes in the gametes (see Chap. 16). When this type of study was carried out in the fruit fly *Drosophila melanogaster* by Morgan and collaborators (1910 to 1915), how-

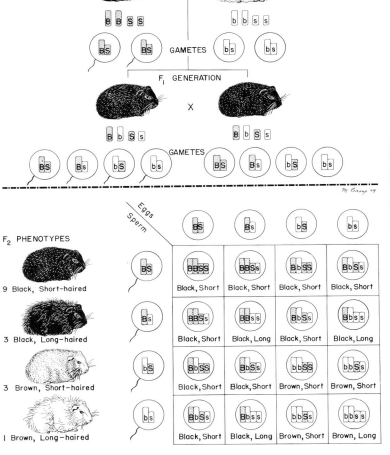

Fig. 17–2. Diagram of a cross between black, short-haired (dominant) and brown, long-haired (recessive) guinea pigs. The independent assortment of genes is evident. (See the text for details.) (From Villee, C.A.: Biology. 7th Ed. Philadelphia, W.B. Saunders Co., 1977, p. 665. Copyright © 1977 by W.B. Saunders. Reprinted by permission of CBS College Publishing.)

ever, it became evident that the law of independent assortment was not universally applicable and that in certain crosses of two or more allelic pairs of genes, free segregation was limited. In each case there was a marked tendency for parental combinations to remain linked and for a lesser proportion of new combinations to be produced. In *Drosophila* there are only four pairs of chromosomes, and this increases the chances for genes to occupy loci in the same chromosome.

If two genes *A* and *B*, with the corresponding alleles *a* and *b*, are in the same chromosome, only two classes of gametes will be obtained, either *AB* or *ab*. Figure 17–3, *left*, illustrates the mechanism of meiosis and the formation of the gametes in this hybrid. The coexistence of two or more genes in the same chromosome is called *linkage*.

After studying a considerable number of different crosses in *Drosophila*, Morgan reached the conclusion that all genes of this fly were clustered into four linked groups corresponding to the four pairs of chromosomes.

Linkage May Be Broken by Recombination during Meiotic Prophase

Further studies along these lines revealed that linkage is not absolute and that it is frequently broken. This concept is explained in the diagram of Figure 17–3, *right*, in which there is a *recombination* between the genes *A* and *B* (i.e., *crossing-over*) at meiosis. In this case four kinds of gametes are formed, in two of which the two genes are recombined. It is evident that in this example the linkage is broken in a certain proportion of the gametes by the interchange of

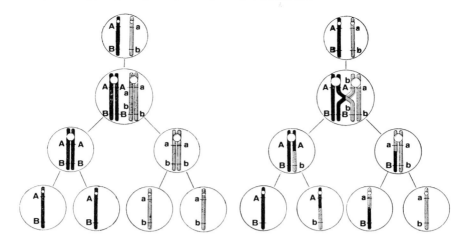

Fig. 17–3.   *Left,* diagram of the segregation of two pairs of allelic genes localized on the same pair of chromosomes without crossing-over. The result is two types of gametes. AB and ab. A case of linkage. *Right,* diagram of the segregation of two pairs of allelic genes on the same chromosome between which crossing-over takes place during meiosis. Four types of gametes result: AB, ab, Ab, aB. A case of linkage with crossing-over.

segments of the homologous chromatids (see Fig. 16–9).

We have seen in Chapter 16 that *genetic crossing-over* or *recombination* takes place between the DNA molecules and is not visible microscopically. The *chiasmata* observed at diplonema morphologically represent in some way the sites of these molecular events, however. Thus there should be a correlation between the two phenomena. It is also known that the frequency of recombination of two linked genes is a function of the distance that separates them along the chromosome. When two genes are close to one another, the probability of crossing-over is less than when they are far apart. If the distance between genes is estimated by linkage analysis, it is possible to construct a map indicating the relative position of each gene along the chromosome.

An estimate of the possible number of new recombinations can be obtained by counting the number of chiasmata during meiosis. The so-called *recombination index* is calculated by adding to the number of bivalents the number of chiasmata detected in the same cell at diplonema. In a species with a higher index, the possibility of new recombinations is higher, and this implies a greater possibility of variation.

There are several other pieces of evidence to support the relationship between chiasmata and crossing-over. The number of chiasmata is related to the length of the chromosomes in the bivalent. The presence of one chiasma reduces the possibility of another occurring in the immediate vicinity. This phenomenon has been called *positive interference*. The chiasma frequency is roughly constant in a given species but can be modified by genetic or environmental action. Chromosome pairing, chiasma formation, and crossing over are under genetic control. Lines of rye with high and low chiasma frequencies have been obtained by inbreeding. Among environmental factors that may affect the number of chiasmata are temperature, radiation, chemicals, and nutrition (see Henderson, 1969 in the Additional Readings at the end of this chapter).

Neurospora—Ideal for Studying Recombination and Gene Expression

Among the organisms studied in genetics, the mold *Neurospora* occupies a special place. The advantage of this material is twofold:

1. It is possible to identify and to follow the fate of each of the four chromatids present in the bivalent meiotic chromosome and thus to determine whether the crossing-over involves two, three, or all four chromatids.
2. It is possible to make a close correlation between genetic constitution and biochemical expression of genes.

As Figure 17–4 shows, the four cells resulting from the two meiotic divisions undergo a mitotic

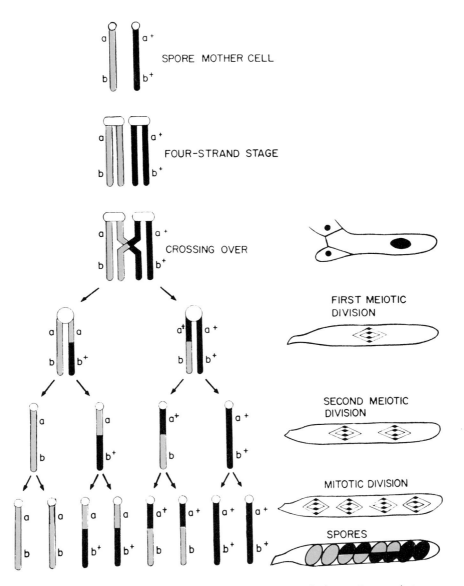

Fig. 17–4. Diagram of the formation of ascospores in *Neurospora crassa*. A single crossing-over between genes a and b, the behavior of one pair of chromosomes during the first and second meiotic divisions, and the division by mitosis of each of the four products are shown. The presence of a single chromatid in each spore is indicated diagrammatically.

division, which gives rise to eight haploid ascospores. Each of these ascospores can be isolated by dissection and cultured separately, giving rise to haploid individuals having the genetic constitution carried in each of the four original chromatids of the bivalent chromosome.

Figure 17–4 indicates a single crossing-over between genes *a* and *b* and the resulting products. Analysis of the eight ascospores shows that only two of the chromatids interchange segments while the other two remain intact.

SUMMARY:

Fundamentals of Genetics

Mendel (1865) discovered the laws of heredity by studying crosses between peas having pairs of *contrasting* characteristics (i.e., allelic). In a cross between parents with yellow seeds and parents with green seeds he found that in F_1 all the *hybrids* had yellow seeds (dominant gene). In the F_2 generation 75% of the plants had yellow seeds and 25% had green seeds (recessive gene). Mendel postulated that the genes are transmitted without mixing (i.e., via the *law of segregation*). He demonstrated that in F_2 there are 1 dominant homozygous, 2 heterozygous, and 1 recessive homozygous offspring. The *genotypic* segregation is 1:2:1 although the *phenotypic* proportion is 3:1. The law of segregation can be explained in terms of the behavior of chromosomes during meiosis (see Chapter 16). At times there is incomplete dominance (i.e., *intermediary heredity*).

The behavior of two or more pairs of allelic genes follows the *law of independent assortment*. Genes that lie in different chromosomes are independently distributed during meiosis. For two pairs of alleles the phenotypic proportion is 9:3:3:1 (Fig. 17–2).

Studies by Morgan and collaborators (1910 to 1915) demonstrated that the law of independent assortment may be limited. In *Drosophila* it was found that all genes were clustered into four linked groups corresponding to the four chromosome pairs. *Linkage* is not absolute and may be broken by *recombination* (Fig. 17–3) during the meiotic prophase. There is a correspondence between the number of chiasmata at diplonema and the number of recombinations *(crossing-over)* taking place at the molecular level. The distance between genes in a chromosome may be measured in *units of recombination*, and *genetic maps* that represent the relative order of genes along the chromosomes may be constructed. The number of chiasmata (and recombinations) is related to the length of the chromosome. A *positive interference* reduces the possibility of one chiasma occurring near another. In a species, chromosome pairing, chiasmata, and crossing-over are rather constant and are under genetic control. The mold *Neurospora* is ideal for the study of recombination and of its relationship to the biochemical expression of genes. After the two meiotic divisions, there is a mitotic one, and eight ascospores are formed. Each one is haploid and contains a single chromatid in which the genetic recombination may be studied by separation and culturing of each of the ascospores.

17–2. CHROMOSOMAL CHANGES AND CYTOGENETICS

The normal functioning of the genetic system is maintained by the constancy of the hereditary material carried in the chromosomes. We mentioned in Chapter 1 that each species has a diploid number of chromosomes which are in homologous pairs, and in Chapter 18, we discuss the normal human karyotype in detail. Under various conditions there may be changes in the karyotype, with different genetic consequences. Chromosomes may *change in number* maintaining a normal structure, or they may have *structural alterations*, frequently called *chromosomal aberrations*. Such structural changes may ap-

pear spontaneously, or they may be caused by the action of ionizing radiation or by chemicals. Sometimes these changes may have a genetic component. For example, in the hereditary disease *xeroderma pigmentosum*, chromosomes are extremely sensitive to ultraviolet radiation.

Euploidy—A Change in the Number of Chromosome Sets; Aneuploidy—A Loss or Gain of Chromosomes

There are two main kinds of change in the number of chromosomes (Table 17–1). In *euploids*, the change is in the haploid number, but the entire set of chromosomes is kept balanced, whereas in *aneuploids* there is a loss or gain of one or more chromosomes, causing the set to become unbalanced. In humans aneuploidy may cause severe alterations of the phenotype (see Chap. 18).

Some exceptional plants and animals have a *monoploid* (or haploid) chromosome set. In these organisms meiosis is irregular because of the absence of homologous chromosomes. As a result, gametes with varying numbers of chromosomes may be formed.

Polyploidy. A plant or animal that has more than two haploid sets of chromosomes is called a *polyploid* (Fig. 17–5). This change is common in nature, especially in the flowering plants. A diploid organism has two similar genomes; a triploid has three; a tetraploid has four; and so on. Polyploids may originate either by reduplication of the chromosome number in somatic tissue with suppression of cytokinesis or by formation of gametes with an unreduced number of chromosomes.

Meiosis in a triploid is more irregular than in a tetraploid. In general, polyploids of uneven number are sterile because the gametes have a more unbalanced number of chromosomes.

The scarcity of polyploids among animals is due to the mechanism by which sex is determined. If polyploidy occurs, the genic balance between the sex chromosomes and the autosomes is disturbed, and the race or species may disappear because of sterility.

Several species of amphibians show spontaneous polyploidy. In such polyploids the DNA content is correspondingly increased.[1] Polyploids are useful in the study of the expression of genes in multiple dosage. Studies of this type

TABLE 17–1.   CHROMOSOME COMPLEMENTS IN EUPLOIDS AND ANEUPLOIDS

Type	Formula	Complement*
EUPLOIDS		
Haploid	n	(ABCD)
Diploid	2n	(ABCD)(ABCD)
Triploid	3n	(ABCD)(ABCD)(ABCD)
Tetraploid	4n	(ABCD)(ABCD)(ABCD)(ABCD)
Autotetraploid	4n	(ABCD)(ABCD)(ABCD)(ABCD)
Allotetraploid	4n	(ABCD)(ABCD)(A'B'C'D')(A'B'C'D')
ANEUPLOIDS		
Monosomic	2n + 1	(ABCD)(ABC)
Trisomic	2n + 1	(ABCD)(ABCD)(B)
Tetrasomic	2n + 2	(ABCD)(ABCD)(B)(B)
Double trisomic	2n + 1 + 1	(ABCD)(ABCD)(AC)
Nullisomic	2n − 2	(ABC)(ABC)

*A, B, C, D are nonhomologous chromosomes.

have been made on amphibians for serum albumin, hemoglobin, and various enzymes.[2]

Polyploidy has been induced experimentally by temperature shock. It is also possible to induce polyploidy with substances such as colchicine, acenaphthene, heteroauxin, and veratrine.

These substances inhibit the formation of the spindle, and thus cell division is not completed. After a time the cells recover their normal activity, but have double the number of chromosomes. From the standpoint of pure and applied scientific work innumerable possibilities are offered by the experimental production of polyploids.

Allopolyploidy. This is a chromosomal variation produced in crosses between two species having different sets of chromosomes. The resulting hybrid has a different number of chromosomes than the parents: For example, the Argentine, black *Sorghum (S. almum)* is an allotetraploid (2n = 4x = 40) originated in nature by an interspecific cross between *S. halepense* (2n = 4x = 40) and *S. sudanense* (2n = 2x = 20) (Fig. 17–6). In this case, the fertilization occurred between one abnormal diploid gamete of *S. sudanense* and a normal gamete of the other

species. In most cases crosses between distantly related species produce sterile diploid hybrids.

Aneuploidy—The Result of Nondisjunction at Meiosis or Mitosis

Aneuploidy is produced by a failure in the separation of chromosomes during meiosis, called *nondisjunction. Monosomic* individuals have lost one of the chromosomes of the karyotype, and in *trisomics* there is a gain of one chromosome. These changes are discussed in Chapter 18 on the human karyotype.

Aneuploidy may also occur in somatic cells in which the *nondisjunction* occurs at mitosis. This process may produce individual cells in various tissues or parts of the body that have different chromosome numbers, e.g., in mosaic individuals, variegations, and gynandromorphs.

In cultures of normal cells, chromosomal variations are common, particularly after several transplants. These changes may lead to malignancy, but this is not the case in all cultures. The cytogenetic analysis of mammalian tumors has led investigators to consider them as altered

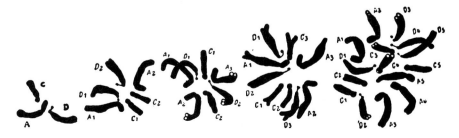

Fig. 17–5. Polyploid series in the plant *Crepis.* (After Nawashin.)

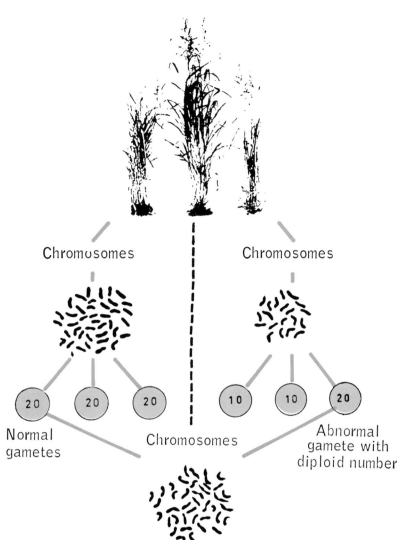

Fig. 17–6. The origin of the Argentine *Sorghum almum* by crossing *Sorghum halepense* (2n = 4x = 40) wth a diploid *Sorghum* (2n = 2n = 20) in which the fertilization occurred between one gamete not reduced (x = 20) and a normal gamete of *S. halepense*. Somatic chromosomes of the parents and the hybrid allopolyploid are illustrated. (After Saez and Nuñez, 1949.)

karyotypes, which are genetically and cytologically unstable.

Chromosomal Aberrations—Structural Alterations; Gene Mutations—Molecular Changes

To understand the mechanism by which chromosomal aberrations are produced the student should recall the concept of the single DNA molecule structure of chromatids (Section 13–4). According to this concept, in the G_1 phase there is a single DNA duplex per chromatid. It is only after DNA duplication, in the S phase, that each chromosome contains two chromatids and two DNA molecules. In a chromosomal aberration there is an alteration in the structural organization of the chromosome. For example, a break in the chromatid may lead to a loss of material, called a *deletion*, and in cases in which there are two breaks, a *translocation*, an *inversion*, or an *exchange between sister chromatids* may be produced. In all these cases the chromosome change can be observed microscopically, especially at the stages of chromosome condensation.

Chromosomal aberrations should be differentiated from *gene mutations* (also called *point mutations*), in which the changes are produced at the molecular level. In these, the alteration resides in the genetic code (i.e., in the sequence of DNA bases) and can be detected only by its genetic or molecular expression (see Section 19–2).

Radiation and Chemical Mutagens Act Mainly on the DNA Molecule

Both gene mutations and chromosomal aberrations may occur spontaneously, but their frequency increases by the action of so-called ionizing radiation (i.e., x-rays, β-rays, γ-rays, and ultraviolet light, as well as accelerated particles, such as fast neutrons, protons, and so forth). Other *mutagenic agents* are chemical substances, viruses, mycoplasms, and temperature changes.

In various organisms it has been demonstrated that the number of mutations induced by radiation is proportional to the dose. In the case of *Drosophila*, the relation is a linear function in the range of 25 to 9000 roentgen units (R).*

The effects of radiation are cumulative over long periods of time. For example, 0.1 R per day for 10 years is enough to increase the mutation rate to about 150% of the spontaneous level.

In contrast to gene mutations, chromosomal aberrations increase exponentially. More chromosomal aberrations are produced by continuous than by intermittent treatment.

A low dose of radiation may not be enough to cause fractures in one chromosome. As the dose increases, the number of breaks increases, and "aberrant" fusions become more and more likely. If the dose is intermittent or of low intensity, there is a greater chance that the broken ends of the chromosome will rejoin, or "heal," in the original chromosome structure before a second break can cause aberration.

The type of chromosomal aberration depends on the period of the cell cycle at the time of irradiation. A *chromatid break* occurs if the cell was in the G_2 phase, and a *chromosomal break* (i.e., two chromatids) occurs if the cell was in the G_1 phase, which is before DNA duplication. Irradiation generally produces localized lesions that may stabilize and form the so-called gaps (Fig. 17–7). After some time, such gaps may be repaired.[3] Such repair may result in complete restitution of the original structure (true repair) (see Howard-Flanders, 1981 in the Additional Readings at the end of this chapter). If there is no repair, the lesion becomes stabilized and cytologically visible. The process of restitution is inhibited by cold, cyanide, and dinitrophenol. Oxygen increases the number of fractures and chromosomal interchanges.

When a chromosome breaks, the two fragments may either reunite or remain separated permanently. If, instead of reconstituting at once, the broken ends reduplicate to form two chromatids, the fragments having the kinetochore migrate to different poles, and the fragments without a kinetochore (acentric) are eliminated in the cytoplasm (Fig. 17–8A). Sometimes the sister fragments reduplicate and unite, forming a dicentric chromosome and another acentric chromosome. During mitosis, the kinetochores of the dicentric chromosome move to opposite poles and form a "bridge" between the daughter nuclei that finally breaks at some point (Fig. 17–8B).

In humans, double or multiple chromosomal breaks are induced by acute exposures (e.g., heavy medical radiations, nuclear accidents, or nuclear warfare), whereas single breaks are produced at low doses. Chromosomal aberrations have been observed in blood cultures of humans who have had radiation treatment or injections of radioactive substances. The genetic effect of radiation has been studied in spaceflights. After Geminis III and IV, no increase in chromosomal aberrations was found in the blood cells of the astronauts who made the flights (see Brinkley and Hittleman, 1975 in the Additional Readings at the end of this chapter).

Germ Cell Mutations Are Transmittable and Somatic Are Not

Mutations in *somatic cells* are not transmitted from generation to generation but may be cumulative and produce severe changes in the individual, depending on the type of cell affected and the time at which the mutation occurs. Radiation can affect tissues that undergo mitosis, as well as tissues in which cell division no longer takes place. If mutation occurs during early embryonic development, a large number of cells are affected. The majority of mutated genes are

---

*A roentgen (R) is the amount of radiation sufficient to produce 2×10^9 ion pairs per cm.[3]

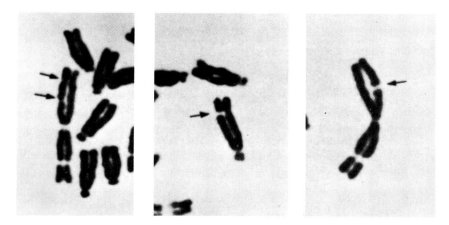

Fig. 17–7. Chromatid gaps and breaks *(arrows)* induced by radiation. (From Evans, H.J.: Radiation Research. Amsterdam, North-Holland Publishing Co., 1967.)

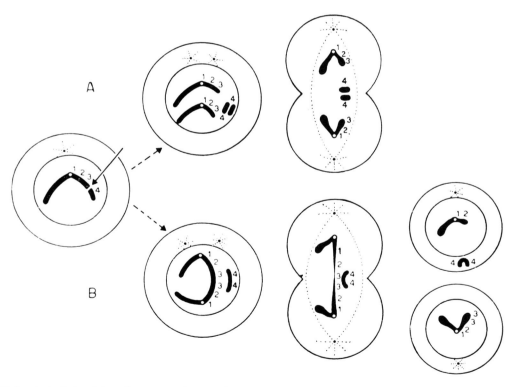

Fig. 17–8. A single break in a chromosome between loci 3 and 4. *A,* the two parts of the broken chromosome reduplicate. The fragments with the kinetochore go to opposite poles. The fragments without the kinetochore remain in the equatorial region and are eliminated. *B,* the two parts of the broken chromosome reduplicate and the broken ends unite, forming a chromosome with two kinetochores and another chromosome without a kinetochore. During mitosis the two kinetochores move to opposite poles and the chromosome section between them breaks. The two daughter cells receive chromosomes with different constitutions. The fragment without a kinetochore is eliminated in the cytoplasm. (From Stern, C: Principles of Human Genetics. 3rd Ed. San Francisco, W.H. Freeman, 1973.)

recessive and thus have no effect as long as the individual is heterozygous.

It is probable that some cases of cancer produced in irradiated individuals are caused by somatic mutation.

In contrast to somatic mutation, mutations in the germ cells may be transmitted to the offspring. In most cases of mutation caused by irradiation, however, both somatic and germ cells are frequently affected.

Even the lowest doses of radiation are genetically harmful, and the effects are dangerous to all organisms, from the simplest to humans. In microorganisms and plants, however, a few useful mutations may be obtained by irradiation. For example, irradiation has been used to produce new antibiotics and plants with high economic value.

Intra- and Interchromosomal Aberrations

In the previous section, we mentioned that in chromosomal aberrations there is an alteration of the structural organization that can be detected under the microscope. Aberrations may be *intrachromosomal*, when the alteration is within a single chromosome; or *interchromosomal*, when it involves the intervention of two or more chromosomes (Fig. 17–9). We shall now briefly mention the main types of chromosomal aberrations.

Deficiency or Deletion. This aberration involves the loss of chromosomal material and may be either *terminal* (at the end of a chromosome) or *intercalary* (within the chromosome). The former aberration originates from a single break at the G_1 phase and the latter from two breaks (Fig. 17–9A). Some intercalary deletions may result in origination of both a ring chromosome and an acentric rod, which is eliminated.

In heterozygous deficiency, one chromosome is normal, but its homologue is deficient. Animals with a homozygous deficiency usually do not survive to an adult stage because a complete set of genes is lacking. This suggests that most genes are indispensable, at least in a single series of alleles, to the development of a viable organism. Deficiencies are important in cytogenetic investigations of gene location for determination of the presence and position of unmated genes.

Duplication. Duplication occurs when a segment of the chromosome is represented two or more times (Fig. 17–9B). The duplicated fragment may be free, with a kinetochore of its own, or it may be incorporated into a chromosomal segment of the normal complement. If the fragment includes the kinetochore, it may be incorporated as a small extra chromosome. In general, duplications are less deleterious to the individual than deficiencies.

Inversion. An inversion is a chromosomal aberration in which a segment is inverted 180°. Inversions are called pericentric when the segment includes the kinetochore, and paracentric if the kinetochore is located outside the segment (Fig. 17–9C). In these aberrations there is a typical configuration at pachynema consisting of a loop that allows the pairing of the inverted segment. In the paracentric inversion, acentric and dicentric chromatids are formed.

Translocation. This interchromosomal aberration is called *reciprocal* when segments are exchanged between nonhomologous chromosomes (Fig. 17–9D). Translocations may be *homozygotic*, if involving segments of two chromosomes of a pair, or *heterozygotic* if only one chromosome of a pair is translocated.

When both chromosomes are broken near the kinetochore, a new chromosome may originate by *centric fusion* or *Robertsonian translocation* (Fig. 17–9E). The fusion creates a metacentric chromosome with two arms in the form of a V, and a small fragment, which tends to be eliminated. This phenomenon has occurred in the phylogeny of many species and has led to a reduction in the number of chromosomes. This mechanism has played a part in the evolution of various species.

Centric fusion has occurred during the phylogeny of *Drosophila*, grasshoppers, reptiles, birds, mammals, and other groups. It is a process that establishes a new type of chromosome and reduces the somatic chromosome number of the species. In Chapter 18 we mention the importance of reciprocal translocations in human cytogenetics.

Isochromosomes. A new type of chromosome may arise from a break (i.e., a misdivision) at the kinetochore. As Figure 17–10 shows, the two resultant telocentric chromosomes may open up to produce chromosomes with two identical arms (i.e., isochromosomes). This type of chromosome has been produced in irradiated

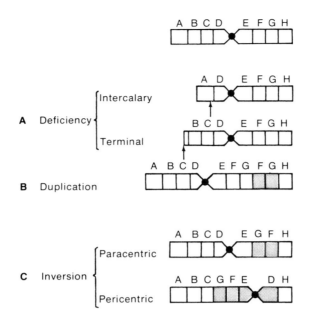

Interchromosomal aberrations

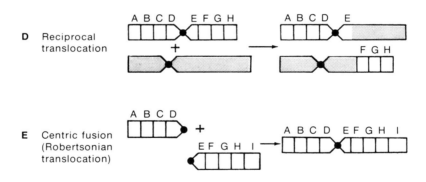

Fig. 17–9. Diagram showing some of the most frequent chromosomal aberrations. (See the text for details.)

material. At meiosis they may pair with themselves or with a normal homologue.

Sister Chromatid Exchanges Increase in Diseases with Altered DNA Repair

A *sister chromatid exchange* is an interchange of DNA between homologous sister chromatids in a chromosome, presumably involving DNA breakage followed by fusion. Sister chromatid exchanges are difficult to find using common cytological methods because the chromatids are morphologically identical. Such chromatid exchanges were first described in studies in which

[3]H-thymidine was added during a replicating cycle, followed by another cycle in a nonradioactive medium. As we mentioned in Chapter 14, this method demonstrated that the duplication of DNA is semiconservative. In a few mitoses observed by radioautography, however, an alteration of the labeling along the chromatids was found, and this was interpreted as the result of chromatid exchange caused by the radioactivity of the tritiated label used.

Analysis of this phenomenon has been greatly facilitated by the use of *bromodeoxyuridine* (BrdU), a thymidine analogue that can be incorporated into the DNA of replicating cells in-

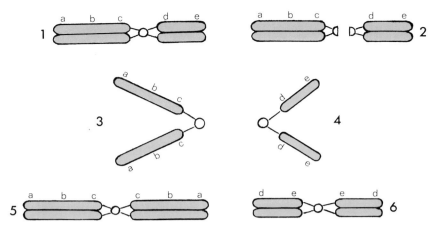

Fig. 17–10. Formation of isochromosomes. *1*, original chromosome; *2*, misdivision of the kinetochore at the beginning of mitotic anaphase; *3* and *4*, the chromatids unfold into two isochromosomes; *5* and *6*, in the next division, two complete isochromosomes are present. Note that in each isochromosome, the arms exhibit similar genetic constitution.

stead of the original base. If a fluorescent dye (Hoechst 33528) is added after the BrdU incorporation, the fluorescence of the segments that contain BrdU is greatly diminished in comparison with those with the original base.[4] If Giemsa stain is used, the BrdU will also diminish its staining.

In using this technique, however, researchers have been unable to discover whether the chromatid exchange can occur spontaneously or whether it is induced by the BrdU. It has, though, been of great help in differentiating the various inherited diseases characterized by chromosome fragility, which have an increased frequency of sister chromatid exchanges and a tendency to have associated neoplasia. Some of these diseases (e.g., Bloom's syndrome, Fanconi's anemia, and ataxia telangiectasia) are presumably related to defects in DNA repair.

Sister chromatid exchange has also been important in studying the effect of mutagens on the chromosomes. Various mutagenic drugs that are alkylating agents, such as *mitomycin C* and nitrogen mustard, produce a great number of breaks and chromatid exchanges (Fig. 17–11). The intimate association of sister chromatid exchange with mutagenesis and carcinogenesis may have important medical implications (see Latt, 1981; Gebhart, 1981 in the Additional Readings at the end of this chapter).

A certain number of chromatid exchanges are probably spontaneous in normal chromosomes. Their frequency varies between 0.15 and 0.4 per chromosome and apparently increases with the

size of the replicon (i.e., the length of the DNA that is involved in DNA duplication; see Section 14–2).[5]

17–3 CHROMOSOMES PLAY A FUNDAMENTAL ROLE IN EVOLUTION

The development of comparative cytology and cytogenetics has brought about great progress toward an understanding of evolution. McClung and Navashin were the first to emphasize the importance of cytogenetics to taxonomy and to the study of evolution by comparing genomes of related species. Systematics has been greatly advanced by cytogenetic investigation, which now provides many of the best methods of elucidating correlations between different taxonomic categories. In general, families, genera, and species are characterized by different genetic systems.

The study of the karyotypes of different species has revealed interesting facts about both the plant and animal kingdoms. It has been demonstrated that individuals in wild populations are, to some extent, heterogeneous cytologically and genetically. In some cases, even if the genes are identical, they may be ordered in a different way owing to alterations of the chromosomal segments. These changes have an important bearing on the evolution of species.

The majority of plant species originate from an abrupt and rapid change in nature, and aneuploidy and polyploidy are the prime sources of

Fig. 17–11. Cultured cell exposed to the mutagenic chemical agent 8-methoxypsoralen and to bromodeoxyuridine. The regions stained only lightly with Giemsa are those in which the DNA was synthesized in the presence of BrdU. Observe the numerous sister chromatid exchanges in each chromosome. (Courtesy of S. Latt.)

variation. In the animal kingdom polyploidy is not so important. Among vertebrates, different species of fishes have a different number of chromosomes. Amphibians are generally characterized by a specific number for each family. Reptiles and birds have large chromosomes (macrochromosomes) and small chromosomes (microchromosomes) that serve to differentiate them cytologically.

Matthey distinguished between the basic chromosome number and the number of chromosomal arms, also called the fundamental number (FN). According to this concept, the metacentric chromosome has *two* arms and acrocentric and telocentric chromosomes have *one*. This is an important distinction in a group having both acrocentric and metacentric chromosomes, and the number of arms in each of the different species can be compared.

Another method used to study the cytogenetics of evolution is the application of measurements of total chromosomal area and DNA content.

With regard to the absolute size of their chromosomes, mammals and birds constitute two independent groups. These two orders have different DNA contents and different sex-determining mechanisms. Speciation depends more on chromosomal rearrangements and mutation of individual genes than on changes in the total amount of genetic content.

Two opposite changes in the number (and configuration) of chromosomes are of particular importance in evolution. In *centric fusion*, a process that leads to a decrease in chromosome number, two acrocentric chromosomes join together to produce a metacentric chromosome. In *dissociation*, or *fission*, a process that leads to an increase in chromosome number, a metacentric (commonly large) and a small supernumerary metacentric fragment become translocated, so that two acrocentric or submetacentric chromosomes are produced.

Fusion and dissociation are the main mechanisms by which the chromosome number can be decreased and increased during evolution of the majority of animals and in some groups of plants (Fig. 17–9).

Studies of somatic and polytene chromosomes in several hundred species of *Drosophila* have elucidated the formation and evolution of this genus, which has been thoroughly analyzed from genetic, ecological, and geographical standpoints.

Observation of chromosomal organization and of the different karyotypes in the individual, the species, genera, and the major systematic groups indicates that a chromosomal mechanism is involved in the process of evolution.

Evolution, however, is very complex and should be considered from the different biochemical, cytological, genetic, ecological, and experimental aspects. All these methods and approaches should be used to analyze the intricate relationships between groups of organisms, particularly those that show marked variations. (These considerations are beyond the scope of this book.)

The Evolution of Chromosomes in Primates

In recent years a great deal of work on evolution has focused on the possible cytogenetic relationship between the great apes (chimpanzee, gorilla, and orangutan) and the human species. It was found that these primates have 48 chromosomes, and attempts were made to correlate the 24 chromosome pairs with the 23 pairs in humans. This analysis was greatly facilitated by the use of the banding techniques that permit one to study the inner structure of each chromosome (see Section 18–1).

These comparative studies have demonstrated: (1) 13 pairs of chromosomes in humans are identical with 13 pairs in the chimpanzee. (2) Chromosome number 2 in humans has resulted from the centric fusion (or Robertsonian translocation; see Fig. 17–9E) of two chromosomes present in hominoid apes.[6,7] (3) The other chromosomes differ in the occurrence of nine

pericentric inversions (Fig. 17–9C) and two additions (Fig. 17–9B) of chromatin material (see Dutrillaux, 1981 in the Additional Readings at the end of this chapter).

In summary, the evolution of chromosomes in primates consists in modifications of the morphology in the form of fusion, fission, inversion, and reciprocal translocation studied earlier in this chapter. At present, pericentric inversion appears to be the main structural difference between individual chromosomes of the great apes and humans.

It has been postulated that evolution of the human karyotype has occurred by a series of pericentric inversions, which permitted genetic isolation of small breeding groups and selection of favorable gene combinations that gave rise to *Homo sapiens.*

At this point let us emphasize again the importance of efficient reproduction in evolution. Reproduction can occur only between individuals that have undergone the same changes in chromosomes (i.e., that are homozygous for the change); in fact, in hybrids there is diminution of fertility. The genetic barrier, which becomes established by this kind of "sexual closure" within a given population, could permit the origin of a new species among the descendants. At present, the homologies in the morphology of chromosomes are being correlated with mapping of genes in similar chromosomes of humans and the great apes (see Pearson, 1977; Jones, 1977; Jeffreys and Barrie, 1981 in the Additional Readings at the end of this chapter). Researchers in the molecular biology field are also mapping mitochondrial DNA after cleavage by restriction enzymes (see Chap. 19). This permits the establishment of genealogical relationships between closely related species such as human and apes and suggests possible evolutionary trees.[8] Analysis of the sequence of mitochondrial DNA is having an important impact in taxonomy.

SUMMARY:
Chromosomal Aberrations, Action of Mutagens, and Cytogenetics and Evolution

Changes in the number and structure of chromosomes may occur spontaneously or experimentally by the action of radiation or chemicals. The number of chromosomes is generally constant for plant and animal species. *Chromosomal changes in number* are of two main kinds: in *euploids* the set is kept balanced; in *aneuploids* there is a loss or gain of one or more chromosomes. In exceptional cases there are *haploid* organisms. In plants, *polyploids* (triploid, tetraploid, etc.) are rather common. They originate by reduplication without cytokinesis. In animals, polyploids are scarce, because sex is frequently determined by a pair of different chromosomes (XY). Polyploidy can be induced by colchicine; this substance is used in agriculture to improve certain plant species. *Allopolyploidy* consists of the formation of a hybrid with different sets of chromosomes. Sometimes a diploid gamete fertilizes a normal haploid gamete, producing a new species with a triploid number.

Among the aneuploid organisms there are the *trisomic* (i.e., with three similar chromosomes instead of a normal pair) and the *monosomic* (missing one member of a pair of homologous chromosomes). These two conditions are important in humans and may arise by nondisjunctional division.

Chromosomal aberrations, also called *structural alterations*, involve a change in the molecular organization of the chromosome. One or two breaks are produced at the level of the chromatin fiber, and the effect depends on whether it occurs in the G_1 phase (one chromatid) or in the G_2 (two chromatids). Chromosomal aberrations should be differentiated from *gene* or *point mutations*, which occur at a molecular level.

Ionizing radiation—x-rays, γ-rays, β-rays, fast neutrons, slow neutrons, and ultraviolet light—can produce point mutations or chromosomal aberrations. The number of mutations increases proportionally with the dose of x-radiation. The effect of radiation is cumulative. An exposure of cultured cells to 20 R (roentgen units) is sufficient to produce one chromosome break per cell. In contrast to mutations, which increase proportionally with the dose, chromosomal aberrations increase exponentially with the dose. Breakage can be followed by "healing" of the broken end (see Section 14–2). (The breaks may be at the chromosome or the chromatid level.) Chromatid breaks are produced in cells irradiated after the S phase. If two breaks are produced, translocation, inversions, and large deletions may be induced. Dicentric chromosomes may be produced, which form a bridge at anaphase. In the human, heavy medical radiation, nuclear accidents, or radioactive substances may produce chromosomal aberrations. In the production of new antibiotics and plants, irradiation is being used to economic advantage. *Somatic mutations* are not transmitted from one generation to another, whereas germ cell mutations may be passed to the offspring.

Some of the main aberrations are: (1) *Deficiency* or *deletion*, in which a part (either *interstitial* or *terminal*) of the chromosome is missing. The parts of the chromosome lacking the kinetochore generally are lost. Deficiency may be *heterozygous* (one chromosome is normal) or *homozygous* (both chromosomes are deficient). The latter generally do not survive. (2) *Duplication*, in which a chromosome segment is represented two or more times (tandem duplication). (3) *Translocation*, in which there is an exchange of segments between nonhomologous chromosomes (*reciprocal* type) or between different parts of the same chromosome (simple type). Sometimes *centric* fusion occurs when the two chromosomes are broken near the kinetochore and form a metacentric (V-shaped) chromosome. (4) *Inversion*, in which there is breakage of a segment, followed by its fusion in a reverse position. It is *pericentric* if it includes the kinetochore, and *paracentric* if the kinetochore is outside. *Isochromosomes* may arise from a break at the kinetochore, resulting in two chromosomes with identical arms.

Sister chromatid exchange consists of the interchange of DNA segments between sister chromatids in a chromosome. First described in studies using ³H-thymidine, it was interpreted as the result of radiation from ³H. This aberration can be studied with bromodeoxyuridine (BrdU), which is incorporated instead of thymidine and which produces changes in fluorescence (by Hoechst 33528) or in the Giemsa stain. This technique permits the detection of diseases characterized by chromosome fragility and enables the study of the action of mutagenic drugs which increase the number of breaks.

Cytogenetic studies have provided excellent methods for establishing taxonomic interrelationships and have, thereby, contributed to studies of *evolution* and *systematics*. One of the most frequent causes of evolution is a change in the order of genes as a result of chromosomal aberrations. In plants, aneuploidy and polyploidy are frequent sources of variation, whereas in mammals and birds, speciation depends more on chromosomal rearrangement and point mutations.

In the study of evolution, the number of chromosomes, the characteristics of the karyotype, the total chromosomal area, and the content of DNA are investigated. The presence of metacentric chromosomes may, in some cases, result from fusion of two acrocentric chromosomes. The contrasting phenomenon (i.e., *dissociation*) may lead to an increase in chromosome number. The problem of *evolution* is very complex and beyond the scope of this book; however, knowledge of cytogenetics is most fundamental to its understanding. One example of the importance of cytogenetics in evolution is given in a comparison of the chromosome pairs of the primates and humans.

REFERENCES

1. Becak, W.: Genetics, *61*:183, 1969.
2. Becak, W., Becak, M.L., and Rebello, M.N.: Chromosoma, *22*:192, 1967.
3. Evans, H.J.: Radiation Research. Amsterdam, North-Holland, 1967.
4. Latt, S.A.: Proc. Natl. Acad. Sci. USA, *70*:3395, 1973.
5. Cleaver, J.E.: Exp. Cell Res., *13*:27, 1981.

6. de Grouchy, J., Turleau, C., Roubin, M., and Chavin-Colin, F.: *In* Chromosome Identification. Edited by T. Caspersson and L. Zech. New York, Academic Press, 1973, p. 127.
7. Lejeune, J., Dutrillaux, B., Rethoré, M., and Prieur, M.: Chromosoma, *43*:423, 1973.
8. Ferris, S.D., et al.: Proc. Natl. Acad. Sci. USA, 78:242, 1981.

ADDITIONAL READINGS

Brinkley, B.R., and Hittleman, W.N.: Ultrastructure of mammalian chromosome aberrations. Int. Rev. Cytol., *42*:49, 1975.

Darlington, G.D.: A diagram of evolution. Nature, *276*:447, 1978.

de Grouchy, J., Turleau, C., and Finaz, C.: Chromosomal phylogeny in the primates. Annu. Rev. Genet., *12*:289, 1978.

Dutrillaux, B.: New chromosome techniques. *In* Molecular Structure of Human Chromosomes. Edited by J.J. Yunis. New York, Academic Press, Inc., 1977, p. 233.

Dutrillaux, B.: Chromosomal evolution in primates. Hum. Genet., *48*:251, 1979.

Dutrillaux, B.: Les chromosomes des primates. La Recherche, *12*:1246, 1981.

Evans, H.J.: Radiation Research. Amsterdam, North-Holland Publishing Co., 1967.

Gebhart, E.: Sister chromatid exchange and structural chromosome aberration in mutagenicity testing. Hum. Genet., *58*:235, 1981.

Henderson, S.A.: Chromosome pairing, chiasmata, and crossing over. *In* Handbook of Molecular Cytology. Edited by A. Lima-de-Faría. Amsterdam, North-Holland Publishing Co., 1969, p. 327.

Howard-Flanders, P.: Inducible repair of DNA. Nature, *245*:72, 1981.

Jeffreys, A.J., and Barrie, P.A.: Sequence variation of nuclear DNA in man and primates. Philos. Trans. R. Soc. Lond., *292*:133, 1981.

John, B., and Lewis, K.R.: The meiotic mechanism. In Readings in Genetics and Evolution. London, Oxford University Press, 1973, p. 2.

Jones, K.W.: Repetitive DNA in primate evolution. *In* Molecular Structure of Human Chromosomes. Edited by J.J. Yunis. New York, Academic Press, Inc., 1977, p. 295.

Latt, S.A.: Sister chromatid exchange formation. Annu. Rev. Genet., *15*:11, 1981.

Latt, S.A., and Schreck, R.R.: Sister chromatid exchange analysis. Hum. Genet., *32*:297, 1980.

Latt, S.A., Allen, J.W., and Stetten, G.: In vitro and in vivo analysis of chromosome structure, replication, and repair using BrdU-33258 Hoechst techniques. *In* International Cell Biology. New York, Rockefeller University Press, 1977.

Lewis, K.R., and John, B.: The Matter of Mendelian Heredity. London, J. & A. Churchill, Ltd., 1964.

Pearson, P.L.: Banding patterns, chromosome polymorphism, and primate evolution. In Molecular Structure of Human Chromosomes. Edited by J.J. Yunis. New York, Academic Press, Inc. 1977.

Seeberg, E.: Chromosome Damage and Repair. New York, Plenum, 1982.

Taylor, J.H.: Meiosis. *In* Encyclopedia of Plant Physiology. Vol. 18, Berlin, Springer-Verlag, 1967, p. 344.

Wolff, S.: Sister chromatid exchange. Annu. Rev. Genet., *11*:183, 1977.

Yunis, J.J.: Molecular Structure of Human Chromosomes. New York, Academic Press, Inc., 1977.

Yunis, J.J., and Prakash, O.: The origin of man: A chromosomal pictorial legacy. Science, *215*:1525, 1982.

HUMAN CYTOGENETICS

In the last two decades great advances have been made in the identification of human chromosomes in the karyotype and in the localization of genes. Such studies have had important biological and medical implications in light of the discovery that many congenital diseases and syndromes are related to chromosomal aberrations. Today, *human cytogenetics* has become a specialized science in itself, the wide-ranging interests of which far exceed the limits of a text on cell biology. Here we present to the student an elementary account that can be supplemented by some of the general references in the Additional Readings at the end of this chapter.

The first step in this field was made by Tjio and Levan in 1956,[1] with the final demonstration that the correct *diploid number* of human chromosomes is 46 (44 autosomes + XY in the male and 44 + XX in the female) (Fig. 18–1). The field of human cytogenetics, however, started to receive great attention three years later with the discovery by Lejeune and coworkers of a *trisomy* (i.e., an extra chromosome) in patients affected by *mongolism* or *Down's syndrome*.[2] This finding led to rapid advances, with the identification in the same year of a series of aberrations of the sex chromosomes, such as *Klinefelter's syndrome* with XXY and *Turner's syndrome* with XO (i.e., without Y),[3,4] and later on a series of autosomal aberrations was discovered.

After 1968 a new era was initiated with the demonstration by Caspersson and coworkers of chromosome banding using a fluorescent dye (quinacrine mustard).[5] This was followed by the development of a number of banding techniques which, by demonstrating a substructure in chromosomes, have permitted a more precise identification not only of individual chromosomes but also of their parts. These methods have increased the precision of cytogenetic diagnosis by allowing the study of finer chromosomal aberrations, such as deletions, translocations, inver-

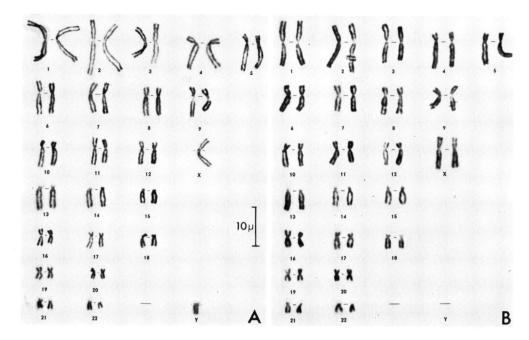

Fig. 18–1.   Human male *(A)* and female *(B)* karyotypes from a blood culture. (Courtesy of M. Drets.)

sions, and so forth, of individual chromosomes (see Section 17–2). The impact of such technical advances has been so great that from the point of view of clinical application, the field can be divided historically into two major periods: one before the discovery of the banding techniques and the other after. These advances have permitted the observation of new chromosomal defects involving almost every chromosome of the human karyotype. (See McKusick and Ruddle, 1977 in the Additional Readings at the end of this chapter.)

In this chapter, in addition to discussing the banding techniques, we also study the sex chromosomes and sex chromatin in relation to sex determination.

In recent years considerable progress has been made in the study of the genetic map of human chromosomes, thanks to modern methods of molecular biology. These include somatic cell genetics, gene transfer, and nucleic acid hybridization. A short account of the progress made will be included at the end of this chapter. (Further details will be given in Chapters 19 to 23.)

18–1   THE NORMAL HUMAN KARYOTYPE

The human karyotype has been studied in tissue cultures of fibroblasts, bone marrow, skin,

and peripheral blood; colchicine and hypotonic solutions are used to block mitosis at metaphase and to separate the chromosomes. An important technical advance has been the introduction of phytohemagglutinin, which induces lymphocytes to transform into lymphoblast-like cells that start to divide 48 to 72 hours after exposure. The strong mitogenic properties of this substance have allowed the development of microtechniques that employ small amounts of blood. Spreading the cells on a slide causes them to burst and to display all the chromosomes, which are usually studied in metaphase. A higher resolution of the chromosome structure has been achieved through the use of banding techniques to study mitosis in prophase and prometaphase.[6]

Karyotype Preparation—Chromosomal Ordering by Size and Kinetochore Position

A *karyotype* of human metaphase chromosomes is usually obtained from microphotographs. The individual chromosomes are cut out of the microphotograph and then lined up by size with their respective partners. The technique can be improved by determining the so-called *centromeric index*, which is the ratio of the lengths of the long and short arms of the chromosome. More recently, a system has been introduced that involves a computer-controlled microscope and several accessories that permit (1) scanning of slides, (2) location of cells in metaphase, (3) counting chromosomes, and (4) transmission of digitally expressed images for computation and storage. All these steps, which can be carried out automatically, may help in making karyotypes more rapidly and in determining chromosomal aberrations.[7]

Table 18–1 shows the classification of human chromosomes now generally used. The 23 pairs are disposed in seven groups (A through G) decreasing in size and having the characteristics described. For example, group A consists of pairs 1, 2, and 3, large almost metacentric chromosomes; group B comprises pairs 4 and 5, large submetacentric chromosomes, and so forth. The X chromosome is in group C (pairs 6 through 12), medium-sized submetacentric chromosomes; chromosome Y is in group G, together with pairs 21 and 22, small acrocentric chromosomes. It is important to remember that pairs 21 and 22 have satellites that correspond to nucleolar organizers, while the Y chromosome lacks satellites. Furthermore, group D (pairs 13, 14, and 15) also contains acrocentric chromosomes with satellites that correspond to nucleolar organizers (see Fig. 18–2).

Banding Techniques Reveal Structural Details of Chromosomes

The main banding techniques are identified by letters, i.e., Q, G, C, R, T, and so forth, which are related to the method used or the results obtained.

Q Banding. We mentioned at the beginning of this chapter that the introduction of certain fluorescent dyes (i.e., proflavine and quinacrine) and especially an alkylating derivative, *quina-crine mustard*, led to the discovery of a substructure along the length of the chromosomes. The bands stained with quinacrine were named *Q bands.*

Later on, other techniques, which involve a prestain treatment that tends to denature the DNA (such as exposure to alkali, increased temperature, or formamide), followed by various staining techniques, were used.

G Banding. One of the best known techniques is that using the Giemsa stain (originally used to stain blood smears). The bands stained with Giemsa were designated *G bands.*[8] Both Q and G bands correlate with the *chromomeres* observed in leptotene and pachytene chromosomes during meiosis.

On the basis of the banding pattern, three major types of chromatin can be identified in chromosomes: (1) *centromeric or kinetochoric constitutive heterochromatin*, (2) *intercalary heterochromatin*, and (3) *euchromatin* (see Section 13–5). The Q and G banding patterns are generally similar and correspond to the second type, intercalary heterochromatin.

During the cell cycle, G bands replicate in the second half of the S phase, while the interbands replicate earlier. The G bands are richer in adenine-thymidine, but both the bands and the interbands are active in transcription producing RNA.[9]

C Banding. Pretreatment of chromosomes with procedures that denature DNA, followed by Giemsa staining, results in an intense staining of the kinetochoric region corresponding to the localization of *constitutive heterochromatin*. This differential staining is the result of the rapid reannealing of satellite DNA, which is located mainly in the kinetochoric region. Satellite DNA renatures easily because it is composed of short, highly repetitive nucleotide sequences (see Section 23–2). In addition to the perikinetochoric

TABLE 18–1. CHARACTERISTICS OF THE CHROMOSOMES IN THE HUMAN KARYOTYPE

Group	Pairs	Description
A	1–3	Large almost metacentric chromosomes
B	4–5	Large submetacentric chromosomes
C	6–12 + X	Medium-sized submetacentric chromosomes
D	13–15	Large acrocentric chromosomes with satellites
E	16–18	No. 16, metacentric; Nos. 17–18, small submetacentric chromosomes
F	19–20	Small metacentric chromosomes
G	21–22 + Y	Short acrocentric chromosomes with satellites. (The Y chromosome belongs to this group but has no satellites.)

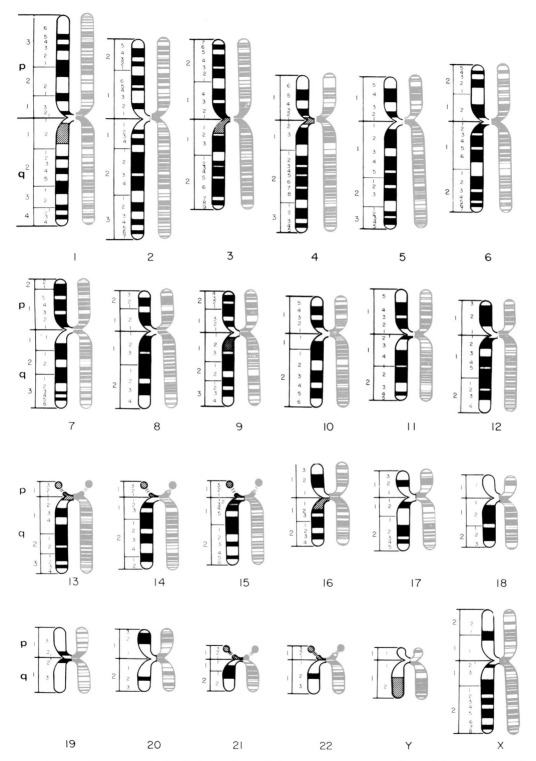

Fig. 18–2.   Human karyotype studied with the banding techniques. Chromosomes are disposed following the Paris conference (see Table 18–1). *Left* chromatid, in black, represents banding pattern at mid-metaphase; *right* chromatid, in red, represents G-banding pattern at late prophase. (Courtesy of J.J. Yunis.)

regions, the secondary constrictions of chromosomes 1, 9, and 16 and the distal segment of the long arm of the Y chromosome are stained. This method has provided the interesting information that there may be a *polymorphism* in human chromosomes. In fact, the size of the C bands may vary from person to person, and this variation may be due to differing amounts of *satellite DNA*.

R Banding. With minor variations in pretreatment of the chromosomes, a banding pattern that is the reverse of that found with the Q and G band techniques is obtained.

T Banding. Another modification has allowed the staining of certain telomeric regions of the chromosomes.

Other banding techniques use the Feulgen stain (F bands), and one selectively stains the *nucleolar organizers* (N bands). These are localized in the satellites of chromosomes 13, 14, 15, 21, and 22 (Fig. 18–2).

Banding Provides New Features for Identification of Human Chromosomes

The recognition of chromosome bands has elucidated new morphological features for use in the identification of human chromosomes and chromosomal parts. In addition to the telomere, kinetochore, and arms, special landmarks (i.e., well-defined bands) were selected to subdivide the arms into regions. The regions are designated 1, 2, 3, and 4, moving from the kinetochore toward the telomere. Using the nomenclature established at the Paris conference in 1971, any particular band or segment of a chromosome can be identified according to the following parameters: (1) chromosome number, (2) an arm symbol (p = short arm, q = long), and (3) the region and band numbers (Fig. 18–3). (For example, 6p23 indicates band number 3 in region 2 of the short arm of chromosome 6.) (For further details, see Sanchez and Yunis, 1977 in the Additional Readings at the end of this chapter.)

With the use of the banding techniques, some 320 bands per haploid set can be identified at metaphase. However, if mitosis is followed through late prophase and prometaphase, up to 1256 bands can be recognized.[6] In the idiogram of Figure 18–2, the left chromatid shows the bands observed at mid-metaphase, and the right

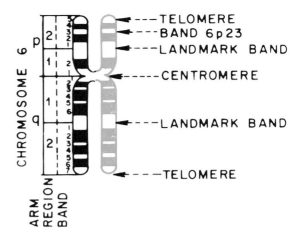

Fig. 18–3. Nomenclature set at the Paris conference (1971) to identify any particular region and band in a chromosome. *p,* short arm; *q,* long arm. Landmark bands divide each arm into regions which are composed of bands. (From Sanchez, O., and Yunis, J.J.: New chromosome techniques and their medical applications. *In* New Chromosome Syndromes. New York, Academic Press, Inc., 1977, p. 1.)

chromatid represents the G banding in late prophase.

Banding techniques, especially G banding, have become important tools in the analysis of mammalian, avian, reptilian, and amphibian chromosomes; distinct G bands have not been found in plant chromosomes, however. This absence has been attributed to the higher compactness of plant chromosomes, with more DNA content per unit length.[8] This assumption has not been proven, however, and we still do not know why plant chromosomes lack G bands.[9]

SUMMARY:

The Human Karyotype

The correct diploid number of human chromosomes (i.e., 44 autosomes + XY in the male, and 44 + XX in the female) was established in 1956. Rapid advances were made in human cytogenetics and several cases of aneuploidy were discovered (mongolism, Turner's and Klinefelter's syndromes, and so forth). After 1968 a new era was begun with the use of the *banding* techniques, which permitted a more precise identification of individual chromosomes and their parts. Thus, chromosomal aberrations were demonstrated using these techniques.

In karyotyping cells in tissue culture, blast-like transformation by phytohemagglutinin and hypotonic treatment with the use of colchicine to arrest mitosis at metaphase have been important technical advances. Recently chromosomes at prophase and prometaphase have been studied with banding techniques, and thus higher resolution has been achieved. The karyotype is generally made by cutting the microphotographs of individual chromosomes and lining them up by size with their respective partners. The *centromeric index* refers to the ratio of the lengths of the long and short arms of the chromosome. Computer-controlled microscopy is now being used in karyotyping.

The 22 pairs of autosomes are in seven groups (A through G) with differing morphology (i.e., metacentric, submetacentric, acrocentric with satellites, and others) (see Table 18–1). The X chromosome is submetacentric and belongs to group C. The Y chromosome falls into group G, as a small acrocentric, but it lacks satellites.

The first banding technique (Q banding) used fluorescent dyes (quinacrine) and identified the so-called Q bands. The other techniques generally involve a prestain treatment followed by various stains. G banding uses Giemsa stain. With G banding, three major types of chromatin can be recognized (euchromatin, centromeric, and intercalary heterochromatin). C banding stains specifically centromeric heterochromatin. R banding gives a pattern that is the reverse of that of Q and G banding. T banding stains telomeres. With banding, the chromosomes show special regions, subregions, and bands. Some 320 bands per haploid set have been identified, and at prometaphase, as many as 1256 bands.

18–2 SEX CHROMOSOMES AND SEX DETERMINATION

The well-known fact that male and female individuals are statistically found in about equal numbers suggests that sex is determined by a hereditary mechanism. There is physiological and cytological proof that sex is determined as soon as the egg is fertilized and that it depends on the gametes. Among the physiological evidence is the finding that identical twins—which originate from a single zygote—are always of the same sex. Cytological evidence was first obtained by McClung, who demonstrated that the karyotype of a cell is composed not only of common chromosomes *(autosomes)* but also of one or more special chromosomes that are distinguished from the autosomes by their morphological characteristics and behavior. These were called *accessory chromosomes, allosomes, heterochromosomes,* or *sex chromosomes.*

The majority of organisms have a pair of sex chromosomes, which, in the course of evolution, have been specialized for sex determination. One of the sexes has a pair of identical sex chromosomes (XX); the other may have a single sex chromosome, which may be unpaired (XO) or paired with a Y chromosome (XY).

In the XY pair there is a *homologous region* in which recombination may take place during meiosis and a *differential region* that is unpaired (Fig. 18–4).

In the human, the gametes are not identical with respect to the sex chromosomes. The male is heterozygous (XY) and produces two types of spermatozoa in similar proportions, i.e., one carries the X chromosome, and the other, the Y. The female, being homozygous (XX), produces only one kind of gamete (the ovum). Therefore, only two combinations of gametes are possible at fertilization, and the result is 50% males and 50% females (Fig. 16–1). This type of sex determination is found in all mammals and in certain insects, such as *Drosophila.* In other vertebrates and certain invertebrates, the female is heterogametic, and the male homogametic. For example, in birds and reptiles it is the female that has the heterozygous-ZW pair of chromosomes. In some animals the sex chromosomes cannot be distinguished from the others. In such cases the sex-determining genes are confined to a region in a pair of chromosomes.

In the human the Y chromosome contains an essential testis-determining factor and determines the male sex.

The chromosomal sex-determining mechanism evolved very early in vertebrate evolution, probably more than 300 million years ago. It is thought that the distinct ZW and XY chromosome pairs may have resulted from restrictions in the recombination of homologous chromosomes. The condensation of certain regions of the chromosome probably led to the deterioration of major genes in the Y and W chromosomes, leaving the alleles intact in the X and Z chromosomes. Such an alteration could have resulted from the production of repeated sequences with the generation of satellite DNA (see Sing et al., 1980 in the Additional Readings at the end of this chapter).

X and Y Sex Chromatin in Interphase Nuclei

In 1949 Barr and Bertram made the important discovery that in the interphase nucleus of females there is a small chromatin body (i.e., a chromocenter) which is lacking in males.[10] This was called *sex chromatin* or the *Barr body,* and

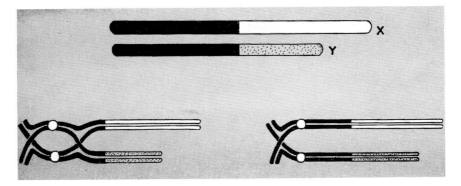

Fig. 18–4.   Diagram of a pair of XY sex chromosomes of a mammal. In black, the pairing or homologous segment; in white, the differential segments of the X chromosome; stippled, the differential segments of the Y chromosome. The configuration of the bivalent depends on the position of the chiasmata, which is produced only in the homologous segment.

more recently, after the 1971 Paris conference, *X chromatin.*

The X chromatin can be found as a small body in different positions within the nucleus. For example, in nerve cells it may be near the nucleolus (Fig. 18–5A), in the nucleoplasm, or near the nuclear envelope (Fig. 18–5B and C). In cells of the oral mucosa it is generally attached to the nuclear envelope (Fig. 18–5F), and in neutrophil leukocytes it may appear as a small rod called the *drumstick* (Fig. 18–5D).

The study of sex chromatin has a wide range

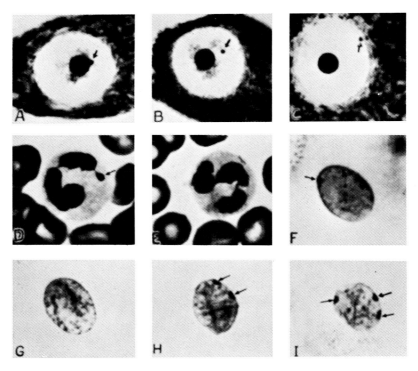

Fig. 18–5.   Sex chromatin in a nerve cell of a female cat. *A,* near the nucleolus; *B,* in the nucleoplasm; *C,* under the nuclear membrane (from M.L. Barr); *D,* normal leukocyte with a drumstick nuclear appendage from a human female, ×1800; *E,* same as *D,* in a male. (Remember that 90% of females also lack the drumstick, as in the male. ×1800. *F,* one sex chromatin corpuscle *(arrow)* in a nucleus from an oral smear. ×2000; *G,* same as *F,* from a male. Notice the lack of sex chromatin. ×1800. *H,* nucleus from the XXX female with two sex chromatin bodies. Vaginal smear. ×2000. *I,* similar, from an XXXX female. The three Barr bodies are indicated by *arrows.* ×2000. (From Barr, M.L., and Carr, D.H.: Nuclear sex. *In* Chromosomes in Medicine. Edited by J.L. Hamerton. Medical Advisory Committee of the National Spastics Society in association with Wm. Heinemann, Little Club Clinics in Developmental Medicine, No. 5, 1963.)

of medical applications and offers the possibility of relating the origin of certain congenital diseases to chromosome anomalies. Among these applications is the diagnosis of sex in intersexual states in postnatal and even in fetal life.

The frequency with which sex chromatin can be detected in the female varies from tissue to tissue. In nervous tissue the frequency may be 85%, whereas in whole mounts of amniotic or chorionic epithelium it may be as high as 96%. In oral smears the frequency may vary between 20 and 50% in normal females (see Hamerton, 1971 in the Additional Readings at the end of this chapter).

X Chromatin and X Chromosome Inactivation

The relationship between sex chromatin and sex chromosomes has been elucidated. Sex chromatin is derived from only one of the two X chromosomes; the other X is euchromatic and is decondensed at interphase. The number of corpuscles of sex chromatin at interphase is equal to $nX - 1$. This means that there is one Barr body fewer than the number of X chromosomes. This relationship between sex chromatin and sex chromosomes is particularly evident in some humans who have an abnormal number of sex chromosomes (see Table 18–2).

The differential behavior of the two X chromosomes in the female led Lyon (see Lyon, 1972 in the Additional Readings at the end of this chapter) to the so-called *inactive X hypothesis,* according to which: (1) only one of the X chromosomes is genetically active; (2) the X undergoing condensation may be of either maternal or paternal origin, and the decision by which X becomes inactive is random; and (3) the inactivation occurs early in embryonic life and remains fixed. It is now admitted that inactivation of the X chromosome is not complete and that only a part of the X chromosome condenses into a Barr corpuscle; this condensed chromosome is said to contain *facultative heterochromatin,* as opposed to the *constitutive heterochromatin* found in other chromosomes. Inactivation of the X chromosome takes place even in 3X and 4X individuals, in which only one X remains uncondensed (Fig. 18–5H and I).

The fact that X inactivation occurs at random has been demonstrated in human diseases linked to the X chromosome. The *Lesch-Nyhan syndrome,* in which a deficiency of one enzyme of the purine metabolism (i.e., hypoxanthine-guanine phosphoribosyl transferase) produces mental retardation and increased uric acid levels, results from a recessive mutation in the X chromosome. This is shown as follows: If fibroblasts of these patients are cultured in vitro, two types of cell clones are obtained. Half the clones

TABLE 18–2.   SEX ANEUPLOIDS IN HUMANS

X \ Y	O	Y	YY	Sex Chromatin
X	Monosomic XO Turner's syndrome 2X − 1 2n = 45	Disomic XY Normal 2X 2n = 46	XYY	0
XX	Disomic XX Normal 2X 2n = 46	Trisomic XXY Klinefelter's syndrome 2X + 2 2n = 47	Tetrasomic XXYY Klinefelter's syndrome 2X + 2 2n = 48	1
XXX	Trisomic XXX Metafemale 2X + 1 2n = 47	Tetrasomic XXXY Klinefelter's syndrome 2X + 2 2n = 48		2
XXXX	Tetrasomic XXXX Metafemale 2X + 2 2n = 48	Pentasomic XXXXY Klinefelter's syndrome 2X + 3 2n = 49		3
Phenotype	♀	♂	♂	

contain the enzyme, whereas the other half (in which the X carrying the normal gene is condensed) lack the enzyme.

The inactive X chromosome is known to begin replication in the latest part of the S phase of the cell cycle, when the autosomes and the active X chromosome have already replicated. As Figure 18–6 shows, this can be easily observed in cells treated with 5-bromodeoxyuridine followed by staining with Giemsa stain or by the fluorescent dye Hoechst 33258.

All the genes on the condensed X chromosome are inactive with the exception of those present in the short arm. Early in embryogenesis, once the inactivation is established, it is irreversibly maintained, except in the germ cell line, in which reactivation occurs at a specific stage of germ cell development. In the human embryo inactivation starts in the late blastocyst about the 16th day of life; the inactivated X chromosome then becomes condensed in the somatic cells (Fig. 18–7).

Most of our present knowledge about developmental regulation of the X chromosome originates from the mouse embryo. In the zygote the maternal X^m is active while the paternal X^p is inactive (sperm). During the segmentation period until the 8- to 16-cell stage (morula) both X^m and X^p are active. Inactivation starts with the formation of the trophoectoderm; at this stage, however, both X chromosomes are functional in the inner cell mass. At the next developmental stage both Xs are still active in the embryonic ectoderm (or epiblast). By 6.5 days, however, all cells of the somatic layers (i.e., ectoderm, mesoderm, and definitive endoderm) show X inactivation.

In germ cells inactivation occurs at a later period, when they enter the genital ridges. Coinciding with the entry into meiotic prophase there is reactivation of the X chromosome and both Xs continue to function throughout ovogenesis (see Martin, 1982 in the Additional Readings at the end of this chapter).

Recent studies have been aimed at understanding the molecular mechanism of inactivation. It has been suggested that if the primary change is in DNA, this could be accomplished by an increased methylation of CG bases. Some experimental support for this hypothesis comes from work in which treatment with 5-azacytidine (which leads to hypomethylation) has produced reactivation of some X-chromosome genes. This treatment, however, does not produce the reaction of the whole X chromosome.[11,12]

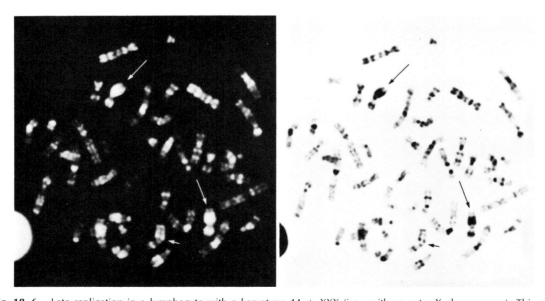

Fig. 18–6. Late replication in a lymphocyte with a karyotype 44 + XXX (i.e., with an extra X chromosome). This cell initiated replication (S phase) in 5-bromodeoxyuridine and then completed it in deoxythymidine. The late replicating X chromosomes have strong fluorescence *(left)* and an intense stain *(right)*. The early-replicating X chromosome is indicated by a short arrow, and the two late replicating X chromosomes by long arrows. (Courtesy of S.A. Latt, H.F. Willard, and P.S. Gerald, 1978.)

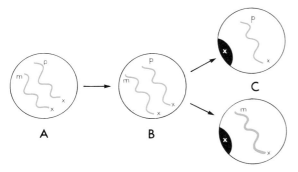

Fig. 18–7. Diagram of the evolution of XX chromosomes. *A,* in the zygote both the paternal *(p)* and maternal *(m)* X are euchromatic. *B,* in the early blastocyst the same is true as in *A. C,* in late blastocyst 50% of the cells have a maternal heterochromatic X chromosome, and the other 50% have the paternal heterochromatic X chromosome.

Y Chromatin—The Heterochromatic Region of the Y Chromosome

The fluorescence method for the study of Q banding led to the demonstration that a large portion of the Y chromosome is heterochromatic and appears, at interphase, as a strongly fluorescent body, the so-called Y *chromatin* (Fig. 18–8). While the X chromatin is 1.0 to 1.2 μm in diameter, the Y chromatin is only about 0.7 μm in diameter and generally is not attached to the nuclear envelope. The number of Y chromatin bodies is identical to the number of Y chromosomes. For example, individuals with XYY have two Y chromatin bodies.

The human Y chromosome consists of two segments: one that does not fluoresce and is ge-

netically active and another that fluoresces and is genetically inactive and contains a highly repetitive (satellite) DNA. It is this latter segment that may be polymorphic. The size of the long arm of the Y chromosome, where this heterochromatic segment is located, may vary considerably among individuals and subpopulations. For example, it is very large in the Arab populations of North Africa and very small in an American sect, the *Amish,* who are descendants of a Swiss and German group that emigrated in the eighteenth century and who have generally intermarried (see Sanchez and Yunis, 1977; Bühler, 1980 in the Additional Readings at the end of this chapter).

Sex is Determined by Sex Chromosomes, but Hormones Influence Differentiation

Although the primary determination of sex is made at fertilization (Fig. 16–1) the embryo acquires its definite sex characteristics by a more complex mechanism. An epigenetic factor (i.e., hormonal) may assume control of the genetic determination during development, thereby changing the phenotypic expression of the sex. Among vertebrates a condition of bisexuality may exist. For example, male amphibians have a rudimentary ovary (Bidder's organ) and vestigial oviducts.

In the human embryo until the sixth week the gonads and the primordia of the urogenital tract are identical in males and females. At this time the gonad has already been invaded by the primary XX or XY germ cells. At this point, a gene (or set of genes) present in the Y chromosome causes the undifferentiated gonad to differentiate into a testis, and the absence of this gene allows the gonad to become an ovary. The development into a testis starts as soon as the *gonocytes* (i.e., primordial germ cells) from the yolk sac have finished their migration into the *gonadal ridge.* Gonocytes of the male (XY) migrate deeper into the gonadal blastema, and female gonocytes (XX) remain at the periphery, forming a thick cortical layer.

H-Y Antigen. It is known that the Y chromosome in mammals is a strong male inducer and directs the organogenesis of the testis. This is done by inducing the production of a protein (or proteins) that becomes localized at the plasma membrane and determines, by cell rec-

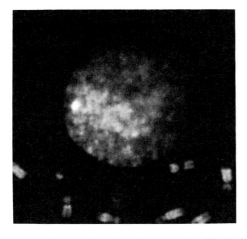

Fig. 18–8. Staining, by quinacrine mustard, of the Y chromosome in a human interphase cell. The strongly fluorescent body is clearly distinguished within the nucleus. (Courtesy of T. Caspersson.)

ognition, the formation of the seminiferous tubule. It has been proposed that in mammals a Y-linked determining gene induces the synthesis of this *H-Y antigen*, which is found in various cell types in males but is absent in females. This antigen is also a histocompatibility factor (H) that is found, for example, in the skin of male mice and that determines the rejection of female skin transplants.

Action of Testosterone. The human gonad differentiates into a definite testicle at the seventh week, whereas the female gonad begins to develop into an ovary between the eighth and ninth weeks of development. At the time of differentiation an important epigenetic factor is the production of androgens by somatic cells in the embryonic male gonad.

In mammals administration of *testosterone* (the male hormone) to the mother produces in the fetus a shift in the differentiation of XX genitalia to a male type, producing what is called *masculine pseudohermaphroditism*. This hormone acting locally accelerates the development of the testis, whereas in the female the absence of the hormone permits the slower development characteristic of the ovary (see Jost, 1970; Ohno, 1976a and b in the Additional Readings at the end of this chapter).

SUMMARY:
Sex Chromosomes and Sex Determination

Sex determination is transmitted from one generation to the next by a hereditary mechanism. In many species, as in the human, the gametes are not identical with respect to the sex chromosomes. One of the sexes is heterozygous; the other is homozygous. Upon fertilization the result is 50% males and 50% females. In most cases a pair of sex chromosomes is involved in sex determination (XX in females, XY or XO in males). Frequently the male is heterogametic, but in certain cases (birds, fishes) the female is heterogametic. In the human and most mammals the male character is determined by the Y chromosome. Sex chromosomes have a homologous region in which recombination may occur and a differential region that is related to sex determination.

X chromatin (i.e., sex chromatin or the Barr body) is a small chromatin body observed in the interphase nucleus of females and has a medical application in the diagnosis of intersexual states. Sex can be determined in the fetus by the study of smears of amniotic epithelium. Sex chromatin is derived from only one of the two X chromosomes, which becomes heteropyknotic. The number of chromatin bodies is nX − 1. It is thought that the X chromosome forming the sex chromatin is genetically inactive—the inactivation occurring early in embryonic life (i.e., *facultative heterochromatin*). This inactivation occurs at random, as can be demonstrated in cell clones from an individual with a hereditary disease involving an enzyme deficiency in purine metabolism. Both in somatic and germ cells, inactivation of the X chromosome has been studied in the mouse embryo. Some investigations support the view that inactivation is accomplished by an increased methylation of GC bases.

Y chromatin is a strongly fluorescent body observed in the interphase nucleus of males by using Q banding techniques. Individuals with XYY have two Y chromatins. This body is generally not attached to the nuclear envelope and represents a large heterochromatic portion of the Y chromosome. This part is genetically inactive and is found only in humans and gorillas and it contains repetitive (satellite) DNA.

Although sex is determined primarily by the sex chromosomes, during embryonic development an epigenetic factor (i.e., hormonal) may be important in sex differentiation.

Until the sixth week of embryonic life, both male and female gonads are identical. At this time they are invaded by XX or XY germ cells. The testicle differentiates in the seventh week, whereas the ovary begins to develop one or two weeks later. The early differentiation in the male depends on a local production of androgen. In the female the lack of androgen results in the slower development of the ovary.

The organogenesis of the testis is directed by the Y chromosome, which carries the gene for the H-Y antigen and induces its production. The H-Y antigen is a protein of the cell membrane that is involved in cell recognition and the formation of the seminiferous tubules. The H-Y antigen is absent in females. The same antigen (histocompatibility factor) is found in the skin of male mice, where it determines the rejection of female skin transplants. Testosterone given to a mother may shift differentiation in the fetus from an XX to a male type (*masculine pseudohermaphroditism*).

18–3 HUMAN CHROMOSOMAL ABNORMALITIES

The study of chromosomal abnormalities in humans can be divided into periods before and after the development of the banding techniques. In the earlier period it was found that 0.5% of all newborns have gross chromosomal abnormalities, about half of which are in the sex chromosomes and the rest of which are in the autosomes.

These abnormalities generally consist of aneuploidy, such as monosomy or trisomy. Structural aberrations; such as translocation, deficiency, duplication, and other more complex alterations, have also been observed (see Section 17–2).

Diagnosis of chromosomal aberrations can be made even before a child is born. This is done by removing amniotic fluid by puncturing the mother's abdomen, a procedure called *amniocentesis,* which does not entail risks to the fetus (Fig. 18–9). The amniotic fluid contains cells shed from the fetus and which can be cultured.

Amniocentesis, performed at mid-trimester of gestation, makes possible detection of chromosomal abnormalities such as Down's syndrome. This technique is useful for women who are carriers of translocations in which the risk of having an affected child is high. It should also be used in older mothers. The technique permits antenatal determinations, by karyotype, of fetal sex, which may be important in cases of sex-linked diseases. Amniocentesis also reveals abnormalities involving errors in metabolism and defects in the neural tube of the embryo (by analysis of *α-fetoprotein*). Other genetic abnormalities such as those involving the hemoglobin molecule (i.e., sickle cell anemia, β-thalassemia) can be revealed only by removing fetal blood (see Fuchs, 1980 in the Additional Readings at the end of this chapter).

At present nearly 200 inborn disorders, many

of which are extremely rare, can be detected in utero. More recently a new technique of prenatal diagnosis has been developed which is based on obtaining a *biopsy of the chorionic villi* present at an early stage in embryonic development. This technique allows prenatal testing during the first 3 months of pregnancy, a time at which amniocentesis is not possible. The tissue obtained by a small catheter is made of rapidly dividing embryo cells that can be analyzed readily for chromosomal and biochemical defects (see Kolata, 1983 in the Additional Readings at the end of this chapter).

Aneuploidy—Caused by Nondisjunction of Chromosomes

In aneuploidy, a frequent chromosomal defect, the genetic message contained in each chromosome is maintained intact. The alteration is quantitative; it resides in a disequilibrium established by the excess (trisomy) or deficit (monosomy) in the amount of genetic material. Such a *dosage effect* may be dangerous to the organism and may produce severe anatomical and functional anomalies (i.e., malformations). It will be shown later that trisomy produces changes that are characteristic for each chromosome present in excess; however, in all instances of trisomy the tendency is toward involution of the

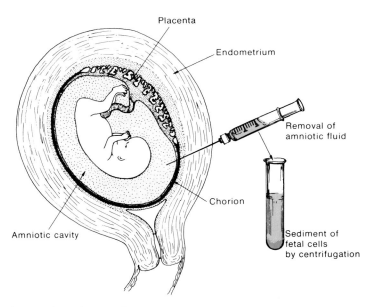

Fig. 18–9. Diagram showing the technique of amniocentesis.

nervous system, resulting in a more or less severe mental defect.

Another important consequence of aneuploidy is spontaneous abortion. Aneuploidy of one of the larger chromosomes is generally lethal, being more severe than an equivalent alteration in a smaller chromosome. Malformation, mental retardation, sterility, and spontaneous abortion operate as strong selective mechanisms tending to eliminate from the general population those individuals carrying deleterious genetic imbalances.

Aneuploidy originates by the mechanism called *nondisjunction*. The immediate cause of nondisjunction is the lagging, during anaphase, of one sister chromatid, which during telophase remains in one of the cells together with the other sister chromatid (Fig. 18–10). This change gives rise to a cell line that lacks one chromosome or has one chromosome in excess for that pair (monosomy and trisomy).

Nondisjunction may occur during *meiosis*, giving rise to an aneuploid *ovum*. This, when fertilized by a normal spermatozoon, results in a zygote and later an organism in which all cells are aneuploid. In some cases the alteration may be in the male gamete or in both, thus producing more complex types of abnormalities. If nondisjunction occurs in a *mitotic division*, which precedes the formation of germ cells, the effects are similar to those of *meiotic nondisjunction*. If mitotic nondisjunction occurs during embryonic development, a *mosaic* of different cell lines is produced.

Reciprocal Translocations—Identified by Banding

We mentioned in Section 17–2 the various structural aberrations that are found in human chromosomes (i.e., translocation, deficiency, and duplication) (see Fig. 17–9). The most common of these is the so-called *reciprocal translocation*, in which chromosome segments are exchanged between nonhomologous chromosomes. Some of these translocations may result in a chromosome having a length and configuration that may be easily identified. In other cases, however, the apparent morphology changes very little. Banding techniques are of fundamental importance in detecting such cases, as is clearly illustrated in Figure 18–11, in which there is a *balanced translocation* between a segment of the long arm of chromosome 2 and a segment of the short arm of chromosome 6 (a 2/6 translocation).

Sex Chromosome Abnormalities in Human Syndromes

Table 18–2 shows some of the most common abnormalities found in sex chromosomes to-

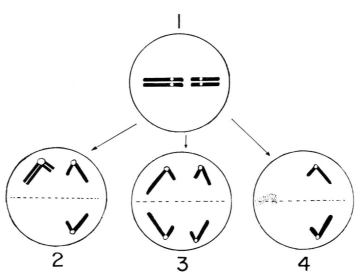

Fig. 18–10. Mitotic nondisjunction and chromosome loss. *1,* normal metaphase. *2,* nondisjunction anaphase giving rise to monosomic and trisomic nuclei. *3,* normal anaphase. *4,* a chromosome loss results in two monosomic nuclei.

A   B

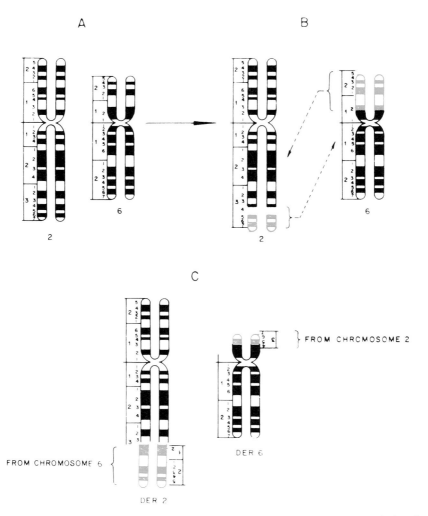

C

Fig. 18–11.   Schematic representation of a balanced translocation between chromosomes 2 and 6. The banding technique facilitates the recognition of the two segments that were translocated. The translocated segments are in red. (From Sanchez, O., and Yunis, J.J.: New chromosome techniques and their medical application. *In* New Chromosome Syndromes. New York, Academic Press, Inc., 1977, p. 1.)

gether with the clinical syndromes and an indication of the presence or absence of sex chromatin. Descriptions of the most important aberrations of this type follow.

Klinefelter's Syndrome. Most affected individuals are practically normal except for minor phenotypic anomalies. They have small testes, frequent gynecomastia (enlarged breasts), tendency to tallness, obesity, and underdevelopment of secondary sex characteristics. Spermatogenesis does not occur, thereby resulting in complete infertility. These individuals have a positive sex chromatin and 47 chromosomes (44 autosomes + XXY).

Males with 48 chromosomes (44 autosomes +

XXXY) and two Barr corpuscles have also been described. These individuals have features of Klinefelter's syndrome and are mentally retarded. Persons with 49 chromosomes (44 autosomes + XXXXY) have also been reported. They display extensive skeletal anomalies, extreme hypogenitalism, and severe mental deficiency. These persons have three X chromatin bodies.

XYY Syndrome. Males having two Y chromosomes have been identified in the past in maximum security institutions.[13,14] It was proposed that such individuals had a strong tendency toward antisocial behavior and aggression (see Hook, 1973 in the Additional Readings at the end of this chapter). More recently, XYY individuals have been found in the normal population in a proportion of 1 in 650 male infants, suggesting that the correlation with violence may not be as strong as was previously thought.

Turner's Syndrome (Gonadal Dysgenesis). Patients with Turner's syndrome usually have a female appearance with short stature, webbed neck (folds of skin extended from the mastoid to the shoulders), and generally infantile internal sexual organs. The ovary does not develop and shows complete absence of germ cells. As a result of this ovarian dysgenesis, menstruation does not occur, and secondary sexual characteristics do not develop. The karyotype shows 45 chromosomes (44 autosomes + X), and there is no sex chromatin. In this syndrome there is little difference in the gonads up to the third month of gestation; the ovaries contain approximately the normal number of germ cells. Later on there is a rapid atresia of the germ cells, leading to their virtual disappearance after puberty. It is probable that the lack of one X chromosome determines the progressive ovarian atresia.

Females with X Polysomy. Triplo-X constitution (47 chromosomes: 44 autosomes + XXX) was detected in phenotypically near-normal females. A number of these females are mentally subnormal or psychotic, and some of them menstruate. Two sex chromatin bodies are found in cells from these women. A few severely retarded patients with three corpuscles of sex chromatin and 48 chromosomes (44 autosomes + XXXX) have been found. These persons are also called metafemales (Table 18–2).

Mixed Chromosomal Aberrations. Klinefelter's syndrome may be found combined with mongolism. Such a person has 48 chromosomes: 45 autosomes (including a trisomic pair 21) + XXY.

Mosaics. Sometimes chromosomal aberrations are produced during development of the embryo. One interesting example is induced by the loss of one Y chromosome in the first division of the zygote. This may result in twins, of which one has *normal* male characters and the other has Turner's syndrome.

Sometimes chromosomal aberrations due to nondisjunction are produced during development of the embryo. These individuals possess different chromosomal complements from cell to cell within the same tissue or between tissues,

TABLE 18–3. SEX CHROMOSOME MOSAICS

	Clinical Syndrome	Sex Chromatin
Females		
XO/XY	Turner	—
XO/XX	Turner	—
XO/XYY	Turner	—
XO/XXX	Variable	—
Males		
XX/XXY	Klinefelter	+
XY/XXY	Klinefelter	+
XXXY/XXXXY	Small gonads and immature sexual characteristics, mental disorder	3 +
XO/XY	Hermaphrodite	

depending on the embryonic stage at which nondisjunction occurred. Table 18–3 illustrates the most frequently found mosaics of the sex chromosomes in females and males.

Sex Chromosomes and True Hermaphroditism. A true hermaphrodite is an individual with both ovarian and testicular tissue. The two types of gonadal tissue may be separated or in close proximity (ovotestis).

Mongolism—The Best Known Autosomal Abnormality

Before the development of the banding techniques about 12 classic syndromes had been described, the majority of which were caused by an extra chromosome or the loss of a chromosome segment. Except for the X monosomy (Turner's syndrome), any other monosomy is believed to be inviable (i.e., leads to death of the embryo). This also holds true for all trisomies other than those of chromosomes 13, 18, or 21.

21-Trisomy (Mongolism or Down's syndrome). Among the most important autosomal aberrations is *mongolism*, which is characterized by multiple malformations, mental retardation, and markedly defective development of the central nervous system. It was discovered that the mongoloid has an extra chromosome. Pair 21 is trisomic instead of normal. This aberration probably originates from nondisjunction of pair 21 during meiosis.

The extra chromosome of pair 21 in some cases may become attached to another autosome (translocation), usually to pair 22.

The phenotype of a mongoloid is recognizable

at birth in most cases. The face of such a patient has a special moon-like aspect, with oblique palpebral fissures, increased separation between the eyes, and a skin fold (epicanthus) at the inner part of the eyes. The nose is flattened, the ears are malformed, the mouth is constantly open, and the tongue protrudes.

Mongolism is the most common congenital disease and is present in more than 0.1% of births. Its frequency increases as the mother's age exceeds 35. The occurrence of this trisomy is sporadic, and in general there is no recurrence in the family. However, in the rarer cases of mongolism by translocation, the disease may affect siblings and may appear in successive generations. Fortunately, this "translocation trisomy" represents only 3 to 4% of all cases of mongolism. In this type there is no change in frequency with the age of the mother, and when the aberration is properly determined by karyotype analysis, the parents should be warned of a repetition of this defect.

21-Monosomy. Complete deletion of one of the chromosomes in pair 21 is apparently lethal, but there is a syndrome in which a large part of one is lacking. Children with this condition have a morphological aspect which is, to some extent, the opposite of mongolism. The nose is prominent, the distance between eyes shorter than normal, and the ears are large and the muscles contracted. It seems that in trisomy and monosomy of the 21st, the phenotype shifts to one or the other side of normal.

18-Trisomy. In this case the child is small and weak, the head is laterally flattened, and the helix of the ear scarcely developed. The hands are short and the digital imprints are rather simple. These children are very retarded mentally and usually die before one year of age.

18-Monosomy. This is the opposite syndrome in which a partial deletion of one chromosome of the pair occurs.[15] The ears are voluminous, the fingers long, and the digital imprints are complex and convoluted.

13-Trisomy. Multiple and severe body malformations, as well as profound mental deficiency, are characteristic features. The head is small and the eyes are often small, or absent. Harelip, cleft palate, and malformations of the brain are frequent. The internal organs are also severely malformed, and in most cases death occurs soon after birth.

Structural Aberrations May Cause Other Syndromes

One of the most remarkable phenotypes associated with a chromosome structural change is the condition known as *cri du chat* syndrome.[16] The affected baby has a strange mewing cry, multiple malformations, and mental retardation associated with partial deletion of the short arm of chromosome 5.

Another type of patient showing facial alteration and skeletal and ophthalmologic abnormalities, along with profound mental retardation, is the carrier of a deletion of the long arm of chromosome 18.

Banding Techniques Have Detected More Than 30 New Syndromes

The use of the banding techniques has permitted the detection of more than 30 chromosomal syndromes that result from a partial trisomy or deletion from each of almost all the chromosomes. In Figure 18–12, syndromes of total or partial trisomy of various chromosomes are indicated by dotted lines, and total or partial monosomy, by solid lines. Such alterations have been found in children with mental retardation or minor congenital defects and also in some with normal intelligence and minor anomalies that were earlier considered deviations of normality.[17]

The technical advances have permitted the identification of chromosome defects even when there are few clinical symptoms. Furthermore, these methods have allowed the study of cases of balanced chromosomal translocations (Fig. 18–11), which, as carriers of this type of syndrome, are important to detect. In such cases, transmission can be prevented by proper counseling and prenatal diagnosis. (Consideration of these syndromes, although of considerable medical importance, is beyond the scope of this book; see the Additional Readings at the end of this chapter.)

Chromosome Aberrations in Cancer, Fragile Sites and Oncogenes

Before the use of banding techniques a variety of chromosome changes were observed in human cancers; however, an association be-

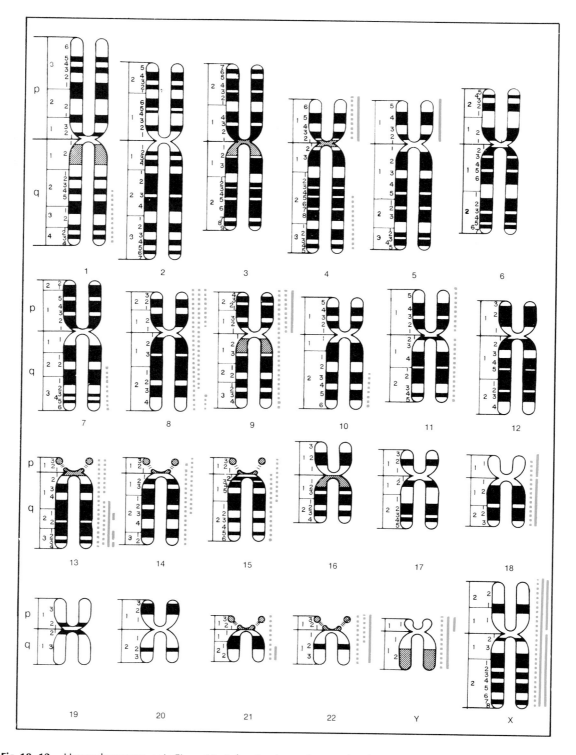

Fig. 18–12. Human karyotype, as in Figure 18–2 showing the syndromes of total or partial trisomy of various chromosomes (indicated by dotted lines) and total or partial monosomy (solid lines). (Courtesy of J.J. Yunis.)

tween them and the specific tumor types could not be established. This situation changed in 1960, when Nowell and Hungerdorf discovered that in *chronic myeloid leukemia* (a cancer of myeloid cells) there was a specific change in the so called *Philadelphia chromosome* (Ph[1]). This aberration involves a balanced translocation between chromosomes 9 and 22. The Ph[1] chromosome is acquired with the disease and is not found in cultured fibroblasts of the patients nor in myeloid cells of an identical twin. The use of banding techniques allowed the discovery of several specific chromosome aberrations in different types of cancer. For example, in Burkitt's lymphoma (a cancer of lymphatic cells), a piece of chromosome number 8 is translocated into number 14 (see Fig. 22–14); in retinoblastoma (a cancer from the retina), there is a deletion in the long arm of chromosome 13; and in some lung cancers, a deletion occurs in chromosome 3. These and many other findings in different types of cancers led to a reconsideration of the role of chromosome aberrations in the origin and development of tumors (see Rowley, 1983 in the Additional Readings at the end of this chapter).

Advances in cytogenetics and studies done at the molecular level have led to the establishment of a close relationship between chromosomal aberrations and certain types of cancer. At the same time, advances have been made in the characterization of special sites having a higher probability of breakage in chromosomes—the so called *fragile sites*—as well as on the localization of certain genes—the *oncogenes*—that have the potential of inducing a *cancerous transformation* (see Section 22–3).

Fragile sites are chromosome regions, comprising both chromatids, that take a paler stain than the rest of the chromosome. They are always present in the same regions of the chromosome and are transmitted by simple mendelian heredity. Fragility can be expressed by the formation of gaps in which a break can be produced; this may originate a deletion or a translocation. Fragile sites probably represent chromosome segments in which abnormal coiling during cell division facilitates the breakage of the chromosome. Such fragile sites are characteristic of leukemias and lymphomas but have also been found in solid tumors, such as carcinomas of the lung, ovary, and meningiomas.

It is apparent that the break points that occur in human cancers and lead to the production of chromosomal aberrations are topographically related to the position of *oncogenes*. Although this special type of gene, which induces cancer, is studied in detail in Section 22–3, a few concepts are mentioned here because of the possible relationship of oncogenes with chromosomal aberrations and fragile sites in cancer cells.

Oncogenes were first discovered in certain retroviruses of avian, rodent, feline, or primate origin. (A retrovirus is an RNA virus that can synthesize DNA by the enzyme *reverse transcriptase*, Section 22–2.) In contrast to *viral oncogenes*, which can induce transformation if they invade a cell, *cellular oncogenes* are present in normal eukaryotic cells and occupy constant positions in the genome. These genes, which coexist in the normal cell, can be activated by different mechanisms; for example, by mutation caused by a carcinogenic agent (e.g., radiations or certain chemicals), or by a change in position in the genome, such as that resulting from a translocation (see Fig. 22–14). Such changes could increase the synthesis of products originated from the oncogene, which can then induce the cancerous transformation (see Section 22–3).

The following could be a hypothetical series of events taking place during a transformation with chromosomal aberration: (1) the presence of fragile segments could be a predisposing condition for cancer; (2) at these points, a breakage occurs; (3) the breakage leads to a chromosome rearrangement that changes the position of the oncogene; (4) the oncogene is activated, and the synthesis of its products determines cancerous transformation with uncontrolled cell multiplication.

(The changes that occur in the cell surface and in the cytoskeleton by transformation were studied in Sections 5–4 and 6–5.)

SUMMARY:
Chromosomal Abnormalities

In the early period of human cytogenetics about 0.5% of newborns were found to have gross chromosomal abnormalities. With the development of the banding techniques, that finding was duplicated and some 30 syndromes and several malignancies have been shown to be associated with chromosomal defects. Abnormalities may consist of aneuploidy (monosomy or trisomy), in which case the genetic message in the chromosome is intact, but there is a *dosage effect* that may produce severe anatomical and functional anomalies (mental defect in trisomy).

Aneuploidy is produced by nondisjunction, which may occur at meiosis or mitosis. In nondisjunction one sister chromatid does not reach the pole and remains with the other sister chromatid (one cell is monosomic, the other trisomic). Several structural aberrations (i.e., translocation, deficiency, or duplication) are found in human chromosomes. In the *reciprocal* translocation segments are exchanged between nonhomologous chromosomes. The banding techniques are very helpful in detecting cases of balanced translocation (Fig. 18–11).

Aberrations of the sex chromosomes lead to the production of several syndromes. In *Klinefelter's* syndrome there are 47 chromosomes (44 + XXY), and the sex chromatin is positive. The individual has male characteristics, but has a small testis and underdeveloped secondary sex characteristics. Cases with 48 and 49 chromosomes (44 + XXXY and 44 + XXXXY) and having 2 and 3 corpuscles of sex chromatin, respectively, have been described (Table 18–2).

In Turner's syndrome the individual has a female appearance with dwarfism. The ovaries do not develop; there are 45 chromosomes (44 + X), and sex chromatin is lacking. Examples of numerical autosomal aberrations are; *mongolism*, generally caused by a trisomy of the 21st pair (sometimes the extra chromosome is translocated upon another autosome); *monosomy of the 21st pair*, which produces individuals in whom the appearance of the face is approximately opposite to that of a mongoloid; *trisomy* and *monosomy* of the 18th pair; and *trisomy* of the 13th pair. Many chromosomal abnormalities have profound alterations of the CNS and may produce lethal malformations.

The 30 chromosomal syndromes discovered with the banding techniques are the result of either partial trisomy or a deletion in each of almost all the chromosomes. These alterations are found in children with mental retardation or minor congenital defects and also in others with minor deviations from normality. Special consideration is given to balanced translocations because their transmission can be prevented by prenatal diagnosis or proper counseling.

Several types of cancer show chromosomal defects. The best known is the balanced translocation 9/22 (*Philadelphia chromosome*) found in *chronic myelogenous leukemia*. This chromosome is acquired with the disease. In *retinoblastoma* there is a deletion of the long arm of chromosome 13.

18–4 HUMAN CHROMOSOMES AND THE GENETIC MAP

In humans it has been calculated that there may be some 50,000 genes. Despite the complexity involved in a such huge number, in re-

cent years important advances have been made in the localization of genes to a particular chromosome or even to a chromosomal region (e.g., arm or band).

Before the era of molecular biology, the presence of a gene was inferred, as in Mendel's experiments, by the existence of two alternatives of a given trait (see Section 17–1). In humans, more than 1200 genes have been identified by this method and by the occurrence of diseases resulting from mutations.

Daltonism and Hemophilia—The Best Known Sex-Linked Diseases

The easiest of the genes to identify are the sex-linked genes, which are present either in the X chromosome or in the nonhomologous region of the Y chromosome (Fig. 18–4). At present more than 100 genes have been assigned to the X chromosome.

In 1911 Wilson assigned the specific gene for *daltonism* (red-green color blindness) to the X chromosome. This mutation is expressed in males (8% of whom have daltonism) by a recessive mechanism, but it is very rare in females. This is because males have only one X chromosome and therefore only one allele for this gene. As a result, any mutation on this chromosome will produce a phenotypic change. On the other hand, females have two copies of the same gene, and in the heterozygous condition color blindness (a recessive gene) is not expressed. However, women are genetic *carriers* and can transmit color blindness to their progeny.

Hemophilia (a disease with a defect in blood clotting) is also inherited by an X-linked recessive gene. This disease is transmitted by females but is expressed in males (Fig. 18–13). The few cases of daltonism or hemophilia in females are due to a homozygous condition in which both X chromosomes are altered at the same locus.

Somatic Cell Genetics and Chromosome Mapping

At present, the most widely used technique for gene mapping is based on *somatic cell genetics*, which involves the transfer of genes by a parasexual mechanism. This implies the fusion or hybridization between two somatic cells, i.e.,

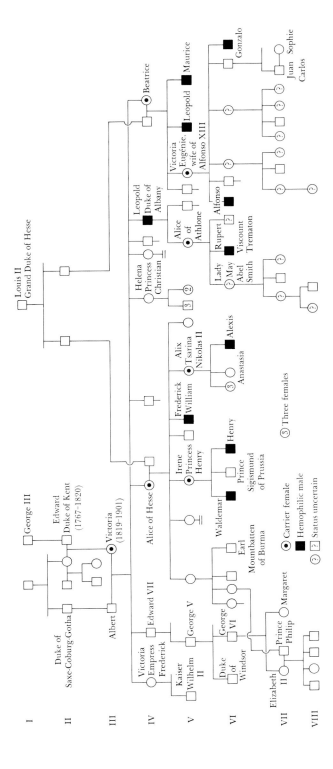

Fig. 18–13. Pedigree of the descendants of Queen Victoria showing carriers and affected males who possessed the X-linked gene conferring the disease hemophilia. (From McKusick, V.A.: Human Genetics, 2nd Ed. Englewood Cliffs, New Jersey, Prentice-Hall, Inc., 1969, p. 56. © 1969 by Prentice-Hall, Inc. Reprinted by permission of Prentice-Hall, Inc.)

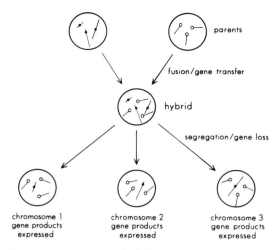

Fig. 18–14. Diagram of a hybrid produced by fusion between two cells having 3 chromosomes each. Observe that, after cell divisions, chromosomes from one parent are lost and finally only one remains in the hybrid. The gene products of the remaining chromosome can be studied by isoenzyme analysis or by nucleic acid hybridization. (Courtesy of F.H. Ruddle.)

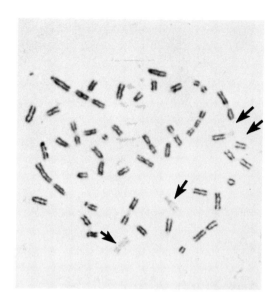

Fig. 18–15. Karyotype of mouse × human hybrid cell stained with alkaline Giemsa. Mouse chromosomes are more deeply stained. There are two human chromosomes with a gray color *(arrows)* and a mouse × human translocation *(double arrow)*. (Courtesy of F.H. Ruddle.)

a mechanism that does not involve sexual fertilization (see Section 23–3).

There are several ways by which the gene transfer from a donor cell to a recipient cell can be carried out:

(1) In *somatic cell hybridization* the whole genome of a cell is transferred into another. Experimentally a cultured human cell is fused with a mouse cell, by the action of fusogenic agents such as Sendai virus or polyethylene glycol (see Fig. 23–4). In subsequent cell divisions these hybrid cells tend to lose human chromosomes (Fig. 18–14). The chromosomes that are retained may be recognized by differences in morphology, especially with the use of banding techniques. Furthermore, as Figure 18–15 shows, by staining with alkaline Giemsa, it is possible to differentiate the more grayish human chromosomes from those of the mouse cell.

(2) In *microcell-mediated gene transfer*, one or a few chromosomes can be separated from the donor and integrated with the recipient cell. The technique consists of a prolonged treatment with colchicine, which produces the formation of micronuclei containing one or a few chromosomes. Then the cells are submitted to *cytochalasin B*, and upon centrifugation, the micronuclei, surrounded by a rim of cytoplasm and the plasma membrane, are separated (see Fig. 23–6).

Thereafter the micronuclei are fused to the recipient cell by Sendai virus.

(3) In *chromosome-mediated gene transfer*, metaphase chromosomes are separated and precipitated with calcium phosphate. Upon addition to recipient cells, chromosomes may be transferred to a cultured cell in a certain proportion. This method facilitates regional mapping of chromosomes.

(4) In *DNA-mediated gene transfer*, a still finer level of resolution is achieved. Fragments of genetically active DNA are first integrated in a plasmid and cloned using the techniques of genetic engineering that will be described in Section 19–2. Then the DNA fragments are incorporated into the recipient cell.

Selection of Hybrids and Gene Mapping

The genes incorporated by any of the above methods are transcriptionally and translationally active (in other words, they are able to produce messenger RNA and the corresponding specific proteins). The products of gene activity can be then used as markers for the localization of the gene in the corresponding chromosome.

This process can be understood more easily in the experiment described in Figure 18–14, in which three chromosomes are involved. After the fusion and the subsequent loss of chromosomes, clones of the cells containing chromosome 1, 2, or 3 are produced. The presence of specific gene products for each chromosome is demonstrated by a sensitive method that involves the separation of isoenzymes by electrophoresis. The human and mouse isoenzymes may have different electrophoretic behavior due to differences in charged amino acids. Other biochemical markers may be structural proteins or surface antigens.

Another useful experiment consists in selecting hybrid cells that retain a specific human chromosome. This may be done using mutated mouse cells, which are unable to grow in a particular medium unless their genetic enzyme deficiency is compensated by the human gene. Figure 18–16 shows that a cell uses two biochemical pathways to produce the nucleotides needed for DNA replication. The *de novo* pathway, in which sugars and amino acids are used, can be blocked with the aminopterin contained in the HAT medium (which contains hypoxanthine, aminopterin and thymidine). Hypoxanthine and thymidine are used for the synthesis of nucleotides by the *salvage pathway*, provided that the enzymes thymidine kinase (TK) and hypoxanthineguanine phosphoribosyl transferase (HGPRT) are present. When TK-deficient or HGPRT-deficient mouse cells are hybridized with human cells, the genes contained in the human chromosomes can complement the deficiency, compensating for the defect. Eventually clones of cells that contain one chromosome containing the TK gene (present in chromosome 17) can be obtained.

Once this is achieved, it is possible by gel electrophoresis to identify isoenzymes and other proteins encoded by this human chromosome.

To select other chromosomes, other mutants that require particular nutrients can be used. Eventually a bank of cell lines could be obtained that contains each individual human chromosome. It would then be possible to detect in each cell line the gene products that are encoded in the particular chromosome that was isolated.

Nucleic Acid Hybridization and Gene Mapping

Genes present in multiple copies can be detected directly by *in situ hybridization*. The ex-periment consists of the use of an RNA complementary to the genes, which can then be hybridized on a section or on a spread of the cell containing those genes. For example, the genes for the 18S and 28S ribosomal RNAs can be localized easily in the satellites of chromosomes 13, 14, 15, 21, and 22, which function as nucleolar organizers (Fig. 18–2). The 5S ribosomal RNA is localized in the distal end of the long arm of chromosome 1.

With the techniques of genetic engineering (see Section 19–2), it is possible to purify a definite messenger RNA and then to produce a *labeled complementary DNA* (cDNA) by the use of reverse transcriptase. This cDNA is then used to hybridize with the recipient hybrid cell. These techniques have led to the mapping of several important human genes (see D'Eustachio and Ruddle, 1980 in the Additional Readings at the end of this chapter).

Progress in Human Gene Mapping

Since 1970, with the development of somatic cell genetics and the techniques of banding, rapid progress has been achieved in mapping human genes. At the 1979 conference in Edinburgh, some 230 genes mapped in autosomes and 112 in the X chromosome were mentioned. According to F. H. Ruddle about three genes per month are mapped in the human species. At present structural genes have been assigned to each of the 22 autosomes, and in chromosome 1 some 28 genes have been mapped. Moreover, several genes have been localized in definite regions of chromosomes.

For example, the genes of the ABO blood groups are at the distal end of the long arm of chromosome 9; the Rh blood factor gene is in the short arm of chromosome 1; and the histone genes are in the short arm of chromosome 7. The loci for hemophilia and daltonism are in the distal part of the long arm of the X chromosome (see McKusick and Ruddle, 1977 in the Additional Readings at the end of this chapter). From these studies some general conclusions about the organization of the human genome can be drawn. One of the most important is that the genes for the different enzymes of a metabolic pathway (e.g., glycolysis, Krebs cycle, biosynthesis of nucleic acids) are mapped in different chromosomes. This implies that the expression

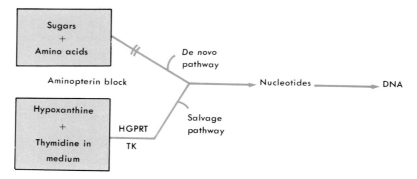

Fig. 18–16. The nucleotides required for DNA replication may come from two biochemical pathways. The *de novo* pathway is blocked by the drug aminopterin present in the HAT medium. Hypoxanthine (a purine precursor) and thymidine (a pyrimidine precursor) are utilized through the salvage pathway provided that the enzymes thymidine kinase (TK) and hypoxanthine-guanine phosphoribosyl transferase (HGPRT) are present.

of such genes must be coordinated by special interchromosomal control mechanisms within the cell nucleus.

SUMMARY:

The Human Genetic Map

In the human there may be some 50,000 structural genes. Before the times of molecular genetics, the presence of a gene was inferred by the existence of two alternatives of a given trait (as in Mendel's experiments). By this method some 1200 genes were identified in humans. The easiest to identify are the sex-linked genes (X or Y chromosome). The first of these to be assigned to the X chromosome was *daltonism*. *Hemophilia* is also X-linked (transmitted by females but expressed in males). More than 100 genes have been assigned to the X chromosome. The Y chromosome carries the testis-determining factor with the H-Y antigen.

Some of the methods used for gene mapping are linkage analysis (Section 17–1) (following a certain variation in a family), somatic cell hybridization, and nucleic acid hybridization. Human genes can be transferred to a mouse cell by somatic cell hybridization, by microcell transfer of chromosomes, as well as by chromosome- or DNA-mediated gene transfer. Clones of hybrid cells tend to lose human chromosomes and those remaining express their function by different isoenzymes or messenger RNAs. A selection (HAT) system may be used to segregate mutations and clones of cells.

In situ hybridization uses labeled RNA to localize a specific chromosome region. Genes with multiple copies are easily localized; for example, 18S and 28S ribosomal RNA are in the satellites of chromosomes 13, 14, 15, 21, and 22 (nucleolar organizers). The 5S ribosomal RNA is in chromosome 1.

At present, structural genes have been assigned to the 22 autosomes and sex chromosomes. Twenty-eight genes have been assigned to the large chromosome 1. The ABO blood group is in chromosome 9, the Rh blood factor in 1, and the histone genes in 7. The exact localization in a chromosomal arm or band has been achieved for certain genes.

REFERENCES

1. Tjio, J.H., and Levan, A.: Hereditas, *42*:1, 1956.
2. Lejeune, J., Turpin, R, and Gautier, M.M.: Ann. Génét., *1*:41, 1959.
3. Jacobs, P.A., and Strong, J.A.: Nature, *183*:302, 1959.
4. Ford, C.E., et al.: Lancet, *1*:711, 1959.
5. Caspersson, T., Zech, L., Johansson, C., and Modest, E.J.: Chromosoma, *30*:215, 1970.
6. Yunis, J.J.: Science, *191*:1268, 1976.
7. Holmquist, G., Gray, M., Porter, T., and Jordan, J.: Cell, *31*:121, 1982.
8. Anderson, L.K., Stack, S.M., and Mitchell, J.B.: Exp. Cell Res., *138*:433, 1982.
9. Greilhuter, D.: Annu. Rev. Genet., *12*:25, 1978.
10. Barr, M.L., and Bertram, E.G.: Nature, *163*:676, 1949.
11. Mohandes, T., Sparkes, R.S., and Shapiro, L.J.: Science, *211*:393, 1981.
12. Johnson, M.H.: Nature, *296*:493, 1982.
13. Jacobs, P.A., et al.: Nature, *208*:1351, 1965.
14. Price, W.W., and Whatmore, P.B.: Nature, *213*:815, 1967.
15. Grouchy, J., Bonnette, J., and Salmon, C.: Ann. Génét., *9*:19, 1966.
16. Lejeune, J., Lafourcade, J., Berger, R., and Rethoré, M.O.: Ann. Génét., *8*:11, 1965.
17. Lewandowski, R., and Yunis, J.J.: Am. J. Dis. Child., *129*:515, 1975.

ADDITIONAL READINGS

Borgaonkar, D.S.: Chromosomal Variation in Man. 3rd Ed. New York, Alan R. Liss. Inc., 1980.
Bühler, E.M.: A synopsis of the human Y chromosome. Hum. Genet., 55:145, 1980.
Comings, D.E.: Prenatal diagnosis and the "new genetics." Hum. Genet., 32:453, 1980.
Comings, D.E.: Mechanism of chromosome banding and implications for chromosome structure. Annu. Rev. Genet., *12*:25, 1978.
D'Eustachio, P., and Ruddle, F.H.: Somatic cell genetics

and the development of the immune system. Curr. Top. Dev. Biol., *14*:59, 1980.

Dutrillaux, B.: New chromosome technique. *In* Molecular Structure of Human Chromosomes. Edited by J.J. Yunis. New York, Academic Press, Inc., 1977, p. 233.

Ford, C.E.: Twenty years of human cytogenetics. Hereditas, *86*:5, 1977.

Fuchs, F.: Genetic amniocentesis. Sci. Am., *242*:47, 1980.

Greagan, R.P., and Ruddle, F.H.: New approaches to human gene mapping by somatic cell genetics. *In* Molecular Structure of Human Chromosomes. Edited by J.J. Yunis. New York, Academic Press, Inc., 1977.

Hamerton, J.L.: Human Cytogenetics. Vol. 1; p. 422; Vol. 2, p. 474. New York, Academic Press Inc., 1971.

Harnden, D.: Human cytogenetic nomenclature. Hum. Genet., *59*:269, 1982.

Harris, H.: The Principles of Human Biochemical Genetics. 2nd Ed. Amsterdam, Elsevier Publishing Co., 1975.

Hook, E.B.: Behavioral implication of the human XYY genotype. Science, *179*:139, 1973.

Jost, A.D.: Development of sexual characteristics. Sci. J. (Lond.), *6*:67, 1970.

Jukes, T.H.: Mitochondrial codes and evolution. Nature, *301*:19, 1983.

Kolata, G.: First trimester prenatal diagnosis. Science, *221*:1031, 1983.

Krumlauf, R., Jean Pierre, M., and Young, B.D.: Construction of genomic libraries from specific human chromosomes. Proc. Natl. Acad. Sci. USA, *79*:2971, 1982.

Latt, S.A., and Darlington, G.J.: Prenatal diagnosis. Cell Biological approaches. Methods in cell biology. Vol. 26. New York, Academic Press, Inc., 1982.

Levan, A., Levan, G., and Mitelman, F.: Chromosomes and cancer. Hereditas, *86*:15, 1977.

Lewandowski, R.C., and Yunis, J.J.: Phenotypic mapping in man. *In* New Chromosomal Syndrome. Edited by J.J. Yunis. New York, Academic Press, Inc., 1977.

Lyon, M.F.: X chromosome inactivation and development patterns in mammals. Biol. Rev., *47*:1, 1972.

Martin, G.R.: X-chromosome inactivation in mammals. Cell, *29*:721, 1982.

McKusick, V.A.: The mapping of human chromosomes. Sci. Am., *224*:104, 1971.

McKusick, V.A., and Ruddle, F.H.: The status of the gene map of the human chromosomes. Science, *390*:405, 1977.

Ohno, S.: Major regulatory genes for mammalian sexual development. Cell, *315*:321, 1976a.

Ohno, S.: La differenciation sexuelle. La Recherche, *7*(65):5, 1976b.

Puck, T.T.: Some new developments in genetic analysis of somatic mammalian cells. *In* Control of Cellular Division and Development: Part B. 1981, p. 393.

Puck, T.T., and Kao, F.T.: Somatic cell genetics and its applications to medicine. Annu. Rev. Genet., *16*:225, 1982.

Rothwell, N.V.: Human Genetics. Englewood Cliffs, New Jersey, Prentice-Hall, Inc., 1977.

Rowley, J.D.: Chromosome abnormalities in human leukemia. Annu. Rev. Genet., *14*:17, 1980.

Rowley, J.D.: Human oncogene locations and chromosome aberrations. Nature, *301*:290, 1983.

Sanchez, O., and Yunis, J.J.: New chromosome techniques and their medical application. *In* New Chromosome Syndromes. Edited by J.J. Yunis. New York, Academic Press, Inc., 1977, p. 1.

Sing, L., Purdom, I.F., and Jones, K.W.: Sex chromosome–associated satellite DNA. Evolution and conservation. Chromosoma, *79*:137, 1980.

Yunis, J.J.: Molecular Structure of Human Chromosomes. New York, Academic Press, Inc., 1977.

Yunis, J.J., Tsai, M.Y., and Willey, A.M.: Molecular Organization and Function of the Human Genome. New York, Academic Press, Inc., 1977.

THE GENETIC CODE AND GENETIC ENGINEERING

Deoxyribonucleic acid (DNA) is the genetic material of cells, carrying information in a coded form from cell to cell and from parent to offspring. When a gene is active, or expressed, its information is copied first into another nucleic acid, *ribonucleic acid* (RNA), which in turn directs the synthesis of the ultimate gene products, the specific *proteins*. RNA is also the genetic material of some viruses that contain RNA instead of DNA. These concepts, which constitute *the central dogma of molecular biology*, were summarized by Francis Crick in the following diagram:

$$\text{REPLICATION } \overset{\displaystyle\frown}{\underset{\displaystyle\smile}{}} \text{ DNA} \overset{\text{TRANSCRIPTION}}{\rightleftharpoons} \text{RNA} \xrightarrow{\text{TRANSLATION}} \text{PROTEIN}$$

The diagram shows that there are three main steps in the flow of genetic information from the genome (DNA): (1) *replication* of the DNA molecule, and thus of its genetic information, by a template mechanism; (2) *transcription* of this information into RNA molecules; and (3) *translation* of this information into the various protein components of a cell—for example, the enzymes.

The student should remember that in molecular biology the word *transcription* is used as a synonym for RNA synthesis, and *translation* as a synonym for protein synthesis.

As the diagram above shows, transcription can be reversed; i.e., RNA sometimes is copied into DNA, which can then become integrated in the genome. This "reverse transcription" occurs during the life cycle of some RNA viruses (called retroviruses). Because most animal cells contain latent retroviruses in their genome (Section 22–3), reinsertion of DNA copied from RNA (cDNA) is not uncommon. Translation of RNA into protein, however, is unidirectional and cannot be reversed.

In this chapter we will analyze the way in which DNA (which is made of four types of nucleotides) codes the information for proteins (which are made of 20 amino acids). We shall see that in the *genetic code,* each amino acid is specified by a group of three nucleotides. The genetic code is universal; i.e., it is the same for all living organisms. Deciphering the genetic code was one of the most important scientific achievements of this century; we will briefly review here the way in which this was done. In order to understand this section it is essential that the student be familiar with the structure of nucleic acids described in Section 2–2.

We will also consider two important topics: (1) DNA sequencing and (2) the isolation of eukaryotic genes using "genetic engineering" techniques by which eukaryotic genes can be grown in *Escherichia coli*. These subjects are included because studies using these techniques have produced a revolution in our knowledge of how genes are organized.

19–1 THE GENETIC CODE

Genes Are Made of DNA

By 1910, long after Mendel discovered that genes were inherited as discrete factors, the work of T.H. Morgan, E.B. Wilson, and others suggested that the mechanisms of heredity could be explained in terms of chromosomal behavior and therefore that genes were located in the chromosomes (see Chapter 13). In 1924 Feulgen developed his well-known histochemical reaction and showed that chromosomes contain DNA. For a long time, however, it was thought that the chromosomal proteins would turn out to be the genetic material, mainly because the composition of DNA (four nucleotides) was considered far too simple to code for the great diversity of proteins. The cytophotometric studies on Feulgen-stained cells by Pollister, Mirsky, and others in 1948 (Table 13–1) showed that eukaryotic nuclei have a *constant* amount of DNA and drew attention to the role of DNA in heredity. By then, however, the demonstration that DNA is the genetic material had already been made in experiments with microorganisms.

As Figure 19–1 shows, in 1928 Griffith found that a nonpathogenic strain of a bacterium called pneumococcus would kill mice if injected together with a dead (heated) culture of a patho-

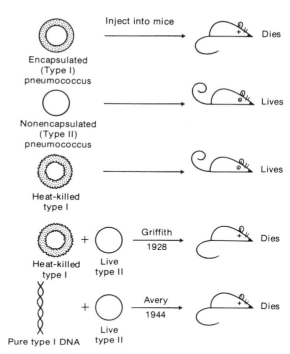

Fig. 19–1. Demonstration by *transformation* experiments with pneumococci that DNA is the genetic material. Type I pneumococcus has an extracellular polysaccharide capsule, which makes it virulent and lethal for mice. Type II lacks one of the enzymes required for making the polysaccharide and is not lethal. Both heated extracts and highly purified type I DNA are able to *transform* type II pneumococci so that they too can make an extracellular capsule and become lethal.

genic strain of pneumococcus. A nonpathogenic bacterium had been somehow *transformed* into a virulent strain. In 1944 Avery showed that a preparation of purified pneumococcus DNA had the transformation capacity, thus proving that DNA and not other cellular components was the genetic material. The virulent strains of pneumococcus have an extracellular polysaccharide capsule that provides resistance to the immunological reaction of the mouse. This capsule is absent in the nonpathogenic strains. In Avery's experiment the DNA transferred the gene coding for the enzyme *UDP-glucose dehydrogenase*, required for the synthesis of the polysaccharide capsule, which is absent in nonvirulent strains. This gene becomes incorporated into the chromosome of the nonvirulent strain and *transforms* it.

Most genes code for proteins; i.e., *one gene* codes for *one polypeptide chain*. Not all genes do so, however; some code only for RNAs, such as ribosomal and transfer RNAs. Although DNA is the genetic material in most organisms, some viruses code their information in RNA. In both cases the same four-letter code is used to store the information for protein synthesis.

Three Nucleotides Code for One Amino Acid

The *codons*—i.e., the hereditary units that contain the information coding for one amino acid—consist of three nucleotides (a triplet). This information is first *transcribed* into the messenger RNA, which has a sequence of bases complementary to the DNA from which it is copied. DNA and mRNA have only 4 different bases (A, G, C, T in DNA, and A, G, C, U in RNA) whereas proteins contain 20 different amino acids (see Fig. 2–13).

Deciphering how cells translate a four-letter code into one of 20 components was a major scientific achievement.

The code is read in groups of three bases, each triplet representing one amino acid (Fig. 19–2). Three is the minimum number of bases needed to code for 20 amino acids; the possible per-

mutations of the four bases are $4^3 = 64$. If the genetic code consisted of doublets, the number of codons would be insufficient ($4^2 = 16$), and if groups of four bases were utilized, the possibilities would be many more than necessary ($4^4 = 256$).

The length of the coding portion of a gene depends on the length of the message to be translated, i.e., the number of amino acids in the protein. For example, a sequence of 1500 nucleotides may contain 500 codons that code for a protein having 500 amino acids. The message is read in groups of three from a *fixed starting point*. The initial amino acid of a protein is determined by a special *initiation codon* (discussed later in this chapter).

The sequence of triplets determines the sequence of amino acids in a protein. Amino acids, however, cannot by themselves recognize a given triplet of mRNA; in order to do this, each amino acid is first attached to an *adaptor* molecule, which is called *transfer* RNA (tRNA), as shown schematically in Figure 19–2. Protein synthesis is described in detail in Chapter 21; here it is enough to note that every tRNA mole-

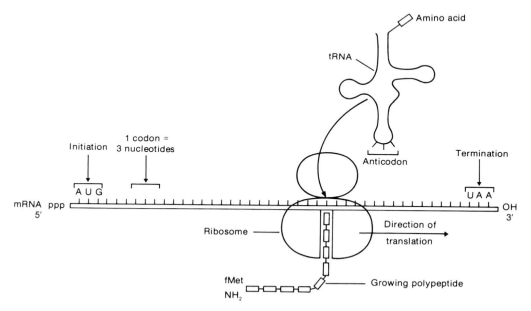

Fig. 19–2. Schematic representation of some of the components required for protein synthesis. mRNA is copied from the DNA, and groups of three nucleotides (codons) code for one amino acid, starting from a fixed starting point (AUG). The codons on the mRNA are recognized by tRNA, which acts as an adaptor with an *anticodon* triplet that reads the mRNA by complementary base pairing and carries a specific amino acid attached to one of its ends. Ribosomes are required to bring all the components of the protein synthesis machinery in place. Note that mRNA is read in the 5′ to 3′ direction and that proteins are synthesized starting from the NH₂ terminus. The 5′ end of mRNA starts in triphosphate in prokaryotes (ppp), and the 3′ end always ends in a hydroxyl group (OH). The initiator amino acid is fMet, which is coded by the first AUG codon in the mRNA. UAA is one of the three possible termination codons that indicate the end of a protein.

cule has an *amino acid attachment site* and a site for the recognition of the triplets in mRNA *(anticodon)*. Each tRNA has an *anticodon* of three nucleotides that can base-pair with the complementary *codon* of mRNA. The translation of the message into protein occurs in the *ribosomes*, which ensure the ordered interaction of all the components involved in protein synthesis.

Artificial mRNAs Were Used to Decipher the Genetic Code

The breakthrough that led to the solution of the genetic code came in 1961 when Nirenberg and Mattaei discovered that synthetic polyribonucleotides used as artificial mRNAs stimulated the incorporation of amino acids into polypeptides in a cell-free protein-synthesizing system. The first RNA used was polyuridylic acid (poly U), and the result was synthesis of polyphenylalanine (a polypeptide made only of phenylalanine). Thus it was deduced that the codon for phenylalanine was coded exclusively by uridine bases (UUU).

The cell-free protein-synthesizing system was an extract of *E. coli* that had been broken open and from which the cell walls had been removed by centrifugation. This extract contained the ribosomes, tRNA, aminoacyl-tRNA synthetases, and other factors required for protein synthesis. Radioactive amino acids were incorporated into polypeptides after the addition of ATP, GTP and exogenous mRNA. (The endogenous mRNAs from *E. coli* are short-lived and are degraded after preincubation of the extract at 37°C for a few minutes.) In addition to poly U, other homopolymers were tested. Poly A stimulated the uptake of lysine (codon AAA), and poly C, that of proline (codon CCC). (Poly G forms complex, triple-stranded structures in solution and is therefore unable to function as a synthetic mRNA.)

The base composition of many codons was deduced using polynucleotides formed by two bases. Thus a random copolymer of U and G contains eight triplets (UUU, UUG, UGU, GUU, UGG, GUG, GGU, and GGG) and directs the synthesis of several amino acids (Phe, Leu, Cys, Val, Trp, Val, Gly, Gly; see Table 19–1).

TABLE 19–1. THE GENETIC CODE

1st Base		2nd Base								3rd Base
		U		C		A		G		
U	UUU	Phe	UCU	Ser	UAU	Tyr	UGU	Cys		U
	UUC	Phe	UCC	Ser	UAC	Tyr	UGC	Cys		C
	UUA	Leu	UCA	Ser	UAA[6]	Term. (ochre)	UGA[2,6]	Term. (opal)		A
	UUG	Leu	UCG	Ser	UAG	Term. (amber)	UGG	Trp		G
C	CUU[4]	Leu	CCU	Pro	CAU	His	CGU	Arg		U
	CUC[4]	Leu	CCC	Pro	CAC	His	CGC	Arg		C
	CUA[4]	Leu	CCA	Pro	CAA	Gln	CGA	Arg		A
	CUG[4]	Leu	CCG	Pro	CAG	Gln	CGG[5]	Arg		G
A	AUU	Ile	ACU	Thr	AAU	Asn	AGU	Ser		U
	AUC	Ile	ACC	Thr	AAC	Asn	AGC	Ser		C
	AUA[1]	Ile	ACA	Thr	AAA	Lys	AGA[3]	Arg		A
	AUG	Met	ACG	Thr	AAG	Lys	AGG[3]	Arg		G
	AUG	fMet								
G	GUU	Val	GCU	Ala	GAU	Asp	GGU	Gly		U
	GUC	Val	GCC	Ala	GAC	Asp	GGC	Gly		C
	GUA	Val	GCA	Ala	GAA	Glu	GGA	Gly		A
	GUG	Val	GCG	Ala	GAG	Glu	GGG	Gly		G

[1]Met and initiator Met in human and yeast mitochondria.
[2]Trp in human, neurospora and yeast mitochondria.
[3]Termination in human mitochondria.
[4]Thr in yeast mitochondria.
[5]Trp in plant mitochondria.
[6]Gln in paramecium nuclear DNA.

The use of copolymers allowed the determination of the *composition* of codons but not the order in which the nucleotides were placed in the codon; for example, UUG could not be distinguished from UGU or GUU by this method. The recognition of the *sequence* of codons was later made possible by the use of trinucleotide templates of known base composition. In 1964 Leder and Nirenberg found that trinucleotides induced the binding of the specific AA-tRNAs (aminoacyl-tRNAs) to ribosomes. The complexes formed when ribosomes are incubated together with the various triplets and ^{14}C-AA-tRNAs can be easily detected because they are retained by nitrocellulose filters, while unbound AA-tRNA passes through the filter. The short triplets can be made using standard organic chemistry methods of synthesis, and about 50 codons were soon deciphered using the binding assay. For example, UUG induced the binding of leucine-tRNA, UGU bound cysteine-tRNA, and GUU, valine-tRNA.

Another method that allowed the recognition of the *sequence* of codons was developed in the laboratory of Khorana. With chemical synthetic and enzymatic methods, it was possible to produce long polyribonucleotides with alternating doublets or triplets of known sequence. These synthetic mRNAs were then used to direct protein synthesis in a cell-free system. When an alternating doublet was used, a polypeptide chain made of two alternating amino acids was formed. For example, with the alternation of U and G the result was:

GUG   UGU   GUG   UGU   GUG
Val – Cys – Val – Cys – Val

This showed that GUG and UGU code for valine (Val) and cysteine (Cys) and also confirmed that each codon is a triplet. When the template consisted of a group of three alternating bases, various polypeptide chains containing only one type of amino acid each were obtained. For example, the following alternation of UUGs was used:

UUG   UUG   UUG   UUG   UUG   UUG

Because this message can be read in three "reading frames," each one out of phase by one base, three homopolypeptide chains result as follows:

UUG   UUG   UUG   UUG   UUG
Leu – Leu – Leu – Leu – Leu

U UGU   UGU   UGU   UGU   UG
    Cys – Cys – Cys – Cys

UU GUU   GUU   GUU   GUU   G
      Val – Val – Val – Val

Synonyms in the Code—61 Codons for 20 Amino Acids

By 1966 all 64 possible codons had been deciphered. As shown in Table 19–1, 61 codons correspond to amino acids, and three represent signals for the termination of polypeptide chains. Because there are only 20 amino acids, it is clear that several triplets can code for the same amino acid; i.e., some triplets are synonyms. (This fact is sometimes called *degeneracy of the genetic code*.) For example, proline is encoded by CCU, CCA, CCG, and CCC. Note that in most cases the synonymous codons differ only in the base occupying the third position of the triplet and that the first two bases of the codon are more important in coding. As a result, mutations that change the third base frequently go unnoticed, since they may not change the amino acid composition of the protein, and are then called *silent* mutations.

DNA sequencing studies have confirmed that all the possible codons are utilized in vivo. The use of all possible sequences probably has advantages in that it minimizes the effect of harmful mutations. If many codons had been left without a meaning, mutations giving rise to them would interfere severely with the normal sequence of protein synthesis.

The use of 61 codons to code for 20 amino acids prompted an additional question: Is each triplet recognized by a special tRNA molecule? It was found that there are fewer than 61 tRNA species and that one tRNA can recognize more than one codon (although always for the same amino acid). This is because the base of the tRNA anticodon that base-pairs with the third base of the codon (the one that is less important in coding) can have some degree of "wobble," i.e., a certain amount of movement that allows this base to establish hydrogen bonds with bases other than the normal complementary ones. For example, G in this position of the anticodon can base-pair with U or C in the mRNA, while U can interact with A or C in the third position of the codon.

AUG—The Initiation Codon; UAG, UAA, and UGA—Termination Codons

Figures 19–2 and 19–3 show that mRNA is translated in the 5′→3′ direction (which is also the direction of RNA synthesis). The polypeptide chain is always assembled sequentially, starting from the end bearing the NH_2-terminus.

The *initiation signal* for the synthesis of a protein is the AUG codon. This codon has a dual function. When the AUG is at the beginning of a bacterial message *(starting codon)*, it will code for the incorporation of N-formylmethionine (fMet), and AUG in any other position of the message will code for normal methionine (Met). There is a special fMet-tRNA that is different from the tRNA for normal methionine. In order to bind to ribosomes, fMet-tRNA requires protein (initiation factor 2), while Met-tRNA uses the same elongation factor used by all other tRNAs. In eukaryotes there is also a special initiation tRNA, but it directs the incorporation of methionine instead of formylmethionine.

Natural mRNAs always start translation at AUG, but this is not an absolute requirement for *artificial* mRNAs in in vitro systems, as in the case of the homopolymers discussed above.

The *termination signal* is provided by the codons UAG, UAA, and UGA. When the ribosome reaches a termination codon, the completed polypeptide chain is released. Unlike all other codons, UAG, UAA, and UGA are not recognized by special tRNAs, but by special proteins, the *releasing factors* of protein synthesis (see Section 21–6).

The Genetic Code Is Universal, Except for Mitochondria

Although most of our knowledge about the genetic code initially came from experiments with *E. coli*, similar results have been obtained with other systems, such as amphibian, mammalian, and plant tissue. It may be said that the genetic code is universal, i.e., there is a single code for all living organisms.

This universality provides strong evidence that life on earth started only once. When the first living forms appeared some 3 billion years ago, the genetic code was established and it has not changed since then throughout the evolution of living organisms. Once the initial code was established, there were strong selective pressures to maintain it invariant because the change in a single codon assignment would change amino acids in a great many proteins at the same time and these multiple mutations would in all likelihood be lethal.

This selective pressure has been less strict in mitochondrial DNA. Mitochondria code for only a few proteins and have their own protein synthesis machinery. Changes in the assignments of a few codons have arisen (Table 19–1), but the overall code has been maintained. The mitochondrial code will be analyzed in detail later in this chapter. In the ciliated protozoan Paramecium, the termination codons UGA and UAA have been found to code for the amino acid glutamine, even in non-mitochondrial genes. These are exceptions, however, and do not change the

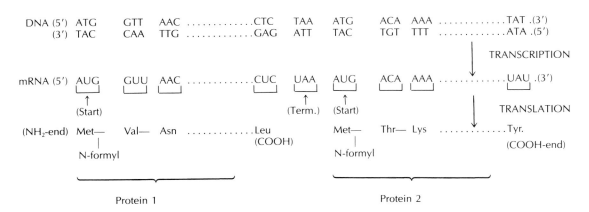

Fig. 19–3. Transcription and translation of a hypothetical prokaryotic mRNA that has two start and two stop signals in the same molecule (polycistronic mRNA). Eukaryotic mRNAs always have only one AUG start per molecule (monocistronic). Note that transcription and translation take place in a 5′ to 3′ direction and that protein synthesis starts at the amino terminus and ends at the carboxyl end.

fact that all living organisms share essentially the same code.

Mutations Produce Amino Acid Changes

In addition to biochemical in vitro systems, genetics was of fundamental importance in deciphering the genetic code. A mutation is a change in the DNA sequence, and this change is reflected in the sequence of the corresponding RNA or protein. Two main types of mutations can be distinguished: (1) *point mutations*, which involve a single nucleotide and (2) changes involving longer segments of DNA. Among the latter there are *deletions* of whole genes, *translocations* to other chromosomes, or *insertions* of pieces of DNA into another gene. In bacteria, as well as in eukaryotes, spontaneous mutations most frequently are caused by *transposons*, which are segments of DNA that have a tendency to jump around in the genome (Fig. 22–11). The frequency of mutations can be increased by ionizing radiation, ultraviolet light (Section 14–2), and chemical agents.

It was the analysis of point (single-base) mutations that was most helpful in explaining how the genetic code works. In fact, the first direct proof that genes code for proteins was provided by Vernon Ingram in 1957 when he showed that the mutation that causes the human disease sickle-cell anemia produces the incorporation of a valine instead of a glutamic acid in hemoglobin (codon change GAA to GUA).

We can distinguish four main types of point mutations:

1. *Silent mutations* change a nucleotide but not the amino acid sequence because they affect the third position of the codon, which is usually less important in coding. Let us consider the following imaginary mRNA:

3	6	9	12	15	18
AUG	UAU	CCA	UAU	CCA	UAG
fMet	Tyr	Pro	Tyr	Pro	Term

An A to U change in position 9 will give:

		↓			
AUG	UAU	CCU	UAU	CCA	UAG
fMet	Tyr	Pro	Tyr	Pro	Term

This is a silent mutation because it leaves the protein sequence unchanged.

2. *Missense mutations* change the meaning of

a codon, changing one amino acid into another. For example, a C to U change at position 7 will give:

		↓			
AUG	UAU	UCA	UAU	CCA	UAG
fMet	Tyr	Ser	Tyr	Pro	Term

3. *Nonsense mutations* arise when a codon for an amino acid is mutated into a termination codon (UAG, UAA, or UGA), resulting in the production of a shorter protein. For example, a U to G change at position 12 will give:

			↓		
AUG	UAU	CCA	UAG	CCA	UAG
fMet	Tyr	Pro	Term	—	—

4. *Frameshift mutations* arise from the insertion or deletion of individual nucleotides and cause the rest of the message downstream of the mutation to be read out of phase, producing an incorrect protein from then on. For example, an insertion of an extra G between nucleotides 6 and 7 of our imaginary mRNA would give:

		↓			
AUG	UAU	GCC	AUA	UCC	AUA
fMet	Tyr	Ala	Ile	Ser	Ile

Nonsense and frameshift mutations rarely go unnoticed by the organism, because the shorter or abnormal proteins are usually inactive. In missense mutations, on the other hand, the change of a single amino acid is frequently compatible with some degree of biological activity. Silent mutations, of course, always go unnoticed.

Suppressor Mutations—tRNA Mutations

The phenotypic effects of point mutations can sometimes be reversed by a second mutation in a different gene, which is called a *suppressor* mutation. There are several types of suppressor mutations, but the most interesting are tRNA mutations that act as *nonsense* (chain termination) *suppressors*. These mutations substitute a base in the anticodon of a tRNA so that it will now incorporate an amino acid at the codon that has mutated to a chain termination codon. In the example shown in Figure 19–4 a mutation in the tyrosine tRNA anticodon causes it to read a UAG (chain termination) codon and

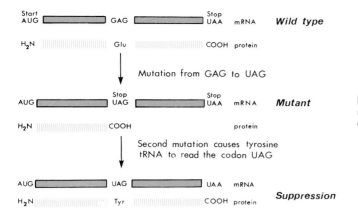

insert tyrosine at its site, thus allowing completion of the polypeptide chain, although with one amino acid change. The tRNA mutations that suppress UAG codons are sometimes called *amber* suppressors for historical reasons (due to the aspect of the plaques of the T_4 bacteriophage mutants in which the phenomenon was initially found; UAA suppressors are called *ochre* suppressors, and those for UGA codons are called *opal* suppressors.

An interesting case is that of mutant tRNAs that can suppress frameshift mutations caused by insertions of single bases. *Frameshift suppressor* tRNAs have an extra base in the anticodon (for example CCCC instead of CCC) and read a four-letter word instead of a triplet, thereby restoring the correct reading frame.

SUMMARY
The Genetic Code

The flow of genetic information is as shown in the following diagram:

$$\text{replication} \underset{(1)}{\overset{}{\curvearrowleft}} \text{DNA} \underset{(2)}{\overset{\text{transcription}}{\rightleftarrows}} \text{RNA} \xrightarrow[(3)]{\text{translation}} \text{PROTEIN}$$

Step (3) is undirectional, but step (2) can be reversed (RNA→DNA) by means of the enzyme *reverse transcriptase*. The information contained in DNA and in RNA is based on four nucleotides, while proteins are made by 20 amino acids. Cells can translate this four-letter code into one of 20 components.

The demonstration that DNA is the informational molecule came mainly from the experiments of Avery, in which purified DNA from virulent pneumococci penetrated the non-virulent type and produced its *transformation* (Fig. 19–1).

The genetic code is read in groups of three nucleotides. These units of genetic information are called the *codons*, i.e., the triplets of bases that spec-

ify single amino acids. The genetic code consists of 4^3 = 64 codons, and the length of a gene is related to number of amino acids in the protein (e.g., 1500 nucleotides constitute 500 codons coding for 500 amino acids). The DNA code is transcribed into an RNA code which is complementary to one DNA strand, and on the ribosomes the message is read by an adaptor molecule, the transfer RNA (tRNA), which carries the amino acid (Fig. 19–2).

The genetic code was deciphered by the use of synthetic mRNAs made of a single type of nucleotide, of polynucleotides containing two or three types of bases, and of trinucleotide templates of known base composition. Chemical synthesis and enzymatic methods produced alternating doublets or triplets forming polyribonucleotides that permitted further studies of the code. Of the 64 codons, 3 represent termination signals, and 61 code for the 20 amino acids (Table 19–1). Thus some triplets are synonyms (i.e., the code is degenerate). The number of tRNAs that carry the anticodons is fewer than 61. This discrepancy is explained by the wobble of the third base in the anticodon of tRNA, which can base-pair with more than one base.

The initial or starting codon is AUG, which codes for N-formylmethionine in prokaryotes. The termination or nonsense codons are UAG, UAA, and UGA. The genetic code is universal, i.e., there is a single code for all living organisms, except for mitochondrial DNA and ciliate protozoans, in which changes in the meaning of some codons have been observed. The universal code strongly suggests that life started only once on earth.

Changes in the base sequence of DNA produce gene mutations. These may involve deletion or insertion of long DNA segments or changes in single bases *(point mutations)*. In *missense mutations* one amino acid is replaced by another (e.g., sickle cell anemia). In *silent mutations* the mutation does not result in a change in amino acid, and is frequently a change in the third position of the codon. *Frameshift* mutations arise through insertion or deletion of single bases, which causes the message to be read out of phase. *Chain termination* or *nonsense mutations* occur when mutation of a codon results in a termination codon, and, therefore in the synthesis of shorter protein. Point mutation can sometimes be reversed by *suppressor mutations* (Fig. 19–4).

19–2   DNA SEQUENCING AND GENETIC ENGINEERING

DNA Sequencing Confirmed the Code

The genetic code can also be deciphered by comparing the amino acid sequence of proteins with the nucleotide sequences of the genes that code for them. However, the methods for sequencing nucleic acids had not yet been developed in the 1960s, and, as we have seen, the code was solved using cell-free systems with artificial mRNAs. Methods that allow very rapid sequencing of DNA have been developed recently and have produced a revolution in molecular biology. The complete nucleotide sequences of living organisms such as bacteriophage φX174 (a virus that infects *E. coli*), SV40 (a virus that infects monkey cells), bacteriophage λ (a virus that infects *E. coli*), and Epstein-Barr virus (a human herpes virus that produces tumors) have been established.

The nucleotide sequences obtained by the newer methods confirmed the genetic code and established that all of the possible codons are used in vivo. These method also provided completely unexpected insights into the way genes are organized. Because of its great importance, we will briefly analyze here how DNA is sequenced and the main conclusions derived from the φX174 and mitochondrial DNA sequencing work.

There are two DNA sequencing methods, both based on the generation of DNA fragments of different lengths which start at a fixed point and terminate at specific nucleotides. The DNA fragments are separated by size on polyacrylamide gels, and the nucleotide sequence is read directly from the gel. The DNA fragments can be produced by chemical cleavage (Maxam and Gilbert method) or by enzymatic copying of single-stranded DNA (Sanger method).

Figure 19–5 outlines the steps in Sanger's chain-termination method. Single-stranded DNA is copied using *DNA polymerase I* from *E. coli*. This enzyme will not work on single-stranded DNA unless a short primer is annealed to it, producing a stretch of double-stranded DNA (Fig. 19–5). The primer has on its deoxyribose a free 3′OH group to which the next nucleotide can be attached; this provides a fixed starting point. The short primers are made synthetically or generated using restriction endonucleases, which are enzymes that cut DNA at specific sites. The primer is then extended using DNA polymerase and radioactive nucleotides. The chains are terminated at specific bases by addition of a 2′, 3′-dideoxynucleotide. Once a 2′,3′-dideoxynucleotide is incorporated, DNA elongation stops because it lacks a 3′-OH group in the sugar moiety, thus preventing the attachment of the next nucleotide. Four parallel reactions are performed, each containing the four common deoxynucleotides plus a small amount of one chain-terminating agent (either ddATP, ddGTP, ddCTP, or ddTTP). These produce DNA chains of various lengths that start at a unique site and end at specific bases.

To read the sequence, the four reactions are electrophoresed on a polyacrylamide gel that separates the radioactive fragments according to size. The radioautography of one such gel is shown in Figure 19–6. The fragments are visualized as a series of bands, each differing in length by one nucleotide. Each of the four tracks in the gel indicates chains terminating at one of the four nucleotides. The DNA sequence is read directly from the gel. For example, the band in position 30 in Figure 19–6 is under the track labeled C and therefore represents a cytosine in the DNA sequence.

Figure 19–6 is a good example of a sequencing gel, and the student is advised to read the sequence unassisted and compare the results with the sequence indicated in the figure legend. The sequence between positions 30 and 70 can be read rather easily.

The number of nucleotides that can be sequenced with these rapid methods is limited only by the resolution of the polyacrylamide gels; in a good experiment 400 nucleotides (more if several experiments are done simultaneously) can be sequenced in only one day's work. These methods, together with new genetic engineering methods that allow the presentation of large amounts of purified genes, have led to important advances in our understanding of how genes are organized, as exemplified by the work with φX174.

(1) Primer (restriction fragment) is annealed to single-stranded DNA template.

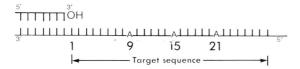

(2) Extension with DNA polymerase and four ^{32}P nucleoside triphosphates. Termination occurs at specific nucleotides because a small amount of one of the four dideoxynucleotides is added to each reaction mixture. For example, ddTTP stops the copy at T.

(3) Gel

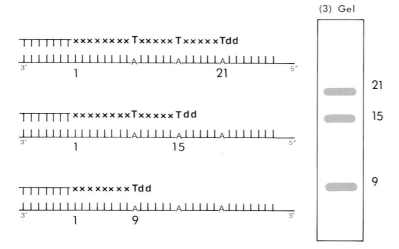

Fig. 19–5. Steps in Sanger's chain-termination sequencing method. (1) Single-stranded DNA s copied by DNA polymerase I from *E. coli.* (2) Extension of DNA copy is terminated by addition of 2', 3'-dideoxynucleotide which lacks the 3'-OH group and prevents the attachment of further nucleotides. (3) So that the sequence can be read, the labeled DNA is denatured and electrophoresed in a polyacrylamide gel (see Fig. 19–6).

The 5375-Nucleotide Sequence of φX174 Revealed Overlapping Genes

φX174 is a small virus having a circular chromosome of single-stranded DNA. After it infects *E. coli*, the complementary strand is synthesized, and the now double-stranded chromosome replicates to produce more phage progeny. The sequence of the 5375 nucleotides of φX174, established by Fred Sanger, Bart Barrell, and colleagues, unexpectedly revealed that sometimes genes overlap, producing two proteins from the same stretch of DNA.

φX174 codes for 10 proteins (Fig. 19–7), and it was estimated from the molecular weight of the proteins that more than 6000 bases would be required to code for them. The reason for the discrepancy between the size of the genome (5375) and the amount of DNA required to code for the ten proteins was explained when the complete nucleotide sequence became known.

Three of the genes of φX174 (genes E, B, and K in Fig. 19–7) overlap with other genes. In these regions, two proteins with different amino acid sequences are coded by the same stretch of DNA. This can be achieved because the proteins are encoded in different reading frames.

Figure 19–8 shows the nucleotide sequence at the start of gene E, which is completely contained within gene D. There is a ribosome recognition site and an ATG initiation codon present within the D gene, which are read by the ribosomes one nucleotide out of frame, thus producing two proteins from a single segment of mRNA. The fact that genes E and D overlap was confirmed by sequencing mutations that generate chain termination codons when the E frame is read, producing a shorter E protein while not significantly affecting the D gene (see Fig. 19–8, bottom line).

The most impressive case of genetic coding economy is found in gene K (see Fig. 19–7), which overlaps with genes A and C. In this region, four nucleotides are read in the three possible translational phases:

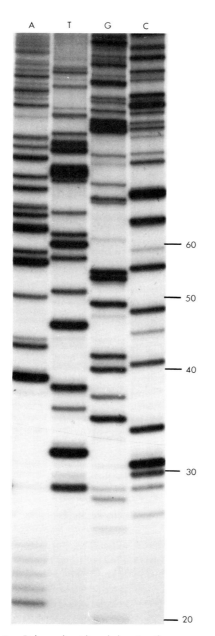

Fig. 19–6. Polyacrylamide gel showing the sequence of a segment of bacteriophage φX174. Each channel represents DNA chains terminated at a specific nucleotide. The sequence can be read clearly (except for weaker G bands at positions 33, 47, 52 and 99). The sequence is:

GAAAAAGCGT	CCTGCGTGTA	GCGAACTGCG	ATGGGCATAC
20	30	40	50
TGTAACCATA	AGGCCACGTA	TTTTGCAAGC	TATTTAACTG
60	70	80	90

(Courtesy of W. Barnes.)

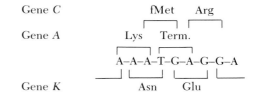

A sequence of 140 base pairs in which two protein coding genes overlap one nucleotide out of phase has also been found in simian virus 40 (SV40). It is possible that overlapping genes are peculiar to small viruses, which can store only a limited amount of DNA within the viral capsid and which are under a strong selective pressure to use their DNA as efficiently as possible. Cells from vertebrates contain large amounts of DNA and may not be under the same constraints as viruses.

The deciphering of the nucleotide sequence in φX174 and SV40 and its comparison to the amino acid sequence of some of their proteins provided a beautiful confirmation of the genetic code and showed that all 64 possible codons are utilized in protein synthesis.

Some Codons Are Used Differently in Mitochondrial DNA

As discussed in Chapter 11, mitochondria can in some ways be considered endosymbionts, which arose some 2 billion years ago when a primitive cell engulfed a bacteria-like organism that had developed the ability to respire. These cells became the ancestors of all eukaryotes (fungi, plants, and animals).

When in 1981 Bart Barrell, Fred Sanger, and their colleagues published the complete nucleotide sequence of human mitochondrial DNA, a number of novel features of gene organization became apparent, including that the genetic code was not, after all, universal.

The 16,569 base pairs of the circular human mitochondrial DNA molecule are so tightly packed with genes that there are few (and sometimes no) unused nucleotides between them. As Figure 19–9 shows, the molecule contains the genes for two ribosomal RNAs, 22 tRNAs (indicated in Figure 19–9 by the name of the amino acid they recognize), five proteins of the inner mitochondrial membrane (3 subunits of cytochrome *c* oxidase, cytochrome *b*, and 1 subunit of the ATPase), and eight unassigned reading

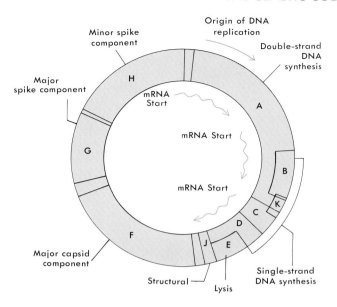

Fig. 19–7. Genetic map of bacteriophage ϕX174. The position of the ten protein coding genes (A through K) is indicated. Note that genes E, B, and K are overlapping.

frames (URFs), which presumably code for as yet unidentified proteins. The protein-coding genes are tightly packed and usually flanked by tRNAs, which provide signals for processing (cleaving) the RNAs. Both strands are entirely transcribed into RNA, each starting at a single promoter, and the RNAs are subsequently cut at the tRNA sites, giving rise to the specific mRNAs. The mRNAs are polyadenylated (a tail of A's is added; see Section 20–2), and the economy of nucleotide usage is so great that many mRNAs terminate in U or UA, so that the termination codon UAA arises only after a tail of A's is added to the mRNA.

The vast majority of the mitochondrial components are coded for in the nucleus and must be imported into the mitochondria. Only the ribosomal RNAs are synthesized in the mitochondria; the ribosomal proteins must be imported after they are synthesized in the cytoplasm. All of the matrix proteins, including the Krebs cycle enzymes, are imported from the cytoplasm too. The five proteins that are coded in mitochondrial DNA are highly hydrophobic proteins of the inner mitochondrial membrane, but they represent only some subunits of oligomeric enzymes, the rest of the subunits being made under the direction of nuclear genes. (Cytochrome *c* oxidase has seven subunits, of which three are coded in mitochondria; the ATPase complex has roughly 10 subunits, of which only one is coded in mitochondria; and the cyto-

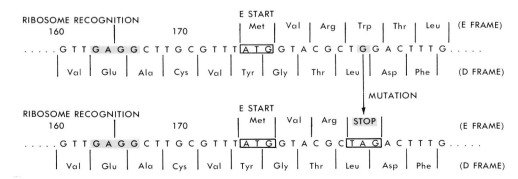

Fig. 19–8. Overlap of D and E genes of ϕX174. In the midst of the D-gene sequence there is a ribosome recognition signal and then an ATG initiation codon in a new reading frame, which begins the E gene. The E-gene frame was identified by the discovery of the mutation of a G to an A in a mutant virus *(bottom line)*, changing the codon for the amino acid tryptophan (Trp) to a premature termination codon, TAG, that stops E-gene translation. The reading frame for D gene is not affected by the mutation because both CTG and CTA specify leucine (Leu). Because the DNA sequence is indicated, T is used (in an RNA sequence U would be used instead). (Redrawn from Fiddes, J.C.: Sci. Am., *237*(6):54, 1977.)

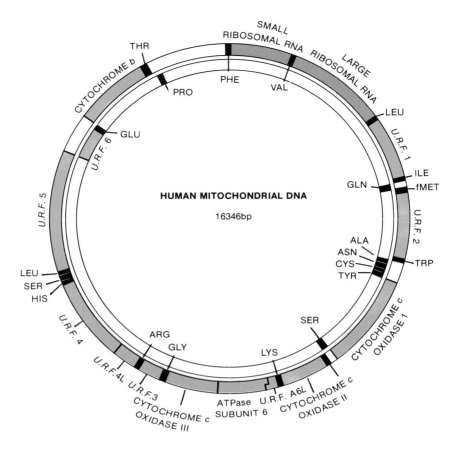

Fig. 19–9. Map of the genes in human mitochondrial DNA. The coding potential of both DNA strands is shown. URF, unassigned reading frames. The position of tRNA genes is indicated by the name of the amino acid they recognize. See the text for details. (Redrawn from Grivell, L.A.: Sci. Am., *248*(3):114, 1983.)

chrome bc_1 complex has 6 to 9 subunits, of which only cytochrome *b* is made in mitochondria.) Thus mitochondrial DNA has retained only a few genes, and it is thought that the vast majority of the genome of the primitive free-living "endosymbiont" was transferred into the nucleus of the "host" cell during evolution.

Although there must have been strong selective pressure to keep the mammalian mitochondrial genome small, in other organisms this apparently has not been the case. Yeast mitochondrial DNA is roughly 75,000 nucleotides long, even though it contains about the same number of genes as its human counterpart. The difference lies in the long stretches of spacer DNA between genes in the yeast; also, the protein-coding regions of yeast are interrupted by multiple intervening sequences or introns (see Section 20–2) that must be spliced out before a mature mRNA is produced. Plant mitochondria

have still larger genomes (250,000 nucleotides in the case of maize) and are known to have exchanged sequences with chloroplast DNA.

The most remarkable finding came when the mitochondrial DNA sequence was compared to the amino acid sequence of the proteins it codes for, which had been previously determined. It turned out that in human mitochondrial DNA, the codon UGA coded for tryptophan (instead of termination), AUA for methionine (instead of isoleucine) and AGA and AGG for termination of protein synthesis (instead of arginine). Even more remarkable, some codons of the mitochondrial DNAs from other organisms (yeast, *Neurospora*, plants) also differ from the universal code, but none of these mitochondrial DNAs have codes identical to each other (Table 19–2). These divergences from the original code arose in mitochondria presumably because they synthesize only a very small number of proteins. In

TABLE 19–2. CHANGES IN CODON USAGE IN MITOCHONDRIAL DNAs COMPARED TO THE UNIVERSAL (PROKARYOTIC AND NUCLEAR GENES) GENETIC CODE

	UGA	CUG CUC CUA CUU	AUA	AGG AGA	CGG
Universal	Termi-nation	Leu	Ile	Arg	Arg
Yeast	Trp	Thr	Met	Arg	Arg
Human	Trp	Leu	Met	Termi-nation	Arg
Neurospora	Trp	Leu	Ile	Arg	Arg
Maize	Termi-nation	Leu	Ile	Arg	Trp

cells that make thousands of proteins, changes in code words could not be possible because they would produce multiple lethal mutations.

If most mitochondrial proteins are made by the nuclear genetic system, why should there be any mitochondrial DNA at all? It has been suggested that the inner membrane proteins coded in mitochondria must be made "in situ" because they are highly hydrophobic and thus too insoluble to be imported from the cytoplasm. This theory is perhaps unlikely; subunit 9 of the ATPase complex, which is highly hydrophobic and coded by mitochondrial DNA in yeast, is coded in the nucleus in *Neurospora* and mammals.

Another view is that mitochondrial DNA is an evolutionary relic (see Grivell, 1983 in the Additional Readings at the end of this chapter). In the early stages of evolution there were few barriers to the transfer of genes from the endosymbiont to the genome of the host, and most of the genes were transferred to the nucleus. Later on, perhaps as a result of the appearance of a new mitochondrial genetic code, the transfer became more difficult and finally impossible, and a few genes became isolated within mitochondria.

Mitochondria have some "prokaryotic-like" features; for example, the protein synthesis machinery is sensitive to the antibiotic chloramphenicol (Chapter 11). They also have some "eukaryotic-like" characteristics, such as polyadenylated mRNAs and, in yeast, frequent intervening sequences interrupting genes. However, they have their own genetic code dif-

ferent from the universal one shared by prokaryotes and eukaryotes. Thus at this stage of evolution the mitochondrial genetic system has acquired so many special features that it should be considered unique.

Recombinant DNA Allows the Study of Complex Genomes

The studies on φX174 and SV40 were possible because their chromosomes are small and can be obtained in pure form. A eukaryotic chromatid, however, contains a single DNA molecule that is very long and consequently is much more difficult to study. This problem has now been overcome by the development of recombinant DNA techniques (known as *genetic engineering*), which allow short segments of eukaryotic DNA to be excised and ligated into small bacterial chromosomes (plasmids), which can then be grown in large amounts in *E. coli*. Although genetic engineering is mainly a technological achievement, it will be considered here because it is now widely used and has practical applications in industry and medicine in the large-scale production of valuable proteins.

Restriction Endonucleases Recognize Specific DNA Sequences

Recombinant DNA is based on the use of *restriction endonucleases*, which are enzymes that recognize specific nucleotide sequences and cut DNA (Fig. 19–10). There are two main types of restriction endonucleases. Class I enzymes recognize a specific sequence but cut the DNA at nonspecific sites, which can be hundreds of nucleotides away, and they are therefore not as useful for genetic engineering. Class II enzymes provide a molecular scalpel for cutting DNA at the specific recognition site, and they are most useful in DNA research. So far, more than 80 class II enzymes have been isolated.

The restriction endonucleases are named according to the microorganisms from which they are isolated (see Table 19–3). For example, Eco RI is the enzyme isolated from *E. coli* containing the drug-resistance plasmid RI. Class II restriction enzymes usually recognize nucleotide sequences four or six nucleotides long (Table 19–3). Those that recognize four nucleotides produce shorter fragments (on average, 250

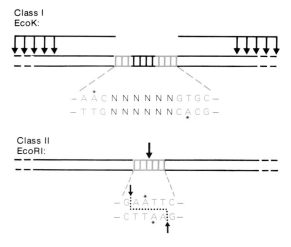

Class I
EcoK:

– A A C N N N N N N G T G C –
– T T G N N N N N N C A C G –

Class II
EcoRI:

– G A A T T C –
– C T T A A G –

Fig. 19–10. The two main types of restriction endonucleases. Both recognize specific nucleotide sequences, but class I endonucleases "slide along" the DNA before cutting once (at any of the positions indicated by the *arrows*), whereas class II endonucleases cut within the recognition sequence (*arrows*). EcoK (class I) can cut several hundred nucleotides away (in both directions) from the recognition sites, which consist of two blocks of specific nucleotides separated by six nucleotides of any sequence (*N*). EcoRI produces staggered cuts, resulting in "sticky ends," as indicated by the dotted line. The asterisks indicate nucleotides that may become methylated ("modified") in vivo, thus preventing cutting by the restriction enzymes.

bases long) than those that recognize six (on average, 4000 bases long). An important feature of these recognition sequences is that usually they are symmetrical. For example, the following recognition sequence has an axis of symmetry from which the nucleotide sequence reads the same on both strands in the 5′ to 3′ direction:

$$5' \quad G \quad A \quad A \mid T \quad T \quad C \quad 3'$$
$$3' \quad C \quad T \quad T \mid A \quad A \quad G \quad 5'$$

Some enzymes cut in the middle of the recognition sequence (producing blunt-ended fragments, see Hae III in Table 19–3), but in many other enzymes the breaks in the two strands are separated by several nucleotides (see Table 19–3). Such breaks result in single-stranded segments at the ends of the cleaved DNA that can base-pair with each other. These *"sticky" ends* can anneal with any DNA fragment cut by the same enzyme, thus providing a very good tool for genetic engineering.

Most bacteria contain restriction endonucleases whose function is to provide protection against foreign DNA. If foreign DNA (for example, from a bacteriophage) invades the cells, it is cleaved by the restriction enzymes. Bacteria protect their own DNA from being cleaved by having *modification enzymes*, which methylate DNA in the same DNA sequences recognized by their particular restriction enzyme, thus rendering it resistant to cleavage (indicated by asterisks in Fig. 19–10). The *restriction-modification phenomenon* was first discovered in the 1950s in studies of the behavior of two strains of *E. coli* (called K and B) that have different restriction enzymes. It was found that bacteriophage λ (lambda) grown in *E. coli* K could infect *E. coli* B only very poorly (while being very active in *E. coli* K), and vice versa. We now know that the DNA grown in each bacterial type is methylated at different sites. If the DNA is not methylated at the proper sites, it is cleaved by the restriction enzymes as it enters the heterologous cells. The study of this apparently esoteric phenomenon of *restriction* of bacteriophage growth led to the discovery of restriction enzymes and to a revolution in modern biology (see Arber, 1975 in the Additional Readings at the end of the chapter). This is a good example of how the beneficial results of scientific research cannot be predicted beforehand.

Eukaryotic Genes Can Be Introduced into Plasmids and Cloned in *E. coli*

Bacteria sometimes contain small plasmids (i.e., small circles of DNA which replicate autonomously) in addition to their chromosome.

TABLE 19–3.   RECOGNITION SEQUENCES OF SOME RESTRICTION ENDONUCLEASES

Name	Recognition Sequence	End after Cleavage*		Source
Eco RI	$\downarrow$ —G A A T T C— —C T T A A G— $\uparrow$	— G — C T T A A	A A T T C— 	*E. coli* containing drug-resistance plasmid RI
Hind III	$\downarrow$ — A A G C T T— — T T C G A A— $\uparrow$	— A — T T C G A	A G C T T— A —	*Hemophilus influenzae,* serotype D
Bam I	$\downarrow$ —G G A T C C— —C C T A G G— $\uparrow$	— G — C C T A G	G A T C C— G —	*Bacillus amylo-liquefaciens*
Hae III	$\downarrow$ — G G C C — — C C G G— $\uparrow$	— G G — C C	C C — G G —	*Hemophilus aegyptius*

* Note Eco RI, Hind III and Bam I produce "sticky" ends, and Hae III "blunt" ends.

Some of the best known plasmids are the resistance factors (R factors), which carry the genes for resistance to various antibiotics. In 1973 it was shown that fragments of foreign DNA could be introduced in vitro into plasmids with the help of restriction endonucleases and that these recombinant molecules could replicate when introduced into *E. coli*. These experiments started a new era in molecular biology.

Figure 19–11 shows the steps involved in the production of a recombinant DNA. The circular plasmid is cut open at a single site with a restriction enzyme. Eukaryotic DNA is also cleaved with the same enzyme (which generates sticky ends). The two DNAs are allowed to anneal to each other at their sticky ends and are then joined together with the enzyme *DNA ligase*. The recombinant plasmids are next introduced into *E. coli* made permeable to DNA by incubation in calcium chloride. Afterwards, the cells that have taken up a plasmid molecule are selected by the use of antibiotics. For example, if the plasmid carries the gene for tetracycline resistance, those cells that do not have such a plasmid can be killed by adding tetracycline to the culture medium. Once an *E. coli* colony is obtained, millions of copies of the eukaryotic DNA segment can be grown. Because all these copies are derived from an *individual* hybrid molecule, this procedure is called *gene cloning* (see Cohen, 1975; Maniatis et al., 1982).

In addition to plasmids, other *vectors* can be used to propagate cloned DNA fragments. One of the most commonly used is bacteriophage λ,

a longtime favorite of microbial geneticists. Phages have been constructed that have single restriction sites for certain enzymes and that are able to accept foreign inserts of 15,000 to 20,000 nucleotides. Plasmids with such long inserts tend to grow poorly, and so phage λ is preferred for cloning large DNA fragments. Furthermore, years of genetic and biochemical experiments with phage λ have shown that it is possible to assemble this bacteriophage in vitro using a mixture of λ DNA and protein extracts. This mixture produces active virus, which can then inject its DNA into *E. coli* as in normal infection. This *"in vitro packaging"* technique is much more efficient (more recombinant clones obtained per μg of DNA) than simple transformation of DNA into *E. coli* cells made permeable with calcium chloride with plasmid vectors (see Murray, 1983). For this reason eukaryotic gene libraries are usually constructed by insertion of random segments of DNA into bacteriophage λ. Then the desired gene is recognized through use of a specific radioactive hybridization probe, for example, globin cDNA (Section 20–2). Because the globin gene is present in a single copy per mammalian genome, 50,000 to 200,000 λ clones have to be screened (10,000 or more λ plaques can be grown per petri dish). Part of the bacteriophage plaques are transferred to a nitrocellulose filter paper, the proteins are removed and the DNA denatured by treatment with alkali, and the DNA bound to the filter is hybridized with the radioactive globin probe. After washing, plaques containing a globin gene can

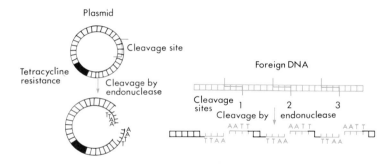

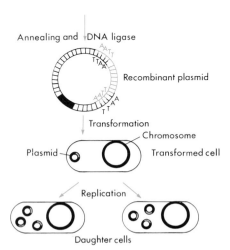

Fig. 19–11. Production of a recombinant DNA molecule by genetic engineering. (Courtesy of S. Cohen.)

teriophage plaques are transferred to a nitrocellulose filter paper, the proteins are removed and the DNA denatured by treatment with alkali, and the DNA bound to the filter is hybridized with the radioactive globin probe. After washing, plaques containing a globin gene can be detected by autoradiography. Once the gene has been identified, the recombinant bacteriophage is grown, and the globin gene is cut out from the recombinant λ with restriction enzymes and cloned into plasmid vectors which are easier to grow in milligram amounts.

Another vector that is much used is bacteriophage M13 (Fig. 19–12), a single-stranded bacteriophage (like φX174). Infected *E. coli* cells secrete single-stranded virus into the culture medium. The DNA is very easy to isolate by simply centrifuging the bacteria out and extracting the supernatant with phenol to remove the viral proteins. This method is excellent for preparing the single-stranded templates required for Sanger's dideoxynucleotide sequencing method (Fig. 19–5).

Yeast cells can also be transformed efficiently

with plasmids, but in this case the prokaryotic plasmids must have a eukaryotic origin of replication in order to multiply (Fig. 14–22). These *"shuttle vectors"* can be grown both in yeast and in *E. coli* and are very advantageous because genes can be introduced into a eukaryote so that their regulation and expression can be studied. However, when milligram amounts of the purified DNA are required, it may be then grown in *E. coli*. These yeast vectors have been essential for the identification of centromeric and telomeric DNA sequences (see Figs. 13–21 and 13–22).

Genetic Engineering and Industrial Production of Valuable Proteins

Recombinant DNA made possible the use of bacteria to produce proteins of medical importance, giving rise to a whole new industry. Expression of eukaryotic genes in high yields in bacteria requires that these genes be hooked up to bacterial promoters (i.e., start signals for RNA polymerase) and ribosomal binding sites (for cor-

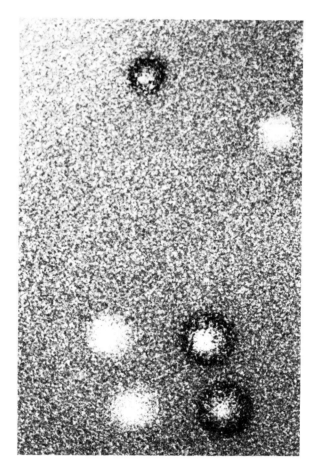

Fig. 19–12. Bacteriophage M13 plaques seen as clear regions on a lawn of *E. coli*. Of the six colonies seen, three have a dark ring (blue in real life), and three are white. The M13 vectors carry an *E. coli* β-galactosidase gene within them and give a blue color in indicator plates resulting from the activity of the enzyme. A restriction site appropriate for cloning other DNA fragments is contained with the β-galactosidase gene. The three white colonies are due to insertions of a segment of foreign DNA, in this case frog DNA, which inactivates the enzyme. The blue/white selection is very useful for choosing which phages have acquired a piece of foreign DNA. Each plaque results from the descendants of a single M13 phage (a clone). (Courtesy of T. Bürglin.)

rect protein synthesis). This requirement has led to painstaking constructions of hybrid genes of great ingenuity. Although these constructs are laborious, when they work well, they produce enormous amounts of protein (i.e., up to 40% of the total cell protein recovered from bacterial cultures). The human growth hormone clones from a leading company in the field produce 200,000 molecules of hormone per bacterium, and each ml of culture contains 5×10^9 bacteria! In some cases it has been necessary to synthesize

whole genes in order to obtain high efficiency expression. This was done for human insulin, whose A and B chains were made *de novo* chemically using polynucleotide synthesis methods and were inserted into high efficiency expression vectors (see Gait, 1979 in the Additional Readings at the end of this chapter). This insulin is now marketed throughout the world.

Many proteins have attracted the attention of the new genetic engineering companies. Among them are the following:

1. *Human insulin.* Some patients cannot tolerate pig insulin, which has a slightly different amino acid sequence than human insulin. Recombinant DNA insulin should eventually become cheaper than that extracted from animal pancreas, and is now available.

2. *Interferon.* This protein has indications as an antiviral agent. The initial hopes of an antitumoral effect now seem less certain. Several types of interferon are available.

3. *Growth hormone.* The human one is useful in the treatment of hypopituitary dwarfs and may prove useful in the treatment of bone fractures, skin burns, and bleeding ulcers of the digestive tract. The human hormone is marketed in the U.S. The bovine hormone is expected to yield bigger cattle and thus more beef.

4. *Vaccine production.* Any antigenic protein can potentially be made. A vaccine against human hepatitis B has already been approved for use. Vaccines made by genetic engineering offer the advantage that the microbial strains from which the proteins are extracted do not contain complete viruses, and thus there is no risk of accidental inoculation with live virus. Cloning directly into vaccinia virus DNA holds great promise, although vaccines so produced are not yet in the market. Recombinant vaccinia viruses containing, for example, a gene from genital herpes virus within its DNA, can multiply and can subsequently be inoculated into humans. As the vaccinia virus produces a mild infection, it should express some of the herpes virus protein and thus produce immunity. This is in a way very similar to what Jenner did over 100 years age when he introduced the first

vaccination scheme, which eventually led to the extinction of smallpox.

5. *Plant genetics.* Though not yet a reality, the potential exists for developing plant strains that can fix nitrogen (by introduction of nitrogen fixation genes into bacteria that live in the roots of nonleguminous plants). Other possible strains might carry genes that confer resistance to disease, insects, or excessive salt in the ground, or might bear fruits or grain with more of certain amino acids in their storage proteins.

Although discussion of this brave new world of commercial exploitation of molecular biology does not really belong in this textbook, we mention it here so that the reader can get a feeling for the long way we have come since the first code word (poly U, polyphenylalanine) was deciphered in 1961.

A New Type of Genetic Analysis Has Become Possible

Genetic engineering has also been invaluable in the study of how genes are controlled in higher organisms. Not only can genes be grown in large quantities and sequenced, but their *expression* can also be studied after reintroduction into eukaryotic cells, thus allowing a new type of genetics.

In its usual form a genetic experiment begins with the observation of mutant individuals, which in eukaryotic organisms are usually recognized by some phenotype abnormality—for example, white eyes in a fly. Using the in vitro manipulations that are now possible, mutations can be introduced into eukaryotic genes that have been cloned in plasmids (Fig. 19–13). For example, a number of nucleotides can be deleted by treating the DNA with a restriction endonuclease, or a particular change can be induced in a single nucleotide by a chemical mutagen. The mutation can be characterized precisely by the new methods for determining DNA sequences. Then the biological effect of the mutation can be tested in a eukaryotic gene expression system (for example, by microinjecting it into the nucleus of living cells), and its function can be analyzed. One can determine, for example, which are the regions containing the ini-

tiation for termination signals for RNA synthesis (Fig. 19–13). Experiments of this type, to be discussed in later chapters, have been essential in showing how eukaryotic genes work (for example, Fig. 21–22). This constitutes a new type of genetics (so-called "reverse genetics") in which the gene is first cloned, then mutated in a chemically defined way, and finally tested for biological activity.

Many types of functional assays for cloned genes are now available. The first was microinjection of DNA into the nucleus of frog oocytes (Fig. 19–13). Later, in vitro transcription systems were developed that could transcribe cloned eukaryotic DNA. Genes can also be transplanted into cultured somatic cells, which will ingest DNA from the culture medium (if supplied as a coprecipitate with calcium phosphate) and integrate it into the genome in a low proportion of cells. Alternatively, cultured somatic cells can be microinjected directly into the nucleus with cloned DNA. Finally, it has now become possible to inject cloned genes into mouse and *Drosophila* embryos (Figs. 22–20 and 22–21), where they become integrated into the germ line, giving rise to new strains of animals that can transmit to their progeny their newly acquired genes. All these systems have been invaluable for analyzing the effect on gene activity of mutations introduced in vitro in cloned DNAs.

SUMMARY:
DNA Sequencing and Genetic Engineering

Methods have been developed that allow the rapid sequencing of DNA (Fig. 19–5) and that have produced a revolution in molecular biology. All the rapid sequencing methods are based on the production of DNA fragments of different lengths which start at a fixed point and terminate at specific nucleotides. The sequence is read on polyacrylamide gels that separate the DNAs by size (Fig. 19–6).

The first complete organism to be sequenced was the small, single-stranded bacteriophage ϕX174 which is 5375 nucleotides long. Unexpectedly it was found that sometimes genes can overlap. In fact, about 6000 nucleotides would be needed to code for the 10 proteins produced by this virus if the overlaps did not exist. In certain regions the same segment of DNA codes for two protein sequences by use of a different reading frame. In gene K, which overlaps with genes A and C, four nucleotides are read in the three possible translational phases. Sequencing of DNA and comparison to the amino acid sequence of the correspond-

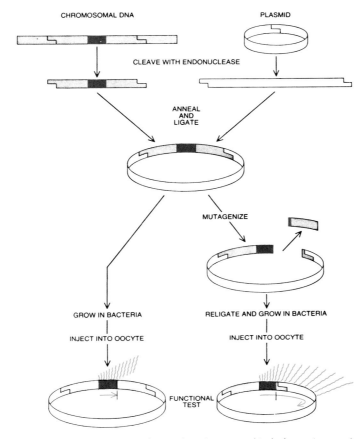

Fig. 19–13.   Gene cloning and microinjection can be combined in a new kind of genetic experiment. Chromosomal DNA bearing a gene under study is recombined with a bacterial plasmid by recombinant-DNA techniques: the DNA's are cleaved with an endonuclease (a restriction enzyme) that provides "sticky ends," so that the gene can be inserted into the plasmid. A sequence of nucleotides at the end of the gene under study is then deleted from some of the recombinant plasmids; the precise extent of the deletion can be determined by nucleotide-sequencing methods. The plasmids are cloned in bacteria to provide large amounts of the original gene *(left)* and the mutated gene *(right)*, which are injected into oocytes for a functional test. In this case the original gene is transcribed into RNA molecules *(color)* of the proper size. The mutated gene is not: transcription continues beyond the end of gene into plasmid DNA. Presumably deletion eliminated the termination signal recognized by RNA polymerase. (From De Robertis, E.M., and Gurdon, J.B.: Sci. Am., *242:*(6):74, 1979.)

ing proteins confirmed the genetic code and showed that all 64 possible codons are used in vivo.

The sequencing of the 16,569 nucleotides of human mitochondrial DNA (Fig. 19–9) also proved very informative. Only a few genes are coded within mitochondria: 2 ribosomal RNAs, 22 tRNAs, 5 hydrophobic inner membrane proteins, and 8 still unassigned reading frames (URFs). Although the genes are tightly packed in human mitochondria, yeasts have five times more mitochondrial DNA but a smilar number of mitochondrial genes. Yeast mitochondrial genes have multiple introns, which are absent in human mitochondrial DNA.

To everyone's surprise it was found that mitochondria have a different genetic code from that of prokaryotic or nuclear genes. Some codons are used differently, and these changes vary among species (Table 19–2). These divergences from the original code arose only in mitochondrial DNA,

probably because it codes for only a few proteins. If mitochondria arose from a free-living endosymbiont engulfed by cells a long time ago, then most of its genes were transferred into the nucleus of the host cell. The appearance of a new code may have been a substantial obstacle for further transfers, thus leaving a few genes isolated permanently within mitochondria.

The techniques of genetic engineering are based on the use of *restriction endonucleases* from bacteria. These enzymes recognize specific DNA nucleotide sequences and provide a molecular scalpel for cutting DNA at these sites (Table 19–3). Some restriction enzymes produce "sticky" ends which can anneal with any DNA fragment cut by the same enzyme The normal function of the bacterial endonucleases is protection against invasion by foreign DNA (e.g., a bacteriophage). The DNA of the bacterium is rendered resistant to these endonucleaves by a *restriction-modification* process in which methylation of the specific DNA sequences is produced.

Genetic engineering techniques permit the cloning (i.e., the replication of a single molecule) of genes by insertion of the genes into plasmids (Fig. 19–11) or into bacteriophage λ. Plasmids are small circles of DNA that replicate autonomously, and many of them carry genes for resistance to antibiotics. The recombinant plasmids carrying eukaryotic genes are introduced into *E. coli* and are then selected by the use of the proper antibiotic. Colonies can be obtained in which millions of copies of the same molecule are produced (gene clones). These techniques make possible the use of bacteria to produce proteins of medical importance (e.g., human insulin or the growth hormone) and to analyze how genes work.

ADDITIONAL READINGS

Anderson, S., et al.: Sequence and organization of the mitochondrial genome. Nature, *290*:465, 1981.

Arber, W.: DNA modification and restriction. Prog. Nucleic Acid Res. Mol. Biol., *14*:1, 1975.

Cohen, S.N.: The manipulation of genes. Sci. Am., *233*:24, 1975.

Cold Spring Harbor Laboratory: The genetic code. Cold Spring Harbor Symp. Quant. Biol., Vol. 31, 1966.

Crick, F.H.C.: Central dogma of molecular biology. Nature, *227*:561, 1970.

Crick, F.H.C.: The origin of the genetic code. J. Mol. Biol., *38*:367, 1968.

Crick, F.H.C.: The genetic code, III. Sci. Am., *215*(4):55, 1966.

De Robertis, E.M., and Gurdon, J.B.: Gene transplantation and the analysis of development. Sci. Am., *241*(6):74, 1979.

Fiddes, J.C.: The nucleotide sequence of a viral DNA. Sci. Am., *237*(6):54, 1977.

Gait, M.J.: Synthetic genes for human insulin. Nature, *277*:429, 1979.

Gilbert, W., and Villa-Komaroff, L.: Useful proteins from recombinant bacteria. Sci. Am., *242*:74, 1980.

Godson, G.N., Barrell, B.G., Staden, R., and Fiddes, J.C.: Nucleotide sequence of bacteriophage G4 DNA. Nature, *276*:236, 1978.

Goeddel, D.V, et al.: Direct expression in *E. coli* of a DNA sequence coding for human growth hormone. Nature, *281*:544, 1979.

Grivell, L.A.: Mitochondrial DNA. Sci. Am., *248*(3):78, 1983. (Highly recommended.)

Itakura, K.: Synthesis of genes. Trends Biochem. Sci.,*5*:114, 1980.

Maniatis, T., Fritsch, E.F., and Sambrook, J.: Molecular Cloning. Cold Spring Harbor, New York, Cold Spring Harbor Laboratory, 1982. (This laboratory manual is the cloner's bible.)

Maxam, A.M., and Gilbert, W.: A new method for sequencing DNA. Proc. Natl. Acad. Sci. USA, *74*:560, 1977.

Murray, N.E.: Phase λ and molecular cloning. *In* The Bacteriophage Lambda. Vol. II. Cold Spring Harbor, New York, Cold Spring Harbor Laboratory, 1983.

Nathans, D., and Smith, H.O.: Restriction endonucleases in the analysis and restructuring of DNA molecules. Annu. Rev. Biochem., *44*:273, 1975.

Roberts, R.J.: Restriction endonucleases. CRC Crit. Rev. Biochem., *4*:123, 1976.

Sanger, F. et al.: Nucleotide sequence of bacteriophage φX174 DNA. Nature, *265*:687, 1977.

Sanger, F. Coulson, A., Hong G. and Petersen, G.B.: Nucleotide sequence of bacteriophage λ. J. Mol. Biol., *162*:729, 1982.

Sanger, F., Nicklen, S., and Coulson, A.R.: DNA sequencing with chain-terminating inhibitors. Proc. Natl. Acad. Sci. USA, *74*:5463, 1977.

Watson, J.D.: Molecular Biology of the Gene. Menlo Park, California, W.A. Benjamin, Inc., 1975.

Wigler, M., et al.: Transformation of mammalian cells with genes of prokaryotes and eukaryotes. Cell, *16*:777, 1979.

Willamson, R.: Genetic Engineering. Vols. 1, 2, 3. London, Academic Press, 1981. (A series of useful reviews.)

Ycas, M.: The Biological Code. New York, John Wiley & Sons, Inc., 1969.

TRANSCRIPTION AND PROCESSING OF RNA

The flow of genetic information follows the pathway:

$$\text{DNA} \underset{\text{transcription}}{\rightleftharpoons} \text{RNA} \xrightarrow{\text{translation}} \text{Protein}$$

The information coded in DNA cannot act directly as a template for protein synthesis but must be first transcribed into messenger RNA (mRNA). In addition to this informational RNA, other RNAs are required for the very complex process of translating a four-base code into a sequence of 20 amino acids. These other nucleic acids are ribosomal RNA (rRNA) and transfer RNA (tRNA), which account for most of the RNA contained in cells (Table 20–1). In eukaryotes other RNAs are present as well, such as heterogeneous nuclear RNA (hnRNA), which consists of mRNA precursors, small nuclear RNAs (snRNAs), and small cytoplasmic RNAs (scRNAs).

In this chapter we analyze how the information contained in DNA is *transcribed* into RNA. RNA is synthesized by the enzyme RNA polymerase. This step is important in gene regulation because the rate of expression of a particular gene is in general controlled by the frequency

TABLE 20–1. RNA MOLECULES IN *E. COLI*

Type	Relative Amount (%)	Sedimentation Coefficient	Number of Nucleotides
Ribosomal RNA (rRNA)	80	23S	3700
		16S	1700
		5S	120
Transfer RNA (tRNA)	15	4S	~75
Messenger RNA (mRNA)	5	Heterogeneous	

with which RNA polymerase starts the transcription of that gene.

The transcripts synthesized by RNA polymerase are frequently not the final products utilized by the cell. To become functionally competent the primary transcripts must be modified by a series of chemical alterations known as *processing*. There are four main types of modifications in the processing of RNA: (1) *cleavage of large precursor RNAs* into smaller RNAs, which involves the removal of extra segments from the beginning or end of the precursor; (2) *splicing*, in which *intervening sequences* or *introns* (extra segments of nucleotides inserted within the genes themselves) are excised and the molecule religated, (3) *terminal addition of nucleotides*, such as poly A segments at the 3' end and *"cap"* nucleotides at the 5' end of eukaryotic mRNAs; and (4) *nucleoside modifications* such as the methylations that are common in transfer and ribosomal RNAs.

The processing of each major RNA species has its own peculiarities. mRNA and snRNA are analyzed here and rRNA, 5S RNA, and tRNA in the following chapter. RNA synthesis and processing differ in many respects in prokaryotes and eukaryotes and are therefore treated separately but compared and contrasted whenever possible.

20–1   TRANSCRIPTION IN PROKARYOTES AND EUKARYOTES

Genetic experiments led François Jacob and Sydney Brenner to propose that in bacteria there must exist a short-lived "messenger" of the information coded in DNA.[1-3] In time, it was found that such messenger RNAs indeed existed (see Fig. 21–1). The average life span of mRNA in *E. coli* is about 2 minutes, after which the molecules are broken down by ribonucleases.

The length of mRNA molecules is heterogeneous, reflecting the variable length of the polypeptide chains for which the mRNA codes. The average protein length is between 300 and 500 amino acids, which are encoded by 900 to 1500 nucleotides. The coding region for a single protein is also called a cistron. In bacteria, however, mRNAs frequently code for several proteins, and the molecules are much longer. As Figure 20–1 shows, bacterial *polycistronic* mRNAs contain several protein synthesis start (AUG) and stop (UAG, UAA, UGA) codons in the same RNA molecule. In eukaryotes, on the other hand, mRNAs always have a single start codon per molecule and are thus *monocistronic* (see Fig. 20–12).

Prokaryotic RNAs Are Transcribed by a Single RNA Polymerase

In prokaryotes all types of RNAs are transcribed by the same enzyme. Using DNA as the template, this RNA polymerase catalyzes the formation of RNA from the four ribonucleoside triphosphates. The enzyme copies the base sequence of one of the DNA strands (called the *coding* strand) according to the Watson-Crick base-pairing rules. Because only one DNA strand is used as a template, transcription is said to be *asymmetrical*. The polymerization reaction is as follows:

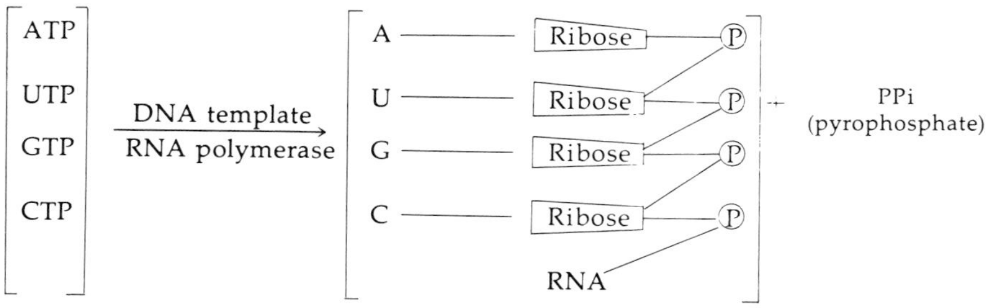

In addition to copying the nucleotide sequence of the DNA template precisely, RNA polymerase is able to recognize a variety of genetic signals on the chromosome, such as the signals for starting and stopping RNA synthesis at precise sites.

There are an estimated 3000 genes in *E. coli* (the E. coli genome has 3.4 million base pairs,

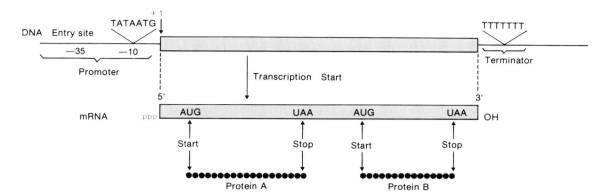

Fig. 20–1. Idealized prokaryotic protein-coding gene. Note the promoter, terminator, and the polycistronic mRNA. Compare with the organization of the eukaryotic gene shown in Figure 20–12.

and the average gene is about 1000 nucleotides long), and they are all transcribed by the same enzyme. Considering these multiple functions, it is not surprising that RNA polymerase is a large and complex enzyme. A complete molecule or *holoenzyme* consists of several polypeptides: two α of 40,000 daltons, one β of 155,000, one β′ of 165,000, and one σ of 90,000. The total molecular weight of the enzyme is therefore 490,000.[4] The sigma (σ) factor is loosely bound to the rest of the enzyme and can be separated by physical means. The ββ′α₂ enzyme (without sigma) is called the *core polymerase*. The sigma factor is required for the enzyme to recognize the correct start signals on the DNA. Figure 20–2 shows that as soon as the RNA chain is

started, the sigma factor is released from the core enzyme and can be used in the transcription of other RNA molecules.[5,6,7]

The following three stages are commonly distinguished in the transcription by RNA polymerase: (1) binding to promoters and chain initiation, (2) elongation, and (3) termination.

The start signals on DNA are called *promoters* and represent the initial binding site for the RNA polymerase. This binding is a crucial step in the regulation of gene expression. As Figure 20–3 shows, prokaryotic promoters have the common sequence TATAATG (or slight variations of it) located some ten nucleotides before the end of the mRNA.[8] This AT-rich region probably favors local separation of the DNA strands,

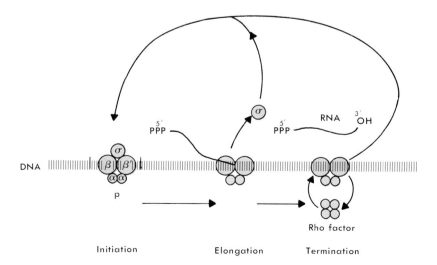

Fig. 20–2. Diagram showing the three stages of transcription by RNA polymerase. At *initiation* the enzyme composed of subunits β, β′, 2α, and σ is at the site of the promoter (*p*). During *elongation* the enzyme transcribes the gene and subunit σ is eliminated. *Termination* occurs at the termination site and the termination factor *rho* causes the release of the transcribed RNA molecule.

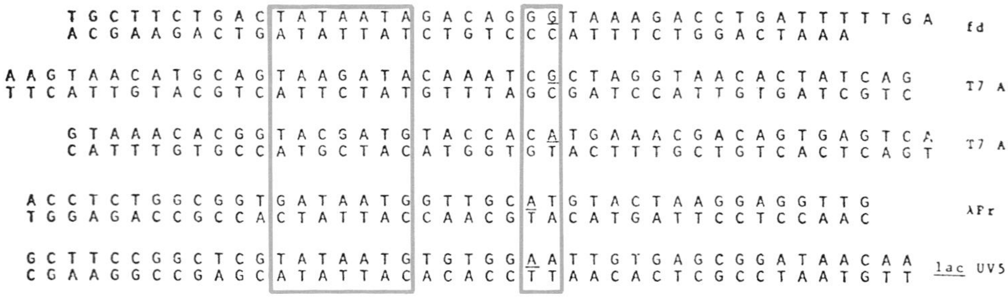

```
T G C T T C T G A C|T A T A A T A|G A C A G G|G|T A A A G A C C T G A T T T T T G A   fd
A C G A A G A C T G|A T A T T A T|C T G T C C|C|A T T T C T G G A C T A A A

A A G T A A C A T G C A G|T A A G A T A|C A A A T|C G|C T A G G T A A C A C T A T C A G   T7 A2
T T C A T T G T A C G T C|A T T C T A T|G T T T A G|C G|G A T C C A T T G T G A T C G T C

G T A A A C A C G|T A C G A T G|T A C C A C|A T|G A A A C G A C A G T G A G T C A   T7 A3
C A T T T G T G C C|A T G C T A C|A T G G T|G T|A C T T T G C T G T C A C T C A G T

A C C T C T G G C G G T|G A T A A T G|G T T G C|A T|G T A C T A A G G A G G T T G   λPr
T G G A G A C C G C C A|C T A T T A C|C A A C G|T A|C A T G A T T C C T C C A A C

G C T T C C G G C T C G|T A T A A T G|T G T G G A|A|T T G T G A G C G G A T A A C A A   lac UV5
C G A A G G C C G A G C|A T A T T A C|A C A C C|T T|A A C A C T C G C C T A A T G T T
```

Fig. 20–3. Promoter sequences that are protected by RNA polymerase from digestion by DNAse. Observe that all promoters have a common sequence, TATAATG, or a close derivative of it. This sequence is situated about ten bases before the start of the mRNA. The base in which transcription starts is outlined by the *small box*.

a step required for RNA polymerase to gain access to the DNA bases. In addition to the TATAATG sequence, the enzyme also recognizes a DNA region located about 35 bases before the start of the mRNA (Fig. 20–1). Single-base changes in this region can inactivate promoters, but different promoters have no obvious common sequences in this region. Individual variations among promoters are not surprising because different genes are expressed with varying efficiencies; that is, some promoters are "stronger" than others. *E. coli* has proteins called *repressors*, which can bind to specific sequences of DNA (called *operators*), turning off the expression of a gene or set of genes. Repressors work by binding to a DNA site that overlaps with the promoter, thereby preventing the binding of RNA polymerase (see Fig. 22–2).

During elongation, RNA polymerase copies the DNA sequence accurately, progressing at a rate of 30 nucleotides per second. Only one strand of the DNA is transcribed. Elongation of the RNA proceeds only in the 5′ to 3′ direction, which, as we have seen, is the same direction used in DNA replication and protein synthesis.

Termination of transcription occurs when the enzyme arrives at a stop signal in the DNA. A *termination factor* or *rho factor* (a protein with ATPase activity) causes the release of the completed RNA molecules (Fig. 20–2).[9] Transcription terminates on a region of four to eight consecutive U residues, as shown in Figure 20–4. In addition to a row of U's, other features are important in prokaryotic termination,[10,11] particularly the ability of the nascent transcript to form a hairpin loop before termination[12] (see Fig. 22–5) and a GC-rich region preceding the row of T's in the DNA. Gene expression in *E. coli*

can be regulated at the level of transcription termination, as is discussed later for the tryptophan operon (Fig. 22–4).

In Prokaryotes Transcription Is Coupled to Protein Synthesis

Because prokaryotes do not have a nuclear membrane, the ribosomes are free to attach to the mRNA molecules as they are being synthesized, as can be beautifully visualized in Figure 20–5. The bacteria were treated with the enzyme lysozyme to render them osmotically sensitive, then gently broken open, and spread for electron microscopy.[13,14] RNA polymerase can sometimes be seen as a small granule attached to the DNA template. Electron micrographs show that prokaryotic mRNAs are translated into protein at one end while the other end is still being transcribed. The half-life of these mRNAs is so brief that sometimes degradation starts even before mRNA synthesis is completed. This is possible because both RNA synthesis and protein synthesis start at the 5′ end of the mRNA. In bacteria, any change in the rate of mRNA synthesis will be followed a few minutes later by a change in the synthesis of the corresponding protein.

Eukaryotes Have Three RNA Polymerases

Eukaryotic transcription differs in several respects from that of prokaryotes. Figure 20–6 illustrates one of the most obvious differences, which is that in higher cells RNA is synthesized within the nucleus. The nuclear membrane introduces a barrier between transcription and translation, and before the mRNA can reach the

```
....-A-G-G-U-A-A-U-A-G-U-U-A-G-A-G-C-C-U-G-C-A-U-A-A-C-G-G-U-U-U-C-G-G-G-A-U-U-U-U-U-A-OH       λ 6S

....-U-U-U-G-U-U-G-C-C-G-C-C-U-U-A-G-A-A-C-G-C-U-C-G-G-U-U-G-C-C-G-G-G-C-U-U-U-U-U-U-A-OH       λ oop

....-C-A-A-U-C-A-G-A-U-A-C-C-C-A-G-C-C-U-A-A-U-G-A-G-C-G-C-C-C-G-C-G-G-G-C-U-U-U-U-U-U-U-(U)-OH   trp
                                                                                                leader

....-G-U-C-A-U-A-A-G-G-G-C-C-U-U-U-G-A-A-G-U-U-A-C-C-G-C-U-U-C-A-A-G-G-G-C-U-U-U-U-U-U-(A)-(A)-OH  φ80 M₃
```

Fig. 20—4. Stop signals for RNA polymerase in four prokaryotic RNAs. Observe the presence of a GC-rich region (underlined) followed by five to eight consecutive U residues.

ribosomes, it must be exported to the cytoplasm. Furthermore, eukaryotic mRNAs are much more stable than those found in prokaryotes, with half-lives that can be measured in hours or days instead of minutes. In addition, these mRNAs are generally monocistronic, and they have special types of posttranscriptional modifications, such as addition of a methyl-G "cap" at the 5' end, a tail of poly A at the 3' end, and splicing of intervening sequences, all of which are not present in E. coli mRNAs.

We have seen that in E. coli all the genes of the chromosome are transcribed by the same enzyme. In higher organisms three different nuclear RNA polymerases have been identified, each of which transcribes a different class of genes. Figure 20–7 shows how the three enzymes can be separated by ion-exchange chromatography. Nuclear extracts of all eukaryotic cells show these three peaks of activity. The enzymes have been classified as I, II, and III according to their order of elution from the column. The most useful criterion for distinguishing the polymerases is their sensitivity to inhibition by α-amanitin, the toxin of the poisonous mushroom Amanita phalloides. This toxin is of medical interest because it is the main cause of lethal intoxication by mushrooms in humans.

Table 20–2 summarizes the properties of eukaryotic RNA polymerases, and Figure 20–8 experimentally confirms what is stated there. RNA polymerase I is not affected by α-amanitin and is localized in the nucleolus. This enzyme transcribes the large ribosomal RNAs (18S, 28S, and 5.8S). RNA polymerase II is completely inhibited by low concentrations of α-amanitin and is responsible for the synthesis of mRNA and U-snRNAs which are small (200–100 nucleotides), uridine-rich, nuclear RNAs (see below).[15] RNA polymerase III is inhibited by high concentrations of α-amanitin[16] and synthesizes tRNA, 5S RNA, and other small RNAs such as 7S scRNA, which is part of the signal recognition particle

(SRP) involved in protein secretion (see Fig. 20–28) The eukaryotic RNA polymerases are very complex enzymes, each one having between six and ten protein subunits. The protein subunits of each enzyme type are different, except for two small polypeptides of 29,000 and 19,000 daltons that are common to all three RNA polymerases.[17] The three types of eukaryotic RNA polymerases differ in their molecular structure and in their mechanism of regulation.

The DNA in mitochondria and chloroplasts is transcribed by yet another enzyme. In contrast to the nuclear enzymes, mitochondrial RNA polymerase is a simple enzyme formed by a single peptide. It is not inhibited by α-amanitin but rather by rifampicin, an antibiotic that inhibits prokaryotic RNA polymerase. This is another property of mitochondria that suggests their origin was prokaryotic.

Transcription May Be Visualized by Electron Microscopy

Oscar Miller has devised a remarkable technique by which transcription complexes can be visualized directly in the electron microscope.[13] In this method, isolated nuclei are lysed gently, and the chromosomal material is centrifuged on top of an electron microscope specimen holder. Transcribing genes are spread together with their attached RNA polymerases and growing RNA chains. Figure 20–9 shows ribosomal RNA genes during transcription, obtained by spreading nucleoli of frog cells. Several conclusions about eukaryotic transcription can be drawn from this type of study:[18,19]

1. The template for transcription is a complex of DNA and protein that has the characteristic beaded appearance of chromatin nucleosomes (see Fig. 14–26).

2. Eukaryotic RNA synthesis starts and ends at precise sites on the DNA (the promoter and terminator sequences). Between these two sites the RNA chains gradually in-

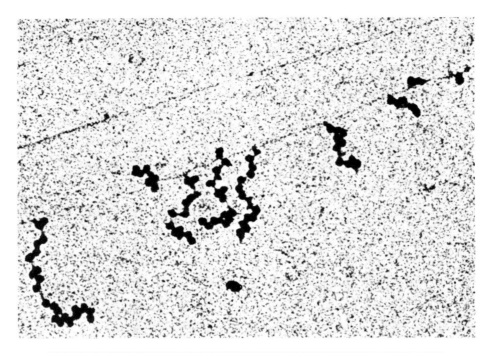

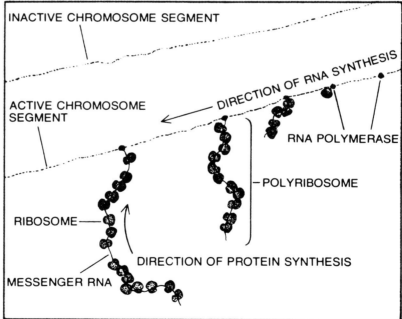

Fig. 20–5. Electron micrograph of the transcription and translation processes in bacteria. Two stretches of DNA, one naked, the other with nascent mRNA arranged at right angles, are observed. The *bottom* diagram facilitates the interpretation of the electron micrograph. (Courtesy of O.L. Miller, Jr.)

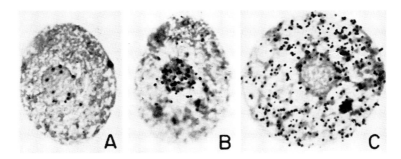

Fig. 20–6. *A,* autoradiograph of a *Tetrahymena* cell (ciliate protozoon) incubated in ³H-cytidine for 1.5 to 12 minutes. Notice that all labeled RNA is restricted to the nucleus. *B,* the same as in *A,* after 35 minutes RNA begins to enter the cytoplasm. *C,* the same, incubated for 12 minutes in ³H-cytidine and then for 88 minutes in a nonradioactive medium. Notice that while the nucleus has lost all labeled RNA, the cytoplasm is heavily labeled. (Courtesy of D.M. Prescott.)

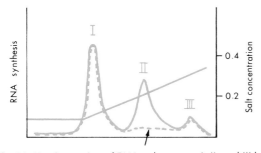

Fig. 20–7. Separation of RNA polymerases I, II, and III by chromatography in DEAE-Sephadex. The peaks are eluted by increasing the salt concentration. *Solid line,* without α-amanitin; *dotted line,* assays performed with 0.1 µg/ml α-amanitin showing selective inhibition of polymerase II.

crease in length. As a result, genes that are very actively transcribed show the so-called "Christmas tree" configuration as shown in Figure 20–9 in rRNA genes and in Figure 20–21 for a protein-coding gene.

3. At any one time, only a very small fraction of the chromatin is being transcribed.

4. In fully transcribing genes the RNA polymerase molecules are packed very close together, with one polymerase every 200 base pairs of DNA or less. Because the elongation rate of RNA synthesis is 20 to 30 nucleotides per second, it follows that a maximally expressed gene can complete one RNA chain at least every 10 seconds.

5. The growing or *nascent* RNA becomes associated with protein as it is being transcribed, producing ribonucleoprotein particles (RNPs) rather than free RNA. Processing of the initial transcripts into smaller RNAs can also start before the RNA chains are completed (see Fig. 20–21).

6. In eukaryotes the nuclear envelope introduces a barrier between transcription and protein synthesis. The mRNA must be transported into the cytoplasm before it can be utilized.

SUMMARY:

TRANSCRIPTION IN PROKARYOTES AND EUKARYOTES

RNA is synthesized by RNA polymerase. The primary transcripts thus produced must usually be processed further before they can be used by the cell. Types of processing include: (1) cleavage of large precursor RNAs, (2) splicing, (3) terminal addition of nucleotides, and (4) nucleoside modifications.

Prokaryotic mRNAs are very short-lived, and are often translated into protein at one end while the other end is still being transcribed (Fig. 20–5). The length of these mRNAs is heterogeneous; some mRNAs (polycistronic mRNAs) code for several proteins (Fig. 20–1). All prokaryotic RNAs are transcribed by the same enzyme, which is a large and complex multi-subunit molecule consisting of several polypeptides. One of these peptides, the

TABLE 20–2.   PROPERTIES AND FUNCTIONS OF EUKARYOTIC RNA POLYMERASES

Enzyme	Localization	Gene Transcripts	Inhibition by α-Amanitin
I	Nucleolus	5.8S, 18S and 28S rRNAs	Insensitive
II	Nucleoplasm	mRNA, U-snRNA	Sensitive to low concentration (0.1 µg/ml)
III	Nucleoplasm	tRNA, 5S RNA, 7S scRNA	Sensitive to high concentration (10 µg/ml)

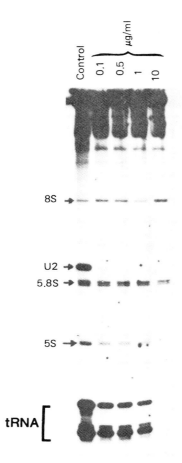

Fig. 20–8. Differential inhibition of the three eukaryotic RNA polymerases by α-amanitin. Frog oocytes were injected into the nucleus with ^{32}P-labeled GTP and, in order to increase U2-snRNA and tRNA expression, a bacteriophage λ recombinant containing *Xenopus* U2-snRNA and tRNA genes. The other bands—5.8S ribosomal RNA, its precursor 8S RNA, and 5S RNA—come from endogenous oocyte transcripts. Note that a low concentration of α-amanitin (0.1 µg/ml) abolishes U2-snRNA transcription, which is carried out by RNA polymerase II (U-snRNA genes are described in detail later in this chapter). RNA polymerase III transcripts (5S RNA and tRNA) require much higher concentrations of α-amanitin for inhibition (10 µg/ml). RNA polymerase I transcripts (5.8S rRNA and a precursor of it called 8S RNA) are not affected at all. (Courtesy of I.W. Mattaj and R. Zeller.)

sigma factor, is loosely bound to the core polymerase and is required for the enzyme to recognize the correct start signals on the DNA. The transcription process involves three stages: (1) binding to promoters and chain initiation, (2) elongation, and (3) chain termination.

Eukaryotic transcription differs in many respects from that of prokaryotes. Transcription takes place in the nucleus; eukaryotic mRNAs are considerably more stable than those of prokaryotes and are monocistronic (i.e., there is only one protein synthesis start site per mRNA). Three nuclear RNA polymerases have been identified, each of which transcribes a different class of genes. These polymerases can be distinguished by their order of elution from ion-exchange columns and by their sensitivity to inhibition by α-amanitin (Table 20–2 and Fig. 20–8). A maximally expressed gene can complete one RNA chain at least every 10 seconds, and the growing chain becomes associated with protein as it is being transcribed. The mRNAs must be transported into the cytoplasm before they can be utilized.

20–2 EUKARYOTIC MESSENGER RNA BIOSYNTHESIS

The origin and fate of messenger RNA in eukaryotic cells is more complex than in bacteria. The formation of a functionally active mRNA is the consequence of a complex series of steps that comprise: (1) the actual transcription of DNA into mRNA precursors, (2) the intranuclear processing of these precursors, and (3) the transport of the mRNAs into the cytoplasm and their association with ribosomes to initiate the process of translation of protein synthesis.

mRNA Has Noncoding Regions

One of the main features of prokaryotic mRNA is its very short half-life, on the order of minutes. In contrast, mostly metabolically stable mRNAs are found in eukaryotes. The best example of a long-lived mRNA is that in the mammalian reticulocyte, an immature red blood cell that has lost its nucleus but still retains the ribosomes and other components of the protein-synthesizing machinery. Reticulocytes contain mRNA that continues to produce hemoglobin for many hours and even days after the nucleus (and therefore the globin genes) has been lost.

Because they are stable, it has been possible to purify several eukaryotic mRNAs, taking advantage of differentiated tissues that synthesize large amounts of a given protein and therefore accumulate considerable amounts of specific mRNAs. Examples of such tissues are mammalian reticulocytes, in which 90% of the protein synthesized is α- and β-globin, the protein subunits of hemoglobin; chick oviduct, which produces ovalbumin; and the silk gland of the moth *Bombyx mori*, which secretes silk fibroin. The availability of pure mRNAs has allowed the study of their structure. In addition, *radioactive*

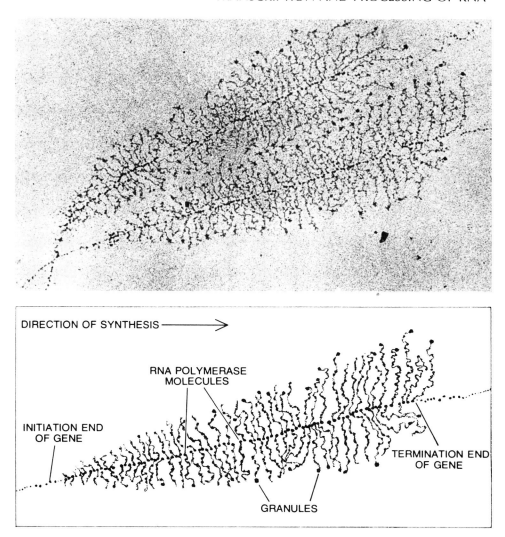

DIRECTION OF SYNTHESIS ⟶

RNA POLYMERASE
MOLECULES

INITIATION END
OF GENE

TERMINATION END
OF GENE

GRANULES

Fig. 20–9. *Above,* electron micrograph showing two nucleolar genes in the process of transcription of the ribosomal RNA. *Below,* labels are self-explanatory. ×35,000. (Courtesy of O.L. Miller and B.R. Beatty.)

probes for specific genes can be prepared from the mRNA by a variety of methods, such as using the enzyme *reverse transcriptase* to copy a labeled *complementary DNA;* this probe in turn allows one to study the genes at the DNA level, for example, by isolating clones from a genomic recombinant DNA library.

Here we analyze the structure of globin mRNA, because certain features which are common to all mRNAs emerged from studies on globin.

When the RNAs extracted from rabbit reticulocyte polysomes are centrifuged in a sucrose gradient, globin mRNA sediments as a small 9S peak that can be separated from ribosomal and

transfer RNAs (i.e., 28S, 18S, 5S, and 4S) by repeated cycles of centrifugation. The estimated length of the 9S mRNA is between 650 and 700 nucleotides. Of these, only 589 are coded by the DNA[20]; the rest are accounted for by the poly A tail that is attached to the mRNA after transcription is completed (see later discussion). The mRNA molecule is much longer than is required to code for the globin protein, which has 146 amino acids and is therefore coded by only 438 nucleotides. The fate of the extra ("nontranslated") nucleotides is shown in Figure 20–10.

In addition to the sequence coding for protein, all eukaryotic mRNAs have two regions that are not translated into protein. The 5' non-

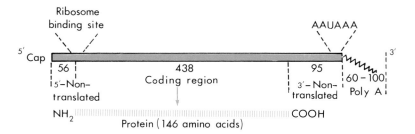

Fig. 20–10. Rabbit globin mRNA. Note that eukaryotic mRNAs have two noncoding regions, a 5′ cap, and a 3′ poly A tail. The number of nucleotides in each section is indicated.

coding region, which is located before the start of the region coding for protein, has about 50 nucleotides in most mRNAs[21] and contains the ribosome binding site (Fig. 20–10). The function of the 3′ noncoding region in protein synthesis is unknown, even though it is about twice the size of the 5′ noncoding region. Some mRNAs have a very large 3′ noncoding region; in ovalbumin mRNA it is 600 nucleotides long.

The noncoding regions are thought to be functionally important because their sequence has tended to be conserved through evolution (i.e., globin mRNAs from different species have similar noncoding sequences).[21,22] In addition, all mammalian polyadenylated mRNAs have the sequence AAUAAA close to the start of the poly A sequence.[22] This is a signal for the addition of poly A some 10 to 20 nucleotides downstream from it.[23]

The 5′ End of Eukaryotic mRNA Is Blocked by 7-Methylguanosine

Most eukaryotic mRNAs have their starting ends blocked by the posttranscriptional addition of a "cap" of 7-methylguanosine. Normally the phosphodiester bonds of an RNA molecule go from the 3′ position of the ribose of one nucleotide to the 5′ position of the ribose of the next one. The linkage of the cap to mRNA is different, because the riboses of 7-methylguanosine and the terminal nucleotide are linked by a 5′ to 5′ triphosphate bridge, as shown in Figure 20–11. A cap amounts to a reversed GTP added to the end of the mRNA. The cap is added at the very first nucleotide transcribed by RNA polymerase II and thus is an indication of the site of transcription initiation. Capping is so rapid that it is thought to take place simultaneously with transcription initiation.

The cap structure m[7]-G[5′]ppp[5′]Xp has no free phosphates, and the mRNA is thus protected against attack by phosphatases and other nu-

cleases.[24] Bacterial mRNA is not capped and therefore starts with pppXp. The cap is important in mRNA translation; when it is removed, mRNA is unable to form an initiation complex with 30S ribosomal subunits. Similarly, cap analogues, such as m[7]-G-5′-monophosphate, are potent inhibitors of protein synthesis initiation. Bacterial mRNAs are translated poorly by eukaryotic ribosomes, but if they are artificially capped in vitro, they become very active templates for mammalian ribosomes. The dependence of protein synthesis on caps is not absolute, however, some mRNAs (e.g., poliovirus mRNA) do not have caps but are still translated normally.

The 3′ End of Eukaryotic mRNA— A Poly A Segment

Most mRNAs contain a sequence of polyadenylic acid attached to their 3′ end. This poly A segment is 100 to 200 nucleotides long and is added sequentially to the mRNA after transcription is completed by the action of a poly A synthetase that uses ATP as substrate (Fig. 20–12). Bacterial mRNAs do not have poly A.

The finding of poly A tails in mRNA represented an advance in molecular biology because it made possible several methods for the rapid separation of mRNA from other contaminant RNAs. Poly A + RNA will bind to poly U or polydeoxythymidine (oligo dT) columns. Not all mRNAs are polyadenylated; up to 15% of mRNAs in polysomes lack poly A. Among these, the most prominent are the histone mRNAs.

Poly A has a role in promoting mRNA stability.[25] When globin mRNA is microinjected into living *Xenopus* oocytes, it is translated into globin. Fully adenylated globin mRNA is very stable in oocytes, as shown by the fact that it is translated continuously for up to 2 weeks after injection, a period during which each globin mRNA molecule can produce 100,000 protein molecules.[26] It has been found that if the poly

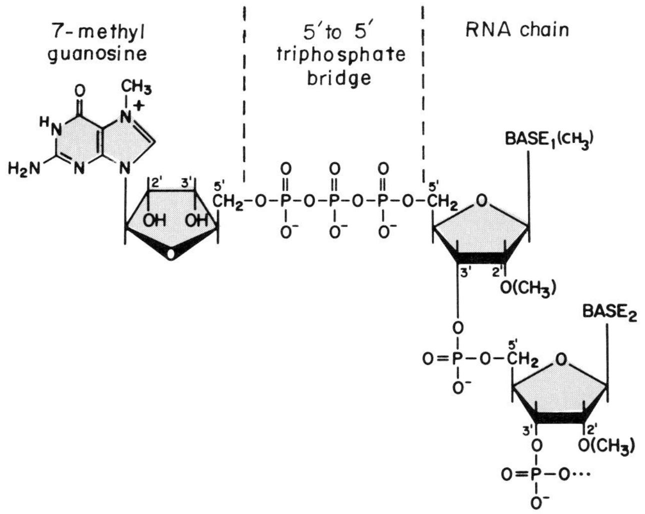

7-methyl
guanosine

5′ to 5′
triphosphate
bridge

RNA chain

Fig. 20–11. Structure of the 5′ cap of messenger RNA. Observe that the linkage of the 7-methylguanosine to the terminal nucleotide is by a 5′ to 5′ triphosphate bridge.

A segment is removed from globin mRNA (using the enzyme polynucleotide phosphorylase, which degrades RNA from the 3′ end), the half-life of the mRNA in oocytes is greatly decreased. This effect is reversible, because when poly A is added back (using an RNA-adenyltransferase from *E. coli*), the stability of the mRNA is restored.[24]

Eukaryotic Genes Are Frequently Interrupted by Intervening Sequences

When the first cloned eukaryotic genes became available in 1977, it was unexpectedly found that the information for mRNA was found in noncontiguous segments of the genome. Figure 20–12 shows that most genes are interrupted by insertions of noncoding DNA, which are later excised at the RNA level and the molecule precisely religated in a process called RNA *splicing*. The extra segments that are removed are called *introns* or *intervening sequences*, and the segments that constitute the mature mRNA are called *exons*. (Exons are not synonymous with protein-coding regions because the 5′ and 3′ nontranslated mRNA sequences are also part of exons.)

Introns have been found in most eukaryotic

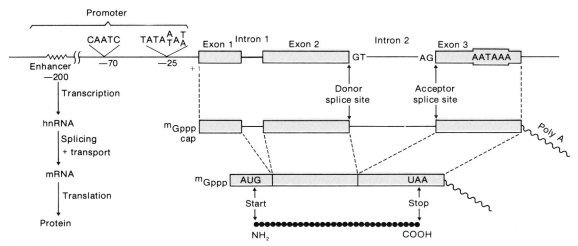

Fig. 20–12. An idealized eukaryotic protein-coding gene. Because it contains two introns, it could correspond to a globin gene. Note the three elements of the promoter (TATA sequence, CAAT sequence, and enhancer), the introns and exons, the polyadenylation signal, and the monocistronic mRNA. Compare with the prokaryotic gene shown in Figure 20–1.

protein-coding genes, such as globin,[28,29] ovalbumin, immunoglobulin, and collagen. All mRNA-coding genes are spliced by the same mechanism, but there are also introns that are removed by different mechanisms in some tRNAs,[30] *Tetrahymena* ribosomal DNA[31] and fungal mitochondrial DNA.[32] Introns are not a special feature of eukaryotes, and splicing has been recently detected in bacteriophages (see below), but introns are much less frequent in prokaryotes.

Globin genes have two introns, one of them being rather short, as shown in the idealized gene of Figure 20–12. The long intron 2 is about 600 base pairs long, i.e., about the same length as the whole mature mRNA; it can be visualized in Figure 20–13B. The electron micrograph shows a cloned segment of mouse DNA that was hybridized to globin mRNA. When RNA hybridizes to the DNA, a "bubble" is seen in which one strand is an RNA-DNA hybrid, and the other is single-stranded DNA. Globin mRNA hybridizes to two discontinuous regions of the genomic DNA, while the intervening sequence, which is not present in the 9S mRNA, remains as a loop of double-stranded DNA (see an interpretation of the structure in the line drawing of Fig. 20–13B).

Intervening sequences are transcribed into precursor RNAs that are larger than the mature mRNAs. To obtain a functional mRNA these internal sequences must be precisely excised and the molecule religated because an error of even a single nucleotide would cause the rest of the message to be read out of phase.

The β-globin gene is not transcribed initially as a 9S molecule, but rather as a precursor that sediments at 15S in sucrose gradients. This sedimentation value represents a molecule about twice as long as globin mRNA, and this extra length is due to the presence of the intervening sequences[33] as can be visualized in Figure 20–13A. Mature 9S mRNA is the result of removing the intervening sequence, a process that the cell does rapidly; the 15S precursor has a half-life of less than two minutes.

Other genes have many intervening sequences; e.g., ovalbumin has 7 introns, albumin 14, vitellogenin 28, and collagen more than 40. Interestingly, the precursor RNAs containing introns are nuclear and do not exit into the cytoplasm until the very last intervening sequence

is removed. Thus the presence of introns might provide a signal indicating that a particular RNA cannot be released to the cytoplasm, which is important, because there it would be translated into an incorrect protein product. Thus splicing is completed in the nucleus before the mRNA molecule is transported into the cytoplasm.

Some protein-coding genes do not contain any introns—for example, the histone and interferon genes—but they constitute a minority of the mammalian genes.

Splice Junctions Have Conserved Sequences That Are Recognized by U1-snRNPs

The splicing mechanism has enormous precision, but we do not yet know how it is achieved. When the nucleotide sequences of a great many intron-exon junctions were examined, it was found that some nucleotides are strongly conserved.[34,35] Figure 20–14 shows the intron-exon boundaries for the seven introns of the ovalbumin gene, and it can be seen that the intron always starts with GT and ends with AG (see also Fig. 20–12). Other flanking nucleotides are also conserved:

5′ exon ag ↑ GTaag . . . intron . . . cAG ↑ g . . . 3′ exon

but these are not present in every gene whereas the GT . . . AG rule applies in every gene. Mutations in the conserved GT and AG nucleotides abolish splicing. In humans, a mutation that changes G for A at the splicing junction of β-globin (AU instead of GU) leads to the production of unspliceable RNA precursors and results in a severe hereditary disease called β_0 thalassemia in which β-globin is absent from red blood cells.

An additional conserved region was detected in yeast introns, which contain the 7-nucleotide conserved sequence TACTAAC roughly 40 nucleotides before the 3′ end of the intron. This sequence is required for splicing, because when the TACTAAC sequence is mutated, yeast introns cannot be spliced[36,37]. Higher organisms are different from yeast in that there is no strict conservation of the seven nucleotide region, but there usually is a somewhat changed version of the TACTAAC sequence, which varies from intron to intron. Although these sequences may vary, they have the same function in higher organisms as in yeast, namely providing the

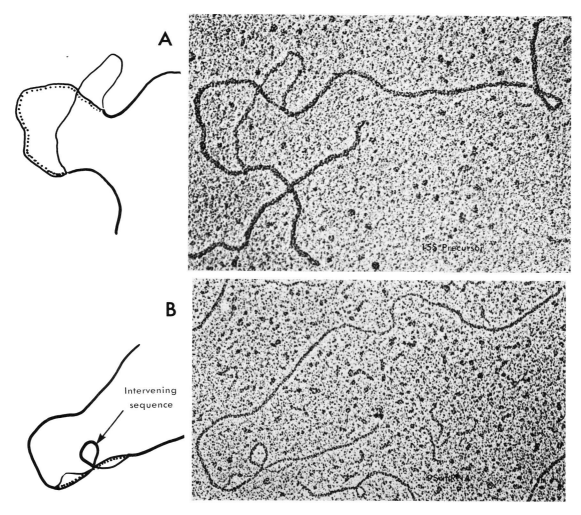

Fig. 20–13. Visualization of the intervening sequence of the globin gene under the EM. A cloned segment of DNA containing the mouse β-globin gene was hybridized with A, 15S globin mRNA precursors; B, mature 9S globin mRNA, and the hybridized molecules spread for observation under the EM. The hybridized RNA is represented by a dotted line in the diagram. The 15S precursor hybridizes in a continuous way, showing that the intervening sequence is transcribed into RNA. Note in B that the mature globin mRNA hybridizes to two discontinuous regions of the DNA, and the intervening sequence remains as a loop of double-stranded DNA. It was from electron micrographs such as these that intervening sequences were first discovered. (Courtesy of P. Leder.)

branching point for splicing intermediates as explained below.

Figure 20–15 shows the three conserved regions in intron: 1) the 5′ end of the intron, 2) the TACTAAC sequence in yeast (or modified versions of it in higher animals and plants) and 3) the 3′ end of the intron. Each one must be recognized precisely, and this is done by U-snRNPs, which are *small nuclear ribonucleoprotein particles* consisting of 7 to 8 different proteins complexed with one molecule of a U-snRNA (a family of uridine-rich small nuclear RNAs, described at the end of this chapter).

Several years ago it was noted that the sequence of the 5′ end of U1-snRNA was complementary, i.e., could base pair with the conserved nucleotides at the 5′ end of introns.[38,39] Subsequent experiments clearly showed that this RNA-RNA interaction is essential for splicing.[40,41] A different U-snRNP, containing the U2-snRNA was found to be bound by the second conserved sequence indicated in Fig. 20–15.[42] The third sequence (3′ end of the intron) is also recognized by an U-snRNP, but it is not yet certain which one, although U5-snRNA is suspected.

Using cell extracts which are able to splice

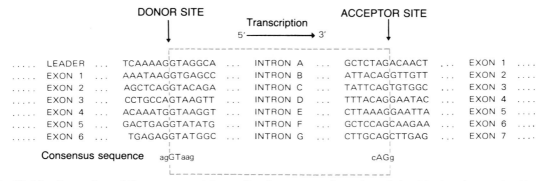

Fig. 20–14. Comparison of the sequences at the ovalbumin gene exon-intron boundaries. The invariant nucleotides GT . . . intron . . . AG are indicated as part of the "consensus sequence." The DNA sequence is shown here; in RNA the T's are replaced by U's.

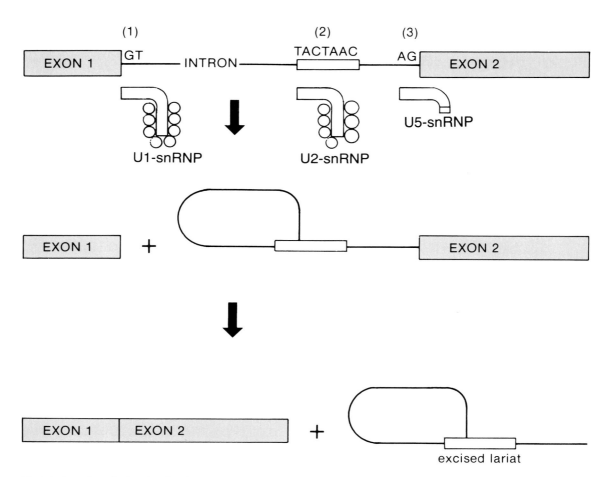

Fig. 20–15. Branched molecules (lariats) are intermediates in pre-mRNA splicing. The three regions of the precursor RNA that must be precisely recognized are numbered. U1-snRNP is bound to the first one, and U2-snRNP to the second one, via RNA-RNA base pairing interactions. The third sequence is also recognized by a U-snRNP and although it is not yet certain which one it is, U5-snRNP is suspected. Each snRNP contains one snRNP and seven or eight different proteins (small circles). In the first stage a concerted reaction cleaves the 5' end of the intron and covalently joins it to an A residue in the intron, creating a lariat. In a second reaction, the 3' end of the intron is cleaved and the two exons are ligated together. All these reactions take place in a large 60S complex called the spliceosome.

radioactive mRNA precursors, it has been possible to identify intermediates in the splicing reaction. In a first step, the appropriate regions of the pre-mRNA are recognized by U1 and U2 snRNPs, and the 5′ end of the intron is cleaved, and joined covalently to an A residue within the conserved region 40 nucleotides before the 3′ end of the intron (Fig. 20–15). This forms a *branch point* in the pre-mRNA, formed between the terminal G residue of the intron joined by a 2′-5′ phosphodiester bond to the ribose of the A nucleotide.[43,44] In a second step (Fig. 20–15), splicing is completed by cleavage at the 3′ splice site followed by ligation of the two exons. The excised intron remains as an excised lariat (called this way because it resembles the lasso employed by cowboys). The lariat is then rapidly degraded by cellular enzymes.

This series of highly precise reactions is carried out by a complex cellular machinery. The splicing intermediates containing the excised 5′ exon and the lariat intron still joined to the 3′ exon (see middle section of Fig. 20–15) sediment together in a very large macromolecular complex which sediments between 50 and 60 S and has been called the spliceosome (see Brody and Abelson, 1985; Grabowski et al., 1985). Spliceosomes are comparable in size to ribosomal subunits, and contain U1, U2 and U5 snRNPs as well as other as yet uncharacterized components of the pre-messenger RNA splicing machinery.

Tetrahymena rDNA Can Splice Itself in the Absence of Proteins

In vitro studies have shown that mRNA splicing requires U1- and U2-snRNPs and in addition other components that are protease-sensitive. In the case of tRNA splicing (see Fig. 21–30), the splicing enzymes (an endonuclease and a ligase) are proteins that have been obtained in pure form. In the case of the ribosomal genes of the ciliate protozoon *Tetrahymena* the situation is quite different because splicing can take place in the absence of proteins.

Tetrahymena rDNA contains an intervening sequence whose splice junctions are different from those conserved in mRNA-coding genes. Splicing of rRNA precursors can take place in *Tetrahymena* nuclear extracts. Cech and his colleagues decided to purify the enzymes that catalyzed this reaction. Using standard biochemical

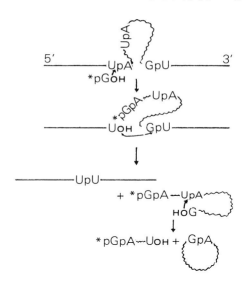

Fig. 20–16. Self-splicing of *Tetrahymena* ribosomal RNA precursors. The reaction does not require any proteins. A guanosine donor is introduced at the 5′ end of the intervening sequence. Cleavage of the 3′ end is immediately followed to ligation. The excised intervening sequence then undergoes a cyclization reaction. (Data from Zaug, A., Grabowski, P.J., and Cech., T.R.: Nature, *301*:578, 1983.)

fractionation methods, they found that the splicing activity always purified together with the rRNA precursor and that it was resistant to proteases and to boiling in the presence of the ionic detergent sodium dodecyl sulfate, a most unusual behavior for a protein. They then did an extraction with phenol, a procedure in which RNA stays in the aqueous phase and protein goes to the phenolic one. Once again, the splicing activity was in what by now had to be a pure RNA fraction. To eliminate completely the possibility that a *Tetrahymena* protein was bound tightly to the rRNA precursor and could not be separated from it, they proceeded to clone the rDNA in an *E. coli* plasmid containing a strong promoter. The rRNA precursor transcribed in vitro with *E. coli* RNA polymerase was still able to splice itself, even though there were no other *Tetrahymena* components in this reaction.

The reaction (Fig. 20–16) only requires magnesium, ATP, a guanosine donor (GMP, GDP, or GTP), and the rDNA precursor.[45,46] The secondary structure of the RNA itself is responsible for this chemical reaction. This structure has been called a *ribozyme*, i.e., a molecule in which the activation energy for a chemical reaction is lowered by RNA (and not by proteins as in enzymes). The energy of the ATP is used to cut and incorporate a G residue at the 5′ end of the intervening sequence (Fig. 20–16). Somehow this stored energy is used to complete excision and ligation in a concerted reaction. The excised intervening sequence then circularizes itself, giving a circle and a 15-nucleotide linear RNA (Fig. 20–16).

More recently, Zaug and Cech (1986) have been able to demonstrate that the excised circular intron has catalytic activity even after its self-splicing is completed. When they added polycytidilic acid, the molecules were progressively converted into both longer and shorter poly C molecules. The kinetics of the reaction can be explained by the Michaelis-Menten equation (Chapter 2), and each intron molecule can catalyze many polymerization reactions. Thus the self-splicing intron is clearly an enzyme, and therefore enzymes are not always proteins (see Cech, 1986).

In yeast mitochondria splicing takes place by a mechanism in which the secondary structure of the intron is also very important[32] and which also yields circular intervening sequences after splicing. Interestingly, fungal mitochondrial introns and the *Tetrahymena* rDNA intron can be folded into similar secondary structures.[32] Mitochondrial introns are also self-splicing RNAs.

The existence of self-splicing RNAs has interesting evolutionary implications. The splicing mechanism could indeed be ancient because it can be a property of RNA itself. Splicing enzymes may have arisen later in order to perfect a preexisting mechanism.

Splicing in Prokaryotes

Until recently, intervening sequences were thought to be a special feature of eukaryotic cells, until two genes containing an intron were discovered in bacteriophage T4. These are the genes coding for thymidylate synthase (Chu et al. 1986) and for ribonucleotide reductase (Sjöberg et al., 1986), two enzymes utilized for the rapid biosynthesis of deoxyribonucleotides for DNA replication after bacteriophage T4 infects its host, *Escherichia coli*. The nucleotide sequences at the exon/intron borders of these genes do not conform to the rules found in eukaryotic introns (Fig. 20–14). The bacteriophage introns, however, can base pair into a secondary structure similar to that found in the Tetrahy-

mena rRNA and in some mitochondrial introns from yeast and neurospora. The T4 introns, like the tetrahymena counterparts, are able to undergo self-splicing in vitro, in the absence of proteins. The energy and molecular information required for this reaction are stored in the secondary structure of the RNA precursor molecule itself.

Because bacteriophage T4 replicates within *E. coli*, these results show that at least in some cases RNA splicing exists in prokaryotic cells. On the other hand, many prokaryotic genes have been sequenced and no others have been found to contain introns. Thus it is clear that intervening sequences are rare in prokaryotic cells. It seems that the splicing mechanism could have existed in primitive cells before eukaryotes and prokaryotes evolved, but that introns may have been gradually lost as the prokaryotic genomes became streamlined for rapid DNA replication.

Why Have Splicing?

To have genes in pieces that must then be precisely spliced together at the RNA level seems a rather complicated way of organizing gene expression. However, the phenomenon is so widespread and so conserved throughout eukaryotic evolution that one would like to believe it serves some useful purpose, even though we might not yet know what it is.

Intervening sequences have been in place for a very long time. Consider the oxygen-binding proteins α-globin, β-globin, and myoglobin. All have a similar three-dimensional structure and carry a heme group that transports oxygen. The amino acid sequence, however, has changed much among the three proteins. It turns out that α-globin, β-globin, and myoglobin all have two introns, placed in analogous positions. Thus it appears that the two introns were present in the common ancestral gene and have remained in place for many million years of evolution.

An additional example is provided by the enzyme triosephosphate isomerase, which is involved in glycolysis and evolved to its optimal enzymatic activity before prokaryotes and eukaryotes diverged 1.5 billion years ago. This enzyme has eight introns in maize and six in humans, of which five are in identical positions.[47] This places the antiquity of introns at least 1000 million years back, when plants and animals diverged. The same enzyme from bacteria and yeast does not contain introns, but it does in the fungus *Aspergillus* (reviewed by Gilbert et al., 1986). The observations support the view that the original gene consisted of several exons since the beginning of evolution when the first triosephosphate isomerase was assembled. Bacteria and yeast probably lost the introns in the course of evolution because their genomes specialized in rapid DNA synthesis and cellular division and lost most of their nonessential DNA. The introns, however, remained in the slowly replicating cells of complex organisms, as birthmarks of their origin.

Proteins frequently are made of different functional domains, which sometimes appear as globular regions in the three-dimensional structure of proteins. The most clear example is provided by the immunoglobulin heavy chains, which have a "variable" domain, a "hinge" region, and three globular heavy chain "constant" domains (see Fig. 23–29). In the DNA each domain is encoded in an individual exon (see Fig. 23–32) and separated from the others by an intervening sequence. There are many other examples in which functional domains are encoded in different exons, and which may explain split genes.

It has been suggested[48] that splicing accelerates evolution enormously because it allows new proteins to evolve by the joining of functional segments of preexisting genes. Evolution in this manner should be much more efficient than by individual amino acids mutating at random. Figure 20–17 shows how splicing enables new genes to be built from bits of old ones. If two functional domains A and B in two different DNA molecules are to be combined without splicing, as in Figure 20–17, *1*, then recombination must occur between two nonhomologous regions of DNA in such a way that the message is still read in phase. The correct recombination between two given nucleotides will be very rare indeed. In Figure 20–17, *2*, in which splicing takes place, recombination need not be in phase and the target for recombination is much larger. If one assumes a target of, say, 1000 nucleotides in each molecule (Fig. 20–17, *2*), recombination with splicing is $1000 \times 1000 = 10^6$ times more probable than recombination without splicing between individual nucleotides in both DNA molecules ($1 \times 1 = 1$).

There may be many other beneficial (or det-

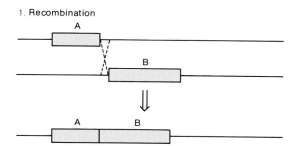

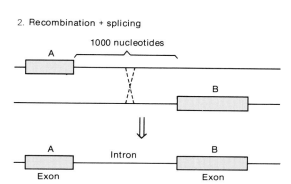

Fig. 20–17. The splicing mechanism facilitates building of new genes from functional bits of old ones. See text for details. Based on thoughts of Walter Gilbert.

rimental) consequences of having introns, but at present they are difficult to see. It is perhaps fitting to remember here that no molecular biologist had the faintest suspicion that splicing existed until split genes were stumbled upon in 1977.

An RNA Polymerase II Promoter Consists of a TATA Box, a CAAT Box, and an Enhancer

Having examined the nature and organization of eukaryotic mRNAs, we can now return to their transcription by RNA polymerase II. The three main elements that compose a polymerase II promoter—the TATA box, the CAAT box, and the enhancer—are shown in Figure 20–12.

The conserved sequence $\text{TATA}^A_T A^T_A$ (also called the Hogness-Goldberg box) occurs 20 to 30 nucleotides before the start of transcription of the vast majority of protein-coding genes. In prokaryotes there is a similar AT-rich region, but it is located slightly closer (-10) to the mRNA start. The TATA box is important for transcription; its main function seems to be to position precisely the start of transcription.[49] Some TATA

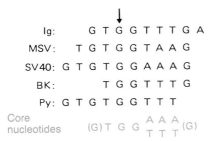

Fig. 20–18. Enhancer "core" sequences. Mutations in the arrowed nucleotide abolish SV40 enhancer activity. Ig, immunoglobin heavy chain; MSV, Moloney sarcoma virus; SV40, simian virus 40; Py, polyoma virus.

box mutants will still start transcription, but at multiple sites and downstream from the normal start site. The TATA box sequence is bound by a specific protein or transcription factor.

Many genes (but not all) have the conserved sequence CAAT between positions -70 and -80. When this sequence is removed by deletion, the level of transcription is drastically reduced, but the remaining transcripts start at the correct site.[50,51] This region probably determines the efficiency with which a gene can be expressed. The CAAT box is also bound by a special transcription factor protein in living cells.

The *enhancer* is a segment of DNA that can increase expression of some genes and that has some puzzling characteristics. Enhancers were first discovered when a segment of SV40 DNA was cloned in a plasmid also containing a rabbit globin gene.[52] After transformation into cultured cells, globin expression was greatly stimulated. What was puzzling was that the stimulation persisted whether the enhancer DNA was introduced in the inverse orientation, or positioned either before or after the globin gene. Thus enhancers can work at a distance from a gene and regardless of the orientation. Enhancers have been found in a variety of viral and cellular genes, and a potential "core sequence" has been identified[53] by sequence comparisons and mutagenesis (Fig. 20–18). The enhancer itself is much larger than this conserved core, and in SV40 it has been mapped to a 72-base-pair region normally located 100 to 170 nucleotides before the "early" promoter. The mechanism of enhancement is unknown at present, but enhancers are hypersensitive to digestion by DNAse I, a characteristic that has been interpreted as indicating that there are proteins bound to them that impede the usual nucleo-

some configuration. After many studies, enhancers are now considered to be regions of DNA that provide binding sites for multiple proteins (called transcription factors) which are then recognized by RNA polymerase II. While prokaryotic RNA polymerase can bind directly to DNA, eukaryotic RNA polymerase II recognizes proteins bound to DNA, but not naked DNA. For a particular promoter to be active, a certain combination of transcription factors will have to be bound to it, and all the required proteins will be present only in specific tissues. Expression of different repertoires of transcription factors is at present the best explanation of the tissue-specific transcription of genes.

One of the most exciting findings about enhancers is that they explain why a particular promoter is activated after the immunoglobulin constant chain region is translocated close to a certain (out of several hundred) variable chain region (see Fig. 23–30). The first intron of the constant region contains an enhancer which then activates the promoter closest to it, i.e., the one corresponding to that particular translocated variable chain region.[54,55] The immunoglobulin enhancers are tissue-specific; i.e., they work only when transformed into lymphocytes, and that fact raises the possibility that other tissue-specific enhancers and their corresponding binding proteins might play a role in switching on and off entire sets of genes during cell differentiation.

3′ End Maturation Involves Cleavage of Longer Precursors Followed by Polyadenylation

When RNA polymerase II reaches the end of a gene, it does not terminate transcription abruptly, but rather goes on transcribing the DNA as shown in Figure 20–19. In the case of mouse β-globin, it has been shown that transcription goes on for at least 1500 nucleotides past the gene.[56] It is not yet known whether precise termination sites exist or whether the RNA polymerase II molecules gradually fall off the DNA at multiple sites.

As Figure 20–19 shows, in order to produce a polyadenylated mRNA the transcript is cut and polyadenylated while it is still being transcribed. The signal for the addition of poly A is AAUAAA,[22,23] which is present in all mRNAs

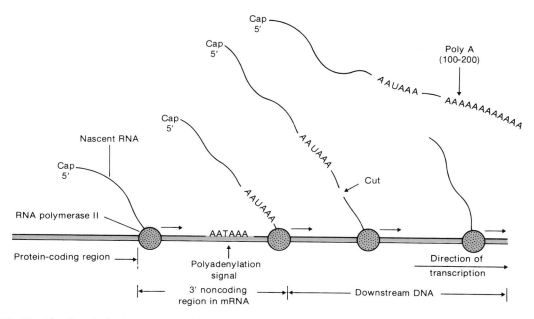

Fig. 20–19. The 3' end of mRNA is matured by cutting and polyadenylation while still being transcribed. The conserved AAUAAA sequence is recognized by a U-snRNP (not shown) and then cut. Note that the 5' ends of the pre-mRNAs already have a cap. Capping is an early event, perhaps simultaneous with transcription initiation.

about 15 nucleotides before the poly A tail. This signal is recognized by a U-snRNP before cleavage (see Hashimoto and Steitz, 1985). U-snRNAs are widely used for recognizing processing signals in pre-mRNAs. As we shall see below, some transcriptional units have multiple polyadenylation signals, of which only one is chosen. These alternative polyadenylation sites represent an important control step in eukaryotic gene regulation (see Figs. 20–22, 20–23, and 23–32).

RNA polymerase II reads through beyond the gene also in non-polyadenylated mRNAs. As Figure 20–20 shows, histone mRNA transcription continues past the gene for about 100 nucleotides. The polymerases fall off the DNA

gradually. All histone mRNAs have a conserved hairpin stem and loop structure at their 3' end which, together with other downstream conserved nucleotides (Fig. 20–20), are necessary for proper cleavage (see Hentschel and Birnstiel, 1981). *Xenopus* oocytes will mature the 3' ends of sea urchin histone mRNA precursors injected into their nucleus. However, the 3' maturation of sea urchin H3 mRNA can be greatly increased by injection of extracts from sea urchin embryos. With this assay a 12S fraction was isolated[57] which turned out to be a small nuclear RNP. The RNA contained in this snRNP is 70 nucleotides long, is called U7-snRNA, and is complementary to the conserved nucleotides at the end

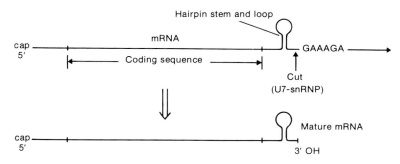

Fig. 20–20. Maturation of the 3' end of histone mRNAs. (After studies of Max Birnstiel.)

of histone mRNA. The injection of the RNA alone also results in the stimulation of 3' cleavage of H3 mRNA. Presumably the injected snRNA binds to oocyte proteins, and the snRNP complex then participates in the reaction that trims the 3' end of the histone mRNA (see Birnstiel et al., 1985).

RNA polymerase II is unique in its lack of precise termination after transcription of a gene is completed. Prokaryotic RNA polymerase, as well as eukaryotic RNA polymerases I and III, terminate at precise sites, which in all cases involve a row of T's in the noncoding strand of the DNA (Fig. 20–1).

Three-Quarters of the Heterogeneous Nuclear RNA (hnRNA) Is Destroyed within the Nucleus

Heterogeneous nuclear RNA was discovered in 1962 when Scherrer and Darnell labeled cultured cells for 5 minutes with ^{3}H uridine. Most of the labeled RNAs were large and heterogeneous in size. hnRNA has a very short half-life, on the order of minutes. When long labeling times are used (several hours), most of the label is found in rRNA and tRNA, which are stable and accumulate. We now know that hnRNA, like all RNA polymerase II transcripts, is capped at the 5' end and is sometimes polyadenylated at the 3' end. hnRNA is much longer than mature mRNAs. The average length of eukaryotic mRNA is 1800 nucleotides, while that of hnRNA is 4000. Some hnRNAs can be 20,000 or more nucleotides long (Darnell, 1983).

Nucleic acid hybridization studies have shown that hnRNA contains more sequences than mRNA. It is estimated that 6 to 20% of the information available in the genome is represented in hnRNA, while only 1% is represented in cytoplasmic mRNA.

It is now clear that hnRNA consists of mRNA precursors. At least some of the hnRNAs are early transcripts from which the intervening sequences have not yet been removed, such as the 15S globin precursor (Fig. 20–13A). Other transcripts may represent the products of RNA polymerases that continue to transcribe after the ends of genes (Fig. 20–19).

Three-quarters of the hnRNA never exits the nucleus and is degraded. One can measure the proportion of caps and poly A tails; in cyto-

plasmic mRNA there is roughly one cap for each poly A tail. In nuclear hnRNA, however, there are four caps for each poly A tail, which suggests that many of the molecules never become polyadenylated. One possibility that has been considered is that correct polyadenylation may be a decisive factor in determining whether a particular transcript will become stable and processed into mature mRNA.

hnRNA and mRNA Are Associated with Proteins

Eukaryotic mRNAs in living cells are not present as naked RNA strands but in association with proteins, forming ribonucleoprotein particles. Although cells have many different mRNAs, only between two and five proteins are associated with mRNPs, and therefore the same proteins are bound to many kinds of RNA. Globin mRNA is associated with two proteins of 78,000 and 52,000 daltons. The 78,000 protein is associated specifically with the poly A segment of globin and of other mRNAs.[58,59]

When the poly A tail of cytoplasmic mRNPs is digested with small amounts of micrococcal nuclease (see Fig. 13–14 for a similar experiment with chromatin), a repeating pattern of organization appears, in which the RNA bands are multiples of 27 nucleotides.[60] When the 78,000 poly A-binding protein is purified and mixed with poly A in vitro, the 27-nucleotide repeat can be reconstituted, suggesting that the repeat is caused by this protein. The 78,000 protein is absent from nuclear poly A and present only in the cytoplasm.[61] It is conceivable that this protein might have a role in mRNA transport by binding to the mRNA and retaining it in the cytoplasm.

Heterogeneous nuclear RNA is also bound to a specific class of proteins that are nuclear (Fig. 13–9) and bind to the nascent transcripts while they are being transcribed. These hnRNA-binding proteins are basic and have molecular weights between 30,000 and 40,000. The proteins associate with each other forming a particle that sediments at about 30S.[62,63] These particles attach to hnRNA giving a beads-on-a-string configuration. The protein particles do not bind randomly and seem to prefer certain regions of the nascent RNA.[64,65] Figure 20–21 shows an RNA polymerase II transcriptional unit spread for

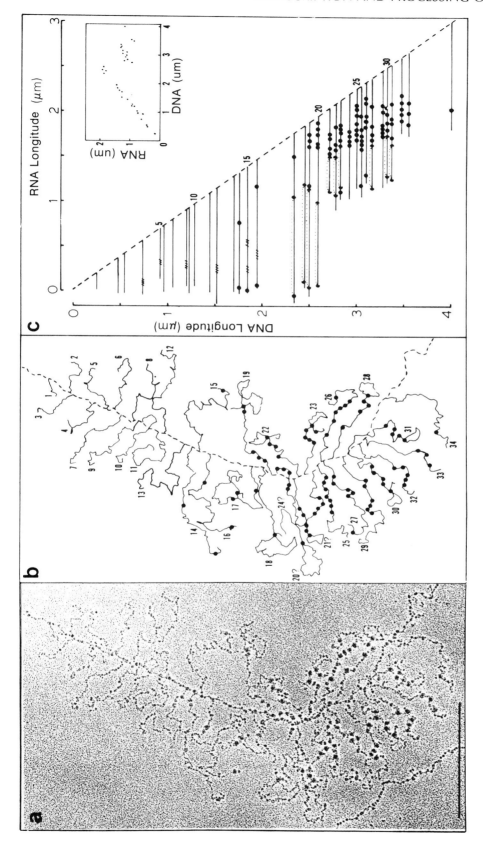

Fig. 20–21. Correlation of hnRNP structure and nascent transcript cleavage in an RNA polymerase II transcriptional unit. The nascent mRNA molecules with their associated protein particles were spread for electronmicroscopy. In Panel B the RNA molecules were traced and the 30S protein particles represented by black dots, while the DNA and associated proteins are indicated by a broken line, with the promoter towards the top. In panel C the position of the transcripts with respect to the DNA molecule (0 corresponds to the promoter) is plotted as a function of the length of the RNA molecules. Note that the distribution of protein molecules is non-random and that cleavage of the RNA molecules occurs at a precise position (perhaps due to a processing event as the one shown in Fig. 20–19). (From Beyer, A., Bouton, A., and Miller, O., Jr.: Cell, 26:155, 1981. Copyright © 1981 by MIT Press.)

electron microscopy with 30S particles bound to it. The diagram in panel C suggests that the binding is not random and also that in some transcriptional units there are abrupt changes in fibril length, indicating that cleavage occurs at specific sites during transcription.[66] These cleavages could be analogous, for example, to those indicated in Figure 20–19. Sometimes loops can be seen in the nascent transcripts, suggesting that the secondary structure is also not random[66] (see transcripts 18, 19, and 20 in Fig. 20–21B).

Alternative RNA Processing Can Regulate Gene Expression

The highly complicated RNA-processing mechanisms in eukaryotes are used to control gene expression. It would be surprising indeed if this had turned out not to be the case. In some cases the choice of the poly A site determines the protein product, e.g., calcitonin mRNA or the membrane-bound or secreted forms of immunoglobulins (see Fig. 23–32) In other cases the choice of both polyadenylation and splicing sites is affected, e.g., in adenovirus (see below) and in the *Drosophila* myosin gene.[67]

Calcitonin is a peptide hormone produced by the C cells of the thyroid, which controls blood calcium levels. As Figure 20–22 shows, the gene has several exons that are spliced together and in thyroid cells the mRNA is translated into a single protein of 17,000 daltons. The completed protein is then cleaved by proteases into several peptides, one of which is the calcitonin hormone (3500 daltons). As in many other instances, the specific protein cleavages take place at sites containing two consecutive basic amino acids; i.e., the signal for protein cleavage is Lys-Arg or Arg-Arg or Lys-Lys. The function of the other peptides is not known, but they could be other hormones as well.

In the hypothalamus (a region of the brain that secretes many hormones into the bloodstream) the same gene is processed to produce a different peptide (called CGRP for calcitonin-gene-related-product, probably a neurohormone) instead of calcitonin (see Amara et al., 1982). As Figure 20–22 shows, when polyadenylation site 1 is used in the thyroid, calcitonin is made; when polyadenylation site 2 is used (in the hypothalamus), the calcitonin exon is treated as if it were an intron and removed from the mature mRNA by splicing, and the CGRP neuropeptide is made instead.

We said earlier that introns represent functional units of proteins that are joined together by splicing.[47] It seems from the results with the calcitonin gene that in the endocrine system introns can represent new functional peptide hormones.

The extreme example of how multiple protein products can be made from a single transcriptional unit is provided by adenovirus. This virus infects human cells and has a linear chromosome of 35,000 base pairs. Late in infection (e.g., after 24 hours) essentially all of the RNA transcripts start at a single "late" promoter (Fig. 20–23) and continue until the end of the virus. There are five possible poly A addition sites (Fig. 20–23A). All the mRNAs share three short "leader" exons but differ in their downstream sequences. Not only are there five polyadenylation sites, but for each of them several splices are possible (Fig. 20–23B). In this way 15 to 20 proteins are made from a single transcriptional unit (reviews by Nevins and Chen-Kiang, 1981; Darnell, 1983, 1985).

Adenovirus splicing is submitted to regulation, for the same mRNAs are spliced differently during the first few hours after infection. Early transcripts sometimes start at the "late" promoter, but all of these are polyadenylated at the first poly A site. More important, as shown in Figure 20–23, panel C, these transcripts contain an extra exon between the second and third leader segments. The protein made by this mRNA is only 14,000 daltons, while the same mRNA made late in infection (lacking the extra leader exon) codes for a protein of 52,000. Thus there is alternative splicing of this intron, and some part of the cellular splicing machinery must change in late infected cells, perhaps as a consequence of the synthesis of some gene product expressed early in infection (Nevins, 1982).

These examples were brought up here not with the purpose of overwhelming the student with the complexities of eukaryotic gene expression, but rather to underline how subtle its control mechanisms can be. The detailed mechanisms by which splicing and polyadenylation sites are chosen remain a mystery. Understanding them is a major goal for molecular biologists.

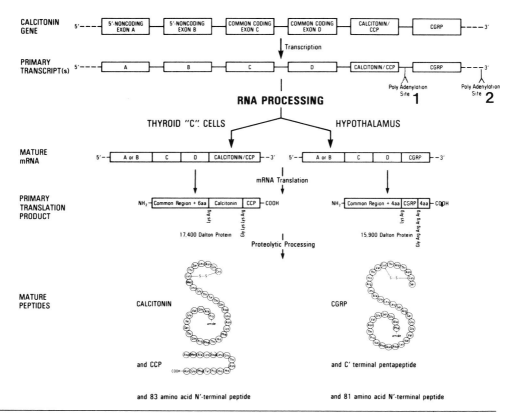

Fig. 20–22. Alternative RNA processing in the calcitonin gene. Two hormones are produced in different tissues from the same gene. (Reprinted by permission from Amara, S., et al.: Nature, 298:240, 1982. Copyright © 1982 Macmillan Journals Limited.)

20–3 snRNPS AND scRNPS

Eukaryotic cells contain small, stable RNAs that are distinct from the main classes of RNA (mRNA, hnRNA, rRNA, and tRNA). These RNAs are associated with proteins forming ribonucleoprotein (RNP) particles, and according to their intracellular localization they are called small nuclear RNAs (snRNAs) or small cytoplasmic RNAs (scRNAs). Figure 20–24 shows the RNAs extracted from whole HeLa cells (lane T), the nuclear fraction (lane N) and the cytoplasm (lane C). Some nuclei break during the biochemical fractionation procedure, and thus there is some cross-contamination of the fractions. Still, it is clear from Figure 20–24 that some RNAs are more concentrated in the nucleus (e.g., U2-RNA) while other RNAs (e.g., Alu 7S RNA) are predominantly cytoplasmic. We now briefly discuss the characteristics of the main small RNAs.

U-snRNPs React with Autoimmune Antibodies

The main type of snRNA is the U family (Table 20–3). These RNAs were named U1 through U7 because they are rich in uridine. They are metabolically stable (same half-life as a ribosome), abundant (there are 1 million U1 molecules per human cell nucleus), and ubiquitous (present in all plants and animals) and have a highly conserved sequence (e.g., human and *Xenopus* U1 have 95% sequence homology). They are transcribed by RNA polymerase II (Fig. 20–8) and start with a trimethylguanosine cap. Messenger RNAs have a monomethylguanosine cap (Fig. 20–11).

U-snRNPs became amenable to study when Lerner and Steitz[68] discovered that patients with autoimmune disease (such as *lupus erythematosus*) sometimes have antibodies that specifically immunoprecipitate U-snRNPs. These human sera are of two types: (1) anti-Sm (after Smith, the first patient in whom the specificity

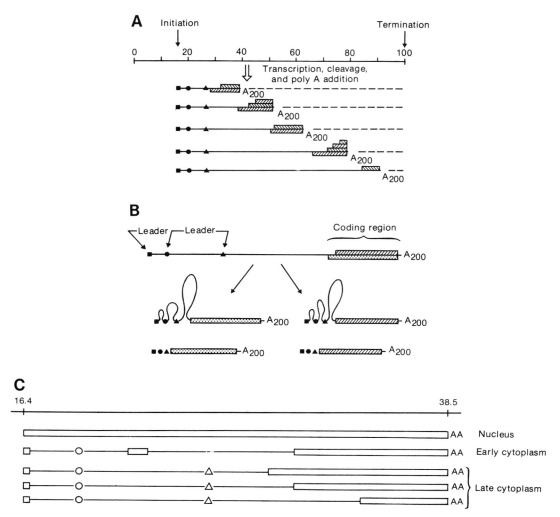

Fig. 20–23. Schematic diagrams representing the selection of a particular mRNA from an adenovirus primary transcript. The square, circle, and triangle represent the three short "leader" exons shared by all adenovirus mRNAs. *A,* selection of a 3' site. One of the five possible 3' sites is selected by cleavage of the still nascent transcript at the appropriate site followed by poly A addition generating a poly A-containing, unspliced precursor as shown. *B,* selection of a 5' site. The poly A molecule formed as described in *A* then serves as the substrate for splicing to select the particular mRNA from the family of mRNAs. *C,* early in infection transcripts are spliced differently, with an extra exon between the second and third leader. (Redrawn from Nevins, J., and Chen-Kiang, S.: Adv. Virus Res., 26:1, 1981.)

was found), which precipitates U1, U2, U4, U5, and U6 snRNPs and (2) anti-U1-RNP, which immunoprecipitates only U1 (Fig. 20–25). The antibodies react with one or more proteins in the snRNP but not with the RNA. Thus U1, U2, U4, U5, and U6 RNPs share at least one common protein (which reacts with Sm antibodies) and U1 has a protein that is not found in the other snRNAs. Figure 20–25 also shows other autoimmune antibodies found in human patients that react with other RNPs.

Although the protein composition is still in-

completely understood, it seems that most U-snRNPs have about seven proteins and one molecule of RNA. Most of the proteins are shared, but others are unique.[69,70] Each snRNP is about three-quarters protein and one-quarter RNA by weight. It was the use of anti-U1 antibodies that demonstrated that U1-snRNP is involved in splicing in the way described in Figure 20–15. U7-snRNP is required for the 3' cleavage of histone mRNA as shown in Fig. 20–16, and another snRNA is involved in polyadenylation. Table 20–3 summarizes our present knowledge

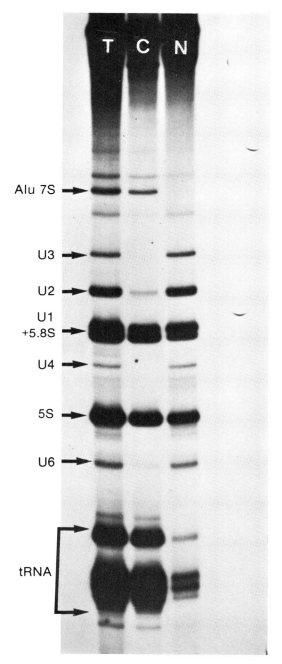

Alu 7S →

U3 →

U2 →

U1
+5.8S →

U4 →

5S →

U6 →

tRNA

Fig. 20–24. Small RNAs from HeLa cells. *T,* total cells; *C,* cytoplasmic fraction; *N,* nuclear fraction. There is some degree of cross-contamination due to the biochemical procedure used to separate nucleus and cytoplasm, but it is clear that some RNAs are predominantly cytoplasmic (Alu 7S, tRNA) whereas others are nuclear (U3, U2, U4, U6). U1 and U5 comigrate with other RNAs and were not resolved by this procedure. HeLa cells were labeled by addition of ^{32}P inorganic phosphate to the culture medium for 18 hours, and then the RNAs were extracted and separated according to size in an 8% polyacrylamide/7 *M* urea gel. (From E.M. De Robertis.)

of snRNAs. Most of the U-snRNAs can be found bound to hnRNA within cells, as part of 60S spliceosome particles.

U3-snRNA is present in nucleoli, where it is tightly bound to 32S rRNA precursor. It is not precipitated by anti-Sm-antibodies, presumably because it is not easily solubilized. U3-snRNA is believed to be involved in rRNA processing. The other U-snRNPs seem to be involved in mRNA processing (Table 20–3).

Microinjected U-snRNAs Migrate into the Nucleus

When naked U-snRNAs are microinjected into the cytoplasm, they migrate into nuclei.[71] Figure 20–26 shows sections of frog oocytes microinjected into the cytoplasm with radioactive U1-RNA or tRNA. As can be seen, U1-RNA becomes nuclear while tRNA remains in the cytoplasm—their normal cellular locations. Soon after microinjection the snRNA becomes bound to about seven snRNA-binding proteins, and then the complex migrates into the nucleus. The snRNP-binding proteins, which can be studied using anti-Sm antibodies, are stored in large amounts in the oocyte cytoplasm. Each oocyte has the equivalent of about 200,000 somatic cells worth of proteins, the vast majority of which are not bound to snRNA.[72] The proteins are stored for use in early development, when large amounts of snRNAs are synthesized at late blastula.[73]

As Figure 20–27 shows, the free oocyte snRNP-binding proteins are stored in the cytoplasm, but when excess pure snRNA is injected, the complex rapidly migrates into the nucleus. When the binding site of the Sm-antigen on the U2-RNA molecule was mutated, the resulting RNAs also lost the ability to migrate into nuclei (see Mattaj and De Robertis, 1985 in the Additional Readings at the end of this chapter). It seems that binding to the RNA exposes a domain of the protein which then acquires karyophilic properties.

This high nuclear affinity of the U-snRNPs could be important in RNA transport. There are sufficient U1-snRNPs in a cell nucleus (10^6) to bind to all the intervening sequences present in hnRNA at any one time. This could in turn provide an anchorage point to the nucleus for those mRNA precursors that have not had all the in-

TABLE 20–3. SOME SMALL RNAS OF HUMAN CELLS

RNA	Length (nucleotides)	Number of Proteins	Number Per Cell	Localization	RNA Polymerase	5' End*	Comments
U3	216	?	3×10^5	Nucleolus	II	m³ cap	rRNA processing?
U2	189	~7	5×10^5	Nucleoplasm	II	m³ cap	binds to splicing branch point
U1	165	~8	1×10^6	Nucleoplasm	II	m³ cap	binds to 5' end of intron
U4	139	~7	1×10^5	Nucleoplasm	II	m³ cap	
U5	118	~7	2×10^5	Nucleoplasm	II	m³ cap	binds to 3' end of intron?
U6	106	?	2×10^5	Nucleoplasm	II	?	Forms complexes with U4 snRNP
U7	70	?	5×10^4	Nucleoplasm	II	m³ cap	Histone mRNA 3' maturation
La	Many, heterogeneous	1 (50K)	?	Nucleoplasm	III	pppX	All RNA polymerase III transcripts, immediately after transcription (tRNA precursors, 5S RNA, 7S RNA)
Alu 7S	280	6	5×10^5	Cytoplasm	III	pppX	Signal recognition particle used in protein secretion
Y1–Y5	60–110	?	$\sim 2 \times 10^5$	Cytoplasm	III	pppX	Part of the Ro antigen

Data mostly from Zieve, G.: Cell, 26:296, 1981; and Lerner, M.R., and Steitz, J.A.: Cell, 25:298, 1981.
*m³ cap stands for a trimethylated 5' cap nucleotide, while pppX stands for a triphosphate at the uncapped 5' end.

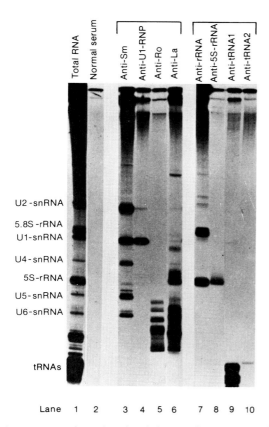

Fig. 20–25. Polyacrylamide gel showing the separation of ribonucleoproteins from a HeLa cell culture. Cells were broken by sonication, and the homogenate was precipitated with serum of patients with autoimmune disease (lanes 3 to 10). The precipitate was extracted with phenol to isolate the RNAs submitted to electrophoresis. Total cell extracts contain many RNAs, none of which is precipitated by normal human serum (lane 2). A whole spectrum of autoimmune antibodies is detected: lane 3, anti-Sm precipitates U-snRNPs containing U2, U1, U4, U5 and U6 snRNAs; lane 4, anti-U1-sRNP precipitates U1-snRNPs; lane 5, anti Ro binds to scRNPs Y1 to Y5; lane 6, anti La binds to a protein that binds to many RNA newly-made polymerase transcripts (mostly tRNA precursors); lane 7, an antibody that binds to naked ribosomal RNAs (all other antibodies in this figure bind to proteins and precipitate RNAs that are bound to them, but do not react with naked RNAs); lane 8, an antibody that reacts specifically with a protein that binds 5S rRNA; lanes 9 and 10, two antibodies which react with aminoacyl-tRNA synthetases and precipitate specific tRNAs. (Courtesy of J.A. Steitz.)

trons removed. This idea is hypothetical, but either this or another mechanism must ensure that hnRNA (which can sometimes have up to 30 introns) will not exit the cell nucleus until the very last intron is spliced out. Only mature mRNA is found in the cytoplasm; this fact is important because if introns were left, aberrant

protein products would be translated in the cytoplasm.

The La Protein Binds to RNA Polymerase III Early Transcripts

The La protein of 50,000 daltons is an abundant (10^7 copies/cell) nuclear protein initially defined by a human autoimmune serum (the name of the patient started with La). It can bind to a variety of RNAs (Fig. 20–25), all of which are RNA polymerase III transcripts shortly after being synthesized. The main RNAs precipitated by La antibodies are tRNA precursors[74] (length 80 to 100 nucleotides), but they also precipitate some 5S RNA, Alu 7s RNA, and Y-scRNAs (described below).

RNA polymerase III transcripts always terminate within a row of U residues. The La protein binds to molecules that have two, three, or more U's at their 3' end. After some time the extra U's are cleaved off the newly-made RNAs, and the RNAs are released from the La protein. This protein might be inolved in RNA polymerase III transcription termination and might also provide a mechanism for retaining some transcripts in the nucleus until they are sufficiently mature to exit into the cytoplasm (many of them undergo methylations and other base modifications in the nucleus, Fig. 21–30).

scRNPs: Alu 7S RNA Is Part of the Signal Recognition Particle Required for Protein Secretion

The only scRNP whose function is known is the one containing the 300-nucleotide Alu 7S RNA (Fig. 20–24). (This 7S RNA is called Alu because it has sequence homologies with repeated sequences of the human genome called the Alu family, see Section 22–2.) As described in Chapter 8 (Fig. 8–11) there is in the cytoplasm a *signal recognition particle* (SRP), which recognizes hydrophobic signal sequences on nascent polypeptides, binds to these ribosomes, and causes a specific arrest of translation. The arrested polysomes can then diffuse to the endoplasmic reticulum, where the polypeptide is then translocated into the lumen of the endoplasmic reticulum. When the SRP particle was purified,[75] it was found to contain six proteins and, unexpectedly, a molecule of 7S RNA (Fig.

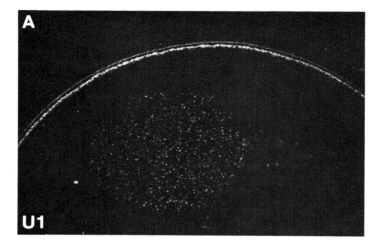

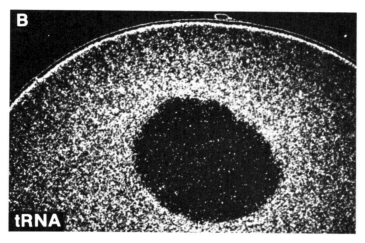

Fig. 20–26. Frog oocytes were injected into the cytoplasm with ³H-labeled U1-RNA *(A)* or with ³H tRNA *(B)*. After 24 hours the oocytes were fixed, sectioned, and covered with photographic emulsion for autoradiography. The silver grains were photographed by dark field microscopy. The white rim at the periphery of the oocyte consists of pigment granules, not autoradiographic grains. Note that U1-RNA accumulates in the nucleus. (From De Robertis, E.M.: Cell, *32*:1022, 1983. Copyright © 1983 MIT Press.)

20–28). Both the RNA and the proteins are necessary for the function of SRP in protein synthesis.[76]

The other major scRNP is called *Ro*, after an antibody that precipitates five RNAs from human cells (Y1 through Y5). Their function is unknown.

Small nuclear RNAs provide an example of how a key discovery can rapidly open up a whole new field of research. Although they were known to exist for 15 years (see review by Busch et al., 1982), snRNAs attracted little attention until it was found that they could be purified with autoimmune antibodies.[68] Now research on snRNPs and scRNPs is flourishing.

SUMMARY:

Eukaryotic mRNA, snRNA, and scRNA

In addition to the protein-coding sequence, all eukaryotic mRNAs have 5' and 3' regions that are not transcribed into protein. Most eukaryotic mRNAs have their starting ends blocked by a 7-methylguanosine cap. This structure protects the mRNA against attack by nucleases, and is necessary for the formation of the initiation complex with 30S ribosomal subunits. Most eukaryotic mRNAs also contain a poly A tail, which helps prevent degradation of the RNA.

Frequently eukaryotic genes are interrupted by *intervening sequences or introns* that do not code for protein (Fig. 20–12). The sequences retained in the mature mRNA are called *exons*. The introns are removed and the molecule is religated in a process called *RNA splicing*.

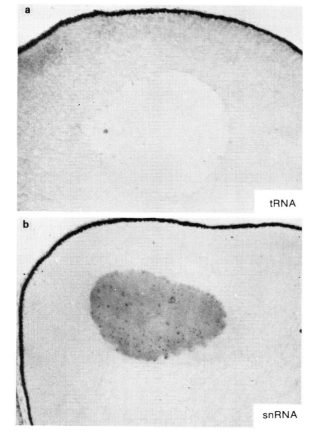

tRNA

snRNA

Fig. 20–27. Nuclear migration of the Sm antigen after microinjection of purified U-snRNA into frog oocytes. The snRNP-binding proteins are stockpiled free of RNA in the oocyte cytoplasm for use in early development. Microinjection of tRNA (a) does not change their localization, but when pure snRNA is injected into the cytoplasm, the complex of protein and snRNA translocates into the nucleus (b). The oocyte sections were stained with Sm antibodies coupled to horseradish peroxidase; the staining represents enzyme activity. (From Zeller, R., Nyfenegger, T., and De Robertis, E.M.: Cell, 32:425, 1982. Copyright © 1982 MIT Press.)

The splice junctions have conserved sequences, so that the intron always starts in GU and ends in AG. UsnRNAs recognize conserved nucleotide sequences involved in splicing by base-pairing as shown in Figure 20–15. The splicing reaction takes place in two steps (Fig. 20–15): first the 5′ end of the exon is cleaved and joined covalently to an A residue about 40 nucleotides before the end of the intron, forming a branched lariat molecule; in a second step the 3′ end is cleaved and the exons are precisely ligated together.

One intron, that of *Tetrahymena* rRNA, has the remarkable property of splicing itself in the absence of any proteins. The secondary structure of the RNA is sufficient to drive the reaction; the intron itself is an enzyme consisting solely of RNA,

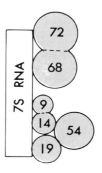

Fig. 20–28. The signal recognition particle (SRP) is made of six proteins (molecular weights indicated in kilodaltons) and one molecule of 7S RNA. (Data from Walter, P., and Blobel, G.: Cell, 34:525, 1983.)

also called a ribozyme. This self-splicing of RNAs has also been found in introns from mitochondria from fungi which have a similar secondary structure to that of *Tetrahymena*. Similar self-splicing introns have been found in two genes of bacteriophage T4, a virus that infects *E. coli*. The splicing mechanisms arose early in evolution as self-splicing reactions, then were made more efficient by splicing protein enzymes that evolved for this purpose.

Splicing is thought to enormously accelerate evolution because it permits the building of new genes by the joining of parts of old ones. In general, exons represent functional domains of proteins. In addition, alternative RNA processing can be used to control gene expression (Figs. 20–22 and 20–23).

RNA polymerase II transcription initiation requires a TATA$_\mathrm{T}^\mathrm{A}$A$_\mathrm{A}^\mathrm{T}$ conserved sequence about 20 nucleotides before the start of transcription (cap site), frequently a CAAT sequence at about −70, and an enhancer. Enhancers are DNA segments that stimulate transcription at a distance and regardless of their orientation in the DNA molecule. Some enhancers stimulate even if downstream from the initiation site, which explains the choice of promoter in rearranged immunoglobulin genes.

Transcription by RNA polymerase II continues past the gene, and the long RNA molecules are then cut and polyadenylated some 15 nucleotides after the poly A addition signal AAUAAA (Fig. 20–19). Histone mRNAs, which are not polyadenylated, are also cleaved at the 3′ end, and in this case U7-snRNP, containing an RNA 70 nucleotides long, is required (Fig. 20–20).

Heterogeneous nuclear RNA (hnRNA) consists of long, unspliced mRNA precursors. It has a short half-life (minutes), and much of it is never polyadenylated and is destroyed within the nucleus. As it is being transcribed, hnRNA-binding proteins attach to it, forming 30S particles that bind in a nonrandom fashion (Fig. 20–21). After transport to the cytoplasm, a new set of proteins binds to mRNA, in particular a 78,000 protein that binds to the poly A tail and confers a 27-nucleotide periodicity to the poly A when partially digested with nucleases.

The main small nuclear RNAs are the seven U-snRNAs, which are uridine-rich, stable, and abundant (Table 20–3). U1 and U2 are involved in splicing and U3 in nucleolar rRNA processing, and it is suspected that the others may also be involved in RNA processing. Each U-RNA is bound to seven or eight proteins, and the snRNP has a strong affinity for the nucleus (Fig. 20–26).

The main small cytoplasmic RNP is a complex of Alu 7S RNA and six proteins (Fig. 20–28), constituting the *signal recognition particle* (SRP) required for protein export into the ER.

REFERENCES

1. Jacob, F., and Monod, J.: J. Mol. Biol., 3:318, 1961.
2. Brenner, S., Jacob, F., and Meselson, M.: Nature, 190:576, 1961.
3. Gros, F., et al.: Nature, 190:581, 1961.
4. Chaberlin, M.J.: *In* RNA Polymerase. Edited by R. Losick and M. Chamberlin. Cold Spring Harbor, New York, Cold Spring Harbor Laboratory, 1976.
5. Burgess, R.R., Travers, A., Dunn, J., and Bautz, E.K.F.: Nature, 221:5175, 1969.
6. Travers, A., and Burgess, R.R.: Nature, 222:537, 1969.
7. Burgess R.R.: *In* RNA Polymerase. Edited by R. Losick and M. Chamberlin. Cold Spring Harbor, New York, Cold Spring Harbor Laboratory, 1976, p. 69.
8. Pribnow, W.: J. Mol. Biol., 99:419, 1975.
9. Roberts, J.W.: *In* RNA Polymerase. Edited by R. Losick and M. Chamberlin. Cold Spring Harbor, New York, Cold Spring Harbor Laboratory, 1976, p. 247.
10. Rosenberg, M., and Court, D.: Annu. Rev. Genet., 13:314, 1979.
11. Stroyhowski, T., and Yanofsky, C.: Nature, 298:39, 1982.
12. Oxender, D.L., Zurawski, G., and Yanofsky, C.: Proc. Natl. Acad. Sci. USA, 76:5524, 1979.
13. Miller, O.L., Jr., and Beatty, B.R.: Science, 164:995, 1969.
14. Miller, O.L., Jr., Hamkalo, B.A., and Thomas, C.A.: Cold Spring Harbor Symp. Quant. Biol., 35:505, 1970.
15. Reeder, R.H., and Roeder, R.G.: J. Mol. Biol., 67:433, 1972.
16. Weinmann, R., and Roeder, R.G.: Proc. Natl. Acad. Sci. USA, 71:1790, 1974.
17. Sklar, V.F., Schwartz, L.B., and Roeder, R.G.: Proc. Natl. Acad. Sci. USA, 72:348, 1975.
18. McKnight, S.L., and Miller, O.L., Jr.: Cell, 8:305, 1976.
19. McKnight, S.L., Bustin, M., and Miller, O.L., Jr.: Cold Spring Harbor Symp. Quant. Biol., 42:741, 1977.
20. Estratiadis, A., Kafatos, F.C., and Maniatis, T.: Cell, 10:571, 1977.
21. Baralle, F.E., and Brownlee, G.G.: Nature, 274:84, 1978.
22. Proudfoot, N.J., and Brownlee, G.G.: Nature, 263:211, 1976.
23. Fitzgerald, M., and Schenk, T.: Cell, 24:251, 1981.
24. Shatkin, A.J.: Cell, 9:645, 1976.
25. Zeevi, M., Nevins, J.R., and Darnell, J.E.: Cell, 26:39, 1981.
26. Gurdon, J.B., Lingrel, J.B., and Marbaix, G.: J. Mol. Biol., 80:539, 1973.
27. Marbaix, G., Huez, G., and Soreq, H.: Trends in Biochem. Sci., 2:106, 1977.
28. Tilghman, S.M., et al.: Proc. Natl. Acad. Sci. USA, 75:725, 1978.
29. Jeffreys, A.J., and Flavell, R.A.: Cell, 12:1097, 1977.
30. Knapp, G., et al.: Cell, 14:221, 1978.
31. Cech, T.R., Zaug, A.J., Grabowski, P.J., and Brehm, S.L.: The Cell Nucleus, 10:179, 1982.
32. Davies, R.W., et al.: Nature, 300:719, 1982.
33. Tilghman, S.M., et al.: Proc. Natl. Acad. Sci. USA, 75:1309, 1978.
34. Breatnach, R., and Chambon, P.: Annu. Rev. Biochem., 50:349, 1981.
35. Mount, S.M.: Nucleic Acids Res., 10:459, 1982.
36. Langford, C.J., and Gallwitz, D.: Cell, 33:519, 1983.
37. Pikielny, C., Teem, J., and Rosbash, M.: Cell, 34:395, 1983.
38. Lerner, M.R., et al.: Nature, 283:220, 1980.
39. Rogers, J., and Wall, R.: Proc. Natl. Acad. Sci. USA, 77:1877, 1980.
40. Mount, S.M., et al.: Cell, 33:509, 1983.
41. Krämer et al.: Cell, 38:299, 1984.
42. Black, D., Chabot, B., and Steitz, J.A.: Cell, 42:737, 1985.
43. Ruskin, B., Krainer, A.R., Maniatis, T., and Green, M.R.: Cell, 38:317, 1984.
44. Konarska, M.M., Grabowski, P.J., Padgett, R.A., and Sharp, P.: Nature, 313:555, 1985.
45. Kruger, K., et al.: Cell, 31:147, 1982.
46. Zaug, A.J., Grabowski, P.J., and Cech, T.R.: Nature, 301:378, 1983.
47. Marchionni, M., and Gilbert, W.: Cell, 46:1694, 1986.
48. Gilbert, W.: Nature, 271:501, 1978.
49. Grosschedl, R., and Birnstiel, M.: Proc. Natl. Acad. Sci. USA, 77:1432, 1980.
50. McKnight, S.L., Gavis, E.R., Kingsbury, R., and Axel, R.: Cell, 25:385, 1981.
51. Grosveld, G.L., De Baer, E., Shewmaker, C.K., and Flavell, R.A.: Nature, 295:120, 1982.
52. Banerji, J., Rusconi, S., and Schaffner, W.: Cell, 27:299, 1981.
53. Weiher, H., Konig, M., and Gruss, P.: Science, 219:626, 1983.
54. Banerji, J., Olson, L., and Schaffner, W.: Cell, 33:729, 1983.
55. Gillies, S.D., Morrison, S.L., Oi, V.T., and Tonegawa, S.: Cell, 33:717, 1983.
56. Hoffer, E., and Darnell, J.E.: Cell, 23:585, 1983.
57. Stunnenberg, E., and Birnstiel, M.L.: Proc. Natl. Acad. Sci. USA, 79:6201, 1982.
58. Blobel, G.: Proc. Natl. Acad. Sci. USA, 70:924, 1973.
59. Greenberg, J.R.: Nucleic Acids Res., 8:5685, 1980.
60. Baer, B.W., and Kornberg, R.D.: Proc. Natl Acad. Sci. USA, 77:1890, 1980.
61. Baer B.W., and Kornberg, R.D.: J. Cell Biol., 96:717, 1983.
62. Beyer, A.L., Christensen, M.E., Walker, B.W., and Le Stourgeon, W.M.: Cell, 11:127, 1977.
63. Martin, T., et al.: Cold Spring Harbor Symp. Quant. Biol., 38:921, 1974.
64. Beyer, A.L., Miller, O.L., Jr., and McKnight, S.L.: Cell, 20:75, 1980.
65. Laird, C.D., and Chooi, W.Y.: Chromosoma, 58:193, 1976.
66. Beyer, A.L., Bouton, A.H., and Miller, O.L.: Cell, 26:155, 1981.
67. Rozek, C.E., and Davidson, N.: Cell, 32:23, 1983.
68. Lerner, M.R., and Steitz, J.A.: Proc. Natl. Acad. Sci. USA, 76:5495, 1979.

69. Hinterberger, M., Pettersson, I., and Steitz, J.A.: J. Biol. Chem., 258:2604, 1983.
70. Kinlaw, L.S., Dusing-Schwartz, S.K., and Berget, S.: Mol. Cell Biol., 2:1159, 1982.
71. De Robertis, E.M., Lienhard, S., and Parisot, R.F.: Nature, 295:572, 1982.
72. Zeller, R., Nyffenegger, T., and De Robertis, E.M.: Cell, 32:425, 1983.
73. Forbes, D., Kirschner, M., and Newport, J.: J. Cell Biol., 97:62, 1983.
74. Rinke, J., and Steitz, J.A.: Cell, 29:149, 1982.
75. Walter, P., and Blobel, G.: Nature, 299:691, 1982.
76. Walter, P., and Blobel, G.: Cell, 34:525, 1983.

ADDITIONAL READINGS

Abelson, J.: RNA processing and the intervening sequence problem. Annu. Rev. Biochem., 48:1035, 1979.

Amara, S.G., et al.: Alternative RNA processing in calcitonin gene expression generates mRNAs encoding different polypeptide products. Nature, 298:240, 1982.

Beyer, A.L., Bouton, A.H., and Miller, O.L., Jr.. Correlation of hnRNP structure and nascent transcript cleavage. Cell, 26:155, 1981.

Birnstiel, M.L., Büslinger, M., and Strub, K.: Transcription, termination and 3′ processing: The end is in site! Cell, 41:349, 1985.

Black, D.L., Chabot, B., and Steitz, J.A.: U2 as well as U1 small nuclear ribonucleoproteins are involved in premRNA splicing. Cell, 42:737, 1985.

Brody, E., and Abelson, J.: The "spliceosome": yeast premessenger RNA associates with a 405 complex in a splicing dependent reaction. Science, 228:963, 1985.

Busch, H., Reddy, R., Rothblum, L., and Choi, Y.C.: snRNAs, snRNPs and RNA processing. Annu. Rev. Biochem., 51:617, 1982.

Cech, T.R.: RNA splicing: Three themes with variations. Cell, 34:713, 1983.

Cech, T.R.: The generality of self-splicing RNA: relationship to nuclear mRNA splicing. Cell, 44:207, 1986.

Cech, T.R.: RNA as an enzyme. Scient. Am., 255(5):64, 1986.

Chabot, B., et al.: The 3′ splice site in pre-messenger RNA is recognized by a small nuclear ribonucleoprotein. Science, 230:1344, 1986.

Chambon, P.: Eukaryotic nuclear RNA polymerases. Annu. Rev. Biochem., 44:613, 1975.

Chu, F.K., Maley, G.F., West, D.K., Befort, M., and Maley, F.: Characterization of the intron in the phage T4 Thymidilate synthase gene and evidence for its self-excision from the primary transcript. Cell, 45:157, 1986.

Crick, F.: Split genes and RNA splicing. Science, 204:264, 1979.

Darnell, J.E.: Variety in the level of gene control in eukaryotic cells. Nature, 297:365, 1982.

Darnell, J.E.: RNA. Scient. Am., 253:68, 1985.

De Robertis, E.M.: Nucleocytoplasmic segregation of proteins and RNAs. Cell, 32:1022, 1983.

Galli, D., Hoffstetter H., Stunnenberg, H.G., and Birnstiel, M.L.: Biochemical complementation with RNA in the *Xenopus* oocyte: A small RNA is required for the generation of 3′ histone mRNA termini. Cell, 34:823, 1983.

Gilbert, W.: Starting and stopping sequences for the RNA polymerase. In RNA Polymerase. Edited by R. Losick and M. Chamberlin. Cold Spring Harbor, New York, Cold Spring Harbor Laboratory, 1976, pp. 193–205.

Gilbert, W., Marchionni, M., and McKnight, G.: On the antiquity of introns. Cell, 46:151, 1986.

Gluzman, Y., and Shenk, T.: Enhancers and Gene Expression. Cold Spring Harbor, New York, Cold Spring Harbor Laboratory, 1983.

Grabowski, P., Seiler, S., and Sharp, P.A.: A multicomponent complex is involved in the splicing of messenger RNA precursors. Cell, 42:345, 1985.

Hardin, J.A., et al.: Antibodies from patients with connective tissue disease bind specific subsets of cellular RNA-protein particles. J. Clin. Invest., 70:141, 1982.

Hashimoto, C., and Steitz, J.A.: A small nuclear ribonucleoprotein associates with the AAUAAA polyadenylation signal in vitro. Cell, 45:581, 1986.

Hentschel, C.C., and Birnstiel, M.L.: The organization and expression of histone gene families. Cell, 25:301, 1986.

Keller, W.: The RNA lariat: a new ring to the splicing of mRNA precursors. Cell, 39:423, 1984.

Lerner, M.R., and Steitz, J.A.: Snurps and Scyrps. Cell, 25:298, 1981.

Lerner, M.R., et al.: Are snRNPs involved in splicing? Nature, 283:220, 1980.

Mattaj, I.W., and De Robertis, E.M.: Nuclear segregation of U2 snRNA requires binding of specific snRNP proteins. Cell, 40:118, 1985.

McKnight, S.L., and Kingsbury, R.: Transcriptional control signals of a eukaryotic protein-coding gene. Science, 217:316, 1982.

Miller, O.L., Jr.: The nucleolus, chromosomes and visualization of genetic activity. J. Cell Biol., 91:155, 1981.

Nevins, J.R.: Adenovirus gene expression: Control at multiple steps of mRNA biogenesis. Cell, 28:1, 1982.

Nevins, J.R., and Chen-Kiang, S.: Processing of adenovirus nuclear RNA to mRNA. Adv. in Virus Res., 26:1, 1981.

Padgett, R.A., et al.: Lariat RNAs as intermediates and products in the splicing of messenger RNA precursors. Science, 225:898, 1984.

Proudfoot, N.: The end of the message. Nature, 298:516, 1982.

Reanney, D.: RNA splicing and polynucleotide evolution. Nature, 277:598, 1979.

Roberts, J.W.: Transcription termination and its control in *E. coli*. In RNA Polymerase. Edited by R. Losick and M. Chamberlin. Cold Spring Harbor, New York, Cold Spring Harbor Laboratory, 1976, pp. 247–271.

Roeder, R.G.: Eukaryotic nuclear RNA polymerases. In RNA Polymerase. Edited by R. Losick, and M. Chamberlin. Cold Spring Harbor, New York, Cold Spring Harbor Laboratory, 1976, pp. 285–329.

Ruskin, B., Krainer, A.R., Maniatis, T., and Green, M.R.: Excision of an intact intron as a novel lariat structure during pre-mRNA splicing in vitro. Cell, 38:317, 1984.

Serfling, E., Jasin, M., and Schaffner, W.: Enhancers and eukaryotic gene transcription. Trends in Genetics, 1:224, 1985.

Sharp, P.A.: Speculations on RNA splicing. Cell, 23:643, 1981.

Sjöberg, B.M., et al.: The bacteriophage T4 gene for the small subunit of ribonucleotide reductase contains an intron. The EMBO J., 5:2031, 1986.

Skoglund, U., et al.: Visualization of the formation and transport of a specific hnRNP particle. Cell, 34:847, 1983.

Walter, P. and Blobel, G.: Signal recognition particle contains a 7S RNA essential for protein translocation across the endoplasmic reticulum. Nature, 299:691, 1982.

Zaug, A.J., and Cech, T.R.: The intervening sequence RNA of tetrahymena is an enzyme. Science, 231:470, 1986.

CHAPTER 21

RIBOSOMES, THE NUCLEOLUS, AND PROTEIN SYNTHESIS

In this chapter we will discuss protein synthesis, the last step in the flow of genetic information. The translation of the four-base code into a 20-amino acid protein sequence involves many cellular components; in addition to mRNA, these include the ribosomes (which in eukaryotes consist of 4 RNAs and 70 protein molecules assembled together at the nucleolus),

about 50 tRNA molecules, and many soluble enzymes. We will first examine these components individually and then study how they interact to produce specific proteins.

Along the way, we will study the regulation of transcription and processing of some fascinating genes which we have not yet considered: rRNA, 5S RNA, and tRNA.

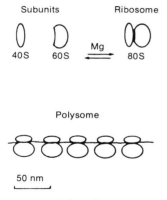

Subunits   Ribosome

40S   60S   Mg   80S

Polysome

50 nm

Fig. 21–1. Diagram of the subunit structure of the ribosome and the influence of Mg. A polyribosome formed by five ribosomes is indicated. The filament uniting the ribosomes represents messenger RNA. The sedimentation constants (S) of the different particles are indicated.

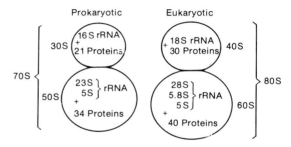

Fig. 21–2. The components of prokaryotic and eukaryotic ribosomal subunits.

21–1   RIBOSOMES

Ribosomes were first observed by Palade in the electron microscope (EM) as dense particles or granules. Upon isolation they were shown to contain approximately equal amounts of RNA and protein. Ribosomes are found in all cells and provide a scaffold for the ordered interaction of all the molecules involved in protein synthesis.

Cells devote considerable effort to the production of these essential organelles. An *E. coli* cell contains 10,000 ribosomes, each one with a molecular weight of about 3 million. Ribosomes represent 25% of the total mass of these bacterial cells. Mammalian cultured cells contain 10 million ribosomes per cell, each of which is about twice as large as a prokaryotic ribosome. Ribosomal biogenesis in eukaryotes provides a wonderful example of how products from different parts of the cell some together at the nucleolus for assembly.

Ribosomes Have Two Subunits

The ribosome is a spheroidal particle of 23 nm and is composed of a large and a small subunit (Fig. 21–1). Eukaryotic ribosomes sediment in sucrose gradients with a *sedimentation coefficient* of 80S, and in the absence of Mg^{++} these ribosomes dissociate reversibly into subunits of 40S and 60S. Prokaryotic ribosomes are smaller and sediment at 70S; they have subunits of 30S and 50S (Fig. 21–2). (Note that the values of the sedimentation coefficients are not additive be-

cause they depend on factors such as the shape of the particles.) Ribosomes are also found in the mitochondria and chloroplasts of eukaryotic cells. They are always smaller than the 80S cytoplasmic ribosomes and are comparable to prokaryotic ribosomes in both size and sensitivity to antibiotics, although the sedimentation values vary somewhat in different phyla.

During protein synthesis several ribosomes become attached to one mRNA molecule (Fig. 21–1), forming a *polyribosome* or *polysome*. In this way a single mRNA molecule can be translated by several ribosomes at the same time. The mRNA is located in the gap between the two ribosomal subunits, as a result of which the ribosome protects a stretch of 25 nucleotides of the mRNA from degradation by ribonuclease.[1] It is possible that the nascent peptide chain grows through a channel or groove in the large ribosomal subunit; this is thought to be the case because ribosomes protect a segment of 30 to 40 amino acids from degradation by proteolytic enzymes (Fig. 21–3).[2]

Ribosomal RNAs Are the Most Abundant Cellular RNAs

The major constituents of ribosomes are RNA and proteins present in approximately equal amounts. The positive protein charges are not sufficient to balance the many negative charges in the phosphates of the RNA, and for this reason ribosomes are strongly negative and bind cations and basic dyes. Ribosomal RNA (rRNA) generally represents more than 80% of the RNA present in cells.

Prokaryotic ribosomes contain three RNA molecules: 16S rRNA in the small subunit, and 23S and 5S in the large subunit (Fig. 21–2). In eukaryotes there are four rRNAs: 18S in the

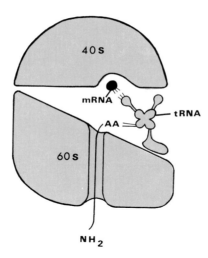

Fig. 21–3. Diagram of a ribosome showing the two subunits and the probable position of the mRNA and the tRNA. The nascent polypeptide chain passes through a kind of tunnel within the subunit. (Courtesy of D.D. Sabatini and G. Blobel.)

small subunit, and 28S, 5.8S, and 5S in the large subunit. The 28S, 5.8S, and 18S rRNAs are synthesized in the nucleolus by cleavage of a single precursor RNA, while 5S RNA is synthesized outside the nucleolus. Eukaryotic ribosomes and rRNAs are much larger than their prokaryotic counterparts, as shown in Table 21–1.

Ribosomal RNA has a high degree of secondary structure; about 70% of it is double-stranded and helical, due to base-pairing. These double-stranded regions are formed by "hairpin loops" between complementary regions of the same linear RNA molecule as shown in Figure 21–4. The various ribosomal proteins adhere at specific points on these loops and stems.[3] Ribosomal RNA provides a three-dimensional matrix to which the various enzymes of the protein synthesis machinery bind in an orderly fashion.

In addition to maintaining ribosome structure, ribosomal RNA also participates in protein synthesis by virtue of its base-pairing properties. The 3' end of 16S RNA has a sequence complementary to the ribosome binding site of most prokaryotic mRNAs, which is indicated in Figure 21–4 and Table 21–3 (Shine and Dalgarno, 1974). The interaction of 16S RNA and mRNA helps the 30S subunit recognize the starting end of the mRNA when binding to it. Similarly, 5S RNA has a sequence complementary to the tetranucleotide TψCG (ψ stands for pseudouridine), which is present in all tRNAs and which is essential for the binding of tRNA to ribosomes. It is also suspected that ribosomal RNAs may also provide some of the catalytic activities required for protein synthesis.

21–2 RIBOSOMAL PROTEINS

The *E. coli* ribosome contains 21 proteins in the small subunit (designated, in order of decreasing size S1 through S21) and 34 in the large (L1 through L34) (Table 21–1). All the proteins are different with the exception of one that is present in both subunits (S20 and L26). Therefore, the total number of proteins in one ribosome is 54. With the exception of S1 (molecular weight 65,000), they range between 7000 and 32,000 in molecular weight. Most of them are rich in basic amino acids.

All the ribosomal proteins have been isolated, and specific antibodies against them have been produced. Various immunological and chemical cross-linking procedures have made possible the construction of maps of the topographical distribution of ribosomal proteins within the subunits (Fig. 21–5).

TABLE 21–1. MOLECULAR WEIGHTS OF RIBOSOMAL COMPONENTS

	Prokaryotes			Eukaryotes		
Ribosomal Particle	70S	50S	30S	80S	60S	40S
Particle MW (RNA + Protein)	2,700,000	1,800,000	900,000	4,500,000	3,000,000	1,500,000
RNA MW		23S 1,100,000	16S 600,000		28S 1,700,000	18S 700,000
		5S 36,000			5.8S 50,000	
					5S 36,000	
Number of Proteins	54	34	21	70	40	30

Compiled from data of Van Holde, K. F., and Hill, W. E., and of Wool, I. G., and Stoffler, G. *In* Ribosomes. Edited by M. Nomura, A. Tissières, and P. Lengyel. Cold Spring Harbor, New York, Cold Spring Harbor Laboratory, 1974.

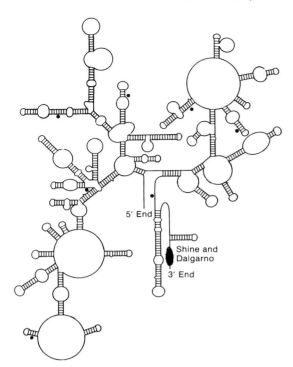

Fig. 21–4. Secondary structure of *E. coli* 16S rRNA. Note that 70% of the RNA is base-paired, forming hairpin loops. A similar folding pattern is present in other bacterial species. Black dots are present every 200 nucleotides from the 5' end. The Shine and Dalgarno sequence, which base-pairs to the ribosome-binding site on the mRNA (Table 21–3), is indicated. (Data from Noller, H.F., and Woese, C.R.: Science, *212*:403, 1981.)

Ribosomes Can Be Dissociated and Self-Assembled

Ribosomal proteins can be dissociated from the ribosome and then added back to reconstitute active ribosomes.[4] For example, ribosomal subunits centrifuged in 5 *M* cesium chloride lose 30 to 40% of their proteins, the so-called *split proteins* (Fig. 21–6). When the split proteins are returned to medium containing the inactive ribosome *cores* and incubated at 37°C, active ribosomes can be reconstituted. In a later series of experiments, it was possible to reconstitute functional ribosomes from naked RNA and proteins.[5]

Reconstitution experiments are useful because they make it possible to identify the function of individual ribosomal proteins. These experiments, in which one ribosomal protein is omitted (or modified) at a time, have shown that certain ribosomal proteins require prior attachment of other proteins in order to become in-

corporated in a stepwise manner. Of particular interest was the identification of "initial binding proteins," which bind at specific sites on naked rRNA and without which the other proteins cannot bind. As we shall see below, the initial binding proteins are crucial in the control of ribosomal protein synthesis.

It is remarkable that such a complex organelle as a ribosome can assemble spontaneously by simple physicochemical interactions; this is another good illustration of the principle of self-assembly of biological structures which, as we have seen already, also applies to viruses and biological membranes (Chapter 2).

Synthesis of Ribosomal Proteins Is Controlled at the Translational Level

Coordinating the synthesis of 54 ribosomal proteins and three rRNAs represents a formidable problem for *E. coli*. An excess of proteins would be wasteful, and if some of them were missing, ribosomes would not assemble properly. Recent studies suggest that control is achieved mainly though blocking the translation of ribosomal protein mRNAs when there is an excess of free ribosomal proteins.

In *E. coli* mRNAs tend to be *polycistronic*, i.e., they contain the information for several proteins in a single mRNA molecule (Fig. 20–1). A transcriptional unit for multiple proteins is called an *operon* (Section 22–1). *E. coli* has six operons that contain genes for ribosomal proteins. Figure 21–7 shows the mRNA transcribed from the longest of these operons, which codes for 11 ribosomal proteins. It has been shown that protein L4 can inhibit the translation of several of these proteins when added to an in vitro protein synthesizing system.[6,7] This inhibition can be overcome by addition of purified 23S rRNA, which binds protein L4. Furthermore, in vivo experiments in which the gene for protein L4 was cloned in a plasmid showed that the induction of protein L4 overproduction greatly reduced the synthesis of all 11 ribosomal proteins present in this operon.[8]

As Figure 21–7 shows, it seems that protein L4 blocks translation by binding to its own mRNA. The protein can also bind to 23S rRNA, and, in fact, striking homologies in sequence and secondary structure are shared by stretches of the polycistronic mRNA and 23S rRNA.[9] Reg-

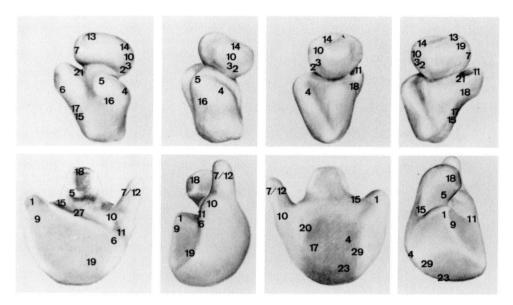

Fig. 21–5. Top row, four views of an *E. coli* small ribosomal subunit model; bottom row, similar views of the large ribosomal subunit. The numbers on the three-dimensional model indicate the topography of individual ribosomal proteins as determined by immuno-electron microscopy. (Courtesy of Dr. M. Stöffler.)

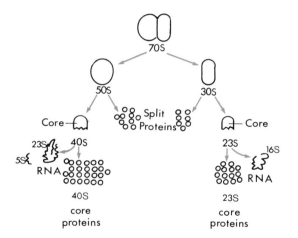

Fig. 21–6. Diagram representing the stepwise dismantling of the two subunits of a 70S ribosome. Note that the proteins may be separated into split and core proteins. The 50S subunit contains 23S and 5S RNAs, and the 30S subunit has 16S RNA.

viewed by Dean and Nomura, 1982, and Nomura, 1986), and the striking common feature is that they all are among the "initial binding proteins" that had been identified by the in vitro assembly studies of ribosomes, which can bind directly to rRNA without the aid of additional proteins.

It is possible that such a translational regulation also exists in eukaryotes. It has been found that frog early embryos contain the mRNAs for all 70 ribosomal proteins, but the translation of most of them (except for four that are translated at all times) is inhibited until mid-blastula; i.e., until ribosomal RNA synthesis is switched on in the nucleolus.[10]

Prokaryotic and Eukaryotic Ribosomes—Little Homology but the Same Basic Function

Eukaryotic ribosomes do not differ functionally from those in prokaryotes in a fundamental way; they perform the same functions, by the same set of chemical reactions. The genetic code is the same for all living organisms, and eukaryotic ribosomes are able to translate bacterial mRNAs efficiently, provided that a "cap" is added enzymatically.[11] However, eukaryotic ribosomes are much larger than bacterial ones (Table 21–1), and most of their proteins are dif-

ulation works as follows: When sufficient rRNA is available, the levels of free L4 protein are low and translation can proceed (Fig. 21–7B); when there is insufficient free rRNA, then L4 binds to its own mRNA and blocks its own synthesis ("*autogenous regulation*") and that of the other ten proteins as well (Fig. 21–7A).

Similar properties have been found for proteins of the other ribosomal protein operons (re-

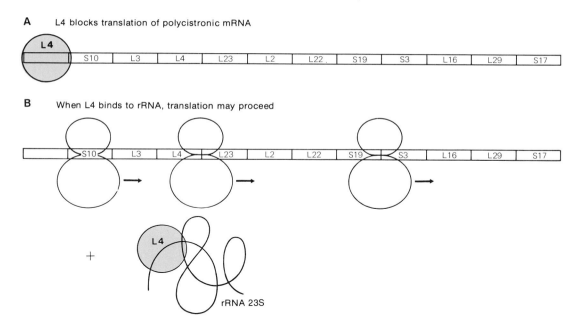

Fig. 21–7. Autogenous regulation of the translation of a polycistronic mRNA coding for 11 ribosomal proteins. The size of protein L4, ribosomes, and rRNA are not up to scale. See description in text.

ferent. Several antibiotics, such as *chloramphenicol*, inhibit bacterial but not eukaryotic ribosomes (this is the basis of the use of many antibiotics in medical treatment). Protein synthesis by eukaryotic ribosomes is inhibited by *cycloheximide*.

Ribosomes from mitochondria and chloroplasts resemble those in bacteria. They are inhibited by chloramphenicol, and hybrid ribosomes containing one bacterial subunit and one subunit from chloroplast ribosomes are fully active in protein synthesis. In eukaryotes hybrid ribosomes with subunits from both plants and mammals are active in protein synthesis, but hybrids are inactive if one subunit is derived from bacteria. Some similarities must also exist, for ribosome reconstitution experiments have shown that proteins L7 and L12 from *E. coli* can replace the homologous proteins in mammalian ribosomes. These studies suggest that there is little homology between ribosomes of prokaryotes and eukaryotes, although the basic function in all organisms remains the same.

21–3   THE NUCLEOLUS AND rRNA BIOGENESIS

The synthesis of a eukaryotic ribosome is a complex phenomenon, involving several regions of the cell. The 18S, 5.8S, and 28S RNAs are synthesized as part of a much longer precursor molecule in the nucleolus, 5S RNA is synthesized on the chromosomes outside the nucleolus, and the 70 ribosomal proteins are synthesized in the cytoplasm. All these components migrate to the nucleolus, where they are assembled into ribosomal subunits and transported to the cytoplasm. Ribosome biogenesis is a striking example of coordination at the cellular and molecular levels.

Nucleoli Are Rich in RNA

Cells contain in their nucleus one or more prominent bodies called nucleoli. Their ease of visualization attracted the attention of early cytologists, and by the end of the 19th century a relationship between the size of the nucleolus and the synthetic activity of the cell was postulated. It was found that nucleoli were small or absent in cells exhibiting little protein synthesis (sperm cells, blastomeres, muscle cells, and so forth), whereas they were large in oocytes, neurons, and secretory cells—those in which protein synthesis is a prominent feature.

In 1934 Barbara McClintock showed that the nucleoli, which disperse during mitosis, are reformed at telophase at specific chromosomal lo-

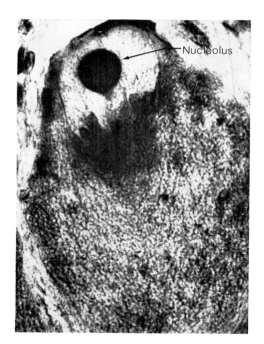

Fig. 21–8. Nerve cell photographed with ultraviolet light at 260 nm. The regions in the nucleus and cytoplasm that absorb the ultraviolet light contain nucleic acid. Note the prominent nucleolus, which is rich in RNA. (From H. Hydén.)

cations, which she called *"nucleolar organizers."* The nucleolar organizers can be readily recognized on mitotic chromosomes because chromosomes are usually thinner at this site, constituting what is called a *secondary constriction* (see Fig. 13–18). The nucleolar organizer was later shown to contain the genes for 18S, 5.8S, and 28S rRNA.

By 1940 Caspersson, using ultraviolet absorption spectra, had shown that the nucleolus and the cytoplasm contain high concentrations of RNA, as depicted in Figure 21–8. Brachet shortly afterwards showed that treatment of tissue sections with ribonuclease abolished the high affinity that nucleoli have for basic dyes, thus confirming that nucleoli contain RNA.

Electron Microscopy of the Nucleolus Shows a Fibrillar and a Granular Zone

The EM makes it possible to distinguish two characteristic components on sections of most nucleoli: a *granular zone* and a *fibrillar zone*, illustrated in Figure 21–9. The granular zone consists of granules 15 to 20 nm in diameter and

frequently occupies the peripheral region of the nucleolus. The fibrillar zone consists of fine fibers 5 to 10 nm in diameter and is located in a more central region of the nucleolus.

When cells are labeled with ³H-uridine for 5 minutes followed by a chase with nonradioactive uridine and examined by autoradiography at the EM level, it is evident that ³H-uridine is incorporated first into the fibrillar zone, and appears only later in the granular area.[12,13] This suggests the relationship:

nucleolar DNA → fibrillar zone → granular zone

The fibrillar area contains the rRNA genes that uncoil from the chromosome and penetrate the nucleolus, together with the nascent RNA molecules attached to it. When spread for observations in the EM, it has the appearance shown in Figure 21–10. The granular area represents ribosome precursor particles at various stages of assembly and processing (Fig. 21–10*B*). Once the ribosomal subunits are mature, they are released from the nucleolus and exit into the cytoplasm through the nuclear pores (Fig. 21–9).

During mitosis the nucleoli undergo cyclic changes. Nucleoli are formed around the DNA loop that extends from the nucleolar organizer. There may be several nucleoli per cell, but frequently they tend to fuse into one or a few nucleoli at this stage. During late prophase the DNA loop containing the rRNA genes gradually retracts and coils into the nucleolar organizer of the corresponding chromosome. The nucleolar organizer region is one of the last to undergo condensation, thus producing a secondary constriction on the chromosome. The fibrillar and granular components are gradually dispersed into the nucleoplasm. After the cell divides, during telophase, the nucleolar organizer DNA uncoils and the nucleolus is reassembled.

The Nucleolar Organizer Contains the Genes Coding for 18S, 5.8S, and 28S rRNA

Direct evidence that the nucleolus is responsible for the synthesis of rRNA was obtained in 1964, when it was discovered that an anucleolate mutant of the South African frog *Xenopus laevis* was unable to synthesize rRNA.[14] Figure 21–11 shows that diploid cells of the wild-type *Xenopus* have two nucleoli (2-*nu*); the heterozygous mutant has only one nucleolus (1-*nu*); the homo-

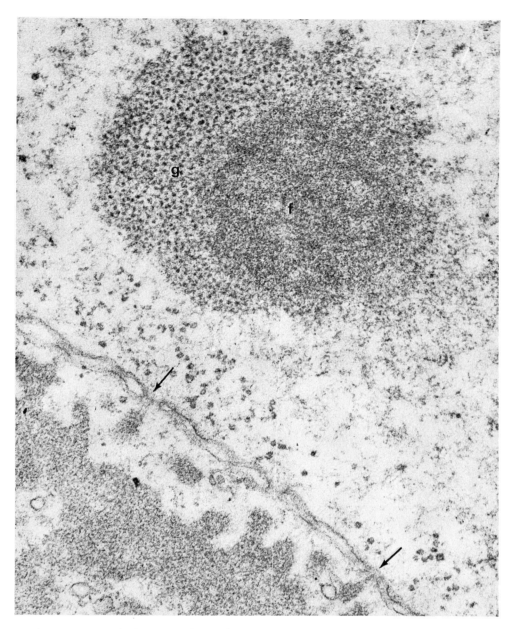

Fig. 21–9. Electron micrograph of an oocyte of *Rana clamitans* showing one of the peripheral nucleoli (containing amplified DNA) with a fibrillar central portion and a granular peripheral portion. Arrows indicate material entering the cytoplasm through the nuclear pores. ×70,000. (Courtesy of O.L. Miller.)

zygous mutant is anucleolate (0-*nu*). When two 1-*nu* heterozygotes are crossed, 25% of the progeny are 0-*nu*. This condition is lethal, and the tadpoles die after one week. Up to this stage the embryos rely on maternal ribosomes inherited from the egg cytoplasm (one *Xenopus* egg contains 10^{12} ribosomes). The 0-*nu* embryos do not synthesize 18S, 28S, or 5.8S rRNA, and DNA-RNA hybridization studies showed that their DNA lacks the genes for rRNA.[15] In other words, 0-*nu* embryos have a deletion of the genes coding for the large ribosomal RNAs.

0-*nu* embryos are able to continue synthesizing 5S rRNA,[16] however, which shows that these genes are not located in the nucleolar organizer. As we shall see later, 5S RNA genes are located at the telomeres of *Xenopus* chromosomes.

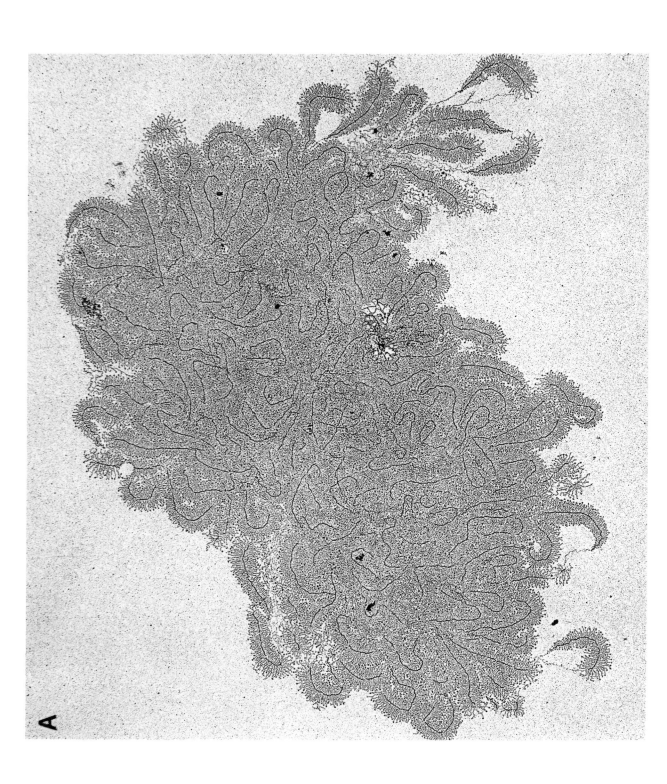

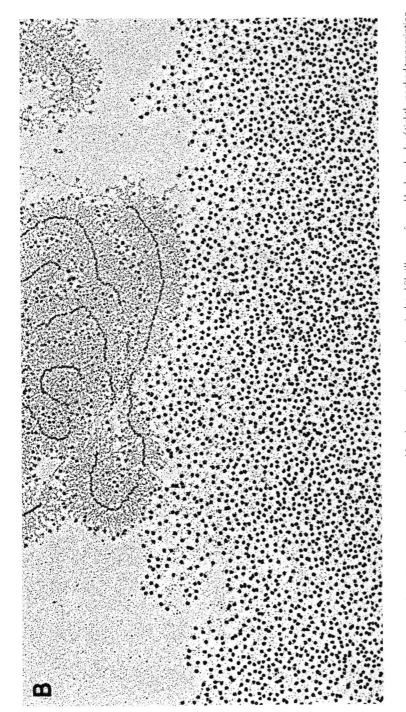

Fig. 21–10. Nucleoli from *Rana pipiens* oocytes spread for electron microscopy. *A*, an isolated fibrillar core formed by hundreds of tightly packed transcription units ("Christmas trees"). *B*, a nucleolus that has been spread with the granular part still attached to it. The granular part represents ribosome subunit precursors at various stages of maturation. (Courtesy of M.F. Trendelenburg.)

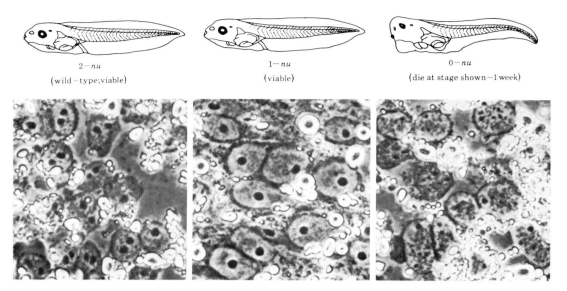

2−nu
(wild−type;viable)

1−nu
(viable)

0−nu
(die at stage shown−1 week)

Fig. 21–11. Number of nucleoli in mutants of *Xenopus laevis*. The wild-type tadpole has two nucleoli (2-*nu*), the heterozygous mutant has one nucleolus (1-*nu*), and the homozygous anucleolate mutant has none (0-*nu*). The 0-*nu* tadpoles do not synthesize ribosomal RNA. (Courtesy of J.B. Gurdon.)

Ribosomal Genes Are Tandemly Repeated and Separated by Spacer DNA

All organisms have multiple rRNA genes. In the case of *Xenopus*, each nucleolar organizer contains 450 rRNA genes. These genes are *tandemly repeated* along the DNA molecule (in a head to tail arrangement) and are separated from each other by stretches of *spacer* DNA, which is not transcribed. These linear repeats of genes can be visualized very clearly in Figure 21–12, which shows a nucleolar organizer spread for electron microscopy.[17,18,19] These rRNA genes are being actively transcribed, and the nascent RNA chains are spread perpendicularly to the DNA axis. Each gene is transcribed into a long RNA molecule (which varies in size from 40S to 45S, according to the species), which will eventually be processed giving rise to 18S, 28S, and 5.8S RNA. Because each gene has a fixed initiation site (promoter) and a fixed termination site, the transcripts adopt the characteristic "Christmas tree" configuration. Nucleolar rRNA genes are transcribed by RNA polymerase I (the α-amanitin-resistant enzyme). The polymerase I molecules (about 100 per gene) can be visualized at the origin of each nascent RNA chain (Figs. 21–12 and 21–15).

The spacer can be divided into a *nontranscribed spacer* and a *transcribed spacer*, which

is copied into RNA but does not give rise to mature rRNA. Studies with restriction endonucleases have shown that although the length of the gene (transcribed segment) is constant, the length of the nontranscribed spacer (Fig. 21–12) can vary considerably between adjacent repeats[20] and between different *Xenopus* individuals.[21] The nontranscribed spacer is formed by short repeated sequences 60 to 81 base pairs in length, and differences in the spacer length are due to variation in the number of these repeating units. It is thought that the spacer heterogeneity might arise from unequal crossing-over events (see Fig. 22–9).

The rDNA Spacer Acts as a Sink for Transcription Factors

The *Xenopus* rDNA genes have been cloned and sequenced and their expression extensively studied by microinjection into *Xenopus* oocyte nuclei. The detailed studies of Ron Reeder have led to the important conclusion that gene spacers are not inert segments of DNA, but rather have a role in the control of transcription.

Deletion studies have shown that rDNA transcription can start with only 140 nucleotides preceding the gene, which must therefore contain the rDNA promoter.[22,23] As Figure 21–13 shows, the promoter contains between positions −72

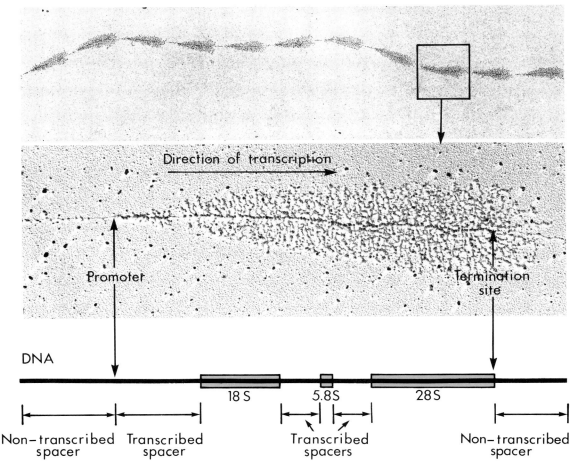

Fig. 21–12. Ribosomal RNA genes are tandemly repeated (head to tail) and separated from each other by nontranscribed spacers. A nucleolar organizer was spread for electron microscopy, and 11 consecutive rRNA genes can be seen. Note the "Christmas tree" or "fern leaf" configuration of the nascent RNA precursor molecules. An RNA polymerase I molecule is present at the base of each RNA molecule. The long precursor molecule contains 18S, 5.8S and 28S rRNAs and is associated with proteins forming a nucleo-protein complex. *Bottom,* The genetic map of an rDNA unit. (Courtesy of U. Scheer, M.F. Trendelenburg, and W.W. Franke.)

and −114 a region necessary for transcription which provides a binding site for a protein factor. The protein itself has not been purified, but is known to bind there on the basis of DNase protection studies (see Fig. 21–23). This protein is known to be species-specific (e.g., mouse cells will transcribe mouse but not human rDNA). This transcription factor must be bound to the promoter before RNA polymerase I can bind (Fig. 21–13) and transcribe the gene.

When the promoter sequence was compared to that of the spacer, two main findings were made (Fig. 21–14):

1. The promoter itself can be duplicated a few times in the spacer. This duplication presumably serves to attract RNA polymerase

I molecules to the region. Sometimes transcription can start from these duplicated promoters, but a row of T nucleotides located 225 nucleotides before the gene (Fig. 21–13) acts as a *failsafe terminator* and stops them.[24]

2. The main component of the spacer, the 60 to 81 base pair (bp) repeat, has a striking nucleotide sequence similarity with the transcription factor binding site (a region of 42 bp within the repeat is 80 to 90% homologous to the promoter binding site). Thus the rDNA spacer is made mostly of repeated binding sites for a transcription factor (Fig. 21–14). This was shown by microinjection experiments in which rDNA

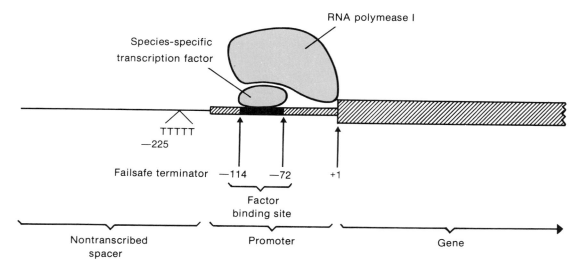

Fig. 21–13. Diagram of the *Xenopus* rDNA promoter.

genes with varying amounts of spacer DNA in front of them were coinjected together with rDNA genes containing the -140 promoter but no spacer DNA. The results showed that whatever gene had most 60 to 81 bp repeats was transcribed more efficiently in this competition assay. (This effect does not require any duplicated promoters present in the spacer.)

The results suggest that the spacer provides multiple functional binding sites that attract factors needed to activate the promoter (Busby and Reeder, 1983). A surprising result was that if the 60 to 81 bp repeats are placed within the same circular plasmid molecule but downstream from the gene, or in opposite orientation to it, they are still able to stimulate transcription. This finding is analogous to what was observed for RNA polymerase II "enhancers" (Section 20–2).

This new role of spacers as "sinks" or storage areas for gene-specific binding proteins could

help explain earlier EM observations that rDNA spacers have a smooth appearance, instead of beads-on-a-string nucleosomes as most of the chromatin does. Figure 21–15 shows a plasmid containing an rDNA gene and spacer which was microinjected into *Xenopus* oocytes and then spread for electron microscopy.[25] Within the same DNA molecule, the plasmid region is covered with nucleosomes while the spacer is "smooth." This smooth appearance could be due to proteins specifically binding to the spacer region.

RNA Polymerase I Remains Bound to the Nucleolar Organizer During Mitosis

During mitosis chromatin condenses maximally. The nucleolar organizer is one of the last sites to achieve condensation, thus leaving a secondary constriction. One might think that chromosome condensation would induce most non-

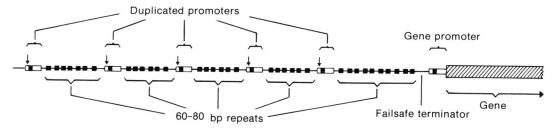

Fig. 21–14. The rDNA spacer of *Xenopus laevis* is made mostly of 60 to 81 bp repeats that are homologous to a region of the promoter, as indicated by the black boxes. In addition, the spacer contains some duplications of the entire promoter and a terminator to stop any transcripts that might start in the spacer. (Courtesy of R. Reeder.)

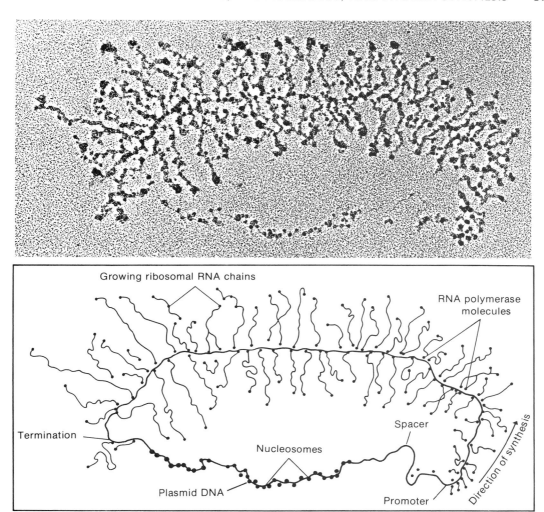

Fig. 21–15. A plasmid molecule containing a *Xenopus* rDNA gene and spacer was injected into frog oocyte nuclei and then spread for electron microscopy. Note that the gene transcripts have a typical Christmas tree appearance, that the plasmid is covered with nucleosomes, but that the spacer DNA has a smooth appearance. (Courtesy of M.F. Trendelenburg.)

essential DNA-binding proteins to be removed from the chromosomes. This is not the case for RNA polymerase I, however. Figure 21–16 shows that RNA polymerase I molecules remain bound to the nucleolar organizer during metaphase and anaphase (see Scheer and Rose, 1984). During this period there is no RNA synthesis (Fig. 14–7), and so the enzyme molecules must remain in an inactive state. The presence of specific proteins bound to the nucleolar organizer might explain why the chromosome has a secondary constriction at this site. Presumably other gene-specific proteins remain bound to DNA throughout the cell cycle as well, RNA polymerase I being only the first example discovered.

Ribosomal DNA is Highly Amplified in Oocytes

Gene amplification is the process by which one set of genes is replicated selectively while the rest of the genome remains constant. The clearest example of gene amplification is seen in the rDNA of amphibian oocytes.

Xenopus eggs are very large cells (1.3 mm in diameter) that accumulate large numbers of ribosomes (10^{12}) for use during early development Anucleolate embryos can survive throughout the first week of development using these maternal ribosomes (Fig. 21–11). Oocytes must synthesize all these ribosomes using the genes contained in a single nucleus (oocyte nuclei have a

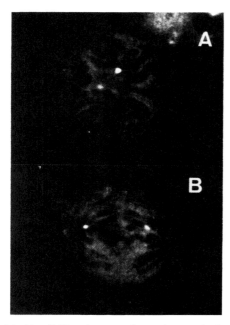

Fig. 21–16. RNA polymerase I remains attached to the rRNA genes during mitosis. Immunostaining of mitotic cells with a specific anti-RNA polymerase I antibody. *A*, metaphase cell showing a single fluorescent body (both chromatids are not resolved). *B*, anaphase cell with a fluorescent nucleolar organizer on each chromosome set. The rat cell line has a single nucleolar organizer because rat rRNA genes are located on the X chromosome, and the cells were taken from a male. (Courtesy of Uli Scheer.)

tetraploid amount of DNA). A somatic cell has 10^7 ribosomes; if it divides every 24 hours, it must be making 10^7 new ribosomes per day. At the same rate, a somatic cell nucleus would need 10^5 days to make the 10^{12} ribosomes present in an oocyte. Even allowing for the fact that an oocyte nucleus has twice the amount of DNA as a diploid cell, the somatic cell would still require more than 100 years to synthesize 10^{12} ribosomes from a single nucleus. Oocytes achieve this synthesis by amplifying the number of rDNA genes 1000-fold (see Table 21–2).

Amplification takes place in very small oocytes in the early stages of meiotic prophase. During pachytene, excess DNA begins to accumulate on one side of the nucleus, forming a Feulgen-positive cap, as shown in Figure 21–17A and B. This cap incorporates ^{3}H-thymidine intensely (while the chromosomes do not), and by the end of the amplification process it contains 25 pg of DNA, equivalent to 2,000,000 rRNA genes (Table 21–2).

The newly synthesized DNA is indeed rDNA.

This can be shown by hybridizing ^{3}H-ribosomal RNA to cytological preparations of *Xenopus* oocytes that have been treated previously with alkali to denature the DNA, a technique called *in situ* hybridization. The hybridized material can then be located by radioautography. Figure 21–17C shows that rRNA hybridizes to the cap of amplified DNA.

As oocytes grow, the extra DNA is accommodated in 1000 to 1500 extra chromosomal nucleoli, which can be seen in Figure 21–18. The rDNA in amplified nucleoli consists of circles of about 450 tandem gene repeats. The circles result from the way in which amplified DNA is replicated ("rolling circles"[26]), and can be visualized when amplified nucleoli are disaggregated to low ionic strength solutions.[27] Amplified rDNA is used only to make ribosomes during oogenesis; it is not inherited by the embryo and it gradually disappears during the first embryonic cleavages.

The initial demonstration that the amplified material is rDNA was dependent on the purification of these genes as a "satellite band" in cesium chloride gradients.[28,29] *Xenopus* rDNA is more GC rich (70%) than bulk DNA (41%) and thus has a higher buoyant density in CsCl gradients when sheared DNA is centrifuged to equilibrium. As Figure 21–19 shows, rDNA forms a satellite band that can be detected by hybridizing individual fractions from the gradient with ^{3}H RNA. The key observation was that oocyte DNA contains more hybridizing material than somatic DNA (Fig. 21–19, lower panel), and thus rDNA is selectively amplified.

Unlike the genes coding for the large ribosomal RNAs, 5S genes are not amplified during oogenesis. Oocytes meet the increased demands for ribosomes by activating all 20,000 5S genes (see below).

The amplification of specific genes is not a common event. rDNA is amplified in oocytes of amphibians, some beetles and spiders, as well as in the macronucleus of ciliate protozoons such as *Tetrahymena* and *Stylonichia*. Gene amplification also occurs in the DNA puffs of *Sciarid* dipterans (see Fig. 23–27) and in *Drosophila* chorion (eggshell) genes (Section 23–4). In all these cases the amplified material is not passed on to future cell generations.

TABLE 21–2.   RIBOSOMAL GENES OF *XENOPUS LAEVIS* IN SOMATIC CELLS AND AFTER AMPLIFICATION IN OOCYTES

Cell Type	Nuclear DNA (pg)	Number of Chromosome Sets per Nucleus	Chromosomal DNA (pg per Nucleus)	Ribosomal DNA (pg per Nucleus)	Number of rRNA Genes	Number of Nucleoli
Somatic	6	2	6	0.012	900	2
Oocyte (after amplification)	37	4	12	25	2,000,000	1000–1500

*Oocytes are in meiotic prophase and therefore contain a tetraploid amount of DNA; ribosomal DNA is amplified to 1000 times more than expected for that amount of chromosomal DNA. (Modified from Gurdon, J.B.: The Control of Gene Expression in Animal Development. Oxford, Oxford University Press, 1974.)

Ribosomal RNAs Undergo Complex Processing in the Nucleolus

The biogenesis of rRNA provides the clearest example of RNA processing. As shown in Figure 21–12, rRNA genes are transcribed into a long precursor RNA which must be cleaved into 18S, 28S, and 5.8S RNA. In the cleavage process about 50% of the precursor RNA is degraded within the nucleus.

Figure 21–20 shows the RNAs that become labeled six hours after microinjection of ^{32}P-GTP into *Xenopus* oocytes. Note that the predominating nuclear RNAs are the primary transcription product (which is 40S in *Xenopus*, but is longer, 45S, in HeLa cells), a 32S rRNA precursor, and 28S rRNA. Most of the 28S rRNA is found in the cytoplasm by this time. 18S rRNA is found predominantly in the cytoplasm because its processing is faster than 28S rRNA and the small ribosomal subunits are rapidly released into the cytoplasm.

The steps involved in rRNA processing have been worked out using HeLa cells pulse-labeled with ^{3}H uridine, from which the nucleoli can be isolated by cell fractionation procedures.[30,31] From these studies it was concluded that:

1. The first ribosomal RNA in HeLa cells is a large 45S molecule of 14,000 nucleotides. Other organisms have shorter precursors (40S for *Xenopus*). Within this molecule the rRNAs are separated by stretches of spacer RNA, and the order of transcription is: 5' end − 18S − 5.8S − 28S − 3' end (Fig. 21–12). On a fully active gene about 100 RNA polymerases are transcribing simultaneously on the ribosomal DNA gene.

2. In the nucleolus, 45S RNA is rapidly methylated, even before transcription is completed. Methylations occur mostly on the ribose moiety (producing 2'-O-methylribose) and occur only in the 18S (46 methylations) and 28S (71 methyl groups) sequences that will be conserved, while those segments that will be degraded remain nonmethylated.

3. 45S RNA has a lifetime of about 15 minutes

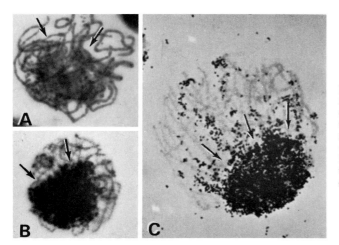

Fig. 21–17. *Xenopus* oocytes during the period of nucleolar DNA synthesis. *A,* at late pachynema, the excess DNA begins to accumulate as granules around the nucleolus *(arrows). B,* later on, the excess DNA appears as a dense cap *(arrows). A* and *B,* Feulgen reaction. ×1700 *C,* large pachynema oocyte hybridized in situ with a probe for rDNA. Observe that the silver grains are deposited mainly on the cap of excess DNA. (See the description in the text.) ×1200. (Courtesy of J.G. Gall.)

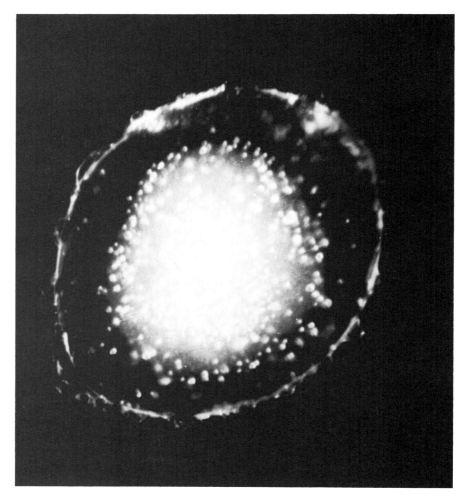

Fig. 21–18.   A manually isolated *Xenopus* oocyte nucleus showing the amplified nucleoli as bright spots on the periphery of the nucleoplasm. Oocytes have a giant nucleus (0.4 mm in diameter), which is sometimes called the *germinal vesicle*. The nuclear membrane is swollen because the nucleus was placed in a hypotonic solution. (From De Robertis, E.M., et al.: Cell, *23*:89, 1981.)

and is then cleaved into smaller components, according to the following general pattern (also indicated in Figure 21–21):

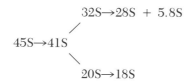

4. 20S RNA is rapidly processed into 18S rRNA, and presumably for this reason the small ribosomal subunits appear in the cytoplasm before the large ribosomal subunits, which have a slower RNA processing (Fig. 21–20).

5. 32S RNA remains in the nucleolus for about 40 minutes and is then cleaved into 28S RNA and 5.8S RNA. These remain in the nucleolus for another 30 minutes before entering the cytoplasm as part of the large ribosomal subunit.

The ribosomal subunits are essentially complete when they exit the nucleus, but minor maturation steps still take place in the cytoplasm, in order to ensure that ribosomes do not engage in protein synthesis within the nucleus.

About half the 45S molecule is lost by the successive degradations. This degradation occurs in the regions that are nonmethylated and have a higher content of GC. Thus the process-

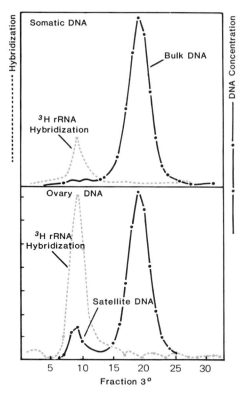

Fig. 21–19. Demonstration that rDNA is selectively amplified in *Xenopus* oocytes. Somatic and oocyte DNA were sheared and centrifuged to equilibrium in CsCl gradients. Individual fractions were then hybridized to ³H rRNA *(dotted line)*. Note that ovary DNA has more rDNA than somatic DNA, and that rDNA behaves as a satellite DNA of buoyant density different from bulk DNA. DNA concentration was determined by measuring the optical density of each fraction. (After studies of Gall, J.G.: Proc. Natl. Acad. Sci. USA, 60:553, 1968.)

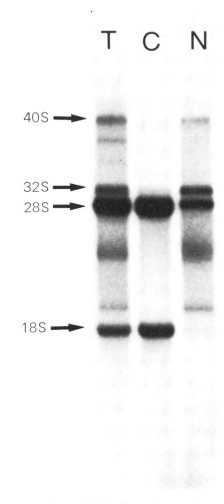

Fig. 21–20. RNAs labeled after a 6-hour pulse in *Xenopus* oocytes. rRNA processing in oocytes is much slower than in cultured cells. *T*, whole oocytes; *N*, hand-isolated nuclei; *C*, cytoplasmic fraction. Note that 40S and 32S precursors are nuclear and that 18S rRNA exits into the cytoplasm rapidly. The RNAs were separated according to size by agarose gel electrophoresis. (From A.E. Carrasco and E.M. DeRobertis, unpublished results.)

ing of the ribosomal RNA leads to an increase in methyl groups and to a decrease in GC content.

All these processing steps do not take place on naked RNA, but rather on RNA-protein complexes. Ribosomal proteins bind to rRNA at the nucleolus, and electron microscopic studies on "Christmas tree" spreads stained with antibodies against specific ribosomal proteins have shown that the initial ribosomal proteins bind before the synthesis of 45S rRNA is completed.[32] In addition, a small nuclear ribonucleoprotein (U3snRNP) is tightly bound to the 32S precursor and is thought to be involved in its processing (Section 20–3).

The nucleolus is the cellular site at which all the ribosomal components are assembled together into ribosomal subunits.

21–4   5S RNA BIOSYNTHESIS

The last ribosomal component that remains to be considered is 5S RNA. The control of the expression of 5S RNA genes has been extensively studied by Donald D. Brown and Robert G. Roeder, and these genes are at present the best understood eukaryotic genes. They merit special study because many of the regulatory features that have emerged from them apply to other eukaryotic genes as well.

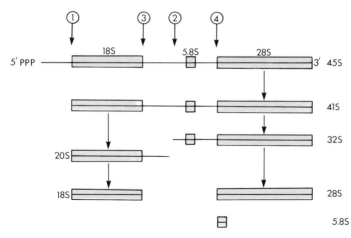

Fig. 21–21. Diagram showing the processing of ribosomal RNA in HeLa cell nucleoli. The order of the various rRNAs in the 45S precursor and the four main sites of processing are indicated.

5S Genes—Tandem Repeats with Spacers, but Outside the Nucleolar Organizer

The genes coding for 5S RNA are not located in the nucleolus and therefore are not deleted in the *Xenopus* anucleolate mutant. The 5S genes are present in multiple copies; *Xenopus* has about 20,000 5S genes per haploid genome, which are located in the tips (telomeres) of most chromosomes, as shown by in situ hybridization of ³H-5S-RNA to chromosome preparations.[33] In *Drosophila* all the 5S genes are located in a single cluster in one of the autosomes.[34]

As with rDNA, the 5S genes are arranged in tandem repeats of the genes separated by segments of spacer DNA. Most of the spacer is formed by repeating segments of a short sequence of 15 base pairs.[35] This sequence is very rich in AT, which explains why 5S DNA migrates as a satellite band lighter than bulk DNA in cesium chloride gradients.[36] Adjacent 5S repeats can be heterogeneous in length owing to variation in the numbers of 15 base pair subrepeats in each spacer,[37] which probably arise by unequal crossing-over events. Tandem repeats of genes separated by spacers seem to be a general feature of the organization of the eukaryotic genome.

Somatic- and Oocyte-Type 5S Genes

Not all 5S genes are identical. The 5S RNA contained in ribosomes from somatic cells differs in about six nucleotides from the sequence of the predominant 5S RNA found in oocytes.[38] There are only 400 somatic-type genes, which produce more than 95% of the 5S RNA present in somatic cells. In oocytes, *both* oocyte-type and somatic-type genes are active, but because the majority of the genes (19,600) are of the oocyte type, most of the 5S RNA produced is of the oocyte type. There are two main questions concerning the expression of this gene: (1) How is a 5S gene transcribed? (2) How are 5S oocyte-type genes repressed in somatic cells, so that only 2% of the genes make 95% of the RNA?

The existence of a large number of 5S genes active only in oogenesis is the way in which *Xenopus* copes with the challenge of making 10^{12} ribosomes for each egg cell without resorting to amplification, as happens with the large rRNA genes.

5S Gene Transcription Requires Binding of a Protein Factor to the Middle of the Gene

Cloned 5S genes are correctly transcribed by RNA polymerase III (the enzyme sensitive to high levels of α-amanitin) when added to an in vitro extract from *Xenopus* oocyte nuclei.[39] A systematic series of deletion mutants was prepared by recombinant DNA techniques. The deletions came in from both ends of the gene, as shown in Figure 21–22.[40,41] The astounding result was that deletions that removed large parts of the 120-nucleotide coding region (from both sides) could still be efficiently transcribed. A region between nucleotides 50 and 83 was found to be essential for transcription, and called the internal control region or ICR.

Progress toward explaining this remarkable finding came from a different set of observations.[42] While 5S genes were actively transcribed in ovary homogenates, extracts from ma-

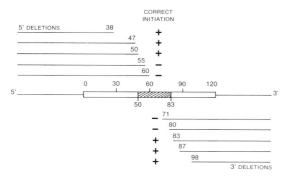

Fig. 21–22. Deletions that include parts of the 5S gene can still be transcribed in oocyte nuclear extracts. Only deletions that include the +50 to +83 internal control region (hatched area) are not transcribed. The numbers indicate the nucleotides at which the deletions stop. (After experiments of D. Brown and colleagues.)

ture eggs (i.e., laid by frogs into water after hormone stimulation) transcribed them very poorly. By adding ovary protein fractions to egg extracts, it was possible to reconstitute 5S transcription. Stimulation of 5S transcription in egg extracts was used as an assay to purify biochemically from oocytes a protein (MW 40,000) called transcription factor III A (TFIII A), which is essential and specific for 5S gene transcription.

Was TFIII A related to the internal control region? To find out, the *"footprint assay"* shown in Figure 21–23 was used. Using end-labeled DNA fragments containing the 5S gene, it was possible to show that pure TFIII A protein protected the middle of the gene from digestion by

DNase I.[42] This experiment showed that TFIII A binds to essentially the same region that had been found to be necessary for transcription in the deletion studies.

Thus 5S gene transcription works as follows: one molecule of a transcription factor specific for an internal region of the gene binds to the DNA, which permits RNA polymerase III to recognize this DNA as a promoter (Fig. 21–24). RNA polymerase III (which is a huge molecule of 700,000 daltons and 10 subunits) will then reach out from this site and start upstream at position +1. The nucleotide sequence at the starting site is not crucial; what is essential is to have an intact TFIII A binding site.

TFIII A has DNA-Binding Fingers

The TFIII A gene has been recently cloned and sequenced (Ginsberg et al., 1984). The amino-terminal three quarters of the TFIII A protein were found to be made of nine repeated domains. Each domain has 30 amino acids and is held in a folded conformation (called a DNA-binding "finger") by a zinc ion, which binds to two cysteines and two histidine residues that are conserved at characteristic intervals. Each finger binds to the major groove of the internal control region DNA at five base intervals, i.e., every half turn of the double helix (Miller et al., 1985; Rhodes and Klug, 1986). Similar DNA-binding "fingers" have been found in other DNA-bind-

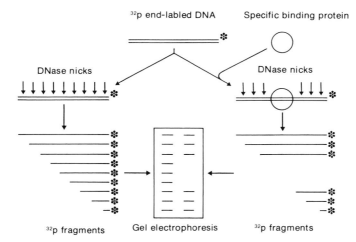

Fig. 21–23. "Footprinting" a protein-binding site on DNA. An end-labeled DNA fragment is digested lightly with DNase I (so that not more than one cut per molecule is produced) in the presence or absence of DNA-binding protein. The DNA is then denatured, separated on a polyacrylamide gel according to size, and autoradiographed. The cut site is indicated by an arrow; the DNA end that was labelled radioactively is indicated by an asterisk. The protein-binding site is seen as a gap ("footprint") on the ladder of DNA fragments. (Courtesy of R. Laskey.)

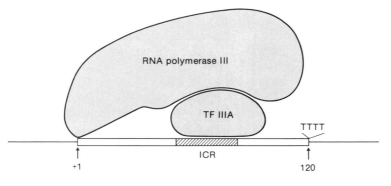

Fig. 21–24. Transcription of a 5S gene requires a transcription factor bound to the middle of the gene, which is then recognized by RNA polymerase III.

ing proteins, such as a gene controlling segment formation in fruit fly development (Rosenberg et al., 1986) and steroid hormone receptors.

TFIII A seems to be the first example of a family of DNA-binding proteins which offers many combinatorial possibilities for the recognition of specific sequences. This is because each DNA-binding finger has slightly different amino acid sequence, and therefore can recognize different nucleotide sequences every half turn of the helix. By introducing different types of DNA binding fingers in different sequence and combination, a large number of proteins recognizing different DNA sequences can be created.

TFIII A Also Binds to 5S RNA, Forming a 7S Storage Particle in Young Oocytes

The purification of TFIII A turned out to be rather easy because this protein is present in enormous amounts in young *Xenopus* oocytes (15% of the total protein) (Fig. 21–25). It had been known for some time that young, previtellogenic oocytes of *Xenopus laevis* synthesize a large amount of 5S RNA, which is stored bound to a protein of 40,000 daltons, forming a 7S RNP particle.[43] The 7S particle stores the 5S RNA until the amplified nucleoli make enough large rRNAs and the 5S RNA can be incorporated into ribosomes. Because this protein had the same molecular weight as TFIII A, the question arose whether they could be the same protein. This turned out to be the case;[44,45] TFIII A is a protein that can bind to a gene and also to the product (5S RNA) of the gene.

This finding suggested a simple feedback model for the regulation of 5S genes: when there is excess 5S RNA, it binds to the transcription factor, which is then unable to bind to DNA, so that 5S RNA transcription is switched off. Fur-

thermore, when TFIII A binds to 5S RNA, it exits from the nucleus and is translocated into the cytoplasm (Fig. 21–25) and is thus removed from the nucleus where the genes are located.[46]

The feedback mechanism must be more complicated than this, however, because TFIII A forms stable transcription complexes[47] with 5S DNA, which can be transcribed hundreds of times by RNA polymerase III while remaining bound to the DNA of the internal control region. These stable complexes can be revealed in "order of addition" experiments in which two different genes are added to a transcription extract at different times. The results show that whichever gene captures TFIII A first will be preferentially transcribed.

How can a 5S gene with a protein bound to

Fig. 21–25. The 7S RNP is cytoplasmic. Immunostaining of a *Xenopus* ovary with anti-TFIII A antibody. Note intense staining on the cytoplasm of small oocytes. As oocytes grow, the concentration of protein decreases. Because the vast majority of TFIII A is bound to 5S RNA in the 7S particle, the staining reflects the distribution of 7S RNP. The amounts of free TFIII A in the nucleus are too low to be detected by this technique. (From Mattaj, I.W., et al.: J. Cell Biol., 97:1261, 1983.)

its middle be copied into RNA hundreds of times without the protein ever being displaced? Detailed studies have shown that TFIII A DNA binding fingers bind mainly to the noncoding strand of the gene (Sakonju and Brown, 1982). This means that the two strands can be opened during transcription and the coding strand copied without dislodging the factor bound to the other strand. Furthermore, the noncoding strand has the same sequence as 5S RNA, and perhaps TFIII A might be bound to common structures in both RNA and DNA.

How Are Oocyte-Type 5S Genes Switched Off in Somatic Cells?

It is perhaps relatively easy to envisage how all 20,000 5S genes are switched on in oocytes. The amount of TFIII A available in young oocytes is so enormous that every gene can bind a TFIII A molecule. It is still not known, however, why only 400 somatic genes are used in somatic cells.

Perhaps an inkling is provided by studies showing that in vitro TFIII A has a fourfold higher affinity for somatic genes than for oocyte-type genes (Bogenhagen et al. 1982). When 5S genes are microinjected into developing embryos (rather than tested in in vitro), somatic-type genes are transcribed 50 to 100 times better than oocyte-type genes. This difference is not enough to explain preference for somatic genes, but it should be kept in mind that somatic cells have minute amounts of TFIII A (by comparison with oocytes) and that the somatic and oocyte genes must therefore compete for a limited amount of factor. It seems possible that TFIII A may bind only once every cell cycle, when DNA is newly replicated and for a brief period devoid of histones. (TFIII A will not bind to inactive oocyte genes in somatic cell chromatin, although it will readily bind to naked DNA.[48,49]) Because TFIII A must bind before histones, more subtle mechanisms could conceivably be involved in the preference for somatic genes: for example, if somatic genes replicated earlier in S phase, they could mop up most of the available TFIII A before the oocyte-type genes replicate. Of course, at this point we cannot exclude that there may be other proteins involved in switching off 5S genes in somatic cells, of which we are still not aware.

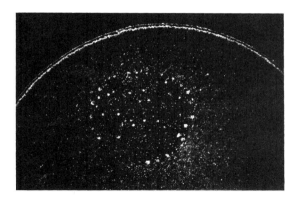

Fig. 21–26. [3]H 5S RNA microinjected into the cytoplasm migrates to and concentrates in the amplified nucleoli of *Xenopus* oocytes. The autoradiograph was photographed by dark-field microscopy. (From De Robertis, E.M., Lienhard, S., and Parisot, R.: Nature, *295*:572, 1982.)

Although their developmental regulation is still incompletely understood, 5S genes are the best studied eukaryotic genes. It is already clear that much of the preference for somatic-type genes over oocyte-type 5S genes will be explained by the differences in affinity of TFIII A for the two types of genes and in differences in the concentration of TFIII A in oocytes and somatic cells. This view is strongly supported by an experiment in which pure TFIII A protein was microinjected in large amounts into developing *Xenopus* embryos, causing a great increase in the transcription of oocyte-type genes, which are normally turned off in somatic cells (Brown and Schliessel, 1985). 5S genes were discussed in detail here because many of the findings (positive transcription factors, stable transcriptional complexes, internal control regions) pioneered with 5S genes are now cropping up in many other types of genes. Their value goes beyond that of a small, 120-nucleotide RNA; they have heavily influenced the way we now think about genes.

5S RNA Is Incorporated into Ribosomes in the Nucleolus

Synthesized by RNA polymerase III outside of the nucleolus, 5S RNA must somehow be transported to the nucleolus, where it becomes incorporated into the large ribosomal subunit. Figure 21–26 is an autoradiograph of a *Xenopus* oocyte whose cytoplasm had been microinjected with purified [3]H-5S RNA 8 hours previously. It may be seen that 5S RNA concentrates in the amplified nucleoli, which presumably provide specific binding sites for it (in the way of large ribosomal subunit precursors). It is worth noting that this nucleolar concentration takes place in the absence of a membrane (nucleoli do not have membranes). A similar migration is followed by the 70 ribosomal proteins, which are synthesized in the cytoplasm and must reach the nucleolus, where ribosome assembly takes place.

SUMMARY:
Ribosome Biogenesis

Ribosomes are spheroidal particles that contain approximately equal amounts of RNA and protein. They are found in all cells and serve as a scaffold for the ordered interaction of the numerous molecules involved in protein synthesis. Eukaryotic ribosomes have a sedimentation coefficient of 80S and large and small subunits of 60S and 40S, respectively. The small subunit contains 18S rRNA and 30 proteins, and the large one 28S, 5.8S, and 5S rRNA and 40 proteins (Fig. 21–2).

An *E. coli* ribosome contains 54 small proteins. These can be dissociated from the ribosome and then added back to reconstitute active particles. The complete assembly of ribosomes from naked RNA and proteins can also be demonstrated, illustrating the principle of biological self-assembly. The synthesis of ribosomal proteins is controlled at the translational level: some of the ribosomal proteins that bind directly to rRNA can also bind to similar structures in their own mRNA and block translation (Fig. 21–7).

Prokaryotic and eukaryotic ribosomes are made up of different RNA and protein molecules, although they are functionally very similar, performing the same operations by the same series of chemical reactions.

Biogenesis of ribosomes is the result of the co-ordinated assembly of several molecular products that converge upon the nucleolus. The genes coding for 18S, 28S and 5.8S RNA are present in the nucleolar organizer; the 5S genes are located in other regions of the genome, and the ribosomal proteins are manufactured in the cytoplasm. The 5S gene products and the proteins migrate toward the nucleolus where they are processed and assembled into active ribosomes.

Direct evidence of the function of the nucleolus as a ribosome factory was obtained with the discovery of an anucleolate mutant of *Xenopus laevis* (Fig. 21–11) which cannot synthesize 18S, 5.8S, and 28S rRNAs. In *Xenopus* there are 450 rRNA genes tandemly repeated (Fig. 21–12) in each nucleolar organizer. In the same species the 5S RNA genes are located at the telomeres of most chromosomes. The rDNA genes are separated by spacer DNA, made mainly by repeated segments

that have a length of 60 to 81 base pairs (Fig. 21–14). The spacer has an important function in rDNA expression because the repeated segment is homologous to a region of the rDNA promoter that binds a transcription factor (Fig. 21–13). Thus the rDNA spacer provides multiple binding sites for specific protein factors.

Gene amplification is the process by which a set of genes is selectively replicated. The rDNA in the amphibian oocyte undergoes this process to accumulate in the egg the huge numbers of ribosomes (10^{12}) that are used in the first stages of development. During pachytene there is an active replication of the nucleolar organizers, and the rDNA is amplified 1000-fold. In a *Xenopus* egg 25 pg of extra DNA with 2,000,000 rRNA genes is accommodated by between 1000 and 1500 nucleoli (Table 21–2, Fig. 21–18). This amplified DNA is then lost during development. The 5S rDNA genes are not amplified during oogenesis, and thus all 20,000 genes become active to meet the need for ribosomes.

Processing of rRNA occurs in the nucleolus. As in the case of other RNAs, longer transcripts are made which are then partially degraded. There is a single precursor for 18S, 28S, and 5.8S rRNAs, which has 14,000 nucleotides and a sedimentation coefficient of 45S.

Within rDNA the order of transcription is:
5′ end − 18S − 5.8S − 28S − 3′ end
and each rRNA is separated by a spacer. Some 100 RNA polymerases are active on a single rDNA cistron. The 45S precursor is methylated in the regions to be conserved, and its lifetime is about 15 minutes. It then is cleaved as follows:

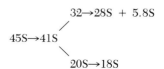

Most nucleoli have a characteristic fibrillar zone, containing the rDNA that uncoils from the chromosome and penetrates the nucleolus along with the 45S RNA molecules attached to it, and a granular zone, representing ribosome precursor particles at various stages of assembly and processing. Once the ribosomal subunits are mature, they are released from the nucleolus and exit into the cytoplasm through the nuclear pores.

5S RNA is synthesized from 20,000 genes in oocytes but only from 400 somatic genes in other cells, which differ slightly in sequence (six nucleotides out of 120). Deletion studies on cloned genes (Fig. 21–22) showed that an *internal control region* in the middle of the gene is essential for transcription. A transcription factor (called TFIII A) binds to this region and enables RNA polymerase III to start transcription at an upstream site (Fig. 21–24). TFIII A has nine "fingers," or small domains held together by zinc ions, which bind to DNA every half turn of the DNA helix. TFIII A, in addition to binding to 5S genes, can also bind to their product, 5S RNA. Young oocytes have massive amounts of TFIII A (15% of total protein), most of it bound to 5S RNA, forming a 7S storage particle. The complex between TFIII A and a 5S gene is stable, and the gene can be transcribed hundreds of times without the protein being displaced, possibly because TFIII A binds to the noncoding strand of 5S RNA, which has a sequence similar to that of 5S RNA. The protein might recognize analogous structural features in 5S DNA and RNA.

In oocytes all 20,000 5S genes are switched on because the massive amounts of transcription factor available can saturate every gene. It is not yet known why oocyte-type genes are switched off in somatic cells, although some ideas are emerging.

21–5 TRANSFER RNA

tRNAs Are Adaptor Molecules

Transfer RNAs are a group of small RNAs (between 75 and 85 nucleotides in length) that have the important role of serving as molecular adaptors during protein synthesis. Because the 20 amino acids have a shape that is not complementary in any way to the mRNA nucleotide triplets, they cannot recognize the codons by themselves. Transfer RNA has a triplet of nucleotides called an *anticodon* that can establish hydrogen bonds with the codon in mRNA; it also can have the amino acid corresponding to that particular codon attached to one of its ends. This entire structure is called an *aminoacyl-tRNA*, and it enables a particular amino acid to be brought to the ribosome in response to the appropriate codon.

Accurate protein synthesis depends on the correct amino acid being attached to the tRNA molecule. This important activity is performed by specific enzymes called *aminoacyl-tRNA synthetases* or *activating* enzymes. If a tRNA is mischarged, it will cause the incorporation of an incorrect amino acid into the protein chain.

In *E. coli* there are about 50 different tRNAs that represent 10 to 15% of the total RNA in the cell. These tRNAs utilize 61 codons from among the 64 possible triplets of the genetic code. Some tRNAs are able to recognize more than one codon because the base of the anticodon that recognizes the third base of the codon in mRNA (which is less important in coding) may have a certain degree of flexibility or "wobble" that en-

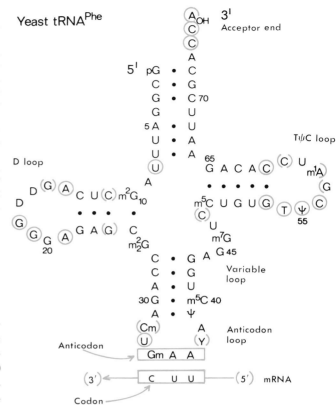

Fig. 21–27. Diagram of the cloverleaf model of transfer RNA showing the amino acid acceptor end (CCA), the anticodon loop with the three bases that read the genetic message; the D loop, which contains several dihydrouridines (D); the TψC loop, which is thought to interact with the 5S ribosomal RNA; and a variable loop. Bases that are invariant in all tRNAs are indicated by circles; bases that tend to be moderately conserved (e.g., always a purine or pyrimidine) are indicated by parentheses. These constant bases interact with each other via hydrogen bonds, producing the three-dimensional folding of the molecule (Fig. 21–28).

ables it to base-pair with more than one nucleotide (see Chapter 19). The three codons for termination of protein synthesis are exceptional in that they are recognized by protein factors called *releasing factors*, instead of by tRNA. There are 20 *isoacceptor tRNA families*, one for each amino acid, and for each one of these families there is one aminoacyl-tRNA synthetase.

tRNAs Have a Common Cloverleaf Secondary Structure

All tRNAs share common features, of which the most notable is the folding into a "cloverleaf" structure as shown in Figure 21–27. Other common features include the sequence CCA at the

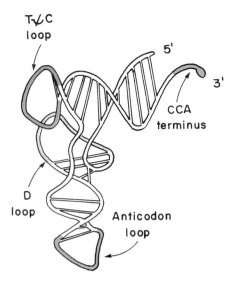

Fig. 21–28. Tertiary structure of yeast phenylalanine transfer RNA showing a compact configuration with an L form. The positions of the various parts shown in Figure 21–27 are indicated. Note that the D and TψC loop are hydrogen bonded together. (Redrawn from Rich, A., and Kim, S.H.: The three-dimensional structure of transfer RNA. Sci. Am., *238*(1):52, 1978.)

3′ end (which attaches covalently to the corresponding amino acid), the conserved nucleotides indicated by circles in Figure 21–22, and unusual bases such as pseudouridine (ψ), inosinic acid, methylcytosine, methylguanine, ribothymidine (T), and others. These unusual bases are modified posttranscriptionally (for example, by the action of methylating enzymes).

The cloverleaf structure consists of a series of loops separated by short stems of helical double-stranded regions between four and seven complementary bases long. Figure 21–27 shows the principal elements of this structure, including (1) the *acceptor end* (CCA), at which the specific amino acid becomes attached; (2) the *anticodon loop*, directly opposite from the acceptor end and having the three bases that recognize and form hydrogen bonds with the mRNA codon; (3) the *D loop*, which contains dihydrouridine; and (4) the *T loop*, which has the conserved sequence TψCG (for which it is named) and is thought to interact with a complementary region of 5S ribosomal RNA during protein synthesis. There is also a highly variable region (variable loop), which differs greatly in length in the different tRNAs.

The tertiary structure of tRNA has been de-

termined by x-ray diffraction studies. As Figure 21–28 shows, tRNA has a compact configuration in which the side arms are folded together, rather than the open configuration (shown for clarity) in Figure 21–27. The molecule has the shape of an L, with the anticodon at one end of the L and the acceptor at the other. The D and T loops are hydrogen-bonded together at the corner of the structure, and the nucleotides involved in these crucial hydrogen bonds tend to be highly conserved in different tRNAs (circled nucleotides in Fig. 21–27). All tRNAs share these characteristics of folding, with the exception of mitochondrial tRNAs, which sometimes do not follow these rules.

The Fidelity of Protein Synthesis Depends on the Correct Aminoacylation of tRNA

The tRNA molecule is directly involved in two events that determine the fidelity of protein synthesis. The first event is the selection of the correct amino acid by the aminoacyl-tRNA synthetase for incorporation into an aminoacyl-tRNA. The second event occurs in the ribosome and involves the selection of the correct tRNA in response to an mRNA codon.

For each amino acid there is a specific aminoacyl-tRNA synthetase that activates the carboxyl group of that amino acid for covalent bonding with the adenylic acid residue of the CCA end of the tRNA. The first step of this reaction requires ATP:

$$AA + ATP \underset{\text{synthetase}}{\overset{\text{aminoacyl}}{\rightleftarrows}} AA{\sim}AMP + P{\sim}P$$

The AA~AMP intermediate formed remains bound to the enzyme until the proper tRNA arrives, and at that moment the formation of the AA~tRNA complex occurs:

$$AA{\sim} AMP + tRNA \underset{\text{synthetase}}{\overset{\text{aminoacyl}}{\rightleftarrows}} AA{\sim}tRNA + AMP$$

To perform these two functions, the activating enzyme has two sites: one to recognize the amino acid and another to recognize the specific tRNA.

The bond between the amino acid and the CCA end of the tRNA is a high-energy one which stores the energy from the ATP used in the first reaction. The energy thus stored is used in the ribosome to drive the formation of the peptide bond.

The considerable accuracy of the tRNA synthetases depends on this two-step reaction. The AA~AMP must remain attached to the enzyme until the tRNA arrives; if an incorrect amino acid were selected initially, during this period it would have a higher chance of falling off because it would have a decreased affinity for the enzyme. Thus there is a double check of the amino acid, first when it initially becomes bound to the enzyme, and second during the period when it must remain attached. This "proofreading" process is an essential feature in the translation of the genetic code.

tRNA Genes Have Two Internal Promoter Regions

Deletion studies on cloned tRNA genes have shown that nucleotide sequences within the gene are essential for transcription by RNA polymerase III,[50-52] as had been found previously for 5S genes. By comparing nucleotide sequences from many tRNA genes and other genes transcribed by RNA polymerase III (such as Alu 7S RNA, Fig. 20–28), and mutating them in vitro in various ways, it has been found that the nucleotide sequences indicated in Figure 21–29 are required for transcription (Galli et al., 1981). They include two separate regions ("A and B boxes") of about 10 nucleotides each that contain many of the nucleotides that are constant in all tRNAs (Figs. 21–27 and 21–29). It is believed that there are two transcription factors that must bind to these sequences[53] before a tRNA gene can be recognized by RNA polymerase III. So far only one of these protein factors, the one that binds to box B, has been isolated.

tRNA Processing—Long Precursors with Complex Posttranscriptional Modifications

As in the case of the ribosomal RNAs, tRNA is not transcribed directly into its final form but rather into a longer precursor. For example, *E. coli* tyrosine tRNA is synthesized as a precursor having 39 extra nucleotides at the 5' end and 3 at the 3' end. These nucleotides must be removed before an active tRNA molecule is obtained. Other *E. coli* tRNAs are synthesized as very long precursors that contain several tRNAs (as many as 5 to 7) in a single RNA chain. These precursor tRNAs are cleaved by a specific enzyme, *ribonuclease P* (P for processing). Ribonuclease P is an interesting enzyme because it has two components or subunits: one is a protein, the other an RNA of about 300 nucleotides in *E. coli*. The RNA moiety is the catalytic subunit of the enzyme, and it can precisely cleave tRNA precursors even without the protein subunit, although somewhat slower (Guerrier-Takada et al., 1983). In fact this finding constituted the first identification of an RNA enzyme.

Figure 21–30 shows the precursor molecules transcribed from a cloned eukaryotic tRNA gene microinjected into frog oocytes. The initial transcript has a 5' leader, a 3' trailer, and an intervening sequence in addition to the coding region. The ends of the molecule are trimmed

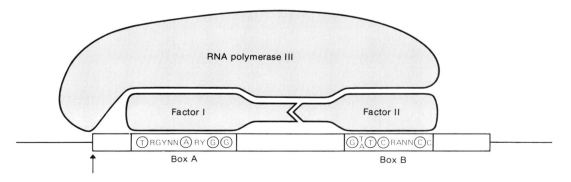

Fig. 21–29. Hypothetical model for the transcription of tRNA genes. Two internal control regions, box A and box B, are required for transcription. Their consensus sequence is indicated, and the nucleotides that are conserved in all tRNA genes (see Fig. 21–27) are circled. Protein transcription factors bind to each box, and the tRNA gene is recognized by RNA polymerase III. (From studies of the laboratories of M. Birnstiel, B. Hall, R. Cortese, and D. Söll.)

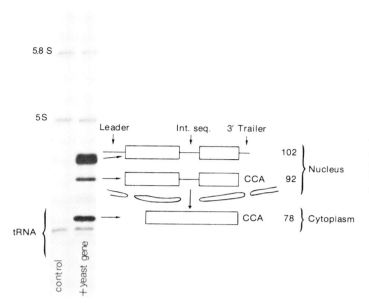

Fig. 21–30. RNAs synthesized in frog oocytes microinjected into the nucleus with a cloned tRNA^{Tyr} gene. The structure and intracellular location of the different precursor bands are indicated. (From De Robertis, E.M., and Nishikura, K.)

first, and then the molecule is spliced and religated. The tRNA precursors are present in the nucleus, while the mature tRNA exits into the cytoplasm.[54,55] Intervening sequences have been found in several eukaryotic tRNA genes, always inserted just after the anticodon.[56] The tRNA intervening sequences, which vary from 10 to 60 nucleotides in length, are removed by two enzymes, a splicing endonuclease and a ligase,[57,58] which are different from the enzymes used for mRNA splicing. tRNA introns do not have conserved sequences at their ends; rather, the splicing enzymes seem to recognize the conserved general shape of the folded tRNA molecule.

Other posttranscriptional modifications important in tRNA function are the base modifications (such as methylations), which lead to the unusual bases found in tRNA, and the terminal addition of CCA. The 3' end of tRNA always terminates with CCA, and its integrty is essential because it is at this end that the amino acid is attached. The 3' end can be degraded in vivo, and cells have an enzyme capable of terminal addition of CCA, which regenerates the active end. In fact, in eukaryotes most tRNA genes do not contain in their DNA the information for the CAA end, which is added posttranscriptionally (Fig. 21–31).

A possible function for the 5' leader sequence became apparent when the order in which the various base modifications appear was examined (Fig. 21–31). The RNA precursors having a leader sequence have particular base modifications added, while other modifications are not present at all and can only be added after the leader sequence is removed. In other words, the precursor RNA with a 5' leader sequence is a substrate for some base modification enzymes, and the RNA without it is a substrate for others.

Intervening sequences are also important in controlling base modifications: when the intervening sequence was deleted precisely from the gene shown in Figure 21–31 using genetic engineering methods[59] and the gene transformed back into yeast cells, it was observed that the i^{6}A modification, that normally is found close to the splicing point, cannot be introduced into the tRNA molecules, and that the resulting tRNA had an impaired function. Thus an intervening sequence is required to obtain a fully functional tRNA. It seems that the extra segments of RNA present in tRNA initial transcripts may introduce a precise order in the intricate processing events involved in tRNA biosynthesis.

21–6 PROTEIN SYNTHESIS

When not engaged in protein synthesis, most ribosomal subunits exist as a cytoplasmic pool of free, exchangeable subunits, rather than as complete ribosomes (Fig. 21–32). The ribosomes are separated into subunits at the end of protein synthesis by a *dissociation factor* that binds to

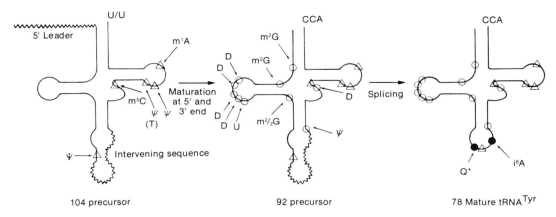

Fig. 21–31.   Diagram summarizing the sequential addition of base modifications to tRNA^{Tyr} precursors in oocytes. $\triangle$, modified bases added at the early precursor stage; $\bigcirc$, modified bases detected at 92-nucleotide stage; $\bullet$, modifications added after splicing of the intervening sequence. (From Nishikura, K., and De Robertis, E.M.: J. Mol. Biol., *145*:405, 1981.)

the 30S subunit. This dissociation factor has been identified as an initiation factor (IF$_3$, discussed below) which is also required for the binding of mRNA to the 30S subunit. Thus the same event that dissociates ribosomes enables the 30S subunit to start a new round of protein synthesis.

Initiation of Synthesis Requires a Specific Initiator tRNA

The first step in protein synthesis is the binding of the small ribosomal subunit to the *ribo-*

some binding site on the mRNA molecule; this site contains the AUG codon at which protein synthesis starts. Ribosome binding sites in prokaryotic mRNAs have between three and eight bases that are complementary to the sequence at the 3' end (tail) of the 16S RNA of the small ribosomal subunit (Fig. 21–4). This complementary region occurs about ten nucleotides before the start of the protein and has been found in all prokaryotic mRNAs studied to date (Table 21–3). It is thought that the end of the 16S RNA hybridizes to the mRNA sequences, thus align-

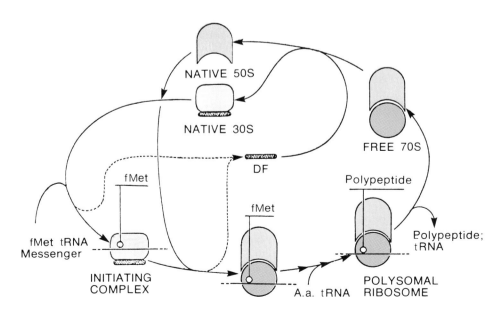

Fig. 21–32.   A model of the ribosome-polysome cycle in which the action of a dissociation factor *(DF)* is indicated. DF is identical to initiation factor 3. (See the text for details.) (From Subramanian, A.R., Ron, E.Z. and Davis, B.D. Natl. Acad. Sci. USA, *61*:761, 1968.)

TABLE 21–3. RIBOSOME BINDING SITES IN BACTERIOPHAGE G₄*

Gene	Nucleotide Sequence
A	A-T-C-A-A-A-C- [G-G-A-G-G] -C-T-T-T-T-C- [A-T-G]
B	A-G-G-C-T- [A-G-G] -G-A-A-T-A-A-A-G-A-A- [A-T-G]
K	C-A-A-G-T-A-C- [G-G-A] -T-A-T-T-T-C-T-G- [A-T-G]
C	G-T-G-G-A-C-T-G-C-T-G-G-T- [G-G-A] -A-A- [A-T-G]
D	A-C-C-A-C-A- [A-A-G-G-A] -A-A-C-T-G-A-A- [A-T-G]
E	C-T-A-T-C- [G-A-G-G] -C-T-T-G-C-G-T-A-T- [A-T-G]
J	C-T-T-T- [T-A-A-G-G-A-G] -T-T-A-T-G-T-A- [A-T-G]
F	C-T-A-T-T- [T-A-A-G-G-A] -T-A-C-A-A-A-A- [A-T-G]
G	A-G-C-C-A-A- [A-A-G-G-A] -C-T-A-A-C-A-T- [A-T-G]
H	T-G-A-A-A- [T-A-A-G-G-A] -T-T-A-T-C-C-T- [A-T-G]
16S rRNA 3' end	$_{HO}$A-U-U-C-C-U-C-C-A-C-U-A-G

*The boxes indicate the initiation codon (ATG) and the bases that are complementary to the 3' end of 16S ribosomal RNA. The DNA sequence is depicted, and therefore T is used instead of U. G₄ is a bacteriophage closely related to φX174. (Data from Godson, G.N., Barrell, B.G., Staden, R., and Fiddes, J.C., Nature, 276:236, 1978.)

ing the initiation codon in the correct position to start protein synthesis. This region of prokaryotic mRNAs (Table 21–3) is also called the Shine and Dalgarno sequence, after their discoverers.

The mRNA start signal is complex and involves other sequences in addition to the AUG codon. These additional sequences differentiate the starting AUG codon from the other AUG codons located within the message. In prokaryotes, the secondary structure of the mRNA (hairpin loops) may be important in the correct selection of the starting AUG (Fig. 2–9). In eukaryotes, the AUG used to initiate translation is usually the first one present in the mRNA sequence, consistent with a model in which the ribosome scans the mRNA starting at the 5' end until it finds an AUG. The starting AUG in eu-

karyotes tends to have in many cases, but not all, conserved nucleotides flanking it. The best consensus for eukaryotic mRNAs is ^a_gccAUGg, of which the most conserved feature is a purine (A or G) at position −3 before the AUG (Kozak, 1986).

In bacteria, the first AA-tRNA to initiate protein synthesis is always a formyl (—CHO) derivative of methionine, called *fMet-tRNA*. Because in protein synthesis the peptide chain always grows sequentially from the free terminal —NH₂ group toward the —COOH end, the role of formylmethionine tRNA is to ensure that proteins are synthesized in this direction. In fMet-tRNA the —NH₂ group is blocked by the formyl group, leaving only the —COOH available to react with the —NH₂ of the second amino acid. In this way synthesis proceeds in the correct sequence. The first amino acid is later separated from the protein by a hydrolytic enzyme.

The ribosome is able to discriminate between an AUG codon within as opposed to at the beginning of the mRNA. When this codon appears at the start of the mRNA, fMet is incorporated, but when an AUG appears in the middle of a message, normal methionine is incorporated instead. It has been learned that there are two different tRNAs that respond to the codon AUG, an initiator fMet-tRNA and one for normal methionine. Methionine is formylated only after it has been bound to this special initiator tRNA.

In eukaryotic cells synthesis is initiated by a special initiator Met-tRNA, but in this case the methionine is not formylated. As in prokaryotes, however, the initial methionine is generally split off from the finished polypeptide.

IF₁, IF₂, and IF₃—Protein Factors That Initiate Protein Synthesis

The small ribosomal subunits in *E. coli* also require three protein factors to start protein synthesis. These *initiation factors* are loosely associated with the 30S subunits and can be isolated by washing ribosomes with salt solutions (Table 21–4). All three factors are essential for initiation when natural mRNAs are used as templates, but they are not required with artificial mRNAs such as poly U.

IF₃ binds to the 30S ribosomal subunit and is required for its binding to the mRNA starting site. IF₃ also functions as a ribosome dissociation

TABLE 21–4. PROTEIN FACTORS IN *E. COLI* PROTEIN SYNTHESIS

Phase	Factor	Source	Function
1° Initiation	IF$_3$	High salt 30S	Dissociation of ribosomal subunits
	IF$_1$ IF$_2$ } GTP IF$_3$	Ribosomal wash	Binding of mRNA and initiator tRNA to 30S subunit
2° Elongation	Tu Ts } T + GTP	Supernatant fraction	Binding of aminoacyl-tRNA Tu-GTP complex to ribosome
	Peptidyl transferase	50S ribosomal subunit	Peptidyl transfer from peptidyl-tRNA to aminoacyl-tRNA
	G + GTP	Supernatant fraction	Translocation of peptidyl-tRNA; release of free tRNA
3° Termination	R$_1$ R$_2$	Supernatant fraction	Release of protein at UAA, UAG, or UGA codons

factor, as diagrammed in Figure 21–32. IF$_2$ binds the initiator fMet-tRNA and carries it to the ribosome (together with GTP) in response to the first AUG codon. IF$_1$ participates in the interaction between IF$_2$ and the initiator tRNA. The upper part of Figure 21–33 shows the main steps of protein synthesis initiation in *E. coli*. Once the large ribosomal subunit binds to the small subunit, the initiation factors are released and can be reutilized.

Eukaryotic IF$_2$ Is Phosphorylated by a Cyclic AMP–Dependent Protein Kinase

Eukaryotic initiation factor 2 (eIF$_2$, similar in function to the *E. coli* IF$_2$) can become phosphorylated under certain circumstances. Phosphorylated eIF$_2$ is inactive in protein synthesis, and the degree of phosphorylation can be regulated by cyclic AMP in a cascade mechanism very similar to the one indicated in Figure 2–29.

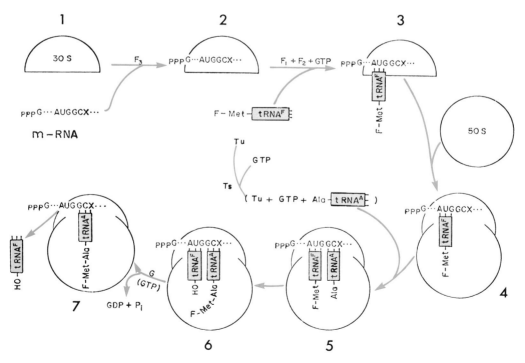

Fig. 21–33. General diagram of the initiation steps in protein synthesis involving 30S and 50S ribosomal subunits, messenger RNA, the initiation factors F$_1$ and F$_1$ + F$_2$, GTP, the elongation factors T and G, formylmethionine-tRNA (fMet-tRNA) and alanine-tRNA. *1*, isolated 30S; *2*, binding of m-RNA and IF$_3$ to the 30S; *3*, binding of IF$_1$ + IF$_2$ + GTP and fMet-tRNA to make the initiation complex; *4*, binding of the 50S subunit to make the complete 70S ribosome; *5*, binding of the second aminoacyl-tRNA; *6*, synthesis of the first peptide bond; *7*, liberation of the free tRNA after translocation.

The eIF$_2$ phosphorylation provides a mechanism by which the rate of protein synthesis can be regulated in eukaryotes.

Chain Elongation Requires Soluble Factors

Once protein synthesis has been initiated, additional factors are required for the elongation of the peptide chain. The elongation factors Tu, Ts, and G are soluble proteins that can be isolated from bacterial cellular supernatants.

EFTu (Tu stands for temperature-unstable factor) forms a complex with an aminoacyl-tRNA and GTP and brings them to the ribosome (Fig. 21–33). In contrast with IF$_2$, which is used only to transport fMet-tRNA to the ribosome, EFTu is used to carry all other aminoacyl-tRNAs to the site of chain elongation. Once the AA-tRNA is in place, the GTP is hydrolyzed and EFTu is released from the ribosome.

EFTs (Ts stands for temperature-stable factor) catalyzes the formation of the complex between EFTu, AA-tRNA, and GTP (Fig. 21–33).

The EFG factor, also called *translocase*, is involved in the translocation process that occurs when the ribosome moves from one codon to the next along the mRNA. G factor binds GTP (hence its name) and carries it to the ribosome, where it is hydrolyzed to GDP and inorganic phosphate. The energy released is used for the translocation process and for the release of the deacylated tRNA used to translate the previous codon.

The energy for peptide bond formation is stored in the AA~tRNA and comes from an ATP molecule used during the aminoacylation of tRNA. Thus protein synthesis is expensive in terms of energy: for each amino acid incorporated, one ATP and two GTP molecules are required.

The large ribosomal subunit contains the enzyme *peptidyltransferase*, or *peptide synthetase*, which catalyzes the formation of the peptide bond. Another function of the 50S subunit is to provide two binding sites for two tRNA molecules; the *aminoacyl* or *acceptor site* and the *peptidyl* or *donor site* (Fig. 21–34).

The stepwise growth of the polypeptide chain involves (1) the entrance of an aminoacyl-tRNA into the aminoacyl site; (2) the formation of a peptide bond and consequent ejection of the tRNA that was occupying the peptidyl site; and

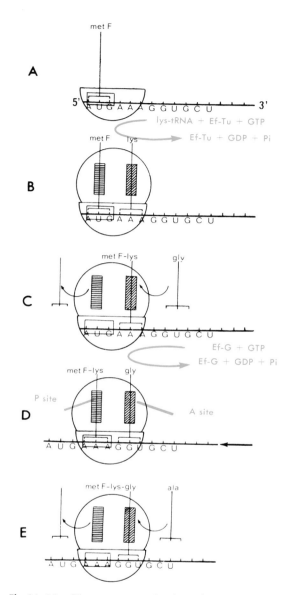

Fig. 21–34. Diagram representing the early stages of translation of mRNA (5'–3'). The *initiation site* in the 30S subunit is indicated by a rectangle. The *aminoacyl* site is indicated by oblique stripes and the *peptidyl* site by horizontal stripes. *A*, initiation complex in which fMet-tRNA binds to the first codon in mRNA (AUG). *B*, the 70S ribosome has been formed, and the second aminoacyl-tRNA (lys-tRNA) binds the second codon (AAA). The function of Ef-Tu and GTP is indicated. *C*, the tRNA is eliminated from the peptidyl site, and the first peptide bond is formed. The function of EF-G and GTP is indicated. *D*, translocation of the mRNA and of the peptidyl-tRNA has occurred, and a new aminoacyl-tRNA (gly-tRNA) binds to the third codon (GGU). *E*, the molecular events of *C* are now repeated. (Adapted from Ochoa, S.: Molecular mechanisms of control of protein biosynthesis. Eur. J. Cell Biol., 26:212, 1981.)

(3) the translocation of the tRNA (now carrying the peptide chain) from the aminoacyl to the peptidyl site. This process should be coupled with the simultaneous movement of the mRNA to place the following codon in position (Fig. 21–34). This translocation, in which the ribosome moves along the mRNA in the 5' to 3' direction, requires the G factor and GTP.

The velocity with which these coordinated processes occurs can be illustrated by the fact that it takes only about 1 minute to construct a hemoglobin chain consisting of 150 amino acids.

Chain Termination Requires Releasing Factors

The termination of the polypeptide chain occurs when the 70S ribosome carrying the peptidyl-tRNA reaches the termination codon located at the end of each cistron. Chain termination leads to the release of the free polypeptide and tRNA, and to the dissociation of the 70S ribosome into 30S and 50S subunits.

The chain termination codons are UAA UGA, and UAG (Table 20–1). Unlike all other triplets, these codons are not recognized by tRNA but rather by two specific proteins, the *releasing factors* R_1 and R_2. R_1 is specific for UAG and UAA, and R_2 is specific for UAA and UGA.

Table 21–4 summarizes the protein factors that are used in the various stages of protein synthesis.

Polysomes Consist of Several Ribosomes Attached to the Same mRNA

During protein synthesis each mRNA molecule is translated by several ribosomes simultaneously. The number of ribosomes in a *polysome* or *polyribosome* depends on the length of the mRNA. A fully active mRNA has a ribosome every 80 nucleotides, and longer mRNAs therefore have larger polysomes.

When eukaryotic cells are lysed and centrifuged on a sucrose gradient, the absorbance at 260 nm (which reflects mainly the location of the ribosomal RNAs, the most abundant nucleic acids in the cell) has the distribution shown in Figure 21–35. The polysomes appear as a series of peaks and are larger than the 80S monomer ribosomes, and therefore sediment closer to the bottom of the centrifuge tube.

If each fraction from such a sucrose gradient

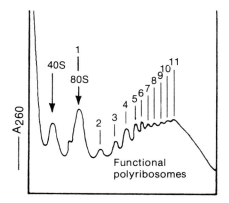

Fig. 21–35. Distribution of ribosomes and polysomes on a sucrose gradient. The numbers of ribosomes per polysome are indicated.

is allowed to complete protein synthesis in the presence of radioactive amino acids, as in Figure 21–36, then the longer a protein is, the higher the number of ribosomes that can translate simultaneously its mRNA. In the example shown in Figure 21–36, light chain immunoglobulin mRNA (MW 24,000) has 6 to 8 ribosomes, while heavy chain immunoglobulin mRNA from the same cells (MW 50,000) has an average of 12 or more ribosomes on it.

The best way to visualize polysomes is to spread them for electron microscopy. Figure 21–37 shows a molecule of silk moth fibroin mRNA with its associated ribosomes. Each ribosome has a nascent protein chain attached to it which lengthens in the direction of protein synthesis (5' to 3'). Fibroin mRNA is very long, and each polysome may contain up to 80 ribosomes; this means that there can be up to 80 polypeptides translated simultaneously from the same RNA molecule.

We have already seen that in prokaryotes protein synthesis starts even before mRNA synthesis is completed, because there is no nuclear envelope interposed between the genome and the ribosomes. Bacterial polysomes are therefore visualized with the nascent mRNA still attached to the DNA via the RNA polymerase (Fig. 20–5).

Antibiotics and Toxins Are Useful in Molecular Biology and Medicine

Antibiotics are useful tools in the study of protein synthesis because many of them work at

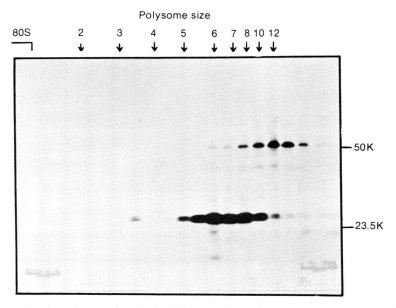

Fig. 21–36.   Polysome size depends on the length of mRNA. A myeloma cell that produces immunoglobulins was separated in a sucrose gradient as in Figure 21–35 and the proteins synthesized by the different fractions separated in a polyacrylamide gel. The number of ribosomes per polysome is indicated at the top of the figure. Note that the immunoglobulin light chain, the lower protein, is made on smaller polysomes of about six ribosomes, while the heavy chain, the upper protein, is made in polysomes of ten or more ribosomes. (Courtesy of B. Mechler.)

very precise steps of the process. For example, *puromycin* is an analogue of AA-tRNA that can bind to the A site of the ribosome. The peptidyl transferase uses it as a substrate, forming a polypeptide-puromycin chain, which is released. *Fusidic acid* blocks the translocation induced by G factor. *Streptomycin* causes misreading of the genetic message, and the errors accumulated in the proteins kill the bacteria. *Tetracycline* inhibits AA-tRNA binding. *Sparsomycin* inhibits

peptidyl transferase. *Chloramphenicol* inhibits protein synthesis of bacteria, chloroplasts, and mitochondria, but not the general cytoplasmic ribosomal system. *Cycloheximide*, on the other hand, affects eukaryotic cells.

Many antibiotics (such as chloramphenicol, streptomycin, and tetracycline) inhibit bacterial ribosomes but not eukaryotic ribosomes. This is the molecular basis for the use of these antibiotics in medical treatment (Table 21–5).

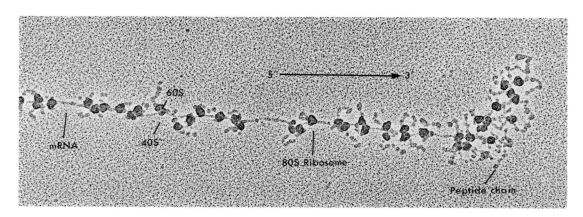

Fig. 21–37.   Silk fibroin mRNA visualized during protein synthesis. A nascent peptide can be observed emerging from each ribosome and increasing in length in the 5′ to 3′ direction. The nascent peptide may be visualized in this case because silk fibroin is a fibrous protein and remains extended, whereas most other proteins adopt a globular conformation as they are being synthesized. (Courtesy of O.L. Miller, Jr.)

TABLE 21–5. MODE OF ACTION OF SOME ANTIBIOTICS USED IN MEDICAL PRACTICE

Antibiotic	Action on Bacterial Cells
Streptomycin	Misreading of mRNA
Tetracycline	Binds to small ribosomal subunit and inhibits binding of AA-tRNA
Chloramphenicol	Inhibits peptidyl transferase in large ribosomal subunit
Rifampicin	Inhibits bacterial RNA polymerase
Penicillin and ampicillin	Inhibit cell wall formation

Other drugs are useful for the study of RNA synthesis. *Actinomycin D* inhibits RNA synthesis by intercalating between the DNA bases, especially in GC-rich regions. *Rifampicin* works by binding to bacterial RNA polymerase and inhibiting initiation of transcription. Rifampicin does not inhibit eukaryotic RNA polymerases and is therefore also used in medical treatment.

Newly Made Proteins are Targeted to Their Final Cellular Destination

Cells devote a very considerable proportion of their constituents to the machinery for protein synthesis. In *E. coli* ribosomes constitute 25% of the total cell mass, and EFTu is the most abundant protein in the cell. Eukaryotic cells not only contain some 10^7 ribosomes, but also have evolved complicated cellular organelles in order to ensure that proteins reach their appropriate destination. Some proteins are translocated during synthesis into the lumen of the rough endoplasmic reticulum (Chapter 8), from which they are channeled via coated vesicles into the Golgi apparatus, where they are glycosylated and targeted to secretory granules, membranes, or lysosomes (Chapter 9). Some proteins remain in the cytoplasm, while others migrate specifically into the nucleus (Chapter 11). Thus a considerable part of the cell is concerned with the synthesis and traffic of proteins. But the effort is worth it, for once all these proteins are assembled into their proper places, they provide the delicate biochemical machinery that keeps cells feeding, locomoting, multiplying, and *alive*.

SUMMARY:
Transfer RNA and Protein Synthesis

Transfer RNAs are small RNAs that serve as molecular adaptors during protein synthesis. Each

tRNA has an anticodon that can hydrogen-bond with a codon triplet in the mRNA and has the corresponding amino acid attached to one of its ends (Fig. 21–27). Faithful protein synthesis depends on the correct charging of these so-called aminoacyl-tRNAs by specific aminoacyl-tRNA synthetases. Some tRNAs recognize more than one codon (third-base "wobble"). There is an isoacceptor tRNA family for each amino acid, and an aminoacyl-tRNA synthetase for each of the 20 amino acids.

All tRNAs share common features including a "cloverleaf" secondary structure, a CCA acceptor end for amino acid attachment, an anticodon loop, and D and T loops that have been conserved throughout evolution. The tertiary structure resembles an L, with the anticodon at one end and the amino acid attachment site at the other. Proper aminoacylation of the tRNA molecule is essential for accurate protein synthesis.

tRNA transcription by RNA polymerase III requires two internal regions within the gene, which contain the conserved nucleotides. Two transcription factors are thought to bind at these sites and permit transcription (Fig. 21–29).

tRNA, like most other RNAs, must be processed from a longer precursor into its final form. Types of posttranscriptional processing include the removal of 5' and 3' extensions, the methylation of some bases, the terminal addition of CCA, and the excision of intervening sequences in some eukaryotic tRNAs.

The first step in protein synthesis is the binding of the small ribosomal subunit to the mRNA binding site, which contains an AUG codon to signal the start of translation, as well as additional sequences that differentiate the starting AUG codon from others within the message. In prokaryotes, the first AA-tRNA is always fMet-tRNA. In eukaryotes, synthesis is initiated by an unformylated initiator Met-tRNA. In both cases the initial methionine is removed from the finished polypeptide. Three protein initiation factors associated with the 30S subunit are also required for synthesis initiation in *E. coli*. Once the large ribosomal subunit binds to the small subunit, the initiation factors are released.

Other protein factors are required for polypeptide chain elongation. These factors are free in the cytoplasm and involved in bringing the correct AA-tRNAs to the ribosome and in the translocation process that occurs when the ribosome moves from one codon to the next. The large ribosomal subunit contains the enzyme that catalyzes the formation of the peptide bond, and provides aminoacyl and peptidyl sites that can accommodate the two tRNA molecules whose amino acids are being added to the protein chain.

Chain termination occurs when the 70S ribosome reaches a termination codon, which is recognized by one of two protein releasing factors.

During protein synthesis each mRNA molecule

is translated by several ribosomes simultaneously, forming a structure called a polysome. This greatly increases the speed with which a long polypeptide can be transcribed. A fully active mRNA molecule has a ribosome every 80 nucleotides.

REFERENCES

1. Nonomura, Y., Blobel, G., and Sabatini, D.D.: J. Mol. Biol., 60:303, 1971.
2. Blobel, G., and Sabatini, D.D.: J. Cell Biol., 45:130, 1970.
3. Noller, H.F., and Woese, C.R.: Science, 212:403, 1981.
4. Nomura, M., and Traub, P.: J. Mol. Biol., 34:609, 1968.
5. Nomura, M., and Held, W.A.: In Ribosomes. Edited by M. Nomura, A. Tissières, and P. Lengyel. Cold Spring Harbor, New York, Cold Spring Harbor Laboratory, 1974, p. 193.
6. Yates, J.L., and Nomura, M.: Cell, 21:517, 1980.
7. Dean, D., Yates, J.L., and Nomura, M.: Cell, 24:415, 1981.
8. Lindahl, L., and Zengel, J.: In Advances in Genetics. Vol. 21. Edited by E.W. Caspari. New York, Academic Press Inc., 1982, p. 53.
9. Olins, P.O., and Nomura, M.: Nucleic Acids Res., 9:1757, 1981.
10. Pierandrei-Amaldi, P., et al.: Cell, 30:163, 1982.
11. Paterson, B.M., and Rosenberg, M.: Nature, 279:692, 1979.
12. Granboulan, N., and Granboulan, P.: Exp. Cell Res., 38:604, 1965.
13. Macgregor, H.C.: J. Cell Sci., 2:147, 1967.
14. Brown, D.D., and Gurdon, J.B.: Proc. Natl. Acad. Sci. USA, 51:139, 1964.
15. Wallace, H., and Birnstiel, M.L.: Biochim. Biophys. Acta, 114:296, 1966.
16. Miller, L.: J. Cell Biol., 59:624, 1973.
17. Miller, O.L., Jr., and Beatty, B.R.: Science, 164:955, 1969.
18. Scheer, U., Trendelenburg, M.F., and Franke, W.W.: J. Cell Biol., 69:465, 1976.
19. Trendelenburg, M.F., and McKinnell, R.G.: Differentiation, 15:73, 1979.
20. Wellauer, P.K., Dawid, I.B., Brown, D.D., and Reeder, R.H.: J. Mol. Biol., 105:461, 1976.
21. Reeder, R.H., Brown, D.D., Wellauer P.K., and Dawid, I.B.: J. Mol. Biol., 105:507, 1976.
22. Moss, T.: Nature, 302:223, 1983.
23. Busby, S., and Reeder, R.: Cell, 34:989, 1983.
24. Labhart, P., and Reeder, R.: Cell, 37:285, 1983.
25. Trendelenburg, M., and Gurdon, J.: Nature, 276:292, 1978.
26. Hourcade, D., Dressler D., and Wolfron, J.: Cold Spring Harbor Symp. Quant. Biol., 38:537, 1973.
27. Miller, O.L.: Natl. Cancer Inst. Monogr., 23:53, 1966.
28. Brown, D.D., and Dawid, I.: Science, 160:272, 1968.
29. Gall, J.G.: Proc. Natl. Acad. Sci. USA, 60:553, 1968.
30. Weinberg, R.A., Loening, U., Willems, M., and Penman, S.: Proc. Natl. Acad. Sci. USA, 58:1088, 1967.
31. Maden, B.E.H.: Trends Biochem. Sci., 1(3):196, 1977.
32. Chooi, W.Y., and Leiby, K.R.: Proc. Natl. Acad. Sci. USA, 78:4823, 1981.
33. Pardue, M.L., Brown, D.D., and Birnstiel, M.L.: Chromosoma, 42:191, 1973.
34. Artavanis, S., et al.: Cell, 12:1057, 1977.
35. Miller, J.R., et al.: Cell, 13:717, 1978.
36. Brown, D.D., Wensink, P.C., and Jordan, E.: Proc. Natl. Acad. Sci. USA, 68:3175, 1971.

37. Carrol, D., and Brown, D.D.: Cell, 7:477, 1976.
38. Ford, P.J., and Southern, E.M.: Nature New Biol., 241:7, 1973.
39. Birkenmeier, E.H., Brown, D.D., and Jordan, E.: Cell, 15:1077, 1978.
40. Sakonju, S., Bogenhagen, D., and Brown, D.D.: Cell, 19:13, 1980.
41. Bogenhagen, D., Sakonju, S., and Brown, D.D.: Cell, 19:27, 1980.
42. Engelke, D.R., Shastry, B.S., and Roeder, R.G.: Cell, 19:717, 1980.
43. Picard, B., Le Maire, M., Wegnez, M., and Denis, H.: Eur. J. Biochem., 109:359, 1980.
44. Pelham, H., and Brown, D.D.: Proc. Natl. Acad. Sci. USA, 77:4170, 1980.
45. Honda, B.M., and Roeder, R.G.: Cell, 22:119, 1980.
46. Mattaj, I.W., Lienhard, S., Zeller, R., and De Robertis, E.M.: J. Cell Biol., 97:1261, 1983.
47. Bogenhagen, D.F., Wormington, W.M., and Brown, D.D.: Cell, 28:413, 1982.
48. Gottesfeld, J., and Bloomer, L.S.: Cell, 28:781, 1982.
49. Gurdon, J.B., Dingwall, L., Laskey, R.A., and Korn, L.J.: Nature, 299:652, 1982.
50. Hoffstetter, H., Kressmann, A., and Birnstiel, M.L.: Cell, 24:573, 1981.
51. Ciliberto, G., Castagnoli, L., Melton, D.A., and Cortese, R.: Proc. Natl. Acad. Sci. USA, 79:1198, 1982.
52. Sharp, S., et al.: Proc. Natl. Acad. Sci. USA, 78:6657, 1981.
53. Klemenz, R., Stillmann, D.J., and Geiduschek, E.P.: Proc. Natl. Acad. Sci. USA, 79:6191, 1982.
54. De Robertis, E.M., and Olson, M.V.: Nature, 278:137, 1979.
55. Melton, D.A., De Robertis, E.M., and Cortese, R.: Nature, 284:143, 1980.
56. Knapp, G., et al.: Cell, 14:221, 1978.
57. Greer, C.L., Peebles, C.L., Gegenheimer, P., and Abelson J.: Cell, 32:525, 1983.
58. Filipowicz, W., and Shatkin, A.J.: Cell, 32:547, 1983.
59. Johnson, P.F., and Abelson, J.: Nature, 302:268, 1983.

ADDITIONAL READINGS

Allison, D., Goh, S.H., and Hall, B.D.: The promoter sequence of a yeast tRNA[tyr] gene. Cell, 34:655, 1983.
Apirion, D.: Processing of RNA. CRC Press Inc., Boca Raton, Florida, 1983.
Bogenhagen, D.F., Wormington, W.M., and Brown, D.D.: Stable transcription complexes of Xenopus 5S RNA genes: A means to maintain the differentiated state. Cell, 28:413, 1982.
Brimacombe, R., Stoffler, G., and Wittman, H.G.: Ribosome structure. Annu. Rev. Biochem., 47:217, 1978.
Brown, D.D.: Gene expression in eukaryotes. Science, 211:667, 1981.
Brown, D.D.: The role of stable transcription complexes that repress and activate eukaryotic genes. Cell, 37:359, 1984. (highly recommended).
Brown, D.D., and Schliessel, M.S.: A positive transcription factor controls the differential expression of two 5S RNA genes. Cell, 42:759, 1985.
Busby, S., and Reeder, R.H.: Spacer sequences regulate transcription of ribosomal gene plasmids injected into Xenopus embryos. Cell, 34:989, 1983.
Clark, B.: The elongation step of protein biosynthesis. Trends Biochem. Sci., 5:207, 1980.
Dean, D., and Nomura, M.: Genetics and regulation of

ribosomal protein synthesis in *E. coli*. Cell Nucleus, *12*:185, 1982.

Engelke, D.R., Shastry, B.S., and Roeder, R.G.: Specific interaction of a purified transcription factor with an internal control region of 5S RNA genes. Cell, *19*:717, 1980.

Galli, G., Hoffstetter, H., and Birnstiel, M.L.: Two conserved sequence blocks within eukaryotic tRNA genes are major promoter elements. Nature, *294*:626, 1981.

Ginsberg, A.M., King, B.O., and Roeder, R.G.: *Xenopus* 5S gene transcription factor TF III A: Characterization of a cDNA clone and measurement of RNA levels throughout development. Cell, *39*:479, 1984.

Guerrier-Takada, C., et al.: The RNA moiety of ribonuclease P is the catalytic subunit of the enzyme. Cell, *35*:849, 1983.

Hunt, T.: The initiation of protein synthesis. Trends Biochem. Sci., *5*:178, 1980.

Kozak, M.: Point mutations define a sequence flanking the AUG initiator codon that modulates translation by eukaryotic ribosomes. Cell, *44*:283, 1986.

Labhart P., and Reeder, R.: Characterization of three sites of RNA 3′ end formation in the Xenopus ribosomal gene spacer. Cell, *45*:431, 1986.

Lake, J.A.: The ribosome. Sci. Am., *245*(2):84, 1981.

Lipmann, F.: Polypeptide chain elongation in protein synthesis. Science, *164*:1024, 1969.

Miller, O.L., Jr.: The nucleolus, chromosomes and visualization of genetic activity. J. Cell Biol., *91*:15S, 1981.

Miller, J., McLachlan, A., and Klug, A.: Repetitive zinc-binding domains in the protein transcription factor IIIA from Xenopus oocytes. The EMBO J., *4*:1609, 1985.

Nishikura, K., and De Robertis, E.M.: RNA processing in microinjected *Xenopus* oocytes: Sequential of base modifications in a spliced transfer RNA. J. Mol. Biol., *145*:405, 1981.

Nomura, M.: Organization of bacterial genes in ribosomal components. Cell, *9*:633, 1976.

Nomura, M.: Assembly of bacterial ribosomes. Science, *179*:864, 1973.

Nomura, M.: The control of ribosome synthesis. Sci. Am., *250*(1):102, 1986.

Ochoa, S.: Molecular mechanisms of control of protein biosynthesis. Eur. J. Cell Biol., *26*:212, 1981.

Ochoa, S.: Regulation of protein synthesis. CRC Crit. Rev. Biochem., 7:7, 1979.

Pierandrei-Amaldi, P., et al. Expression of ribosomal protein genes in *Xenopus laevis* development. Cell, *30*:163, 1982.

Prince, J.B., Gutell, R.R., and Garret, R.A.: A consensus model of the *E. coli* ribosome. Trends Biochem Sci., 8:359, 1983.

Reeder, R.H.: Enhancers and ribosomal gene spacers. Cell, 38:349, 1984.

Reeder, R.H.: Ribosomes from eukaryotes: Genetics. *In* Ribosomes. Edited by M. Nomura, A. Tissières, and P. Lengyel. Cold Spring Harbor, New York, Cold Spring Harbor Laboratory, 1974, pp. 489–518.

Reeder, R.H., Roan, J., and Dunnaway, M.: Spacer regulation of *Xenopus* ribosomal gene transcription: Competition in oocytes. Cell, *35*:449, 1983.

Rhodes, D., and Klug, A.: An underlying repeat in some transcriptional control sequences corresponding to half a double helical turn of DNA. Cell, *46*:123, 1986.

Rich, A., and Kim, S.H.: The three-dimensional structure of transfer RNA. Sci. Am., *238*(1):52, 1978.

Rosenberg, U.B., et al.: Structural homology of the Drosophila Kruppel gene with Xenopus transcription factor III A. Nature, *319*:336, 1986.

Sakonju, S., and Brown, D.D.: Contact prints between a positive transcription factor and the *Xenopus* 5S RNA gene. Cell, *31*:395, 1982.

Scheer, U., and Rose, K.M.: Localization of RNA polymerase I in interphase cells and mitotic chromosomes by light and electron microscopic immunocyto-chemistry. Proc. Natl. Acad. Sci. USA, *81*:1431, 1984.

Shine, J., and Dalgarno, L.: The 3′ terminal sequence of *E. coli* 16S ribosomal RNA: Complementarity to ribosome binding sites. Proc. Natl. Acad. Sci. USA, *71*:1342, 1974.

Smith, J.D.: Transcription and processing of transfer RNA precursors. Prog. Nucleic Acid Res., *16*:25, 1976.

The mechanisms of protein synthesis. Cold Spring Harbor Symp. Quant. Biol., *34*:1969.

Watson, J.D.: Molecular Biology of the Gene. Menlo Park, California, W.A. Benjamin Inc., 1975.

Weissbach, H., and Ochoa, S.: Soluble factors required for eukaryotic protein synthesis. Annu. Rev. Biochem., *45*:191, 1976.

GENE REGULATION

In previous chapters we have seen that the information encoded in DNA is transcribed into RNA and then translated into protein. In this chapter we analyze the ways in which a cell regulates the expression of its genes. The ability to switch genes on and off is of fundamental importance because it enables cells to respond to a changing environment and is the basis for cell differentiation. Gene regulation in eukaryotes is achieved in many ways, some of which have already been considered. For example, in Chapters 20 and 21 we examined the importance of RNA processing as well as the transcriptional regulation of the 5S rRNA, the large rRNA and tRNA genes. In Chapter 23, aspects of cell differentiation will be analyzed, such as gene amplification, transposition of silent genes into expression loci, and gene rearrangements during antibody production. In this chapter we discuss in some detail how the eukaryotic genome is organized into highly repeated, moderately repeated, and single-copy DNA. We see that the genome is constantly submitted to rearrangements mediated by transposons and retroviruses, which can copy RNA into DNA and reinsert it into the genome. Valuable information on how genes are controlled at the transcriptional level has been obtained from studies on polytene and lampbrush chromsomes, in which the transcription process can be seen directly in the optical microscope; these findings will be presented in Sections 22–4 and 22–5.

22–1 GENE REGULATION IN PROKARYOTES

The rate of expression of bacterial genes is controlled mainly at the level of mRNA synthesis (transcription). Regulation can occur at both the initiation and the termination of mRNA transcription. Because bacteria obtain their food from the medium that immediately surrounds them, their gene regulation mechanisms are designed to adapt quickly to changes in the environment.

Eukaryotes, however, have a much larger and more complex genome than prokaryotes; furthermore, the cells of higher organisms are surrounded by a constant internal milieu. The ability of such cells to respond to hormones and to impulses from the nervous system, for example, is thus comparatively more important than the ability to respond rapidly to the presence of certain nutrients.

The Lac Operon Is Induced 1000-fold by Lactose

The cell is an energy-efficient unit that makes only the proteins it needs. *E. coli* has about 3000 protein-coding genes, but only a subset of them is expressed at any one time. The best illustration of this efficiency is provided by the enzymes involved in the utilization of the disaccharide lactose: the synthesis of these enzymes may be induced up to 1000-fold in response to the addition of lactose to the culture medium. Regulation by *enzyme induction* is found in many other bacterial *catabolic systems* that degrade sugars, amino acids, and lipids. In these systems the availability of a substrate stimulates the production of the enzymes involved in its degradation.

The three enzymes required for the utilization of lactose are β-*galactosidase, lac permease,* and *transacetylase,* and the synthesis of all of them is regulated coordinately as a unit called an *operon* (Fig. 22–1).[1] An operon is a group of genes that are next to each other in the DNA and that can be controlled (i.e., turned on and off) in a unified manner. The *lac* operon consists of the genes *z, y,* and *a,* which code for the three enzymes, and two elements involved in their regulation, the *promoter (p)* and the *operator (o).* The genes for the three enzymes are always transcribed together into a single *polycistronic lac* mRNA, which explains why they are always expressed together (Fig. 22–1).

The expression of the *lac* operon is regulated by the *lac repressor,* a protein having four iden-

tical subunits of 40,000 daltons each (Fig. 22–2). In an *E. coli* cell there are about ten *lac* repressor molecules, which are coded by the regulatory gene *i.* The repressor binds strongly and specifically to a short DNA segment called the *operator,* which is located very close to the start of the β-galactosidase gene. As Figure 22–2 shows, a repressor molecule bound at the operator site prevents the transcription of *lac* mRNA.

The lac repressor binds to the following 21 base-pair sequence of operator DNA.

A A T T G T	G	A	G	C	G	G	A	T	A	A C A A T T
				•						
T T A A C A	C	T	C	G	C	C	T	A	T	T G T T A A

This sequence contains regions of twofold symmetry. Some sequences (boxed areas) on the left side of the operator are also present on the right side but on the opposite strand. Similar symmetrical regions have been found in other operators and in DNA-binding proteins (such as restriction enzymes) and probably reflect the fact that most proteins that recognize specific DNA sequences are multimeric. The two symmetrical sites represent recognition sites for different subunits of the repressor.

The Inducer Binds to the Repressor Protein

The affinity of repressor binding to the operator is regulated by the *inducer,* a small molecule that can bind to the repressor. The natural inducer of the lactose operon is *allolactose,* a metabolite of lactose; however, the analog IPTG (isopropylthiogalactoside) is a more powerful inducer that is preferred in laboratory experiments. Each subunit of the repressor has one binding site for the inducer and, upon binding it, undergoes a conformational change by which it becomes unable to bind to the operator. In this way the presence of the inducer permits the transcription of the *lac* operon, which is no longer blocked by the repressor (Fig. 22–2).

The effect of this conformational change is dramatic. While in the absence of lactose *E. coli* cells have an average of only three molecules of β-galactosidase enzyme per cell, after induction of the lac operon 3000 molecules of β-galactosidase are present in each cell, representing 3 percent of the total protein. Furthermore, these adaptations occur very rapidly; most bacterial mRNAs have a half-life of only a few minutes,

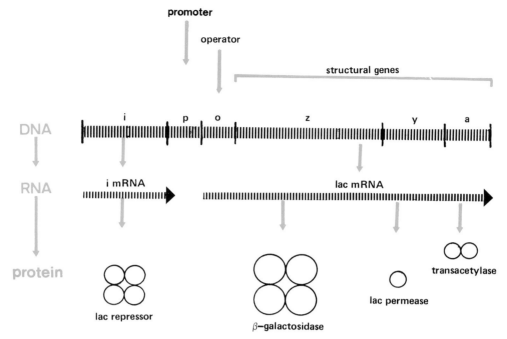

Fig. 22–1.   Diagram representing the *lac operon. i,* the regulatory gene that produces the *i mRNA* that codes for the *lac repressor,* a protein with four subunits. The *promoter (p)* is the region of attachment of RNA polymerase; the *operator (o)* is the region where the repressor binds. The *i gene,* the promoter, and the operator are regulatory elements. The structural genes, *z, y,* and *a,* produce a polycistronic lac mRNA, and the three proteins indicated below.

NO INDUCER

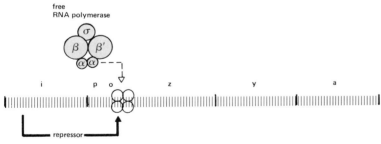

INDUCTION

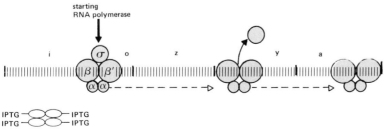

Fig. 22–2.   Diagram representing the regulation of the lactose operon in the absence and in the presence of the inducer. In the absence of the inducer, the *lac* repressor binds tightly to the operator, interfering with the transcription of the structural genes. The binding sites for the repressor and for RNA polymerase overlap, and RNA polymerase is therefore unable to bind when the repressor occupies the operator. The repressor is a tetramer with subunits of 40,000 daltons. When the inducer (IPTG) is present, it binds to the repressor, eliciting a conformational change that prevents its binding to the operator; as a consequence, RNA polymerase is free to transcribe the structural genes. Observe that *E. coli* RNA polymerase has five subunits of different sizes: σ (95,000 daltons); β/β' (about 160,000); and two α (40,000). Note that after transcription has started, the sigma (σ) subunit is released and only the core enzyme remains bound to the DNA.

and therefore any increase or decrease in the rate of mRNA synthesis is rapidly reflected in the rate of protein synthesis.

Repressors Bind Within Promoters and Prevent Attachment of RNA Polymerase

The *promoter (p)* is the DNA segment to which RNA polymerase binds when initiating transcription. Figure 22–3 shows the nucleotide sequence of the *lac* promoter region. Two sections are particularly important for RNA polymerase binding: (1) a TATGTTG sequence located 6 to 12 bases before the transcription starting site, which is conserved in most prokaryotic promoters, and (2) a region-located about 35 bases before the beginning of the mRNA, which is known to be important because mutations in it severely inhibit *lac* expression.

RNA polymerase binds to a region of about 80 nucleotides of DNA, and Figure 22–2 shows that the RNA polymerase binding site in fact overlaps with the region covered by the repressor (that is, the operator). In vitro studies have shown that a repressor bound to the operator blocks the binding of RNA polymerase. The way in which repressors work is therefore very simple: they bind within the promoter sites and prevent the attachment of RNA polymerase.

Lac Transcription Is Also Subject to Positive Control by the CAP-cAMP Complex

Cyclic AMP (cAMP) has a regulatory role in bacteria, activating inducible operons at the level of transcription. *E. coli* has a cAMP receptor protein that is able to bind cAMP with high affinity. This protein is also known as "ca-

tabolite gene *activator protein.*" CAP is a dimeric protein which, when complexed with cAMP, is able to bind to a specific site within the *lac* promoter region (Fig. 22–3). RNA polymerase will recognize the *lac* promoter only if the CAP-cAMP complex is already bound to it. Therefore, in the *lac* operon, in addition to the *negative control* provided by the repressor there is also *positive control* provided by the CAP-cAMP complex.

The positive control mechanism is also required for the expression of many other inducible operons, such as those involved in the utilization of maltose, galactose, and arabinose. However, cAMP is not necessary for the synthesis of those enzymes required for the utilization of glucose as an energy source. This mechanism is highly beneficial in *E. coli* because its cAMP levels vary according to the available food source. Bacteria growing in the presence of glucose have a lower cAMP content than those grown in a poorer energy source, such as lactose. When the intracellular cAMP level is low (i.e., when glucose is available), the CAP protein does not bind to the promoter region, and the *lac* operon will not be turned on even in the presence of lactose. As a result, if *E. coli* is grown in the presence of both glucose and lactose, it will utilize only the glucose. This makes good sense because glucose is the richest and most efficient energy source. It is clear that this mechanism of positive control enables *E. coli* to adapt more efficiently to the changing environment of its natural habitat, the human intestine.

The *trp* Operon: Regulation of Transcription at Initiation and Termination

The *tryptophan operon* codes for the five enzymes that are required for the synthesis of tryp-

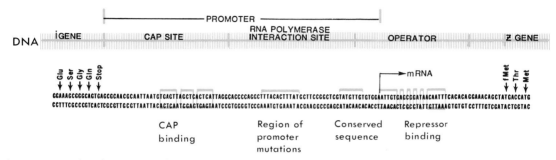

Fig. 22–3. Nucleotide sequence of the *lac* promoter, showing its relation to the overlapping operator sequence. Note that the repressor and CAP protein bind at regions of symmetry. The sequence TATGTTG, which is conserved in prokaryotic promoters, and the region where several promoter mutations occur are indicated. (Data from Dickson, R.C., Abelson, J., Barnes, W.M., and Reznikoff, W.S.: Genetic regulation: The lac control region. Science, *187*:27, 1975. Copyright 1975 by the American Association for the Advancement of Science.)

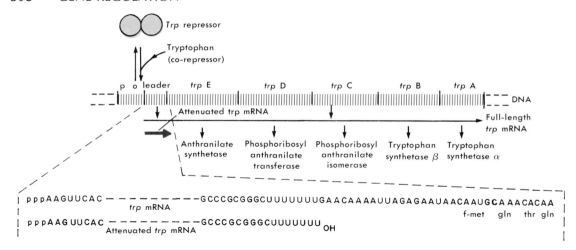

Fig. 22–4.   The tryptophan operon of *E. coli* is regulated at transcription initiation (by the *trp* repressor) and at transcription termination. The leader RNA has a termination sequence called the attenuator, and two types of mRNA can be produced: a short leader RNA or a full-length mRNA that codes for the five enzymes of the operon. The proportion of the two RNAs is regulated by the tryptophan concentration in the cell. Note that as with many other terminators, the attenuated mRNA ends in a row of U's. (Data from Bertrand et al., *Science, 189:*22, 1975.)

tophan (Fig 22–4). It has been known for many years that the expression of this operon is regulated by the availability of tryptophan in the culture medium, and that the presence of tryptophan *represses* the synthesis of *trp* enzymes. Regulation by *enzyme repression* is also found in many other bacterial *anabolic systems* involved in the synthesis of amino acids or nucleic acid precursors, in which the synthesis of the enzymes of a metabolic pathway is selectively inhibited by the end product of that metabolic chain. This is another mechanism that enables *E. coli* to synthesize enzymes only when they are required.

Enzyme repression can also be explained on the basis of a mode similar to that depicted for enzyme induction in the *lac* operon. In enzyme repression the regulatory gene produces a repressor protein which is normally in the inactive form. The repressor, upon binding with a metabolite called a *co-repressor* (in this case the amino acid tryptophan), undergoes a conformational change that enables it to bind to the operator and thereby inhibit the binding of RNA polymerase to the *trp* promoter (Fig. 22–4). The affinity of the repressor for binding to the operator is normally low, but it is increased by the action of the co-repressor. (This is in contrast to the situation in the *lac* operon, in which the repressor is active on its own and loses affinity for the operator when bound to the inducer.)

The tryptophan operon also has a second mechanism of regulation, which operates at the level of *termination of RNA synthesis.*[2] Normally 90% of the RNA polymerase molecules that start transcription at the promoter terminate at a nearby *attenuator* site, yielding a short RNA (140 nucleotides) that does not code for *trp* enzymes (Fig. 22–4). Only 10% or less of the molecules continue on to produce a full-length *trp* mRNA. Termination at the attenuator site is regulated by the level of tryptophan in the cells. When the tryptophan level is low, the proportion of RNA molecules yielding full-length *trp* mRNA increases. When it is high, most of the mRNAs terminate prematurely. Other repressible operons, such as those for histidine *(his)* and phenylalanine *(phe)* biosynthesis, also show similar regulation at the level of RNA termination.

The mechanism by which the metabolic end products regulate transcription termination became apparent when the nucleotide sequence of the short "leader" RNAs became known. The three leader RNAs of the *trp, his,* and *phe* operons code for short peptides between 14 and 16 amino acids long. Their striking feature is that many of these amino acids are the same as the one whose synthesis is directed by the corresponding operon; for example, the *his* operon leader has seven histidine codons in a row.[3]

In these operons, transcription termination is regulated by the ribosomes, which in prokaryotes translate nascent RNAs simultaneously with transcription. The natural tendency of RNA polymerase is to terminate transcription prematurely at the attenuator site, but when the cells are starved for that amino acid, this termination signal is ignored. This is because a ribosome remains attached to the leader RNA (instead of completing the synthesis of the short peptide) due to the insufficient concentration of Trp (or His, or Phe).[4]

Transcription termination in *E. coli* requires a hairpin loop on the RNA followed by a row of U's (Section 20–1). As Figure 22–5 shows, the secondary structure of the "leader" RNA is such that the presence of a ribosome stalled within it when the cells are starved for Trp stops the formation of the hairpin loop required for termination. In this way, the lack of tryptophan stimulates synthesis of the long mRNA that codes for the five enzymes required for its own biosynthesis.

In summary, *E. coli* uses two distinct mechanisms to ensure that the *trp* enzymes are not synthesized unless tryptophan is required. The use of two mechanisms that operate at different levels of tryptophan probably allows a finer control of *trp* enzyme synthesis over a wider range of tryptophan concentrations that would be possible with the repression mechanism alone.

Enzymatic Activity Control Provides the Fine Tuning of Metabolism

Enzyme induction and repression provide a coarse control of metabolism by switching off the synthesis of enzymes when they are not re-quired. However, as we discussed in Chapter 2–6, cells also have fine controls that operate directly at the level of enzyme activity. The most common mechanism is feedback inhibition, in which the end product of a metabolic pathway acts as an allosteric inhibitor of the first enzyme of its metabolic chain. This process is diagrammed in Figure 2–26. Whereas gene induction and repression save valuable energy by preventing the synthesis of unnecessary enzymes, control of enzyme activity by allosteric regulation allows an almost instantaneous fine tuning of catalytic activity.

22–2 GENE ORGANIZATION IN EUKARYOTES

In Chapter 13 we discussed the C-value paradox, i.e., that most eukaryotes have a very large amount of DNA when compared to *E. coli*, but even organisms of the same general level of morphological complexity (which presumably have similar genetic requirements) nevertheless have vastly different amounts of DNA. The classic example is provided by salamanders, which can have huge genome sizes, say 10 to 20 times larger than that of a frog or a human, although there is no reason to believe that life for a salamander is any more sophisticated than for other vertebrates. This paradox began to be explained when it was found by DNA renaturation studies that eukaryotic DNA contains repeated sequences.

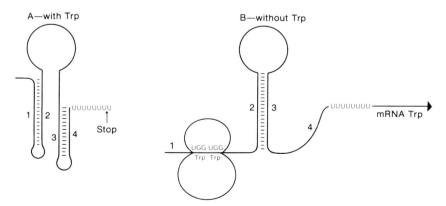

Fig. 22–5. Control of *trp* operon expression by transcription termination. *A*, in the presence of tryptophan, a loop required for termination (formed by segments 3 and 4) is formed and leader RNA is made. *B*, in Trp starvation a ribosome becomes stalled at two consecutive Trp codons, segment 2 base-pairs with segment 3, and transcription of full-length Trp mRNA can proceed.

TABLE 22–1. GENOME SIZE AND REPETITIVE DNA IN VARIOUS ORGANISMS

	Species	Base Pairs in Haploid Genome	Single Copy (%)	Moderately Repetitive (%)	Highly Repetitive (%)
Bacterium	*E. coli*	3.4×10^6	100	—	—
Fruit fly	*D. melanogaster*	1.4×10^8	74	13	13
Frog	*Xenopus laevis*	3.1×10^9	54	41	6
Mouse	*M. musculus*	2.7×10^9	70	20	10
Cow	*B. domesticus*	3.2×19^9	60	38	2

Highly Repetitive, Moderately Repetitive, and Single-Copy DNA

Table 22–1 shows that the genomes of eukaryotic cells contain much more DNA than *E. coli*. For example, a bull spermatozoon (haploid) contains 3.2×10^9 base pairs, which corresponds to a 700-fold increase in information relative to the size of the *E. coli* genome. Human cells contain a similar amount of DNA.

As we have seen in Figure 2–6, the size of a genome can be estimated by denaturing the DNA and allowing it to reanneal: the larger the genome, the longer it will take homologous sequences to reanneal. The speed of reassociation is measured in *Cot* units or *concentration* × *time* (expressed in nucleotide moles × seconds/liter). Britten and Kohne measured the reassociation rate of eukaryotic DNA and found that the curves did not have a simple sigmoid shape as had been found with *E. coli* DNA, but rather had multiple components.[5] Figure 22–6 shows that about 60% of bovine DNA reassociates very slowly (*Cot* between 10^2 and 10^4); this is the *single-copy DNA*. A second component of about 38% of the DNA renatures more quickly (*Cot*

between 10^{-2} and 10); this is the *moderately repetitive DNA*. An additional component of 2% of bovine DNA renatures too quickly to be distinguished in this experiment and constitutes the *highly repetitive DNA* (Table 22–1). The proportion of reassociated DNA can be measured by several procedures, but the one used often is hydroxylapatite chromatography, which retains double-stranded DNA but not single-stranded DNA.

Figure 22–7 shows the components of mouse DNA measured by renaturation kinetics, but plotted in a different way. Ten percent of the total was found as short nucleotide sequences that were repeated in about 1 million copies. Another 20% of the genome consisted of repeated sequences of 1000 to 100,000 copies; the remaining 70% was represented in only one copy.

The three classes of DNA contain the following elements:

1. *Highly repetitive DNA* contains short, simple sequences that are generally species-specific (*satellite DNAs*).

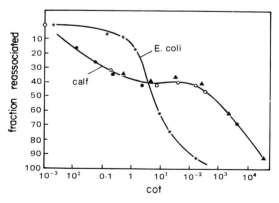

Fig. 22–6. Reassociation of calf thymus DNA compared to that of *E. coli* DNA. Calf DNA has several components, but *E. coli* gives a smooth curve. (From Britten, R.J., and Kohne, D.E.: Repeated sequences in DNA. Science, *161*:529, 1968.)

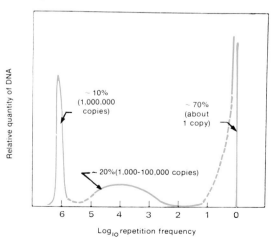

Fig. 22–7. Frequency of repeated DNA sequences present in DNA from a mouse. (See the text for details.) (From Britten, R.J., and Kohne, D.E.: Repeated sequences in DNA. Science, *161*:529, 1968.)

2. *Moderately repetitive DNA* contains (a) moderately repetitive genes arranged in tandem arrays (5S genes, rDNA, histone genes), (b) chromosome functional elements such as telomeres and centromeres and (c) mobile genetic elements or *transposons*, which are able to "jump" into other places of the genome and which are the main constituent of the middle-repetitive sequences.

3. *Single-copy DNA* contains protein-coding genes with their intron-exon structure and also a variety of *pseudogenes* and *processed* genes.

We will now consider these three classes of DNA in more detail. As we will see, the eukaryotic genome is very flexible, and segments of DNA are being rearranged almost constantly. This plasticity of the genome is probably very important in accelerating evolution.

Satellite DNAs Contain the Most Highly Repetitive Sequences

The most highly repetitive DNAs in eukaryons are also called *satellite* DNAs because they can frequently be separated from the bulk of the DNA by equilibrium centrifugation in cesium chloride, as shown in Figure 22–8. Cesium chloride gradients separate DNAs according to their buoyant densities, which in turn depend on their base composition; DNA rich in AT is less dense than GC-rich DNA. All eukaryotes contain satellite DNAs, which are present in many thousands or millions of copies per cell.[6] The heterochromatin regions located near the centromeres have been shown to contain satellite DNA in a variety of organisms. This fact has been established by hybridizing labeled nucleic acids in situ to chromosome preparations (Fig. 13–28).[7] The centromeric heterochromatin is particularly obvious in *Drosophila virilis*, a species with large heterochromatic segments flanking the centromeres of mitotic chromosomes and with approximately half its DNA devoted to three main satellite DNAs. *D. melanogaster* has smaller heterochromatic regions and consequently less satellite DNA.

In general, satellite DNAs are found in the pericentromeric regions of all the chromosomes of a cell, but sometimes preferential localizations

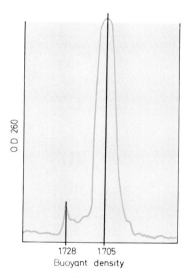

Fig. 22–8. Density gradient in cesium chloride of the DNA extracted from cells of the salamander. Observe the main peak of DNA with a buoyant density of 1.705 and the small satellite peak of higher density (1.728). DNA concentration was measured by its optical density (O.D.) at 260 nm. From MacGregor, H.C., and Kezer, J.: Chromosoma, 33:167–182, 1971.)

can be observed,[8] as in human chromosome Y, which is mostly heterochromatic and has a special satellite DNA found only in male cells.[9]

Satellite DNAs are formed by short nucleotide sequences that are tandemly repeated (head to tail) in DNA. The three satellites of *D. virilis* have a very simple base composition, each resulting from repeating units of seven nucleotides.

Satellite	Nucleotides
I	—ACAAACT—
II	—ATAAACT—
III	—ACAAATT—

All three basic sequences are related, differing by only one nucleotide. Because the repeating sequence is so short, it is repeated many times. For example, *D. virilis* has 12 million copies per haploid genome of the 7-nucleotide repeat unit of satellite I (25% of the total DNA content).[10]

The function of the highly repetitive DNAs is unknown, and most of them are not transcribed into RNA. From an evolutionary standpoint, satellite DNAs tend to change rapidly; there is little selective pressure to keep the sequences constant, and as a result satellites vary widely among species.

A major problem in the study of the evolution

of repeated DNAs is to explain how they evolve together. In other words, how is it that millions of repeats in the genome change their sequence in a coordinated manner and not independently of each other? It has been suggested that this could occur by unequal crossing-over events, which are particularly frequent in repeating sequences, and as shown in Figure 22–9, lead to duplication or deletion of repeating units.[11,12]

A second mechanism, called *gene conversion*, might help explain how sequences that are not tandemly repeated but rather dispersed throughout the genome might evolve together. Genetic experiments in yeast have shown that nonreciprocal exchanges might occur between genes on different chromosomes by mechanisms analogous to recombination.[13,14] An example of gene conversion is provided by the mating-type switch in yeast (Fig. 23–28).

Moderately Repetitive DNA—rDNA, 5S DNA, Histone Genes, Telomere and Centromere DNA

In Chapter 21 we saw that some genes are present in multiple copies. For example, the frog *Xenopus laevis* has 450 ribosomal RNA (rRNA) genes and 20,000 5S RNA genes. These genes are clustered together in the chromosomes, separated from each other by segments of *spacer* DNA. The spacers themselves consist of tandem repeats of shorter DNA sequences.

Most protein-coding genes (such as hemoglobin and ovalbumin) are present only once per haploid genome, but some are reiterated. The genes coding for the 5 histone proteins are repeated about 400 times in the sea urchin and between 10 and 20 times in humans.

Cells have five histone proteins. The genes for all five are arranged in repeating clusters, separated by segments of spacer DNA of a higher AT content than the histone genes. The sea urchin histone genes have been cloned in *E. coli* plasmids. In Figure 22–10 one such set of genes has been partially denatured to reveal the AT-rich spacers and the five genes. Histone genes tend to be clustered in a few chromosome loci, where many of the five-gene units are tandemly repeated (head to tail). Sea urchins synthesize enormous amounts of histone mRNA during early development in order to keep pace with the rapid DNA replication. The 400 tandemly repeated gene units are turned on during this period. Later on in somatic cells, however, individual groups of genes interspersed within the genome are used for histone synthesis. In *Drosophila* all the histone genes are located in a single chromosome band. The fact that the histone genes are clustered together suggests that their expression is regulated in a coordinated manner.

Tandem repeats of genes, alternating with stretches of spacer DNA, seem to be a central feature in the organization of the eukaryotic genome.

Centromere (Fig. 13–21) and telomere (Fig. 13–22) DNA has been cloned from yeast cells (Chapter 13); sequences from different chromosomes cross-hybridize with each other, which places them in the moderately repetitive category.

Transposons Are the Main Component of Moderately Repetitive DNA

Segments of DNA that can "jump" to new places in the genome were first discovered by

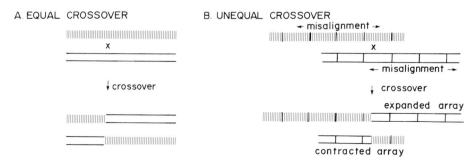

Fig. 22–9. Possible mechanism for the evolution of repetitive DNAs by unequal crossing-over. *A*, in equal crossing-over the DNA molecules are correctly aligned. *B*, in unequal crossing-over tandem repeats (shown by the vertical lines) become misaligned. This misalignment results in recombinant molecules having duplications or deletions of repeating units. By this mechanism mutations in repeating sequences can spread throughout the genome.

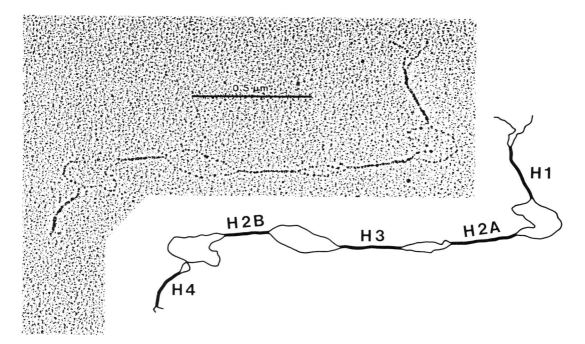

Fig. 22–10. Electron micrograph and diagram of a partially denatured DNA molecule containing the five histone genes. The genes are separated by spacers that are AT-rich and therefore denatured at a lower temperature than the gene regions. This molecule was cloned in *E. coli* and cut with a restriction enzyme. In the cell, many units of these five genes are tandemly repeated. (Courtesy of R. Portmann and M.L. Birnstiel.)

Barbara McClintock in the late 1940s. She studied strains of Indian corn (maize) that had cobs with kernels of different colors. On the basis of cytogenetic observations, she proposed that the light-colored kernels were caused by a segment of DNA that jumped into the gene coding for the pigmented appearance, thus inactivating it.[15] This correct proposal was met with complete disbelief by the scientific community and ignored for the next 20 years, until similar findings were made in *E coli*.

Transposable elements or *transposons* are stretches of DNA that usually have repeated DNA segments at their ends and can jump within the genome. (They may have either direct or inverted repeats, depending on the element.) In the hypothetical transposon shown in Figure 22–11, two "long" repeats flank the gene coding for a recombination enzyme called *transposase*, which specifically recognizes the long terminal repeats and helps them to insert into a new place in the genome. Sometimes an isolated repeat ("solo" sequence; see Fig. 22–11) is left in the DNA at the place where a transposon used to be inserted.

The telltale sign of a transposon is that it is flanked on both sides by short, direct repeats of the DNA sequences into which it is inserted. This duplication is due to the mechanism of insertion, shown in Figure 22–12, in which staggered cuts are made by the transposase and subsequently repaired. The number of repeated nucleotides is usually uneven and fixed for a given transposon (e.g., 5, 7, or 9 nucleotides). As we shall see later, transposons have striking similarities with a group of RNA viruses called *retroviruses*. The similarities are so extensive that one view is that retroviruses are highly successful transposons that have learned to jump from animal to animal.

In *Drosophila melanogaster* the majority of the moderately repetitive DNA is formed by 10 to 15 families of transposable elements. Each one may be present in 20 to 80 copies in the genome. We know that these sequences tend to jump around because in strains of fruit flies from different parts of the world, the transposons are found in different sites on the chromosomes. Circular DNA molecules that may correspond to transposons in the process of being inserted in the chromosomes have been isolated from *Drosophila* nuclei (see Fig. 22–11). The majority

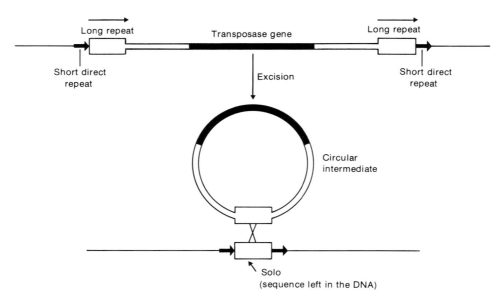

Fig. 22–11.   Excision of a hypothetical transposon, leaving behind a "solo" sequence in the DNA. The circular molecule can jump into a different place in the genome. The transposons *copia* from *Drosophila* and *Ty-1* from yeast behave in this way.

of *Drosophila* spontaneous mutations are caused by transposons jumping into a gene. The first mutant isolated in 1910, a white-eyed fly, turned out to have a transposable element inserted within the *white* gene, which, despite its name, codes for the enzyme that makes the *red* eye pigment.

In humans the most abundant mobile element is called the *Alu* family. In the human genome there are about 300,000 copies of a sequence of 300 nucleotides that contains a site for cutting by the restriction enzyme *Alu* I. The sequence is flanked by direct nucleotide repeats generated as shown in Figure 22–12.[16] *Alu* sequences are interspersed in the human genome, one every 10,000 nucleotides of DNA or less. Many of them are located within introns. Alu sequences make up 3 to 6% of the human genome.

The constant shuffling of Alu sequences may have important consequences. For example, deletions of nonessential DNA regions flanked by Alu sequences are known to occur frequently,[17] and recombination between Alu sequences in introns from different genes might lead to new genes created by exon reshuffling. Transposons may also lead to gene duplications. Similarly, any stretch of DNA flanked by Alu sequences may be moved to other positions in the genome. Because only 1% of the mammalian DNA is thought to code for sequences expressed as proteins, most of the remaining 99% might tolerate insertions and deletions and, in fact, may be derived from, and maintained by, a continuous series of such events (Sharp, 1983; Schmid and Jelinek, 1982).

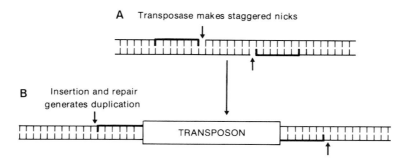

Fig. 22–12.   Transposons are flanked by duplications of the target DNA. *A,* taggered cuts are made in the DNA by the insertion enzyme (transposase). *B,* the duplication results from filling-in (repair) of the gaps generated in DNA.

Single-Copy Genes and Pseudogenes

Single-copy DNA makes up 60 to 70% of the mammalian genome. It contains most protein-coding genes and their introns. Genes present two or three times in the genome are also considered essentially single-copy because they reassociate together with this DNA fraction. Frequently, protein-coding genes constitute small multigene families, with some genes being expressed preferentially in some tissues or at certain developmental stages. The genes coding for the various forms of actins, tubulins and globins are examples of multigene families.

How do cells make new genes? From what we now know from eukaryotic genome organization, it is clear that cells do not start from scratch (i.e., a random sequence of DNA). One of the main mechanisms is *exon shuffling*, by which functional units of two existing genes are recombined, generating a new one (Fig. 20–17). Most new genes start with a duplication event, and then the second gene copy can be mutated to produce a different protein. This happened with α- and β-globins, which evolved from a common ancestral gene. Transposable elements may help establish the initial duplication required to make a new gene.

The enormous genomic flexibility introduced by transposable elements should be of fundamental importance in evolution. Sometimes duplicated genes do not succeed in making a functional product, and they become *pseudogenes*, or dead by-products of evolution. These pseudogenes, which are relatively frequent in the mammalian genome, can be identified because they have undergone mutations that render them unable to produce functional mRNA. For example, a pseudogene might contain premature chain termination codons, insertions, or deletions that cause the message to be read out of frame. Most frequently the promoter of pseudogenes is inactive. Because there is no selective pressure to maintain a functional gene product, in due course pseudogenes accumulate multiple mutations and eventually become random sequence DNA.

Processed Genes Arise by Reverse Transcription of mRNA and Insertion in the Genome

The genome also contains gene-like sequences that cannot have arisen from gene du-

plication. These are called *processed genes* and are considered to come from processed mRNA fragments that have been copied by reverse transcriptase into DNA and reinserted into the genome, usually flanked by the short, direct repeats that are the hallmarks of transposases. Processed genes have been found for immunoglobulins,[18] globin[19] and β-tubulin[20] and analysis of their sequence bears testimony that they came from mRNA. Pseudogenes can be distinguished from normal genes by their DNA sequence if one of the following signs is found: (1) introns have been removed precisely, as when mRNA is spliced; (2) a run of polydeoxy-A is found at the 3′ end, resulting from copying the poly A tail (which is normally not encoded in the genome); and (3) the sequence ends abruptly at the 5′ end of the mRNA. In a pseudogene for α-tubulin, all three characteristics were found.[21]

Processed genes are probably innocuous because they lack promoters and cannot be transcribed. On the other hand, because they code for protein-coding reading frames, they could be useful as potential building blocks for new proteins. Alternatively, they might simply be the scars left in the DNA by the activities of the retroviruses that all mammals have dormant in their genome.

22–3 RETROVIRUSES AND CELLULAR ONCOGENES

About Twenty Genes Can Produce Cancer in Humans, Acting on the Same Pathway of Cell Proliferation

Only a small subset of genes is responsible for the induction of the uncontrolled cell proliferation that leads to cancer. These genes were first identified from oncogenic RNA viruses called *retroviruses*. In the life cycle of these viruses genetic information is copied in the *retrograde* direction (RNA → DNA) by reverse transcriptase and then inserted in the genome (hence the name).

In 1910 Rous sarcoma virus was shown to produce tumors in chickens. As Figure 22–13 shows, a retrovirus has only three genes required for its life cycle and has terminal repeats at its ends, which resemble very much those found at the ends of transposons (Fig. 22–11).

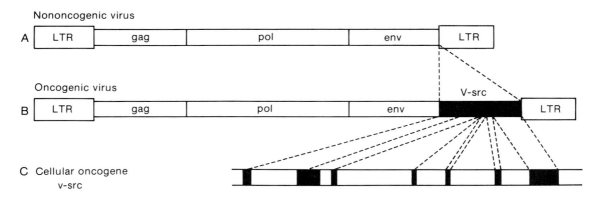

Fig. 22–13. Structure of *(A)* a retrovirus, *(B)* a retrovirus that has picked up oncogene *v-src* (Rous sarcoma virus), and *(C)* the cellular version of the same oncogene *(c-src),* containing multiple introns. The genes of the virus are *gag* (for group-specific antigen), *pol* (for reverse transcriptase *polymerase*), and *env* (for the *env*elope glycoprotein). LTRs, or long terminal repeats, are involved in integration into the genome and transposition. A single gene product *src* can produce sarcomas in chickens.

Some retroviruses, the oncogenic ones, carry an extra gene that induces cancer in animals. Many oncogenic viruses have been isolated, each one having one oncogene, and these have been shown to be able to carry over 20 different oncogenes (Table 22–2). Many independently isolated viruses have been found to carry the same gene, so it is believed that not many more cancer-producing genes remain to be discovered.

The *src* gene of Rous sarcoma virus (Fig. 22–13) codes for a plasma membrane-bound protein kinase that specifically phosphorylates the amino acid *tyrosine* (most phosphorylations normally occur in serine or threonine) on proteins on the cell membrane (one of them being *vinculin* at the cell adhesion plaques; see Figure 6–22). In 1968 Huebner and Todaro predicted that normal cells contained enemies within in the form of oncogenes similar to the viral ones, which could produce cancer when activated. When cDNA probes became available from the *viral src* gene, it was found that all cells contain a gene that is homologous in *v-src*, but that differs in that it has multiple exons and introns, *(c-src*, Fig. 22–13). Retroviruses pick up new genes by recombining with them at the RNA level, so that the introns have already been removed by mRNA splicing. This is analogous to what we saw in Section 22–2 for processed genes.

All 20 viral oncogenes derive from cellular genes in normal cells. The normal cellular version of the gene is called a proto-oncogene. Retroviruses pick up into their genome sequences from the mRNA population that will increase viral production (sometimes by increasing cell proliferation). These viruses can transmit genes from one species to another,[22] thus breaking evolutionary barriers (Bishop, 1983).

Another agent that stimulates proliferation, epidermal growth factor (Chapter 14), like *src*, increases phosphorylation of membrane proteins at tyrosines, and so it is thought that this tyrosine phosphorylation somehow controls cell proliferation. Cancer cells (also called transformed cells) contain ten times more phosphotyrosine than normal cells. About half of the oncogenes code for *Tyr* protein kinases (Table 22–2), and in all cases the proteins are integral components of the cell membrane, with a large extracellular region, a hydrophobic transmembrane domain, and an intracellular domain that has the tyrosine kinase activity. The function of these proteins in normal cells seems to be as receptors that transduce the signal provided by growth factor polypeptides bound to their extracellular domain, which stimulate their intracellular tyrosine kinase activity.

The *erb*-B oncogene has been found to be a truncated version of the gene coding for epidermal growth factor receptor. The tyrosine kinase domain in the *erb*-B viral oncogene, is intact, but its catalytic activity is increased because of a shorter extracellular domain. In the case of the *sis* oncogene (Table 22–2), the increased proliferation is induced in a different way: the *sis* gene codes for the platelet-derived growth factor (PDGF, Section 14–1) polypeptide. When this gene is overexpressed, the increase in PDGF

TABLE 22–2.   ONCOGENES ARE NAMED AFTER THE VIRAL STRAIN FROM WHICH THEY WERE ISOLATED

Oncogene*	Virus Strain	Probable Animal Origin of c Gene	Mode of Action
rel	Avian reticuloendotheliosis virus-T	Turkey	?
src	Rous sarcoma virus	Chicken	Tyr protein kinase
myb	Avian myeloblastosis virus strain BAI-A	Chicken	Nuclear Protein
myc	Avian myelocytomatosis virus MC29	Chicken	Nuclear Protein
erb-A	Avian erythroblastosis virus gene A	Chicken	Nuclear protein, DNA-binding region similar to glucocorticoid receptor
erb-B	Avian erythroblastosis virus gene B	Chicken	Tyr protein kinase, gene for epidermal growth factor receptor
fps	Fujinami sarcoma virus	Chicken	Tyr protein kinase
yes	Y73 avian sarcoma virus	Chicken	Tyr protein kinase
ros	UR2 avian sarcoma virus	Chicken	Tyr protein kinase
mos	Moloney murine sarcoma virus	Mouse	?
ras	Rasheed rat sarcoma virus	Rat	GTP-binding protein in plasma membrane
abl	Abelson murine leukemia virus	Mouse	Tyr protein kinase
fes	Snyder-Theilen feline sarcoma virus	Cat	Tyr protein kinase
fms	McDonough feline sarcoma virus	Cat	?
sis	Simian sarcoma virus	Woolly monkey	Platelet-derived growth factor gene

*The viral oncogenes are preceded by a *v* (e.g., *v-src*) and their cellular counterparts by a *c* (e.g., *c-src*).

stimulates cell proliferation (Waterfield et al., 1983). Even oncogenes that code for proteins that function inside the nucleus may be controlled by the growth factor pathway, as in the case of the *myc* and *fos* genes, whose transcription is greatly stimulated by adding PDGF to cultured cells (Kelly et al., 1983). One of the main conclusions of research with tumor viruses has been that most oncogenes seem to be related in one way or another with the same pathway of regulation of cell proliferation by protein growth factors. This represents great progress, and gives hope that one day soon we will learn how to control this common pathway and with it the proliferation of cancer cells.

Human Cancers Have Mutated Oncogenes

The finding that the same oncogenes present in viruses of animal origin also are responsible for human cancer came from a different line of studies. If cultured cells are treated with DNA from human tumors, on occasion colonies that resemble cancer cells ("transformed cells") and have increased proliferation are obtained. Nontumor human DNA produces very few colonies. The test cell used was a mouse cell line called 3T3, which is itself on the verge of transformation, but is still unable to grow without attachment to a plastic surface, i.e., when suspended in agar. Transformed cells, on the other hand, are attachment-independent and grow happily even if suspended in agar.

Mouse 3T3 cells were treated with DNA (by addition of DNA precipitated with calcium phosphate to the culture medium) from human bladder, colon, or lung carcinomas, and transformed colonies were obtained. When DNA was taken from these colonies and 3T3 cells were transformed again with it, DNA consisting mostly of mouse DNA with very little human DNA was obtained. These human sequences were then cloned (for example, by screening recombinant phages with Alu sequence probes specific for human DNA). The finding was that the genes so isolated were indeed cellular oncogenes, homologous to the ones found in retroviruses. In the case of colon and lung carcinomas, the same oncogene *(c-ras)* was isolated (Cooper, 1982; Weinberg, 1982).

Human Cancer Arises by Multiple Mechanisms, Including Chromosome Translocations

Although retroviruses were of enormous importance in identification of the oncogenes, no direct evidence indicates that retroviruses are actually responsible for human cancer.[23] Tumors are clonal (i.e., derived from a single cell), and it is known that mutagens are in general potent carcinogens, the classic example being tobacco smoke condensate. In general, cancer is due to *increased* activity of a normal gene product encoded by an oncogene, which causes cells to proliferate without control, invading the whole organism. There are multiple ways in which oncogenes have been found to become activated:

1. By insertion of a transposable element containing a strong promoter in front of the oncogene (found in animal models infected with some retroviruses).
2. By point mutations that change a single amino acid in the protein sequence and change its enzymatic activity. The best known is amino acid 12 of *ras*, which is activated by a number of point mutations that change it into a number of different amino acids.
3. By an increase in the number of copies of an oncogene by DNA amplification (leading to special, homogeneously staining regions on chromosomes), as discussed in detail in Chapter 23–4 under DNA amplification.
4. By mutations in two different oncogenes. It has been shown in DNA transformation studies that frequently two cooperating oncogenes are required to produce transformation (Land et al., 1983). This finding is in line with what we know from how tumors arise in humans; statistical studies

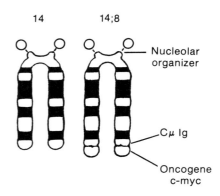

Fig. 22–14. In Burkitt's lymphoma the translocation of a small fragment of chromosome 8, containing an oncogene *(c-myc)*, to chromosome 14 at the immunoglobulin µ heavy chain locus (Cµlg), activates oncogene expression and produces cancer.

have shown that cancer is not a one-step phenomenon, but rather results from multiple mutations that are accumulated throughout our lifetime.

5. By chromosome translocations. This is perhaps the most amazing of all mechanisms, because it explains, in molecular terms, long-standing observations. It had been known for many years that some cancers carry specific chromosome translocations. The most famous one is the *"Philadelphia chromosome,"* a small chromosome resulting from a translocation of a fragment of chromosome 9 into chromosome 22. A Philadelphia chromosome is found almost invariably in cells from chronic myelogenous leukemia, one of the most frequent cancers of the white blood cells in the adult. The chromosome 9 fragment carries the *c-abl* oncogene at the translocation site, which is expressed more actively in the Philadelphia chromosome than at its normal location (Rowley, 1983).

Figure 22–14 shows the translocation that is found in Burkitt's lymphoma. This type of cancer spreads in infectious form in Africa and correlates with a herpesvirus called Epstein-Barr virus (herpesviruses are DNA viruses, not retroviruses). In Caucasians the same virus produces a more benign disease called *infectious mononucleosis.*

In Burkitt's lymphoma cells a translocation of a chromosome 8 fragment containing oncogene *c-myc* takes place in the immunoglobulin heavy chain locus on chromosome 14.[24,25] Immunoglobulin genes undergo DNA rearrangements during differentiation (Fig. 23–30) in lymphoid cells, and it could well be that the virus simply increases the frequency of abnormal rearrangements. Interestingly, in mouse tumors of antibody-producing cells (called myelomas) induced by a completely different method (injection of mineral oil) the *c-myc* oncogene is also translocated either into the heavy chain or into the light chain immunoglobulin genes. In all cases the translocation increases the expression of the *myc* oncogene and results in cancer (Rowley, 1983; Croce and Klein, 1985).

One might envisage that in the future it will be possible to diagnose in individual patients which oncogenes have been turned on by the cancerous transformation. Perhaps one day we will have a treatment for each type of oncogene. This is now merely a hope, but the rational pursuit of anticancerous agents such as inhibitors of tyrosine protein kinases is already under way.

SUMMARY:

Gene Organization and Regulation

Prokaryotes

Regulation by enzyme induction occurs in many catabolic systems, in which the availability of a substrate stimulates the production of the enzymes involved in its degradation. The synthesis of the three enzymes required for lactose utilization, for example, is coordinately regulated by the presence of lactose in the culture medium. *Lac* operon expression is regulated by the *lac* repressor, which binds to a DNA segment called the operator and thus prevents RNA polymerase from binding to the promoter (Fig. 22–2). The affinity of repressor binding to the operator is regulated by the inducer (a derivative of lactose). When bound to the repressor, the inducer causes a conformational change in the repressor that prevents it from binding to the operator.

cAMP also has a regulatory role in bacteria. *E. coli* also contains a protein, CAP, which when complexed with cAMP is able to bind to a specific site within the *lac* promoter region. RNA polymerase recognizes the *lac* promoter only if the CAP-cAMP complex is already bound to it. This positive control mechanism is also required for the expression of many other inducible operons.

Regulation by enzyme repression occurs in many anabolic systems, in which the production of the enzymes of a synthetic pathway is selectively inhibited by the end product of that metabolic chain. The *trp* operon, for example, is repressed by the presence of tryptophan in the medium. In enzyme repression the regulatory gene produces a repressor protein that is normally inactive. Upon binding with a co-repressor, the repressor undergoes a conformational change that enables it to bind to the operator (Fig. 22–4).

The trp operon is also regulated at the level of termination of RNA synthesis. When the Trp level in the cell is low, the proportion of RNA molecules yielding *trp* mRNA is high; when the Trp level in the cell is high, most of the mRNAs terminate prematurely at an attenuator site. Termination at the attenuator site can be stopped by a ribosome stalled in the RNA in conditions of Trp starvation (Fig. 22–5).

Eukaryotes

Eukaryotes contain far more DNA than prokaryotes, and much of their genome is repeated. The discovery of repeated sequences was a consequence of the fact that separated complementary strands can recognize each other and reassociate; the greater the number of repeated sequences, the faster strand reassociation will occur. Three kinds of DNA can be distinguished by this method:

1. *Highly repetitive DNAs*, or satellite DNAs, can be separated from the rest of the DNA by centrifugation in CsCl gradients. Satellite DNAs are frequently located in heterochromatic regions of the chromosome and are formed by short nucleotide sequences that are tandemly repeated many times. Their function is not known, and their sequences vary widely among species.

2. *Moderately repetitive DNAs* include tandemly repeated genes such as 5S genes, rDNA, and histone genes. Tandemly repeated genes alternating with stretches of spacer DNA are a characteristic feature of the eukaryotic genome. Other types of moderately repetitive DNAs are telomeres, centromeres, and *transposons* (which are mobile genetic elements that make up the bulk of moderately repetitive DNA). Transposons are flanked by terminal repeats of DNA (Fig. 22–11) and jump into new sites in DNA with the help of specific transposases. Transposons are the most frequent cause of spontaneous mutation. In human the largest family, called Alu, has 300,000 members dispersed along the genome, which constitute 3 to 6% of the genome. Transposons give great flexibility to the eukaryotic genome.

3. *Single-copy DNA* includes most protein-coding genes and also *pseudogenes*, which are duplicated genes that have become inactivated by mutations and are considered dead by-products of evolution. In *processed genes*, gene-like sequences arise from mRNA copied into DNA by reverse transcriptase and inserted back into the genome.

Considerable progress in the study of cancer has been made using *retroviruses*, which are RNA viruses similar in many ways to transposons, and *transformation* of cultured cells with DNA. Twenty *oncogenes* have been identified. Oncogenes are normally expressed in cells, but their *increased* expression leads to cancer. Many of them are protein kinases that phosphorylate tyrosines, one is similar to a growth factor polypeptide, and another one resembles growth factor receptor (Table 22–2). Multiple mechanisms lead to activation of oncogenes in human cancer, one of them being their translocation to other chromosomal sites, where their expression increases (Fig. 22–14).

22–4 POLYTENE CHROMOSOMES

Polytene chromosomes provided the first evidence that eukaryotic gene activity is regulated at the level of RNA synthesis. They constitute a valuable material for the study of gene regulation because their gene transcription can be visualized directly in the microscope.

Polytene Chromosomes Have a Thousand DNA Molecules Aligned Side by Side

Some cells of dipteran (flies, mosquitoes, midges) larvae become very large and have a high DNA content. The most prominent ones are located in the salivary gland (Fig. 22–15), but other cells from the gut, fat body, and malpighian tubules of the larva also become *polytenic*. Polyteny is achieved by replication of the DNA several times without nuclear division (*endomitosis*); and the resulting daughter chromatids do not separate and remain aligned side by side. (Polyteny differs from polyploidy, in which there is also excess DNA per nucleus, but in which the new chromosomes are separate from each other.)

A polytene chromosome of *Drosophila* salivary glands has about 1000 DNA molecules arranged side by side which arise from 10 rounds of DNA replication ($2^{10} = 1024$). Other dipteran species have more DNA molecules per polytene chromosome; for example, *Chironomus* has 16,000.

In polytene cells the chromosomes are visible during interphase, and the *chromomeres* (regions in which the chromatin is more tightly coiled) alternate with regions where the DNA fibers are folded more loosely. The alignment of many chromomeres gives polytene chromosomes their characteristic morphology, in which a series of dark *bands* alternates with clear zones

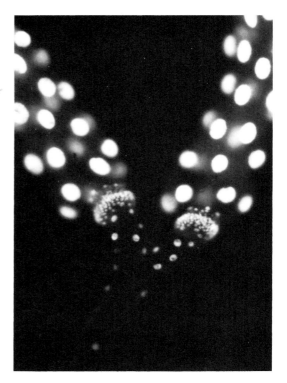

Fig. 22–15. Salivary glands of a *Drosophila melanogaster* larva showing variable degrees of polyteny. The cells were stained with ethidium bromide, a fluorescent compound that binds to DNA. Note that the secretory cells *(top)* are larger and contain more DNA than those of the excretory duct *(bottom)*. The rings of small nuclei located between the excretory ducts and the secretory cells are the *imaginal discs*, which are diploid and destined to produce the salivary glands of the adult fly. (Courtesy of M. Jamrich.)

called *interbands* (Fig. 22–16). There are about 5000 bands in the *Drosophila* genome. They have characteristic morphology and positions, which permit detailed chromosome mapping.

An additional characteristic of polytene chromosomes is that the maternal and paternal homologues remain associated side by side, in what is called *somatic pairing* (Fig. 22–17). This permits the identification of deletions, inversions, and duplications as regions looped out of the chromosomes (see inversion loop on right arm of chromosome 3 in Fig. 22–17). The pericentromeric heterochromatin of all the *Drosophila* chromosomes coalesces in a *chromocenter*, where the chromosomes are joined together (Fig. 22–17). The satellite DNAs of the chromocenter are underreplicated with respect to the rest of the chromosome (i.e., they undergo fewer rounds of replication).

Polytene cells are unable to undergo mitosis

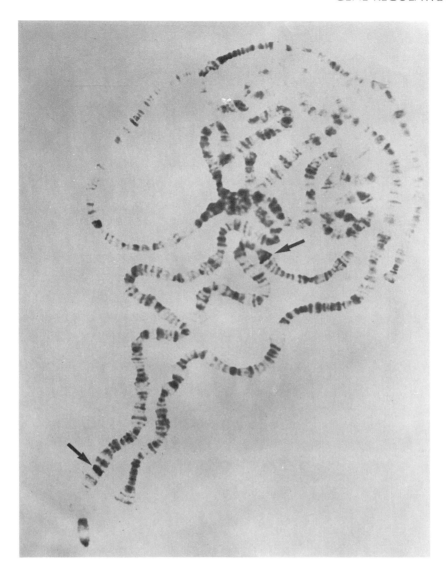

Fig. 22–16. In situ hybridization of a cloned *white* gene to polytene chromosomes of *Drosophila melanogaster*. The chromosomes of a larval salivary gland cell were squashed between coverslip and slide, the coverslip was removed, and the chromosomes were submitted to in situ hybridization and finally stained with Giemsa stain. Note the bands, interbands, and puffs. All the chromosomes are joined together at the *chromocenter*.

  In this new method of hybridization, the DNA on the chromosomes was denatured and then incubated with cloned *white* DNA that had been labeled with deoxynucleotides that have a covalently bound *biotin* molecule. (*Biotin* is a vitamin whose detection with antibodies and a specific binding protein called *avidin* is extremely efficient.) After washing away the nonbound probe, the biotin on the hybridized DNA was detected using an antibiotin antibody coupled to horseradish peroxidase. Two black bands of peroxidase activity were detected and are indicated with *arrows*. The lower hybridization band is at the normal position of the *white* gene; the other white gene is in a new location, where it was introduced by transposon-mediated gene transfer (Fig. 22–20). (Courtesy of W. Gehring.)

and are destined to die. Not all the cells in a dipteran larva have polytene chromosomes; those destined to produce the adult structures after metamorphosis *(imaginal discs)* remain diploid. Figure 22–15 shows a ring of small, diploid imaginal cells that will give rise to the adult salivary glands.

Polytene chromosomes have become even more important with the advent of recombinant DNA techniques because they make it possible to map any DNA segment to specific chromosomal loci by in situ hybridization. Polytene chromosomes are very suitable for in situ hybridization because their 1000 DNA molecules

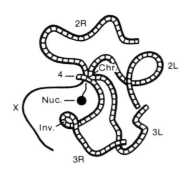
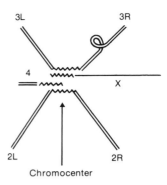

Fig. 22–17. In *Drosophila,* maternal and paternal homologues remain paired and joined together at the *chromocenter* (pericentromeric heterochromatin). *Left,* drawing of squashed chromosomes by M. White, 1942. *Right,* diagrammatic interpretation. This fly was heterozygous for an inversion in the right arm of chromosome 3, as denoted by the inversion loop. The autosomes 2 and 3 have right *(R)* and left *(L)* arms. *Chr,* chromocenter; *Inv.,* inversion; *Nuc,* nucleolus.

are aligned side by side, thereby greatly facilitating the detection of single-copy genes (Fig. 22–16).

Puffs Are Sites of Intense Gene Transcription

One of the most remarkable characteristics of polytene chromosomes is that it is possible to visualize in them the genetic activity of specific chromosomal sites at local enlargements called *puffs* (Fig. 22–18). A puff can be considered a band in which the DNA unfolds into open loops as a consequence of intense gene transcription.

In 1952 W. Beerman compared the polytene chromosomes of different tissues of *Chironomus* larvae and showed that although the pattern of bands and interbands was similar in all tissues, the distribution of puffs differed from one tissue to another.[26] He correctly interpreted the puffing of polytene chromosomes as an expression of intense gene transcription.

In salivary glands the appearance of some puffs has been correlated with the production of

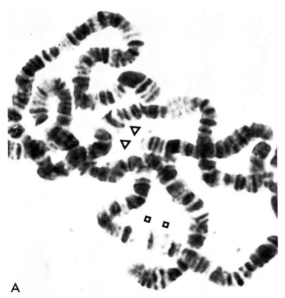
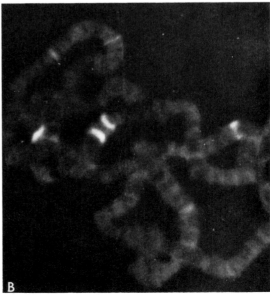

Fig. 22–18. Localization of RNA polymerase II in heat-shock puffs. Larvae were placed at 37°C for 45 minutes before preparation of chromosomes. *A,* orcein stain; *B,* immunofluorescence. Triangles indicate the largest heat-shock puffs, 87 A and 87 C. The squares indicate pre-existing puffs. RNA polymerase II accumulates in the heat-shock puffs, while it disappears from other regions of the chromosome (even from the pre-existing puffs). (Courtesy of E.K.F. Bautz.)

specific proteins which are secreted in large amounts in the larval saliva (see Grossbach, 1977). For example, *Chironomus* has at the base of the salivary glands four specialized cells that contain cytoplasmic granules of a special secretory protein. The gene for this protein is located in a distinct puff that appears only in the four specialized cells. These results clearly show that cell specialization results from variable gene transcription.

Puffs Can Be Induced by Ecdysone and by Heat Shock

Puffing is a cyclic and reversible phenomenon; at definite times, and in different tissues of the larvae, puffs may appear, grow, and disappear. Puff formation can be studied experimentally using factors that will induce their formation. The steroid hormone *ecdysone*, which induces molting in insects, will induce the formation of specific puffs when injected into larvae or when added to salivary glands in culture.

Puffing can also be induced by temperature shock (Fig. 22–18). When *Drosophila* larvae, normally grown at 25° C, are exposed to a temperature of 37° C, a series of specific genes is activated, while most other genes are repressed. Five minutes after heat shock nine new puffs are seen on the giant chromosomes of the salivary glands. These puffs are active in RNA synthesis, as can be demonstrated by ^{3}H-uridine labeling and radioautography (Fig. 22–19b and c). The newly made RNA is released from the puffs (Fig. 22–19d and e) and accumulates in the cytoplasm. New kinds of mRNA can be isolated from heat-treated cells, and these mRNAs give rise to eight specific heat-shock proteins (Ashburner and Bonner, 1979). It is clear, once again, that in this experimental situation the induction of specific proteins is due to an increased rate of transcription of individual genes (Pelham and Bienz, 1982).

One of the earliest events in the induction of heat-shock puffs is the accumulation of RNA polymerase II in the puff loci. RNA polymerase can be located on polytene chromosome preparations by immunofluorescence using specific antibodies.[27] When cells are placed at 37° C, RNA polymerase II rapidly accumulates in the heat-shock puffs, while it disappears from other

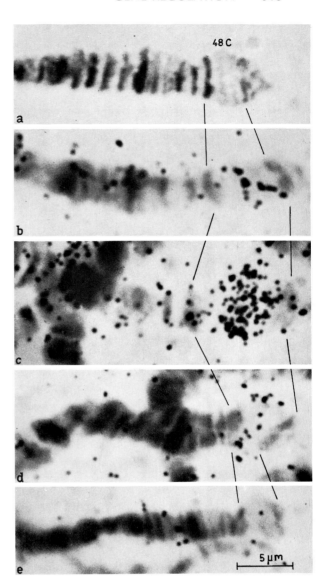

Fig. 22–19. Experiment demonstrating how ^{3}H-uridine is incorporated into puff 48 C of *Drosophila hydei* during heat. shock. *a,* control; *b,*after 3 minutes of temperature shock (a few silver grains have been deposited); *c,* 15 minutes after temperature shock. Observe the greatly increased incorporation; *d,* 15 minutes after injection of actinomycin D in larvae having puffs as in *c; e,* 30 minutes after injection of the antibiotic in animals with puffs as shown in *c.* These results indicate that in 15 to 30 minutes most of the synthesized RNA has been eliminated from the puff. (Courtesy of H.D. Berendes.)

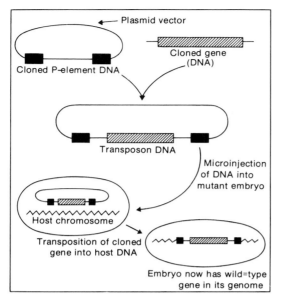

Fig. 22–20. Gene transfer into *Drosophila* chromosomes using the *P-element* transposon.

regions of the chromosome, as shown in Fig. 22–18.

Microinjection of a Transposon Permits the Transfer of Genes into Different Sites in the *Drosophila* Chromosome

An important question in gene regulation is, what is the effect of the surrounding chromosomal regions on the activity of genes? It could be, for example, that chromatin is organized in long looped domains of DNA, some of which would be active in one tissue and not in others (Chapter 13). In *Drosophila* one can now directly test these questions because Allan Spradling and Gerry Rubin have devised a technique by which cloned genes can be introduced with high frequency into the chromosomes.[28,29]

The procedure is based on using a *Drosophila* transposon, called the *P-element*, to carry cloned genes into the chromosome, as shown in Figure 22–20. The recombinant DNA is introduced by by microinjection into the posterior pole of an early embryo (Fig. 22–21), which is the site where the germ cells form. Injection at this site maximizes the chances that integration will be into the germ line, and thus the recombinant DNA will be inherited by the progeny of the fly. Genetic markers are used to select the flies in which integration has taken place. For example,

white-eyed flies are used as recipients if the injected gene is the one coding for red eye pigment (the gene is called *white*, even though its normal function is to produce red eyes). Some of the resulting flies may have red eyes, and these are then bred. Some individuals are then screened by in situ hybridization of the injected cloned gene to denatured DNA on squashed polytene chromosomes to see at which location the new gene has been inserted. In the example shown in Figure 22–16 two bands can be seen hybridizing to the *white* gene: one was introduced with the P-element, and the other is at the normal *white* locus (which was a mutated inactive gene in the recipient white-eyed flies, but is able to hybridize because it still contains gene sequences).

Several genes have been introduced in this way at multiple sites in the chromosomes.[30–32] These studies have shown that when a gene is inserted, in general it is active in the correct tissue-specific way, regardless of the chromosomal environment it is in. (There is one important exception, however; genes integrated in or near the pericentromeric heterochromatin tend to be inactivated.) Thus these studies suggest that long chromosomal domains which could be inactivated in some tissues are not a decisive factor in gene regulation during cell differentiation in *Drosophila*.

22–5   LAMPBRUSH CHROMOSOMES

Lampbrush Chromosomes Are Found in Oocytes at the Diplotene Stage of Meiosis

Lampbrush chromosomes were first observed by Flemming in 1882 and were described in detail in shark oocytes by Ruckert in 1892. He coined the name because the chromosomes look like the brushes used in those times to clean the chimneys of oil lamps. It is only more recently, however, that the structure of these chromosomes has been interpreted in functional terms, largely through the work of Gall and Callan.

Lampbrush chromosomes occur at the diplotene stage of meiotic prophase in oocytes of all animal species, in spermatocytes of several species, and even in the giant nucleus of the unicellular alga *Acetabularia*.[33] Lampbrush chromosomes have many fine lateral projections

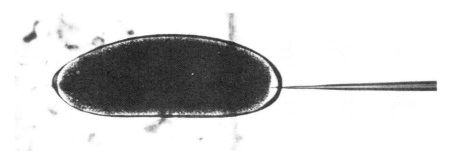

Fig. 22–21. Injection of DNA into a *Drosophila* embryo. The length of the embryo is approximately 0.5 mm. The tip of the needle has been introduced into the posterior pole of the embryo, the site of germ-cell formation. (Courtesy of G. Rubin and A. Spradling.)

(Fig. 22–22), giving them the characteristic "hairy" appearance. They are best visualized in salamander oocytes because they have a high DNA content and therefore very large chromosomes.

Because lampbrush chromosomes are found in meiotic prophase, they are present in the form of *bivalents* in which the maternal and paternal chromosomes are held together by chiasmata at those sites where crossing-over has previously occurred (Fig. 22–22A). Each bivalent has four chromatids, two in each homologue. The axis of each homologue consists of a row of granules or chromomeres (best seen in Figure 22–22D and Figure 22–24A), from which lateral loops extend. The loops are always symmetrical, each chromosome having two of them, one for each chromatid (two additional symmetrical loops will be found on the other homologue because cells in prophase of meiosis have a 4C DNA content—equivalent to a tetraploid cell) (Fig. 22–22A).

The loops can be distinguished by size, thickness, and other morphologic characteristics. Each loop appears at a constant position in the chromosome, and detailed chromosome maps can be drawn. There are about 10,000 loops per chromosome set. Each loop has an axis formed by a single DNA molecule that is unfolded from the chromosome as a result of intense RNA synthesis. About 5 to 10% of the DNA is in the lateral loops, the rest being tightly condensed in the chromomeres of the chromosome axis, which are transcriptionally inactive.

Lampbrush Chromosomes—the Unineme Theory and the C-Value Paradox

We know that each loop has just one DNA molecule because of two observations: (1) as

shown in Figure 22–23, they are extremely rapidly fragmented into linear fragments by digestion with DNase, and (2) sometimes lampbrush chromosomes become stretched during laboratory preparation and the base of the loop accidentally opens up, in what is called a *double bridge*. Figure 22–24 shows one such double-bridge loop; it may be observed that the chromosome fragments are still connected by two individual filaments, each one corresponding to a linear DNA molecule that extends throughout the length of the chromosome. This may now seem a trivial thing, but in the 1950s it was a major finding because it showed that a single thread of DNA runs through each chromatid and led to the elaboration of the *unineme* view of chromosome structure (Chapter 13).

The loop is covered by a matrix that consists of RNA transcripts with hnRNA-binding proteins attached to them. In general the ribonucleoprotein matrix is asymmetrical, being thicker at one end of the loop than at the other (Fig. 22–24). RNA synthesis starts at the thinner end and progresses toward the thicker end. Preparations spread for electron microscopy show the typical "Christmas tree" images with nascent ribonucleoprotein chains attached perpendicularly to the DNA axis.[34] Many of the loops correspond to a single transcriptional unit and show a continuous thin-thick matrix (as in Fig. 22–22B and Fig. 22–24). Other loops contain several units of transcription.[35]

Some transcriptional units in lampbrush chromosomes are extremely long; the one shown in Figure 22–24 is over 100 μm in length (a useful rule to remember is that 1 μm of DNA = 3000 base pairs). Why they are so enormous is a mystery. Even more puzzling is the observation that

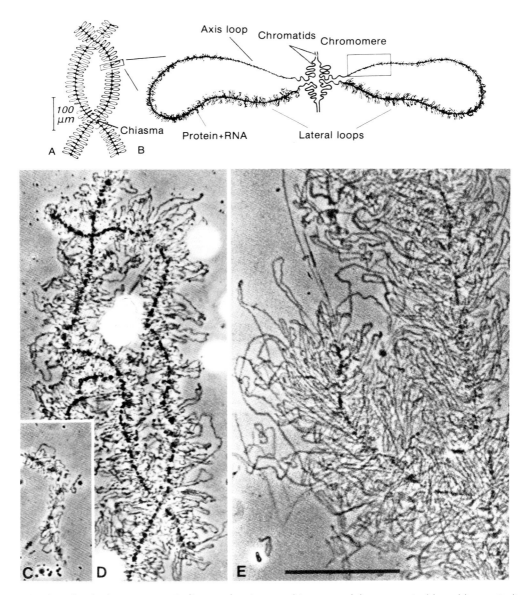

Fig. 22–22. Lampbrush chromosomes. *A,* diagram showing two chiasmata and the symmetrical lateral loops. *B,* diagram of region boxed in A, showing two lateral loops and the folding of the chromatids in the chromomere. *C, D,* and *E* are phase contrast photographs of lampbrush chromosomes of *Xenopus, Triturus,* and *Necturus,* respectively, all at the same magnification (bar = 50 μm), illustrating that loop size increases with the C-value (see Table 22–3). (*A* and *B* courtesy of O.L. Miller, Jr.: *C, D,* and *E* courtesy of U. Scheer.)

the length of the loops increases with the C-value of the species (Table 22–3 and Figure 22–22*C, D* and *E*). As a result, a salamander has transcriptional units 10 times longer than those of a frog, although both must code for similar gene products. We do not have an answer to this paradox, but one possibility is that it might be connected with inefficient termination of transcription in oocytes.

The Long Lampbrush Transcripts May Be Due to Failure of Termination

Lampbrush chromosomes are a very good material for in situ hybridization of cloned DNA to RNA because the thousands of nascent RNA molecules aligned side by side along the loop greatly amplify the hybridization signal (Fig. 22–24).[36] Studies with this technique have

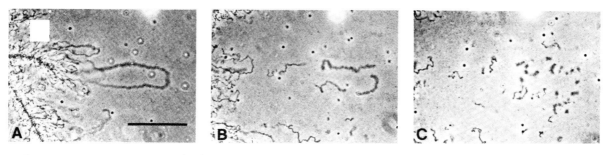

Fig. 22–23. Successive stages in the digestion of a newt lampbrush chromosome loop (chromatid) by pancreatic DNase I. Analysis of the kinetics of digestion indicates that the DNA axis of the loop consists of a single Watson-Crick double helix. Bar = 50 μm. (Courtesy of J.G. Gall.)

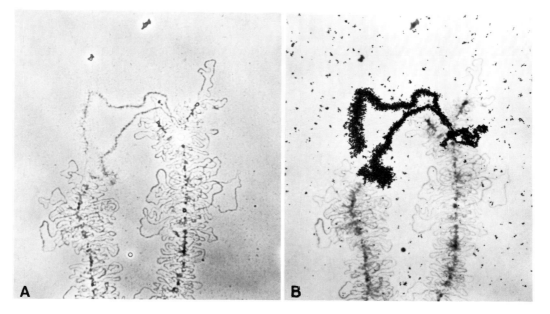

Fig. 22–24. *A,* phase contrast photograph of a long loop of newt chromosomes. *B,*same region after hybridization with a segment of cloned DNA and autoradiography. Note the thin-thick gradient of hybridization to the nascent RNA on the loop, which is over 100 μm long and has a single transcriptional unit. The loop was stretched during preparation and opened at its base, forming a *double bridge* on the chromosome. (Courtesy of Malalo Diaz and Joe Gall.)

TABLE 22–3. SIZE OF THE LAMPBRUSH LATERAL LOOPS INCREASES WITH THE GENOME SIZE

	C-value (pg of DNA)	C-value Increase	Lampbrush Loop Size
Xenopus laevis (South African clawed frog)	3.1	1	5–10 μm
Triturus cristatus (common European newt)	23	7.4	30–50 μm
Necturus maculosus (American "mudpuppy" salamander)	78	25.1	>100 μm

Data from Scheer, U., and Sommerville, J.: Exp. Cell Res., *139*:410, 1982. (See Fig. 22–22 for photographs of Xenopus, Triturus and Necturus lampbrush chromosomes.)

shown that in some loops simple-sequence satellite DNAs are transcribed,[37,38] which is unusual because satellite DNAs are normally not expressed (Section 22–2).

Joseph Gall and collaborators found that in the American newt *Notophthalmus viridescens*, clusters of the five histone genes are tandemly repeated in the genome, but separated by about 50,000 base pairs of repeats of a 225-base-pair satellite DNA (Figure 22–25).[39] The satellite DNA runs in the same orientation over long stretches of the chromosome, periodically interrupted by the histone gene clusters (Fig. 22–25, *IIIB*). When single-stranded DNA from this satellite was hybridized to the histone loops (Fig. 22–25, *II*), it was found that hybridization was discontinuous. Each loop had several thin-

thick gradients (transcriptional units), and a given strand of the satellite DNA always hybridized to the transcriptional unit of a given orientation, as indicated in Figure 22–25, *III*. (The opposite strand would hybridize to those on the other orientation.) The histone gene clusters have genes pointing in both orientations, each one with its own promoter (Fig. 22–25, *IIIC*). The hybridization results are due to initiation of transcription at a histone gene promoter that then fails to terminate normally. Enormous transcripts are produced which result from read-through into the satellite DNA. Eventually transcription stops when the next transcriptional unit is reached. This failure of termination may be generalized throughout the lampbrush chromosome and may explain the

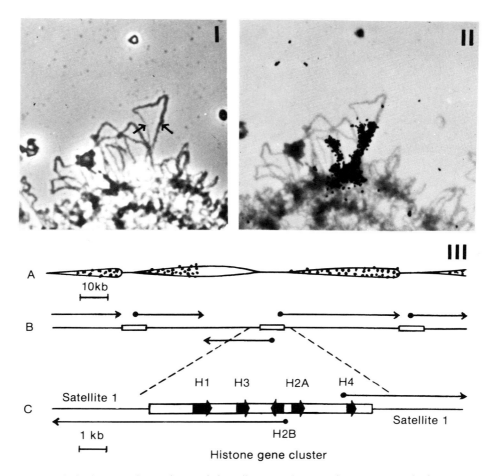

Fig. 22–25. In situ hybridization of a single-stranded satellite DNA fragment from a newt to the histone gene locus. *I*, phase contrast photograph before hybridization; *arrows* indicate places where the polarity of the matrix changes. *II*, hybridization after autoradiography. *III*, interpretation of the results as failure of termination. Transcription initiates at a histone gene promoter but reads through into the flanking satellite DNA. Transcripts in a certain orientation hybridize only to one DNA strand of the satellite DNA. See the text for details. (Courtesy of M. Diaz and J. Gall.)

long loops. (For a beautiful example of how genetic engineering and classical cytology are now coming together, see Diaz et al., 1981.)

The question remains, what is the function of these long transcripts? The majority of them are degraded in the nucleus, but presumably some of the RNAs have some function in preparing an oocyte for the difficult trip that the egg undertakes after fertilization: the development of a new organism.

Summary:
Polytene and Lampbrush Chromosomes

Polytene chromosomes provided the first evidence that eukaryotic gene activity is regulated at the level of RNA synthesis. When dipteran chromosomes become polytenic, the DNA replicates by endomitosis, and the resulting daughter chromatids remain aligned side by side. These chromatids are visible during interphase and have a characteristic morphology of dark bands and alternating interbands. Within these chromosomes it is possible to observe the genetic activity of specific loci at local enlargements called puffs, which represent DNA undergoing intense gene transcription. Puff distribution varies from one tissue to another and can be induced experimentally, indicating that cell specialization results from variable gene transcription.

Lampbrush chromosomes occur at the diplotene stage of meiotic prophase of all animal species. Highly condensed chromomeres form the chromosome axis, from which loops of DNA extend laterally as a result of intense RNA synthesis. Each loop has an axis formed by a single DNA molecule which is covered by a matrix of nascent RNA with hnRNA-binding proteins attached to it. Transcriptional units in oocytes may be extremely long—up to 100 μm (Fig. 22–22). Some experiments suggest that the long transcriptional units of lampbrush chromosomes may be due to a failure of transcription termination.

REFERENCES

1. Jacob, F., and Monod, J.: J. Mol. Biol., 3:318, 1961.
2. Bertrand, K., et al.: Science, 189:22, 1975.
3. Johnston, H., Barnes, W.M., Chumley, F.G., and Roth, J.: Proc. Natl. Acad. Sci. USA, 77:508, 1979.
4. Stroynowski, I. and Yanofsky, C.: Nature, 298:34, 1982.
5. Britten, R.J., and Kohne, D.E.: Science, 161:529, 1968.
6. Skinner, D.M.: Bioscience, 27:790, 1978.
7. Pardue, M.L., and Gall, J.G.: Science, 168:1356, 1970.
8. Appels, R., and Peacock, W.J.: Int. Rev. Cytol. (Suppl.), 8:70, 1978.
9. Cooke, H.J., and McKay, R.: Cell, 13:453, 1978.
10. Gall, J.G., and Atherton, D.D.: J. Mol. Biol., 85:633, 1974.
11. Petes, T.D.: Cell, 19:765, 1980.
12. Szostak, J.W., and Wu, R.: Nature, 284:426, 1980.
13. Klein, H.L., and Petes, T.D.: Nature, 289:144, 1981.
14. Ollo, R., and Rougeon, R.: Cell, 32:515, 1983.
15. McClintock, B.: Cold Spring Harbor Symp. Quant. Biol., 16:13, 1951.
16. Schmid, C.W.: Nature, 284:372, 1980.
17. Calabretta, B. et al.: Nature, 296:219, 1982.
18. Hallis, G.F., et al.: Nature, 296:321, 1982.
19. Flavell, R.A.: Nature, 295:370, 1982.
20. Wilde, C.D., et al.: Nature, 297:83, 1982.
21. Lemischka, I., and Sharp, P.: Nature, 300:330, 1982.
22. Busslinger, M., Rusconi, S., and Birnstiel, M.C.: EMBO J., 1:27, 1982.
23. Cairns, J.: Cancer: Science and Society. San Francisco, W.H. Freeman and Co., 1978.
24. Dalla Favera, R., et al.: Proc. Natl. Acad. Sci. USA, 70:7824, 1982.
25. Dalla Favera, R., et al.: Science, 219:963, 1983.
26. Beerman, W.: Chromosoma, 5:139, 1952.
27. Jamrich, M., Greenleaf, A.L., and Bautz, E.K.F.: Proc. Natl. Acad. Sci. USA, 74:2079, 1977.
28. Spradling, A.C., and Rubin, G.M.: Science, 218:341, 1982.
29. Rubin, G.M., and Spradling, A.C.: Science, 218:348, 1982.
30. Scholnick, S.B., Morgan, B.A., and Hirsh, J.: Cell, 34:37, 1983.
31. Spradling, A.C., and Rubin, G.M.: Cell, 34:47, 1983.
32. Goldberg, D.A., Posakony, J.W., and Maniatis, T.: Cell, 34:59, 1983.
33. Spring, H., Scheer, V., Franke, W.W., and Trendelenburg, M.F.: Chromosoma, 50:25, 1975.
34. Hamkalo, B.A., and Miller, O.L.: Annu. Rev. Biochem., 42:376, 1973.
35. Scheer, V., Franke, W.W., Trendelenburg, M., and Spring, H.: J. Cell Sci., 22:503, 1976.
36. Old, R.W., Callan, H.G., and Gross, K.W.: J. Cell Sci., 27:57, 1977.
37. Varley, J.M., Macgregor, H.C., and Erba, H.P.: Nature, 283:686, 1980.
38. Diaz, M.O., Barsacchi-Pilone, G., Mahon, K.A., and Gall, J.G.: Cell, 24:649, 1981.
39. Gall, J.G., et al.: Chromosoma, 84:159, 1981.

ADDITIONAL READINGS

Adhya, S., and Gargas, S.: How cyclic AMP and its receptor protein act in E. coli. Cell, 29:287, 1982.

Appels, R., and Peacock, W.J.: The arrangement and evolution of highly repeated (satellite) DNA sequences, with special reference to Drosophila. Int. Rev. Cytol. (Suppl.), 8:70, 1978.

Ashburner, M., and Bonner, J.J.: The induction of gene activity in Drosophila by heat shock. Cell, 17:241, 1979.

Beckwith, J., and Zipser, D.: The Lactose Operon. Cold Spring Harbor, New York, Cold Spring Harbor Laboratory, 1968.

Bishop, J.M.: Cancer genes come of age. Cell, 32:1018, 1983.

Bishop, J.M.: Oncogenes. Ann. Rev. Biochem., 53:301, 1983.

Bishop, J.M.: Oncogenes. Sci. Am., 246(3):69, 1982.

Britten, R.J., and Kohne, D.E.: Repeated sequences in DNA. Science, 161:529, 1968.

Brown, D.D.: Gene expression in eukaryotes. Science, 211:667, 1981.

Calos, M.P., and Miller, J.M.: Transposable elements. Cell, 20:579, 1980.

Cooper, G.M.: Cellular transforming genes. Science, 217:801, 1982.

Croce, C.M., and Klein, G.: Chromosome translocations and human cancer. Sci. Am., 252(3):54, 1984.

Daneholt, B.: Transcription in polytene chromosomes. Cell, 4:1,1975.

Davidson, E.H., and Britten, R.J.: Organization, transcription, and regulation in the animal genome. Q. Rev. Biol., 48:565, 1973.

D'Eustachio, P., and Ruddle, F.H.: Somatic cell genetics and gene families. Science, 220:919, 1983.

Diaz, M.O., Barsacchi-Pilone, G., Mahon, K.A., and Gall, J.G.: Transcripts from both strands of a satellite DNA occur in lampbrush chromosome loops of the newt *Notophthalmus*. Cell, 24:649, 1981. (Highly recommended.)

Flavell, A.J., and Ish-Horowicz, D.: Extrachromosomal circular copies of the eukaryotic transposable element copia in cultured *Drosophila* cells. Nature, 292:591, 1981.

Gall, J.G., and Callan, H.G.: ³H-uridine incorporation in lampbrush chromosomes. Proc. Natl. Acad. Sci. USA, 48:562, 1962.

Gall, J.G., Cohen, E.H., and Atherton, D.D.: The satellite DNAs of *Drosophila virilis*. Cold Spring Harbor Symp. Quant. Biol., 38:417, 1973.

Gall, J.G.: Chromosome structure and C-value paradox. J. Cell. Biol., 91:35, 1981. (Highly recommended.)

Gall, J.G., Diaz, M.O., Stephenson, E.C., and Mahon, K.A.: The transcription unit of lampbrush chromosomes. *In* Gene Structure and Regulation in Development. New York, Alan R. Liss, Inc., 1983, p. 137.

Grossbach, U.: The salivary gland of *Chironomus:* A model system for the study of cell differentiation. *In* Results and Problems in Cell Differentiation. Vol. 8. Edited by W. Beerman. Berlin, Springer-Verlag, 1977, p. 147.

Hallis, G.F., et al.: Processed genes: A dispersed human immunoglobulin gene bearing evidence of mRNA-type processing. Nature, 296:321, 1982.

Hunter, T.: Phosphotyrosine, a new protein modification. Trends in Biochem. Sci., 246:249, 1982.

Hunter, T., and Cooper, J.A.: Protein-tyrosine kinases. Ann. Rev. Biochem., 54:897, 1985.

Hynes, R.O.: Cellular location of viral transforming proteins. Cell, 21:601, 1980.

Jacob, F., and Monod, J.: Genetic regulatory mechanisms in the synthesis of proteins. J. Mol. Biol., 3:318, 1961. (Highly recommended.)

Kelly, K., Cochran, B., Stiles, C.D., and Leder, P.: Cell-specific regulation of the c-myc gene by lymphocyte mitogens and platelet-derived growth factor. Cell, 35:604, 1983.

Lamb, M.M., and Daneholt, B.: Characterization of active transcription units in Balbiani rings of *Chironomus tentans*. Cell, 17:835, 1979.

Land, H., Parada, L.F., and Weinberg, R.A.: Tumorigenic conversion of primary embryo fibroblasts requires at least two cooperating oncogenes. Nature, 304:596, 1983.

Lewin, B.: Genes. New York, John Wiley and Sons, Inc., 1985.

Maniatis, T., and Ptashne, M.: A DNA operator-repressor system. Sci. Am., 234(1):64, 1976.

Miller, J.H., and Reznikoff, W.S.: The Operon. Cold Spring Harbor, New York, Cold Spring Harbor Laboratory, 1978.

Pastan, I., and Perlman, R.: Cyclic adenosine monophosphate in bacteria. Science, 169:339, 1970.

Pelham, H.R.B., and Bienz, M.: A synthetic heat-shock promoter element of 13 nucleotides confers heat-inducibility. EMBO J., 1:1473, 1982.

Rowley, J.D.: Human oncogene locations and chromosome aberrations. Nature, 301:290, 1983.

Rubin, G.M., and Spradling, A.C.: Genetic transformation of *Drosophila* with transposable element vectors. Science, 918:348, 1982.

Scheer, U., and Sommerville, J.: Sizes of chromosome loops and hnRNA molecules in oocytes of amphibia of different genome sizes. Exp. Cell Res., 139:410, 1982.

Schmid, C.W., and Jelinek, W.R.: The Alu family of dispersed repetitive sequences. Science, 216:1065, 1982.

Sharp, P.A.: Conversion of RNA to DNA in mammals: *Alu*-like elements and pseudogenes. Nature, 301:471, 1983.

Smith, G.P.: Evolution of repeated DNA sequences by unequal crossing over. Science, 191:528, 1976.

Spradling, A.C., and Rubin, G.M.: The effect of chromosomal position on the expression of the *Drosophila* xanthine dehydrogenase. Cell, 34:47, 1983.

Stroynowski, I., and Yanofsky, C.: Transcript secondary structures regulate transcription termination at the attenuator of *S. marcescens* tryptophan operon. Nature, 298:34, 1982.

Swift, H.: The organization of genetic material in eukaryotes: Progress and prospects. Cold Spring Harbon Symp. Quant. Biol., 38:963, 1973.

Waterfield, M., et al.: Platelet-derived growth factor is structurally related to the putative transforming protein p28[sis] of simian sarcoma virus. Nature, 304:35, 1983.

Weinberg, R.A.: Fewer and fewer oncogenes. Cell, 30:3, 1982.

Yanofsky, C.: Attenuation in the control of expression of bacterial operons. Nature, 289:751, 1981.

CELL DIFFERENTIATION

It is known that the protoplasm of different parts of the embryo is somewhat different. The initial differences in the protoplasmic regions may be supposed to affect the activity of genes. The genes will then in turn affect the protoplasm, which will start a new series of reciprocal reactions. In this way we can picture to ourselves the gradual elaboration and differentiation of the various regions of the embryo.

T.H. Morgan   Embryology and Genetics, 1934

Cell differentiation is the process by which stable differences arise between cells. All higher organisms develop from a single cell, the fertilized ovum, which gives rise to the various tissues and organs. The question of how an apparently structureless egg converts itself into a complex and highly organized embryo has interested scientists since the time of Aristotle 2000 years ago, and still remains one of the major unanswered questions of biology.

In most animal species females produce large unfertilized eggs that contain most of the materials and nutrients required to form an embryo. Development is triggered by *fertilization*. The sperm contributes a small, condensed nucleus *(male pronucleus)*, which rapidly enlarges in the egg cytoplasm, fuses with the *female pronucleus*, and finally divides. The fertilized egg then undergoes a series of very rapid cycles consisting of DNA synthesis followed by cell divisions. These divisions are called *cleavage*, because, unlike normal cell division, the cytoplasm is partitioned without growth. Then the cells form a hollow sphere *(blastula)* in which tissues are not yet evident (Fig. 23–1). Some of the cells then invaginate in a series of cell movement known as *gastrulation*, and the first signs or morphological differentiation appear (Figs. 23–1 and 23–2). These complex changes take place in a comparatively short time. For example, in the South African frog *Xenopus laevis*, a swimming tadpole containing most differentiated tissues (such as blood, nerve, eye, muscle, and so forth) hatches only 72 hours after fertilization.

In molecular terms, cell differentiation means *variable gene activity* in different cells of the same organism. Cell specialization involves the preferential synthesis of some specific proteins (e.g., hemoglobin in erythrocytes, antibodies in plasma cells, and ovalbumin in oviduct). Each eukaryotic cell expresses only a small percentage of the genes it contains, and cells of different tissues express different sets of genes. It is thus clear that cell differentiation will be understood in detail only when the mechanisms of gene regulation in higher cells are understood. Much of the current research in eukaryotic molecular biology is directed toward understanding how some genes are expressed in a tissue-specific way.

Not all genes, however, are tissue-specific. Some genes—called *housekeeping genes*—are expected to be active in all types of cells. Examples of housekeeping genes are those required for building membranes, ribosomes, mitochondria, and the glycolytic enzymes, which are components common to all types of cells. The genes that are expressed differentially, such as globin, ovalbumin, and immunoglobulins, are examples of what in the developmental biology jargon are called *luxury functions*.

Although an understanding of gene regulation will be essential for understanding how the differentiated state is maintained in adult tissues, it may turn out that a better knowledge of cytoplasmic functions is also important for clarifying how the initial differences between cells are established in early embryos. We shall see that all the nuclei in an early embryo are genetically equivalent, and thus the initial differences between cells must reside in the cytoplasm they inherit. It is believed that the cytoplasm of most eggs contains *cytoplasmic determinants* of development which at some point become unequally distributed among the cells of an embryo and subsequently change the activity of genes. The mechanism by which these determinants are segregated unequally in cleaving embryos is not known, and understanding

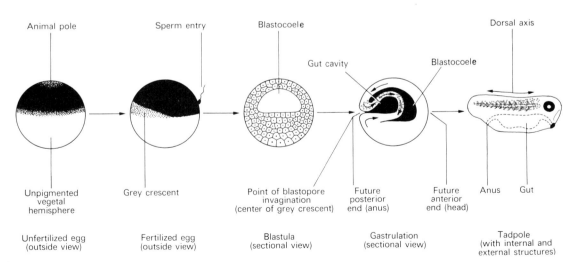

Fig. 23–1. Development of *Xenopus laevis*. During gastrulation cells invaginate at the blastopore, and extensive morphogenetic movements of cells occur *(arrows)*. The blastopore always forms at the center of a pigmented area called the *gray crescent*, which appears in eggs opposite to the point of sperm entry shortly after fertilization. Thus the orientation of the dorsal axis of the embryo, which forms on the blastopore side, is determined by the point of sperm entry. (From J.B. Gurdon, *Gene Expression during Cell Differentiation*. Carolina Biology Readers. Vol. 25. Burlington, North Carolina, Carolina Biological Supply Co., 1978.)

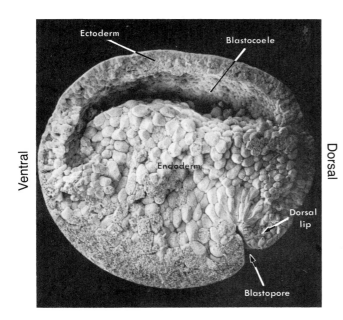

Fig. 23–2. Scanning electron micrograph of a frog embryo during gastrulation. At this stage the embryo has about 25,000 cells that have divided in only 9 hours of development, and the dorsoventral axis of the embryo has been determined. The cells that invaginate at the blastopore will give rise to the mesoderm, and the blastopore itself will give rise to the anus. After extensive cell migrations, the side of the embryo containing the blastopore will give rise to dorsal structures, and the opposite side to ventral structures. The dorsal lip of the blastopore is also called the *organizer* because the invaginated cells produce a diffusible substance (of unknown chemical nature) which later induces the ectoderm cells to differentiate into the neural cord. ×65. (Courtesy of J. Herkovits.)

this segregation will probably be as important as understanding how they modify gene activity.

In this chapter we analyze the possible mechanisms by which the initial differences between cells arise in early embryos. After analyzing the main characteristics of cell differentiation, we then discuss the cytoplasmic determinants of embryonic cells. Next we turn to *nucleocytoplasmic interactions*. A large part of our knowledge of the mechanisms by which differentiation is established and maintained comes from experiments in which nuclei are experimentally placed in a foreign cytoplasm. In the final part of this chapter we analyze the levels at which differential gene expression is controlled in early development.

23–1 GENERAL CHARACTERISTICS OF CELL DIFFERENTIATION

The Differentiated State Is Stable

One of the principal characteristics of cell differentiation in higher cells is that once established, the differentiated state is very stable and can persist throughout many cell generations. For example, a neuron will persist as such throughout the lifetime of an individual, and a cell committed to become a skin cell will gradually keratinize and eventually die. These persistent changes are very different from the type

of regulation involved in enzyme induction and repression in bacteria, which is specially designed to respond rapidly to changes in the environment. As we saw in the previous chapter, E. coli responds to the addition of lactose to the culture medium by synthesizing β-galactosidase a few minutes later, and when lactose is exhausted, enzyme synthesis is stopped with equal speed.

Another important characteristic of cell differentiation is that it is induced in the organism by various stimuli, but once it has been established, it can persist even in the absence of the initial stimulus. For example, differentiated cloned cell lines, such as steroid-secreting cell lines and neuroblastoma cell lines (which differentiate into neurons), are able to grow indefinitely in vitro.[1,2] Without this type of data it could have been argued that differentiation persists in the organism because the stimulus that has given rise to it also persists. In vitro, although such a stimulus cannot be present, the differentiated state is maintained over many cell generations.

Determination Can Precede Morphological Differentiation

In many cases before it is possible to recognize morphologically that a cell has differentiated, there is a period during which the cell is already committed to a particular change. After this *de-*

termination has been made, the cell will differentiate along a specific pathway even if several cell generations intervene before overt morphological differentiation.

The best example of determination is provided by the *imaginal discs of Drosophila* (Imago is an old name for the adult fly, and thus *imaginal disc* means the discs that give rise to the adult). The discs are groups of cells that are present in the larva in an undifferentiated form, but that upon metamorphosis will give rise to legs, wings, antennae, and so forth. Each disc is predetermined to become a particular type of adult structure. This commitment is very stable and heritable, as shown by the disc transplantation experiments carried out by Hadorn.[3,4] Discs transplanted into the abdomen of adult flies remain undifferentiated and multiply as solid masses of tissue in the abdominal cavity. Fragments of these transplanted discs can be transplanted serially into adult flies without giving rise to differentiated tissues. The transplanted discs can, however, be stimulated to differentiate by transplantation into a larva that will undergo metamorphosis, thus providing the appropriate hormonal conditions for differentiation. Discs transplanted serially in the abdomens of adult flies for 9 years (1800 cell generations) were still able, after metamorphosis, to differentiate into structures belonging to the original disc (leg, wing, or antenna).[5]

Cell Differentiation Results from Stepwise Decisions that Are Genetically Controlled

One might imagine that there would be little hope of understanding the enormously complicated patterns of cell arrangements involved in the development of, say, a fly or a frog. Matters can be simplified, however, with the use of genetics, which allows the analysis of the effect of single genes (via mutations) on development. In fact, modern genetics began in 1910 when Thomas Hunt Morgan, who until then had been an experimental embryologist studying frog and invertebrate embryos, discovered the famous white-eyed fly.[6] Morgan realized that development could not be understood before understanding how genes worked, and so he began studying genes by obtaining and analyzing mutants. He chose *Drosophila* because the generation time of this fruit fly is only 14 days. After more than 70 years of research on *Drosophila* genetics, a number of developmental mutants have been found (see Gehring, 1985 in the Additional Readings at the end of this chapter).

The most interesting mutations are those that can, by inactivating a single gene, change one segment of the body into a different one. These are called *homeotic* mutations. Adult flies have a head (made originally of six segments), three thoracic segments (pro-, meso- and metathorax), and eight abdominal segments. Each thoracic segment has a pair of legs, and the second thoracic segment also has a pair of wings. The third thoracic segment does not have wings in flies (which belong to the diptera or two-winged insects) but instead a pair of *halteres*, which are small, drumstick-shaped stumps used for equilibrium during flight. Presumably a gene arose during evolution that changed the second pair of wings into halteres in diptera. Figure 23–3 shows a mutant four-winged fly, generated by a mutation in the so-called *bithorax* locus. A single mutation changes a haltere into a wing. The bithorax locus is a complex of many genes. Some of these genes suppress the formation of legs in the abdominal segments, which suggests that insects evolved from ancestors such as millipedes with a pair of legs in each segment (Lewis, 1978). Another homeotic mutation called *antennapedia* transforms the antennae into legs.

During early *Drosophila* development the cells first become determined to a particular segment,[7,8] and then the cells in each segment subdivide further. These cellular decisions are controlled by the switching on of single *selector genes* in groups of cells[9] (groups of about 50 cells, called *polyclones*, are determined simultaneously). Segments are divided into anterior and posterior portions, or *compartments*, and subsequently into dorsal and ventral compartments.

The existence of compartments can be shown by experiments involving mitotic recombination, such as the one shown in Fig. 23–4. If a mutant heterozygous fly is irradiated with x-rays at a given time of development, somatic recombination can be induced in an individual cell so that one of its daughters becomes homozygous for the mutant and its descendants generate a clone of genetically marked cells. In Figure 23–4 the marker used was a thermosensitive succinate dehydrogenase enzyme, and homozygous mu-

Fig. 23–3. A four-winged fly. Normal flies have only two wings and a pair of halteres, or equilibrium organs, instead of the second pair of wings present in other insects. In this fly a mutation in the bithorax locus was introduced, and the development of the second pair of wings was no longer repressed. Thus a single locus can control the development of the wing. Development is thought to proceed through an orderly activation of selector genes such as this one, which in turn will modify the activity of entire sets of other genes. (Courtesy of E.B. Lewis, who pioneered the study of *Drosophila* developmental genetics.)

tant clones of cells were revealed by histochemical staining of tissues after heating to 47°C.[10,11] The white clone in Figure 23–4 cannot cross the border between the anterior and posterior parts of the imaginal disc that will give rise to the adult wing, which runs roughly vertically through the middle of the disc. After analysis of many such specimens, it has been concluded that each segment has an anterior and a posterior compartment, whose boundaries the cells cannot cross after determination. There is reason to believe that the cells in different compartments differ in their surface properties[12] and that cells with similar surface properties attach to each other within a compartment.

In this view of development, largely elaborated by the Spanish geneticist A. Garcia Bellido, the formation of complicated patterns such as those found in the tissues of a fly arises from a series of binary decisions made by the *selector* genes. Some genes—e.g., *ultrabithorax* and *an-*

tennapedia—determine very early on in embryogenesis the segmental pattern, and then other selector genes subdivide these segments into further compartments.

The genes for *ultrabithorax* and *antennapedia* have now been cloned through the use of recombinant DNA techniques. They have very long RNA transcripts (80 kilobases for bithorax, 100 kilobases for antennapedia), which are spliced differently depending on the stage of embryogenesis. Usng the cloned genes as a probe it has been possible to hybridize them to RNA transcripts on histological sections of embryos.[13] Figure 23–5 shows how this powerful method of in situ hybridization allows one to locate in which tissues particular genes are expressed. It was found that very early in development *ultrabithorax* is expressed in the third thoracic segment and the eight abdominal ones, and that later on it becomes localized predominantly on the third thoracic segment (see Akam, 1983) as

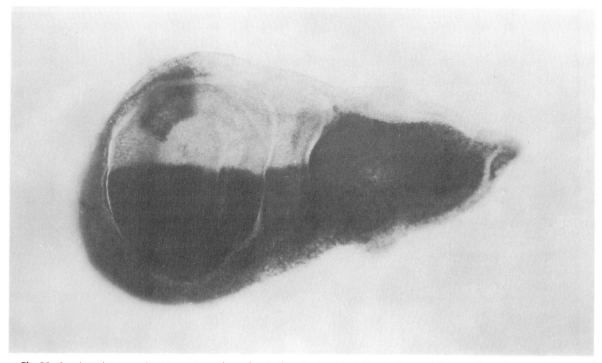

Fig. 23–4. Anterior-posterior compartment boundary in the wing imaginal disc of a *Drosophila* larva. Somatic recombination was induced in early development by x-ray irradiation of a *Drosophila* embryo that was heterozygous for a chromosomal mutation that produces a thermosensitive succinate dehydrogenase enzyme. After histochemical staining, the descendants of a cell that became a homozygous mutant can be seen as a white patch. The cells do not cross a border (vertical line about the center of the disc), even though no physical barrier exists between the two types of cells. This type of experiment demonstrates that cells are determined to become part of a certain compartment early in development. (Courtesy of P.A. Lawrence, who developed this cell marking technique.)

predicted by the genetics. Similarly, *antennapedia* is expressed initially in the second thoracic segment as well as all segments posterior to it; but later becomes localized preferentially at the second thoracic segment, as shown in Figure 23–5 (Hafen et al., 1983). That *antennapedia* should be expressed at the second thoracic segment was also predicted from the genetics, because when in the mutant flies an antenna changes into a leg, the leg has special characteristics unique to the second pair of legs (and thus the normal function of the gene is somehow involved in the determination of the second thoracic segment). In situ hybridization methods have been extremely useful for mapping the regions of the embryo in which a number of genes involved in segmentation, homeotic transformations, compartments and dorsoventral polarity are transcribed.

Many of the genes involved in early development of *Drosophila* were unexpectedly found to share a short (180 nucleotide) region of sequence homology, which was called the homeobox (because it was first found in homeotic genes such as *antennapedia* and *ultrabithorax*). The homeobox codes for a conserved region of 60 amino acids which is usually located close to the carboxyl end of the various proteins involved in the control of gene activity in development (McGinnis et al., 1984, reviewed by Gehring, 1985). The discovery of the homeobox was a breakthrough because this short stretch of sequence is so highly conserved that it can be used as a probe to isolate new genes that also contain it. In Drosophila, about 20 different genes contain homeoboxes, of which about 15 have been studied in some detail. All of them have been found to control the determination of segmentation, except one that is expressed in dorsal cells and is involved in the establishment of the dorso-ventral polarity of the embryo (Doyle et al., 1986).

The junior author of this book works in embryology of the frog *Xenopus*, and decided to

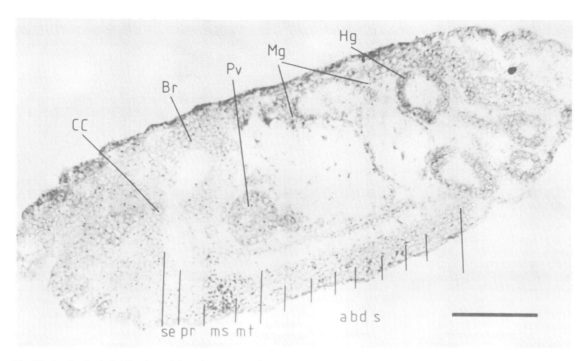

Fig. 23–5. In situ hybridization of cloned *antennapedia* ^{3}H-DNA to RNA transcripts in a histological section of a *Drosophila* embryo. This method, which allows the study of gene expression in different tissues, is based on the use of frozen sections that are digested with the protease pronase to enhance the accessibility of the mRNA on the section to the ^{3}H-DNA probe. After hybridization and washing, the sections are dipped in autoradiographic emulsion, and radioactivity is detected as silver grains. At this stage of development *antennapedia* transcripts are found mostly in the second thoracic segment (ms) ventral nervous tissue. Weaker labeling can also be observed in the abdominal ventral nervous system, following a segmented pattern. *pr*, prothorax; *ms*, mesothorax; *mt*, metathorax; *abds*, the eight abdominal segments; *br*, brain; *pv*, proventriculus; *mg*, midgut; *hg*, hindgut. (Courtesy of M. Levine, E. Hafen and W. Gehring.)

test whether homeotic-like genes could be isolated from vertebrate DNA. Using the Drosophila homeobox as a probe, we cloned Xenopus genes that contain the very conserved homeobox region (Carrasco et al., 1985). Figure 23–6 shows the nucleotide sequence, and the amino acid sequence translated from it, of two Xenopus homeoboxes. The amino acid sequence is highly conserved between the two different genes, while the nucleotide sequence may contain silent changes, usually in the codon third position, which do not change the protein sequence (because the genetic code has synonyms). This suggests that the 60 amino acid protein segment has an essential function which cannot be changed by mutation easily without consequences. The Xenopus homeobox indicated in the top row of Figure 23–5 is identical to the *Drosophila antennapedia* homeobox in 59 out of its 60 amino acids.

Subsequent studies by other research groups have shown that all vertebrates contain about 20 to 30 different homeobox-containing genes, and many of these have been cloned, particularly from mice and humans. As in Drosophila, these genes are expressed during vertebrate embryogenesis, frequently in the nervous tissue, and sometimes in a spatially restricted way (for example, one homeobox-containing gene is expressed only in certain cervical and thoracic-segments of the spinal cord of the mouse; see Awgulewitsch et al., 1986). Thus the vertebrate homeobox-containing genes resemble their *Drosophila* counterparts in many respects, and it seems likely that they will turn out to have similar functions. This gives hope that it will be possible in the near future to understand vertebrate development using the genes identified with the help of the molecular genetics of the fly. Only a few years ago there was little hope of understanding the enormously complicated cellular events involved in early vertebrate development and now, in an unexpected development, genes that probably control the switch-

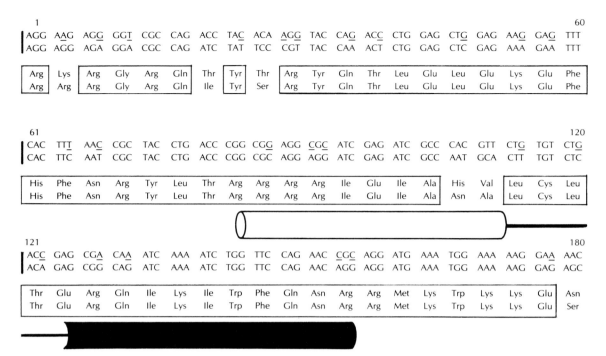

Fig. 23–6.   Sequence of two *X. laevis* homeoboxes. Silent nucleotide changes that do not change amino acids are underlined. The position of the helix-turn-helix motif present in prokaryotic DNA-binding proteins is indicated by a white cylinder and a black cylinder under the sequence. The presence of these two helices suggests that the homeobox is a DNA-binding domain. The conserved region comprises only 60 amino acids, always located close to the carboxy-terminus of the protein. (From E.M. De Robertis, et al., Cold Spring Harbor Symp. Quant. Biol. 50:271, 1985.)

ing on and off of sets of other genes during embryogenesis have been isolated (reviewed by De Robertis et al., 1985).

What is the function of the homeobox? When the homeobox amino sequence was compared by computer methods to a collection of all the proteins sequenced to date, it was found that the only proteins of a similar sequence of amino acids were the yeast mating-type proteins (see section 23–4 and Fig. 23–28). These proteins are known to determine the mating type of yeast in a stable way, i.e., a type of cell differentiation. The mating type proteins control sets of other genes in yeast by binding directly to DNA in their promoter region. Thus the homology found by using computers suggests that the homeobox-containing proteins are DNA binding proteins (Shepherd et al., 1984). Recent studies have shown that the homeobox is indeed a sequence-specific DNA binding domain.

SUMMARY
Cell Differentiation

The differences between cells in various tissues of a multicellular organism are stable and heritable.

Although differentiation may be induced in the embryo by a certain stimulus, it will persist even in the absence of this initial stimulus in tissue culture.

Frequently cells become *determined* or *committed* to become a certain cell type but do not differentiate morphologically. In experiments with fruit flies it was shown that this determined state sometimes lasts for years.

The fruit fly *Drosophila melanogaster* is suitable for genetic studies, and an interesting type of developmental mutant has been isolated. These mutations define the homeotic genes, which when inactivated change one segment of the fly into another (e.g., haltere into wings or an antenna into a leg). Fly development may be conceived as a series of binary decisions in which *selector* genes are switched on or off, giving rise first to the different segments of the fly and then to smaller compartments within each segment. Each selector gene modifies the activity of entire sets of other genes.

The *Drosophila* homeotic genes have been cloned and found to be expressed in a tissue-specific way as predicted from the genetics (Fig. 23–5). A conserved protein-coding segment of 180 nucleotides, the homeobox, has been found in several *Drosophila* homeotic genes. Using the homeobox as a probe, it has been possible to isolate, by nucleic acid hybridization, a large number of homeobox-containing genes. These genes code for DNA-binding proteins, which are involved in controlling the early differentiation of the embryo. Homeobox-containing genes have been isolated from vertebrates and this may provide a key to understanding development.

23–2 LOCALIZATION OF CYTOPLASMIC DETERMINANTS IN EGGS

We now turn to how the initial differences between cells of the early embryo arise. Our thinking on this matter is still heavily influenced by the findings of the experimental embryologists of the past century (wonderfully reviewed by Wilson, 1928). At the time embryology was the very center of biological research, and in particular of heredity. With the help of the light microscope and crude dissection instruments such as glass needles, embryologists described normal development and modified it experimentally, drawing major conclusions. After the turn of the century genetics moved into the limelight, and recently researchers in genetic engineering have made the main contributions to the study of biology. Today, however, embryologists are closing the circle as, armed with the powerful tools of molecular genetics, they return to the questions posed a century ago. Let us now briefly examine some of these questions.

Experimental Embryology 100 Years Ago

By 1860 it was already clear that the egg was a single cell, but the role of the sperm remained a mystery until O. Hertwig in 1875 showed that the sea urchin sperm donated a nucleus, which later fused with the female one. Thus it seemed that the nucleus would contain the hereditary material. In his epoch-making study on *Ascaris* fertilization, Van Beneden (1883) showed that each parent contributed an equal number of chromosomes to each offspring. He was able to show this because *Ascaris* is a nematode (roundworm) with an extremely low haploid chromosome number (either one or two, depending on the variety). Figure 23–7 shows clearly that each parent contributes two chromosomes. This may now seem a trivial finding, but at the time it meant that one could begin to understand heredity. It should be kept in mind that although Mendel's work dated from 1865, it was forgotten

by the scientific community until it was rediscovered in 1900 (independently by De Vries, Correns, and Tschermack). It quickly became clear that the behavior of Mendel's genes could be equated to the behavior of the chromosomes seen under the light microscope.

The studies with *Ascaris* led Weissmann to propose in 1883 his theory of heredity. As Figure 23–8 shows, he proposed that it is the germ cells that carry heredity and that the body or *soma* is a mere offshoot of the germ cells whose main function is to carry the germ cells (i.e., the chicken is the egg's way of making more eggs). We now know that in most organisms the germ cells are kept separate from other cells from early on in development; this part of Weissmann's ideas has been fully confirmed. If we could follow the history of our own germ cells back in evolution, we would eventually be led to the origin of life on earth. At the time this theory of heredity was of great importance, because it clearly precluded the view of those who, like Lamarck, believed that evolution resulted from the inheritance of characteristics acquired by individuals (the soma) during their lifetime. Even Darwin's original theory (1859) on the origin of the species accommodated inheritance of acquired characteristics, but this erroneous view was dispelled by the work of the early embryologists.

Although Weissmann's views on heredity were essentially correct, his theory on development turned out to be wrong. He proposed that the fertilized egg had all the information to make an individual, but that with each successive division part of the information was lost (except from the germ cells). Weissmann thought that in the end some cells retain only genetic material to make a specific type of tissue, and lose all other genetic material. He was misled by what happens in *Ascaris* development, in which in fact large segments of chromatin are lost from somatic cells, as will be explained later on (see Fig. 23–13).

It turned out, however, that *Ascaris* is an exception, and in almost all other animals DNA is not lost during development. Although wrong, Weissmann's ideas on development stimulated a great deal of research and in this sense were highly beneficial. The fact remains that the germ cells are set apart very early during embryogenesis in most organisms (see Figs. 23–13 and

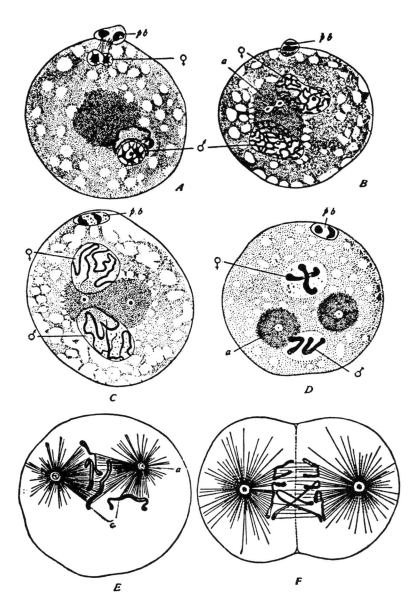

Fig. 23–7. Fertilization of the egg of *Ascaris megalocephala* var. *bivalens*, illustrating Van Beneden's demonstration that each parent contributes an equal number of chromosomes. A, the sperm has entered the egg and its nucleus has swollen (♂). The female nucleus is completing the second meiotic division and is eliminating the second polar body (pb). Each nucleus contains two chromosomes (the diploid number for this variety of *Ascaris* is four). B, both pronuclei (♀, ♂) have swollen; the centrosphere *(a)* contains the dividing centrioles, C, chromosomes start condensing. D, two chromosomes clearly visible in each nucleus. E, first division; the nuclear membranes dissolve, and the chromosomes align in a common metaphase plate. F, first cleavage anaphase (only three chromosomes shown on this section). In *Ascaris*, as in frogs, the pronuclei do not fuse before first cleavage. In other species, such as sea urchins, the membranes of the ♂ and ♀ pronuclei fuse before the first cleavage. (Drawing by T. Boveri, 1888.)

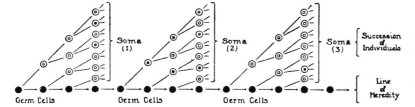

Fig. 23–8. The Weissman theory of heredity. In each generation the germ cell gives rise on the one hand to the body or soma and on the other to more germ cells, which carry the line of heredity. (From E.B. Wilson, 1897.)

23–33). Why this is so is still unexplained, and many scientists wonder whether the germ cell DNA might not need to be especially protected from rearrangements.

The Genome Remains Constant during Early Development—Nuclear Transplantation

Experimental embryology began in 1888 when Roux killed one of the cells of the frog embryo at the two-cell stage with a hot needle. He observed that the other cell gave rise to a half-embryo, usually a right or a left half. He interpreted this finding in light of Weissmann's ideas and proposed that development gradually limited the potential of a cell. In 1892, however, Driesch separated the two cells of a sea urchin embryo and found that both could give rise to complete, although smaller, embryos. Driesch's finding sparked heated controversies, and eventually it became clear that if the dead blastomere is removed from the frog embryo, it too could give rise to a complete tadpole. We now know that in the Roux experiment leaving the dead cell in situ interfered with the invagination of cells during gastrulation.

Many experiments have since showed that the genome is not irreversibly altered during development. Figure 23–9 shows how Spemann constricted a frog egg with a loop of thin (human baby) hair. The nuclei remained in one half of the egg and continued to divide. After the 8- or 16-cell stage was reached, the loop was loosened allowing a nucleus to pass to the other half. Because this half also developed into a normal (but smaller) tadpole, the experiment showed that the 8- or 16-cell nuclei contain all the information required for making the tissues of a tadpole.

Rigorous proof that during embryonic differentiation there is no loss of genetic information came from the classical nuclear transplantation experiments performed by J.B. Gurdon in the frog egg. *Xenopus laevis* unfertilized eggs can be irradiated with ultraviolet light to destroy the

endogenous nucleus and can then be injected with a single *Xenopus* diploid nucleus. Nuclei obtained from *Xenopus* tadpole intestinal cells (which are clearly differentiated cells, having a "brush border" of microvilli) are able to sustain development of normal adult frogs which are fertile.[14] This demonstrates that the intestine cells retained all the genes required for the complete life cycle of a frog, including the formation of germ cells. Not all transplants are successful, the main limiting factor being the rapidity (90 minutes) with which the transplanted nucleus must replicate its DNA before first cleavage. The success rate can be increased by using "partial" blastulae (which arise frequently in nuclear transplantation experiments), in which the nucleus escapes the first division, and which can be recognized readily because only one half of the egg cleaves. These nuclei take a longer time to complete their DNA replication and have a much higher success rate when transplanted into new recipient eggs. This technique is called serial nuclear transplantation (Fig. 23–10).

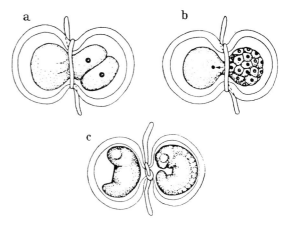

Fig. 23–9. Spemann's demonstration that embryonic nuclei are totipotent until the 16-cell stage. A constricted egg continues to divide only in the half containing the nuclei *(a)*. If the constriction is released at the 16-cell stage, a nucleus may escape to the other side *(b)*. This nucleus can still give a complete embryo, although its development is retarded with respect to the other half *(c)*. (After H. Spemann.)

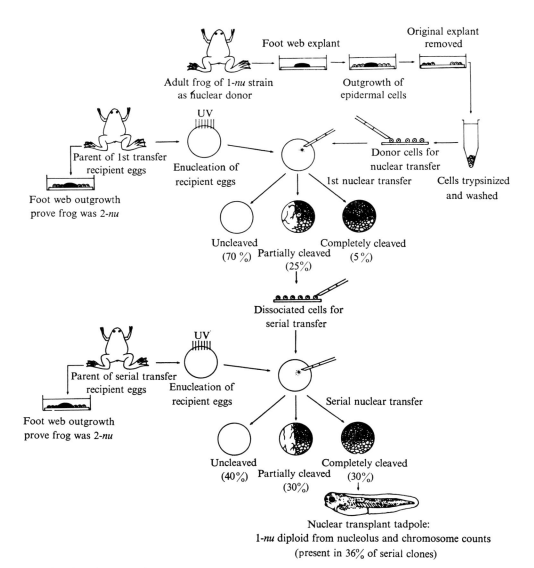

Fig. 23–10. Serial nuclear transplantation in *Xenopus*. Skin cells from the foot web of one-nucleolus frogs (the number of nucleoli provides a genetic marker) were cultured and shown by immunofluorescence to produce keratin, which is a characteristic of differentiated skin cells. Nuclei from these cells were transplanted into eggs whose own nucleus had been killed with ultraviolet light. Nuclei from partial blastulae (see text for details) were then transplanted into new recipient eggs, and swimming tadpoles were obtained. Thus adult skin cells contain all the genes necessary to build a tadpole of the stage indicated. (Courtesy of J.B. Gurdon.)

Development up to the stage at which the tadpole swims has been obtained by serial nuclear transplantation of a variety of adult tissues, such as keratinizing skin cells[15] (Fig. 23–10) and lymphocytes,[16] thus demonstrating that the genes required to make nerve, blood, muscle, cartilage, and other tadpole tissues were not irreversibly inactivated in the donor nuclei. It has not yet been possible to obtain adult frogs using adult tissues, and thus we cannot entirely exclude the possibility that some genes might in fact be lost in some cells.

Similar conclusions have been obtained using plant cells. It is common knowledge that whole plants can be grown from cuttings. In some cases as with carrots, a complete plant can be grown from a single cultured cell.[17]

Nuclear transplantation is also possible in the mouse. Nuclear transplantation permits production of *clones* of genetically identical twins, be-

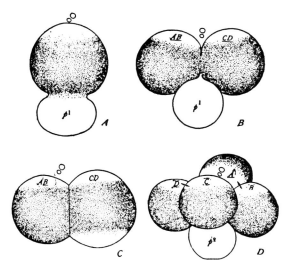

Fig. 23–11. Early cleavage in the mollusc *Dentalium*. *A*, beginning of the first cleavage by formation of the polar lobe (p^1); *B*, later (trefoil) stage of the same cleavage; *C*, resulting two-cell stage, in which the first polar lobe has been resorbed; *D*, second cleavage in progress, with second polar lobe (p^2). The cell that receives the polar lobe cytoplasm will give rise to the mesoderm. If the polar lobe is excised at stage shown in *B*, the embryos lack mesodermic tissues. (From Wilson, E.B.: The Cell in Development and Heredity. 3rd Ed. New York, MacMillan Publishing Co., Inc., 1928.)

cause many individuals can be obtained from the cells of a single frog or mouse embryo. Human and mouse embryos are very similar, and in theory human nuclear transplantation should be feasible. This has not been attempted, fortunately, because of many ethical considerations.

Cytoplasmic Localizations May Determine the Initial Differences between Embryonic Cells

If the information contained in the cleavage nuclei is identical, then the initial differences between cells should reside in the cytoplasm they inherit. In some eggs the segregation of cytoplasmic components, which are thought to be able to affect the activity of genes, is particularly clear. Figure 23–11 shows that in some molluscs, such as *Dentalium* (a sea slug), the cytoplasm of the vegetal pole is extended transiently during the first and second cleavages. This *"polar lobe"* lacks a nucleus and after cell division is incorporated into one of the blastomeres (Fig. 23–11*C*). The cell that inherits the polar lobe cytoplasm eventually gives rise to the mesoderm. E.B. Wilson (1904) removed the polar lobe at the two-cell stage by sucking the

eggs up and down a thin pipette. He found that the lobeless embryos lacked mouth, shell gland, and foot, as well as other mesodermal tissues, and he concluded that the polar lobe cytoplasm contains mesodermal determinants.

Ascidians (phylum Chordata) are particularly convenient organisms for studies on cytoplasmic determinants. Some ascidian eggs have regions of different pigmentation, and it is possible to see that after fertilization these regions undergo extensive movements and eventually become included in the cells that give rise to certain tissues. For example, the ascidian *Styela* has a region of yellow cytoplasm (rich in mitochondria), which eventually gives rise to mesoderm and muscles; a region of clear cytoplasm (which has the contents of the giant oocyte nucleus or germinal vesicle), which becomes ectoderm; and a region of gray cytoplasm (rich in yolk platelets), which becomes endoderm.[18]

If the embryos are compressed so that the yellow cytoplasm is distributed into more cells than usual, the cells that acquire it will give rise to muscle cells, suggesting that this yellow cytoplasm contains determinants for muscle tissue.[19,20]

The segregation of these muscle-inducing substances within the egg can be followed using the experimental trick shown in Figure 23–12. The enzyme acetylcholinesterase is a good marker of muscle differentiation but does not normally appear until the embryo is 9 hours old and has several hundred cells. When developing embryos are placed in sea water containing cytochalasin B (an inhibitor of actin microfilaments), the cells no longer divide. The nuclei, however, continue to multiply, and the acetylcholinesterase activity appears at the normal time if these cleavage-arrested embryos are incubated for several hours. Figure 23–12 shows that the potential to produce acetylcholinesterase, which was present in the unfertilized egg (Fig. 23–12*A*), becomes progressively segregated into subsets of cells during cleavage (e.g., into only two cells at the eight-cell stage, Fig. 23–12*D*). This is precisely the behavior expected of cytoplasmic determinants (see Whittaker, 1979).

Germ Cell Determinants

The best example of *determinants* in development is provided by the germ plasm. Amphib-

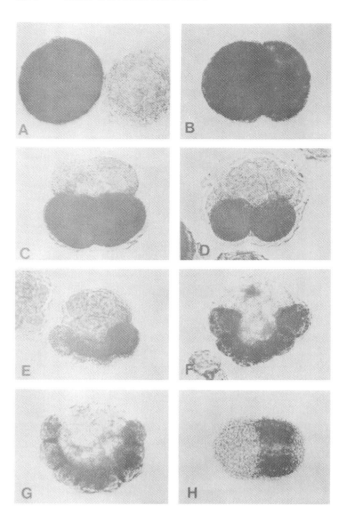

Fig. 23–12. Acetylcholinesterase (AchE) development in cleavage-arrested embryos. To stop cytokinesis embryos were placed in cytochalasin B at following stages: *A*, 1-cell; *B*, 2-cell; *C*, 4-cell; *D*, 8-cell; *E*, 16-cell; *F*, 32-cell; *G*, 64-cell. The embryos were kept for 15 to 16 hours after fertilization at 18°C, before being stained histochemically for the muscle enzyme AchE. *H*, AchE in a 9-hour, non-arrested control embryo. The enzyme stains the region of the tail muscles of the future swimming larva.

ian and other eggs contain in their vegetal (yolky) pole a specialized region of cytoplasm that can be recognized morphologically by the presence of special granules. *Drosophila* eggs have an equivalent region located in the posterior pole of the egg (which is therefore called the *pole-plasm*). This cytoplasm has the property of inducing germ cell formation, i.e., those cells that contain the germ plasm will eventually become the germ cells of the new organism. When the posterior poles of eggs are irradiated with ultraviolet light, sterile (but otherwise normal) animals are obtained. If UV-treated eggs are injected with pole-plasm of normal eggs, however, fertile flies are obtained.[21] Furthermore, if cytoplasm containing the germ cell determinants is injected into the anterior part of a *Drosophila* egg, germ cells develop in an anterior position.[22]

As we mentioned earlier, the germ cells are set apart early in development in a great many organisms, and perhaps the clearest example is the nematode *Ascaris megalocephala*. The painstaking studies of Boveri (Fig. 23–13) showed that the future germ cells can be recognized because the chromosomes remain intact, while somatic cells undergo a process called *chromatin diminution*. *Ascaris* germs cells have only two large chromosomes (some Ascaris varieties have a haploid chromosome number of only one), which have a diffuse centromere (holocentric chromosomes with multiple microtubule attachment sites). In the cells that will become somatic, the central part of the chromosomes breaks into multiple small chromosomes (Fig. 23–13a and b), and the ends (which are heterochromatic and do not have microtubule attachment sites) remain as visible chromatin masses in the cytoplasm of the daughter cells (Fig.

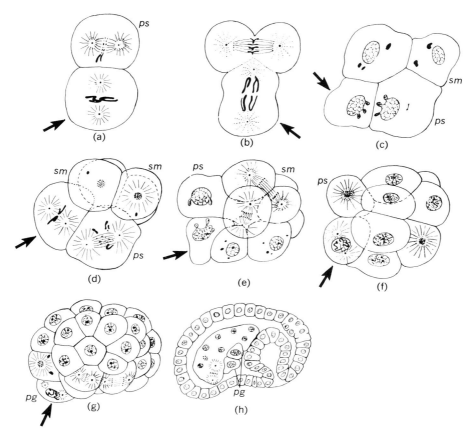

Fig. 23–13. Chromosomal diminution and determination of primordial germ cells in *Ascaris megalocephala*. *ps*, primordial somatic cell, yet to undergo diminution; *sm*, somatic cell which has undergone diminution. *arrow*, germ-line stem pg cell. *(a)* Second cleavage in progress. In the primordial somatic cell chromosome diminution is in progress. *(b)* Later stage, elimination-chromatin at equator of upper spindle. *(c)* 4-cell stage showing eliminated chromatin in the cytoplasm of the upper two cells. *(d)* Third cleavage in progress, second diminution at ps. *(e)* 10-cell embryo showing mitosis of somatic cells with diminished nuclei each containing many small chromosomes. *(f)* 12-cell embryo. *(g)* About 32 cells, fourth diminution in progress, leaving primordial germ cell (pg) (in prophase). *(h)* Gastrula completed with two primordial germ cells. (From the studies of T. Boveri, 1899.)

23–13*c*, from which they are eventually lost. The germ line does not undergo this chromatin diminution process. At each cleavage division a single cell retains its chromosomes intact (the germ-line stem cell). Thus at the fifth cleavage the embryo has 31 somatic cells and one germ cell (Fig. 23–13*g*).

Boveri reasoned that there must be a cytoplasmic substance that becomes segregated into a single cell of the embryo and protects its chromosomes from chromatin diminution. He then proceeded to centrifuge *Ascaris* eggs to redistribute this hypothetical cytoplasmic material. Figure 23–14 shows that centrifuged embryos can have multiple germ cell stem cells instead of a single one. This was one of the earliest (1910), and surely one of the most beautiful,

demonstrations that the egg cytoplasm has components that distribute unequally and determine the fate of cells.

More recently it has been possible with the help of specific antibodies to visualize cytoplasmic components that segregate with the germ cells in nematodes (see Strome and Wood, 1982). Figure 23–15 shows a two-cell embryo of the nematode *Caenorhabditis elegans* stained with an antibody that reacts with granules that are restricted to the germ cell cytoplasm. What is most interesting about this picture is that the granules are localized in one half of the lower cell, even though the chromosomes are still in anaphase. The distribution of the granules does not seem to follow the shape of the mitotic apparatus. Detailed studies have shown that the

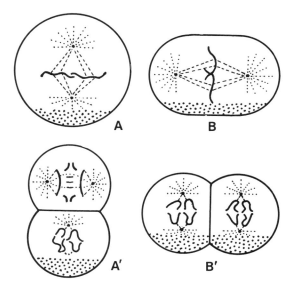

Fig. 23–14. Cytoplasmic control of chromosome diminution in *Ascaris*. *A, A'*, normal eggs; *B, B'*, centrifuged eggs. The shaded area indicates the distribution of a hypothetical cytoplasmic material. In blastomeres containing this material the two large chromosomes remain intact; in blastomeres lacking it the chromosomes undergo diminution (*A'*, top cell).

granules, which are initially distributed throughout the cell, move to one half of it during prophase. The cellular machinery that moves these granules is entirely unknown, but is clearly of great importance for normal development. Understanding how egg cytoplasmic components segregate asymmetrically among blastomeres should be a major goal for embryologists because these movements eventually determine the activity of genes.

The chromatin diminution observed in *Ascaris* is an exceptional phenomenon. In fact, it does not explain cell differentiation even in *Ascaris*, because somatic cells for all tissues undergo a similar diminution. The DNA that is lost consists mostly of repetitive satellite DNA sequences. These sequences must be somehow dispensable in somatic cells, but are necessary for the integrity of germ cells. In the vast majority of plant and animal species DNA is not lost from somatic cells.

Eggs Stockpile Materials for Use During Early Development

Eggs are in general very large cells that stockpile many of the molecules required for early development. For example, a *Xenopus* egg contains about 100,000 times more RNA polymerase, histones, mitochondria, and ribosomes than a normal adult *Xenopus* somatic cell.[23] One reason for accumulating these ready-made materials during oogenesis, rather than making them *de novo* during early embryogenesis is the extraordinarily rapid rate of cell division during cleavage. This rapid pace allows little time for new RNA and protein synthesis, but it is during this period that the first differences between cells are established. Presumably most of the developmentally important substances are made during oogenesis and stored in the egg.

The fate of the different regions of the egg, however, is not fixed beforehand. As Figure 23–1 shows, after fertilization the frog egg cortex contracts toward the sperm entry point, giving an area of different pigmentation called the *gray crescent* at the opposite side. Later on, the gray crescent will give rise to the blastopore, the point at which cells invaginate during gastrulation. The side with the blastopore will become the dorsal part of the embryo. Thus the sperm entry point, which is random, determines the position of the dorsoventral axis of the future embryo.[24,25]

The Mid-Blastula Transition

The first cleavage of the *Xenopus* egg takes place 90 minutes after fertilization (Fig. 23–16). The 11 subsequent cleavages take place synchronously every 35 minutes.[26] This compares with a doubling time of 24 hours for an adult frog somatic cell. In order to achieve this extraordinarily short cell cycle, cleaving embryos increase the number of origins of replication in their DNA (which shortens the S phase) and omit the G1 and G2 phases of the cell cycle, so that the end of DNA synthesis is immediately followed by mitosis.[27] During this initial phase of development there is no transcription of the DNA. After the 12th cleavage, or 4000-cell stage, the cell cycles become longer and asynchronous, the cells become motile, and RNA synthesis starts (Fig. 23–16). This turning point in development is called the *mid-blastula transition*.

Several types of RNA start to be expressed simultaneously at the mid-blastula transition (mRNA, snRNA, tRNA, 5SRNA, and rRNA),

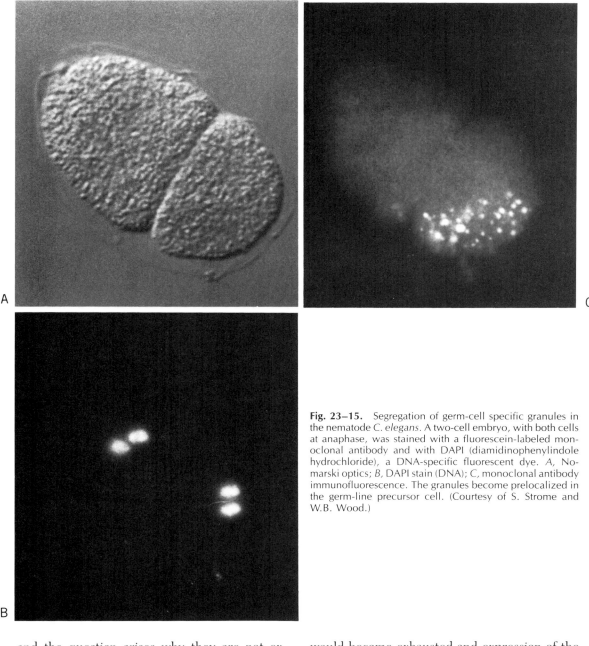

Fig. 23–15. Segregation of germ-cell specific granules in the nematode *C. elegans*. A two-cell embryo, with both cells at anaphase, was stained with a fluorescein-labeled monoclonal antibody and with DAPI (diamidinophenylindole hydrochloride), a DNA-specific fluorescent dye. *A*, Nomarski optics; *B*, DAPI stain (DNA); *C*, monoclonal antibody immunofluorescence. The granules become prelocalized in the germ-line precursor cell. (Courtesy of S. Strome and W.B. Wood.)

and the question arises why they are not expressed earlier. The experimental evidence suggests that there is a critical ratio of nuclei to egg cytoplasm before transcription can start. For example, polyspermic fertilization or the injection of large amounts of extra DNA will induce the mid-blastula transition earlier.[26] Thus it seems that *Xenopus* eggs may have a substance that binds to chromatin and turns it off transcriptionally. The amount of this substance would be sufficient to block up to 4000 nuclei, but then would become exhausted and expression of the genome would start. Needless to say, the identification and purification of this substance is being actively pursued by embryologists.

Cell Interactions Become Increasingly Important as Development Advances

Although cytoplasmic determinants laid down in the egg during oogenesis are undoubtedly very important in establishing early differences

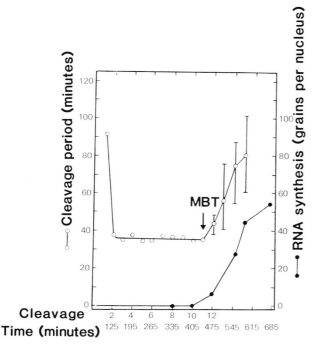

Fig. 23–16. Cell cycle length and RNA synthesis during early *Xenopus* development. The first cleavage takes place 90 minutes after fertilization and the 11 subsequent ones every 35 minutes. When the mid-blastula transition (MBT) is reached at the 12th cleavage (4000-cell stage), the cell cycle becomes longer and asynchronous, and RNA synthesis starts. (Data from Newport, J., and Kirschner, M.: Cell, *30*:675, 1982.)

between cells, they cannot entirely explain development. For example, there is no evidence of cytoplasmic localization in mammalian eggs. Furthermore, as development advances, *cell interactions* become increasingly important. At gastrulation, extensive cell movements and migrations occur, and different types of cells interact with each other in the phenomenon known as *embryonic induction*.

The cells that invaginate at the blastopore form a layer of cells called the *mesoderm*, which, as Figure 23–17 shows, soon divides into a central region or *notochord* and two lateral regions (which later give rise to the *somites*, which in turn give rise to muscle and other mesodermic derivatives). The notochord induces the overlying ectoderm to become neural tissue (Fig. 23–17). Later on, other inductions take place; for example, the optic vesicles (an outgrowth of the brain) induce the nearby ectoderm to become eye lens cells. These inductions are mediated by diffusible substances, but despite numerous attempts to isolate them, their chemical nature remains unknown.

The influence of neighboring cells on cell differentiation can sometimes be quite dramatic. *Teratomas* are tumors of the germ cells, which are the most frequent tumors of the human testis or ovary. When they are proliferating rapidly,

they remain relatively undifferentiated and are highly malignant, but sometimes they have patches of several tissues, such as teeth, hair, muscle, and nerves. In some ways teratomas are like disorganized embryos. In fact, in mice teratomas can be induced experimentally by transplantation of early embryos into ectopic positions (liver, kidney, or under the testicular capsule).[28] A highly malignant teratoma had been kept in the laboratory for eight years by serial grafting into mice. Figure 23–18 shows that when single cells of this tumor were injected into genetically marked mouse blastocysts, they gave patches of normal tissues in the resulting mice.[29] The injected cells[30] were able to contribute to organs coming from the endo-, ecto- and mesoderm. Thus, through interaction with a normal embryonic environment, a normal pathway of cell differentiation was restored to these eight-year-old cancer cells.

SUMMARY
Cytoplasmic Determinants

Nuclear transplantation experiments in the frog (Fig. 23–9) have shown that the genome remains constant during cell differentiation. This can be concluded because nuclei from differentiated intestinal tadpole cells can give rise to fertile adult frogs after transplantation into enucleated frog

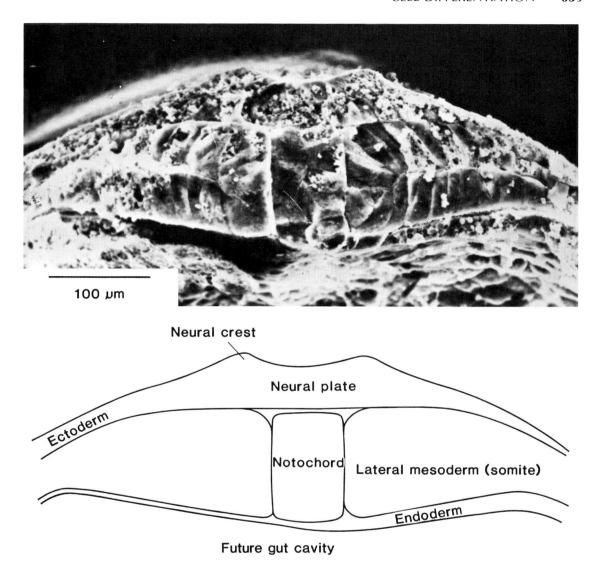

Future gut cavity

Fig. 23–17. Scanning electron micrograph of a *Xenopus laevis* neurula. The embryo was fixed and cut transversely in half with a razor blade. The notochord is the central part of the mesoderm, and it induces a thickening of the overlying ectoderm called the neural plate. The rims of the neural plate (neural crest) will eventually fold and meet, forming the neural tube or central nervous system. The lateral mesoderm will give rise to the *somites*, from which muscles and other mesodermic tissues derive. The bar is 100 μm long. (From J. Marthy and E.M. DeRobertis.)

eggs. If the nuclei remain genetically identical, what causes the initial differences between cells in an embryo? Work with ascidian eggs has suggested that the cytoplasm contains substances (called *determinants*) that become unequally distributed among embryonic cells and cause them to follow a particular differentiation pathway (Fig. 23–12).

The best example of cytoplasmic determinants is provided by the granules present in germ cells. When they are centrifuged (Fig. 23–14) or transplanted into different positions, they will induce the formation of germ cells in a different position. In nematode embryos these granules can be followed with monoclonal antibodies and have been

observed to move to one side of the cytoplasm during prophase, so that only one of the daughter cells inherits the cytoplasmic determinants (Fig. 23–15).

The rate of cell division in early embryos is usually very fast; the first 12 divisions in the frog *Xenopus* take place every 35 minutes. Most of the materials for early development, such as ribosomes, mitochondria, histones, and RNA and DNA polymerases, are stockpiled in frog eggs, each one containing the equivalent of about 100,000 somatic cells.

During this period of rapid division there is no synthesis of RNA, but when the 4000-cell stage

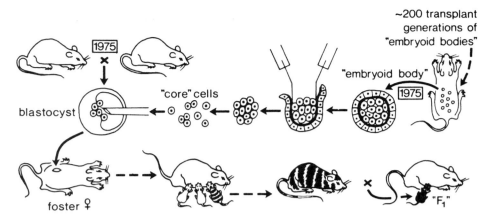

Fig. 23–18. Malignant teratoma cells can be restored to normal development by injection of single intact cells into blastocysts. The highly malignant cells were kept for 8 years by transplantation into the peritoneal cavity ("ascitis tumor"). Under these culture conditions the teratoma cells form aggregates of "embryoid bodies." Single cells from the central part of the embryoid bodies were injected into blastocysts. The resulting mice had coat-color mosaicism and several other genetic markers from the donor cells. Production of F₁ progeny showed that the mosaic mouse also had normal sperm derived from the malignant teratoma cell. (Courtesy of K. Illmensee.)

(called mid-blastula transition) is reached, synthesis of most types of RNA starts simultaneously.

Intracellular cytoplasmic determinants are important in development, but cannot by themselves explain it. As development progresses, cell–cell interactions become increasingly important, as shown by the experiment in Figure 23–18.

23–3 NUCLEOCYTOPLASMIC INTERACTIONS

Nucleus and cytoplasm are interdependent; one cannot survive without the other. The cytoplasm provides most of the energy for the cell through oxidative phosphorylation (in mitochondria) and anaerobic glycolysis, and the cytoplasmic ribosomes contain most of the "machinery" for protein synthesis. The nucleus provides mRNA and also supplies the other important RNA molecules (rRNA and tRNA).

That the nucleus is necessary for cell survival was first found by Waller in 1852, who cut nerves in half and found that the part of the nerve cells containing the nucleus survived, while the axons degenerated. Experiments with protozoa then showed that enucleated fragments are able to sustain most cellular activities; e.g., they can form a cellulose membrane and carry on photosynthesis (plant cells), react to stimuli and ingest food (amoebae), activate cilia (ciliated cells), undergo cytoplasmic streaming, and so forth. These cells generally survive for only a limited time, however, and cannot multiply. An extreme

case is the unicellular marine alga *Acetabularia*, which can survive for many weeks and can even form a specialized "cap" structure (which is normally used to store spores) in the absence of a nucleus.[31]

The interdependence between nucleus and cytoplasm is best shown by experiments involving cell fusion and by microinjection of nuclei into frog oocytes.

Red Blood Cell Nuclei Can Be Reactivated by Cell Fusion

Cells can be fused through the use of inactivated *Sendai virus* (a member of the parainfluenza viruses) and other agents that affect membrane structure, such as polyethyleneglycol and lysolecithin. Through these techniques a nucleus can be placed in a different cytoplasmic environment.

The initial product of the fusion of two different cells is a *heterokaryon* (i.e., a single cell containing nuclei of two types), as shown in Figure 23–19. Eventually, both nuclei might enter mitosis synchronously, form a single metaphase plate, divide, and produce a *hybrd cell line* (also known as a *synkaryon*). The cells of a hybrid cell line have a single nucleus containing chromosomes from both parental nuclei.

In 1965 H. Harris[32] found that chick erythrocyte nuclei are reactivated when fused to HeLa cells (an undifferentiated cell line so

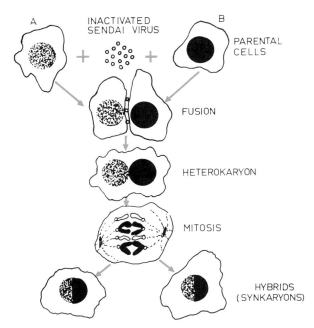

Fig. 23–19. Cell fusion with Sendai virus results first in the production of a *heterokaryon* with two nuclei, one from each parent. A hybrid cell line (synkaryon) arises when the two nuclei undergo mitosis synchronously. The cells of the hybrid cell line have one nucleus that contains chromosomes from both parents and are capable of multiplying in cell culture. (Modified from Ringertz, N.R., and Savage, R.E.: Cell Hybrids. Academic Press, Inc., New York, 1976.)

named because it was derived from the uterine carcinoma cells of a woman named *H*enrietta *L*acks). These heterokaryons are of interest because the erythrocyte nucleus does not normally synthesize RNA or DNA. Chick erythrocytes are terminally differentiated cells that have a highly condensed nucleus and are destined to die (Fig. 23–20, inset). (Mammalian erythrocytes normally eliminate their nuclei during red blood cell maturation, whereas the red cells of birds, amphibians, and reptiles retain their nuclei, which become inactivated.) When fused to HeLa cells, the chick erythrocyte nucleus increases 20 times in volume, disperses its chromatin, resumes RNA synthesis, develops a nucleolus, and eventually replicates its DNA. This process is accompanied by the uptake of large amounts of human nuclear proteins[33] which are thought to reactivate the erythrocyte nucleus.

These experiments clearly show that the synthesis of macromolecules in a nucleus is controlled by the cytoplasmic environment. Even though the erythrocytes are terminally differ-

entiated cells, they can resume RNA and DNA synthesis.

Only the cell cytoplasm is required to reactivate the chick erythrocyte nucleus. This was established by fusing erythrocytes to enucleated HeLA cells, which were still able to reactivate the nuclei.[34] Populations of enucleated cultured cells can be obtained by centrifugation after treatment with cytochalasin B, a drug that inhibits actin microfilaments. Cells exposed to this drug cease moving and extrude their nuclei, which remain attached to the rest of the cell by only a cytoplasmic stalk. Figure 23–21 shows the method used to obtain large numbers of enucleated cells.[35] Cultured cells that have been grown attached to a plastic surface are centrifuged in the presence of cytochalasin B, and the nuclei form a pellet, while the cytoplasm of the cells remains attached to the plastic surface. The detached nucleus in the pellet is surrounded by the cell membrane and a small amount of cytoplasm and is called a *karyoplast*. The enucleated cytoplasm is called a *cytoplast*.

Cells enucleated by the cytochalasin method are viable for at least two days after enucleation and perform many cell functions normally. For example, cell movements, pinocytosis, and contact inhibition are unaffected, and if cells are detached from the plastic surface of a petri dish, they can reattach and spread on a new dish.[36] Thus it is evident that many cytoplasmic functions are independent of the cell nucleus.

Because karyoplasts are still surrounded by the cell membrane, it is possible to fuse them to a different cytoplasm by using Sendai virus.[37] *Reconstituted cells*, which arise from the fusion of a karyoplast and a cytoplast, can survive longer than enucleated cytoplasm, can synthesize RNA, and in some cases can undergo cell division.

In a few cases it has been possible to activate latent genes coding for specific proteins by fusion of differentiated cells. For example, albumin-secreting rat hepatoma cells fused with mouse lymphocytes (which do not produce albumin) will occasionally give rise to clones of hybrid cells that are able to produce both rat and mouse albumin.[38] Somatic cell hybridization experiments show that the patterns of nucleic acid synthesis and gene expression by a nucleus can be modified by substances present in the cell cytoplasm. Similar conclusions can be drawn

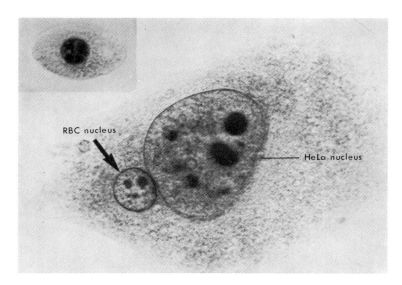

Fig. 23–20. Heterokaryon formed by fusing a chick erythrocyte with a human HeLa cell. The inset shows a normal chick erythrocyte with condensed, inactive chromatin. Three days after fusion the red blood cell nucleus has enlarged, and its chromatin has dispersed. (Courtesy of N.R. Ringertz.)

from work with nuclei transplanted into *Xenopus laevis* oocytes.

Cell Fusion Yields Pure Antibodies of Medical Importance

When animals are injected with a macromolecule of a shape that is recognized as foreign to that individual (e.g., a protein or complex polysaccharide), *antibodies* appear in the serum

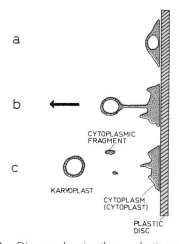

Fig. 23–21. Diagram showing the enucleation of cells adhering to plastic discs. *a*, at the beginning of the experiment the cells are attached to the plastic surface. *b*, the nuclei are pulled out into protrusions by the combined action of cytochalasin and centrifugal force. *c*, the stalks connecting the nuclei are broken by centrifugation, and karyoplasts (nuclei surrounded by a rim of cytoplasm) form a pellet, while the cytoplasts (enucleated cells) remain attached to the plastic disc. (From Ringertz, N.R., and Savage, R.: Cell Hybrids. New York, Academic Press, Inc., 1976.)

several days later. Antibodies are protein molecules that are able to bind specifically to the foreign *antigen*. Many different antibodies appear in the serum of an immunized animal, each one recognizing a different part of the antigen's shape (or different antigens if more than one was injected). Furthermore, individuals of the same species have different immunological responses, so that two antisera directed against the same antigen can in fact be very different. This is a major problem in medicine because, for example, success or failure of an organ transplantation depends on whether the donor and the recipient patient have the correct *histocompatibility antigens* on the cell surface. Clearly, antibodies that can be used throughout the world as standardized diagnostic reagents are highly desirable. These have been made possible through the use of cell fusion techniques to create cell lines that produce only a single type of antibody, which can then be obtained in unlimited amounts.

Antibodies are produced by the lymphocytes. (When producing large amounts of antibody, lymphocytes adopt a special morphology and are also called plasma cells.) Each antibody-producing cell can synthesize only one type of antibody. Lymphocytes do not multiply in culture, but G. Köhler and C. Milstein developed a technique whereby a *single* antibody-producing cell can be propagated indefinitely in culture by hybridization with a tumor cell.

Figure 23–22 shows the way in which these pure "monoclonal" antibodies are made. Initially a mouse is immunized against the desired

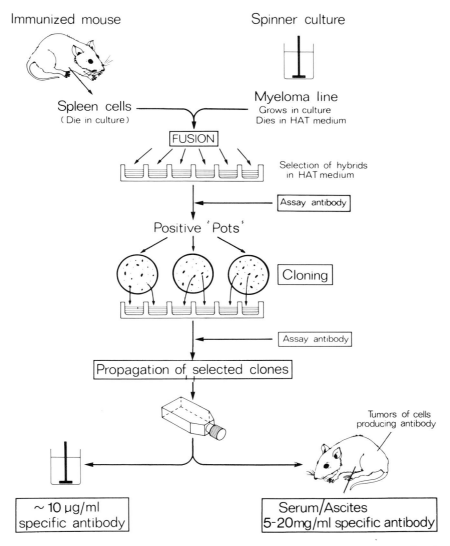

Fig. 23–22. Production of a monoclonal antibody. The hybrid cell line produces only one type of antibody because a single colony (a clone) of cells is propagated. The resulting hybrid cell line can be grown in suspension (spinner) culture, or as a tumor in mice. (Courtesy of C. Milstein and G. Köhle.)

antigen. With Sendai virus, the lymphocytes from the spleen of this mouse are then fused to a mouse cell line derived from plasma cell tumors (called *myelomas* because they invade the bone marrow). The myeloma cell carries a mutation in the enzyme *hypoxanthine-guanine phosphoribosyl transferase* and is therefore unable to grow in HAT medium, as explained in Figure 18–16. However, the resulting hybrid cells are able to grow in HAT medium (Fig. 23–22), and the clones that secrete the desired antibody can be identified and grown. The fusion with the tumor cell "immortalizes" the

spleen lymphocyte, which can then be grown indefinitely in culture or injected into mice, where the cells produce secreting tumors that can be maintained by serial transplantation. Because all of the cells of one clone are derived from a single lymphocyte, a *monoclonal antibody* of high purity is produced.

Monoclonal antibodies have proven very useful as antiviral antibodies (e.g., for diagnosis of influenza virus type and for the treatment of rabies), as diagnostic reagents in clinical biochemistry, for the classification of human histocompatibility antigens, and for the purification on an industrial scale of protein products (such as interferon) made in bacteria by genetic engineering techniques.

Monoclonal antibodies provide a good example of the fact that the practical applications of science cannot be predicted beforehand. Cell fusion was initially developed to study the seemingly specialized cell biology problem of nucleocytoplasmic interactions after fusion of chick erythrocytes with HeLa cells,[32] but eventually produced immunological tools of practical importance to all mankind.

Gene Expression by Somatic Nuclei Is Reprogrammed in *Xenopus* Oocytes

Frog oocytes are growing egg cells, obtained from the abdominal cavity of frogs, which due to their large size (1.2 mm) and tolerance for micromanipulation have been used in numerous microinjection experiments. Aside from their hardiness, oocytes interest developmental biologists because they are destined to become eggs and embryos and already contain most of the molecular machinery necessary for early development. Oocytes are active in RNA synthesis but do not synthesize DNA, and if somatic cell nuclei are injected into them, the transplanted nuclei also show this type of synthetic activity. The mature frog egg (which is laid into pond water), in contrast, does not synthesize RNA but actively synthesizes DNA after penetration by sperm or a microinjection needle. Nuclei transplanted into mature eggs also mimic the behavior of the host cell, actively replicating their DNA but not synthesizing RNA.[39]

The oocyte cytoplasm not only affects the pattern of macromolecular synthesis but is also able to *reprogram* the expression of individual genes in transplanted nuclei. Figure 23–23 shows *Xenopus* oocytes that were injected with a suspension of about 200 HeLa nuclei. The nuclei survive for several weeks inside the oocytes and during that period synthesize substantial amounts of RNA. In the first few days after injection the somatic nuclei enlarge to about ten times their original volume. This change is accompanied by chromatin dispersion and a massive uptake of proteins from the surrounding cytoplasm.[40] The injected nuclei resemble the oocyte's own nucleus morphologically, and sometimes the human chromosomes form structures reminiscent of lampbrush chromosomes (Fig. 23–23E).

The resemblance between the injected and oocyte nuclei is not only morphological. The oocyte cytoplasm reprograms the gene expression of the injected nuclei in such a way that only those genes that are normally active in oocytes are expressed. This can be determined by taking advantage of the fact that the RNAs synthesized by the injected nuclei in the course of several days accumulate˙ in the cytoplasm, where they code for new proteins that can be labeled with radioactive amino acids.[41] In one such experiment, shown in Figure 23–24, *Xenopus* cultured kidney cell nuclei were injected into oocytes of a different amphibian species (the salamander *Pleurodeles*, whose proteins can be readily distinguished from those of *Xenopus* by two-dimensional electrophoresis).[42] Those genes that are normally expressed in kidney cells but not in oocytes became inactive after injection into *Pleurodeles* oocytes. More importantly, some oocyte-active genes that were not expressed by the kidney cell nuclei were activated by the oocyte cytoplasm. The oocyte therefore contains components that can determine that a particular spectrum of protein-coding genes will be active and that others will be inactive.

The work with cell fusion and *Xenopus* oocytes suggests that the cytoplasm of all cells contains components that determine the state of activity of nuclear genes. If these components were distributed asymmetrically among daughter cells, they could play a crucial role in the establishment of cell differentiation.

SUMMARY
Nucleocytoplasmic Interactions

Experiments in which a nucleus is placed in a foreign cytoplasm have provided insights on how genes are switched on and off.

Cell fusion by Sendai virus allows a nucleus to be placed in a different cytoplasm. A single cell with two nuclei is called a *heterokaryon*. Cell nuclei that no longer synthesize RNA or DNA (e.g., chicken erythrocytes) increase in volume, resume RNA synthesis, and may replicate their DNA when fused with HeLa cells. Reactivation is probably related to the entrance of proteins from the cytoplasm. *Karyoplasts* and *cytoplasts* can be obtained by the action of cytochalasin B and can be reconstituted into cells by Sendai virus fusion, in which the nucleus and the cytoplasm come from different cells. Fusion of B lymphocytes with myeloma cell

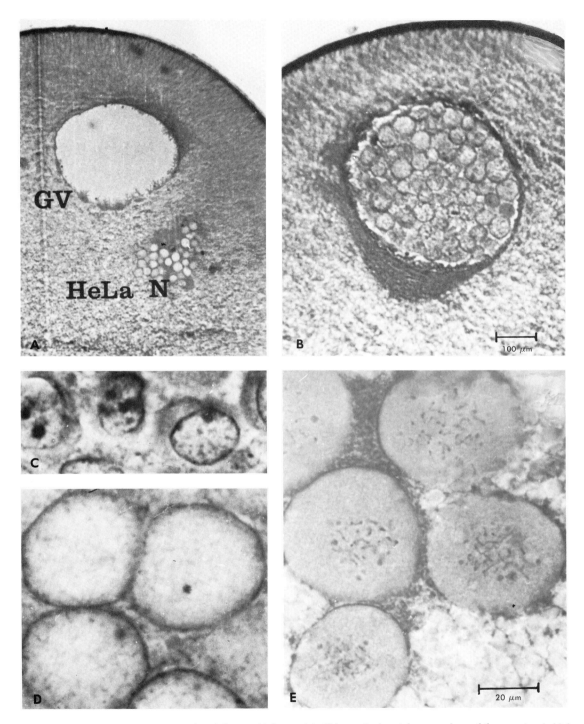

Fig. 23–23. *Xenopus* oocytes injected with human HeLa nuclei. *GV*, germinal vesicle or nucleus of the oocyte. *A*, HeLa nuclei injected into the cytoplasm three days previously. *B*, HeLa nuclei injected into the oocyte nucleus enlarge further. *C*, HeLa nuclei immediately after injection into the cytoplasm. *D*, three days later, the nuclei have swollen more than 10-fold. *E*, one week after injection the HeLa chromosomes become visible, and the HeLa nuclei tend to resemble the nucleus of the oocyte, which is in meiotic prophase (late lampbrush stage). (From E.M. De Robertis.)

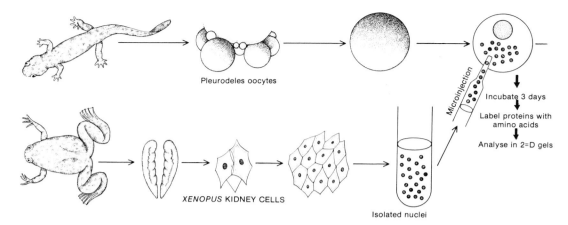

Pleurodeles oocytes

Microinjection

Incubate 3 days

Label proteins with amino acids

Analyse in 2=D gels

XENOPUS KIDNEY CELLS

Isolated nuclei

Fig. 23–24. Microinjection of *Xenopus* somatic nuclei into *Pleurodeles* oocytes. The proteins from these two species can be distinguished in two-dimensional gels. See the text for details. (From De Robertis, E.M., and Gurdon, J.B.: Gene transplantation and the analysis of development. Sci. Am., *241*:74–82, 1979.)

lines permits the production of immortal cell lines that secrete highly specific *monoclonal antibodies* (Fig. 23–22).

Frog oocyte cytoplasm is able to reprogram the expression of genes in transplanted nuclei. This reprogramming ability was found when kidney nuclei of *Xenopus* were injected into oocytes of the salamander *Pleurodeles* (Fig. 23–24). The work with somatic cell fusion and with transplantation of nuclei into oocytes suggests that the cytoplasm can indeed reprogram the activity of nuclear genes.

23–4   MOLECULAR MECHANISMS OF CELL DIFFERENTIATION

There is not a single eukaryotic gene for which we understand for certain how it is expressed in some tissues but not in others. The more we known about eukaryotic gene regulation, however, the more it becomes clear that eukaryotic genes are controlled at multiple levels.

Two main mechanisms can be envisaged: those in which genes are differentially activated (at the transcriptional, post-transcriptional, or translational levels) or those in which the genes themselves are altered (by amplification, DNA rearrangements, and methylation). These mechanisms are known to operate in different cell systems, and we discuss some examples in the following section.

Control at the Level of Transcription

The nuclear transplantation experiments discussed earlier (Fig. 23–10) show that in most cases genes are not irreversibly lost during cell differentiation. Therefore the differences between specialized cells must be explained in terms of variable gene activity. Transcriptional control is probably the most important mechanism. The clearest example is provided by polytene chromosomes, in which transcription can be directly visualized in the form of puffs. As we saw in Chapter 22, specialized cells have distinct patterns of puffing and therefore of transcription. Transcriptional control has been clearly proven also for those genes that code for abundant specialized proteins such as globin, ovalbumin, and silk fibroin.

To make hybridization probes for these genes the mRNA is copied into DNA with reverse transcriptase (producing complementary DNA or cDNA) and then the cDNA is cloned into plasmids. With these hybridization probes, transcripts from the corresponding gene can be detected only in those cells that produce the protein product (e.g., ovalbumin mRNA is detected in oviduct after stimulation with estrogen). Enormous amounts of protein can be produced from the transcriptional induction of a single protein-coding gene because stable mRNA molecules can be translated many times. A fully induced chicken ovalbumin gene makes 17 mRNA molecules per minute[43] (24,500 molecules a day). A single silkworm fibroin gene makes 10^{10} protein molecules in the course of a few days; roughly 10^5 mRNA molecules are made and translated about 10^5 times each. We still do not know, however, the mechanism by which the transcription of these genes is switched on.

Transcriptional control can best be studied by isolation of nuclei from different tissues. The RNA polymerase II molecules that had initiated RNA synthesis in vivo remain attached to the chromatin and can elongate transcription if radioactive ribonucleotides are added (although they cannot initiate new rounds of transcription under these conditions). These labeled RNAs can then be hybridized to specific cloned segments of DNA. The results from such experiments have clearly shown that transcription is tissue-specific (see the review by Darnell, 1982).

Chromatin structure is one possible level at which transcription may be controlled. As discussed in Chapter 13, genes in condensed heterochromatin are inactive, frequently in a very stable form. Nuclease digestion studies suggest that the chromatin of active genes is more susceptible to nuclease attack than that of inactive genes. One cannot tell, however, whether this nuclease sensitivity is merely a consequence of transcription or whether the increased transcription is due to a more open configuration of the chromatin in the active gene.

DNA methylation has also been considered a likely candidate to control transcription. Eukaryotic DNA contains some 5-methylcytosine, which is always next to guanosine (CpG). The enzyme that methylates DNA prefers templates that already have a methyl group on the other strand, and this preference is very interesting because it provides a stable signal on the DNA, which can be transmitted to daughter cells after DNA replication. In many cases inactive genes have been shown to be more methylated than active ones.[44] There are some species (e.g., *Drosophila*), however, in which the DNA is never methylated, and so the significance of DNA methylation is still under debate.

The post-transcriptional processing of the transcripts is also of considerable regulatory importance. Three-fourths of the transcripts made in cells are not polyadenylated and are discarded (Chapter 20), and several instances of alternative splicing pathways are known (Figs. 20–22 and 20–23). A dramatic example is the control of immunoglobulin secretion by RNA processing, which will be discussed later (see Fig. 23–32).

Translational Control

One possibility that has been considered is that cells have mechanisms by which some mRNAs can be translated in a given cell type but not in others. Microinjection studies, however, have shown that living cells can translate a wide variety of mRNAs.

Figure 23–25 shows that microinjected *Xenopus* oocytes can translate SV40 viral mRNAs. In fact, the frog oocytes can also efficiently transcribe and process injected DNAs, copying them into mature mRNAs, provided that the DNA is injected into the nucleus, which contains RNA polymerases and other factors required for transcription. When the living oocytes are used solely as a test tube for translation of mRNA, the mRNAs are introduced into the cytoplasm, which contains the protein synthesis machinery.

Many mRNAs of animal and plant origin are efficiently translated in oocytes (such as those coding for globin, immunoglobulins, thyroglobulin, interferon, collagen, tobacco mosaic virus coat proteins, and many others). This finding suggests that oocytes do not have a mechanism that excludes the translation of certain injected mRNAs.

Globin mRNA in oocytes is very stable and can be continually translated for weeks; translation is extremely efficient, and each mRNA molecule can give rise to up to 100,000 molecules of globin protein.[45] In fact, living oocytes translate mRNA much more efficiently than any cell-free system and have therefore been widely used for assaying mRNA. It is interesting to note that purified mRNA requires great care when handled in vitro to avoid degradation by ribonucleases, which are very active and ubiquitous enzymes, and that homogenates of *Xenopus* eggs and oocytes have a great deal of RNase activity. Injected *living* oocytes, however, do not degrade mRNA.

Living cells can sometimes provide very favorable conditions for biochemical analysis. The oocyte not only will translate the mRNAs, but also will segregate the proteins into the appropriate cellular compartment. The proteins from injected mRNAs coding for secretory proteins will be packed into secretory vesicles and frequently even exported out into the culture medium surrounding the oocyte.[46,47] In this way the

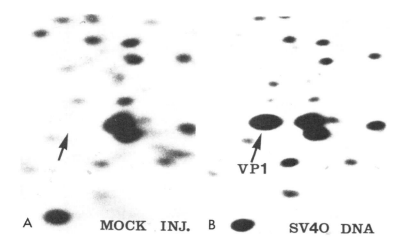

Fig. 23–25. Radioactive proteins synthesized by *Xenopus* oocytes whose nuclei were injected with DNA of SV40 virus. The DNA was transcribed, and the mRNA produced was translated into the viral protein 1 (VP1), which is absent from mock injected controls. (From E.M. De Robertis.)

pathways of cell secretion can be studied—for example, by introduction of mutant mRNAs.

Although there does not seem to be any translational selectivity for injected naked mRNAs, embryologists have long suspected the existence of "masked" mRNAs in eggs, whose translation is activated only after fertilization. Sea urchin eggs increase 10-fold their protein synthesis after fertilization, and this is due to stored maternal mRNA because the increase in protein synthesis is not inhibited by blocking new transcription with actinomycin D. The *kinds* of proteins made by sea urchin eggs before and after fertilization, however, are the same,[48] and this finding cast doubt for some time on whether true translational control existed. In other eggs, however, such as those of the clam *Spisula*, the translation of masked mRNAs for specific proteins after fertilization has been clearly demonstrated.[49] These mRNAs are thought to be bound to proteins that prevent their translation.

The clearest example of translational regulation is provided by the heat shock response of *Xenopus* oocytes. When frog oocytes are exposed to high temperature, the synthesis of a major heat-shock protein of 70,000 daltons is induced while the synthesis of the other endogenous proteins is repressed (Fig. 23–26). As we discussed in Chapter 22, most other cells also induce the synthesis of heat-shock proteins, but this synthesis is controlled by activation of the transcription of the heat-shock genes (see Fig. 22–19). In the case of the frog oocyte, however, synthesis is accomplished by a translational mechanism, i.e., by activation of the translation of stored (but masked) heat-shock mRNA. This

is known because oocytes can still respond in the same way after manual removal of the cell nucleus and because heat-shock mRNA can be detected biochemically in oocytes that do not synthesize the corresponding proteins at the normal temperature (20°C).[50] The biology of the oocyte supports this finding. The oocyte is an enormous cell (200,000 times larger than a somatic cell) with only four times the haploid amount of DNA in its nucleus, so that transcriptional activation (which does occur in *Xenopus* cultured somatic cells) could not provide a significant amount of protein in a short period of time. The stored mRNAs, however, can be made and accumulated during the several months of oogenesis. Further analysis of this mRNA activation in oocytes should provide insights into the still poorly understood mechanisms of translational control during development.

Gene Amplification—A Rare Event

One way of obtaining differential gene activity would be to increase the number of copies of a specific gene. There are three circumstances in which *gene amplification* is known to occur. One is ribosomal DNA amplification observed in the oocytes of amphibians and insects. As we saw in Chapter 21, *Xenopus* oocytes selectively replicate their rDNA genes; a mature oocyte has 2 million copies of them (compared to 900 for a diploid somatic nucleus) and 1000 nucleoli in order to produce the vast number of ribosomes (10^{12}) contained in a single egg (see Table 21–2). The second circumstance of amplification occurs

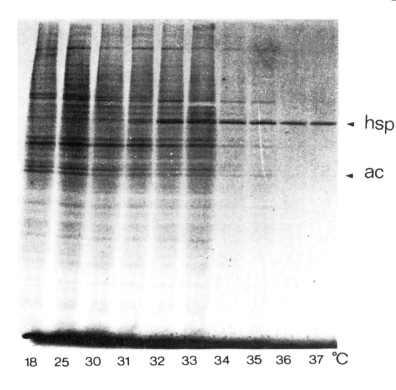

◄ hsp

◄ ac

18 25 30 31 32 33 34 35 36 37 °C

Fig. 23–26. Heat shock response in *Xenopus* oocytes. Oocytes were heat-shocked at the indicated temperature and then labeled with ^{35}S methionine for 1 hour at the same temperature; an autoradiogram of an SDS-polyacrylamide gel is shown. *hsp*, major heat shock protein; *ac*, actin. Identical results are obtained when enucleated oocytes are used, suggesting that the oocyte heat shock activation is at the translational level. (Courtesy of M. Bienz.)

in the DNA replication observed in certain puffs of the dipteran *Rhynchosciara* salivary gland polytene chromosomes[51] (see Glover et al., 1982). In these "DNA" puffs, in addition to an increased RNA synthesis as in all other puffs, there is incorporation of ^{3}H-thymidine. As Figure 23–27 shows, DNA amplification can be visualized by an increase in the amount of DNA in the particular polytene band when the puff collapses after the phase of active RNA synthesis is over. DNA puffs in *Rhynchosciara* code for salivary proteins.[52] The third case is that of the egg shell *(chorion)* proteins of *Drosophila*. The egg chorion is formed by the follicle cells during a period of five hours. It has been found that the follicle cells replicate more of a 90-kilobase segment of DNA containing the chorion protein genes, so that finally it is 16 times more abundant than the rest of the DNA.[53,54]

Although it is clear that selective gene amplification does indeed occur, in all known cases the specially amplified DNA is not passed on to future cell generations. The larval polytene salivary glands die at metamorphosis, the follicle cells die when the egg is laid, and the amplified oocyte rDNA is not inherited by the frog embryo.

The possibility that specific gene amplification occurs for other genes that produce large amounts of protein products has been explored, but with negative results. Nucleic acid hybridization experiments have shown that there is only one copy per haploid genome of globin or chicken ovalbumin genes in all tissues, regardless of whether the gene is preferentially expressed. In other words, a single globin gene, when fully activated, can given rise to all the globin required by a red blood cell. Gene amplification does not seem to be a widespread phenomenon that can explain most cases of cell differentiation.

Drug Resistance Can Induce Gene Amplification

Cells in culture can be forced to amplify genes for certain proteins by selective pressure. Methotrexate is an analogue of folic acid that inhibits the enzyme dihydrofolate reductase (DHFR). The product of DHFR action (tetrahydrofolate) is required for purine and thymidine biosynthesis. Methotrexate is used for the clinical treatment of rapidly dividing tumor cells (especially leukemias), and it has been known for a long time that tumor cells eventually become resistant to the drug. Studies in cell culture have

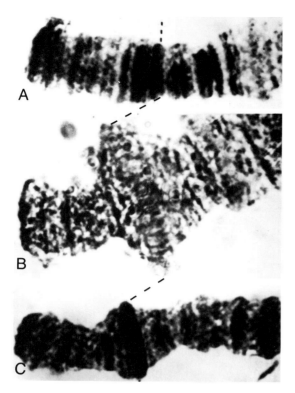

Fig. 23–27. DNA amplification in polytene chromosomes of the dipteran *Rhynchosciara angelae*. The evolution of a "DNA puff" of the salivary gland is shown. *A,* a band of the chromosome is indicated before it becomes active. *B,* the band becomes decondensed and is now called a *puff.* This puff incorporates both ^{3}H uridine and ^{3}H thymidine (commonly puffs make only RNA). *C,* the puffed region becomes inactive again and the DNA collapses into a band that is now much stronger than in *A.* Because the chromosomes were stained with the Feulgen reaction, which is specific for DNA, this experiment shows that the DNA of a specific band was replicated differentially. The puff produces transcripts coding for a protein that is secreted in large amounts in *Rhynchosciara* saliva. (Courtesy of F.J. Lara.)

shown that, after stepwise addition of methotrexate to the culture medium, mutant cells can be obtained which overproduce DHFR, and in fact have increased their number of DHFR genes from 1 to 250 genes per haploid genome.[55] A very long segment of DNA (10^6 base pairs) becomes tandemly repeated, perhaps owing to rare unequal crossing-over events (Fig. 22–9) that might occur spontaneously in cell populations. These duplications could be followed by selection for those cells with increased gene numbers. The stretches of amplified DNA can be seen cytologically as "*homogeneously staining regions*" (HSRs) producing very long arms on some chromosomes.[56] When the selective pres-

sure is removed, the extra sequences tend to be lost, and small, acentric chromatin bodies can be observed in mitotic cells (these tend to come in pairs, perhaps a result of DNA replication, and are called "*double minute*" *chromosomes* or DMs).

Pathologists have observed several chromosomal abnormalities associated with human cancers, and it is not infrequent to find HSRs and DMs in tumors (from patients who have not received drug therapy). As we discussed in Chapter 22, a series of oncogenes has been found in human DNA and some of these oncogenes were found to hybridize strongly to the abnormal chromosomes, which represent, in fact, oncogenes that have been amplified selectively. It is not yet known how frequently this event causes human cancer, nor which is the initial stimulus that selects for the amplification of these genes.

Although the amplification of genes can clearly be "forced" by drugs or other selective agents, there is no evidence that it plays a role in any developmental process leading to normal cell differentiation.

Transposable Genes in Yeast and Trypanosomes

There are cases in which DNA rearrangements cause stable changes in patterns of gene expression—for example, the mating-type switch in yeast. Baker's yeast (*Saccharomyces cerevisiae*) generally propagates as haploid cells, but at a low frequency two cells may fuse and produce diploids that are able to produce spores if life becomes hard (i.e., when food is less plentiful). The cells can fuse, however, only if they are of opposite "sex" or mating type. The mating types are *a* and α. When haploid strains of one mating type are cultured, some cells are transformed into cells of the opposite sex with a certain frequency. If the latter are cultured again, they can switch sex again. Furthermore, these changes are stable and persist over several cell generations.

Genetic and DNA cloning studies have shown that yeast switch their sex in a most unexpected way.[57,58] Figure 23–28 shows that the genetic information for the α or *a* mating types is stored in silent places on yeast chromosomes. These storage sites are called silent *cassettes*, and their expression is activated only when these genes

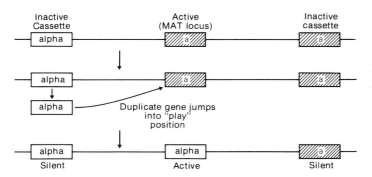

Inactive Cassette | Active (MAT locus) | Inactive cassette

alpha | a | a

alpha | a | a

alpha | Duplicate gene jumps into "play" position

alpha (Silent) | alpha (Active) | a (Silent)

Fig. 23–28. Mating type switching in yeast. The genes for the two mating types, α and a, are stored in inactive "cassettes." They can be copied and inserted into the "play" site or mating type locus (MAT), inducing a switch in the "sex" of the yeast strain.

are transposed into the expressed ("play") position on the "MAT" mating type locus. The silent genes are copied before their transposition (by the mechanism of gene conversion discussed in Chapter 22) and therefore a copy remains at their original position (Fig. 23–28).

Another example of transposition of silent genes into an expression site is provided by the surface antigens of *Trypanosoma brucei*. This flagellate is transmitted to mammals by the tsetse fly and causes sleeping sickness in humans. The trypanosome surface is covered by large amounts of a glycoprotein against which the host produces antibodies. The parasite has adapted by expressing periodically a different surface glycoprotein. Waves of trypanosomes of a new antigenic type appear in the blood of patients at 7- to 10-day intervals, causing a periodical flare-up of the sickness. It has been shown that each trypanosome has more than 100 surface glycoprotein genes, but they are not expressed until transposed into a single "expression site,"[59] in a way analogous to yeast mating type switching.

DNA transpositions of this type are of course extremely attractive as a mechanism for producing stable changes in gene expression. The nuclear transplantation results (Fig. 23–9) demand that if there were any change in the genome, it should be reversible, and these directed transposition mechanisms allow for this. It should be pointed out, however, that studies with restriction enzymes have shown that DNA transpositions do not occur however, near the globin, silk fibroin, or ovalbumin genes in the tissue where they are expressed, and thus in most instances cell specialization can occur without changes at the DNA level.

Gene Rearrangements during Lymphocyte Differentiation—Molecular Immunology

A mammalian organism is always being invaded by foreign substances or *antigens*, many of which (microorganisms, for example) would be lethal if not promptly destroyed. The organism responds by producing *antibody* molecules that will bind to the antigen leading to its subsequent destruction. Figure 23–29 shows the structure of an antibody molecule, which consists of two *light* and two *heavy chains* joined together by disulphide (—S—S—) bridges. Each light or heavy chain consists of two parts: a *variable region*, which differs for each antibody specificity, and a *constant region*, which is the same in many antibody molecules. The composition of the two variable regions (V_L and V_H) will determine the properties of the antibody *combining site*. The constant part of the heavy chain (C_H) determines other properties related to the elimination of the antigen. The heavy chain has three globular protein *domains* (C_{H1}, C_{H2}, C_{H3}) of related amino acid composition which presumably arose in evolution through duplications of a common DNA sequence. At the DNA level each domain (including the *hinge* region, Fig. 23–29) is encoded in a different *exon*, and the introns are then removed by RNA splicing. The antibody molecule provides one of the best examples of how different functional domains of proteins can be brought together by the splicing mechanism, a process that probably accelerates evolution because it allows the assembly of new genes from parts of old ones (Chap. 20).[60]

It has been estimated that a mouse has a repertoire of at least 2×10^7 different antibody combining sites. The whole genome of a mouse can code for only, say, 10^6 genes in all, so it was clear that each antibody gene could not be encoded in the DNA in the usual way. The presence of a variable and a constant region in the same protein chain led Dreyer and Bennet in 1965[61] to make the radical proposal that the variable and constant parts of the protein chains were not encoded in the same contiguous stretch

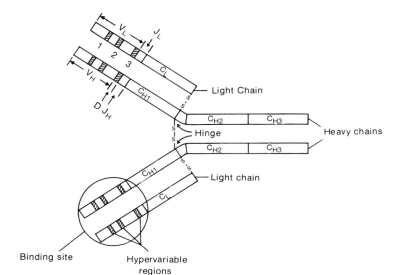

Fig. 23–29. Schematic diagrams of an IgG antibody molecule. C_{H1}, C_{H2}, C_{H3}, and C_L are the heavy (H) and light (L) chain constant region domains. These are regions 100 to 125 amino acids long within the chains that show some homology with each other in amino acid sequence and secondary structure. Each domain (and the hinge region) is encoded in a separate exon; the exons are then spliced together by RNA processing. The parts of the light chain V region coded by the V_L and J_L gene segments are shown, as are the sections of the heavy chain V region coded by V_H, D, and J_H.

of DNA, and that the great variety of antibodies was generated by bringing together one of many possible variable chain sequences and a single constant chain sequence.

When the new tools of DNA cloning and sequencing became available, molecular biologists (such as Tonegawa, Leder, Hood, and others) were able to unravel the way in which these genes are rearranged to generate diversity.

Let us start by analyzing the heavy chain genes (Fig. 23–30): In the mouse germ line there are about 200 variable chain genes, which are located thousands of nucleotides away from the constant chain gene but on the same chromosome. There are eight kinds of constant heavy chain sequences (μ, δ, $\gamma3$, $\gamma1$, $\gamma2b$, $\gamma2a$, ϵ, and α) which give rise to immunoglobulins IgM, IgD, IgG3, IgG1, IgG2b, IgG2a, IgE, and IgA.

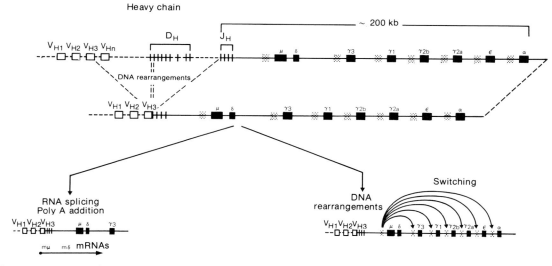

Fig. 23–30. Heavy-chain DNA rearrangements during the development of a B lymphocyte. The first event joins one V segment (out of about 200 variable regions) to a D segment (out of 10 to 50 diversity segments), and this D segment to a J segment (out of 4 possible joining segments). This event determines the type of variable region the cell will express from then on. All B cells express initially the μ heavy chain and δ heavy chains (by differential splicing of long transcripts). After antigen stimulation the cell may switch to producing a different type of immunoglobulin (IgG, IgE, or IgA) with the same variable region. This switch is mediated by a second DNA rearrangement that involves the deletion of the constant heavy chain segments that preceded the expressed one. For example, a cell expressing γ, will delete μ, δ, and $\gamma3$. The DNA regions used for recombination in the heavy chain class switch are indicated by the stippled boxes. The intron-exon structure of each gene is not shown in this diagram. (Courtesy of K. Marcu.)

Antibodies are produced by *B lymphocytes.* They are called *B* because they are initially derived from bone marrow; *T lymphocytes* are derived from the thymus and have other functions in the immune response. When an immature B lymphocyte becomes committed to produce an antibody, one of the V regions becomes joined at the DNA level to two segments of DNA called D (for "diversity") and J (for "joining") (Fig. 23–30). The DNA segments between V, D, and J are deleted and lost after the recombination events. There are at least ten (most likely many more) D segments, which give rise to one of the hypervariable regions of the V chain (indicated in Fig. 23–29). There are four J regions. Because the segments can be recombined at random, we can calculate that with heavy chain rearrangement, at least 8000 heavy chains are possible:

$$\underset{V}{200} \times \underset{D}{10} \times \underset{J}{4} \times \underset{\substack{\text{constant} \\ \text{region}}}{1} = 8000$$

After the heavy chain genes are rearranged in a pre-B cell, the light chain genes (which are on another chromosome) become rearranged in a similar way (except that they lack a D segment). Thus by recombination at least 600 light chains are possible:

$$\underset{V}{150} \times \underset{J}{4} \times \underset{\substack{\text{"Kappa"} \\ \text{constant} \\ \text{region}}}{1} = 600$$

Because the antigen combining site results from the properties of the V_H and V_L regions, by recombination of gene segments at least

$$8000_H \times 600_L = 4,800,000$$

different antibody molecules can be generated.

In fact, the potential for diversity is even greater, because the recombination point between V and J can vary over a range of several nucleotides, producing proteins that have different amino acids at the junction point. Thus the number of possible molecules is brought close to the requirements of antibody diversity (reviewed by Leder, 1982 and Tonegawa, 1985).

Each lymphocyte produces only one type of antibody, which is determined by the early recombination events. During the lifetime of the cell, however, *somatic mutation* can produce additional changes in the V regions, introducing finer adjustments in the affinity of an antibody for its antigen. Somatic mutation has been shown to occur only in the variable regions and not in the constant regions and thus must have a specialized mechanism of its own, which is still incompletely understood.

Control of Immunoglobulin Secretion by Differential RNA Processing

Figure 23–31 shows that after a B lymphocyte has undergone DNA rearrangement, it expresses IgM (and usually IgD as well) on its membrane. This cell will stay quiescent until the binding of specific antigen to its surface induces it to proliferate vigorously and to differentiate further. Proliferation will produce *clones* of cells producing the same type of antibody. The immune system works by the *clonal selection* (by proliferation) of those B cells that have an antibody on their surface that has a good fit with the shape of the antigen.

After a cell has been stimulated by antigen, it starts *secreting* IgM. This switch from membrane-bound to secreted IgM is controlled at the level of RNA processing.[62] As Figure 23–32 shows, membrane-bound form is coded by an mRNA that has two extra exons at its poly-A end. This confers to the resulting protein a highly hydrophobic region on its COOH-terminus, which traps the IgM in the ER membrane, from which it is transferred to the plasma membrane. After stimulation by antigen, however, the transcriptional unit is polyadenylated earlier, eliminating the membrane exons and thus the hydrophobic COOH region (Fig. 23–32). The resulting IgM protein will now be secreted into the surrounding medium.

Many cells co-express IgM and IgD. This co-expression results from very long transcripts (shown in Fig. 23–30), which include the constant regions for both, but all the segments corresponding to IgM (μ) are treated as if they were introns and spliced out. Thus RNA processing is an important level of control in the expression of immunoglobulin genes.

The Heavy-Chain Class Switch Involves Additional DNA Deletions

B lymphocytes stimulated to proliferate by antigen become specialized cells (called *plasma cells*), with an extensive RER and secreting into the bloodstream up to 1000 antibody molecules

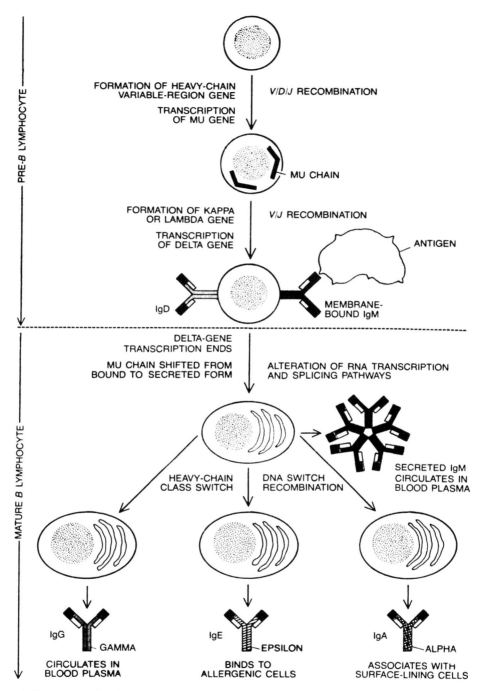

Fig. 23–31.   Differentiation of antibody-producing cells begins with the pre-B lymphocyte. The heavy chain is manufactured first, then the light chain. IgM and IgD are displayed on the cell surface. When a specific antigen is recognized and bound by a surface antibody, the cell is driven further in development: it proliferates to form a clone of mature B lymphocytes, which are specialized to synthesize large amounts of protein. The expression and the arrangement of the heavy-chain constant-region gene may change so that different types of antibody are produced. The variable regions remain the same, and the antibody continues to be directed against the same antigen, but point mutations can accumulate to change the variable regions slightly in the course of maturation. (Courtesy of P. Leder, from Sci. Am. *246*:102, 1982.)

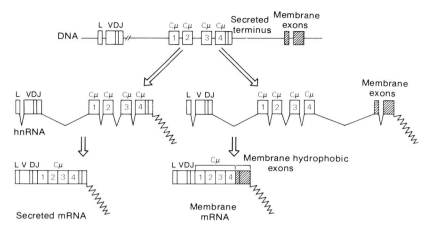

Fig. 23–32. Diagram representing the processing of messenger RNA in IgM that is secreted and in IgM that remains attached to the membrane. Observe that the membranous form has two extra exons at the poly A end, which code for highly hydrophobic amino acids. This hydrophobic region, near the COOH terminal, is used to fix IgM to the membrane. When the production of IgM is stimulated by means of an antigen, these exons are eliminated by choosing an earlier polyadenylation site, and the protein, lacking the hydrophobic region, is secreted. L, signal leader peptide (hydrophobic region required for transport into the ER); V, variable chain (already rearranged at the DNA level by joining to the D and J segments); D, diversity segment; J, joining region; C$_\mu$ 1 through 4, constant domains of μ genes. According to the choice of polyadenylation sites, the different exons are spliced together to produce two mRNAs that differ at their 3′ end. (After the studies of Early et al., Cell *20*, 313–319 1980.)

per second per cell under full expression. The descendants of one clone of cells will have identical variable regions, but some cells will undergo a *heavy-chain class switch*, whereby instead of IgM, other types of antibody, such as IgG, IgE, or IgA, are secreted (Fig. 23–31). The difference lies in the constant region of the heavy chain and confers important properties to the antibodies. For example, IgG can bind *complement* (a series of serum proteins), which will induce killing of microorganisms; IgE binds to receptors on the surface of *mast cells* (a specialized cell that contains large amounts of histamine), where the binding of antigen produces release of histamine and allergic reactions such as asthma; and IgA is present in epithelia.

This switch during lymphocyte differentiation is due to a second deletion at the DNA level of the intervening constant segments that precede the expressed one,[63,64] as explained in Figure 23–30.

The differentiation of a lymphocyte illustrates how intricate the control of cell differentiation can be. It involves initial DNA rearrangements that bring a variable gene next to a constant one, followed by changes in the pattern of RNA processing that change a membrane protein into a secreted one, and a final DNA deletion that brings the variable segment close to a new con-

stant region of different biological properties. The more molecular biologists probe into the genome, the more flexible and changing it seems to be. Although the DNA deletion events present in B lymphocytes differentiation are not applicable to all types of cell specialization (e.g., the nuclear transplantation experiments preclude irreversible losses of DNA), it would be indeed surprising if they were not utilized in other cases of terminal cell differentiation. As for the differential RNA processing mechanisms, first discovered in the immune system, they are now known to occur in many other examples of differential gene expression (Figs. 20–22 and 20–23).

Cell Differentiation in Adult Tissues

Clearly gene expression can be regulated at many levels. We still do not know why a globin gene (or any other gene) is active in red blood cells and inactive in other cells. The methodological advances in cell and molecular biology have been so considerable, however, that we have little doubt that gene control will be understood soon. This knowledge, however, although of great importance, might not be enough. It may turn out that the principles involved in the initial establishment of the differentiated state will be very different from those involved in the maintenance of differential gene expression.

It is possible that the same principles involved in the action of egg cytoplasmic determinants could also apply to *adult cells* (reviewed by Gurdon, 1978). Figure 23–33 shows how this could occur in adult and embryonic cells. In differentiation of adult tissues it is frequently observed that only one of the daughter cells becomes specialized; the other one remains as a *stem cell*, which is able to divide again. Stem cells occur in red blood cell differentiation and could occur in skin and intestinal epithelium, in which the dividing cells are located in certain regions of the epithelium (attached to the basal membrane or at the bottom of the intestinal crypts).

The hypothetical mechanism shown in Figure 23–33b is supported by experimental evidence. During nerve cell differentiation in grasshoppers, some cell divisions result in the formation of a neuron (ganglion cell) and a stem cell (neuroblast), which are always in the same position and morphologically recognizable. By introducing a needle at mitosis, it is possible[65] to rotate the spindle and chromosomes 180 degrees; but despite this rotation, the resulting daughter cells still have the neuron and stem cell in the normal position. Thus the ability to become a neuron does not depend on a particular chromosome set but rather on the type of cytoplasm inherited by the daughter cell.

The idea that the cytoplasm contains determinants that can become unequally distributed in the daughter cells and affect nuclear activity is by no means new. In the 1896 edition of his classic book *The Cell in Development and Heredity*, E.B. Wilson viewed development as follows:

If chromatin be the idioplasm [an old term referring to the genes] in which inheres the sum-total of hereditary forces, and if it be equally distributed at every cell-division, how can its mode of action so vary in different cells as to cause diversity of structure [i.e., differentiation]? Through the influences of this idioplasm [i.e., the genes] the cytoplasm of the egg, or of the blastomeres derived from it, undergoes specific and progressive changes, each change reacting upon the nucleus and thus inciting a new change. These changes differ in different regions of the egg because of preexisting differences, chemical and physical, in the cytoplasmic structure; and these form the conditions under which the nucleus operates.

Thus some of our views on development have changed little over the years. Yet along the way we have learned much about how genes are controlled. The new methods now available have made eukaryotic gene control one of the fastest advancing fields of science. The sense of optimism and excitement of those who believe that, indeed, we will be able to understand soon how genes are controlled during embryonic development is, we think, well justified. We find it fitting to conclude this chapter with a quotation from another great embryologist, Donald D. Brown, who in 1981 wrote: "After all, a few years ago we could not imagine how we could ever isolate a gene."

SUMMARY
Molecular Differentiation

How is a gene expressed in certain tissues and not in others? It seems that not one but multiple mechanisms are used. Activation of *transcription* is probably the most common one, used in many protein-coding genes, e.g., globin in red blood cells, ovalbumin in oviduct, and silk fibroin in the silk gland. Exactly how this is achieved is not known. Chromatin structure and DNA methylation have been explored as possible mechanisms.

Translational control is known to occur in some eggs that have "masked" mRNAs that are translated only after fertilization. The best example of translational control is provided by the heat-shock mRNAs of *Xenopus* oocytes, which are stored but not translated unless the temperature is raised (Fig. 23–26).

In *gene amplification* specific genes are selectively replicated to attain a higher level of expression. Gene amplification happens with rDNA in some oocytes, the DNA puffs of *Rhynocosciara*, and with the chorion (eggshell) genes of *Drosophila*. It is a rare event, and in all cases the amplified DNA is not passed on to future cell generations.

Transposition of genes, from a "silent" site in the chromosome into an "expression" site, where they are actively transcribed, occurs in the yeast mating type switch (Fig. 23–28) and in trypanosomes that change periodically their surface antigens as a defense against the host's immune system. Transposition is known *not* to occur for globin, ovalbumin, and many other genes.

The production of antibodies by B lymphocytes is the best example of how intricate the control of cell differentiation might be. Initially variable region DNA is brought close to a constant region by a DNA deletion. A membrane-bound antibody molecule is initially produced (Fig. 23–31), but upon stimulation with antigen the cell starts *secreting* antibody by a mechanism controlled at the

(a) Propagation of the determined or specialized state

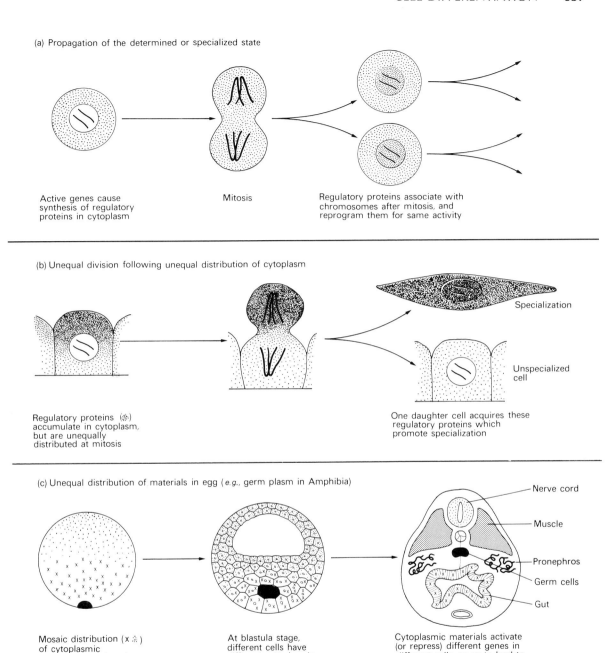

Active genes cause
synthesis of regulatory
proteins in cytoplasm

Mitosis

Regulatory proteins associate with
chromosomes after mitosis, and
reprogram them for same activity

(b) Unequal division following unequal distribution of cytoplasm

Specialization

Unspecialized
cell

Regulatory proteins (※)
accumulate in cytoplasm,
but are unequally
distributed at mitosis

One daughter cell acquires these
regulatory proteins which
promote specialization

(c) Unequal distribution of materials in egg (e.g., germ plasm in Amphibia)

Nerve cord

Muscle

Pronephros

Germ cells

Gut

Mosaic distribution (x ÷)
of cytoplasmic
materials in activated egg

At blastula stage,
different cells have
different cytoplasmic
compositions

Cytoplasmic materials activate
(or repress) different genes in
different cells, so as to lead to
cell differentiation

Fig. 23–33.   Possible mechanism for the cytoplasmic control of cell differentiation. *a,* under normal conditions the daughter cells have the same differentiated state as the parent cell. *b,* the parent's gene-regulating ectoplasmic molecules become unequally distributed, and one of the daughter cells differentiates, while the other remains as a stem cell. *c,* egg cells have unequally distributed determinants that cause embryonic cells to specialize in different ways, as in the case of the germ plasm, which induces germ cell differentiation. (Courtesy of Gurdon, J.B.: Gene Expression During Cell Differentiation. Carolina Biology Readers. Vol. 25. Carolina Biological, 1978.)

RNA processing level (Fig. 23–30). Finally, another DNA deletion might bring the same variable region closer to a new type of constant heavy chain, producing different types of immunoglobulin (IgG, IgE, IgA).

It is possible that in adult tissues mechanisms similar to those described by cytoplasmic determinants in embryos might operate. For example, when cytoplasmic components become unequally distributed among daughter cells, one cell may become differentiated while the other remains as a *stem cell* (Fig. 23–33).

REFERENCES

1. Yasumura, Y., Tashjian, A.H., and Sato, G.H.: Science, *154*:1186, 1966.
2. Augusti-Tocco, G., and Sato, G.: Proc. Natl. Acad. Sci. USA, *64*:311, 1969.
3. Hadorn, E.: *In* The Genetics and Biology of *Drosophila*. Vol. 2c. Edited by M. Ashburner and T.R.F. Wright. New York, Academic Press, Inc., 1978, pp. 555–617.
4. Gehring, W., and Nöthiger, R.: *In* Developmental Systems: Insects. Vol. 2. Edited by S. Counce and C. Waddington. New York, Academic Press, Inc., 1973, pp. 211–290.
5. Gehring, W.J.: *In* The Genetics and Biology of *Drosophila*. Vol. 2c. Edited by M. Ashburner and T.R.F. Wright. New York, Academic Press, Inc., 1978, pp. 511–554.
6. Morgan, T.H.: Science, *32*:100, 1910.
7. Struhl, G.: Nature, *293*:36–41, 1981.
8. Nüsslein-Volhard, C., and Wieschaus, E.: Nature, *287*:795–801, 1980.
9. Garcia Bellido, A.: Am. Zool., *17*:613–629, 1977.
10. Lawrence, P.A.: J. Embryol. Exp. Morphol., *64*:321–332, 1981.
11. Brower, D., Lawrence, P.A., and Wilcox, M.: Dev. Biol., *86*:448, 1981.
12. Kornberg, T.: Proc. Natl. Acad. Sci. USA, *78*:1095, 1981.
13. Levine, M., Hafen, E., Garber, R., and Gehring, W.: EMBO J., *2*:617, 1983.
14. Gurdon, J.B.: J. Embryol. Exp. Morphol., *10*:622, 1962.
15. Gurdon, J.B., Laskey, R.A., and Reeves, O.R.: J. Embryol. Exp. Morphol., *34*:93, 1975.
16. Whabl, M.R., Brun, R.B., and Du Pasquier, L.: Science, *190*:1310, 1975.
17. Steward, F.C.: Proc. R. Soc. Lond. (Biol.), *175*:1, 1970.
18. Conklin, E.: J. Acad. Natl. Sci. Philadelphia, *13*:1, 1905.
19. Whittaker, J.R.: J. Embryol. Exp. Morphol., *55*:343, 1980.
20. Whittaker, J.R.: Dev. Biol., *93*:463, 1982.
21. Okada, M., Kleinman, I.A., and Scheiderman, H.A.: Dev. Biol., *37*:43, 1974.
22. Illmensee, K., and Mahowald, A.: Proc. Natl. Acad. Sci. USA, *71*:1016, 1974.
23. Laskey, R.A., Gurdon, J.B., and Trendelenburg, M.: Br. Soc. Dev. Biol. Symp., *4*:65, 1979.
24. Brachet, J.: Curr. Top. Dev. Biol., *11*:133, 1977.
25. Gerhart, J., et al.: Nature, *292*:511, 1981.
26. Newport, J., and Kirschner, M.: Cell, *30*:675, 1982.
27. Graham, C.F., and Morgan, R.W.: Dev. Biol., *14*:439, 1966.
28. Illmensee, K., and Stevens, L.C.: Sci. Am., *241*:87, 1979.
29. Illmensee, K., and Mintz, B.: Proc. Natl. Acad. Sci. USA, *72*:3585, 1975.
30. Brinster, R.L.: J. Exp. Med., *140*:1049, 1974.
31. Haemmerling, J.: Ann. Rev. Plant Physiol., *14*:65, 1963.
32. Harris, H.: Nature, *206*:583, 1965.
33. Ringertz, N.R., Carlsson, S.A., Ege, T., and Bolund, L.: Proc. Natl. Acad. Sci. USA, *68*:3228, 1971.
34. Ege, T., Zeuthen, J., and Ringertz, N.R.: J. Cell Sci., *1*:65, 1975.
35. Prescott, D.M., Myerson, D., and Wallace, J.: Exp. Cell Res., *71*:480, 1972.
36. Goldman, R.D., Pollack, R., and Hopkins, N.H.: Proc. Natl. Acad. Sci. USA, *70*:750, 1972.
37. Veomett, G., Prescott, D.M., Shay, J., and Porter, K.R.: Proc. Natl. Acad. Sci. USA, *71*:1999, 1974.
38. Malawista, S.E., and Weiss, M.C.: Proc. Natl. Acad. Sci. USA, *74*:1502, 1974.
39. Gurdon, J.B.: J. Embryol. Exp. Morphol., *20*:401, 1968.
40. Gurdon, J.B., De Robertis, E.M., and Partington, G.A.: Nature, *260*:116, 1976.
41. De Robertis, E.M., Partington, G.A., Longthorne, R.F., and Gurdon, J.B.: J. Embryol. Exp. Morphol., *40*:199, 1977.
42. De Robertis, E.M., and Gurdon, J.B.: Proc. Natl. Acad. Sci. USA, *74*:2470, 1977.
43. Palmiter, R.D.: J. Biol. Chem., *248*:8260, 1973.
44. Felsenfeld, G. and McGee, J.: Nature, *296*:602, 1982.
45. Gurdon, J.B., Lingrel, J.B., and Marbaix, G.: J. Mol. Biol., *80*:539, 1973.
46. Lane, C.D.: Cell, *24*:281, 1981.
47. Valle, G., Jones, E.A., and Colman, A.: Nature, *300*:71, 1982.
48. Bradhorst, B.P.: Dev. Biol., *52*:310, 1976.
49. Rosenthal, E.T., Hunt, T., and Ruderman, J.V.: Cell, *20*:487, 1980.
50. Bienz, M., and Gurdon, J.B.: Cell, *29*:811, 1982.
51. Pavan, C., and Da Cunha, A.B.: Annu. Rev. Genet., *3*:425, 1969.
52. Winter, C.E., De Bianchi, A.G., Terra, W.R., and Lara, F.J.: Chromosoma, *61*:193, 1977.
53. Spradling, A.C.: Cell, *27*:193, 1981.
54. Spradling, A.C., and Mahowald, A.P.: Cell, *27*:203, 1981.
55. Schimke, R.T., Kaufman, R.J., Alt, F.W., and Kellems, R.F.: Science, *202*:1051, 1978.
56. Bostock, C.J., and Clark, E.M.: Cell, *19*:709, 1980.
57. Strathern, J.N., and Herskowitz, I.: Cell, *17*:371, 1979.
58. Hicks, J., Strathern, J.N., and Klar, A.J.S.: Nature, *292*:478, 1979.
59. Borst, P., and Cross, G.A.: Cell, *29*:291, 1982.
60. Gilbert, W.: Nature, *271*:501, 1978.
61. Dreyer, W.J., and Bennet, J.C.: Proc. Natl. Acad. Sci. USA, *54*:864, 1965.
62. Early, P., et al.: Cell, *20*:313, 1980.
63. Maki, R., et al.: Proc. Natl. Acad. Sci. USA, *77*:2133, 1980.
64. Maki, R., et al.: Cell, *24*:353, 1981.
65. Carlson, J.G.: Chromosoma, *5*:199, 1952.

ADDITIONAL READINGS

Akam, M.: The location of ultrabithorax transcripts in *Drosophila* tissue sections. EMBO J., *2*:2075, 1983.
Awgulewitsch, A., et al.: Spatial restriction of a mouse homeobox locus within the central nervous system. Nature, *320*:328, 1986.
Briggs, R., King, T.J.: Nucleocytoplasmic interactions in

eggs and embryos. *In* The Cell. Vol. 2. Edited by J. Brachet and A.E. Mirsky. New York, Academic Press, Inc., 1959.

Brown, D.D.: Gene expression in eukaryotes. Science, *211*:667, 1981. (A comprehensive review on the molecular biology of differentiation mechanisms.)

Carrasco, A.E., McGinnis, W., Gehring, W.J., and De Robertis, E.M.: Cloning of an *X. laevis* gene expressed during early embryogenesis coding for a peptide region homologous to *Drosophila* homeotic genes. Cell, 37:409, 1984.

Darnell, J.E.: Variety in the level of gene control in eukaryotic cells. Nature, 297:365, 1982.

Davidson, E.H.: Gene Activity in Early Development. New York, Academic Press, Inc., 1986.

De Robertis, E.M., et al.: The Xenopus Homeoboxes. Cold Spring Harbor Symp. Quant. Biol., 50:271, 1985.

De Robertis, E.M., and Gurdon, J.B.: Gene transplantation and the analysis of development. Sci. Am., *241*:73, 1979.

Doyle, H.J., Harding, K., Hoey, T., and Levine, M.: Transcripts encoded by a homeobox gene are restricted to dorsal tissues in Drosophila embryos. Nature, 323:76, 1986.

Gehring, W.J.: The molecular basis of development. Sci. Am., 253:152, 1985.

Gehring, W.J.: The homeobox: A key to the understanding of development? Cell, *40*:3, 1985.

Gehring, W.J.: Developmental genetics in *Drosophila*. Annu. Rev. Genet., *10*:209, 1976.

Glover, D.M., et al.: Gene amplification in *Rhynchosciara* salivary gland chromosomes. Proc. Natl. Acad. Sci. USA, 79:2947, 1982.

Gurdon, J.B.: Gene Expression during Cell Differentiation. Carolina Biology Readers. Vol. 25. Burlington, North Carolina, Carolina Biological Supply Co., 1978.

Gurdon, J.B., Laskey, R.A., De Robertis, E.M., and Partington, G.A.: Reprogramming of transplanted nuclei in amphibia. Int. Rev. Cytol. (Suppl.), 9:161, 1978.

Hafen, E., Levine, M., and Gehring, W.J.: Transcripts of the homeotic antennapedia gene complex. EMBO J., 2:617, 1983.

Harris, H.: Nucleus and Cytoplasm. Oxford, Clarendon Press, 1974. (Highly recommended.)

Illmensee, K., and Hoppe, P.C.: Nuclear transplantation in *Mus musculus*: Developmental potential of nuclei from preimplantation embryos. Cell, 23:9, 1981.

Köhler, G., and Milstein, C.: Continuous cultures of fused cells secreting antibody of predefined specificity. Nature, *256*:495, 1975.

Lawrence, P.A.: The cellular basis of segmentation in insects. Cell, 26:3, 1981.

Leder, P.: The genetics of antibody diversity. Sci. Am., *246*:72, 1982. (Highly recommended.)

Lewis, E.B.: A gene complex controlling segmentation in *Drosophila*. Nature, *276*:565, 1978.

McGinnis, W., et al.: A homologous protein-coding sequence in *Drosophila* homeotic genes and its conservation in other metazoans. Cell, 37:403, 1984.

Milstein, C.: Monoclonal antibodies. Sci. Am., *243*:66, 1980.

Morata, G., and Lawrence, P.A.: Homeotic genes, compartments and cell determination in *Drosophila*. Nature, *265*:211, 1977.

Morgan, T.H.: Experimental Embryology. New York, Columbia University Press, 1927.

Müller, M.M., Carrasco, A.E., and De Robertis, E.M.: A homeobox-containing gene expressed during oogenesis in *Xenopus*. Cell, 39:157, 1984.

Pollack, R.: Readings in Mammalian Cell Culture. 2nd Ed. Cold Spring Harbor. New York, Cold Spring Harbor Laboratory, 1981.

Ringertz, N.R., and Savage, R.E.: Cell Hybrids. New York, Academic Press, Inc., 1976. (Recommended for study of cell fusion.)

Schneiderman, H.A.: New ways to probe pattern formation and determination in insects. *In* Insect Development. Edited by P.A. Lawrence. Oxford, Blackwell Scientific Publications, Ltd., 1976.

Shepherd, J.C.W., et al.: Fly and frog homeo domains show homologies with yeast mating-type regulatory proteins. Nature, *310*:70, 1984.

Strome, S., and Wood, W.B.: Immunofluorescence visualization of germ-line-specific granules in embryos, larvae and adults of *C. elegans*. Proc. Natl. Acad. Sci. USA, 79:1558, 1982.

Tonegawa, S.: The molecules of the immune system. Sci. Am., *253*:122, 1985.

Whittaker, J.R.: Cytoplasmic determinants of tissue differentiation in the Ascidian egg. *In* Determinants of Spatial Organization. Edited by S. Subtelny and I.R. Konisberg. New York, Academic Press, Inc., 1979.

Wilson, E.B.: The Cell in Development and Heredity. 3rd Ed. New York, MacMillan Pub. Co. Inc., 1928. (The best treatment of cytoplasmic localizations in development.)

CELLULAR AND MOLECULAR
NEUROBIOLOGY

One of the most important functions of living organisms is reacting to an environmental change. Such a change, called a *stimulus*, generally elicits a *response*. In its most basic sense this general property is called *irritability*. For example, a unicellular protozoon may react to stimuli, such as changes in heat or light or the presence of a food particle, by a mechanical response, such as ciliary motion, ameboid movement, etc. *Plants* react by slow responses, which produce differential growth, also called a *tropism*. Irritability reaches its maximal development in animals, and special cells forming the nerve tissue are differentiated to respond rapidly and specifically to stimuli.

In these organisms special physiological *receptors* are differentiated. Receptors are made of special cells or of the distal endings of neurons, specialized to receive a particular stimulus. For example, the receptors of light, touch, taste, pressure, heat, and cold are characterized by their great sensitivity to the specific stimuli.

In an animal, the responses to stimuli vary. The different types of responses are produced in special tissue (e.g., muscles, glands, electric plates, luminous organs), called *effectors*, that are controlled by efferent neurons. The simplest mechanism of nerve action is represented by the so-called monosynaptic reflex. This consists of a neuronal circuit formed by two *neurons*. One neuron is *sensorial* (afferent) and has a receptor at one end to receive the stimulus. At the other end the sensory neuron makes a special contact, called a *synapsis*, with a *motor* (efferent) neuron, which in turn acts on the effector (i.e., muscle).

Figure 24–1 is a simplified diagram of the way in which the information *received* at the receptor is *conducted* along the sensory neuron and then *transmitted* at the synapse. Notice that a new wave of information starts in the second neuron, which finally reaches the effector, where the final response is elicited. In nerves and muscle, information is propagated by a change in the *resting* or *steady potential*. This change originates a *wave of excitation*, which moves along the surface of the cell. In the nerve cell this *propagated* or *action potential* is also known as a *nerve impulse*.

The objective of this chapter is to introduce the cellular and molecular bases of *nerve conduction* and *synaptic transmission*. The ultrastructure of the axon, the biosynthetic proper-

ties of the perikaryon, and the nature of axon flow both in anterograde and retrograde directions are emphasized in the first part of this chapter. Synaptic transmission can be electrical, but it is more frequently chemically driven. Chemical synapses have a complex structure, both at the membranes and within the ending, in which the main components are the synaptic vesicles, the true quantal units of the transmitter. In general terms, in chemical synapses a localized process of neurosecretion takes place that is similar to the production of other neurohumors. Synaptic transmission is the result of the interaction between the neurotransmitter and specific receptor proteins present at the neuronal membrane. This interaction produces a translocation of ions and, in some cases, the production of the secondary messengers cyclic AMP and GMP.

24–1 GENERAL ORGANIZATION OF NERVE FIBERS—AXON TRANSPORT

The neuron is specially differentiated for the conduction and transmission of nerve impulses. After embryonic life, neurons, in general, do not divide, but remain in a permanent interphase throughout the life of the organism. During this time a neuron undergoes an increase in volume and number and complexity of its processes and functional contacts. This fact may be of paramount significance because in addition to conducting and transmitting impulses, nerve cells store instinctive and learned *information* (e.g., conditioned reflexes, memory)—a property that would be best served by a more permanent system of structures than in cells that divide.

Neurons are adapted to their specialized functions by means of different types of outgrowths. The cell body (perikaryon) may emit one or more short outgrowths, or *dendrites*, which carry nerve impulses centripetally, and a longer one, the *axon*, which carries the pulse centrifugally to the next neuron or the effector. An axon is also called a *nerve fiber* when it is wrapped in the sheaths after emerging from the neuron. The axon terminates, ramifying in the telodendrons, or endings. Some neurons *(bipolars)* have only one dendrite and the axon and others *(monopolars)* have only the axon that divides into an afferent and an efferent nerve fiber. *Multipolars*

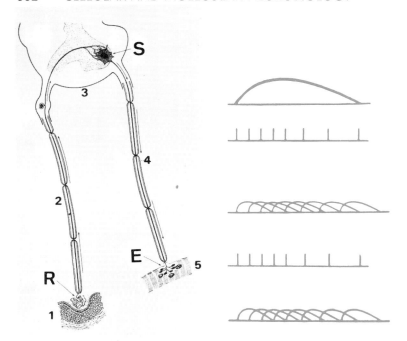

Fig. 24–1. Diagram showing the monosynaptic reflex arc. *Left,* one sensory and one motor neuron with the synaptic function. Notice the receptor and the effector. *Right,* different types of potentials produced at the portions of the reflex arc (1–5), indicated in the figure. *R,* receptor; *S,* synapsis; *E,* effector.

are the most common. In invertebrates most neurons are monopolar.

The perikaryon is characterized by the presence of considerable amounts of basophilic material—the Nissl substance—which is composed of ribosomes and ER. A well-developed Golgi complex is also characteristic of the neuron. The abundance of ribosomes is related to the biosynthetic functions of the perikaryon, which has to maintain a volume of cytoplasm in the outgrowths that may be considerably greater than its own. If a nerve fiber is cut, the distal part degenerates (wallerian degeneration), and the proximal stump may regenerate later on by a growing process that is dependent on the perikaryon. There is also experimental evidence that the axon is continuously growing and being used at the endings.[1]

Axon Ultrastructure—Neurotubules and Neurofilaments

In fixed and stained preparations, observed under the light microscope, the neuronal cytoplasm shows a fine structure made of the so-called *neurofibrils.* These run in all directions and continue into the dendrites and the axon. The electron microscope has revealed that the neurofibrils result from the clumping together of two main fibrous components, the *neurotubules* and the *neurofilaments.*

Neurotubules or *neuronal microtubules* were first discovered in this type of cell.[2] They have characteristics similar to those of other microtubules described in Section 6–2 and are present in all types of cells. In the axon, neurotubules have a parallel arrangement (Fig. 24–2) and in cross section show the typical dotted wall, mainly composed of *tubulin.* These microtubules also have microtubule associated proteins (MAPs) of high molecular weight that in vitro are bound periodically to the wall. In some cases a granular material may be observed inside the microtubules (Fig. 24–3). Unmyelinated axons generally contain a larger number of these structures.

Neurofilaments belong to the system of intermediate filaments of 10 nm described in Section 6–6. They are composed of triplet polymers of 200,000, 150,000, and 70,000 daltons and show cross bridges attached to them. By the use of specific antibodies against each of the polypeptides it is possible to show that while the two polymers of smaller molecular weight are present prenatally, the 200,000-dalton ones appear during postnatal development and may have a more specialized role in the structure of the neurofilament.[3] Furthermore, the core of the filament is made of the 70,000-dalton subunit; the others are at the periphery. In studies of crushed myelinated nerve fibers in which there are wide changes in cross-sectional area during regener-

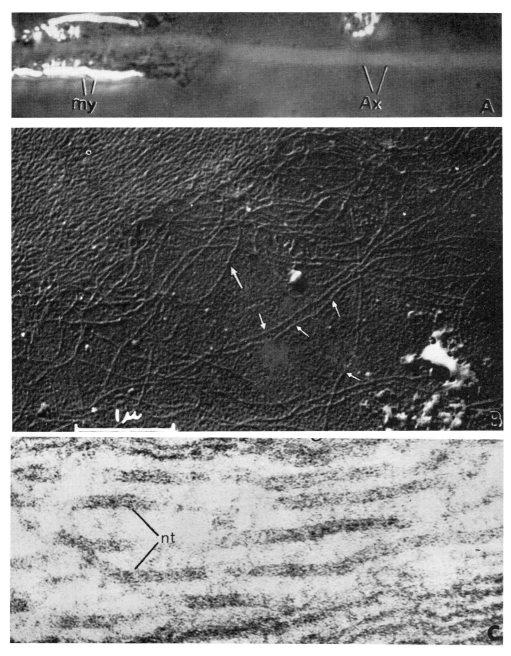

Fig. 24–2. *A,* extruded axon *(Ax)* observed under the polarization microscope. Note its weak positive birefringence. After appropriate treatment, the axon exhibits both intrinsic and form birefringence. *my,* myelin with a strong birefringence. In the normal fiber, the axon birefringence is obscured by that of the myelin sheath. ×260. (From Thornburg, W., and De Robertis, E.: J. Biophys. Biochem. Cytol., *2*:475, 1956.) *B,* fibrillar material (neurotubules and neurofilaments) observed under the EM in an axon extruded from the myelin fibers and compressed. Preparation shadow cast with chromium. The arrows indicate some neurotubules. ×260. (From De Robertis, E., and Franchi, C.M.: J. Exp. Med., *98*:269, 1953.) *C,* section of longitudinally oriented neurotubules *(nt).* ×120,000. (Courtesy of E.L. Rodríguez Echandía, R.S. Piezzi, and E.M. Rodríguez.)

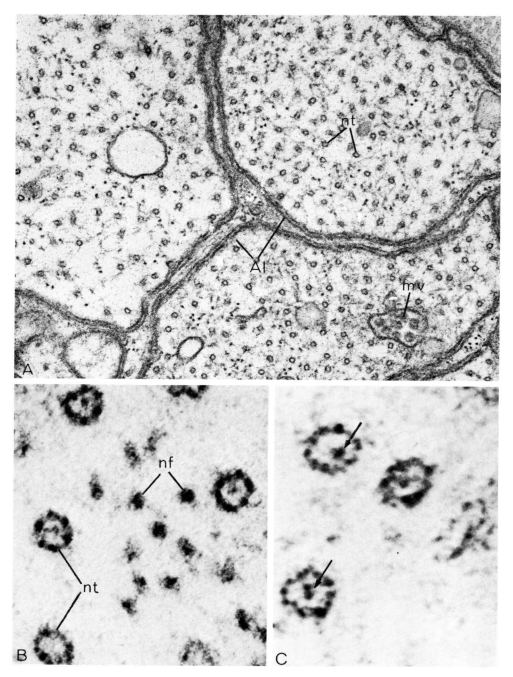

Fig. 24–3.   *A,* cross section of an unmyelinated nerve showing the axolemma *(Al),* neurotubules *(nt),* and a multivascular body *(mv).* ×60,000. *B,* same as *A* at higher magnification. *nf,* neurofilaments; *nt,* neurotubules. ×400,000. *C,* neurotubules containing a dense granule *(arrows).* ×600,000. (Courtesy of E.L. Rodríguez Echandía, R.S. Piezzi, and E.M. Rodríguez.)

ation, it has been observed that the number of neurofilaments changes in parallel with the cross-section while the microtubules do not. This means that the caliber of the nerve fiber is in some way regulated by the concentration of neurofilaments.[4] Recently a biochemical and immunological analysis of the proteins in the neuronal cytoskeleton has been achieved in cultured neurons (see Peng et al., 1986).

In addition to neurotubules and neurofilaments the axon contains membranous organelles represented by mitochondria, smooth ER, and multivesicular bodies.

The use of high voltage electron microscopy has permitted a better knowledge of the axon cytoskeleton. The microtubules and the organelles appear to be suspended in a continuous latticework of fine microtubular filaments (Fig. 24–4).[5] These findings suggest extensive "bridging" between microtubules, neurofilaments, and other cytoplasmic organelles as well as with the plasma membrane. There is evidence that these interactions are mediated by the MAPs. The connection between neurotubules and neurofilaments is apparently done by spherical 15-nm particles from which thin filaments irradiate.[6] We shall see later on that this cytoskeletal structure, by way of cycles of contraction and expansion of the lattice, could regulate the transport of organelles and macromolecules.

Biochemically the axon cytoskeleton contains both stable polymers represented by the neurofilament proteins and soluble polymers composed mainly of tubulin and actin. While the former may conserve the cytoskeletal organization, the other could allow for rapid and reversible changes.[6]

Of the various organelles present in the perikaryon of the neuron, only the smooth ER appears to continue into the axon (see Fig. 24–6). The use of thick sections (1–2 μm), impregnated with heavy metals, and high voltage electron

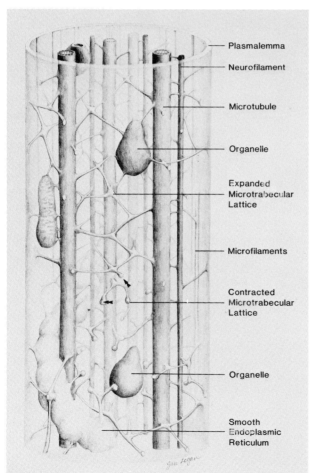

— Plasmalemma

— Neurofilament

— Microtubule

— Organelle

Expanded Microtrabecular Lattice

— Microfilaments

Contracted Microtrabecular Lattice

— Organelle

Smooth Endoplasmic Reticulum

Fig. 24–4. Diagram of the axon ultrastructure as revealed by high voltage electron microscopy. The axon cytoskeleton, in addition to microtubules and neurofilaments (intermediate filaments), shows a microtrabecular lattice in an expanded or a contracted configuration. (Courtesy of M.E. Stearns.)

microscopy, had led to recognition that the smooth ER represents a rather continuous system of parallel and interconnected channels along the length of the axon. This channel system should not be considered as made of rigid tubes, but rather as a dynamic structure that is constantly being remodeled (see Rambourg and Droz, 1980 in the Additional Readings at the end of this chapter).

Macromolecules Are Transported through the Axon

Axons and nerve endings are generally devoid of ribosomes and thus are unable to synthesize significant amounts of protein locally.[7] Most axonal and synaptic proteins must be manufactured in the perikaryon and then transported through the axon to the nerve endings. We will see that the axonal transport may be *anterograde* or *orthograde* (i.e., from the perikaryon to the nerve endings) but also *retrograde* (i.e., in the opposite direction). In both cases the velocity of transport may vary for the different proteins and structural components and may be *fast* or *slow*.

In anterograde transport, macromolecules such as proteins, glycoproteins, and enzymes, which are soluble in the axoplasm or integrate the various axonal and synaptic structures, are first synthesized in the cell body and then find their way into the dendritic arborizations. They may also pass into the axon and nerve endings, at a distance that may vary from millimeters to meters. The proteins present in nerve endings have half-lives ranging from 12 hours to 50 to 100 days; therefore new proteins must reach the nerve endings to compensate for their loss. Experimental work with colchicine and vinblastine has demonstrated that axoplasmic transport is related to the neurotubules. In fact, the integrity of these axonal organelles is critical in maintaining a normal axoplasmic flow. One of the best approaches to the study of axonal transport is the use of radiolabeled precursors for proteins or glycoproteins (e.g., ^{3}H-lysine, ^{3}H-fucose) and the monitoring of the final product by autoradiography.

The transport may be studied at the cellular level by microinjection of the soma of a single neuron with a protein or a glycoprotein precursor. The transport of the protein is then followed by autoradiography. With this method it can be observed that the protein is transported not only into the axon but also into the dendrites (Fig. 24–5). This *dendritic transport* proceeds at about 3 mm/hour and, like the axonic transport, is sensitive to colchicine.[8]

Experimentally the morphology of both anterograde and retrograde transports may be studied on the saphenous nerve submitted to local cooling through the skin for a few hours. Cooling stops both anterograde and retrograde transports. Proximal to the cooling site, vesicles and smooth ER (SER) accumulate, while distally multivesicular bodies and lysosomes accumulate.[9]

Retrograde transport comprises lysosomes, endocytosed proteins, and coated vesicles derived from the exocytosis of synaptic vesicles as well as nerve growth factor (a protein indispensable for the growth of adrenergic neurons), neurotoxins, and neuroviruses.

Anterograde Axonal Flow May Be Fast or Slow

Proteins may be transported in an anterograde direction at different rates, and there are generally two definite groups—those of fast and those of slow transport. The diagram of Figure 24–6 gives a general idea of these two systems.

Fast Axonal Flow. Axonal flow of about 250–400 mm/day is found in proteins that are synthesized in polysomes bound to the ER, are transferred to the Golgi complex, and then penetrate the axon and dendrites. On their way through the Golgi apparatus, some proteins are converted into glycoproteins by the addition of carbohydrates.

Materials that are fast transported are moved through the perikaryon, axon, and dendrites in the form of preassembled membranes and are mainly destined to the axolemma and to the synaptic terminals. These materials include membrane-bound enzymes, synaptic vesicle components, trophic factors, and a large proportion of glycoproteins that may be involved in cell–cell recognition. This fast transport can be compared with the delivery of membranes and secretory proteins in non-neuronal cells. In this case, however, the materials to be delivered become associated with a transport system that extends into the axon and dendrites (see Hammerschlag

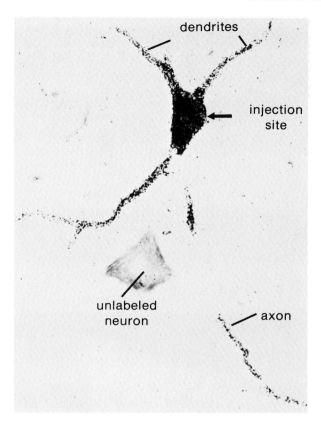

dendrites

injection site

unlabeled neuron

axon

Fig. 24–5. Experiment showing the intracellular transport of proteins. ³H-glycine was injected into the soma of a single motoneuron of the cat spinal cord, and the dendritic transport of the synthesized proteins was followed by autoradiography. *Arrow* shows the site of injection. Another motoneuron shows the lack of labeling. × 250. (Courtesy of G.W. Kreutzberg.)

and Stone, 1982 in the Additional Readings at the end of this chapter).

In autoradiographs of labeled proteins that are undergoing fast transport, the radioactivity is detected in the SER situated at the axon periphery, below the axolemma. These and other studies suggest that the SER constitutes the preferential pathway for the fast axonal transport of membrane proteins. Some proteins go by way of sub-axolemmal tubules, to be incorporated into the axolemma.

The fast moving proteins are thus assigned mainly to the renewal of various membrane components.[10] They participate in the renewal of components of the presynaptic membrane, synaptic vesicles, neurosecretory granules, and mitochondria (Fig. 24–6).

Slow Axonal Flow. In proteins present in the soluble fraction of the nervous tissue, and to a lesser degree in mitochondria, axonal flow is slow. Figure 24–6 shows that these proteins are synthesized mainly by free polysomes and probably bypass the Golgi complex.

The cytoskeletal proteins (see Table 6–1) move by slow axonal flow but in two waves: (1)

tubulin and the proteins integrating the neurofilaments move at a rate of 1 mm/day and, (2) actin, chlatrin, calmodulin, and a number of enzymes move at a rate of 3 to 4 mm/day. This difference implies that in the axon there are two cytoskeletal systems that move separately, i.e., the microtubular-neurofilament network on one side and the actin-microfilament network on the other.[11]

The enzymes related to the synthesis of neurotransmitter, such as choline acetyltransferase and tyrosine hydroxylase, are transported by slow transport. The same mechanism is used by the polypeptide neurotransmitters, such as the enkephalins. Slow transport is arrested if protein synthesis is inhibited, but it continues for 24 hours after section of the axon. The fast and slow moving proteins that enter the axon and nerve endings may compensate for the local breakdown of proteins by proteolytic enzymes, as well as the release of proteins that may be related to the function of synaptic vesicles. A transsynaptic transfer of macromolecules of small magnitude may also occur.

To be kept alive, most neurons require the

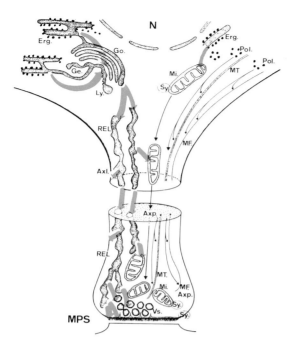

Fig. 24–6. Diagram representing the orthograde axonal transport of macromolecules. *Left, the fast transport; right, the slow transport.* In fast transport the proteins synthesized in the ergastoplasm *(Erg.)* are transferred to the Golgi complex *(Go.* and *Ge.)* where carbohydrates may be added to make glycoprotein. Along the axon the proteins and glycoproteins are transported by way of the smooth endoplasmic reticulum *(REL)*. At the rapid speed of 250 to 400 mm/day such proteins participate in the turnover of macromolecules associated with the axolemma *(Axl.)*, the synaptic vesicles *(Vs.)*, the presynaptic membrane *(MPS)*, and mitochondria *(Mi.)*. In slow transport, indicated by lightly drawn arrows, the proteins synthesized in free ribosomes *(Pol.)* are slowly transported with the axoplasm *(Axp)* along the axon. The soluble proteins, as well as the microtubules *(MT.)* and microfilaments *(MF.)* are transported at the slow speed of 1.5 mm/day, *N*, nucleus; *Ly*, lysosome; *Sy*, synaptic protein. (From Droz, B., Koenig, H.L., and Di Giamberardino, L.: Brain Res., 60:93, 1973.)

"trophic" influence from other neurons; the lack of such influence may produce a *transneuronal degeneration.* Thus, the neurons, besides communicating nerve impulses, are specifically linked by way of special chemical substances. The transneuronal transport of macromolecules was first demonstrated by the injection of labeled amino acids into the posterior chamber of the eye of a mammal. By autoradiography it was possible to follow proteins traversing the primary synapses in the geniculate nucleus and entering into the visual cortex.

As a general conclusion it can be said that axonal transport is related to synaptic function

and to other neuronal activities implying interneuronal communication (see Grafstein and Forman, 1980 in the Additional Readings at the end of this chapter). Little is known about the molecular mechanisms involved in axonal transport. Both anterograde and retrograde transports are energy-dependent, and ATP is indispensable. Calcium ions acting by way of calmodulin are also needed, and transport is blocked by Ca^{++} blockers, such as cobalt, or by drugs acting on Ca^{++} channels (e.g., verapamil). Ca^{++}-sensitive contractile proteins, such as actin and myosin, have been identified in neurons, and the transport is blocked by cytochalasin B, by an anti-α-actinin antiserum, and by a concentration of vanadate that inactivates myosin ATPase.[12] This and other evidence suggests that axon transport is mediated by mechanisms similar to those involved in cell motility, which were studied in Chapter 6 (see Thoenen and Barde, 1980, Lasek et al., 1984 in the Additional Readings at the end of this chapter).

Macromolecules Such as NGF Undergo Retrograde Axonal Transport

Although by 1917 Giuseppe Levi, observing axons of cultured nerve cells, had seen movement of particles in both anterograde and retrograde directions, the latter type of transport has been neglected until recently. By injecting high concentrations of horseradish peroxidase or serum albumin, coupled to a fluorescent dye, in the vicinity of nerve terminals, it was found that proteins that are taken up by endocytosis were transferred through the axon to the corresponding cell bodies. For example, if peroxidase is injected into the visual cortex, at the occipital pole of the brain, it is possible to trace the neurons into the lateral geniculate nucleus.[13] While this technique is important in tracing neuroanatomic connections in the central nervous system and is now widely used, the physiological importance of the retrograde transport was questionable. In fact, peroxidase is not a protein normally found in the nervous system and had to be injected at rather high concentrations.

The physiological importance of this mechanism was recognized when it was found that the *nerve growth factor* (NGF)[14] is transported by a retrograde mechanism. For example, if ^{125}I-labeled NGF is injected into the anterior chamber of the eye of a rat, the radioactivity is taken up by the adrenergic terminals of the iris and transported to the superior cervical ganglia, where the adrenergic perikarya are present. These findings suggested that NGF could act as a retrograde "trophic" messenger between the effector cells and the neurons.

At present the following two types of retrograde transport are postulated: (1) a *nonspecific* type, which depends, as indicated above, on the injection of a macromolecule at high concentration (e.g., peroxidase or serum albumin) and (2) a *specific retrograde axonal transport*, which depends on the binding of the macromolecule to specific binding sites or receptors present at the nerve-ending membrane.[15] The typical example of specific transport is that of the NGF. Other macromolecules, however, such as tetanus and cholera toxins and several lectins, may also be transported by a selective mechanism. Because there is a definite number of receptors at the membrane, with increasing concentration of the ligands, this type of transport reaches a saturation. The transport velocity of NGF in sympathetic axons was estimated at 2 to 3 mm/hr, and in sensory neurons, at 10 to 13 mm/hr. By coupling NGF to peroxidase it has been possible to

follow under the electron microscope the subcellular components that are involved in the retrograde axonal transport. As Figure 24–7 shows, they follow essentially the same route as indicated in Figure 24–6 for fast orthograde transport. The macromolecules are found in the axon in tubules of SER and then in the perikaryon, in tubules, multivesicular bodies, and dense bodies that may correspond to secondary lysosomes.[16]

NGF Function. NGF has been isolated mainly from the male mouse submaxillary gland and is a complex of three subunits (α, β, and γ) with the following composition: $\alpha_2\beta_2\gamma_2$. The molecular weight is about 130,000, and the sedimentation is 7S. This complex contains one or two zinc ions that are important in its stability. The active part of the NGF is a small subunit of 13,260 daltons. Antibodies against NGF produce extensive destruction of sympathetic neurons in neonates and an impairment of the function of neurons in the adult.

If newborn rats are treated with 6-hydroxydopamine (precursor analogue of the neurotransmitter norepinephrine), the adrenergic nerve terminals and the cell bodies of sympathetic neurons are destroyed. The degeneration

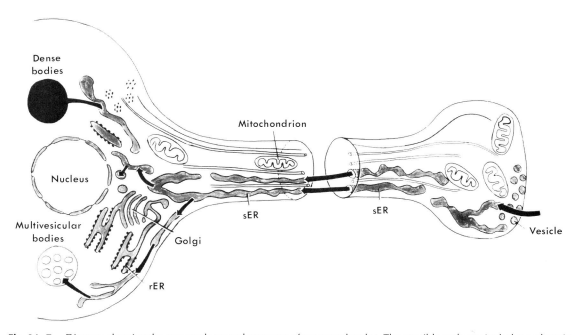

Fig. 24–7. Diagram showing the retrograde axonal transport of macromolecules. The possible pathway includes endocytic vesicles, the smooth endoplasmic reticulum *(sER)*, and multivesicular and dense bodies. Notice that the macromolecules do not penetrate the nucleus. (Courtesy of H. Thoenen.)

of the cell bodies can be prevented with NGF, however.[17] Similar results are obtained by surgical axotomy or by the action of *vinblastine*, which acts on the assembly of microtubules and blocks both anterograde and retrograde axonal transport.

The NGF is indispensable at the early stages of neuronal growth, and with increasing age (after 3 weeks) the neuronal perikaryon becomes resistant to axotomy, to the action of vinblastine, and to 6-hydroxydopamine. In the adult 6-hydroxydopamine affects only the adrenergic nerve endings. The NGF not only produces the growth of the adrenergic neuron but also stimulates the key enzymes involved in the synthesis of the neurotransmitter, *tyrosine hydroxylase* and *dopamine-β-hydroxylase*. In recent years other trophic factors acting on neurons have been described. (For a detailed account on NGF and other neuronotrophic factors, see Thoenen and Barde, 1980; Lindslay et al., 1985 in the Additional Readings at the end of this chapter).

Tetanus Toxin and Some Neurotropic Viruses May Undergo Retrograde and Transsynaptic Transport

Tetanus toxin, one of the most potent neurotoxins, is transported by the axons of the motoneurons to the anterior horn of the spinal cord by a retrograde transport similar to that of NGF. However, the toxin does not remain confined to the perikarya and dendrites of the motoneurons but is also found in the presynaptic nerve terminals that are attached to the neuron.[18] From the medical viewpoint it is interesting that the transsynaptic transfer of the toxin coincides with the first symptoms of local tetanus.

In addition to tetanus toxin, a series of neurotropic viruses (e.g., polio virus) reach the nervous system by binding to the peripheral nerve terminals and then undergoing retrograde transport, possibly followed by transsynaptic transfer. In this way viruses may gain access to the central nervous system despite the blood-brain barrier that blocks entry from the bloodstream.

TABLE 24–1.   PROPERTIES OF NEURONS OF DIFFERENT SIZES (CAT AND RABBIT SAPHENOUS NERVES)

Properties	Group		
	A	**B**	**C**
Diameter of fiber (μm)	20–1	3	—
Conduction velocity (m/sec)	100–5	14–3	2
Duration of action potential (msec)	0.4–0.5	1.2	2.0
Absolute refractory period (msec)	0.4–1.0	1.2	2.0

Modified from Grundfest, H.: Ann. Rev. Physiol., 2:213–242, 1940.

24–2   FUNCTIONS OF THE NERVE FIBER

Nerve Conduction Velocity—Related to Diameter, Myelin, and Internode Distance

Nerve fibers are *nonmyelinated* when wrapped only in Schwann cells without a myelin sheath. *Myelinated* nerve fibers also have a myelin sheath that consists of a multilayer lipoprotein system (see Fig. 4–10). In the autonomic system of vertebrates, most nerve fibers are unmyelinated and are contained within invaginations of the plasma membrane of the Schwann cells (Fig. 24–3). The myelin sheath is interrupted at the *nodes of Ranvier.*

The *internode*, i.e., the distance between successive nodes, is the segment of myelin that is produced and contained within a single Schwann cell. The internode is 0.2 mm in a bull frog fiber of 4 μm, about 1.5 mm in a fiber of 12 μm, and 2.5 mm in one of 15 μm.

Within the internode, obliquitous (conic) *incisures* go across the myelin sheath where the myelin leaflets have a looser disposition. At the node the myelin lamellae are loosely arranged, and a small zone of axon is in direct contact with the extracellular fluid. Myelinated fibers conduct nerve impulses at a much faster rate than unmyelinated fibers. The diameter of the fiber also influences the conduction rate. As shown in Table 24–1, nerve fibers can be classified according to their diameters into groups A, B, and C. C fibers are unmyelinated. The diameter may vary from 20 μm in A fibers to less than 1 μm in C fibers, and the conduction velocity varies from 100 to 2 meters or less per second. The rate of conduction of the nerve impulse follows a linear relationship with the fiber diameter in mammalian myelinated fibers, and it is also related to the internode distance.

The Action Potential—Nondecremental and Propagated as a Depolarization Wave

When a nerve fiber is stimulated, a profound change is produced in the electrical properties of the surface membrane and in the steady potential, and a wave of depolarization is conducted. Here, the subject is discussed briefly as a continuation of the discussion of active transport and membrane potentials in Section 4–3. As shown in Figure 24–8, through the use of intracellular recording, it was demonstrated that upon excitation the resting potential suddenly changes. At the point of stimulation there is not only a depolarization, but also an overshoot, and the potential inside becomes positive.

Hodgkin, Huxley and Katz, using the giant axon of the squid and a direct electrical device called the *voltage clamp*, were able to measure the kinetics of ionic permeability. (In this technique the voltage of the membrane is maintained at a fixed level and the movement of ions is recorded as the current flowing across the membrane.) With this method, later confirmed by the use of radioactive tracers,[19] Hodgkin and Huxley identified two types of ion channels, *Na+ channels* and *K+ channels*, whose opening and closing determined the so-called *action potential*. At the point of stimulation there is a sudden, and several-hundred-fold increase in permeability to Na+, which reaches its peak in 0.1 millisecond. At the end of this period the membrane again becomes essentially impermeable to Na+, but the K+ permeability increases, and this ion leaks out of the cell, repolarizing the nerve fiber. In other words, during the rising phase of the spike Na+ enters through the Na+ channels, and in the descending phase K+ is extruded through the K+ channels (Fig. 24–8). Complete restoration of the ionic balance takes a longer time after the electrical event.

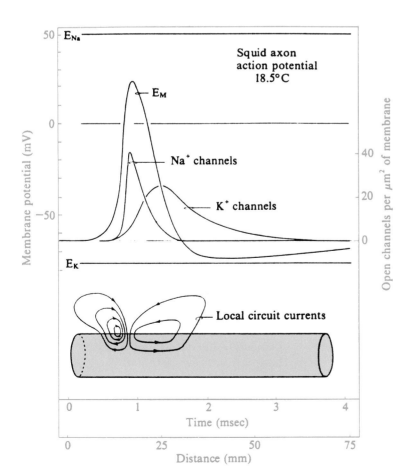

Fig. 24–8. Diagram representing the events in the action potential (E_M) in a squid axon according to the model of Hodgkin and Huxley. Notice the time sequence in milliseconds of the opening and closing of Na+ and K+ channels. The lower part of the diagram shows the local circuit current, which is spreading the depolarization toward the left into the unexcited membrane. E_{Na} and E_K represent the equilibrium potentials for Na and K ions. (From Hillie, B.: Excitability and ionic channels. *In* Basic Neurochemistry. Edited by G.J. Siegel and R.W. Albers. Boston, Little, Brown and Company, 1981. © 1981 by Little, Brown and Company.)

The action potential that develops in the nerve fiber has several other characteristics:

1. The stimulus produces a slight local depolarization in the fiber, which, after reaching a certain *threshold of activation*, produces spikes of the same amplitude. If the intensity of the stimulus is increased, the height of the spike always remains the same but the frequency of spikes increases. This is called an *all-or-none response.*

2. The nerve impulse is *nondecremental*; i.e., the amplitude of the spike does not decrease and is the same all along the course of the nerve fiber. The action potential is thus well adapted to conduction over long distances without losses (see Fig. 24–1).

3. Once a nerve impulse has passed over any point of the fiber, there is a *refractory period* during which it cannot react to another stimulus.

Nerve impulses represent the only language used by the brain to communicate at a distance. The information is coded by a train of uniformly sized action potentials, in which the only variable is the frequency (i.e., the number of impulses in a unit of time). In the experiment shown in Figure 24–1, the stronger the stimulation of the receptor, the faster the train of impulses.

The *propagation* of the nerve impulse is generally explained by the so-called *local circuit theory* (Fig. 24–8). At the point of stimulation the area becomes depolarized (negative outside) and acts as a sink toward which the current flows from the adjacent areas. This wave of depolarization advances along the nerve fiber at a rate of conduction that is characteristic for each fiber (Table 24–1). While this wave of depolarization advances, repolarization is so rapid that only a fraction of the nerve fiber (a few millimeters or centimeters, depending on the conduction rate) is depolarized at a time. In the recovery period sodium leaves the cell by the action of the sodium pump (Section 4–3) and potassium reenters to restore the steady state.

The preceding theory of nerve conduction applies to unmyelinated nerve fibers. In myelinated fibers it is thought that the local circuits occur only at the nodes. According to this so-called *saltatory theory*, at the internode the impulse is conducted electrotonically, and at each node the action potential is boosted to the same height by ionic mechanisms.[20]

Propagated Action Potentials Depend on Na$^+$ and K$^+$ Channels in the Axon Membrane

In Section 4–3 we gave an introduction to the concepts of Na$^+$ and K$^+$ channels, and we described the Na$^+$-K$^+$ pump (Na$^+$-K$^+$ ATPase), the function of which is to extrude Na$^+$ from the axon and to allow the entrance of K$^+$ (Fig. 4–16). These three components are essential to explain the molecular basis of the propagation of action potential along the axonal as well as the muscle membrane.

An ion channel can be envisioned as a hydrophilic channel formed by the interstice between protein subunits embedded in the axon membrane. It is now generally admitted that the ion channel consists of two functional elements: (1) a *selectivity filter*, which determines the kind of ion that will be translocated, and (2) *a gate*, which by opening and closing the channel, regulates the ion flow. In the Na$^+$ and K$^+$ channels of nerve (and muscle) membranes the gating mechanism is electrically driven and is controlled by the membrane potential, without the need of other energy source.

In Figure 24–9 the voltage dependence of Na$^+$ and K$^+$ channels is indicated diagrammatically. In the resting (steady state) condition both the Na$^+$ and K$^+$ channels are closed. With depolarization, the Na$^+$ channel is opened, and during repolarization it closes again, and the K$^+$ channel opens.

Sodium Channel. The Na$^+$ channel is better known than the K$^+$ channel because there are several neurotoxins that affect it in a different manner. The channel has at least three separate binding sites for the neurotoxins. For example, *tetrodotoxin* (TTX), a toxin produced by the Japanese puffer fish, and *saxitoxin* (STX), a shellfish poison, completely block the Na$^+$ channels.[21] Both molecules, labeled with tritium, can be used to determine the number of Na$^+$ channels. There are between 200 and 500 binding sites per μm^2 in the squid axon and in frog skeletal muscle, but many more present at the Ranvier nodes. TTX acts only from the outside and is generally thought to block the selectivity filter located at the axon outer surface.

A second class of Na$^+$ channel blockers are

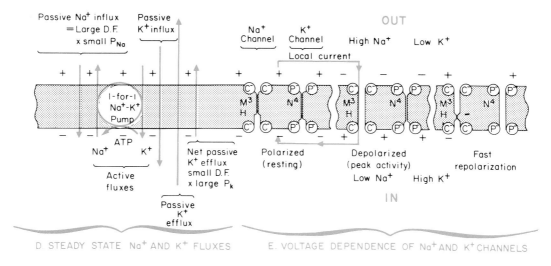

Fig. 24–9. Diagram of the molecular structure of the plasma membrane in relation to the transport of ions during the action potential. The different types of fluxes that maintain the steady state of resting potential are indicated (see Fig. 4–13). With depolarization there is the opening first of Na$^+$ and then of K$^+$ channels. (Courtesy of J.W. Woodbury.)

the *local anesthetics*, such as lidocaine and procaine, of clinical use. These substances are lipid-soluble amines that act by blocking nerve conduction, an effect that is due to the inactivation of the gating mechanism. Other drugs, such as *batrachotoxin, aconitine, veratridine*, and others, either open the Na$^+$ channels or prevent them from closing. Veratridine has been found to bind to a very hydrophobic protein (i.e., proteolipid).[22]

A third type of neurotoxin is *scorpion toxin*, which slows down the inactivation of the channel.

The minimum size of the Na$^+$ channel at the selectivity filter is estimated at 0.3×0.5 nm; thus a sodium ion (0.19 nm), to go through, should be fully or partly dehydrated. From voltage clamp studies it can be demonstrated that a single Na$^+$ channel can translocate 4000 Na$^+$/msec.

The Na$^+$ channel from rat brain has been isolated and purified as a complex having a molecular weight of 316,000 and made of three subunits (α, a large glycoprotein of 270,000, β_1 of 39,000, and β_2 of 37,000 daltons) that form a stoichiometric complex. These isolated Na$^+$ channels can be reconstituted in liposomes, and the three receptor sites for the various neurotoxins can be demonstrated in them.[23]

Potassium Channel. The potassium conductance is selectively inhibited by tetraethylam-

monium ions (TEA). In the squid giant axon, TEA blocks only the K$^+$ channel from the inside. This molecule has the same diameter as the hydrated potassium ion. This and other findings have led to the postulation of a funnel shape for the axonal K$^+$ channel. The large opening of the funnel is toward the interior, and it has a diameter of about 0.8 nm, which is large enough for the potassium ion, as well as TEA. The selectivity filter, which corresponds to the narrow end of the funnel, is only about 0.3 nm in diameter, i.e., large enough to let only the dehydrated potassium ion pass. A hydrophobic site is thought to be present at the large entrance of the channel.[24] The gating mechanism, sensitive to changes in the membrane potential could be located on the inside of the channel. The number of sites for the squid axon has been calculated to be 70/μm^2. In recent times several different types of K$^+$ channels have been described (see Hucho and Schiebler, 1977 in the Additional Readings at the end of this chapter).

Calcium Channel. In addition to the Na$^+$ and K$^+$ channel the Ca^{++} channel is of great importance because, as we have seen in earlier chapters, the entrance of Ca^{++} plays a fundamental role in many cellular functions (e.g., exocytosis, secretion, contraction, etc.). In the neuronal membrane there are a number of Ca^{++} channels that are driven by the membrane potential and are essential in the release of neurotransmitters (see Section 24–3). Ca^{++} channels can be blocked by *verapamil*, or the related drug D-600, and by divalent ions such as Mn^{++}, Co^{++}, Cd^{++}, and Ni^{++} applied from outside. A series of drugs based on a 1,4 dihydropyridine nucleus (e.g., *nifedipine*) are now used as blockers of the Ca^{++} channels and employed therapeutically to control angina as well as hypertension (see Glossman et al., 1982 in the Additional Readings at the end of this chapter).

In Section 24–5 we will see that in addition to these voltage-dependent channels there are others gated by the interaction with specific ligands (i.e., neurotransmitters). These are the receptor-linked channels of which the best known is the nicotinic receptor.

At the Physiological Receptors and Synapses Potentials Are Graded and Nonpropagated

Physiological studies have demonstrated that in addition to the all-or-none response, there is another type of electrical activity in nervous tissue. This is by far the most frequent in the central nervous system and is referred to as a *graded response*. In the graded response the impulse is not *propagated*, and the *amplitude* varies with the intensity of the stimulus. This type of response is characteristic of the physiologic receptors and synapses (see Figure 24–1). Both the *generator potentials* found at the physiological receptors and the *synaptic potentials* are graded responses.

If a peripheral receptor, such as a Pacini corpuscle, or a stretch receptor (neuromuscular spindle) is mechanically stimulated, a local, graded, and decremental potential is recorded, the amplitude and duration of which depend on the intensity and duration of the stimulus. In the Pacini corpuscle it is possible to remove most of the connective lamellae that surround the nerve ending without impairing the generator potential (Fig. 24–10). It appears that in this case the *biological transducer* capable of transforming the mechanical energy (pressure) into the electrical energy (generator potential) is localized at the sensory part of the ending. Probably the mechanical deformation of the ending produces a change in permeability with entrance of ions and partial depolarization. In the Pacini corpuscle it has been observed that the nerve

impulse starts at the first node of Ranvier (Fig. 24–10).[25]

The intensity of the sensory stimulus is reflected in the amplitude of the generator potential and this in turn in the *frequency* of the propagated signal—the stronger the generator potential, the higher the frequency. In this way the information received is coded for conduction along the nerve fiber in the form of a train or volley of impulses (see Figure 24–1).

SUMMARY
General Organization of the Neuron and Function of the Nerve Fiber

The nervous tissue is specially differentiated to respond rapidly and specifically to different stimuli. Protozoa and plants contain no nerve differentiation but have more primitive mechanisms of *irritability*. The simplest mechanism in the nervous tissue is the monosynaptic reflex arc, which is integrated by a sensorial *(afferent)* neuron and a motor *(efferent)* neuron. These two neurons are related, respectively, to a physiological *receptor* and an *effector; they are connected by synapses*. The nerve impulse is *conducted* along the axon of the nerve fibers and is *transmitted* at the level of the synapse and the effector.

Neurons do not divide during postnatal life, but undergo an increase in volume and in the number and complexity of their processes and functional contacts. Neurofibrils can be observed in the axon and dendrites after fixation and staining of the perikaryon. They result from the clumping together of neurotubules and neurofilaments. Neurotubules are similar to the microtubules of other cells (see Chapter 6). Neurofibrils are not involved in nerve conduction, which takes place at the axonal membrane.

The perikaryon of neurons is rich in ribosomes and endoplasmic reticulum (Nissl substance). The great abundance of ribosomes is related to their biosynthetic functions. If the nerve fiber is cut, the distal stump degenerates (wallerian degeneration), and the proximal stump may regenerate later on. There is little or no local protein synthesis in the axon and nerve endings, which are devoid of polysomes. Proteins and glycoproteins are synthesized in the perikaryon and are transported to the axon and nerve endings as well as to the dendrites.

The anterograde axonal transport of macromolecules depends on the integrity of neurotubules. Colchicine and vinblastine, which destroy these organelles, also stop axonal transport. At the cellular level, this process can best be studied with autoradiography.

Axonal transport may be *fast* (250 to 400 mm/day) or *slow* (1 to 1.5 mm per day). Fast transport is found in proteins integrating the axonal membrane, synaptic vesicles, presynaptic membrane,

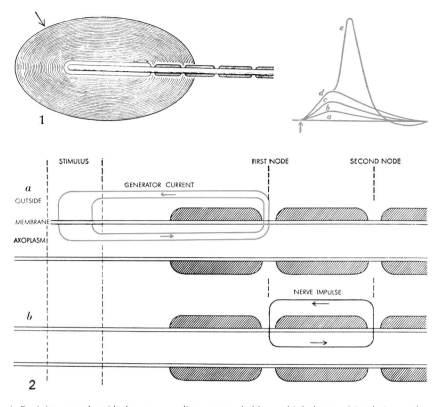

Fig. 24–10. *1,* Pacini corpuscle with the nerve ending surrounded by multiple layers. Stimulation at the point marked by the arrow produces a generator potential *(right),* which increases in amplitude *(a–d)* until it fires an all-or-none nerve impulse *(e). 2,* mechanism of the transducer. The stimulus produces a drop in the resistance of the membrane with ion transfer. Notice the generator current induced by the stimulus *(a)* and the nerve impulse *(b)* originating at the first node. (Courtesy of W.R. Loewenstein.)

and mitochondria, i.e., in membrane-bound proteins. The glycoproteins studied with ^{3}H-fucose are also fast moving. Slow axonal flow is found in soluble proteins (e.g., enzymes, tubulin) that are in the axoplasm or that integrate the neurotubules, neurofilaments, and mitochondria. Few of these proteins reach the nerve ending. The fast and slow moving proteins compensate for the local breakdown and release of proteins taking place within the axon and at the nerve ending.

There is also a retrograde axonal transport at macromolecules that are taken up by the nerve terminals and transported to the perikaryon. This type of transport is *nonspecific* when the macromolecule is an extraneous one and is injected at high concentration (e.g., peroxidase, bovine serum albumin), and *specific* when there are special binding sites at the nerve-terminal membrane. The nerve growth factor is transported in a retrograde direction at 2 to 13 mm/hr through the smooth endoplasmic reticulum of the axon to the perikaryon. Tetanus toxin is also transported by a specific mechanism, but after reaching the perikaryon of the motoneuron, it is transferred transsynaptically to the nerve endings in contact with the motoneu-

ron. It is thought that neurotropic viruses are transported by a similar mechanism.

Nerve fibers are *nonmyelinated* when they are wrapped only in Schwann cells, or they are *myelinated*. Myelin is interrupted at the nodes of Ranvier. The internode distance varies from 0.2 to 2.5 mm and is related to the conduction velocity. The diameter of the nerve fiber is also related to nerve conduction. Nerve conduction is faster with fibers of a larger diameter and which have longer distances between internodes. A, B, and C fibers can be distinguished. Nerve conduction is propagated along the axonal membrane by the *action potential.* This consists of a sudden depolarization with increased permeability to Na$^+$. The membrane potential may depolarize from -90 mV and may overshoot to $+50$ mV. In the ascending phase of the spike there is entrance of Na$^+$. In the descending phase, K$^+$ leaks out. The action potential has a threshold of activation, is an all-or-none response, is nondecremental, and has a refractory period. In unmyelinated fibers, *propagation* is accounted for by the *local circuit theory.* In myelinated fibers conduction is considered *saltatory,* from one node to the other. In the internode the impulse is electrotonically conducted.

Propagated action potentials depend on the presence of Na^+ and K^+ channels in the axon membrane. An ion channel has a kind of selectivity filter that is specific for the ion species and a gating mechanism that is electrically driven and regulates the ion flow. The Na^+ channel is blocked by tetradotoxin, which acts externally at the selectivity filter, and by local anesthetics acting on the gating mechanism. The voltage-dependent gate is explained by conformational changes in protein subunits caused by the rising action potential. The K^+ channel is blocked from inside by tetraethylammonium (TEA). A funnel shape has been postulated for the K^+ channel with a mouth that is 0.8 nm in diameter and that fits the hydrated K^+ or TEA. A selectivity filter 0.3 nm in diameter allows the passage of only the dehydrated K^+. Ca^{++} channels are important in transmitter release and are blocked by verapamil and divalent cations.

In physiological receptors and synapses there are *graded* responses that are *not propagated* (generator and synaptic potentials). A good example of a receptor is the Pacini corpuscle, which is a *biological transducer* capable of transforming mechanical energy into electrical energy.

24–3 SYNAPTIC TRANSMISSION AND STRUCTURE OF THE SYNAPSE

The earliest knowledge of *synapses*, or *synaptic junctions*, came from the discoveries at the turn of the last century of the morphological and physiological organization of the nervous system. The so-called *neuron theory*, established mainly by Cajal, led to the assumption that the functional interactions between nerve cells was by way of contiguities or *functional contacts*.

In 1897, Sherrington coined the name *synapse* to explain the special properties of the reflex arc, which he considered to be dependent on the functional contact between neurons. He attributed to the synapse a valve-like action, which transmits the impulses in only one direction (see Figure 24–1). He discovered some of the fundamental properties of synapses, such as the *synaptic delay* (the delay that the impulse experiences in traversing the junction), the fatigability of the synapse, and the greater sensitivity to reduced oxygen and to anesthetics. He also pointed out that the many synapses situated on the surface of a motoneuron could interact, and that some would have an additive excitatory action, whereas others would be inhibitory and antagonize the excitatory ones.

Synapses can thus be defined as all the regions anatomically differentiated and functionally specialized for the transmission of excitations and inhibitions from one element to the following in an irreversible direction. A more modern definition of the synapse, however, should also include the existence of a complex submicroscopic organization in both the presynaptic and postsynaptic parts of the junction and the presence of a specific neurochemical mechanism in which transmitters, receptor proteins, synthetic and hydrolytic enzymes, and so forth, are involved.

From Figure 24–1, it is clear that the main problem in synaptic transmission consists of finding out by what molecular mechanism the information carried by one neuron is transferred to the next. In other words, how does the code of frequency conducted by one neuron originate a new code of frequency in the following neuron?

Nerve Impulse Transmission Is Mediated Mainly by a Chemical Mechanism

DuBois-Reymond (1877) was the first to suggest that transmission could be either *electrical* or *chemical*. Both types of mechanisms have been observed, but chemical synapses seem to be by far the most frequent.

Electrical transmission was first demonstrated in a giant synapse of the abdominal ganglion of the crayfish cord, and since then, in several other cases. In this type of synapse the membrane contact acts as an efficient rectifier, allowing current to pass relatively easily from the pre- to the postsynaptic element, but not in the reverse direction. The arriving nerve impulse is passed without delay and can depolarize directly and excite the postsynaptic neuron. Here the one-way transmission is due to the valve-like resistance of the contacting synaptic membranes.

In addition to electrical transmission that is unidirectional there are regions of the adjacent neuronal membranes in which there is an *electrotonic coupling* due to the presence of gap junctions (see Section 5–2). In this type of transmission the electrical flux is bidirectional.

Chemical transmission presupposes that a specific chemical transmitter is synthesized and stored at the nerve terminal and is liberated by the nerve impulse. In 1904, Elliot suggested that sympathetic nerves act by liberating adrenalin at the junctions with smooth muscle. Later, von Euler demonstrated that *noradrenalin* was the true adrenergic transmitter. The studies of Dixon (1906) and particularly of Dale (1914) strongly supported chemical transmission in the parasympathetic system. This was finally proved in the heart by Loewi in 1921. Since then, *acetylcholine* has been demonstrated to act in sympathetic ganglia, neuromuscular junctions, and in many central synapses. Modern studies on chemical synaptic transmission have revealed that synapses are the sites of a transducing mechanism in which the electrical signals are converted into chemical signals, and these, in turn, again into electrical signals. It will be shown that active substances of low molecular weight (e.g., acetylcholine, noradrenalin, dopamine, glutamate, γ-aminobutyrate, and others) are produced at the nerve endings and packaged in special containers (i.e., synaptic vesicles) in multimolecular quantities. These packages of the so-called *transmitters* are released when the activation is produced by the nerve impulse. The transmitter, in turn, reversibly reacts with special receptor proteins, present at the postsynaptic membrane. This transmitter-receptor interaction produces a change in permeability to certain ions, thereby creating a *synaptic potential* in the postsynaptic cell.

In the following sections we shall learn that the basic principle of communication in the nervous system consists in the successive change, at each synapse, from an electrical signal to a chemical mechanism and back to another electrical signal.

Synaptic Transmission May Be Excitatory or Inhibitory

Physiological studies on synaptic transmission were greaty improved by the use of microelectrodes, which can be implanted near the synaptic region or intracellularly in the pre- and postsynaptic neuron.

With intracellular recordings in large nerve cells (e.g., motoneurons, pyramidal cells, invertebrate ganglion cells, etc.), it was observed that the arrival of the presynaptic nerve impulse produces a local synaptic potential. *Synaptic potentials*, like the generator potentials, are graded and decremental and do not propagate. They

extend electrotonically only for a short distance with reduction in amplitude.

A typical experiment, shown in Figure 24–11, involves two ganglion cells of *Aplysia* (a marine mollusc), one of which *(P)* acts synaptically upon the other *(F)*. Neuron P is impaled with two microelectrodes, one of which is used for stimulation *(St)* and the other for recording *(R)*. Neuron F is impaled with one microelectrode *(R)* to register the synaptic potential.

Two main types of cells exist, one of which produces an excitatory synaptic potential and the other an inhibitory postsynaptic potential.[26]

Excitatory synapses induce a depolarization of the postsynaptic membrane, which, upon reaching a certain critical level, causes the neuron to discharge an impulse. The *excitatory postsynaptic potential* (EPSP) is due to the action of the transmitter released by the ending (Fig. 24–11,1). This causes a change in permeability of the subsynaptic membrane, allowing the free passage of small ions, such as Na^+, K^+, and Cl^-.

Similarly, *inhibitory synapses* affect the subsynaptic membrane. In these instances the transmitter causes a transient increase in membrane potential, the so-called *inhibitory postsynaptic potential* (IPSP) (Fig. 24–11,2). This hyperpolarizing effect induces a depression of the neuronal excitability and an inhibitory action.

The excitatory or inhibitory action is not dependent exclusively on the type of transmitter substance. For example, acetylcholine is excitatory in the myoneural junction, sympathetic ganglia, and so forth, but inhibitory in the vertebrate heart, in which it reduces the frequency of contraction.

The experiment of Figure 24–12 shows that in the ganglion cells of *Aplysia* the injection of acetylcholine may have an excitatory synaptic effect in certain cells producing depolarization and increased frequency of discharges *(1)* or only a depolarization without firing (i.e., discharge of an action potential) *(2)*. In other cells the same treatment provokes a hyperpolarization and inhibition of spontaneous discharge *(3)*.

These facts indicate that the nature of a synapse depends, in particular, on the mechanism of the receptor (i.e., the chemical reactivity of the membrane in the postsynaptic neuron) and of the associated channel.

The use of intracellular recording has greatly

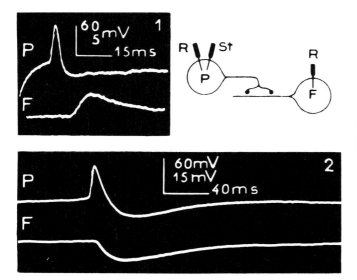

Fig. 24–11. Diagram of the experiment in two ganglion cells of *Aplysia* that are related synaptically. (See the text for details.) *1*, excitatory response. Depolarization of the membrane at *F* after arrival of the action potential from *P*. *2*, inhibitory response. Hyperpolarization of the membrane at *F* after arrival of the action potential from *P*. (Courtesy of L. Tauc and H.M. Gerschenfeld.)

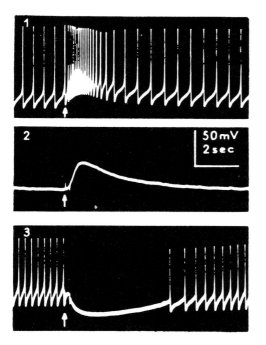

Fig. 24–12. *1*, intracellular recording in a neuron of *Aplysia* that is firing spontaneously. At the point marked by the *arrow*, acetylcholine is added, producing depolarization and increasing the frequency of discharges (excitatory synapse). *2*, same experiment on a neuron, without firing. Only depolarization is produced *3*, the action of acetylcholine induces hyperpolarization and inhibition of spontaneous discharges. (Courtesy of L. Tauc and H.M. Gerschenfeld.)

contributed to the delineation of some of the basic mechanisms by which the code of signals is transmitted from one cell to the next. All the synaptic potentials from the different excitatory and inhibitory endings impinging upon a neuron are integrated and will change the electrical properties of the membrane in a critical zone of the cell of low excitatory threshold, which is called the "pacemaker." In this region, which in the motoneurons is located at the initial segment of the axon, new impulses are fired.[27] In this respect a neuron acts in a way similar to an analogue computer, i.e., integrating all the inhibitory and excitatory messages that impinge on its surface before deciding whether to fire a new impulse.

Several Thousand Synapses May Impinge on a Single Neuron

Morphological studies with the light microscope revealed that the size, shape, and distribution of synapses of different regions of the central and peripheral nervous tissue vary considerably. Synapses are classically categorized as axodendritic, axosomatic, or axo-axonic, according to the relationship of the ending to the postsynaptic component. The endings may have different sizes and shapes (e.g., bud, foot, or button ending, club ending, and calix [cup] ending). *The number* of synapses that terminate on the perikaryon and dendrites varies considerably. In the ciliary ganglion there is a single huge synapse per cell. A large motoneuron of the spinal cord may receive 10,000 synaptic contacts, a pyramidal neuron of the cerebral cortex may have 40,000 synaptic contacts, and a large Purkinje cell of the cerebellum may receive as many as 200,000. These figures give an idea of the extraordinary complexity of the nervous system. This immense number of synapses carries information from numerous other neurons, some of which have an excitatory effect, and others, an inhibitory. Thus the neuron is a real computation center where all the information is integrated and from which new information is sent along the axon by nerve impulses.

The Number of Synapses Is Related to That of the Dendritic Spines

In general the number of synapses relates to the number and length of the dendrites on which most of the synaptic contacts are produced. The dendrites have fine protrusions on the surface, the so-called *spines*, upon which synaptic contacts with the nerve endings are made.

A typical spine is pedunculated with an enlarged tip between 0.5 and 2.0 μm in diameter and a narrow stalk between 0.5 and 1 μm long. The number and size of the dendritic spines, which can be easily studied in neurons stained by the Golgi method, allow one to judge the importance of the synaptic connections of a definite neuron (Fig. 24–13). Such studies are important in some pathological conditions, because the reduction in the number of spines may indicate a degeneration of afferent axons and terminals.

Although, embryologically, the brain is completed in all its details before it is used and its building is under genetic control, nevertheless with performance it can undergo fine changes in microstructure at the synaptic level. The number and size of spines can be modified and this modification is related to the so-called plasticity of the general nervous system.

Synaptic contacts are not completely stable and may change not only during development but also in relation to the influence of the inputs upon a neuron. For example, sensory deprivation produces alteration and reduction in the number of spines.[28] It has been observed that fish reared in community tanks have more dendritic branches and spines in interneurons of the deep tectal layers than those reared without visual-tactile contact with other fish of the same species (Fig. 24–13). The socialized controls not only have more spines but also have shorter stems. These findings suggest that social stimulation induces formation and widening of the spines.[29] Such changes in plasticity may have a profound influence on the effectiveness of the synaptic transmission and thus may be at the base of learning processes.

In fact, one of the most accepted theories of learning is based on the growth and branching of spine synapses. This *synaptic growth* implies the stimulus of protein synthesis and RNA transcription, and the inhibition of these processes may abolish learning and long-term memory. (For more details on this important subject see McGeer, Eccles, and McGeer, 1978 in the Additional Readings at the end of this chapter.)

The Synapse Ultrastructure Suggests Many Types of Functional Contacts

The discovery of the synaptic vesicles by De Robertis and Bennett[30] and of special features of the synaptic membranes[31] with the electron microscope provided new morphological landmarks for the study of synapses. At the synaptic junction both synaptic membranes appear thicker and denser. The synaptic cleft between the synaptic membranes may be about 30 nm wide (up to 50 nm wide at the myoneural junction) and is occupied by a dense cleft material, which may show a system of fine *intersynaptic filaments* of about 5 nm that join both synaptic membranes (Fig. 24–14).

Another system of filaments has been observed to penetrate a variable distance into the postsynaptic cell. This is the so-called subsynaptic web[31] *or postsynaptic density.*[32] The demonstration of the intersynaptic material between the membranes suggests that there is greater adhesion at the junction as was demonstrated by microdissection experiments. In fact, in an isolated cell the endings break their connection with the axons but remain attached to the cell. The discovery of the positions of the synaptic vesicles and the presynaptic and postsynaptic densities has suggested the possible functional polarity of the synapse and the recognition of a series of contacts that could not be demonstrated with histological methods.

Synaptic contacts may be established between

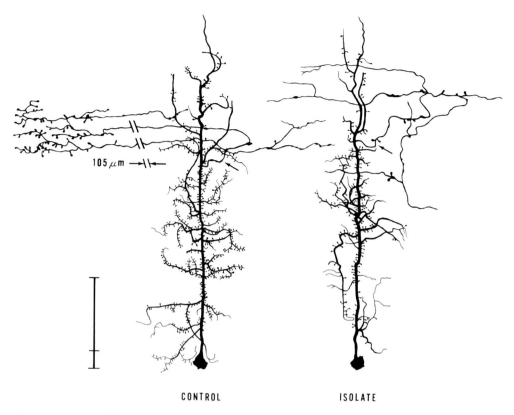

Fig. 24–13.   Interneurons of the optic tectum of the jewel fish stained by the Golgi method. *Control,* numerous dendrites and spines; *Isolate,* a reduction in the dendrites and in the number of spines may be observed. These spines are longer and more slender than in the control. (From Goss, R.G., and Globus, A.: Spine on tectal interneurons in jewel fish is shortened by social stimulation. Science, *200*:786, 1978. Copyright 1978 by the American Association for the Advancement of Science.)

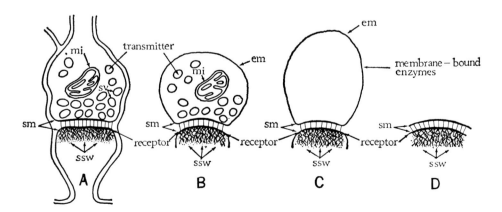

Fig. 24–14.   Diagram showing the systematic dissection of the synaptic region achieved by the use of cell fractionation methods. *A,* typical synapse of the cerebral cortex showing: *mi,* mitochondria; *sv,* synaptic vesicles; *sm,* synaptic membranes; and *ssw,* subsynaptic web. *B,* isolated nerve endings. *em,* nerve-ending membrane. *C,* after the osmotic shock, only the nerve-ending membrane remains. *D,* after treatment with a mild detergent, only the junctional complex remains. (From De Robertis, E.: Molecular biology of synaptic receptors. Science, *171*:963, 1971. Copyright 1971 by the American Association for the Advancement of Science.)

any portions of the neuronal membranes; in fact, the only parts of the neuron that never receive a synapse are the segments of nerve fibers covered by myelin. Furthermore, release of neurotransmitter does not seem to be restricted to axon terminals, and it has been suggested that dendrites may also represent sites of release.

From the morphological viewpoint the following types of synapses can be recognized under the electron microscope:

1. *Axodenritic synapses* (Fig. 24–15*a*), which may be directly on the dendrite but in most cases are established by way of a spine (*axospinous synapses*).

2. *Axosomatic synapses*, which terminate on the perikaryon and some of which may be at the axon-hillock of the neuron (i.e., the portion of the perikaryon from which the axon emerges). These synapses are sometimes referred to as *initial-segment synapses* and are assumed to exert an inhibitory action on the postsynaptic membrane. Such synapses are important because at this strategic point they may influence the firing of the nerve impulse.

3. *Axo-axonic* synapses are those in which a nerve ending terminates on another (Fig. 24–15*d*). This type of synapse appears to be involved in *presynaptic inhibition*, because its effect may produce a reduction in the release of the neurotransmitter.

4. *Serial synapses*, in which axo-axonic synapses are disposed serially, i.e., one process is postsynaptic to one synapse and presynaptic to another.

5. *Reciprocal synapses*, which correspond to the situation in which a neuronal process may be presynaptic at one point and postsynaptic at another (Fig. 24–15*e*).

6. *Somatodendritic synapses*, which correspond to the special type of synapses between the perikaryon and a dendrite (Fig. 24–15*c*).

7. *Dendrodendritic synapses*, which are found in several regions of the central nervous system and in which dendrites contain clusters of synaptic vesicles opposing other dendrites. Some of these contacts may be of the reciprocal type (Fig. 24–15*e*).

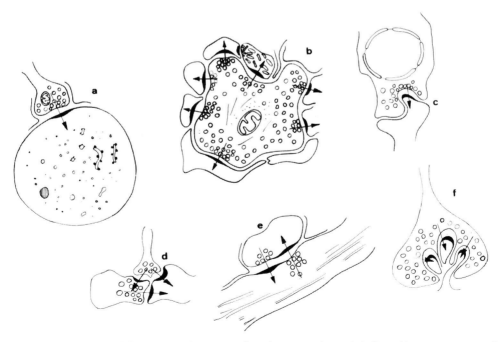

Fig. 24–15.   Diagram showing different types of synapses. The polarity in each case is indicated by an arrow. *a*, exodendritic; *b*, glomerular, showing several axodendritic and one axo-axonic synapse, *c*, somatodendritic; *d*, a triad, in which one terminal synapses with another and with a dendrite; *e*, a reciprocal synapse, which may involve two dendrites; *f*, an invaginated synapse. Notice that the polarity is determined by the synaptic vesicles and the postsynaptic density. (Courtesy of J. Pecci Saavedra, O. Vilar, and A. Pellegrino de Iraldi.)

These and other more complex types of synapses, such as the *glomerular* (Fig. 24–15*b*), the *triad* (Fig. 24–15*d*), and the *invaginated* synapses (Fig. 24–15*f*), suggest that the functional interactions between neurons may be much more complex than previously imagined and that practically all regions of the neuronal membrane may play a synaptic role.

Lectin Receptors and Postsynaptic Densities May Play a Role in the Formation and Maintenance of Synapses

The cleft material and the postsynaptic density may play a role in the formation and maintenance of the synaptic connections.[32] By using plant lectins, such as concanavalin A and others, it has been possible to detect the presence of specific glycoproteins on the neuronal membrane (see Section 5–2). It has been demonstrated that these so-called *lectin* receptors in the postsynaptic membrane are restricted in their mobility. It is thought that the synaptic receptors are linked across the membrane and that the underlying postsynaptic density may be involved in this linkage.

In a study of the synaptogenesis of the Purkinje cells of the cerebellum, three major stages have been recognized:

1. In the first stage (corresponding to day 13 to 14 of incubation in the chick embryo) numerous coated vesicles (Section 10–4) are formed in the Golgi complex. These vesicles migrate to the surface where they form dense plaques and expose glycoproteins. At these sites, the incoming parallel nerve fibers become attached and form contacts.
2. Once these contacts are formed, at the postsynaptic side, pinocytic vesicles are incorporated which carry part of the presynaptic (axonic) membrane. During this stage, which extends to day 15, many membranes of the parallel fibers are incorporated into the cytoplasm of the Purkinje cell.
3. At a later stage (15 to 16 days), the postsynaptic densities appear. These densities seem to be formed by ribosomes that are associated with these structures.

Thus the three stages described may correspond to: (1) *the intermembranous adherence* between the axon terminal and the postsynaptic site, (2) *the recognition* of the correctness of the attachment, probably mediated by the incorporation of axon membranes into the postsynaptic cell, and (3) the *definite formation* of the synapse with its postsynaptic density (see Palacios-Pru et al., 1981 in the Additional Readings at the end of this chapter).

The Presynaptic Membrane Shows Special Projections at Active Zones

From the initial electron microscopic studies of synapses it has become common knowledge that synaptic vesicles tend to converge on special regions of the presynaptic membrane called the *active zones*. At these points there are projections of a dense material that bear a special relationship with the synaptic vesicles. In central synapses these *presynaptic projections* are about 50 nm in diameter and are arranged in precise array on the presynaptic membrane.[33] Figure 24–16 shows that these projections form a hexagonal network and that the spaces in between accommodate the attachment sites of the vesicles. In freeze-fracturing studies the inner leaflet (P face) of the presynaptic membrane shows clusters of 10-nm particles; the most characteristic feature, however, is the presence of small pits in the P face which correspond to protuberances on the outer leaflet (E face). These pits, geometrically arranged, correspond to the attachment sites of the vesicles (Fig. 24–16).[33]

In the neuromuscular junction of vertebrates the presynaptic dense projections appear as long bands that run perpendicular to the axis of the nerve terminal (Fig. 24–17). Facing these bands, on the postsynaptic side, are the typical folds of the postsynaptic membrane. While most of the synaptic vesicles of the neuromuscular junction are not in contact with the presynaptic membrane, those that are on both sides of the projections are in intimate relationship with that membrane. The freeze-etching technique has revealed in the presynaptic membrane a double row of 10-nm particles that are parallel to the dense projections bearing the vesicles. We shall see later that these large intramembrane particles may correspond to calcium channels. Despite the geometric differences (i.e., a *hexagonal grid* in central synapses and *parallel bands* in the myoneural junction), the presynaptic pro-

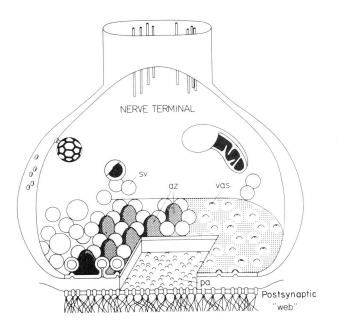

Fig. 24–16. Diagram of a central synapse. *az*, active zone showing the presynaptic projection surrounded by vesicles attached to the membrane. Observe that the vesicles and the attachment sites *(vas)* form a hexagonal pattern. *pa*, postsynaptic intramembrane particles restricted to the area facing the active zones; *sv*, synaptic vesicles. (Courtesy of K. Akert.)

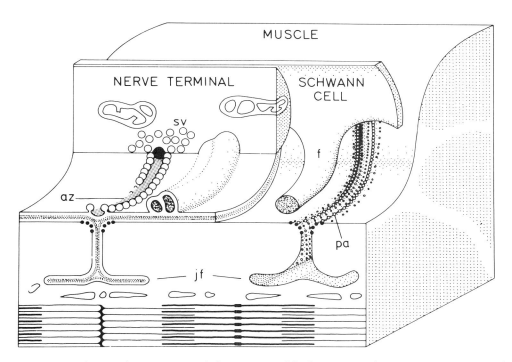

Fig. 24–17. Diagram showing the main structural characteristics of the frog myoneural junction. *az*, active zone showing the attachment of synaptic vesicles on both sides of the presynaptic band; *pa*, particles found in the postsynaptic membrane that may correspond to cholinergic receptors; *sv*, synaptic vesicles; *jf*, junctional folds; *f*, finger of a Schwann cell. Notice that on the right side of the figure the presynaptic membrane and the cleft are not represented. (Courtesy of K. Akert.)

jections on both synapses are essentially the same. There is, however, a quantitative difference; central synapses can accommodate up to eight times more vesicles near the membrane than can the motor end plate. In synapses of photoreceptors there are special presynaptic projections, the *synaptic ribbons*, which appear as flattened bands upon which numerous synaptic vesicles converge.

The function of these various types of presynaptic projections is still unknown, but it seems possible that they act as guides for transporting the vesicles to the sites of release. This hypothesis is supported by the finding that both actin and myosin are present within the nerve ending and are associated with the presynaptic membrane and the vesicles. These proteins could provide the motive force that pulls the vesicles toward the presynaptic membrane.

The Postsynaptic Membrane Shows a Complex Macromolecular Organization

Studies using freeze-etching and negative staining have revealed a fine macromolecular organization in the postsynaptic membrane.

In the myoneural junction, at the entrance of the postsynaptic folds, there are numerous 10-nm membrane particles that have a density on the order of $7500/\mu m^2$. As Figure 24–17 shows, such particles correspond, on the postsynaptic membrane, to regions opposed to presynaptic active zones.[45] These regions correspond to the localization of the cholinergic receptors (see later discussion). In fact, autoradiographic studies with ^3H-α-bungarotoxin (a toxin that binds specifically to nicotinic cholinergic receptors) have demonstrated that these particles are localized in the same region of the postsynaptic membrane. There is good evidence suggesting that these particles represent receptor-channel complexes and that they protrude from the surface of the membrane into the synaptic cleft.[34]

Acetylcholinesterase, the enzyme that hydrolyzes acetylcholine, is more widely distributed into the postsynaptic folds than the receptor. In the postsynaptic membrane of *Torpedo* electroplax, particles with a packing density of 10,000 to $15,000/\mu m^2$ have been observed. These particles have a hexagonal arrangement with a center-to-center distance of 8 to 9 nm.

The precise arrangement between the ace-tylcholine receptors, at the entrance of the folds, and the synaptic vesicles, attached to the presynaptic active zones, is remarkable. It can be hypothesized that this orderly organization ensures that the secreted neurotransmitter rapidly reaches the sites of action at the receptors.

In central synapses a somewhat similar organization has been found. Here there are postsynaptic particles in aggregates confined roughly to the regions opposed to active sites in the presynaptic membrane (Fig. 24–16).

SUMMARY:
Structure of the Synapse

Synapses are regions of contact between neurons, or between a neuron and a nonneuronal cell (i.e., a physiological receptor or effector) at which excitatory or inhibitory actions are transmitted in one direction. They may be *electrical*, in places in which there is a direct contact of membranes, and in such cases the nerve impulse passes from one neuron to the other without delay. Most synapses, however, involve *chemical transmission* based on the release of specific transmitters such as acetylcholine (ACh), noradrenalin, dopamine, and other biogenic amines, as well as amino acids such as glutamate, aspartate, γ-aminobutyrate (GABA), and glycine.

The microphysiological study of synaptic transmission has demonstrated that the nerve impulse stops at the nerve ending, and that at the synapse a *synaptic potential* is generated in the postsynaptic cell. This potential may be *excitatory* (depolarization) or *inhibitory* (hyperpolarization); in both cases, it is graded and does not propagate. In some cases the same transmitter (i.e., ACh) may be excitatory or inhibitory; the different effect depends on the receptor protein with which the transmitter interacts. All the excitatory and inhibitory inputs acting on a neuron are added algebraically, and the neuron will fire new impulses in relation to these inputs as well as to its own spontaneous firing rhythm.

Morphologically, synapses may be *axodendritic*, *axosomatic*, or *axo-axonic*. With the EM, *dendrodendritic*, *serial*, and *reciprocal* synapses may be distinguished by virtue of their presence on one side of synaptic vesicles and the differentiations of the synaptic membranes (i.e., presynaptic projections, intersynaptic filaments, and subsynaptic web).

Several thousand synapses may impinge on a single neuron, and in general, the number of synapses is related to that of the dendritic spines. Synaptic contacts are not completely stable and may change not only during development but also in relation to the stimulation of the neuron (e.g., in sensory deprivation the number of spines and synapses diminishes).

Special lectin receptors are present at the neuronal membrane, and these correspond to glycoproteins. Most of these are mobile within the membrane, but at the postsynaptic membrane they are less mobile. The postsynaptic density may play a role in this change in mobility, as well as in the development of the synapse. In cultured neurons it has been observed that the postsynaptic densities are made by coated vesicles coming from the Golgi, prior to the appearance of synaptic vesicles at the terminal.

A number of synaptic vesicles are attached to the presynaptic membrane at special sites arranged around presynaptic projections. These projections make a hexagonal grid in central synapses or parallel bands in the myoneural junction and may be related to the release of the vesicles.

At the postsynaptic membrane there are particles that in some cases correspond to the localization of receptor proteins and channels. There is a precise arrangement of attachment of synaptic vesicles at the presynaptic membrane, which parallels the position of receptors in the postsynaptic membrane.

24–4 SYNAPTIC VESICLES AND QUANTAL RELEASE OF NEUROTRANSMITTER

Since the discovery of the synaptic vesicles, they have been implicated in the storage, transport, and release of the neurotransmitters. Following the demonstration by Fatt and Katz (1952) of the quantal nature of chemical transmission, De Robertis and Bennett postulated that the synaptic vesicles are involved in the storage of the neurotransmitter.[30] This was proved by the isolation of these organelles from synapses of the central nervous system and the demonstration that acetylcholine was concentrated in this fraction.[35] Later on, other neurotransmitters were localized in the vesicles (see De Robertis, 1964 in the Additional Readings at the end of this chapter), and pure populations of the vesicles were isolated from cholinergic and adrenergic neurons. In this section we will study the morphology and structure of the synaptic vesicles in relation to their possible role in synaptic transmission.

Several Types of Synaptic Vesicles May Be Recognized by Morphology and Cytochemistry

Morphologically five types of synaptic vesicles have been recognized.

Agranular or Electron-Translucent Vesicles. These vesicles have a diameter of 40 to 50 nm with a limiting membrane 4 to 5 nm thick (Fig. 24–18). They are characteristic of peripheral cholinergic synapses and are most frequently found in central synapses. In a highly purified fraction of synaptic vesicles from the brain, it was observed by quick freeze-etchings that they have "knob-like" protrusions on the surface. It has been suggested that these protrusions represent a part of a proton Mg^{++}-ATPase pump. This pump may be necessary in the packaging of the neurotransmitter into the synaptic vesicle (see Stadler and Tsukita, 1984). They are distributed more or less uniformly throughout the axoplasm of the ending, but as mentioned earlier, a certain proportion make contact with the presynaptic membrane at the *active points* of the synapse. At the myoneural junction of the frog these active points are localized on both sides of the presynaptic band (Fig. 24–19). In this type of synapse there are about 1000 synaptic vesicles per μm^3 and a total of about 3×10^5. Of these about 20% are near the membrane and readily available for release of the transmitter upon arrival of the nerve impulse.[36]

Large Dense-Core or Large Granular Vesicles. These may be observed in central, as well as peripheral, synapses (Fig. 24–19). In peripheral cholinergic and adrenergic endings only a few percent of the vesicles are large granular vesicles. It has been postulated that these vesicles may contain neuropeptides.

Flattened or Elliptical Vesicles. These vesicles are found in certain central and peripheral synapses (Fig. 24–19). They have been related to inhibitory synapses[37] and found in a fraction of isolated nerve ending (i.e., the synaptosomes) that is rich in the inhibitory transmitter γ-aminobutyrate.[38]

Small Granulated Vesicles. These vesicles, first described in 1961 by De Robertis and Pellegrino de Iraldi,[39] have a diameter of 40 to 60 nm and contain a dense core and a clearer region or matrix. Through the use of pharmacological agents that release or increase the concentration of catecholamines, and with cytochemical techniques, it has been demonstrated that they contain norepinephrine and, in the case of the pineal nerves, also 5-hydroxytryptamine.[40] Recently, a two-compartment model of these vesicles has been postulated: (1) a loosely bound,

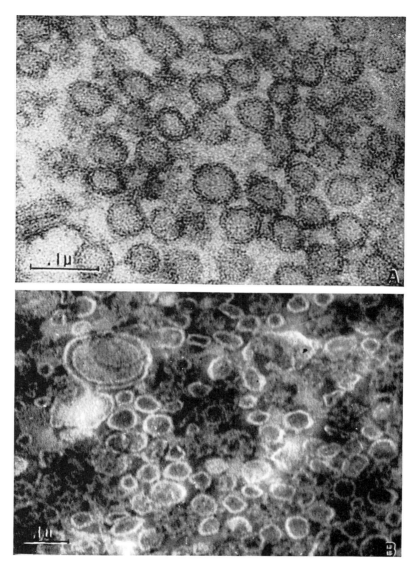

Fig. 24–18. *A,* high resolution electron micrograph of synaptic vesicles in the hypothalamus of a rat, showing the fine structure of the vesicular membrane. ×180,000. *B,* isolated synaptic vesicles from rat brain after osmotic shock of the mitochondrial fraction. Negative staining with phosphotungstate. ×120,000. (From De Robertis, E. et al.: J. Neurochem. *10*:225, 1963.)

more easily releasable pool in the outer compartment or matrix and (2) a tightly bound, more resistant pool in the core compartment (Fig. 24–20; see Pellegrino de Iraldi, 1980 in the Additional Readings at the end of this chapter).

Coated Vesicles. These are found in a small proportion in central synapses and were first recognized as complex vesicles in isolated fractions.[35] Coated vesicles are produced by the process of selective endocytosis and may be involved in the process of membrane recycling.[41] In myoneural junctions labeled with peroxidase

that have produced a considerable amount of transmitter in response to stimulation, it has been found that these vesicles increase in number and contain the label. In Section 10–4 it was mentioned that the coat of these vesicles is made of the protein *clathrin.*

Neuronal Development—The Type of Neurotransmitter and Synaptic Vesicle May Be Medium-Determined

The development of synaptic connections involves the synthesis of a specific neurotransmit-

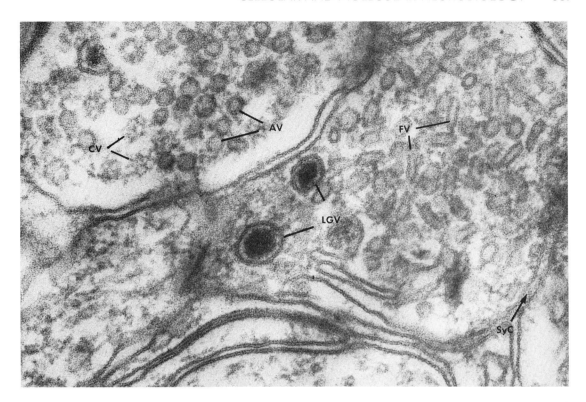

Fig. 24–19. Four types of synaptic vesicles are observed in two nerve endings of the rat hypothalamus. *AV*, agranular vesicle; *CV*, coated vesicle; *FV*, flattened vesicle; *LGV*, large granular vesicle. *SyC* points to the synaptic cleft. × 135,000. (From E. De Robertis, unpublished.)

ter and a special type of vesicle in which the transmitter is stored. In the autonomic system, there are two main types of neurotransmitters, acetylcholine and norepinephrine, contained respectively in clear agranular vesicles and small granulated vesicles. Both the adrenergic and the cholinergic neurons are embryologically derived from special regions of the neural crest. The neuroblasts migrate to the sites that they will occupy in the adult, cease dividing, and differentiate, producing one transmitter or the other.

Several experiments have demonstrated that this differentiation is influenced by the immediate environment. For example, cells of regions of the neural crests that normally provide adrenergic neurons, if transplanted to the rostral regions of the embryo, perform the cholinergic function, producing acetylcholine and clear vesicles; if neural crest cells from the rostral region are implanted in the mid-trunk, they develop into adrenergic neurons with terminals containing granular vesicles.

A more direct experiment consists of culturing

sympathetic neuronal cells from a newborn rat by a method that destroys all the non-neuronal (i.e., glial) cells. In this case, the neurons become adrenergic and produce norepinephrine. If the same neurons are grown in the presence of heart tissue, most of them become cholinergic. This effect can also be obtained by adding the fluid in which heart tissue has been grown to the pure adrenergic cell culture. This is an indication that the heart cells produce a *cholinergic developmental factor* that is able to promote the differentiation of the neuron toward the synthesis of acetylcholine.

These and other experiments have led to the hypothesis that at an early stage neurons are still "plastic" and, under the influence of the medium, can change the neurotransmitter (and the type of vesicle) they produce. Work in which single neurons are cultured in the presence of heart cells has permitted the transformation from adrenergic to cholinergic to be followed. The most interesting finding is that there are neurons that have a dual function, i.e., that be-

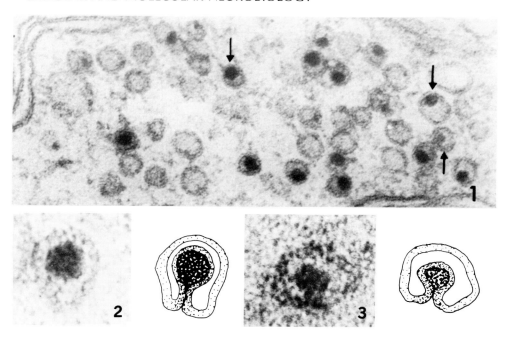

Fig. 24–20. Electron micrographs of sympathetic nerve endings in rat pineal and diagrams illustrating the two-compartment model of the small granulated vesicles. An outer clear compartment or matrix and a core compartment in the invaginated membrane is suggested. (Courtesy of A. Pellegrino de Iraldi.)

have simultaneously as adrenergic and cholinergic and have the corresponding two types of synaptic vesicles. These neurons could correspond to an intermediary stage in the differentiation. The hypothesis has been put forward that the chemical differentiation of a neuron (which determines the choice of neurotransmitter) is influenced by non-neuronal cells by way of the release of special developmental factors that may enter the neuron by retrograde transport. In addition, the influence of the electrical activity of other neurons may modify the response to such developmental factors (see Bunge et al., 1978; Patterson et al., 1978 in the Additional Readings at the end of this chapter).

Synaptosomes, Synaptosomal Membranes, and Synaptic Vesicles May Be Isolated by Cell Fractionation

The application of cell fractionation methods to the study of the central nervous system has permitted a true dissection of the synaptic region, with the isolation of the synaptosome as a structural unit. As Figure 24–14 shows, a typical axodendritic synapse on a spine, under the action of shearing forces, breaks its connections

with the axon and with the dendrite, and the nerve ending, or *synaptosome*, becomes isolated, together with the presynaptic and postsynaptic membranes. The synaptosomal membrane heals, and the synaptosome becomes a structural and functional unit in which many processes related to the metabolism of the neurotransmitters may be studied. The synaptosome is sensitive to changes in the osmotic pressure of the medium, and in a hypotonic solution it undergoes an osmotic shock, with breakage of the membrane and release of the synaptic vesicles, mitochondria, and the entire content of the nerve ending. At this point (Fig. 24–14C) only the nerve-ending membrane remains. By a further treatment with a mild detergent, however, only the *functional complex*, composed of the synaptic membranes and related structures, is found (Fig. 24–14D).[42] Further treatment with detergents leads to the isolation of the subsynaptic web or postsynaptic density. Such studies have demonstrated that the nerve-ending membrane contains important membrane-bound enzymes, such as acetylcholinesterase, N^+-K^+ ATPase, and adenylate cyclase. They have also allowed the study of the binding of synaptic receptors to the neurotransmitters.[43]

TABLE 24–2. CONTENT OF BIOGENIC AMINES IN THE BULK FRACTION (M_1), IN SYNAPTIC VESICLES (M_2), AND IN THE SOLUBLE FRACTION (M_3)

Biogenic Amines	Fraction		
	M_1	M_2	M_3
Acetylcholine	0.55	2.85	1.20
Noradrenaline	0.40	2.56	1.93
Dopamine	0.46	2.46	1.72
Histamine	0.39	2.24	2.27

The crude mitochondrial fraction of the brain was osmotically shocked and then centrifuged. The results are expressed in relative specific concentration. For literature, see De Robertis, 1967 in the Additional Readings at the end of this chapter.

These cell fractionation studies also permitted the isolation of the synaptic vesicles[35] (Fig. 24–18) and demonstrated that they contain the neurotransmitters. In Table 24–2 it is shown that fraction M_2, corresponding to the synaptic vesicles, has the highest content of various neurotransmitters.

Neurotransmitters—Synthesis and Metabolism of Acetylcholine

The number of substances produced by neurons which may be considered as possible neurotransmitters has increased considerably in recent years. Table 24–3 lists more than 30 putative neurotransmitters, but the actual number is 50 or more and is increasing rapidly. Most of them are substances of low molecular weight that belong to the following groups: biogenic amines, amino acids, purine, and neuropeptides. Among this last group, some were known as hormones (i.e., oxytocin, vasopressin), releasing factors, or active substances that are present in both the gut and the brain (e.g., cholecystokinin, substance P, etc.) (see Gainer and Brownstein, 1981 in the Additional Readings at the end of this chapter).

It should be emphasized, however, that only a few of these substances meet all the criteria of and thus can be fully identified as neurotransmitters. Others are considered for the moment as *neuromodulators*—that is, substances that may influence synaptic transmission by different mechanisms.

The main criteria used to define a substance as a neurotransmitter are:

1. It must be present inside the nerve ending.
2. It must be released upon nerve stimulation.
3. Its effects should mimic those produced by nerve stimulation.
4. Its effects should be modified by specific drugs (agonists and antagonists) in the same direction and magnitude as in the in vivo condition.
5. It must be inactivated after its action in synaptic transmission.

Let us consider the acetylcholine (ACh) system as an example. The neurotransmitter ACh is synthesized by way of a combination of choline and acetylcoenzyme A by the enzyme *choline acetyltransferase*. The enzyme is produced in the neuronal body and is transported to the nerve terminal by axonal flow (Fig. 24–21). Acetylcholine is stored mainly within the synaptic vesicles, although some ACh that is soluble in the cytoplasm may be present. After its release into the synaptic cleft, acetylcholine interacts with the cholinergic receptor present in the postsynaptic membrane and is degraded into choline and acetate by *acetylcholinesterase*, a hydrolytic enzyme present in nerve-ending membranes. As Figure 24–21 shows, choline is again taken up by a high-affinity transport mechanism into the nerve terminals. By cell fractionation it can be demonstrated that the three main components of the acetylcholine system (i.e., choline acetyltransferase, acetylcholine, and acetylcholinesterase) are found in the synaptosomes (see Rodríguez de Lores Arnaiz and De Robertis, 1973 in the Additional Readings at the end of this chapter).

Neuropeptides and the Opioid System

In the last decade one of the most important developments in the field of neurotransmitters and neuromodulators has been the discovery of peptides in neurons which are able to participate in the transmission of the nerve impulse. These neuropeptides include molecules that may contain from two (e.g., carnosine) to about 30 amino acids (Table 24–3). They originate as larger precursors from ribosomes in the perikaryon, are stored in vesicles, and are transported to the nerve terminal, where they are finally processed

TABLE 24–3. SUBSTANCES THAT CAN BE CONSIDERED POSSIBLE NEUROTRANSMITTERS OR NEUROMODULATORS

Biogenic amines
 Acetylcholine
 Catecholamines
  Dopamine
  Norepinephrine
  Epinephrine
 5-Hydroxytryptamine (serotonin)
 Other primary amines
  Histamine
  Octopamine
  Phenylethylamine
  Phenylethanolamine
 Polyamines
  Putrescine
  Spermine
  Spermidine

Amino acids
 Glutamic acid
 Aspartic acid
 Glycine
 γ-aminobutyric acid (GABA)
 Taurine
 Proline

Purines
 Adenosine
 ATP, etc

Neuropeptides	*Number of amino acids*
Substance P	11
Carnosine	2
Thyrotropin-releasing hormone (TRH)	3
Neurotensin	13
Somatostatin	14
β-endorphin	31
Enkephalin	5
Angiotensin I	10
Angiotensin II	8
Oxytocin	9
Vasopressin	9
Cholecystokinin	8
Bradykinin	9

as peptides and released. As occurs with the catecholamines and serotonin, peptides are taken up again by the nerve ending or inactivated after degradation by peptidases.

Although some neuropeptides, such as substance P, have been known for a long time, undoubtedly the discovery of the enkephalins by Hughes in 1975 brought the problem of neuropeptides to the forefront of neurobiology. It is interesting that the two opioid pentapeptides—methionine-enkephalin and leucine-enkephalin—having the following structure

Tyr-Gly-Gly-Phen-Met (met-enkephalin)
Tyr-Gly-Gly-Phen-Leu (leu-enkephalin)

were discovered long after the specific receptor for morphine in the nervous system was known.

The isolation of these neuropeptides was facilitated by their property of acting on the receptor to morphine in a bioassay. Experimentally the vas deferens of mice were electrically stimulated, and it was found that these peptides inhibited the contraction. The specificity of the action was demonstrated by the effect of naloxone (a blocker of the morphine receptor) that inhibited the effect of the enkephalins.

The problem of the opioid peptides was complicated by the finding of other larger peptides with a similar function, such as the β-endorphins and dynorphins.

At present three main groups of these opioid peptides are recognized.

1. The *β-endorphin system*, mainly present in the anterior and intermediary lobe of

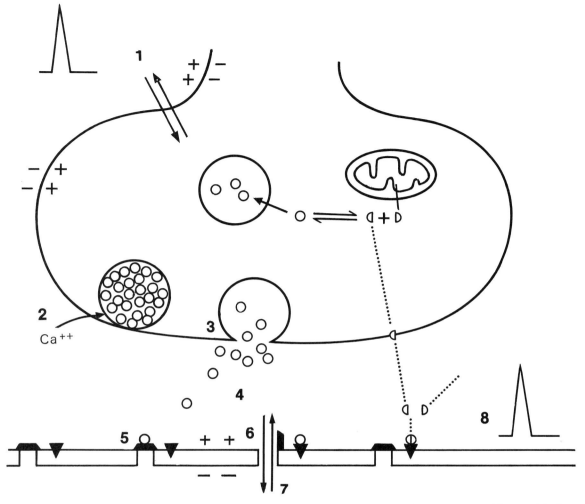

Fig. 24–21.   Diagram of a cholinergic synapse, such as the myoneural junction, showing the components of the acetylcholine system and steps in the mechanism of neurotransmission. *1,* action potential arriving to the nerve ending resulting from ion currents; *2,* Ca^{++} entry into the presynaptic ending; *3,* liberation of the acetylcholine molecules contained in the synaptic vesicle by fusion with the membrane and opening into the synaptic cleft; *4,* diffusion of the neurotransmitter within the cleft; *5,* interaction with the receptor; *6,* opening of the postsynaptic channels; *7,* Na$^+$ and K$^+$ currents through the channel; *8,* interaction of acetylcholine with acetylcholinesterase and hydrolysis, the choline moiety is taken up again; postsynaptic action potential. Within the nerve ending the synthesis of acetylcholine by cholineacetylase is shown. (Courtesy of O. Uchitel.)

the hypophysis, as well as in a limited number of neurons in the *nucleus arcuatus* of the hypothalamus. These neurons have long axons that project to limbic structures, thalamus, locus ceruleus, etc. This system is limited and in it β-endorphin coexists with α-melanotropin (α-melanocyte stimulating hormone, or α-MSH) and β-lipotropin (β-LPH).

2. The *enkephalin system* has a much wider and more diffuse distribution in the CNS and is generally present in neurons having short axons.

3. The *dynorphin system* comprises several neuropeptides having a chain length of 8 to 32 amino acids. The concentration of the dynorphins in the brain is much lower than that of enkephalins, and they are present mainly in the basal nuclei.

Cytochemical Identification of Neuronal Systems—Coexistence of Neurotransmitters

The vast number of neurons present in the central and peripheral nervous system can be differentiated not only by their morphology and synaptic contacts, but by the neurotransmitters they are able to synthesize, store, and release at the nerve endings. In the last few years, powerful methods have been developed to identify the biochemical nature of the neurotransmitters in different neuronal systems. These studies are leading to a neurochemical mapping of neurons based on the neurotransmitters they synthesize, but also on the specific receptors they carry on their membranes.

Identification can be done by *cell fractionation* of different regions of the brain, including the separation of neurons, synaptosomes, synaptic vesicles, and synaptic membranes. Neurons producing catecholamines (norepinephrine, dopamine) or indolamines (serotonin) may be identified by the use of *fluorescence histochemical techniques. Autoradiography* can be used to localize neurons and nerve endings by the administration of labeled neurotransmitters or precursors. For example, Figure 24–22 shows a glomerular synapse from the cerebellum which has been submitted to the uptake of radioactive serotonin.[44] Nowadays the most powerful technique is based on the use of *immunochemical methods* against neurotransmitters (especially neuropeptides) or the specific enzymes involved in their metabolism.

For example, in Figure 3–16 we showed the localization of serotonin in nerve endings by the use of an antiserotonin antiserum.

The use of these methods seemed in some cases to confirm the so-called Dale's rule that each neuron produces a single neurotransmitter. In recent years, however, several exceptions to this rule have been observed, and it is now accepted that two or more neurotransmitters can coexist in the same nerve terminal. One of the first cases of *coexistence of neurotransmitters* was observed in the sympathetic nerves of the pineal gland, whose synaptic vesicles contain both norepinephrine and serotonin (see Pellegrino de Iraldi, 1980 in the Additional Readings at the end of this chapter). More frequent is the case in which a biogenic amine, such as dopamine or serotonin, coexists with neuropeptides. For example, serotonin and substance P (Table 24–3) can coexist in the same terminal and even in the same dense core synaptic vesicle.[45] Another example is the coexistence of the vasoactive intestinal polypeptide (VIP) with acetylcholine in cholinergic neurons; neuropeptides also

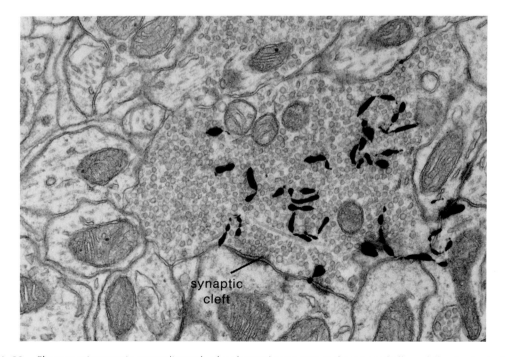

synaptic cleft

Fig. 24–22. Electron microscopic autoradiograph of a glomerular synapse in the rat cerebellum, following topical superfusion with ^{3}H-serotonin. The uptake of serotonin in the nerve ending is demonstrated by the presence of silver grains. ×41,000. (Courtesy of C. Sotelo and A. Beaudet.)

coexist in the cholinergic nerve endings that innervate the adrenal medulla and also in sympathetic neurons and chromaffin cells.

The coexistence of a neurotransmitter with the neuropeptide may modulate the response of the effector cell. For example, the neuropeptide may modify (positively or negatively) the interaction of the neurotransmitter with its receptor. In the nerve ending it is possible to postulate the presence of mixed vesicles containing both substances or of different vesicles either for the neurotransmitter or the neuropeptide. In some cases the release of the neuropeptide or of the neurotransmitter may depend on the intensity of the stimulation. In some peripheral systems it has been found that the neurotransmitter causes a rapid response of short duration, while the neuropeptide produces a more long-lasting effect (see Lundberg and Hökfelt, 1983 in the Additional Readings at the end of this chapter).

Related to the problem of the coexistence of neurotransmitters and the origin of neuropeptides is the fact that several neuropeptides can coexist in a system within the same cell. For example, the *pro-opiocortin system* simultaneously produces several active peptides that function as hormones or neurotransmitters. As Figure 24–23 shows, a single large peptide of 134 amino acids is a common precursor for the adrenocortical hormone (ACTH) and for *β-lipotropin*. In addition, ACTH may originate α-melanotropin and a fragment (corticotropin-like intermediate lobe peptide, or CLIP) in the intermediate lobe of the hypophysis. β-lipotropin

may give rise to γ-LPH, β-MSH, and to the neuropeptides *β-endorphin* and *met-enkephalin* (Table 24–3; see Cooper and Martin, 1982 in the Additional Readings at the end of this chapter).

In addition to the pro-opiocortin system other precursor systems, such as those of the *pro-enkephalin A* and *pro-enkephalin B* can give rise to different neuropeptides.

The Transport of the Neurotransmitter Involves the Synaptic Vesicles

The synthesis of neurotransmitter, although occurring predominantly at the nerve terminal, may take place in the perikaryon. Thus part of the transmitter must be transported to the sites of release by way of synaptic vesicles. The possible flux of the various kinds of vesicles is shown diagrammatically in Figure 24–24. The vesicles are transported by fast axoplasmic flow, possibly with the involvement of neurotubules, which under certain conditions are even seen associated with the vesicles within the nerve terminal. Drugs that depolymerize the neurotubules (colchicine and vinblastine) produce a reduction or blockage of the transport of the vesicles along the axon. In Figure 24–24 it may be observed that while most types of synaptic vesicles are transported by anterograde transport, the coated vesicles, which arise through endocytosis, can travel by retrograde transport toward the perikaryon.

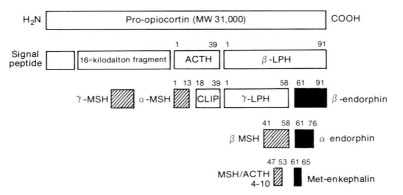

Fig. 24–23. Diagram representing the large precursor of the pro-opiocortin system and the various peptides that are produced by successive proteolysis. It is suggested that the common precursor is cleaved to yield equimolecular amounts of adrenocorticotropin (ACTH) and β-lipotropin (β-LPH), in addition to the signal peptide and a 16-kilodalton fraction. ACTH is cleaved to yield α-melanocyte stimulating hormone (α-MSH) and a corticotropin-like intermediate lobe peptide (CLIP). β-LPH yields γ-LPH, and β-endorphin. This last, together with α-endorphin and Met-enkephalin are opioid peptides (in black). The recurring MSH sequences are indicated by shading. (Courtesy of P.E. Cooper and J.B. Martin.)

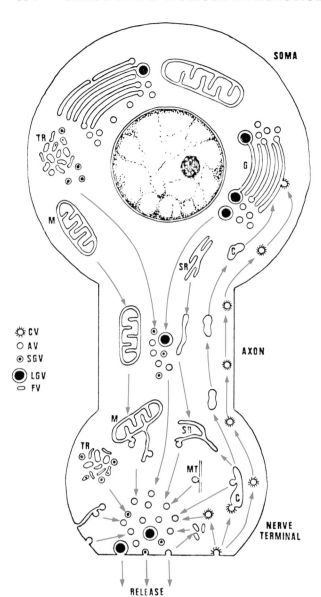

Fig. 24–24. Composite diagram showing possible sites of origin, transport, and release of various types of synaptic vesicles. *AV*, agranular vesicle; *CV*, coated vesicles; *FV*, flattened vesicle; *LGV*, large granular vesicle; *SGV*, small granular vesicle. The SGVs are shown to originate from a tubular reticulum. *(TR). M*, mitochondria; *MT*, microtubule. Note that some coated vesicles are transported by retrograde axon flow. (From Osborne M.P., in Cottrell, G.A., and Usherwood, P.N.R. (eds.): The Synapse. Glasgow, Blackie, 1977.)

Transmitter Release Is Related to the Role of Synaptic Vesicles in Nerve Transmission

The so-called *vesicle hypothesis* of synaptic transmission postulates that at the arrival of the nerve impulse at the nerve ending there is a synchronous release of a large number of quanta (i.e., multimolecular units or packets) of neurotransmitter, which are liberated from the synaptic vesicles into the synaptic cleft. For example, the postsynaptic potential at the myoneural junction may be the result of the simultaneous release of 100 quanta of acetylcho-line, which may represent the liberation of some 2×10^6 molecules of neurotransmitter.[46] In a resting synapse a spontaneous release may occur about every second, producing the so-called *miniature end plate potentials* (mepp), each of which is attributed to the release of single quanta. When the nerve ending is partly depolarized, or the K^+ concentration in the medium is increased, the number of mepp also increases.

Direct estimates of ACh content and the number of vesicles in nerve endings of *Torpedo* electroplax suggest that the content may be very concentrated. In fact, the solution of ACh inside the vesicle is probably isosmotic with plasma (i.e., 0.4 to 0.5 M). Estimates of the ACh released in each quantum have varied in the literature; in the rat diaphragm a fair estimate is 12,000 to 21,000 ACh molecules.[34] The presence of ATP and some binding component within synaptic vesicles has been demonstrated and it is possible that these substances play some role in the ACh concentration within the vesicle.

Since the time of the discovery of the synaptic vesicles, experiments have been carried out to demonstrate their possible role in nerve transmission. Axotomy resulted in an early degeneration of the endings, with clumping and lysis of the vesicles.[47] These alterations coincide with the time in which nerve transmission is interrupted.[48] Electrical stimulation of the cholinergic nerve endings of the adrenal medulla at high frequencies resulted in a reduction in the number of vesicles.[47] This finding has now been confirmed by numerous authors in the myoneural junction and in electric tissue (see Osborne, 1977 in the Additional Readings at the end of this chapter). In *Torpedo* electroplax stimulated to fatigue there is a loss of about 50% of the vesicles;[49] at the same time the presynaptic membrane increases, forming finger-like protrusions, probably corresponding to the integration of the vesicle membranes.[50]

Fixation by quick-freezing is the technique used to show morphological events during the transmission of a single nerve impulse, which takes place in milliseconds time. With this technique it was observed that the number of pits, corresponding to the opening of vesicles on both sides of the active zone of the myoneural junction, increased with the stimulation. Furthermore, by increasing the transmitter release using 4-aminopyridine and rapid freezing, it has been possible to observe synaptic vesicles in the process of opening into the synaptic cleft. These findings imply that the vesicle hypothesis is essentially correct (see Carpenter and Reese, 1981 in the Additional Readings at the end of this chapter).

An interesting finding is that a certain number of synaptic vesicles in the synapse of the electroplax and in the myoneural junction, when fixed in the presence of Ca^{++}, show inside a small granule attached to the membrane, which corresponds to calcium binding. When the synapsis is intensely stimulated, the vesicles that carry this Ca^{++} granule are preferentially released.[51] These findings are related to the role of calcium in transmitter release.

Another role of calcium is demonstrated by the following experiment. If venom of the black widow spider is applied to the neuromuscular junction, an intense release of acetylcholine and depletion of synaptic vesicles is produced. By freeze-fracture, pits in the presynaptic membrane are observed. In the presence of Ca^{++} the venom produces acetylcholine release, but the synaptic vesicles are not depleted. If peroxidase is put in the extracellular medium, it is taken up into the vesicles. This finding implies that the vesicles are being recycled and reused. The presence of Ca^{++} is thus important for the recovery of the synaptic vesicles.[52]

Other experiments have shown that the lack of calcium produces a disruption of the orderly array of intramembranous presynaptic particles present at the active sites of the myoneural junction (see Fig. 24–17).

Transmitter Release Probably Involves Exocytosis with Recycling of Synaptic Vesicles

One of the most discussed problems of transmitter release is whether it involves exclusively a mechanism of exocytosis or whether the synaptic vesicles are reused again and again. In the latter case, the attachment of the vesicle to the releasing sites of the membrane would be temporary, and the vesicle could be recharged with transmitter for reuse. There are arguments that favor both sides, and probably both mechanisms are at work. Quantitative work done on crayfish motor neurons has led to the proposal that there must be vesicle reuse or a mechanism by which incorporation of the vesicle membrane is balanced with an equivalent withdrawal. Related to these mechanisms is the finding that newly synthesized transmitter or transmitter recaptured from the cleft, as occurs in adrenergic synapses, is preferentially released.[53]

In synapses of *Torpedo* electroplax there are apparently two populations of cholinergic vesicles, one with a higher ratio of ACh/ATP, the so-called *supercharged vesicles*, and the other with a lower ratio.[54] It has been suggested that the supercharged vesicles are associated with the presynaptic membrane and upon stimulation are released first.

It has been postulated that newly formed synaptic vesicles are capable of several cycles of secretion and reloading before they undergo exocytosis and are incorporated into the membrane.[52] The existence of a recycling mechanism for the vesicles is now generally accepted, however.[55] It is possible that after several rounds of reuse the vesicle membrane is incorporated into the presynaptic membrane and subsequently recaptured. This could be done by direct endocytosis and formation of new vesicles or by intermediary membranous structures (e.g., by way of the smooth endoplasmic reticulum). Frequently, coated vesicles are formed in this process (Fig. 24–24). Recycling of vesicle membrane is an energy-saving mechanism that avoids the problem of constant synthesis, with all the overwork that this implies for the neuron.

The Depolarization–Transmitter Secretion Coupling Is Mediated by Calcium Ions

The coupling between the depolarization caused by the arrival of the nerve impulse and the secretion of acetylcholine is mediated by the influx of calcium ions. Extracellular Ca^{++} is an absolute requirement for the release of ACh from the nerve terminal; in the absence of calcium, there is no release of ACh. When the nerve impulse arrives at the terminal, there is an increase in Ca^{++} permeability (Fig. 24–21).

The entry of calcium into the ending has been demonstrated by ^{45}Ca and the use of *aequorin*, a substance that fluoresces in the presence of Ca^{++}.[56]

The entry of calcium through the presynaptic membrane presumably takes place through voltage-dependent channels. These calcium channels may be the large intramembrane particles observed by freeze-fracture at the active zones.[57] In the neuromuscular junction and in the other synapses, these particles are present in the vicinity of the sites where the release from the synaptic vesicles is produced. A tentative diagram of the hypothetical mechanism is shown in Figure 24–25 for the myoneural junction. The entry of Ca^{++} through the channels generates a localized increase in Ca^{++} concentration that triggers the opening of the vesicle (see Llinás,

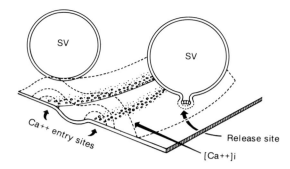

Fig. 24–25. Diagram of the active zone of the myoneural junction (see Fig. 24–17), showing the intramembranal particles, observed by freeze-fracture, as the possible Ca^{++} channels. The voltage-dependent opening of these channels allows Ca^{++} entry and ensures the release of the transmitter from the synaptic vesicle (SV). The broken lines illustrate possible changes in Ca^{++} concentration in the vicinity of the vesicle. (Courtesy of R. Llinás.)

1979 in the Additional Readings at the end of this chapter).

Experimentally the presynaptic entry of Ca^{++} can be favorably studied in the giant synapse of the stellate ganglion of the squid. The dimensions of the presynaptic terminal and of postsynaptic fiber permits the introduction, under microscopic observation, of microelectrodes so that voltage clamp experiments can be carried out. In this preparation it is possible to block the Na^+ conductance of the action potential with tetrodotoxin as well as the K^+ conductance with tetraethylammonium. If a depolarization is produced using the voltage clamp technique, there is an inward Ca^{++} current that induces transmitter release despite the lack of Na^+ and K^+ conductance.

An even more elegant experiment consists in

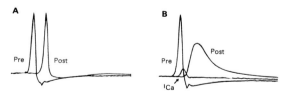

Fig. 24–26. Recordings of action potentials in an experiment carried out in the giant synapse of the stellate ganglion of the squid in which electrodes are implanted in the pre- and postsynaptic portions. In *A*, the presynaptic *(Pre)* and the postsynaptic *(Post)* action potentials are recorded prior to the blockage of the Na^+ and K^+ currents. In *B*, the stored presynaptic action potential is used on the blocked synapse to generate a postsynaptic potential. The calcium current (I_{Ca}) that triggers the release of the transmitter is shown. This I_{Ca} and the postsynaptic potential can be blocked by Cd^{++} or Mn^{++}. See the text for details. (Courtesy of R. Llinás.)

recording and storing the action potential in a digital memory. After blocking of both Na^+ and K^+ conductances, the simulated action potential is introduced into the voltage clamp circuit.[58] As Figure 24–26 shows, in these conditions a Ca^{++} current (I_{Ca}) becomes apparent at the falling phase of the presynaptic action potential. Then the postsynaptic potential is caused by the release of the transmitter and its interaction with the receptor. The Ca^{++} current can be blocked by Mn^{++} or Cd^{++}. These experiments permit a direct measurement of the Ca^{++} current generated by the action potential.

Another important consequence of these studies is that they have revealed the *synaptic delay*, which occurs between the end of the presynaptic action potential and the beginning of the synaptic potential. This can be divided into two portions: (1) the time actually involved in the opening in the Ca^{++} channel and (2) a shorter interval (about 200 μsec) representing release of the transmitter (see Llinás, 1982; Zucker and Lanitò, 1986 in the Additional Readings at the end of this chapter).

While our description of the possible role of synaptic vesicles in nerve transmission is largely in accordance with the vesicular hypothesis (Fig. 24–21), it should be mentioned that alternative hypotheses for the release of the neurotransmitter, in which the vesicles serve only as storage sites and the release is supposedly from acetylcholine free in the cytoplasm, have been postulated (see Israel et al., 1979; Tauc, 1982 in the Additional Readings at the end of this chapter).

Role of Calmodulin

Throughout this chapter the role of Ca^{++} in many activities of the synapse has been mentioned. The effect of Ca^{++} in several cases appears to be mediated by the calcium-binding protein *calmodulin*. For example, calmodulin present in the nerve ending can mediate: (a) the effect of Ca^{++} in the release of the neurotransmitter, (b) the interaction of the synaptic vesicle with the presynaptic membrane, and (c) the phosphorylation of synaptic proteins and of tubulin. All of these functions are inhibited by inhibitors of calmodulin such as trifluoperazine (see De Lorenzo, 1981 in the Additional Readings at the end of this chapter).

Calmodulin is concentrated at the postsynaptic density (subsynaptic web, Fig. 24–14), where it is bound to a major protein of 51,000 daltons. In addition it activates the phosphorylation of this protein by a protein kinase.[59]

SUMMARY:
Synaptic Vesicles and Transmitter Release

The presence of *synaptic vesicles* is the main characteristic of synapses. Most of these structures are spherical and electron-translucent, and have a diameter of 40 to 50 nm and a membrane 4 to 5 nm thick. In the myoneural junction there are some 1000 vesicles per μm^3 and a total of 3×10^5; about 20% of them are attached to the presynaptic membrane and are ready to discharge the transmitter. In many central and peripheral synapses there are *large dense-core* or *granular vesicles*.

Some synapses, thought to be inhibitory, contain *flattened* or *elliptical vesicles*. In sympathetic axons and endings there are small *granulated vesicles*; the granule contains the transmitter (norepinephrine). Another type is the *complex* or *coated vesicle*, which has a shell of chlotrin material surrounding it. These vesicles are formed by endocytosis and may represent a mechanism of membrane recycling.

During synaptic development there is a critical period in which peripheral synapses may change from adrenergic (with small granular vesicles) into cholinergic (with clear vesicles). This change depends on the influence of the surrounding cells, which produce developmental factors. Cultured neurons with dual function and with both types of vesicles have been observed. By cell fractionation methods a true dissection of the synaptic region may be achieved. In this way, synaptosomes, synaptic vesicles, nerve-ending membranes, synaptic membranes, and postsynaptic densities may be isolated. A direct demonstration that synaptic vesicles are the sites of storage of transmitter was obtained after the isolation of these organelles (Table 24–2).

At least 50 putative neurotransmitters and neuromodulators are produced by neurons. They include biogenic amines (acetylcholine, norepinephrine, dopamine, serotonin, etc.), amino acids (glutamic and aspartic acid, GABA, glycine, and others), purines, and numerous neuropeptides. Several criteria are used to define a substance as a neurotransmitter.

In the acetylcholine system, synthesis is produced by the enzyme choline acetyltransferase in the following reaction:

$$Choline + Acetyl\ CoA \rightarrow Acetylcholine$$

After its release, acetylcholine is hydrolyzed by acetylcholinesterase, thereby releasing acetyl groups and choline, which are again taken up by the nerve ending.

Cell fractionation, fluorescence, autoradiography, and especially inmunochemical methods are being used to identify special neuronal systems. Through the use of these techniques, researchers have discovered that neurotransmitters sometimes coexist in the same neuron and even in the same synaptic vesicle. The case of the pro-opiocortin system is a special one.

Transmitter release is related to the role of synaptic vesicles in nerve transmission. The vesicle hypothesis implies that the neurotransmitter is liberated in multimolecular packets of quanta. Each quantum released produces a spontaneous miniature end plate potential. The ACh content of vesicles may be very high. In *Torpedo*, its concentration is about 0.4 to 0.5 M. Each quantum involves some 12,000 to 21,000 ACh molecules, which are packed within a vesicle.

When nerve impulses arrive at the endings, 100 or more quanta are released, thereby producing the synaptic or end plate potential. The relationship between vesicles and quantal release has been demonstrated by a variety of experiments.

The mechanism of transmitter release probably involves reuse of the vesicles, followed by exocytosis and recycling of the membrane vesicle.

The coupling between the depolarization and the secretion of the transmitter is mediated by calcium ions. Ca^{++} penetrates through special channels situated at the active zones of the presynaptic membrane. The synaptic delay involves the time used in the opening of the Ca^{++} channels and in the actual release of the transmitter.

24–5 SYNAPTIC RECEPTORS AND THE PHYSIOLOGICAL RESPONSE

Since the beginning of this century, through the work of Langley, Ehrlich, and others it has been postulated that the transmitter interacts with a specific receptor localized at the cell membrane. For many years, however, the knowledge of synaptic receptors has been principally indirect—based on the final response obtained from this interaction. For example, it was known that acetylcholine produced contraction of skeletal muscle, or the secretion of a gland, and that these two effects could be blocked with curare and atropine, respectively.

In 1955 Nachmanson had already postulated that the "acetylcholine receptor is a protein that upon the binding of ACh undergoes a conformational transition which results in changes in membrane permeability." Figure 24–27 shows that the primary transmitter-receptor interaction is coupled, probably by way of a conformational change, to the translocation of ions across the cell membrane (i.e., the channel or

ionophore) and by several interposed mechanisms can produce a final response. This figure also shows the probable relationship between the primary interaction, the displacement of Ca^{++}, and certain metabolic processes involving cyclic AMP and other substances.

The primary interaction between the neurotransmitter and the receptor can now be studied directly on isolated synaptic membranes through the use of labeled ligands. These may be the corresponding neurotransmitter or drugs that act as *agonists* (i.e., in a way similar to the neurotransmitter) or *antagonists* (i.e., that block the action of the agonists).

The ligand-receptor interaction shows the following characteristics:

1. *Saturability.* Because there is a definite number of receptor sites in the membrane, as the concentration of the ligand is increased, the binding should reach a saturation level, called the maximal binding, or Bmax.
2. *High affinity.* The affinity (i.e., the relative "tightness" of the binding) depends on the dissociation constant determined kinetically or at equilibrium (K_D). The affinity is inversely related to the K_D. (In general agonists have a K_D in the micromolar range, and the antagonists in the nanomolar or picomolar range.)
3. *Reversibility.* The ligand-receptor interaction, in general, does not involve a covalent binding and can be dissociated, either by dilution of the ligand or by the action of other drugs.
4. *Specificity.* The interaction should be specific for the ligand and should be displaced by the corresponding agonists and antagonists.

At a second stage (Fig. 24–27) neurotransmitters alter the excitability of the postsynaptic membrane by changing the membrane potential and/or the membrane resistance. This last effect implies a change in the transmembrane permeability to one or more ions. The effect on the potential depends on which ionic permeability has changed.

The Acetylcholine Receptor Is Coupled to the Translocation of Sodium and Potassium Ions

Taking as an example the interaction of acetylcholine at the myoneural junction (or the elec-

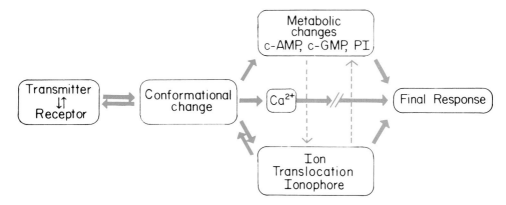

Fig. 24–27. Diagram showing the primary interaction between transmitter and receptor and its consequences. PI = phophatidyl inositol. (See the text for details.)

troplax) the effect in ionic translocation can be expressed as in the following diagram:

$$ACh + Rc \xrightleftharpoons[K_{-1}]{K_1} AChRc \xrightleftharpoons[K_{-2}]{K_2} AChRo$$

in which ACh interacting with the receptor (Rc) forms first a complex, which is in the closed condition, and then, in a second step, goes into the open condition AChRo. (Ic and Io correspond to the closed and open ionophores.)

Here we should stress again the difference between the axonal channels studied in Section 24–1 and those related to this acetylcholine receptor. While in the axon there are separate Na^+ and K^+ channels that are regulated by the membrane potential, in the postsynaptic membrane both ions probably use a common channel, and the gating is driven by the chemical interaction of the neurotransmitter with the receptor protein (see Fig. 24–21).

Experimentally, by the use of fine microelectrodes in the myoneural junction, some of the best evidence of the functioning of receptors at the molecular level has been obtained.[60] It was mentioned before that in this type of preparation it is possible to record the spontaneous miniature end plate potentials (mepp), which have an amplitude of 70 to 100 µV and represent the discharge of single multimolecular quanta of ACh. Katz and Miledi have observed that if a minimal but steady dose of acetylcholine is applied to the myoneural junction, there are minute fluctuations of the membrane potentials that are superimposed upon a small depolarization of the membrane. The amplitude of these fluctuations is about a hundred times smaller than the mepp (i.e., on the order of 0.3 µV) and constitute the so-called *membrane noise* (Fig. 24–28).[60]

In denervated frog muscle, in which the acetylcholine receptors appear beyond the myoneural junction, it has been possible to record the actual opening and closing of individual channels[61] (see Salpeter and Loring, 1985). The current produced is equivalent to the translocation of about 5×10^4 univalent ions. The enormous amplification that is produced at the chemical synapse is readily understandable; in fact, the interaction of one or two ACh molecules with a receptor translocated more than 10,000 ions!

The Change in Membrane Potential Depends on the Ions Translocated

As we studied in Chapter 4, the equilibrium or *resting membrane potential* (RMP) is determined by the concentration gradient of the different ionic species. In a neuron, the RMP varies between -50 and -100 mV and depends on the distribution of Na^+, K^+, and Cl^- across the membrane. Each of these ions has its own equilibrium (E) or *Nernst potential* (Table 24–4).

Under resting conditions, the permeability of K^+ (whose concentration is high inside the cell) predominates, and the RMP is closer to the E_{K^+}. In the myoneural junction or the electroplax, when the ACh ionophore opens, both Na^+ and K^+ move in opposite directions. The new equilibrium reached is intermediary between E_{Na^+}

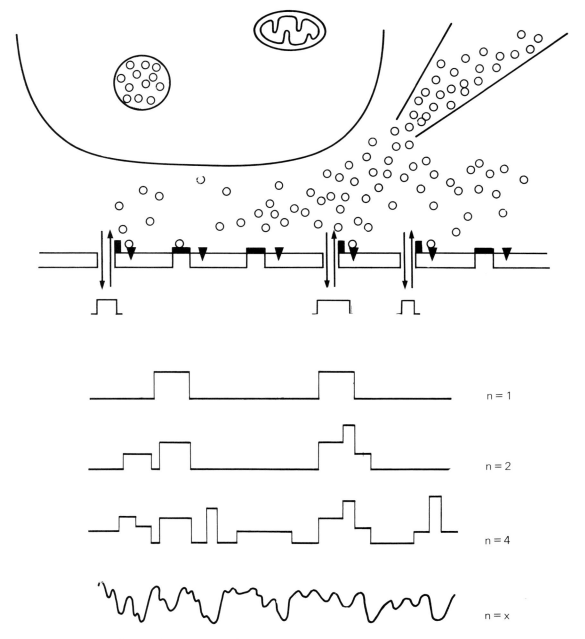

Fig. 24–28. Diagram showing the experiment of Katz and Miledi for the demonstration of the membrane noise. *Top,* from a micropipette a steady dose of acetylcholine is applied to the myoneural junction; *Middle,* the neurotransmitter diffuses and interacts with single receptors opening channels, as in Fig. 24–21, and interacts with acetylcholinesterase. *Bottom,* the opening of channels produces currents of varying lengths of time depending on the kinetics of the opening and closing. n = 1, recording from a single channel opening; n = 2, from two channels; n = 4, from four channels, n = X, when the number of channels increases, the result is the membrane noise. (Courtesy of O. Uchitel.)

TABLE 24–4. CONCENTRATION OF IONS AND NERNST POTENTIAL

Ion	Extracellular Concentration (mM)	Intracellular Concentration (mM)	Nernst Potential (mV)
Na$^+$	145	12	E_{Na^+} +66
K$^+$	4	155	E_{K^+} −97
Ca^{++}	1.5	<10^{-3}	$E_{Ca^{++}}$ > +97
Cl$^-$	120	4	E_{Cl^-} −90

and E_{K^+} (usually − 15 mV), and the membrane is depolarized. In this case the synaptic potentiao is excitatory (Fig. 24–12).

In other cases, when there is an increase in permeability to a single ion, the RMP will be closer to the Nernst potential of the corresponding ion. For example, when the channel selectively translocates Cl$^-$, a hyperpolarization is produced (E_{Cl^-} = − 90 mV). This is what occurs with the GABA receptor, which produces an *inhibitory* effect.

In mollusc neurons (Fig. 24–12) selective permeabilities to Na$^+$, K$^+$, Cl$^-$, and Ca^{++} have been observed. In addition, the permeability may be increased or decreased allowing for numerous combinations (i.e., P_{Na^+} ↑ is excitatory, and P_{Na^+} ↓ is inhibitory; P_{Kes^+} ↑ is inhibitory, and P_{K^+} ↓ is excitatory).

Synaptic Receptors Are Hydrophobic Proteins Embedded in the Lipoprotein Framework of the Membrane

Only in recent years has it been possible to separate synaptic receptor proteins as chemical entities and to analyze, in a more direct way, their interaction with the transmitter. The separation of these substances has been difficult because they are intrinsic proteins (see Table 4–2) that require strong treatments to be separated from the lipoprotein framework of the membrane.

Receptor proteins are highly hydrophobic and intimately related to lipids of the membrane. Two main procedures are used to isolate their proteins: extraction with organic solvents[62] and separation by detergents. Cholinergic receptor proteins have been isolated not only from the electroplax of *Torpedo* and *Electrophorus* tissues, which have the richest cholinergic innervation, but also from skeletal muscle, smooth muscle, and brain. Adrenergic receptor proteins have been separated from brain, from spleen capsule, and from heart. Receptor proteins for the amino acids glutamate and γ-aminobutyrate have also been isolated from muscle of *Crustacea* and from insects that have a double innervation: excitatory (via glutamate) and inhibitory (via γ-aminobutyrate). From synaptic membranes of the cerebral cortex, hydrophobic proteins binding L-glutamate, L-aspartate, and γ-aminobutyrate have been separated (see De Robertis, 1975 in the Additional Readings at the end of this chapter).

After extraction the receptor proteins are generally separated by column chromatography or, in the case of the cholinergic receptor, by affinity chromatography. In the latter technique, the gel used contains an active group that specifically binds the receptor protein and is then eluted by a pulse of the corresponding transmitter.

Figure 24–29,*1* shows the two-step separation of the cholinergic receptor of skeletal muscle. The small second peak of protein is the one having a high affinity for ACh and other cholinergic drugs. In Figure 24–29,*2* this protein has been further purified by affinity chromatography. In this case, the protein peak, which appears after the ACh pulse, is the specific receptor protein fraction. With this procedure a total purification of about 15,000 times has been achieved using rat diaphragm.[63]

The ideal source for the isolation of the acetylcholine receptor (AChR) is the electric tissue (electroplax) of the eel (*Electrophorus electricus*) and of rays. The most commonly used ray is the *Torpedo*, in whose electroplax the density of cholinergic synapses may occupy 50% of the postsynaptic area.

The molecular properties of the AChR have been characterized both in detergent solubilized membranes (see Hucho, 1981; Anholt, 1981 in the Additional Readings at the end of this chapter) and after organic solvent extraction. In the first case the isolated AChR consists of a complex of four glycopeptides with molecular weights of

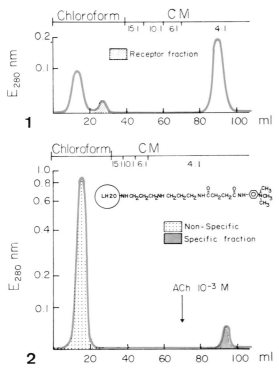

Fig. 24–29. Separation of the cholinergic receptor from rat diaphragm measured by absorption at E_{280} nm; *1*, by conventional chromatography in Sephadex LH20; *2*, by affinity chromatography. Observe that the receptor protein fraction from *1* has been purified further in *2*. The specific receptor fraction appears only after a pulse of 10^{-3} M acetylcholine (ACh). The inset shows the chemical composition of the affinity column with the cholinergic end (p-aminophenyl-trimethylammonium) linked to the gel by a 1,4 nm "arm" (3'-3'-iminobispropylamino). (From Barrantes, F., Arbilla, S., de Carlin, C., and De Robertis, E.: Biochem. Biophys. Res. Commun., 63:194, 1975.)

40,000 (α), 50,000 (β), 57,000 (γ) and 64,000 (δ). The complex has the composition $\alpha_2 \beta\gamma\delta$ and corresponds to a total molecular weight of 250,000. It is a monomer, but in the postsynaptic membranes of *Torpedo*, a large proportion of the receptor may be present as dimers ($\alpha_2\beta\gamma\delta$—S—S—$\delta\gamma\beta\alpha_2$) with an S—S bond between the two δ subunits. Treatment with a thiol reagent results in breakage of the S—S bond and separation of the monomers. The sequence of the various protein subunits has been determined, and the genes producing them have been cloned (see Stevens, 1985, 1986 in the Additional Readings at the end of this chapter).

So far the functional role has been demonstrated only for the α subunit, in which the binding site for acetylcholine (ACh) and for the neurotoxin α-bungarotoxin is located. Similar findings were obtained using extraction with organic solvents of *Torpedo* electroplax. Several proteolipid bands were found, of which only the one, of 39,000 daltons, specifically binds ACh, α-bungarotoxin, and other cholinergic ligands.[64]

The use of high resolution electron microscopy,[65] immunoelectron microscopy, and x-ray and neutron diffraction[66] have provided a three-dimensional view of the AChR in the postsynaptic membrane. The AChR complexes are densely packed in the lipid bilayer (about 10^4 per μm^2), and each of them has a diameter of 8.5 nm (Fig. 24–30A). The complex has a total height of 11 nm and protrudes about 5.5 nm above the lipid bilayer. It traverses the membrane and protrudes slightly on the cytoplasmic side. As Figure 24–30B shows, the extension above the bilayer lies on the synaptic side of the membrane and has a funnel shape with a central well. There are two sites for the binding of α-bungarotoxin which lie on this extension, but not in the central well. When observed in planar arrays a large proportion of the complexes have a dimeric array due to the S—S bridge that binds the δ subunits in each complex (Fig. 24–30C). Information about the position of the subunits in this structure is becoming available, and in general it is thought that the subunit arrangement limits a central ion channel of about 0.7 nm. It is interesting that the monomer has an asymmetric structure with a kind of horseshoe shape in which the position of the two α subunits (α_1 and α_2) has been determined under the EM by the binding of α-bungarotoxin. There is also some evidence about the position of the δ and γ subunits within the monomer (see Zingheim et al., 1982; Stroud, 1983 in the Additional Readings at the end of this chapter).

Reconstitution of the AChR in Model Membranes

A useful method to study the function of the AChR is to reconstitute the purified receptor into liposomes or planar lipid bilayers. Incorporation into lipid vesicles was achieved by extraction of the AChR with cholate, followed by dialysis in a cholate-free medium. The success of this preparation came from the addition of a crude soybean lipid extract that apparently protected the ionophonic properties of the AChR.[67] With this method it has been possible to study

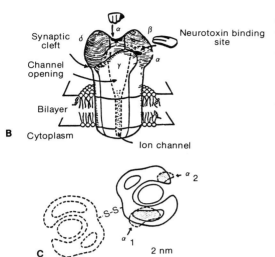

Fig. 24–30. *A,* diagram of the postsynaptic membrane of *Torpedo* showing the densely packed AChR embedded in the lipid bilayer. (From Anholt, R., Lindstrom, J., and Montal, M.: J. Biol. Chem., *256*:4377, 1981.) *B,* three-dimensional structure of the AChR in a lipid bilayer with the neurotoxin binding sites as deduced from x-ray scattering and electron microscopy. The position of the subunits is tentative. (From Kistler, J., et al.: Biophys. J., *37*:371, 1982.) *C,* schematic representation of the subunit topography of the membrane-bound AChR, viewed in axial projection perpendicular to the membrane. Two monomers held by an S—S bond between the δ subunits are shown. The position of α_1 and α_2 subunits has been determined under the EM by the binding of α-bungarotoxin in negative staining, using an averaging image technique. (Courtesy of F.J. Barrantes.)

the *influx* and *outflux* of $^{22}Na^+$, and to stimulate both with cholinergic agonists.

This method has also permitted observation of the phenomenon of *desensitization*, which also occurs in vesicles or microsacs made of natural membranes. Desensitization is a well-known phenomenon in the myoneural junction and consists of a preexposure of the AChR to agonists, after which the AChR changes to a high affinity state and no longer reacts with a physiological response (i.e., the Na^+ fluxes are diminished or even blocked).

Reconstitution may also be carried out in planar bilayers (i.e., black lipid membranes) across a hole in a septum of plastic. This allows the study of the conductance response produced by the addition of cholinergic agonists and its blocking by antagonists. As Figure 24–31 shows, addition of ACh produces an increase in conductance, accompanied by fluctuations, in a bilayer containing the AChR extracted in organic solvent. These fluctuations disappear when the membrane voltage is reduced to zero or the cholinergic blocker α-tubocurarine is added.[68] Similar observations, including the recording of single channels (as in Fig. 24–28) have been observed in detergent-extracted AChR.[69]

Based on the hydrophobic properties of the AChR receptor and on reconstitution experi-ments in lipid bilayers, a tentative model was postulated.[70] The most interesting features of this model were the oligomeric arrangement of the protein subunits and their disposition traversing the postsynaptic membrane.

Figure 24–32 is a schematic representation to explain the molecular mechanisms involved in the activation and inactivation of the AChR in *Electrophorus.*

The receptor is represented in the resting condition, and after the activation with one and then two acetylcholine molecules, which leads to the opening of the channel, desensitization is depicted as a conformational change by which the receptor site enters a state of high affinity and can no longer open the channel. The steps involve different rate constants that have been measured or calculated.[71]

Because the ACh-receptor interaction is reversible, it may easily go back toward the closed position once the ACh has been removed from the site of binding. This reversible effect is facilitated because the receptor molecules are held in place by hydrophobic interactions within the framework of the membrane.

We mentioned that the ionophore of the AChR is permeable both to Na^+ and K^+. In the case of other receptors, the ionophore may be selective for a single ion (e.g., Na^+, K^+, or Cl^-).

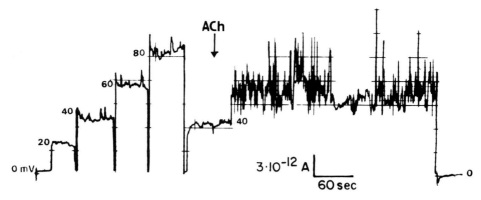

Fig. 24–31. Time-current recording in a black lipid membrane of 2% egg lecithin in decane containing, in the forming solution, 4 μg/ml of cholinergic proteolipid from *Torpedo*. At the time indicated, acetylcholine was applied to the membrane. Observe that the membrane is maintained at various voltages and that the transmitter is applied at 40 mV. (Courtesy of P. Schlieper and E. De Robertis.)

For these channels it is possible to assume that (as in the case of the Na$^+$ and K$^+$ channels in the axon) there is a *gate* that regulates the ion flow and a *selectivity filter* that determines the kind of ion that is translocated.

Based on these considerations models for receptor-ionophore complexes have been postulated in which either the receptor or the ionophore moiety changes. For example, an ionophore with P$_{Cl-}$ may be coupled to an AChR or to a dopamine receptor; and a single receptor may be coupled to different ion-selective ionophores (see Carpenter and Reese, 1981 in the Additional Readings at the end of this chapter).

Neurotransmitter Synthesis and Release—Modulated by Presynaptic Receptors

Until recently only the postsynaptic receptors present on effector cells were known, but it is now recognized that receptors may be localized in other portions of the neuronal membrane and particularly at the nerve terminal. When these *presynaptic receptors* (or *autoreceptors*) are activated, there is a modulation of the secretion by which the synthesis and release of the neurotransmitter can increase or decrease.

The following experiment is illustrative: ^{3}H-norepinephrine is provided to the spleen and accumulates at the sites of storage (i.e., the synaptic vesicles of adrenergic terminals). Then the

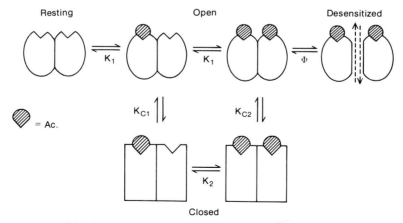

Fig. 24–32. Diagram proposed for the molecular mechanisms involved in the opening and closing of the AChR channel. The interaction of two molecules of acetylcholine produces the open condition with the entrance of Na$^+$ and the exit of K$^+$. The desensitized state of the AChR is shown as a change in conformation which results in a more permanent closing of the channel. K and K$_c$ are various constants. (From Aoshima, H., Cash, D.J., and Hess, G.P.: Biochemistry, *20*:3467, 1981. Copyright 1981 by the American Chemical Society.)

splenic nerve is stimulated, and there is secretion of ^{3}H-norepinephrine. If an adrenergic (α) blocker such as phenoxibenzamine is added, the secretion is increased. This experiment is interpreted as indicating that α-adrenoreceptors, present presynaptically, produce a negative *feedback inhibition* of the secretion, which is blocked by phenoxibenzamine (see Langer, 1977 in the Additional Readings at the end of this chapter).

At present we know that in peripheral as well as in central nerve endings, there are numerous receptors that may produce either a *facilitation* or an *inhibition* of synaptic transmission. The use of cell fractionation, which allows for the separation of the synaptosomal membrane and the postsynaptic membrane (see Fig. 24–14), has shown that various central receptors are located both pre- and postsynaptically.

Many nerve terminals have receptors to their own transmitter (autoreceptors) or to other neurotransmitters. These receptors may explain the well-known phenomenon of *presynaptic inhibition*, by which the transmitter released by one terminal may have an inhibitory action on the release of another transmitter in another nerve ending.

The importance of presynaptic receptors in the modulation of synaptic transmission is becoming recognized in many peripheral and central synapses, and it is possible that they are involved in certain long-lasting effects in nerve transmission.

Long-lasting Synaptic Functions Involve the Use of a Second Messenger

The description given so far of the neurotransmitter-receptor interaction involves the direct translocation of ions. As we have seen in the case of the nicotinic cholinergic receptor of the myoneural junction, this translocation is a very fast phenomenon, occurring in a few milliseconds. In other cases, however, as Figure 24–27 shows, the interaction may produce, in addition, metabolic changes in the neuron with transient but longer-lasting physiological effect. In Chapter 2 a section was dedicated to discussion of the action of hormones with the intervention of specific receptors and the enzyme adenylate cyclase, discovered by Sutherland in 1956. As shown in Figure 2–27, the specific hormone re-

ceptor is coupled within the membrane with adenylate cyclase. When the hormone interacts with the receptor, the enzyme is activated and produces *3'-5' cyclic AMP (cAMP)* from ATP. This nucleotide is considered a *second messenger* that is able to translate extracellular messages into an intracellular response.

In Figure 2–29 the classic example of the so-called *cascade effect* of cAMP on glucogenolysis was indicated. It may be observed that under the action of the hormones, epinephrine or glucagon adenylate cyclase is activated, increasing the levels of cAMP. This increase in cAMP in turn activates *protein kinase*, a phosphorylating enzyme, which converts phosphorylase kinase from an inactive into an active form, which will finally degrade the glycogen.

The first indication that adenylate cyclase could be involved in synaptic function was provided in 1967 by De Robertis et al.[72] with the demonstration that this enzyme was concentrated in synaptosomes and was found mainly in the synaptosomal membrane. After osmotic shock of the synaptosome, the enzyme activity was found to be increased, indicating that the active groups are facing toward the inside of the synaptosomal membrane.[72]

The synaptosome also contains high levels of *cyclic phosphodiesterase*, the enzyme that degrades cAMP into inactive AMP.

At the present time it is thought that several neurotransmitter-receptor interactions that generate slow postsynaptic potentials (i.e., on the order of 100 to 500 msec) involve a second messenger mechanism, with the production of cAMP. Another second messenger, *cyclic GMP*, was found in brain, and it is produced by the enzyme *guanylate cyclase*. In contrast to adenylate cyclase, this enzyme is usually soluble in the cytosol.

The levels of cAMP or cGMP depend not only on the activation of the corresponding enzymes but also on the action of phosphodiesterase, which apparently acts on both nucleotides. For example, if phosphodiesterase is inhibited by methylxanthines (such as caffeine) or other drugs, the levels of these cyclic nucleotides increase. Phosphodiesterase and adenylate cyclase are also regulated by Ca^{++} by way of a *Ca^{++} binding protein* (e.g., calmodulin).

The link between cAMP and synaptic transmission was more firmly established when it was found that the iontophoretic application of either norepinephrine or cAMP upon the Purkinje cells of the cerebellum slowed the firing of these neurons. Furthermore, it was shown the application of norepinephrine produced a large increase of cAMP within these cells.[73]

As Figure 24–33 shows, various neurotransmitters and other regulatory agents such as hormones and prostaglandin E_1, acting on specific receptors, are able to activate either the adenylate or the guanylate system, causing the increase of the corresponding second messenger. (The student should be aware that the same neurotransmitter—e.g., histamine or norepinephrine—acting on different receptors, may activate either cAMP or cGMP.)

A Third Step in Neuronal Regulation—Protein Phosphorylation and Synapsin I

The concept of the second messenger is complemented by that of protein phosphorylation and dephosphorylation; this is a third step in the mechanism of neuronal regulation. In brain there are some 100 neuronal specific proteins that may be phosphorylated by protein kinases.

In most cases the PO_4 is linked to a serine or an asparagine residue, although there is also another class of protein kinases that bind PO_4 to tyrosine.

There are four classes of protein kinases depending on the second messenger that activates the enzyme: cAMP- and cGMP-dependent protein kinases and others that are related to Ca^{++}-calmodulin and to Ca^{++}-diglycerol. In the CNS these enzymes are involved in transmitter biosynthesis, in cytoskeletal proteins, and in the phosphorylation of Ca^{++} and K^+ channels and of receptors.

As shown in Figure 24–33, Ca^{++} may act as a second messenger when these ions are displaced by the depolarization caused by the nerve impulse, but all the other messengers are also able to phosphorylate certain substrates. The phosphorylation system includes phosphoprotein phosphatases that are able to reverse the reaction producing dephosphorylation.

The complete system for phosphorylation-dephosphorylation is present within the synaptosome, and various protein kinases have been purified and characterized. In the case of the *cAMP-dependent protein kinase* it is thought

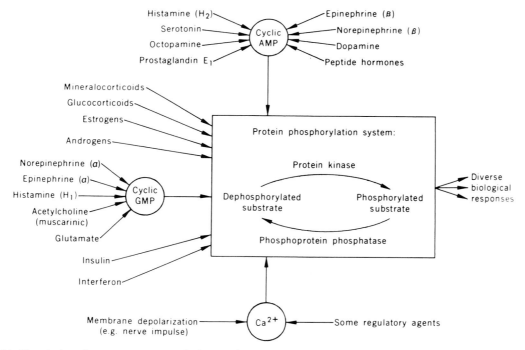

Fig. 24–33.   Action of neurotransmitters and other regulatory agents on specific receptors that are able to activate adenylate cyclase or guanylate cyclase producing cyclic AMP or cyclic GMP. These second messengers and also Ca^{++} are able to act on the phosphorylation system producing diverse physiological responses. (Courtesy of P. Greengard.)

that the cAMP binds to the inhibitory subunit of the kinase causing it to dissociate from the catalytic subunit. In this way the activated enzyme may readily transfer phosphate groups from ATP to the endogenous substrate protein.

Using an antiserum against *cGMP-dependent protein kinase* it was found that this enzyme is highly concentrated in the cerebellum and especially in Purkinje cells, whose dendrites, perikarya (with exception of the nucleus), and axon appear stained by the immunohistochemical reaction.[74]

In recent years, the work of Greengard and collaborators has concentrated on the study of the endogenous substrates for the protein kinases in the nervous tissue. Of the various proteins that are phosphorylated there is one that is remarkable because of its high concentration and its unique localization in axon terminals. This was named *protein I* and today is called synapsin I.[75] This protein is a doublet of two subunits, synapsin 1A and 1B, of 86,000 and 80,000 daltons, respectively, and present in a 1:2 molar proportion. Other characteristics are the basic properties and the elongated shape of these two subunits. Synapsin I is the most prominent substrate, not only for the cAMP-dependent protein kinase, but also for the Ca^{++}-calmodulin-dependent kinase in brain.

Studies on the localization of synapsin I are of great interest and suggest that it may play an important function in synapses.[76,77] It is present in most if not all synapses, including peripheral ones, such as the myoneural junction. Within the nerve terminal, synapsin I is found associated with synaptic vesicles. In the isolated vesicle fraction it represents up to 6% of the total protein. This protein is of the peripheral type, bound to the vesicle by ionic forces, and it can be extracted with solutions of high ionic strength.

The synapsin I molecule has a head and a tail, that attaches to the vesicle and contains two sites of phosphorylation. The cAMP protein kinase acts on the head portion, while the Ca^{++}-calmodulin protein kinase acts on the tail and in this way detaches the synapsin from the vesicle. In resting conditions synapsin I makes a kind of a cage around the vesicle, as can be demonstrated at the EM level using antibodies coupled to ferritin (Fig. 24–34). The cage is supposed to keep the synaptic vesicle in place, inhibiting

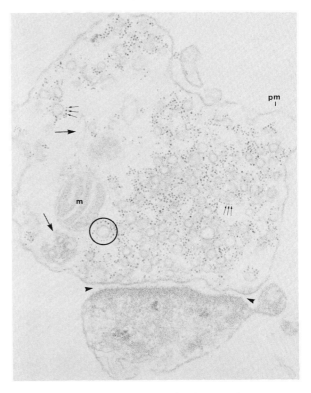

Fig. 24–34. Immunocytochemical localization of synapsin I in an isolated synaptosome incubated first with rabbit antibodies against synapsin I. Subsequently the antibodies bound to synapsin I have been labeled by ferritin-conjugated goat antibodies directed against rabbit IgG. The dark particles visible in the field *(small arrows)* are the iron cores of the ferritin molecules. In the nerve ending the large majority of ferritin particles decorate the cytoplasmic side of synaptic vesicles. The profile of a synaptic vesicle uniformly decorated by ferritin particles is enclosed by a black circle. No ferritin particles are associated with the plasmalemma *(pm)*, with the mitochondria *(m)*, or with other membranous organelles *(large arrows)*. Arrowheads indicate the synaptic cleft. × 140,000. (Modified from De Camilli et al.: J. Cell Biol., 96:1374, 1983; reprinted with permission from the Rockefeller University Press).

transmitter release. When the cage is removed after phosphorylation of synapsin I by the Ca^{++} calmodulin protein kinase, the vesicle can fuse with the presynaptic membrane and the release of the neurotransmitter is produced (see Greengard and De Camilli, 1982 in the Additional Readings at the end of this chapter). The possible function of synapsin I has been demonstrated by Llinás in experiments of microinjection into the presynaptic ending of the giant synapse of the crayfish. The injection of this protein inhibits the release of neurotransmitter; however, when PO_4-synapsin is injected, there is no inhibition. If Ca^{++}-calmodulin protein kinase is injected, there is a large increase in neurotransmitter release.

Learning, Memory, and the Functional Effectiveness of Synapses

We mentioned earlier that in cAMP-mediated synapses the physiological events may last 100 milliseconds or more. There are indications in mollusc ganglion cells and in the sympathetic system of vertebrates that these events may be much longer, lasting for hours. This finding has led to the hypothesis that synaptic events mediated by cyclic nucleotides could be related to long-term changes in the nervous system and could play a role in learning and memory. Furthermore, phosphorylation of regulatory proteins within the nucleus of the neuron could be involved in gene expression by the transcription of RNA from certain DNA genes. This could lead to translation into important proteins, such as enzymes or receptors, and to more permanent changes in the neuron (see Nathanson and Greengard, 1981 in the Additional Readings at the end of this chapter).

In Section 24–3 we mentioned that one of the most accepted theories of learning is based on fine changes in microstructure at the synaptic level which involves the growth and branching of spine synapses. In recent years the work of Kandel and collaborators in the marine snail *Aplysia* has shed light on some of the most elemental forms of learning behavior (see Kandel and Schwartz, 1982 in the Additional Readings at the end of this chapter). The advantage of this species for studying the cellular bases of learning stems from the fact that the neuronal circuitry is rather simple and well known from the viewpoint of structure and the neurotransmitters involved. Kandel and collaborators analyzed the simplest behavior controlled by the abdominal ganglion, which consists in defensive reflexes, i.e., the withdrawal of the gill in response to a tactile stimulus to the siphon. These reflexes can be modified by experience resulting in *habituation*, when the stimulus is repeated, or in *sensitization* (i.e., a stronger response) if another stimulus is simultaneously applied to the tail. These two mechanisms involve a change or modulation in synaptic effectiveness acting at the presynaptic level. In habituation there is a decrease in transmitter release, due to a reduced

influx of Ca^{++} into the presynaptic terminal. In sensitization the opposite phenomenon occurs; there is a sudden increase in Ca^{++} influx which is mediated by cAMP.[78] In this phenomenon a serotoninergic interneuron acting presynaptically on the sensory terminal produces the increase in cAMP and the *facilitation* of the transmitter release. Further studies have revealed that there are fine structural modifications accompanying the functional changes. Thus, the number of active presynaptic zones and of synaptic vesicles attached to them is reduced in habituation and increased in sensitization.[79]

Similar mechanisms are involved in classical conditioning, in which the *conditioned stimulus* must discharge the sensory neuron before the *unconditioned stimulus* activates the serotoninergic neurons producing presynaptic facilitation (as in the case of sensitization). In this case the cAMP produced in the terminal activates a protein kinase which in turn, by phosphorylation of the K^+ channel, prolongs the duration of the action potential and increases the Ca^{++} influx.[80]

These studies reveal that, at the cellular and molecular levels, some of the basic phenomena of learning can be explained on the basis of the regulation of synaptic function. If protein phosphorylation is involved in the general mechanism of memory, the neurons must be programmed to keep phosphorylating the proteins. (This assumption is made because the lifetime of membrane proteins is only 2 to 3 weeks.) Recently it has been suggested that memory can be considered a continuation of neuronal differentiation, a change probably occurring at the synaptic level under the influence of acquired experience. The versatility in memory may depend not on the production of numerous proteins by DNA transcription and translation but on a limited number of receptors and channel proteins that confer special properties to the nerve cells (see Stevens, 1982; Robertson, 1982 in the Additional Readings at the end of this chapter).

SUMMARY:

Receptors and the Physiological Response

The primary interaction between the neurotransmitter and the receptor at the membrane shows the following characteristics: saturability (Bmax), high affinity $\left(\dfrac{1}{K_D}\right)$, reversibility, and specificity.

This interaction produces a conformational change in the protein by which channels for the translocation of ions are opened (ionophoric response). In contrast to the axonal ionophores, in this type of response both Na^+ and K^+ are translocated. In the myoneural junction the application of a minimum dose of ACh produces fine fluctuations in the membrane potential. These are a hundred times smaller (i.e., 0.2 to 0.5 μV) than the mepp (70 to 100 μV) and probably correspond to the opening of individual channels. Each of these elementary fluctuations corresponds to the passage of some 50,000 univalent ions across the membrane. In the extrajunctional receptors of denervated muscle the actual opening and closing of single channels has been observed.

The effect of the transmitter on the membrane potential depends on the ionic species that are translocated. If both Na^+ and K^+ ions are translocated, the potential is decreased (*depolarization*); with Cl^- ions a *hyperpolarization* is produced.

Receptor proteins are localized at the chemosensitive sites of the neuronal membrane. They are in low concentration, highly hydrophobic, and embedded in the membrane.

They may be extracted with organic solvents or by the use of strong detergents and can be purified by column chromatography and affinity chromatography. Receptor proteins from electroplax, skeletal muscle, brain, smooth muscle, and crustacean heart muscle have been isolated. These proteins are either cholinergic, adrenergic, or related to the amino acids glutamate and GABA. The cholinergic receptor from electroplax is about 270,000 daltons and has several subunits ranging between 40,000 and 64,000 daltons. The smallest α subunit carries the receptor site.

The cholinergic receptor proteolipid from the electroplax has been reconstituted in black lipid membranes. In this case of reconstitution, under the influence of ACh, fluctuations of the same order of amplitude as those of the myoneural junction have been observed. An oligomeric model of the cholinergic receptor, in which the receptor and the ionophore traverse the membrane, has been postulated.

Presynaptic receptors or autoreceptors can modulate the secretion of the neurotransmitter at the nerve terminal. These receptors may be important in explaining the phenomena of presynaptic inhibition and facilitation that are found in the nervous system.

In addition to the 1-millisecond events produced by the interaction of ACh with the nicotinic cholinergic receptor, there may be other cases in which these events are longer lasting. The physiological effects involve the production of a second intracellular messenger represented by the cyclic nucleotides cAMP and cGMP. Cyclic AMP is produced by adenylate cyclase, which is concentrated in synaptosomal membranes, and cGMP by guanylate cyclase, which is usually soluble. Several neurotransmitters are known to stimulate adenylate cyclase, and others are known to stimulate guanylate cyclase; the influence of cAMP in synaptic transmission has been demonstrated in certain cases.

A third step in neuronal regulation consists in the phosphorylation of certain proteins by way of protein-kinases. Cyclic AMP, cGMP, and Ca^{++} are all able to stimulate protein kinases. Protein 1, also called synapsin I, is a protein that it is rapidly phosphorylated by a Ca^{++}- or a cAMP-dependent protein kinase. This protein is situated around the synaptic vesicles within the nerve ending.

Some basic mechanisms of learning in *Aplysia*, such as habituation, sensitization, and conditioned reflexes, involve the presynaptic modulation of neurotransmitter release, in which Ca^{++} influx, cAMP, and the phosphorylation of proteins play an important role.

REFERENCES

1. Weiss, P., and Hiscoe, H.B.: J. Exp. Zool., *107*:315, 1948.
2. De Robertis, E., and Franchi, C.M.: J. Exp. Med., *98*:269, 1953.
3. Shaw, G., and Weber, K.: Nature, *298*:277, 1982.
4. Nagele, R.G., and Roisen, F.J.: Brain Res., *253*:31, 1982.
5. Stearns, M.E.: J. Cell Biol., *92*:765, 1982.
6. Morris, J.R., and Lasek, R.J.: J. Cell Biol., *92*:192, 1982.
7. Gilbert, J.M.: Biol. Chem., *247*:6541, 1972.
8. Kreutzberg, G.W., and Schubert, P.: Neuroscience, *7*:537, 1982.
9. Tsukita, S., and Ishikawa, H.: J. Cell Biol., *84*:513, 1980.
10. Di Giamberardino, L., Bennett, G., Koenig, H.L., and Droz, B.: Brain Res., *60*:129, 1973.
11. Garner, J.A., and Lasek, R.J.: J. Cell Biol., *88*:172, 1981.
12. Isenberg, G., Schubert, P., and Kreutzberg, G.W.: Brain Res., *194*:588, 1980.
13. Kristensson, K.: Annu. Rev. Pharmacol. Toxicol., *18*:97, 1978.
14. Levi-Montalcini, R.: Harvey Lect., *60*:219, 1966.
15. Schwab, M.E., and Thoenen, H.: Brain Res., *105*:213, 1976.
16. Schwab, M.E.: Brain Res., *130*:190, 1977.
17. Levi-Montalcini, R., et al.: Proc. Natl. Acad. Sci. USA, *72*:595, 1975.
18. Schwab, M.E., and Thoenen, H.: Brain Res., *105*:213, 1976.
19. Hodgkin, A.L., and Huxley, A.F.: Cold Spring Harbor Symp. Quant. Biol., *17*:43, 1952.
20. Tasaki, I.: Nervous Transmission. Springfield, Illinois, Charles C Thomas, 1953.
21. Conti, F., de Felice, L.J., and Wanke, E.: J. Physiol. (Lond.), *248*:45, 1975.
22. Taylor, R.F.: J. Neurochem., *31*:1199, 1978.
23. Catteral, W.A., et al.: Neuroscience, 7:537, 1982.
24. Armstrong, C.M.: *In* Membranes. Edited by G. Eisenman. New York, Marcel Dekker, Inc., 1975.
25. Loewenstein, W.R.: Sci. Am., *203*:98, 1960.

26. Tauc, L., and Gerschenfeld, H.M.: C.R. Acad. Sci. (Paris), *257*:3076, 1960.
27. Eccles, J.C.: The Physiology of Synapses. Berlin, Springer-Verlag, 1964.
28. Valverde, F.: Exp. Brain Res., *3*:337, 1967.
29. Coss, R.G., and Globus, A.: Science, *200*:787, 1978.
30. De Robertis, E., and Bennett, H.S.: J. Biophys. Biochem. Cytol., *2*:307, 1955.
31. De Robertis, E., Pellegrino de Iraldi, A., Rodríguez de Lores Arnaiz, G., and Salganicoff, L.: Anat. Rec., *139*:220, 1961.
32. Cotman, C.W., and Banker, G.A.: *In* Review of Neurosciences. Vol. 1. Edited by S. Ehrenpreis and I.J. Kopin. New York, Raven Press, 1974, p. 2.
33. Akert, K., Peper, K., and Sandri, C.: *In* Cholinergic Mechanisms. Edited by P.G. Waser. New York, Raven Press, 1975, p. 43.
34. Grohovaz, F., Limbrick, A.R., and Miledi, R.: Proc. R. Soc. Lond. [Biol.], *215*:147, 1982.
35. De Robertis, E., et al.: J. Neurochem., *10*:225, 1963.
36. Hubbard, J.I.: Physiol. Rev., *53*:674, 1973.
37. Uchizono, K.: Nature, *207*:642, 1965.
38. De Robertis, E.: *In* Structure and Function of Inhibitory Neuronal Mechanisms. Edited by C. Von Euler, et al. Oxford, Pergamon Press, 1968, p. 511.
39. Pellegrino de Iraldi, A., and De Robertis, E : Int. J. Neuropharmacol., *2*:231, 1963.
40. Jaim-Etcheverry, J., and Zieher, L.M.: *In* Advances in Cytopharmacology. Vol. 1. New York, Raven Press, 1971, p. 343.
41. Kanaseki, T., and Kadota, K.: J. Cell Biol., *42*:202, 1969.
42. De Robertis, E.: Science, *156*:907, 1967.
43. De Robertis, E.: Science, *171*:963, 1971.
44. Sotelo, C., and Beaudet, A.: Proc. R. Soc. Lond. [Biol.], *206*:133, 1979.
45. Pelletier, G., Steinbusch, H.W.M., and Verhofstad, A.A.J.: Nature, *293*:71, 1981.
46. Katz, B.: Science, *173*:123, 1971.
47. De Robertis, E.: Int. Rev. Cytol., *8*:61, 1958.
48. Pecci Saavedra, J., Vaccarezza, O.L., Reader, I.A., and Pasqualini, E.: Exp. Neurol., *26*:607, 1970.
49. Whittaker, V.P., and Zimmermann, H.: *In* Synaptic Transmission and Neuronal Interaction. Edited by M.V.L. Bennett. New York, Raven Press, 1974.
50. Boyne, A.L., Bohan, T.P., and Williams, T.H.: J. Cell Biol., *67*:814, 1975.
51. Pappas, G.D., and Rose, S.: Brain Res., *103*:362, 1976.
52. Ceccarelli, B., et al.: Eur. J. Cell Biol., *22*(1):189, 1980.
53. Jones, D.G.: Synapses and Synaptosomes: Morphological Aspects. London, Chapman and Hall, 1975.
54. Dowall, M.J., and Zimmermann, H.: Brain Res., *71*:160, 1974.
55. Heuser, J.E., and Reese, T.S.: *In* Synaptic Transmission and Neuronal Interaction. Edited by M.V.L. Bennett. New York, Raven Press, 1974, p. 59.
56. Llinás, R., and Nicholson, C.: Proc. Natl. Acad. Sci. USA, *72*:187, 1975.
57. Pumplin, D.W., Reese, T.S., and Llinás, R.: Proc. Natl. Acad. Sci. USA, *78*:7210, 1981.
58. Llinás, R., Steinberg, I.Z., and Walton, K.: Biophys. J., *33*:289, 1981.
59. Grab, D.J., Carlin, R.K., and Siekevitz, P.: Ann. N.Y. Acad. Sci., *356*:55, 1980.
60. Katz, B., and Miledi, R.: J. Physiol. (Lond.), *224*:665, 1972.
61. Neher, E., and Sakmann, B.: Nature, *260*:799, 1976.
62. De Robertis, E., Fiszer, S., and Soto, E.F.: Science, *158*:928, 1967.
63. Barrantes, F.J., Arbilla, S., De Carlin, M.C.L., and De Robertis, E.: Biochem. Biophys. Res. Commun., *63*:194, 1975.
64. Criado, A., Aguilar, J.S., and De Robertis, E.: J. Recept. Res., *1*:169, 1980.
65. Kistler, J., et al.: Biophys. J., *37*:371, 1982.
66. Wise, D.S., Schoenborn, B.P., and Karlin, A.: J. Biol. Chem., *256*:4124, 1981.
67. Epstein, M., and Racker, E.: J. Biol. Chem., *253*:6660, 1978.
68. Schlieper, P., and De Robertis, E.: Biochem. Biophys. Res. Commun., *75*:886, 1977.
69. Anholt, R., Lindstrom, J., and Montal, M.: J. Biol. Chem., *256*:4377, 1981.
70. De Robertis, E.: Science, *171*:963, 1971.
71. Aoshima, H., Cash, D.J., and Hess, G.P.: Biochemistry, *20*:3467, 1981.
72. De Robertis, E., et al.: J. Biol. Chem., *242*:3487, 1967.
73. Bloom, F.E.: Rev. Physiol. Biochem. Pharmacol., *74*:1, 1975.
74. Ueda, T., and Greengard, P.: J. Biol. Chem., *252*:5155, 1977.
75. De Camilli, P., Cameron, R., and Greengard, P.: J. Cell Biol., *96*:1337, 1983.
76. De Camilli, P., Harris, S.M., Jr., Huttner, W.B., and Greengard, P.: J. Cell Biol., *96*:1355, 1983.
77. Huttner, W.B., Schiebler, W., Greengard, P., and De Camilli, P.: J. Cell Biol., *96*:1374, 1983.
78. Klein, M., Camardo, J., and Kandel, E.R.: Proc. Natl. Acad. Sci. USA, *79*:5713, 1982.
79. Siegebaum, S., Camardo, J., and Kandel, E.R.: Nature, *299*:413, 1982.
80. Bailey, C.H., and Cheu, M.: Science, *220*:91, 1983.

ADDITIONAL READINGS

Adams, R.J.: Organelle movement in axons depends on ATP. Nature, *297*:327, 1982.
Anholt, R.: Reconstitution of acetylcholine receptors in model membranes. Trends Biochem. Sci., *6*:288, 1981.
Barrantes, F.J.: Endogenous chemical receptors: Some physical aspects. Annu. Rev. Biophys. Bioeng., *8*:287, 1979.
Bartfai, T.: Cyclic nucleotides in the central nervous system. Trends Biochem. Sci., *3*:121, 1978.
Briley, M.S., and Changeaux, J.P.: Isolation and purification of the nicotinic receptor and its functional reconstitution into a membrane environment. Int. Rev. Neurobiol., *20*:31, 1977.
Brown, I.R.: Molecular Approaches to Neurobiology. New York, Academic Press, Inc., 1982.
Bunge, R., Johnson, M., and Ross, C.D.: Nature and nurture in development of the autonomic neuron. Science, *199*:1409, 1978.
Burnstock, N.: Purinergic nerves. Pharmacol. Rev., *24*:509, 1972.
Carpenter, D.O., and Reese, T.S.: Chemistry and physiology of synaptic transmission. *In* Basic Neurochemistry. Edited by G.J. Siegel and R.W. Albers. Boston, Little, Brown and Company, 1981, p. 161.
Ceccarelli, B., et al.: Freeze-fracture studies of frog neuromuscular function during intense release of neurotransmitter I and II. J. Cell Biol., *81*:163, 1979.
Cooper, P.E., and Martin, J.B.: Neuroendocrinology and brain peptides. Trends in Neuroscience, *5*:187, 1982.
De Lorenzo, R.J.: The calmodulin hypothesis of neurotransmission. Cell Calcium, *2*:365, 1981.
De Robertis, E.: Synaptic Receptors: Isolation and Molec-

ular Biology. New York, Marcel Dekker, Inc., 1975, pp. 1–387.

De Robertis, E.: Ultrastructure and cytochemistry of the synaptic region: The macromolecular components involved in nerve transmission are being studied. Science, *156*:907, 1967.

De Robertis, E.: Histophysiology of Synapses and Neurosecretion. Oxford, Pergamon Press, 1964.

Gainer, H., and Brownstein, M.J.: Neuropeptides. *In* Basic Neurochemistry. Edited by G.T. Siegel and R.W. Albers. Boston, Little, Brown and Company, 1981, p. 269.

Glossman, H., et al.: Calcium channels direct identification with radioligand binding studies. Trends Pharmacol. Sci., 431, 1982.

Grafstein, B., and Forman, D.S.: Intercellular transport in neurons. Physiol. Rev., *60*:1167, 1980.

Greengard, P., and De Camilli, P.: Protein phosphorylation in neurons. *In* Disorders of the Motor Unit. Edited by D.L. Schotland. New York, John Wiley and Sons, Inc., 1982, p. 441.

Hammerschlag, K., and Stone, G.C.: Membrane delivery by fast axonal transport. Trends Neurosci., 5:12, 1982.

Hille, B.: Excitability and ionic channels. *In* Basic Neurochemistry. Edited by G.T. Siegel and R.W. Albers. Boston, Little, Brown and Company, 1981, p. 95.

Hucho, F.: The nicotinic acetylcholine receptor. Trends in Biochem. Sci., 6:242, 1981.

Hucho, F., and Schiebler, W.: Biochemical investigations on ionic channels in excitable membranes. Mol. Cell. Biochem., *18*:151, 1977.

Israel, M., Dunant, Y., and Manaranche, R.: The present status of the vesicular hypothesis. Prog. Neurobiol., *13*:237, 1979.

Kandel, E.R., and Schwartz, J.H.: Molecular biology of learning: Modulation of transmitter release. Science, *218*:433, 1982.

Langer, S.Z.: Presynaptic receptors and their role in the regulation of transmitter release. Brit. J. Pharmacol., *73*:17, 1981.

Lasek, R.F., Garner, F.A., and Brady, S.T.: Axonal transport of the cytoplasmic matrix. J. Cell Biol., 99:2125, 1984.

Lindslay, R.M., et al.: Differences and similarities in the neurotrophic growth factor requirements of sensory neurons derived from neural crests and neural placode. J. Cell Sci. Suppl. 3:115, 1985.

Llinás. R.: Calcium in synaptic transmission. Sci. Am., *247*:56, 1982.

Llinás, R.: The role of calcium in neuronal function. *In* The Neurosciences. Edited by F.V. Schmitt and F.G. Worden. Cambridge, MIT Press, 1979, p. 555.

Lundberg, J.M., and Hökfelt, T.: Coexistence of peptides and classical neurotransmitters. Trends in Neuroscience, 6:325, 1983.

McGeer, P.L., Eccles, J.C., and McGeer, E.G.: Molecular Neurobiology of the Mammalian Brain. New York, Plenum Press, 1978.

Nathanson, J.A., and Greengard, P.: Cyclic nucleotides and synaptic transmission. *In* Basic Neurochemistry. Edited by G.T. Siegel and R.W. Albers. Boston, Little, Brown & Company, 1981, p. 297.

Nestler, E.J., and Greengard, P.: Protein phosphorylation in the brain. Nature, *305*:583, 1983.

Osborne, M.P.: The role of synaptic vesicles with some observations on vertebrate sensory cells. *In* The Synapse. Edited by G.A. Cottrell and P.N.R. Usherwood. Glasgow, Blackie, 1977.

Palacious-Pru, E.L., Palacios, L., and Mendoza, R.V.: Synatogenetic mechanisms during chick cerebellar cortex development. J. Submicrosc. Cytol., *13*:145, 1981.

Patterson, P.H., Potter, D.D., and Furshpan E.J.: The chemical differentiation of nerve cells. Sci. Am., *239*:38, 1978.

Pellegrino de Iraldi, A.: Compartmentation of synaptic vesicles in autonomic neurons in histochemistry. *In* Cell Biology of Autonomic Neurons. Edited by O. Eränkö, et al. New York, Raven Press, 1980, p. 255.

Peng, I., Benoler, L.L., and Black, M.M.: Biochemical and immunological analysis of cytoskeletal domains of neurons. J. Cell Biol., *102*:252, 1986.

Rambourg, A., and Droz, B.: Smooth endoplasmic reticulum and axonal transport. J. Neurochem., *35*:16, 1980.

Rees, R.P., Bunge, M.P., and Bunge, R.P.: Morphological changes in the neuritic growth cone and target neuron during synaptic junction development in culture. J. Cell Biol., *68*:240, 1976.

Robertson, M.: Learning, forgetting and the cell biology of memory. Nature, *300*:219, 1982.

Rodriguez de Lores Arnaiz, G. and De Robertis, E.: Drugs affecting the synaptic components of the CNS. *In* Fundamentals of Cell Pharmacology. Edited by S. Dikstein. Springfield, Illinois, Charles C Thomas, 1973, p. 280.

Rodriguez de Lores Arnaiz, G., and De Robertis, E.: Properties of isolated nerve endings. *in* Current Topics in Membrane Transport. Vol. 3. New York, Academic Press, Inc., 1972, p. 237.

Salpeter, M.M., and Loring, R.H.: Nicotinic acetylcholine receptors in vertebrate muscle: properties distribution and neural control. Prog. in Neurob., *25*:297, 1985.

Stadler, H., and Tsukita, S.: Synaptic vesicles contain an ATP-dependent proton pump and show "knob-like" protrusions on their surface. EMBO J., 3:3333, 1984.

Stevens, C.F.: AChRs: five-fold symmetry and the γ-subunit. Trends Neurosci., 8:335, 1985.

Stevens, C.F.: Channels as molecular entities. Trends in Neurosc. 9:453, 1986.

Stroud, R.M.: Acetylcholine receptor structure. Neurosci. Commentaries, 3:124, 1983.

The Synapse. Cold Spring Harbor Symp. Quant. Biol., *40*:1975.

Tauc, L.: Non-vesicular release of neurotransmitter. Physiol. Rev., 62:857, 1982.

Thoenen, H., and Barde, Y.A.: Physiology of nerve growth factor. Physiol. Rev., *60*:1284, 1980.

Williams, M.: Protein phosphorylation in mammalian nervous system. Trends Biochem. Sci., 4:25, 1979.

Yamamura, H.I., Enna, S.J., and Kuhar, M.J., eds.: Neurotransmitter Receptor Binding. New York, Raven Press, 1978.

Zingheim, H.P., et al.: Direct visual localization of two toxin recognition sites on an ACh receptor protein. Nature, *299*:81, 1982.

Ziecker R.S., and Lanolò, L.: Mechanism of transmitter release, voltage hypothesis and calcium hypothesis. Science, *231*:574, 1986.

GLOSSARY

Acetyl coenzyme A. A high-energy intermediate in fuel molecule metabolism, bringing suitably modified acetyl groups into the tricarboxylic acid cycle.

Actin. Force-generating cytoplasmic and muscle protein, present in both globular (G-actin) and fibrous (F-actin) forms.

Active site. The location(s) on an enzyme molecule at which the substrate attaches and is chemically modified.

Active transport. Membrane transport that requires the expenditure of energy.

Adenylate cyclase. Membrane-bound enzyme that catalyzes the production of cAMP from ATP.

Aerobic respiration. Process requiring O_2 by which organisms generate energy from ingested organic molecules.

Allele. One of two genes for a given trait that occurs in a specific position on each homologous chromosome.

α-amanitin. Toxin that inhibits the eukaryotic RNA polymerases to different extents and thus is useful for characterizing them.

Ameboid motion. Type of locomotion in which the cell changes shape actively, sending cytoplasmic pseudopodia in the direction of movement.

Amino acid. One of about 20 nitrogen-containing organic acids used by the cell for polymerization into protein molecules.

Aminoacyl-tRNA. tRNA molecule carrying an amino acid.

Aminoacyl-tRNA synthetase. Enzyme responsible for attaching the correct amino acid to the tRNA molecule.

Amniocentesis. Puncture of the uterine wall with a needle for the purpose of obtaining amniotic fluid, which can be analyzed to determine whether the fetus has a genetic abnormality.

Amphipathic. Term used to describe a molecule having both a hydrophobic and a hydrophilic region.

Anaerobic glycolysis. Degradation of glucose into lactic acid in the absence of oxygen, yielding two ATP molecules.

Anaphase. Stage of mitosis during which the daughter chromosomes separate and begin moving toward opposite poles of the cell.

Aneuploidy. Karyotypic abnormality in which a specific chromosome(s) is present in too many or too few copies.

Annulus. Ring-shaped structure enclosing the nuclear pore, consisting of eight surrounding granules and a proteinaceous material in the pore opening.

Anticodon. Triplet of tRNA nucleotides complementary to those of the mRNA codon.

Aster. Microtubule fibers radiating out from each cell pole during the metaphase stage of mitosis.

ATP (adenosine triphosphate). A nucleotide containing high-energy bonds that provides energy for many biochemical cellular processes by undergoing enzymatic hydrolysis.

Autophagy. Digestion of cellular material by the cell's own enzymes; part of the normal regeneration and turnover of eukaryotic cellular components.

Autoradiography. Technique by which emissions from a radioactively labeled microscopic specimen can be visualized by exposure of a photographic emulsion to the labeled tissue.

Axoneme. Axial microtubular component of cilia and flagella that is their essential motile element.

Bacteriophage. Virus that attacks bacterial cells.

Basal body. Structurally identical to the centriole but serving as a cilium or flagellum attachment site; also called a kinetosome.

Birefringence. Property of some substances that affect the polarization of transmitted light in such a way that these substances appear prominent under a polarization microscope.

Blastula. Early embryonic form consisting of a hollow sphere of cells.

cAMP (cyclic adenosine monophosphate). Cyclic nucleotide having a regulatory function in many metabolic processes.

5′ Cap. 7-methyl-guanosine structure added to the beginning of eukaryotic mRNAs.

Catalyst. Substance that facilitates a chemical reaction without being depleted or permanently modified in the process.

Cell coat. Continually renewed outermost layer of an animal cell, in direct contact with the plasma membrane; also called glycocalyx.

Cell fractionation. Technique by which subcellular components are separated and purified on the basis of their physical properties.

Cell wall. Rigid exoskeletal structure enclosing and protecting the contents of most plant and bacterial cells.

Centriole. Hollow cylindrical organelle involved in the organization of the spindle; its wall contains nine microtubule triplets.

Centromere. Site of chromosome attachment to the mitotic spindle.

Chase. Washing and incubation of a "pulsed" tissue sample in medium containing excess unlabeled precursor.

Chiasma. Point of chromosomal interchange that becomes visible when the homologues begin to separate at diplonema.

Chloroplast. Chlorophyll-containing plastid in which photosynthesis occurs.

Chromatid. In the metaphase chromosome, one of two symmetrical structures, each containing a single DNA molecule.

Chromatin. Complex of DNA and proteins.

Chromomere. Granule found along the axis of a decondensed chromosome, in which the DNA is tightly folded.

Chromosome. Structure consisting of chromatin strands, which become extremely compact during cell division.

Cilium. Short, microtubular organelle projecting into the extracellular medium; used for cell locomotion or to create currents in the surrounding fluid.

Cis or **forming face.** That side of the Golgi (dictyosome) closest to the nuclear envelope and the ER.

Cisterna. One of the flattened, membrane-enclosed sacs found in parallel stacks in, for example, the nuclear envelope, the ER and the Golgi complex.

Cleavage. Rapid cycles of DNA synthesis followed by cell division in which the cytoplasm is partitioned without growth, occurring very early in embryonic development.

Cloning. Asexual production of a line of cells or organisms or segments of DNA genetically identical to the original.

Codon. Triplet of consecutive nucleotides specifying a particular amino acid.

Colchicine. Drug that prevents microtubule polymerization and facilitates depolymerization of existing microtubules by binding to tubulin subunits; its derivative Colcemid has similar effects.

Condensing vacuole (prozymogen granules). An immature secretory granule.

Connexon. Single unit of a gap junction, consisting of hexagonally arranged particles surrounding a central channel.

Contact inhibition. The tendency of normal cultured cells to limit their own movement, cell division, and growth.

Crista. Infolding of the inner mitochondrial membrane, containing components for respiratory metabolism and oxidative phosphorylation and providing a relatively large surface area on which these reactions can take place.

CURL (endosome). Compartment where *U*ncoupling of *R*eceptor and *L*igand occurs.

C value. Characteristic DNA content of the cells of a given species.

Cyclosis. Generation of cytoplasmic currents by the action of microfilaments.

Cytochalasin B. Drug that disrupts actin microfilaments without affecting microtubules.

Cytogenetics. The study of inheritance using the methods of cytology.

Cytokinesis. Process of cleavage and separation of the cytoplasm at the final stage of mitosis.

Cytoplasm. Protoplasmic material of the cell not including the nucleus and its contents.

Cytoplasmic matrix. That portion of the cell cytoplasm not contained within the endomembrane system.

Cytoskeleton. Complex cytoplasmic network of microtubules and microfilaments located predominantly near the cell membrane and thought to be responsible for cell shape and movement.

Dalton. Unit of molecular weight; one dalton equals one-twelfth the mass of carbon 12.

Daltonism. Red and green color blindness, a recessive trait known to be X-linked.

Desmosome. Mechanical attachment formed by the thickened plasma membrane of two adjacent cells, containing intermediate filaments.

Determinants. A group of substances that, upon becoming unequally distributed into the cytoplasm of certain groups of cells, commit those cells to a particular course of differentiation.

Diakinesis. The stage of meiosis during which chromosome contraction increases; at the end of this stage the homologues are attached only at the chiasmata.

Dictyosome. A stack of cisternae making up the Golgi complex.

Diplonema. Stage of meiosis during which paired homologues begin to separate.

Disulfide bond. A covalent bond formed between —SH groups of two cysteine residues (—S—S—).

Dynein. Microtubule subfiber protein, having ATPase activity and thought to be essential for movement.

Ecdysone. Insect steroid hormone that can induce chromosomal puffing.

Electron microscope (EM). Instrument that uses a focused beam of electrons to produce an enlarged image of an object.

Endocytosis. Collective term for phenomena involving the surrounding and ingestion of various substances by the plasma membrane.

Endoplasmic reticulum (ER). System of membrane-enclosed cytoplasmic channels involved in cellular transport processes; rough ER has ribosomes attached to its outer surface whereas smooth ER does not.

Enzyme. A protein that acts as a biological catalyst.

Erythrocyte. A red blood cell.

Escherichia coli (E. coli). Easily cultured prokaryotic organism, the best understood and most studied organism.

Euchromatin. Uncondensed, genetically active regions of the chromatin.

Eukaryote. Organism (uni- or multicellular) whose cells contain a true nucleus.

Exocytosis. Energy-requiring process by which secretory granules discharge their contents by fusing with the cell membrane.

Exon. Part of the gene that becomes part of the mature mRNA after the intervening sequences (introns) are spliced out.

Fibronectin. High-molecular-weight glycoprotein component of the cell coat, thought to aid cellular locomotion.

Flagellum. Long microtubular organelle projecting into the extracellular medium used for cell locomotion.

Fluid mosaic model. Model of plasma membrane structure postulating integral proteins embedded in the lipid bilayer but free to move laterally and projecting out of the bilayer at one surface or the other.

Frameshift mutation. Insertion or deletion of a single base(s), causing the message beyond the mutation to be read out of phase.

G_1 period. Interval between the end of mitosis and the start of the S phase.

G_2 period. Interval between the end of the S phase and the start of mitosis.

G_0 state. Condition of a cell whose division has been arrested at G_1 and which is considered to be withdrawn from the cell cycle.

Gap junction. Array of plaque-like connections between the plasma membranes of adjacent cells.

Gene. Hereditary unit specifying the production of a distinct protein (e.g., an enzyme) or RNA.

Gene amplification. Process by which a gene is replicated selectively—for example, in the rDNA of amphibian oocyte—while the rest of the genome remains constant.

Genetic engineering. Array of techniques that facilitate the manipulation and duplication of pieces of DNA for industrial, medical, and research purposes.

Genotype. Genetic constitution of an organism.

GERL (Golgi-ER-Lysosome). Acid phosphatase–rich region of the Golgi maturing face, thought to be involved in the production of lysosomes and/or presecretory granules.

Glyoxysome. Membrane-enclosed organelle found in plant cells and involved mainly in triglyceride metabolism.

Golgi complex. Cytoplasmic membrane system consisting of stacks of flattened sacs and other vesicles involved in secretion processes.

HeLa cell. Cell line originally derived from the uterine carcinoma cells of a patient named Henrietta Lacks.

Heterogeneous nuclear RNAs (hnRNAs). Large nuclear RNAs of variable length, most of which are degraded within the nucleus and many of which represent unspliced mRNA precursors.

Heterokaryon. A single cell containing nuclei of two types, produced by the fusion of two cells.

Heterophagosome. Digestive vacuole resulting from the ingestion of foreign substances by the cell.

Heterozygous. Having two different alleles for a given trait.

Histones. Group of five basic proteins that associate with DNA in the cell nucleus to form chromatin.

Homozygous. Having two identical alleles for a given trait.

Hydrophobic interactions. Exclusion of water by nonpolar groups of a macromolecule, causing compaction with hydrophobic groups inside and hypophilic groups outside.

Inducer. Molecule which, when bound to the repressor, causes a conformational change that prevents the repressor from binding to the operator.

Initiation signal. The AUG codon that specifies the incorporation of N-formylmethionine at the 5′ end of a new protein chain.

In situ hybridization. Technique in which radioactively labeled RNA or DNA is used to identify the specific chromosomal regions or RNA transcripts complementary to it.

Interphase. Interval between eukaryotic cell divisions during which growth and synthetic activities take place.

Intron. Intervening sequence of DNA, located within a gene, that is not included in the mature mRNA.

Isochromosome. Chromosome formed by the joining of two identical chromosome arms.

Karyotype. Characteristic chromosome set of a given species.

Kinetochore. Region of the centromere to which spindle microtubules attach during cell division.

Klinefelter's syndrome. Genetic disorder characterized by an XXY karyotype, outwardly male phenotype, and various anomalies including sterility.

Lampbrush chromosome. Chromosome with highly extended lateral loops of DNA; found in all animal oocytes at the diplonema stage of meiosis.

Leptonema. Stage of meiosis during which the chromosomes condense and become visible.

Locus. Position of an allele on the chromosome.

Lysosome. One of a diverse group of single-membrane organelles containing many kinds of hydrolytic enzymes; involved in intracellular digestion.

Meiosis. Division process occurring during gamete formation that reduces the number of chromosomes in each cell to one half.

Messenger RNA (mRNA). Class of RNA molecules that encode the amino acid sequence of a protein in their nucleotide base sequence and serve as templates for protein synthesis.

Metaphase. Stage of mitosis during which the chromosomes become attached to spindle fibers and arrayed in the equatorial plane of the cell.

Microsome. Small closed vesicle of ER membrane obtained upon differential centrifugation.

Microtubule. Hollow cylindrical fibril involved in intracellular movements.

Microvillus. Plasma membrane differentiation that greatly increases the effective absorption surface of the cell.

Missense mutation. Change that causes a codon to specify a different amino acid.

Mitochondrion. Membrane-bound organelle that generates chemical energy in the form of ATP for use in cellular metabolic processes.

Mitosis. Process of cell division, whereby the genetic material is precisely divided and two new chromosome sets identical to the original are generated.

Mosaic. Tissue or organism whose cells exhibit more than one genotype—for example, as a result of mitotic nondisjunction during embryonic development.

Mutation. Physical or chemical alteration in the genetic material.

Nondisjunction. Failure of one or more homologous chromosome pairs to separate properly during cell division.

Nuclear envelope. Flattened sac or double membrane surrounding the nucleoplasm and genetic material.

Nuclear pore. Interruption in the nuclear envelope that allows the exchange of material between the nucleus and cytoplasm.

Nucleic acid. Macromolecule consisting of repeating units made up of a sugar moiety, a nitrogenous base, and phosphoric acid.

Nucleoid. Nuclear region within a prokaryotic cell, visibly distinct from the cytoplasm but not isolated by a membrane.

Nucleolar organizer. Chromosomal region that contains the genes for ribosomal RNAs and induces formation of the nucleolus.

Nucleolus. Structure within the nucleus, consisting of chromatin and large amounts of RNA; site of rRNA synthesis and ribosome formation.

Nucleoside. Nucleotide minus the phosphoric acid (i.e., the combination of pentose moiety and nitrogenous base).

Nucleosome. Repeating unit of chromatin consisting of 200 base pairs of DNA coiled around a histone octamer (two each of histones H2A, H2B, H3, and H4).

Nucleotide. Monomeric unit of a nucleic acid macromolecule; also used to store and transfer energy.

Nucleus. Membrane-enclosed cell organelle containing the genetic material and various proteins involved in its replication and transcription.

Operator. Short DNA region to which the repressor binds.

Operon. Group of genes (and regions governing their transcription) that occur sequentially in the DNA and can be regulated in a unified manner.

Oxidative phosphorylation. ATP synthesis by phosphorylation of ADP using the energy provided by electron transfer during aerobic respiration.

Pachynema. Stage of meiosis during which two chromatids belonging to different homologues exchange segments of genetic material.

Peroxisome (microbody). Membrane-enclosed organelle containing enzymes for the metabolism of hydrogen peroxide.

Phagocytosis. Ingestion of particulate material by the cell, either as a means of feeding or defense.

Phenotype. Any visible characteristics of an organism that are ultimately an expression of gene activity.

Photosynthesis. Process of trapping the energy of sunlight, transforming it into chemical energy, and storing it in the chemical bonds of carbohydrates.

Pinocytosis. Ingestion of liquids and soluble materials by the cell.

Plasma membrane. Lipoprotein structure serving as a selective barrier between the cell's cytoplasm and its immediate environment.

Plasmid. Small piece of autonomous, extrachromosomal, circular DNA found in some bacterial cells.

Plasmodesmata. Cytoplasmic bridges between adjacent plant cells.

Point mutation. Change in a single base of the DNA molecule.

Poly A. Long polyadenylic acid segment added post-transcriptionally to many eukaryotic mRNAs at the $3'$ end.

Polycistronic mRNA. mRNA molecule that codes for more than one protein.

Polymer. Large molecule made up of repeating structural subunits.

Polysome. Translational unit formed by the attachment of several ribosomes to a single mRNA molecule.

Polytene chromosome. Chromosome (usually studied in the cells of dipteran larvae) in which the DNA is replicated in such a way that the daughter chromatids do not separate, producing instead a characteristic morphology of bands and interbands.

Pore complex. Structure formed by the nuclear pores and annuli.

Primary lysosome. Small storage vesicle formed at the GERL region; a lysosome that has not yet fused with other vesicles.

Prokaryote. Organism that lacks a true nucleus.

Promoter. DNA region to which RNA polymerase binds when initiating transcription.

Prophase. Stage of mitosis during which the chromosomes condense and become visible within the nucleus, followed by the dissolution of the nuclear envelope.

Protein kinase. Type of enzyme whose activity may be stimulated by cAMP and which catalyzes the phosphorylation of proteins.

Puff. Local enlargement of a polytene chromosome in which active gene transcription can be visualized.

Pulse. Incubation of a tissue sample to medium containing a radioactively tagged precursor for a short time.

Purine. Type of nitrogenous base found in nucleic acids consisting of two fused rings consisting of nitrogen and carbon atoms (adenine or guanine).

Pyrimidine. Type of nitrogenous base found in nucleic acids, consisting of a single heterocyclic ring (cytosine, thymine, or uracil).

Reciprocal translocation. Translocation in which small chromosome segments are exchanged between chromosomes of different pairs.

Recombination. Exchange of genetic material between homologue chromosomes.

Replicon. "Replication unit" of eukaryotic DNA, having its own replication initiation site.

Repressor. Protein that can bind strongly and specifically to a short region of DNA preceding the structural gene(s), thus preventing transcription.

Restriction endonuclease. Any of several enzymes capable of recognizing and cutting a specific symmetrical nucleotide sequence in DNA.

Reverse transcriptase. Enzyme that can transcribe RNA into DNA, found in some RNA tumor viruses.

Ribosome. Cytoplasmic granule composed of RNA and protein, at which protein synthesis takes place.

Ribosome binding site. Series of bases on the mRNA molecule, including the AUG codon, to which the small ribosomal subunit binds.

RNA (ribonucleic acid). Single-stranded molecule having a nucleotide subunit structure similar to that of DNA.

RNA polymerase. Enzyme involved in the synthesis of RNA from a DNA template.

Robertsonian translocation. A special type of translocation in which the long arms of two acrocentric chromosomes form a metacentric chromosome; also called centric fusion.

Sarcomere. Structural unit of the myofibril, extending between adjacent Z lines and containing identifiable bands or zones.

Sarcoplasmic reticulum. Specialized component of the endoplasmic reticulum that is found in muscle fibers, consisting of tubules and vesicles connected to the muscle cell membrane.

Satellite DNA. Region of highly repeated DNA (constitutive heterochromatin), which may have a structural role in chromosomes.

Sendai virus. Parainfluenza virus that adheres to the cell surface and facilitates cell fusion.

Sex chromatin. An X chromosome randomly selected and inactivated at an early stage of development in the cells of mammalian females; also called a Barr body.

Sex-linked genes. Genes located on the X chromosome.

Signal hypothesis. Proposal that the mRNA sequence of a secretory protein contains signal codons that direct its synthesis on ER-bound as opposed to free ribosomes.

Signal peptide. Sequence of 15 to 30 hydrophobic amino acids whose function is to establish the initial ribosome-mRNA association with the ER membrane.

Signal recognition particle (SRP). Complex of six proteins and a 7S RNA that binds to nascent proteins having a signal peptide, stopping translation until the ribosomes bind ("dock") to the endoplasmic reticulum.

Silent mutation. Alteration in the DNA molecule that does not produce any amino acid change in the protein synthesized.

Sister chromatids. Identical chromatids belonging to the same homologue.

Spindle. Microtubular structure spanning the cell from pole to pole during metaphase; involved in the correct partitioning of chromosomes to daughter cells.

Splicing. Precise excision of the intervening sequences from an RNA primary transcript, followed by ligation of the message to produce a functional molecule.

Stem cell. Undifferentiated cell capable of giving rise to specialized daughter cells.

SV40. Virus that infects monkey cells and contains a circular DNA molecule.

Symbiont hypothesis. Proposal that mitochondria and chloroplasts might have originated as prokaryotic intracellular parasites that established a symbiotic relationship with the eukaryotic cell.

Synaptonemal complex (SC). Proteinaceous structure, composed of two lateral arms and a medial element, that forms between pairing homologues during zygonema.

Syncytium. Large protoplasmic mass not divided into distinct cells.

Synkaryon. Hybrid cell line generated by a heterokaryon whose nuclei have entered mitosis synchronously and merged.

Synthetic (S) phase. Specific part of interphase during which DNA synthesis occurs.

Telophase. Stage of mitosis during which the chromosomes uncoil and become surrounded by two new nuclear envelopes.

Thylakoid. Flattened vesicle found in chloroplasts, in which the light-dependent reactions of photosynthesis take place.

Tight junction. Structure formed when adjacent cell membranes adhere to each other, creating a barrier to diffusion.

Transcription. Process by which an RNA molecule is polymerized on a DNA template with the aid of various enzymes.

Trans face. The side of the Golgi dictyosome from which secretory vesicles are released toward the plasma membrane.

Transfer RNA (tRNA). Type of RNA molecule that identifies an amino acid in the cytoplasm and transports it to the ribosome.

Transformation. Conversion of normal cultured cells into cancerous cells.

Translation. Process by which a protein is synthesized from amino acids according to specifications encoded in the RNA.

Tropomyosin. Regulatory protein complexed with F-actin in the thin myofilament.

Troponin. Regulatory protein bound to the actin-tropomyosin complex and having Ca^{++}-binding properties.

Tubulin. Principal protein in many types of microtubules, occurring as a dimer whose two subunits (α and β) differ slightly from each other.

Turner's syndrome. Genetic disorder characterized by an XO karyotype, outwardly female phenotype, and various anomalies including sterility and growth retardation.

Unineme theory. Theory that each chromatid represents a single linear DNA molecule with its associated proteins.

Viroid. Virus-like particle that consists of a single RNA molecule without a protein coat and that infects plant cells.

Virus. Microorganism consisting of a DNA or RNA core surrounded by a protein coat capable of replicating only within the cells of a host organism.

X-chromatin. An X chromosome randomly selected and inactivated at an early stage of development in the cells of mammalian females.

Xenopus laevis. Species of South African frog well suited for many kinds of biological studies.

Xeroderma pigmentosum. Human hereditary disease in which cells are unable to repair the damage caused by ultraviolet light.

X-ray diffraction. Scattering of x-rays by the atoms of a crystal in a manner such that the resulting diffraction pattern provides information about the structure and/or identity of the substance.

Zygonema. Stage of meiosis during which the homologous chromosomes align and pair with each other.

Zygote. Product of the union of two gametes, which develops into a new organism.

Zymogen granules. Mature electron-opaque secretory granule that releases its contents into the extracellular medium when the appropriate stimulus is provided.

INDEX